Atlas of the Vascular Plants of Texas

Volume I

24

Sida, Botanical Miscellany
Botanical Research Institute of Texas

Atlas of the Vascular Plants of Texas

Introduction

Dicots

B.L. Turner

Holly Nichols, Geoffrey Denny, Oded Doron

2003

ISBN 1-889878-08-1
ISSN 0833-1475

Sida, Botanical Miscellany, Number 24, Volume 1

Founded by
Wm. F. Mahler and Barney L. Lipscomb, 1987
Editor: Barney Lipscomb
Botanical Research Institute of Texas
509 Pecan Street
Fort Worth, Texas 76102-4060, USA
Contributing Spanish Editor: Félix Llamas
Dpt. Biología Vegetal, Universidad de León
Campus de Vegazana s/n, 24071 León, SPAIN

Atlas of the Vascular Plants of Texas
By B.L. Turner, Holly Nichols, Geoffrey Denny, Oded Doron

Cover
Photograph: Field of Texas bluebonnets
Photo credit: Geoffrey Stanford
Inset photograph: *Lupinus texensis* (bluebonnet) is one of six species of *Lupinus* in the state and all six species are state flowers of Texas
Photo credit: George M. Diggs Jr., Austin College, Sherman, Texas
Cover Design: Rebecca Horn, rhorngraphics, rhorn@airmail.net

Inside cover map
The inside front cover map is provided for easy reference to county names in Texas (courtesy of Jim Henrickson, University of Texas at Austin).

Atlas of the Vascular Plants of Texas
is published with the support of:
Billie Lee Turner and The University of Texas at Austin
and the Botanical Research Institute of Texas

Distribution of copies by:
Botanical Research Institute of Texas
509 Pecan Street
Fort Worth, Texas 76102-4060, USA
Telephone: 1 817 332 4441
Fax: 1 817 332 4112
Website: www.brit.org/sida
Email: sida@brit.org

BRIT
PRESS

Sida, Botanical Miscellany, No. 24
25 Apr 2003

Printed in the United States of America

Atlas of the Vascular Plants of Texas

A Collaborative Project between the Botanical Research Institute of Texas and Professor Emeritus Billie Lee Turner, University of Texas Austin

A Book to Promote

Botanical and Environmental Education

Scientific Exploration and Conservation

In Texas

Table of Contents

Volume 1

Introduction 1

Plant identification and sources of data 2

Arrangement of text 4

Synonymy and author citations 4

Floristic and vegetational regions 4

New varietal and species combinations for taxa occurring in Texas 4

Tabulation of flora 9

Atlas

(Dicots) 11

Adventives in Texas 622

Index 625

New names and new combinations

	Page
Acleisanthes lanceolatus var. **maloneanus** (B.L. Turner) B.L. Turner, comb. nov.	**7**
Acleisanthes megaphyllus (Fowler & B.L. Turner) B.L. Turner, comb. nov.	**7**
Acleisanthes nevadensis (Standley) B.L. Turner, comb. nov.	**8**
Acleisanthes purpusianus var. **marshii** (I.M. Johnst.) B.L. Turner, comb. nov.	**8**
Antennaria parlinii var. **fallax** (Greene) B.L. Turner, comb. nov.	**6**
Arceuthobium vaginatum var. **cryptopodum** (Engelm. in A. Gray) B.L. Turner, comb. nov.	**8**
Arceuthobium vaginatum var. **durangense** (Hawksw. & Wiens) B.L. Turner, comb. nov.	**8**
Astragalus austrinus var. **trichocarpus** (Torrey & A. Gray) B.L. Turner, comb. nov.	**7**
Chamaesyce spurca (M.C. Johnst.) B.L. Turner, comb. nov.	**7**
Chamaesyce triligulata (Wheeler) B.L. Turner, comb. nov.	**7**
Cymopteris filifolius (Mathias, Constance & Theobald) B.L. Turner, comb. nov.	**6**
Cymopteris hallii (A. Gray) B.L. Turner, comb. nov.	**6**
Cymopteris longiradiatus (Mathias et al.) B.L. Turner, comb. nov.	**6**
Dalea arenicola (Wemple) B.L. Turner, comb. nov.	**7**
Dalea wrightii var **warnockii** (Tharp & Barkley) B.L. Turner, comb. nov.	**7**
Dyschoriste linearis var. **cinerascens** (Henrickson & Hilsenbeck) B.L. Turner, comb. nov.	**4**
Dyschoriste linearis var. **decumbens** (A. Gray) B.L. Turner, comb. nov.	**5**
Dyschoriste linearis var. **schiedeana** (Nees) B.L. Turner, comb. nov.	**5**
Gaillardia pulchella var. **drummondii** (Hook.) B.L. Turner, comb. nov.	**6**
Galium mexicanum var. **asperrimum** (A. Gray) B.L. Turner, comb. nov.	**8**
Galium mexicanum var. **flexicum** (Dempster) B.L. Turner, comb. nov.	**8**
Garrya ovata var. **goldmanii** (Woot. & Standl.) B.L. Turner, comb. nov.	**7**
Helianthus praecox var. **runyoni** (Heiser) B.L. Turner, comb. nov.	**6**
Helianthus rigidus var. **subrhomboideus** (Rydb.) B.L. Turner, comb. nov.	**6**
Ipomopsis longiflora var. **australis** (Fletcher & Wagner) B.L. Turner, comb. nov.	**8**
Ipomopsis longiflora var. **neomexicana** (Wilken) B.L. Turner, comb. nov.	**8**
Liatris punctata var. **mucronata** (DC.) B.L. Turner, comb. nov.	**6**
Oenothera pilosella var. **sessilis** (Pennell) B.L. Turner, comb. nov.	**7**
Osmorhiza mexicana Griseb. var. **bipatriata** (Constance & Shan) B.L. Turner, comb. nov.	**6**
Phoradendron villosum var. **coryae** (Trelease) B.L. Turner, comb. nov.	**8**
Portulaca umbraticola var. **lanceolata** (Engelm. in A. Gray) B.L. Turner, comb. nov.	**8**
Sambucus nigra var. **canadensis** (L.) B.L. Turner, comb. nov.	**6**
Sambucus nigra var. **cerulea** (Raf.) B.L. Turner, comb. nov.	**6**
Sideroxylon lanuginosum var. **oblongifolium** (Nutt.) B.L. Turner, comb. nov.	**8**
Sideroxylon lanuginosum var. **rigidum** (A. Gray) B.L. Turner, comb. nov.	**8**
Silphium integrifolium var. **asperrimum** (Hook.) B.L. Turner, comb. nov.	**6**
Silphium integrifolium var. **wrightii** (Perry) B.L. Turner, comb. nov.	**6**

Atlas of the Vascular Plants of Texas

Volume I

"...a man's reach should exceed
his grasp, Or what's a heaven for?"
—*Robert Browning*

Idle thoughts that occurred to the
senior author in the assembly
of these volumes.

Dedicated to all the
plant naturalists
who ever collected
in Texas!

ATLAS OF THE VASCULAR PLANTS OF TEXAS

B.L. Turner

Plant Resources Center
University of Texas
Austin, TX 78713, U.S.A.

Holly Nichols

Geomar
Christian-Albrechts-Universität
Kiel, Germany

Geoffrey Denny

Dept. of Horticultural Sciences
Texas A&M University
College Station, TX 77843, U.S.A.

Oded Doron

University of Texas
Austin, TX 78712, U.S.A.

ABSTRACT

The *Atlas of the Vascular Plants of Texas* provides county-level dot maps for about 6000 taxa of vascular plants native and naturalized in Texas. Some taxa reported for Texas may not be accounted for in the Atlas, when vouchered material of such collections has not been personally studied. Within each of four major categories (ferns, gymnosperms, monocots, and dicots), the taxa are arranged alphabetically. If a species is deemed to have well-defined infraspecific categories in Texas, such are mapped. Distribution records for the Atlas are based primarily on herbarium vouchers from herbaria at the University of Texas-Austin (TEX-LL) and Sul Ross State University (SRSC), where all identifications have been personally verified by the first author. Other records have been taken from published literature from various sources and web-posted records from the "Texas flora" site (Texas A&M). Maps for the family Cactaceae were largely accounted for by A.M. Powell (Sul Ross State University-Alpine). All other distribution records have been manually recorded by Turner. Manually prepared maps were translated into electronic format by the three junior authors. If two or more independent collections were available from a given county, multiple symbols are placed in that county, but only rarely more than five to ten per county, depending upon its size, providing a broad picture of the more heavily collected portions of the state as well as insight into the regions of the state where the species concerned is most likely to be observed. An introduction, index, and maps of Texas counties, floristic regions, and vegetational regions provide additional clarification and assistance in using the Atlas. Thirty-seven new nomenclatural combinations at specific and infraspecific rank are made in Acanthaceae, Apiaceae, Asteraceae, Brassicaceae, Caprifoliaceae, Euphorbiaceae, Fabaceae, Garryaceae, Nyctaginaceae, Onagraceae, Polemoniaceae, Portulacaceae, Rubiaceae, Sapotaceae, and Viscaceae.

RESUMEN

El *Atlas de las Plantas Vasculares de Texas* aporta mapas de puntos a nivel condado de unos 6000 taxa de plantas vasculares nativas y naturalizadas en Texas. Algunos taxa citados para Texas puede que no estén en el Atlas, si los materiales testigos de las recolecciones no han podido estudiarse personalmente. Dentro de las cuatro categorías mayores (helechos, gimnospermas, monocotiledóneas, y dicotiledóneas), los taxa se colocan ordenados alfabéticamente. Si se cree que una especie tiene definidas categorías infraspecíficas en Texas se cartografían. Las citas de distribución para el Atlas se basan en primer lugar en pliegos de los herbarios de la Universidad de Texas-Austin (TEX-LL) y Universidad Sul Ross State (SRSC), en los que todas las identificaciones han sido verificadas personalmente por el primer autor. Otras citas se han tomado de artículos publicados por varias fuentes y citas de la web del sitio "Texas flora" (Texas A&M). Los mapas de la familia Cactaceae han sido mayormente representados por A.M. Powell (Sul Ross State University-Alpine). Todas las demás distribuciones han sido realizadas manualmente por Turner. Los mapas preparados manualmente se traspasaron a formato electrónico por tres autores jóvenes. Si dos o más colecciones independientes estaban disponibles de un condado determinado, se colocaron símbolos múltiples en ese condado, pero rara vez más de cinco a diez por condado, dependiendo de su tamaño, ofreciendo una visión amplia de las partes del estado colectadas con más intensidad así como la inspección de las regiones del estado donde la especie implicada puede observarse más fácilmente. Una introducción, índice, y mapas de los condados de Texas, regiones florísticas, y regiones de vegetación clarifican algunos puntos y ayudan en el uso del Atlas. Se hacen treinta y siete nuevas combinaciones nomenclaturales a nivel específico e infraspecífico en Acanthaceae, Apiaceae, Asteraceae, Brassicaceae, Caprifoliaceae, Euphorbiaceae, Fabaceae, Garryaceae, Nyctaginaceae, Onagraceae, Polemoniaceae, Portulacaceae, Rubiaceae, Sapotaceae, y Viscaceae.

INTRODUCTION

B.L. Turner

Correll and Johnston (1970), in their massive *Manual of the Vascular Plants of Texas*, and Johnston (1988, 1990) provided an excellent overview of the state's vascular flora. Checklists by Hatch et al. (1990) and Jones et al. (1997) also have provided useful information and perspective. Approximately 5,479 taxa were recognized by Correll and Johnston, including 4,839 species, among which were recognized 576 infraspecific taxa, exclusive of formae and named natural hybrids. Altogether then, some 6,055 taxa were said to be persistent elements of the state's flora as of 1970. Relatively few of these were thought to be introduced or persistent weedy species (by rough estimate about 200 or less). Since that time numerous new aliens, mostly persistent weedy species, have been introduced, as might be deduced from the recent Flora of North Central Texas (Diggs et al. 1999). Such newly introduced aliens are likely to increase in diversity and geographic range as urban areas are expanded and new crop and grazing plants introduced, along with their baggage of weeds.

Regardless, published dot-maps of a detailed nature for these 6,000 or so Texas taxa (at least at the county level) are largely lacking, except for those published in selected monographs and/or specialized texts such as my *Legumes of Texas* (Turner 1959), the excellent series of distributional maps for the northern Panhandle of Texas provided by the *Atlas of the Flora of the Great Plains* (Great Plains Flora Association 1976), and the *Wild Orchids of Texas* (Liggio & Liggio 1999).

Impetus for the present Atlas arose largely out of the senior author's concern for an up-to-date, uniformly constructed, authoritative account of the distribution of the state's flora as shown by dot-records positioned among its 254 counties. I am well aware that several government agencies, beginning about 1996, have provided funds for databasing those Texas collections housed at the larger herbaria in the state. But progress on this project has been slow, and while what has been catalogued to date is available on the web, dot-maps showing distribution at the county level are not yet available, nor has there been any concerted effort to arrive at agreement as to accepted formal names and/or their rank (for the taxa concerned). Such databasing, while commendable, has proved a daunting undertaking since, in cataloguing herbarium specimens assembled over the years, one is faced with identification control: taxonomic evaluation of each such sheet entered into the database.

Plant identifications and sources of data

The senior author has been solely responsible for the taxonomic identification, determination of rank, and nomenclature of the taxa treated in the present text. The family Cactaceae was largely accounted for by A.M. Powell (Sul Ross State University, Alpine, Texas), drawing upon his wide knowledge of the family and especially his distributional maps for the state (Powell, in prep.). Maps of the Poaceae were prepared by S.L. Hatch (Texas A&M University, College Station, Texas), who draws from personal and institutional experience with this family. The junior authors have been responsible for the mapping programs, i.e., their computer skills permitted translation of my manually prepared dot-maps (each dot inked by hand, either as a solid dot, circle, or other symbols when infraspecific taxa are depicted) into electronic format. This collaborative enterprise has taken many thousands of hours over a five-year period (1997–2001). Most of the dots are based on specimens at LL, SRSC, and TEX—all of these were annotated by the senior author according to the identifications rendered. The number and positions of the dots within individual counties also represent his judgments. In short, the senior author has examined personally, touched, or "pored over" an estimated several hundred thousand sheets in the preparation of the present Atlas.

I have attempted to be consistent in my evaluation and recognition of specific and infraspecific categories. In this text I advocate a biological species concept, using the varietal category for meaningful infraspecific elements. It will be noted that the term variety is used exclusively for infraspecific categories, the term subspecies treated as a clustering or systematic device to accommodate the classificatory process (see Turner & Nesom [2000] for additional discussion of this matter). Briefly then, varietal taxa, as recognized herein, are believed to be meaningful taxonomic units having morpho-geographical integrity. Most such varieties usually abut each other in distribution (i.e., are largely allopatric) and show some degree of populational intergradation near regions of contact. In any case, positive detection of the latter processes requires much field work and subsequent analysis and need not concern us here. The senior author has inferred intergradation and varietal status based upon his evaluation of herbarium material and/or field-based experience with living populations. Infraspecific elements recognized by other workers which did not meet my standards of varietal recognition have not been accounted for in the present text.

In addition to herbarium collections and monographic listings, we have incorporated into our maps the dot-maps of others with whose taxonomy we agree. The greater part of these were extracted from the Atlas of the Great Plains (1977) and Liggio and Liggio (1999). We have also included the county notations given by Diggs et al. (1999) in their superlative Illustrated Flora of North Central Texas, Harcombe and Neaville (1977), and Nesom and Brown (1998), but only where their systematic views are concordant with ours. In addition, several individuals contributed county records for this or that taxon, most notably Bob O'Kennon for sites about Gillespie County and Michael and Barbara MacRoberts for sites in eastern Texas.

At least a few species, the senior author has "road-logged" sightings of well-marked species, placing a series of abutting dots along the major highways whenever the species being observed occurred at least every ten miles along the routes concerned (see maps for *Cercis occidentalis*, the red bud tree, and the various species of *Yucca*). To some, this may seem an unscientific enterprise, the dots not being vouchered, but I have no doubts as to the accuracy of my observations. Most such sighting exercises were performed along the highways between Austin and Alpine during the years 1997–2000, the senior author making a nearly weekly run between these two cities while on part-time retirement from the University of Texas at Austin.

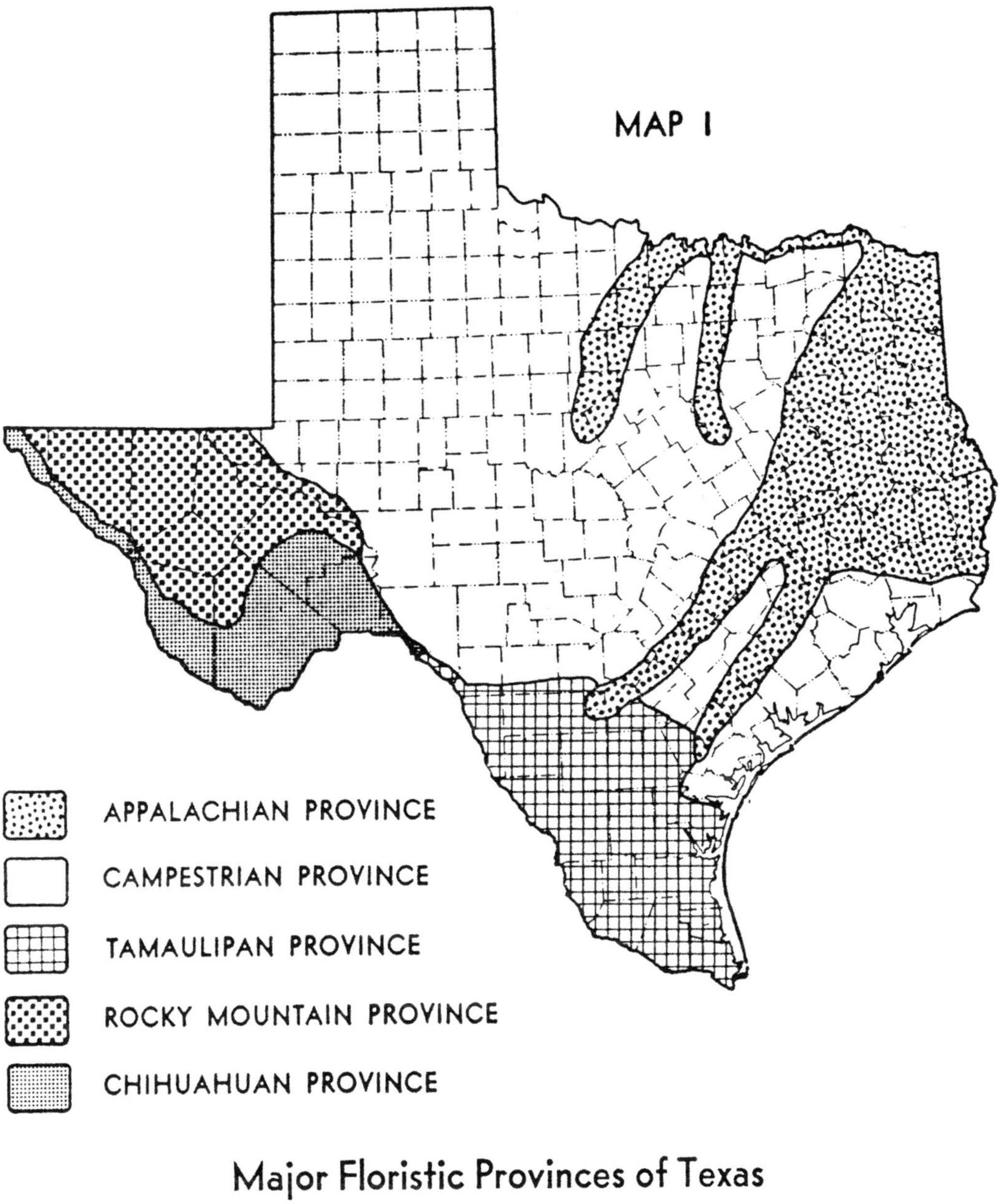
MAP I
APPALACHIAN PROVINCE
CAMPESTRIAN PROVINCE
TAMAULIPAN PROVINCE
ROCKY MOUNTAIN PROVINCE
CHIHUAHUAN PROVINCE
Major Floristic Provinces of Texas

FIG. 1. Floristic regions of Texas (from Turner 1959).

In the positioning of dots within counties, if only a single collection was noted I usually placed the dot in the center of the county concerned (except in the Trans-Pecos region where positions in the large counties concerned might be associated with mountain ranges, or the like). If two or more independent collections were gathered from a given county, multiple dots were placed in that county, but only rarely more than five to ten dots per county, depending upon its size. This is a helpful addition to the present Atlas, because the user is provided with a broad picture of the more heavily collected portions of the state as well as insight into the regions of the state where the species concerned is most likely to be observed. Heavily collected counties, as might be surmised, are those with major academic or botanical institutions housing herbaria, such as Brewster, Bryan, Dallas, El Paso, Lubbock, Tarrant, and Travis counties.

The senior author would not have attempted this tedious enterprise were he not reasonably familiar with most Texas plants and those of adjacent regions (including Mexico), this a result of 52 years of herbarium and field work, beginning in 1948 at Sul Ross State University in Alpine, Texas. Having cut my taxonomic teeth on the family Poaceae (as represented in the Trans-Pecos region), I gravitated to monographic work on the family Asteraceae, then to a treatment of the Legumes of Texas, these several families constituting about 30 percent or more of the state's vascular species. As might be expected, these several families possess formidable taxonomic problems, especially at the infraspecific levels. Fortunately, most of the remaining families have been well-studied by taxonomic gurus, especially trees, shrubs, sedges, and ferns. Additionally, the detailed citations given in the various treatments of genera given in the three-volume accounts of the Flora of Texas (Lundell et al. 1961–1969), along with the county citations provided by diverse generic monographers, have proven very useful in the present compilation.

Finally, it should be noted that a number of plant taxa reported for Texas by others may not be accounted for in this Atlas. This is because I have not personally seen vouchered material of such collections or else such names refer to cultivated plants that have not (as yet) escaped the confines of cultivated tracts.

Arrangement of text

The taxa accounted for in the present text are arranged alphabetically by family, by genus, by species. This precludes the need for an index, except as how one might wish to ascertain synonymies and author citations, these readily available in the several texts alluded to below. Especially useful in this connection are the texts of Jones et al. (1997) and Kartesz (1994). In the few instances where the names used in the present are absent from the aforementioned books, comments regarding such absences are given parenthetically below the scientific names preferred. Plants that appear to be adventive are listed by family just before the index on page 622. If a species is deemed to have well-defined infraspecific categories in Texas, such are mapped; if not, no such categories are mapped even though yet other workers might think such elements exist, as noted above.

Synonymy and author citations

We have not given full or even partial synonymies for the various taxa mapped in this text. Such can be obtained from the recent listings of Jones et al. (1997) and Kartesz (1994), if needed. In some cases we have included synonymous names or comments in brackets below the maps, most of these not given in the above-mentioned texts or where our taxonomic views differ with their compilations.

Floristic and vegetational regions

For those workers unfamiliar with the floristic and vegetational regions of Texas, two maps are provided herein (Fig. 1 and Fig. 2). The floristic regions (or Provinces) (Fig. 1) are those of Turner (1959), the vegetational regions (Fig. 2) from Hatch et al. (1990). Floristic regions are largely the result of climatic factors (mainly rainfall and temperature) working their effect over long periods of time. Vegetational regions are ecosystems which not only reflect the interplay of climate over time but also incorporate yet other variables, mainly elevational and edaphic factors (such as sandy or clay soils, etc.). Distinction between the floristic regions as perceived by Turner (1959) is becoming increasingly blurred, because of anthropogenic effects over the past century—mainly highway construction, which creates corridors for range expansions, especially along the interstate highway systems.

NEW VARIETAL AND SPECIES COMBINATIONS FOR TAXA OCCURRING IN TEXAS

B.L. Turner

In keeping with appropriate nomenclatural practice, especially at the infraspecific level (cf. Turner & Nesom 2000), the following new combinations are proposed for use in the present text.

ACANTHACEAE

I accept the view of Henrickson and Hilsenbeck (1999) that the *Dyschoriste linearis-D. schiedeana* complex constitutes a series of essentially allopatric intergrading populations; the earliest specific name for the complex is *D. linearis*. This perspective necessitates the following new combinations:

Dyschoriste linearis var. **cinerascens** (Henrickson & Hilsenbeck) B.L. Turner, comb. nov. BASIONYM: *Dyschoriste schiedeana* var. *cinerascens* Henrickson & Hilsenbeck, Lundellia 2:82. 1999.

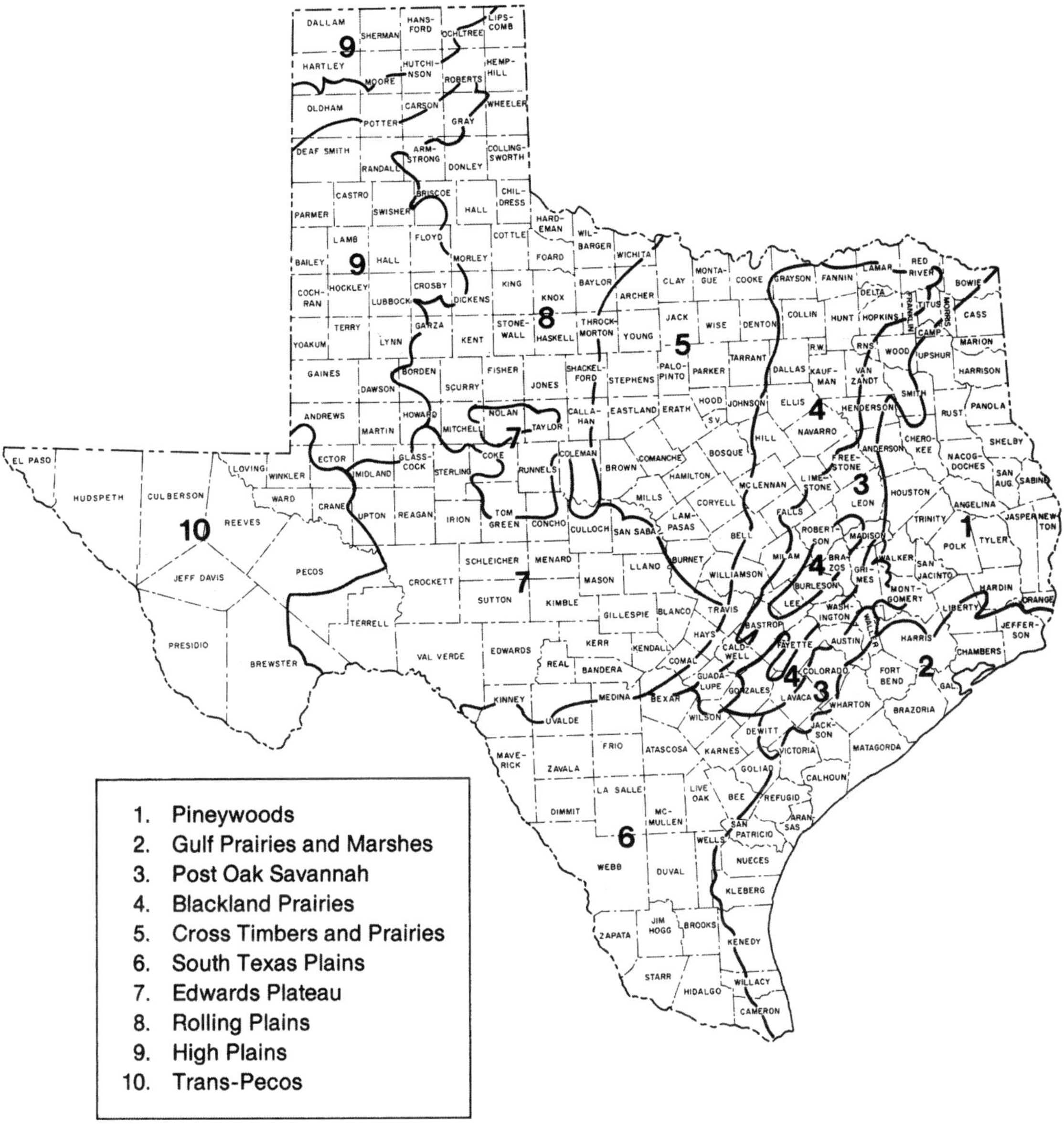
1. Pineywoods
2. Gulf Prairies and Marshes
3. Post Oak Savannah
4. Blackland Prairies
5. Cross Timbers and Prairies
6. South Texas Plains
7. Edwards Plateau
8. Rolling Plains
9. High Plains
10. Trans-Pecos
DALLAM
SHERMAN
HANS-FORD
OCHLTREE
LIPS-COMB
HARTLEY
MOORE
HUTCHI-NSON
ROBERTS
HEMP-HILL
OLDHAM
POTTER
CARSON
GRAY
WHEELER
DEAF SMITH
RANDALL
ARM-STRONG
DONLEY
COLLING-SWORTH
PARMER
CASTRO
SWISHER
BRISCOE
HALL
CHIL-DRESS
BAILEY
LAMB
HALE
FLOYD
MOTLEY
COTTLE
HARD-EMAN
WIL-BARGER
FOARD
WICHITA
COCH-RAN
HOCKLEY
LUBBOCK
CROSBY
DICKENS
KING
KNOX
BAYLOR
ARCHER
CLAY
MONTA-GUE
COOKE
GRAYSON
FANNIN
LAMAR
RED RIVER
BOWIE
YOAKUM
TERRY
LYNN
GARZA
KENT
STONE-WALL
HASKELL
THROCK-MORTON
YOUNG
JACK
WISE
DENTON
COLLIN
HUNT
DELTA
HOPKINS
FRANKLIN
TITUS
MORRIS
CASS
CAMP
MARION
GAINES
DAWSON
BORDEN
SCURRY
FISHER
JONES
SHACKEL-FORD
STEPHENS
PALO-PINTO
PARKER
TARRANT
DALLAS
R.W.
KAUF-MAN
VAN ZANDT
RNS
WOOD
UPSHUR
HARRISON
SMITH
ANDREWS
MARTIN
HOWARD
MITCHELL
NOLAN
TAYLOR
CALLA-HAN
EASTLAND
ERATH
HOOD
JOHNSON
ELLIS
NAVARRO
HENDERSON
RUSK
PANOLA
EL PASO
HUDSPETH
CULBERSON
LOVING
WINKLER
ECTOR
MIDLAND
GLASS-COCK
STERLING
COKE
RUNNELS
COLEMAN
BROWN
COMANCHE
HAMILTON
BOSQUE
HILL
ANDERSON
CHERO-KEE
NACOG-DOCHES
SHELBY
SAN AUG.
SABINE
WARD
CRANE
UPTON
REAGAN
IRION
TOM GREEN
CONCHO
MCCULLOCH
SAN SABA
MILLS
CORYELL
MC LENNAN
LIME-STONE
FREE-STONE
LEON
HOUSTON
ANGELINA
TRINITY
JASPER
NEW-TON
TYLER
REEVES
LAM-PASAS
BELL
FALLS
ROBERT-SON
MADISON
POLK
JEFF DAVIS
PECOS
CROCKETT
SCHLEICHER
MENARD
MASON
LLANO
BURNET
MILAM
WILLIAMSON
BRA-ZOS
GRI-MES
WALKER
SAN JACINTO
MONT-GOMERY
HARDIN
ORANGE
LIBERTY
PRESIDIO
TERRELL
SUTTON
KIMBLE
GILLESPIE
BLANCO
TRAVIS
BURLESON
LEE
WASH-INGTON
WALLER
BASTROP
JEFFER-SON
BREWSTER
VAL VERDE
EDWARDS
KERR
KENDALL
HAYS
CALD-WELL
FAYETTE
AUSTIN
HARRIS
CHAMBERS
REAL
BANDERA
COMAL
GUADA-LUPE
GONZALES
COLORADO
FORT BEND
GAL.
KINNEY
UVALDE
MEDINA
BEXAR
WILSON
LAVACA
WHARTON
BRAZORIA
DEWITT
JACK-SON
MAVE-RICK
ZAVALA
FRIO
ATASCOSA
KARNES
VICTORIA
MATAGORDA
GOLIAD
CALHOUN
DIMMIT
LA SALLE
LIVE OAK
BEE
REFUGIO
MC-MULLEN
SAN PATRICIO
ARAN-SAS
WEBB
DUVAL
JIM WELLS
NUECES
KLEBERG
ZAPATA
JIM HOGG
BROOKS
KENEDY
STARR
HIDALGO
WILLACY
CAMERON

FIG. 2. Vegetational regions of Texas (from Hatch et al. 1990).

Dyschoriste linearis var. **decumbens** (A. Gray) B.L. Turner, comb. nov. BASIONYM: *Calophanes decumbens* A. Gray, Syn. Fl. N. Amer. ed. 1, 2(1):325. 1875.

Dyschoriste linearis var. **schiedeana** (Nees) B.L. Turner, comb. nov. BASIONYM: *Calophanes schiedeanus* Nees in DC., Prodr. 11:111. 1847.

My concept of this highly variable taxon includes material of *Dyschoriste schiedeana* var. *prostrata* Henrickson. As noted by its author, the latter taxon "appears to be a transitional taxon between the larger, more uniformly vestitured var. *decumbens* to the north and the equally small, often equally vestitured var. *schiedeana* to the east." In short, it presumably represents a series of populational intergrades between these two regional infraspecific taxa, those to the east more reasonably positioned within var. *schiedeana*; those to the west and north to var. *decumbens*.

Finally, I can't help but note that Henrickson's concept of *Dyschoriste linearis* var. *sanpatriciensis* Henrickson (the type, itself, from the Gulf Coastal sands of San Patricio County, Texas), is highly abiological in that the five scattered specimens cited by him appear to be but sporatic pubescent forms of a widespread variable *D. l.* var. *linearis*, the sheets concerned occurring sympatrically with the latter, as mapped by Henrickson. Indeed, one of these (*McVaugh 10558*, NY, SRSC), from Val Verde County was annotated by me as an intermediate between var. *linearis* and var. *cinerascens*.

APIACEAE

Cymopteris filifolius (Mathias, Constance & Theobald) B.L. Turner, comb. nov. BASIONYM: *Aletes filifolius* Mathias, Constance & Theobald, Madrono 20:214. 1869.

Cymopteris hallii (A. Gray) B.L. Turner, comb. nov. BASIONYM: *Seseli hallii* A. Gray, Proc. Amer. Acad. Arts 8:288. 1870.

= *Aletes acaulis* (Torrey) Coulter & Rose, N. Amer. Umbell. 26. 1888. not *Cymopteris acaulis* (Pursh) Raf.

Cymopteris longiradiatus (Mathias et al.) B.L. Turner, comb. nov. BASIONYM: *Pseudocymopteris longiradiatus* Mathias, Constance & Theobald, Madrono 20:217. 1965.

Osmorhiza mexicana Griseb. var. **bipatriata** (Constance & Shan) B.L. Turner, comb. nov. BASIONYM: *Osmorhiza bipatriata* Constance & Shan, Univ. Calif. Publ. Bot. 23:121. 1948. According to annotations (herbarium SRSC 1981), P.P. Lowry anticipates treatment of this taxon as a subspecies, this more appropriately treated as a variety.

ASTERACEAE

Antennaria parlinii var. **fallax** (Greene) B.L. Turner, comb. nov. BASIONYM: *Antennaria fallax* Greene, Pittonia 3:321. 1898.

Gaillardia pulchella var. **drummondii** (Hook.) B.L. Turner, comb. nov. BASIONYM: *Gaillardia bicolor* var. *drummondii* Hook., Curtis Bot. Mag. 61:pl. 3368. 1834.

This early name must replace that of *Gaillardia pulchella* var. *picta* (Sweet) A. Gray, as noted in our upcoming monograph of the genus *Gaillardia* (Turner & Watson, in prep.).

Helianthus rigidus var. **subrhomboideus** (Rydb.) B.L. Turner, comb. nov. BASIONYM: *Helianthus subrhomboideus* Rydb., Mem. New York Bot. Gard. 1:419. 1900.

Helianthus praecox var. **runyoni** (Heiser) B.L. Turner, comb. nov. BASIONYM: *Helianthus praecox* subsp. *runyoni* Heiser, Madrono 13:161. 1956.

Heiser (1956) treated these two taxa of *Helianthus* as subspecies, hence the varietal combinations provided here (Turner & Nesom 2001).

Liatris punctata var. **mucronata** (DC.) B.L. Turner, comb. nov. BASIONYM: *Liatris mucronata* DC., Prodr. 5:129. 1836.

This combination was informally proposed in my treatment of the tribe Eupatorieae for Mexico (Turner 1997); the var. *mucronata* is mostly allopatric with the typical variety and intergrades with it, at least in their region of contact in central Texas.

Silphium integrifolium var. **asperrimum** (Hook.) B.L. Turner, comb. nov. BASIONYM: *Silphium asperrimum* Hook., Comp. Bot. Mag. 1:99. 1835.

Silphium integrifolium var. **wrightii** (Perry) B.L. Turner, comb. nov. BASIONYM: *Silphium simpsonii* var. *wrightii* Perry, Rhodora 39:288. 1937.

The above two taxa are allopatric and intergrade near regions of contact, the latter occurring in sandy soils of eastern Texas, this typified (GH!) by a Wright collection from that area made in 1848, the former occurring in calcareous soils of central Texas.

CAPRIFOLIACEAE

Sambucus nigra var. **cerulea** (Raf.) B.L. Turner, comb. nov. BASIONYM: *Sambucus cerulea* Raf., Alsogr. Amer. 48. 1838.

Bolli (1994) treated this taxon as the sole member of his *Sambucus nigra* subsp. *cerulea* (Raf.) Bolli, noting its wide range and intergradation into the more northern entity.

Sambucus nigra var. **canadensis** (L.) B.L. Turner, comb. nov. BASIONYM: *S. canadensis* L., Sp. Pl. 269. 1753.

Such intergradation is not obvious in Texas or the western U.S.A., but Bolli contends that the two taxa intergrade southwards into Mexico. He also noted the misapplication of the name, *Sambucus mexicana* Presl, long accepted in western Texas as the correct name of *S. nigra* var. *cerulea*.

EUPHORBIACEAE

Chamaesyce triligulata (Wheeler) B.L. Turner, comb. nov. BASIONYM: *Euphorbia fendleri* var. *triligulata* Wheeler, Bull. Torrey Bot. Club 45:5. 1936.

Johnston (1975) treated this taxon as a variety of *Euphorbia chaetocalyx* but I take this to be a clean species, there being no intergrades between it and and either *C. fendleri* (Torrey & A. Gray) Small or *C. chaetocalyx* (Boiss.) Woot. & Standl.

Chamaesyce spurca (M.C. Johnst.) B.L. Turner, comb. nov. BASIONYM: *Euphorbia theriaca* var. *spurca* M.C. Johnst., Wrightia 5:138. 1975.

As in the above, I also find this taxon to be relatively clean, showing little if any intergradation with *Chamaesyce therica* Wheeler.

FABACEAE

Astragalus austrinus var. **trichocarpus** (Torrey & A. Gray) B.L. Turner, comb. nov. BASIONYM: *Astragalus nuttallianus* var. *trichocarpus* Torrey & A. Gray, Fl. N. Amer. 1:333. 1838.

Isely (1998) included this taxon as a variety of *Astragalus nuttallianus* DC. but noted that it blended with *A. austrinus* Small, the latter also included within his broad concept of *A. nuttallianus* as a variety.

Dalea arenicola (Wemple) B.L. Turner, comb. nov. BASIONYM: *Petalostemon arenicola* Wemple, Iowa State J. Sci. 45:94. 1970.

I agree with Wemple's treatment of this taxon at the specific level, although Isely (1998), treated this at the varietal level under *Dalea purpurea* Vent., as did Barneby before him.

Dalea wrightii var **warnockii** (Tharp & Barkley) B.L. Turner, comb. nov. BASIONYM: Based upon *Parosela warnockii* Tharp & Barkley, Anal. Esc. Nac. Ciences Biol. 4:284. 1946.

GARRYACEAE

Garrya ovata var. **goldmanii** (Woot. & Standl.) B.L. Turner, comb. nov. BASIONYM: *Garrya goldmanii* Woot. & Standl., Contr. U.S. Natl. Herb. 16:157. 1913.

Dahling (1978) treated this taxon as a subspecies of *Garrya ovata*, neglecting to give it varietal rank.

ONAGRACEAE

Oenothera pilosella var. **sessilis** (Pennell) B.L. Turner, comb. nov. BASIONYM: *Kneiffia sessilis* Pennell, Bull. Torrey Bot. Club 46:366. 1919.

Straley (1978) treated this taxon as a subsp. of *Oenothera pilosella*, neglecting to give it varietal rank.

NYCTAGINACEAE

Levin (2002) proposed a number of new combination in the Nyctaginaceae, this a result of DNA studies on the family. In short, her data suggest that the monotypic genus *Ammocodon* and the small genus *Selinocarpus* are best positioned, on phylogenetic grounds, in an expanded *Acleisanthes*. I have accepted this arrangement but hasten to note that I cannot agree with some of her proposed synonymies and create the following new combinations to accommodate taxa that I would recognize.

Acleisanthes lanceolata var. **maloneana** (B.L. Turner) B.L. Turner, comb. nov. BASIONYM: *Selinocarpus maloneanus* B.L. Turner, Phytologia 75:239. 1953.

Additional collections of this taxon garnered after its original description clearly show that it intergrades into *A. lanceolata*, hence the combination proposed here.

Acleisanthes megaphylla (Fowler & B.L. Turner) B.L. Turner, comb. nov. BASIONYM: *Selinocarpus lanceolatus* var. *megaphyllus* Fowler & B.L. Turner, Phytologia 37:184. 1977.

Turner (1993) elevated this taxon to specific rank, while Levin (2002) treated it as but a leaf variant of *Acleisanthes lanceolata*. Field and herbarium studies strongly suggest that the populations concerned are meaningful biological units that are deserving of some form of nomenclatural recognition, be this varietal or specific. I prefer the latter based upon current observations.

Acleisanthes nevadensis (Standley) B.L. Turner, comb. nov. BASIONYM: *Selinocarpus diffusus* A. Gray subsp. *nevadensis* Standl., Contr. U.S. Natl. Herb. 12:388. 1909.

Levin (2002) treated this geographically isolated taxon as but a synonym of her broad concept of *Acleisanthes diffusa*. Because it is quite isolated from *A. diffusa* proper and does not seem to intergrade with populations of the latter, I recognize it as specifically distinct.

Acleisanthes purpusiana var. **marshii** (I.M. Johnst.) B.L. Turner, comb. nov. Basionym: *Selinocarpus marshii* I.M. Johnst., J. Arnold Arbor. 25:162. 1944.

Levin (2002) did not account for this well-marked Mexican taxon in her overview of *Acleisanthes*, hence the combination here.

POLEMONIACEAE

Wilken (2001) has provided an excellent overview of the *Ipomopsis longiflora* complex, in which three subspecific categories were recognized. As already noted (Turner & Nesom 2000), I would treat these essentially allopatric taxa (presumably to some extent intergrading) as varieties, the nomenclature for which follows: *Ipomopsis longiflora* (Torrey) V. Grant var. *longiflora*.

Ipomopsis longiflora var. **neomexicana** (Wilken) B.L. Turner, comb. nov. Basionym: *I. longiflora* subsp. *neomexicana* Wilken, Madrono 48:116. 2001.

Ipomopsis longiflora var. **australis** (Fletcher & Wagner) B.L. Turner, comb. nov. Basionym: *I. longiflora* subsp. *australis* Fletcher & Wagner, Madrono 31:20. 1984.

PORTULACACEAE

Portulaca umbraticola var. **lanceolata** (Engelm. in A. Gray) B.L. Turner, comb. nov. Basionym: *Portulaca lanceolata* Engelm. in A. Gray, Boston J. Nat. Hist. 6:154. 1850. = *P. umbraticola* subsp. *lanceolata* (Engelm. in A. Gray) Matthews & Ketron, Castanea 56:305. 1991.

RUBIACEAE

Galium mexicanum var. **flexicum** (Dempster) B.L. Turner, comb. nov. Basionym: *Galium mexicanum* subsp. *flexicum* Dempster, Madrono 23:384. 1976.

Galium mexicanum var. **asperrimum** (A. Gray) B.L. Turner, comb. nov. Basionym: *Galium asperrimum* A. Gray, Pl. Fendl. 60. 1849.

Dempster (1976) has provided an excellent account of the *Galium mexicanum* complex, both of the above new combinations applying to Trans-Pecos taxa.

SAPOTACEAE

Sideroxylon lanuginosum var. **oblongifolium** (Nutt.) B.L. Turner, comb. nov. Basionym: *Bumelia oblongifolia* Nutt., Gen. 1:135. 1818.

Sideroxylon lanuginosum var. **rigidum** (A. Gray) B.L. Turner, comb. nov. Basionym: *Bumelia lanuginosa* var. *rigida* A. Gray, Syn. Fl. N. Amer., ed. 2, 2(1):68. 1886.

Cronquist (1945) accepted both of the above varieties as worthy of recognition (albeit within *Bumelia lanuginosa*), as do I. The two taxa intergrade in the western Edwards Plateau of central Texas.

VISCACEAE

Arceuthobium vaginatum var. **cryptopodum** (Engelm. in A. Gray) B.L. Turner, comb. nov. Basionym: *Arceuthobium cryptopodum* Engelm. in A. Gray, Boston J. Nat. Hist. 6:214. 1850.

Arceuthobium vaginatum var. **durangense** (Hawksw. & Wiens) B.L. Turner, comb. nov. Basionym: *Arceuthobium vaginatum* subsp. *durangense* Hawksw. & Wiens, Brittonia 17:230. 1965.

Hawksworth and Wiens (1965) treated both of the above taxa as subspecies, neglecting to accord them varietal rank.

Phoradendron villosum var. **coryae** (Trelease) B.L. Turner, comb. nov. Basionym: *Phoradendron coryae* Trelease, Gen. Phorad. 43:1916.

TABULATION OF FLORA

One of the stimulants of the present study was to provide a base line with which to evaluate the introduction and spread of the non native elements, and at the same time to show the distribution of our native elements as compiled for the year 2000.

Based upon the present tabulation there are about 5,100 species of vascular plants in Texas. Of these, ca. 10 % are believed to be introduced species, most of which have become established (i.e., reproducing or persisting without cultivation). This suggests a native vascular flora of ca. 4,580 species. Correll and Johnston (1970, p.15) reckoned the vascular flora of Texas to total 4,839 species, including established aliens or introductions. The latter figure compares favorably with the present tabulation, the discrepancy (4,839 vs 5,100) no doubt the result of numerous introductions since their study. Indeed, I estimate that ca. 300 new introductants have become established in Texas since the Correll and Johnston study, this over a period of about thirty years. One can only imagine the plethera of new introductions to be expected in the near future. Almost certainly many of the introduced taxa will replace or reduce the populations of our native taxa, such as occurred over much of Eurasia and elsewhere.

The present accounting would not have been possible without the collective efforts of numerous individuals, both professional and amateur, over a 160 year period, but mainly over the past 80 years, following the development of institutional herbaria in the state. We owe a debt of gratitude to those dedicated field workers who tramped the roadsides and out backs collecting natives and aliens alike, always with a sense of joy (I like to think) pleased that something "new" or interesting had come under their scrutiny.

ACKNOWLEDGMENTS

We would like to thank Guy Nesom (BRIT) and Michael and Barbara MacRoberts (LSUS) for their thorough reviews that definitely improved this Atlas; Guy Nesom also helped with the adventive list and index. Robert O'Kennon (BRIT) and George Diggs (Austin College, BRIT) also offered valuable comments. Tom Wendt (TEX) and Rob Maushardt (BRIT) provided important technical and computer support.

REFERENCES

Bolli, R. 1994. Revision of the genus *Sambucus*. Diss. Bot. 22:1–223.

Correll, D.S. and M.C. Johnston. 1970. Manual of the vascular plants of Texas. Texas Research Foundation, Renner.

Cronquist, A. 1945. Studies in the Sapotaceae, III. *Dipholis* and *Bumelia*. J. Arnold Arbor. 26:435–471.

Dahling, G.V. 1978. Systematics of *Garrya*. Contr. Gray Herb. 209:1–104.

Dempster, L.T. 1976. *Galium mexicanum* (Rubiaceae) of Central America and western North America. Madrono 23:378–386.

Diggs, G.M., Jr., B.L. Lipscomb, and R.J. O'Kennon. 1999. Shinners & Mahler's Illustrated Flora of North Central Texas. Sida, Bot. Misc. No. 16.

Great Plains Flora Association (R.L. McGregor, coordinator; T.M. Barkley, editor). 1976. Atlas of the Flora of the Great Plains. Iowa State University Press, Ames.

Harcombe, P.A. and J.E. Neaville. 1977. Vegetation types of Chambers County, Texas. Texas J. Sci. 29:209–234.

Hatch, S.L., Gandhi, K.N., and L.E. Brown. 1990. Checklist of the vascular plants of Texas. Miscell. Publ. 1655, Texas Agric. Exp. Station, Texas A&M Univ. System, College Station.

Hawksworth, F.G. and D. Wiens. 1965. *Arceuthobium* in Mexico. Brittonia 17:213–238.

Henrickson, J. 1999. Studies in the genus *Dyschoriste* (Acanthaceae): 1. Plants of northern Mexico, Texas to Arizona. Lundellia 2:72–99.

Isely, D. 1998. Native and naturalized Leguminosae (Fabaceae) of the United States. Brigham Young Univ., Provo, Utah.

Johnston, M.C. 1975. Studies of the *Euphorbia* species of the Chihuahuan Desert Region and adjacent areas. Wrightia 5:120–143.

Johnston, M.C. 1988. The vascular plants of Texas: A list, updating the manual of the vascular Plants of Texas (ed. 2, 1990). Published by the author, Austin, Texas.

Johnston, M.C. 1990. The vascular plants of Texas: A list, updating the manual of the vascular Plants of Texas (ed. 2). Published by the author, Austin, Texas.

Jones, S.D., J.K. Wipff, and P.M. Montgomery. 1997. Vascular plants of Texas. Univ. of Texas Press, Austin.

Kartesz, J.T. 1994. A synonymized checklist of the vascular flora of the United States, Canada, and Greenland (ed. 2). Vol. 1–Checklist. Timber Press, Portland, Oregon.

Levin, R.A. 2002. Taxonomic status of *Acleisanthes*, *Selinocarpus*, and *Ammocodon* (Nyctaginaceae). Novon 12:58–63.

Liggio, J. and A. O. Liggio. 1999. Wild orchids of Texas. Univ. of Texas Press, Austin.

Lundell, C.L. et al. 1961–1969. Flora of Texas. Volumes 1–3. Texas Research Foundation, George Bantu Co., Inc. Menasha, Wisconsin.

Nesom, G.L. and L.E. Brown. 1998. An annotated checklist of the vascular plants of Walker, Montgomery, and San Jacinto counties, Texas. Phytologia 84:107–153.

Straley, G.B. 1977. Systematics of *Oenothera* sect. *Kneiffia* (Onagraceae). Ann. Missouri Bot. Gard. 64:381–424.

Turner, B.L. 1959. The legumes of Texas. University of Texas Press, Austin.

Turner, B.L. 1993. New species and combinations in *Selinocarpus* (Nyctaginaceae). Phytologia 75:239–242.

Turner, B.L. 1997. *Liatris*, in Comps of Mexico 1. Phytologia Memoirs 11:152.

Turner, B.L. and G.L. Nesom. 2000. Use of variety and subspecies and new varietal combinations for *Styrax platanifolius* (Styracaceae). Sida 19:257–262.

Wilken, D. 2001. A new *Ipomopsis* (Polemoniaceae) from the southeast USA and adjacent Mexico. Madrono 48:116–122.

Anisacanthus linearis

Anisacanthus puberulus

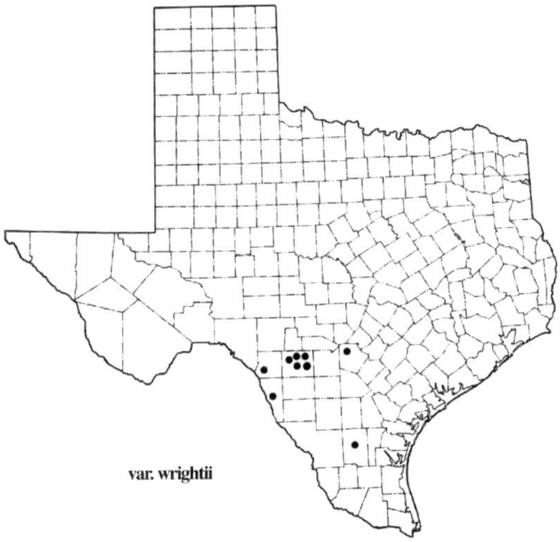
var. wrightii

Anisacanthus quadrifidus

Carlowrightia arizonica

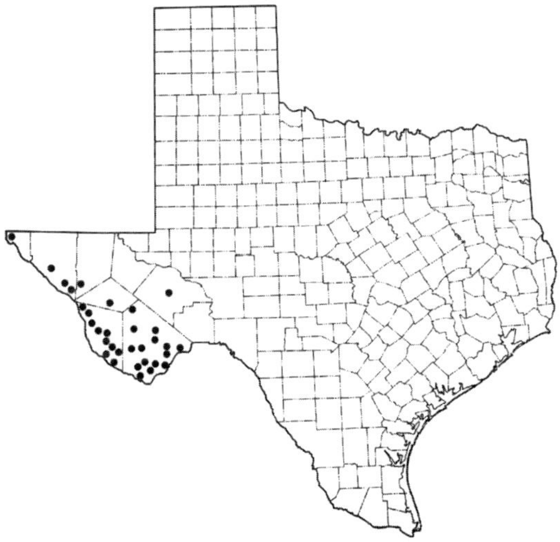

Carlowrightia linearifolia

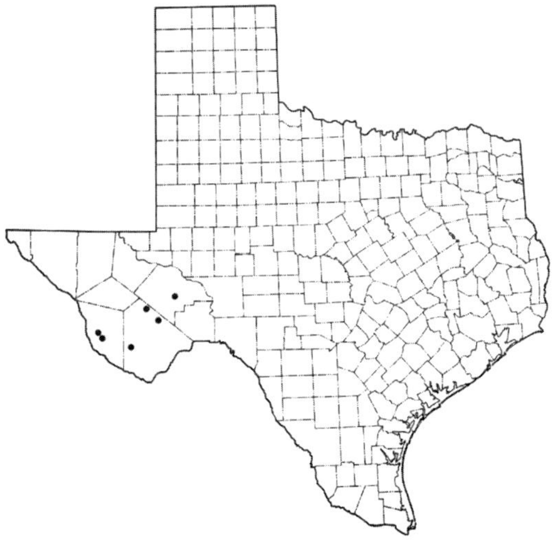

Carlowrightia mexicana

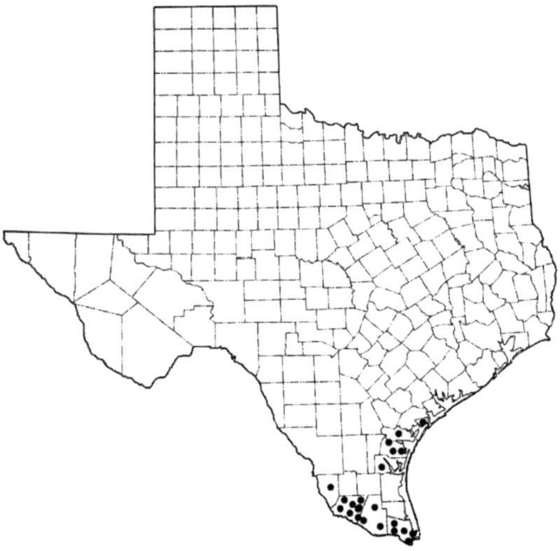

Carlowrightia parviflora

Carlowrightia parvifolia

Carlowrightia serpyllifolia

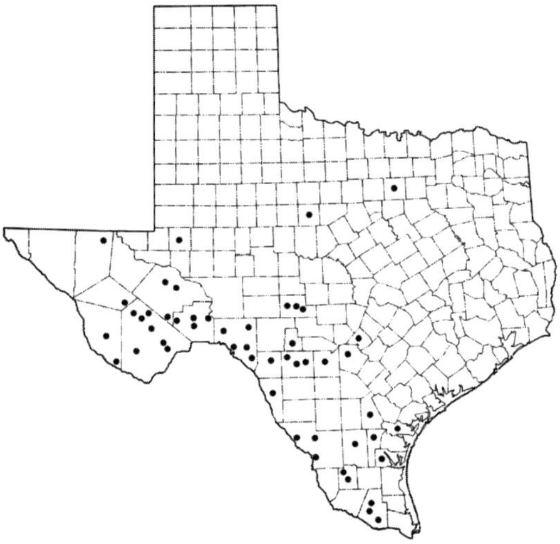

Carlowrightia texana

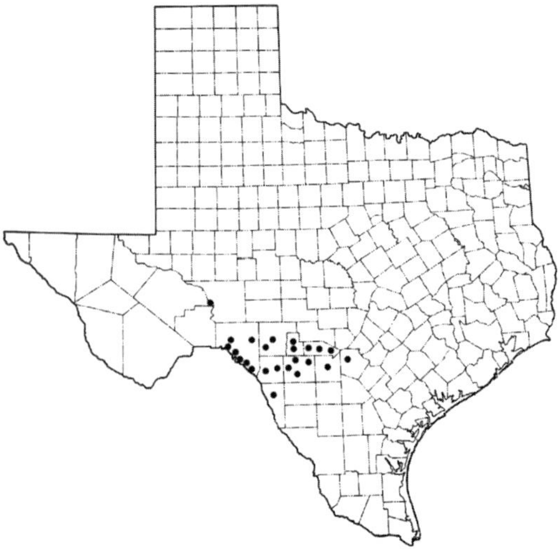

Carlowrightia torreyana

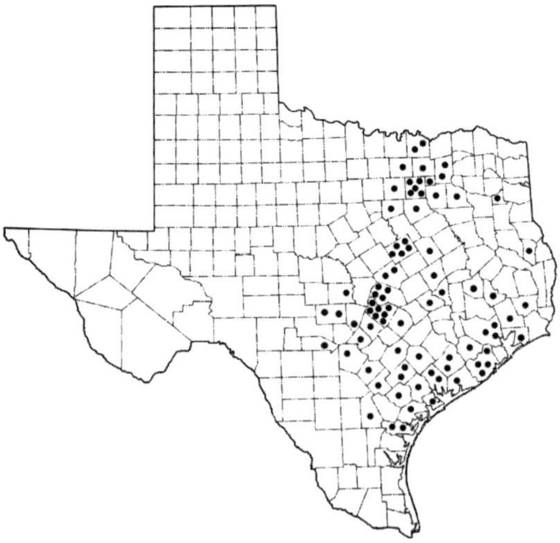

Dicliptera brachiata

Dicliptera sexangularis

Dyschoriste crenulata

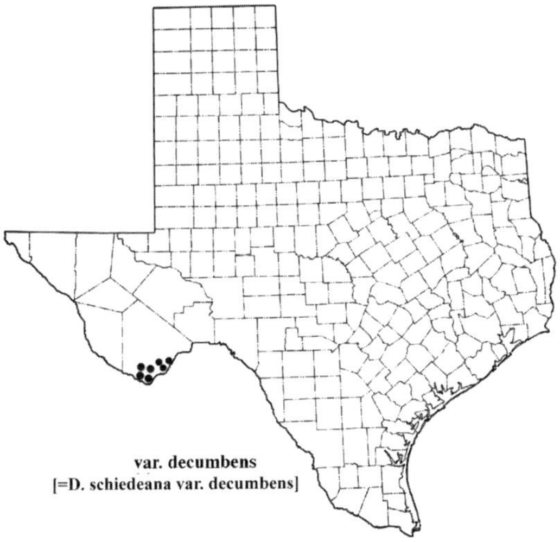
var. decumbens
[=D. schiedeana var. decumbens]

Dyschoriste decumbens

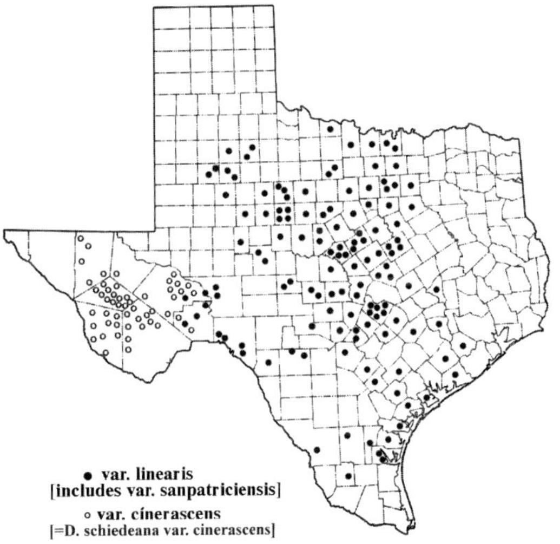
• var. linearis
[includes var. sanpatriciensis]
○ var. cinerascens
[=D. schiedeana var. cinerascens]

Dyschoriste linearis

Elytraria bromoides

Elytraria imbricata

Hygrophila lacustris

Hygrophila polysperma

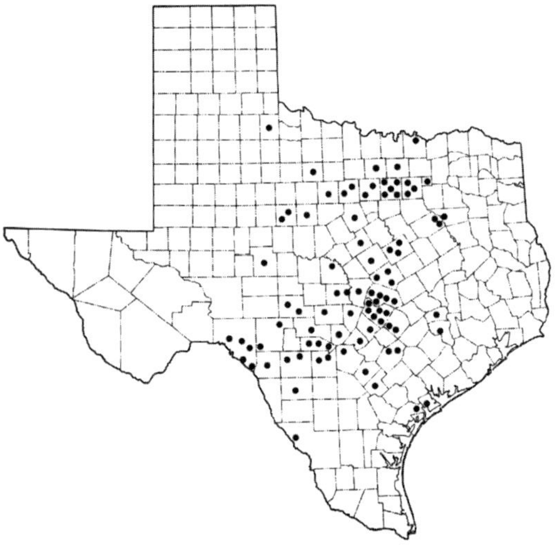

Justicia americana

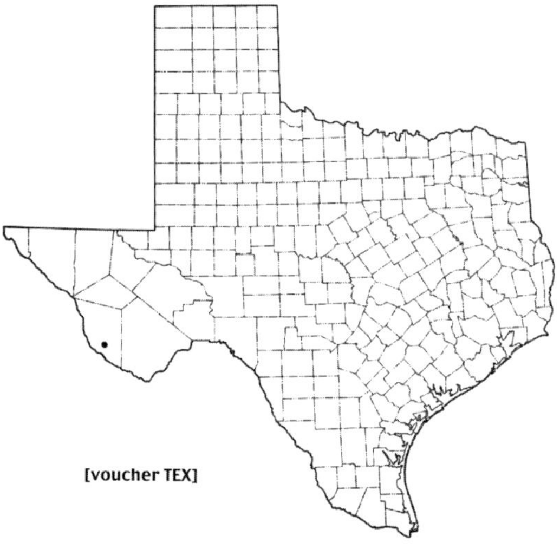
[voucher TEX]

Justicia longii

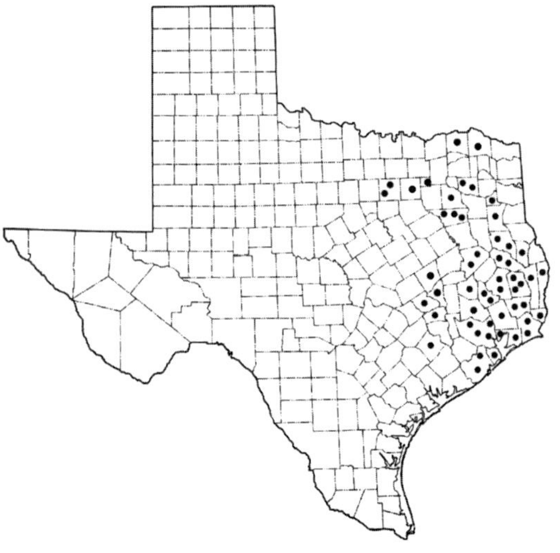

Justicia ovata

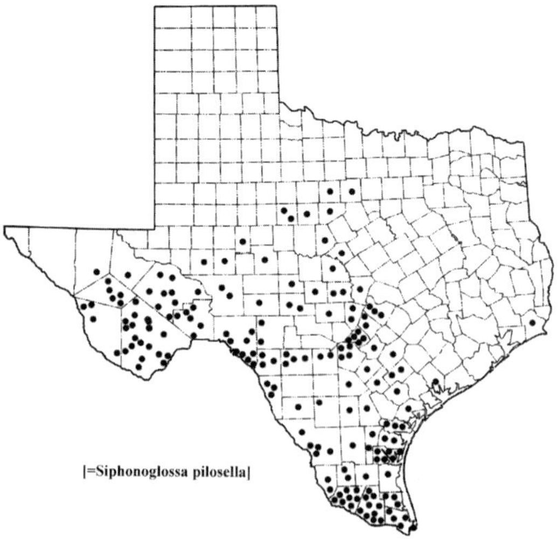
[=Siphonoglossa pilosella]

Justicia pilosella

Justicia runyonii

[= J. greggii of authors]

Justicia turneri

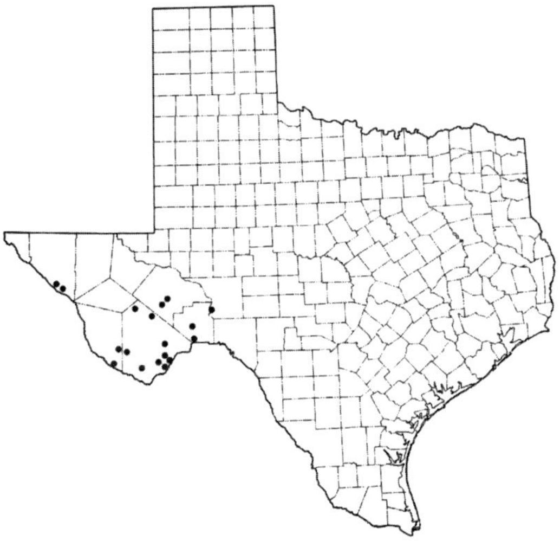

Justicia warnockii

Justicia wrightii

Nomaphila stricta

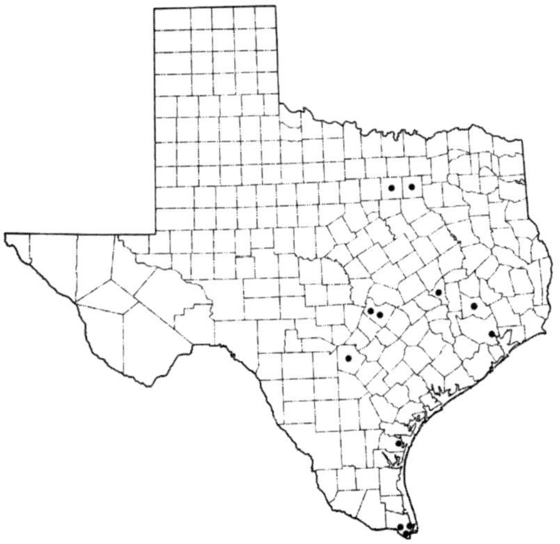

Ruellia brittoniana

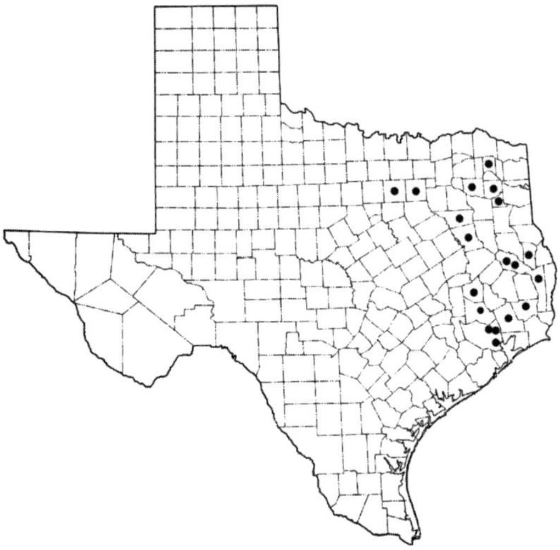

Ruellia caroliniensis

Ruellia corzoi

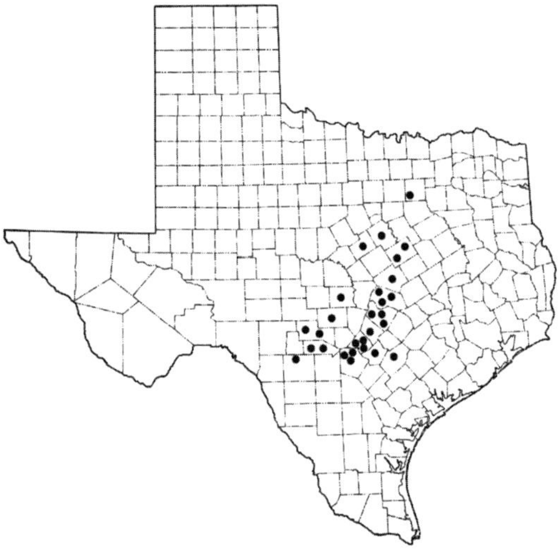

Ruellia drummondiana

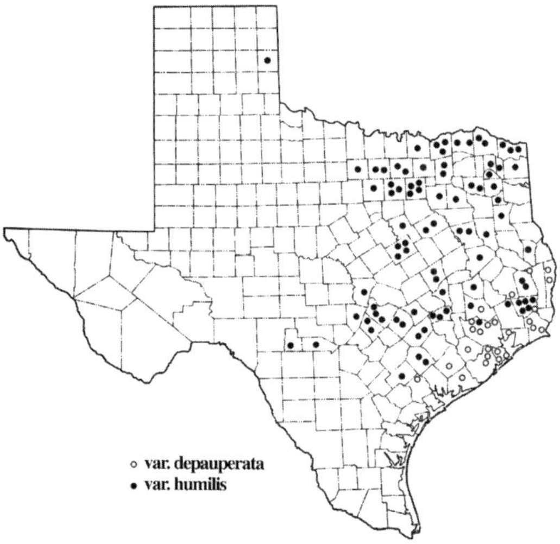
var. depauperata
var. humilis

Ruellia humilis

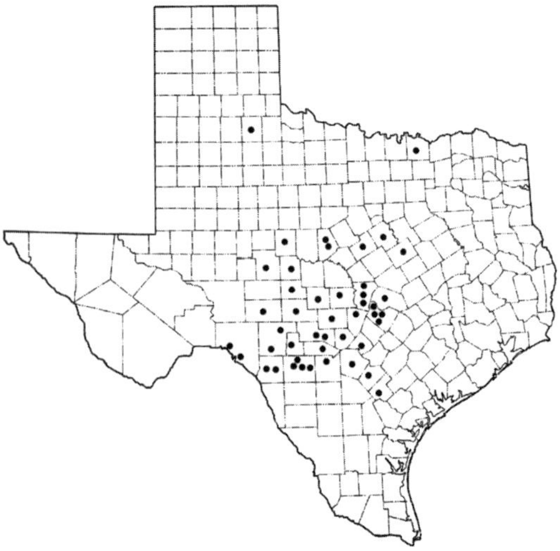

Ruellia metziae

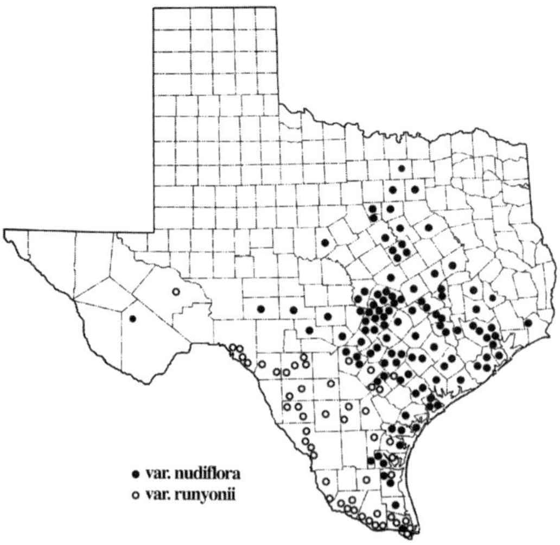
var. nudiflora
var. runyonii

Ruellia nudiflora

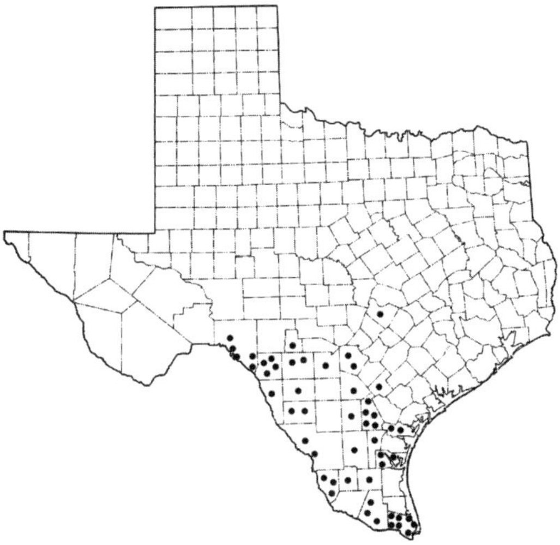

Ruellia occidentalis

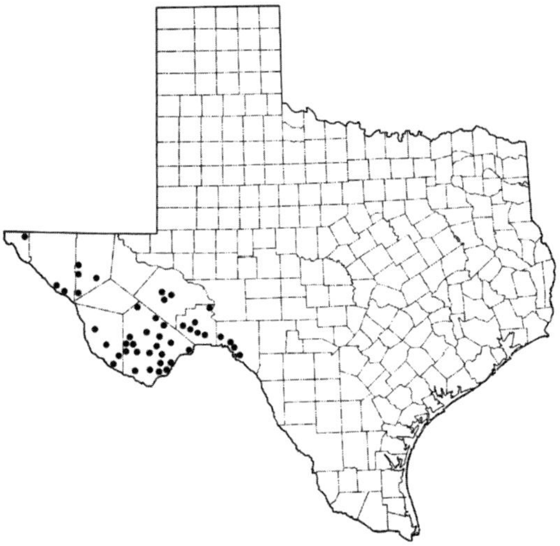

Ruellia parryi

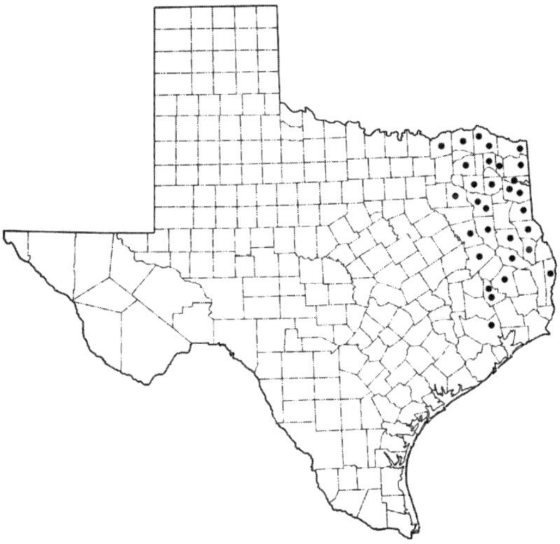

Ruellia pedunculata

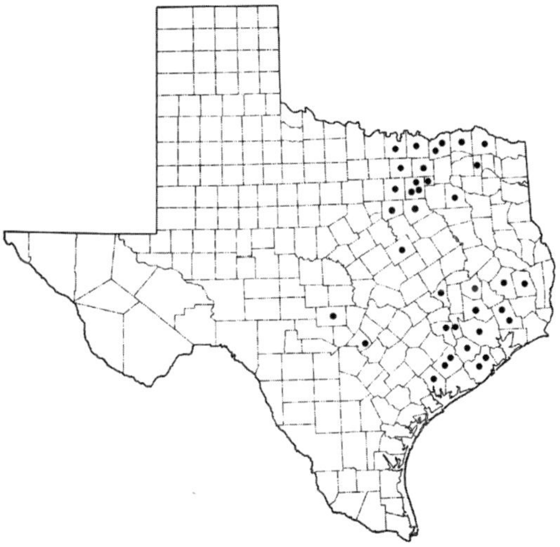

Ruellia strepens

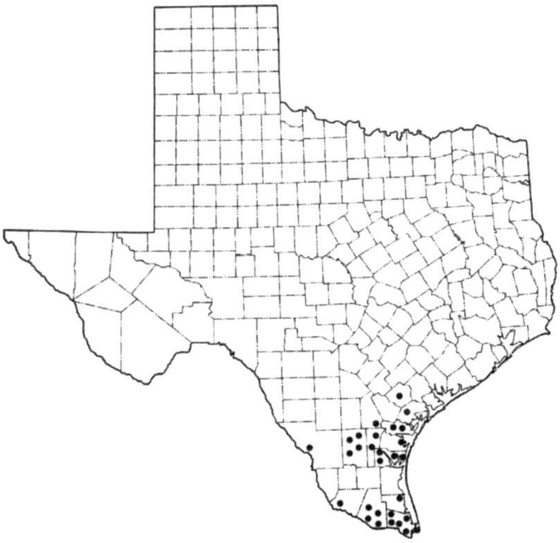

Ruellia yucatana

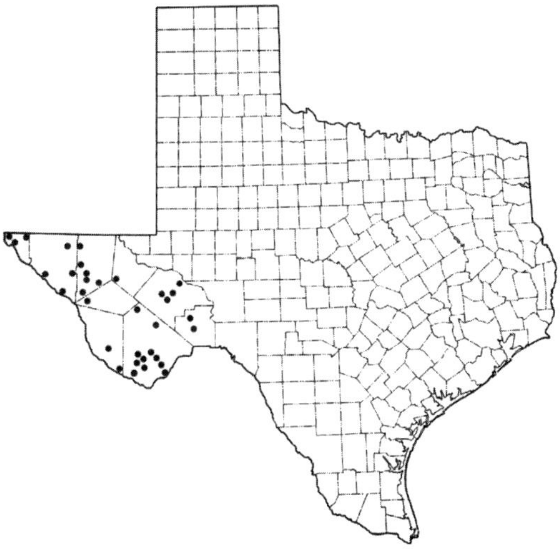

Stenandrium barbatum

Stenandrium dulce

Tetramerium nervosum

Thunbergia alata

Yeatesia platystegia

Yeatesia viridiflora

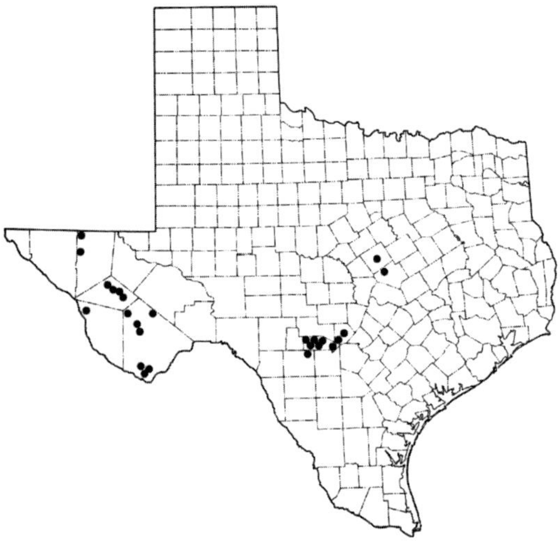

Acer grandidentatum

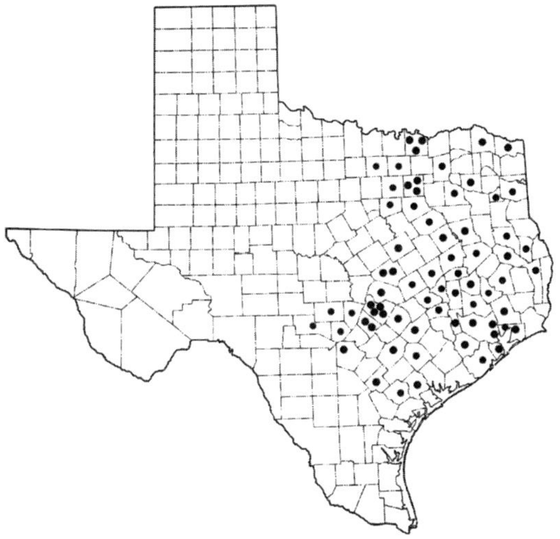

Acer negundo

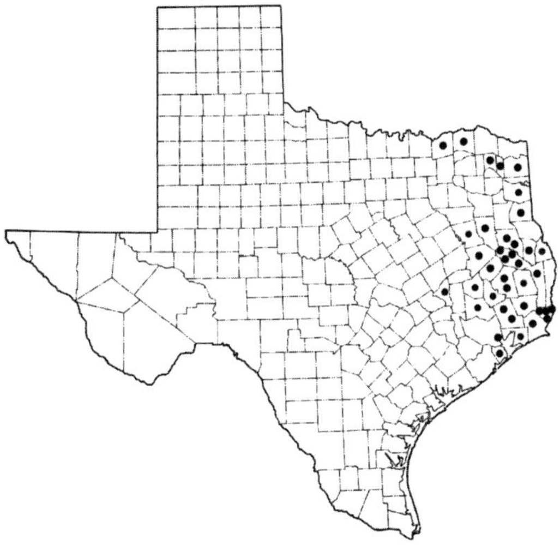

Acer rubrum

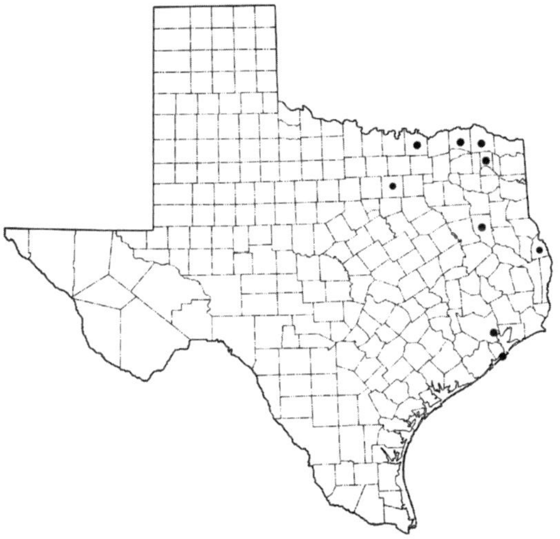

Acer saccharinum

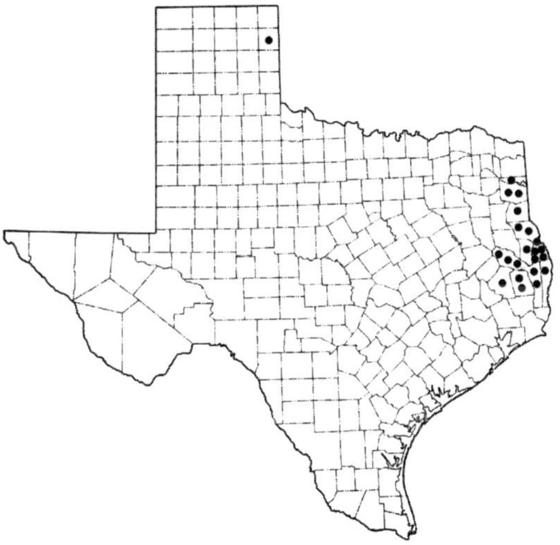

Acer saccharum

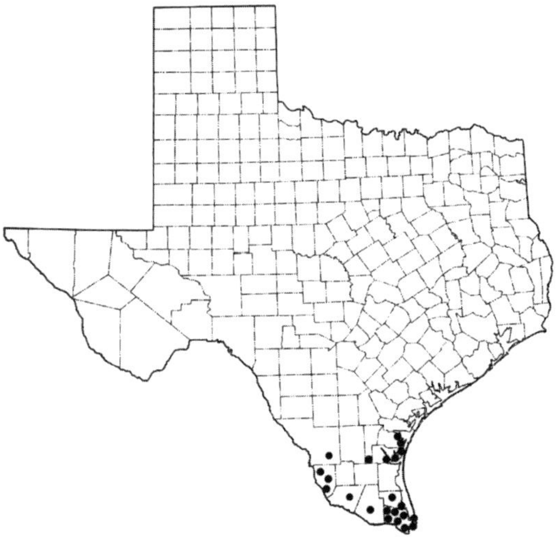

Phaulothamnus spinescens

Sesuvium maritimum

Sesuvium portulacastrum

Sesuvium trianthemoides

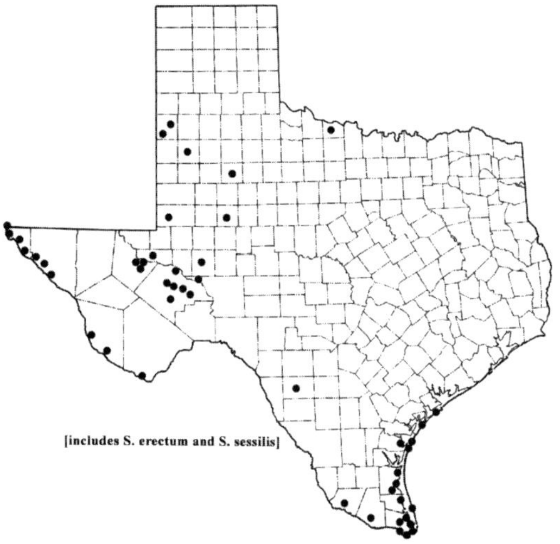
[includes S. erectum and S. sessilis]

Sesuvium verrucosum

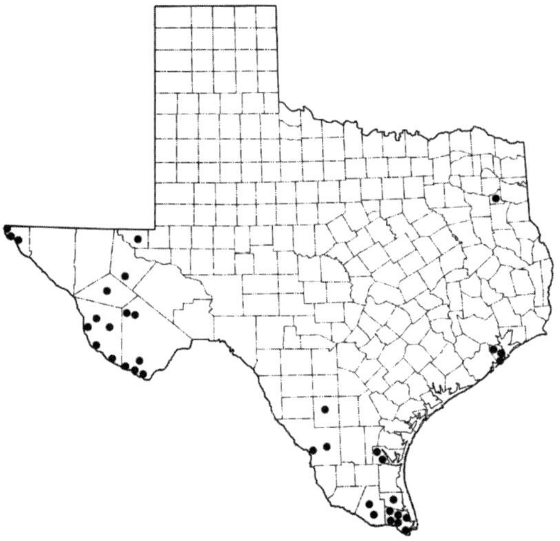

Trianthema portulacastrum

Achyranthes aspera

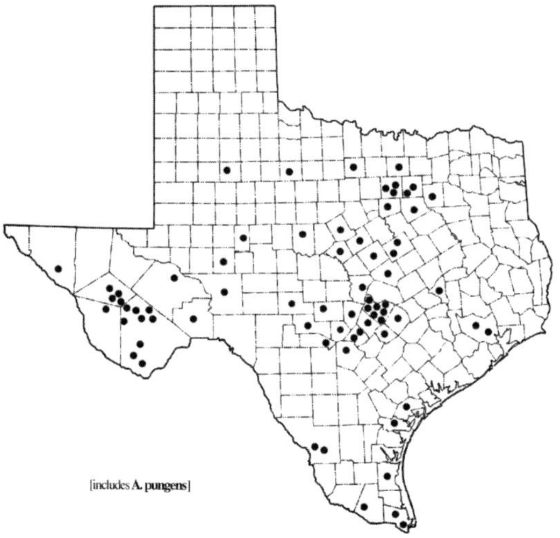
[includes A. pungens]

Alternanthera caracasana

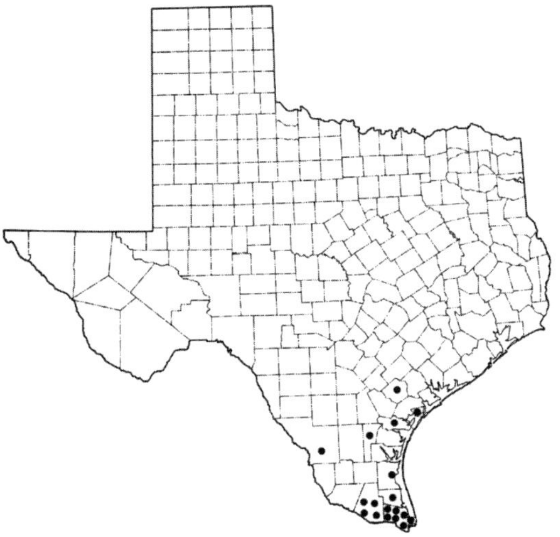

Alternanthera paronichyoides

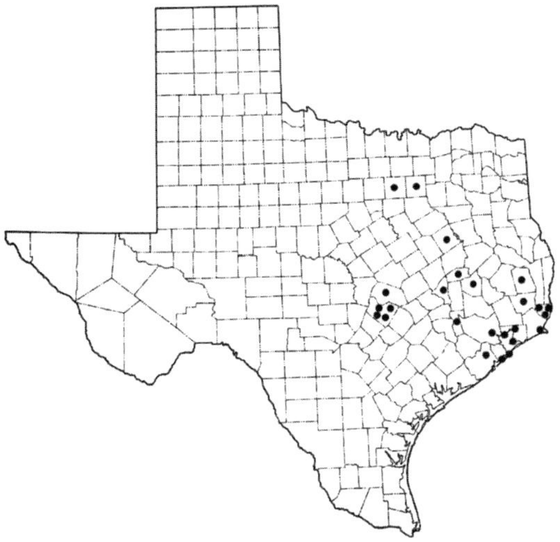

Alternanthera philoxeroides

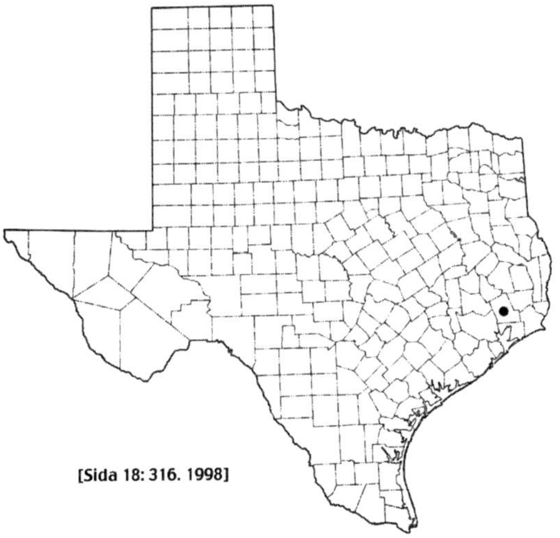
[Sida 18: 316. 1998]

Alternanthera sessilis

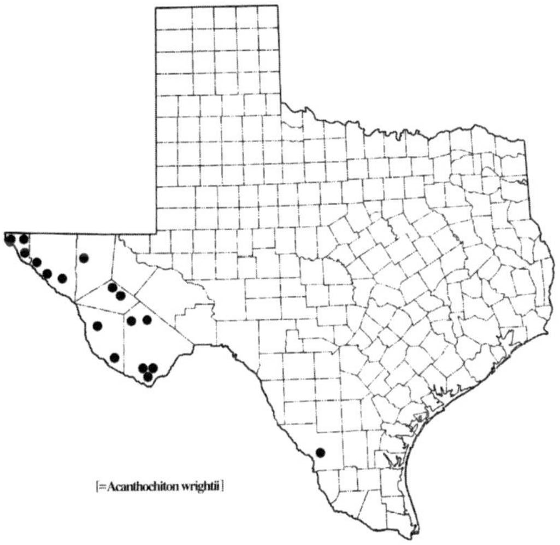
[=Acanthochiton wrightii]

Amaranthus acanthochiton

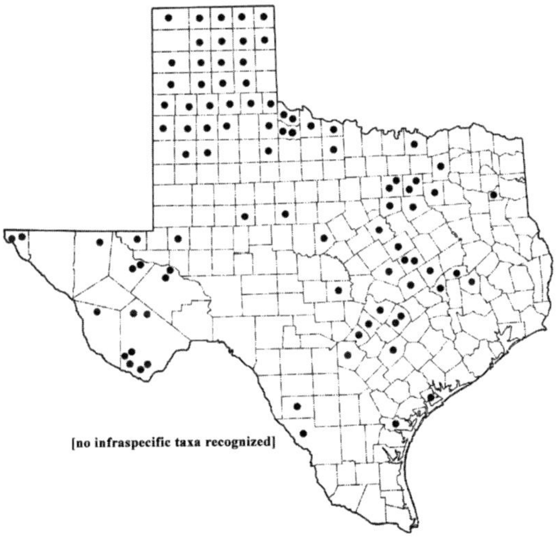
[no infraspecific taxa recognized]

Amaranthus albus

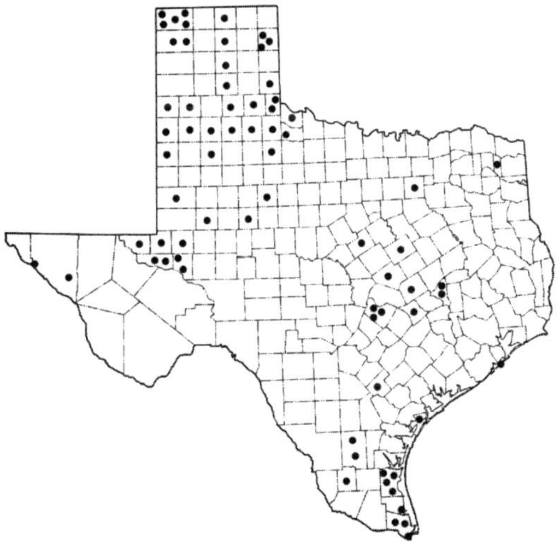

Amaranthus arenicola

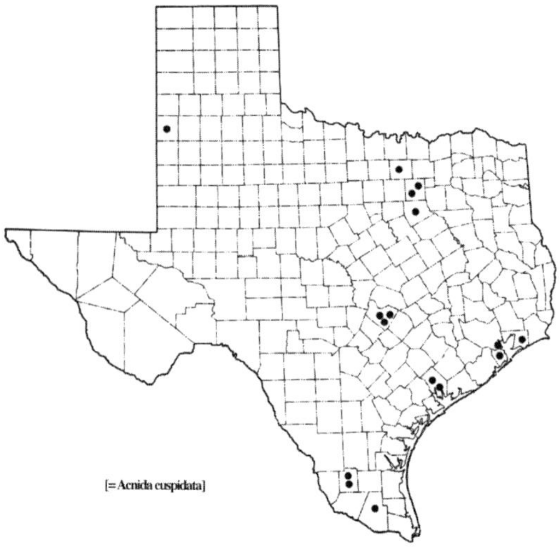
[= Acnida cuspidata]
Amaranthus australis

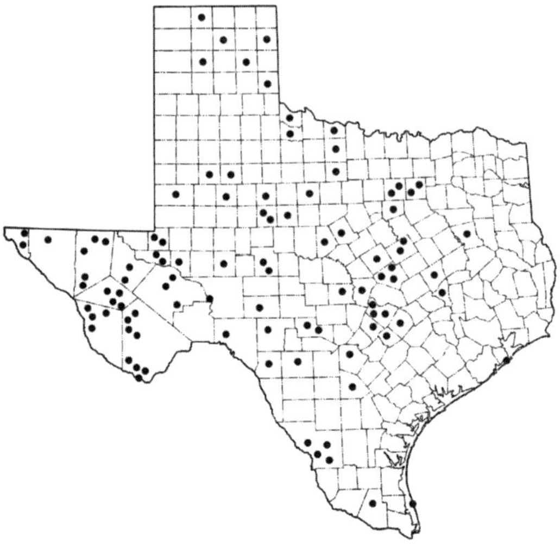
Amaranthus blitoides

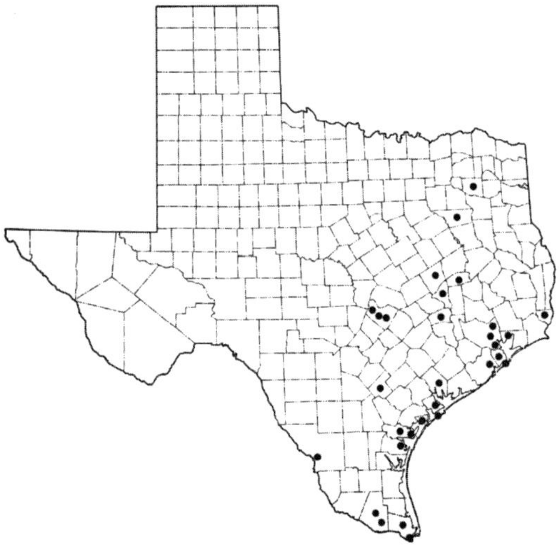
Amaranthus blitum

Amaranthus californicus

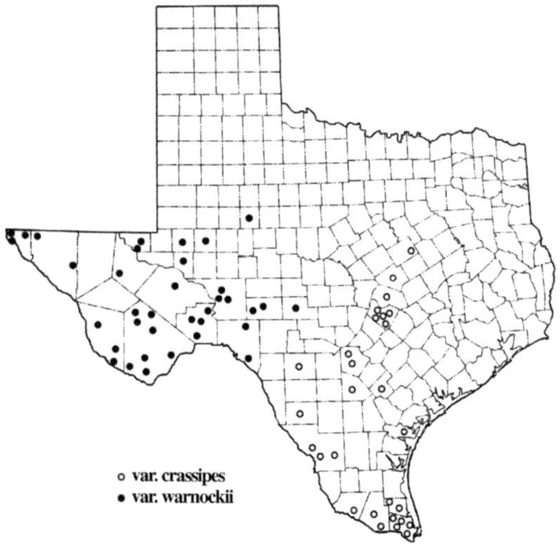
o var. crassipes
• var. warnockii
Amaranthus crassipes

Amaranthus cruentus

Amaranthus fimbriatus

Amaranthus greggii

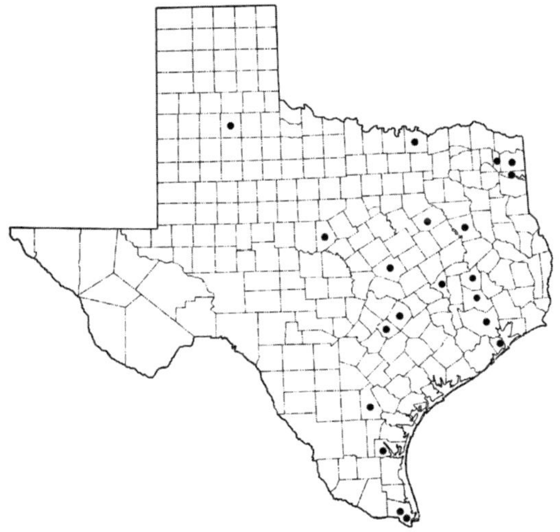

Amaranthus hybridus

Amaranthus hypochondriacus

Amaranthus lividus

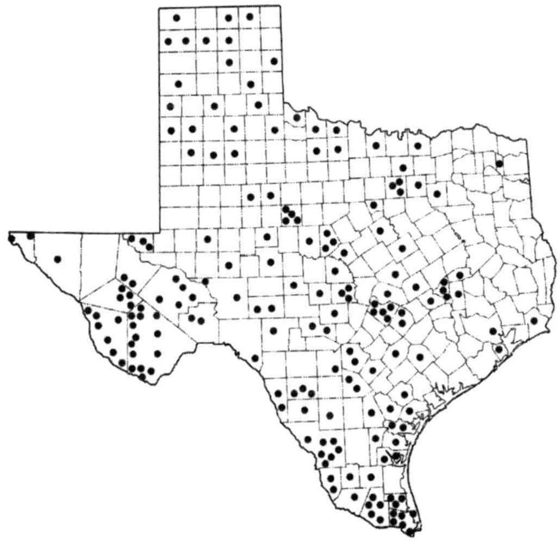

Amaranthus palmeri

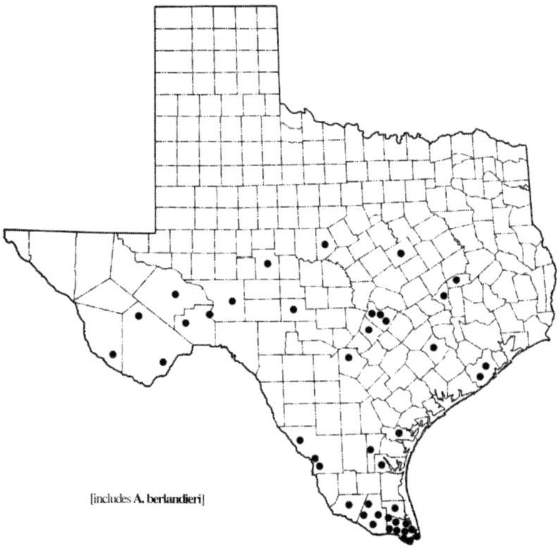
[includes A. berlandieri]

Amaranthus polygonoides

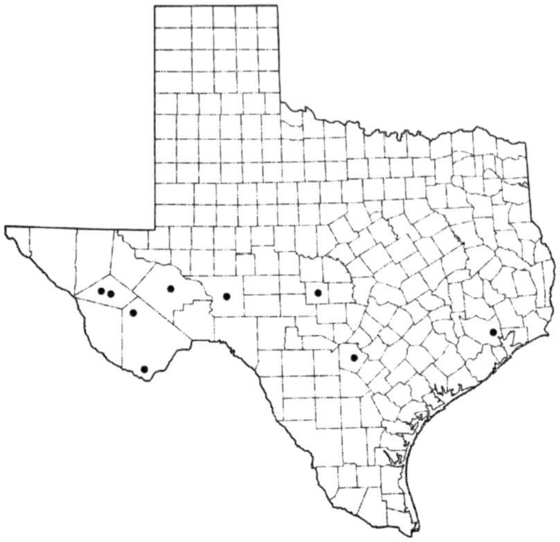

Amaranthus powellii

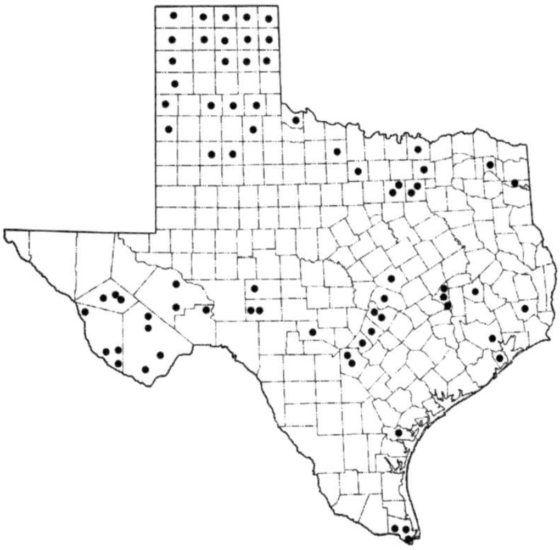

Amaranthus retroflexus

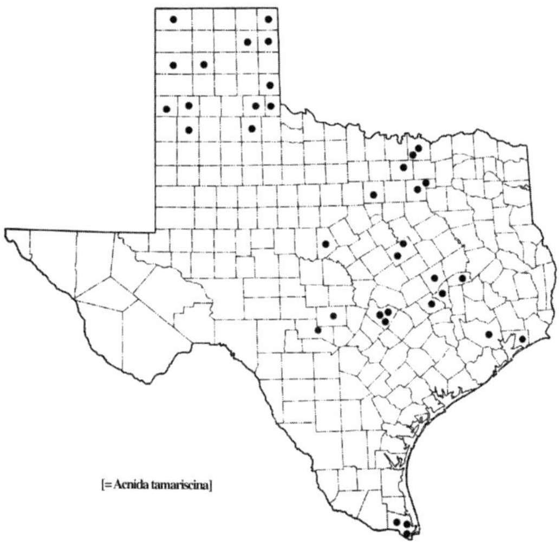
[= Acnida tamariscina]

Amaranthus rudis

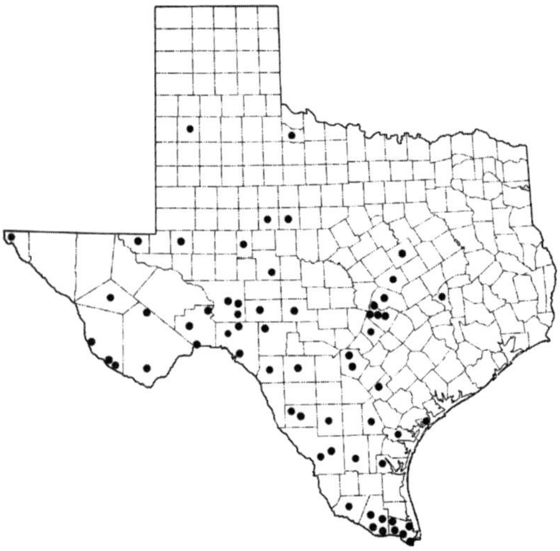

Amaranthus scleropoides

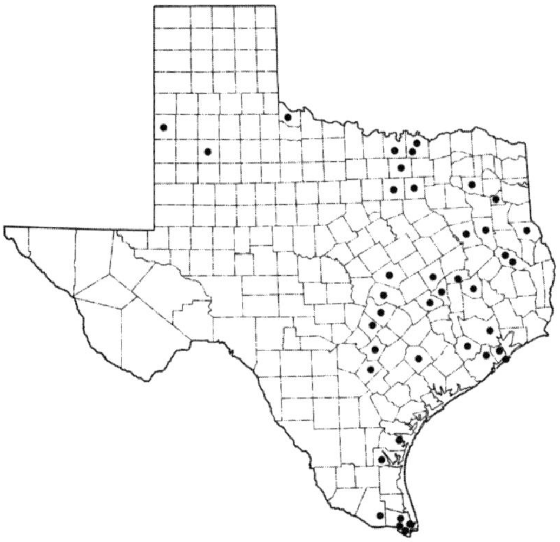

Amaranthus spinosus

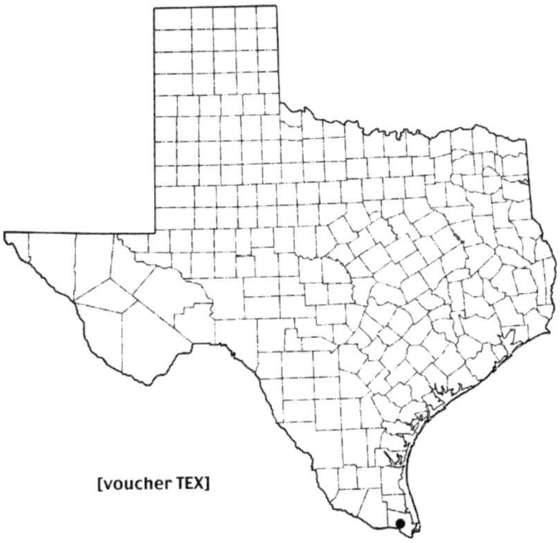
[voucher TEX]

Amaranthus tamaulipensis

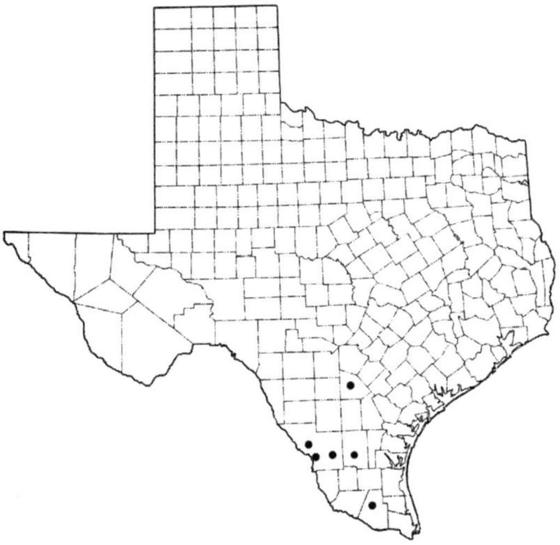

Amaranthus x texensis

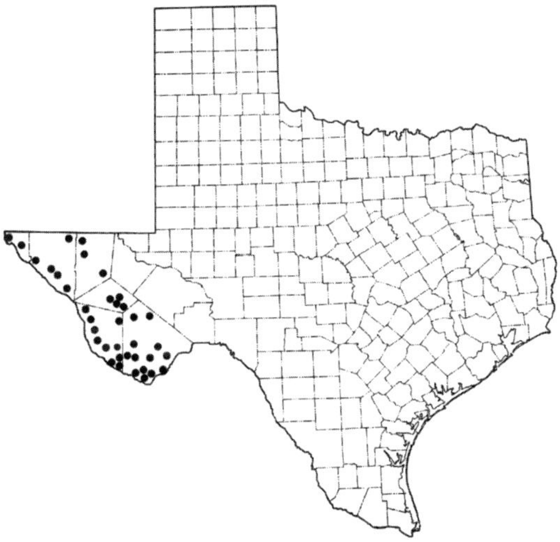

Amaranthus torreyi

Celosia argentea

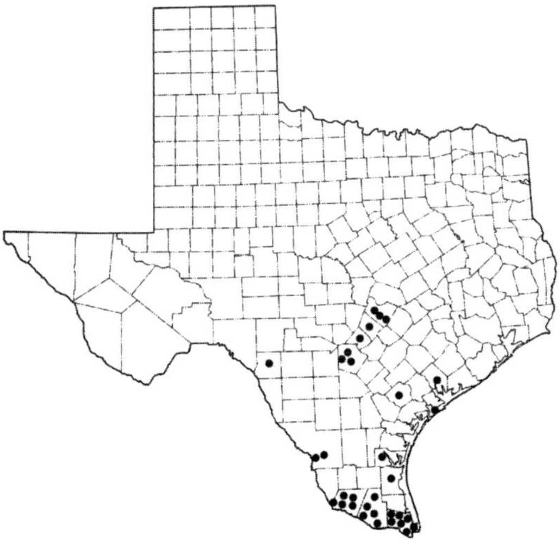

Celosia nitida

Celosia palmeri

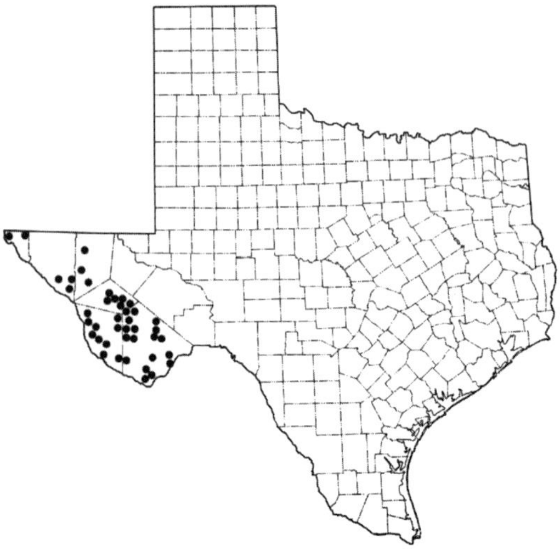

Froelichia arizonica

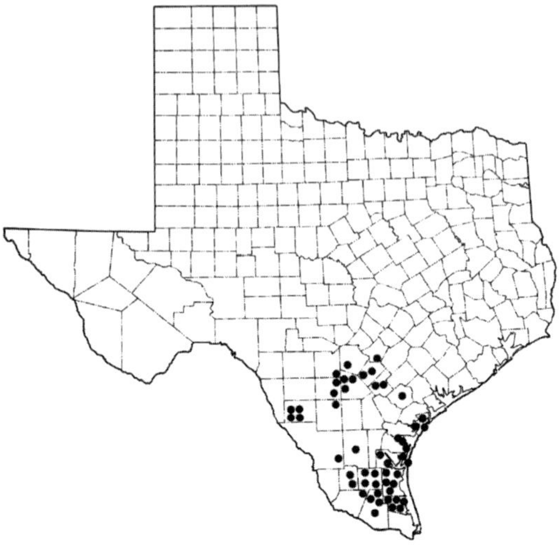

Froelichia drummondii

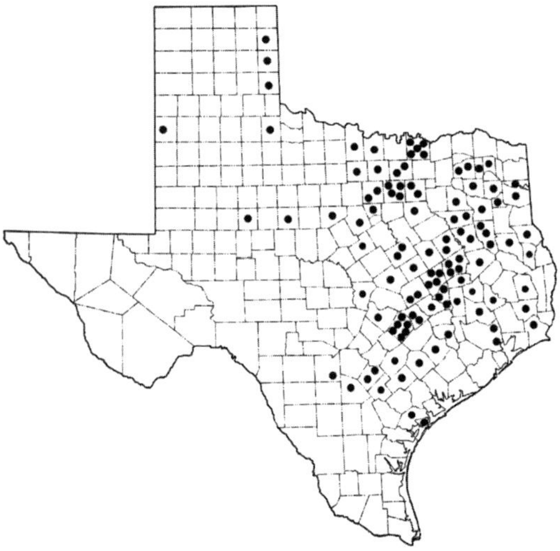

Froelichia floridana

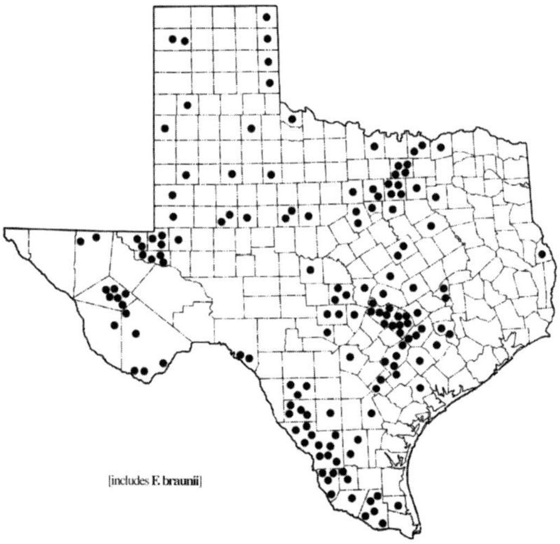
[includes F. braunii]

Froelichia gracilis

Gomphrena caespitosa

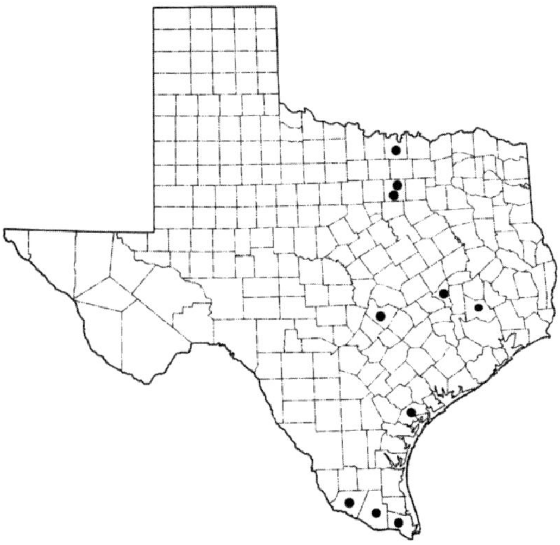

Gomphrena globosa

Gomphrena haageana

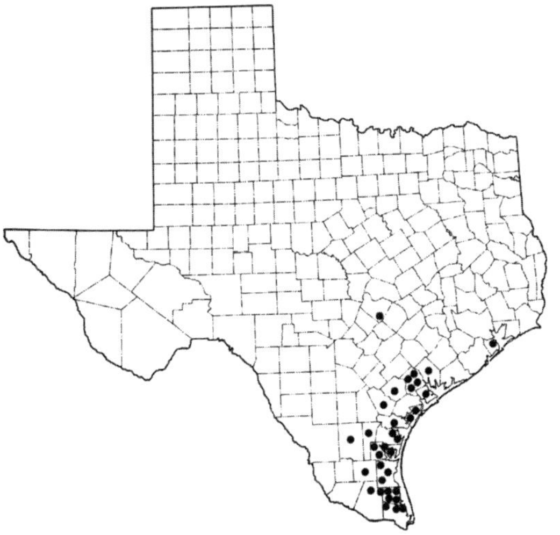

Gomphrena nealleyi

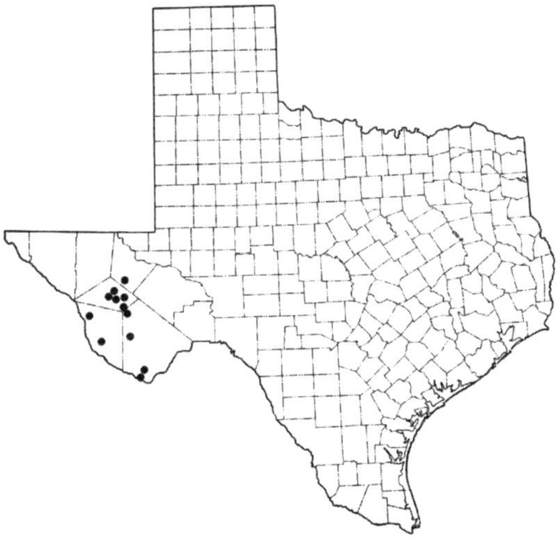

Gomphrena nitida

[includes **G. decumbens** and **G. dispersa**]

Gomphrena serrata

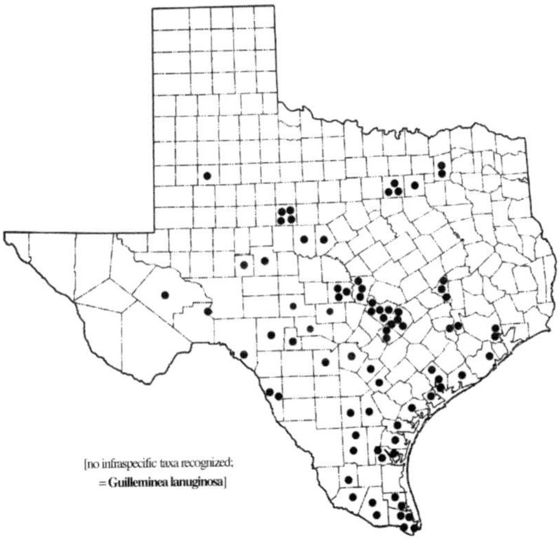
[no infraspecific taxa recognized;
= **Guilleminea lanuginosa**]

Gossypianthus lanuginosus

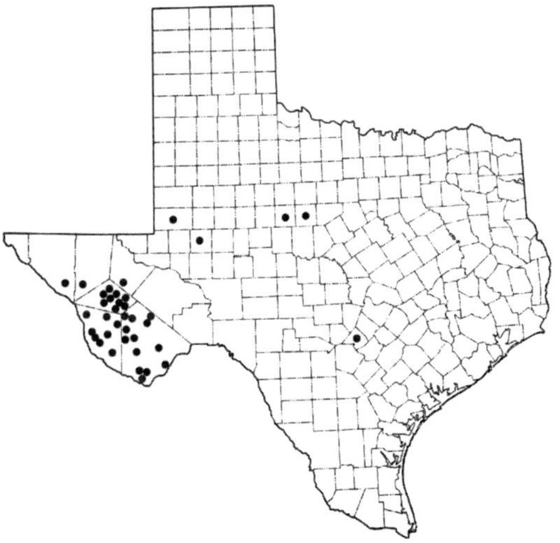

Guilleminea densa

Iresine diffusa

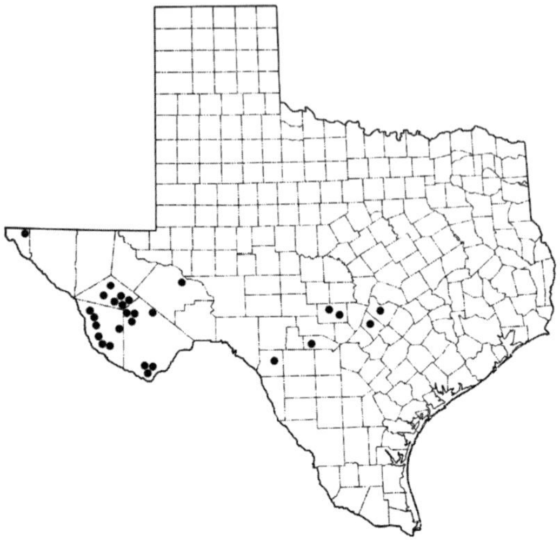

Iresine heterophylla

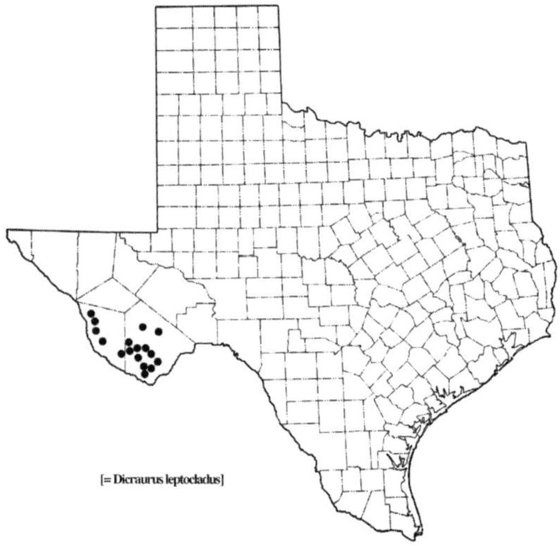
[= Dicraurus leptocladus]

Iresine leptoclada

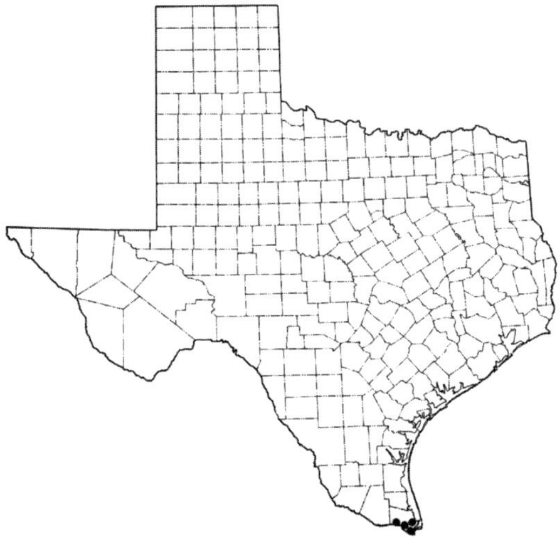

Iresine palmeri

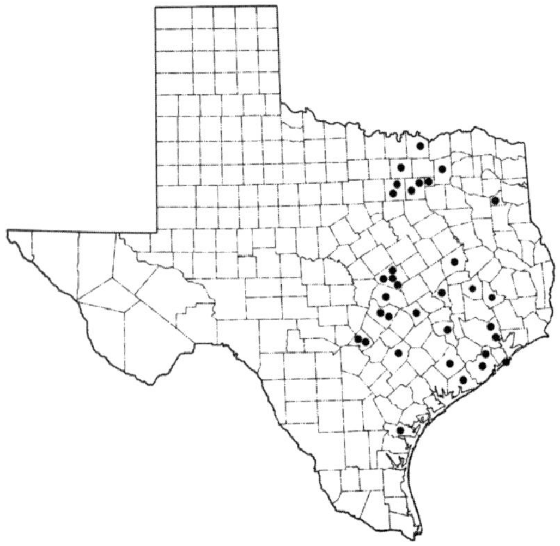

Iresine rhizomatosa

Philoxerus vermicularis

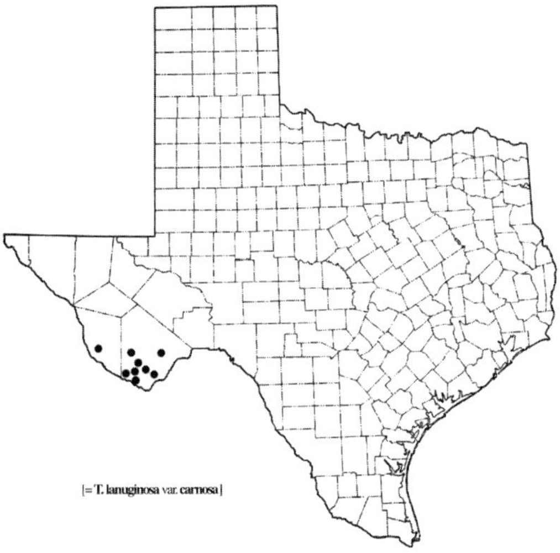
[= T. lanuginosa var. carnosa]

Tidestromia carnosa

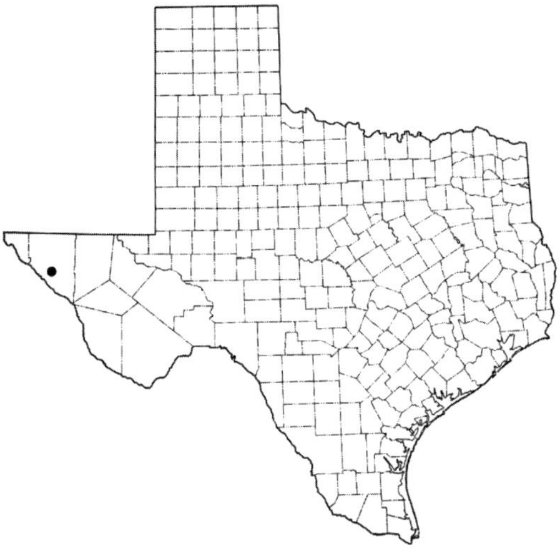

Tidestromia gemmata

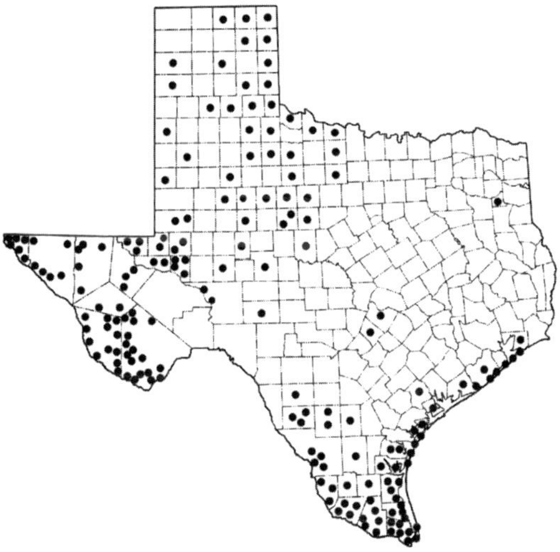

Tidestromia lanuginosa

Tidestromia oblongifolia

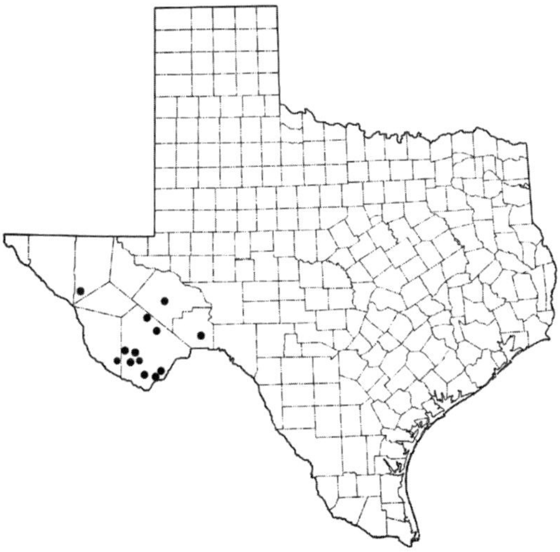

Tidestromia suffruticosa

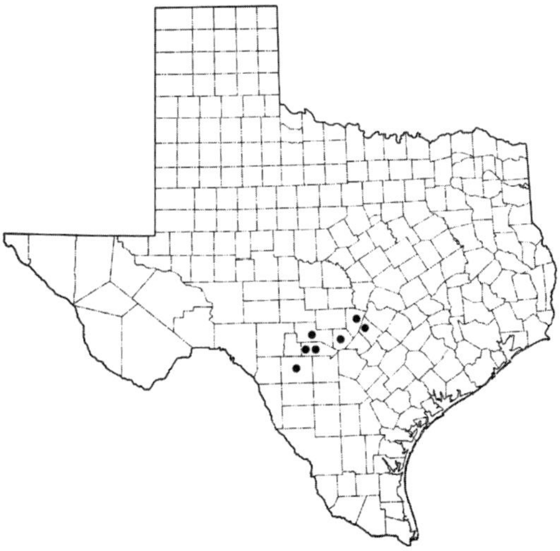

Cotinus obovatus

Pistacia mexicana

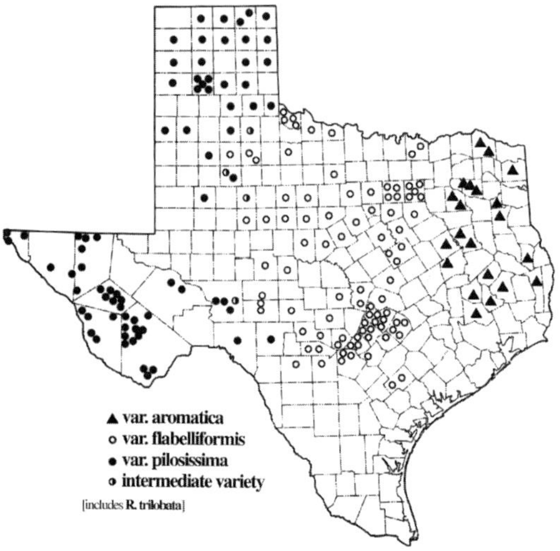
▲ var. aromatica
○ var. flabelliformis
● var. pilosissima
◐ intermediate variety
[includes R. trilobata]

Rhus aromatica

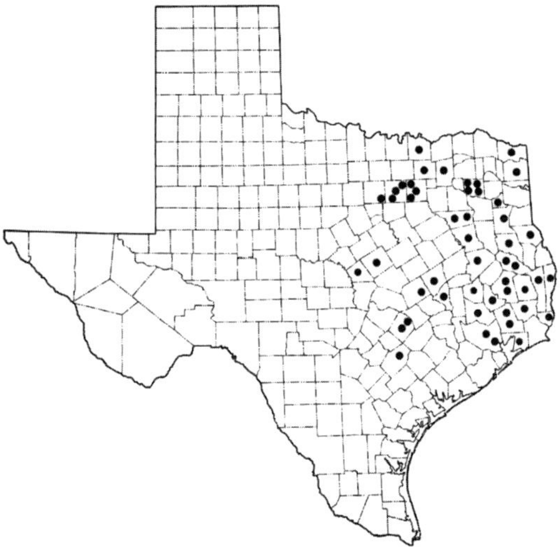

Rhus copallinum

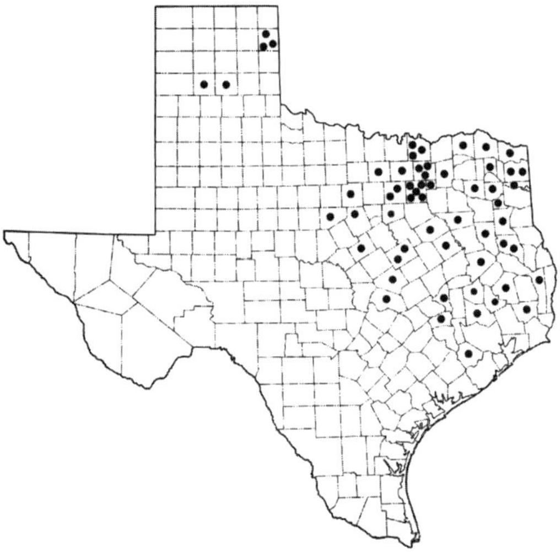

Rhus glabra

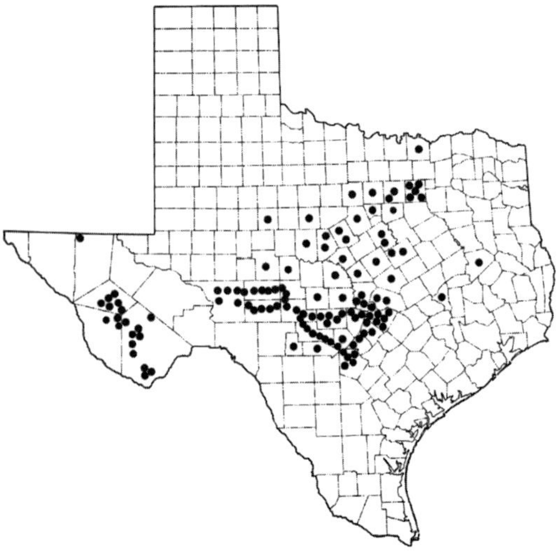

Rhus lanceolata

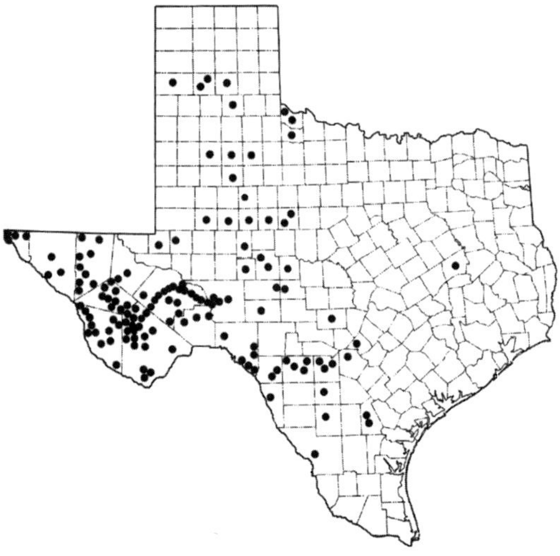

Rhus microphylla

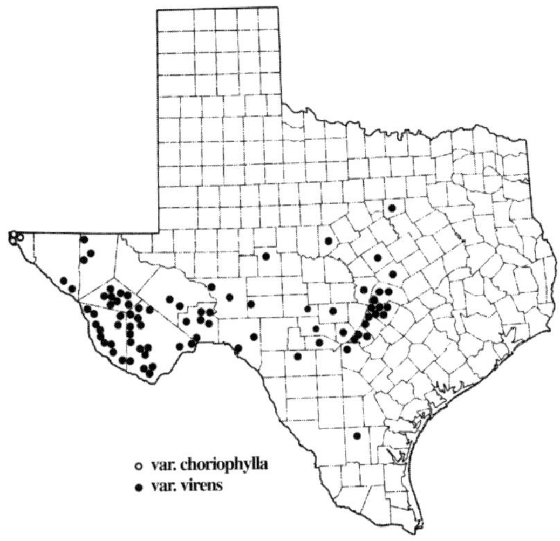
var. choriophylla
var. virens

Rhus virens

Schinus molle

Schinus terebinthifolius

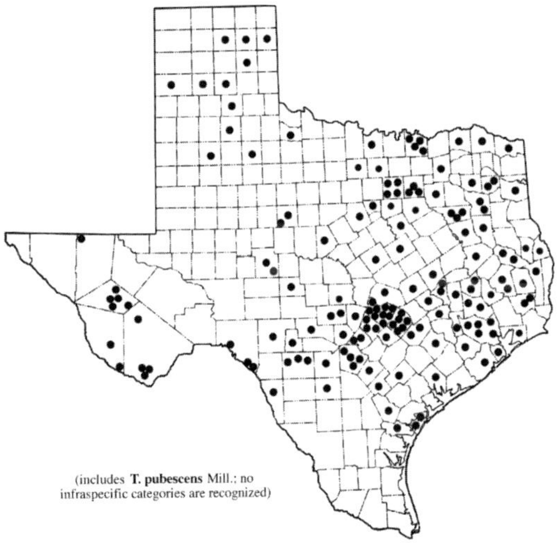
(includes T. pubescens Mill.: no infraspecific categories are recognized)

Toxicodendron radicans

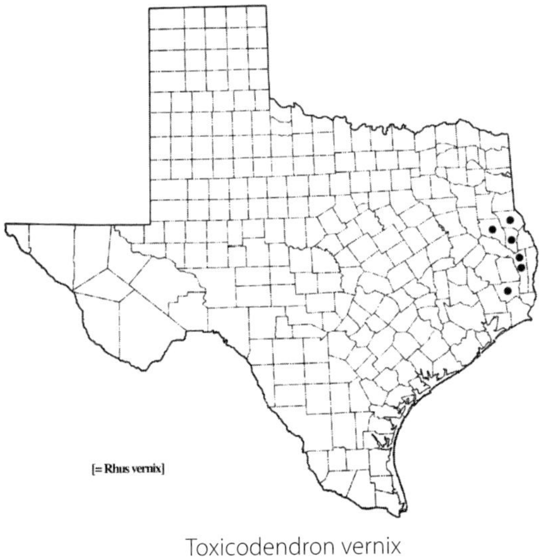
[= Rhus vernix]

Toxicodendron vernix

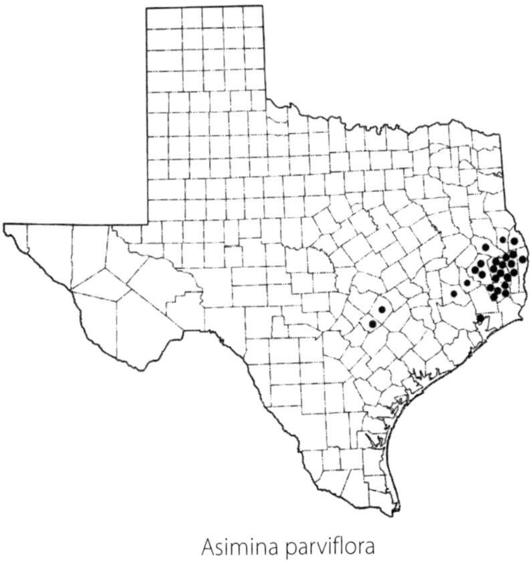

Asimina parviflora

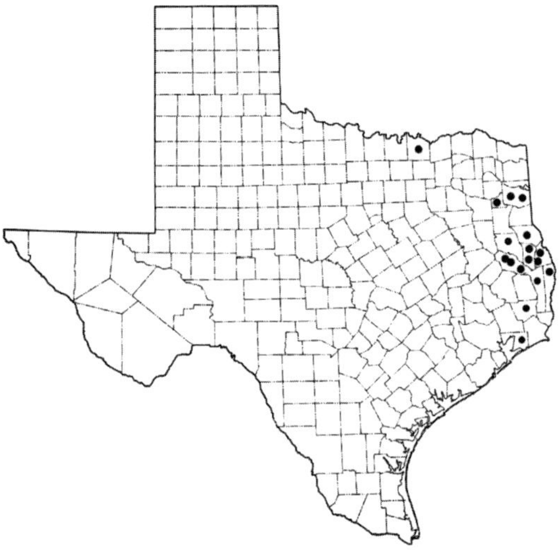

Asimina triloba

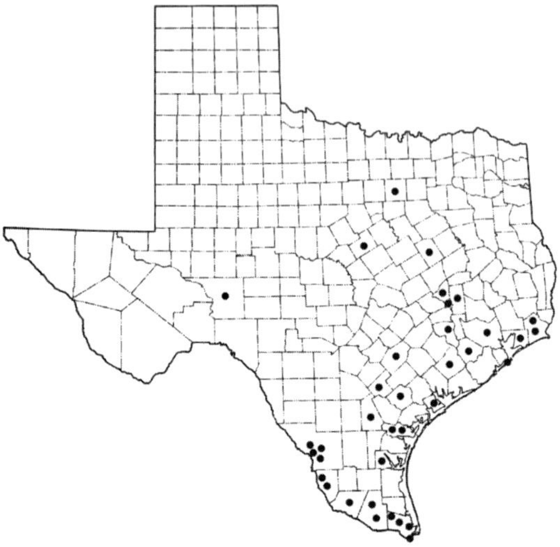

Ammi majus

Ammi visnaga

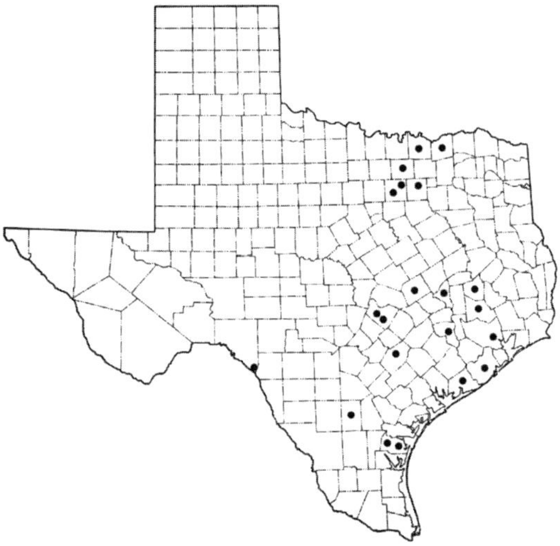

Ammoselinum butleri

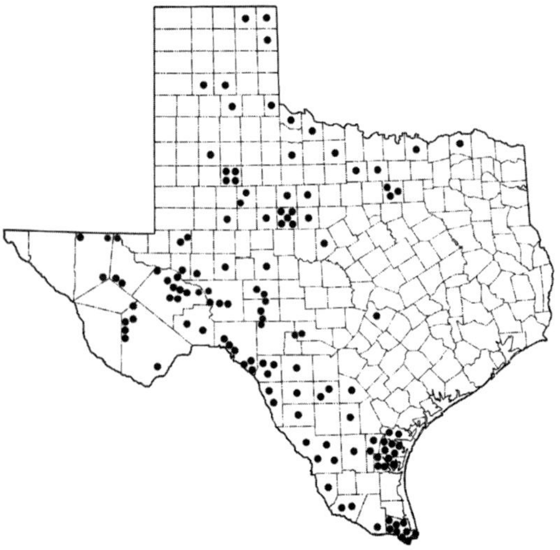

Ammoselinum popei

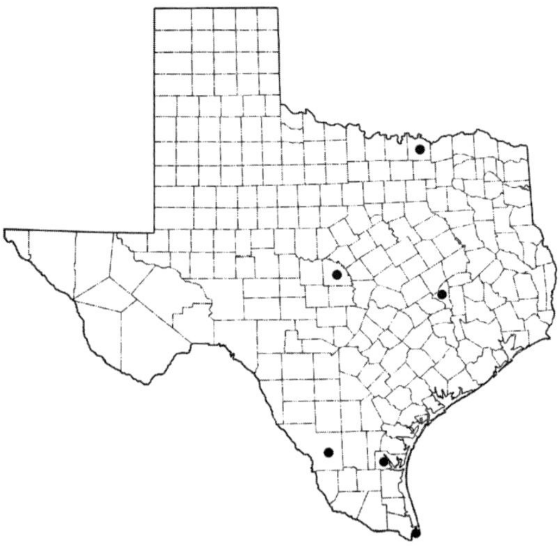

Anethum graveolens

Apium graveolens

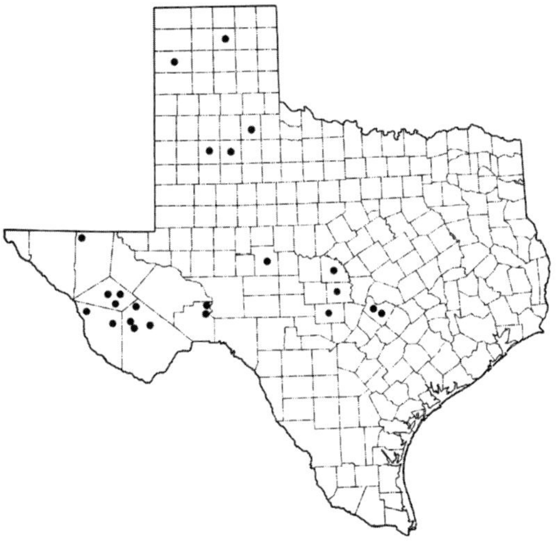

Berula erecta

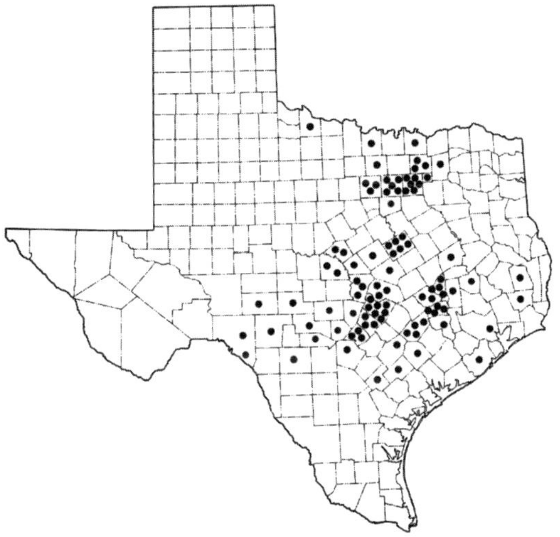

Bifora americana

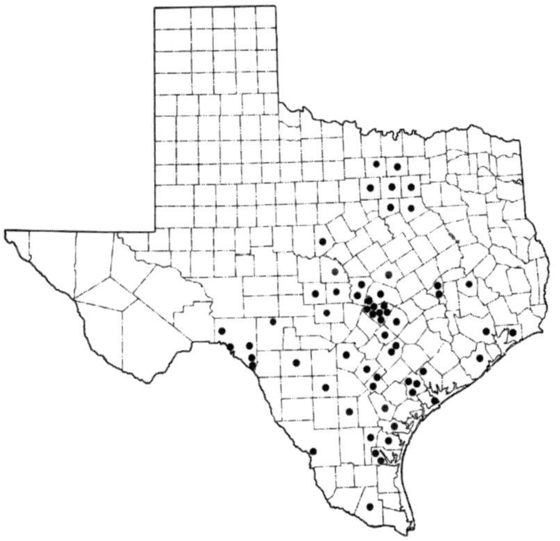

Bowlesia incana

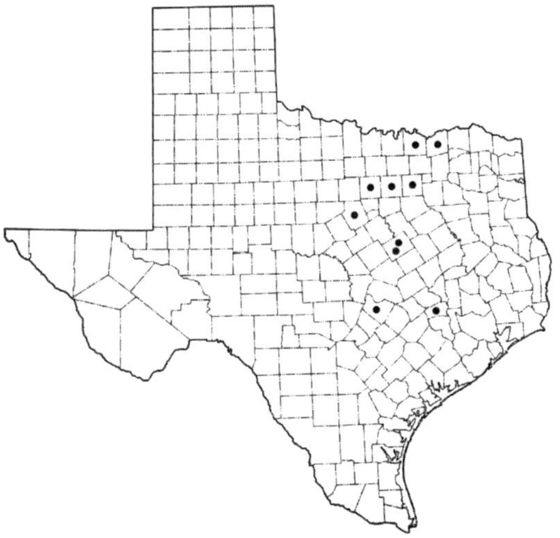

Bupleurum rotundifolium

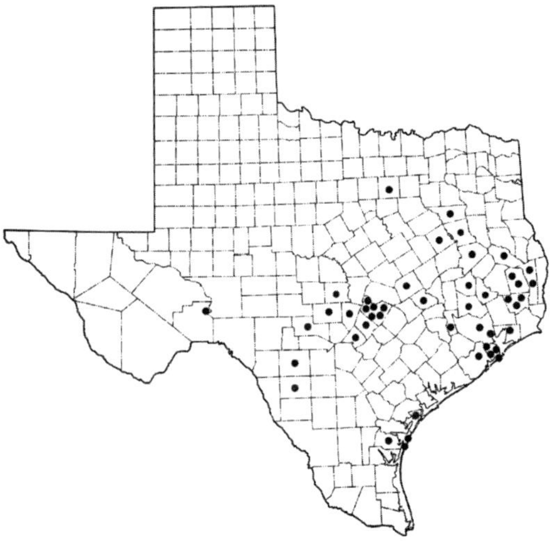

Centella erecta

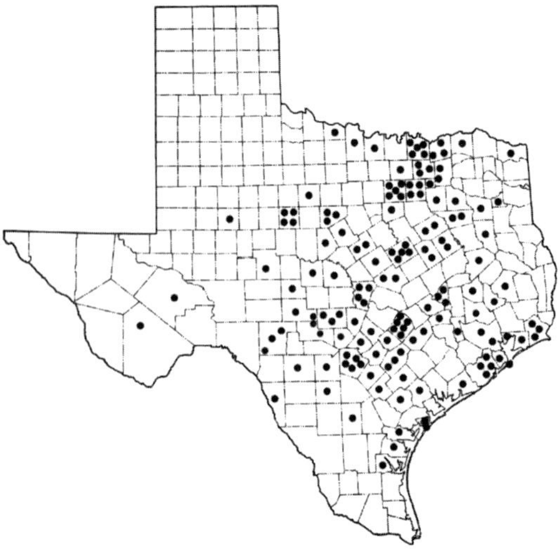

Chaerophyllum tainturieri

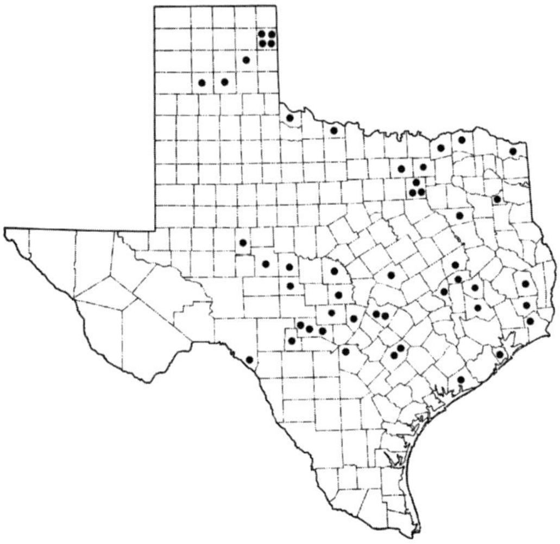

Cicuta maculata

Conium maculatum

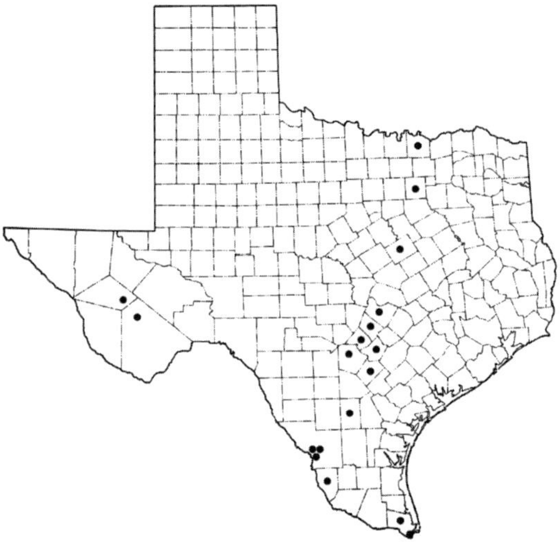

Coriandrum sativum

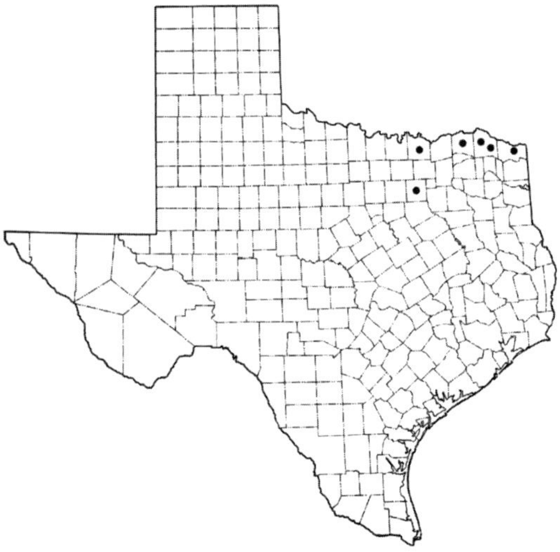

Cryptotaenia canadensis

Cuminum cyminum

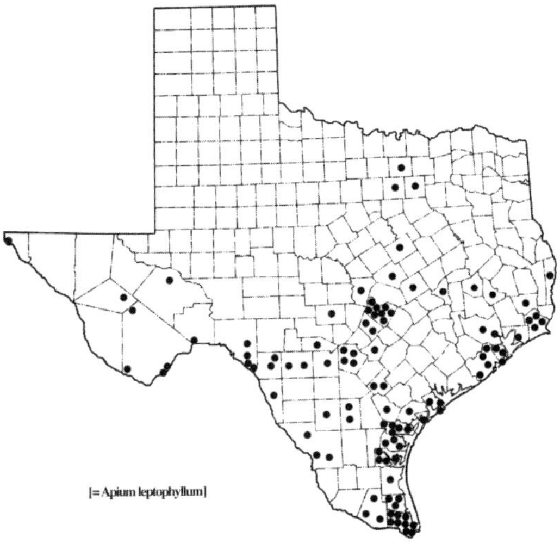
[= Apium leptophyllum]

Cyclospermum leptophyllum

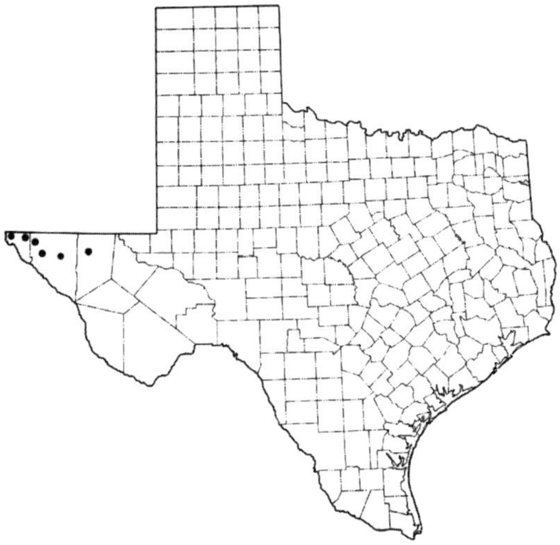

Cymopterus acaulis

Cymopterus bulbosus

[=Aletes filifolius
=Pseudocymopteris filifolius]

Cymopterus filifolius

[=Aletes montanus
=Cymopteris montanus]

Cymopterus hallii

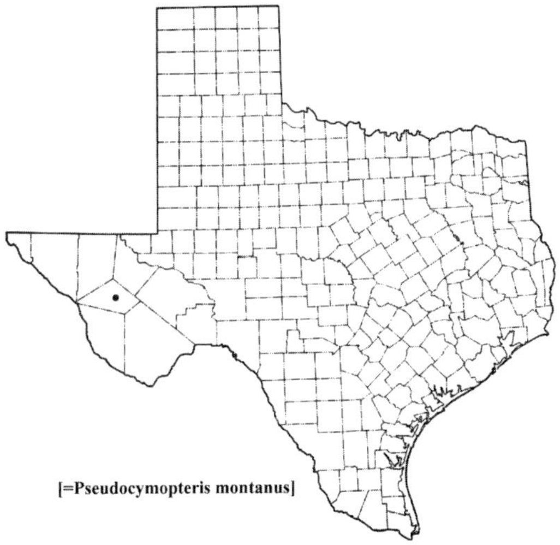
[=Pseudocymopteris montanus]

Cymopterus lemmonii

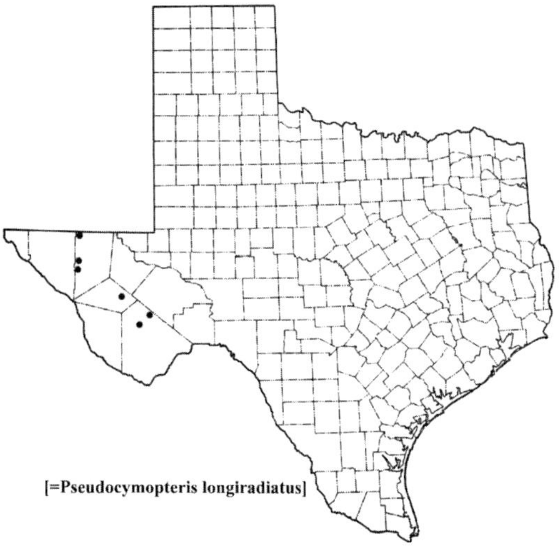
[=Pseudocymopteris longiradiatus]

Cymopterus longiradiatus

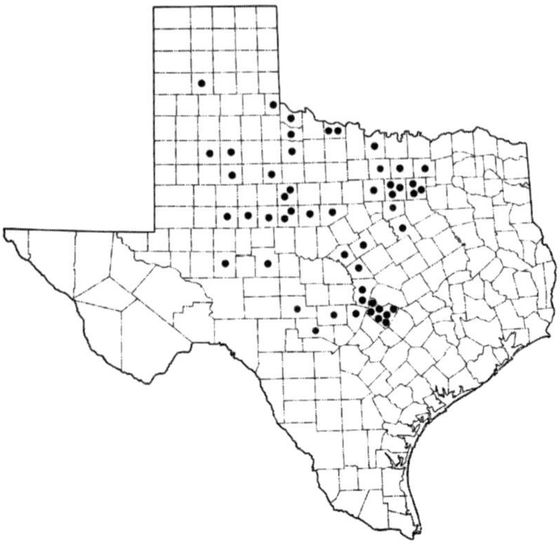

Cymopterus macrorhizus

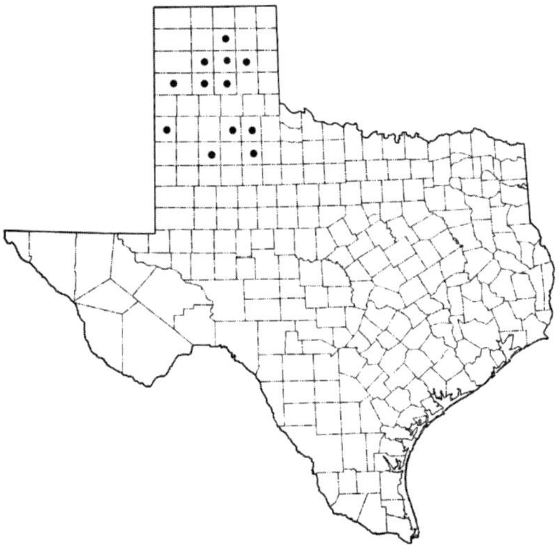

Cymopterus montanus

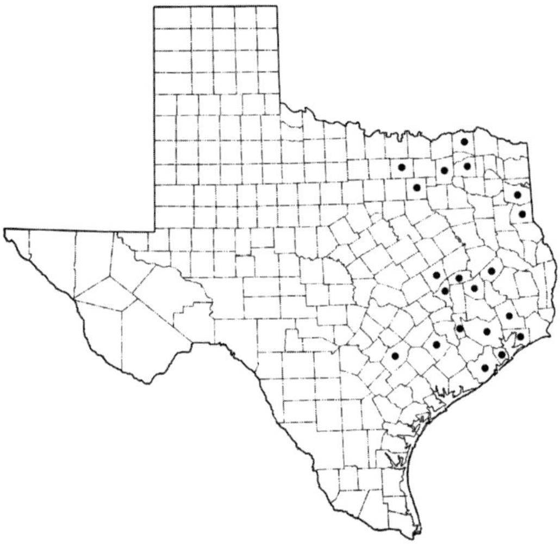

Cynosciadium digitatum

Daucosma laciniata

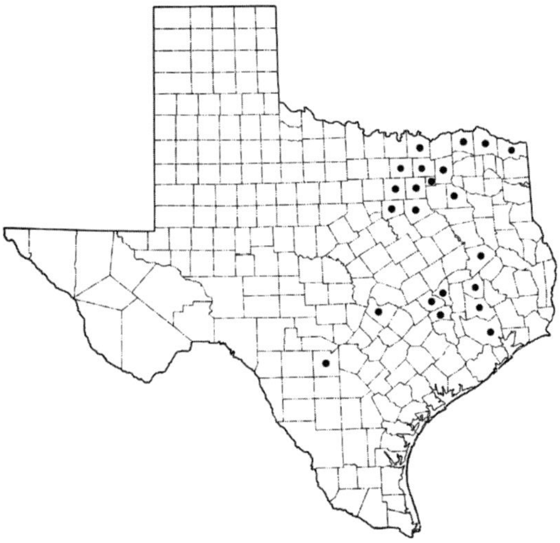

Daucus carota

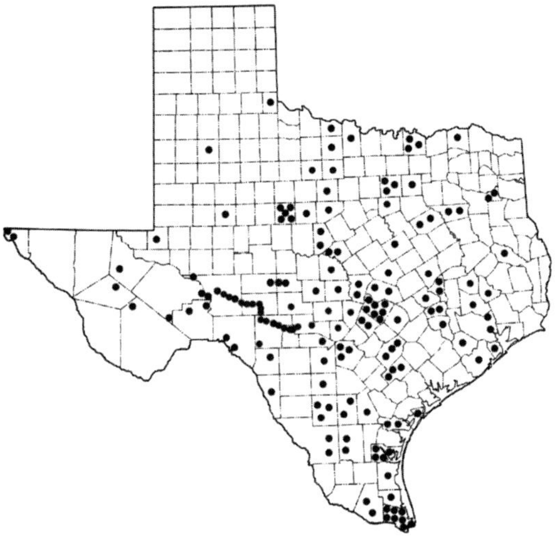
Daucus pusillus

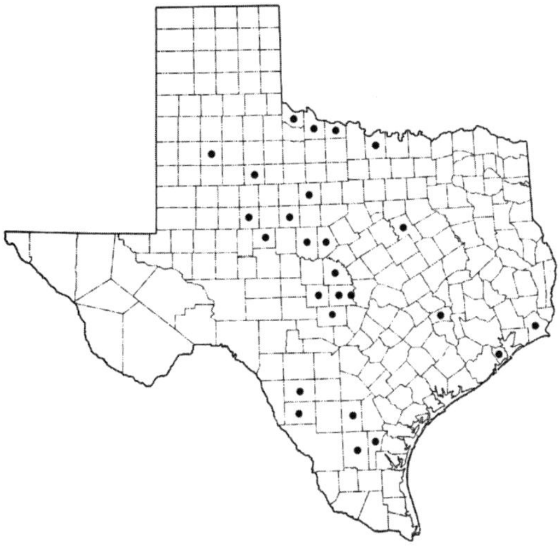
Eryngium diffusum

Eryngium heterophyllum

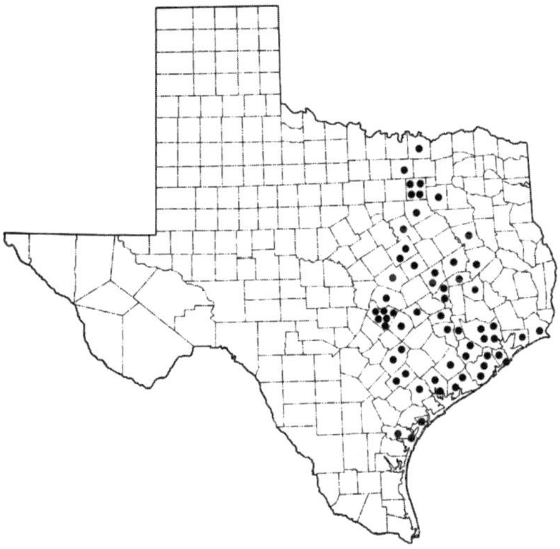
Eryngium hookeri

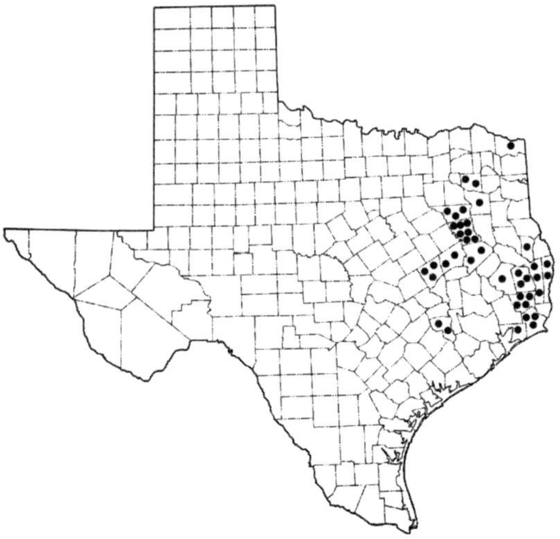
Eryngium integrifolium

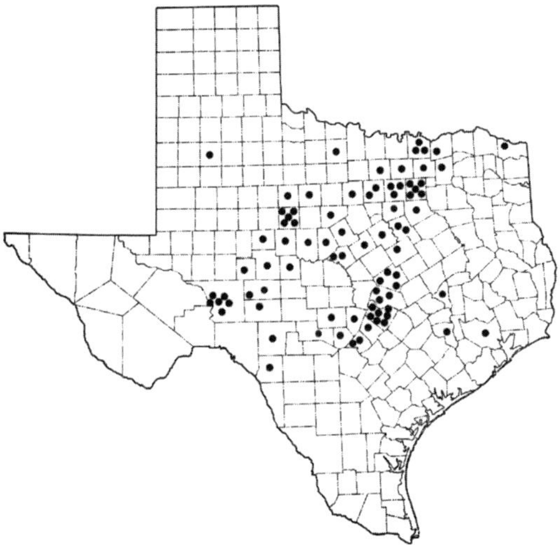
Eryngium leavenworthii

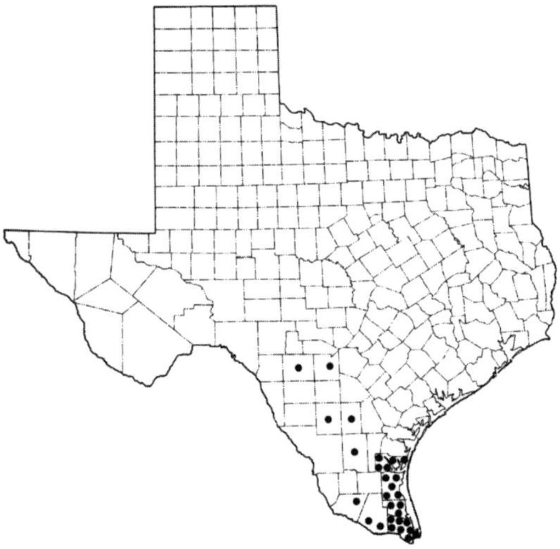

Eryngium nasturtiifolium

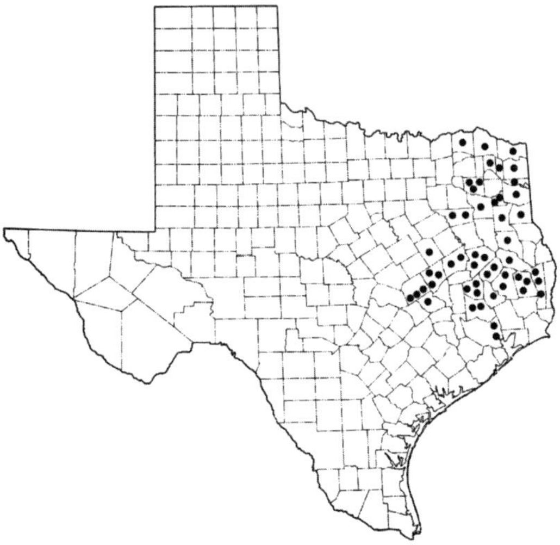

Eryngium prostratum

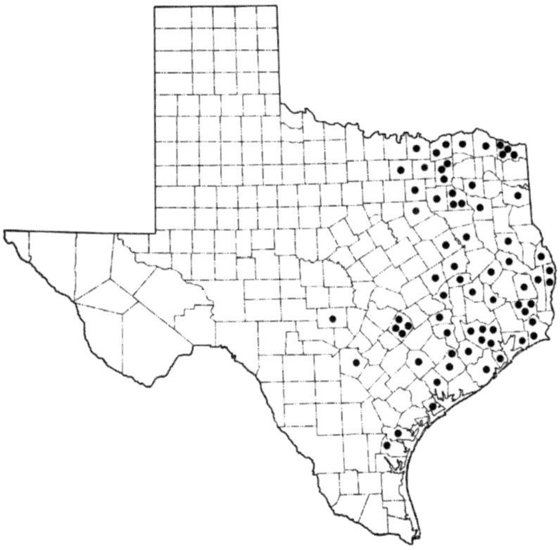

Eryngium yuccifolium

Eurytaenia hinckleyi

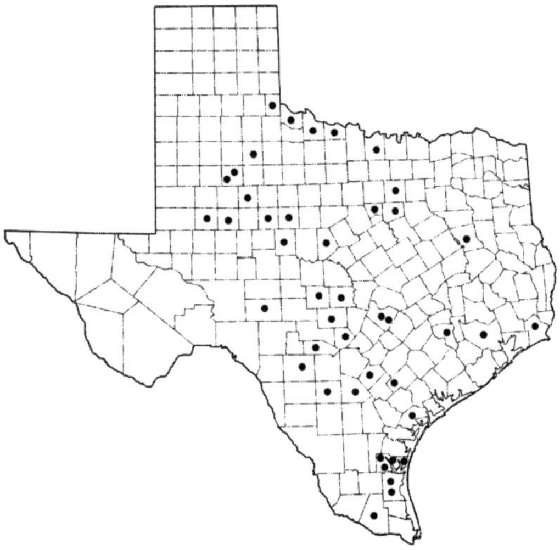

Eurytaenia texana

Foeniculum vulgare

Hydrocotyle bonariensis

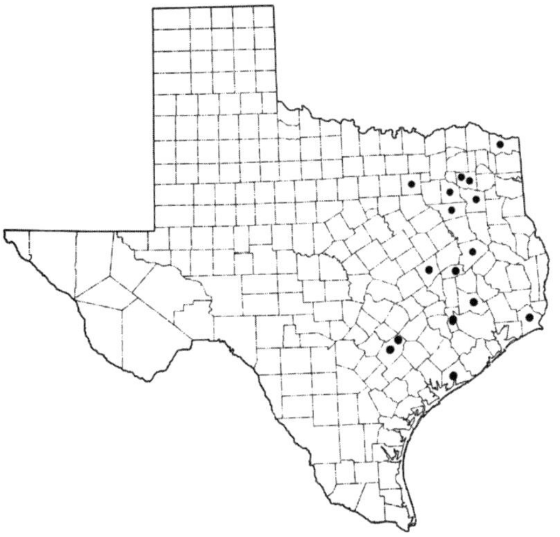

Hydrocotyle ranunculoides

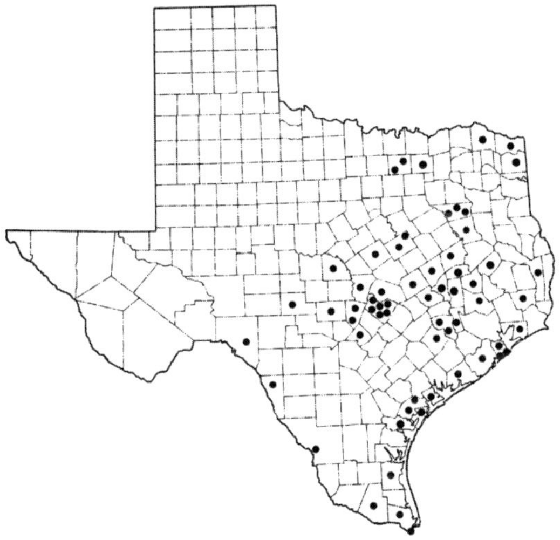

Hydrocotyle umbellata

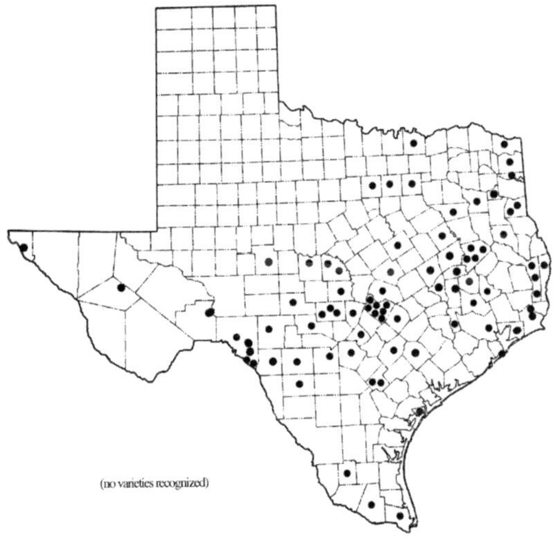
(no varieties recognized)

Hydrocotyle verticillata

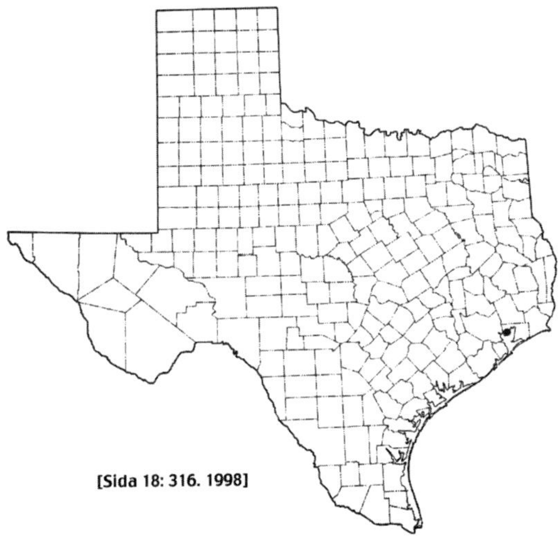
[Sida 18: 316. 1998]

Lilaeopsis chinensis

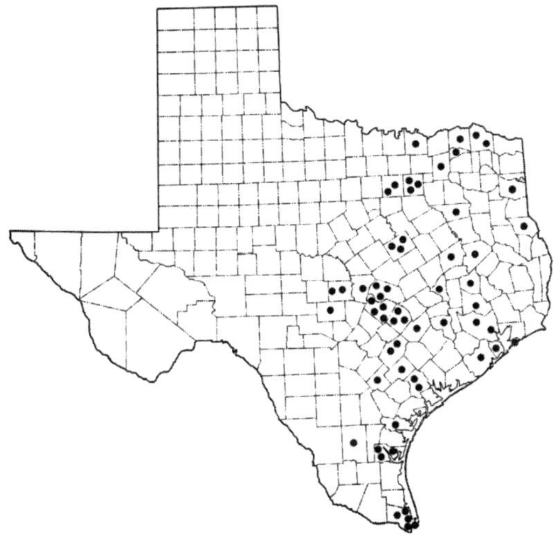

Limnosciadium pinnatum

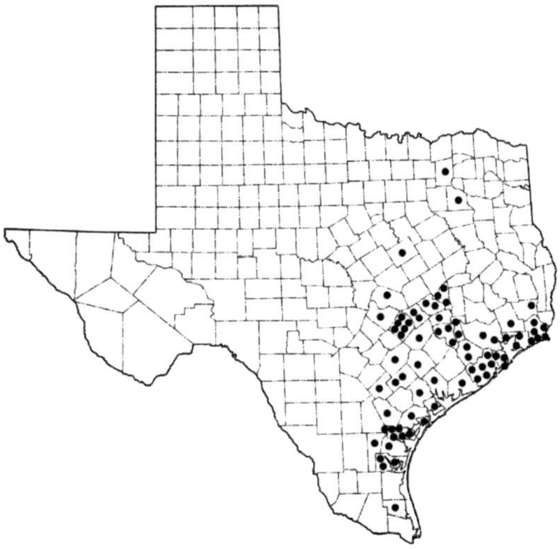

Limnosciadium pumilum

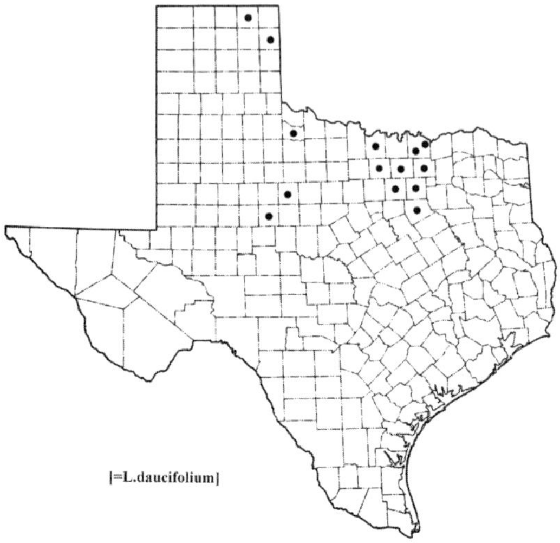
[=L.daucifolium]

Lomatium foeniculaceum

Lomatium orientale

Osmorhiza longistylis

var. bipartriata
[= O. bipatriata
= O. mexicana var. bipatriata
(Const. & Shan) B.L. Turner, comb. nov.]

Osmorhiza mexicana

Oxypolis filiformis

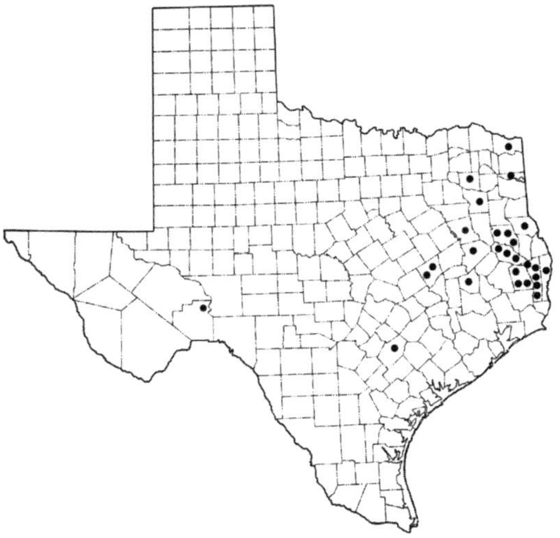

Oxypolis rigidior

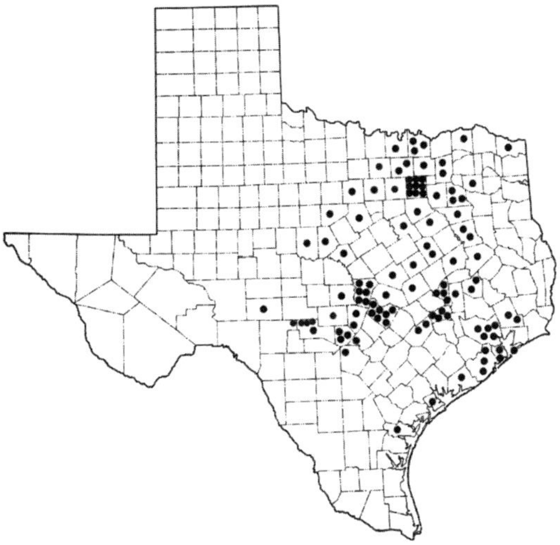

Polytaenia nuttallii

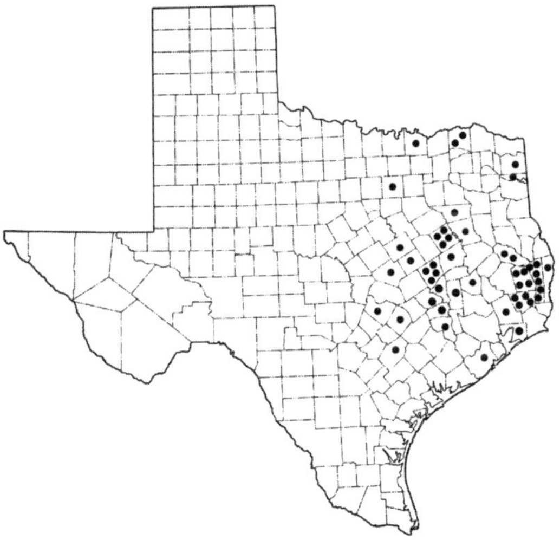

Ptilimnium capillaceum

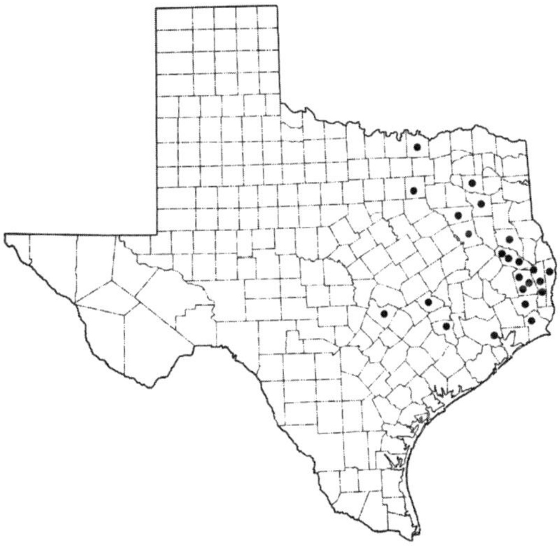

Ptilimnium costatum

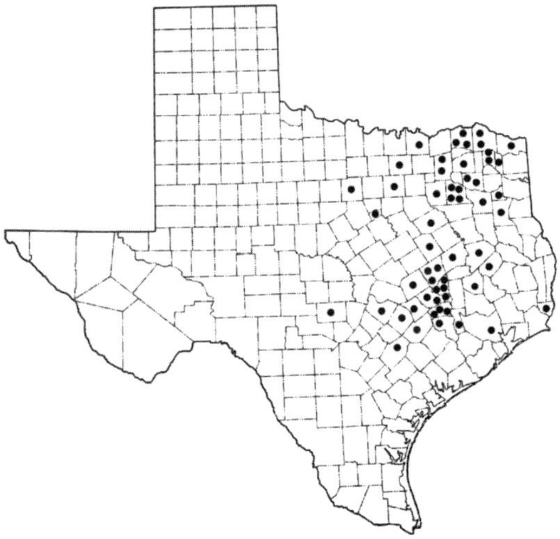

Ptilimnium nuttallii

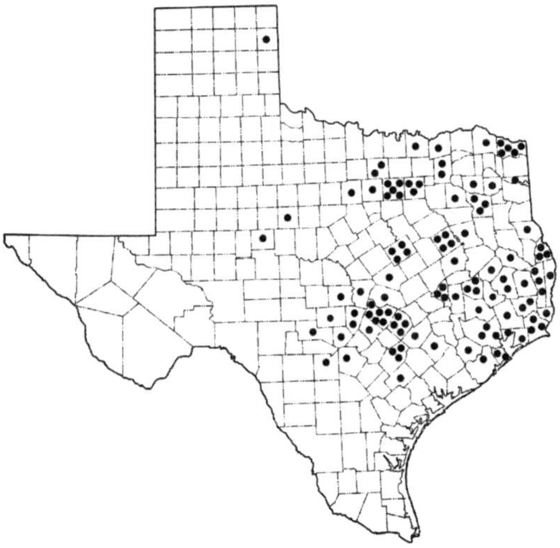

Sanicula canadensis

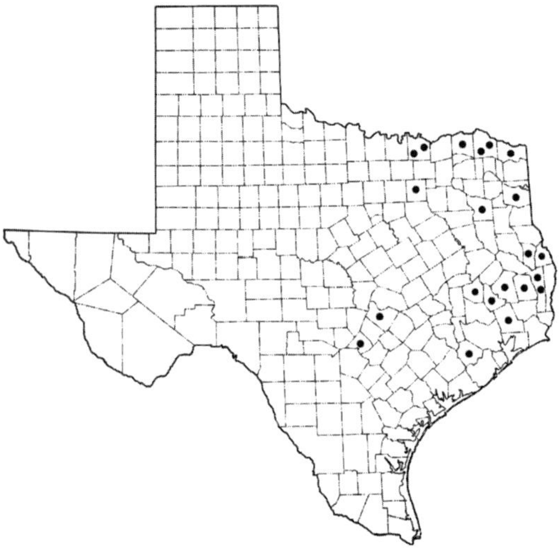

Sanicula odorata

Sanicula smallii

Scandix pecten-veneris

Sium suave

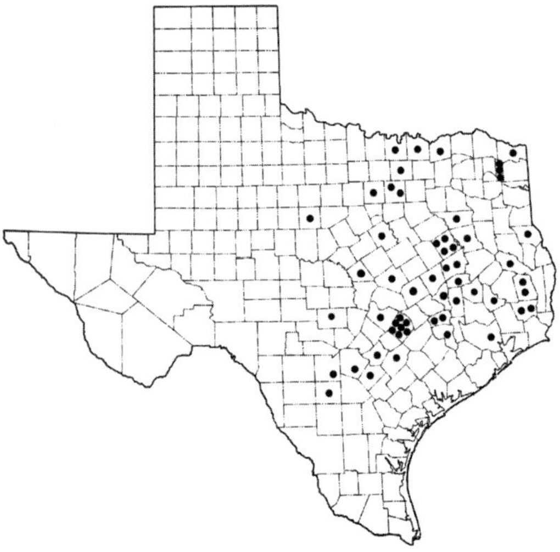

Spermolepis divaricata

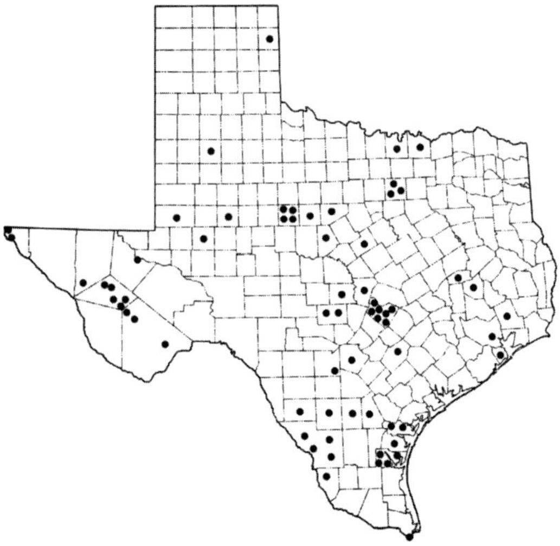

Spermolepis echinata

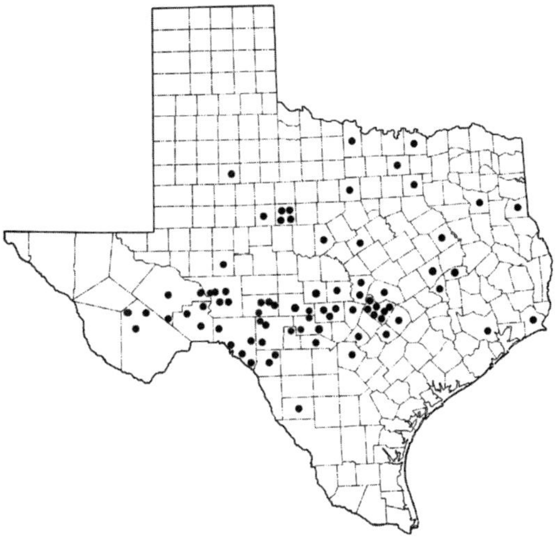

Spermolepis inermis

Taenidia integerrima

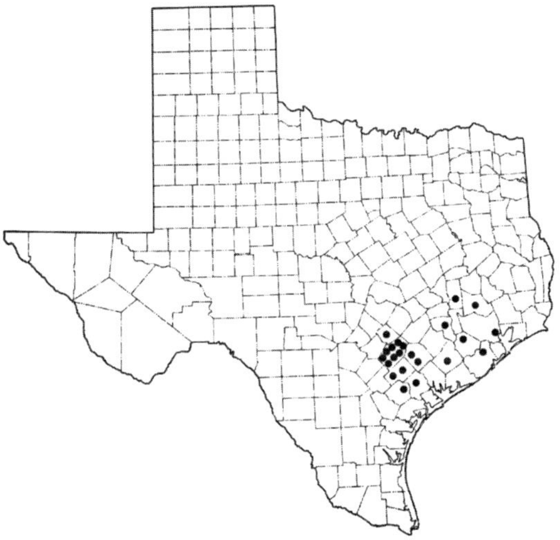

Tauschia texana

Thaspium barbinode

Thaspium trifoliatum

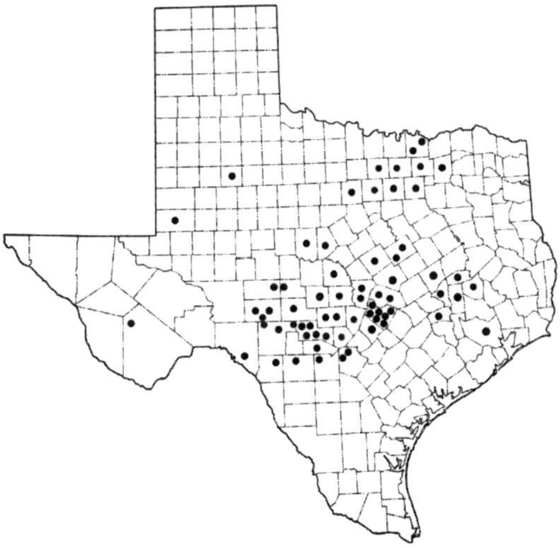

Torilis arvensis

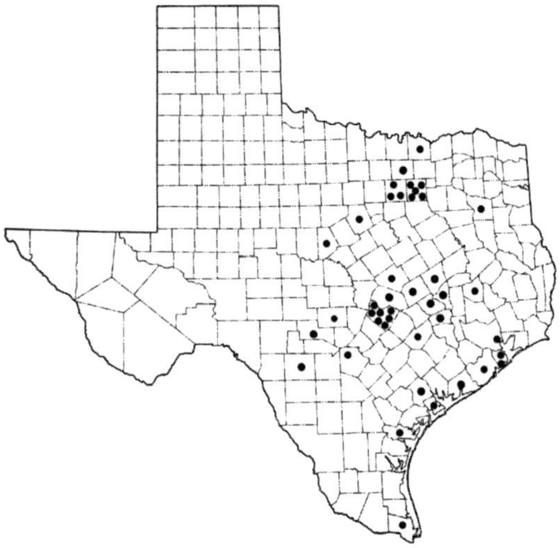

Torilis nodosa

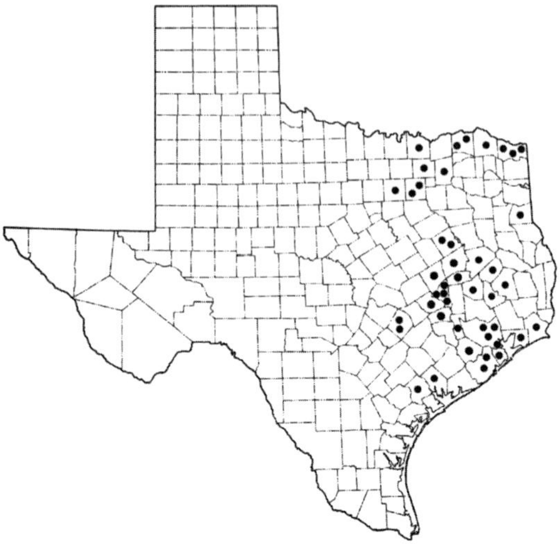

Trepocarpus aethusae

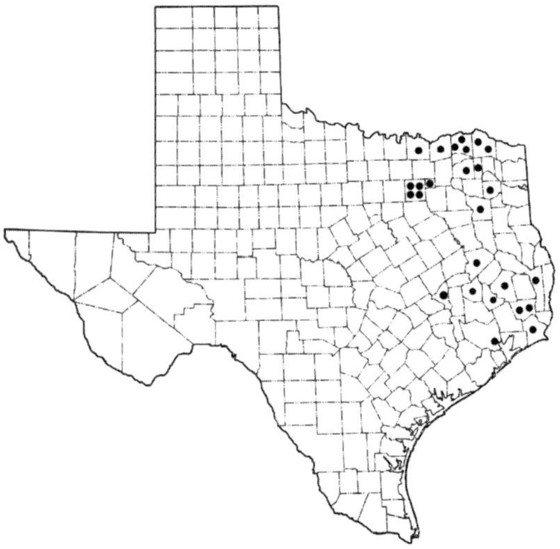

Zizia aurea

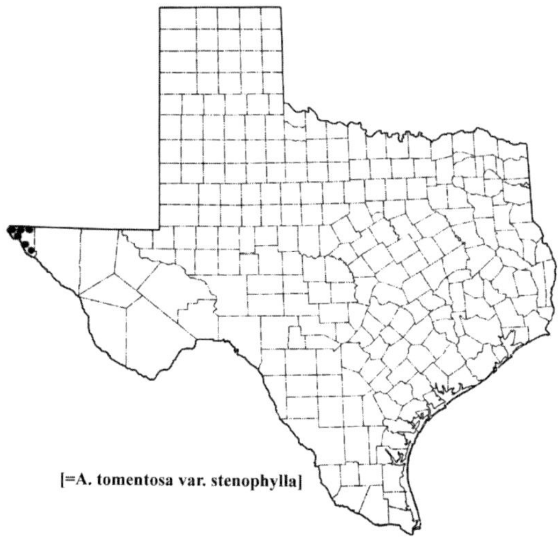
[=A. tomentosa var. stenophylla]

Amsonia arenaria

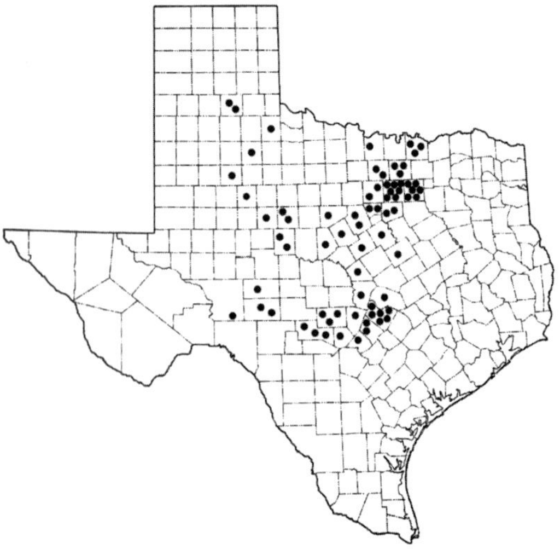

Amsonia ciliata

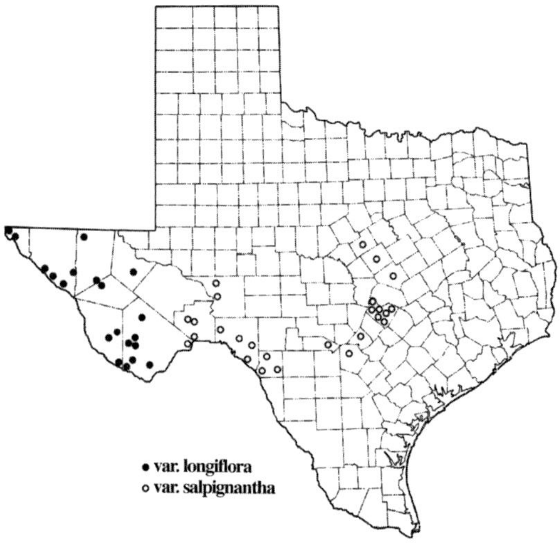
• var. longiflora
○ var. salpignantha

Amsonia longiflora

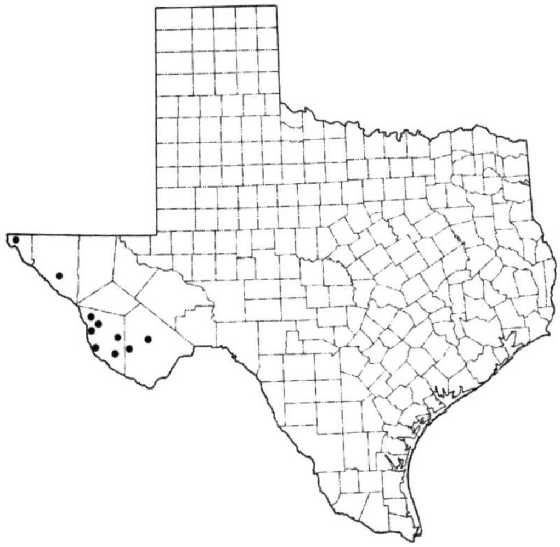

Amsonia palmeri

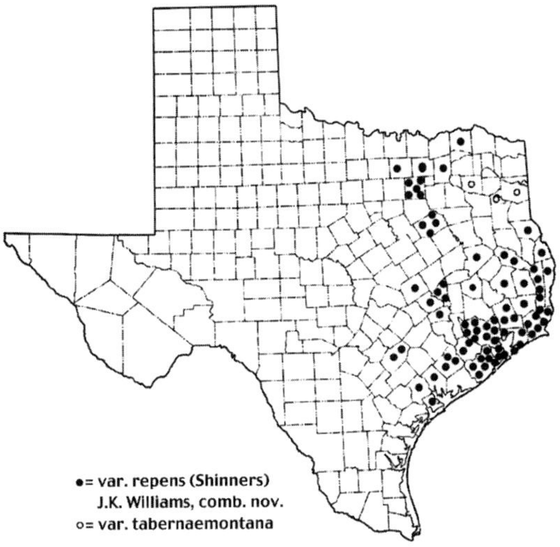
•= var. repens (Shinners) J.K. Williams, comb. nov.
○= var. tabernaemontana

Amsonia tabernaemontana

Amsonia tharpii

Apocynum androsaemifolium

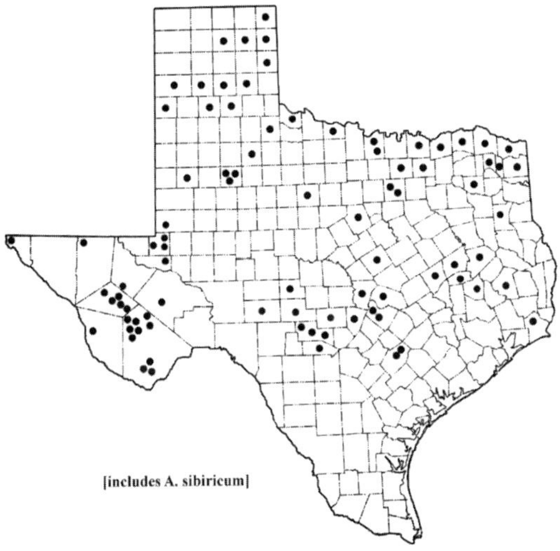
[includes A. sibiricum]

Apocynum cannabinum

Catharanthus roseus

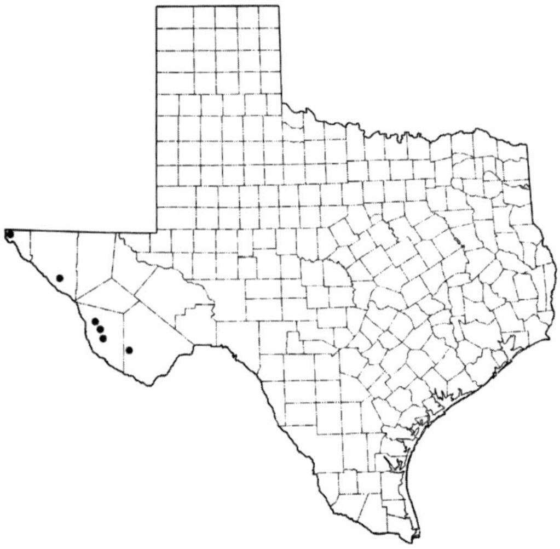

Haplophyton cimicidum

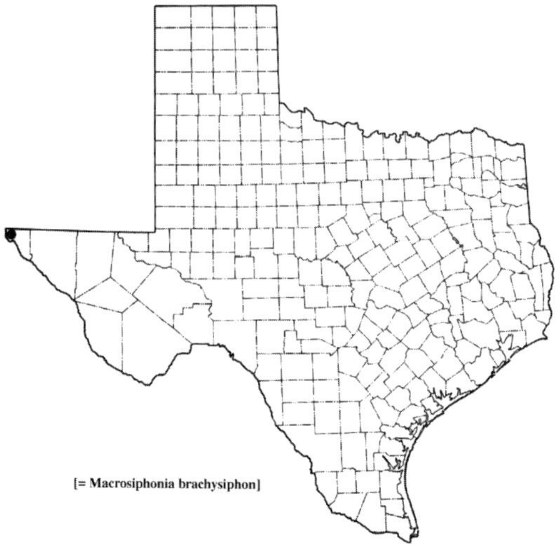
[= Macrosiphonia brachysiphon]

Telosiphonia brachysiphon

[= Macrosiphonia hypoleuca]

Telosiphonia hypoleuca

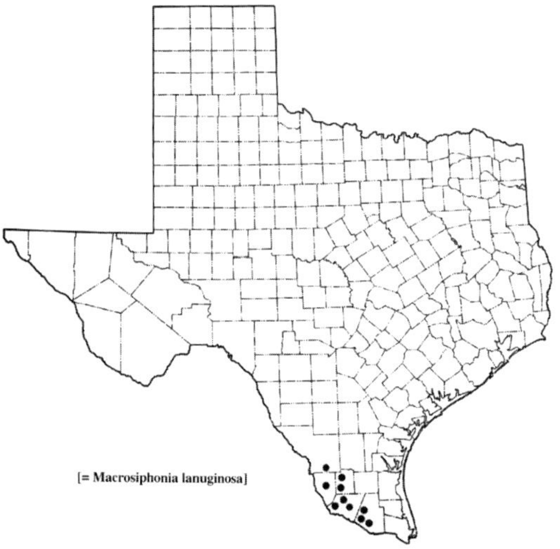
[= Macrosiphonia lanuginosa]

Telosiphonia lanuginosa

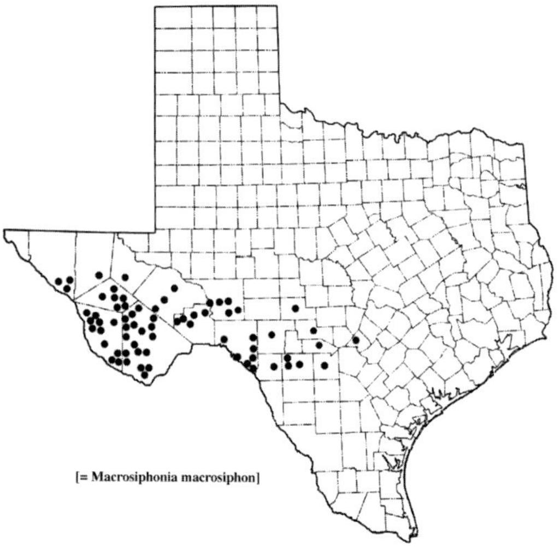
[= Macrosiphonia macrosiphon]

Telosiphonia macrosiphon

Thevetia peruviana

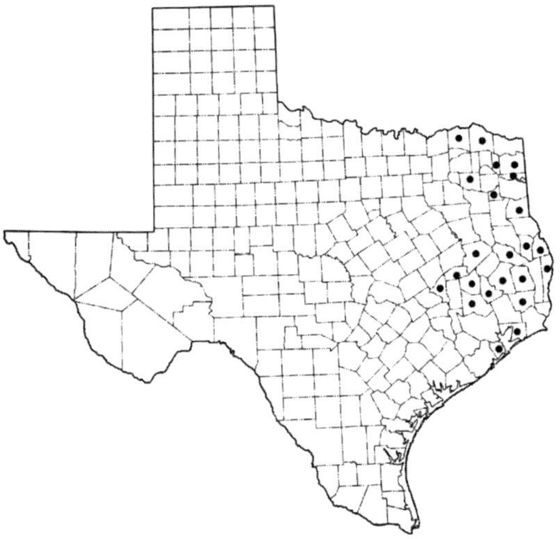

Trachelospermum difforme

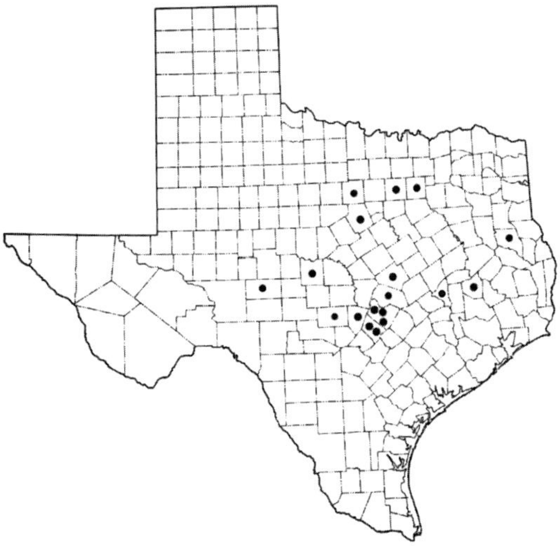

Vinca major

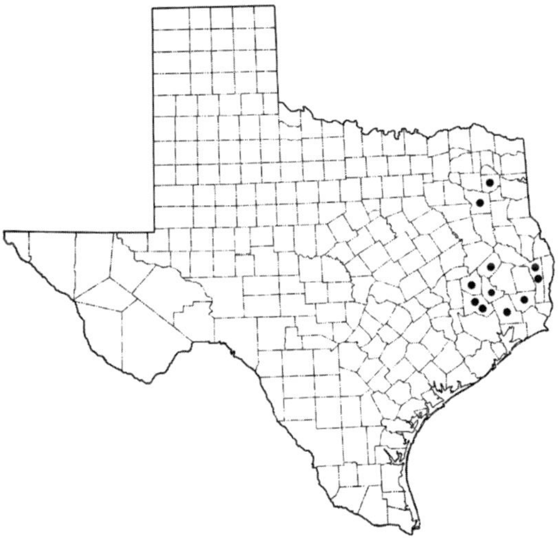

Ilex ambigua

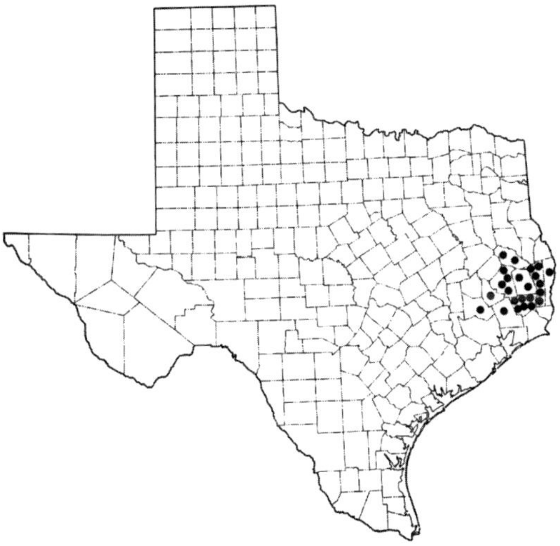

Ilex coriacea

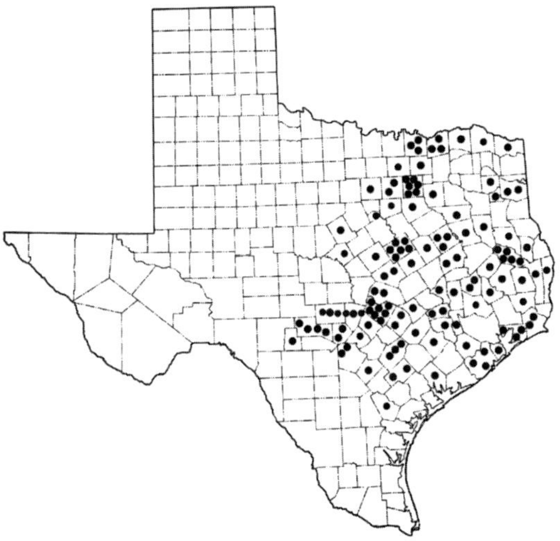

Ilex decidua

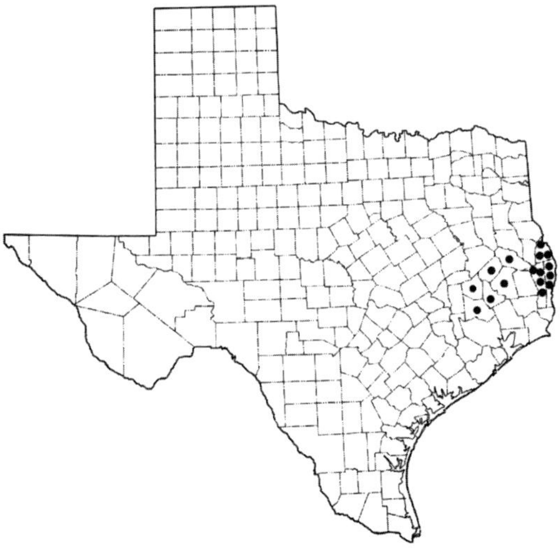

Ilex longipes

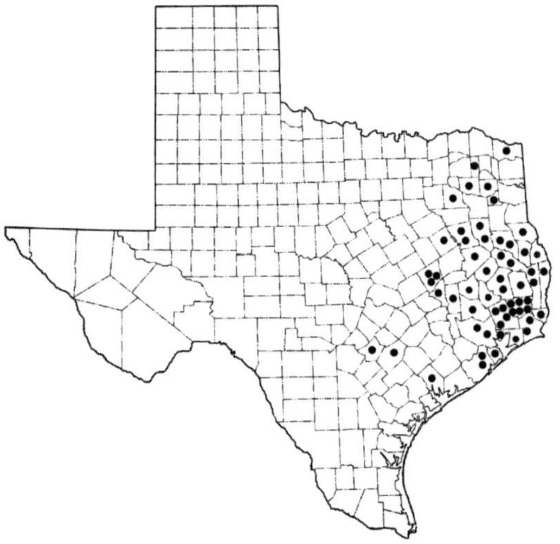

Ilex opaca

Ilex verticillata

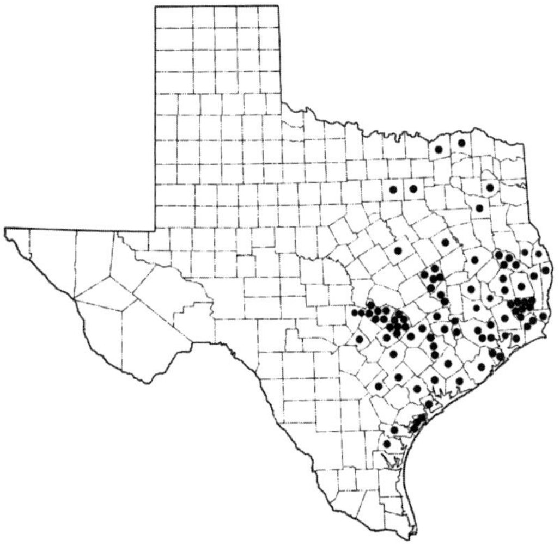

Ilex vomitoria

Aralia racemosa

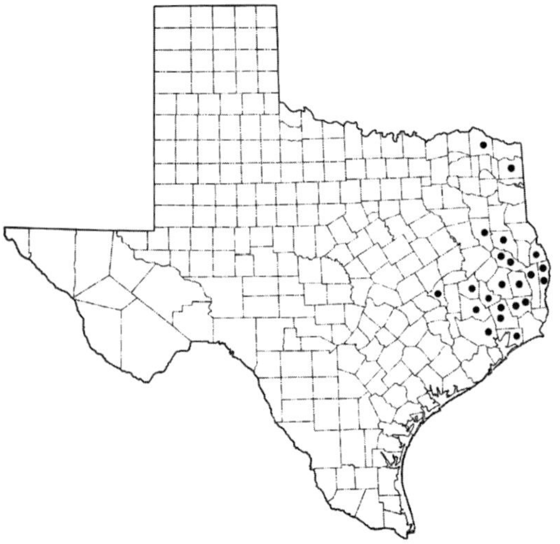

Aralia spinosa

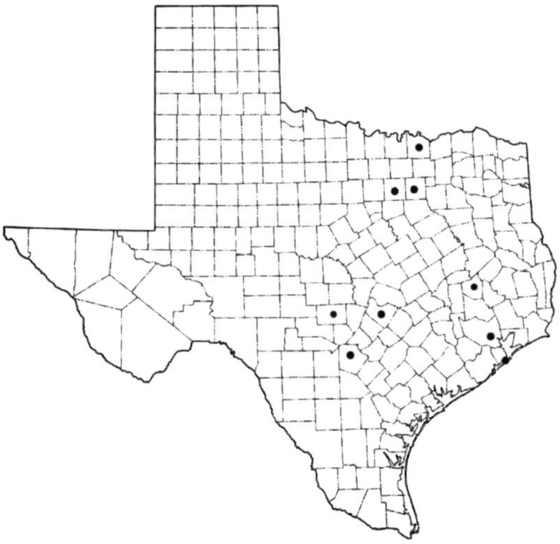

Hedera helix

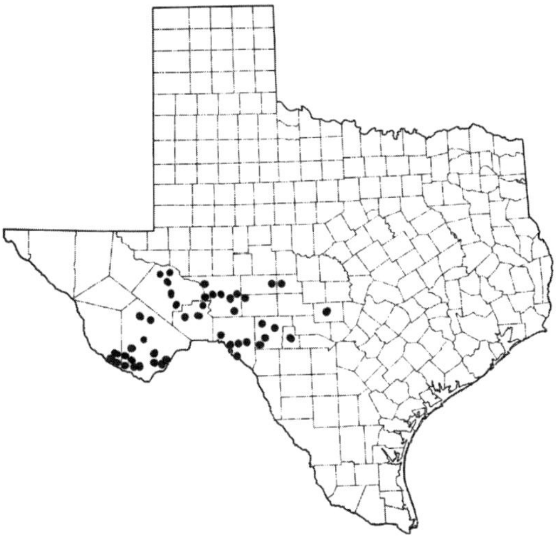

Aristolochia coryi

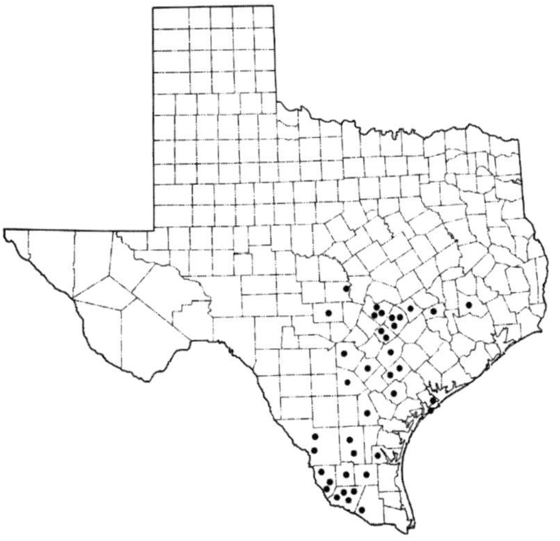

Aristolochia erecta

Aristolochia pentandra

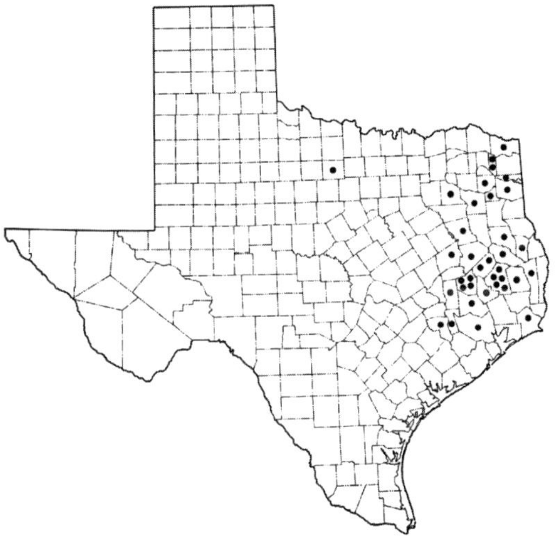

Aristolochia reticulata

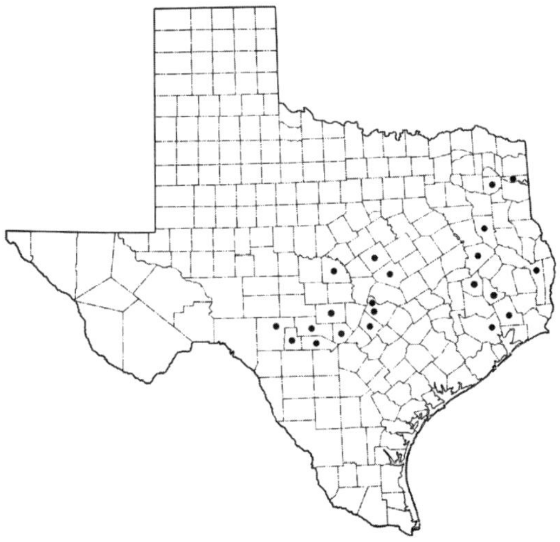

Aristolochia serpentaria

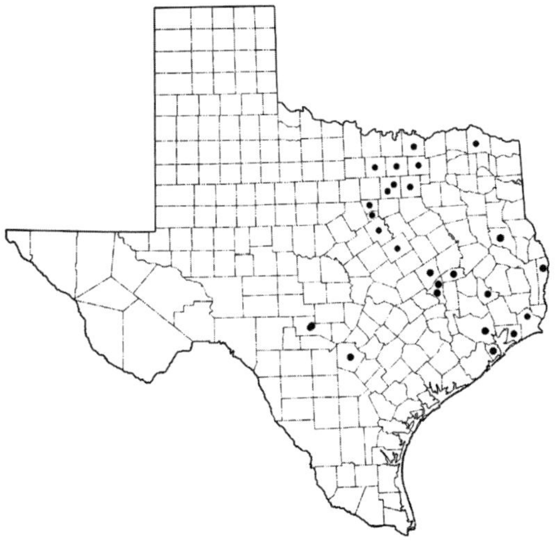

Aristolochia tomentosa

Aristolochia wrightii

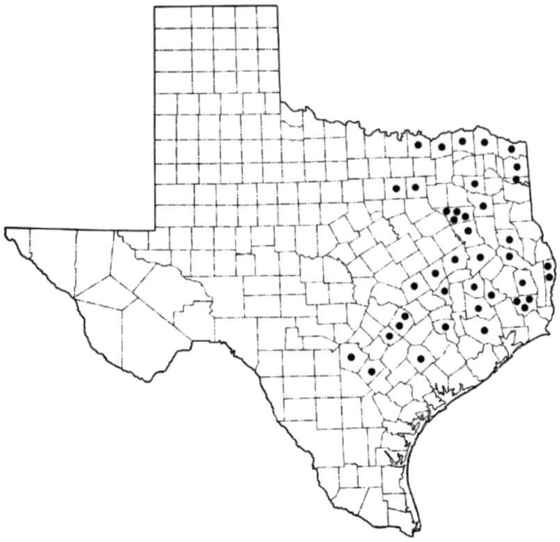

Asclepias amplexicaulis

Asclepias arenaria

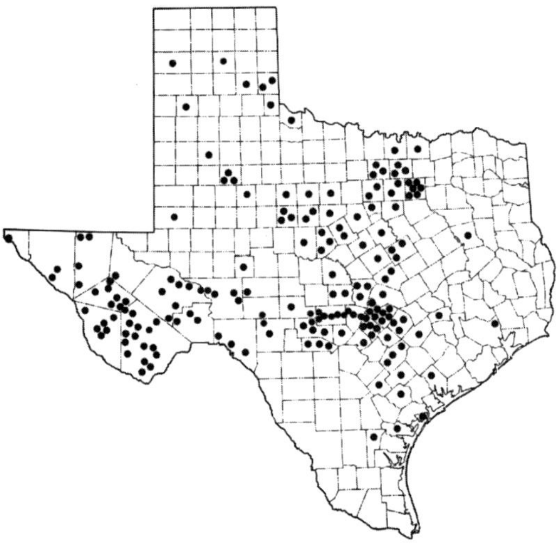

Asclepias asperula

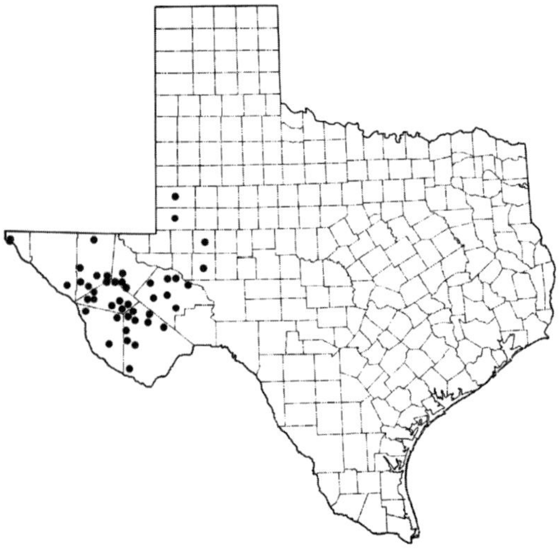

Asclepias brachystephana

Asclepias curassavica

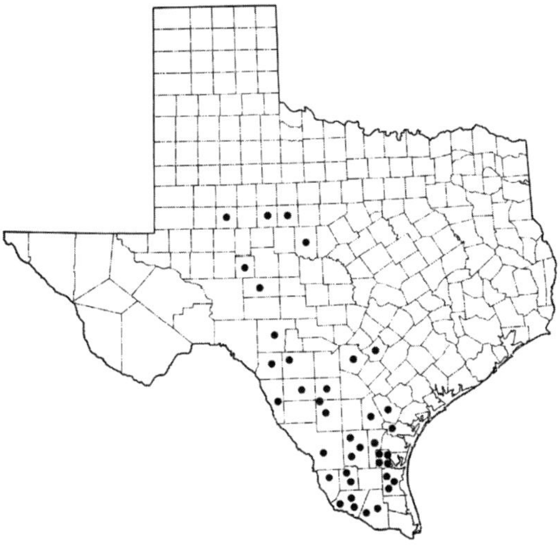

Asclepias emoryi

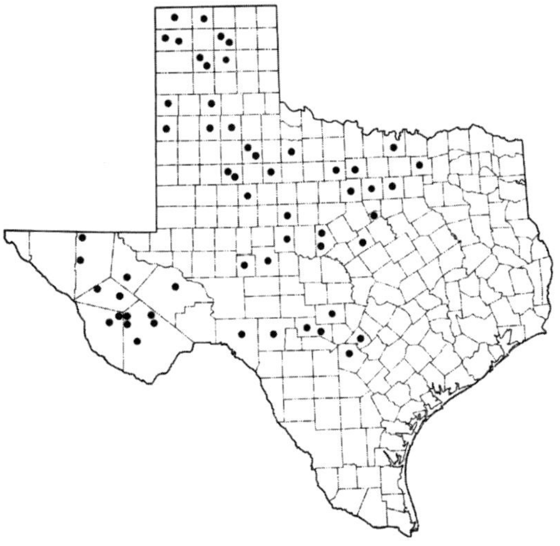

Asclepias engelmanniana

[includes **A. elata**]

Asclepias glaucescens

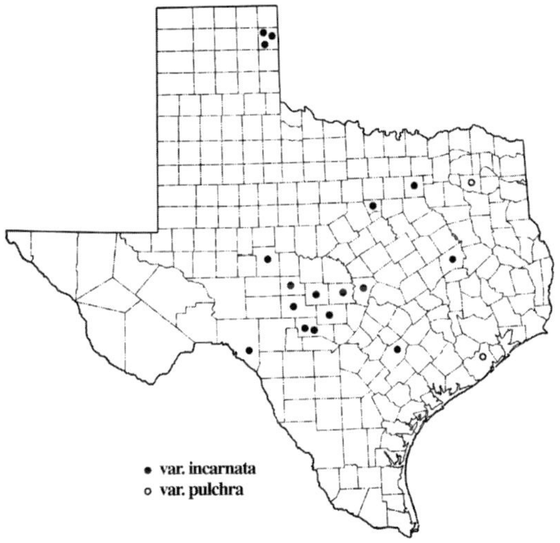
● **var. incarnata**
○ **var. pulchra**

Asclepias incarnata

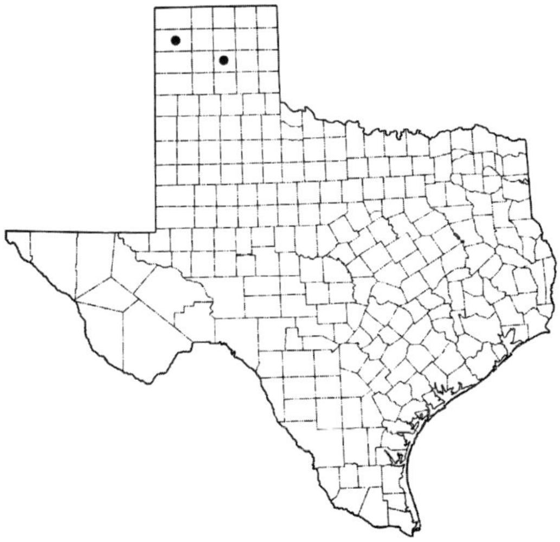

Asclepias involucrata

Asclepias lanceolata

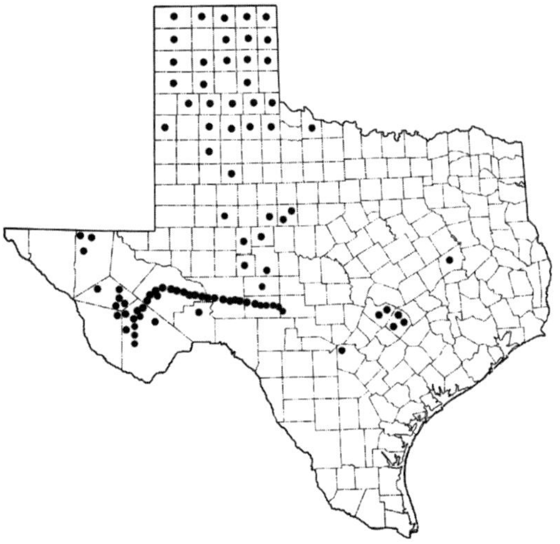

Asclepias latifolia

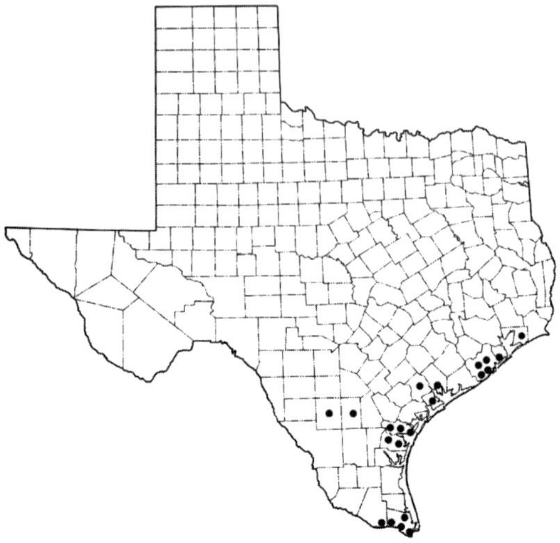

Asclepias linearis

Asclepias longifolia

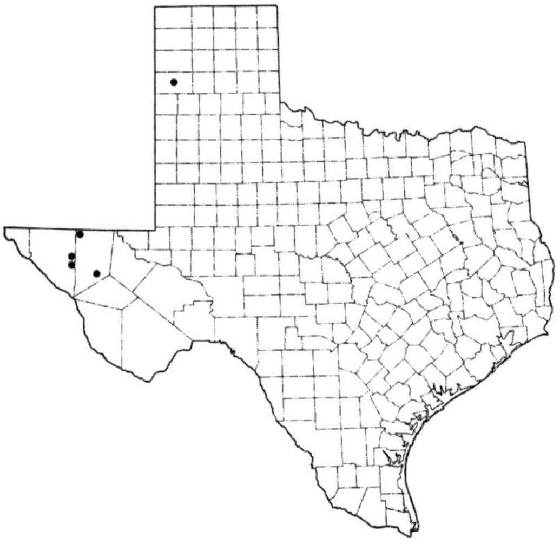

Asclepias macrotis

Asclepias nummularia

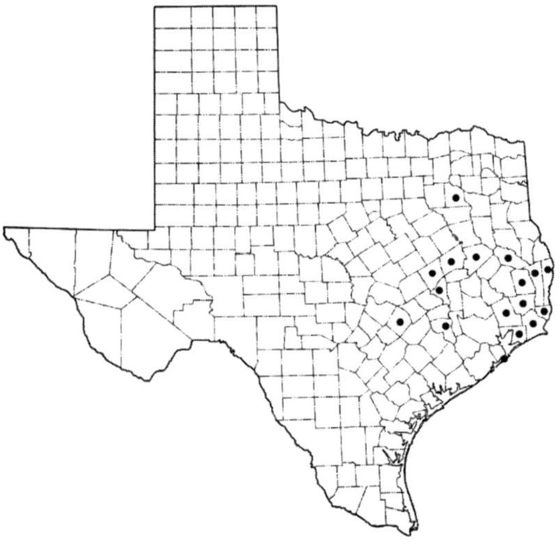

Asclepias obovata

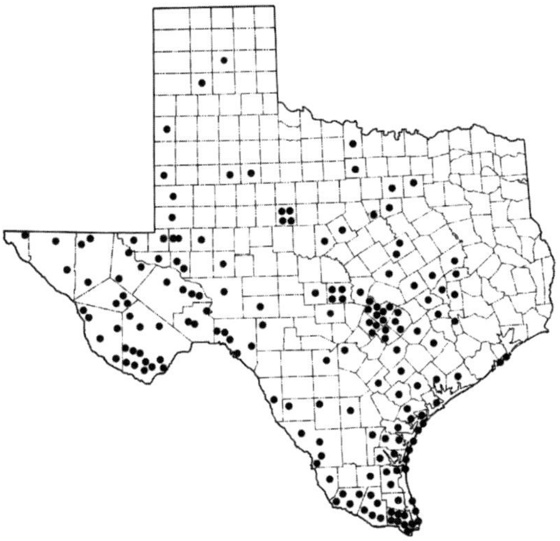

Asclepias oenotheroides

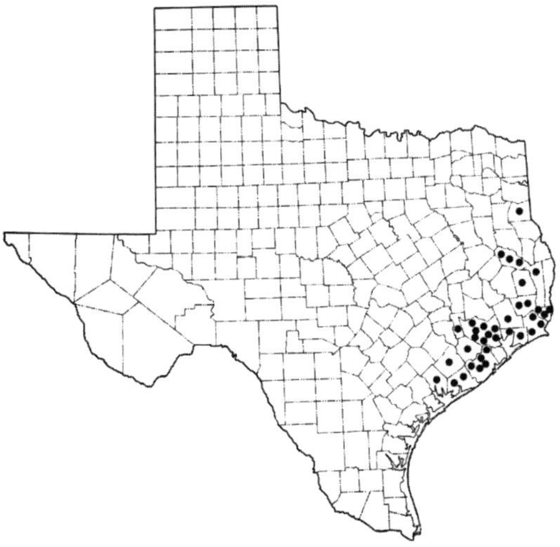

Asclepias perennis

Asclepias prostrata

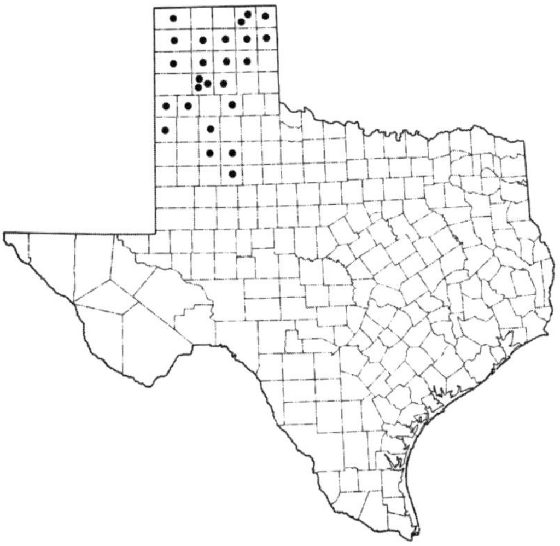

Asclepias pumila

Asclepias purpurascens

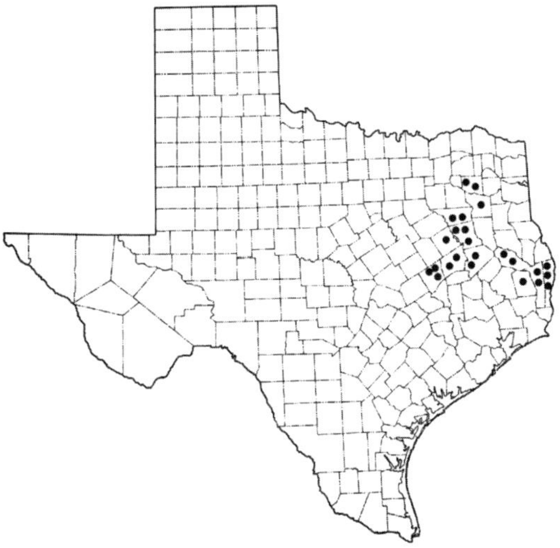

Asclepias rubra

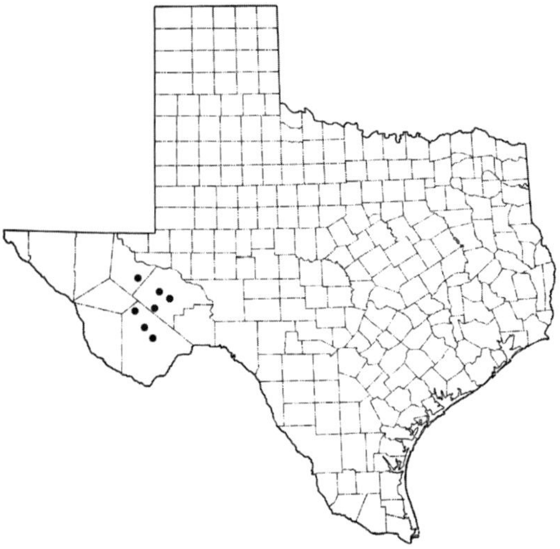

Asclepias scaposa

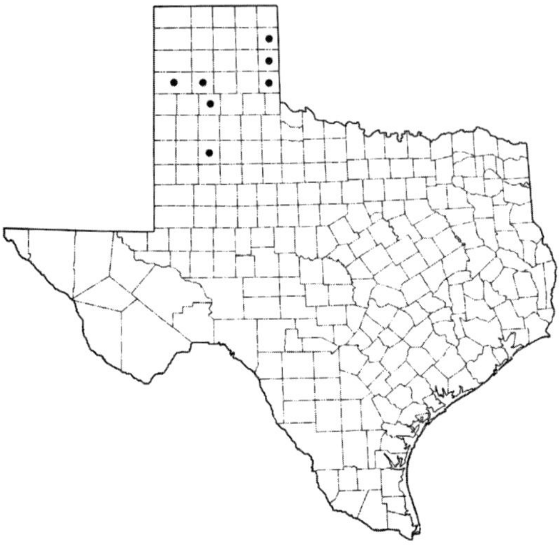

Asclepias speciosa

Asclepias sperryi

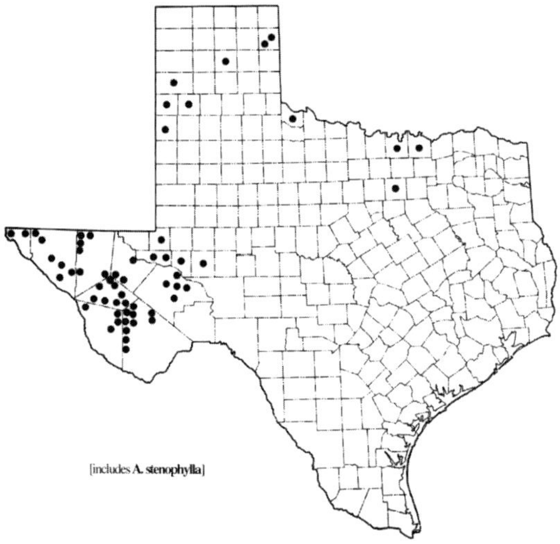
[includes **A. stenophylla**]

Asclepias subverticillata

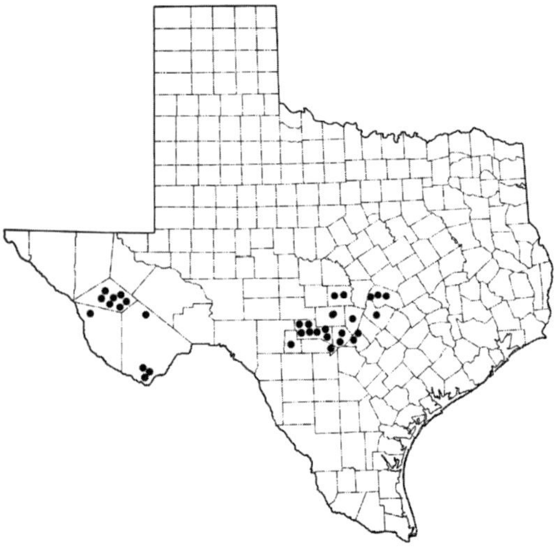

Asclepias texana

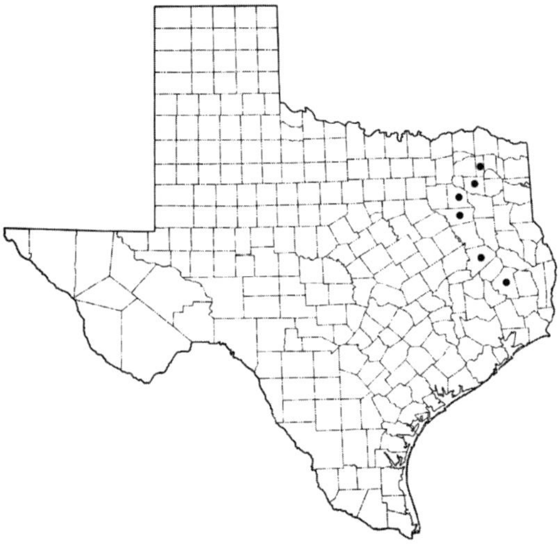

Asclepias tomentosa

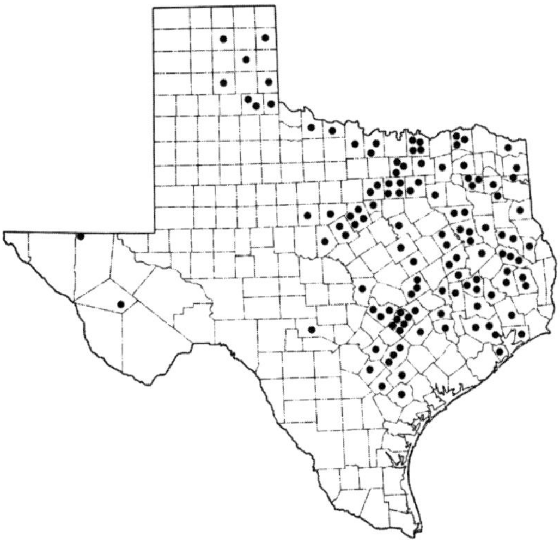

Asclepias tuberosa

Asclepias uncialis

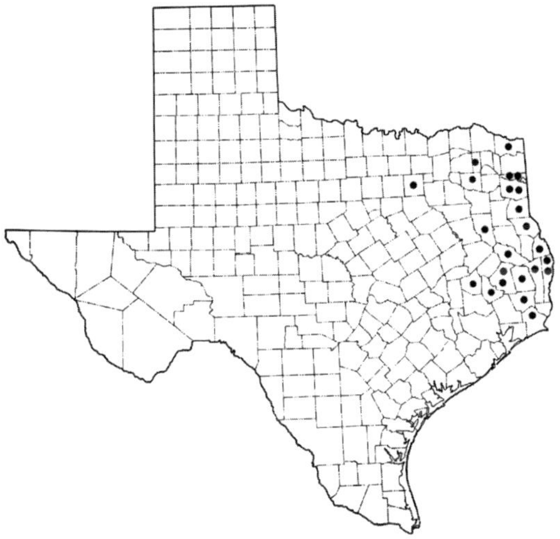

Asclepias variegata

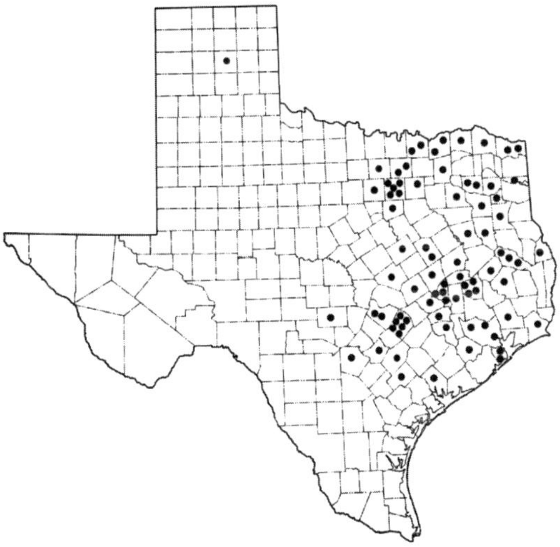

Asclepias verticillata

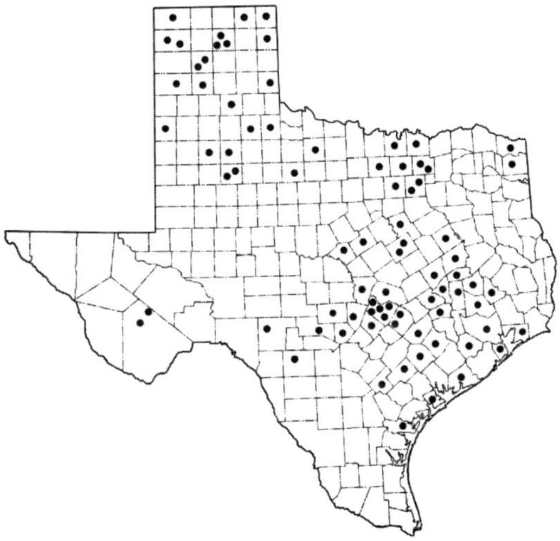

Asclepias viridiflora

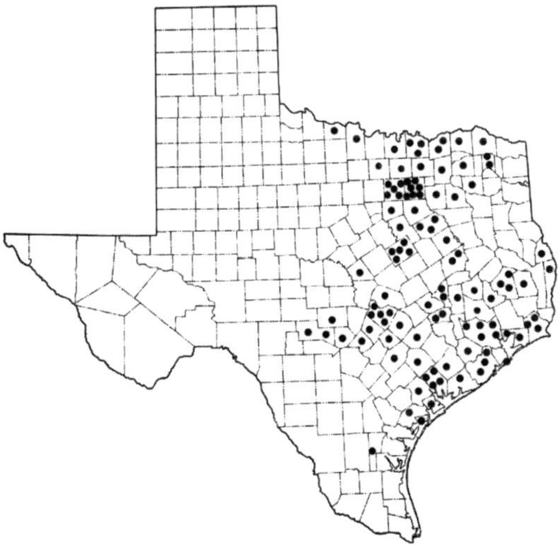

Asclepias viridis

Cynanchum angustifolium

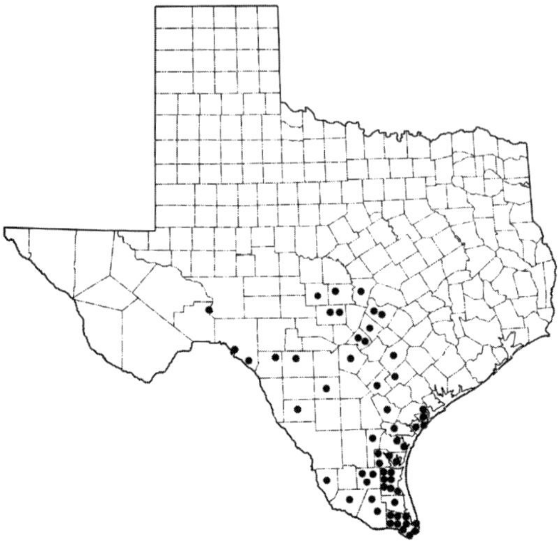

Cynanchum barbigerum

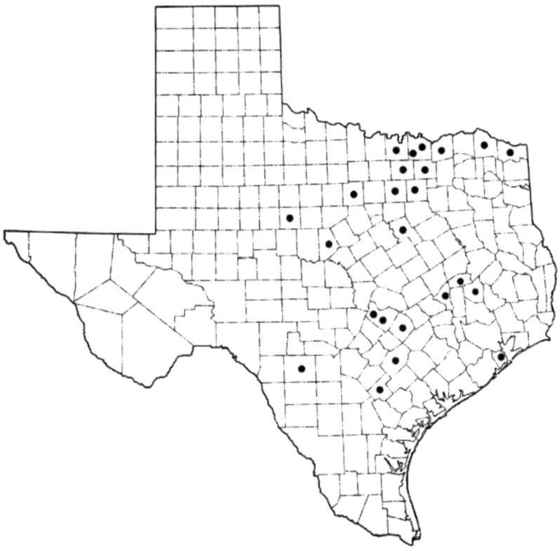

Cynanchum laeve

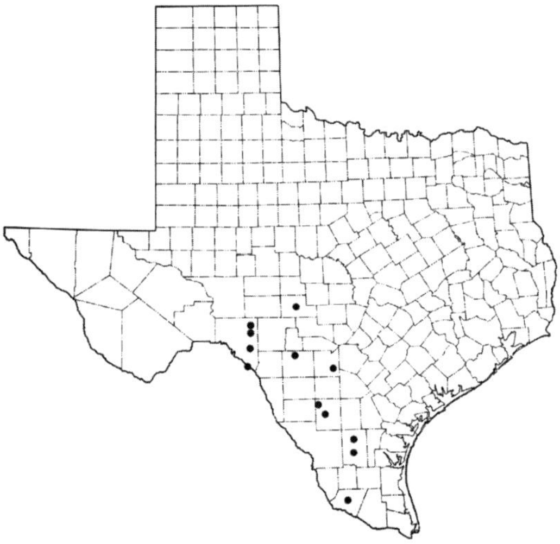

Cynanchum maccartii

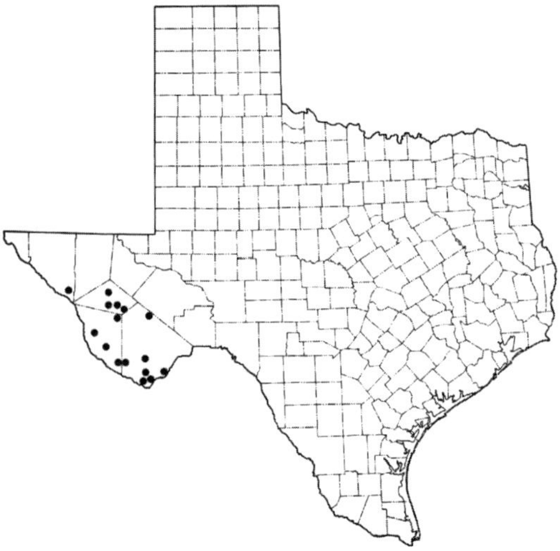

Cynanchum pringlei

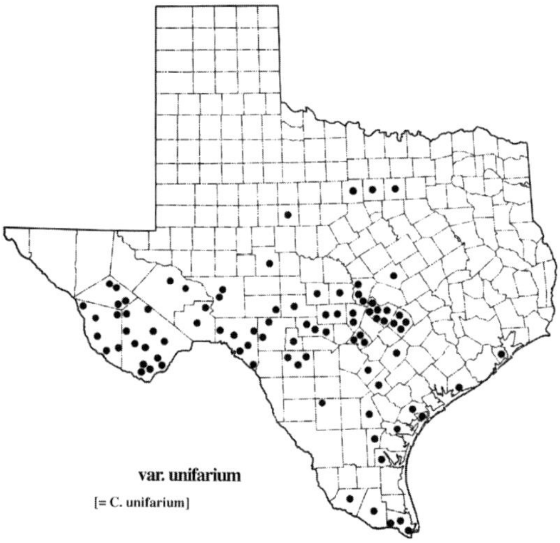
var. unifarium
[= C. unifarium]

Cynanchum racemosum

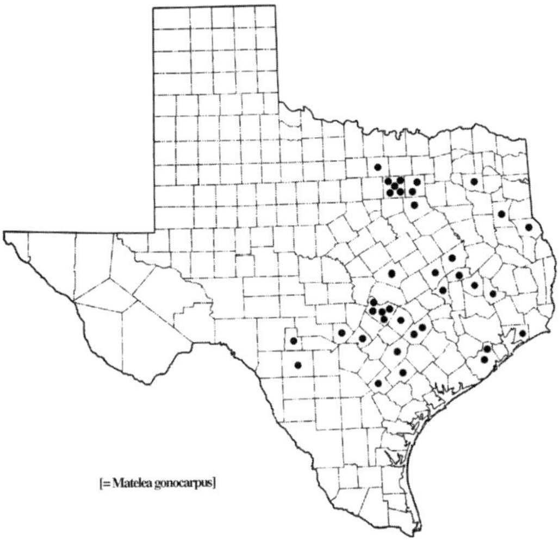
[= Matelea gonocarpus]

Gonolobus gonocarpos

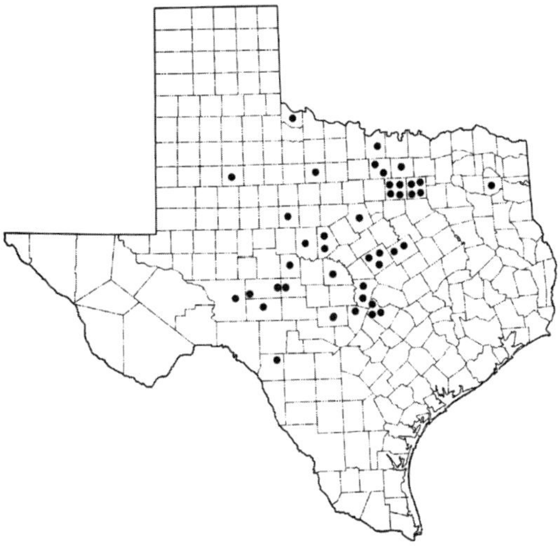

Matelea biflora

Matelea brevicoronata

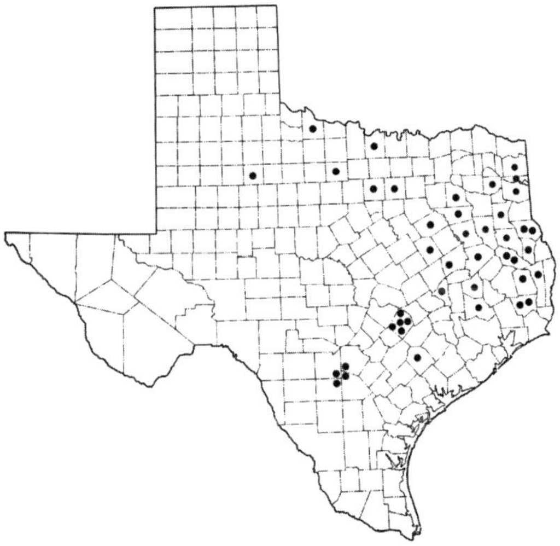

Matelea cynanchoides

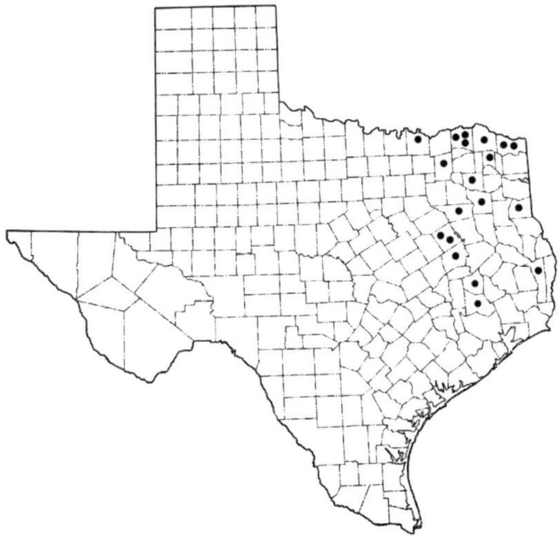

Matelea decipiens

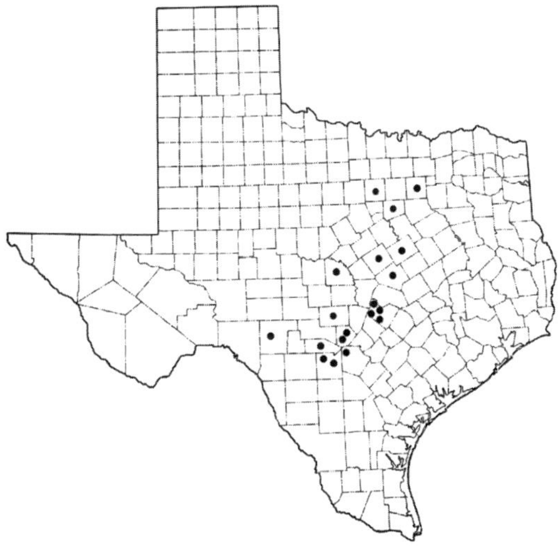

Matelea edwardsensis

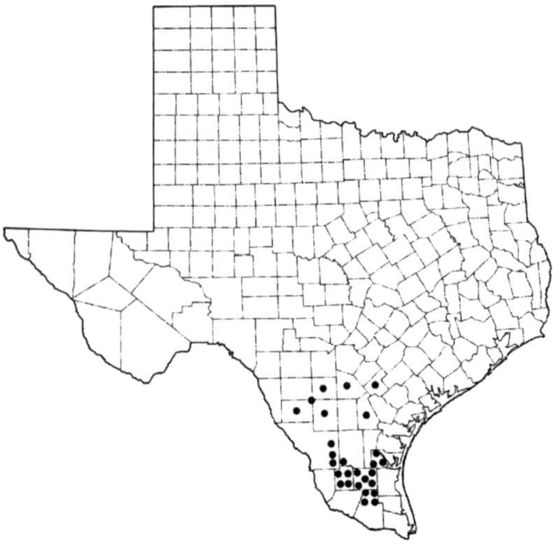

Matelea parviflora

Matelea parvifolia

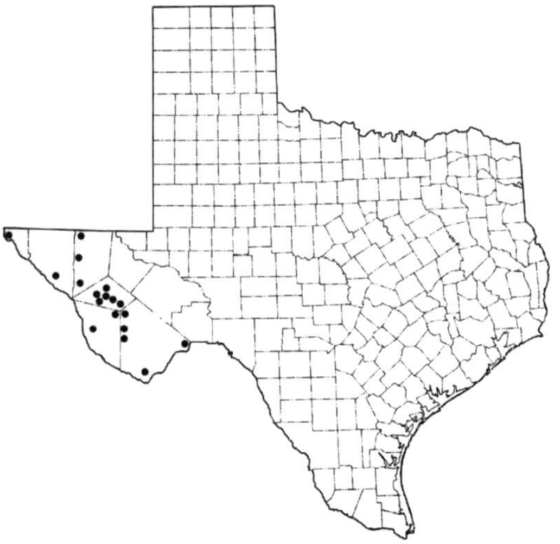

Matelea producta

Matelea radiata

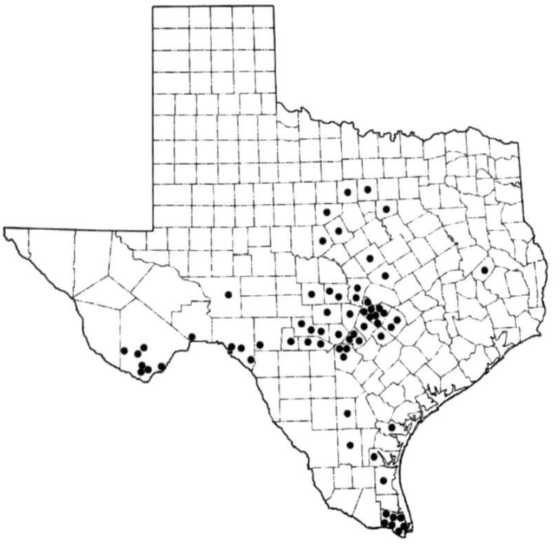

Matelea reticulata

Matelea sagittifolia

Matelea texensis

Periploca graeca

[= Funastrum clausum]

Sarcostemma clausum

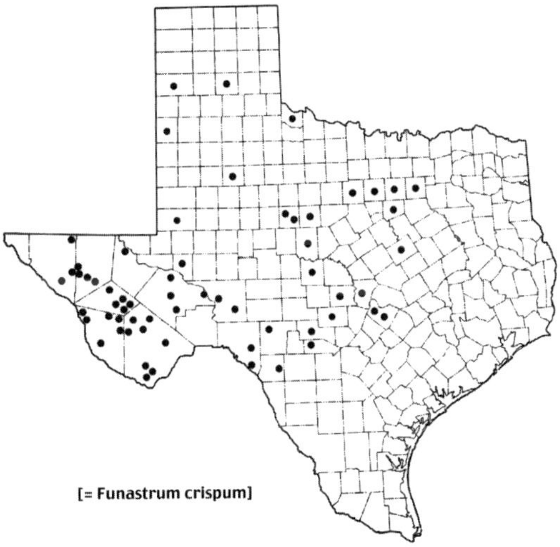
[= Funastrum crispum]

Sarcostemma crispum

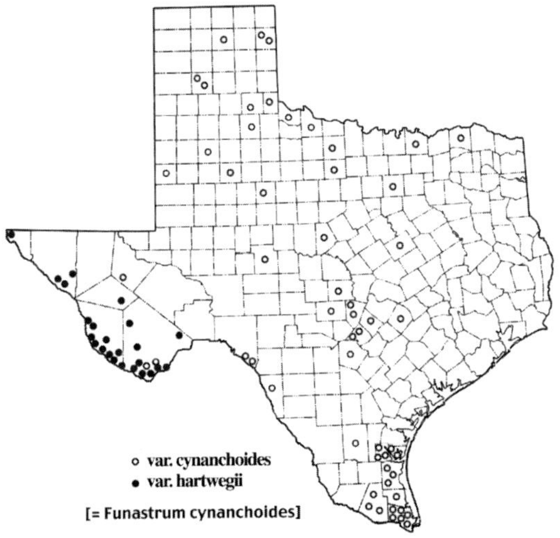
○ var. cynanchoides
● var. hartwegii
[= Funastrum cynanchoides]

Sarcostemma cynanchoides

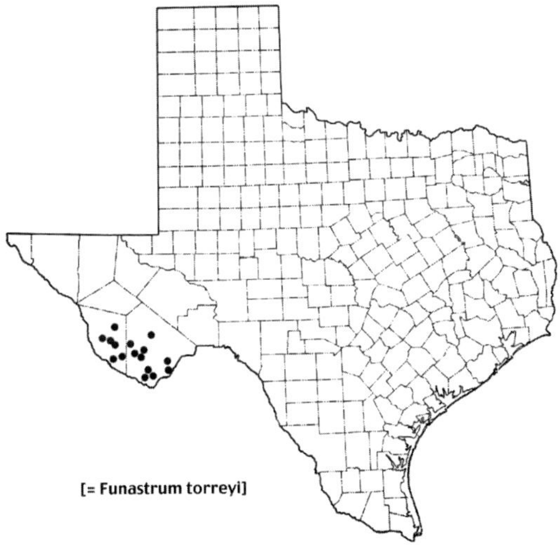
[= Funastrum torreyi]

Sarcostemma torreyi

Acanthospermum australe

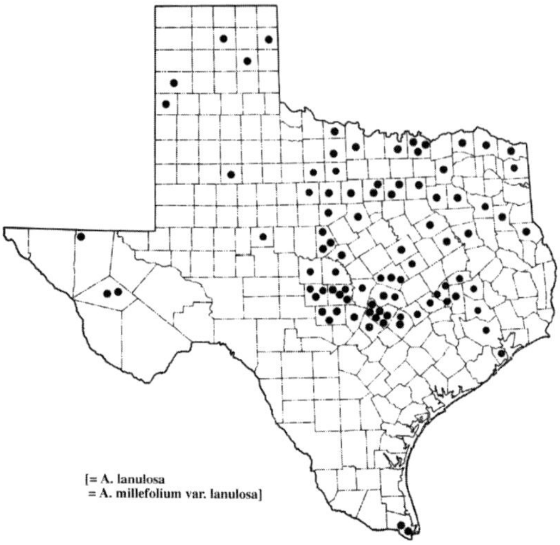
[= A. lanulosa
= A. millefolium var. lanulosa]

Achillea millefolium

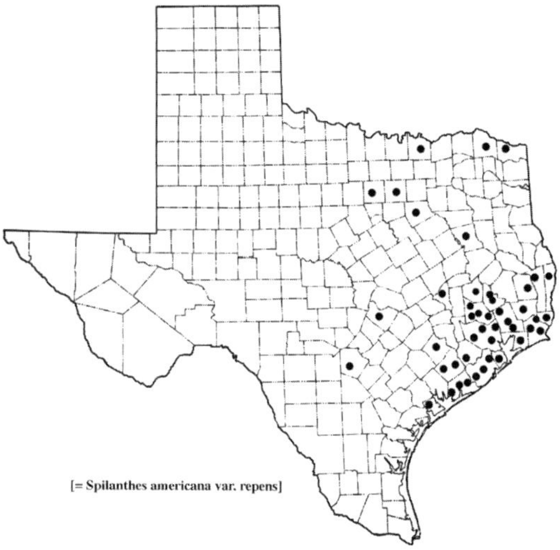
[= Spilanthes americana var. repens]

Acmella oppositifolia

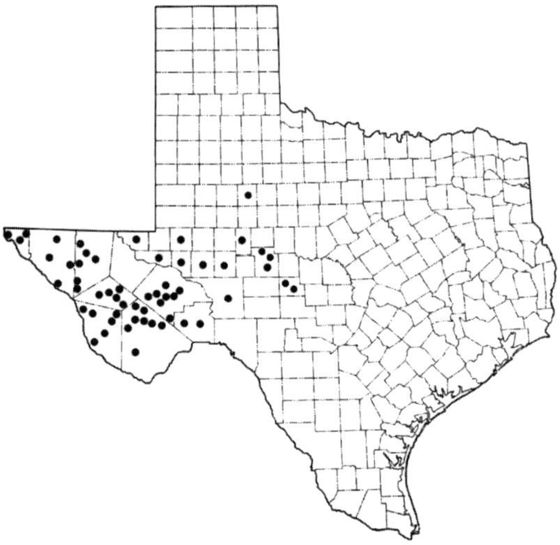

Acourtia nana

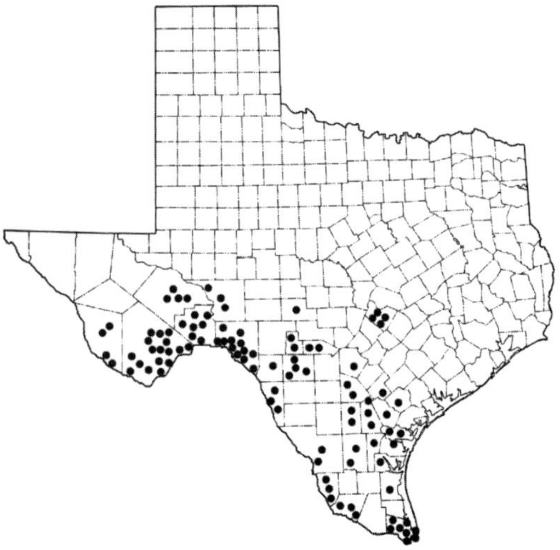

Acourtia runcinata

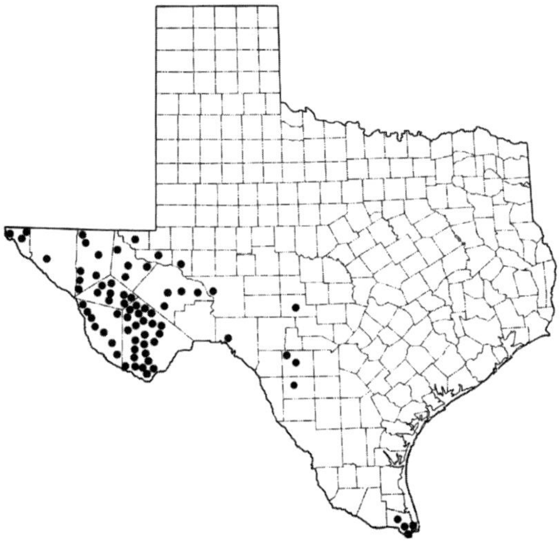

Acourtia wrightii

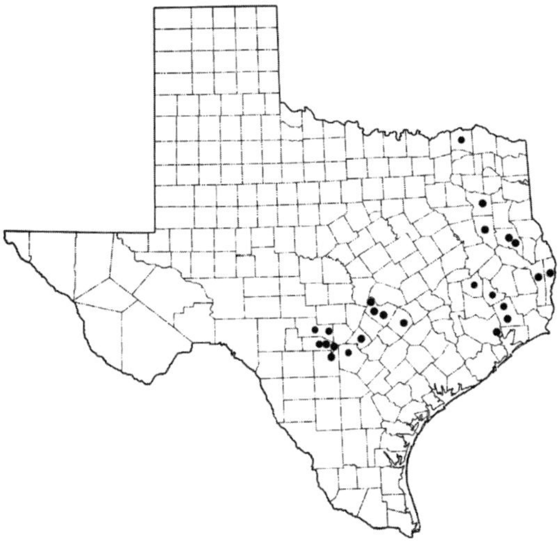

Ageratina altissima

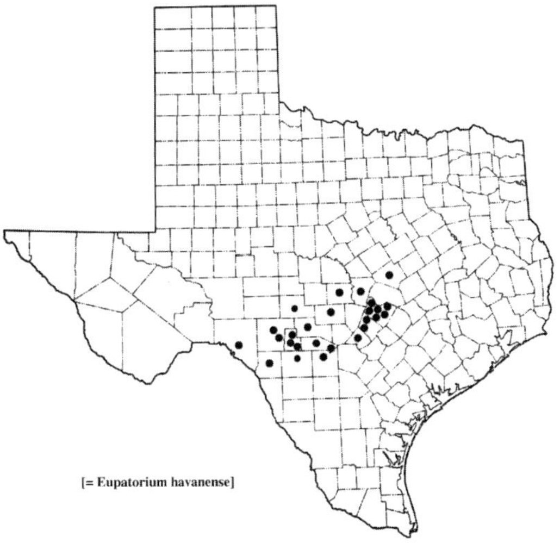
[= Eupatorium havanense]

Ageratina havanensis

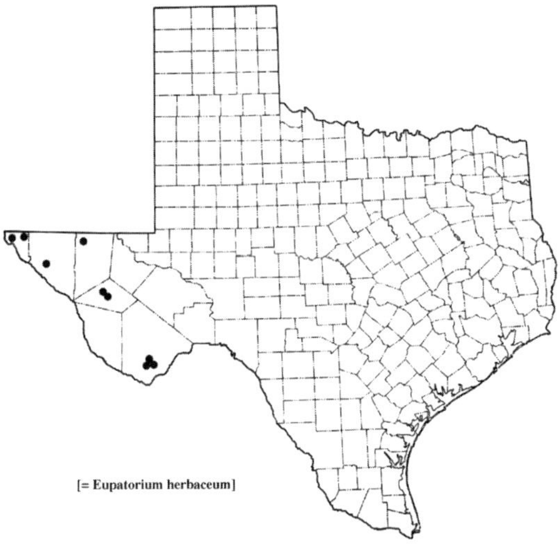
[= Eupatorium herbaceum]

Ageratina herbacea

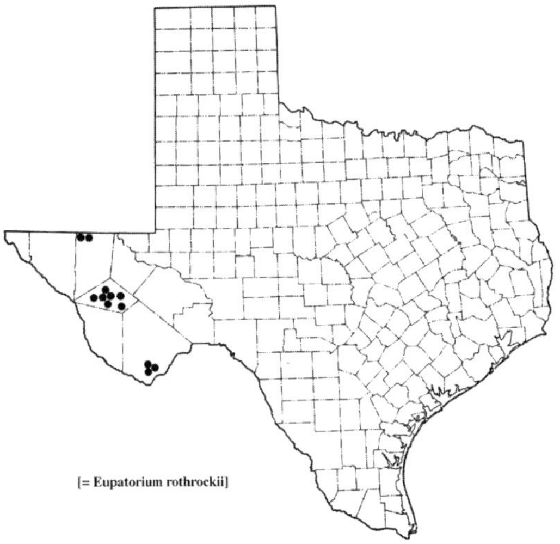
[= Eupatorium rothrockii]

Ageratina rothrockii

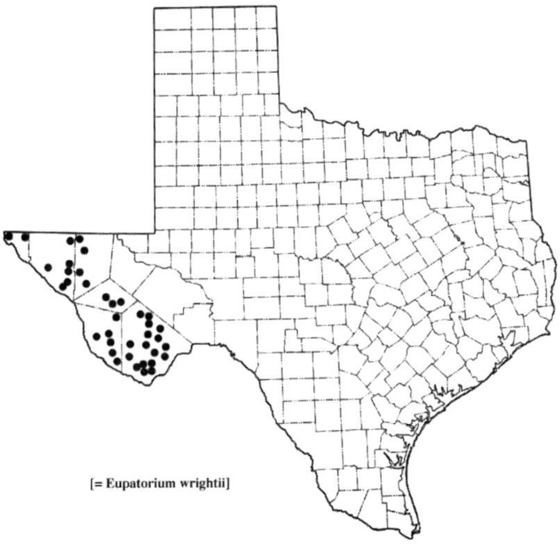
[= Eupatorium wrightii]

Ageratina wrightii

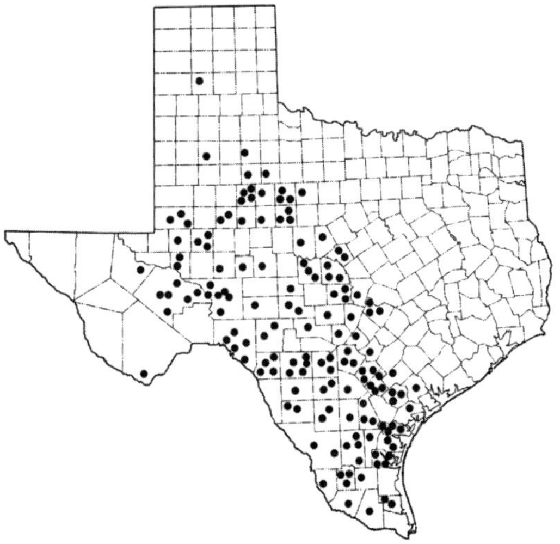

Amblyolepis setigera

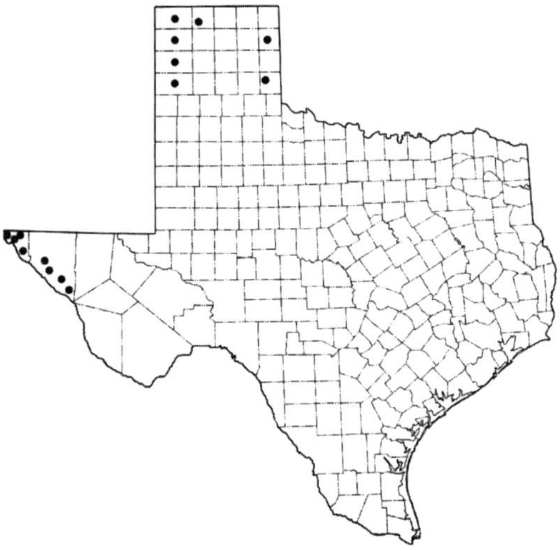

Ambrosia acanthicarpa

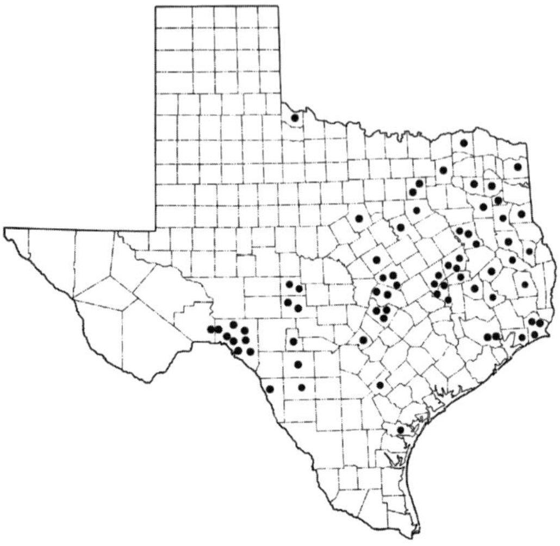

Ambrosia artemisiifolia

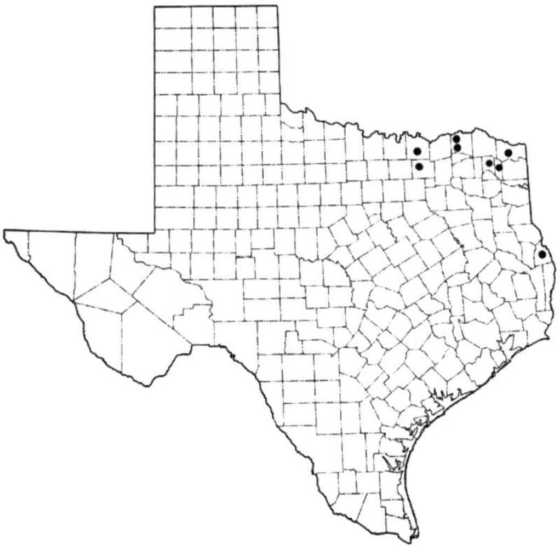

Ambrosia bidentata

Ambrosia cheiranthifolia

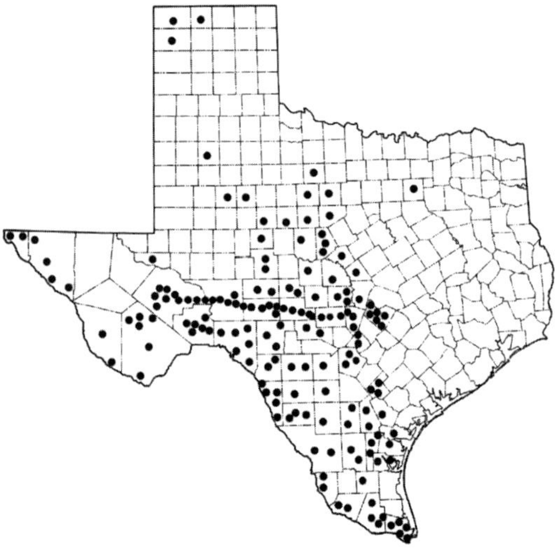

Ambrosia confertiflora

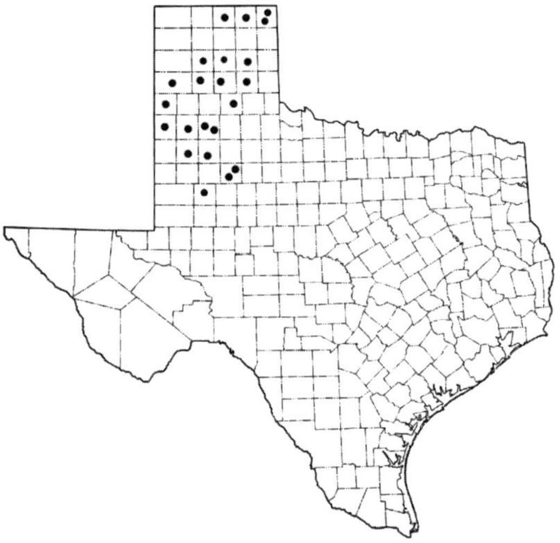

Ambrosia grayi

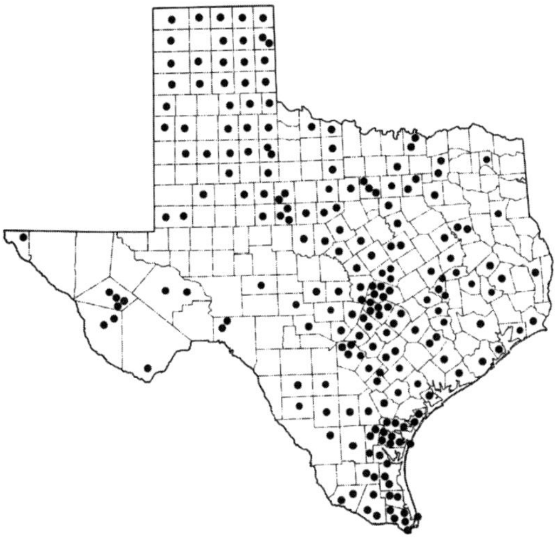

Ambrosia psilostachya

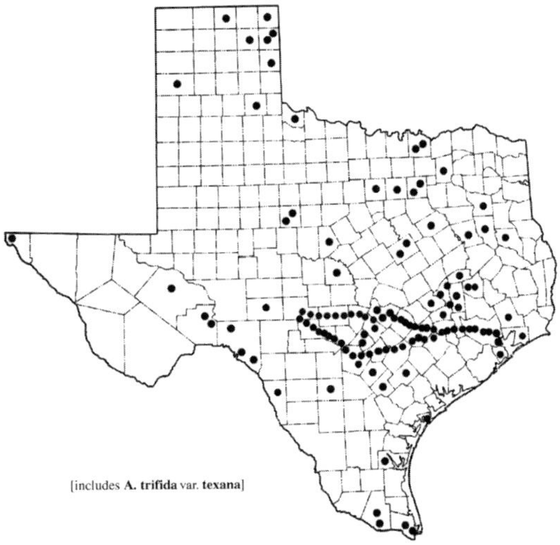
[includes **A. trifida** var. **texana**]

Ambrosia trifida

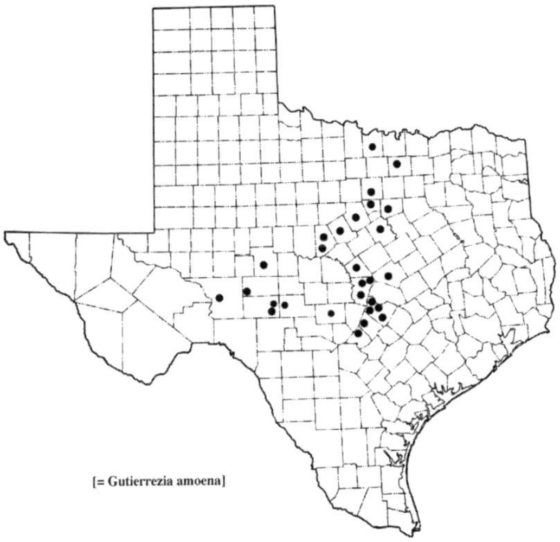
[= **Gutierrezia amoena**]

Amphiachyris amoena

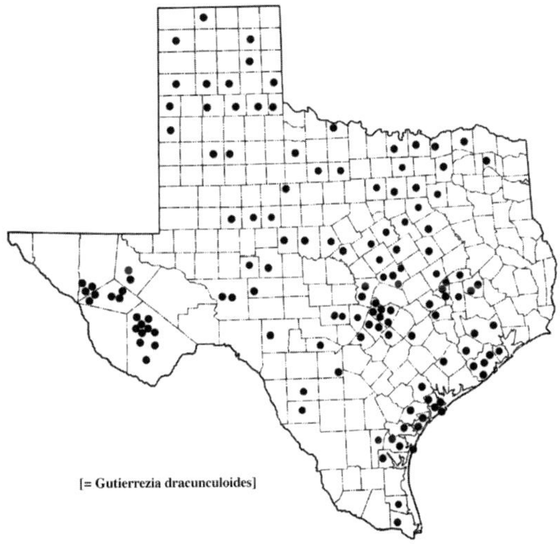
[= **Gutierrezia dracunculoides**]

Amphiachyris dracunculoides

Anaphalis margaritacea

Antennaria marginata

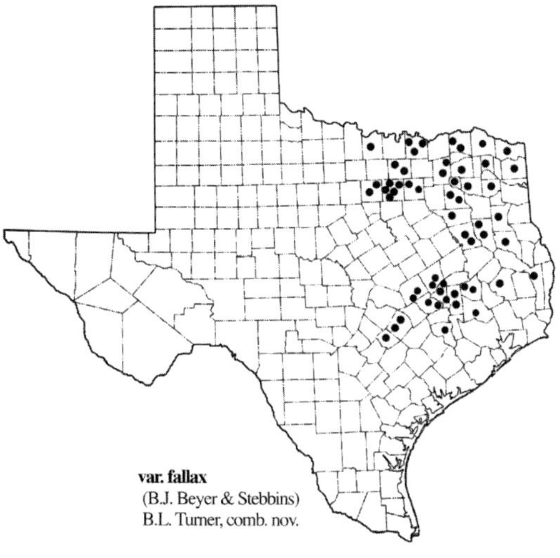
var. fallax
(B.J. Beyer & Stebbins)
B.L. Turner, comb. nov.

Antennaria parlinii

Antennaria parvifolia

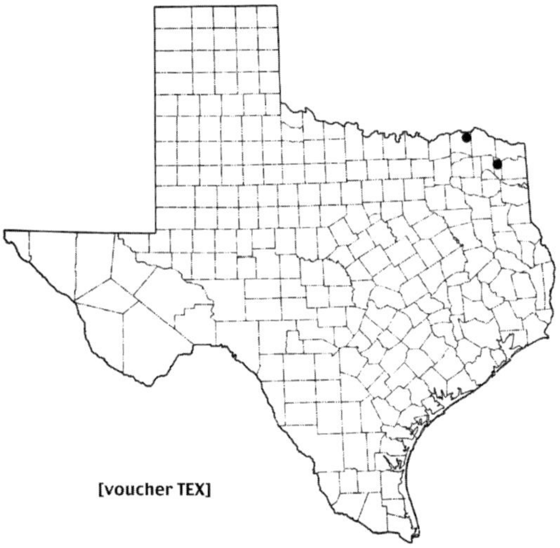
[voucher TEX]

Antennaria plantaginifolia

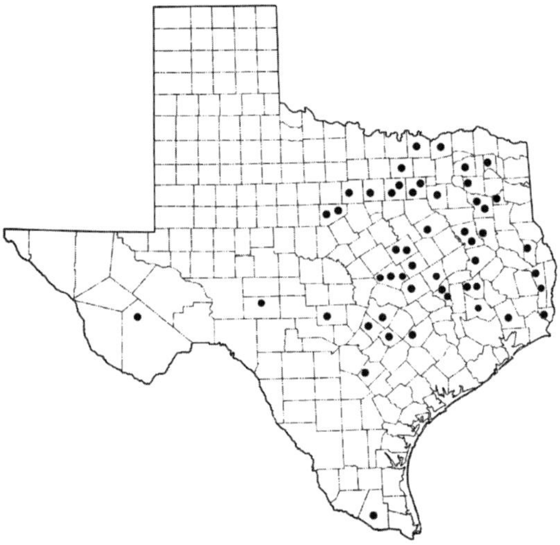

Anthemis cotula

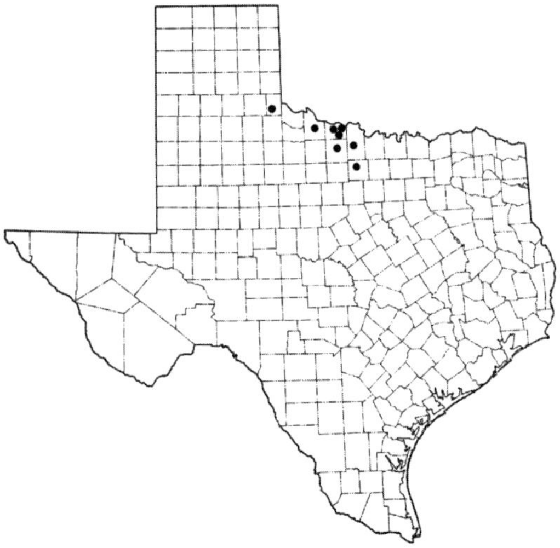

Aphanostephus pilosus

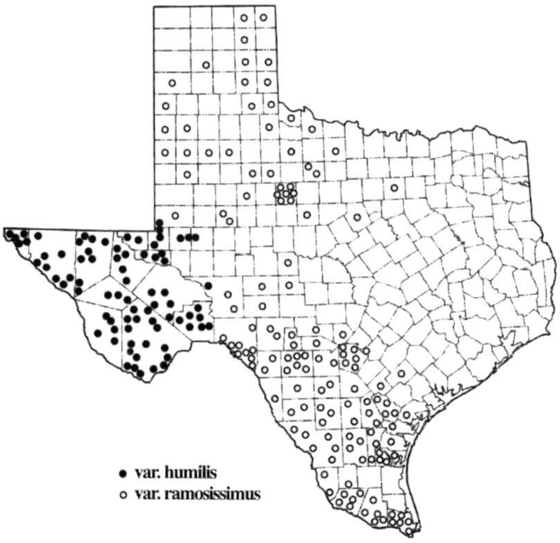
var. humilis
var. ramosissimus

Aphanostephus ramosissimus

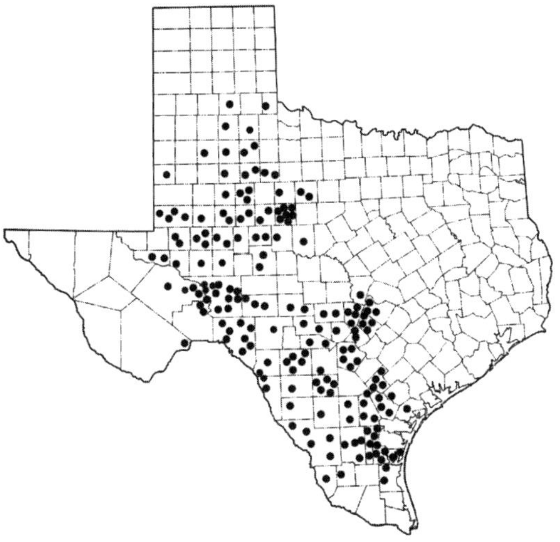

Aphanostephus riddellii

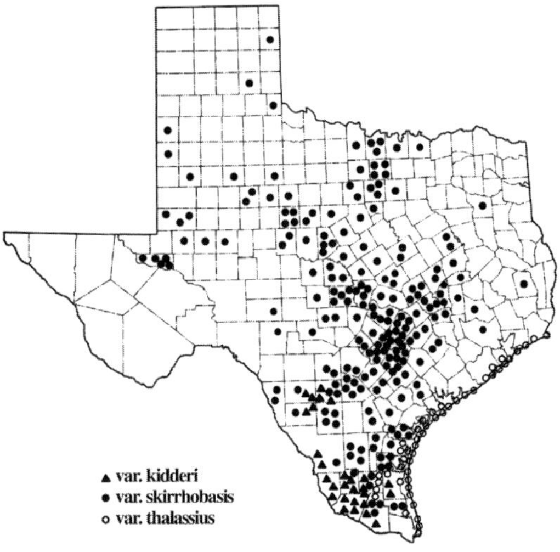
▲ var. kidderi
● var. skirrhobasis
○ var. thalassius

Aphanostephus skirrhobasis

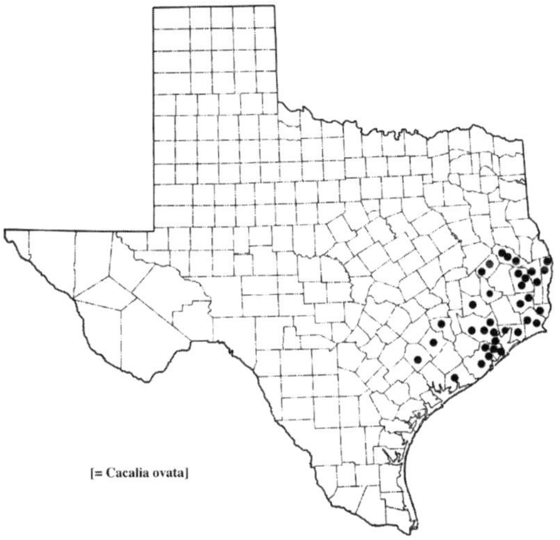
[= Cacalia ovata]

Arnoglossum ovatum

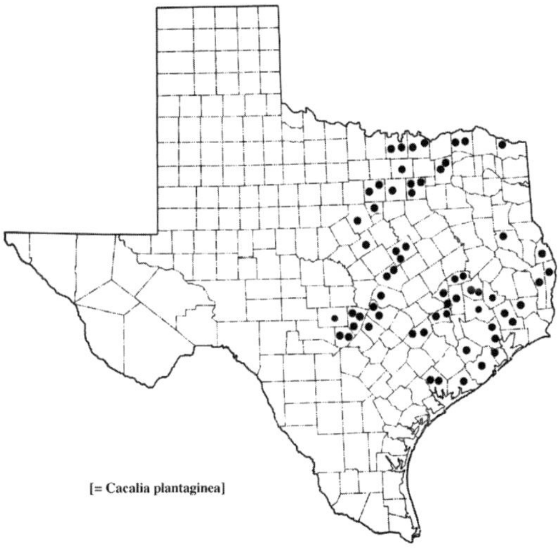
[= Cacalia plantaginea]

Arnoglossum plantagineum

Artemisia bigelovii

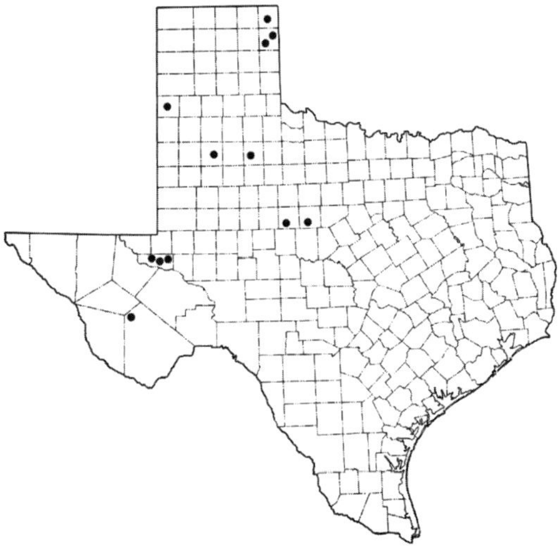

Artemisia campestris

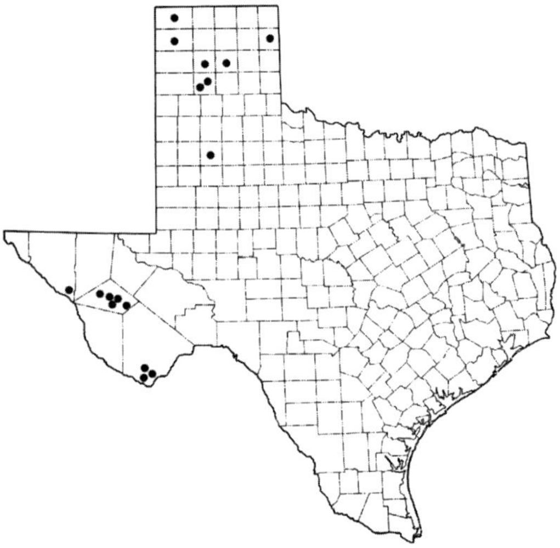

Artemisia carruthii

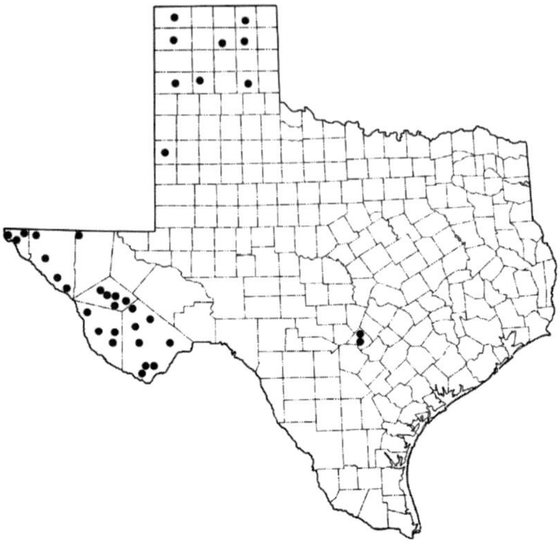

Artemisia dracunculoides

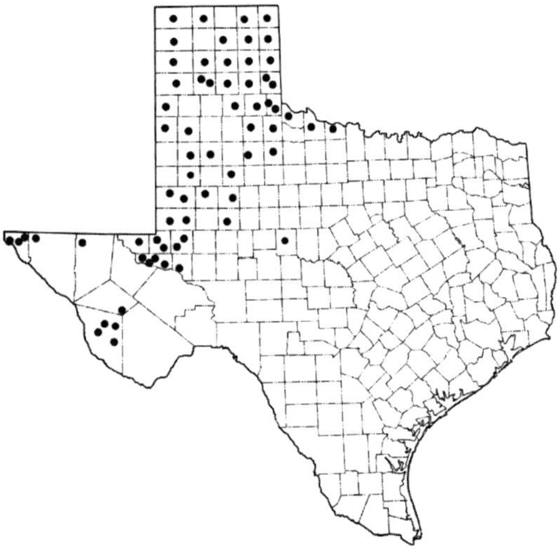

Artemisia filifolia

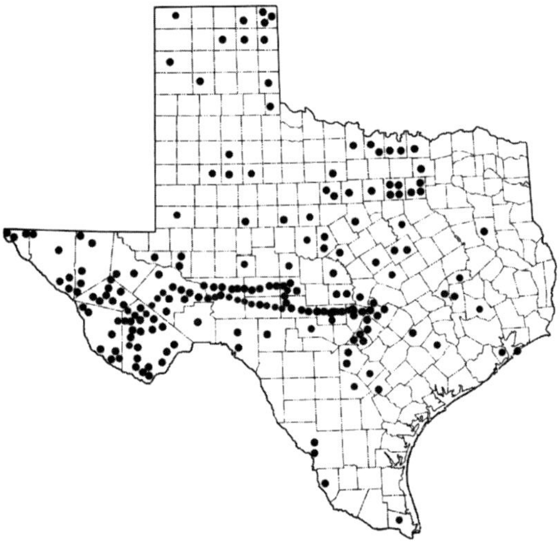

Artemisia ludoviciana

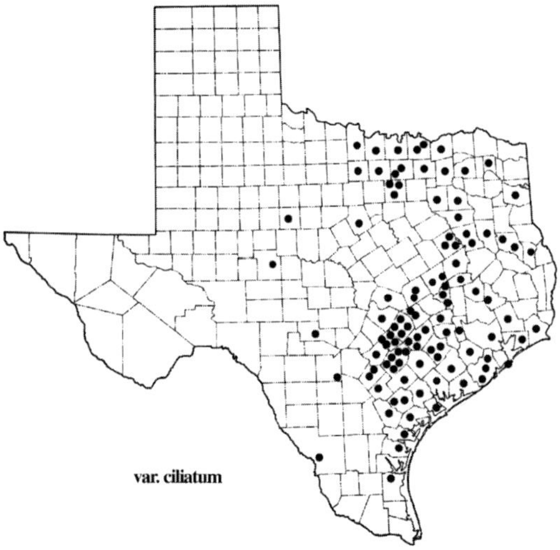
var. ciliatum

Astranthium integrifolium

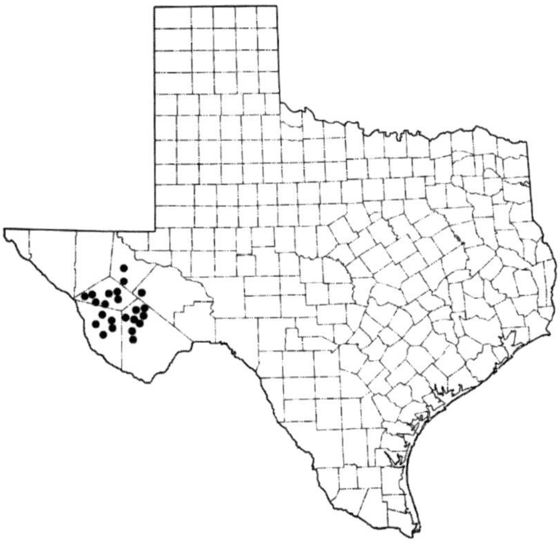

Astranthium robustum

Baccharis bigelovii

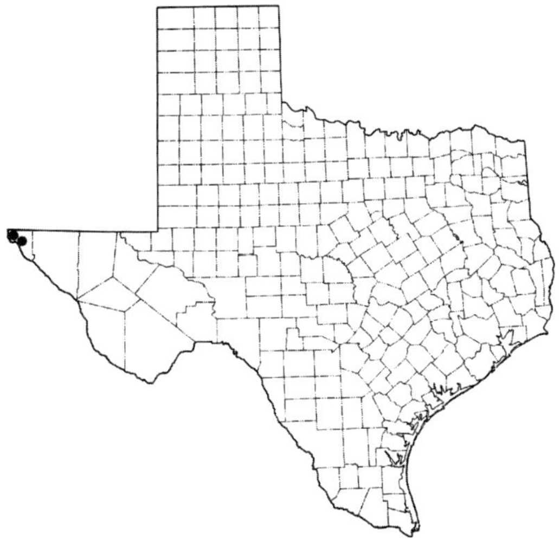

Baccharis brachyphylla

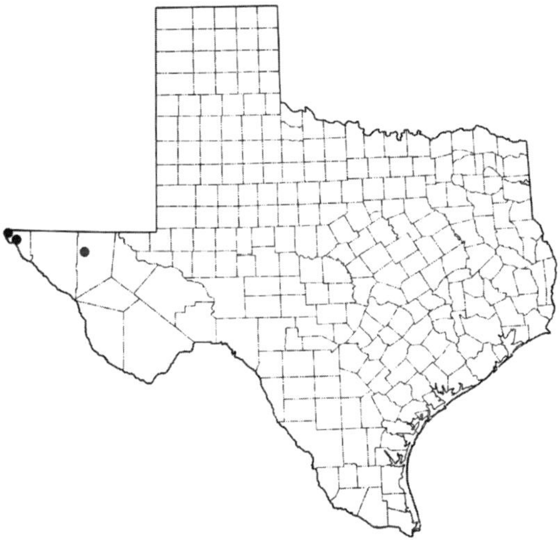

Baccharis emoryi

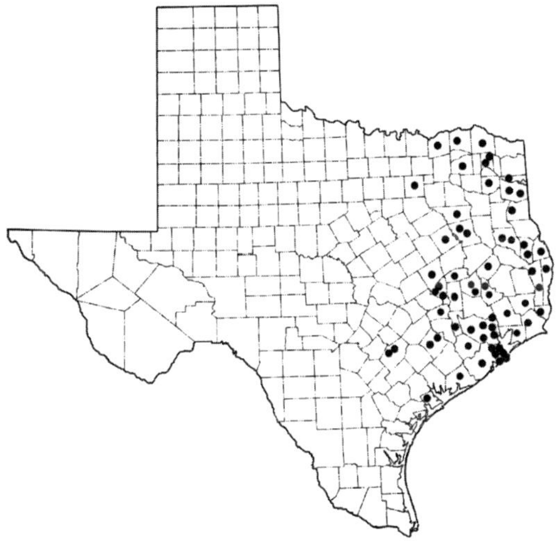

Baccharis halimifolia

Baccharis havardii

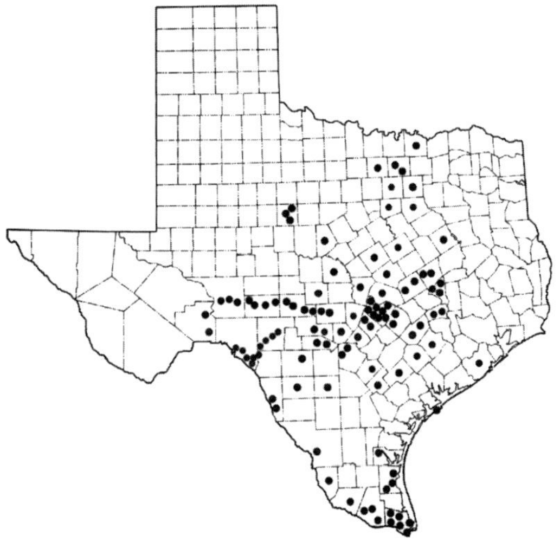

Baccharis neglecta

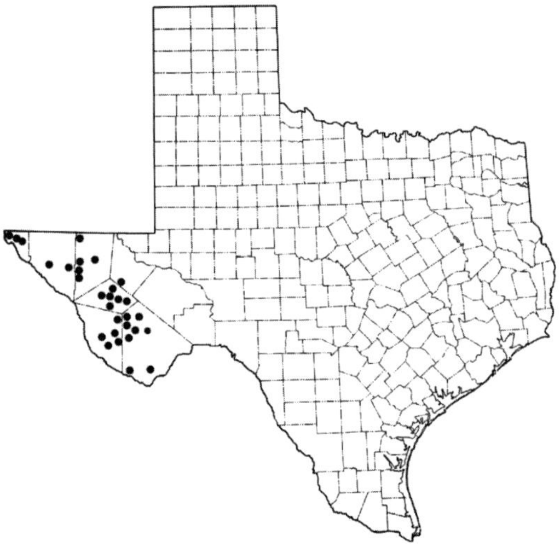

Baccharis pteronioides

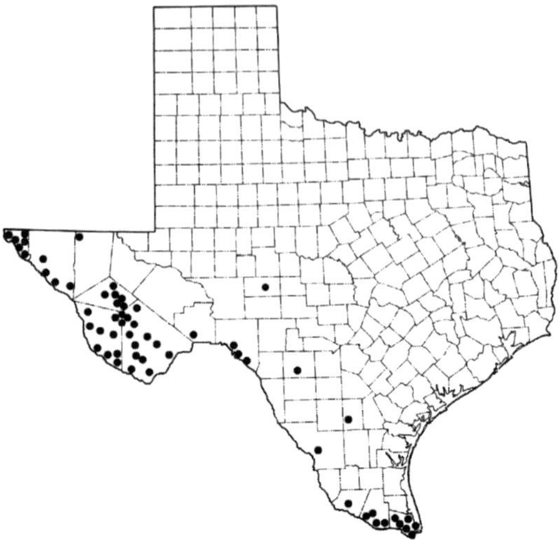

Baccharis salicifolia

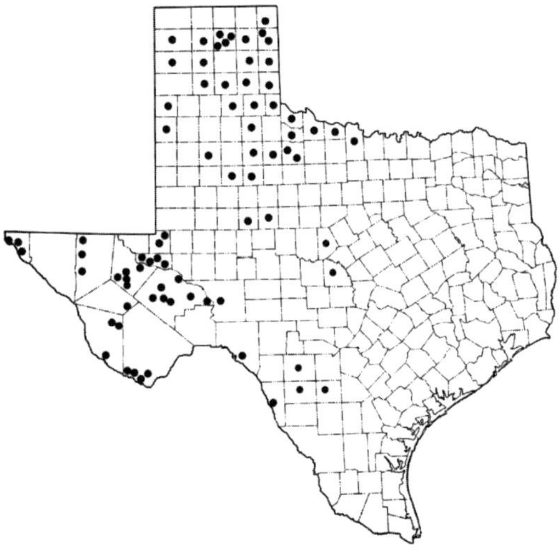

Baccharis salicina

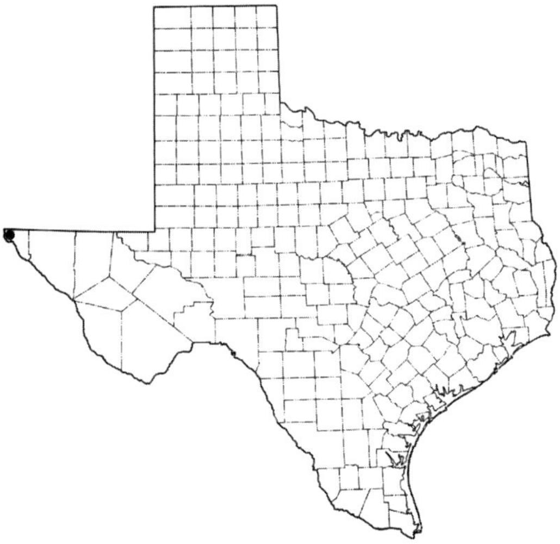

Baccharis sarothroides

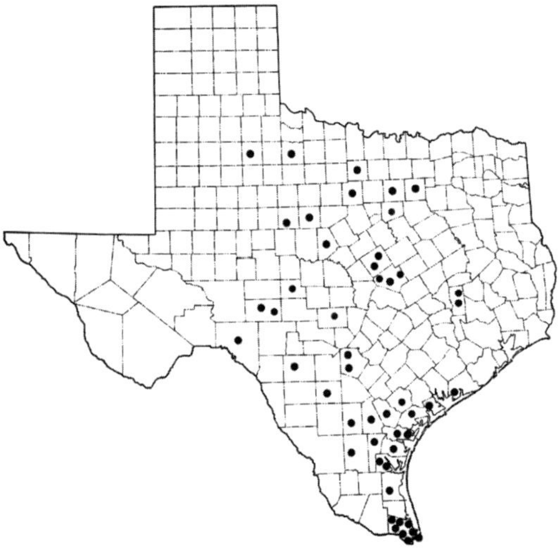

Baccharis texana

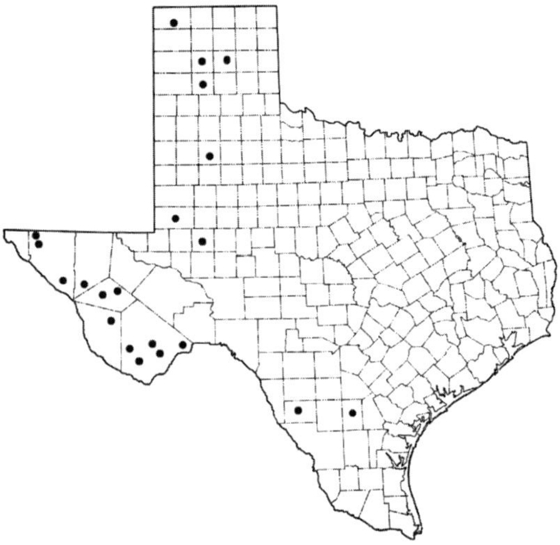

Baccharis wrightii

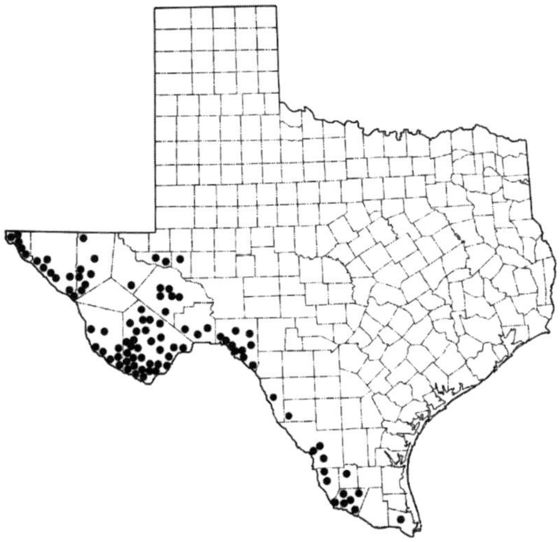

Bahia absinthifolia

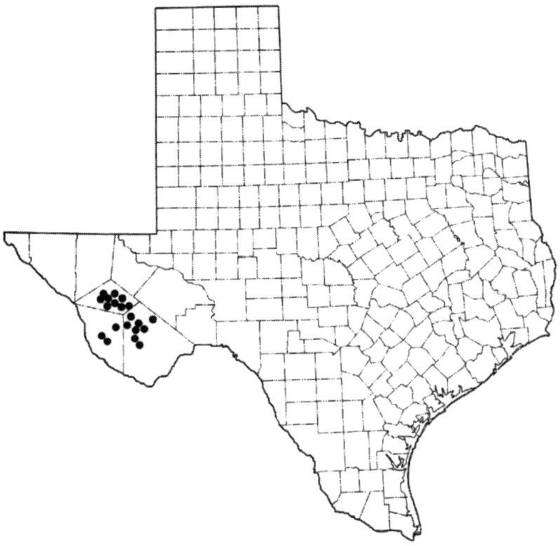

Bahia bigelovii

Bahia dissecta

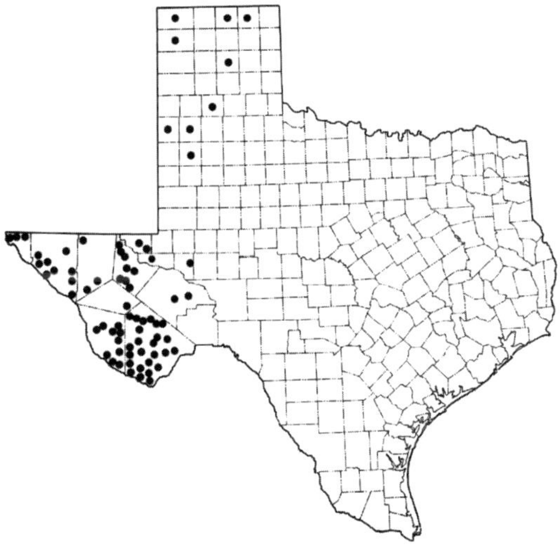

Bahia pedata

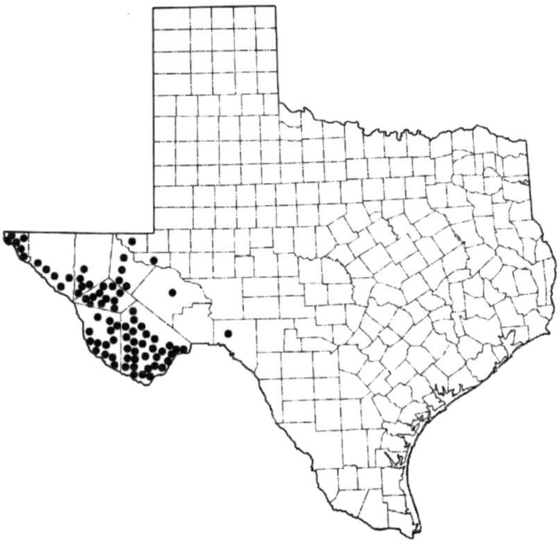

Baileya multiradiata

Bartlettia scaposa

Bebbia juncea

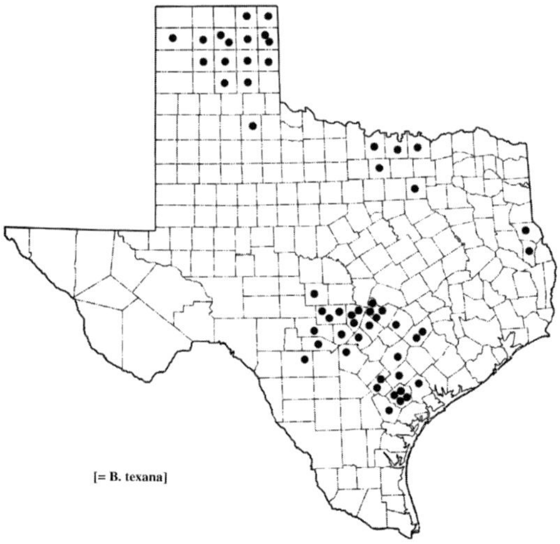
[= B. texana]

Berlandiera betonicifolia

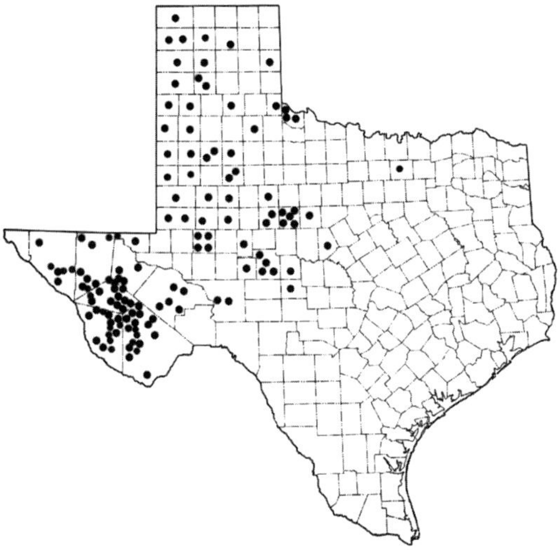

Berlandiera lyrata

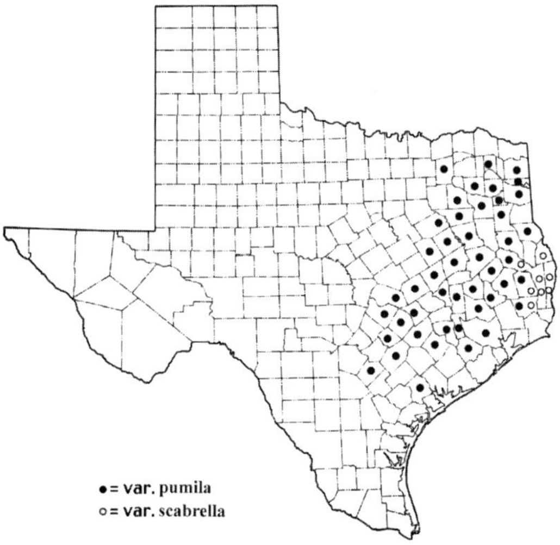
● = var. pumila
○ = var. scabrella

Berlandiera pumila

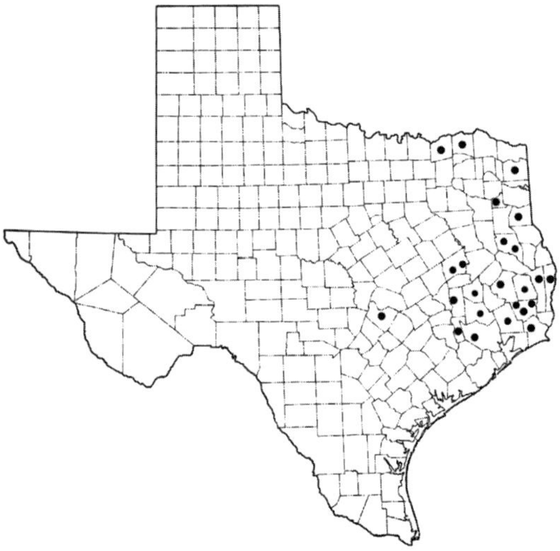

Bidens aristosa

Bidens bigelovii

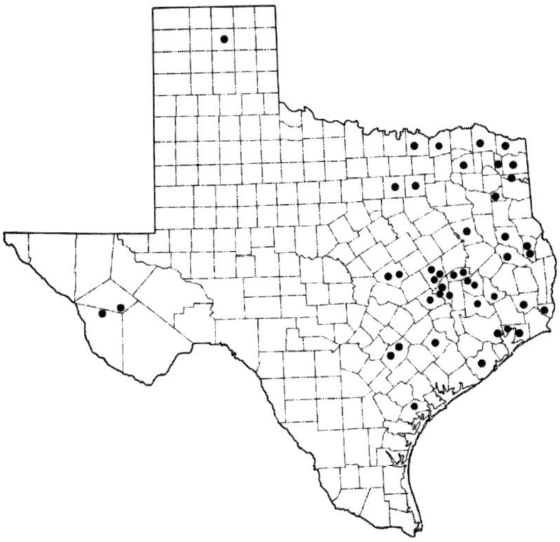

Bidens bipinnata

Bidens cernua

Bidens comosa

Bidens discoidea

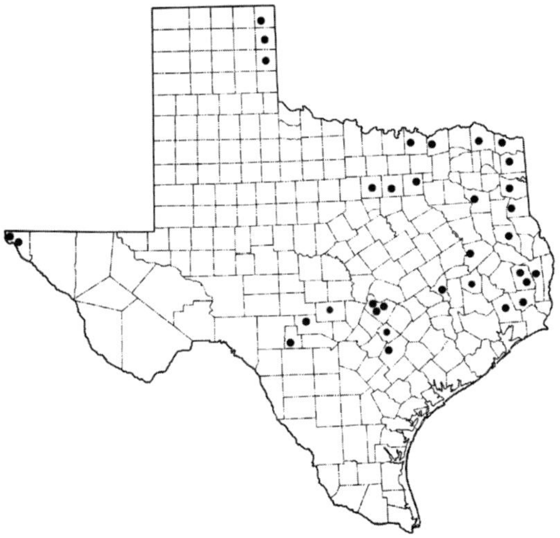

Bidens frondosa

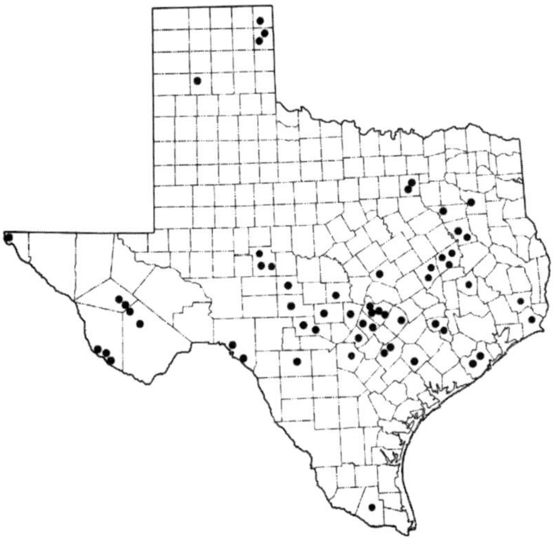

Bidens laevis

Bidens leptocephala

Bidens odorata

Bidens pilosa

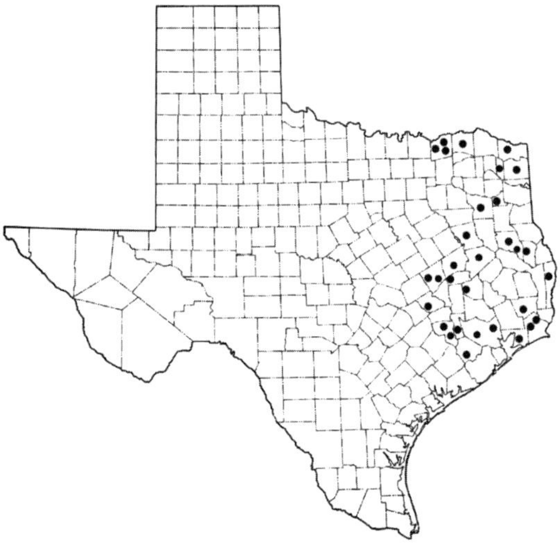
Bidens polylepis

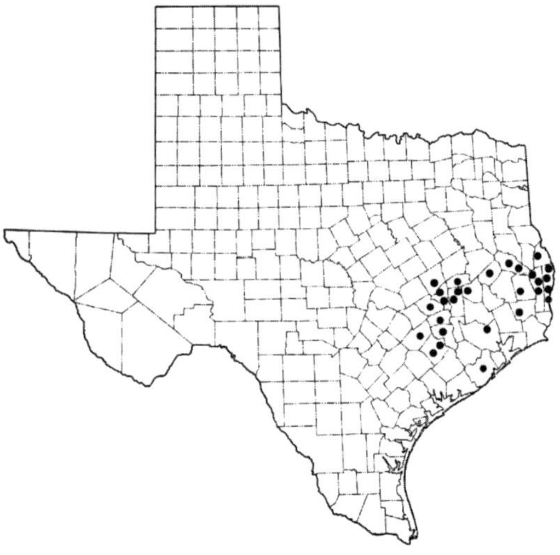
Bigelowia nuttallii

Boltonia asteroides

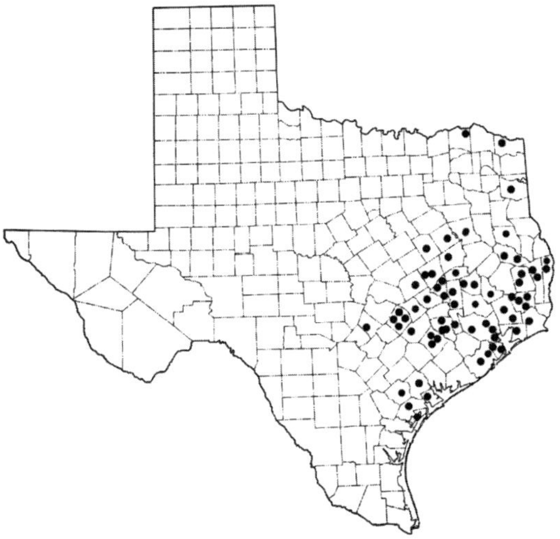

Boltonia diffusa

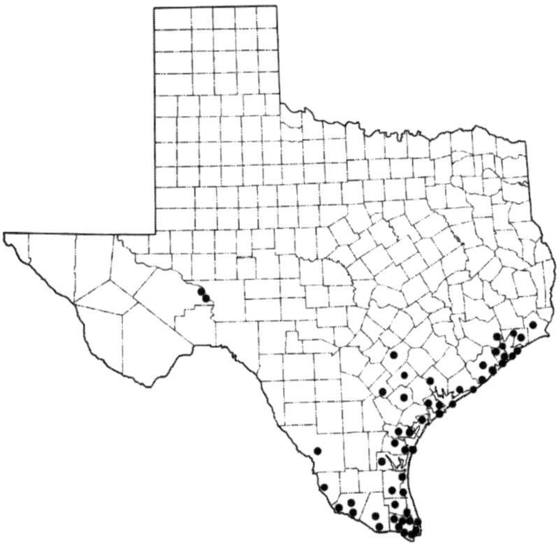

Borrichia frutescens

Brickellia baccharidea

Brickellia brachyphylla

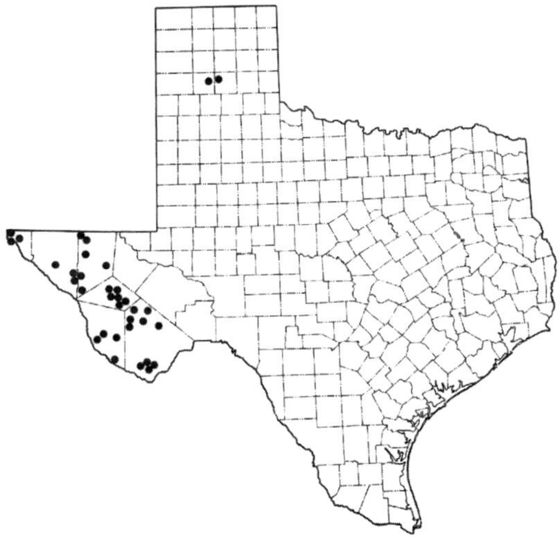

Brickellia californica

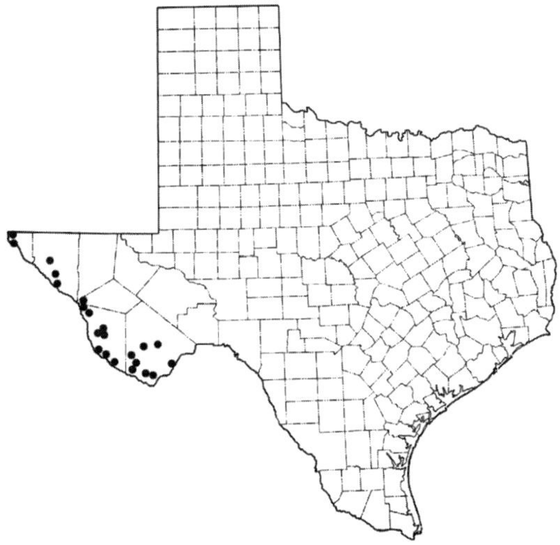

Brickellia coulteri

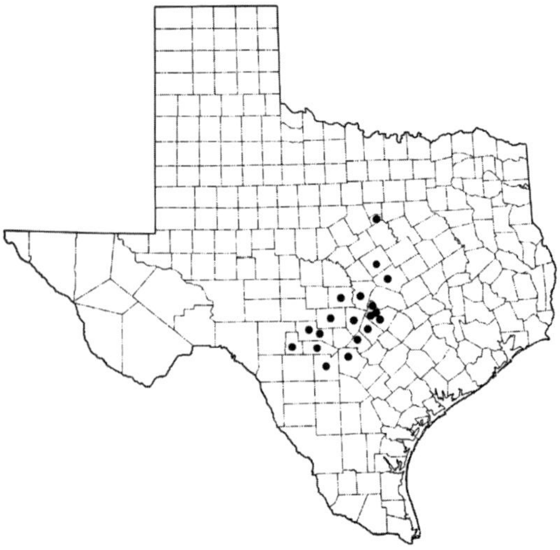

Brickellia cylindracea

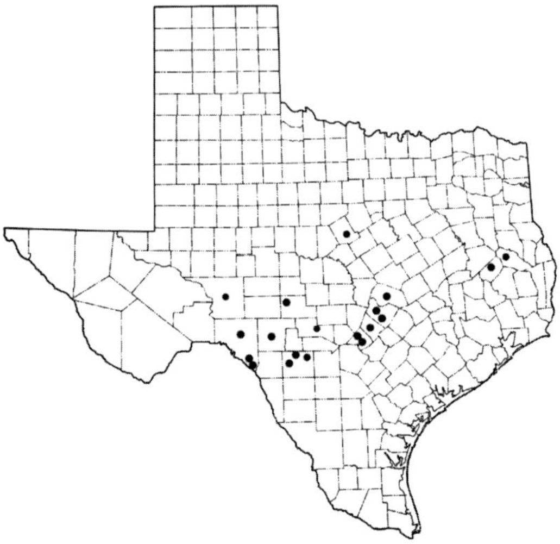

Brickellia dentata

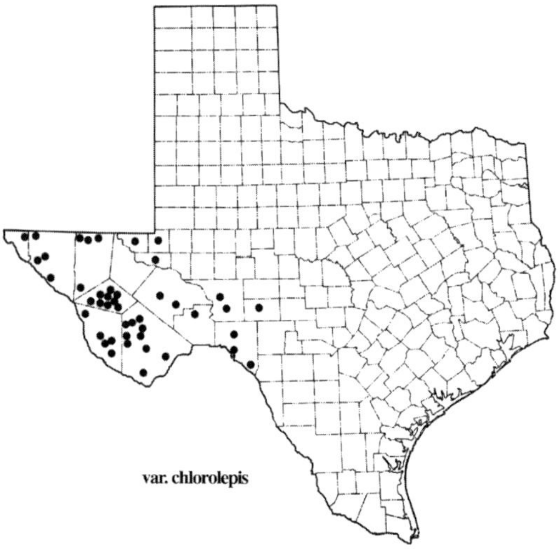
var. chlorolepis

Brickellia eupatorioides

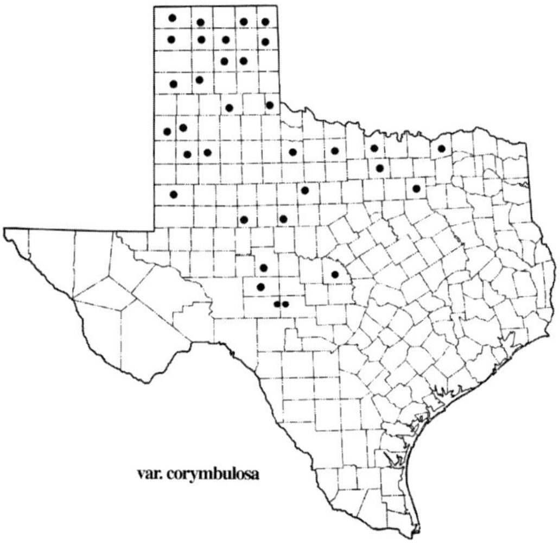
var. corymbulosa

Brickellia eupatorioides

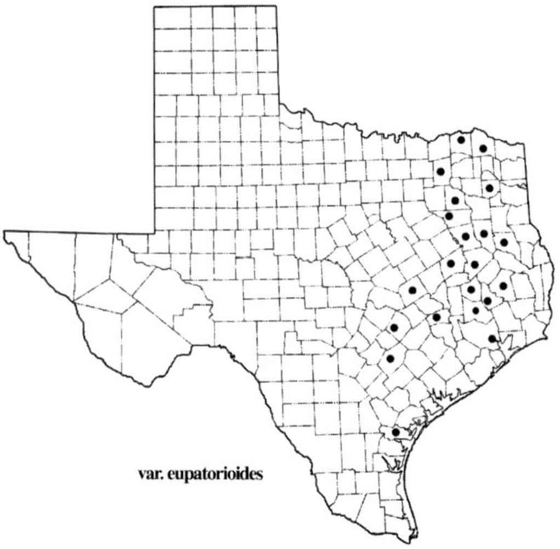
var. eupatorioides

Brickellia eupatorioides

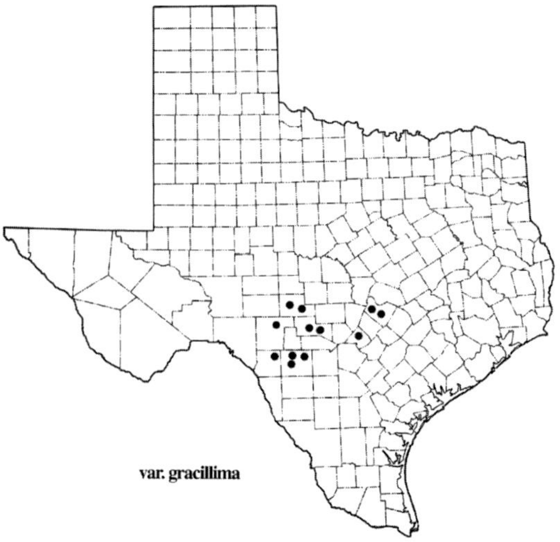
var. gracillima

Brickellia eupatorioides

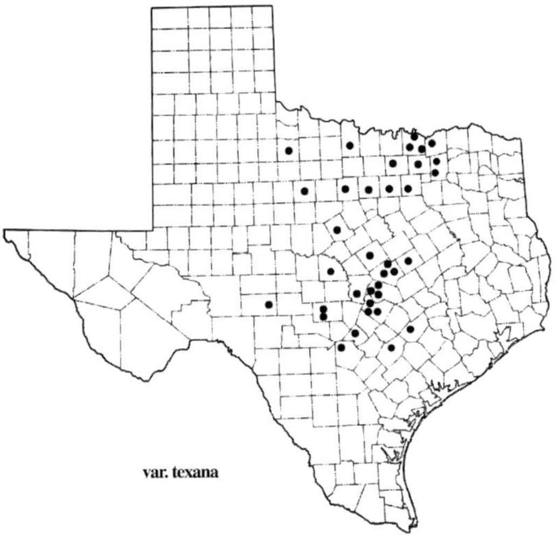
var. texana

Brickellia eupatorioides

Brickellia grandiflora

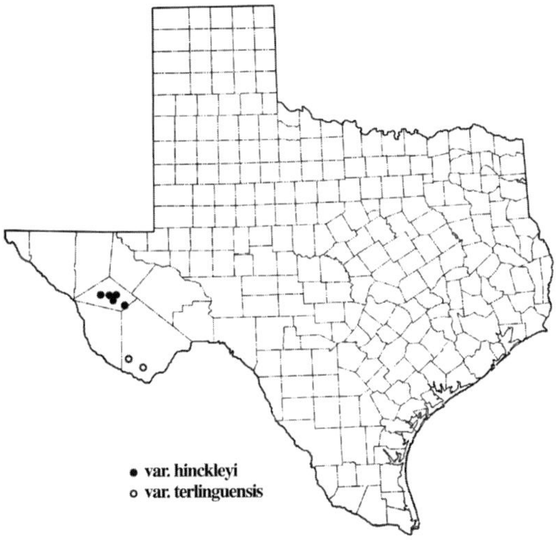
• var. hinckleyi
○ var. terlinguensis

Brickellia hinckleyi

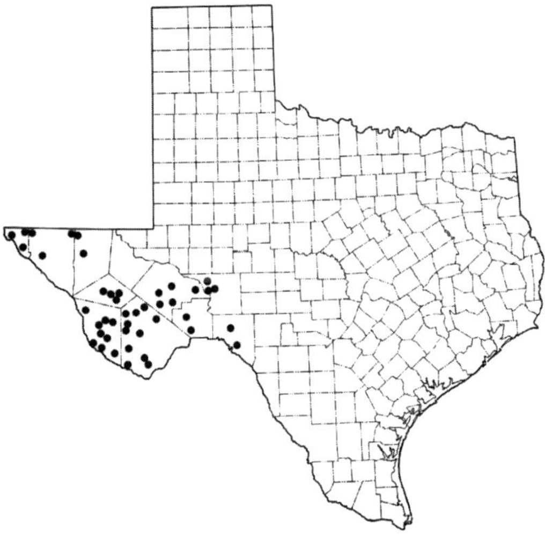

Brickellia laciniata

Brickellia lemmonii

Brickellia parvula

Brickellia venosa

Brickellia veronicifolia

Calycoseris wrightii

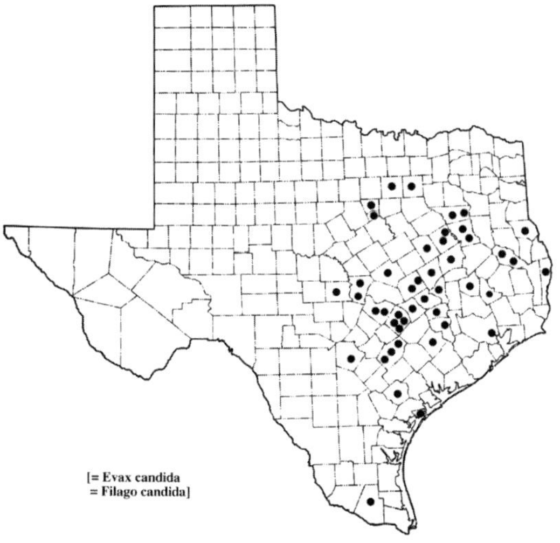
[= Evax candida
= Filago candida]

Calymmandra candida

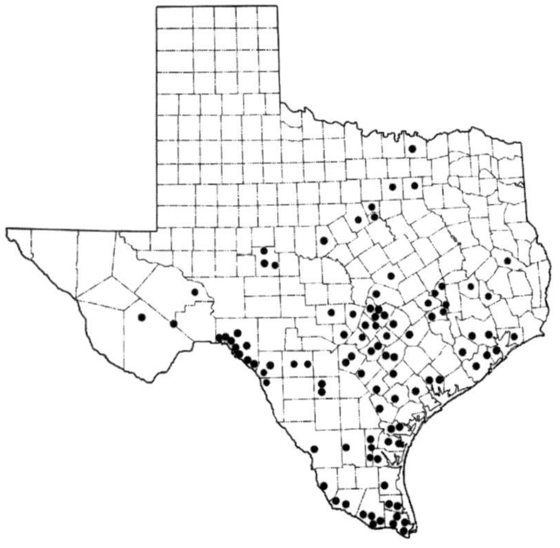

Calyptocarpus vialis

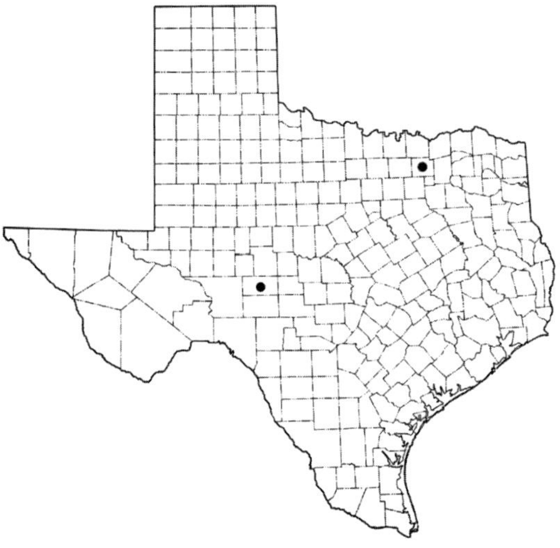

Carduus acanthoides

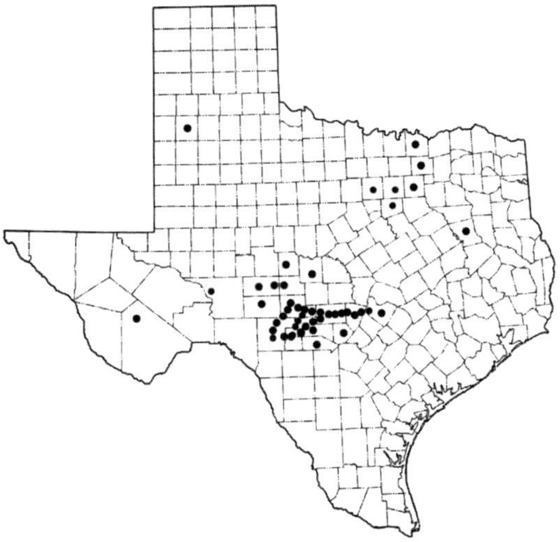

Carduus nutans

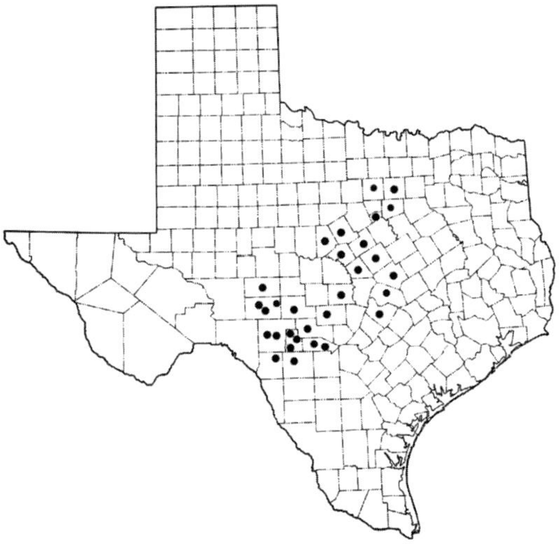

Carduus tenuiflorus

Carminatia tenuiflora

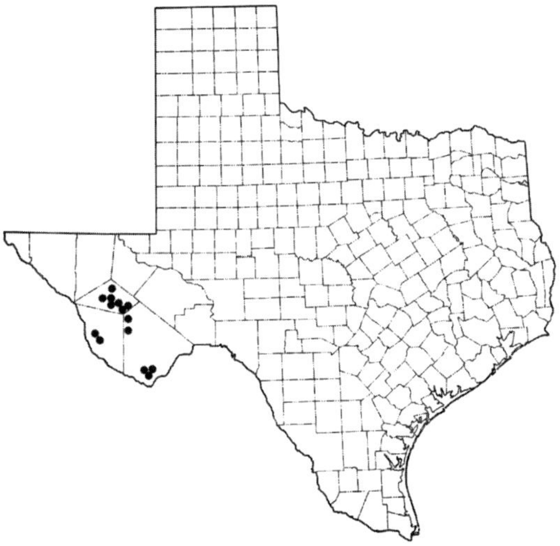

Carphochaete bigelovii

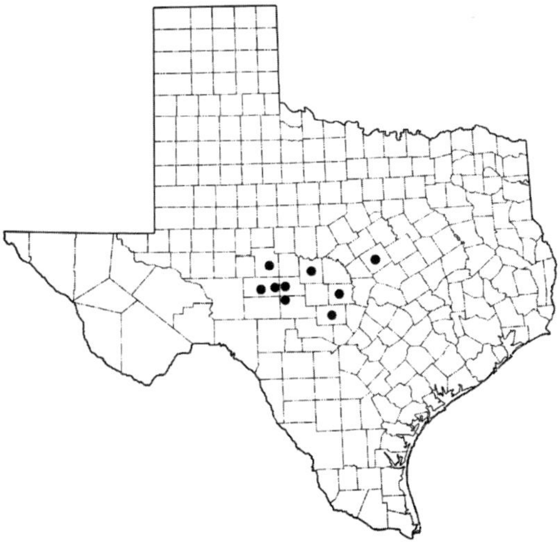

Carthamus lanatus

Carthamus tinctorius

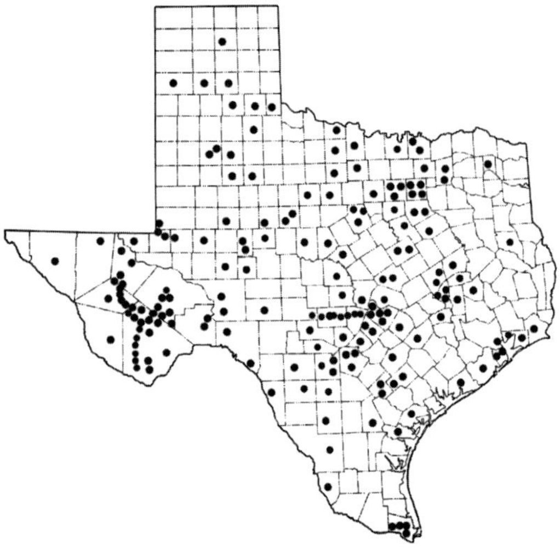

Centaurea americana

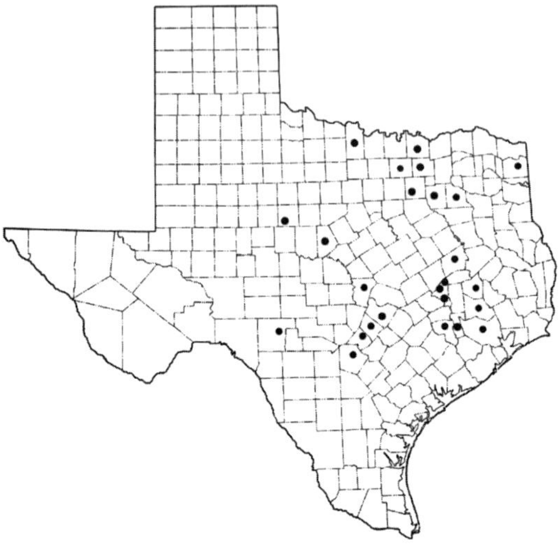

Centaurea cyanus

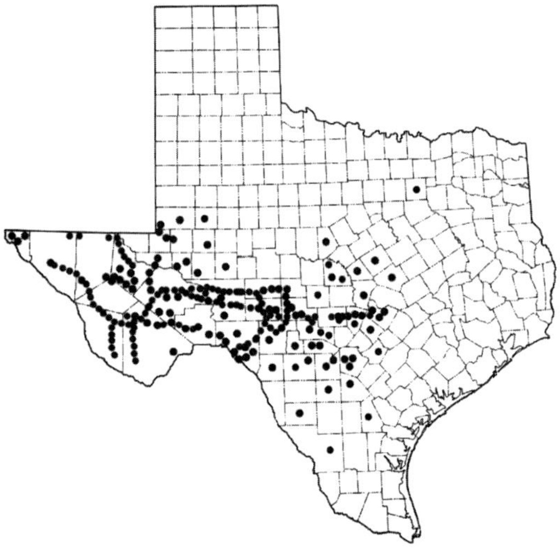

Centaurea melitensis

Centaurea repens

Centaurea solstitialis

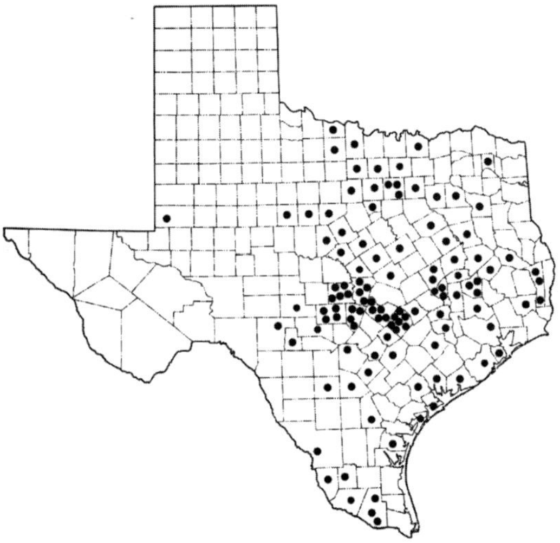

Chaetopappa asteroides

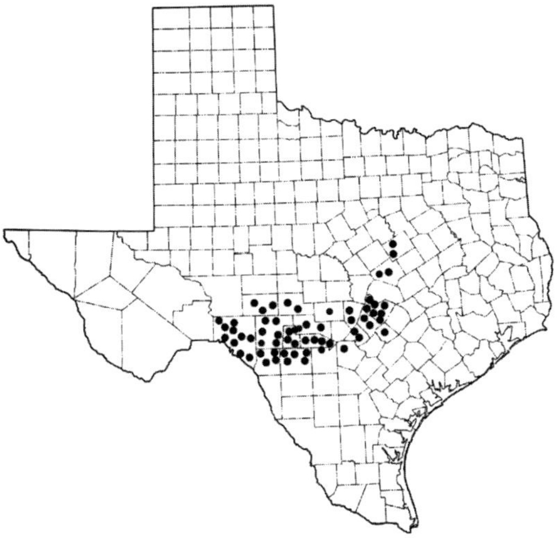

Chaetopappa bellidifolia

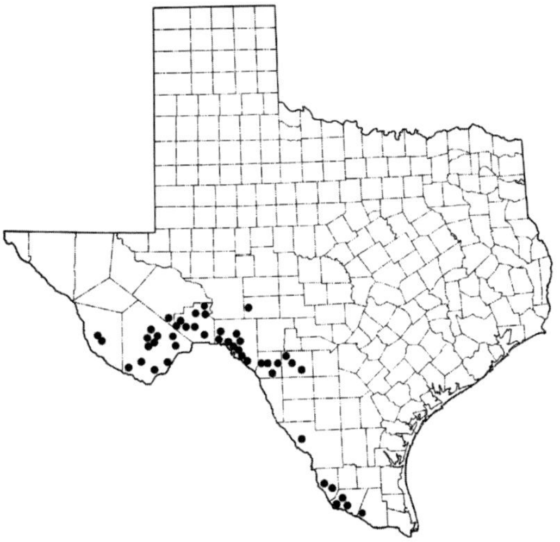

Chaetopappa bellioides

Chaetopappa effusa

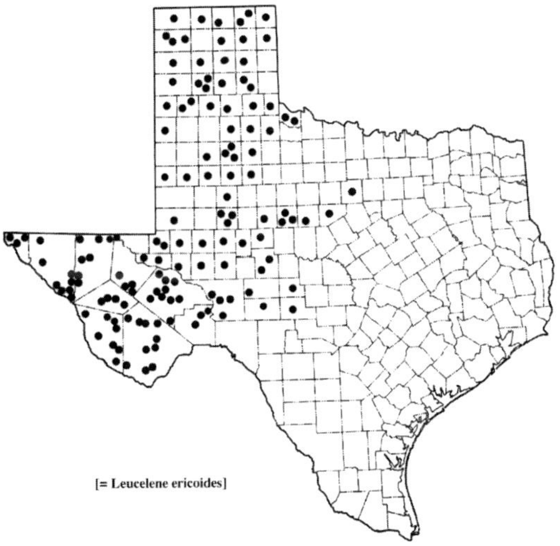
[= Leucelene ericoides]

Chaetopappa ericoides

Chaetopappa hersheyi

Chaetopappa imberbis

[= C. keerlioides]

Chaetopappa parryi

Chamomilla recutita

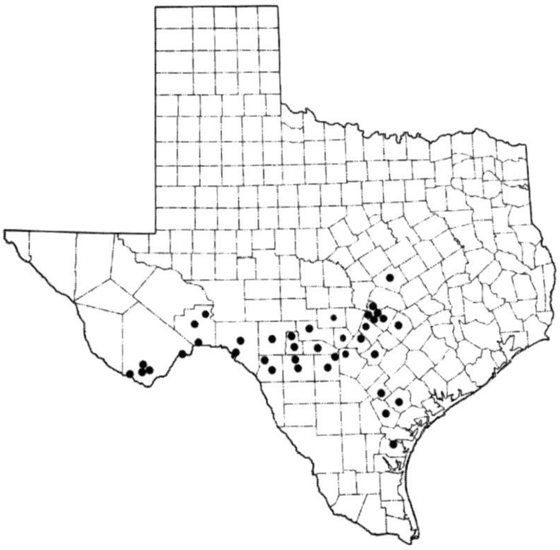

Chaptalia texana

Chaptalia tomentosa

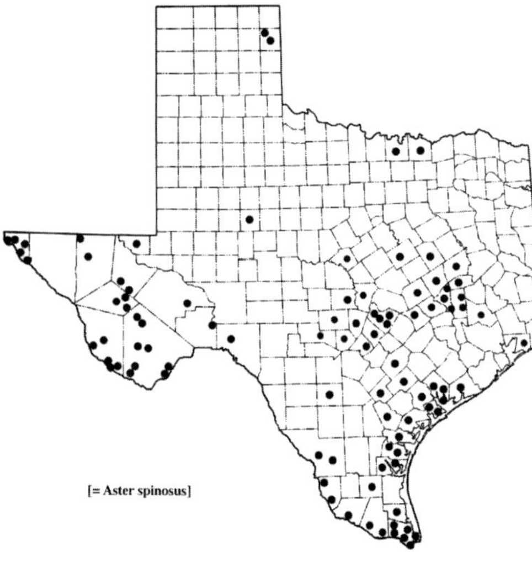
[= Aster spinosus]

Chloracantha spinosa

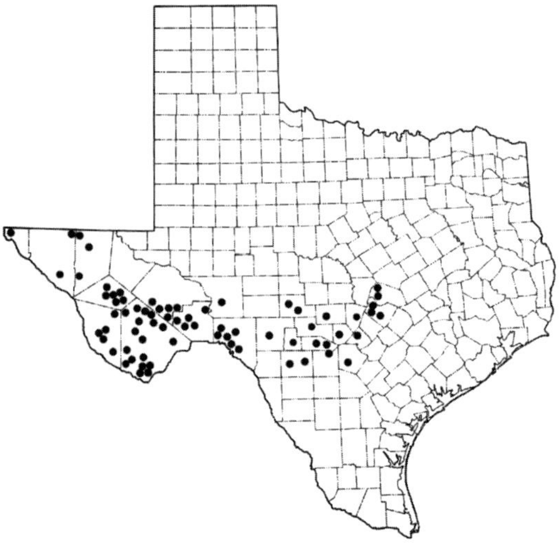

Chrysactinia mexicana

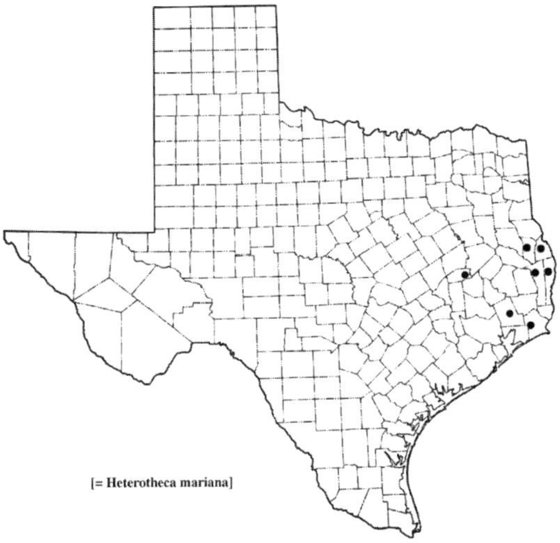
[= Heterotheca mariana]
Chrysopsis mariana

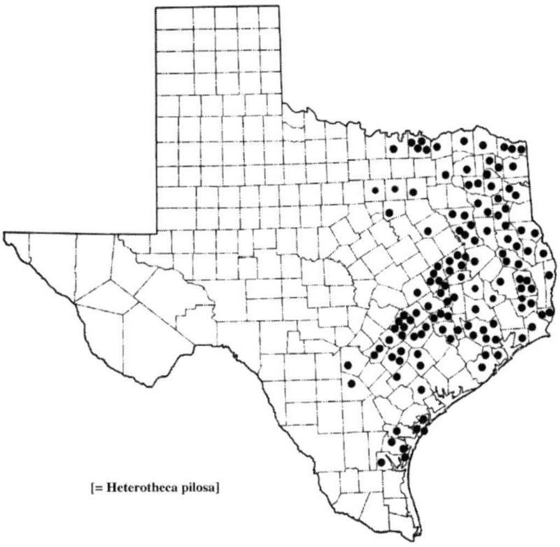
[= Heterotheca pilosa]
Chrysopsis pilosa

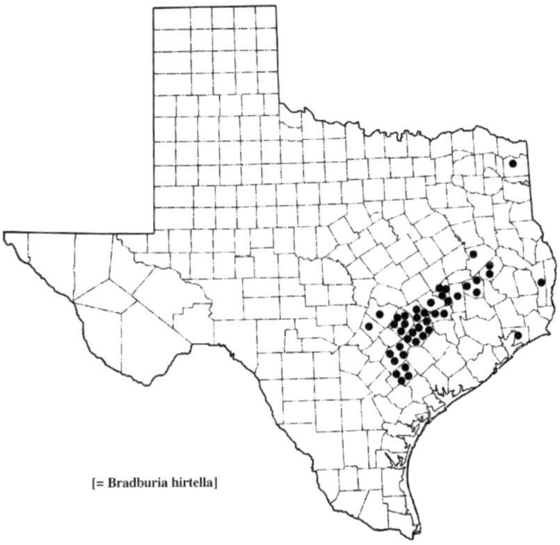
[= Bradburia hirtella]
Chrysopsis texana

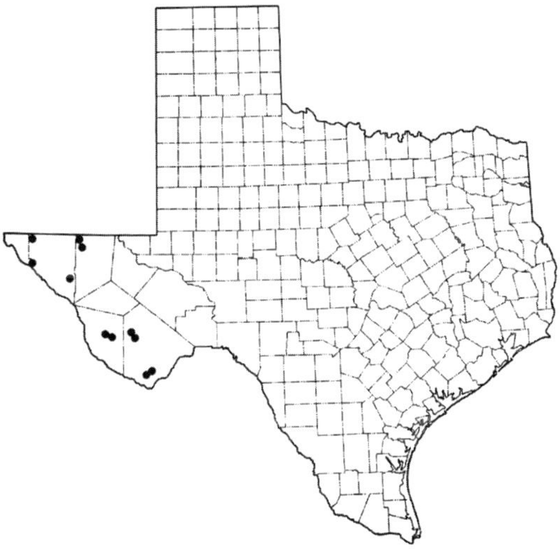
Chrysothamnus baileyi

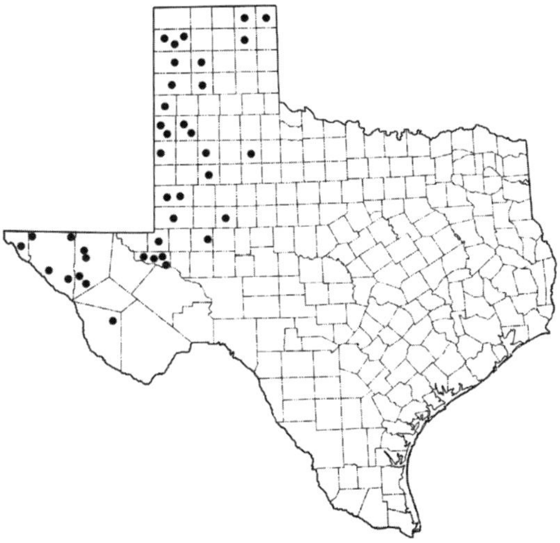
Chrysothamnus pulchellus

Chrysothamnus spathulatus

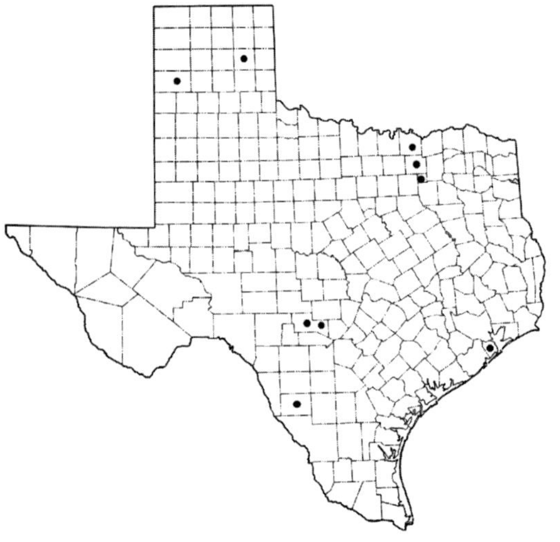

Cichorium intybus

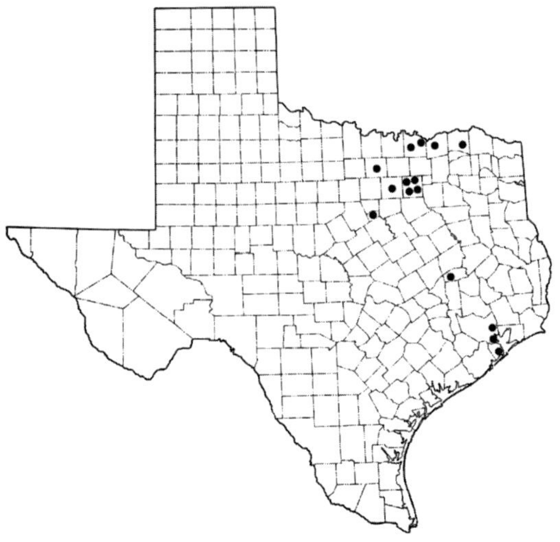

Cirsium altissimum

Cirsium carolinianum

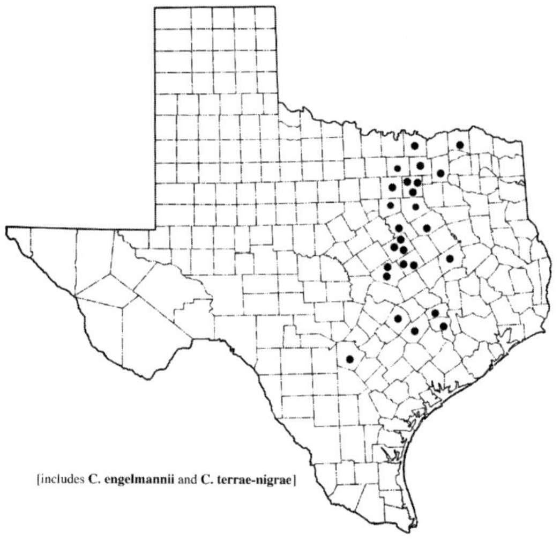
[includes **C. engelmannii** and **C. terrae-nigrae**]

Cirsium discolor

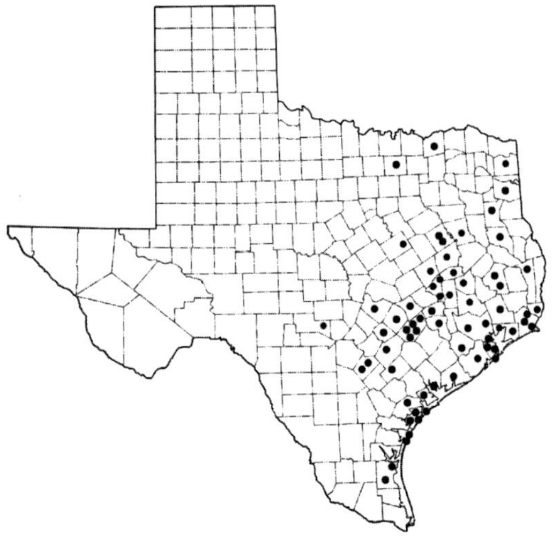

Cirsium horridulum

Cirsium muticum

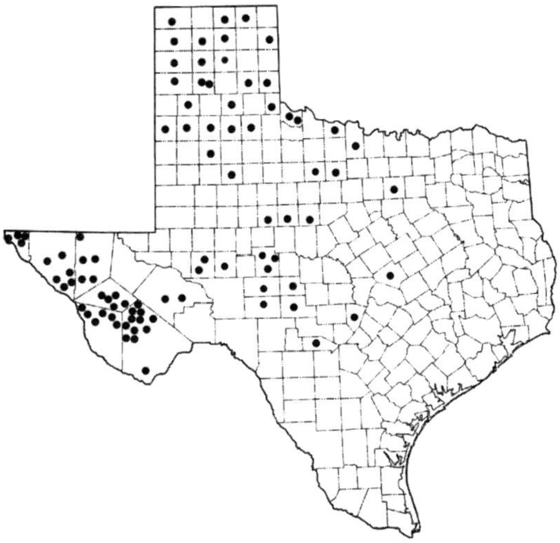

Cirsium ochrocentrum

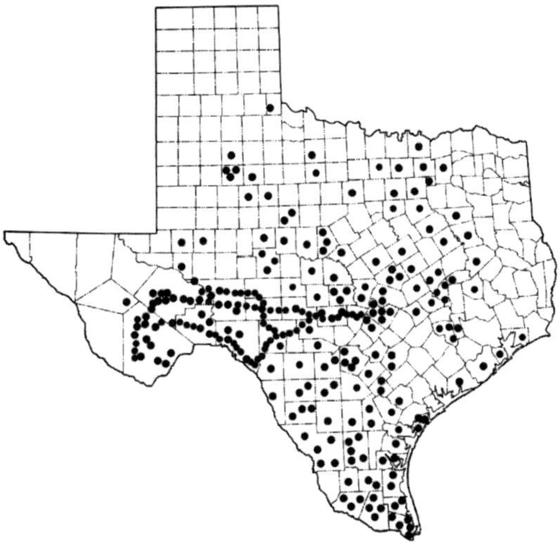

Cirsium texanum

Cirsium turneri

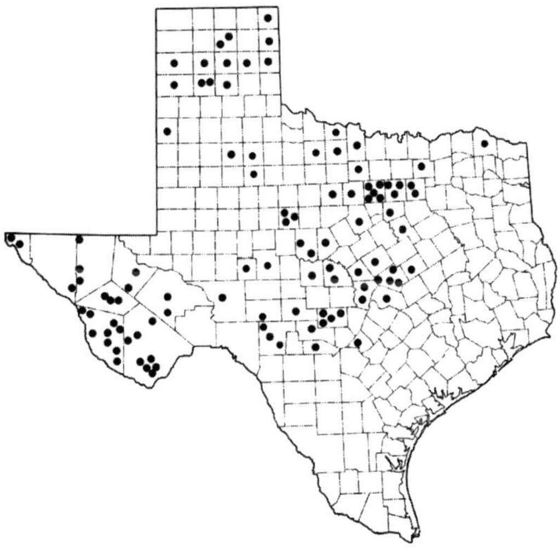

Cirsium undulatum

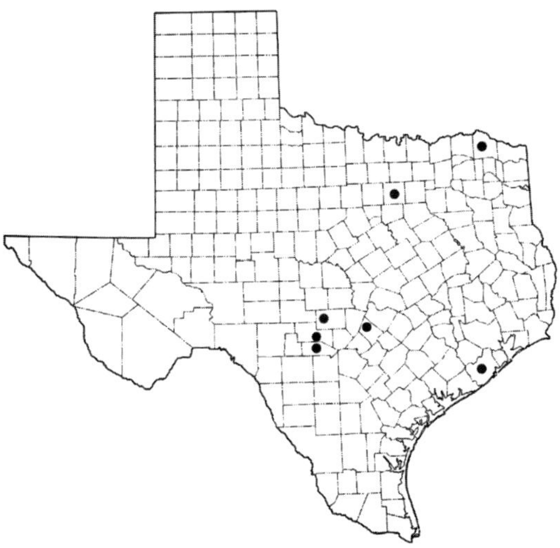

Cirsium vulgare

Clappia suaedifolia

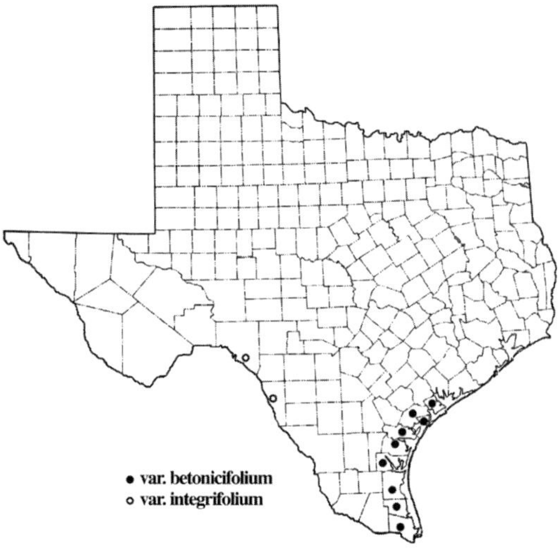
• var. betonicifolium
○ var. integrifolium

Conoclinium betonicifolium

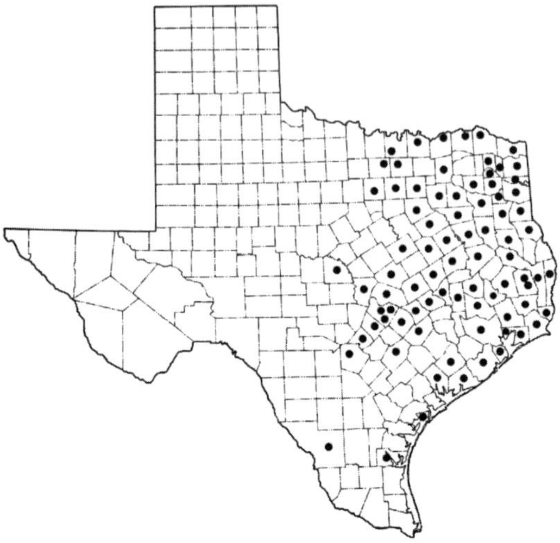

Conoclinium coelestinum

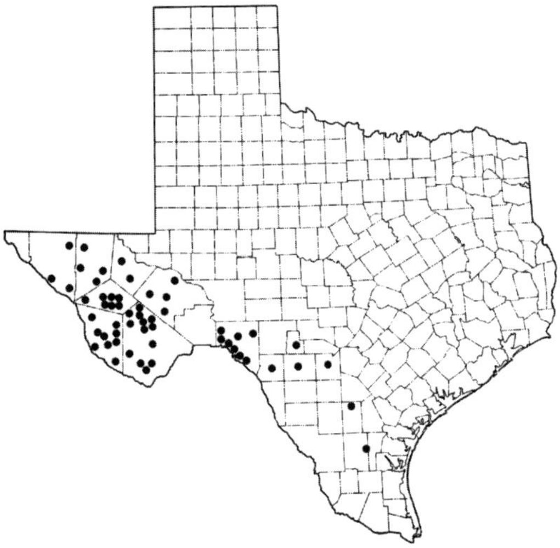

Conoclinium dissectum

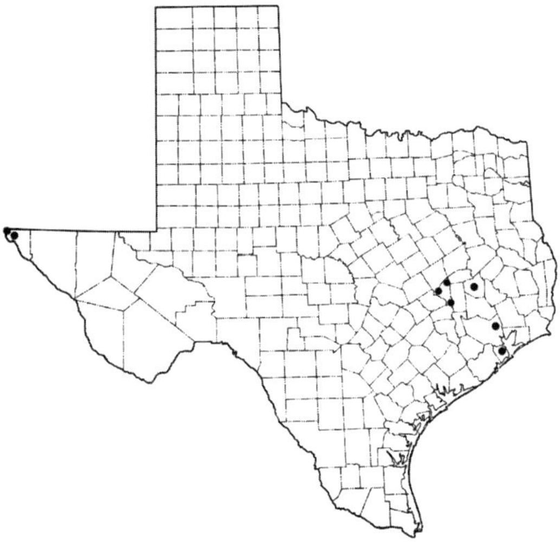

Conyza bonariensis

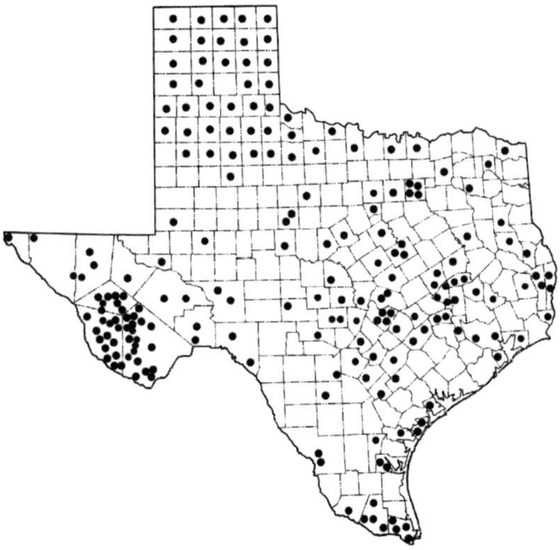

Conyza canadensis

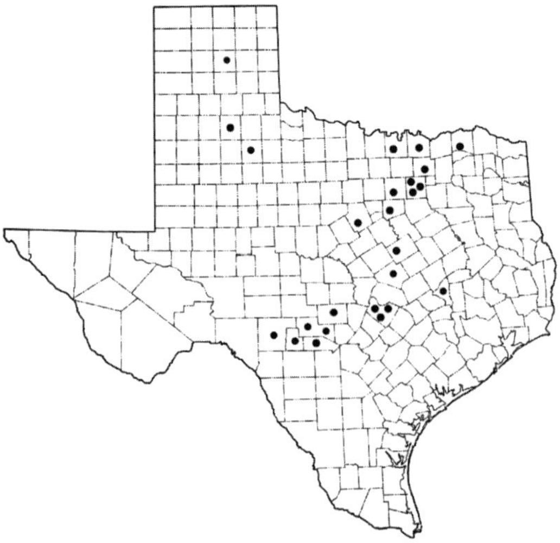

Conyza ramosissima

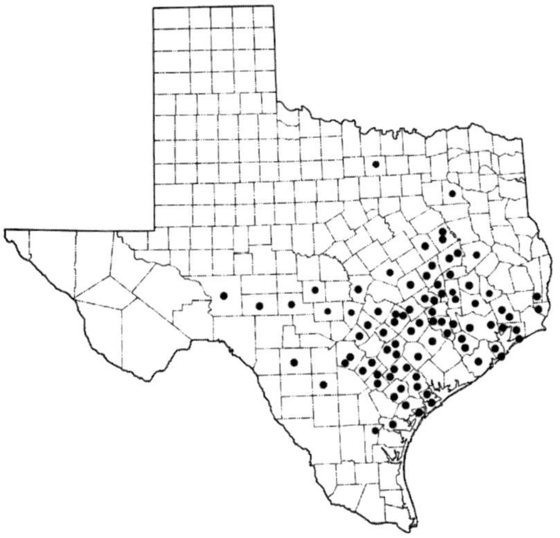

Coreopsis basalis

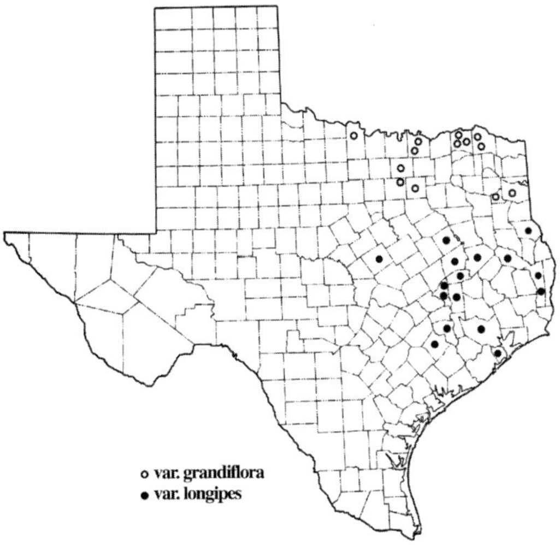
o var. grandiflora
• var. longipes

Coreopsis grandiflora

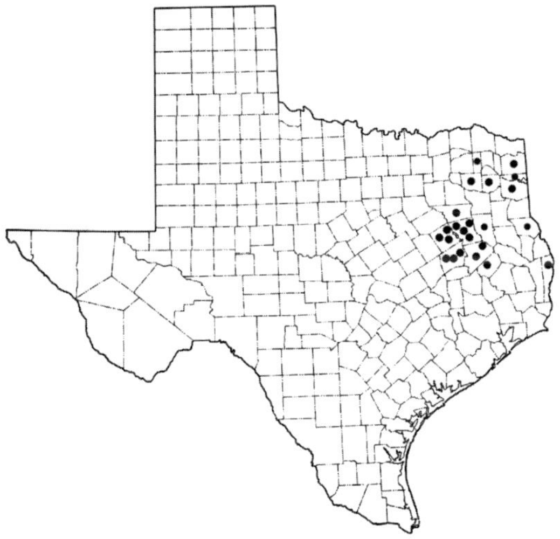

Coreopsis intermedia

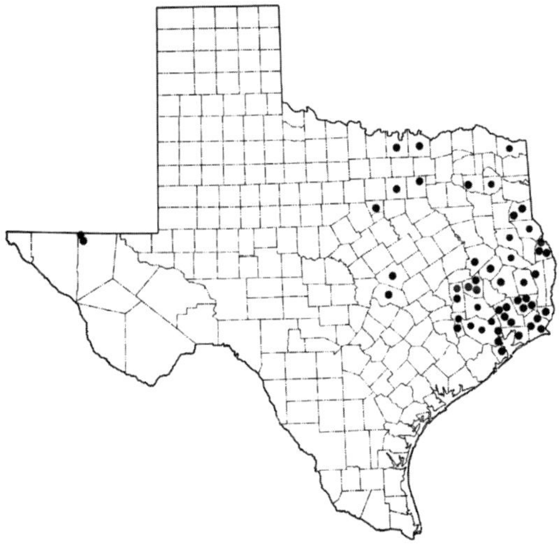

Coreopsis lanceolata

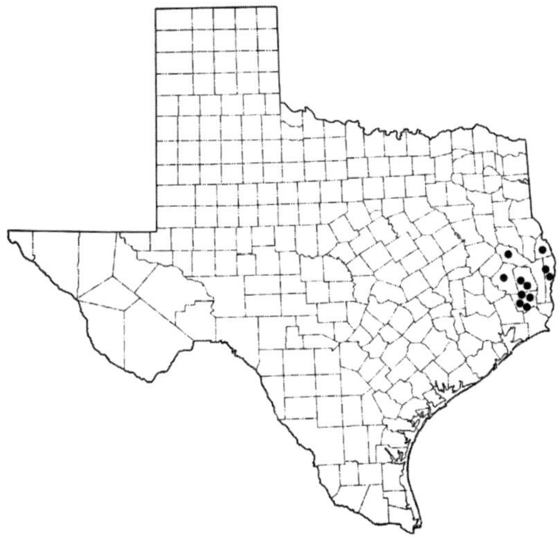

Coreopsis linifolia

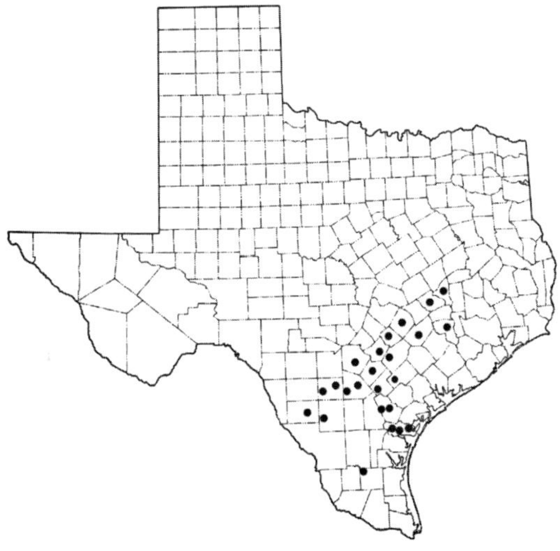

Coreopsis nuecensis

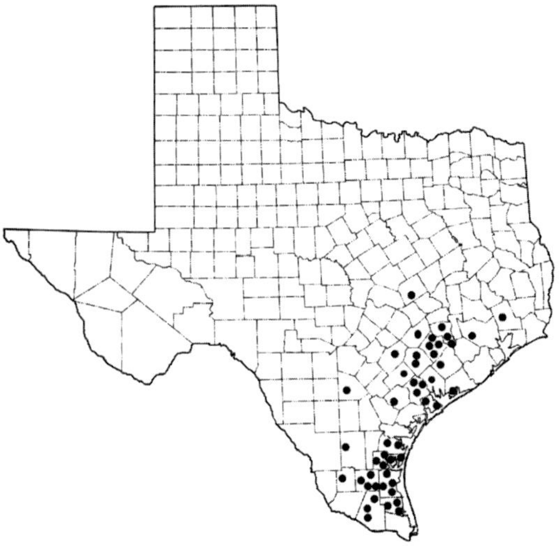

Coreopsis nuecensoides

Coreopsis pubescens

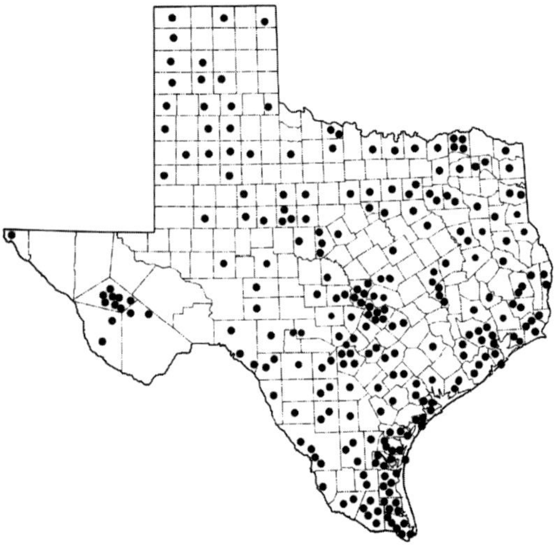

Coreopsis tinctoria

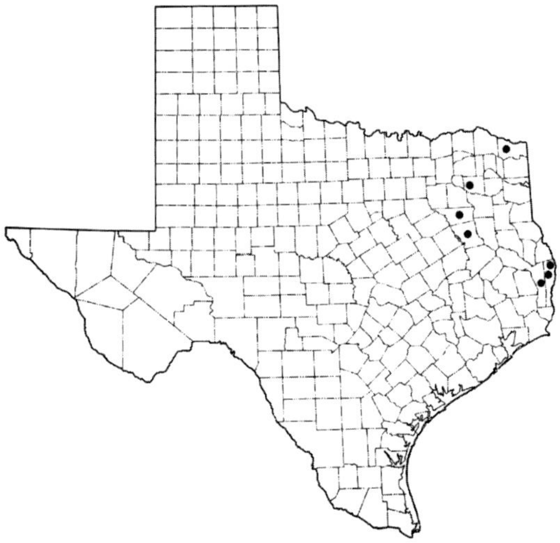

Coreopsis tripteris

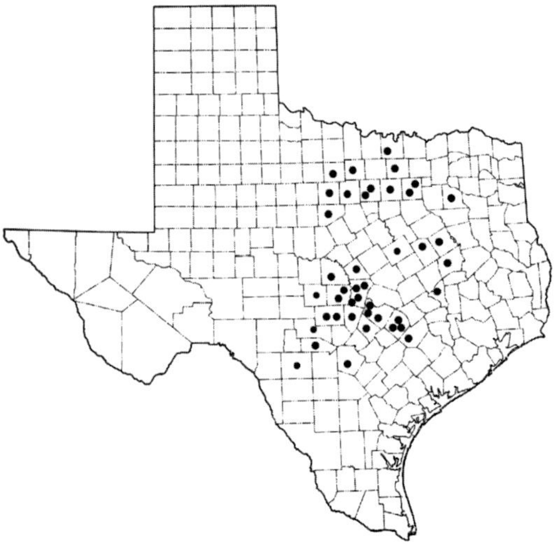

Coreopsis wrightii

Cosmos parviflorus

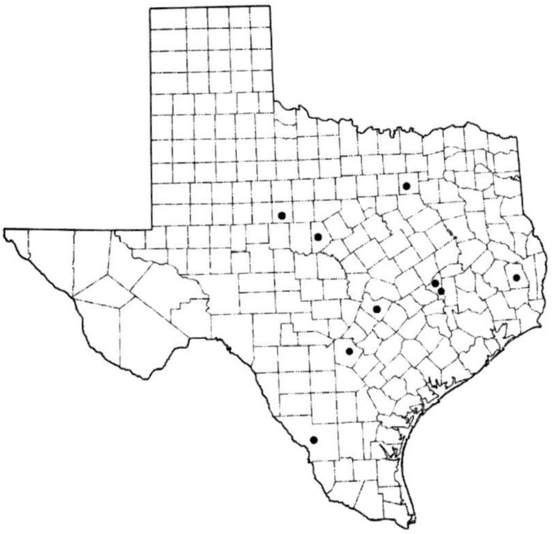

Cosmos sulphureus

Cotula australis

Crepis capillaris

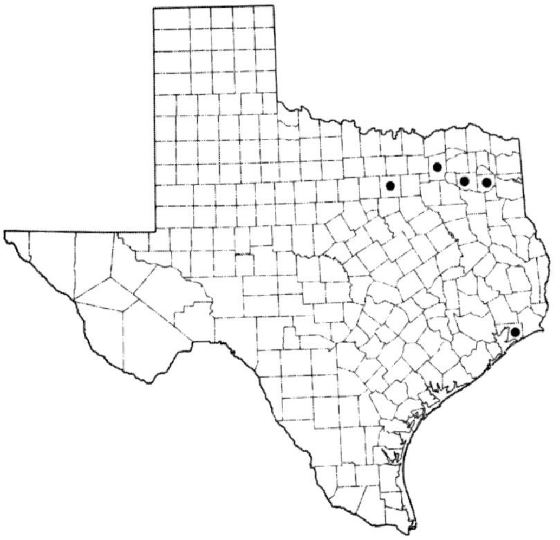

Crepis pulchra

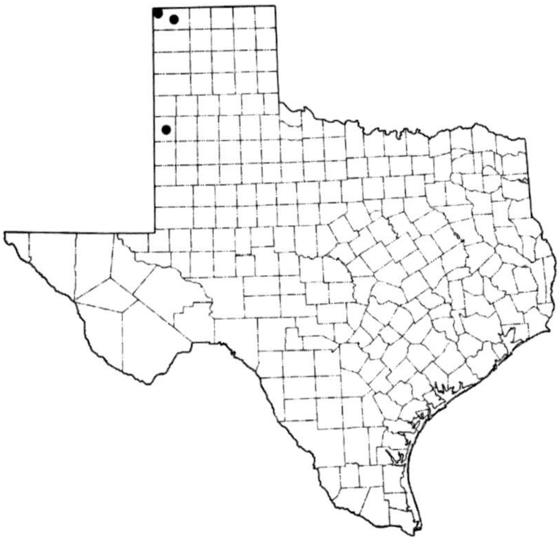

Crepis runcinata

Crepis setosa

Crepis zacintha

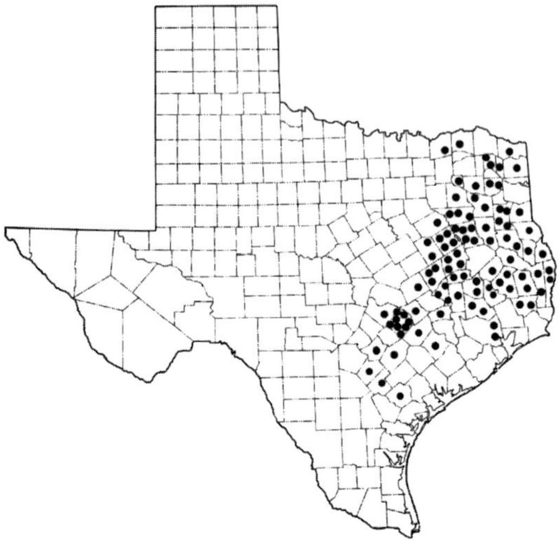

Croptilon divaricatum

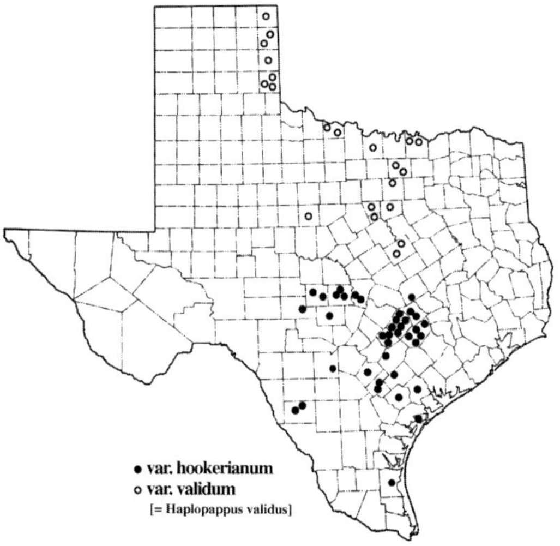
• var. hookerianum
○ var. validum
[= Haplopappus validus]

Croptilon hookerianum

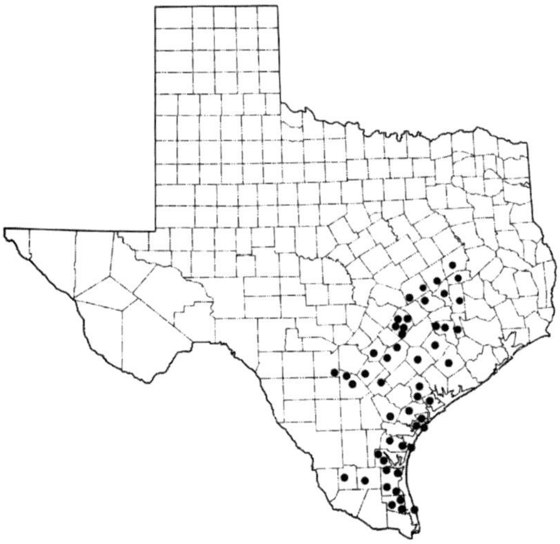

Croptilon rigidifolium

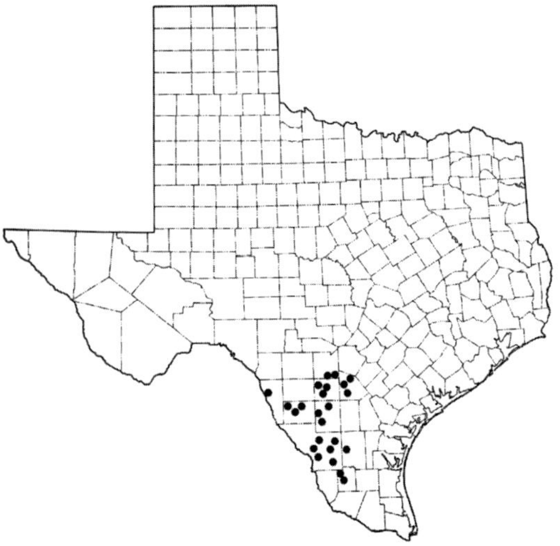

Dichaetophora campestris

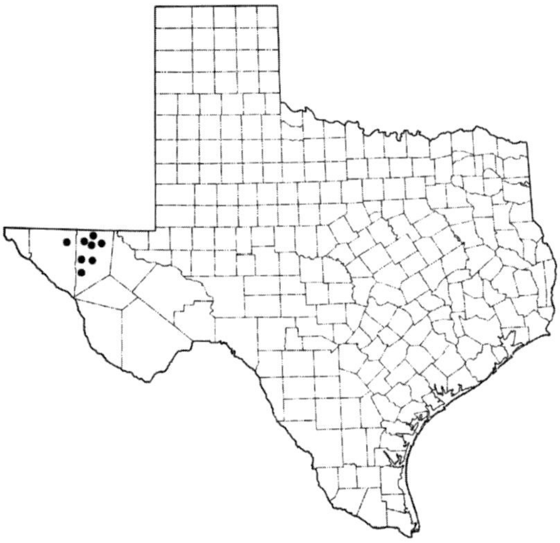

Dicranocarpus parviflorus

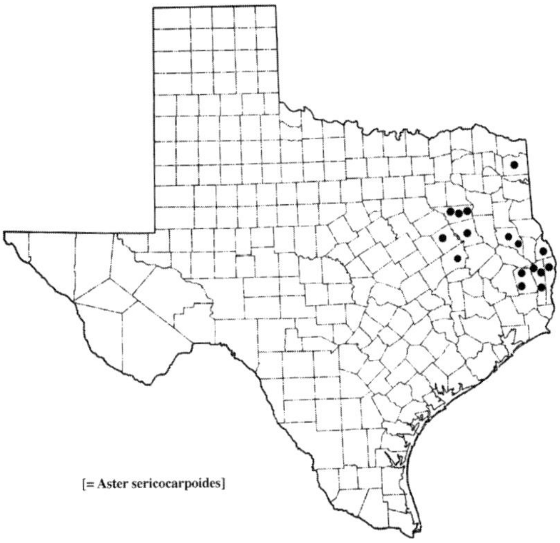
[= Aster sericocarpoides]

Doellingeria sericocarpoides

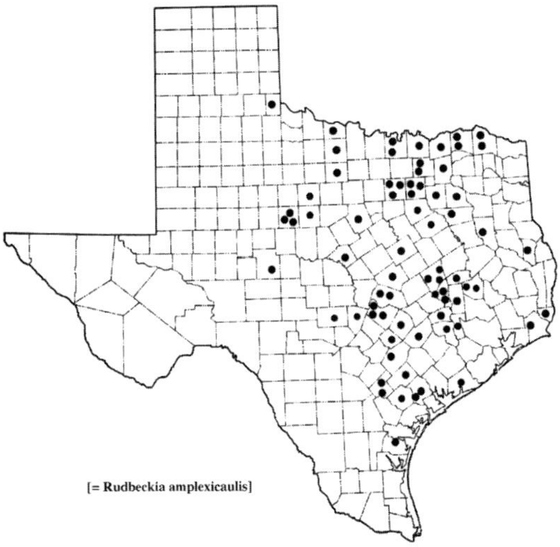
[= Rudbeckia amplexicaulis]

Dracopis amplexicaulis

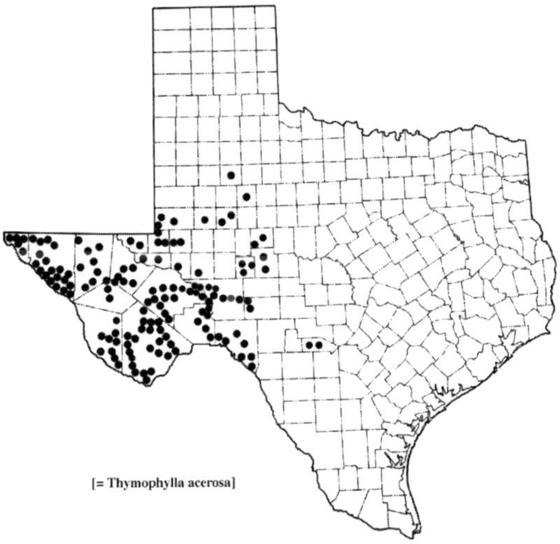
[= Thymophylla acerosa]

Dyssodia acerosa

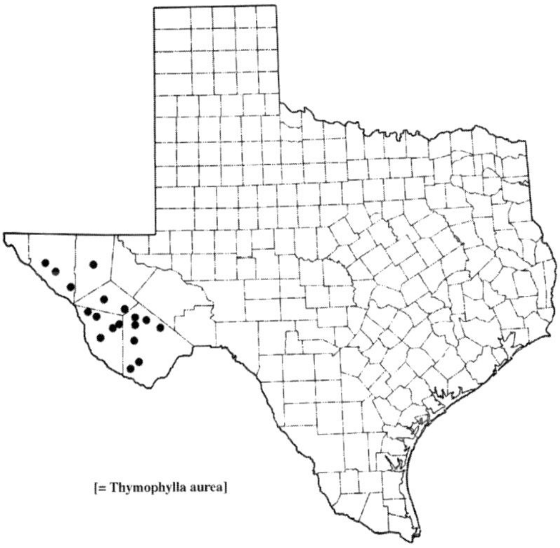
[= Thymophylla aurea]

Dyssodia aurea

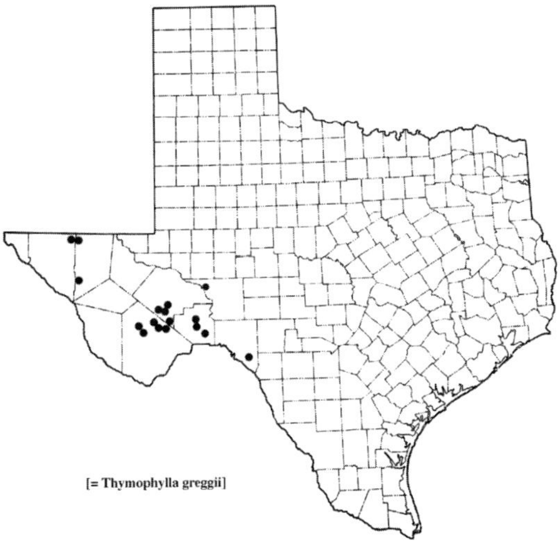
[= Thymophylla greggii]

Dyssodia greggii

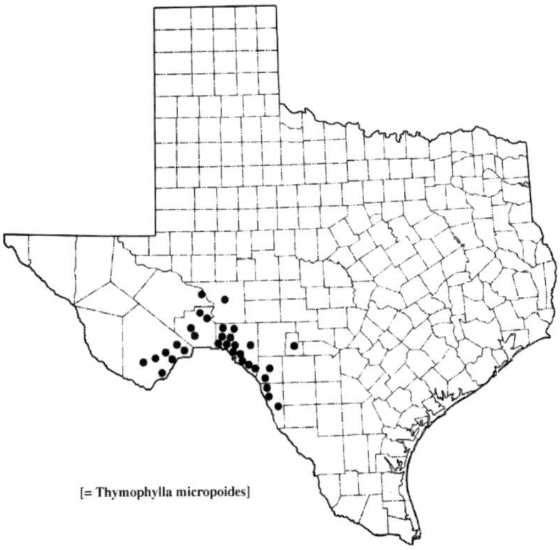
[= Thymophylla micropoides]

Dyssodia micropoides

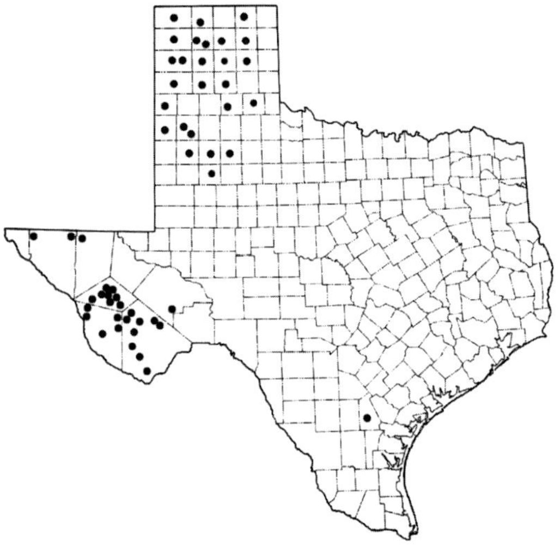

Dyssodia papposa

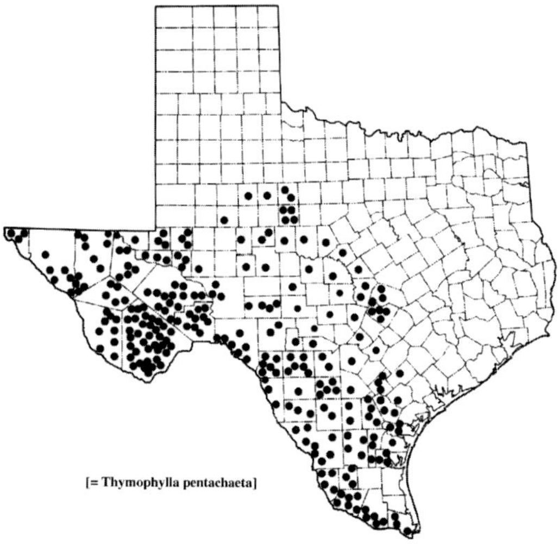
[= Thymophylla pentachaeta]

Dyssodia pentachaeta

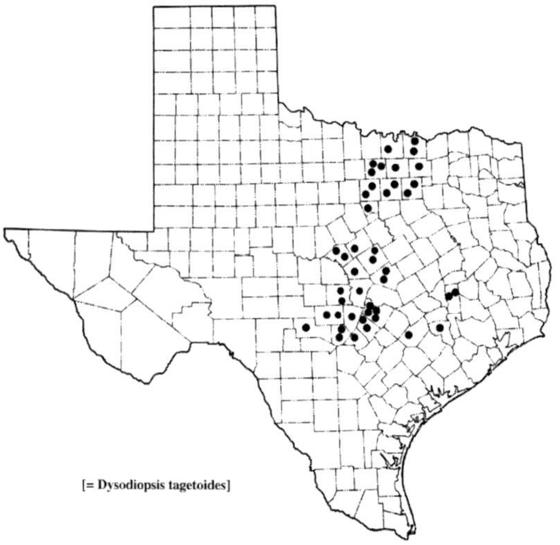
[= Dysodiopsis tagetoides]

Dyssodia tagetoides

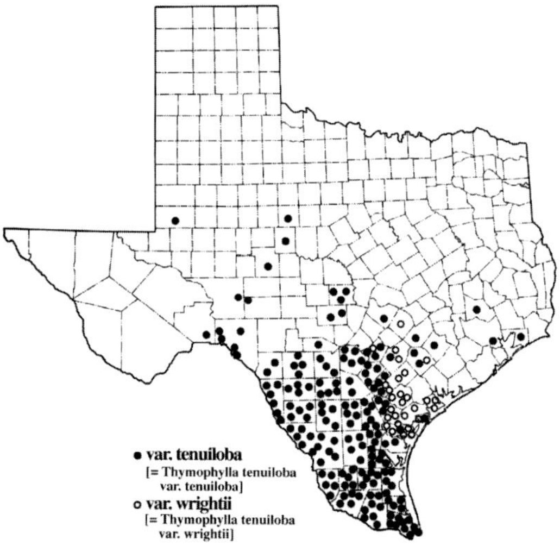
• var. tenuiloba
[= Thymophylla tenuiloba var. tenuiloba]
○ var. wrightii
[= Thymophylla tenuiloba var. wrightii]

Dyssodia tenuiloba

[= Thymophylla tephroleuca]

Dyssodia tephroleuca

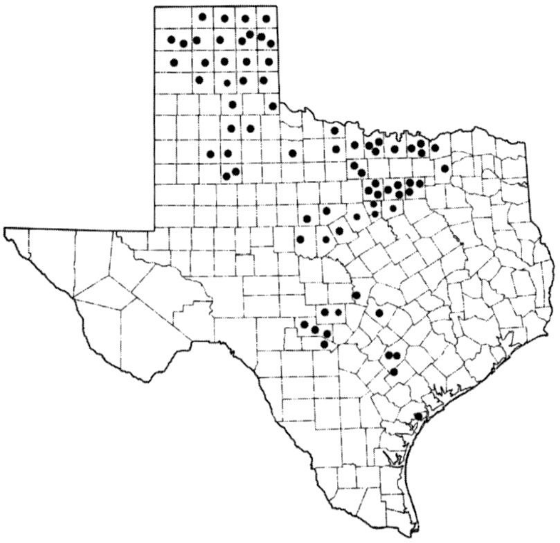

Echinacea angustifolia

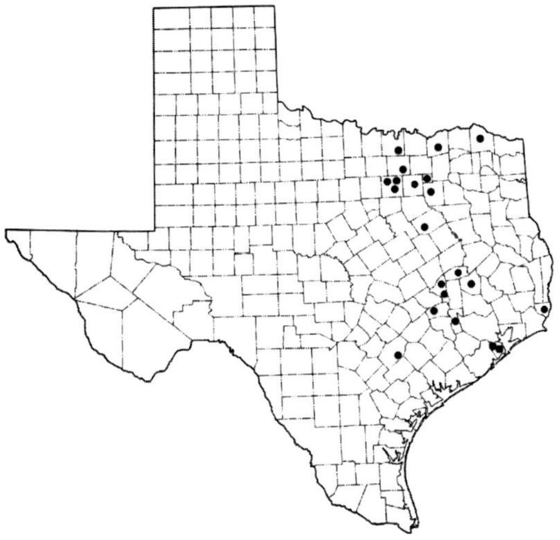

Echinacea atrorubens

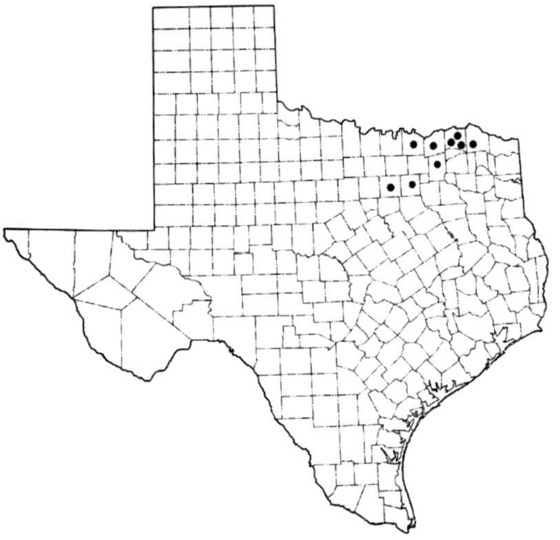

Echinacea pallida

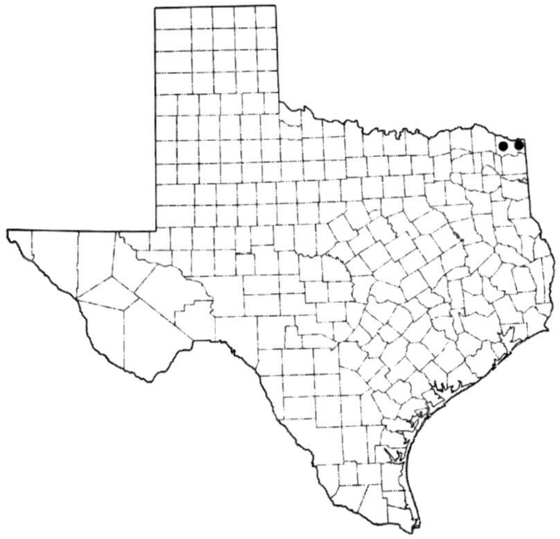

Echinacea purpurea

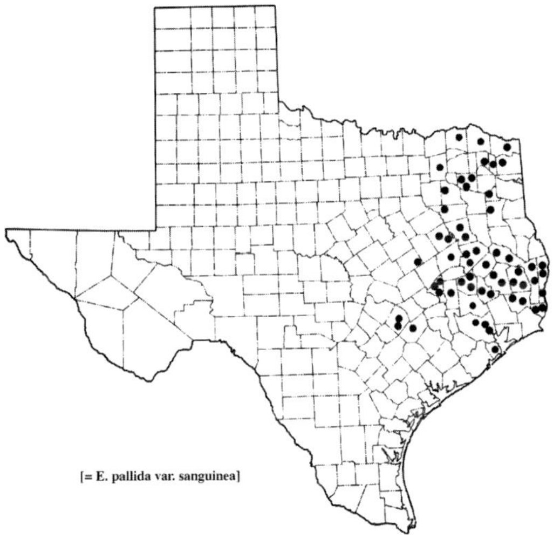
[= E. pallida var. sanguinea]

Echinacea sanguinea

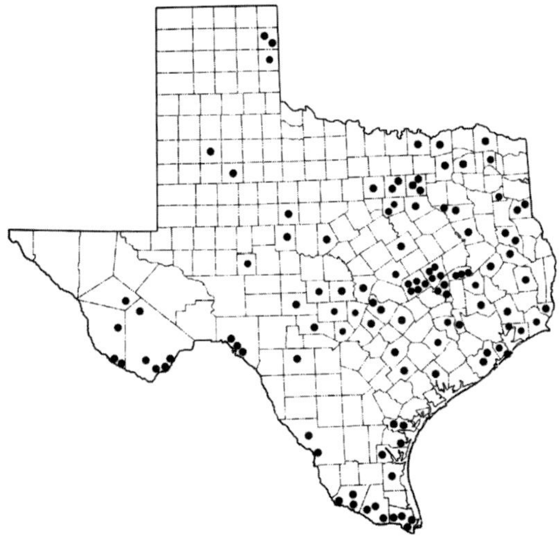

Eclipta prostrata

Egletes viscosa

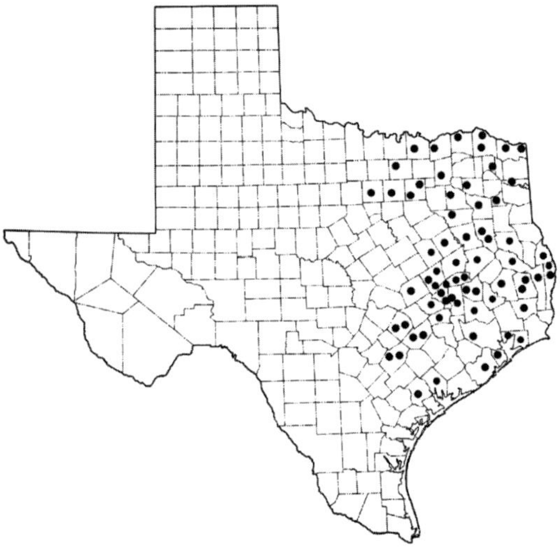

Elephantopus carolinianus

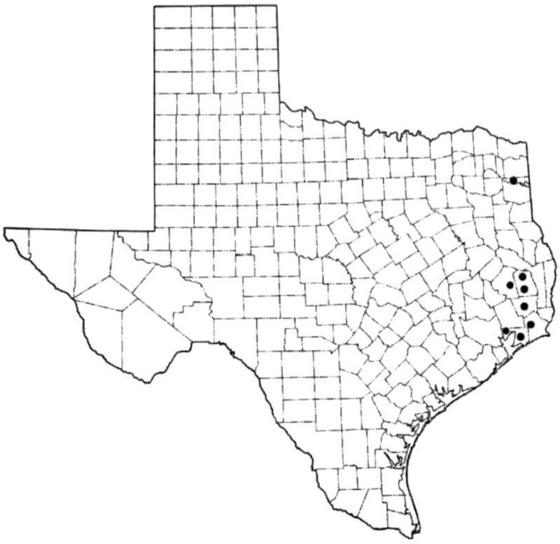

Elephantopus nudatus

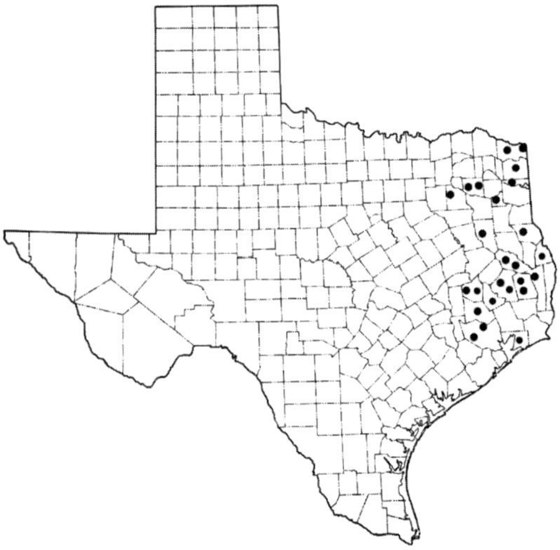

Elephantopus tomentosus

Emilia fosbergii

Encelia scaposa

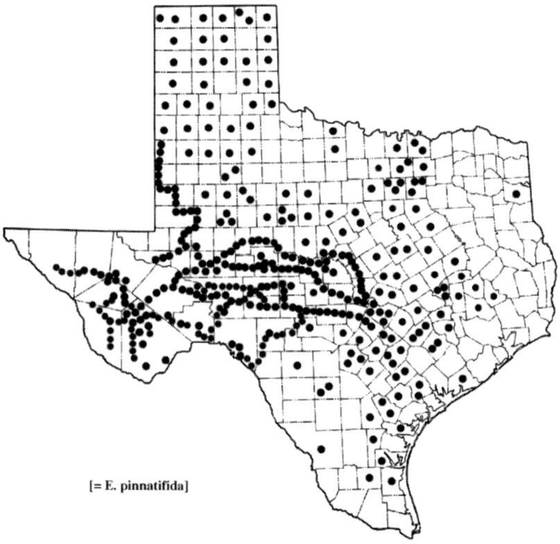
[= E. pinnatifida]

Engelmannia peristenia

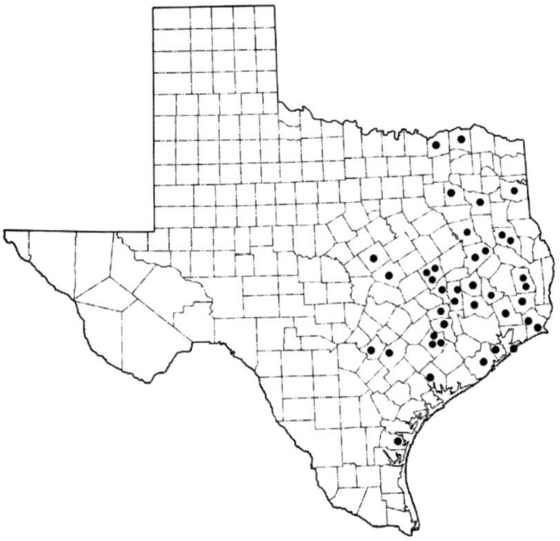

Erechtites hieraciifolia

Ericameria laricifolia

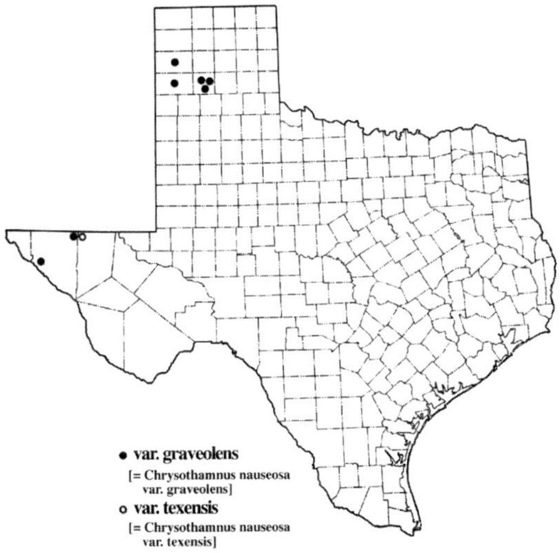
• var. graveolens
[= Chrysothamnus nauseosa var. graveolens]
o var. texensis
[= Chrysothamnus nauseosa var. texensis]

Ericameria nauseosa

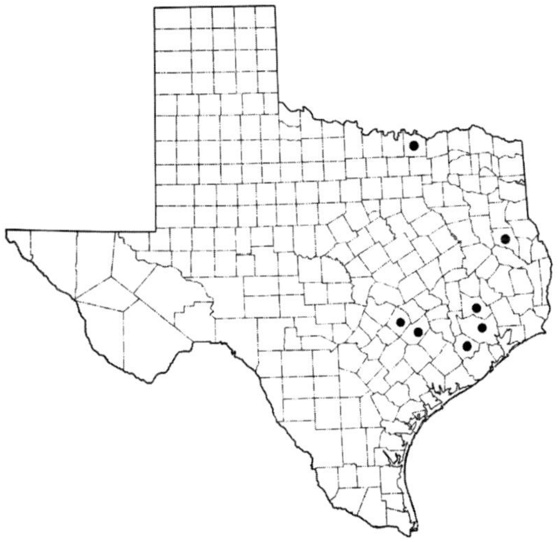

Erigeron annuus

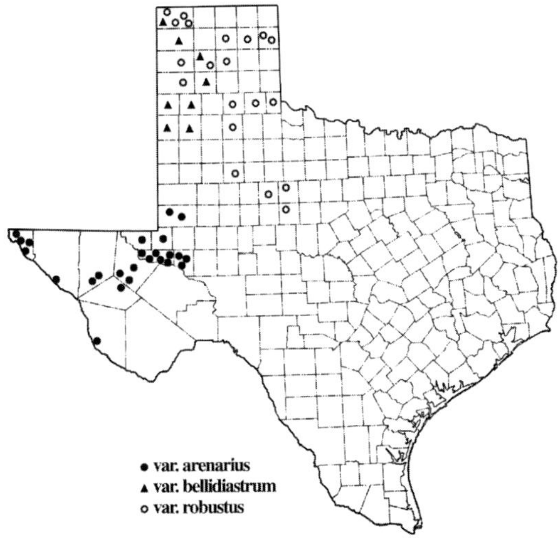
• var. arenarius
▲ var. bellidiastrum
o var. robustus

Erigeron bellidiastrum

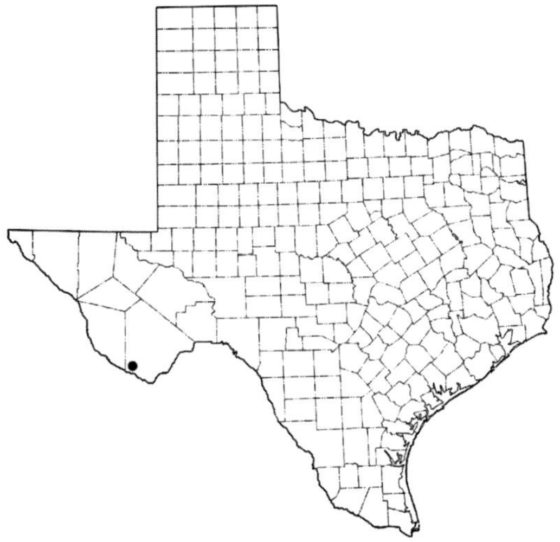

Erigeron bigelovii

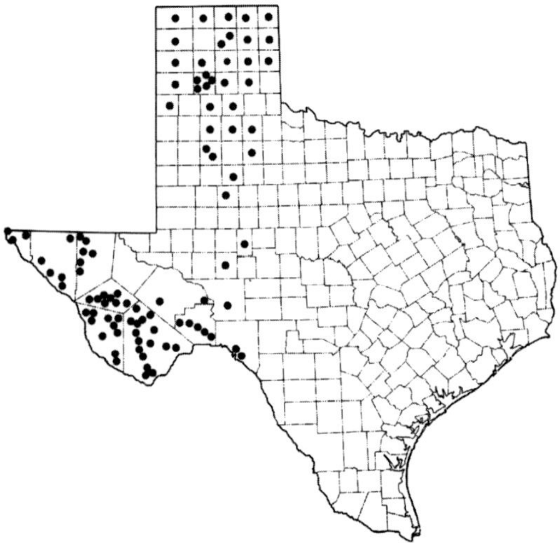

Erigeron colomexicanus

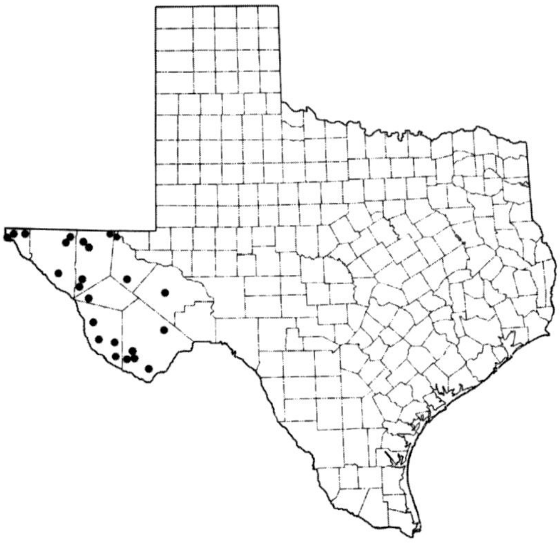

Erigeron divergens

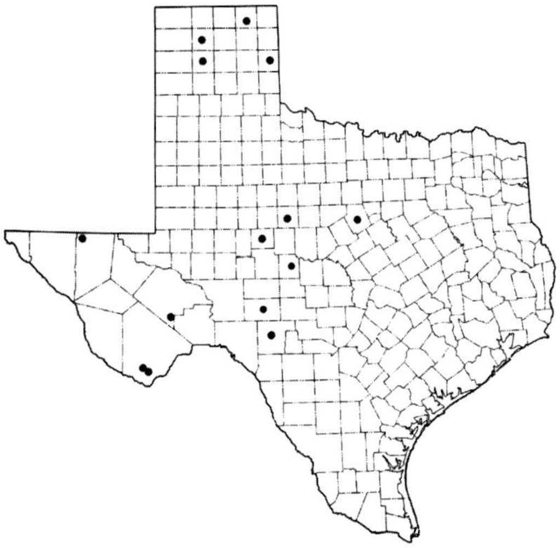

Erigeron flagellaris

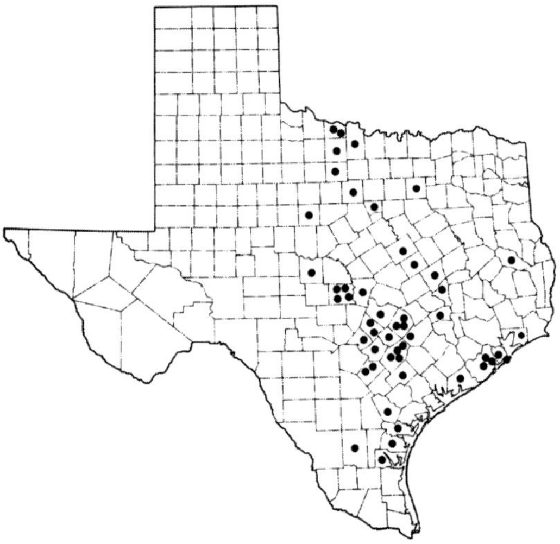

Erigeron geiseri

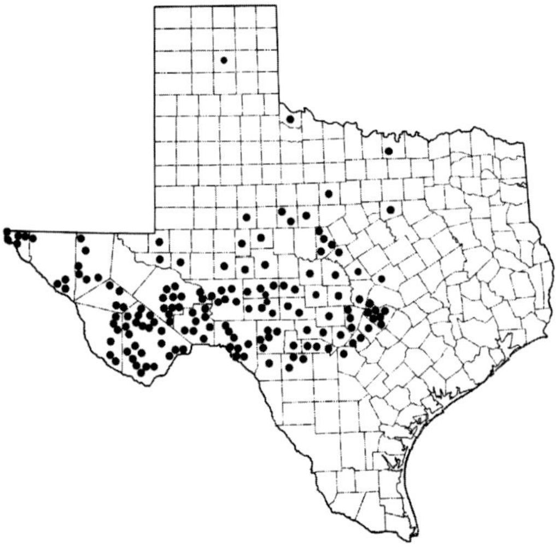

Erigeron modestus

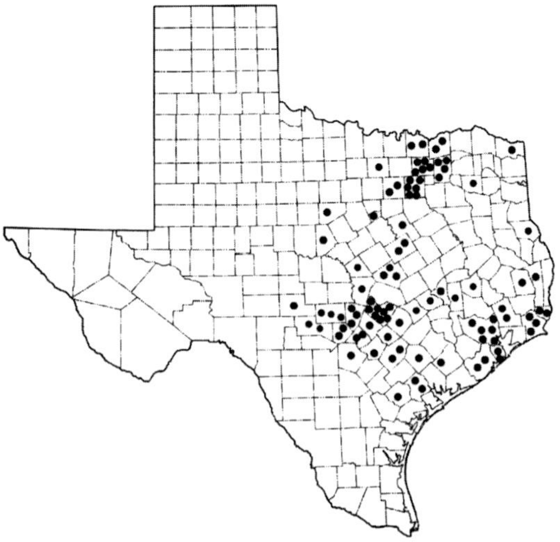

Erigeron philadelphicus

Erigeron procumbens

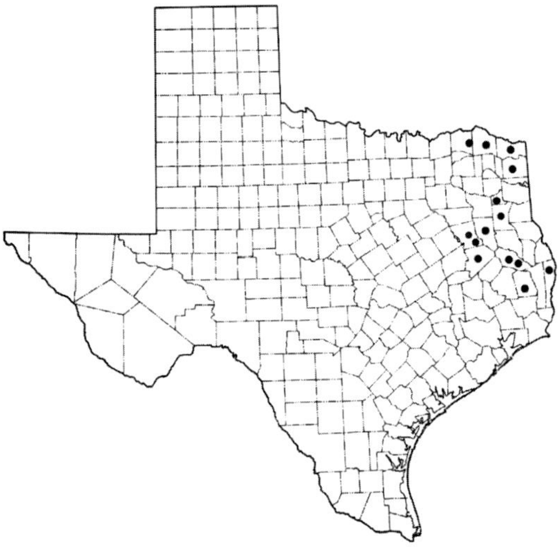
Erigeron pulchellus

Erigeron rybius

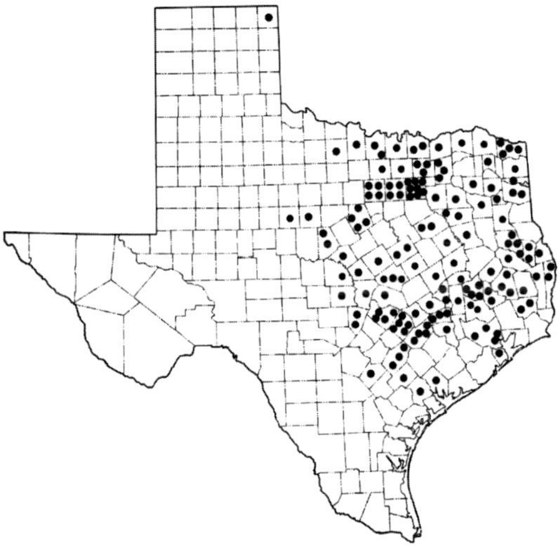
Erigeron strigosus

Erigeron superbus

Erigeron tenellus

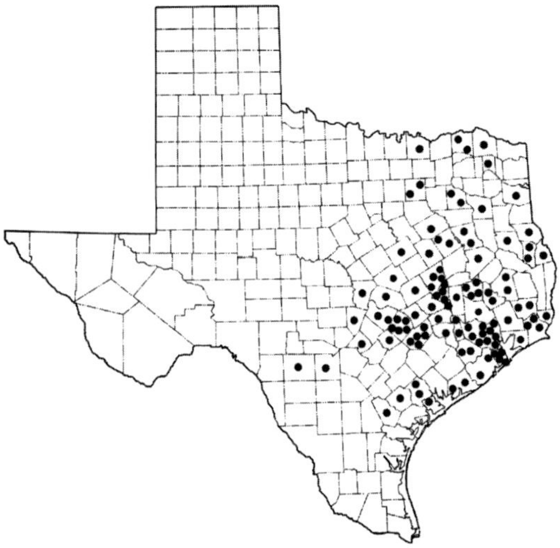

Erigeron tenuis

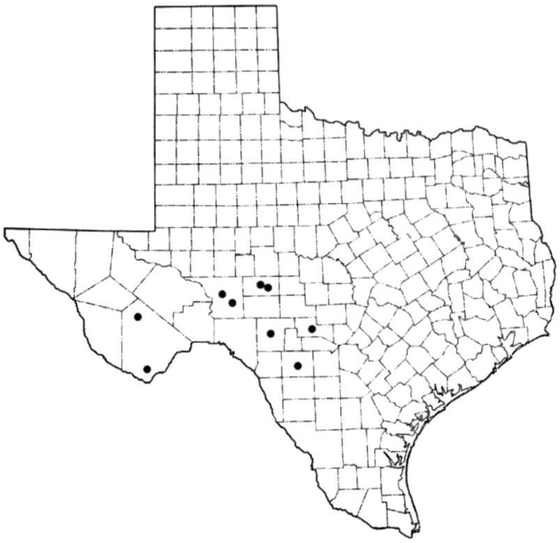

Erigeron versicolor

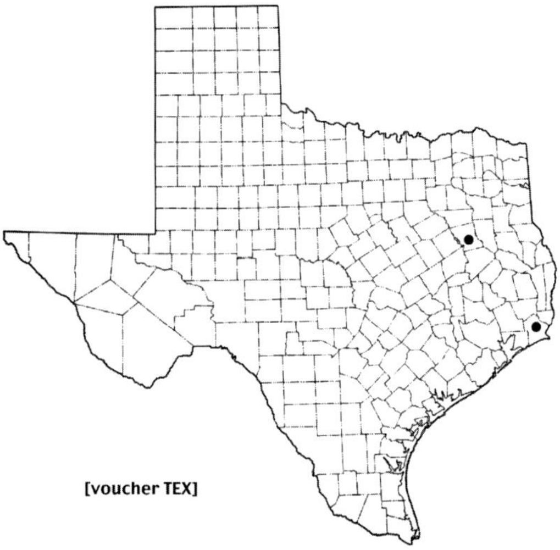
[voucher TEX]

Eupatorium album

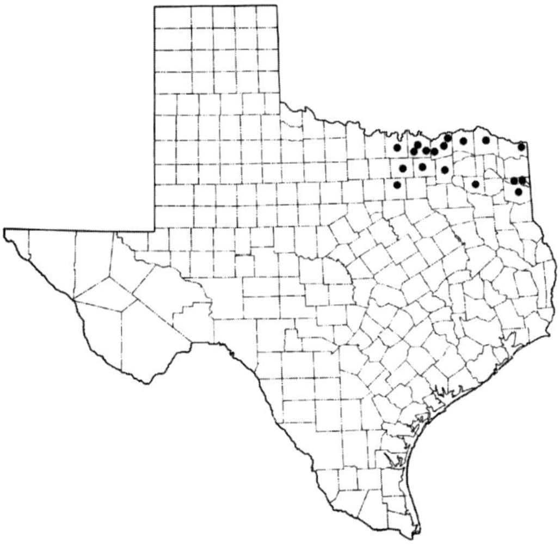

Eupatorium altissimum

[= Tamaulipa azurea]

Eupatorium azureum

[= Chromolaena bigelovii]

Eupatorium bigelovii

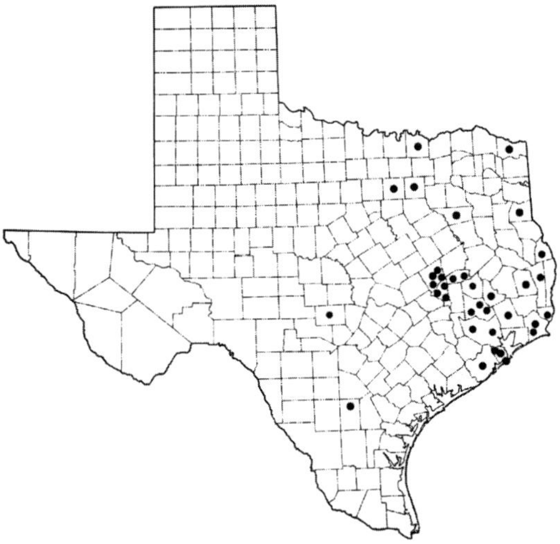

Eupatorium capillifolium

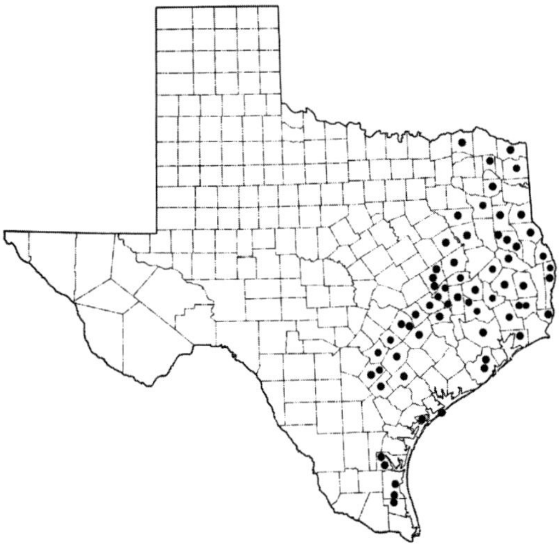

Eupatorium compositifolium

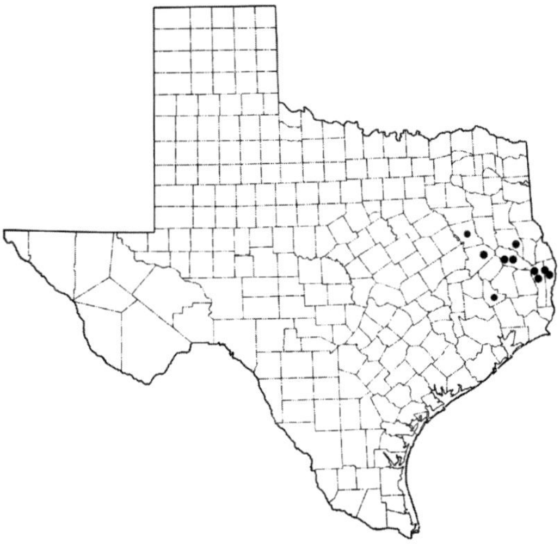

Eupatorium fistulosum

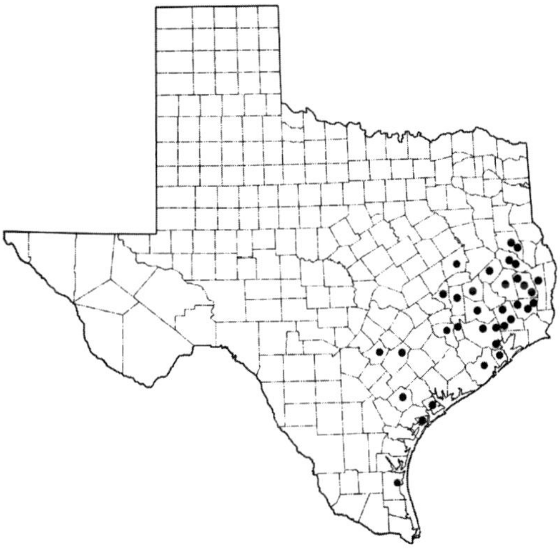

Eupatorium glaucescens

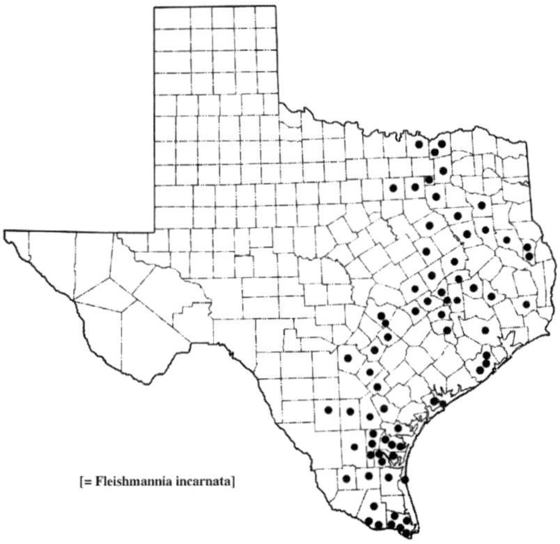
[= Fleishmannia incarnata]

Eupatorium incarnatum

Eupatorium ivifolium

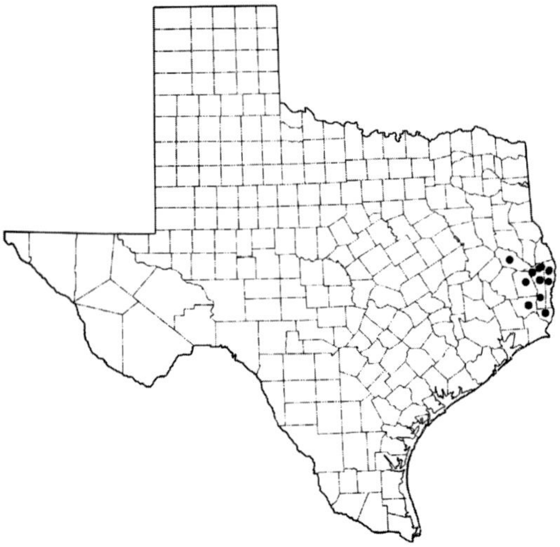

Eupatorium leucolepis

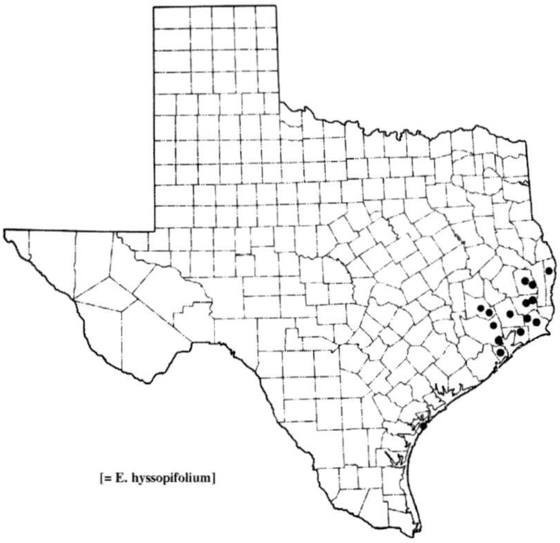
[= E. hyssopifolium]

Eupatorium mohrii

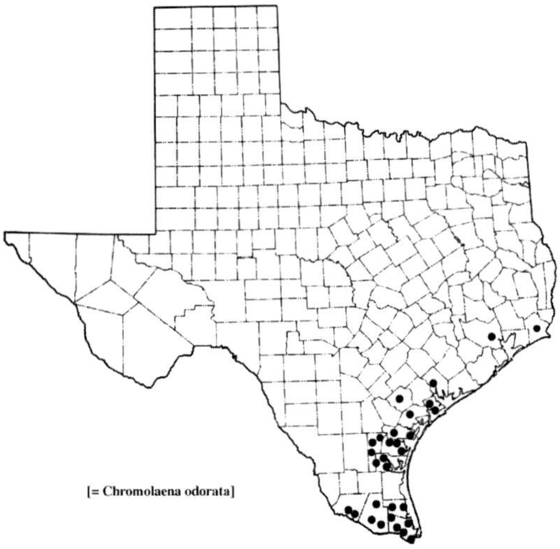
[= Chromolaena odorata]

Eupatorium odoratum

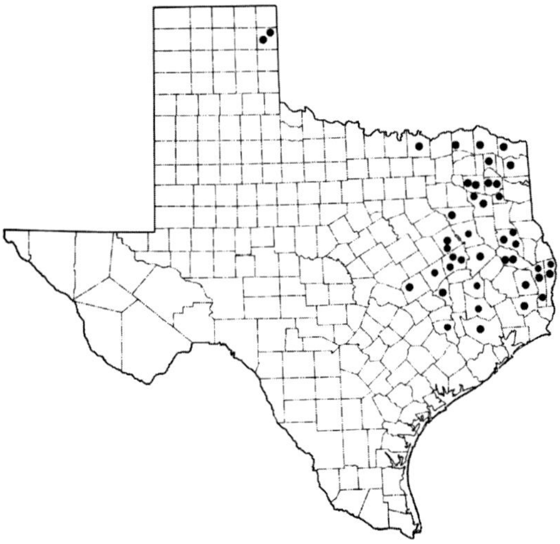

Eupatorium perfoliatum

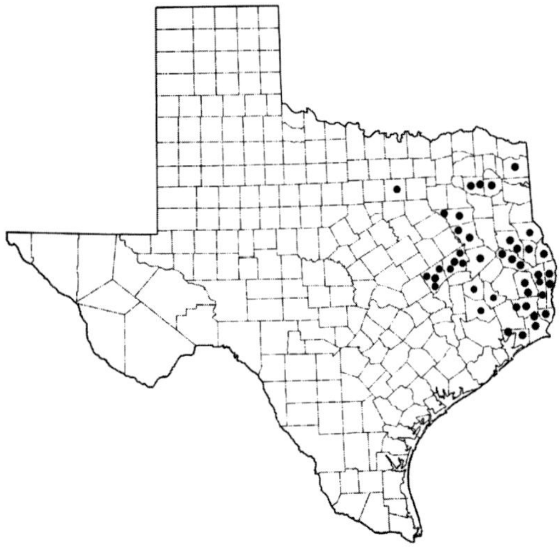

Eupatorium rotundifolium

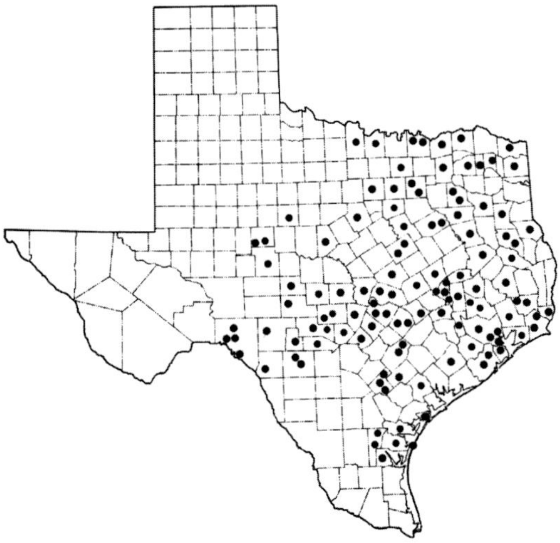

Eupatorium serotinum

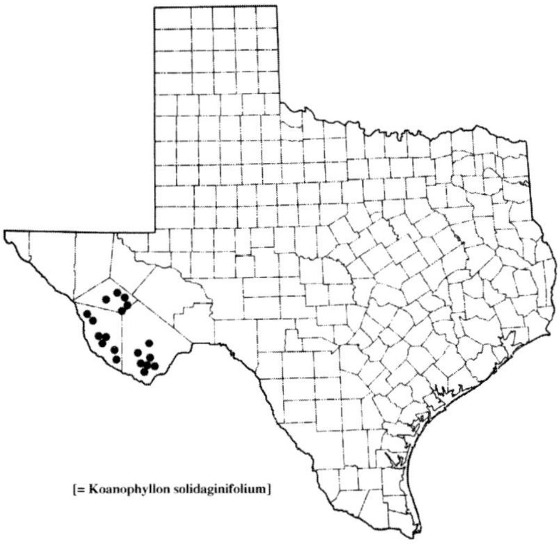
[= Koanophyllon solidaginifolium]

Eupatorium solidaginifolium

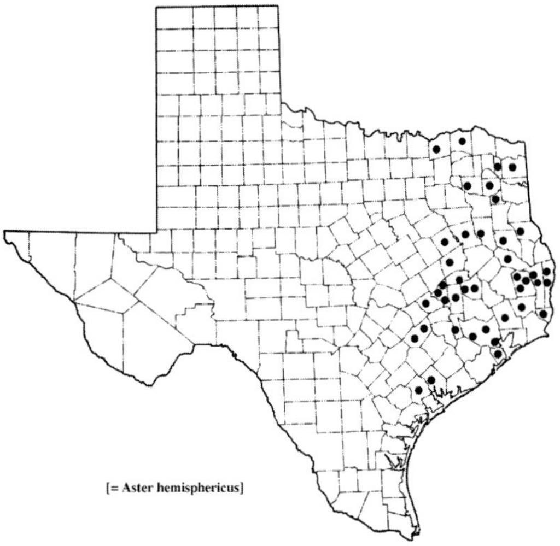
[= Aster hemisphericus]

Eurybia hemispherica

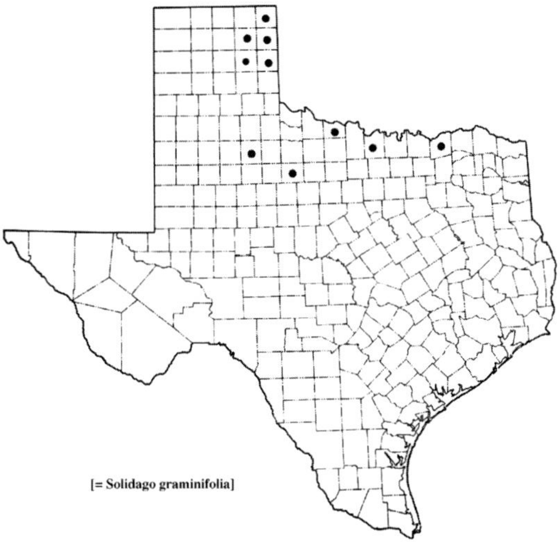
[= Solidago graminifolia]

Euthamia graminifolia

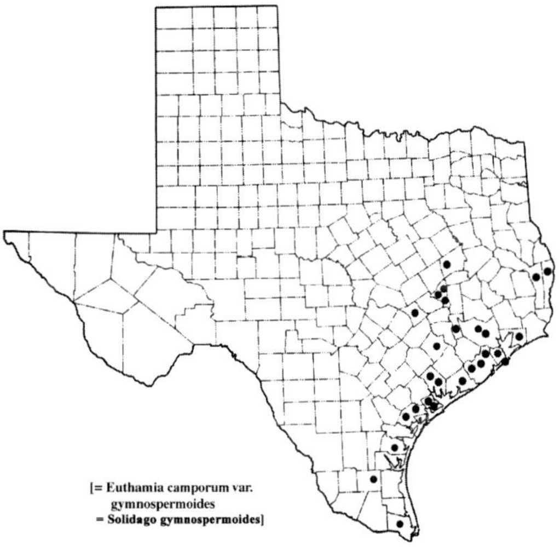
[= Euthamia camporum var.
gymnospermoides
= Solidago gymnospermoides]

Euthamia gymnospermoides

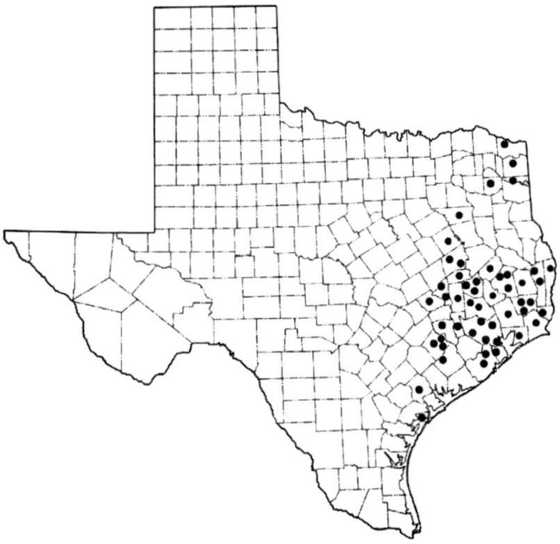

Euthamia leptocephala

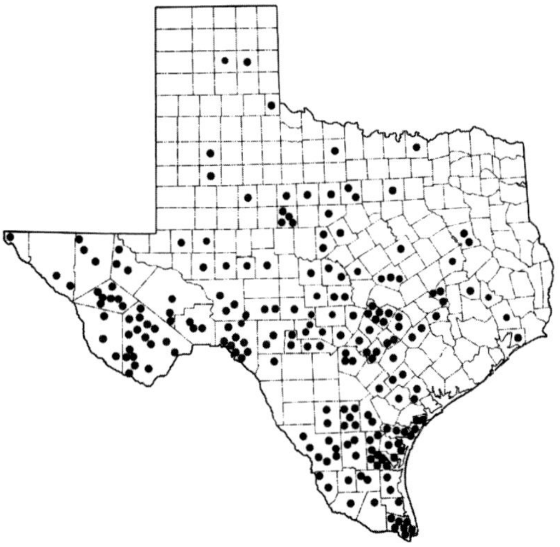

Evax multicaulis

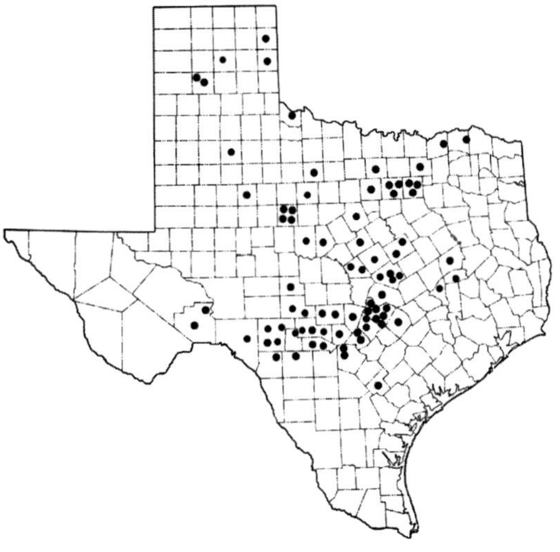

Evax prolifera

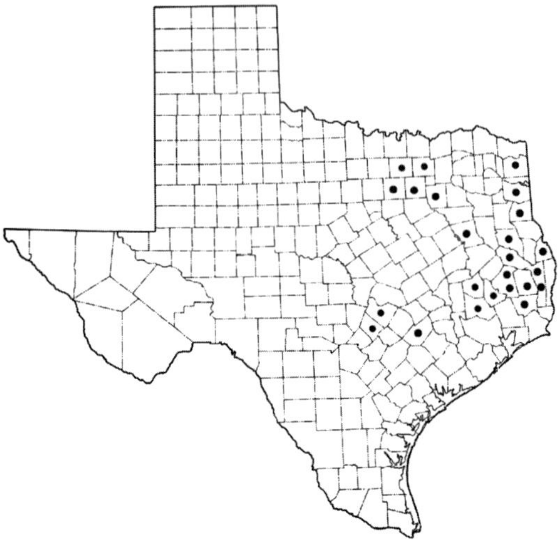

Facelis retusa

Filago californica

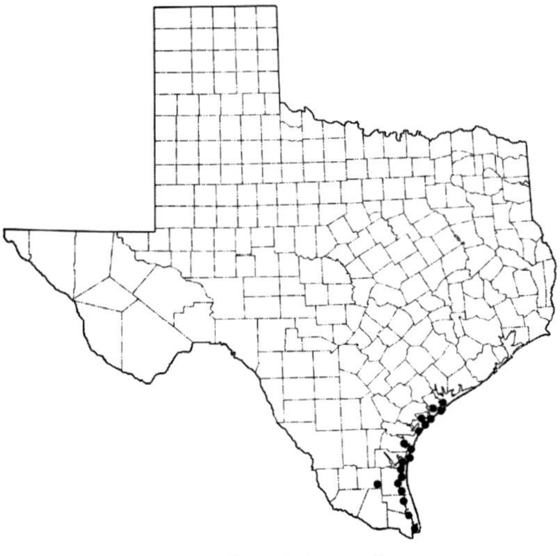

Flaveria brownii

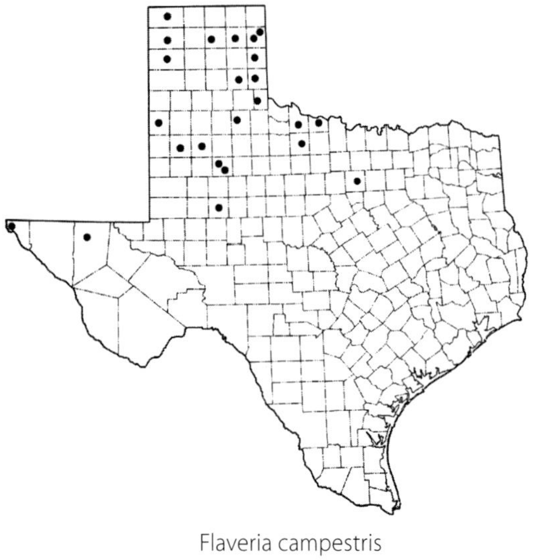

Flaveria campestris

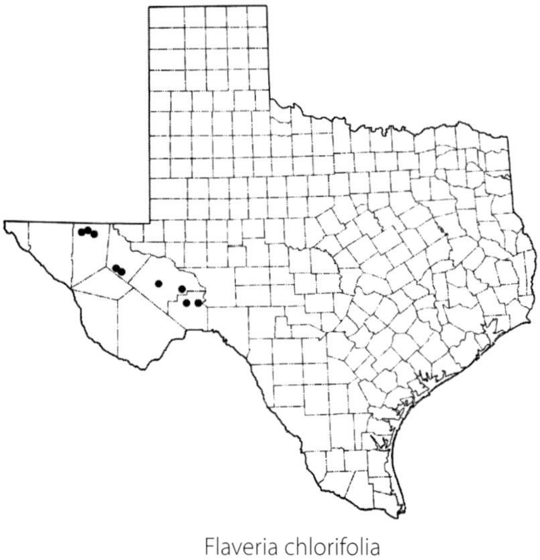

Flaveria chlorifolia

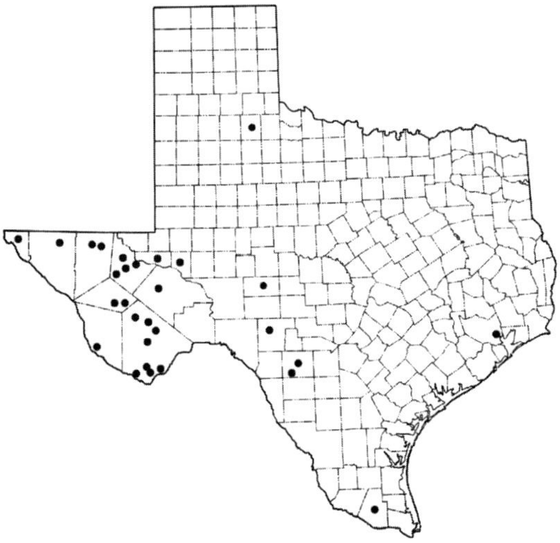

Flaveria trinervia

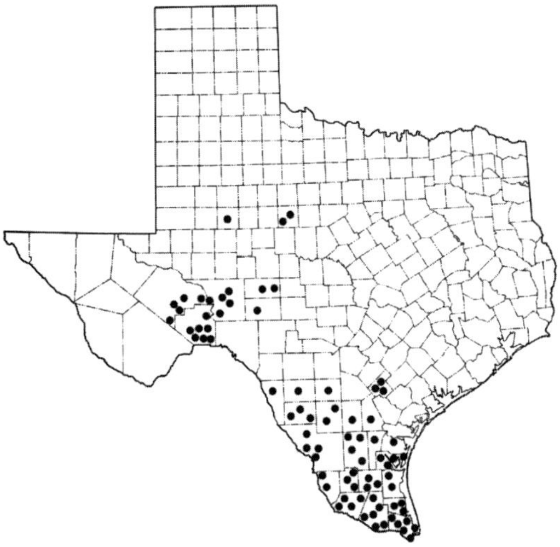

Florestina tripteris

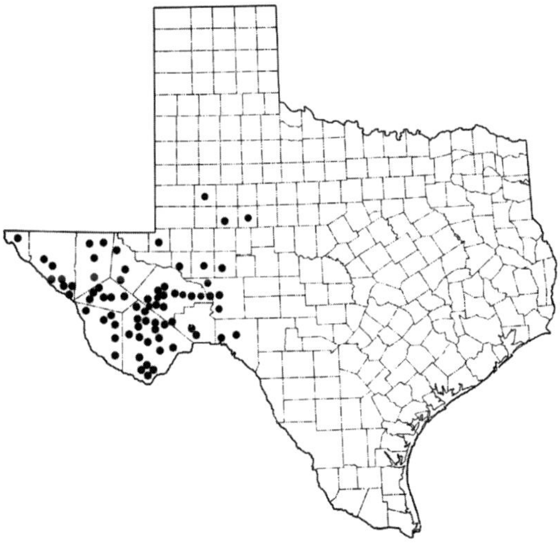

Flourensia cernua

Flyriella parryi

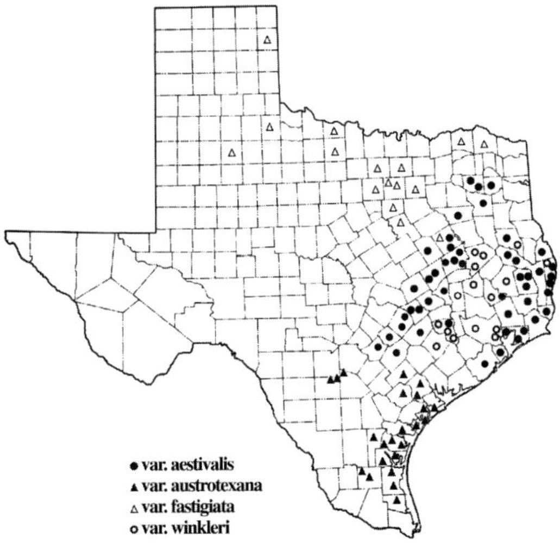
• var. aestivalis
▲ var. austrotexana
△ var. fastigiata
○ var. winkleri

Gaillardia aestivalis

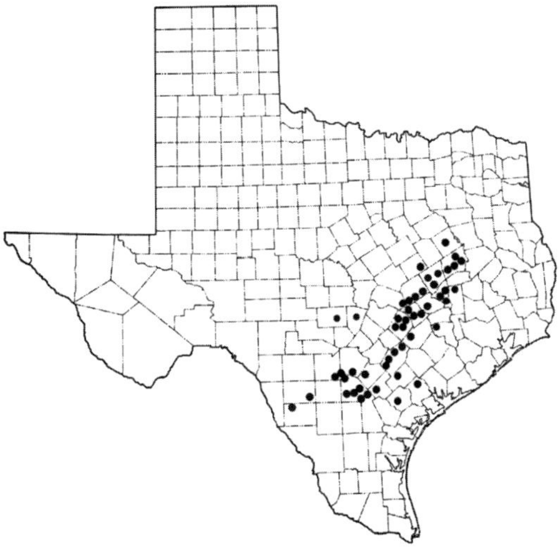

Gaillardia amblyodon

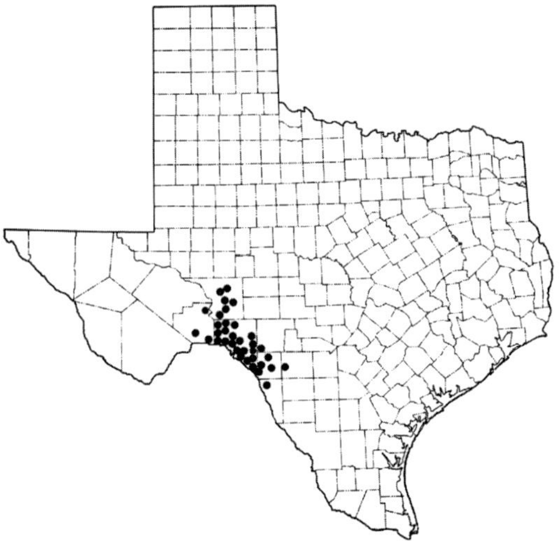

Gaillardia coahuilensis

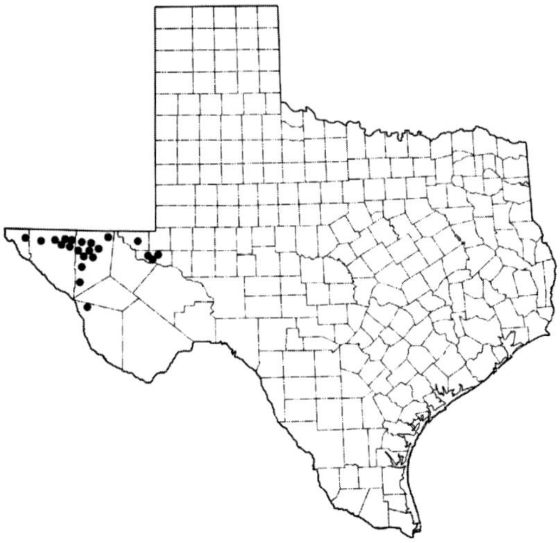

Gaillardia multiceps

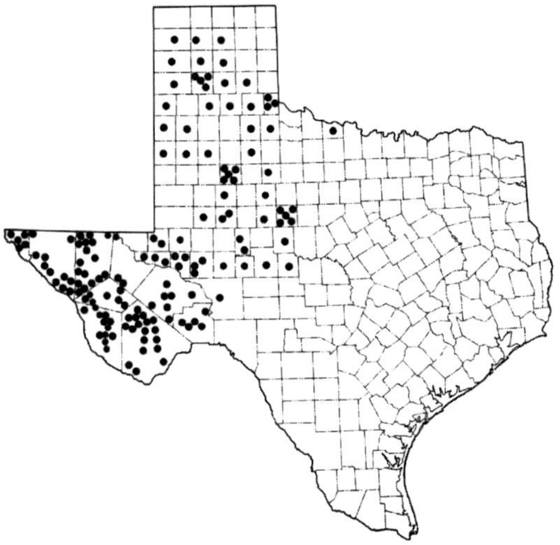

Gaillardia pinnatifida

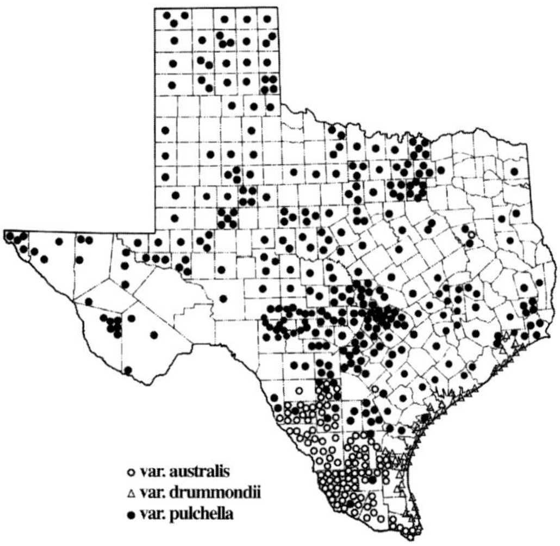
○ var. australis
△ var. drummondii
● var. pulchella

Gaillardia pulchella

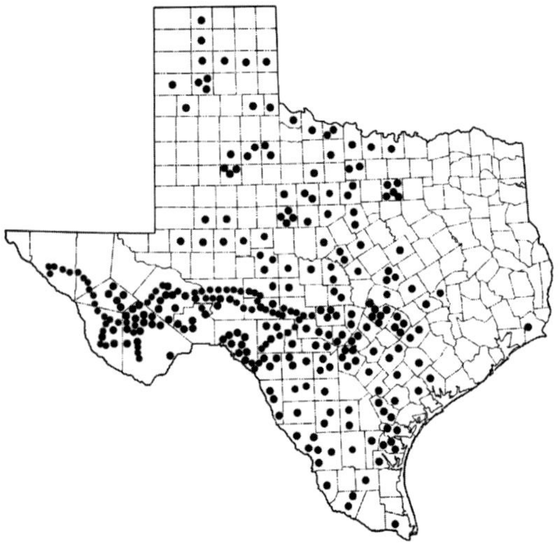

Gaillardia suavis

Galinsoga parviflora

[= Gnaphalium americanum
= G. purpureum var. americanum]

Gamochaeta americana

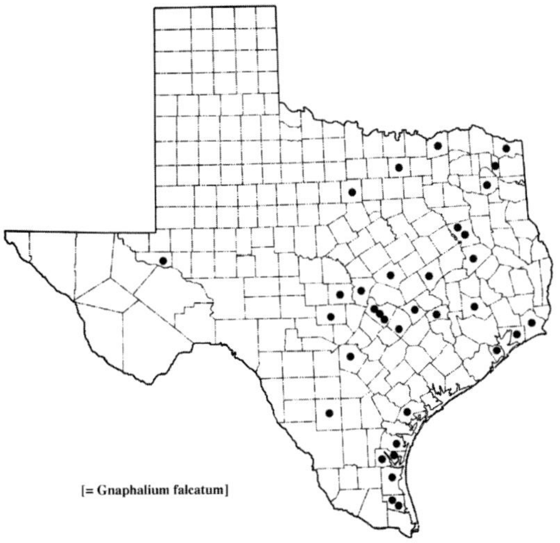
[= Gnaphalium falcatum]

Gamochaeta falcata

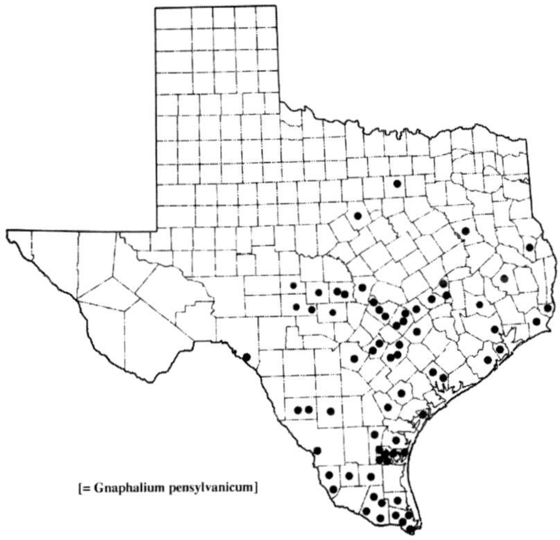
[= Gnaphalium pensylvanicum]

Gamochaeta pensylvanica

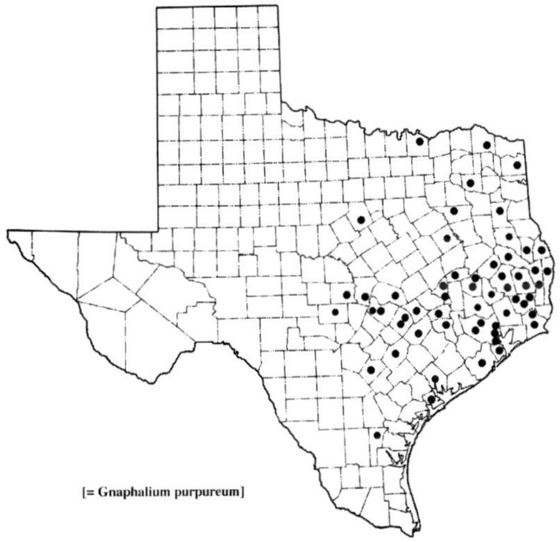
[= Gnaphalium purpureum]

Gamochaeta purpurea

Gamochaeta sphacilata

Gochnatia hypoleuca

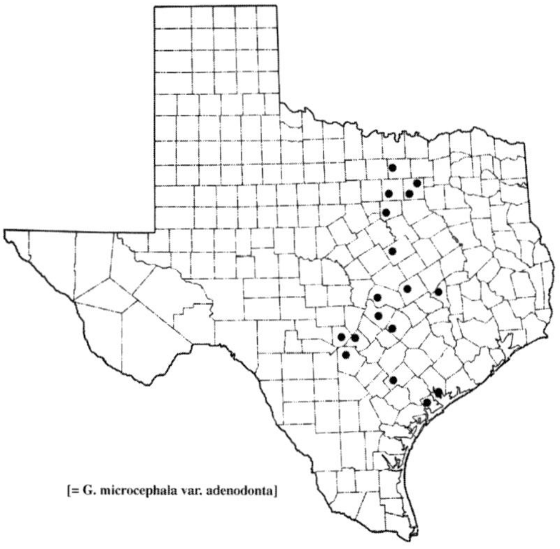
[= G. microcephala var. adenodonta]

Grindelia adenodonta

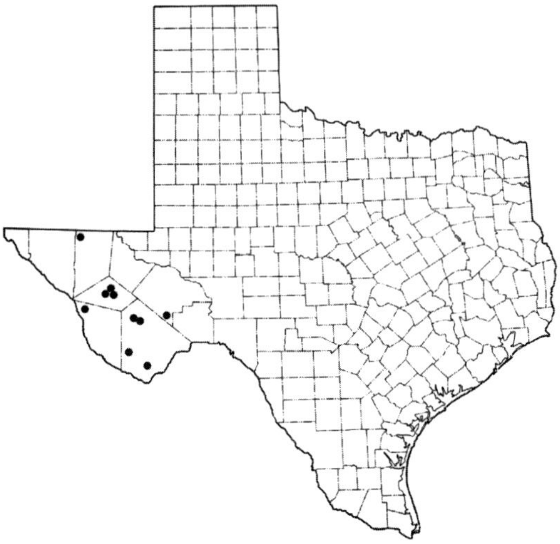

Grindelia arizonica

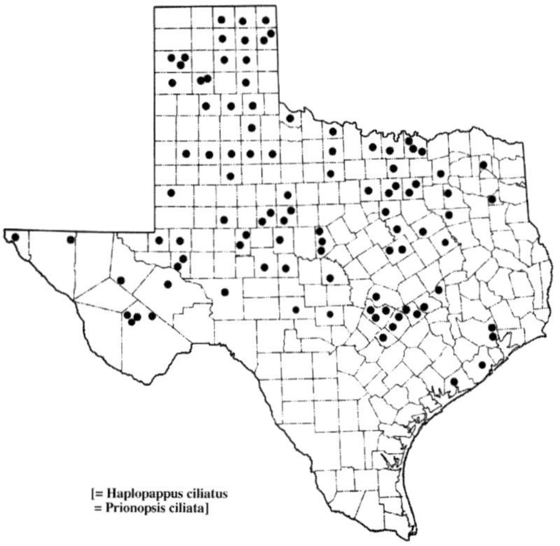
[= Haplopappus ciliatus
= Prionopsis ciliata]

Grindelia ciliata

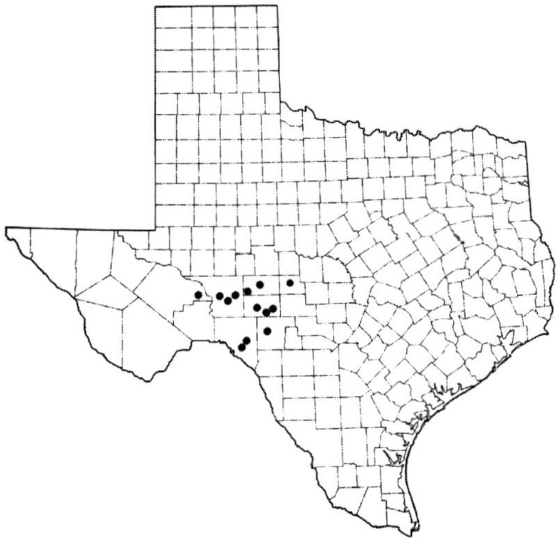

Grindelia grandiflora

Grindelia havardii

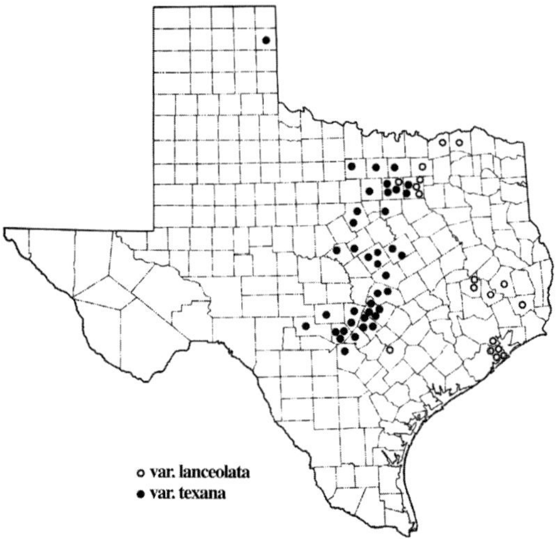
o var. lanceolata
• var. texana

Grindelia lanceolata

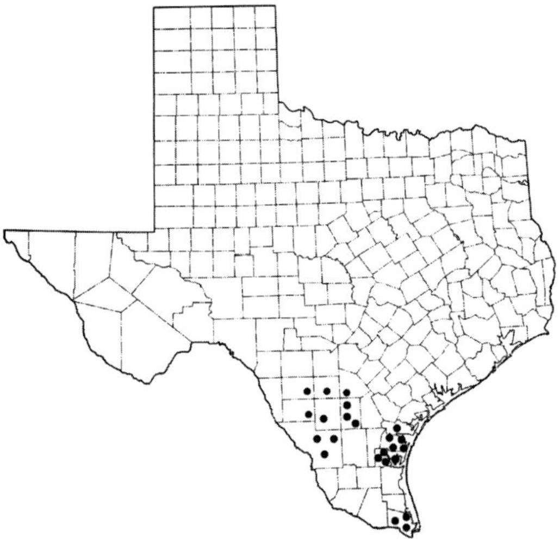

Grindelia microcephala

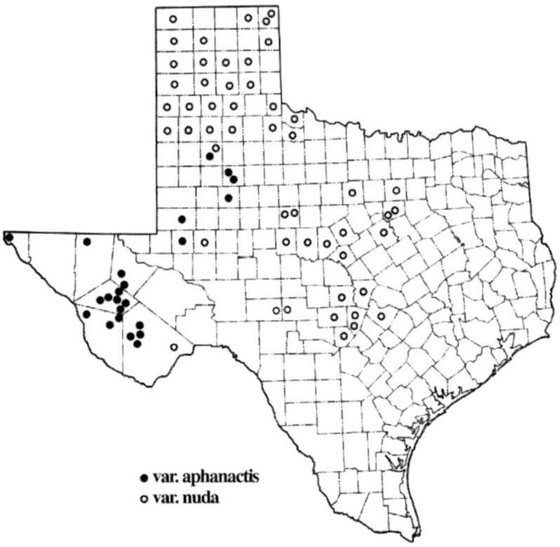
• var. aphanactis
○ var. nuda

Grindelia nuda

Grindelia oolepis

Grindelia oxylepis

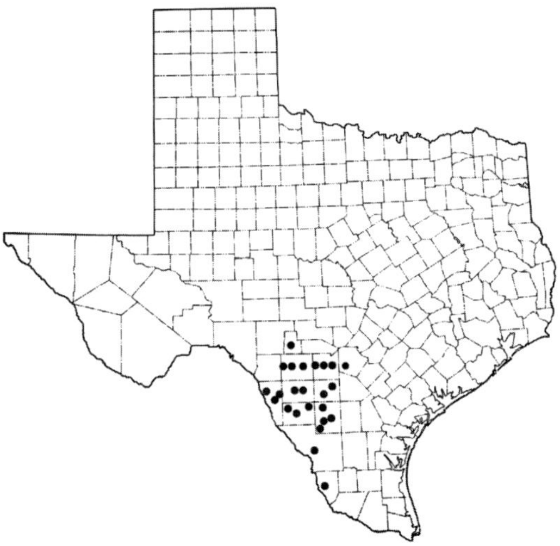

Grindelia pusilla

Grindelia scabra

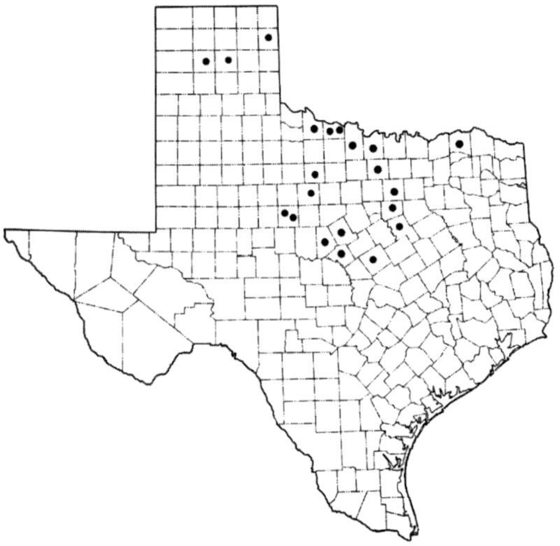

Grindelia squarrosa

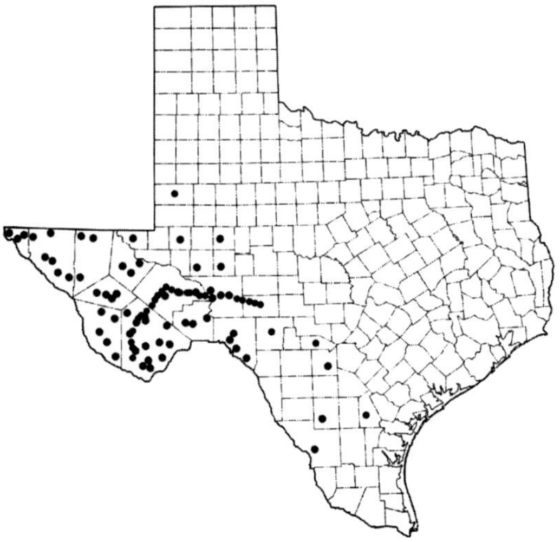

Gutierrezia microcephala

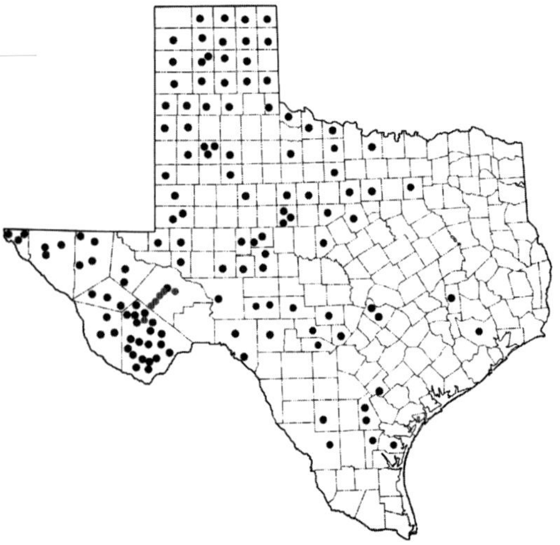

Gutierrezia sarothrae

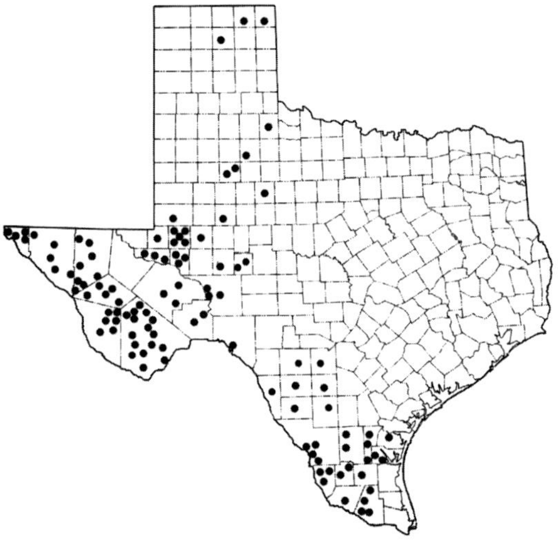

Gutierrezia sphaerocephala

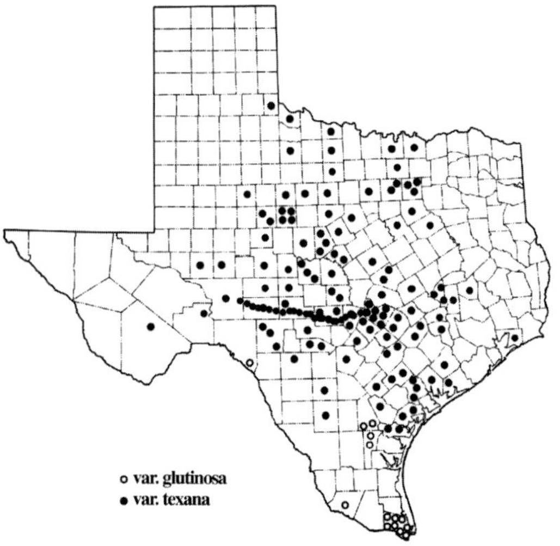
○ var. glutinosa
● var. texana

Gutierrezia texana

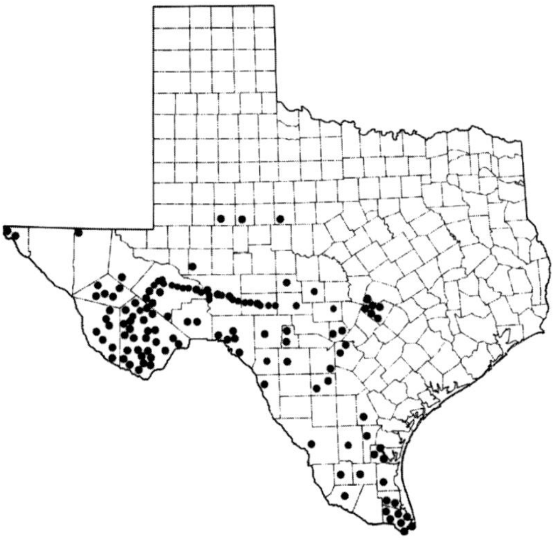

Gymnosperma glutinosum

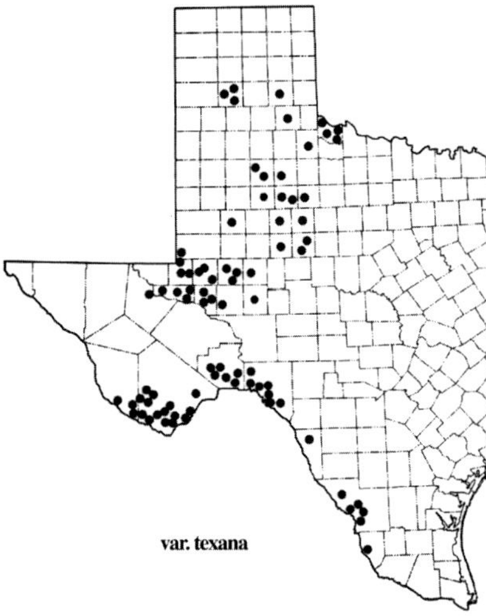
var. texana

Haploesthes greggii

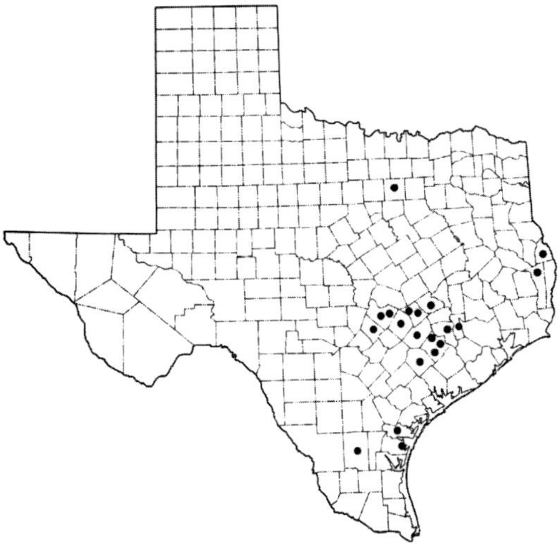

Hedypnois cretica

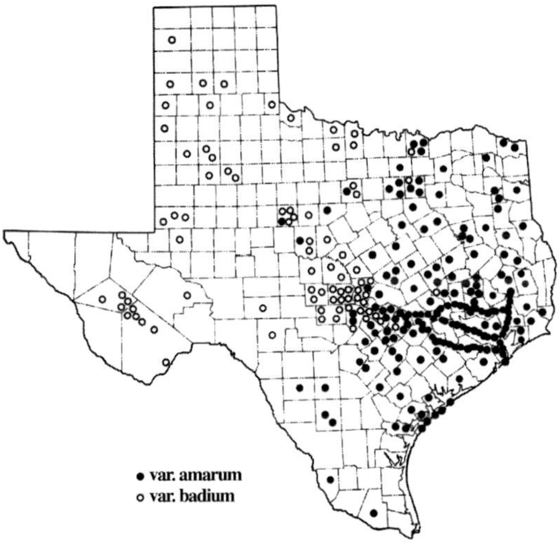
• var. amarum
○ var. badium

Helenium amarum

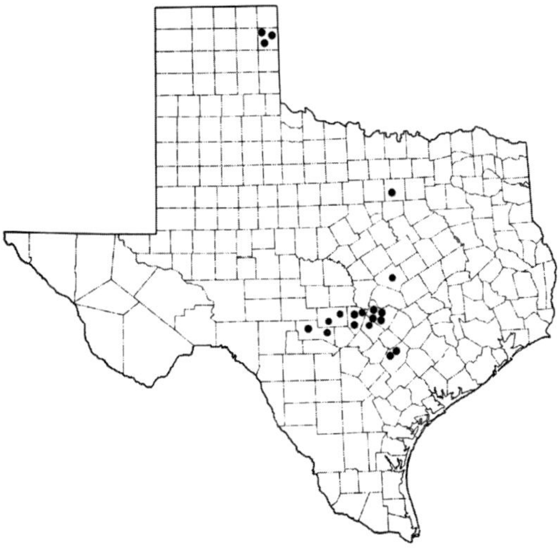

Helenium autumnale

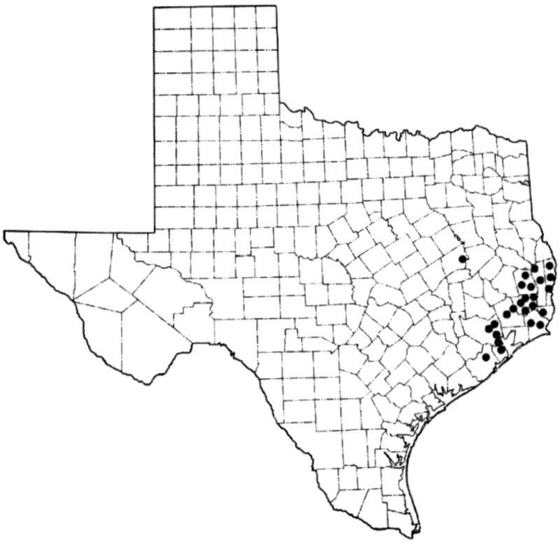

Helenium drummondii

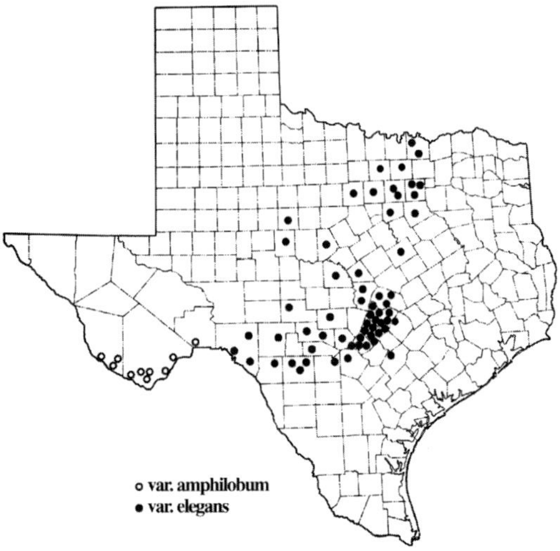
○ var. amphilobum
• var. elegans

Helenium elegans

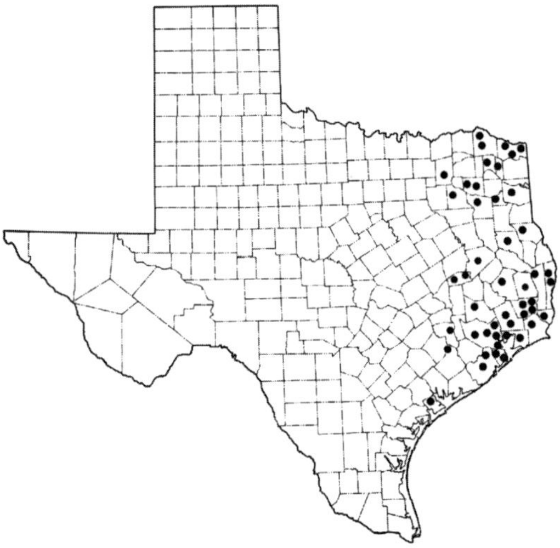

Helenium flexuosum

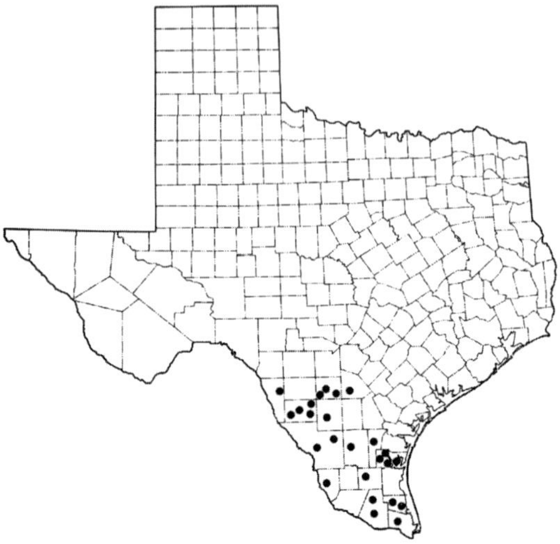

Helenium linifolium

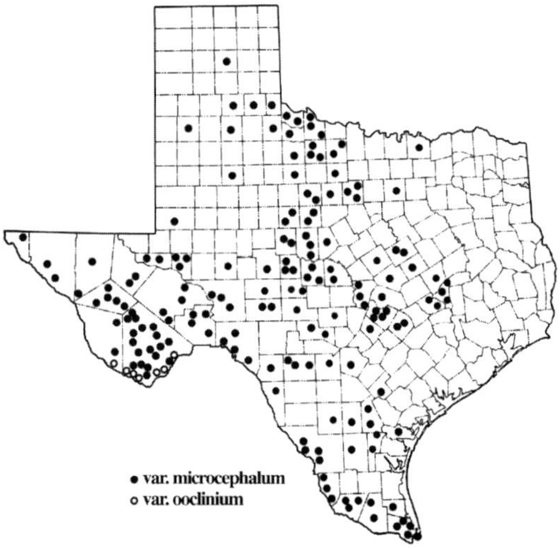
• var. microcephalum
o var. ooclinium

Helenium microcephalum

Helenium quadridentatum

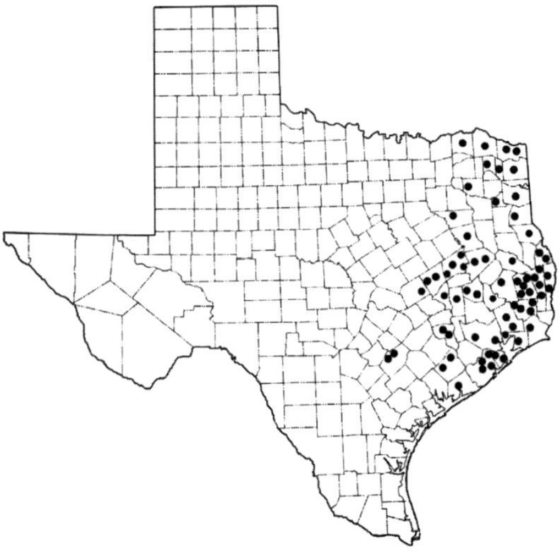

Helianthus angustifolius

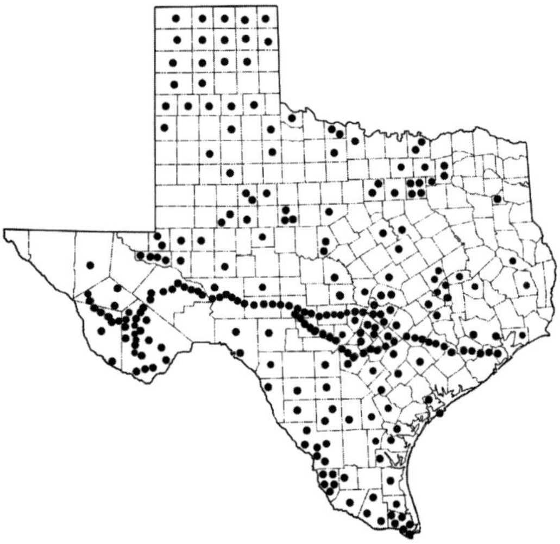

Helianthus annuus

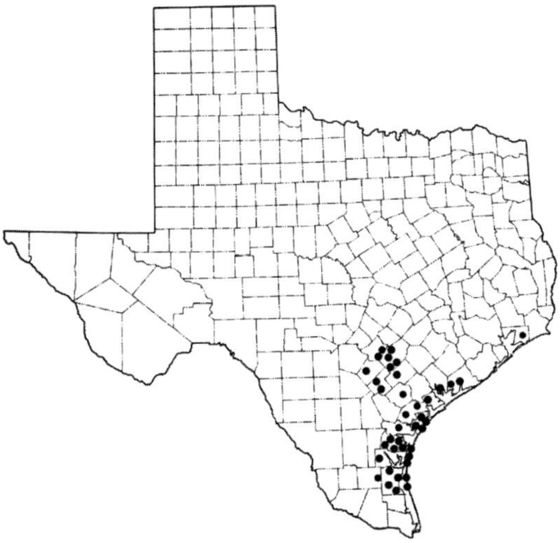

Helianthus argophyllus

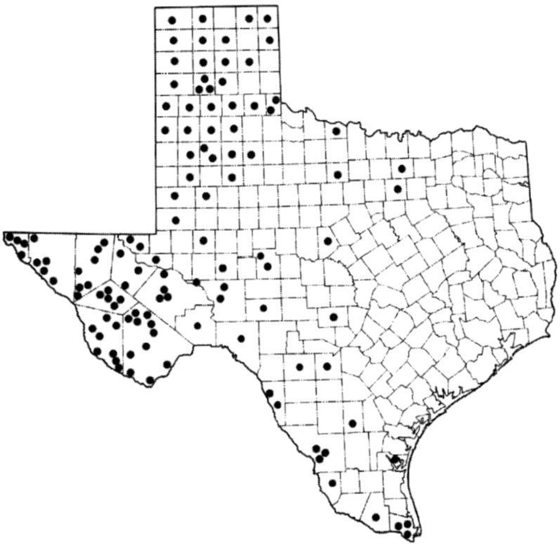

Helianthus ciliaris

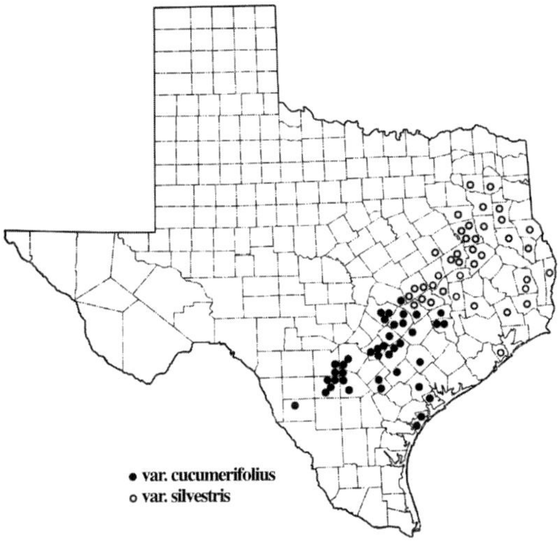
• var. cucumerifolius
○ var. silvestris

Helianthus debilis

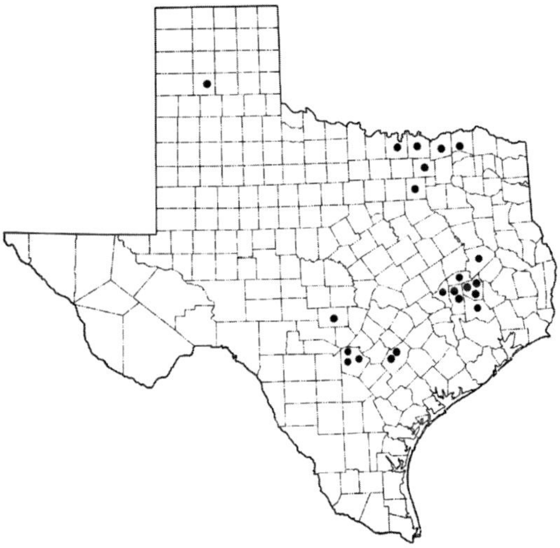

Helianthus grosseserratus

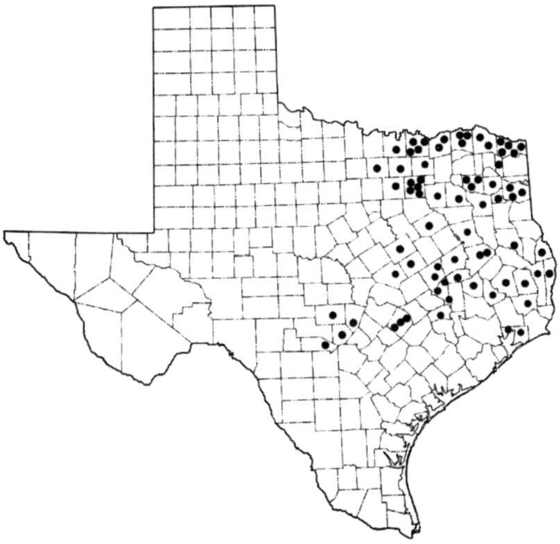

Helianthus hirsutus

Helianthus laciniatus

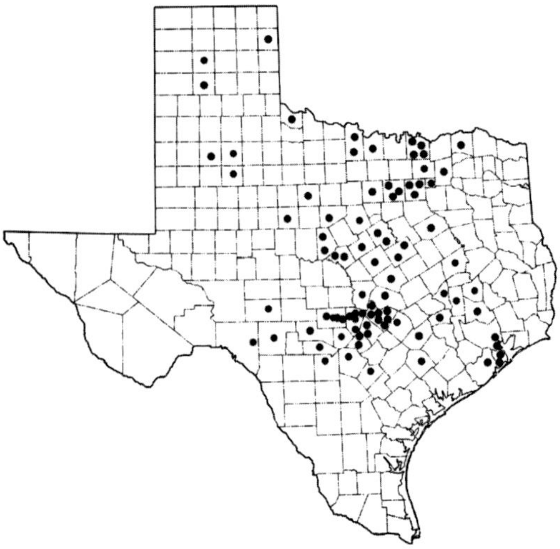

Helianthus maximiliani

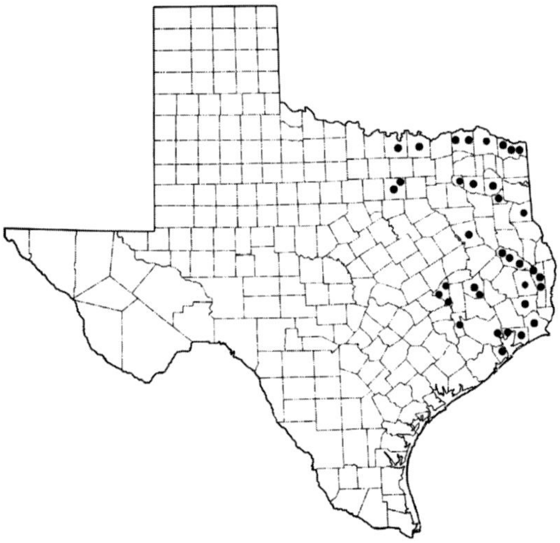

Helianthus mollis

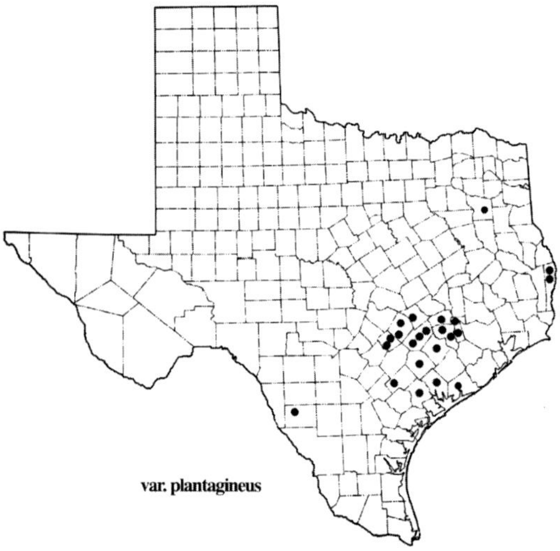
var. plantagineus

Helianthus occidentalis

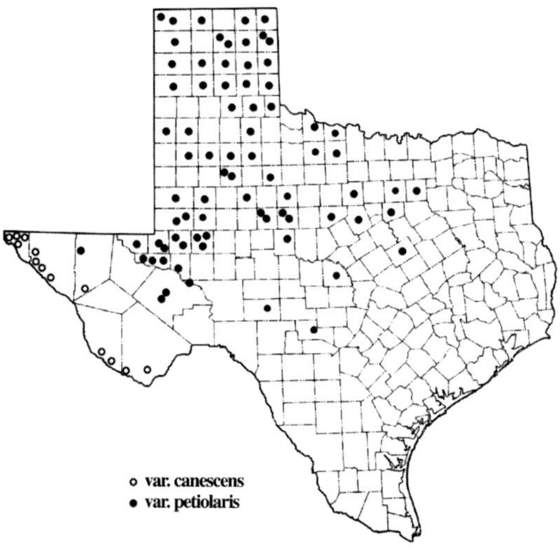
○ var. canescens
● var. petiolaris

Helianthus petiolaris

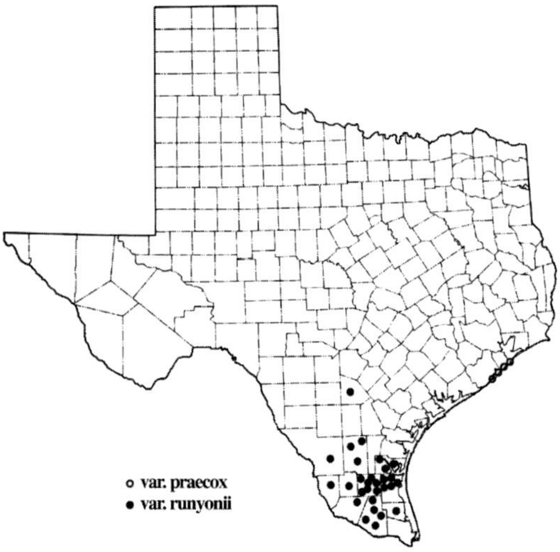
○ var. praecox
● var. runyonii

Helianthus praecox

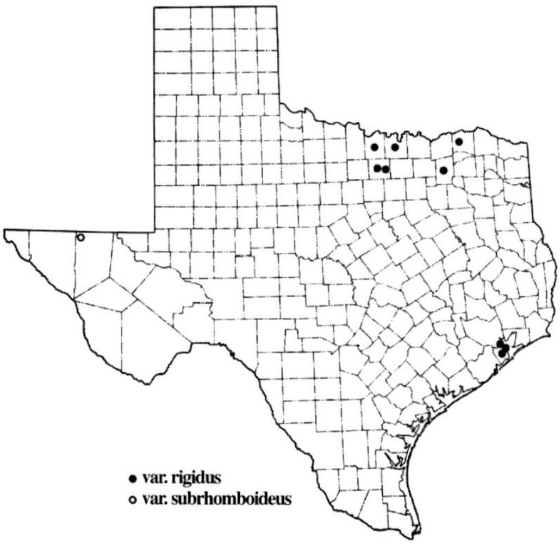
● var. rigidus
○ var. subrhomboideus

Helianthus rigidus

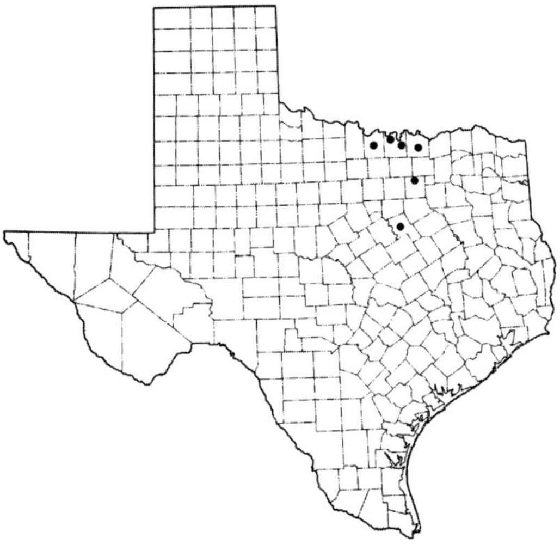

Helianthus salicifolius

Helianthus tuberosus

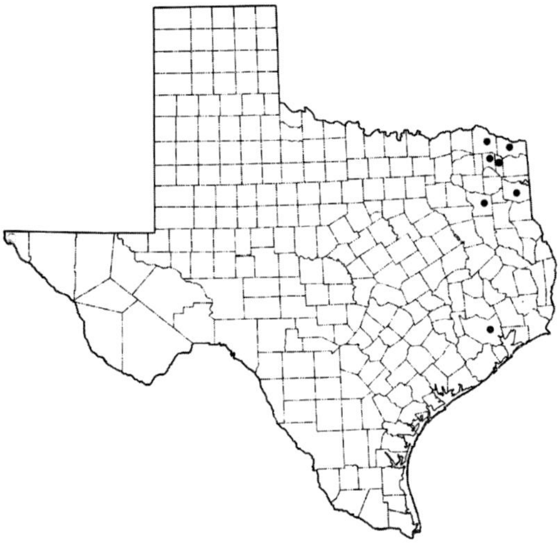

Heliopsis helianthoides

Heliopsis parvifolia

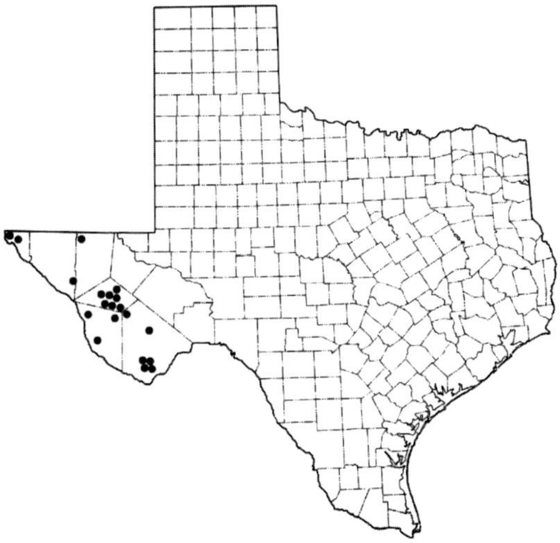

Heterosperma pinnatum

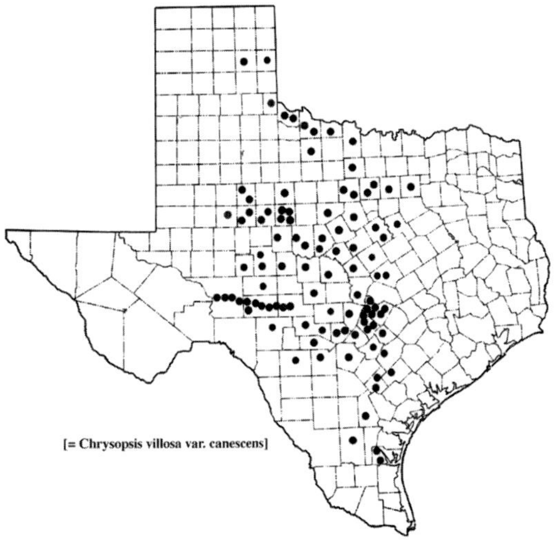
[= Chrysopsis villosa var. canescens]

Heterotheca canescens

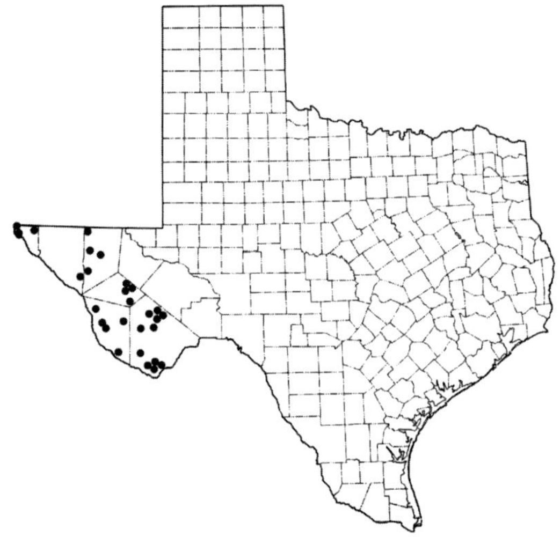

Heterotheca fulcrata

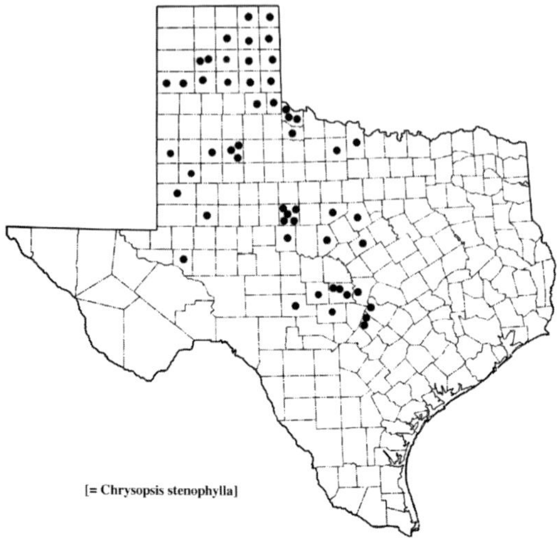
[= Chrysopsis stenophylla]

Heterotheca stenophylla

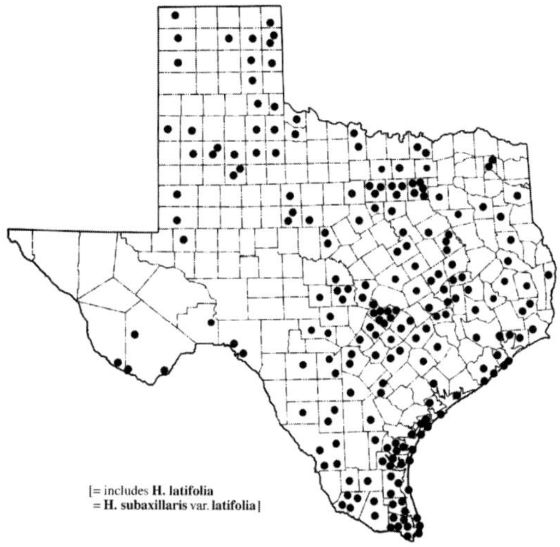
[= includes H. latifolia
= H. subaxillaris var. latifolia]

Heterotheca subaxillaris

Heterotheca viscida

Hieracium carneum

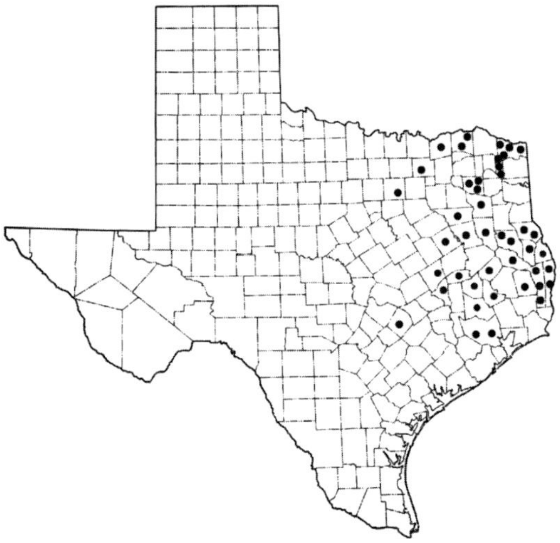

Hieracium gronovii

Hieracium longipilum

Hieracium schultzii

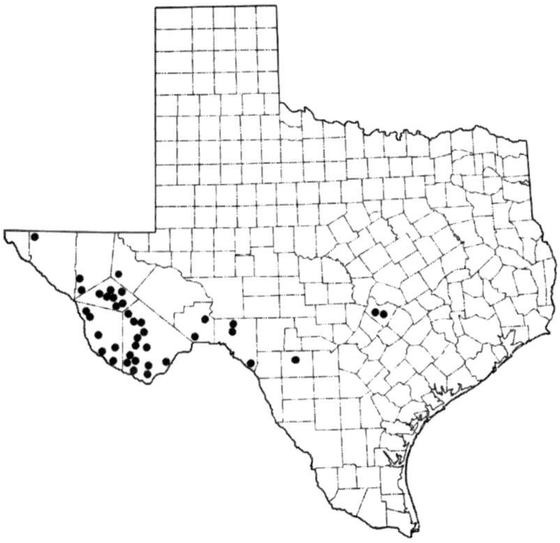

Hymenoclea monogyra

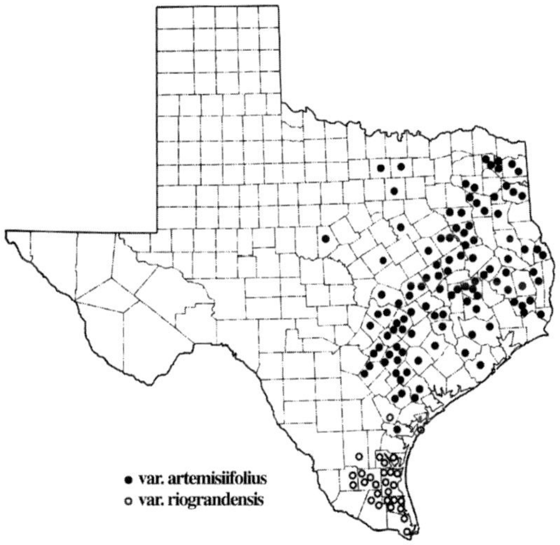
var. artemisiifolius
var. riograndensis

Hymenopappus artemisiifolius

Hymenopappus biennis

Hymenopappus carrizoanus

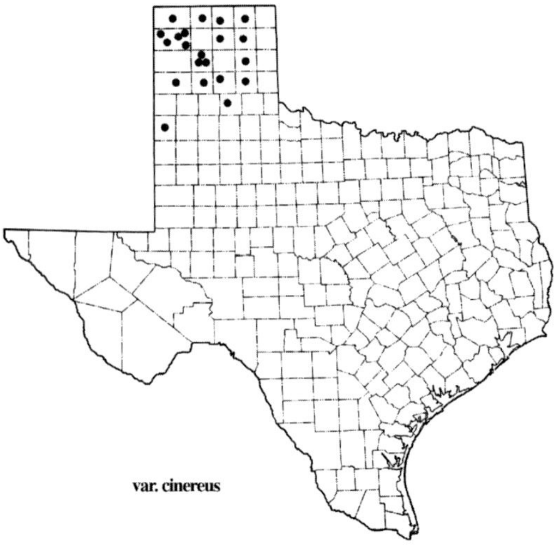
var. cinereus

Hymenopappus filifolius

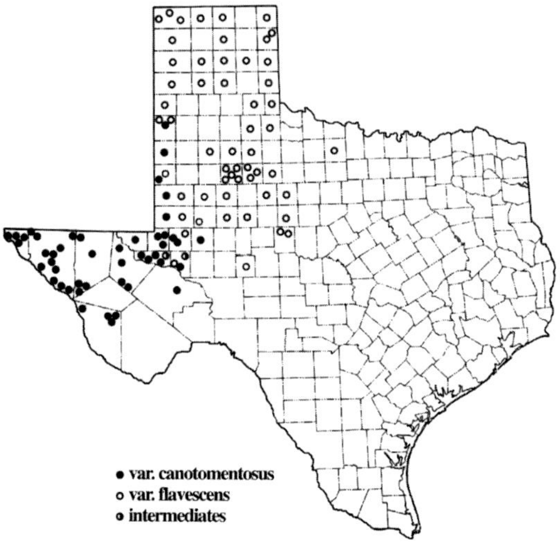
• var. canotomentosus
○ var. flavescens
⊙ intermediates

Hymenopappus flavescens

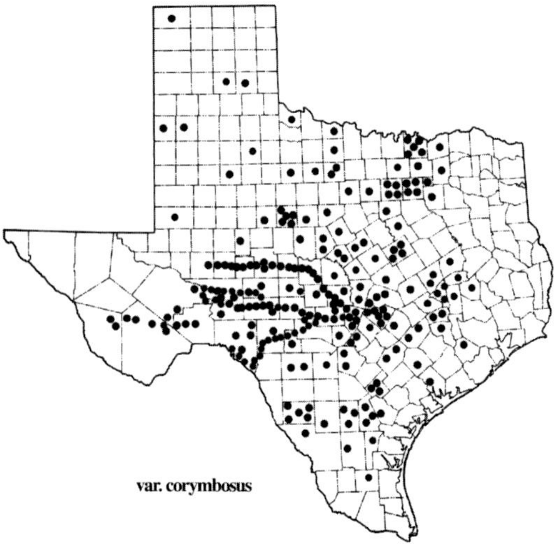
var. corymbosus

Hymenopappus scabiosaeus

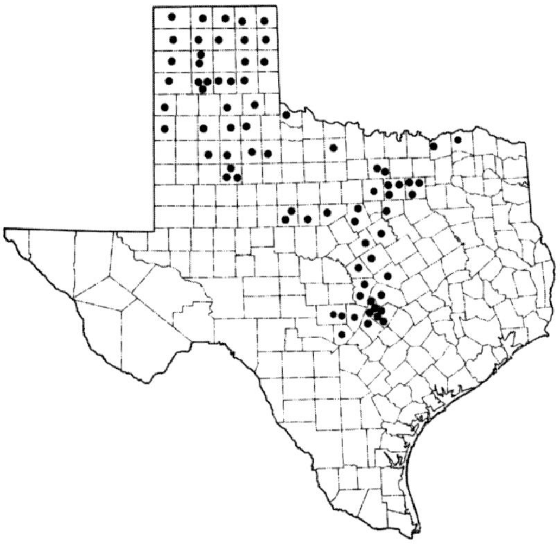

Hymenopappus tenuifolius

Hymenothrix wislizeni

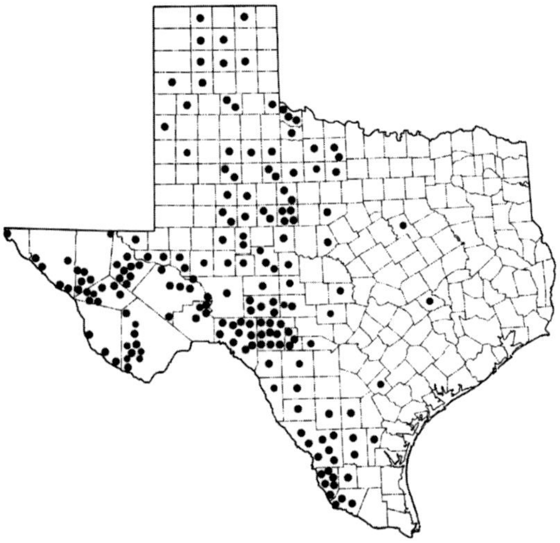

Hymenoxys odorata

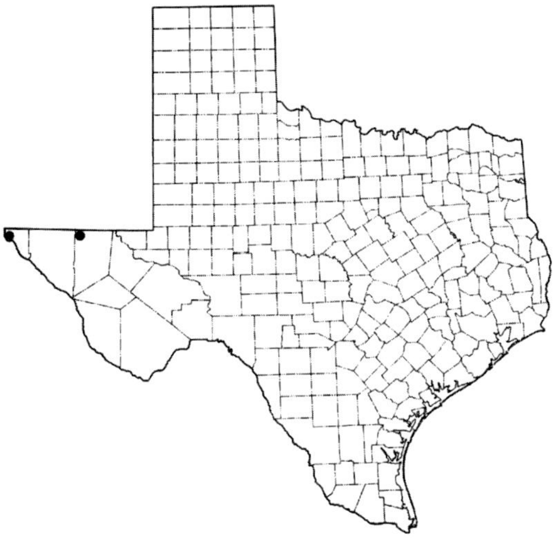

Hymenoxys richardsonii

Hymenoxys texana

[Sida 17: 633. 1997]

Hypochaeris glabra

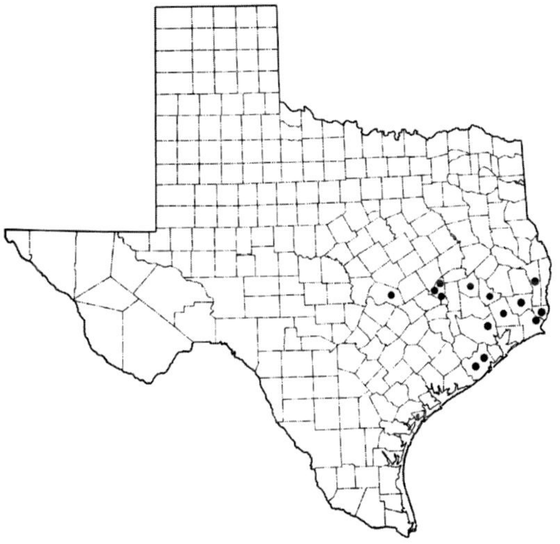

Hypochaeris microcephala

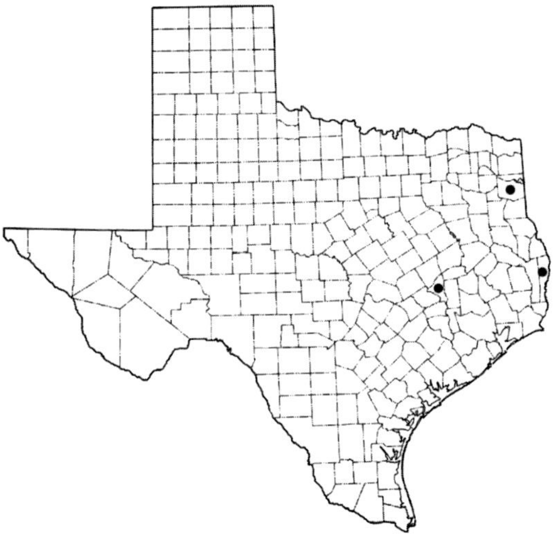

Hypochaeris radicata

Ionactis linariifolius

var. achyranthes

Isocarpha oppositifolia

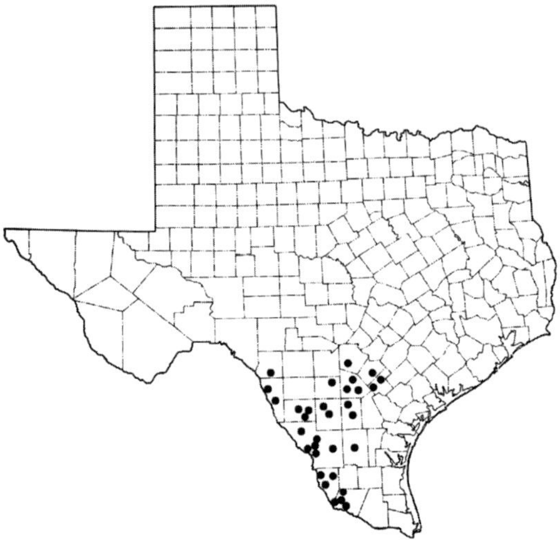

Isocoma coronopifolia

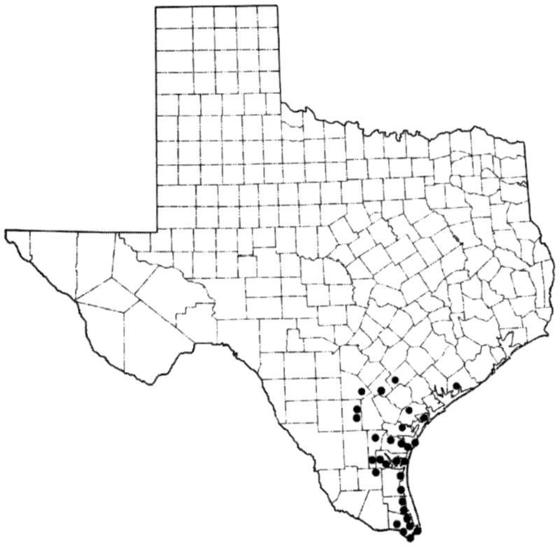

Isocoma drummondii

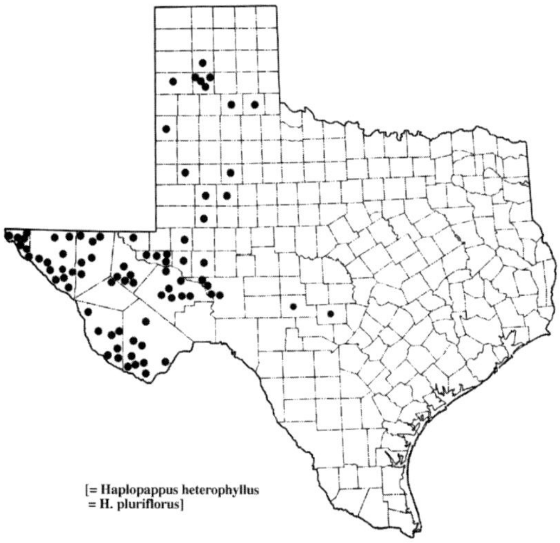
[= Haplopappus heterophyllus
= H. pluriflorus]

Isocoma pluriflora

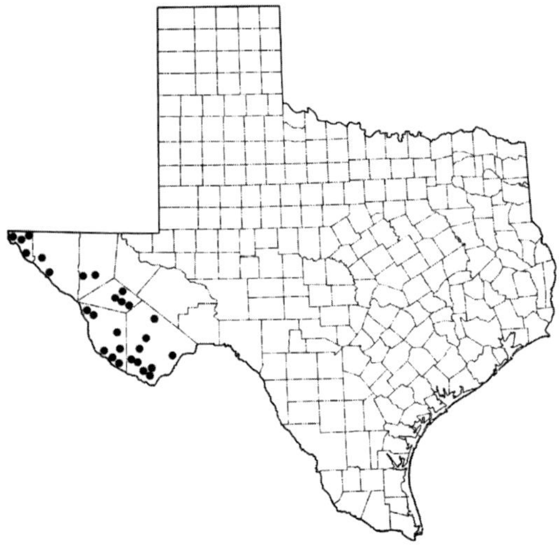

Iva ambrosiifolia

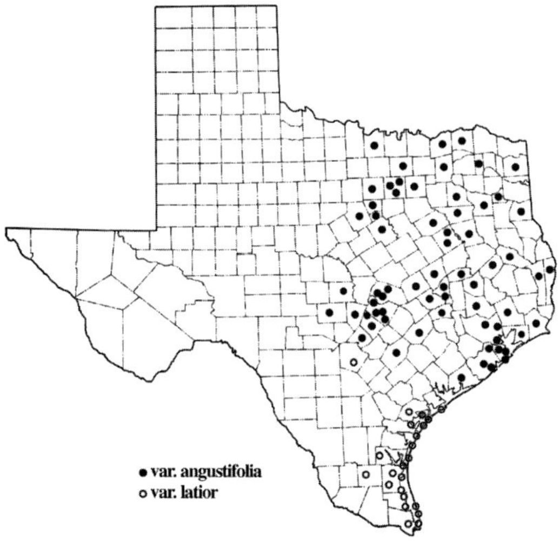
• var. angustifolia
○ var. latior

Iva angustifolia

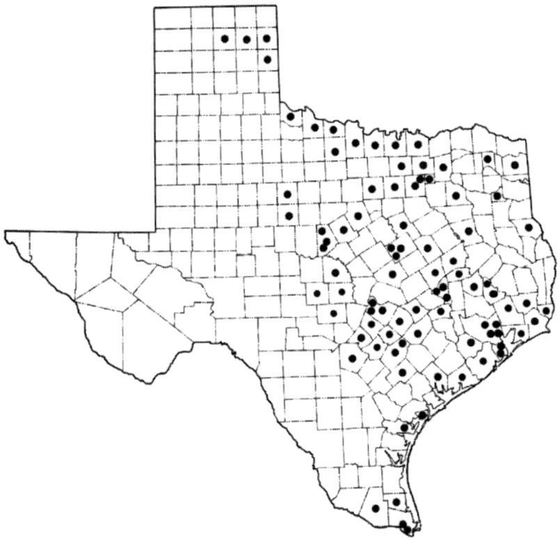

Iva annua

Iva axillaris

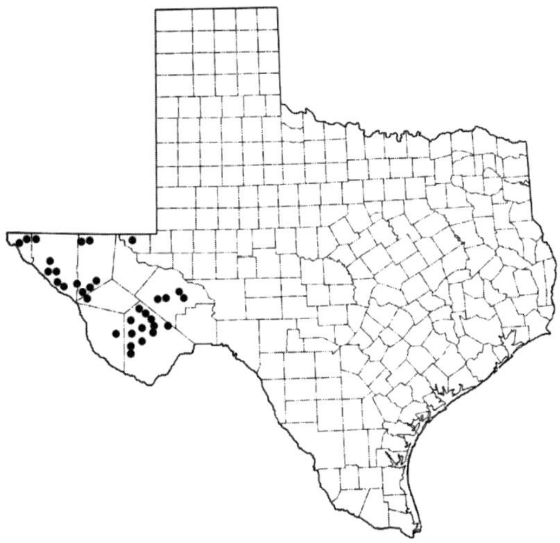

Iva dealbata

Iva frutescens

Iva imbricata

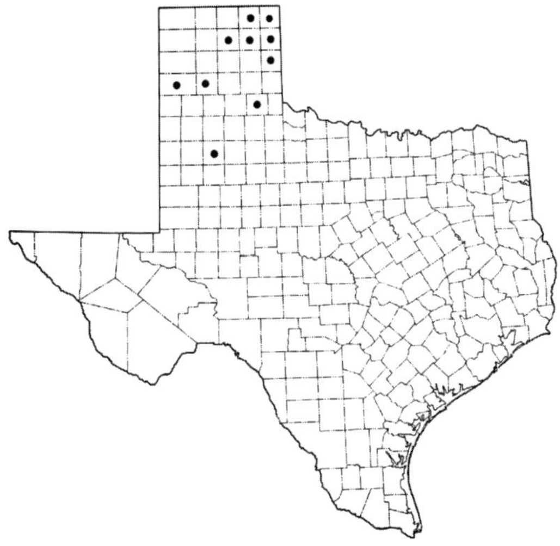

Iva xanthifolia

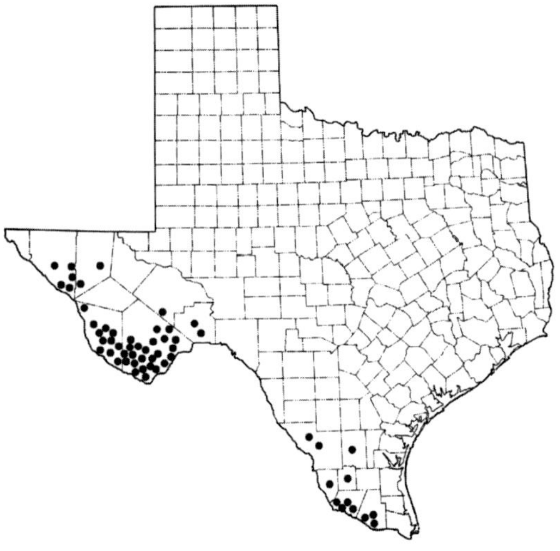

Jefea brevifolia

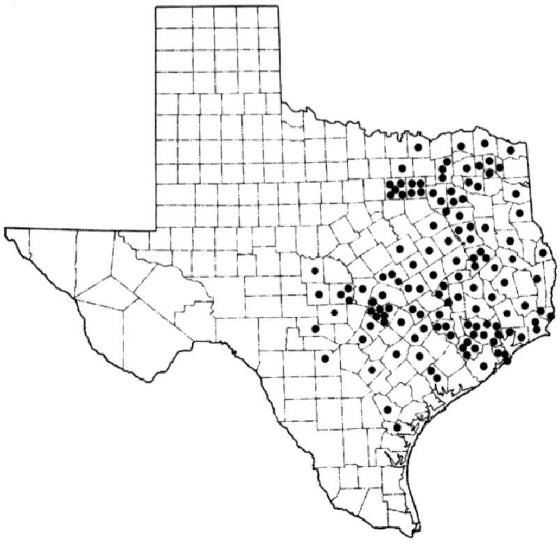

Krigia cespitosa

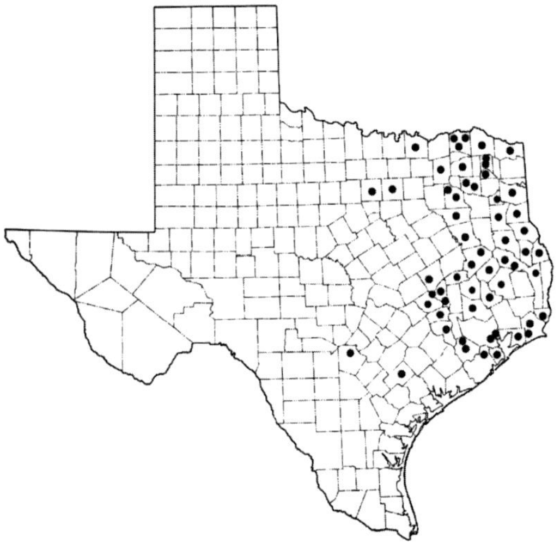

Krigia dandelion

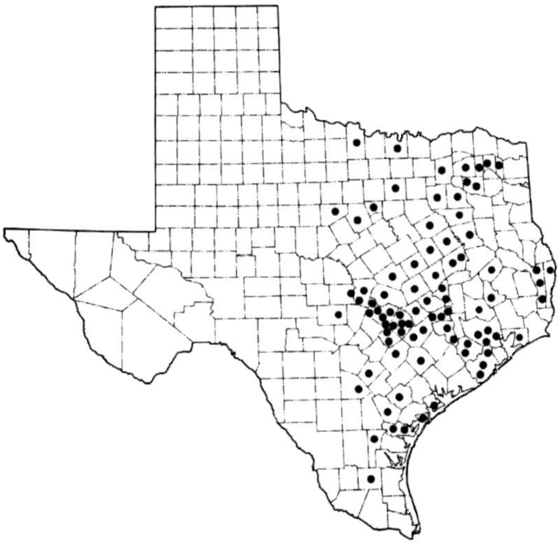

Krigia occidentalis

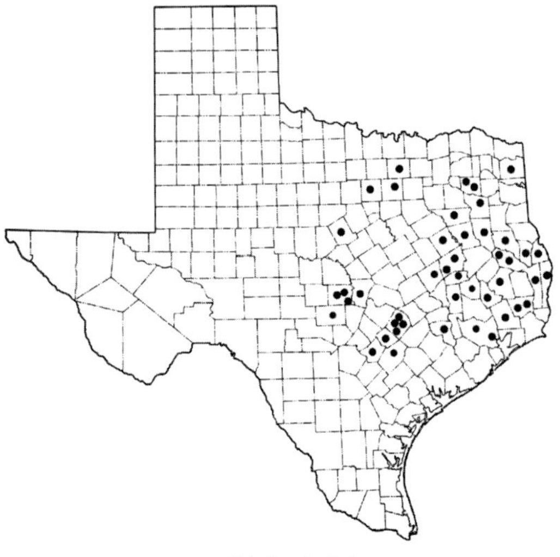

Krigia virginica

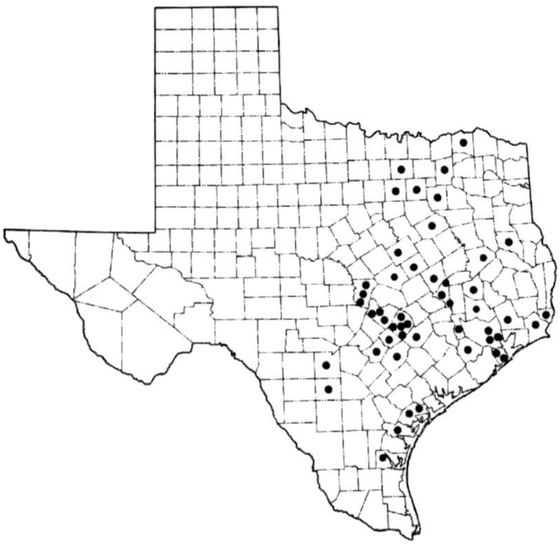

Krigia wrightii

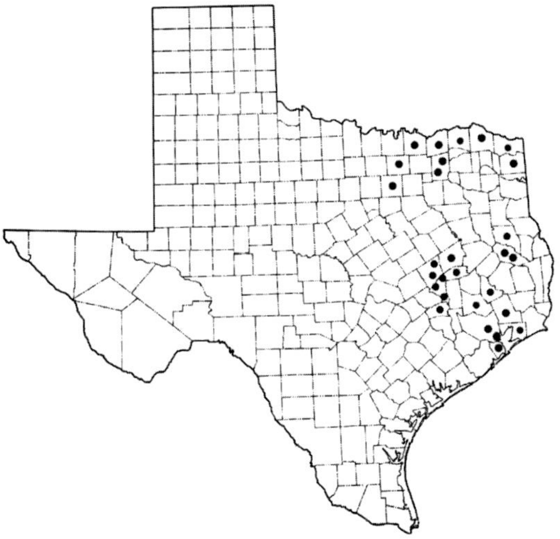

Lactuca canadensis

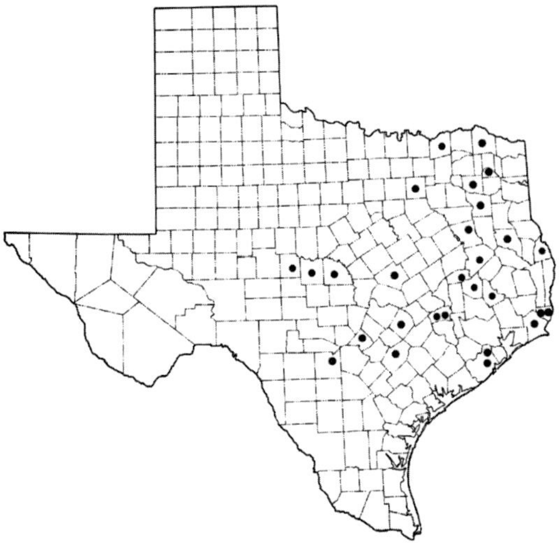

Lactuca floridana

Lactuca graminifolia

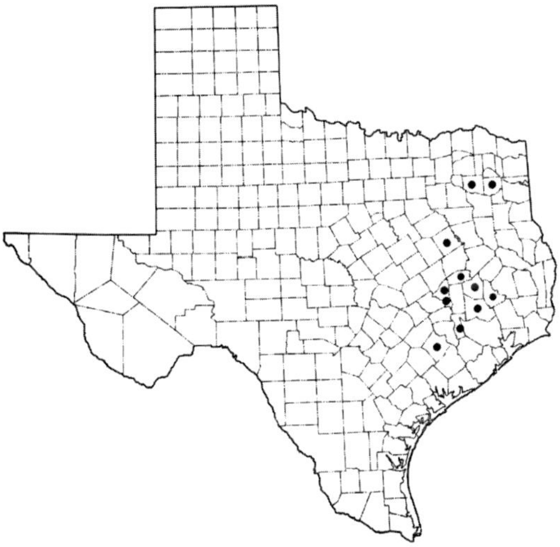

Lactuca hirsuta

Lactuca intybacea

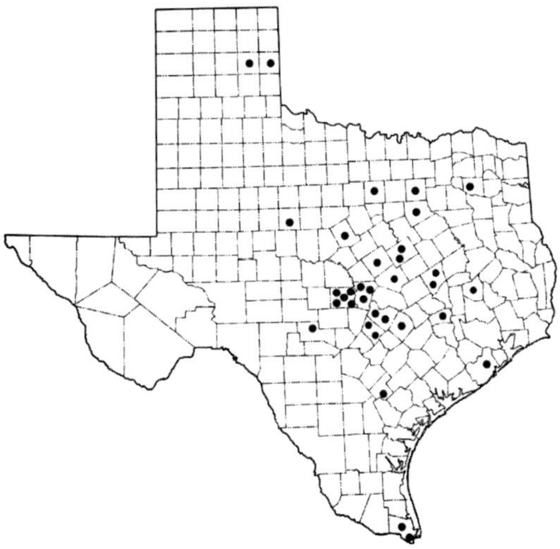

Lactuca ludoviciana

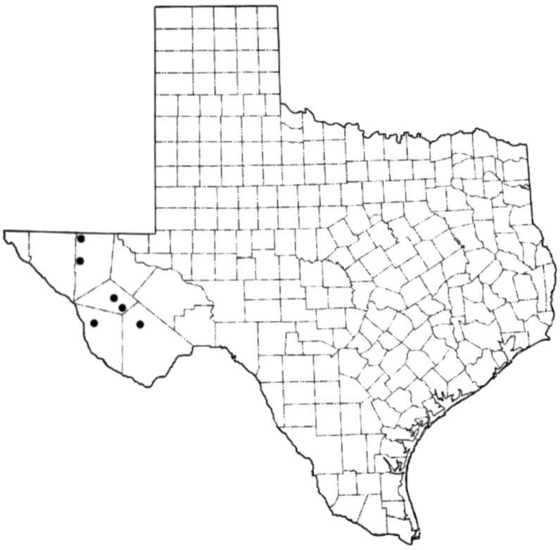

Lactuca pulchella

Lactuca saligna

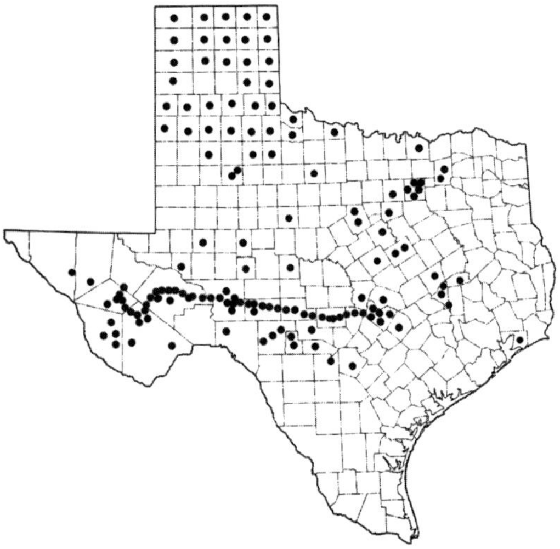

Lactuca serriola

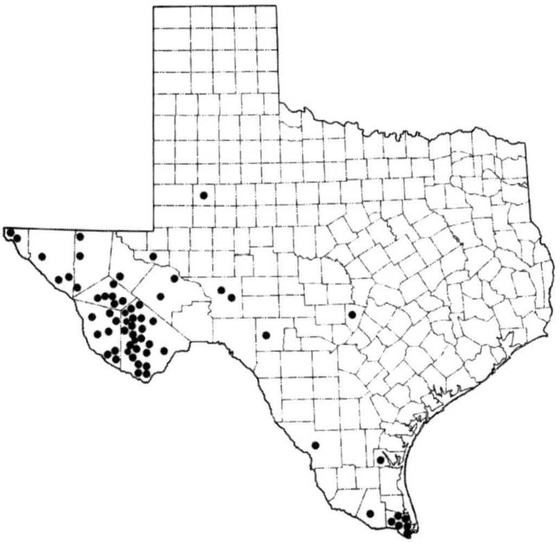

Laennecia coulteri

Laennecia filaginoides

Laennecia sophiifolia

Laennecia turnerorum

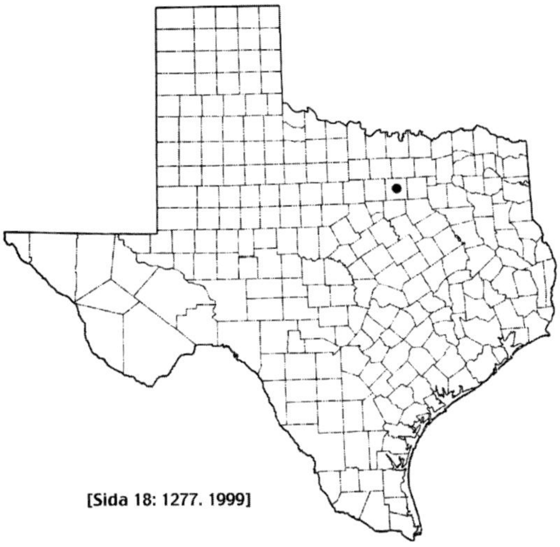
[Sida 18: 1277. 1999]

Lapsana communis

Leontodon hispidus

Leontodon taraxacum

Lepidospartum burgessii

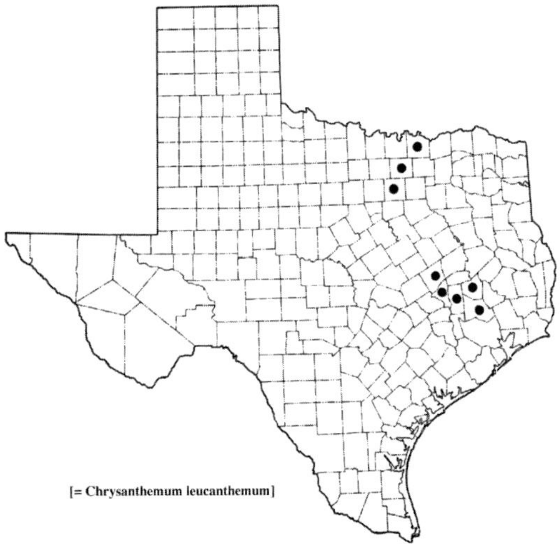
[= Chrysanthemum leucanthemum]

Leucanthemum vulgare

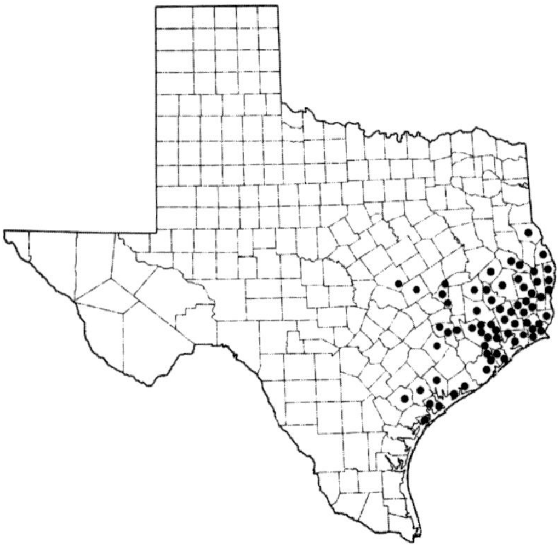

Liatris acidota

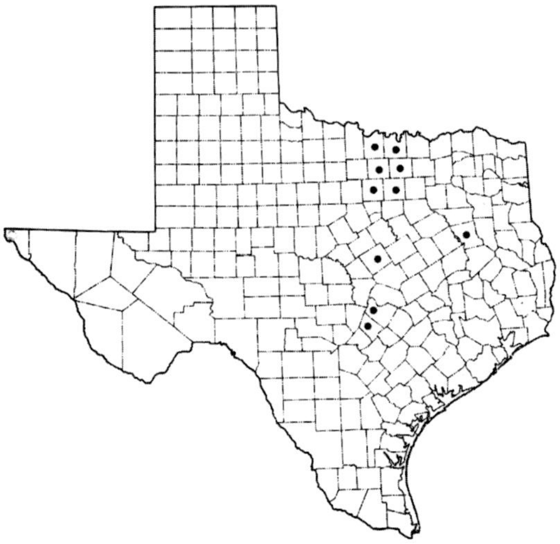

Liatris aestivalis

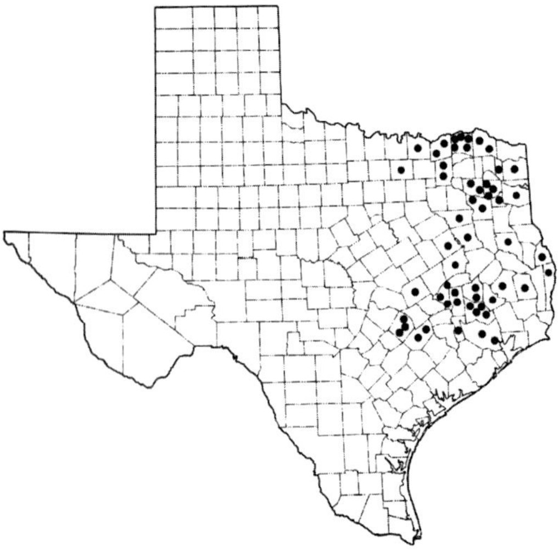

Liatris aspera

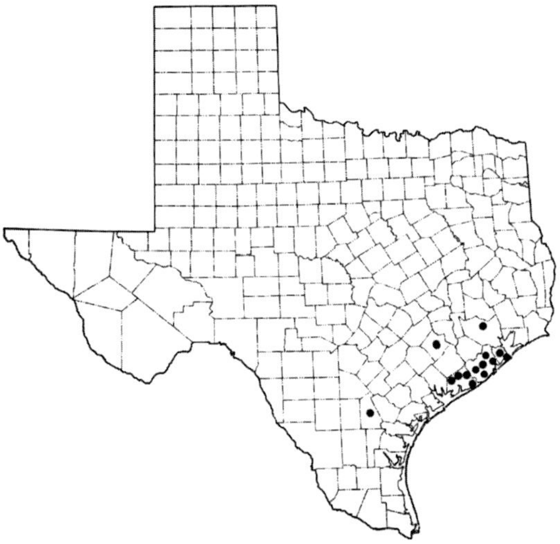

Liatris bracteata

Liatris cymosa

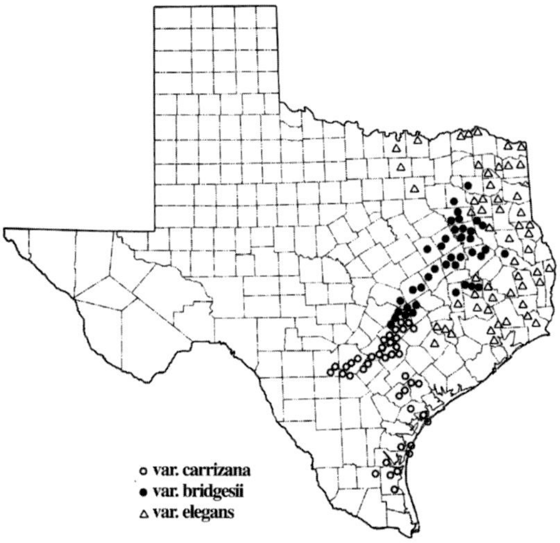
○ var. carrizana
● var. bridgesii
△ var. elegans

Liatris elegans

Liatris glandulosa

Liatris lancifolia

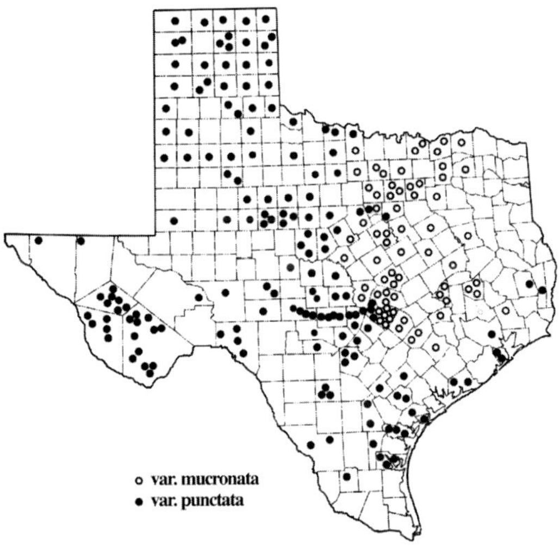
var. mucronata
var. punctata

Liatris punctata

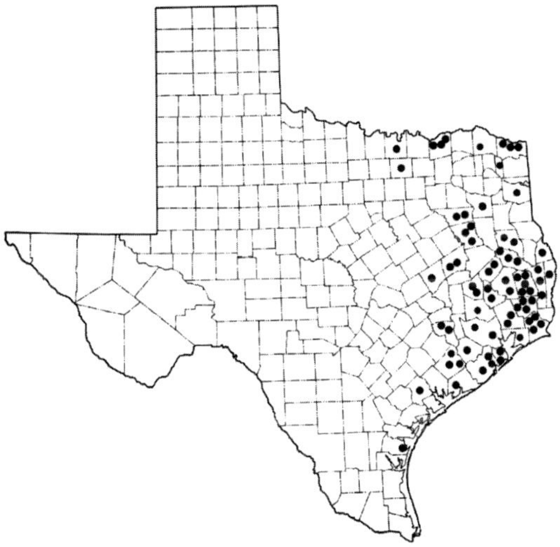

Liatris pycnostachya

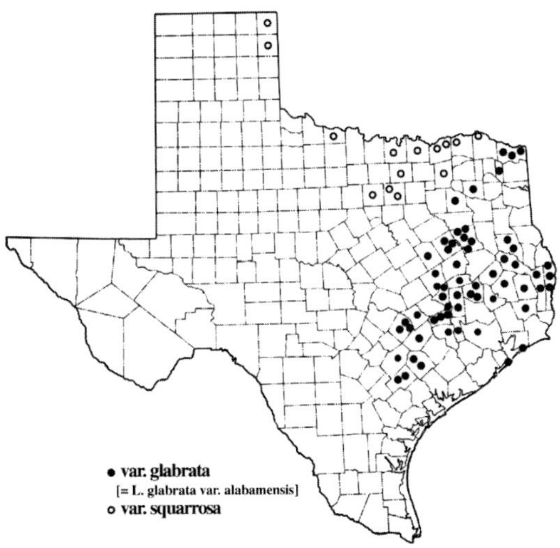
var. glabrata
[= L. glabrata var. alabamensis]
var. squarrosa

Liatris squarrosa

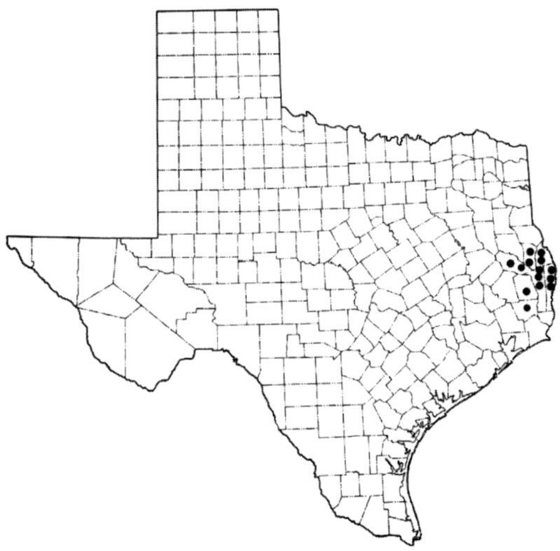

Liatris tenuis

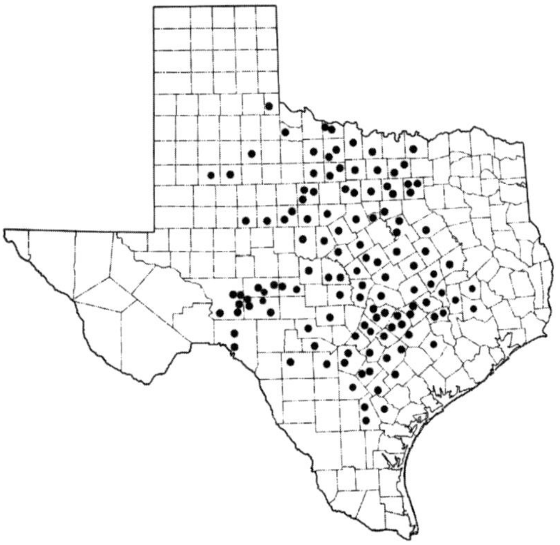

Lindheimera texana

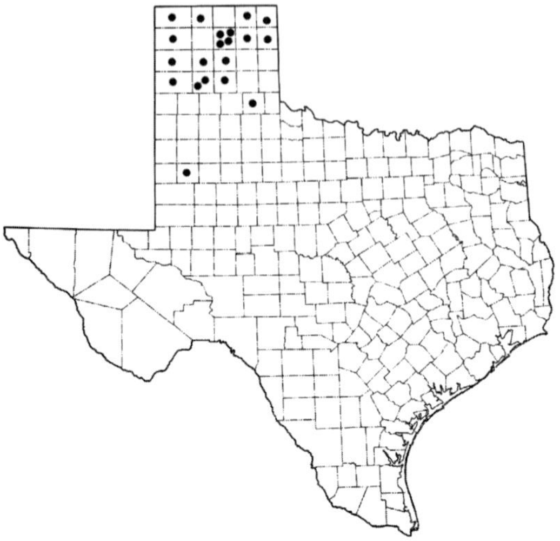

Lygodesmia juncea

Lygodesmia ramosissima

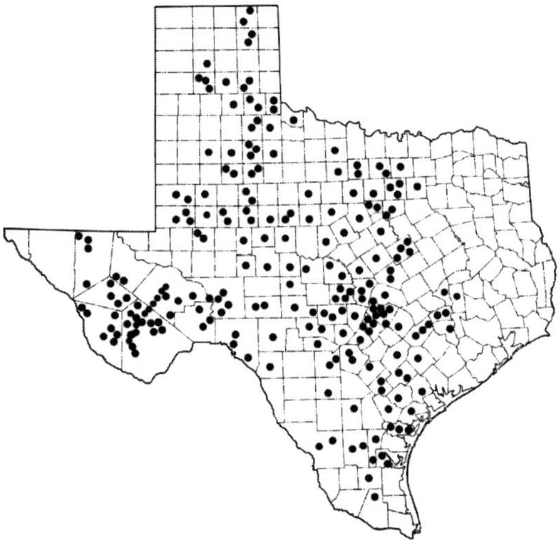

Lygodesmia texana

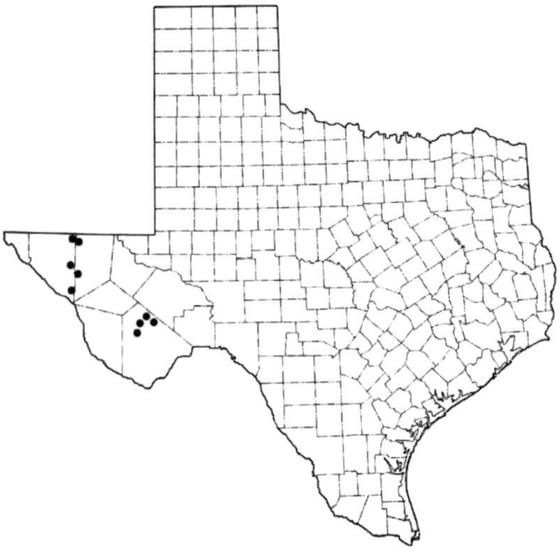

Machaeranthera blephariphylla

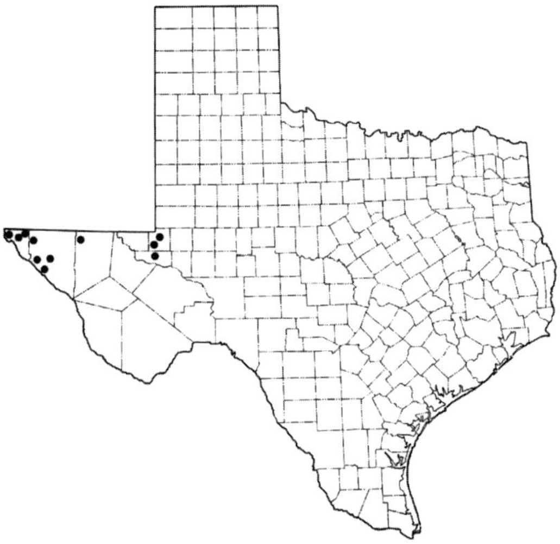

Machaeranthera canescens

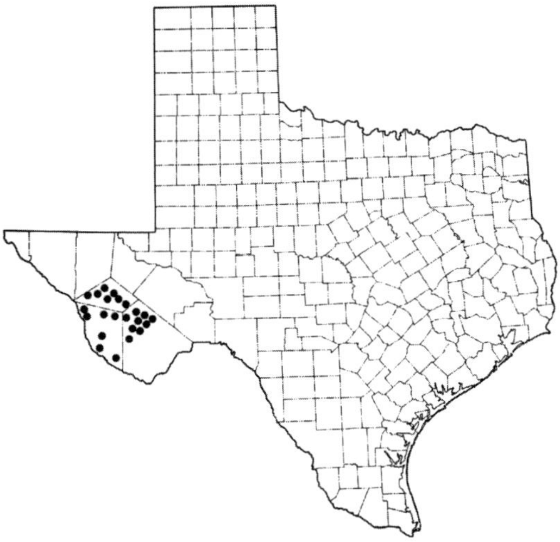

Machaeranthera gracilis

Machaeranthera gypsophila

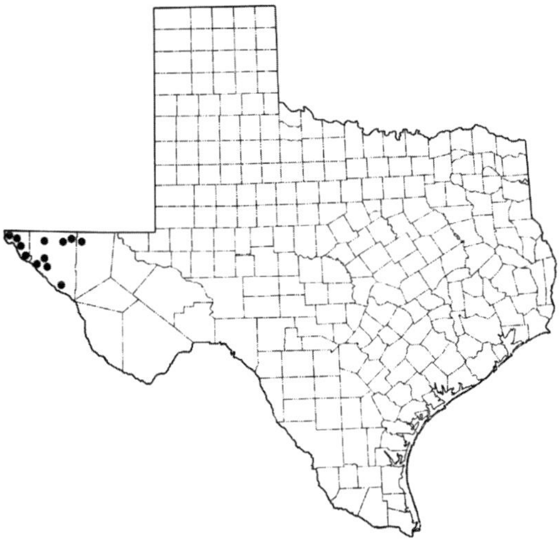

Machaeranthera parviflora

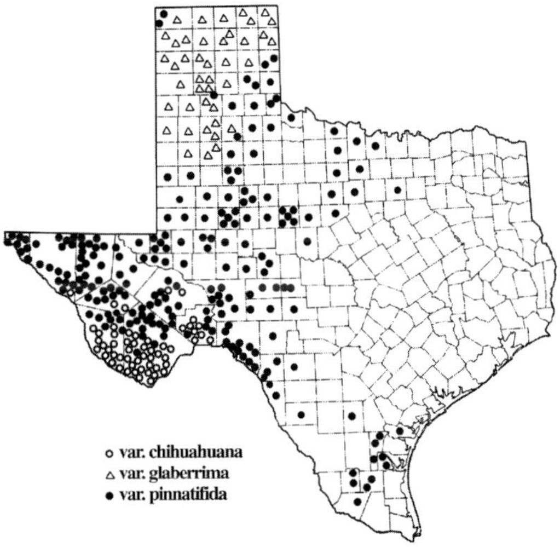
var. chihuahuana
var. glaberrima
var. pinnatifida

Machaeranthera pinnatifida

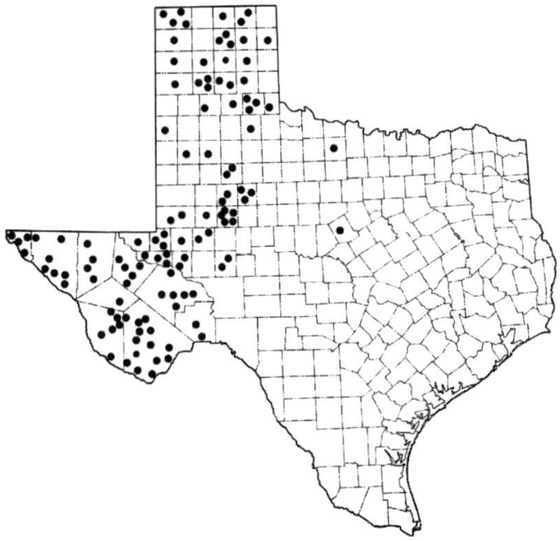

Machaeranthera tanacetifolia

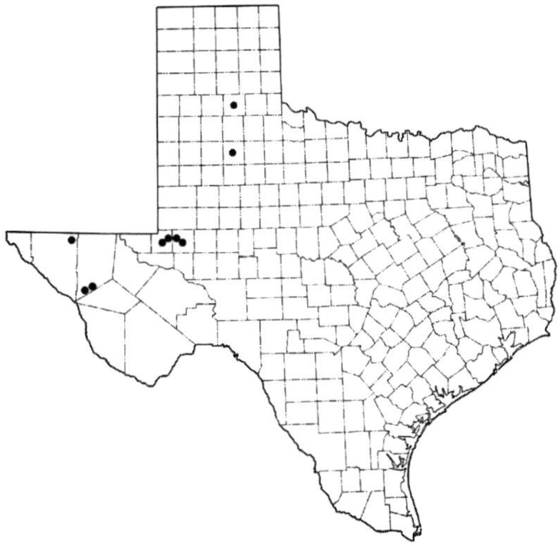

Machaeranthera viscida

Madia elegans

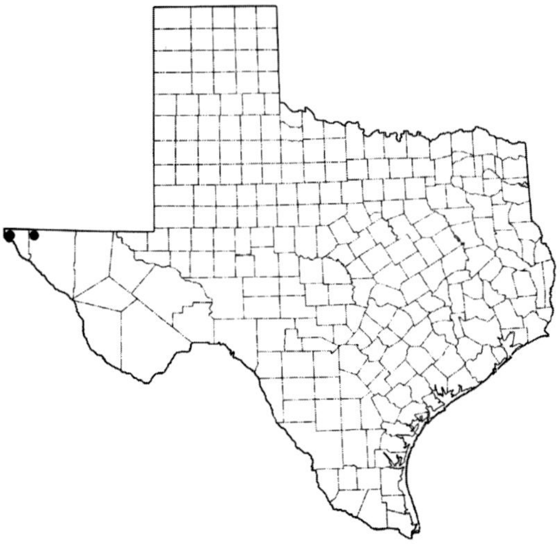
Malacothrix fendleri

Malacothrix sonorae

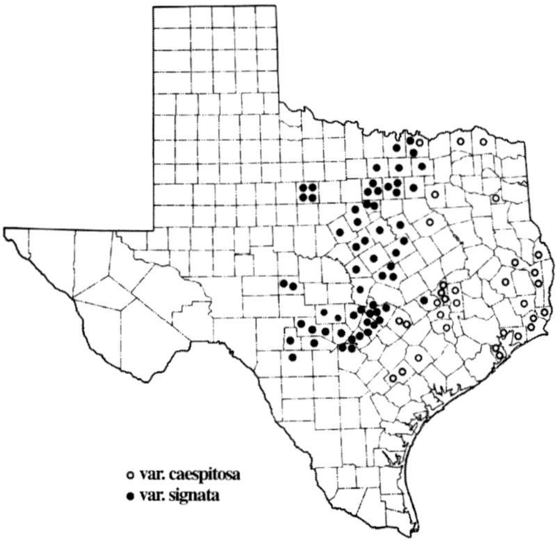
o var. caespitosa
• var. signata
Marshallia caespitosa

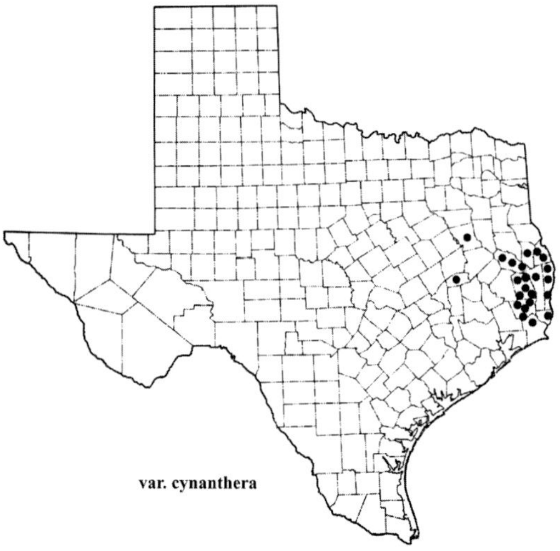
var. cynanthera
Marshallia graminifolia

Matricaria matricarioides

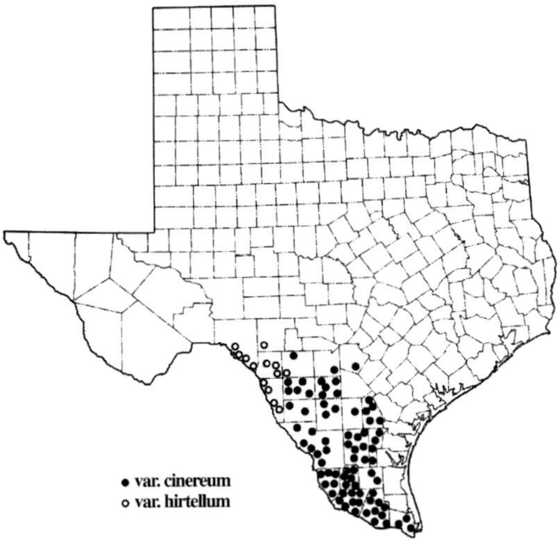
• var. cinereum
o var. hirtellum

Melampodium cinereum

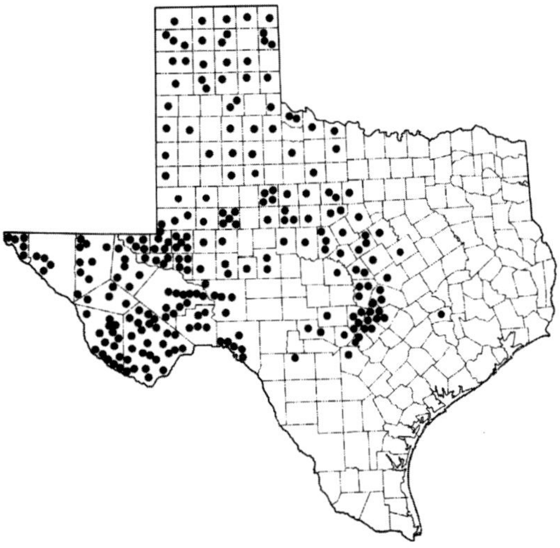

Melampodium leucanthum

Melampodium strigosum

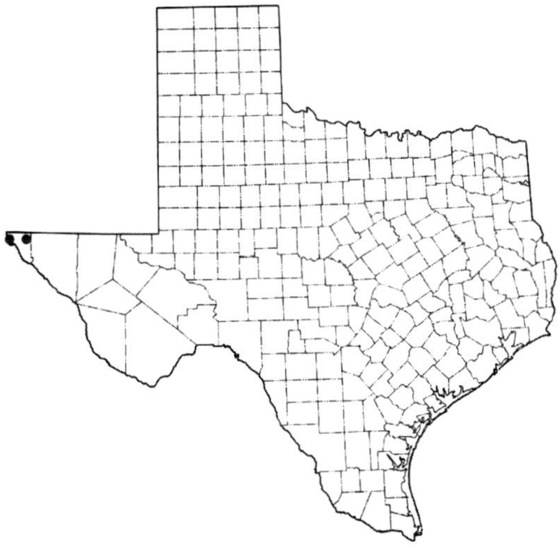

Microseris linearifolia

Mikania cordifolia

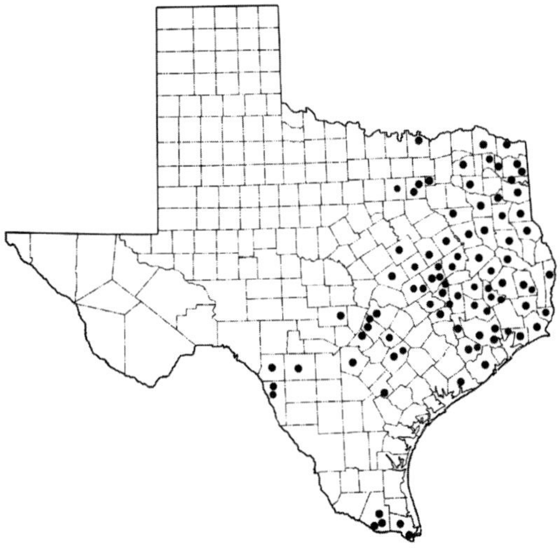

Mikania scandens

Nicolletia edwardsii

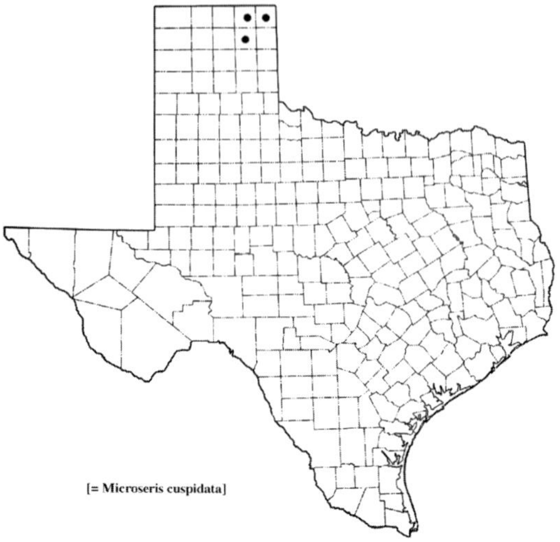
[= Microseris cuspidata]

Nothocalais cuspidata

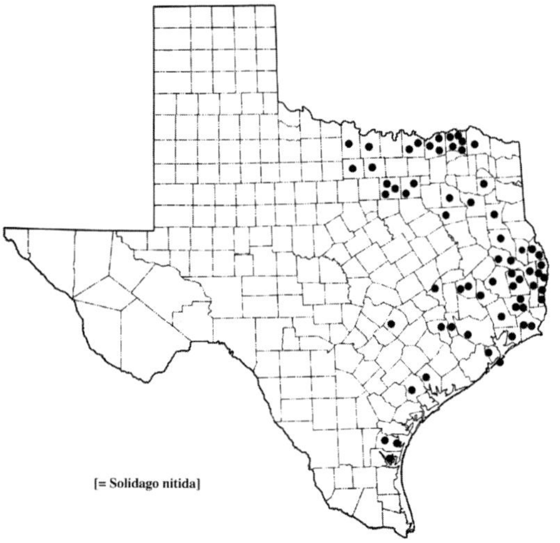
[= Solidago nitida]

Oligoneuron nitidum

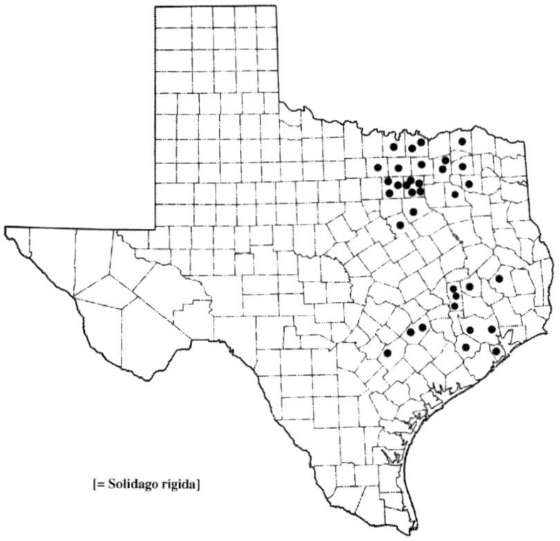
[= Solidago rigida]

Oligoneuron rigidum

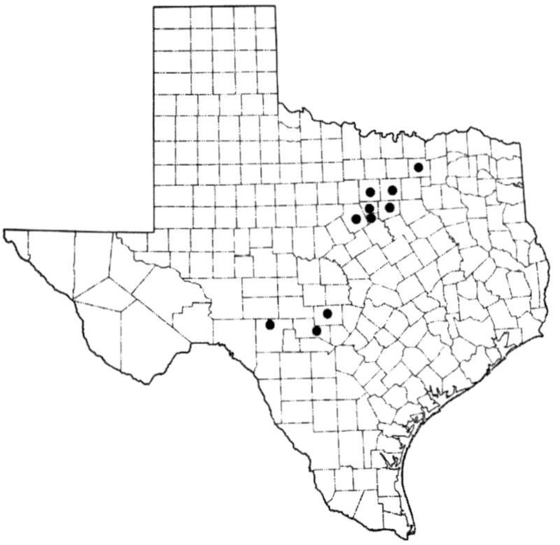

Onopordum acanthium

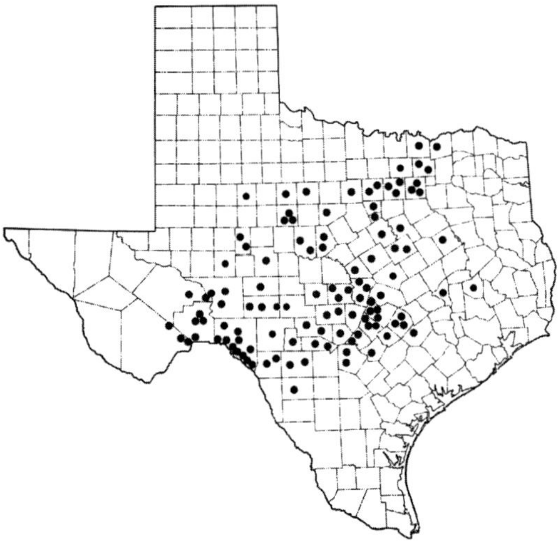

Palafoxia callosa

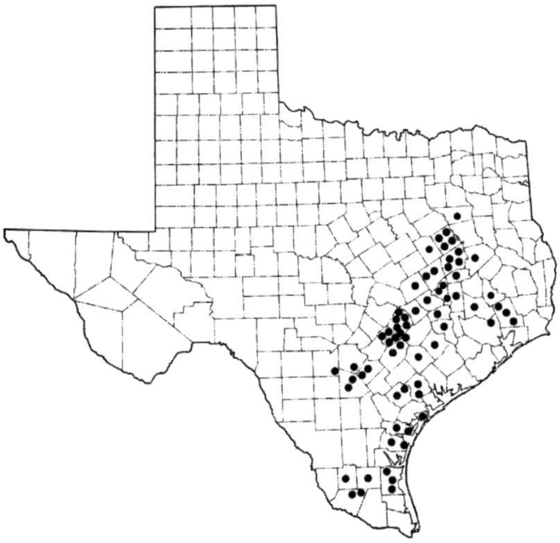

Palafoxia hookeriana

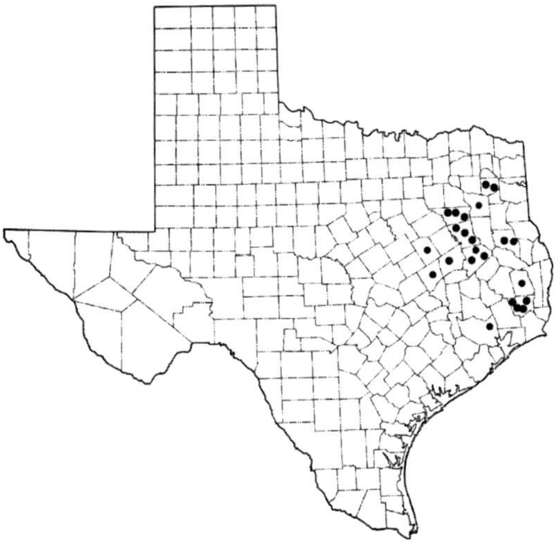

Palafoxia reverchonii

Palafoxia riograndensis

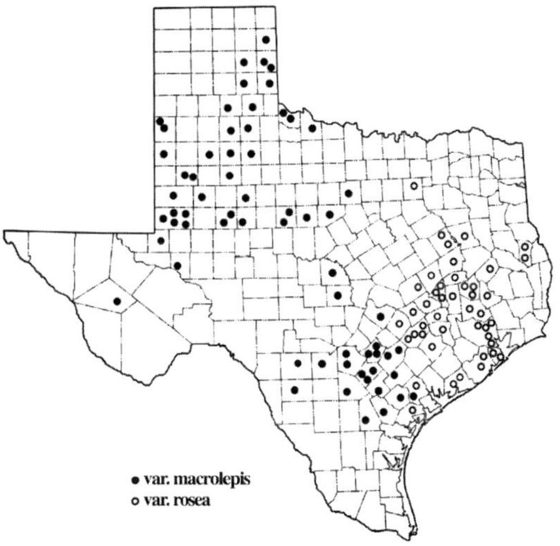
• var. macrolepis
○ var. rosea

Palafoxia rosea

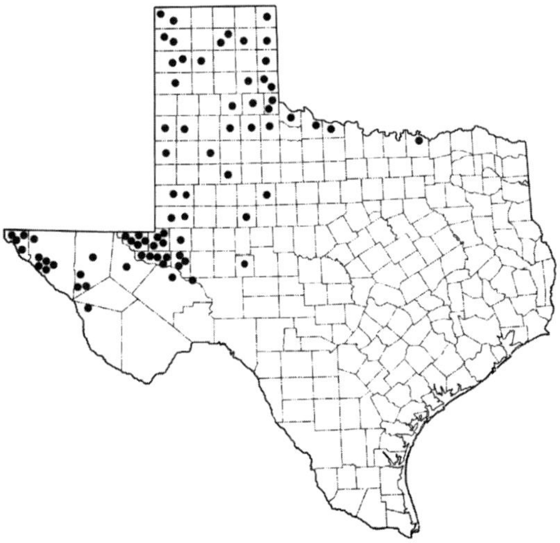

Palafoxia sphacelata

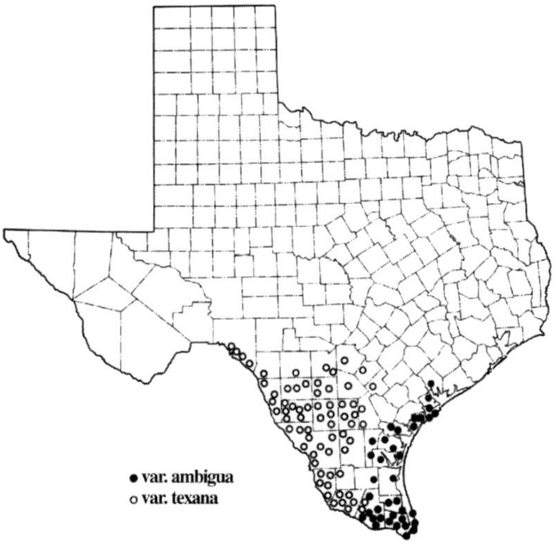
• var. ambigua
○ var. texana

Palafoxia texana

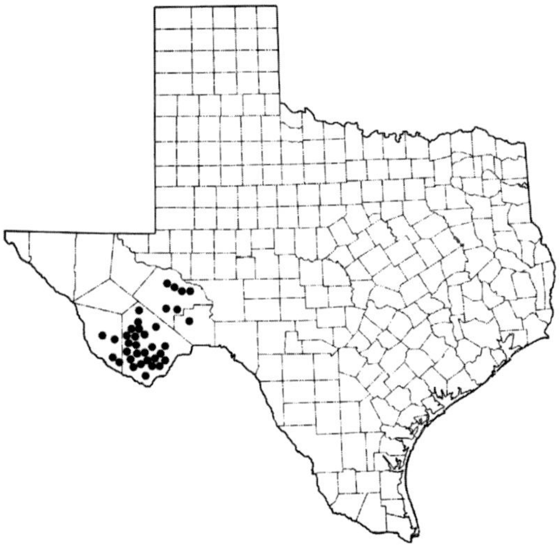

Parthenium argentatum

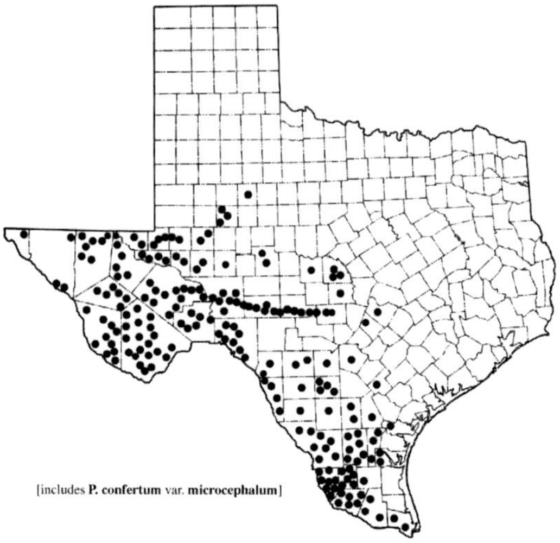
[includes P. confertum var. microcephalum]

Parthenium confertum

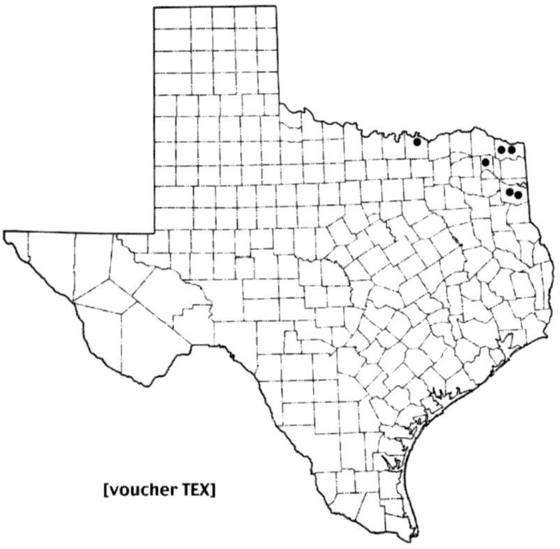
[voucher TEX]

Parthenium hispidum

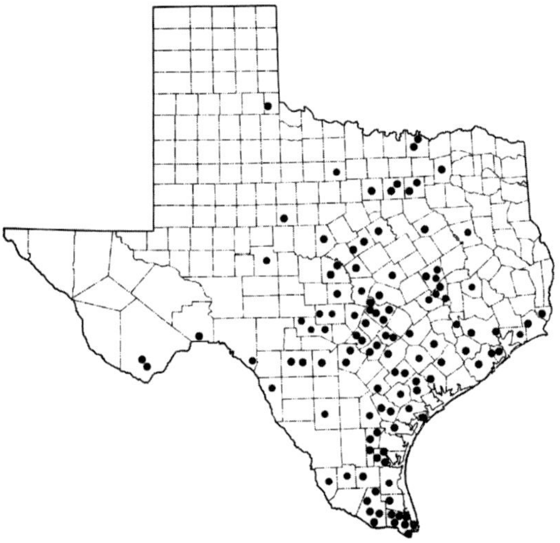

Parthenium hysterophorus

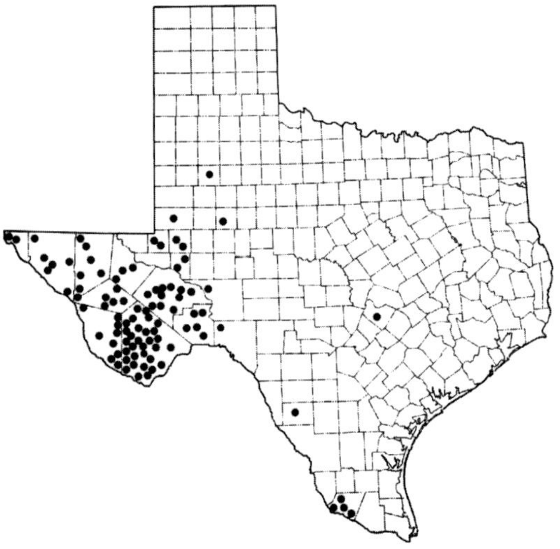

Parthenium incanum

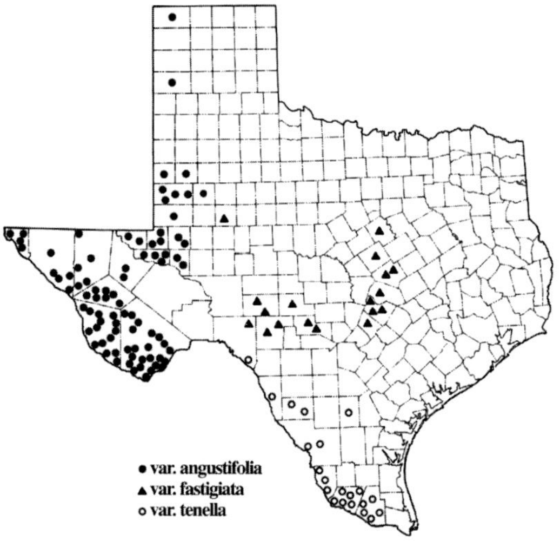
● var. angustifolia
▲ var. fastigiata
○ var. tenella

Pectis angustifolia

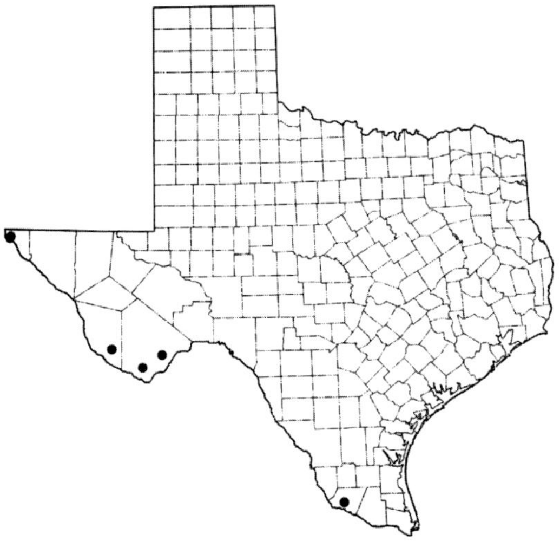

Pectis cylindrica

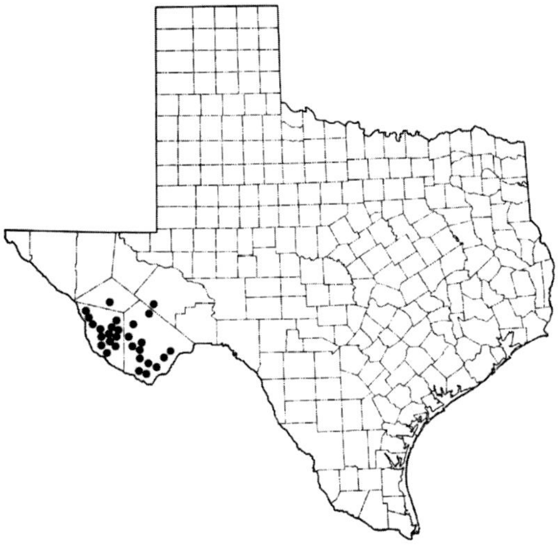

Pectis filipes

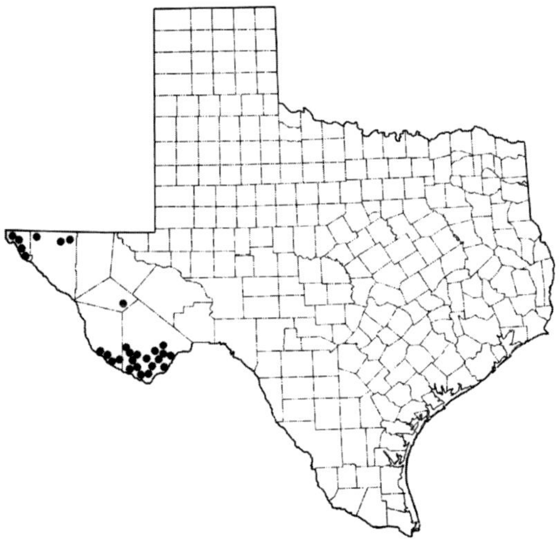

Pectis papposa

Pectis prostrata

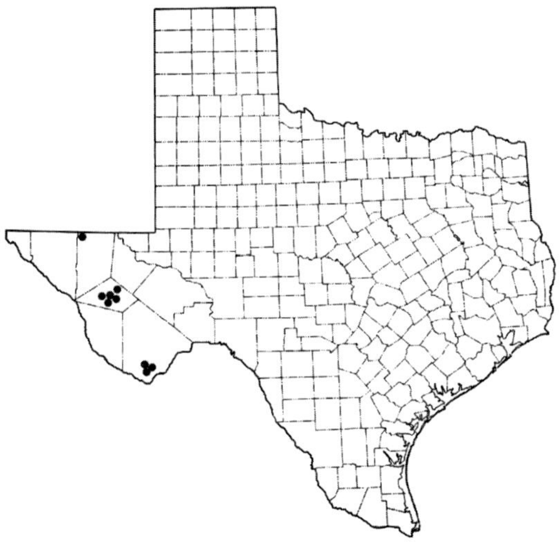

Pericome caudata

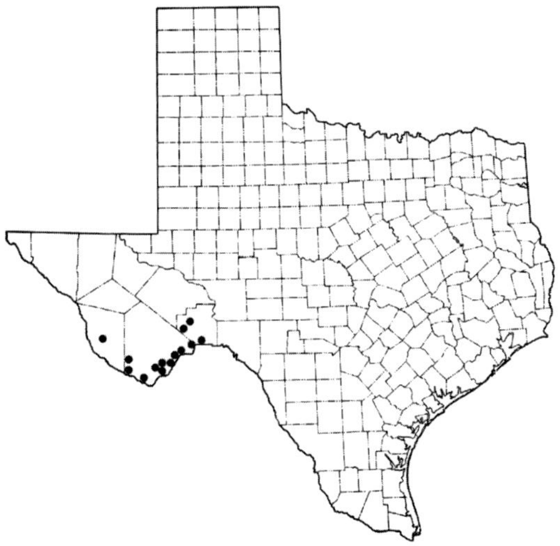

Perityle aglossa

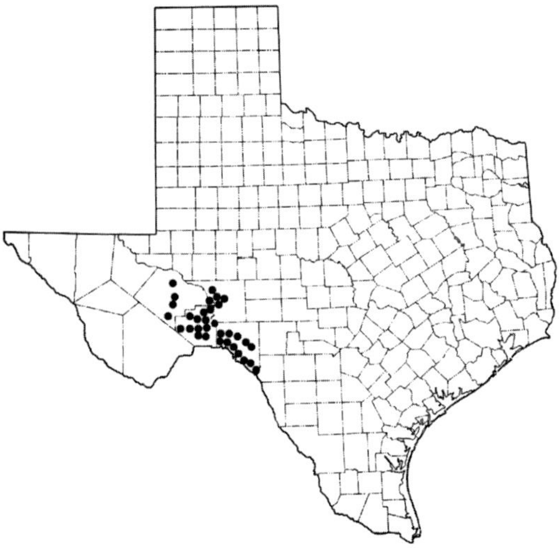
Perityle angustifolia

• var. appressa
○ var. bisetosa
△ var. scalaris
Perityle bisetosa

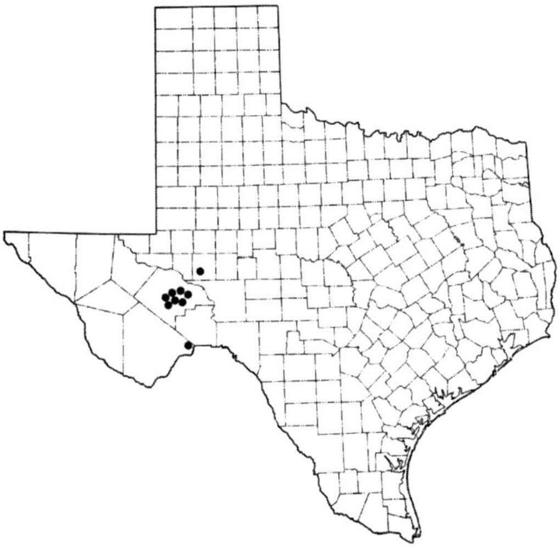
Perityle cinerea

Perityle dissecta

Perityle fosteri

Perityle huecoensis

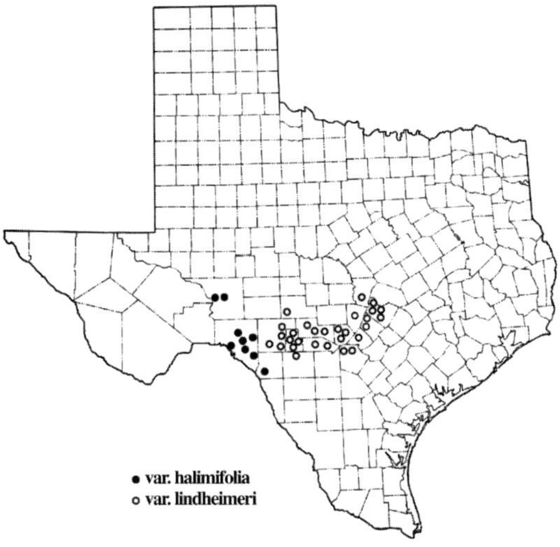
var. halimifolia
var. lindheimeri

Perityle lindheimeri

Perityle microglossa

Perityle parryi

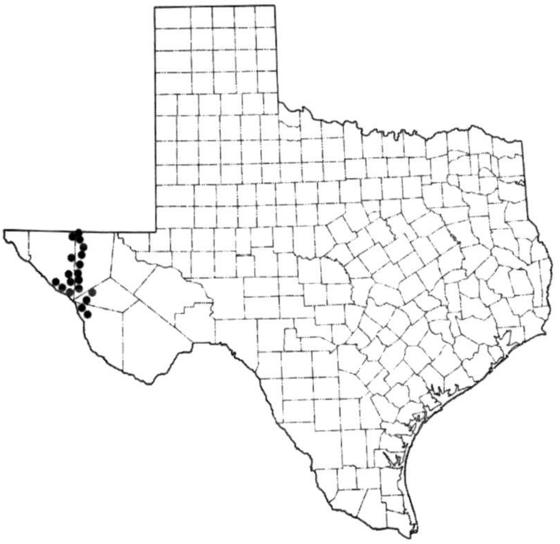

Perityle quinqueflora

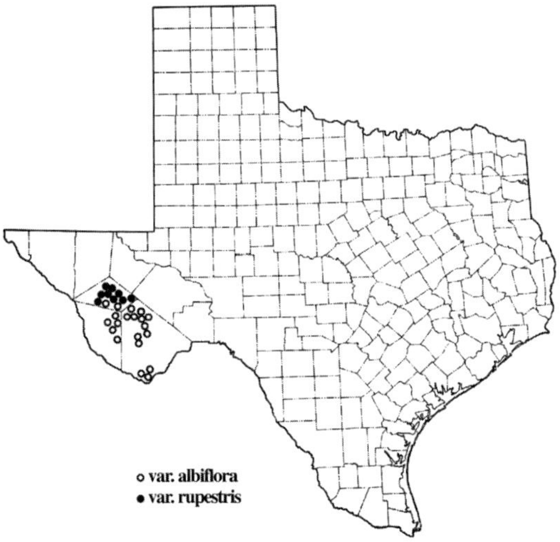
var. albiflora
var. rupestris

Perityle rupestris

Perityle vaseyi

Perityle vitreomontana

Perityle warnockii

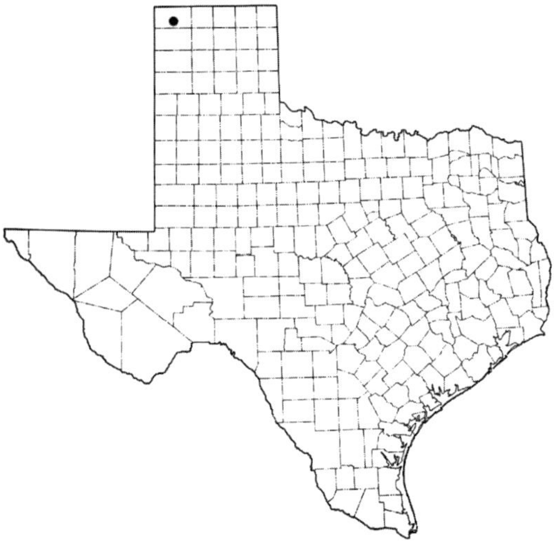
Picradeniopsis oppositifolia

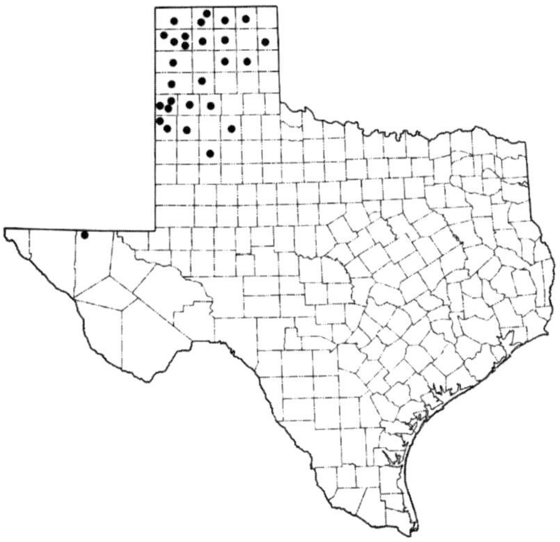
Picradeniopsis woodhousei

Pinaropappus parvus

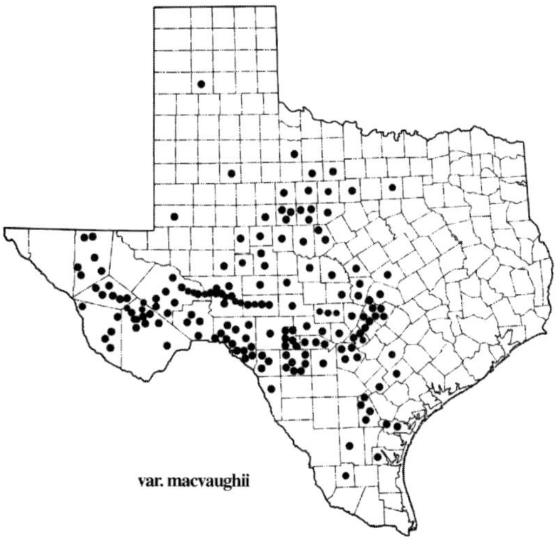
var. macvaughii
Pinaropappus roseus

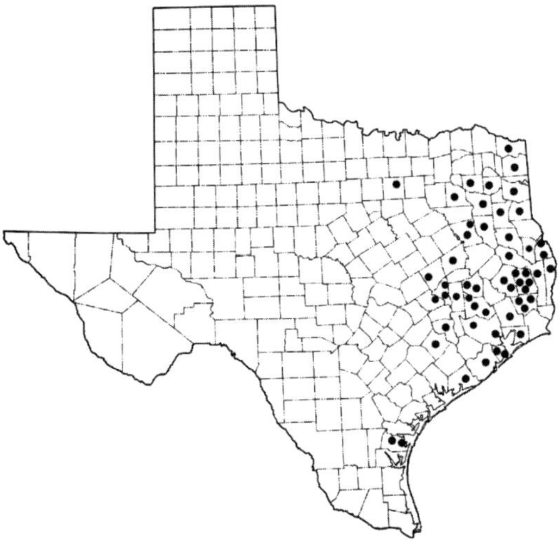

Pityopsis graminifolia

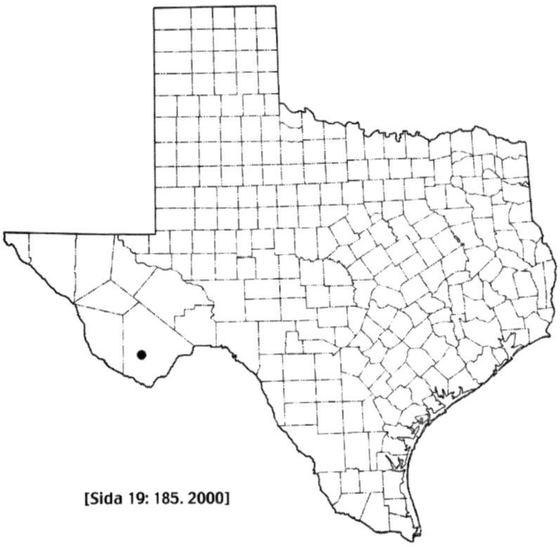
[Sida 19: 185. 2000]

Plateilema palmeri

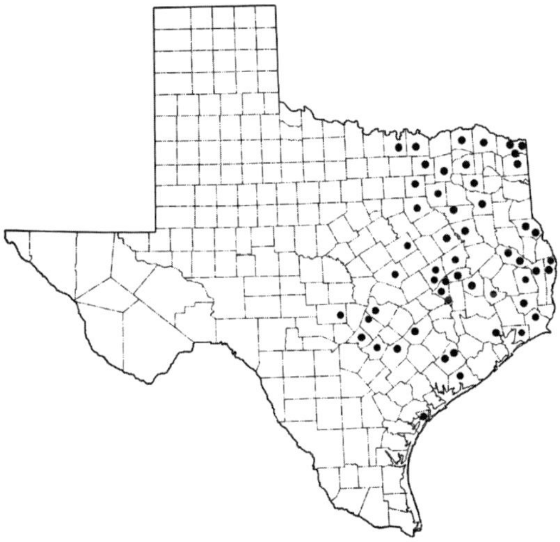

Pluchea camphorata

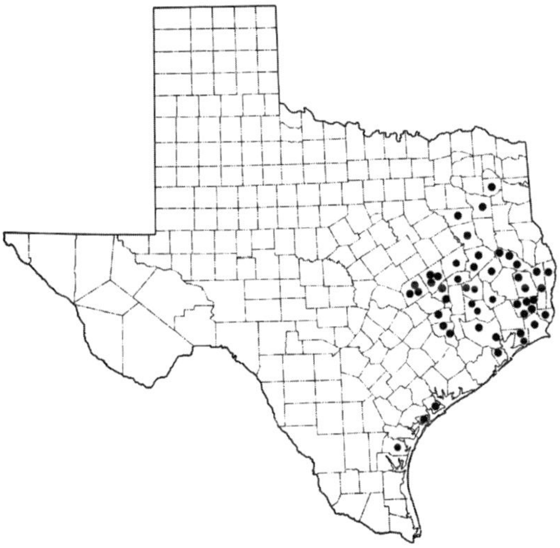

Pluchea foetida

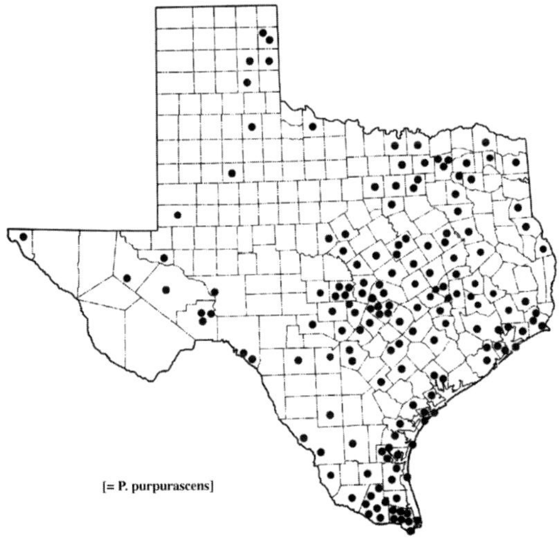
[= P. purpurascens]

Pluchea odorata

Pluchea rosea

Pluchea sericea

Porophyllum gracile

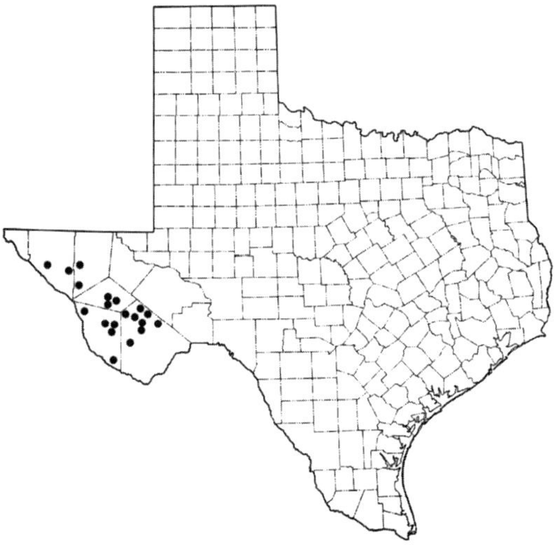

Porophyllum greggii

Porophyllum macrocephalum

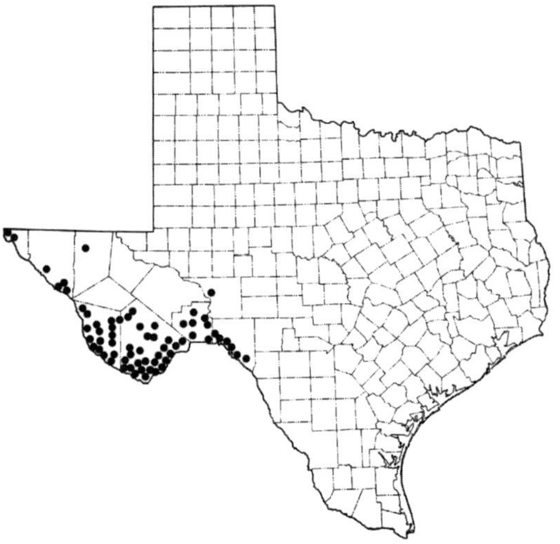

Porophyllum scoparium

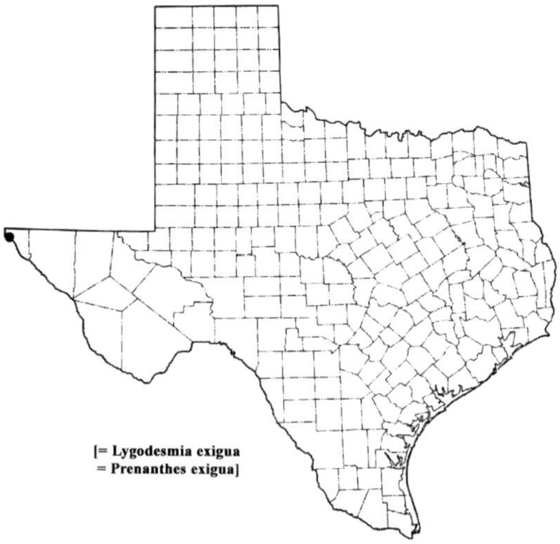
[= Lygodesmia exigua
= Prenanthes exigua]

Prenanthella exigua

Prenanthes altissima

Prenanthes barbata

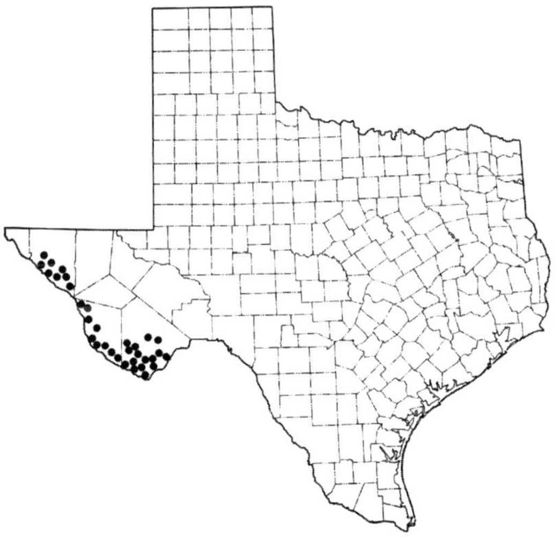

Psathyrotes scaposa

Pseudoclappia arenaria

Pseudoclappia watsonii

Pseudognaphalium arizonicum

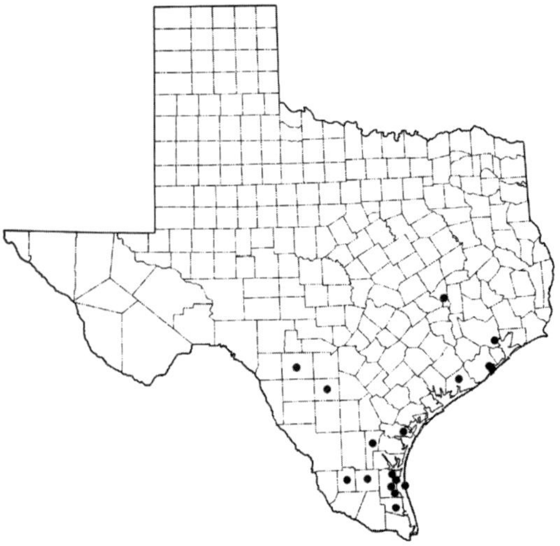
Pseudognaphalium austrotexanum

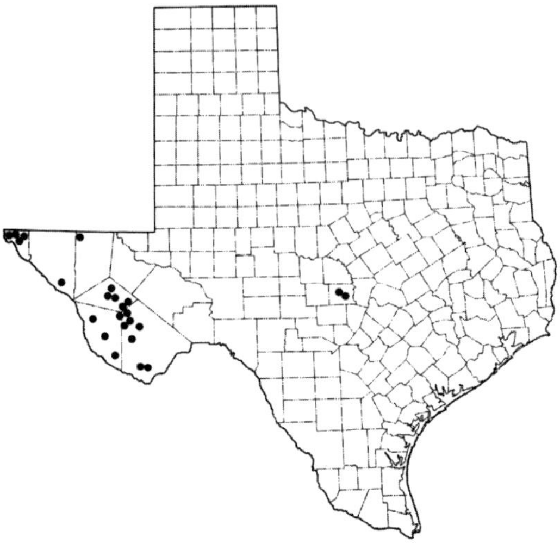
Pseudognaphalium canescens

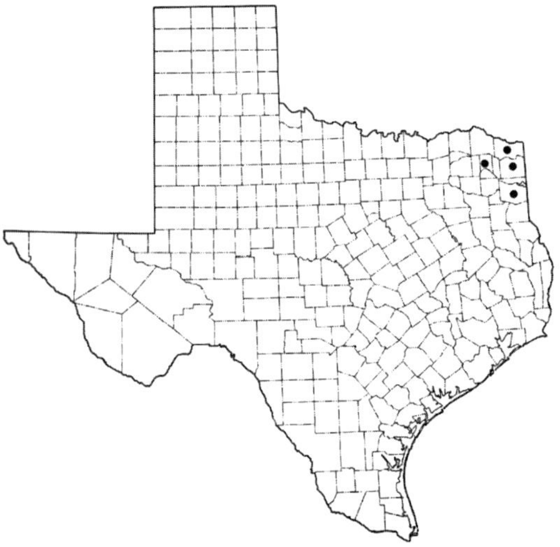
Pseudognaphalium helleri

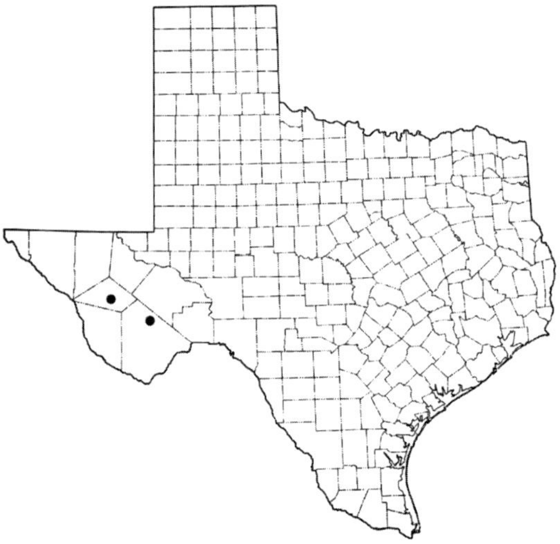
Pseudognaphalium jaliscense

Pseudognaphalium luteoalbum

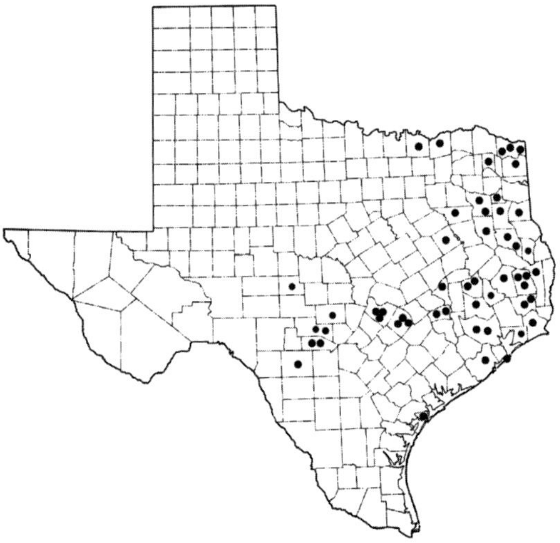
Pseudognaphalium obtusifolium

Pseudognaphalium pringlei

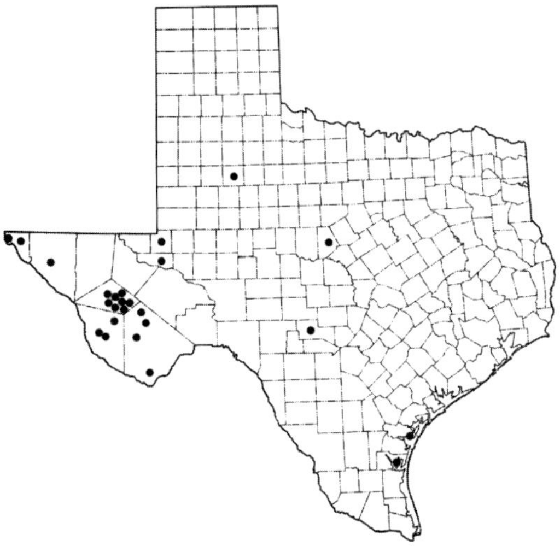

Pseudognaphalium stramineum

Pseudognaphalium viscosum

Pseudogynoxys chenopodioides

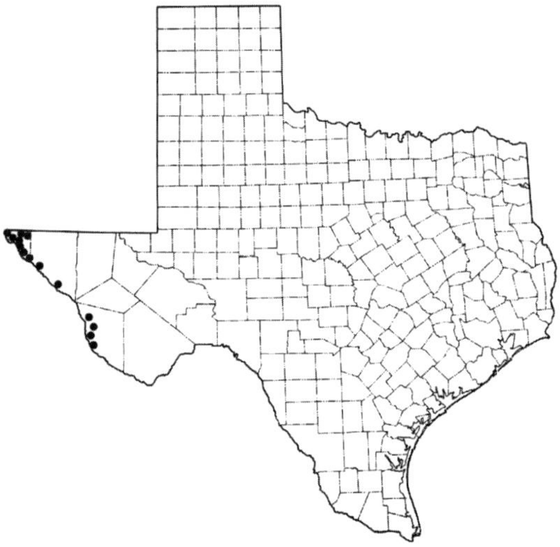

Psilactis asteroides

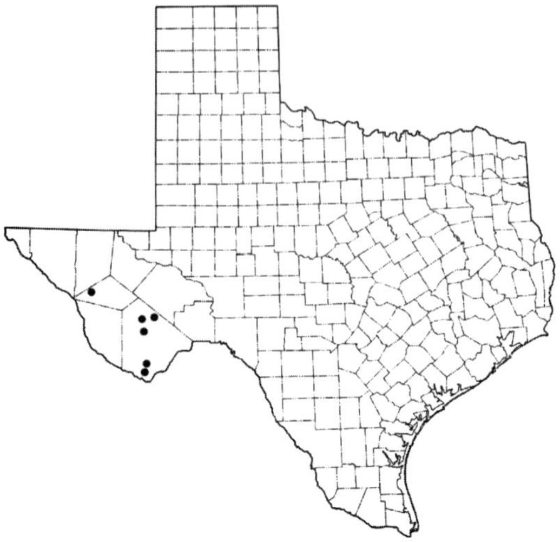

Psilactis brevilingulata

Psilactis heterocarpa

Psilactis tenuis

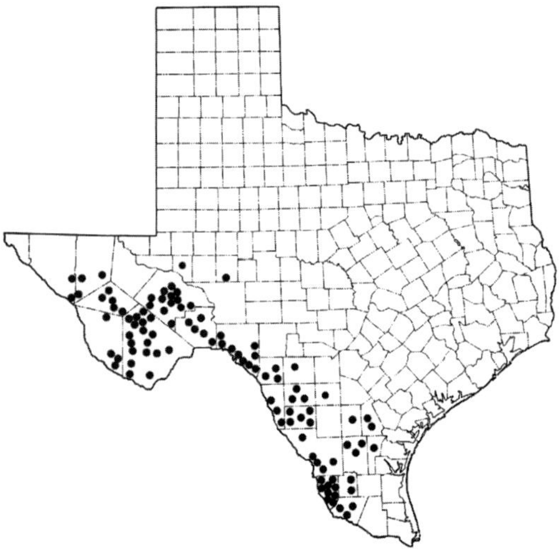

Psilostrophe gnaphalioides

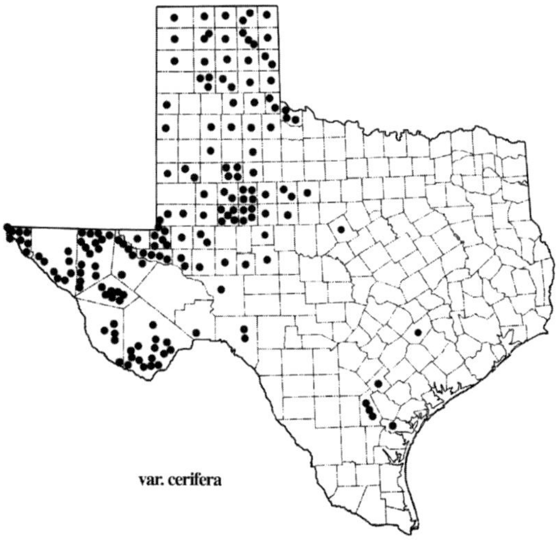
var. cerifera

Psilostrophe tagetina

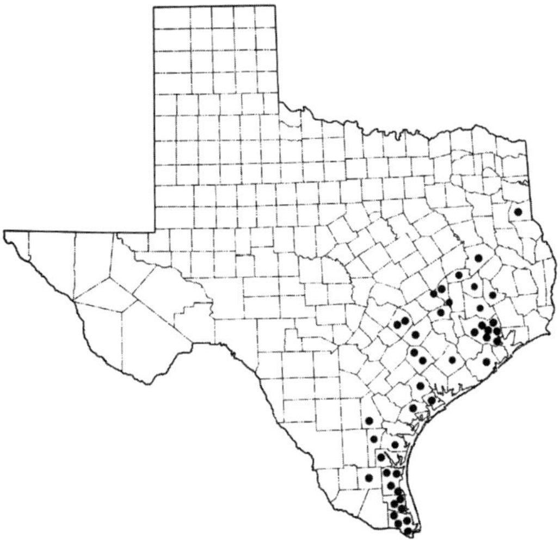

Pterocaulon virgatum

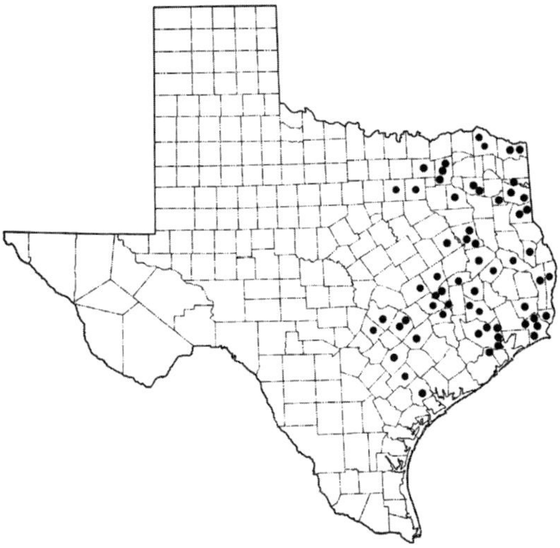

Pyrrhopappus carolinianus

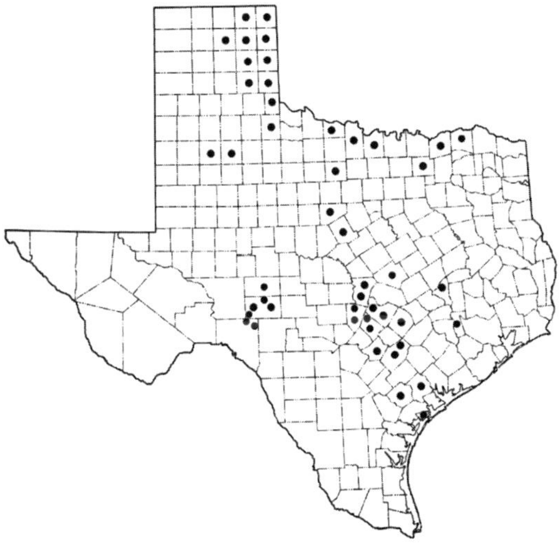

Pyrrhopappus grandiflorus

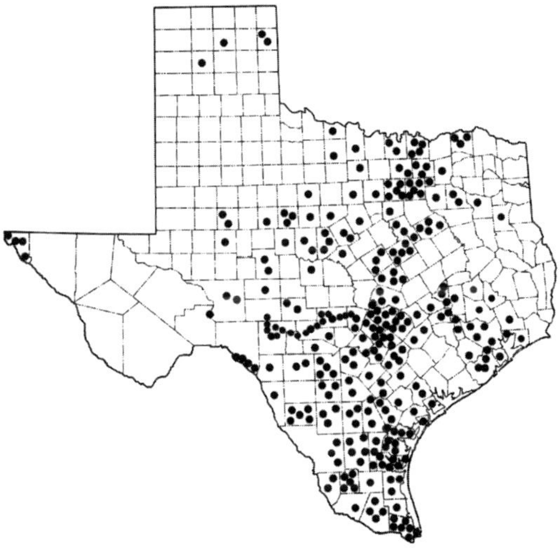

Pyrrhopappus pauciflorus

Rafinesquia neomexicana

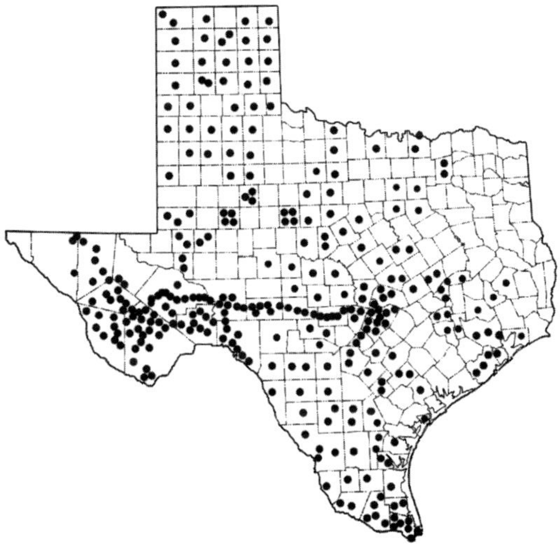

Ratibida columnifera

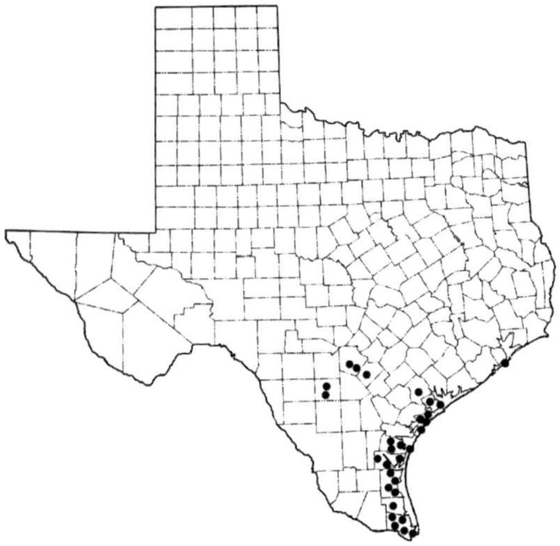

Ratibida peduncularis

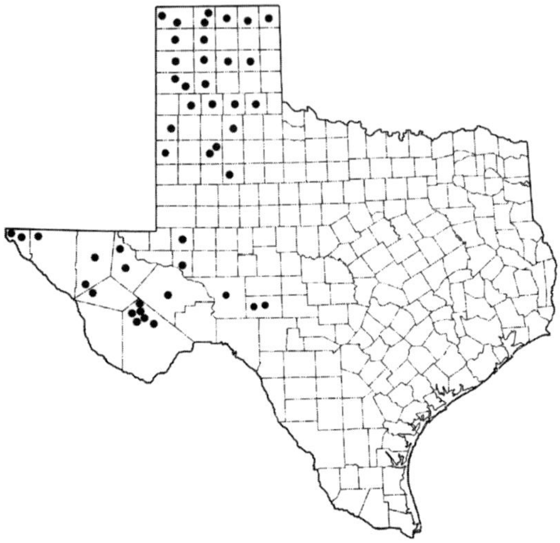

Ratibida tagetes

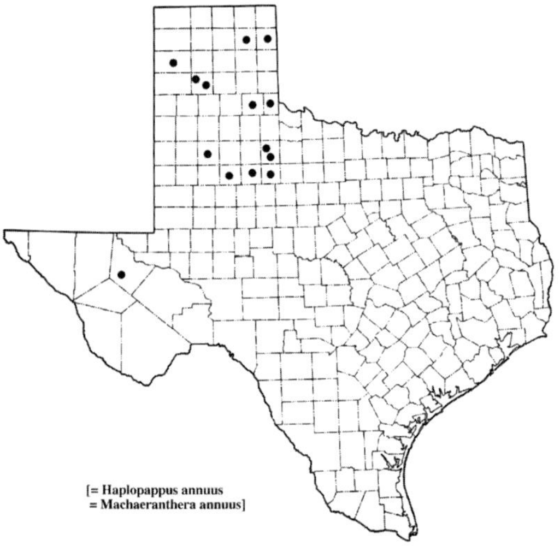
[= Haplopappus annuus
= Machaeranthera annuus]

Rayjacksonia annua

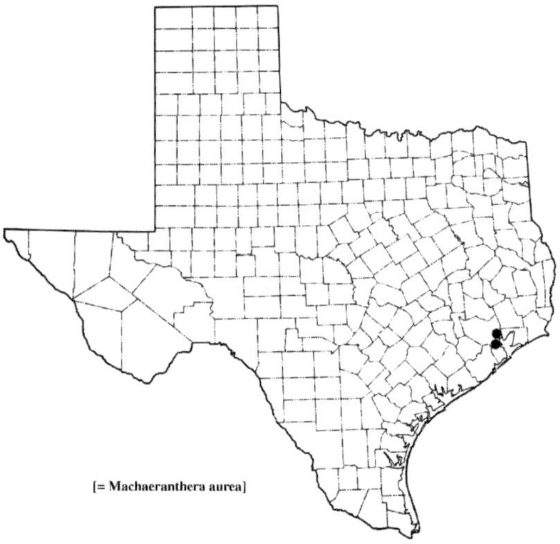
[= Machaeranthera aurea]

Rayjacksonia aurea

[= Machaeranthera phyllocephala]

Rayjacksonia phyllocephala

Rudbeckia fulgida

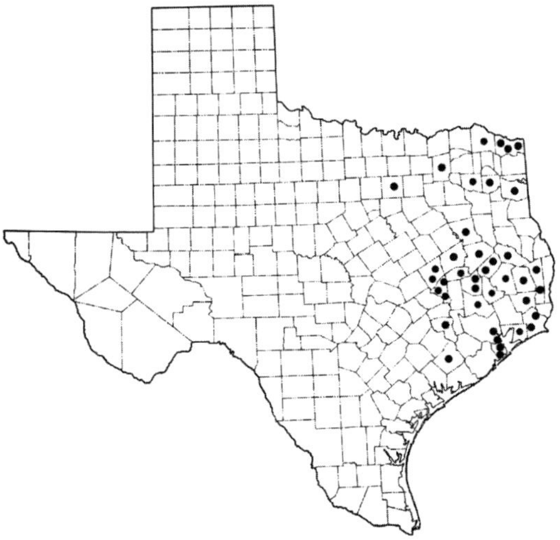

Rudbeckia grandiflora

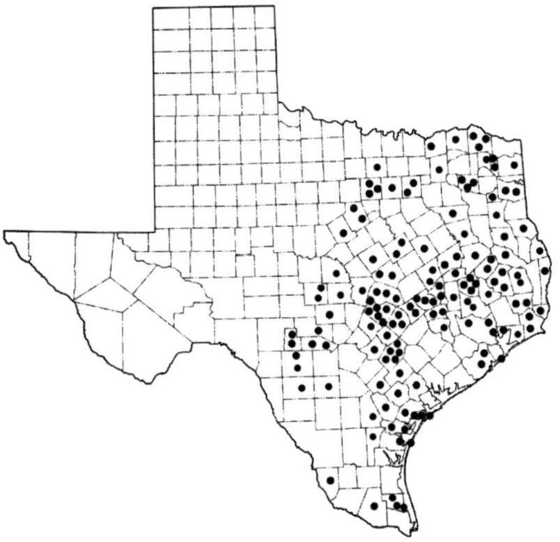

Rudbeckia hirta

Rudbeckia laciniata

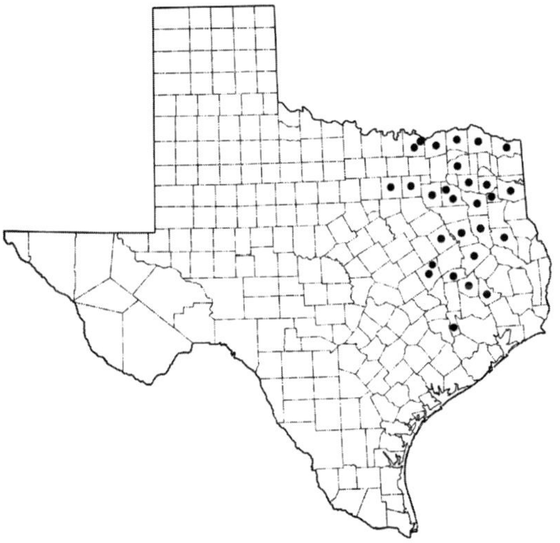

Rudbeckia maxima

Rudbeckia missouriensis

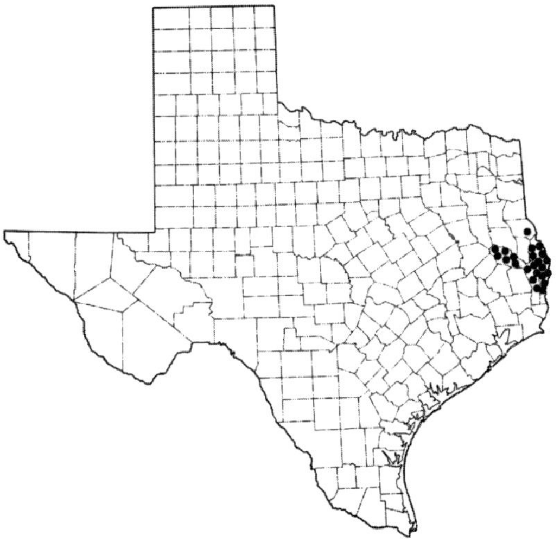

Rudbeckia scabrifolia

Rudbeckia subtomentosa

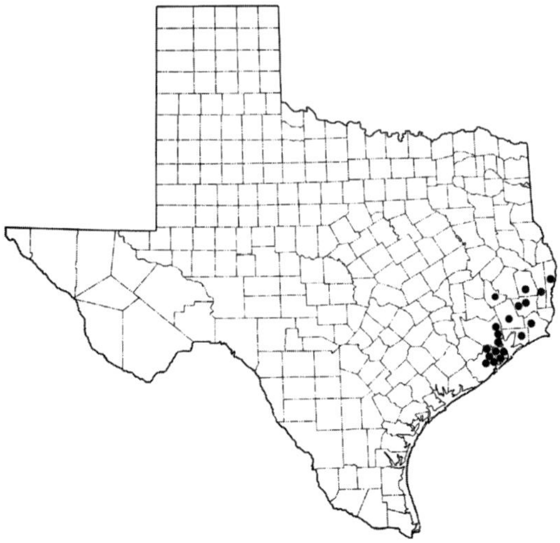
Rudbeckia texana

Rudbeckia triloba

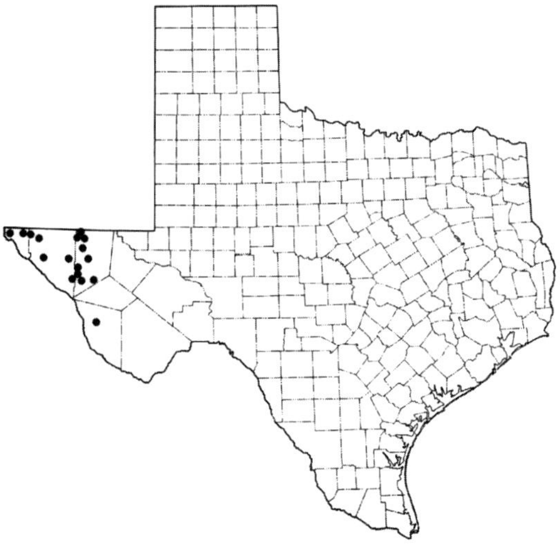
Sanvitalia abertii

Sanvitalia angustifolia

Sanvitalia ocymoides

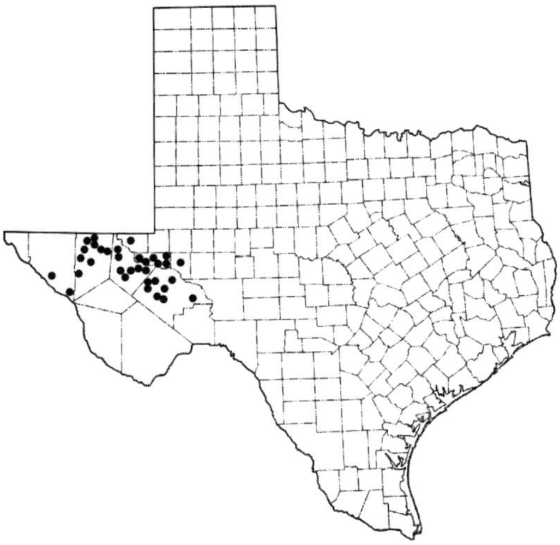

Sartwellia flaveriae

Schkuhria multiflora

var. wislizeni

Schkuhria pinnata

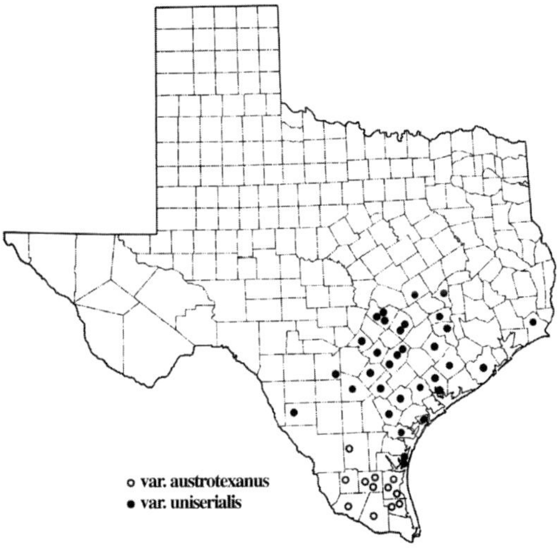
o var. austrotexanus
• var. uniserialis

Sclerocarpus uniserialis

Scorzonela laciniata

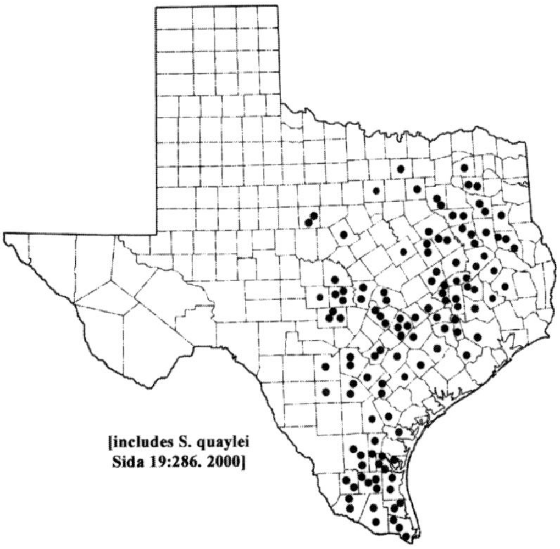
[includes S. quaylei
Sida 19:286. 2000]

Senecio ampullaceus

[=Packera fendleri]
Senecio fendleri

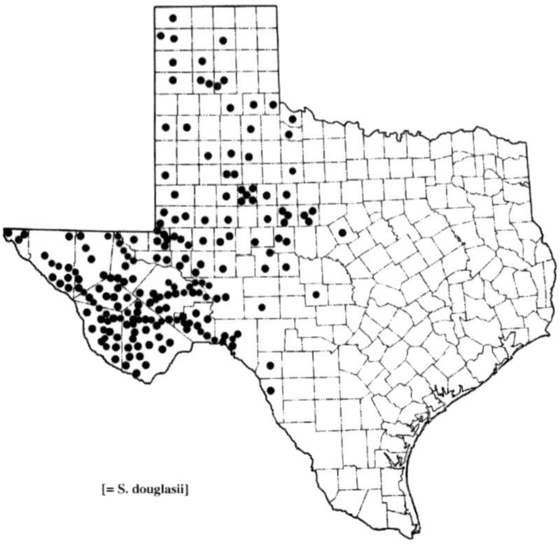
[= S. douglasii]
Senecio flaccidus

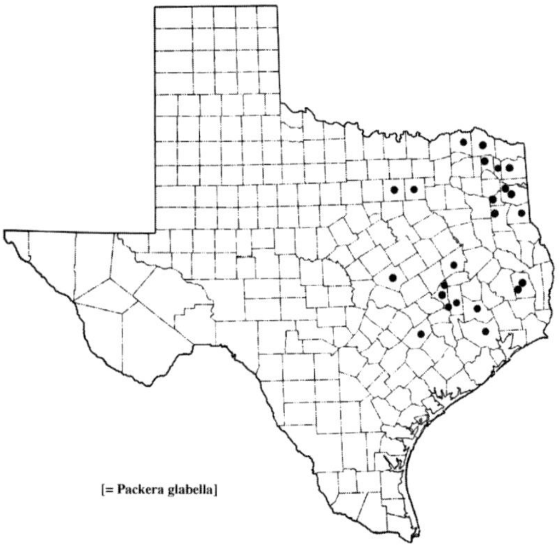
[= Packera glabella]
Senecio glabellus

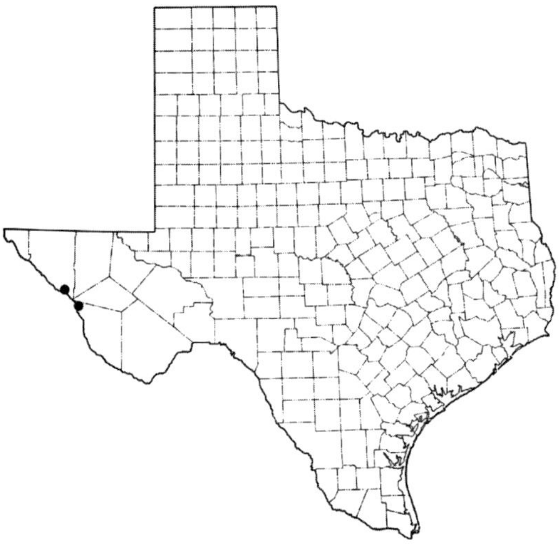
Senecio greggii

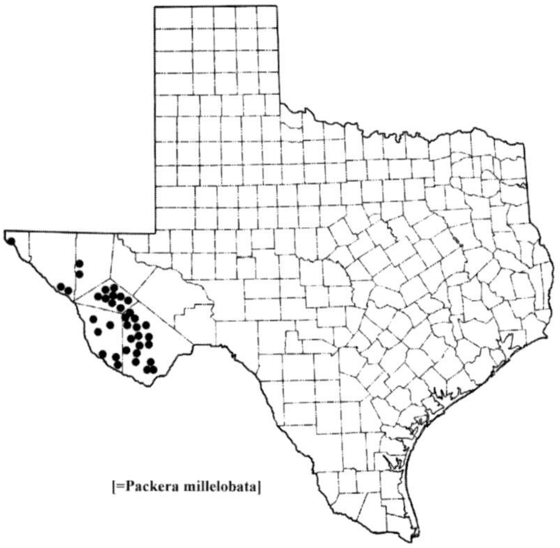
[=Packera millelobata]
Senecio millelobatus

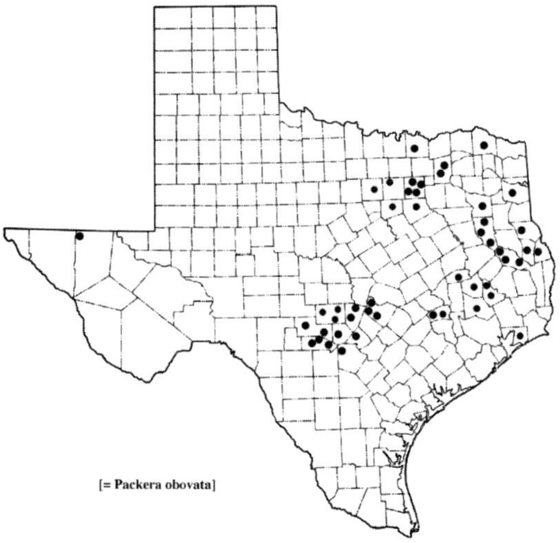
[= Packera obovata]
Senecio obovatus

Senecio parryi

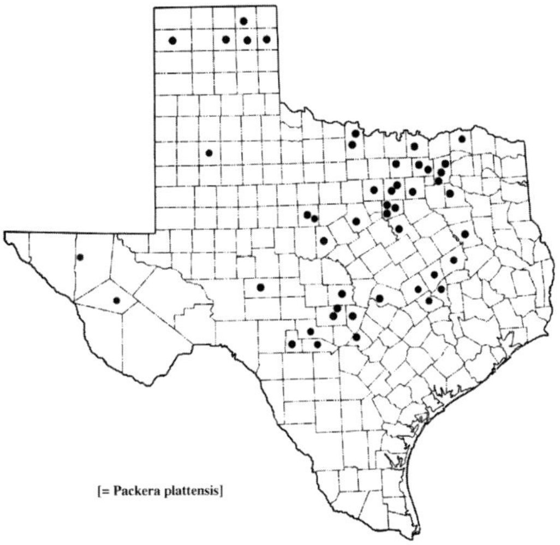
[= Packera plattensis]

Senecio plattensis

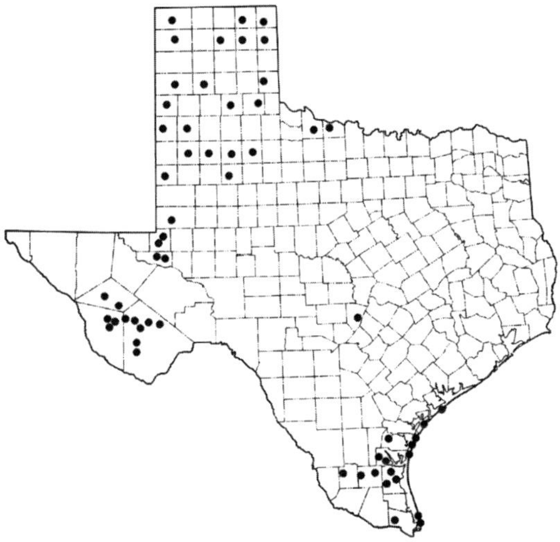

Senecio riddellii

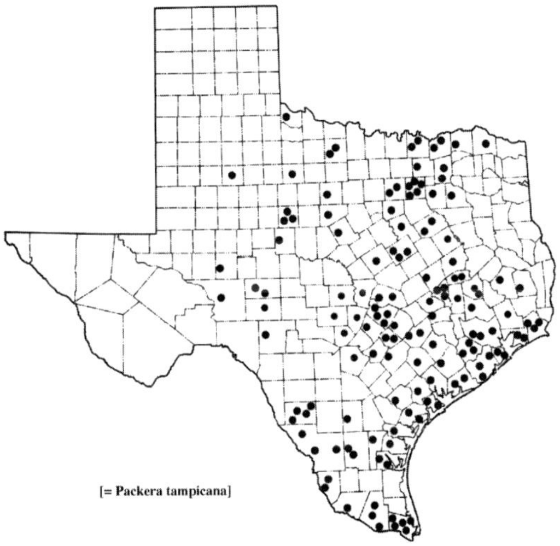
[= Packera tampicana]

Senecio tampicanus

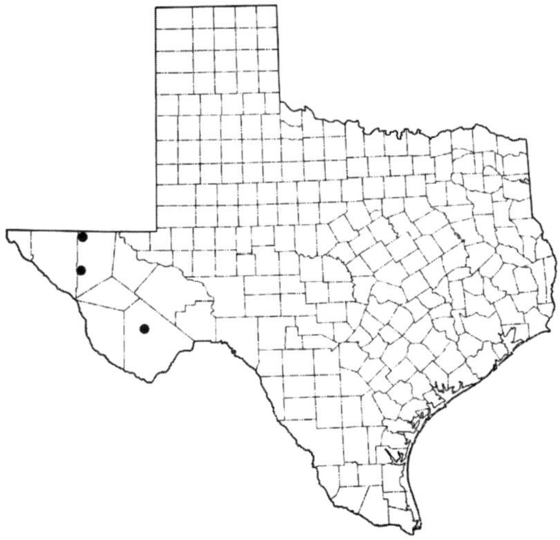

Senecio thurberi

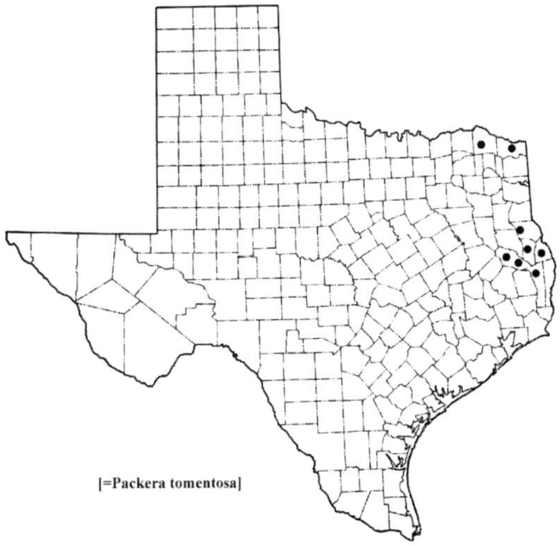
[=Packera tomentosa]

Senecio tomentosus

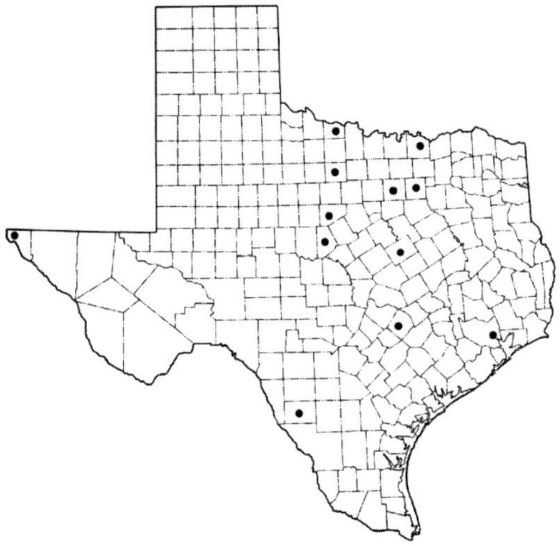
Senecio vulgaris

Senecio warnockii

Senecio wootonii

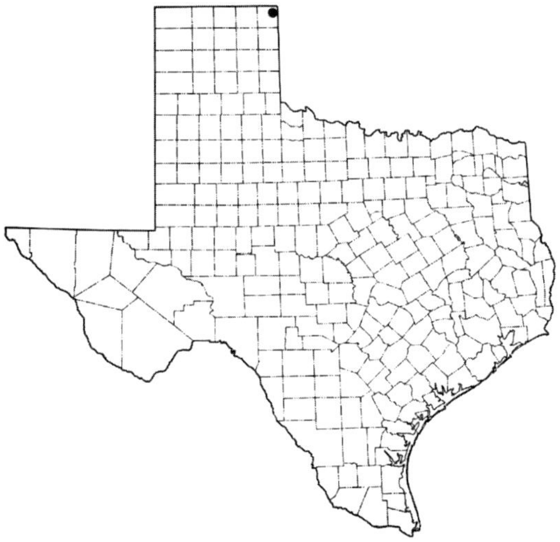
Shinnersoseris rostrata

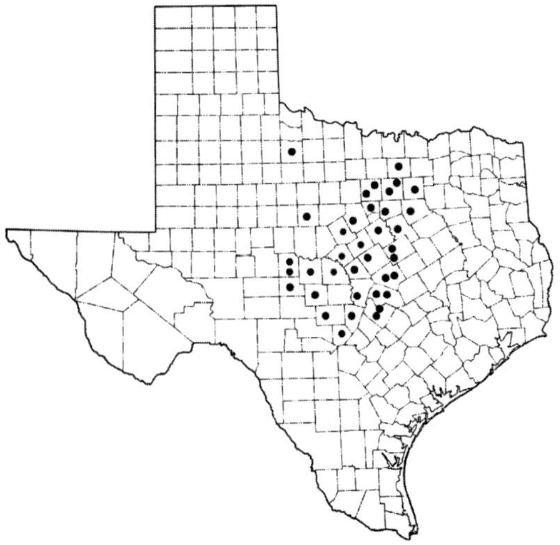
Silphium albiflorum

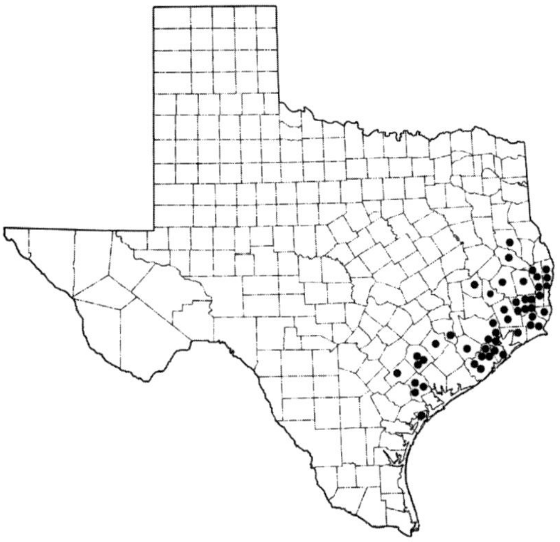
Silphium gracile

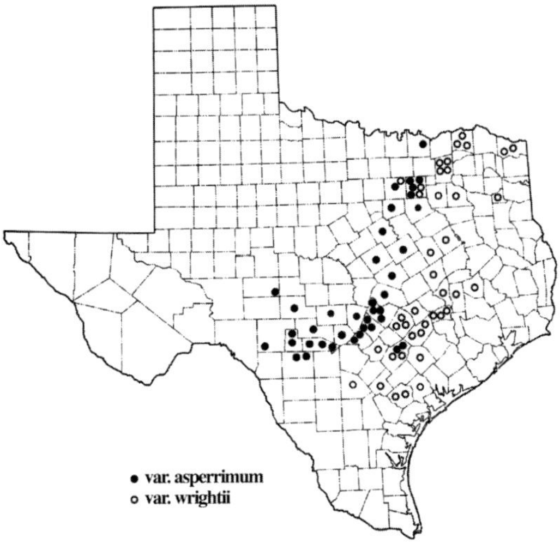
• var. asperrimum
○ var. wrightii

Silphium integrifolium

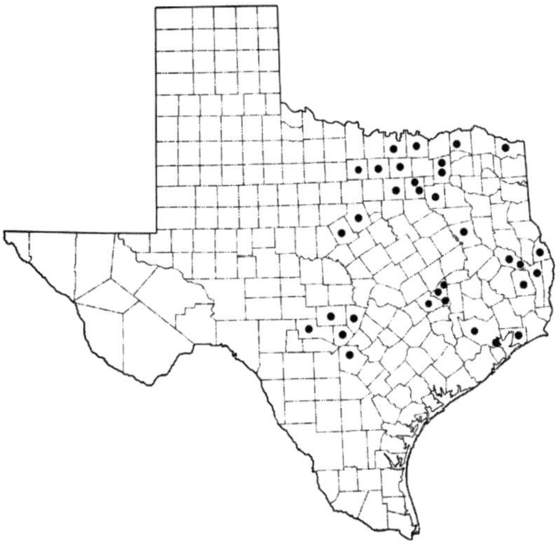

Silphium laciniatum

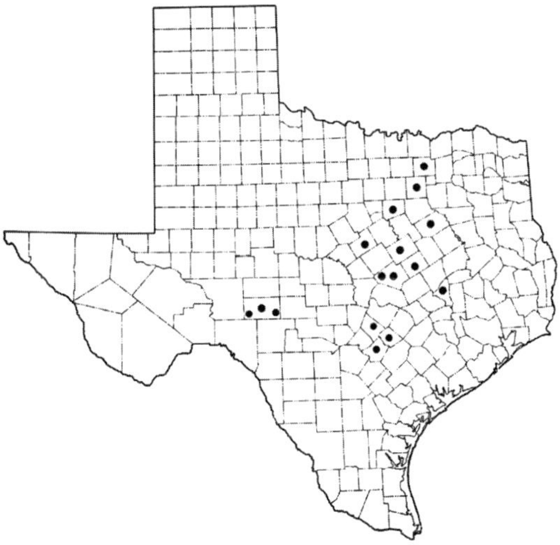

Silybum marianum

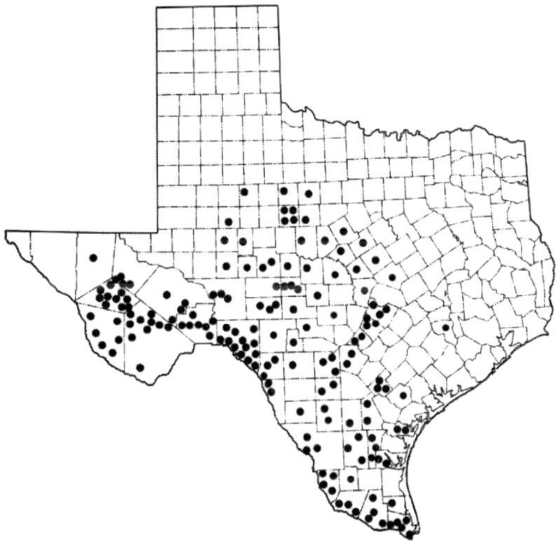

Simsia calva

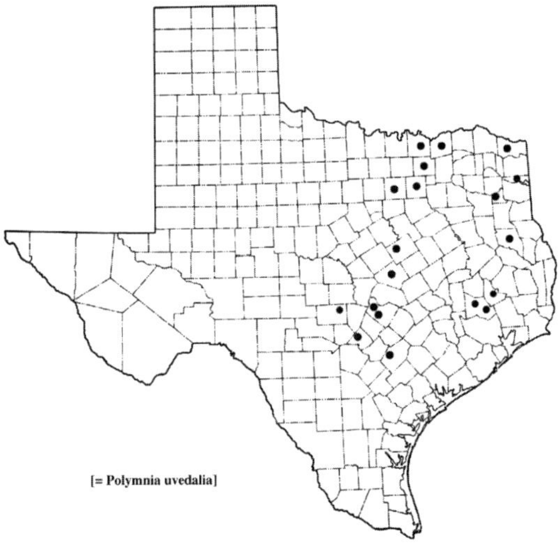
[= Polymnia uvedalia]

Smallanthus uvedalius

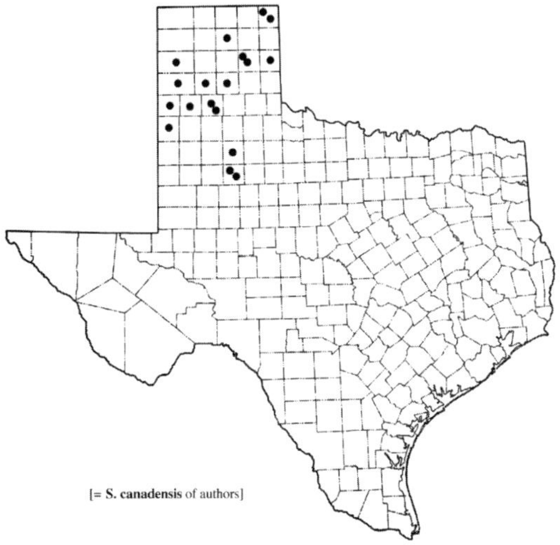
[= S. canadensis of authors]

Solidago altiplanities

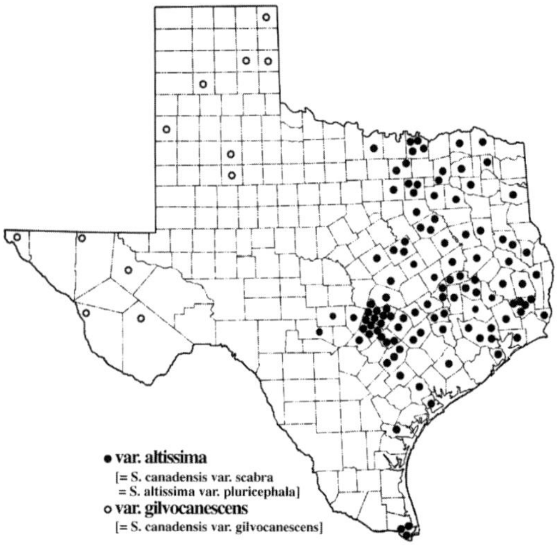
• var. altissima
[= S. canadensis var. scabra
= S. altissima var. pluricephala]
○ var. gilvocanescens
[= S. canadensis var. gilvocanescens]

Solidago altissima

Solidago arguta

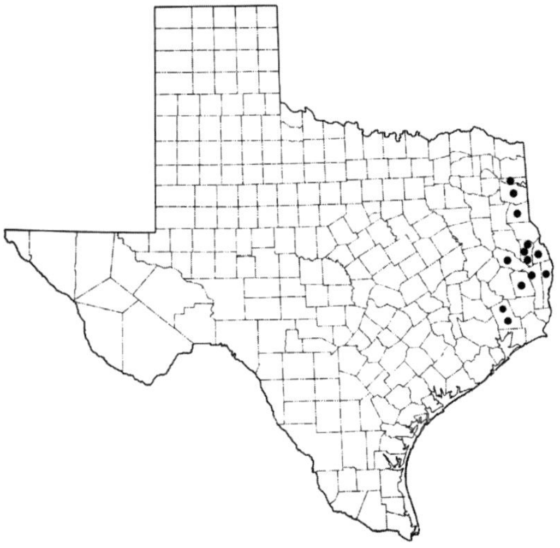

Solidago auriculata

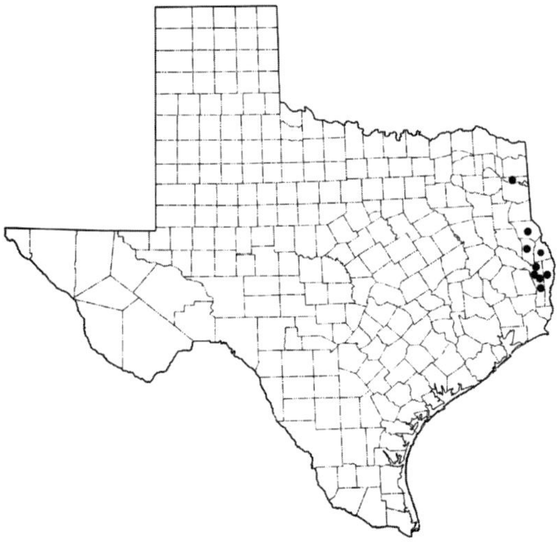

Solidago caesia

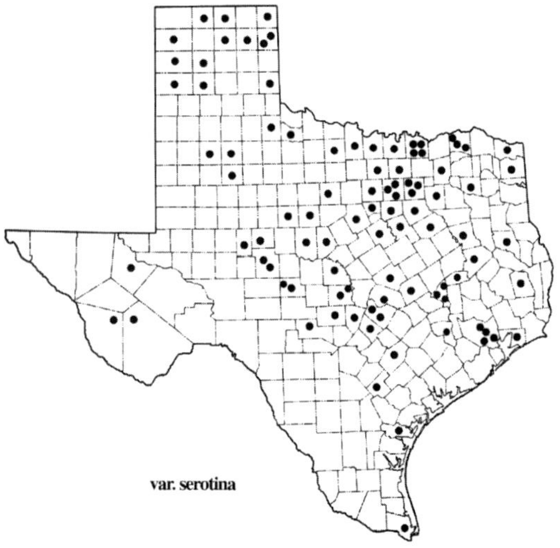
var. serotina

Solidago gigantea

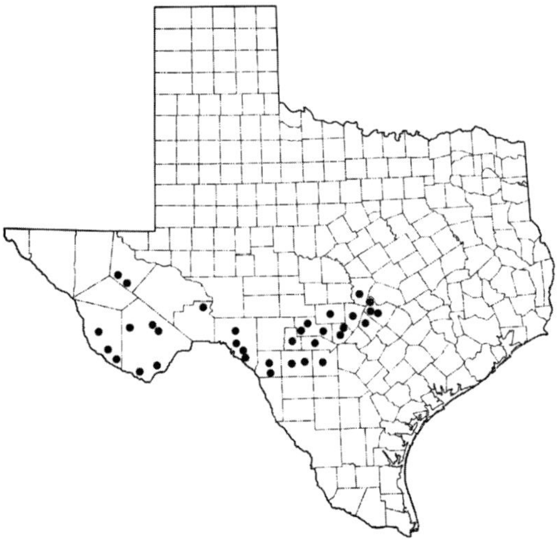

Solidago juliae

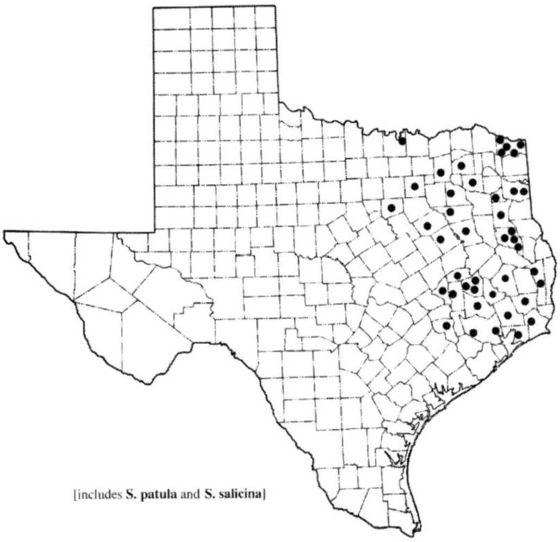
[includes S. patula and S. salicina]

Solidago ludoviciana

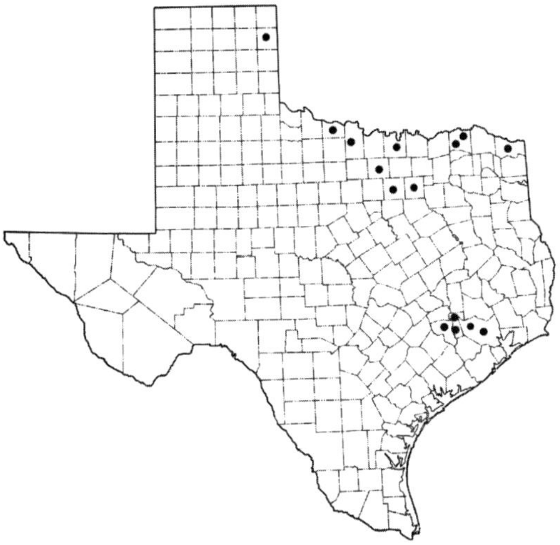

Solidago missouriensis

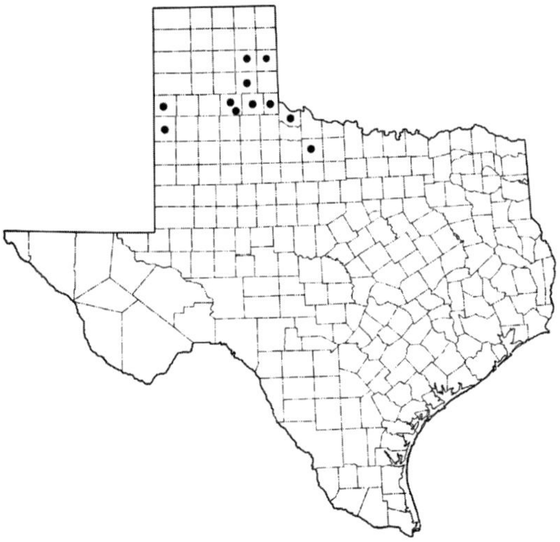

Solidago mollis

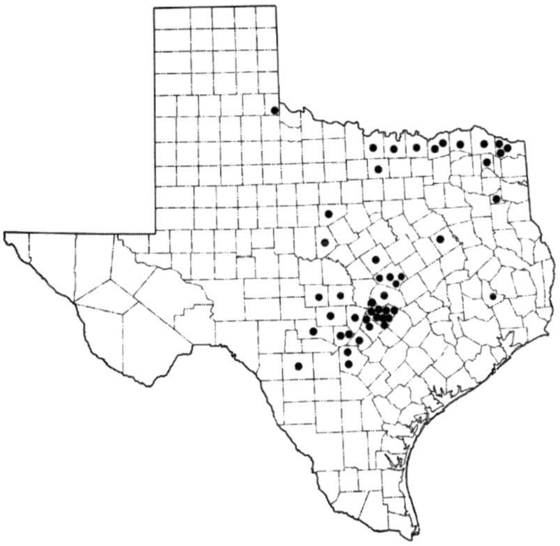

Solidago nemoralis

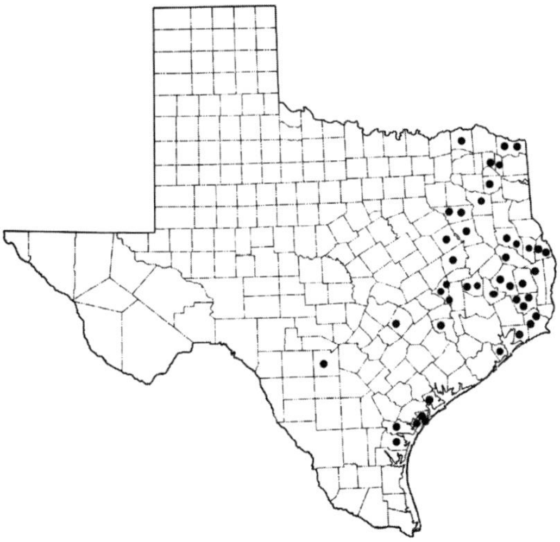

Solidago odora

Solidago patula

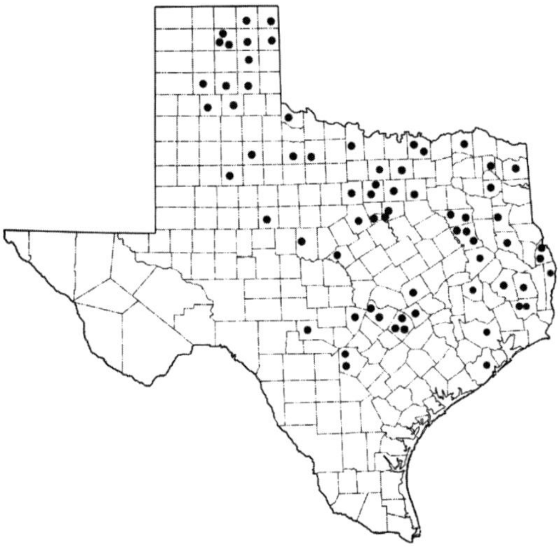

Solidago petiolaris

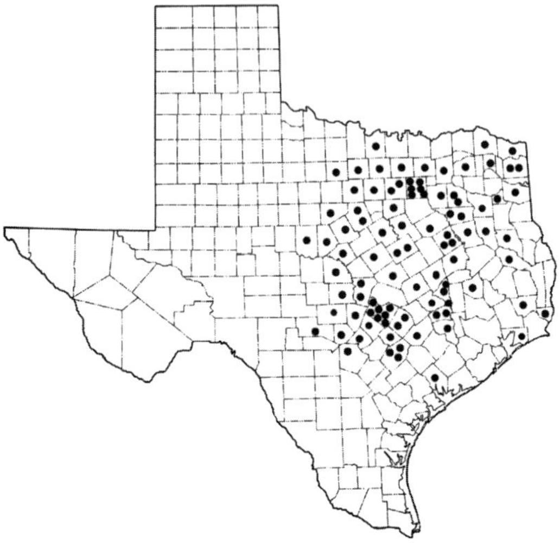

Solidago radula

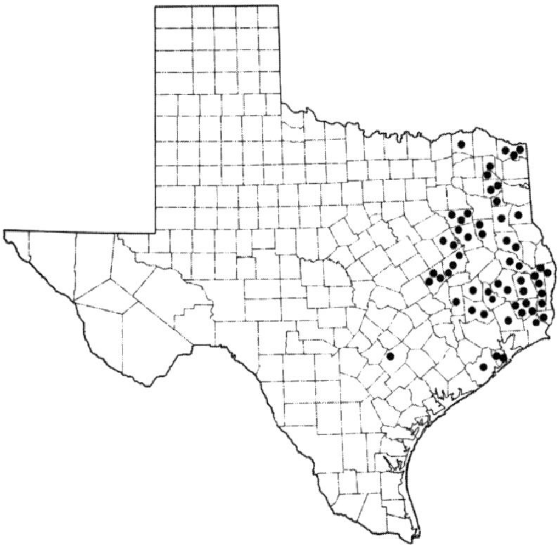

Solidago rugosa

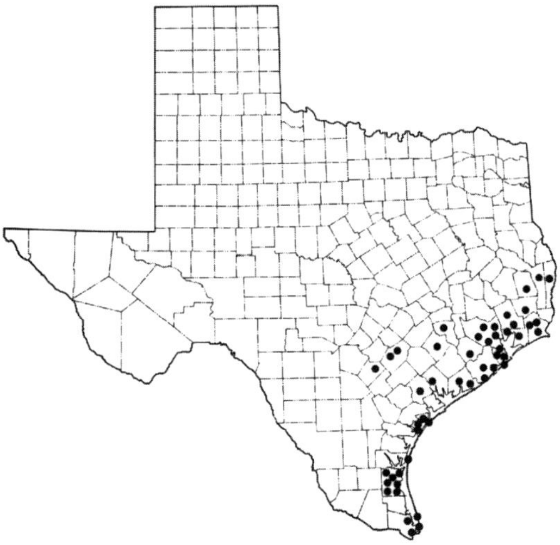

Solidago sempervirens

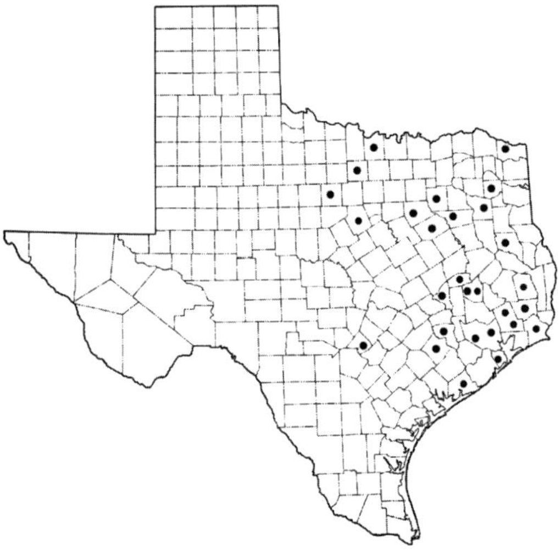

Solidago speciosa

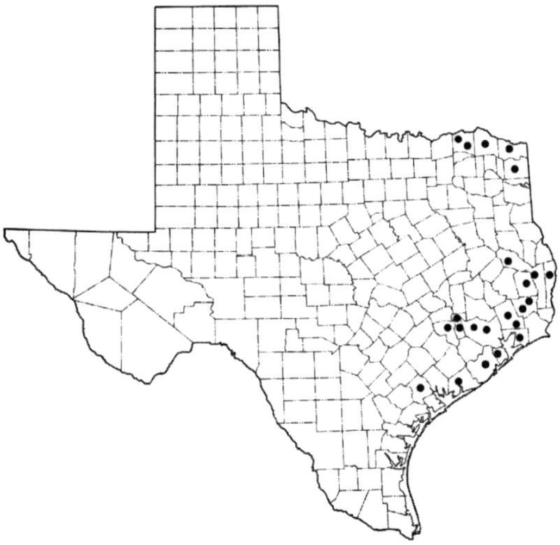

Solidago tortifolia

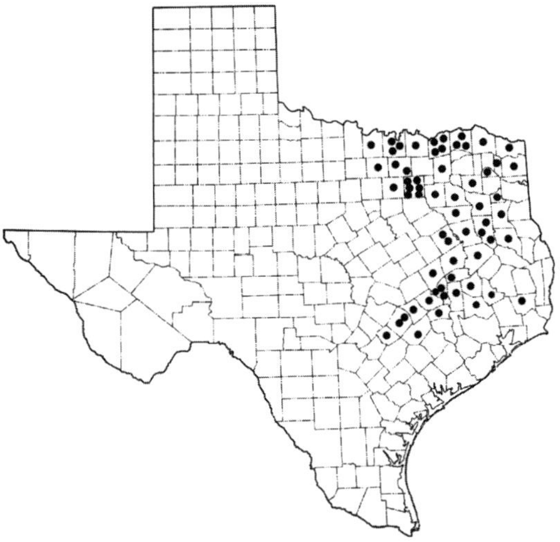

Solidago ulmifolia

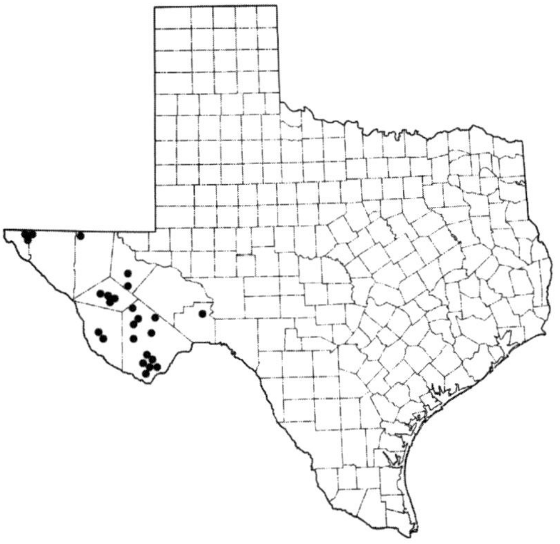

Solidago velutina

Solidago wrightii

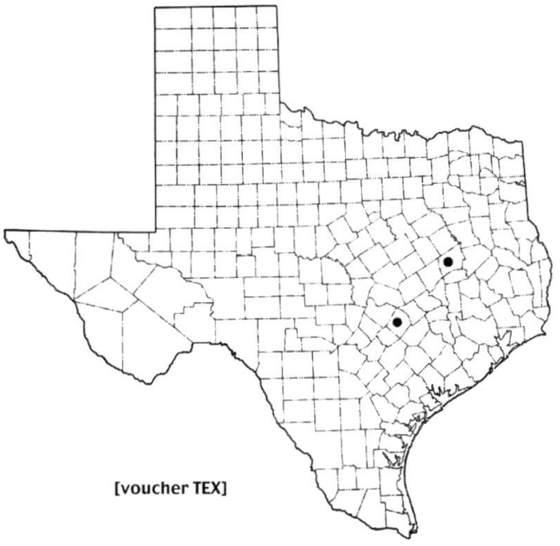
[voucher TEX]

Soliva anthemifolia

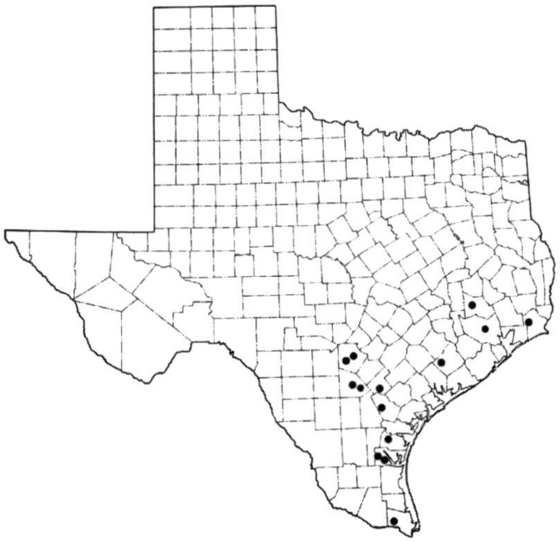

Soliva mutisii

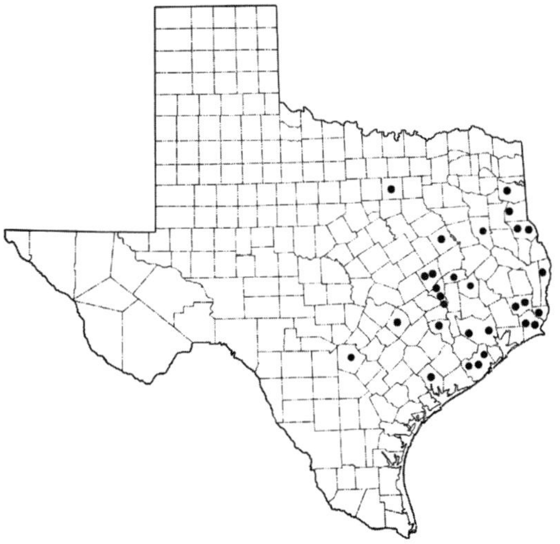

Soliva pterosperma

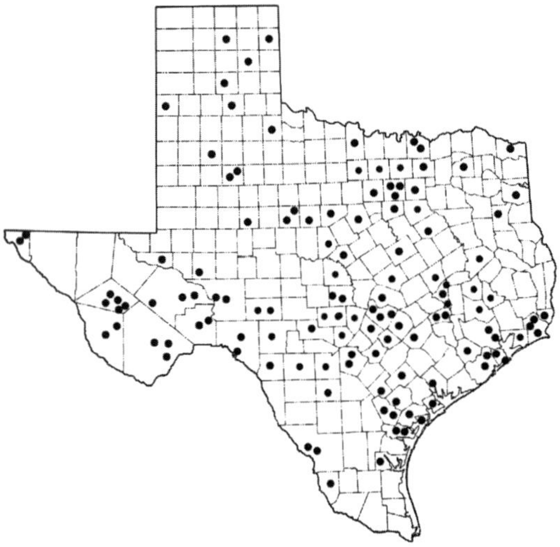

Sonchus asper

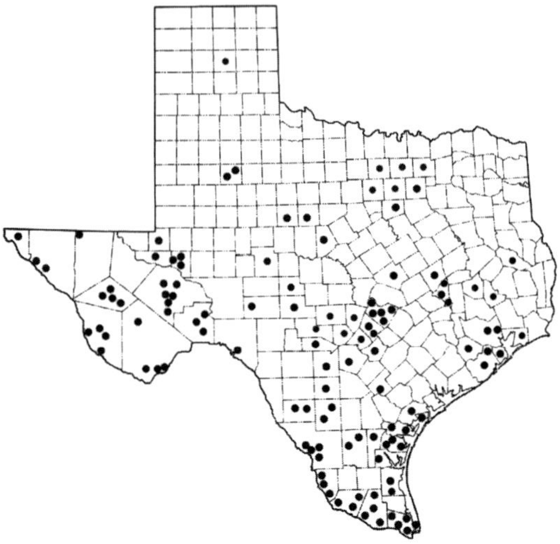

Sonchus oleraceus

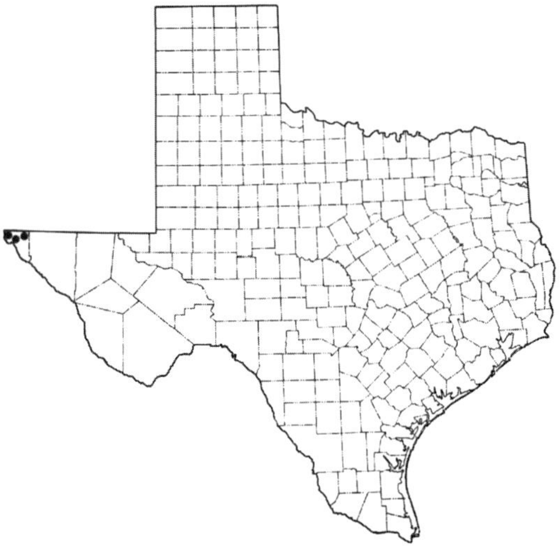

Stephanomeria exigua

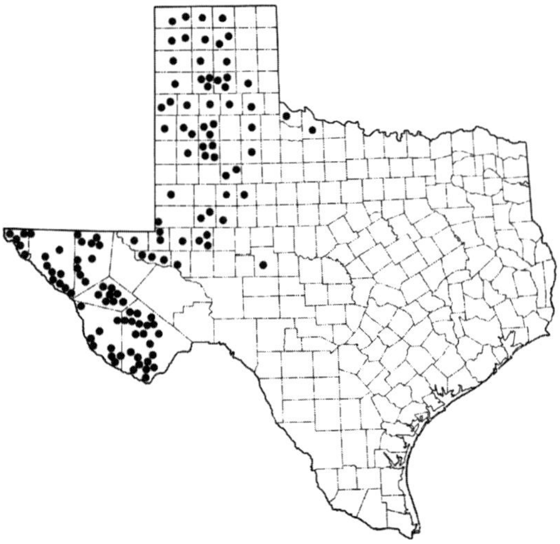

Stephanomeria pauciflora

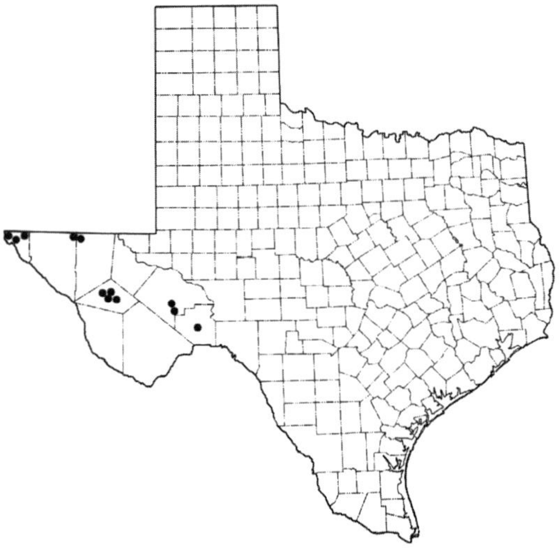

Stephanomeria tenuifolia

Stevia ovata

Stevia serrata

Stevia viscida

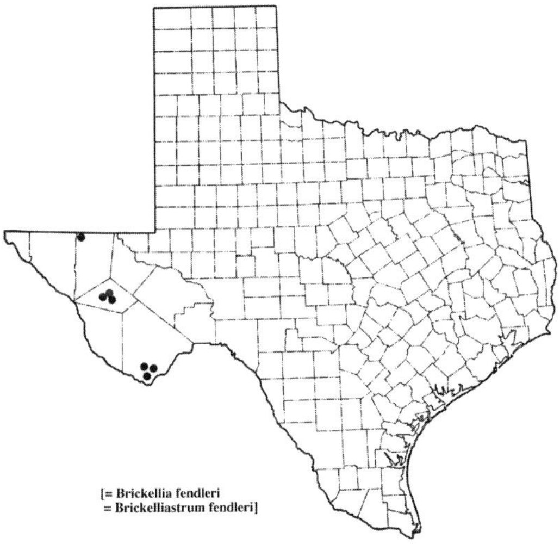
[= Brickellia fendleri
= Brickelliastrum fendleri]

Steviopsis fendleri

Stylocline micropoides

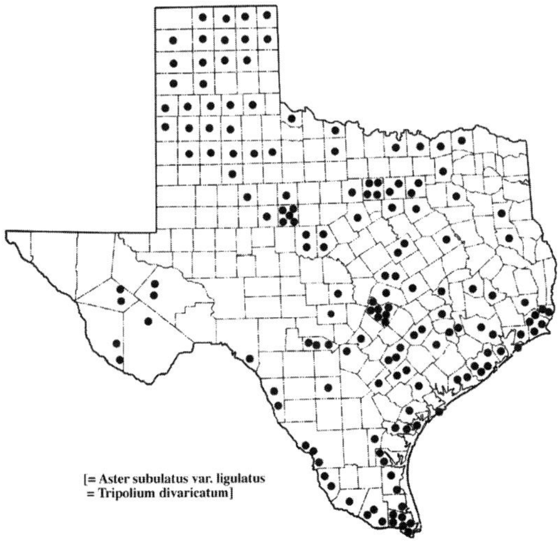
[= Aster subulatus var. ligulatus
= Tripolium divaricatum]

Symphyotrichum divaricatum

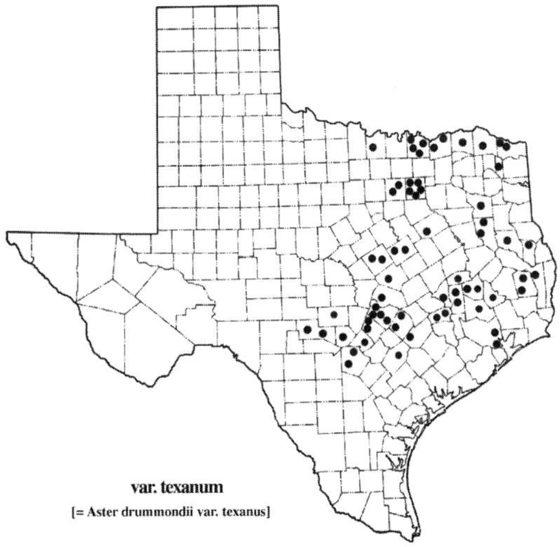
var. texanum
[= Aster drummondii var. texanus]

Symphyotrichum drummondii

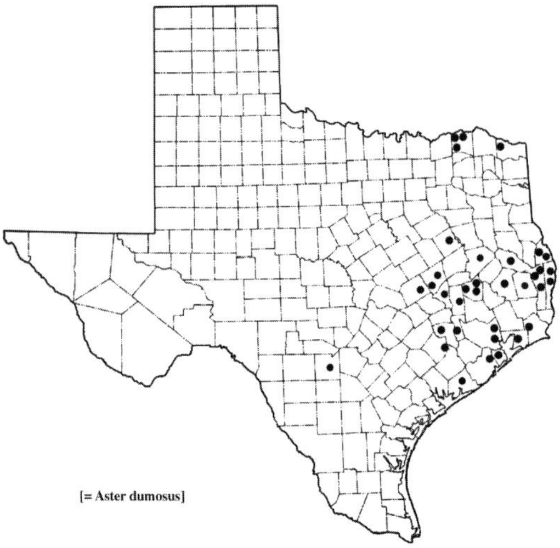
[= Aster dumosus]

Symphyotrichum dumosum

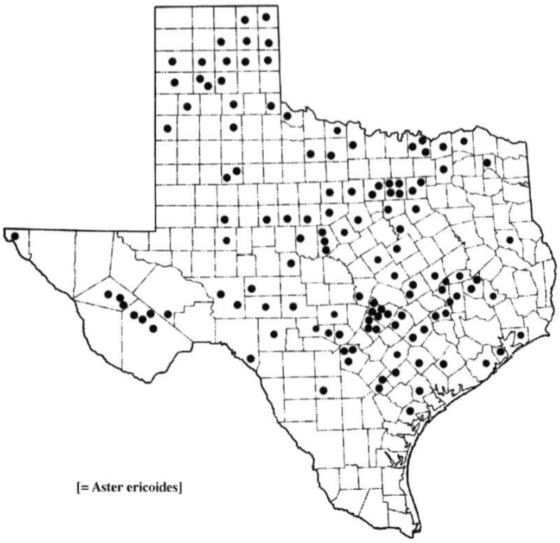
[= Aster ericoides]

Symphyotrichum ericoides

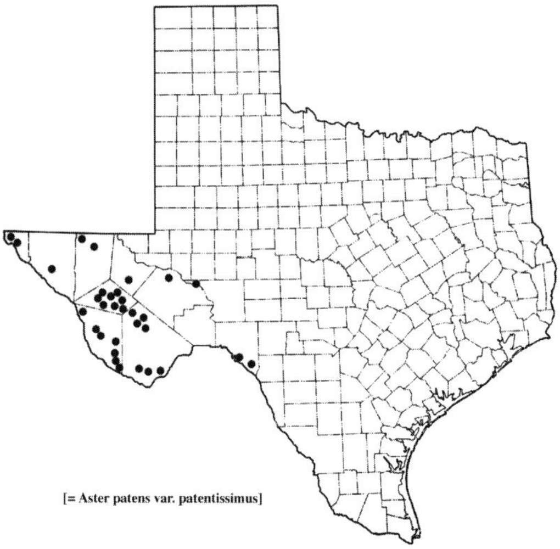
[= Aster patens var. patentissimus]

Symphyotrichum expansum

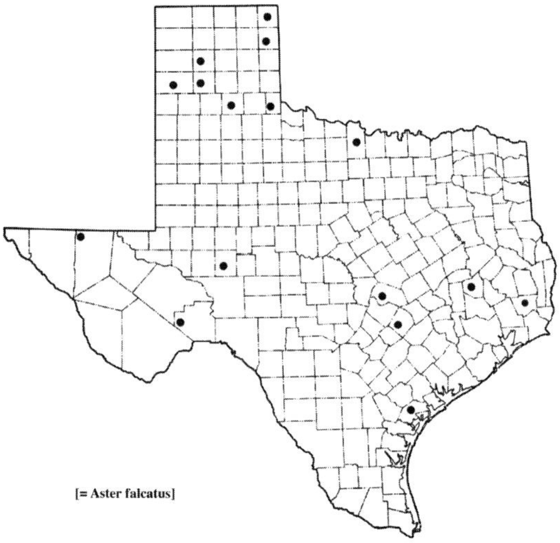
[= Aster falcatus]

Symphyotrichum falcatum

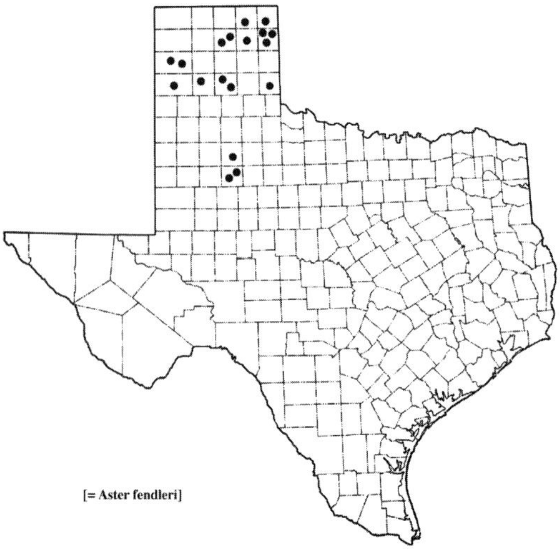
[= Aster fendleri]

Symphyotrichum fendleri

[= Aster laeve]
● var. geyeri
○ var. purpuratum

Symphyotrichum laeve

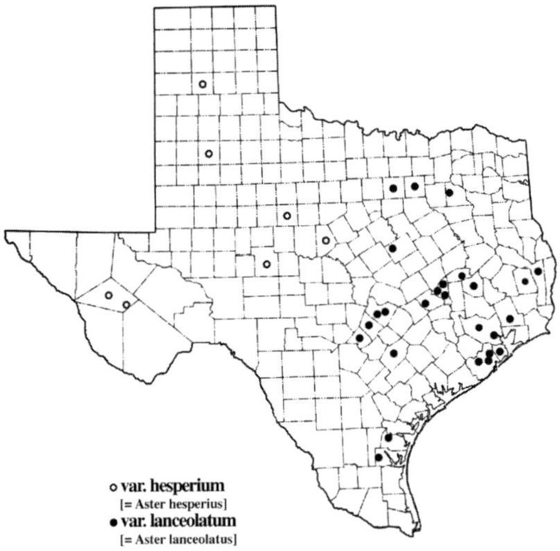
○ var. hesperium
[= Aster hesperius]
● var. lanceolatum
[= Aster lanceolatus]

Symphyotrichum lanceolatum

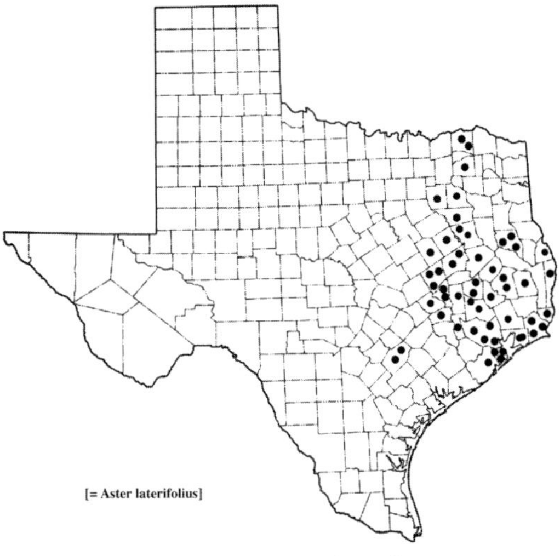
[= Aster laterifolius]

Symphyotrichum lateriflorum

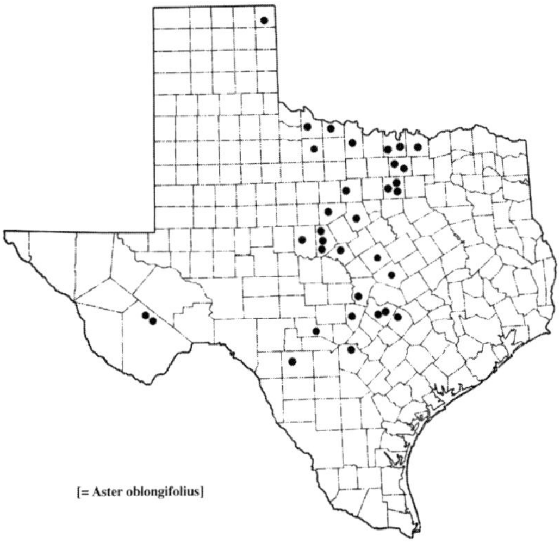
[= Aster oblongifolius]

Symphyotrichum oblongifolium

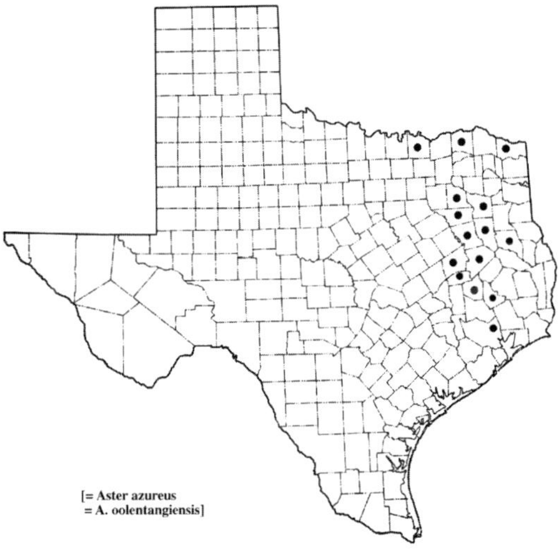
[= Aster azureus
= A. oolentangiensis]

Symphyotrichum oolentangiense

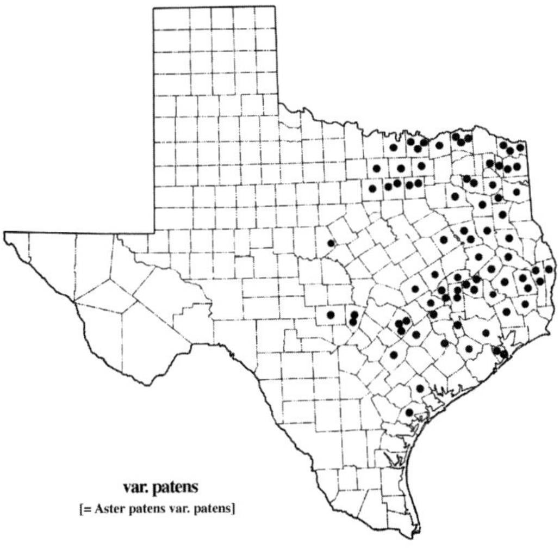
var. patens
[= Aster patens var. patens]

Symphyotrichum patens

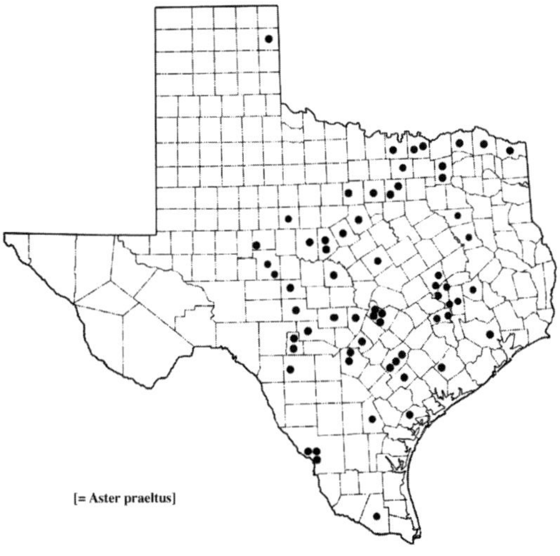
[= Aster praeltus]

Symphyotrichum praealtum

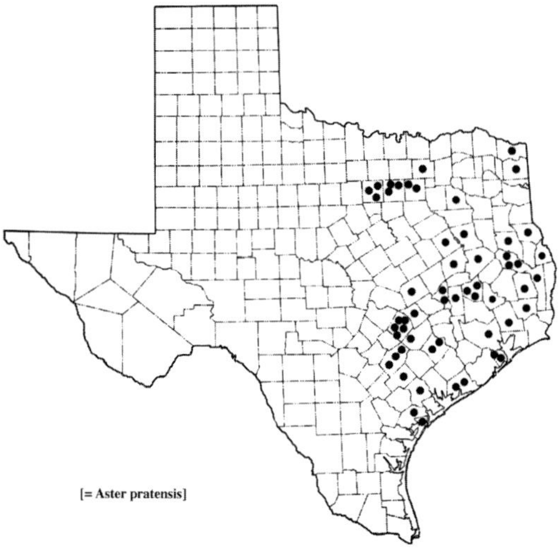
[= Aster pratensis]

Symphyotrichum pratense

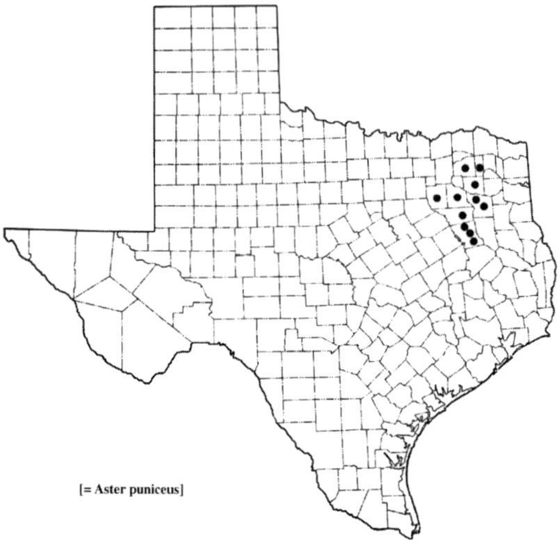
[= Aster puniceus]

Symphyotrichum puniceum

[= Aster sericeus]

Symphyotrichum sericeum

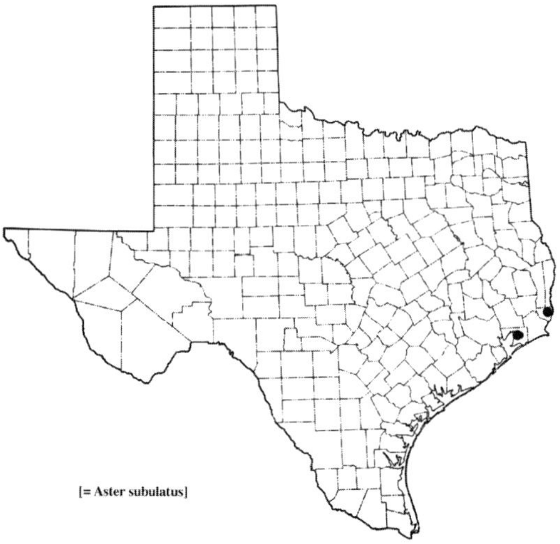
[= Aster subulatus]

Symphyotrichum subulatum

[= Aster tenuifolius]

Symphyotrichum tenuifolium

Tagetes micrantha

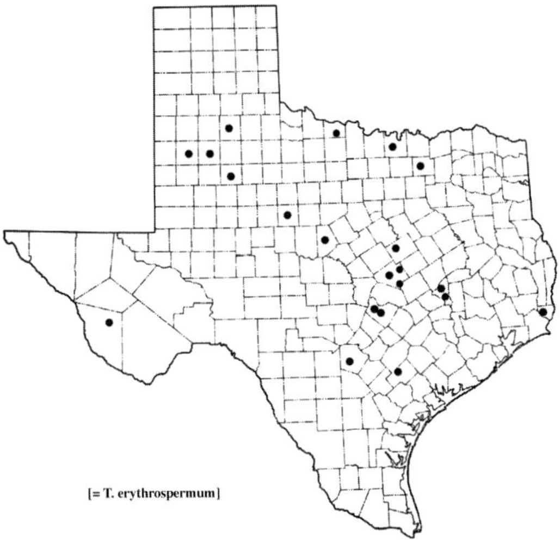
[= T. erythrospermum]

Taraxacum laevigatum

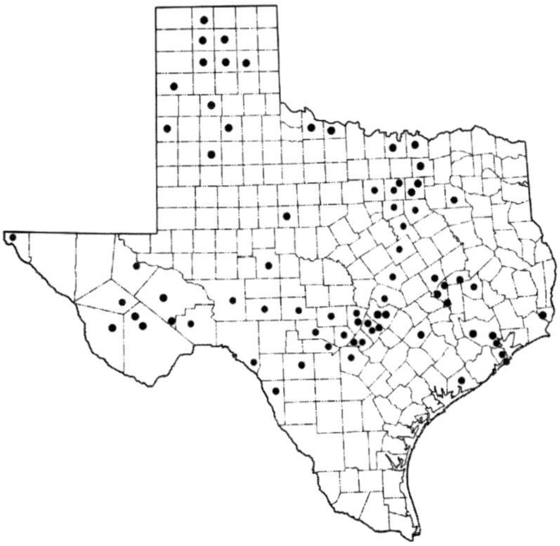

Taraxacum officinale

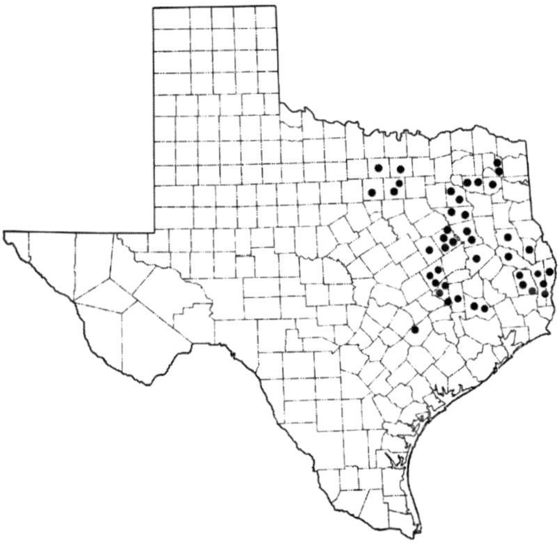

Tetragonotheca ludoviciana

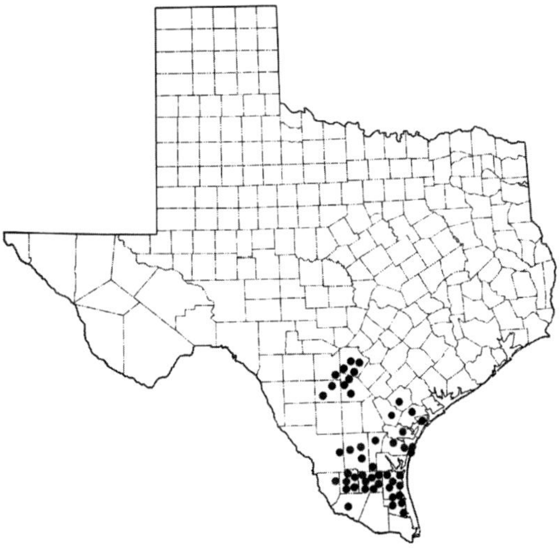

Tetragonotheca repanda

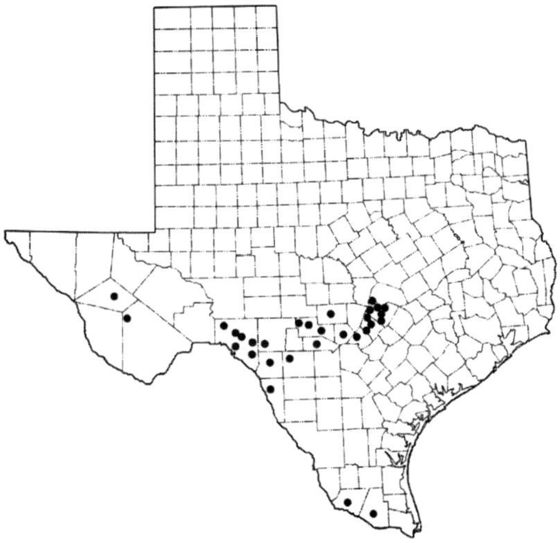

Tetragonotheca texana

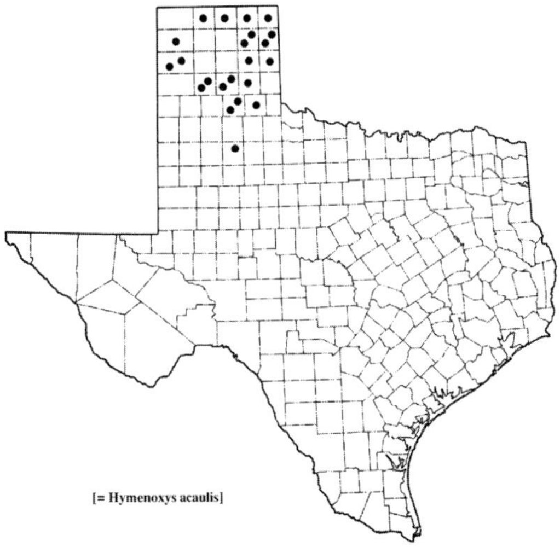
[= Hymenoxys acaulis]

Tetraneuris acaulis

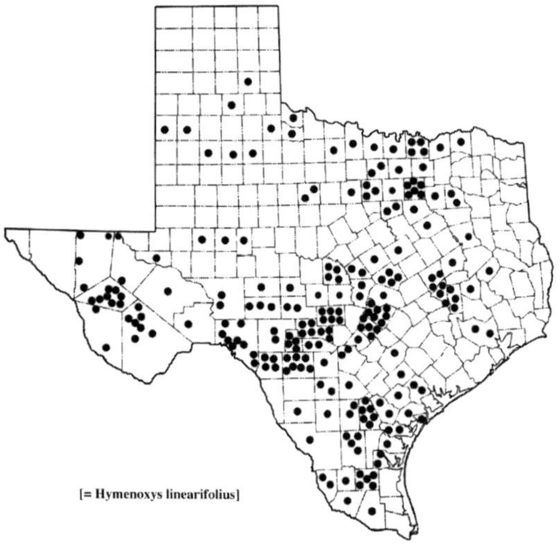
[= Hymenoxys linearifolius]

Tetraneuris linearifolia

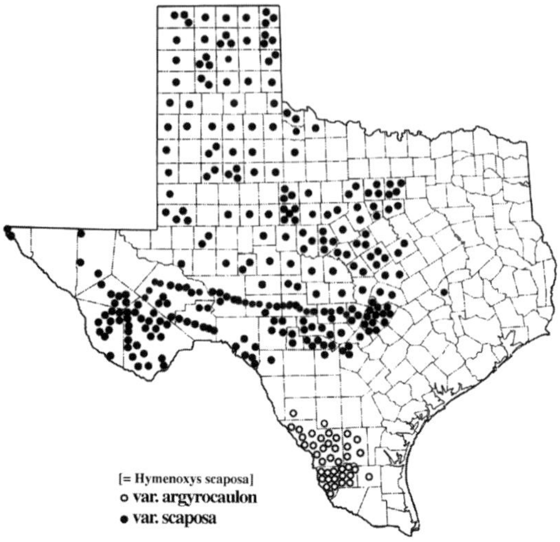
[= Hymenoxys scaposa]
○ var. argyrocaulon
● var. scaposa

Tetraneuris scaposa

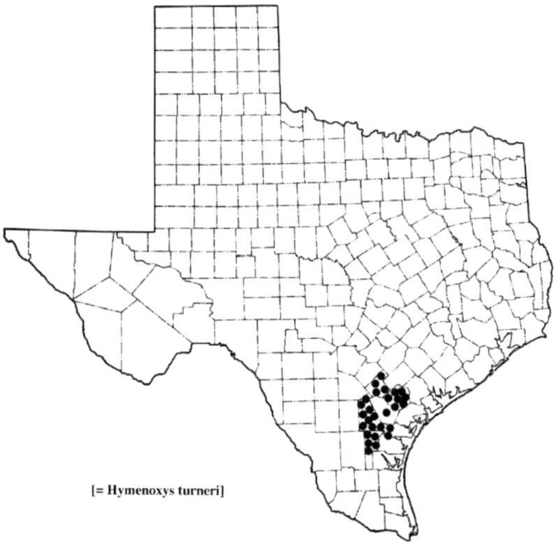
[= Hymenoxys turneri]

Tetraneuris turneri

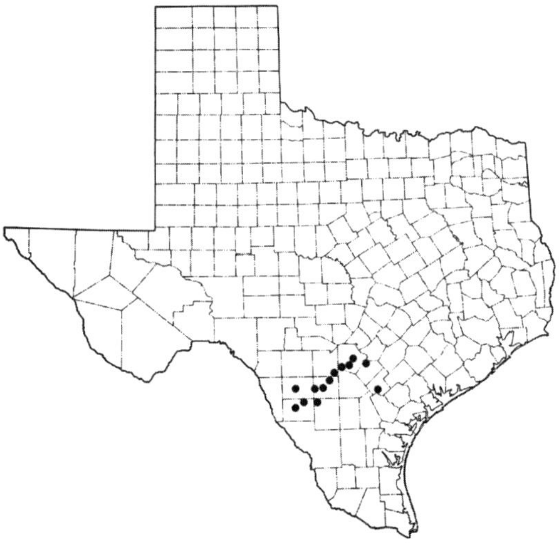

Thelesperma burridgeanum

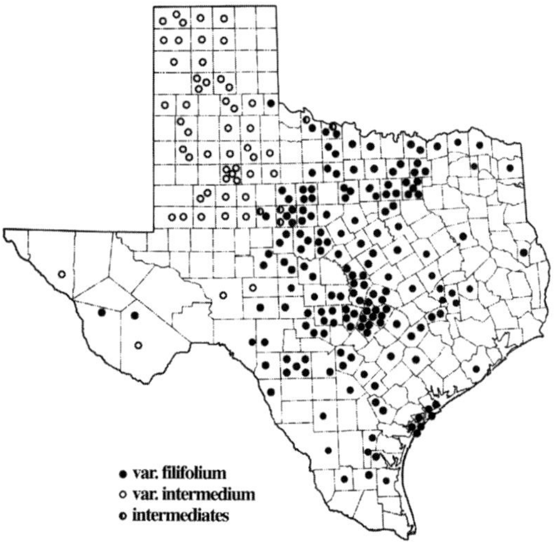
● var. filifolium
○ var. intermedium
◉ intermediates

Thelesperma filifolium

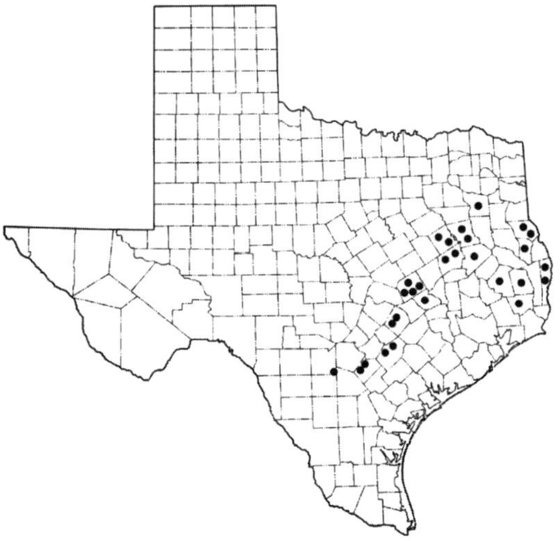

Thelesperma flavodiscum

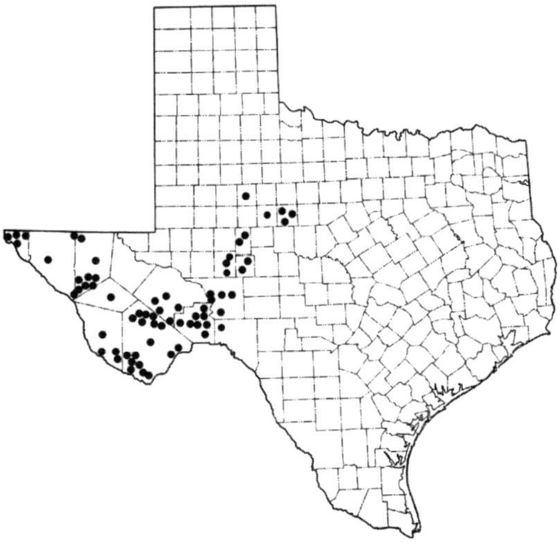

Thelesperma longipes

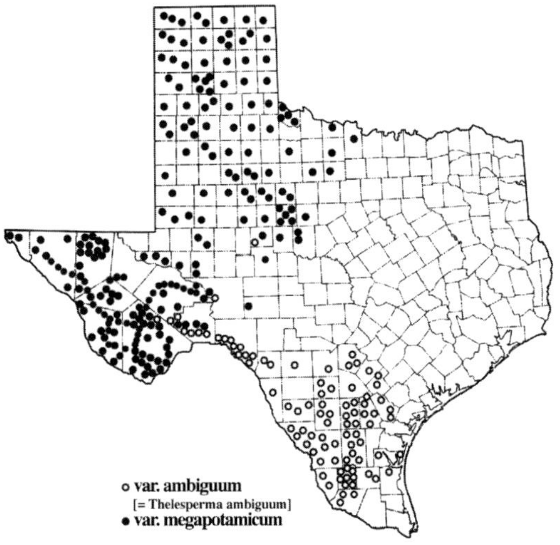
○ var. ambiguum
[= Thelesperma ambiguum]
● var. megapotamicum

Thelesperma megapotamicum

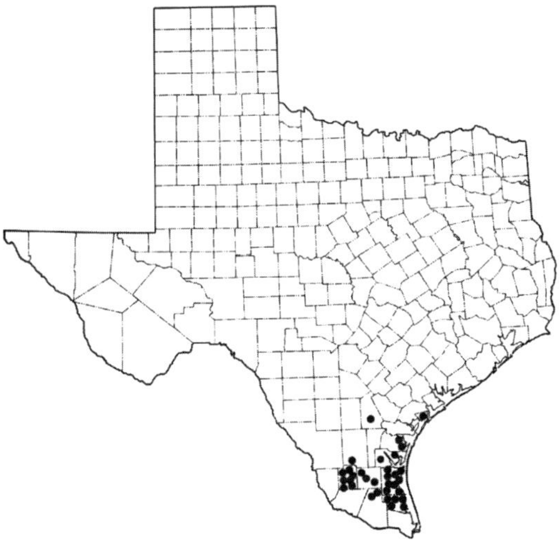

Thelesperma nuecense

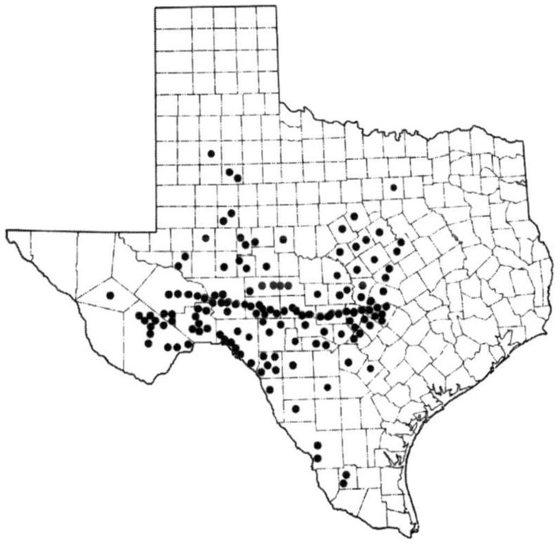

Thelesperma simplicifolium

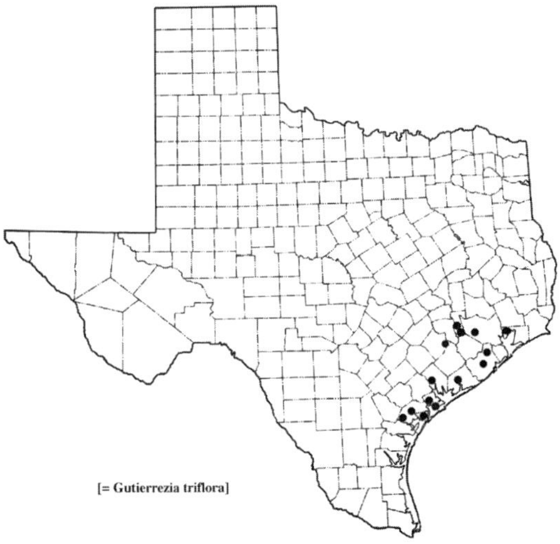
[= Gutierrezia triflora]

Thurovia triflora

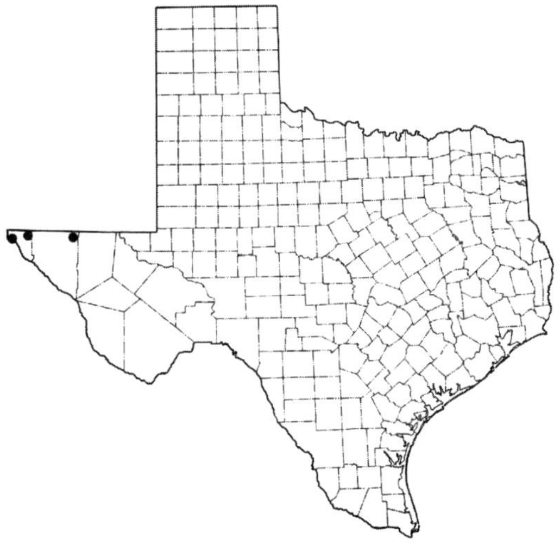

Townsendia annua

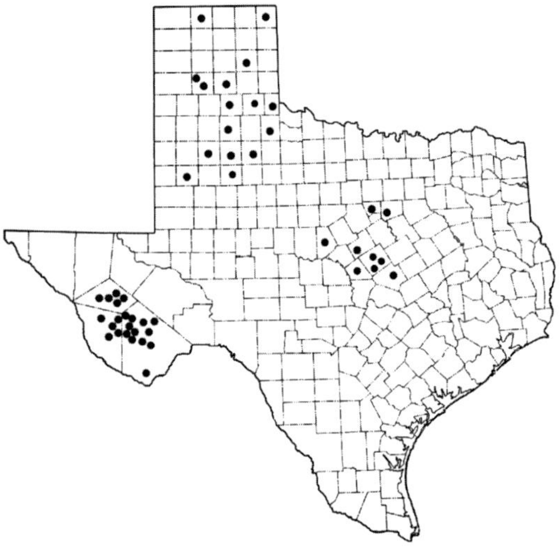

Townsendia exscapa

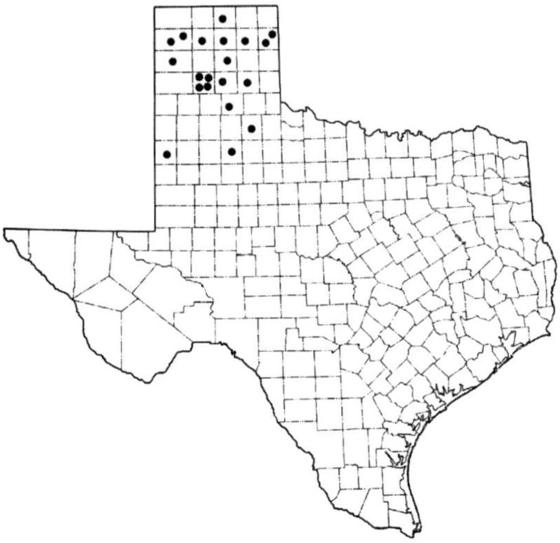

Townsendia texensis

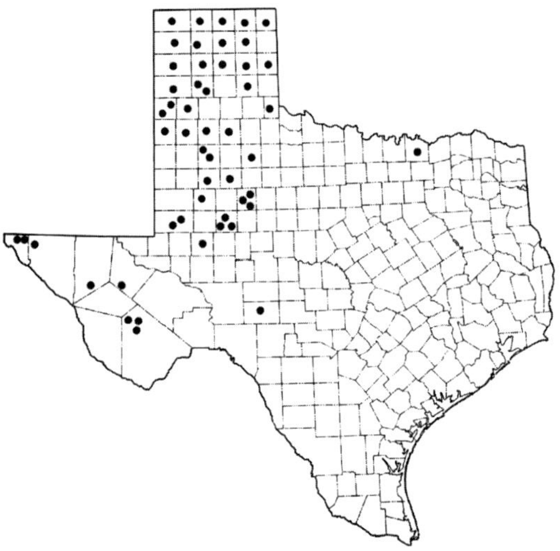

Tragopogon dubius

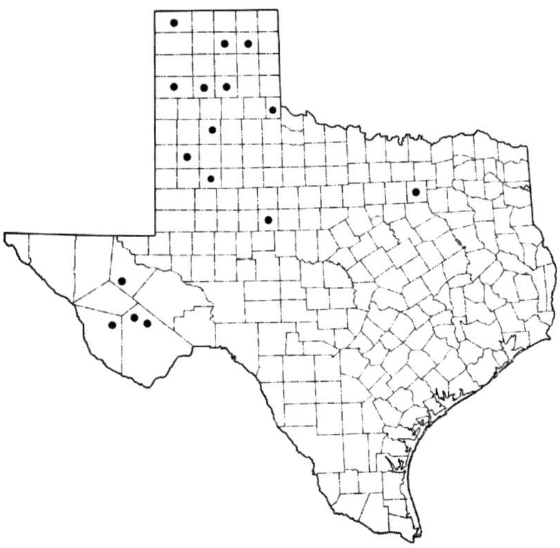

Tragopogon porrifolius

Trichocoronis rivularis

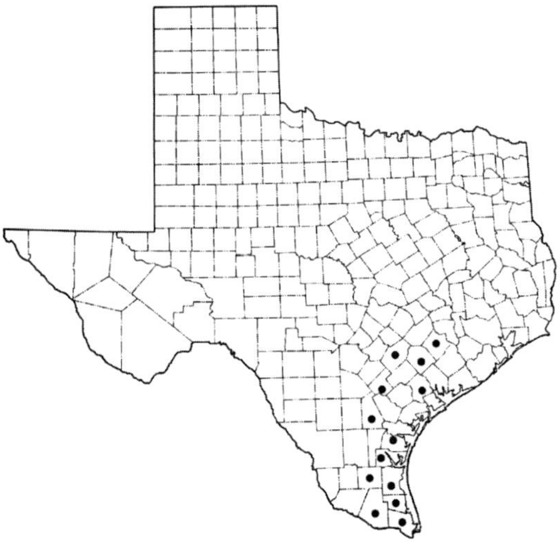

Trichocoronis wrightii

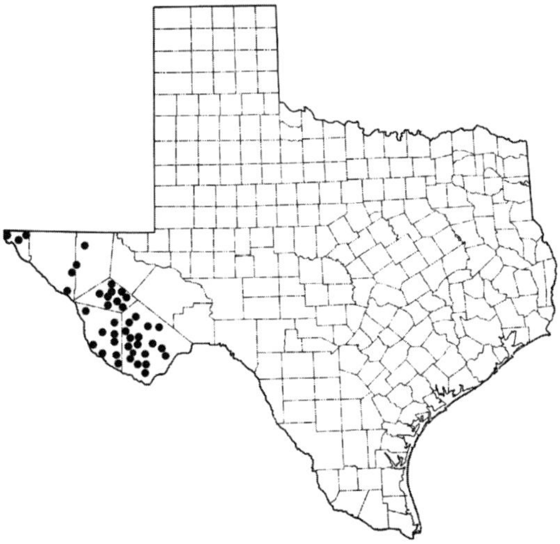
Trixis californica

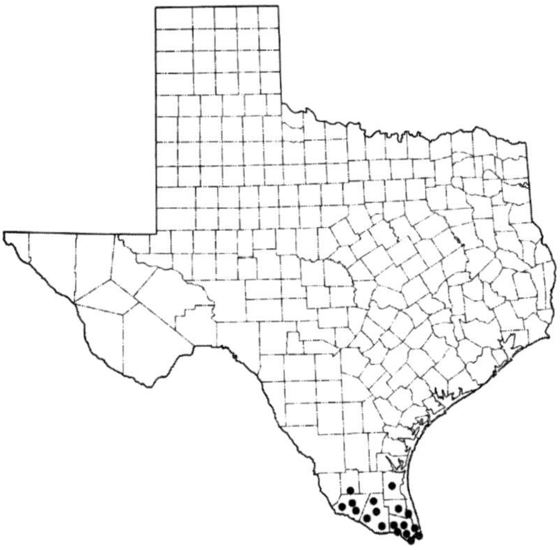
Trixis inula

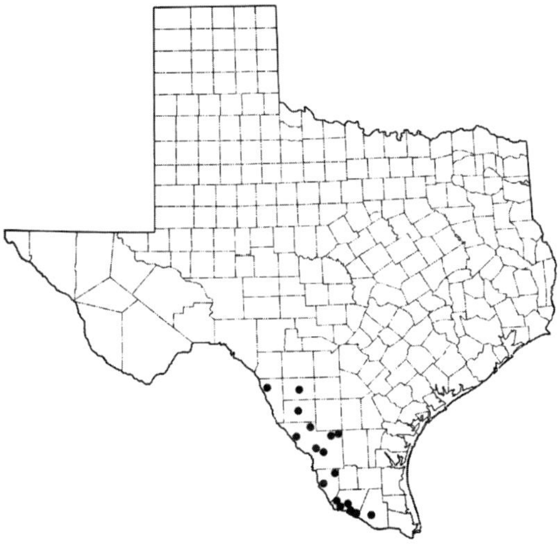
Varilla texana

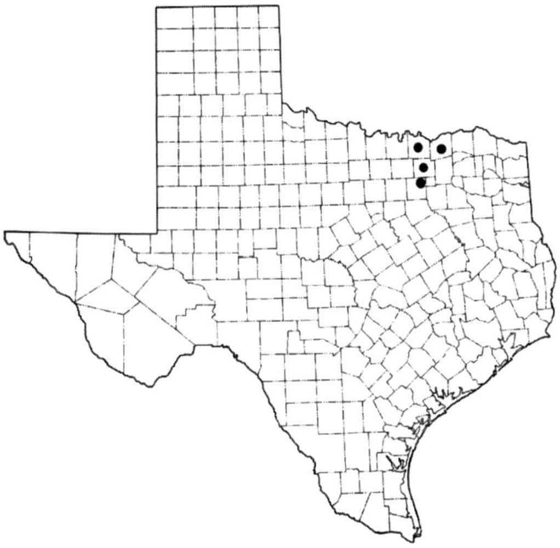
Verbesina alternifolia

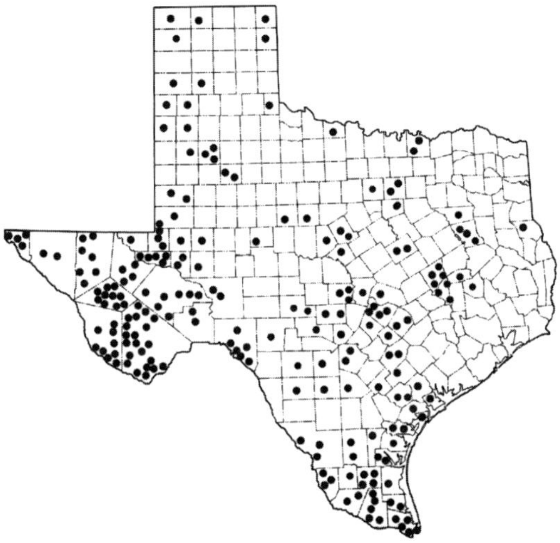
Verbesina encelioides

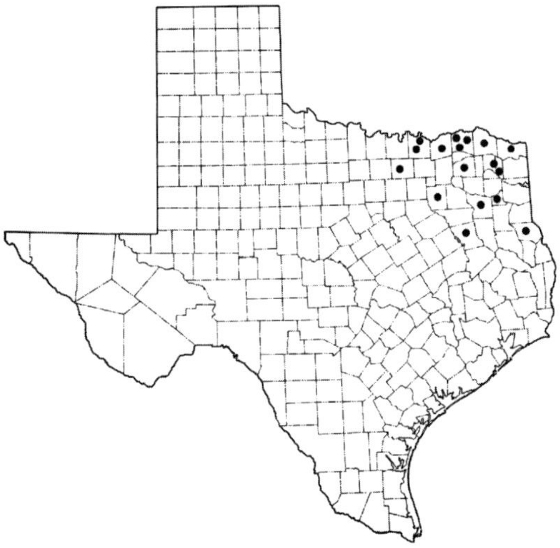
Verbesina helianthoides

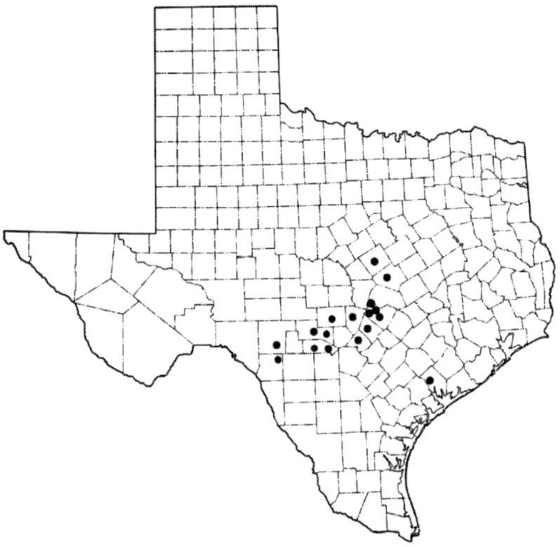
Verbesina lindheimeri

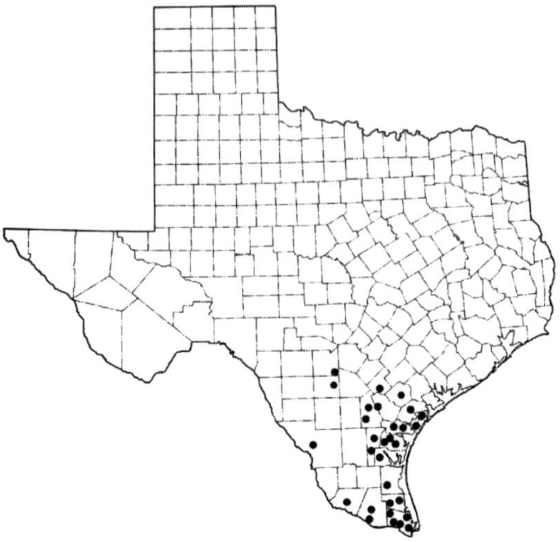
Verbesina microptera

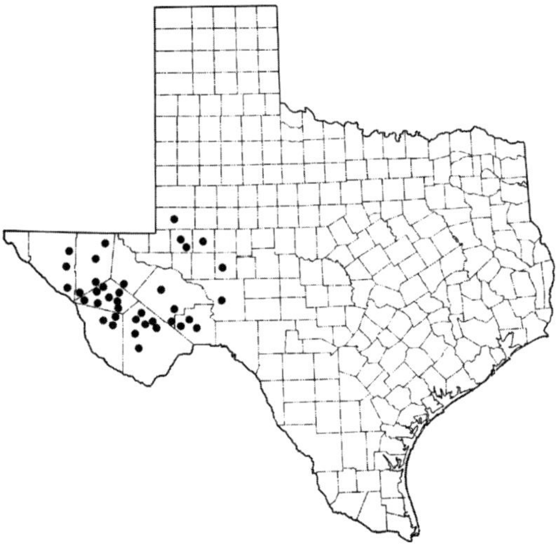
Verbesina nana

Verbesina oreophila

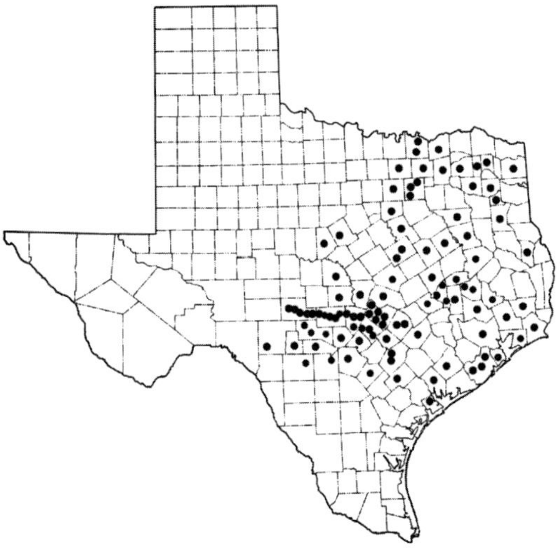
Verbesina virginica

Verbesina walteri

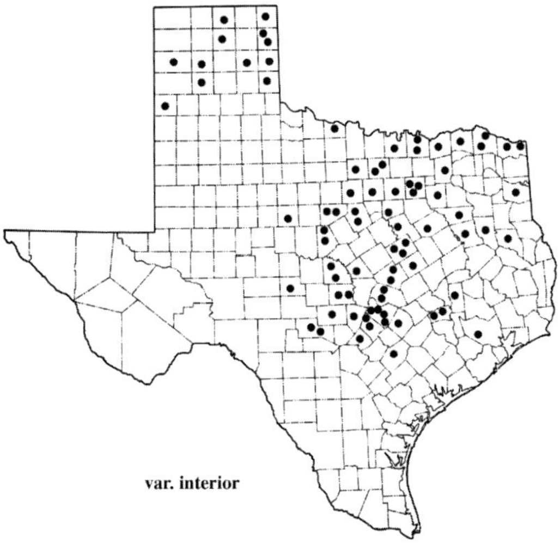
var. interior

Vernonia baldwinii

Vernonia gigantea

[This taxon is a hybrid between
V. baldwinii x V. lindheimeri]

Vernonia guadalupensis

Vernonia larsenii

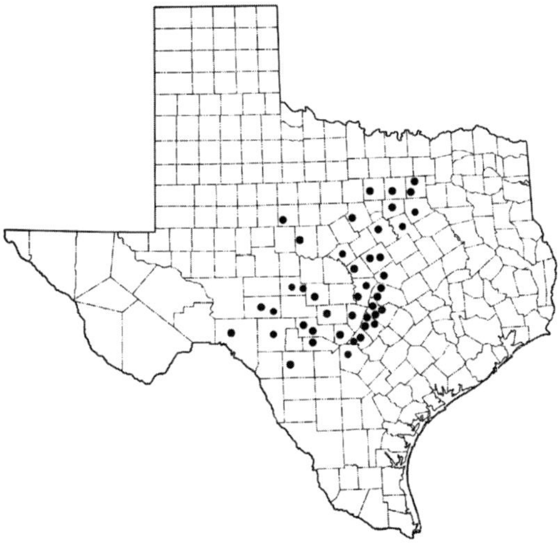

Vernonia lindheimeri

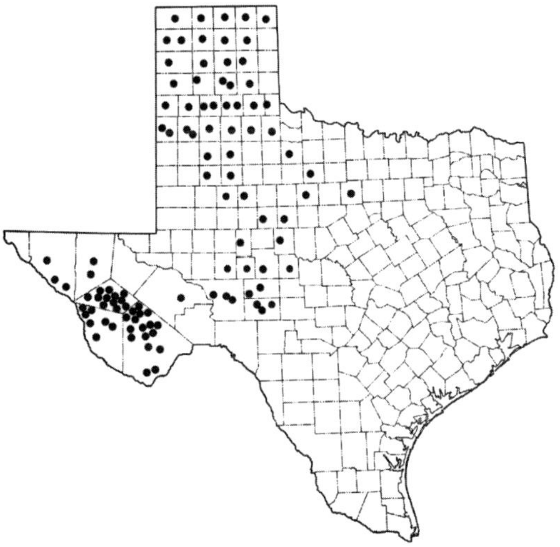

Vernonia marginata

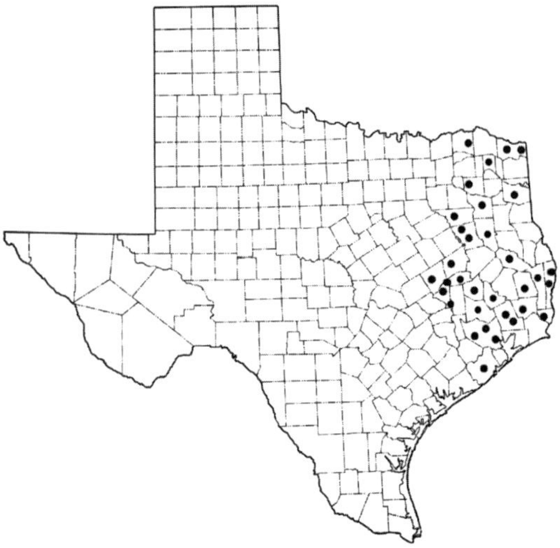

Vernonia missurica

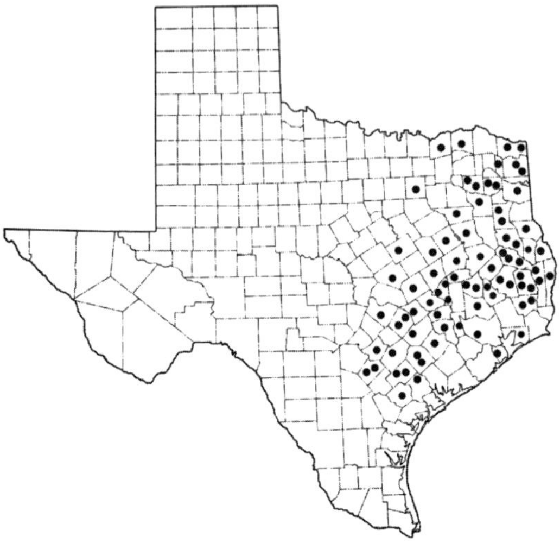

Vernonia texana

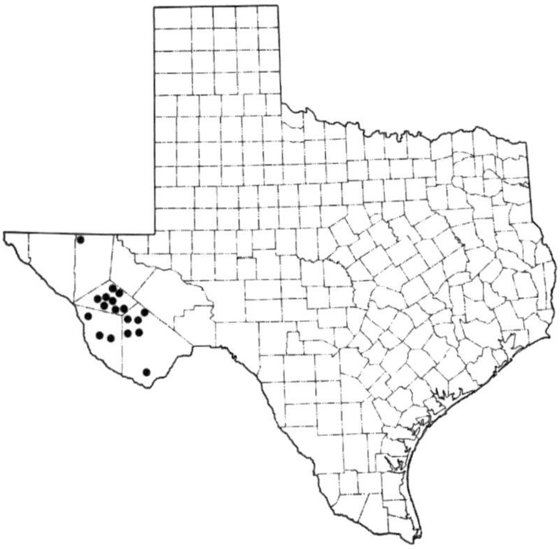

Viguiera cordifolia

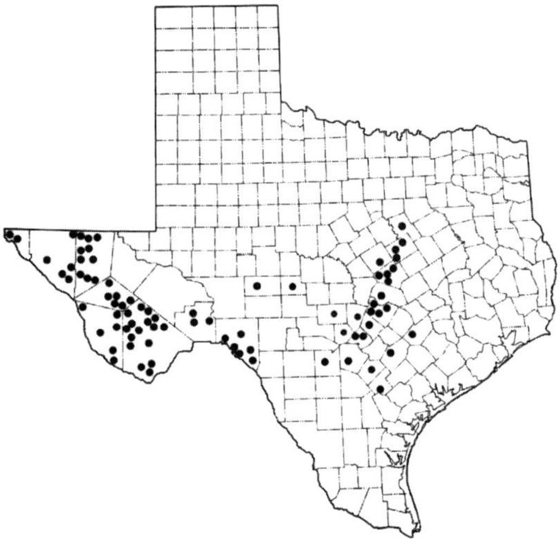

Viguiera dentata

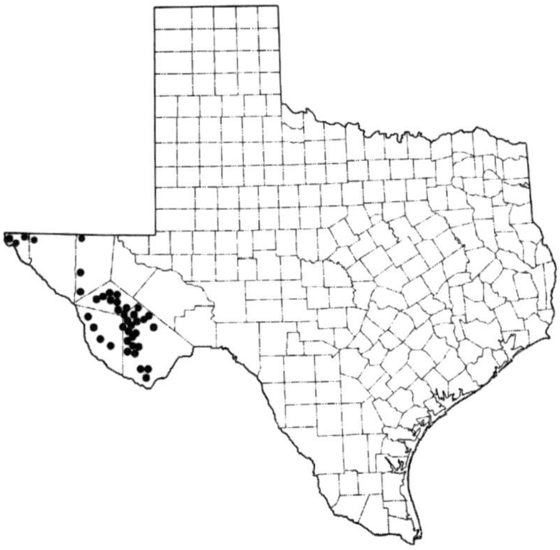

Viguiera longifolia

Viguiera multiflora

Viguiera phenax

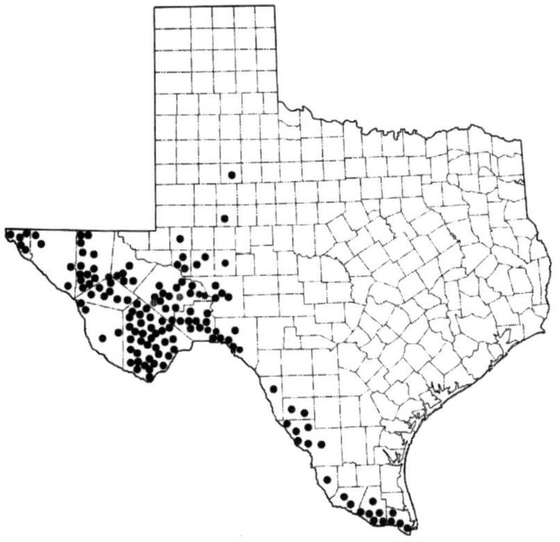

Viguiera stenoloba

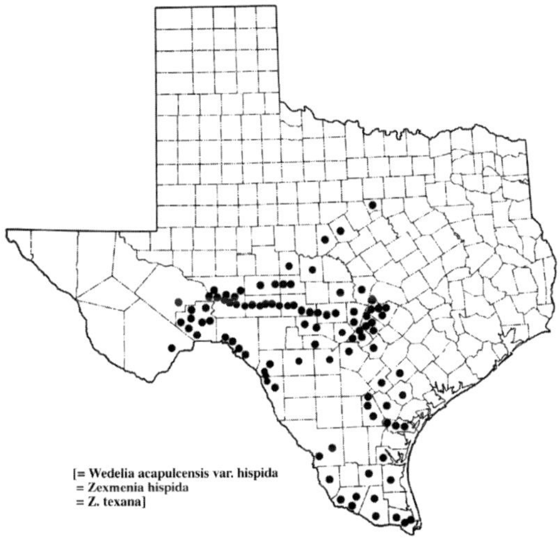
[= Wedelia acapulcensis var. hispida
= Zexmenia hispida
= Z. texana]

Wedelia hispida

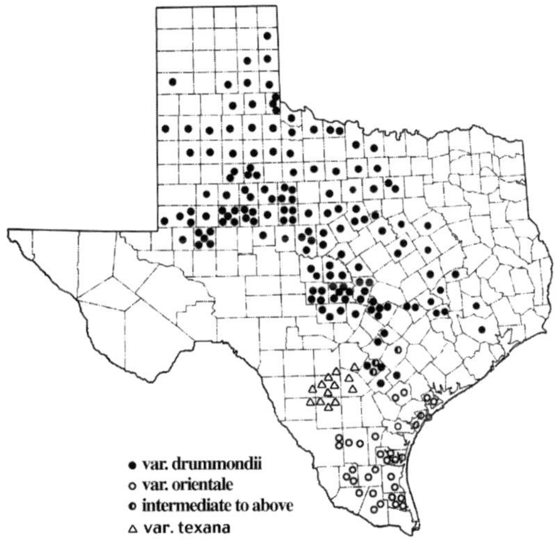
• var. drummondii
○ var. orientale
◉ intermediate to above
△ var. texana

Xanthisma texana

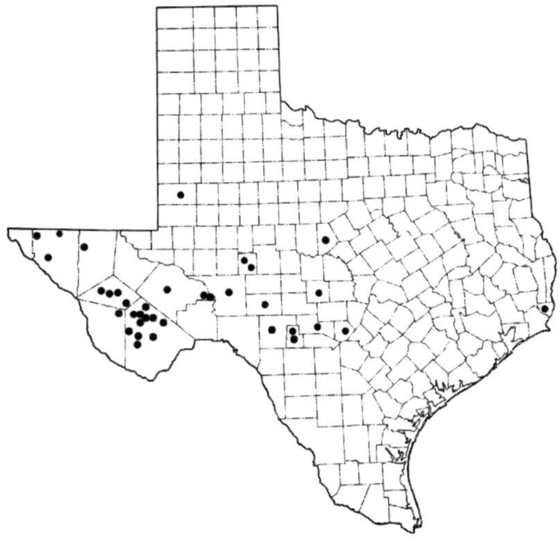

Xanthium spinosum

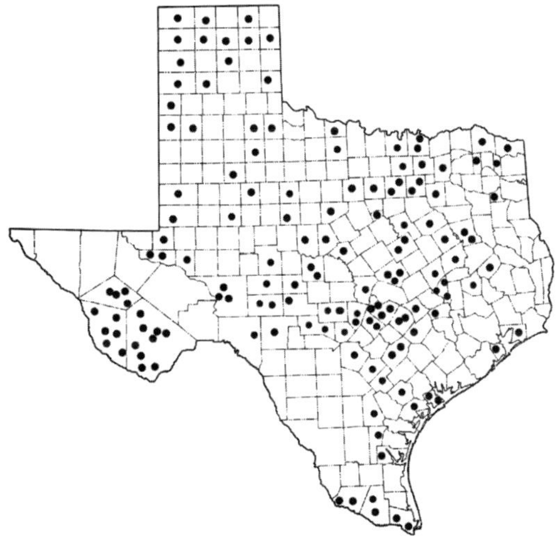

Xanthium strumarium

Xanthocephalum gymnospermoides

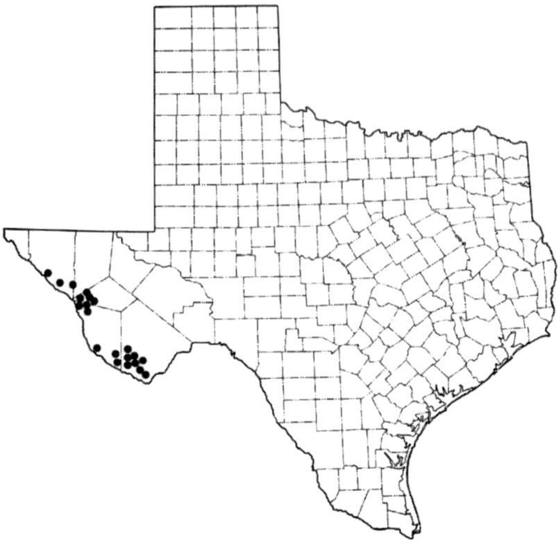

Xylorhiza wrightii

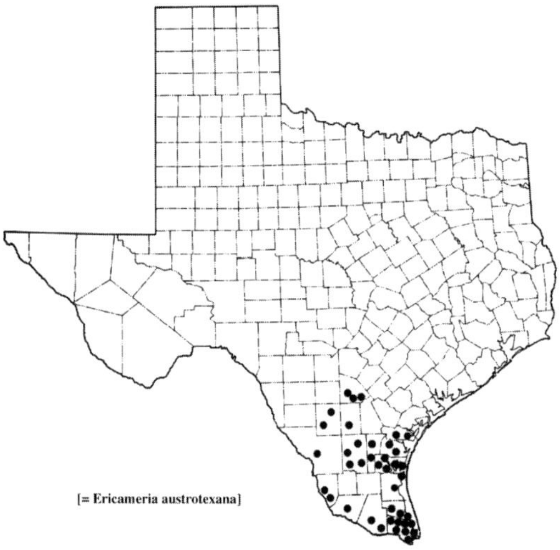
[= Ericameria austrotexana]

Xylothamia palmeri

[= Ericameria triantha]

Xylothamia triantha

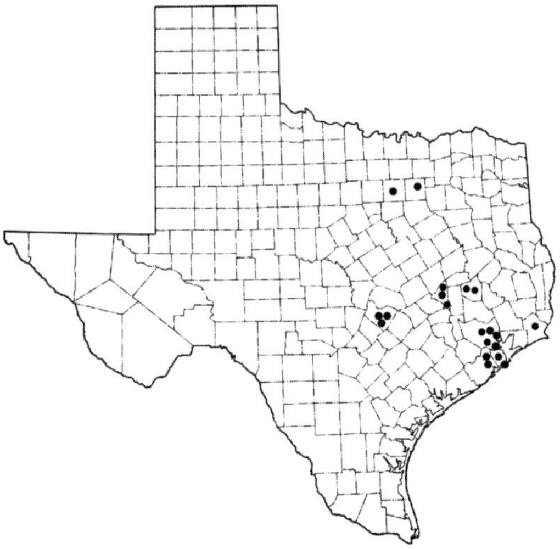

Youngia japonica

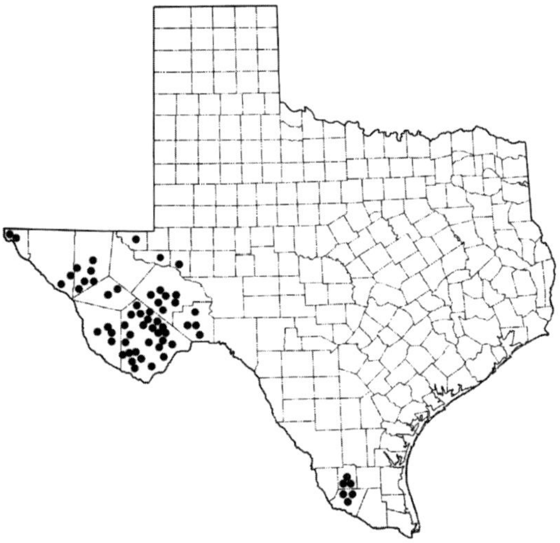

Zinnia acerosa

Zinnia anomala

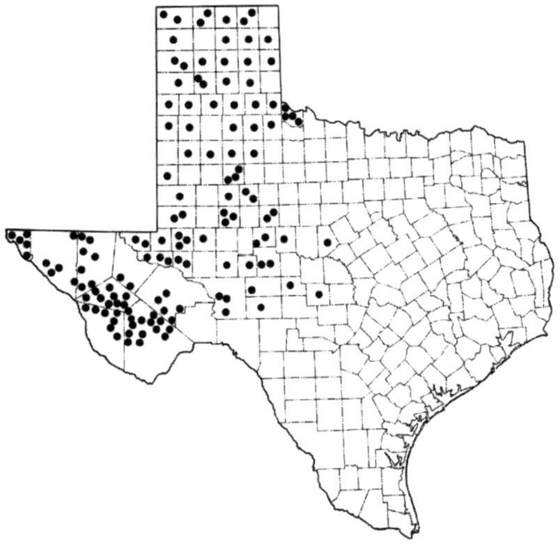

Zinnia grandiflora

Avicennia germinans

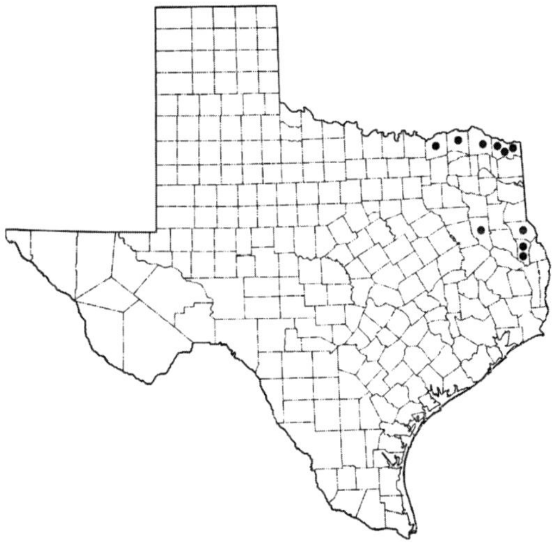

Impatiens capensis

Anredera cordifolia

[includes **A. leptostachys** and **A. scandens**]

Anredera vesicaria

Batis maritima

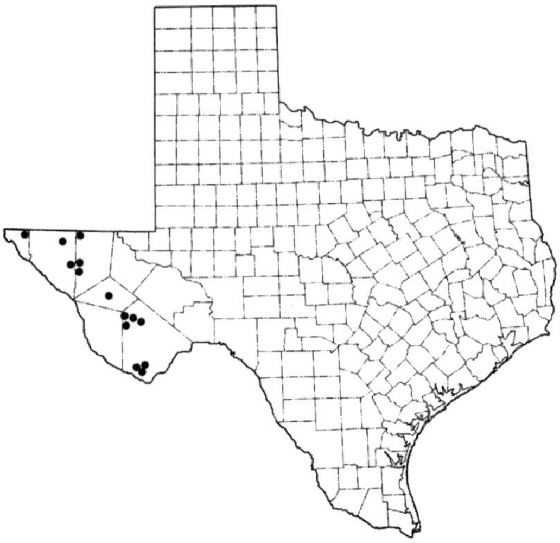

Berberis haematocarpa

Berberis repens

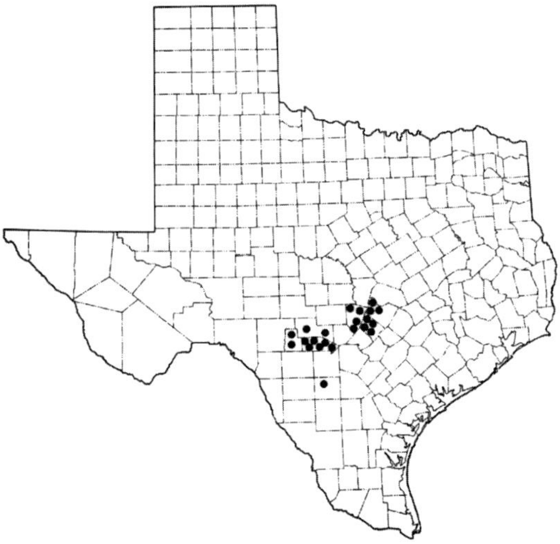

Berberis swaseyi

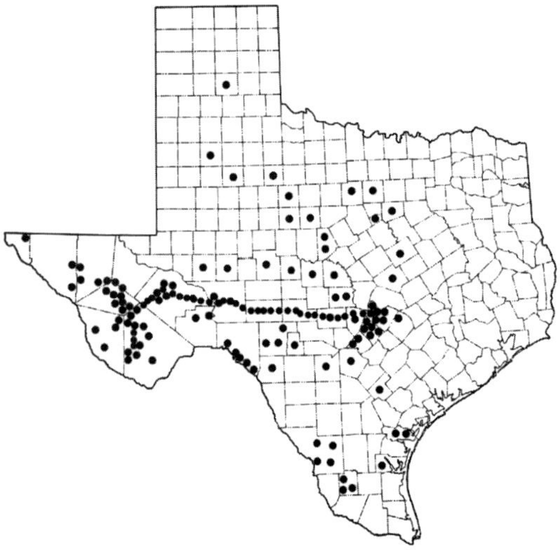

Berberis trifoliata

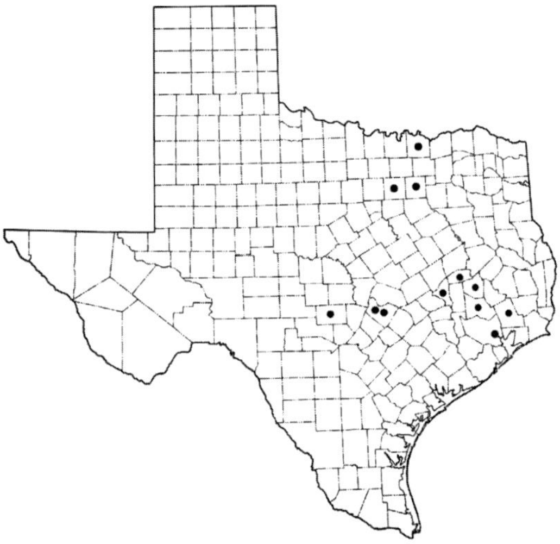

Nandina domestica

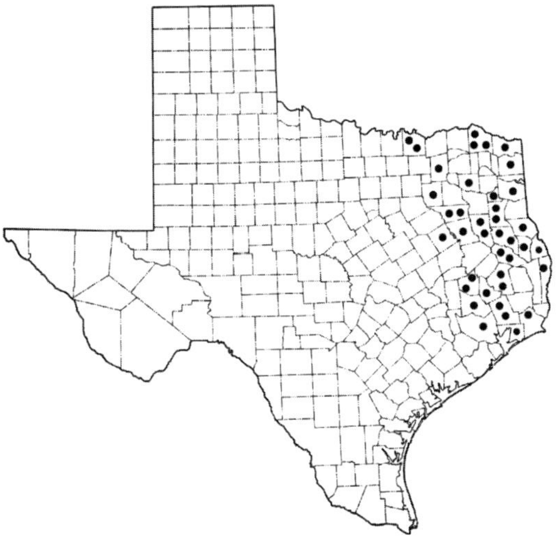

Podophyllum peltatum

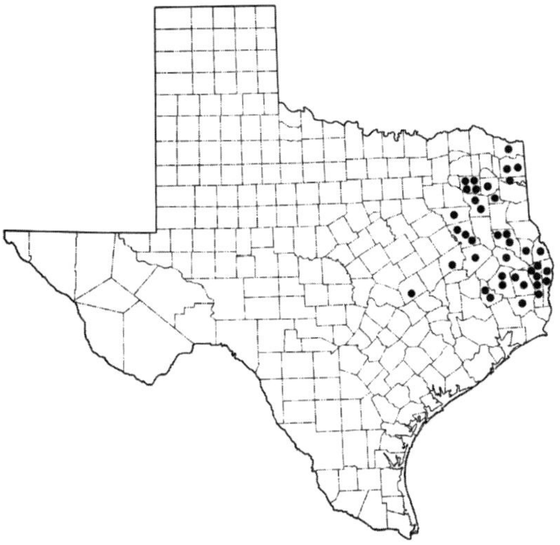

Alnus serrulata

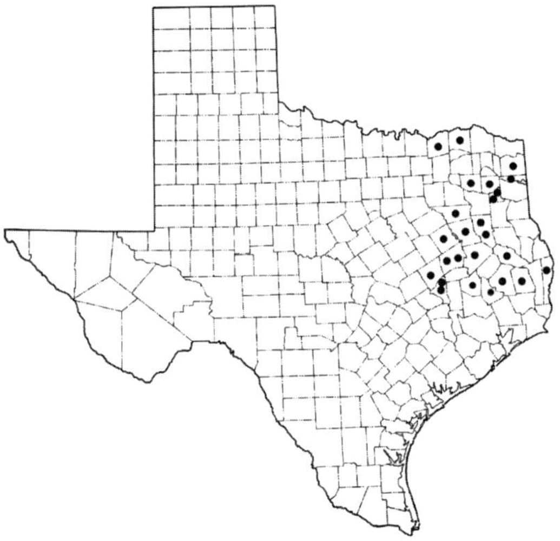

Betula nigra

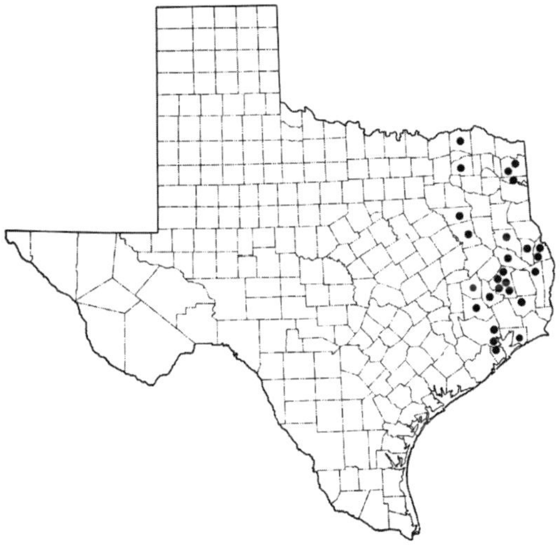

Carpinus caroliniana

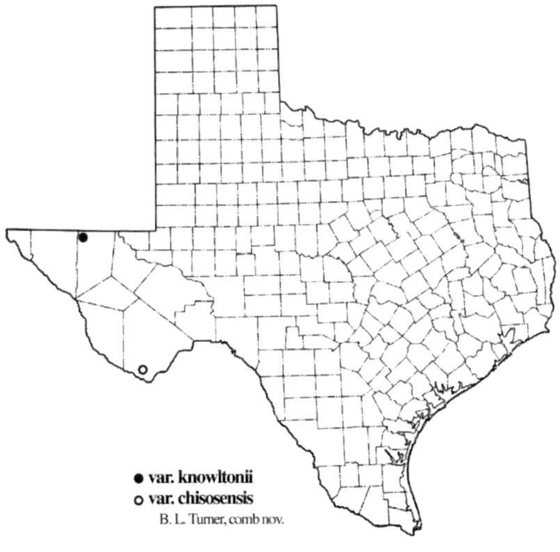
● var. knowltonii
○ var. chisosensis
B. L. Turner, comb nov.

Ostrya knowltonii

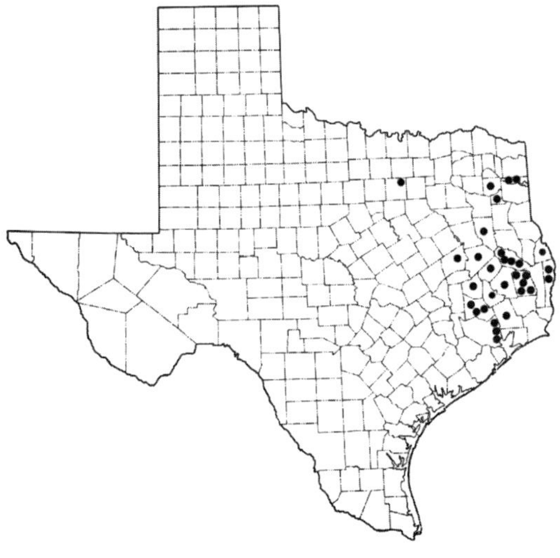

Ostrya virginiana

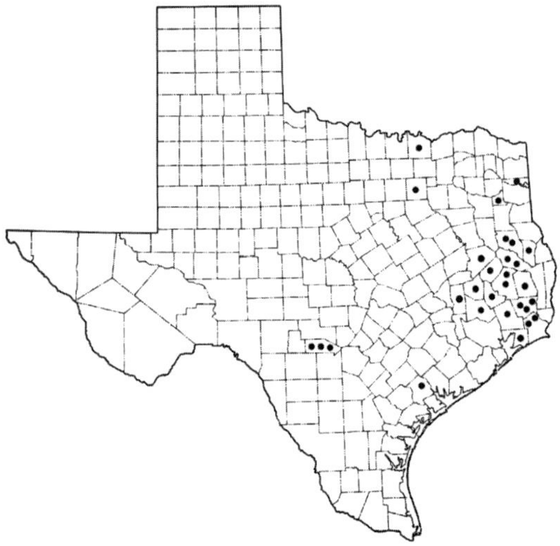

Bignonia capreolata

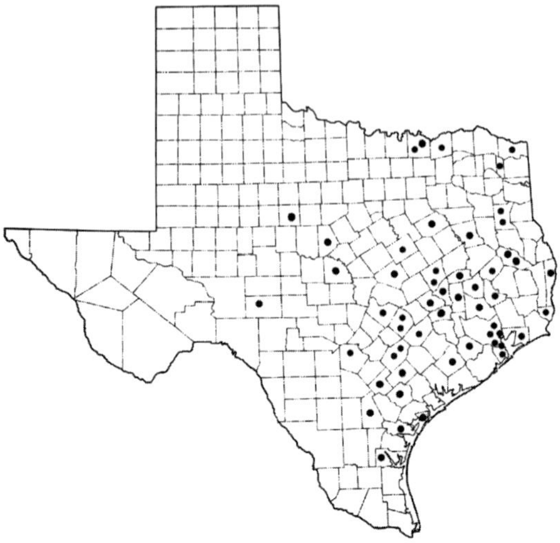

Campsis radicans

Catalpa bignonioides

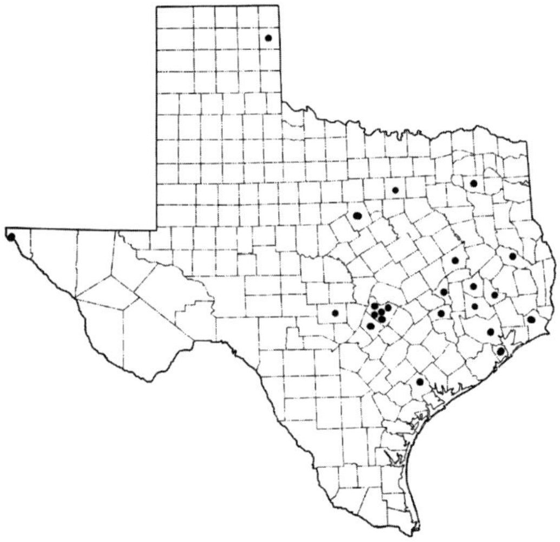

Catalpa speciosa

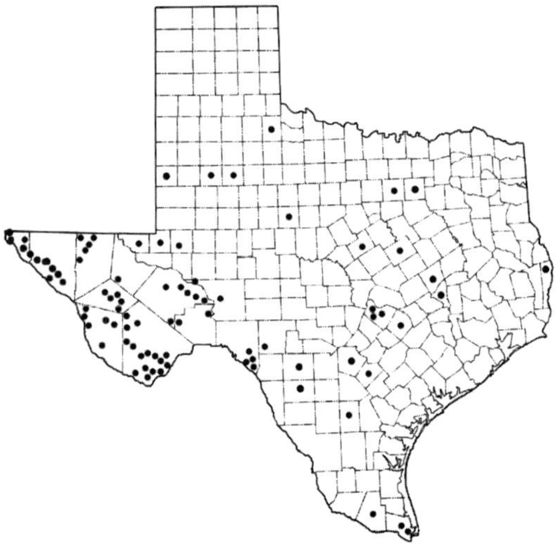

Chilopsis linearis

first collected 1989

Macfadyena unguis-cati

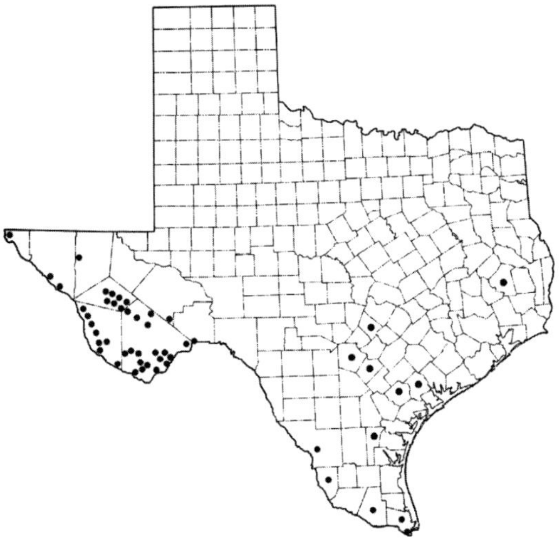

Tecoma stans

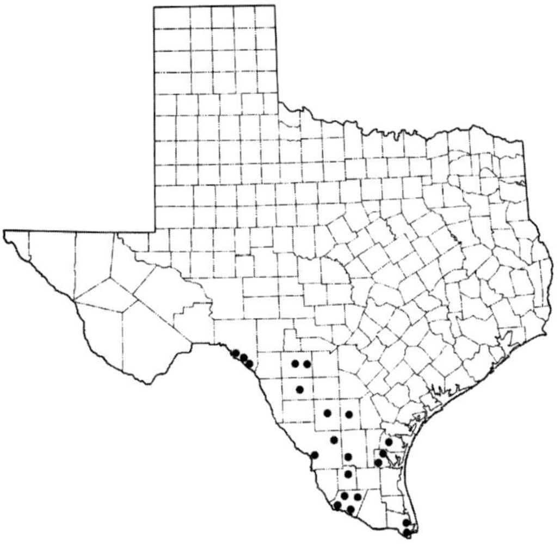

Amoreuxia wrightii

Amsinckia intermedia

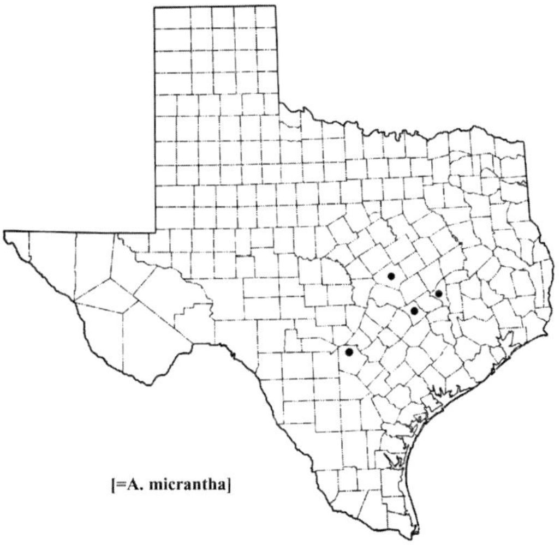
[=A. micrantha]

Amsinckia menziesii

Anchusa azurea

Antiphytum floribundum

Antiphytum heliotropoides

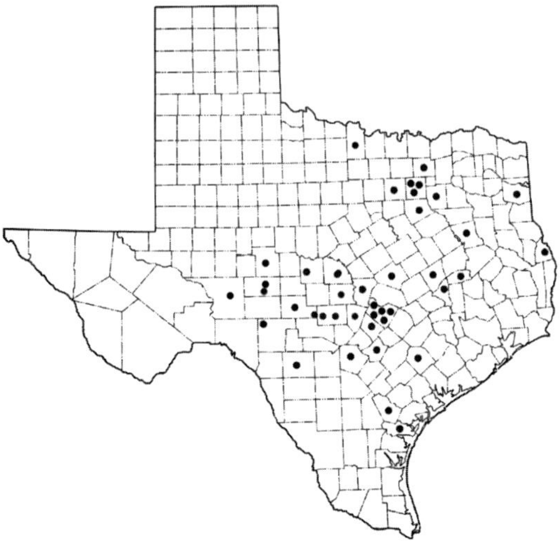

Buglossoides arvensis

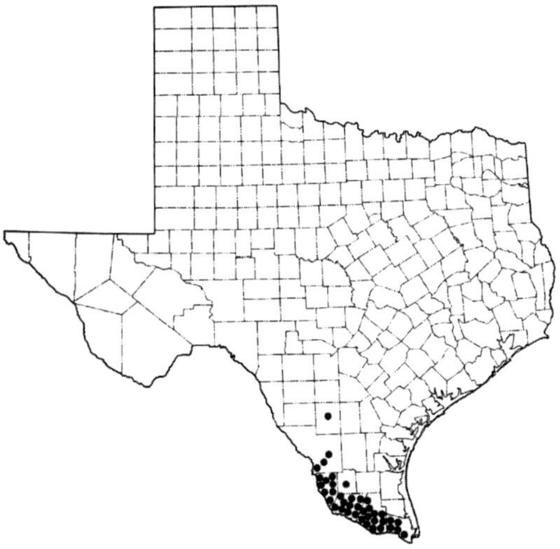

Cordia boissieri

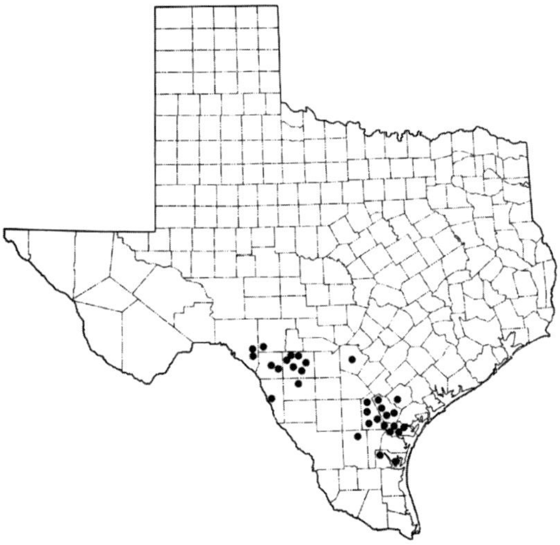

Cordia podocephala

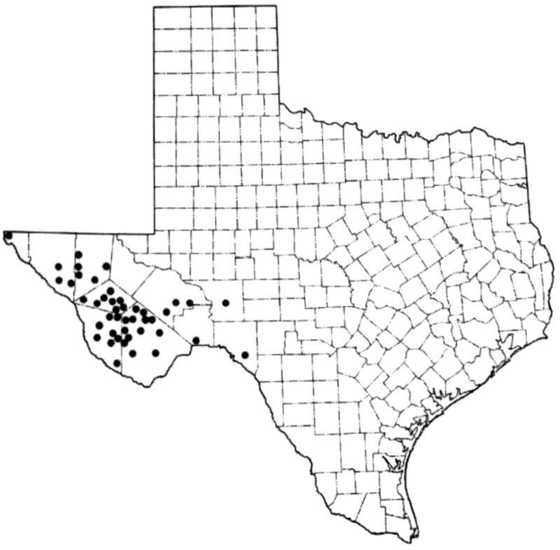

Cryptantha albida

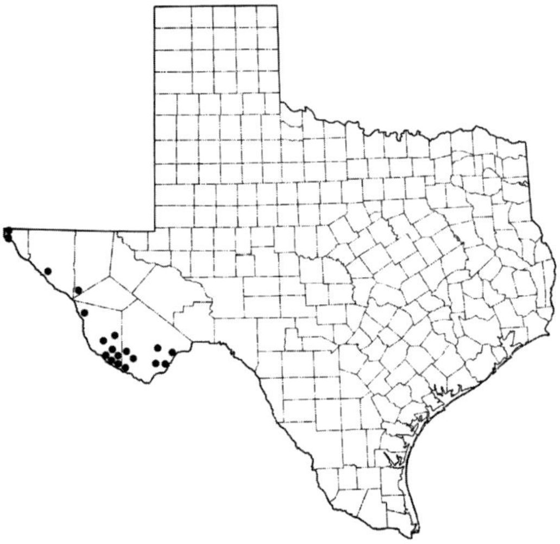

Cryptantha angustifolia

Cryptantha barbigera

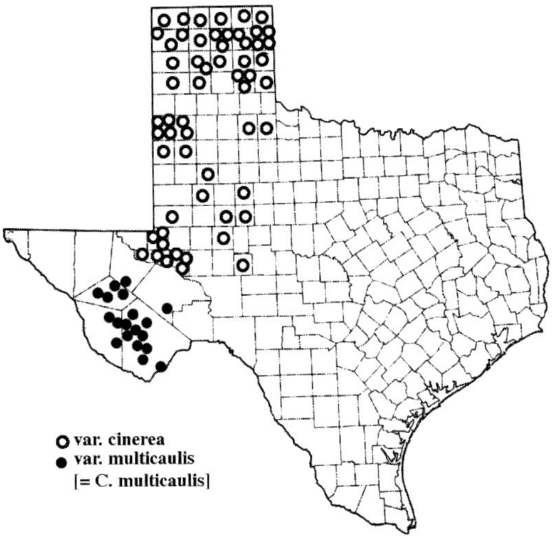
var. cinerea
var. multicaulis
[= C. multicaulis]

Cryptantha cinerea

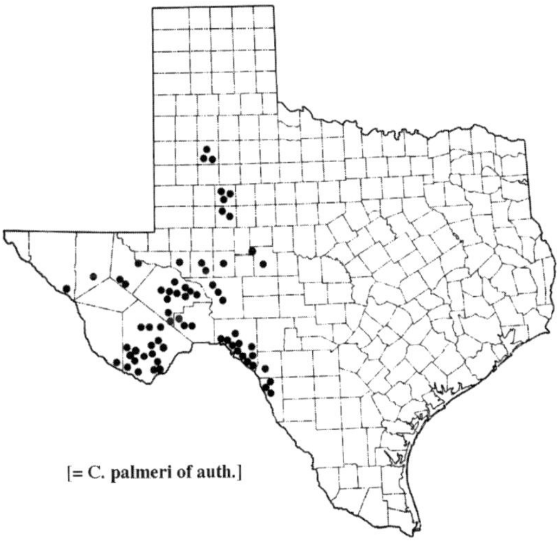
[= C. palmeri of auth.]

Cryptantha coryi

Cryptantha crassipes

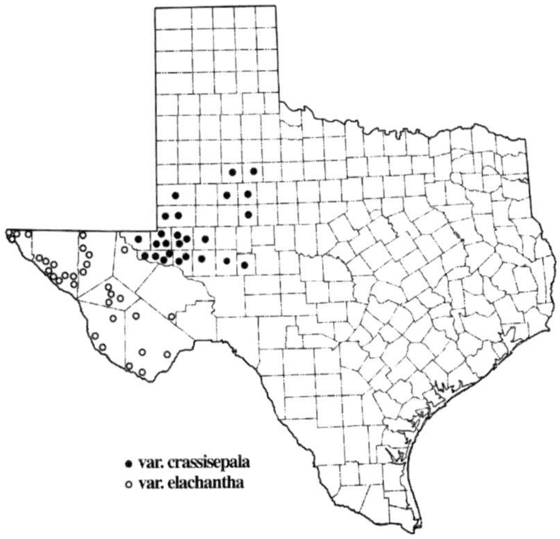
var. crassisepala
var. elachantha

Cryptantha crassisepala

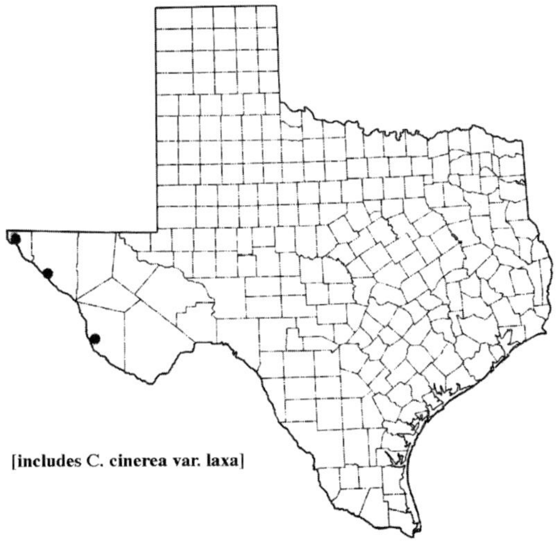
[includes C. cinerea var. laxa]

Cryptantha laxa

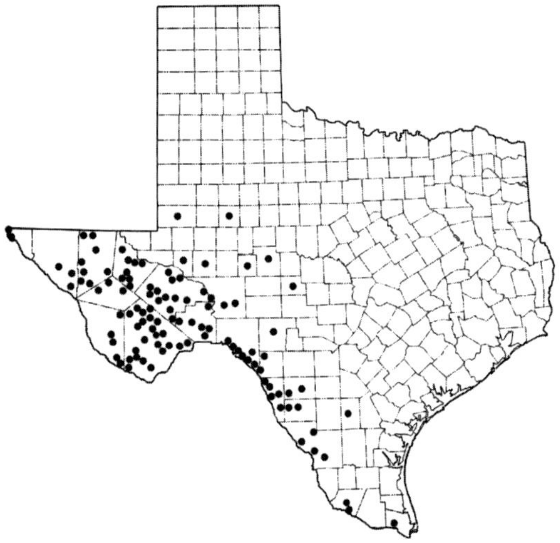

Cryptantha mexicana

Cryptantha micrantha

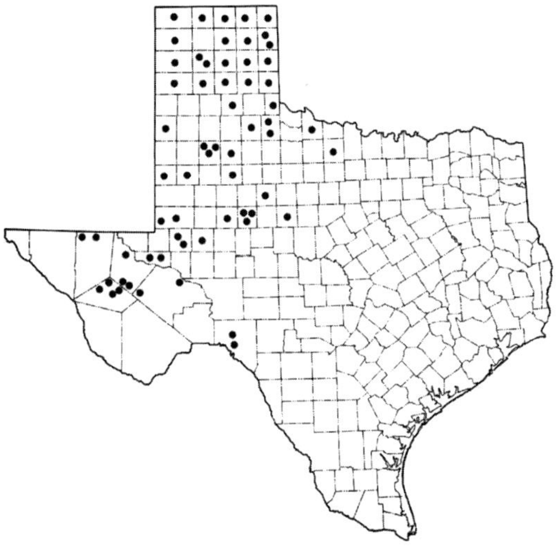

Cryptantha minima

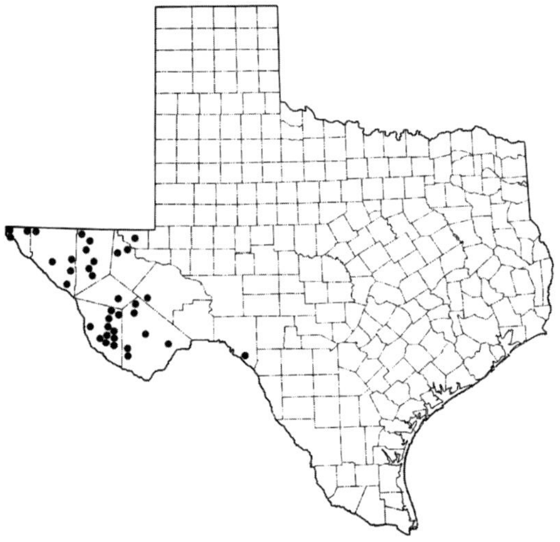

Cryptantha oblata

Cryptantha paysonii

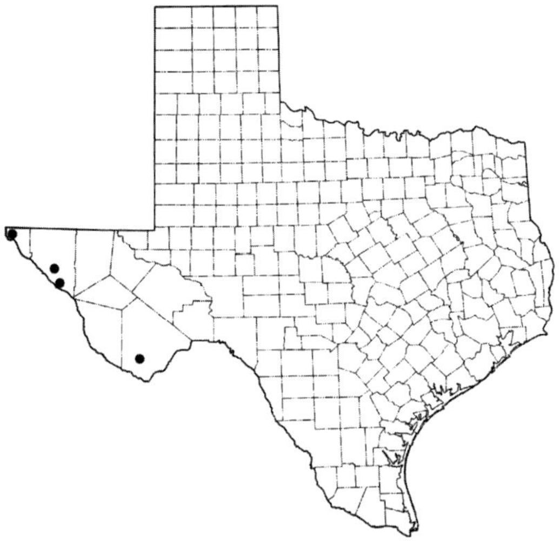

Cryptantha pterocarya

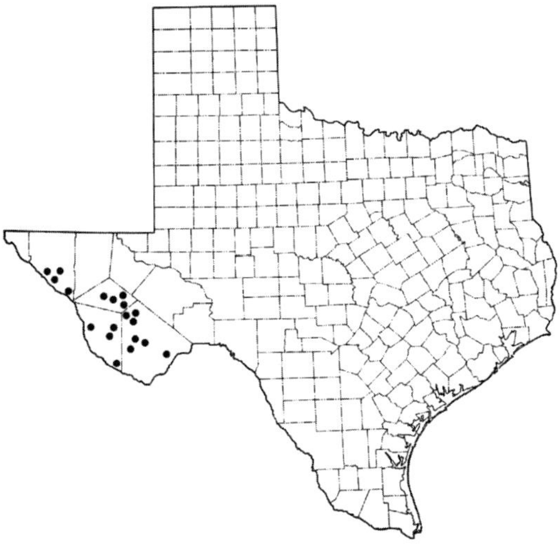

Cryptantha pusilla

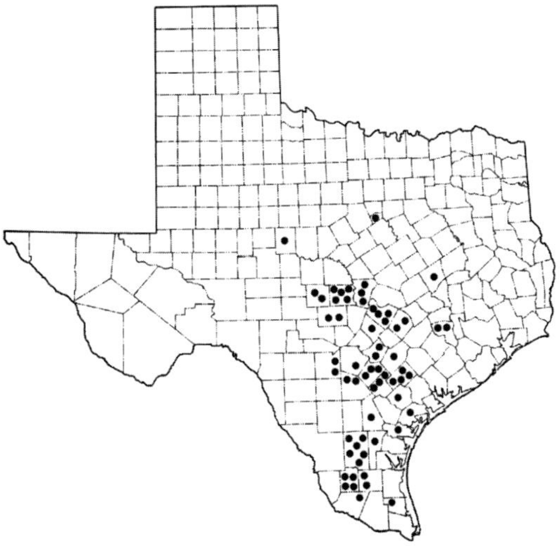

Cryptantha texana

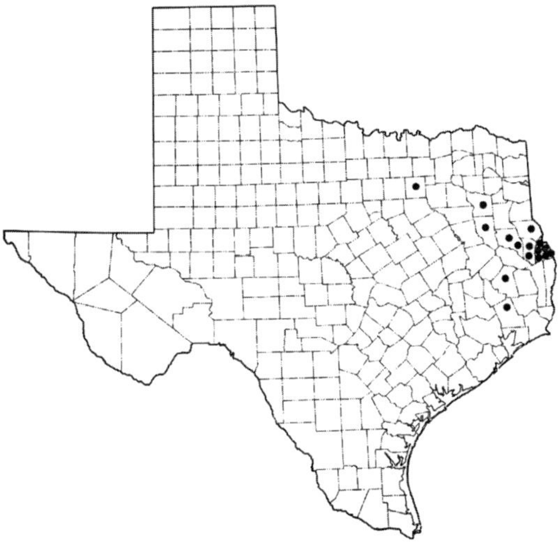

Cynoglossum virginianum

Cynoglossum zeylanicum

Echium vulgare

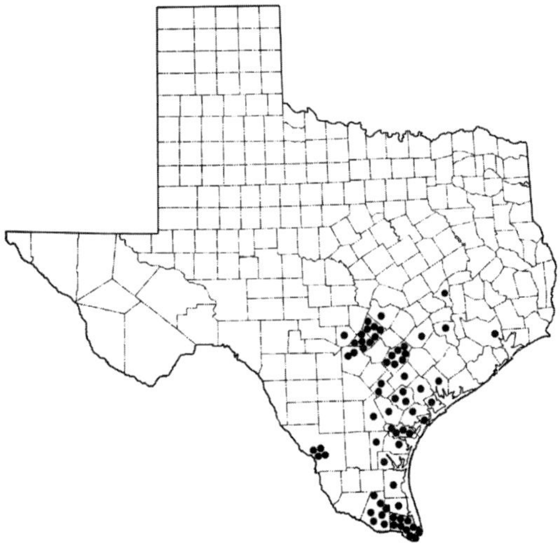

Ehretia anacua

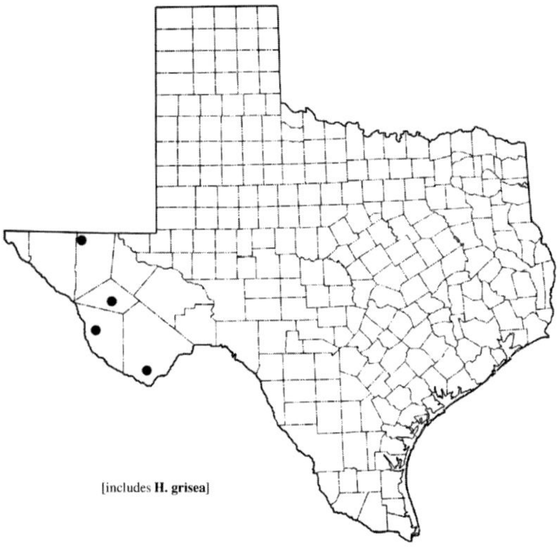
[includes **H. grisea**]

Hackelia besseyi

Hackelia floribunda

Hackelia pinetorum

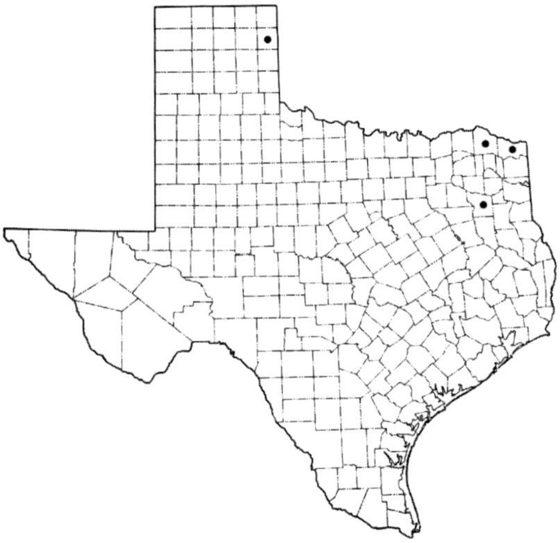

Hackelia virginiana

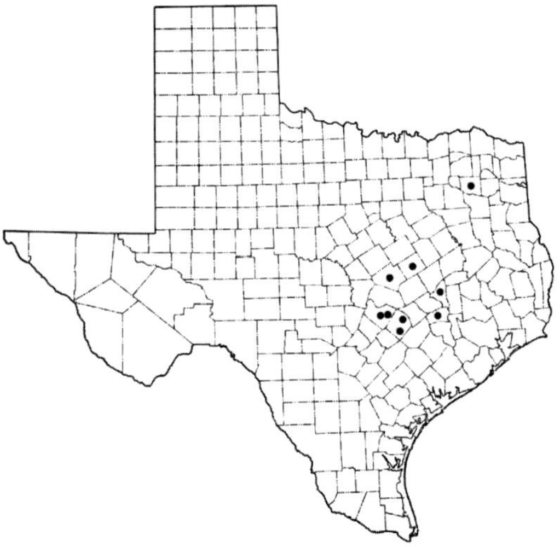

Heliotropium amplexicaule

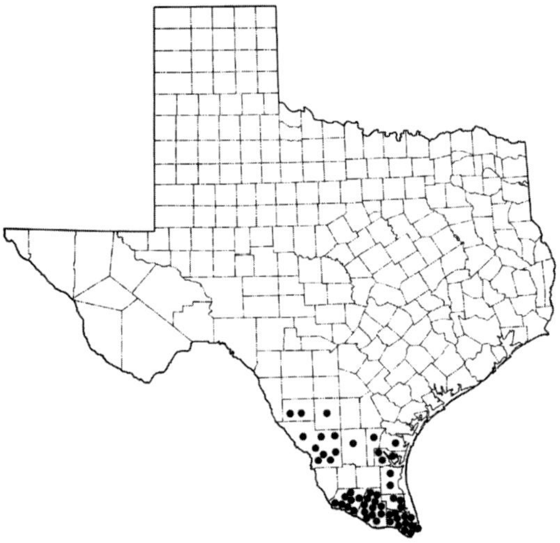

Heliotropium angiospermum

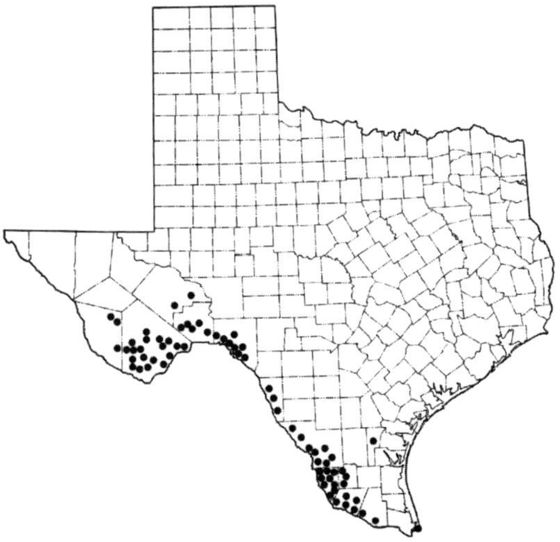

Heliotropium confertifolium

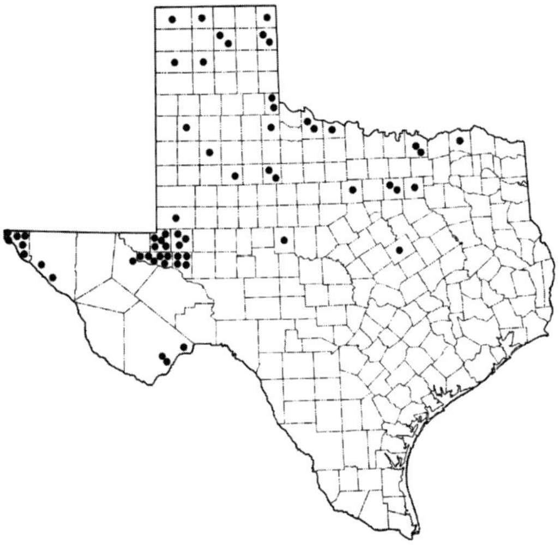

Heliotropium convolvulaceum

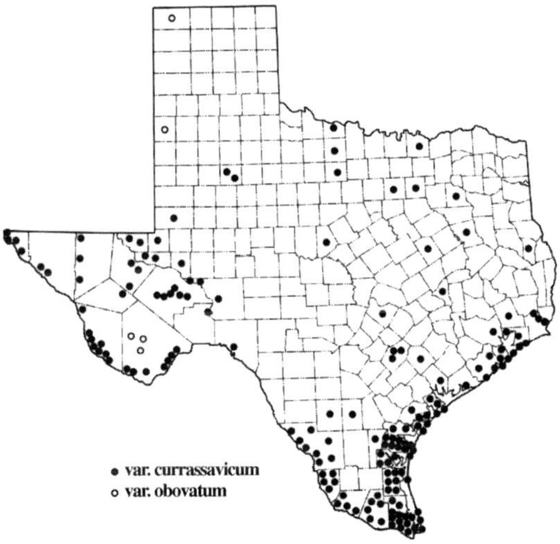
var. currassavicum
var. obovatum

Heliotropium currassavicum

Heliotropium europaeum

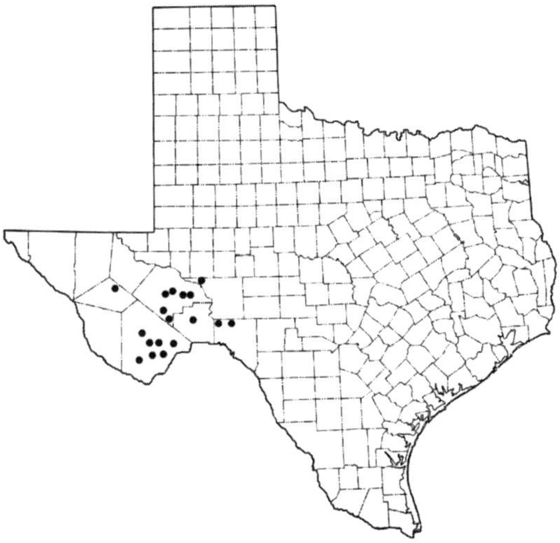

Heliotropium glabriusculum

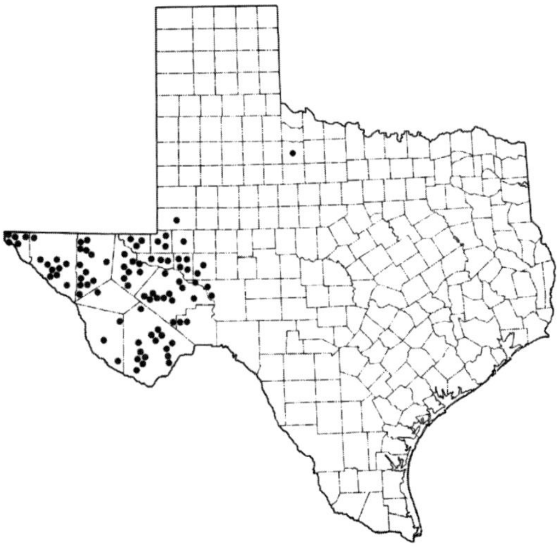

Heliotropium greggii

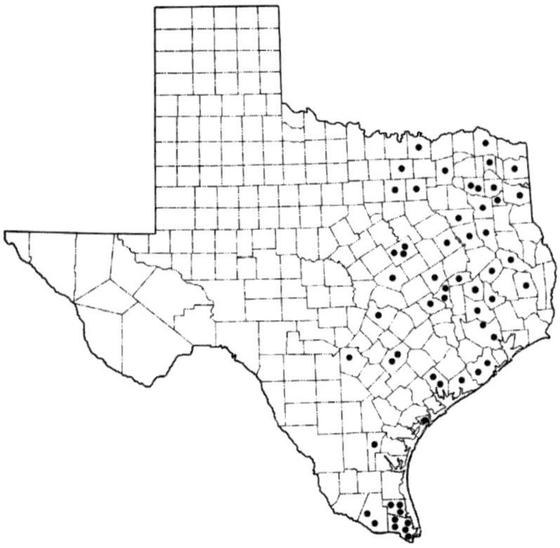

Heliotropium indicum

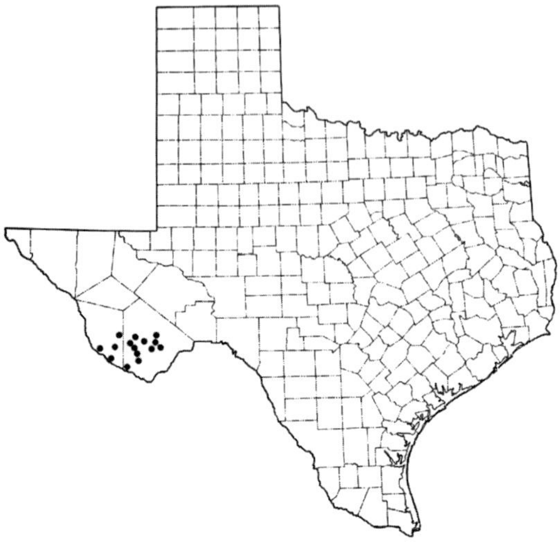

Heliotropium molle

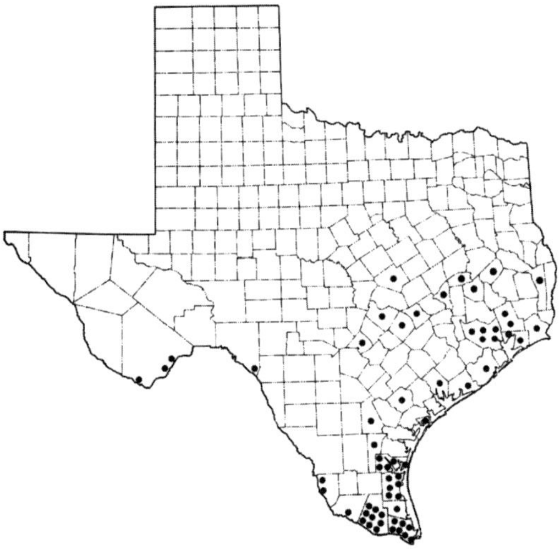

Heliotropium procumbens

Heliotropium racemosum

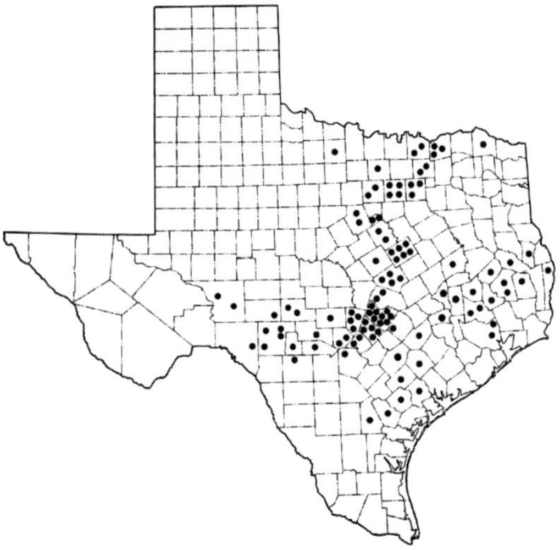

Heliotropium tenellum

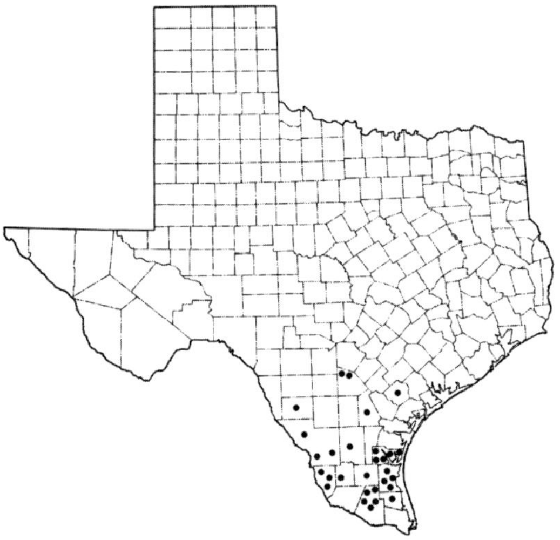

Heliotropium texanum

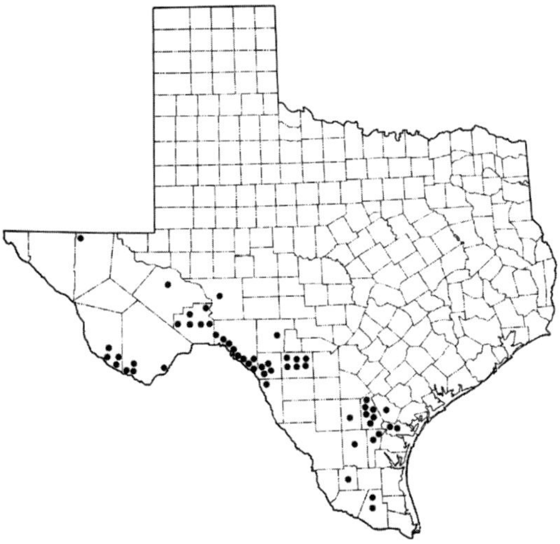

Heliotropium torreyi

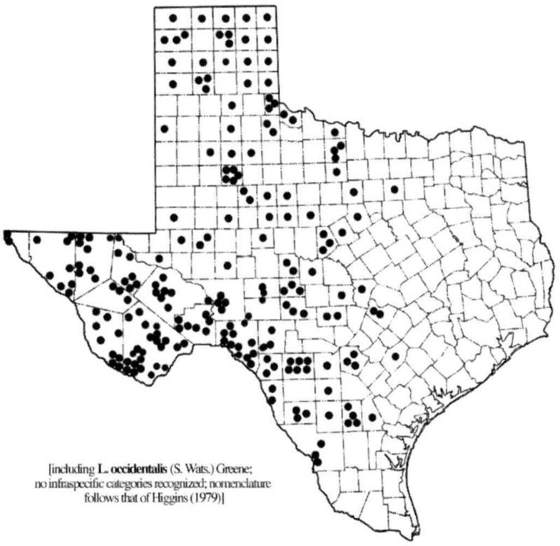
[including **L. occidentalis** (S. Wats.) Greene;
no infraspecific categories recognized; nomenclature
follows that of Higgins (1979)]

Lappula redowskii

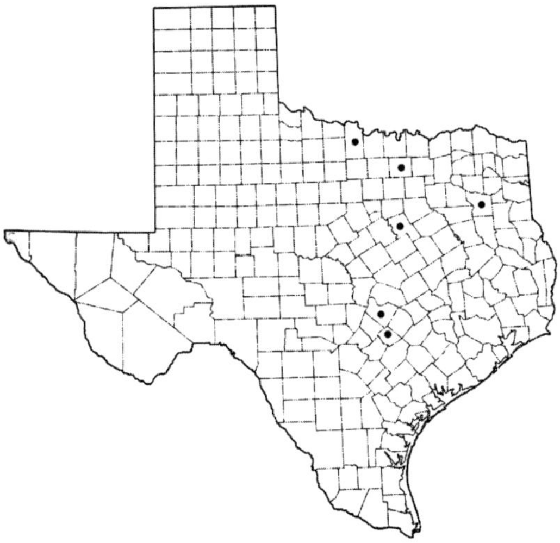

Lithospermum arvense

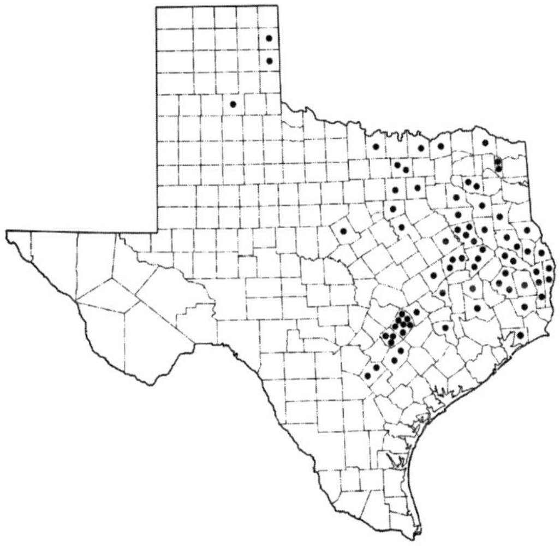

Lithospermum caroliniense

Lithospermum cobrense

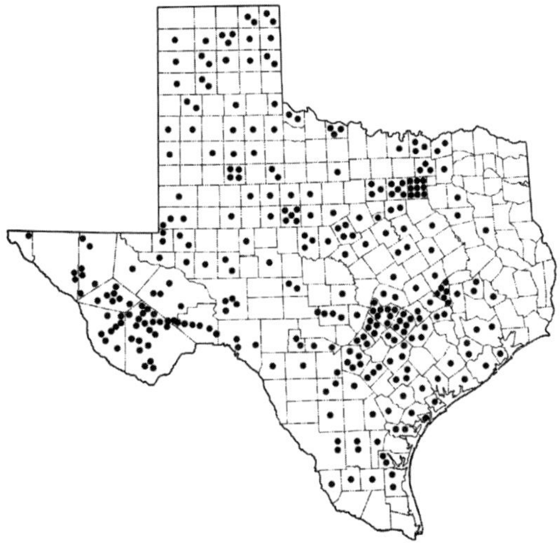

Lithospermum incisum

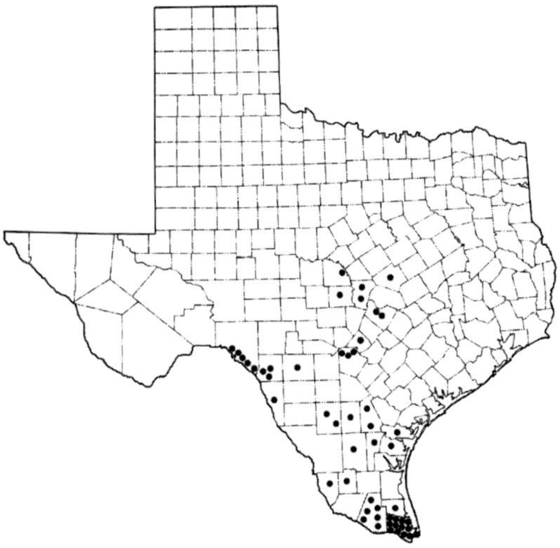

Lithospermum matamorense

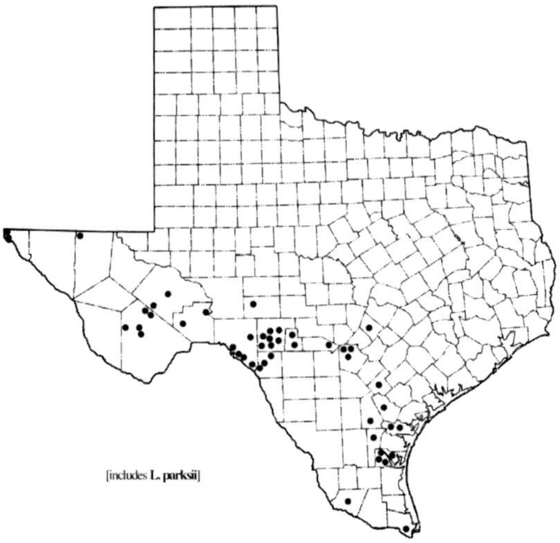
[includes **L. parksii**]

Lithospermum mirabile

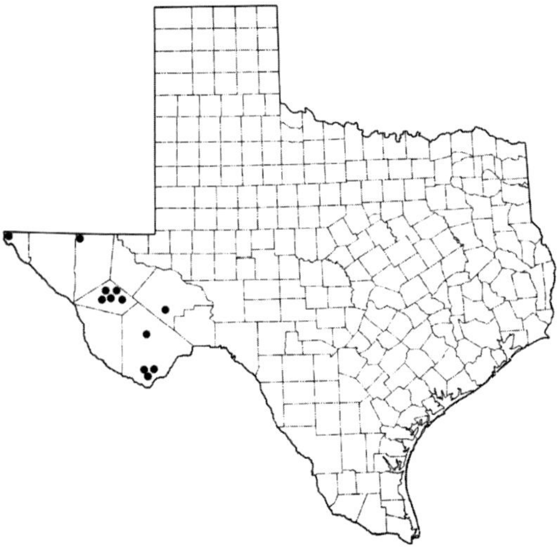

Lithospermum multiflorum

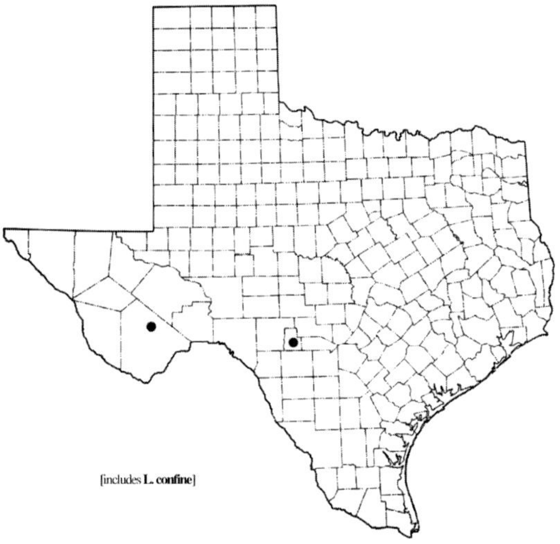
[includes **L. confine**]

Lithospermum obtusifolium

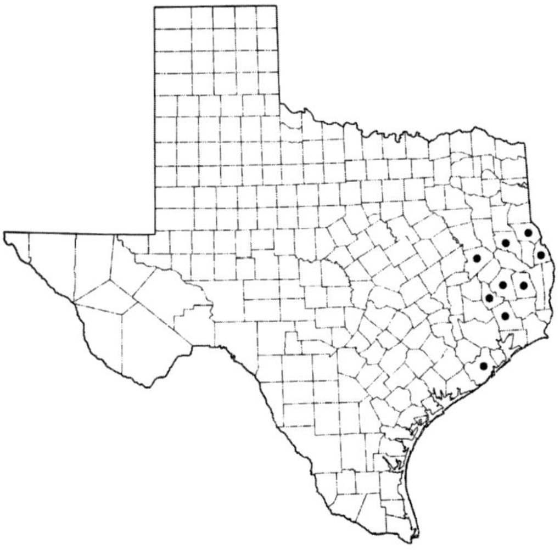

Lithospermum tuberosum

Lithospermum viride

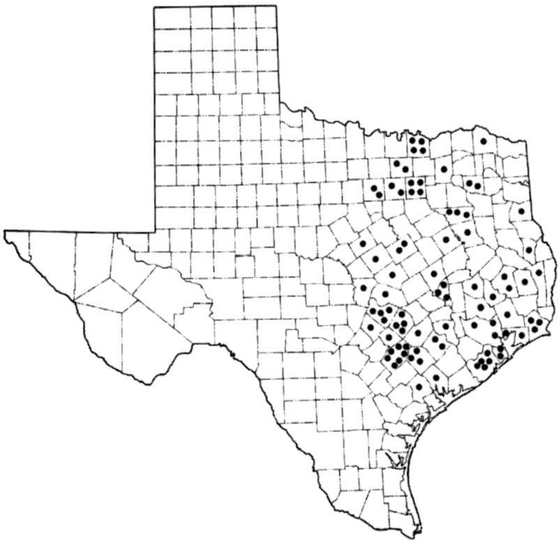

Myosotis macrosperma

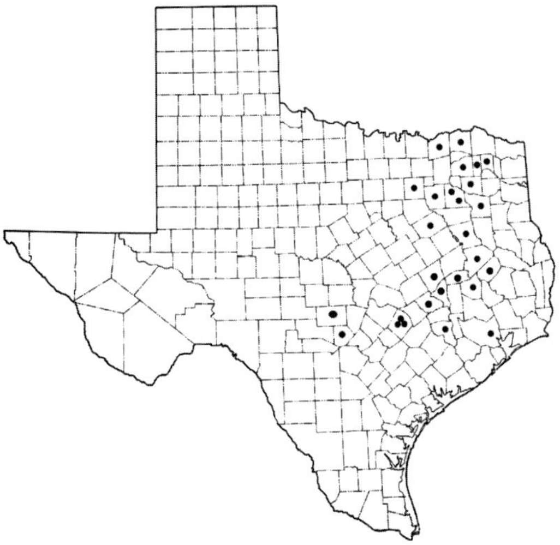

Myosotis verna

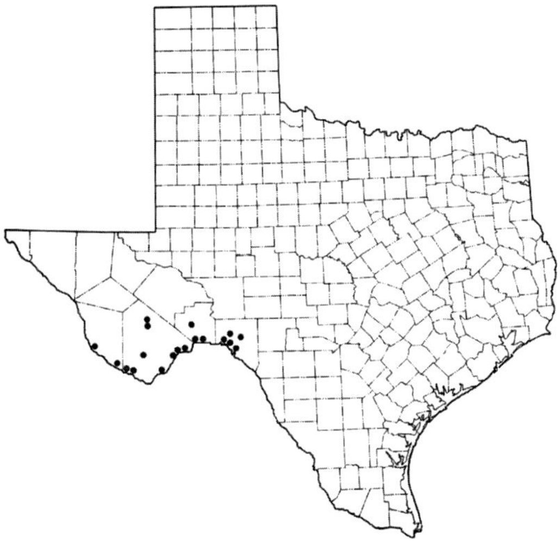

Omphalodes aliena

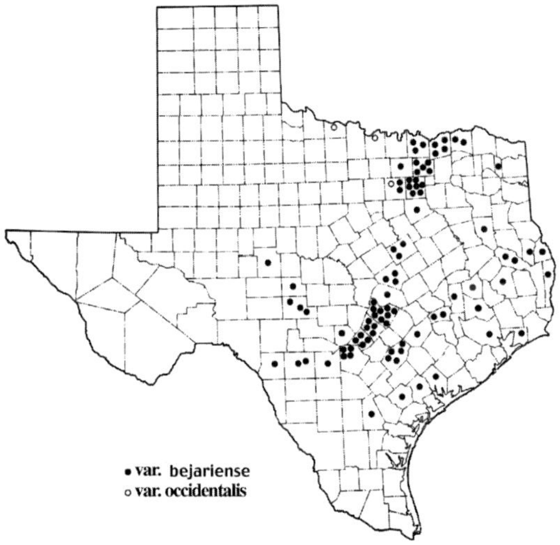
• var. bejariense
○ var. occidentalis

Onosmodium bejariense

Onosmodium helleri

Pectocarya heterocarpa

Pectocarya platycarpa

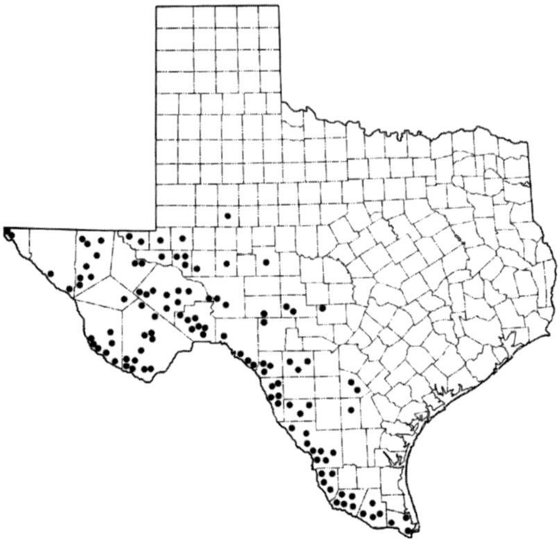

Tiquilia canescens

Tiquilia gossypina

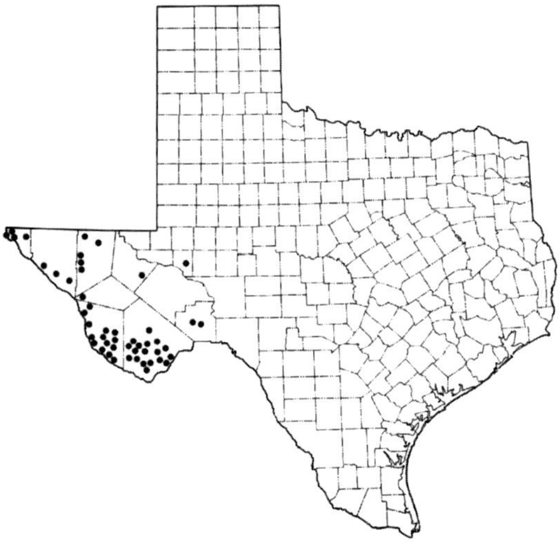

Tiquilia greggii

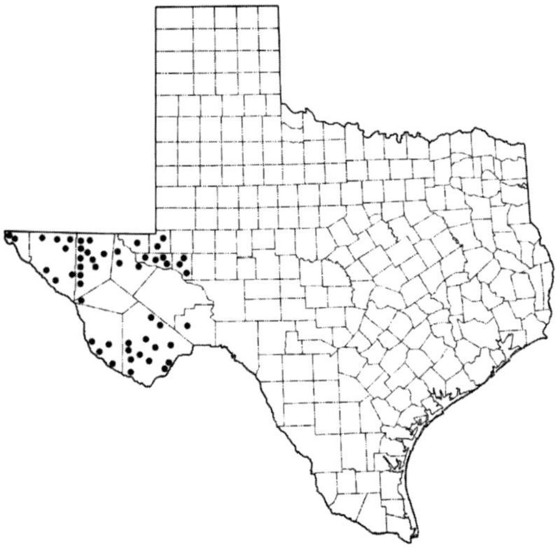

Tiquilia hispidissima

Tiquilia mexicana

Tournefortia volubilis

Arabidopsis thaliana

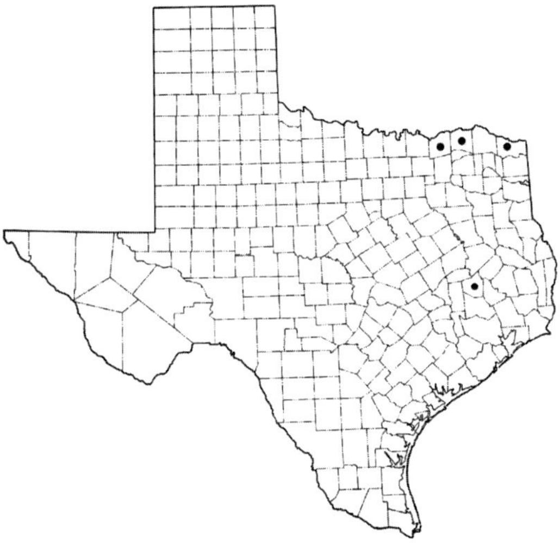

Arabis canadensis

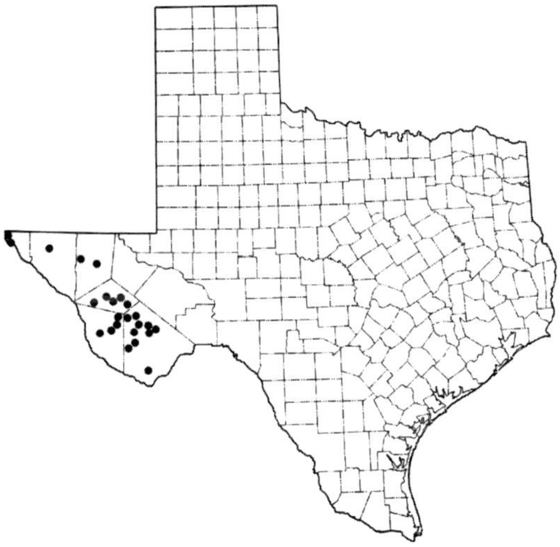

Arabis fendleri

Arabis perennans

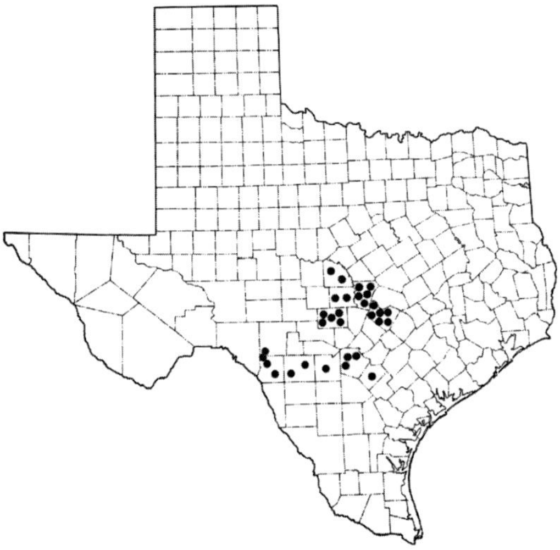

Arabis petiolaris

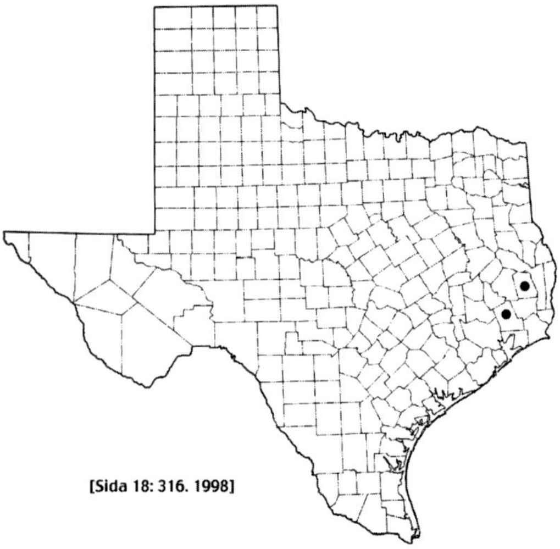
[Sida 18: 316. 1998]
Armoracia lacustris

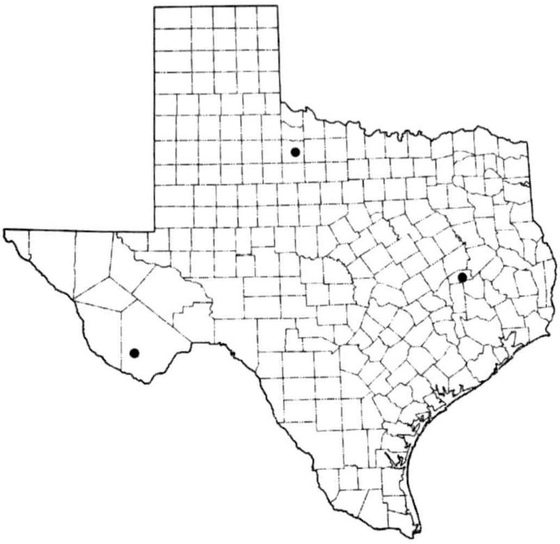
Brassica nigra

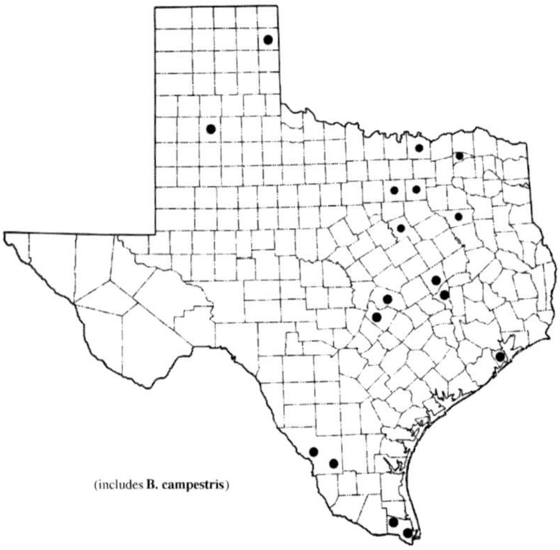
(includes B. campestris)
Brassica rapa

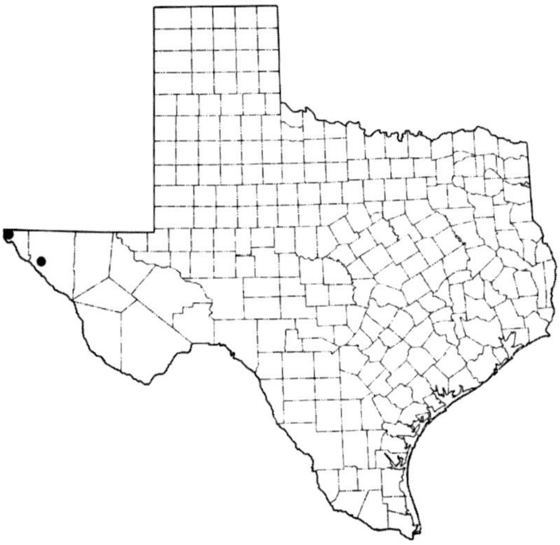
Brassica tournefortii

Cakile constricta

Cakile geniculata

[includes C. fusiformis]

Cakile lanceolata

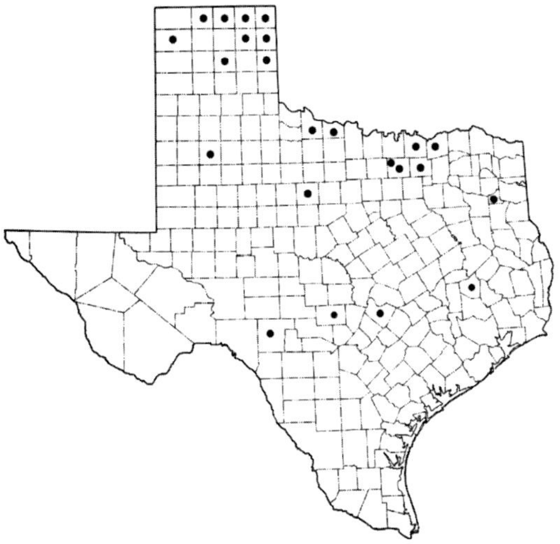

Camelina microcarpa

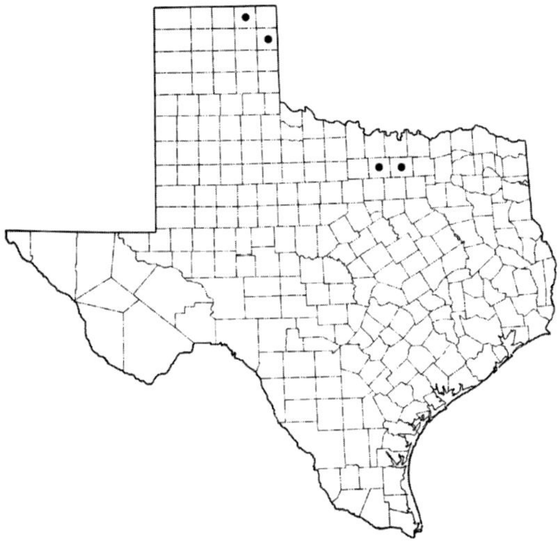

Camelina rumelica

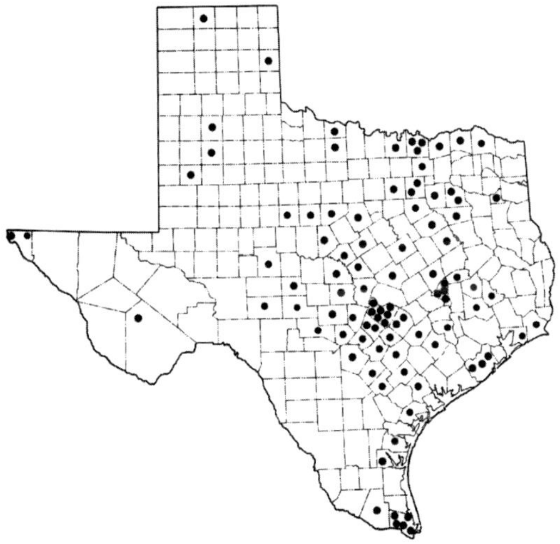

Capsella bursa-pastoris

[= Dentaria laciniata]

Cardamine concatenata

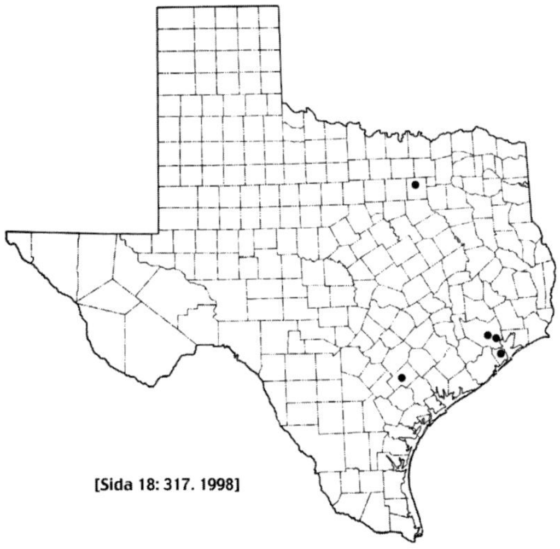
[Sida 18: 317. 1998]

Cardamine debilis

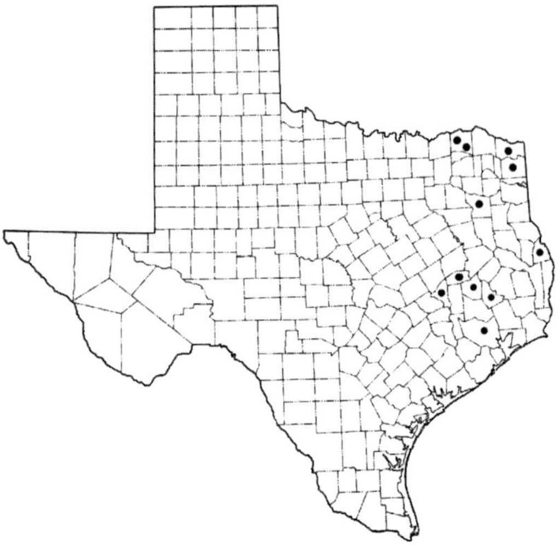

Cardamine hirsuta

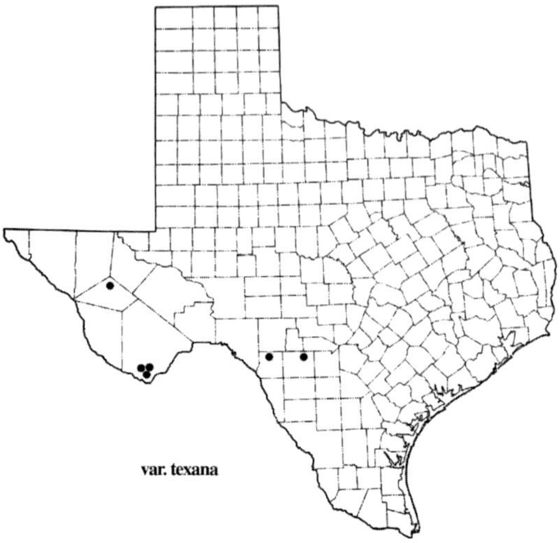
var. texana

Cardamine macrocarpa

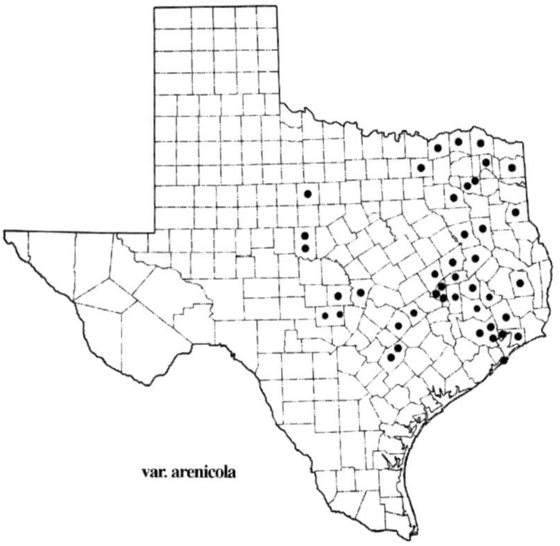
var. arenicola

Cardamine parviflora

Cardamine pensylvanica

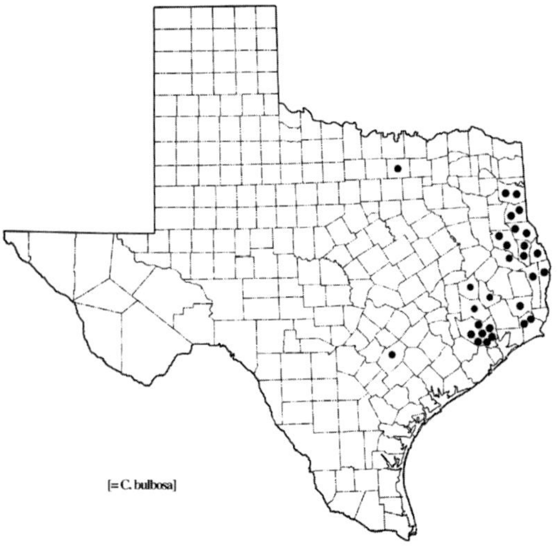
[= C. bulbosa]

Cardamine rhomboidea

Cardaria draba

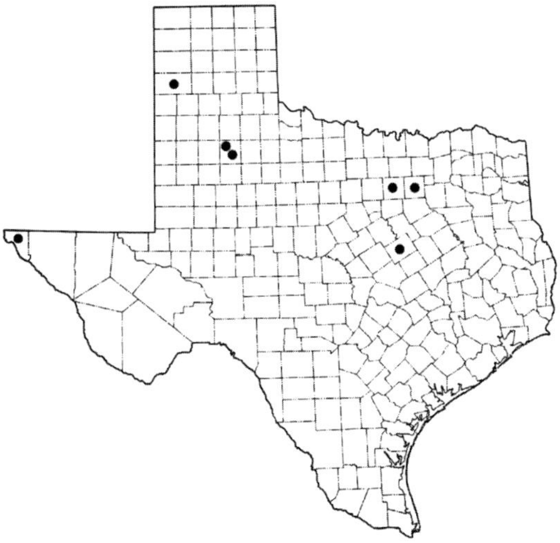
Chorispora tenella

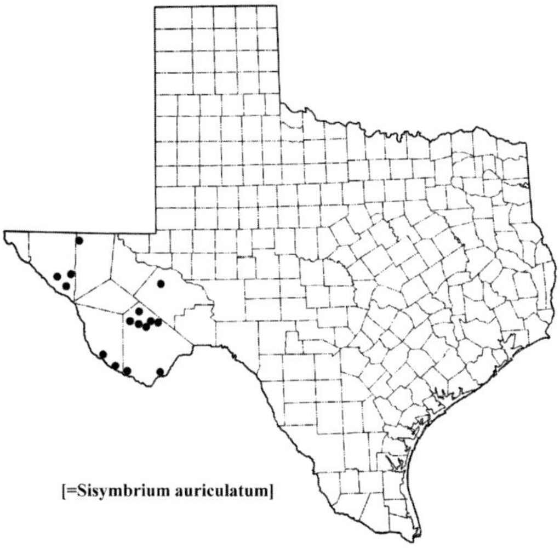
[=Sisymbrium auriculatum]
Coelophragmus auriculatus

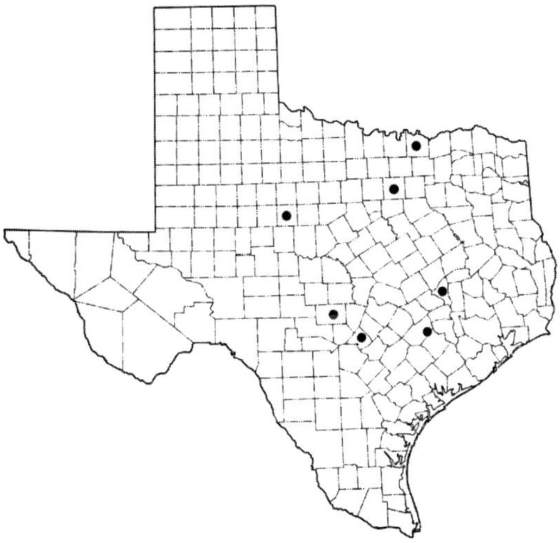
Conringia orientalis

Coronopus didymus

[= D. richardsonii var. viscosa]
Descurainia incana

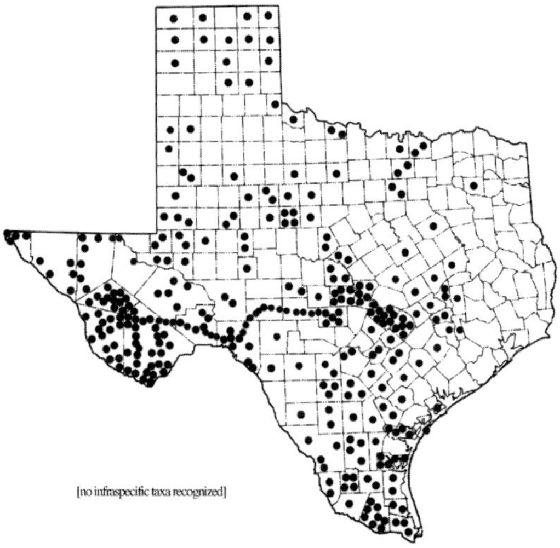
[no infraspecific taxa recognized]
Descurainia pinnata

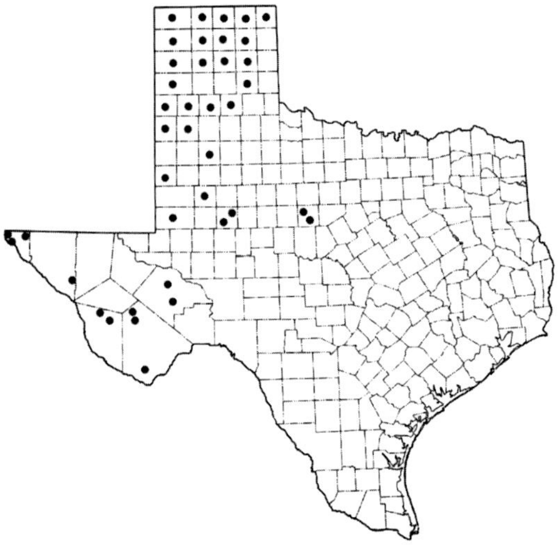

Descurainia sophia

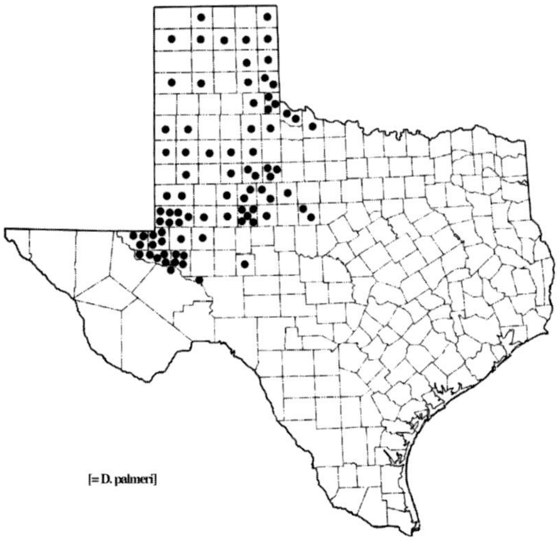
[= D. palmeri]

Dimorphocarpa candicans

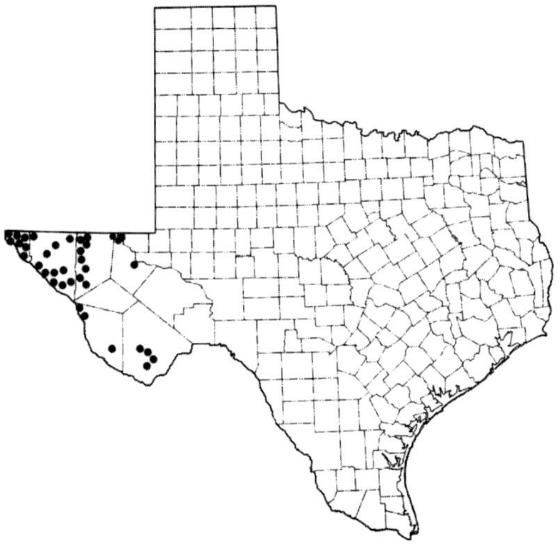

Dimorphocarpa wislizeni

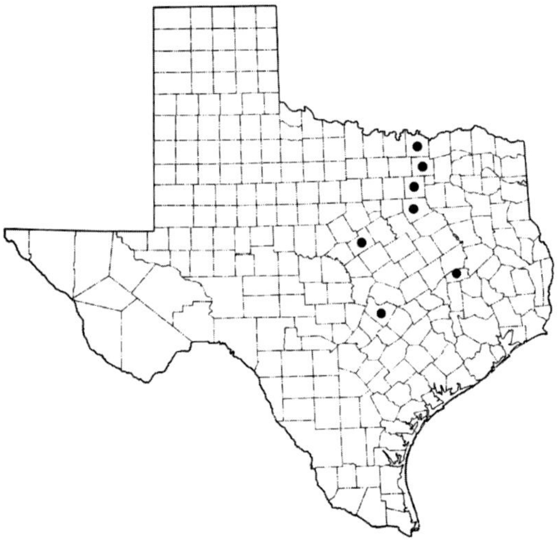

Diplotaxis muralis

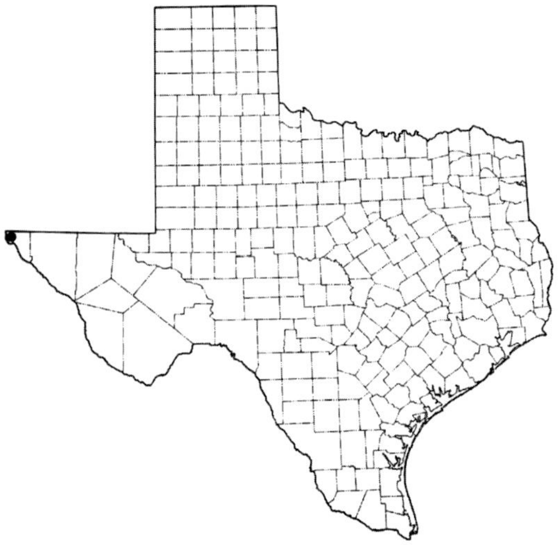

Diplotaxis tenuifolia

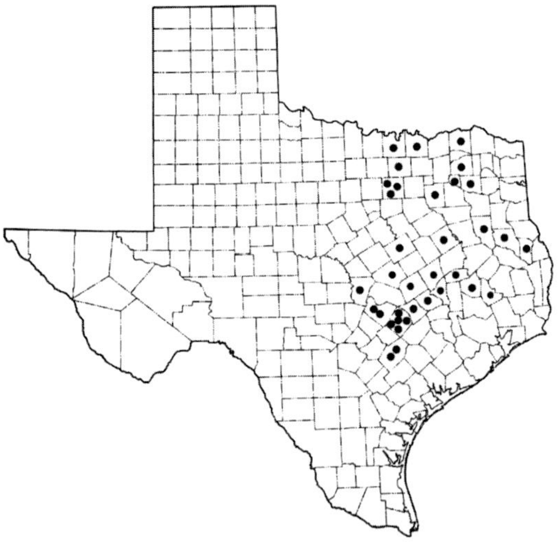

Draba brachycarpa

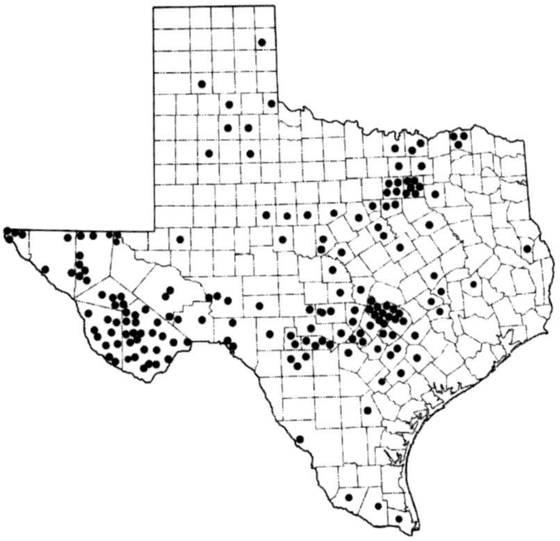

Draba cuneifolia

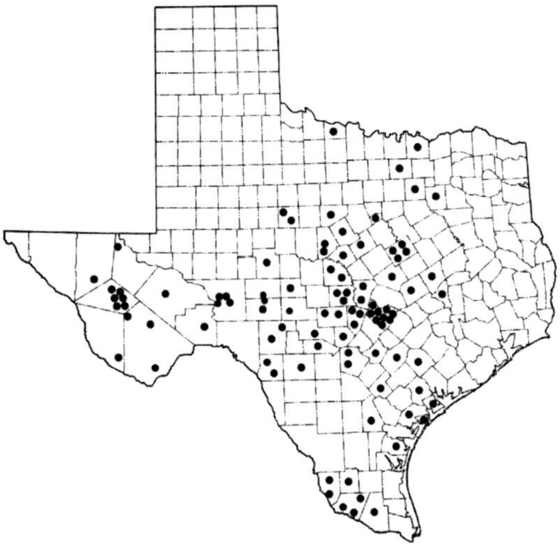

Draba platycarpa

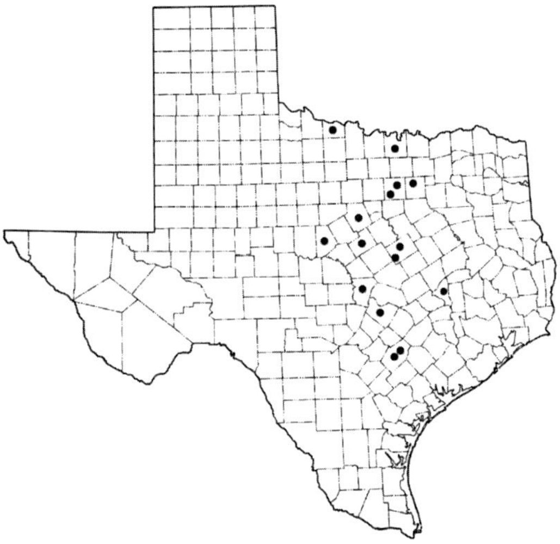

Draba reptans

Draba standleyi

Dryopetalon runcinatum

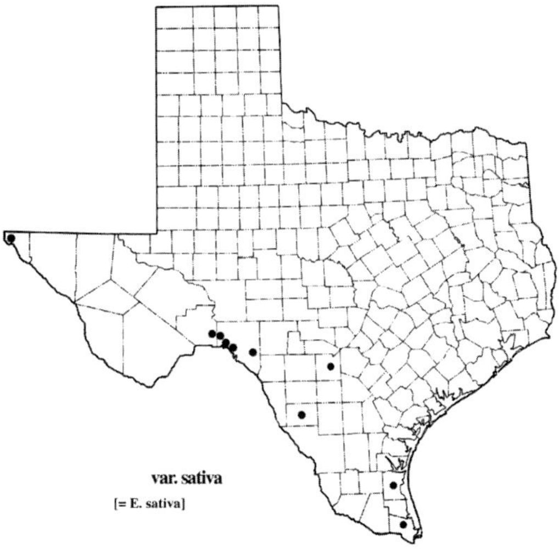
var. sativa
[= E. sativa]

Eruca vesicaria

Erucastrum gallicum

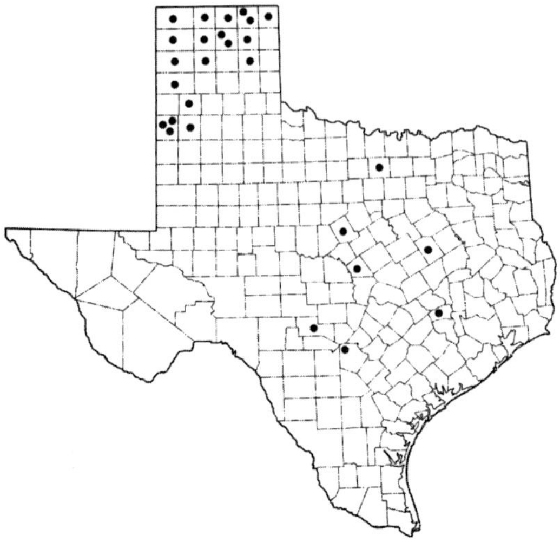

Erysimum asperum

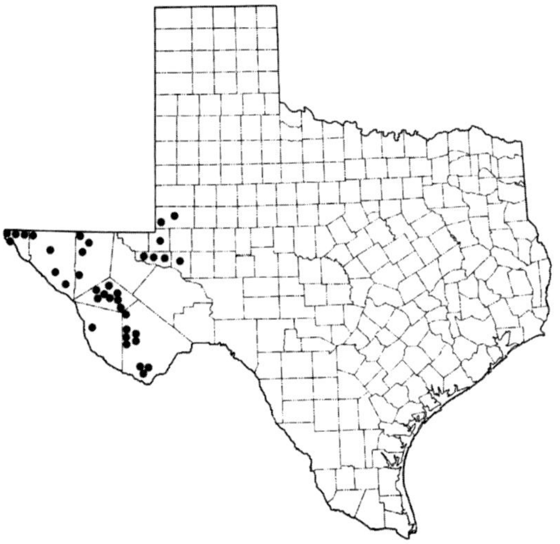

Erysimum capitatum

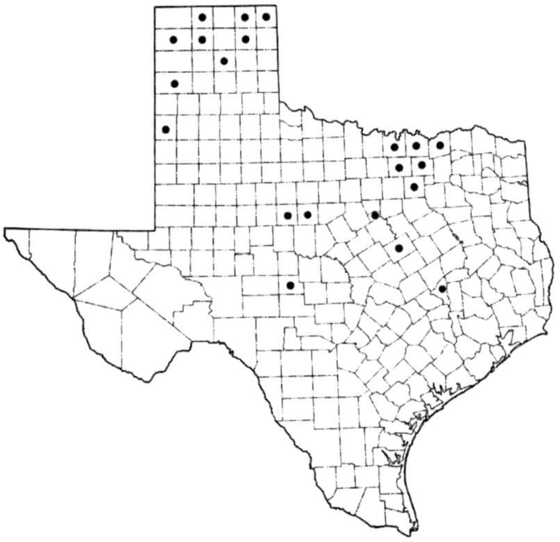

Erysimum repandum

Halimolobos diffusa

Iodanthus pinnatifidus

Leavenworthia texana

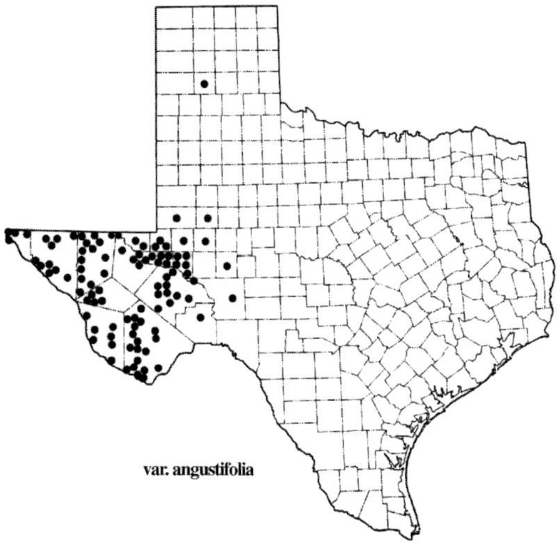
var. angustifolia

Lepidium alyssoides

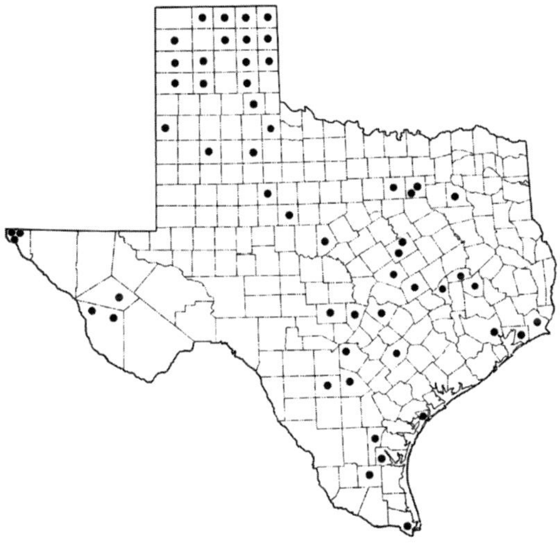

Lepidium densiflorum

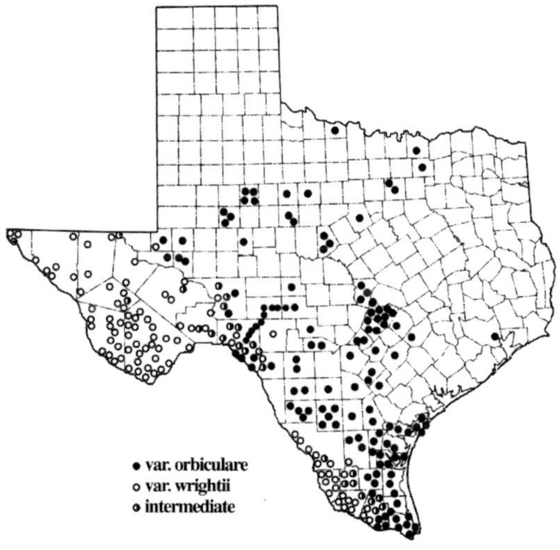
var. orbiculare
var. wrightii
intermediate

Lepidium lasiocarpum

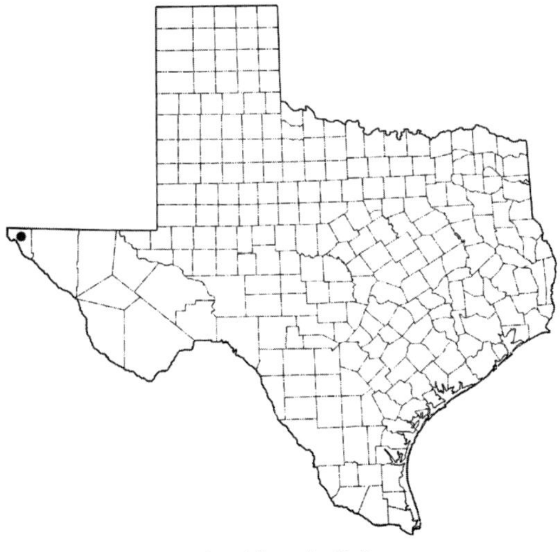

Lepidium latifolium

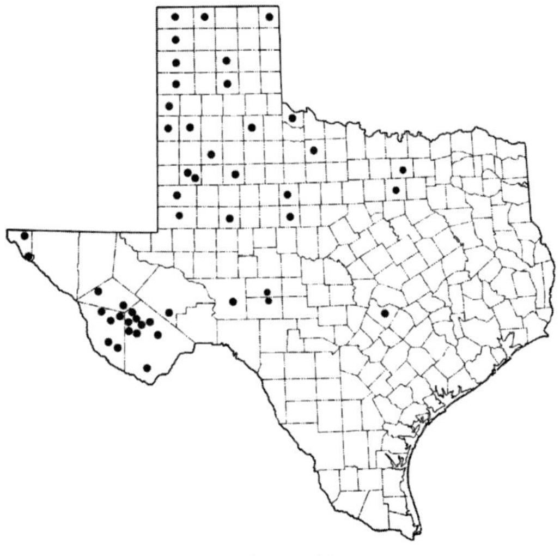

Lepidium oblongum

Lepidium perfoliatum

Lepidium ruderale

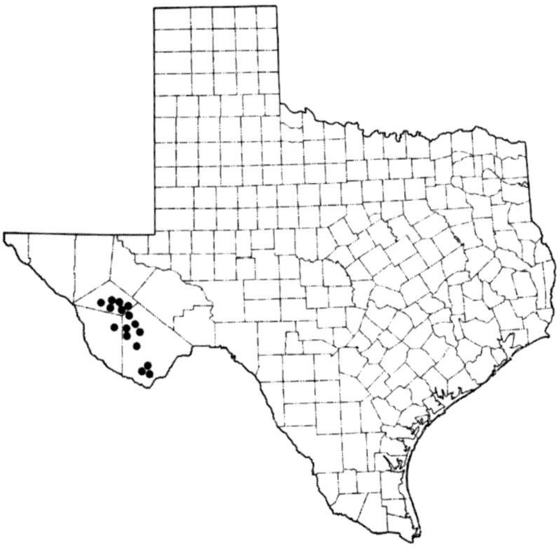

Lepidium sordidum

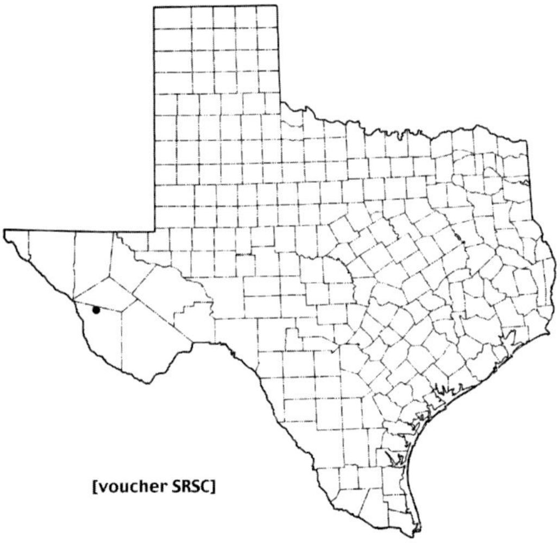
[voucher SRSC]

Lepidium thurberi

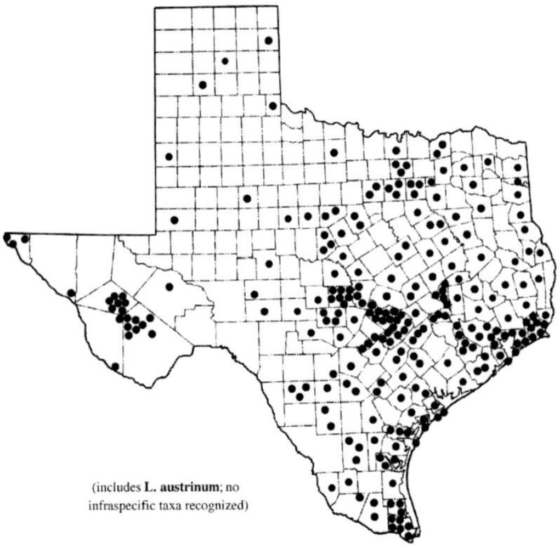
(includes **L. austrinum**; no infraspecific taxa recognized)

Lepidium virginicum

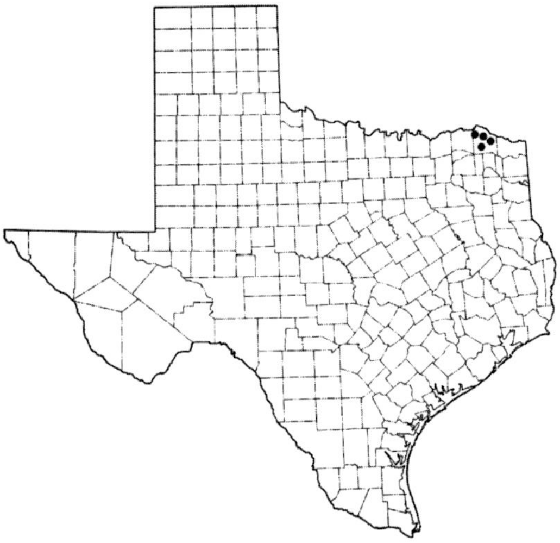

Lesquerella angustifolia

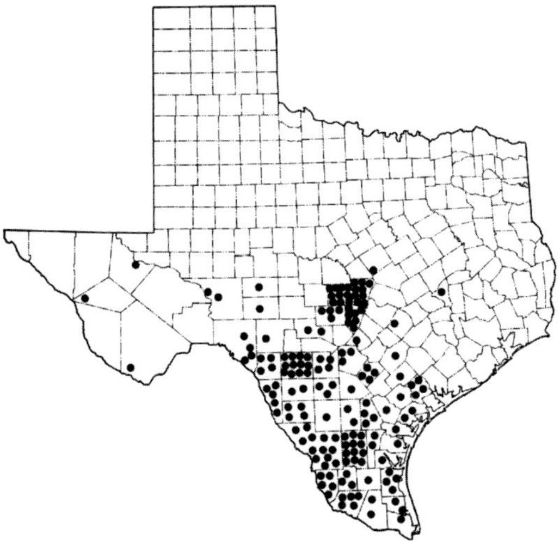

Lesquerella argyraea

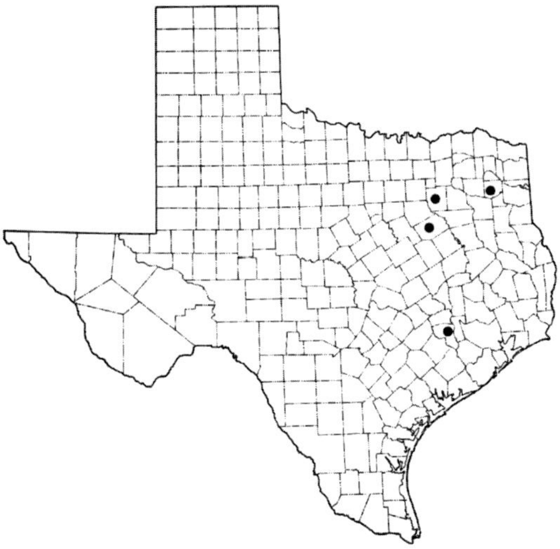

Lesquerella auriculata

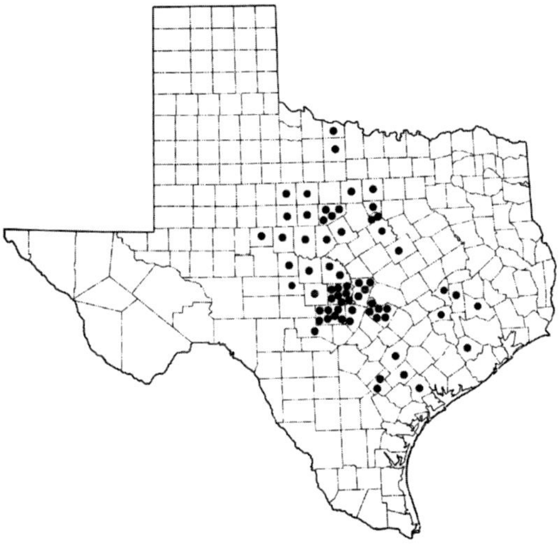

Lesquerella densiflora

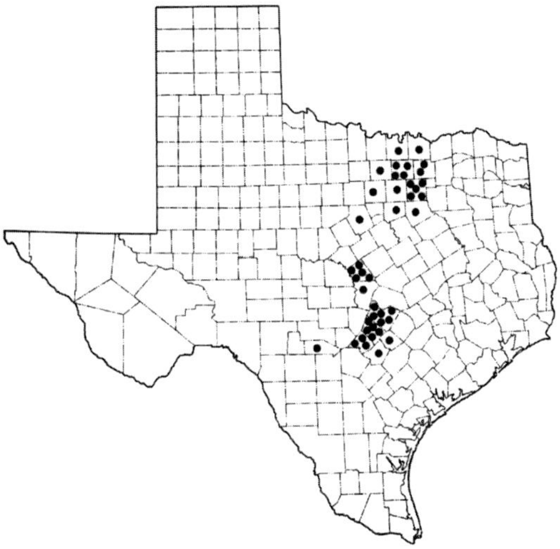

Lesquerella engelmannii

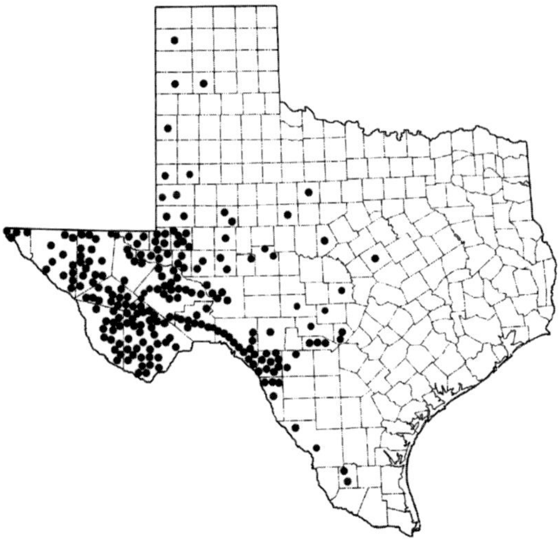

Lesquerella fendleri

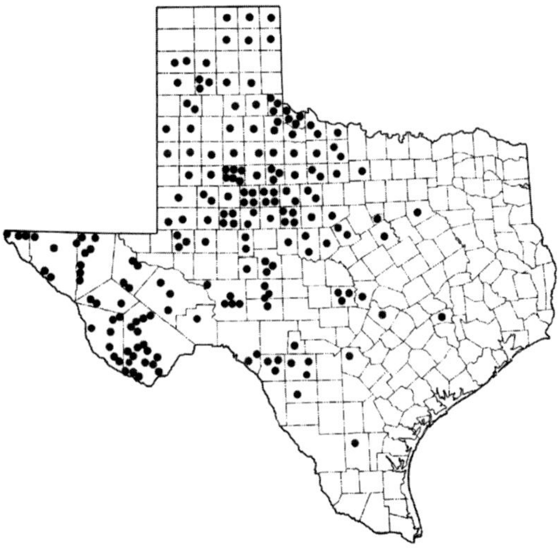

Lesquerella gordonii

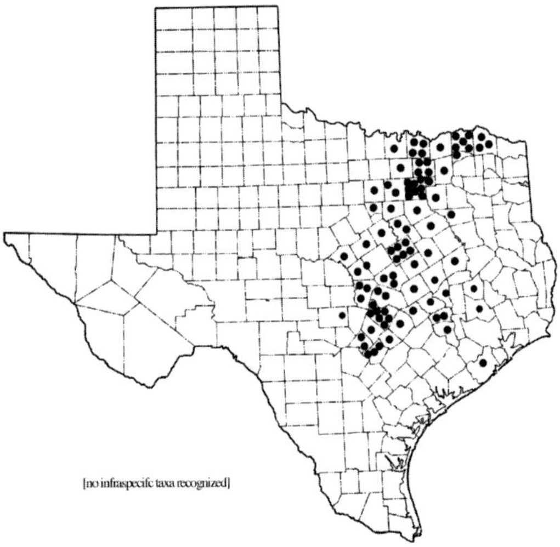
[no infraspecifc taxa recognized]

Lesquerella gracilis

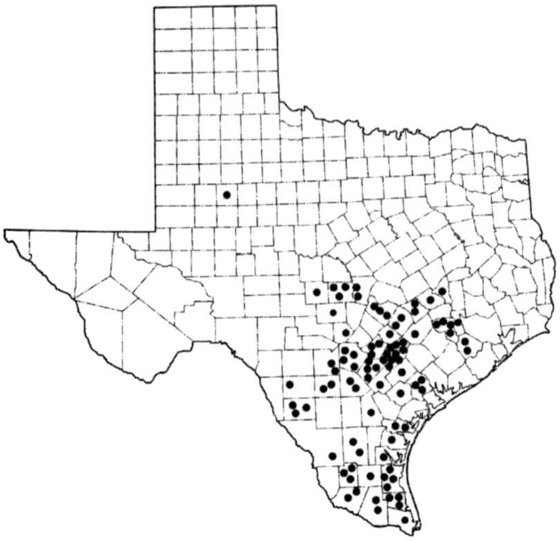

Lesquerella grandiflora

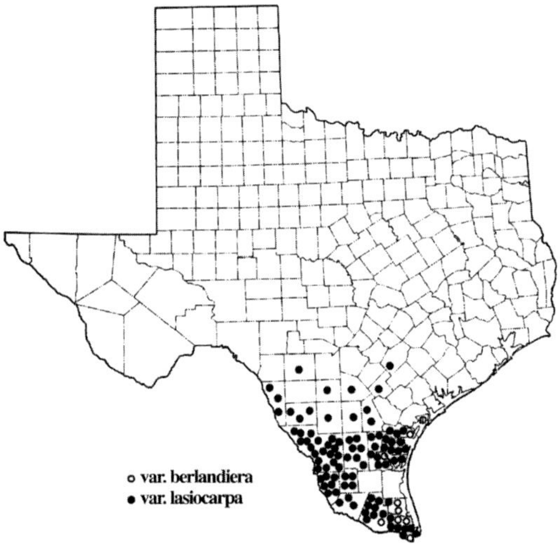
o var. berlandiera
• var. lasiocarpa

Lesquerella lasiocarpa

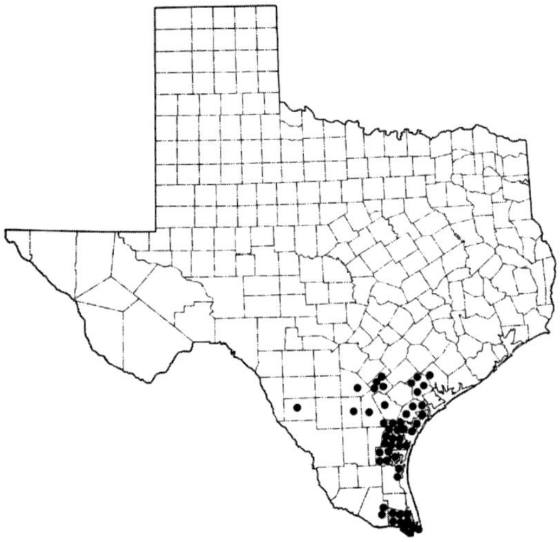

Lesquerella lindheimeri

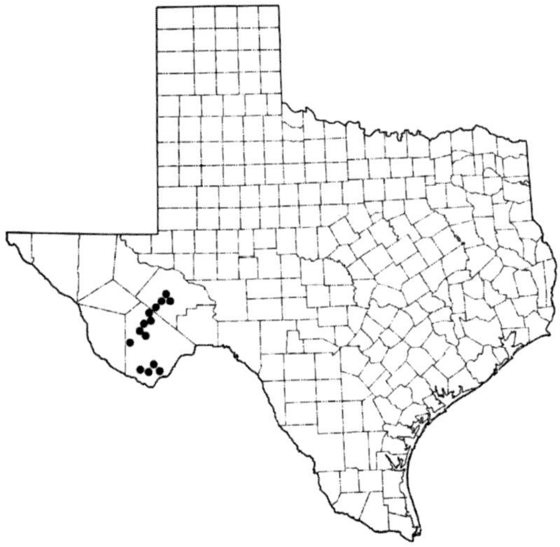

Lesquerella mcvaughiana

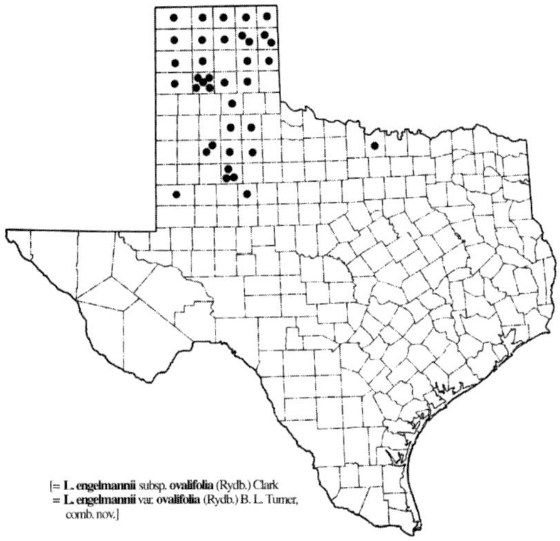
[= L. engelmannii subsp. ovalifolia (Rydb.) Clark
= L. engelmannii var. ovalifolia (Rydb.) B. L. Turner, comb. nov.]

Lesquerella ovalifolia

Lesquerella pallida

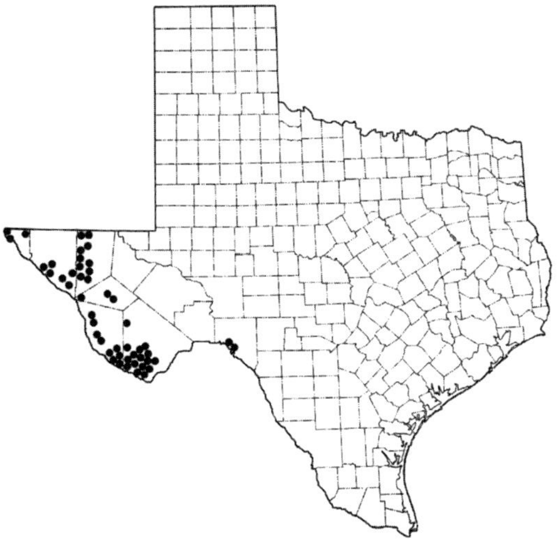

Lesquerella purpurea

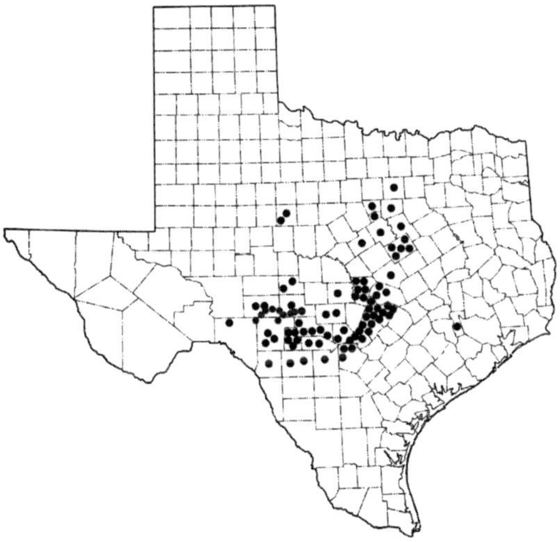

Lesquerella recurvata

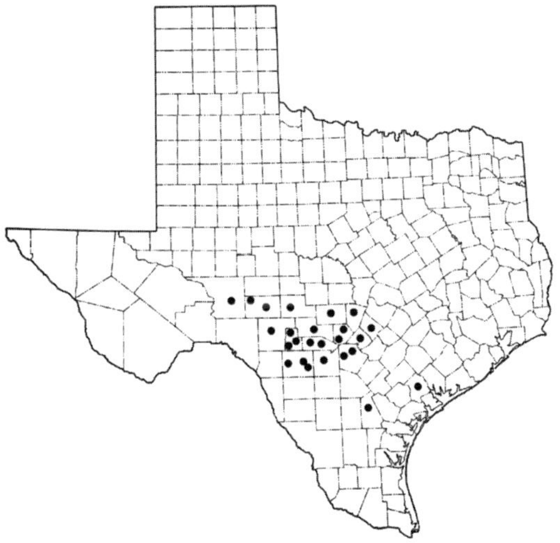

Lesquerella sessilis

Lesquerella thamnophila

Lesquerella valida

Lobularia maritima

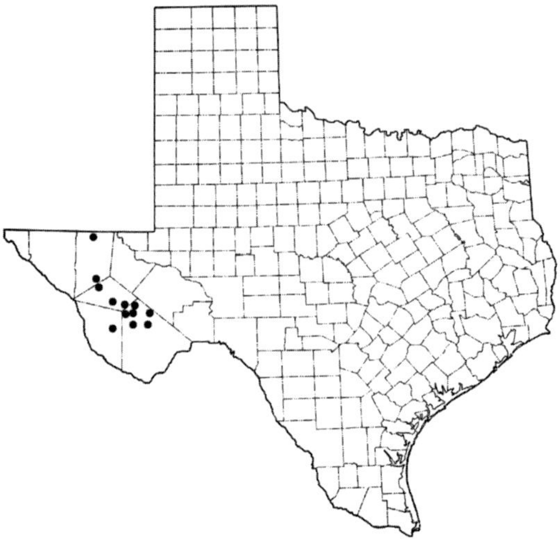
Mancoa pubens

[= M. bicornis]
Matthiola longipetala

Myagrum perfoliatum

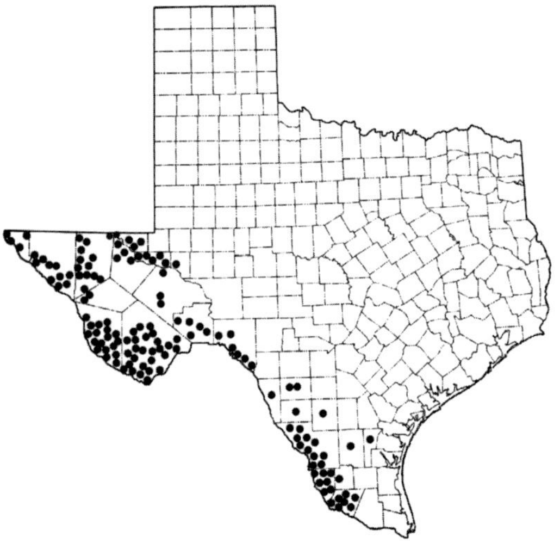
Nerisyrenia camporum

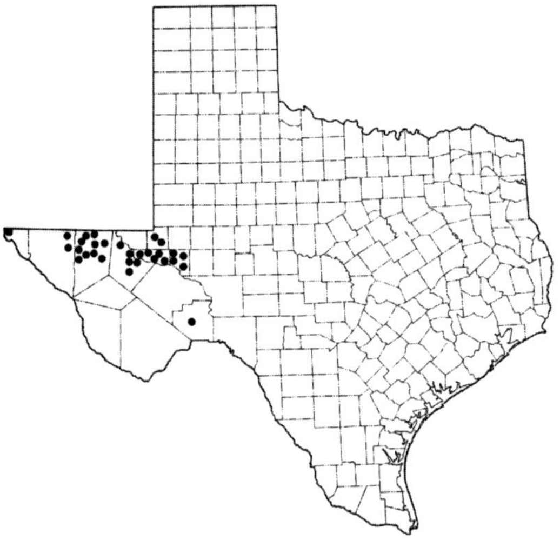
Nerisyrenia linearifolium

Pennellia longifolia

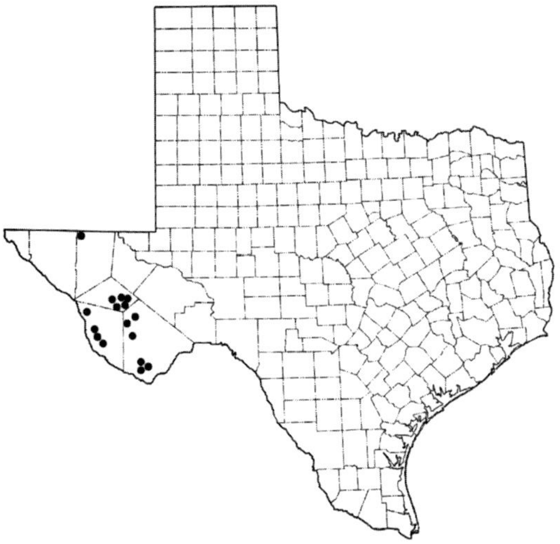

Pennellia micrantha

Raphanus sativus

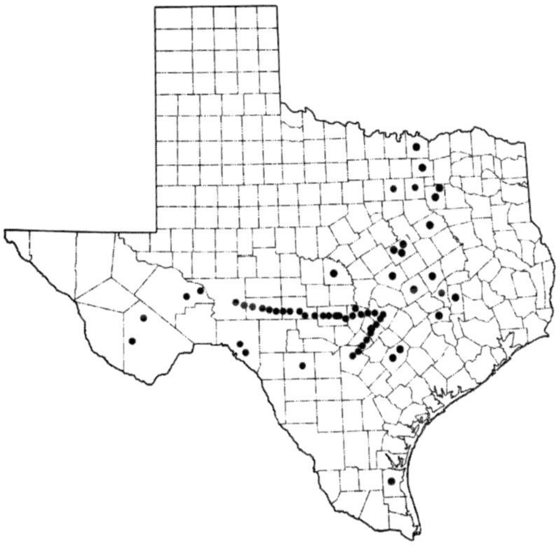

Rapistrum rugosum

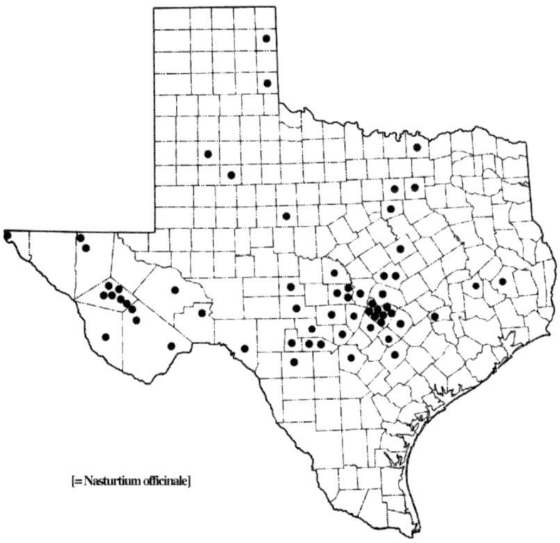
[= Nasturtium officinale]

Rorippa nasturtium-aquaticum

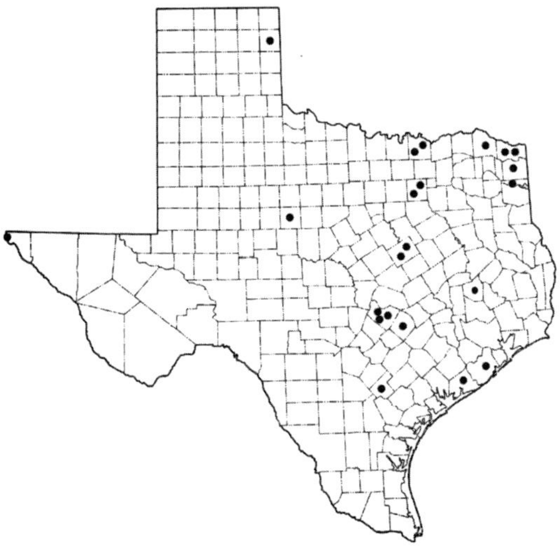

Rorippa palustris

Rorippa ramosa

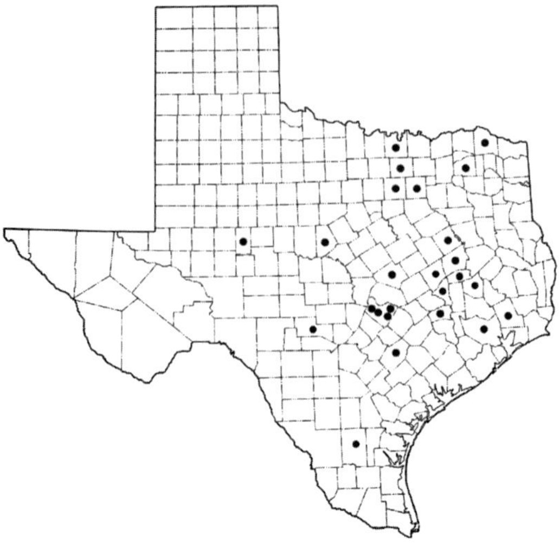

Rorippa sessiliflora

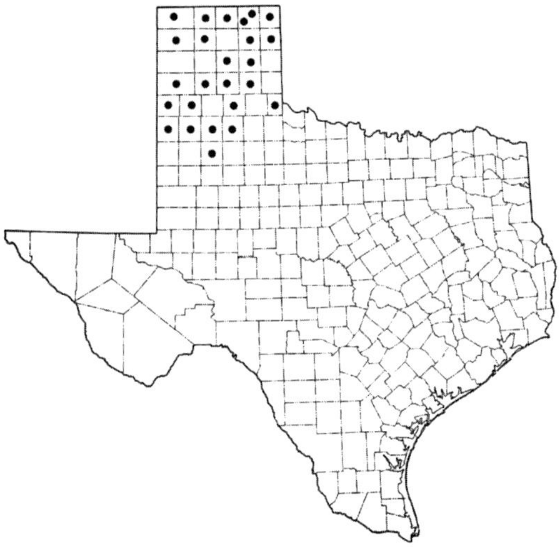

Rorippa sinuata

Rorippa sphaerocarpa

Rorippa tenerrima

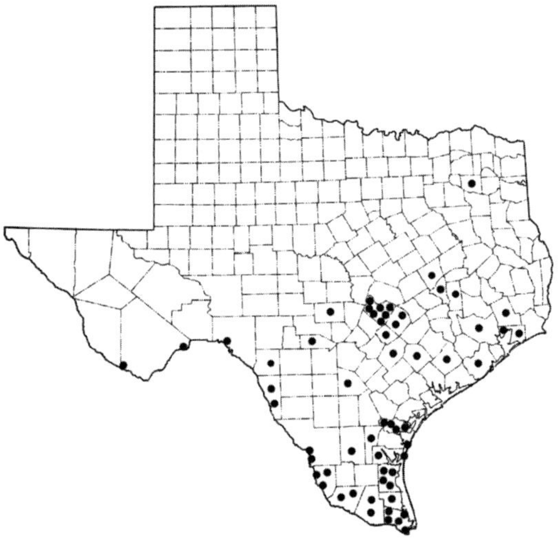

Rorippa teres

Rorippa truncata

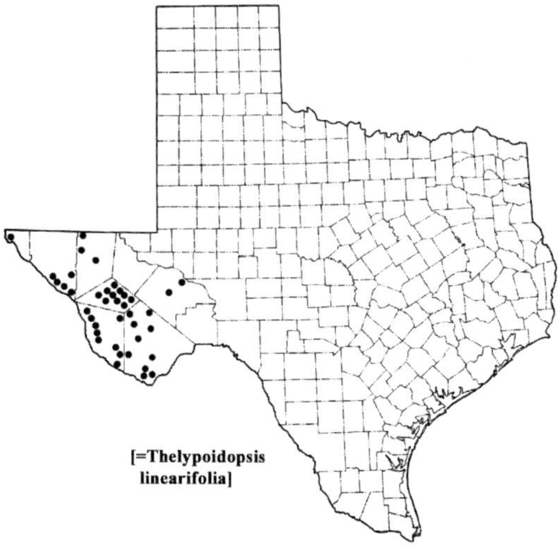
[=Thelypoidopsis linearifolia]

Schoenocrambe linearifolia

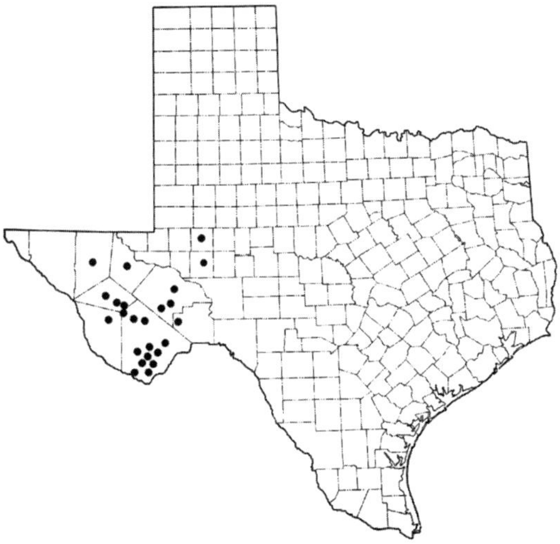

Selenia dissecta

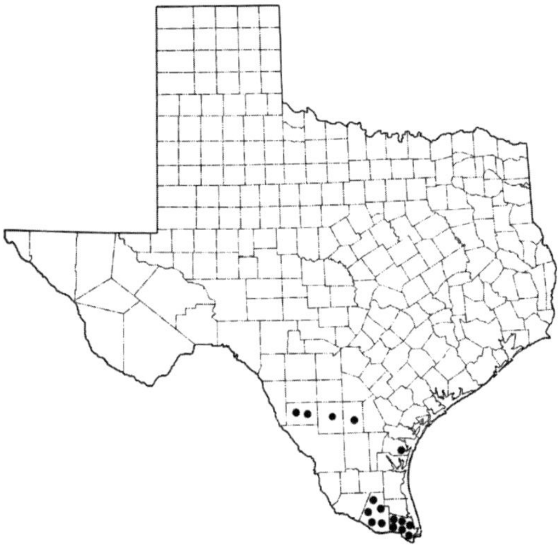

Selenia grandis

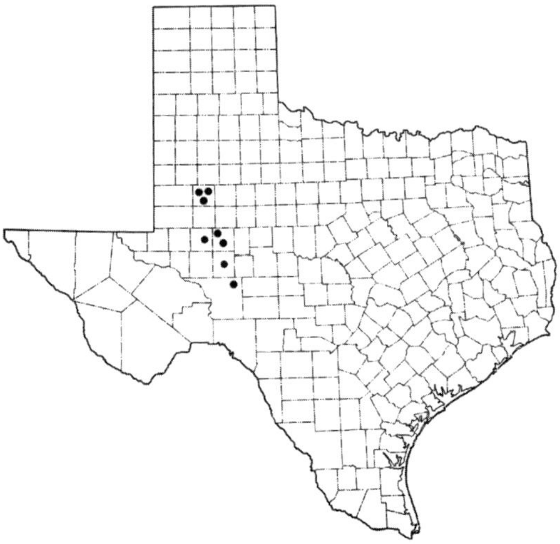

Selenia jonesii

[vouchers, SRSC, TEX]

Sibara grisea

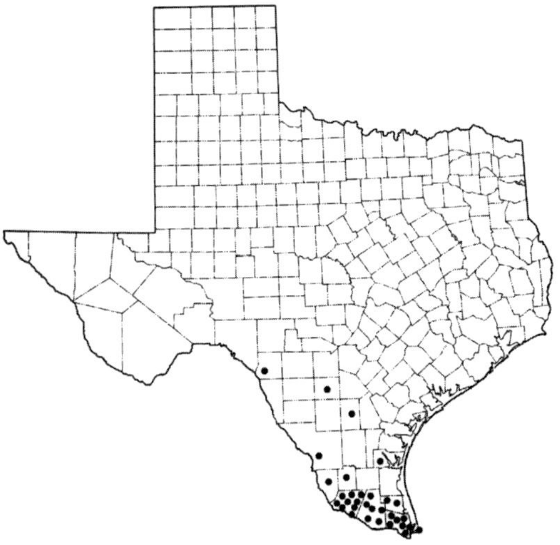

Sibara runcinata

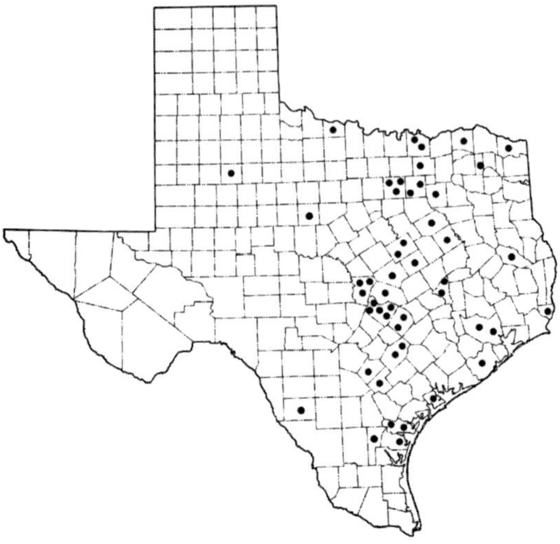

Sibara virginica

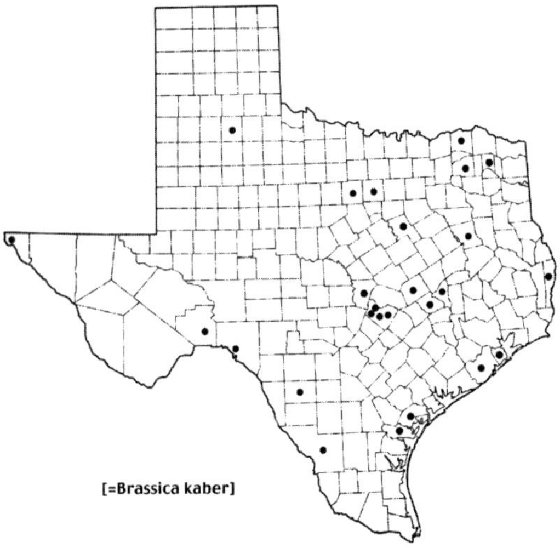
[=Brassica kaber]

Sinapsis arvensis

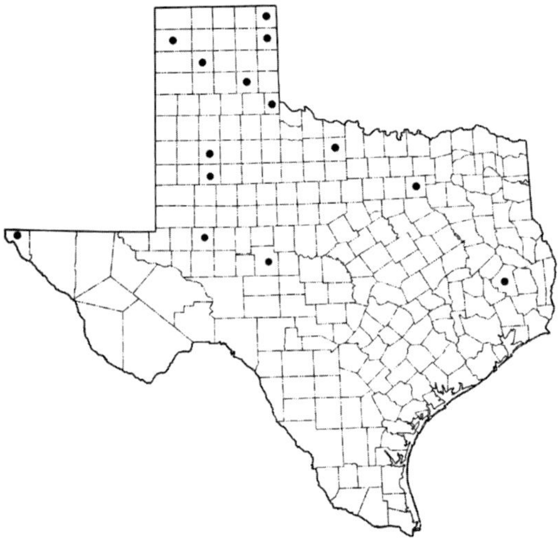

Sisymbrium altissimum

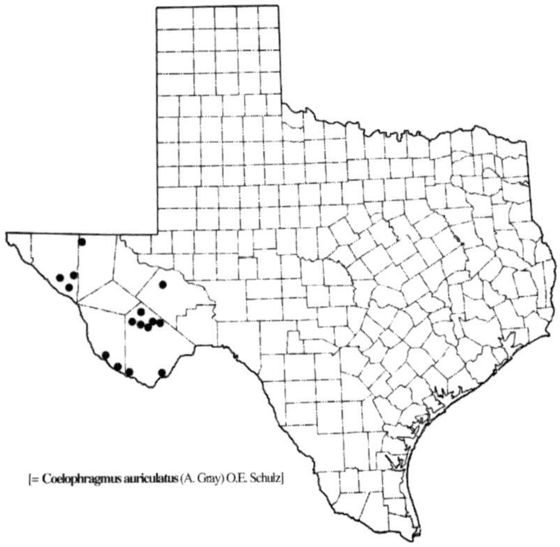
[= Coelophragmus auriculatus (A. Gray) O.E. Schulz]

Sisymbrium auriculatum

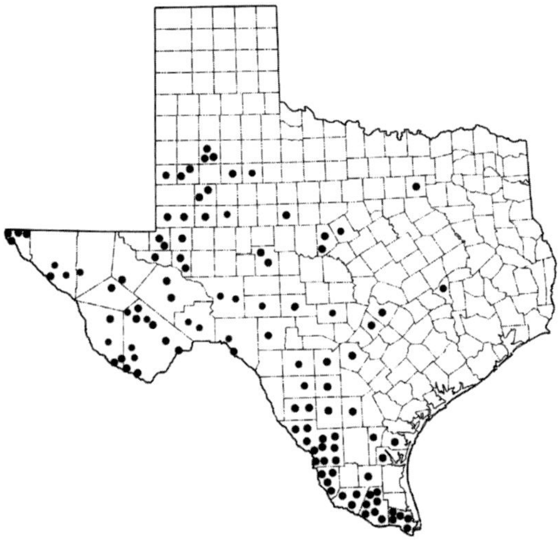

Sisymbrium irio

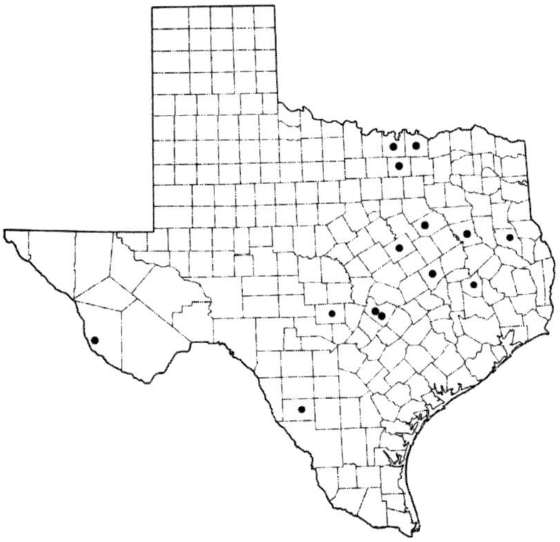

Sisymbrium officinale

Sisymbrium orientale

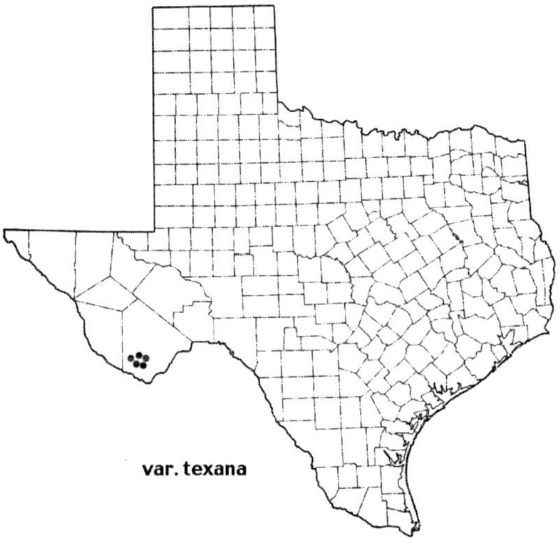
var. texana

Stanleya pinnata

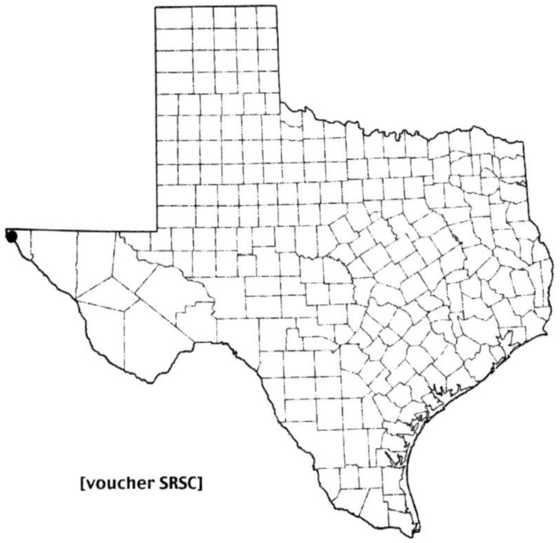
[voucher SRSC]

Streptanthus arizonicus

Streptanthus bracteatus

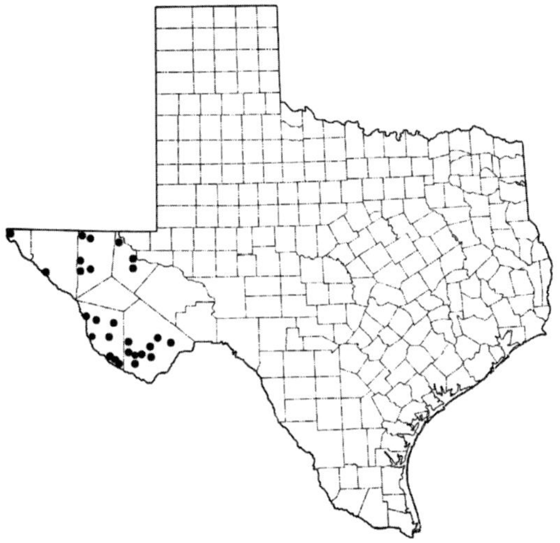

Streptanthus carinatus

Streptanthus cutleri

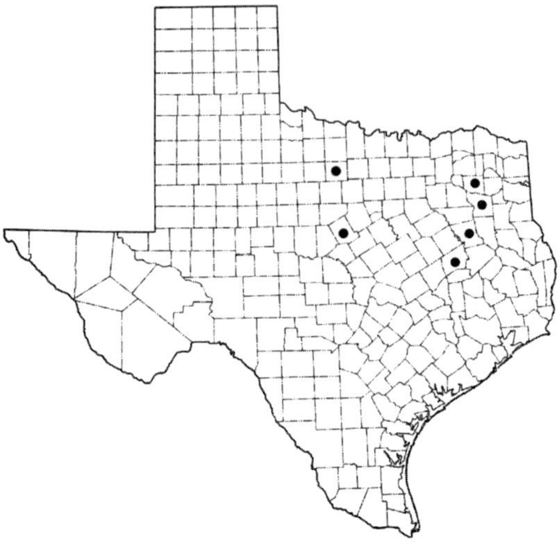

Streptanthus hyacinthoides

Streptanthus maculatus

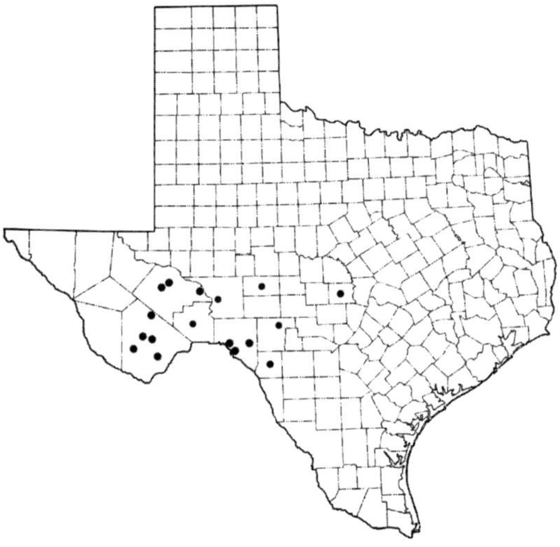

Streptanthus playtcarpus

Streptanthus sparsiflorus

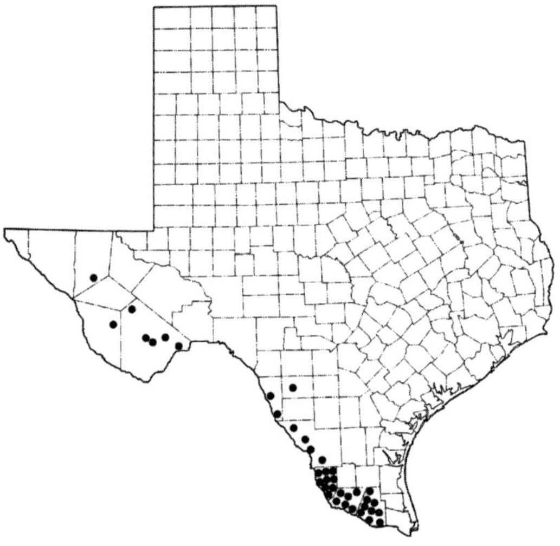

Synthlipsis greggii

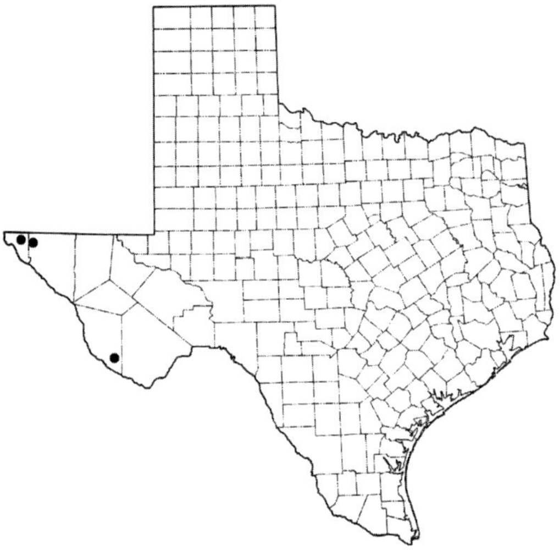

Thelypodiopsis purpusii

[= Sisymbrium shinnersi]

Thelypodiopsis shinnersii

Thelypodiopsis vaseyi

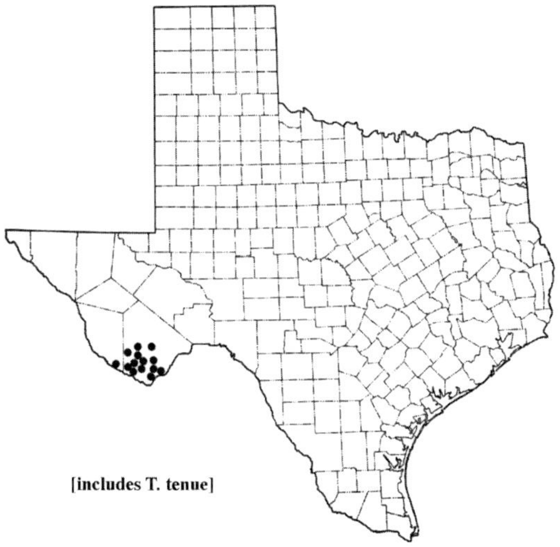
[includes T. tenue]

Thelypodium texanum

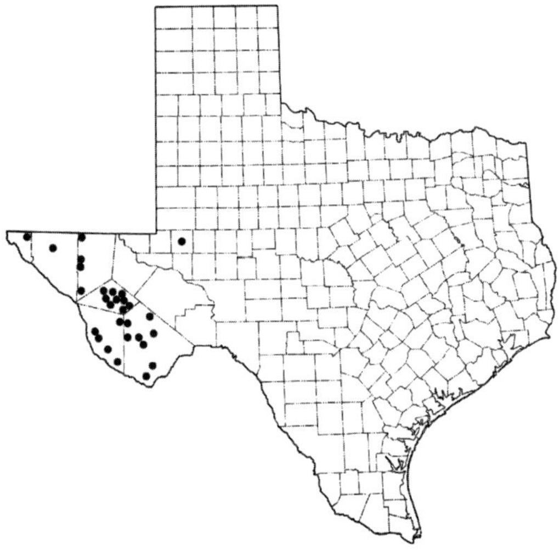

Thelypodium wrightii

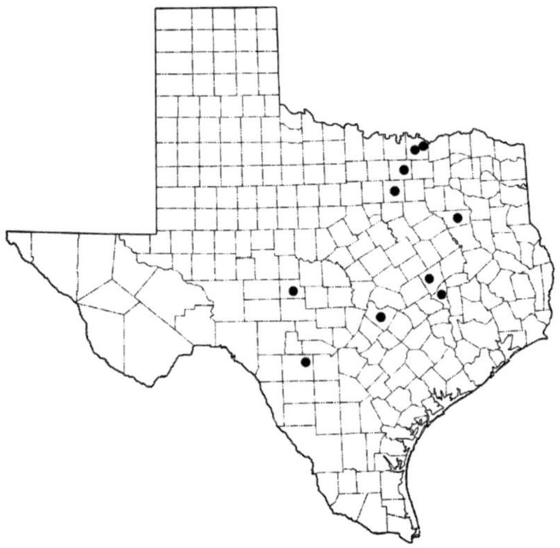

Thlaspi arvense

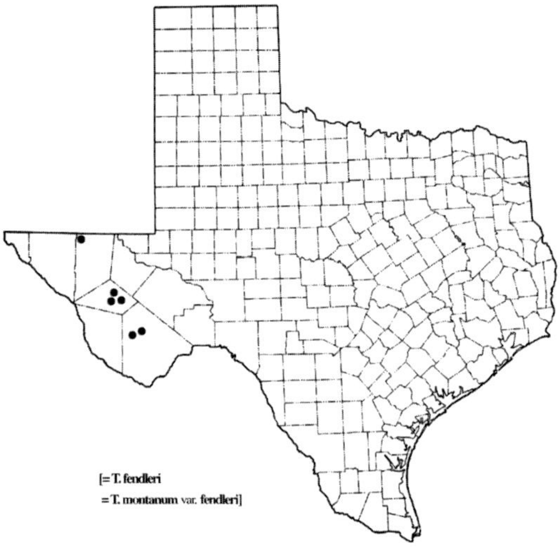
[= T. fendleri
= T. montanum var. fendleri]

Thlaspi fendleri

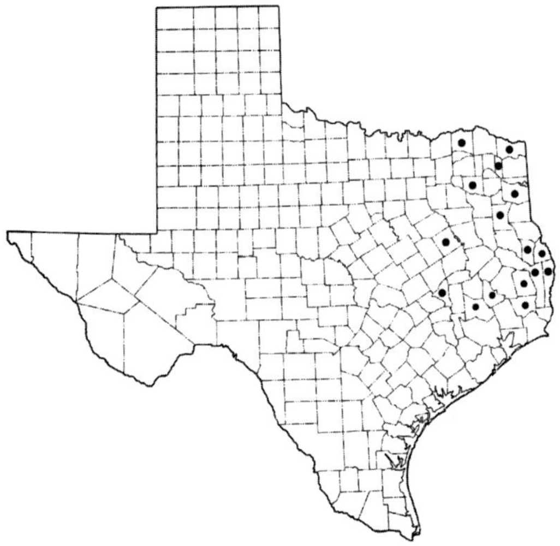

Brasenia schreberi

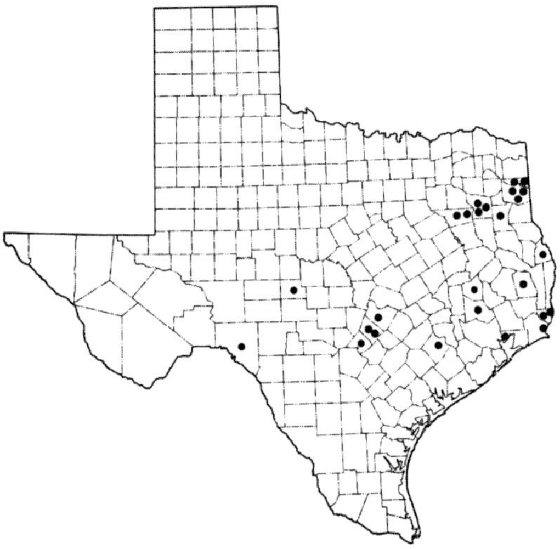

Cabomba caroliniana

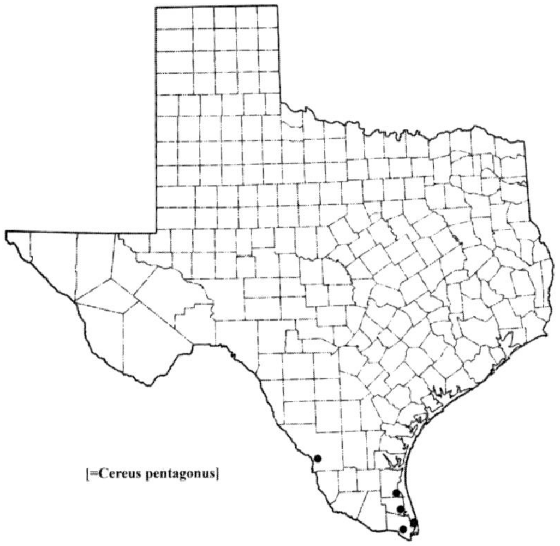
[=Cereus pentagonus]

Acanthocereus pentagonus

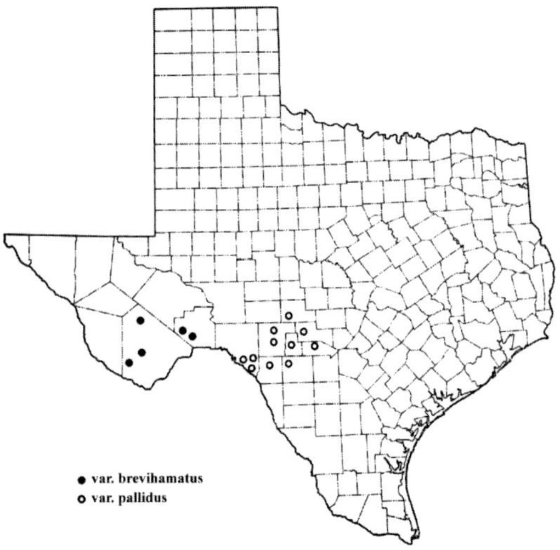
• var. brevihamatus
o var. pallidus

Ancistrocactus brevihamatus

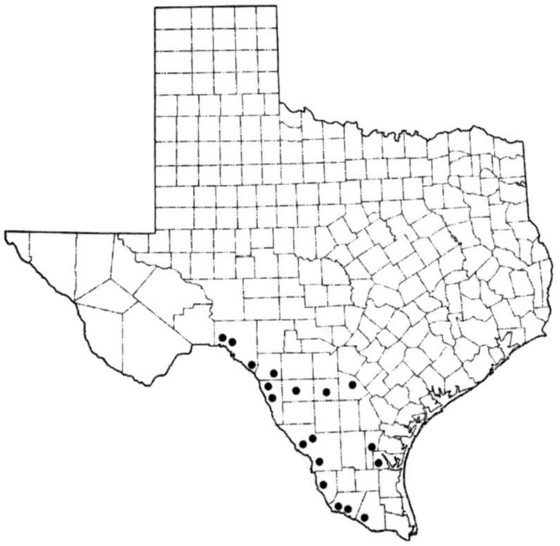

Ancistrocactus scheeri

Ancistrocactus tobuschii

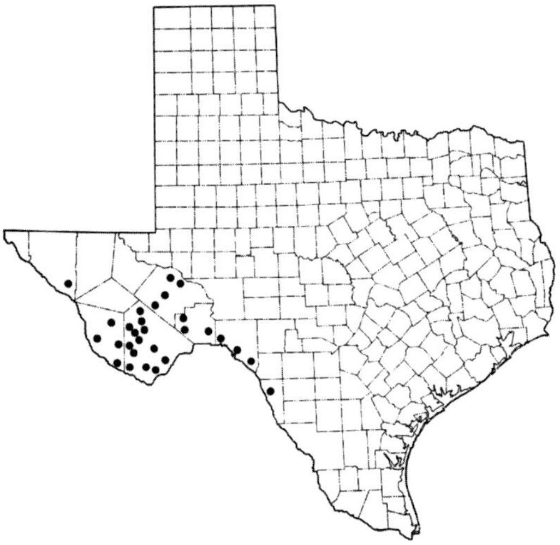

Ariocarpus fissuratus

[=Echinocactus asterias]

Astrophytum asterias

Coryphantha chaffeyi

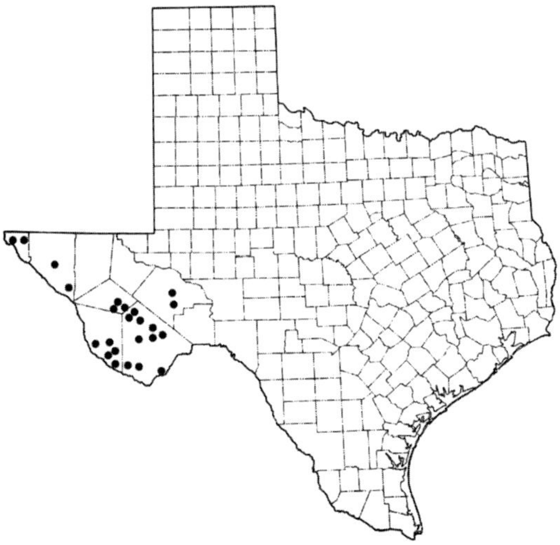

Coryphantha dasyacantha

Coryphantha duncanii

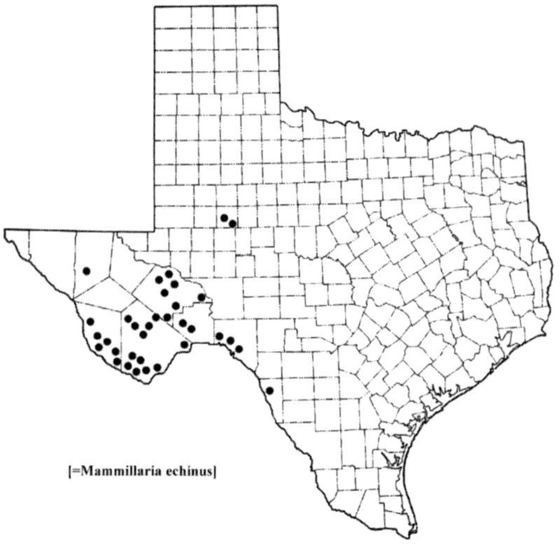
[=Mammillaria echinus]

Coryphantha echinus

Coryphantha hesteri

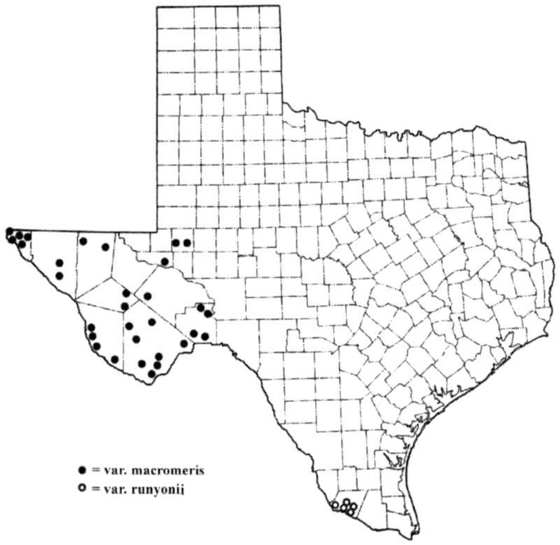
● = var. macromeris
○ = var. runyonii

Coryphantha macromeris

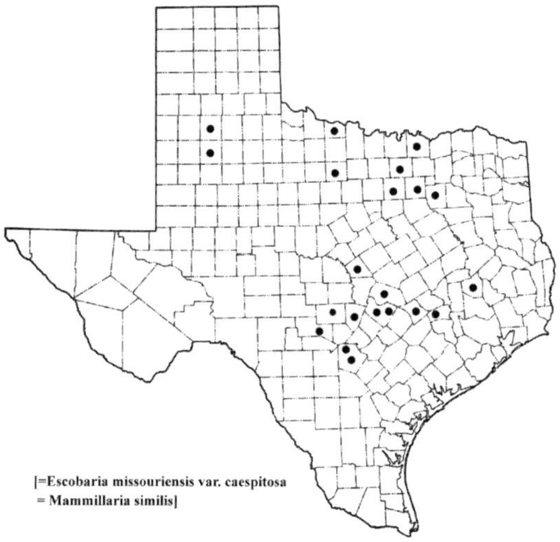
[=Escobaria missouriensis var. caespitosa
= Mammillaria similis]

Coryphantha missouriensis

Coryphantha pottsiana

Coryphantha ramillosa

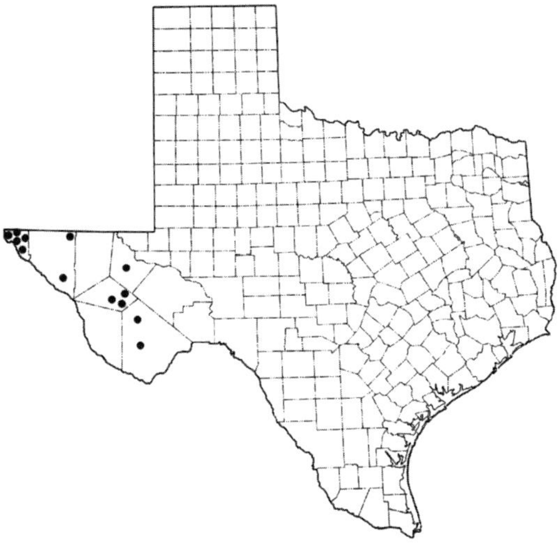

Coryphantha scheeri

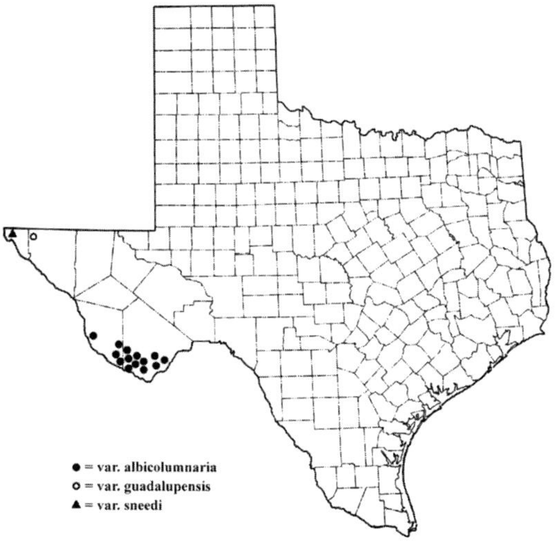
● = var. albicolumnaria
○ = var. guadalupensis
▲ = var. sneedi

Coryphantha sneedii

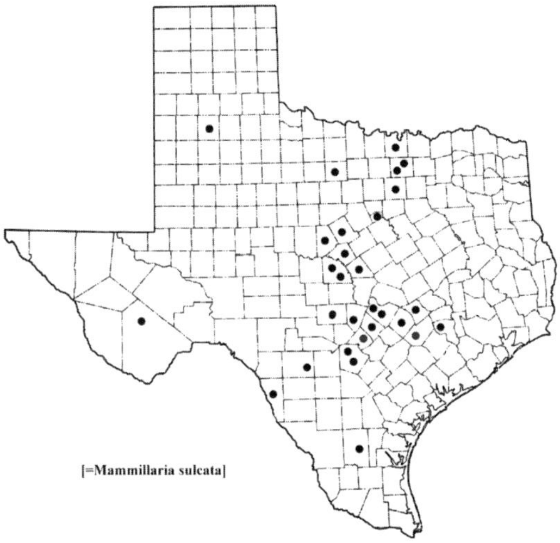
[=Mammillaria sulcata]

Coryphantha sulcata

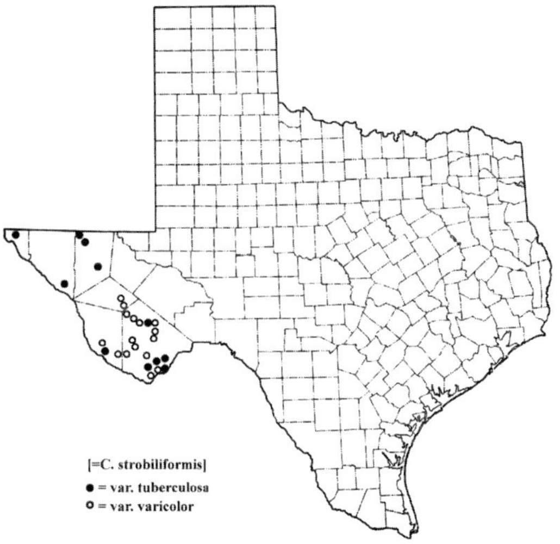
[=C. strobiliformis]
● = var. tuberculosa
○ = var. varicolor

Coryphantha tuberculosa

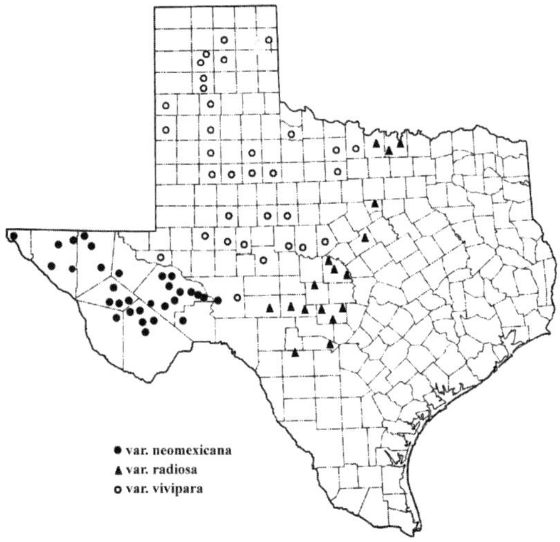
● var. neomexicana
▲ var. radiosa
○ var. vivipara

Coryphantha vivipara

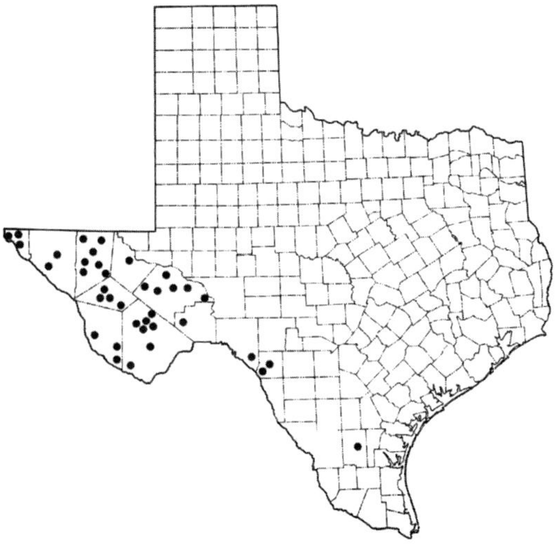

Echinocactus horizonthalonius

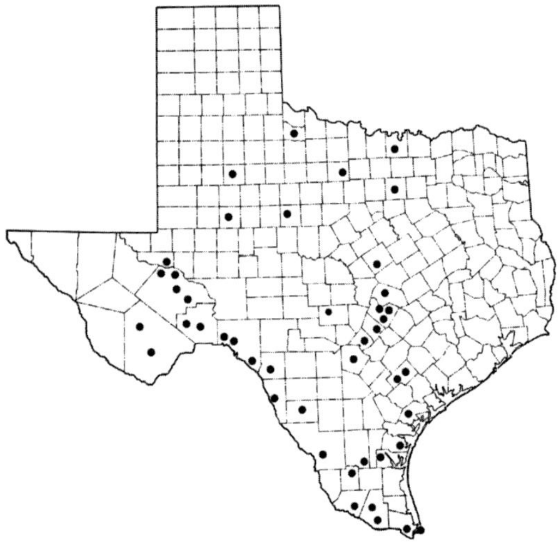

Echinocactus texensis

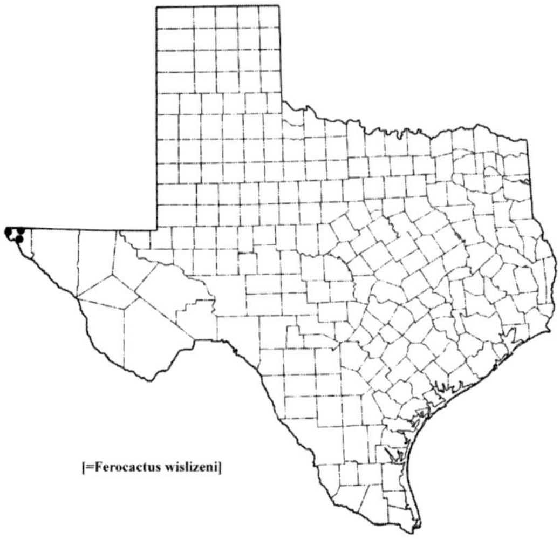
[=Ferocactus wislizeni]

Echinocactus wislizeni

Echinocereus berlandieri

Echinocereus chisoensis

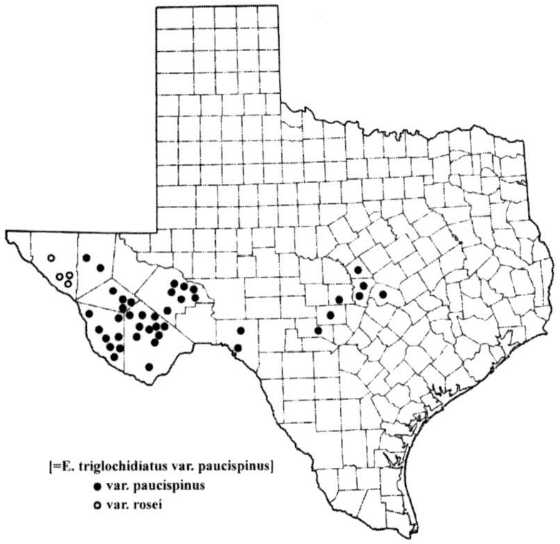
[=E. triglochidiatus var. paucispinus]
● var. paucispinus
○ var. rosei

Echinocereus coccineus

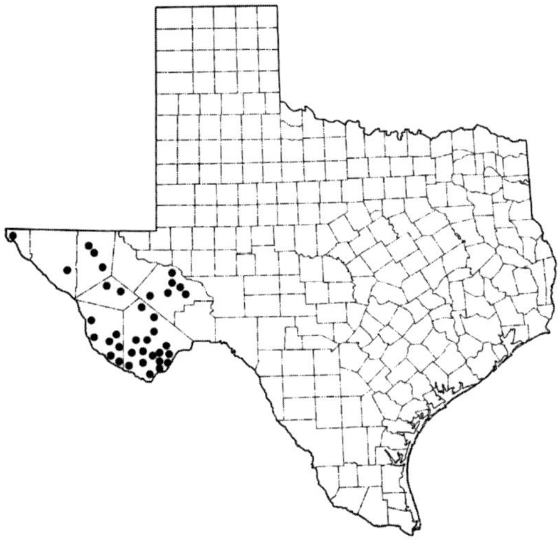

Echinocereus dasyacanthus

Echinocereus davisii

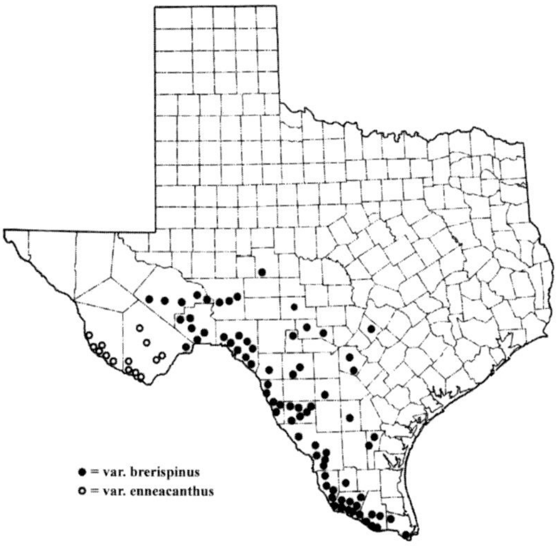
● = var. brerispinus
○ = var. enneacanthus

Echinocereus enneacanthus

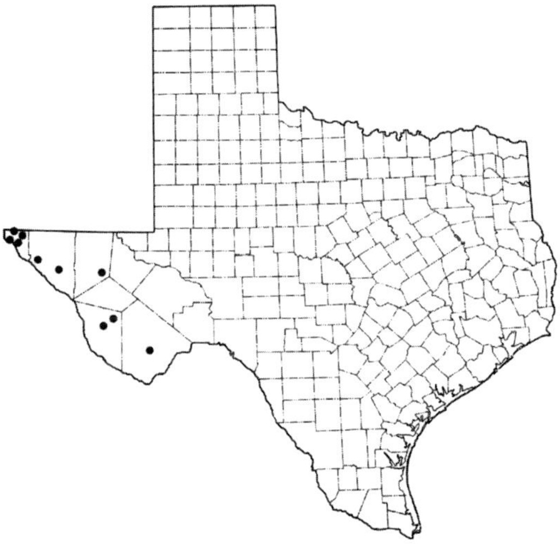

Echinocereus fendleri

Echinocereus fitchii

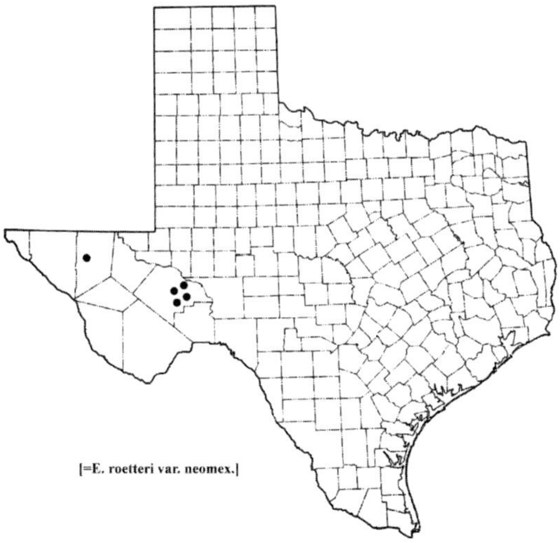
[=E. roetteri var. neomex.]

Echinocereus lloydii

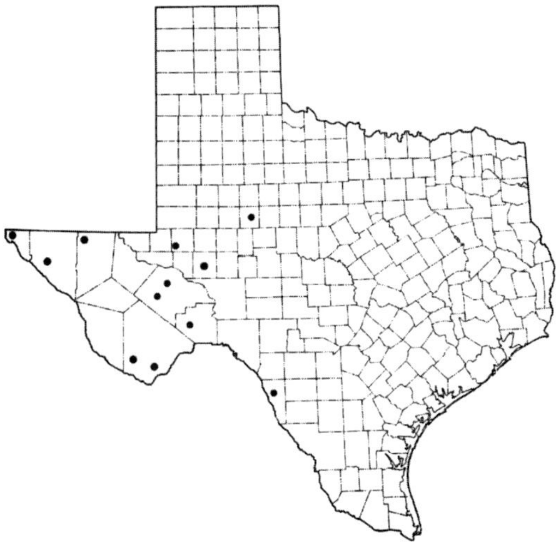

Echinocereus neomexicanus

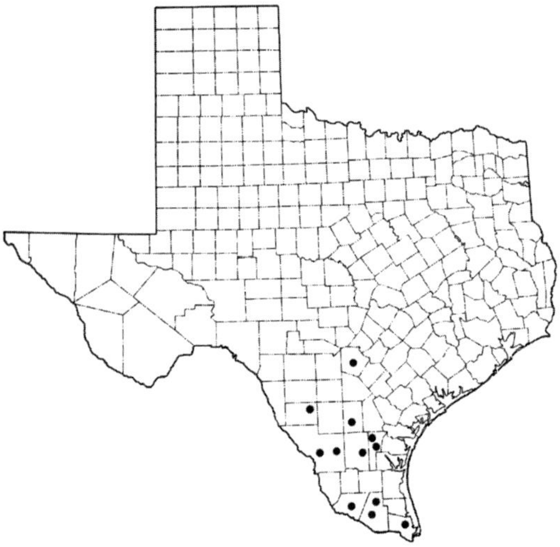

Echinocereus papillosus

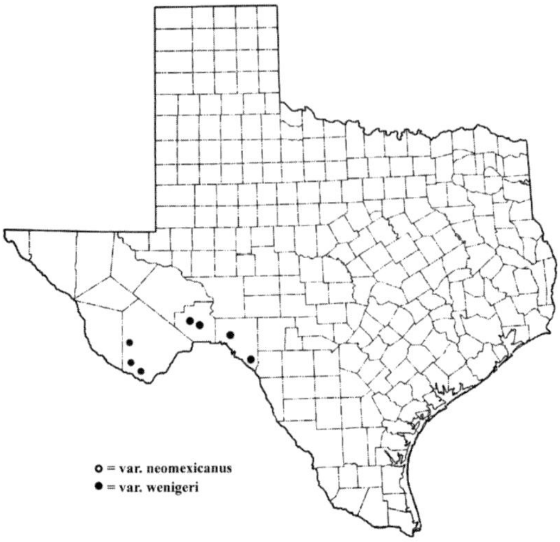
o = var. neomexicanus
● = var. wenigeri

Echinocereus pectinatus

Echinocereus pentalophus

[=Cereus poselgeri = Wilcoxia poselgeri]

Echinocereus poselgeri

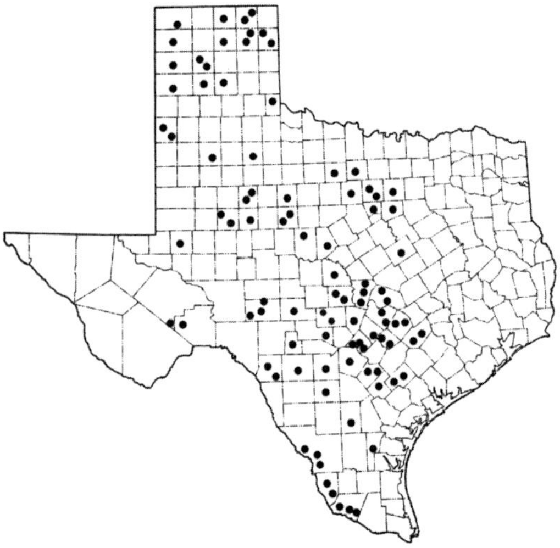

Echinocereus reichenbachii

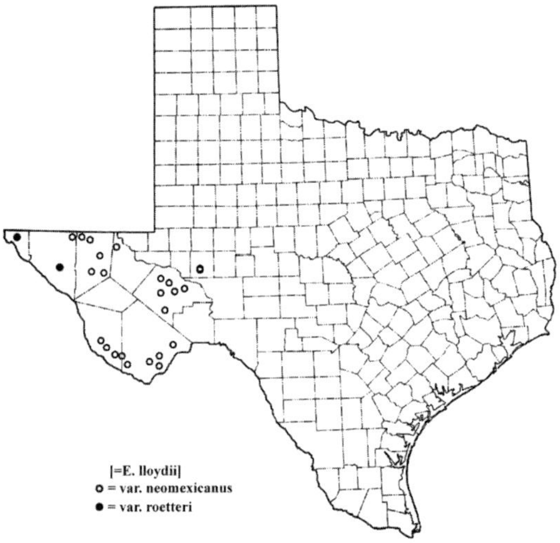
[=E. lloydii]
o = var. neomexicanus
● = var. roetteri

Echinocereus roetteri

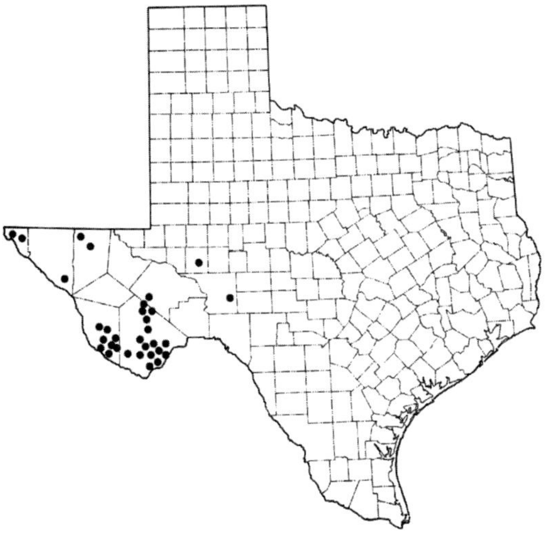

Echinocereus stramineus

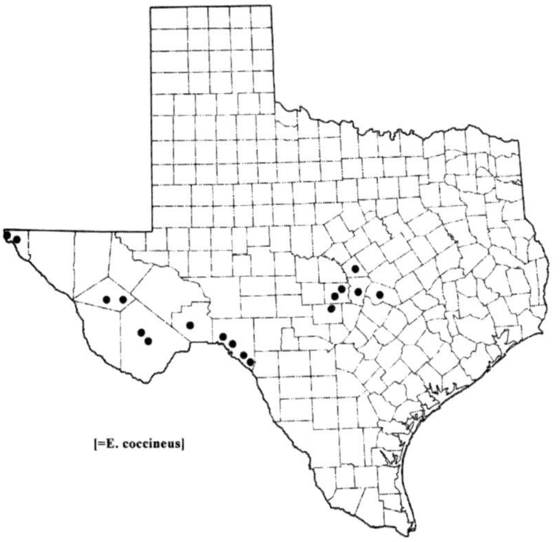
[=E. coccineus]

Echinocereus triglochidiatus

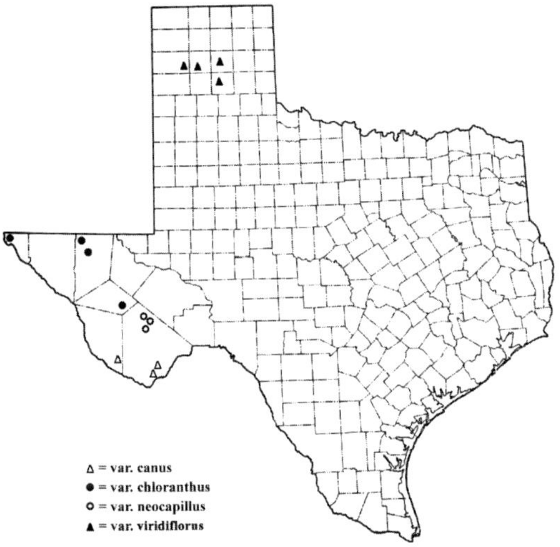
△ = var. canus
● = var. chloranthus
○ = var. neocapillus
▲ = var. viridiflorus

Echinocereus viridiflorus

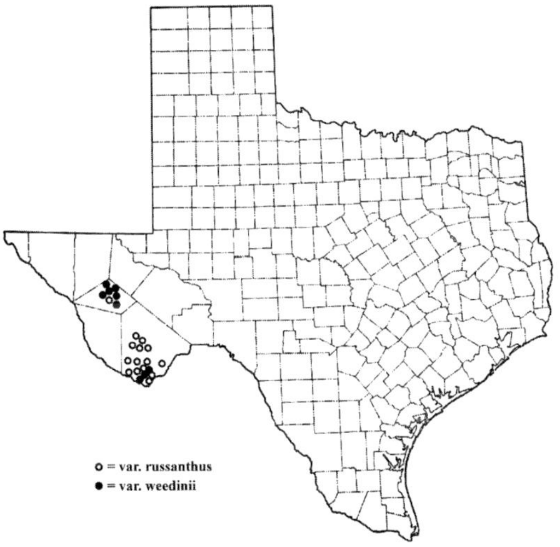
○ = var. russanthus
● = var. weedinii

Echinocereus viridiflorus

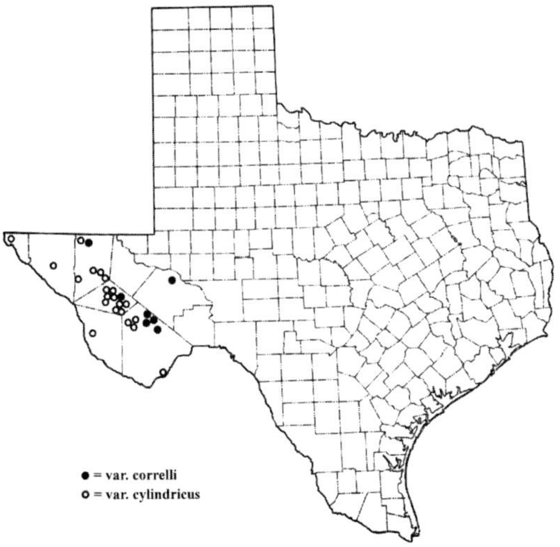
● = var. correlli
○ = var. cylindricus

Echinocereus viridiflorus

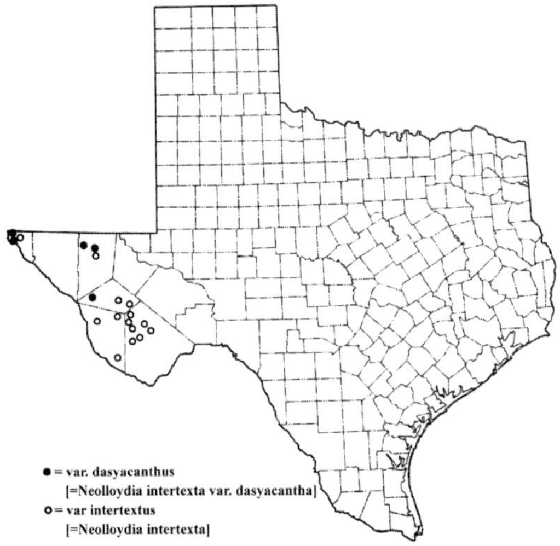
● = var. dasyacanthus
[=Neolloydia intertexta var. dasyacantha]
○ = var intertextus
[=Neolloydia intertexta]

Echinomastus intertextus

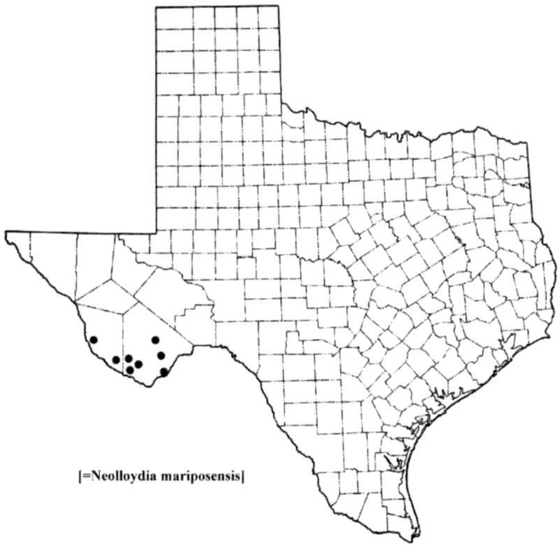
[=Neolloydia mariposensis]

Echinomastus mariposensis

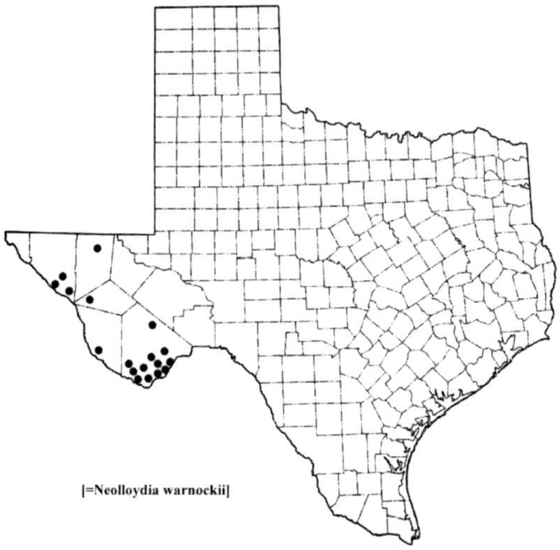
[=Neolloydia warnockii]

Echinomastus warnockii

Epithelantha bokei

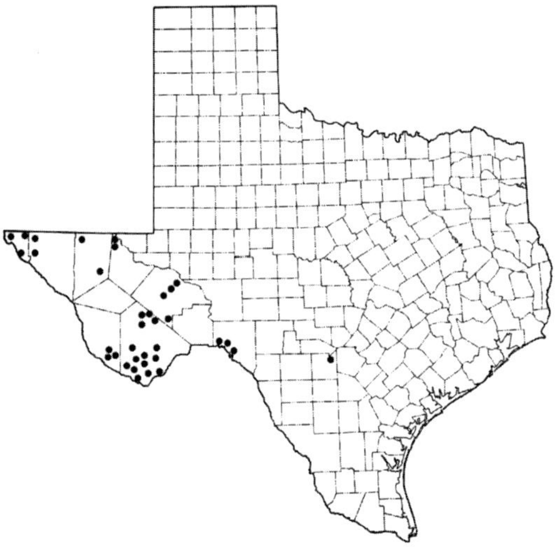

Epithelantha micromeris

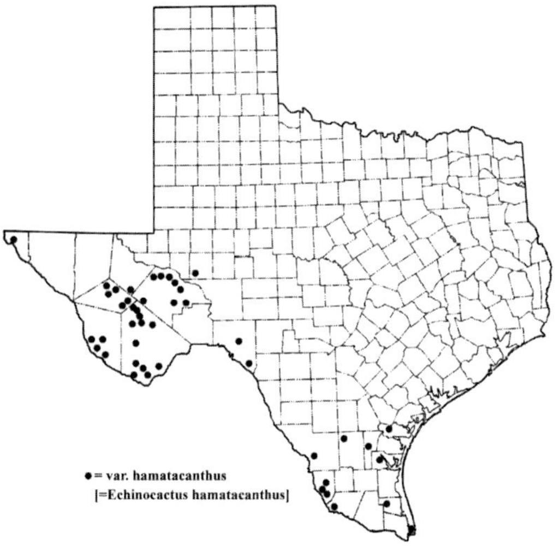
● = var. hamatacanthus
[=Echinocactus hamatacanthus]

Ferocactus hamatacanthus

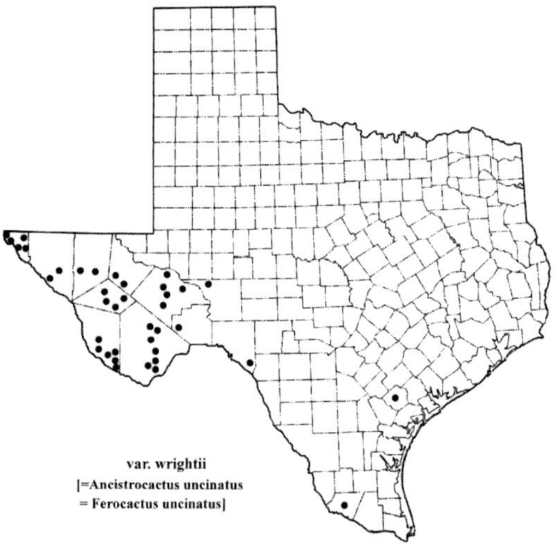
var. wrightii
[=Ancistrocactus uncinatus
= Ferocactus uncinatus]

Glandulicactus uncinatus

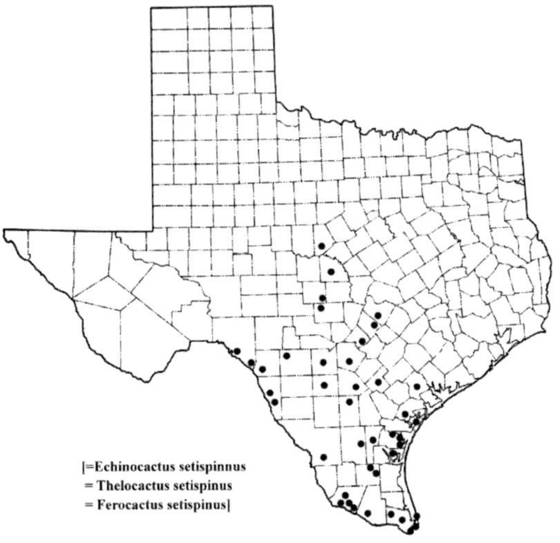
[=Echinocactus setispinnus
= Thelocactus setispinus
= Ferocactus setispinus]

Hamatocactus bicolor

Lophophora williamsii

Mammillaria grahamii

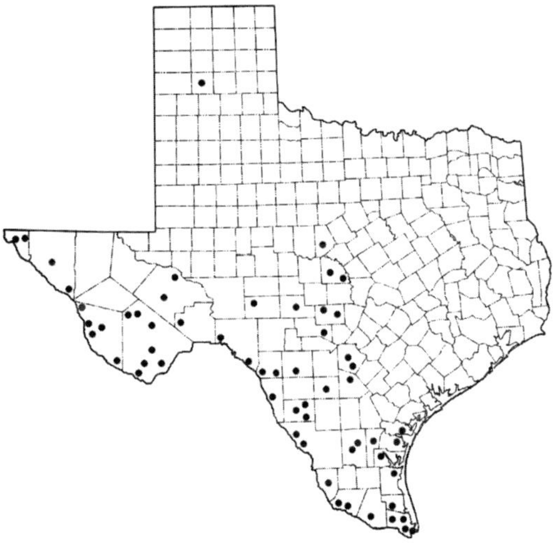

Mammillaria heyderi

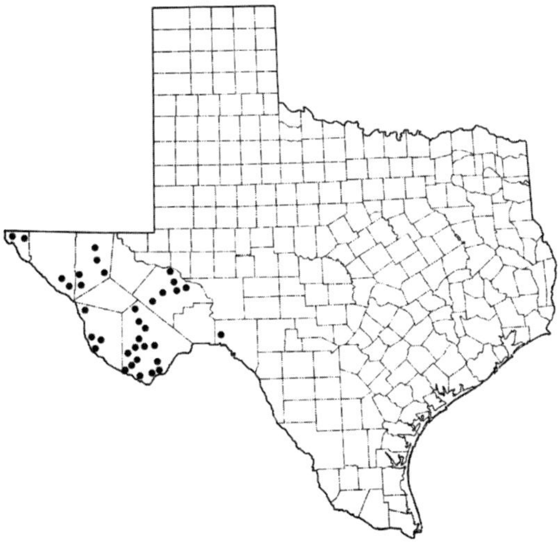

Mammillaria lasiacantha

Mammillaria meicantha

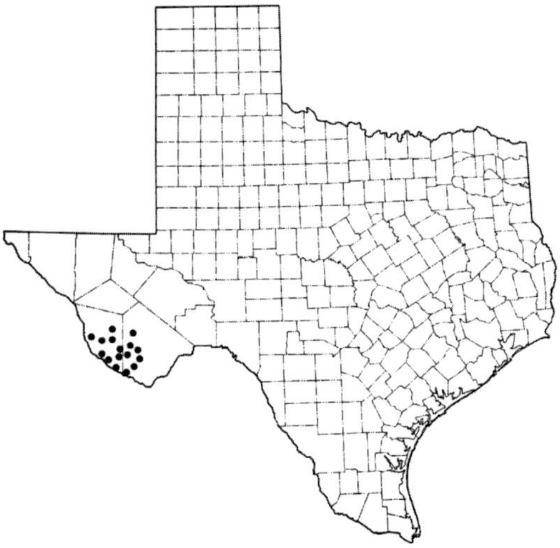

Mammillaria pottsii

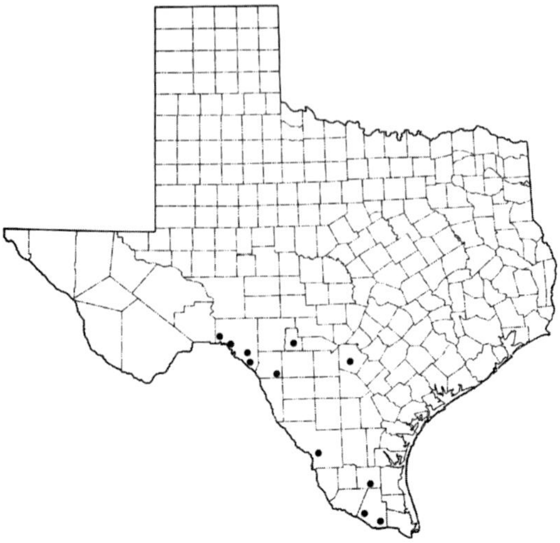

Mammillaria prolifera

Mammillaria wrightii

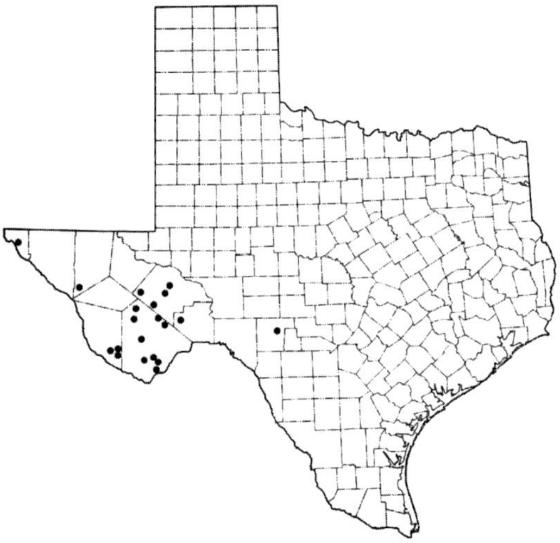

Neolloydia conoidea

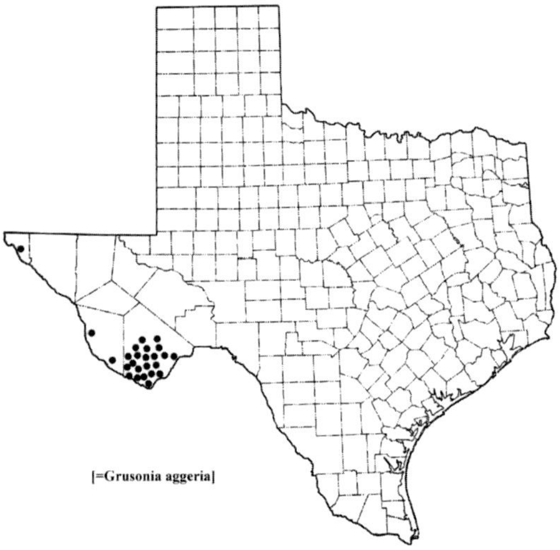
[=Grusonia aggeria]

Opuntia aggeria

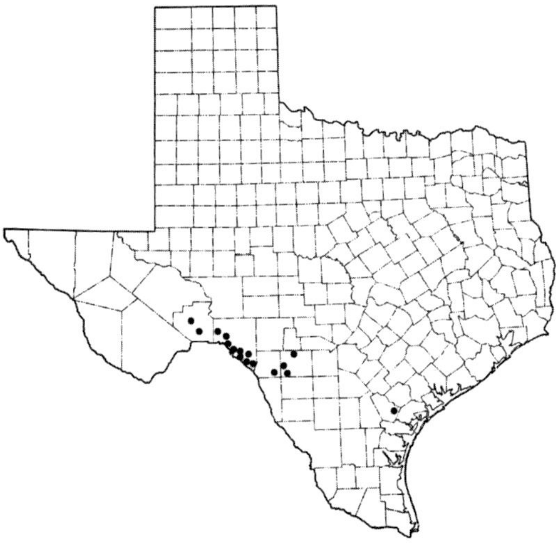

Opuntia atrispina

[=O. macrocentra var. aureispina]

Opuntia aureispina

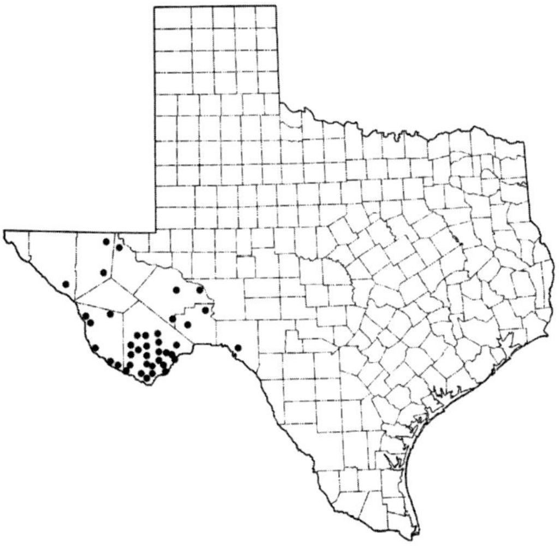

Opuntia camanchica

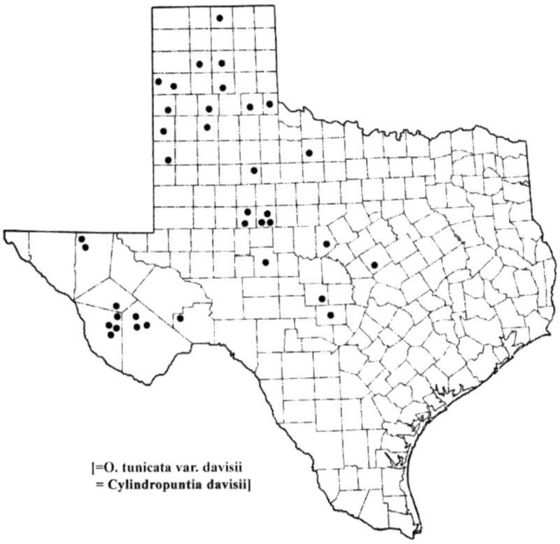
[=O. tunicata var. davisii
= Cylindropuntia davisii]

Opuntia davisii

Opuntia densispina

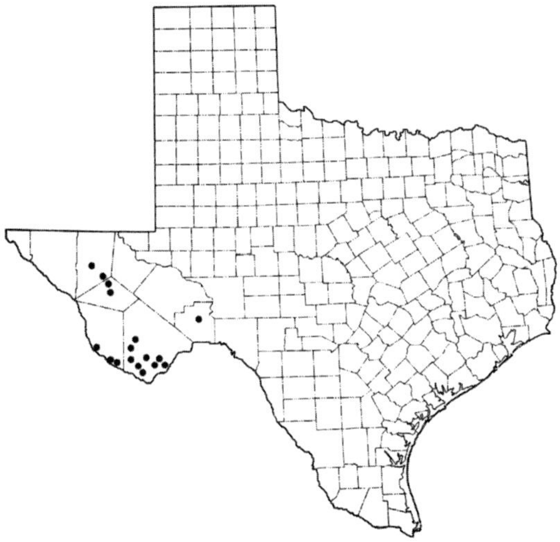

Opuntia dulcis

Opuntia edwardsii

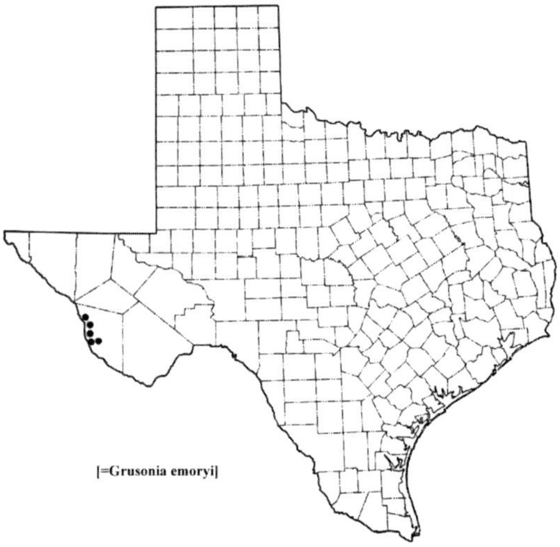
[=Grusonia emoryi]

Opuntia emoryi

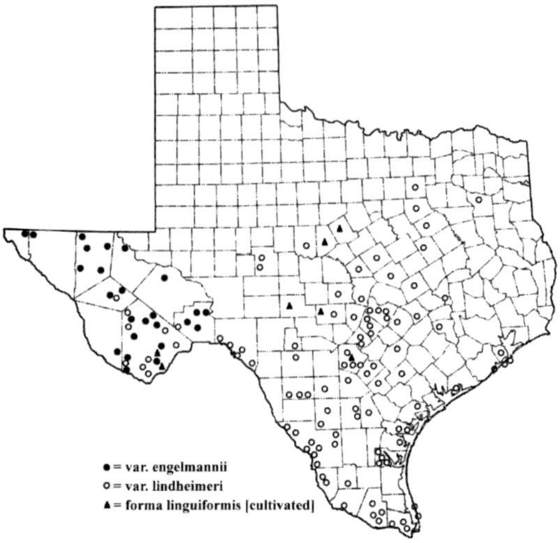
● = var. engelmannii
○ = var. lindheimeri
▲ = forma linguiformis [cultivated]

Opuntia engelmannii

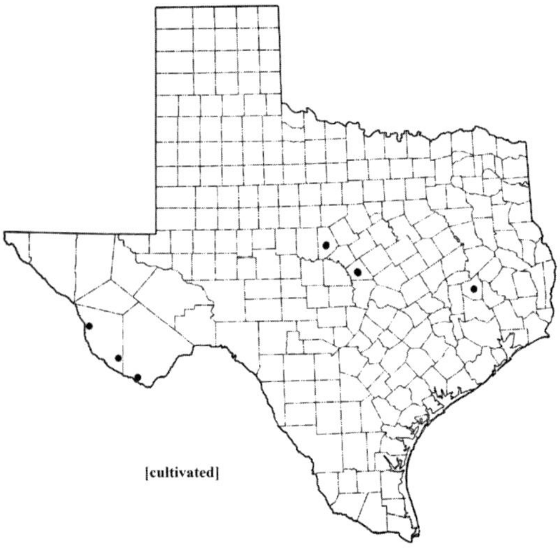
[cultivated]

Opuntia ficus-indica

Opuntia fragilis

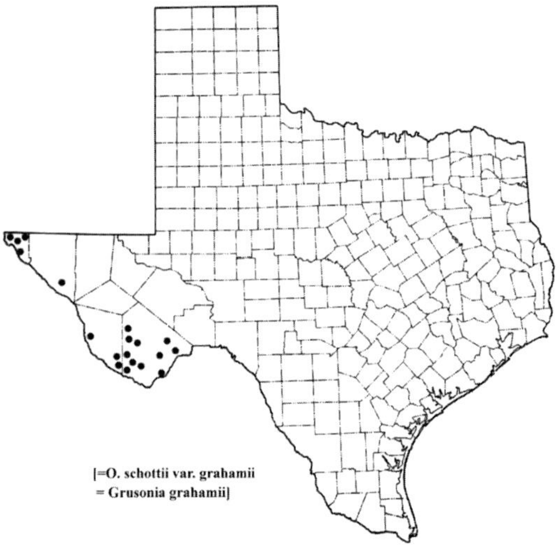
[=O. schottii var. grahamii
= Grusonia grahamii]

Opuntia grahamii

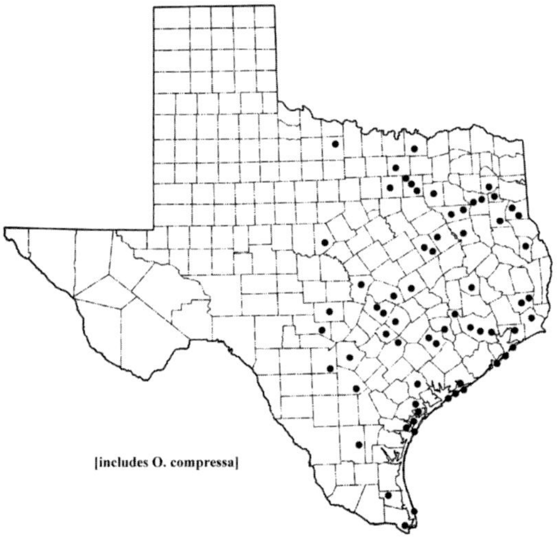
[includes O. compressa]

Opuntia humifusa

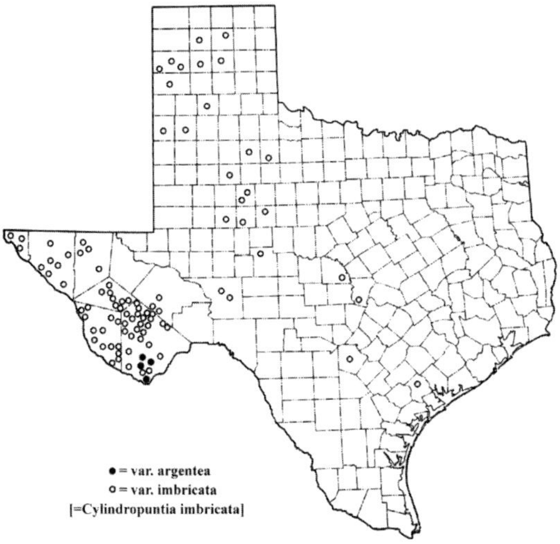
● = var. argentea
○ = var. imbricata
[=Cylindropuntia imbricata]

Opuntia imbricata

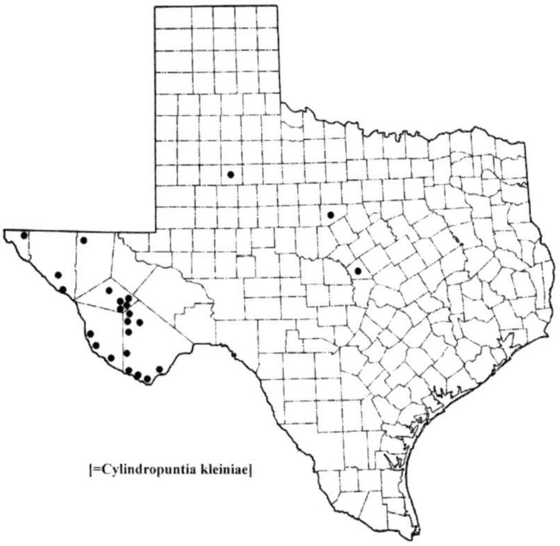
[=Cylindropuntia kleiniae]

Opuntia kleiniae

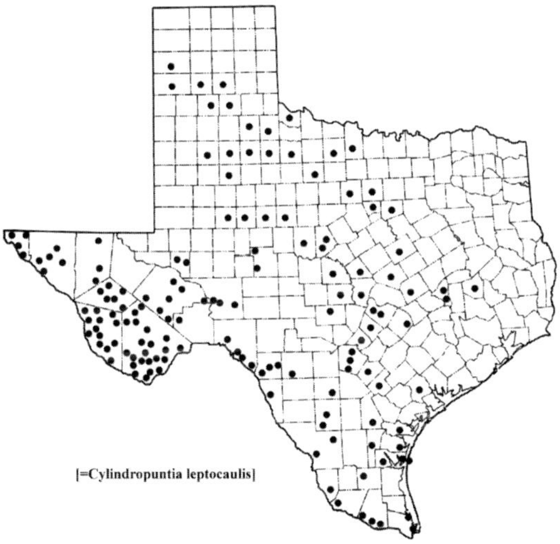
[=Cylindropuntia leptocaulis]

Opuntia leptocaulis

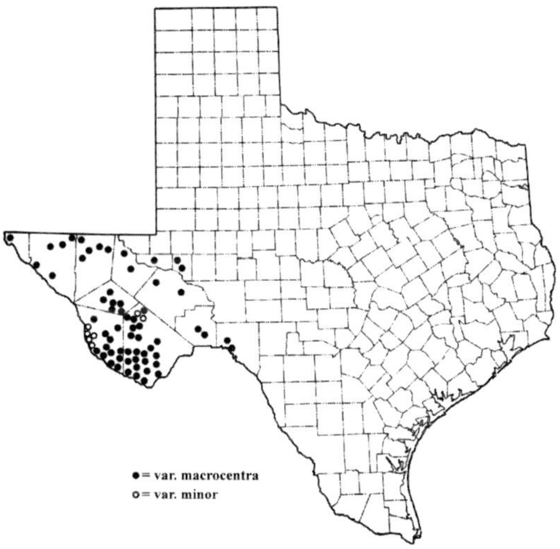
● = var. macrocentra
○ = var. minor

Opuntia macrocentra

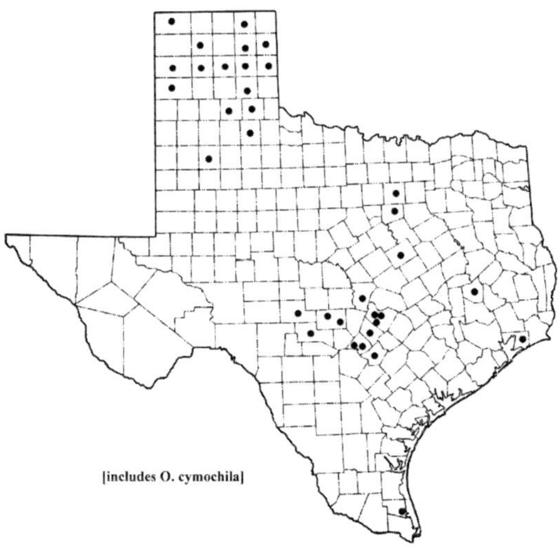
[includes O. cymochila]

Opuntia macrorhiza

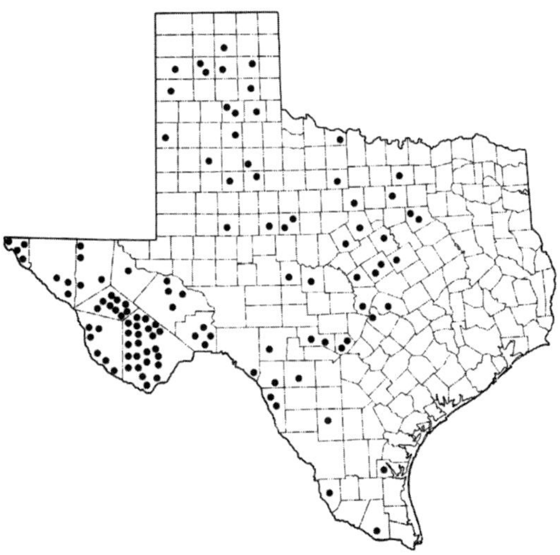

Opuntia phaeacantha

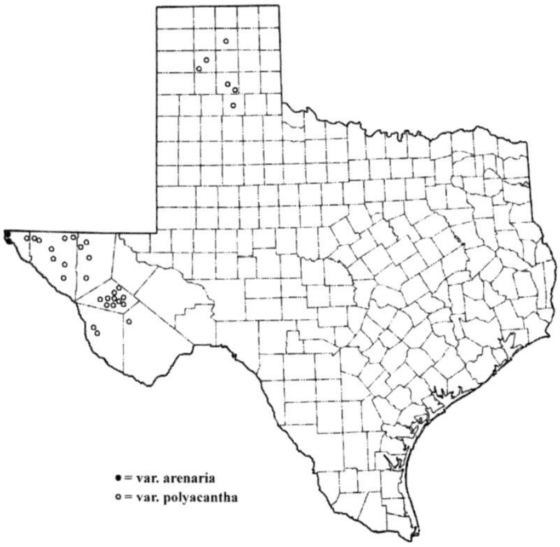
• = var. arenaria
○ = var. polyacantha
Opuntia polyacantha

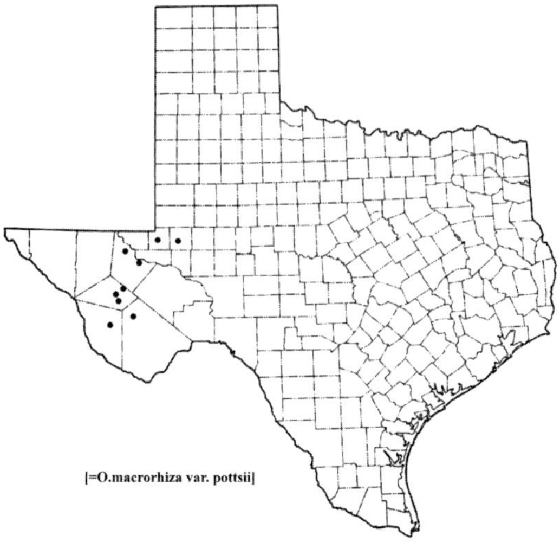
[=O.macrorhiza var. pottsii]
Opuntia pottsii

Opuntia rufida

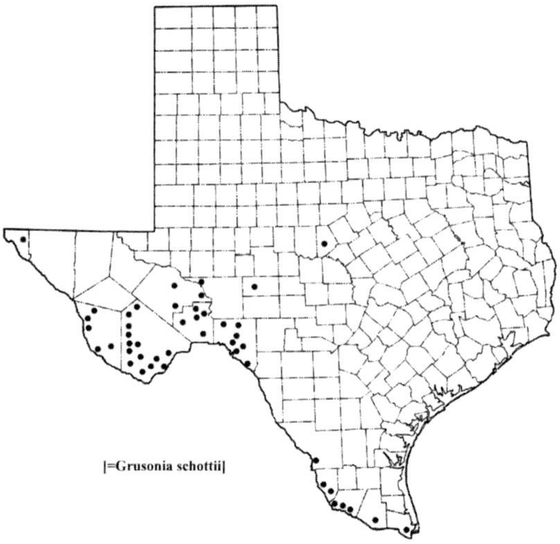
[=Grusonia schottii]
Opuntia schottii

Opuntia spinosibacca

Opuntia stricta

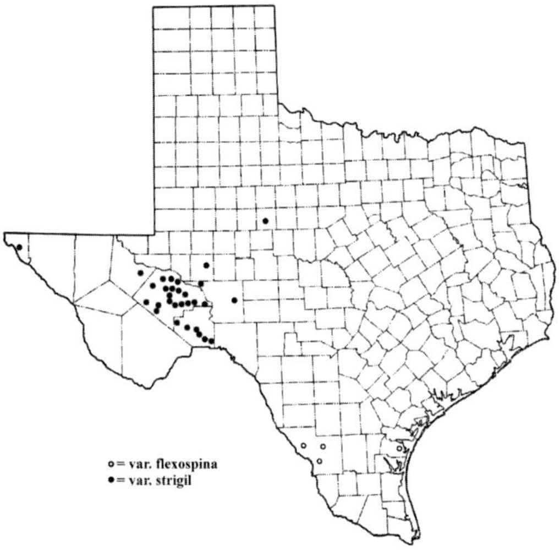
o= var. flexospina
•= var. strigil

Opuntia strigil

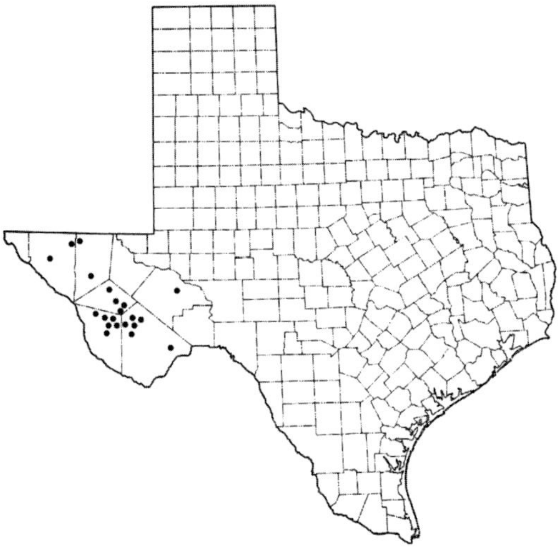

Opuntia tortispina

[=Cylindropuntia tunicata]

Opuntia tunicata

[=Cereus greggii]

Peniocereus greggii

Pereskia aculeata

o= var. bicolor
•= var. flavidispinus

Thelocactus bicolor

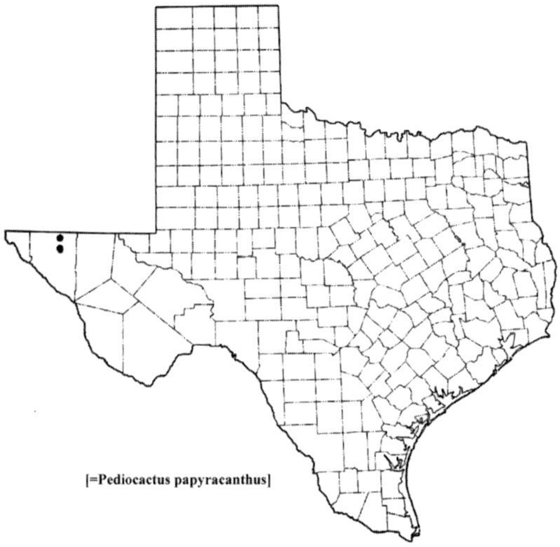
[=Pediocactus papyracanthus]

Toumeya papyracantha

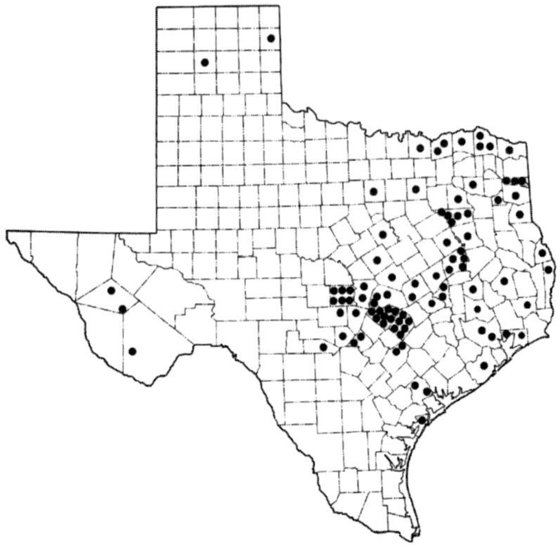

Callitriche heterophylla

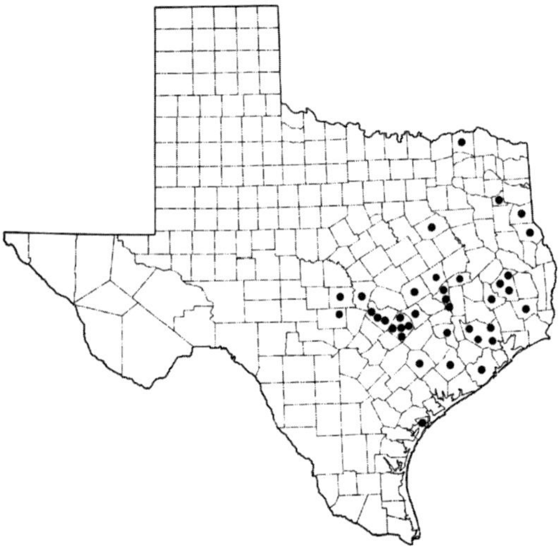

Callitriche nuttallii

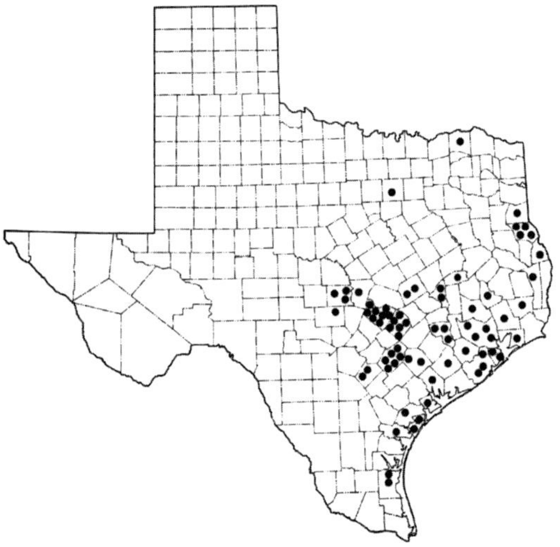

Callitriche peploides

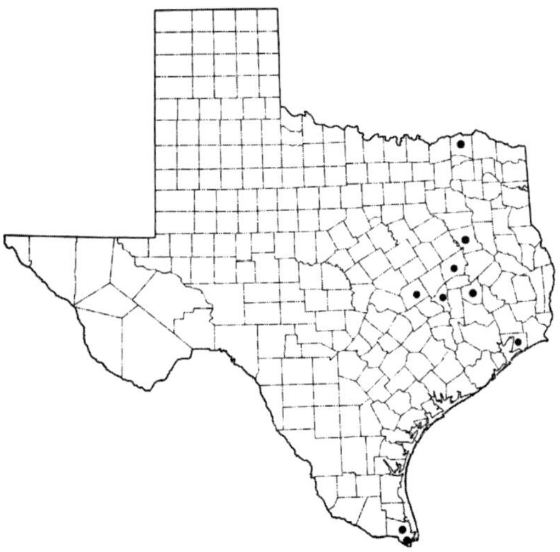

Callitriche terrestris

Campanula rapunculoides_

Campanula reverchonii

Campanula rotundifolia

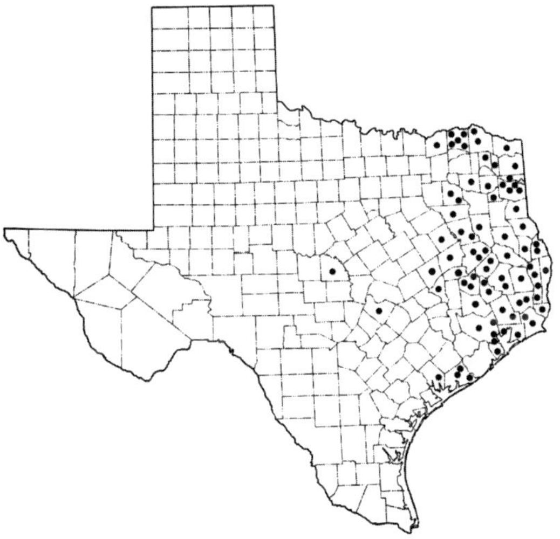

Lobelia appendiculata

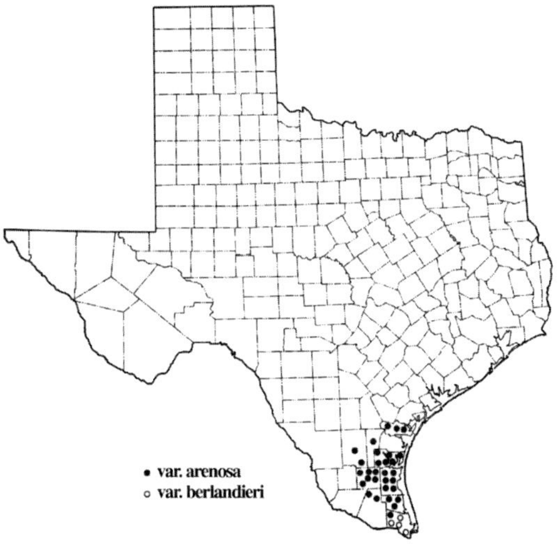
• var. arenosa
○ var. berlandieri

Lobelia berlandieri

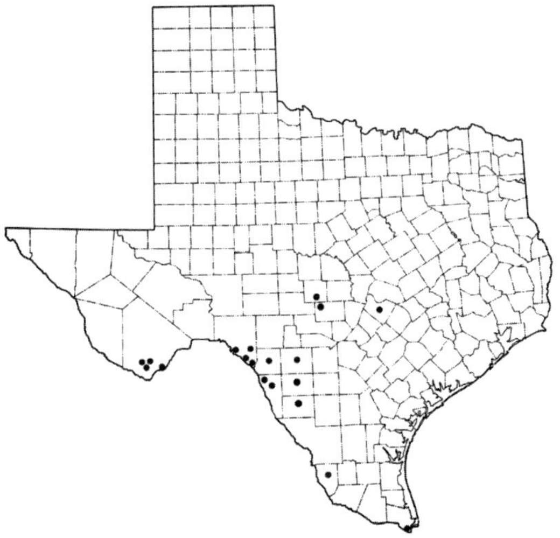

Lobelia brachypoda

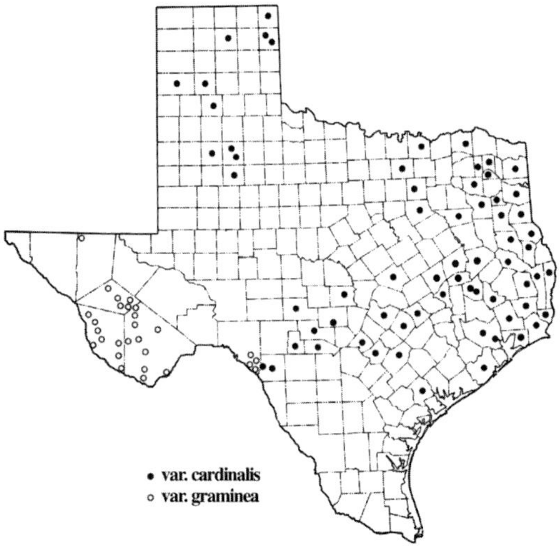
• var. cardinalis
○ var. graminea

Lobelia cardinalis

Lobelia fenestralis

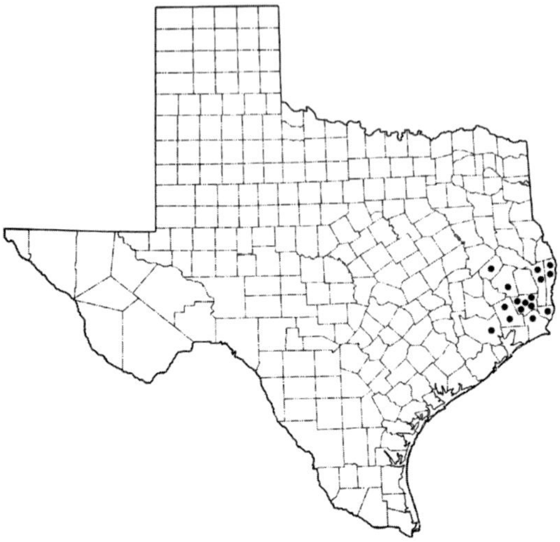

Lobelia flaccidifolia

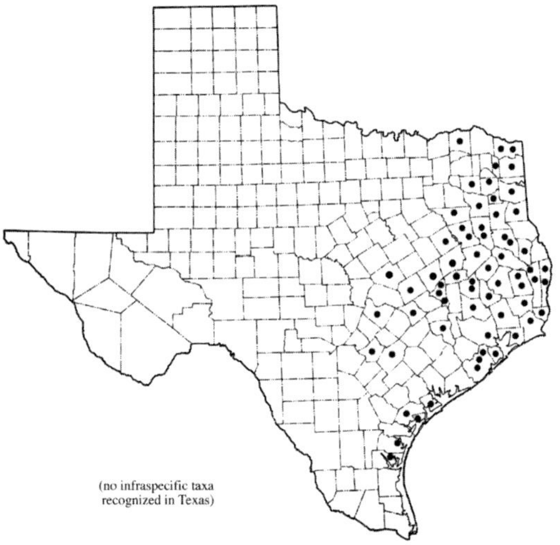
(no infraspecific taxa
recognized in Texas)

Lobelia puberula

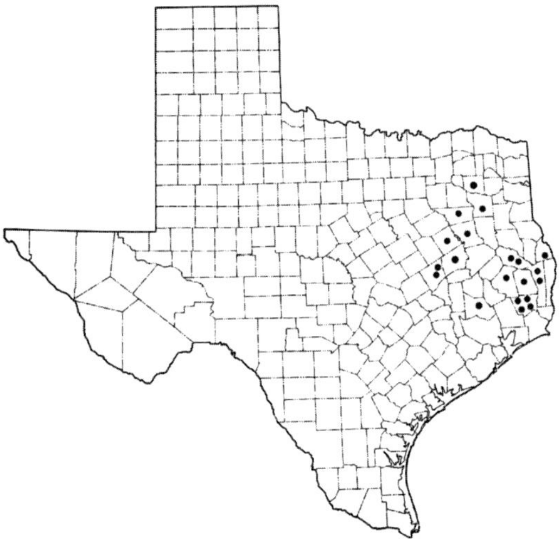

Lobelia reverchonii

Lobelia siphilitica

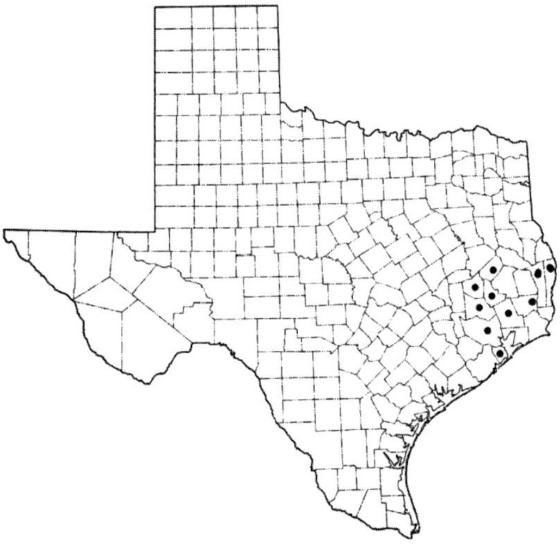

Lobelia spicata

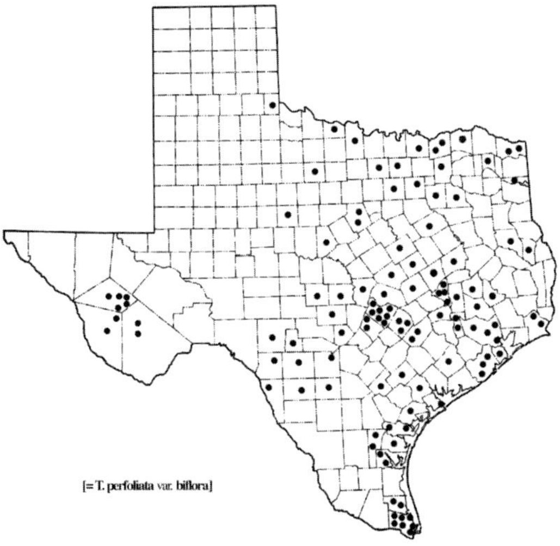
[= T. perfoliata var. biflora]
Triodanus biflora

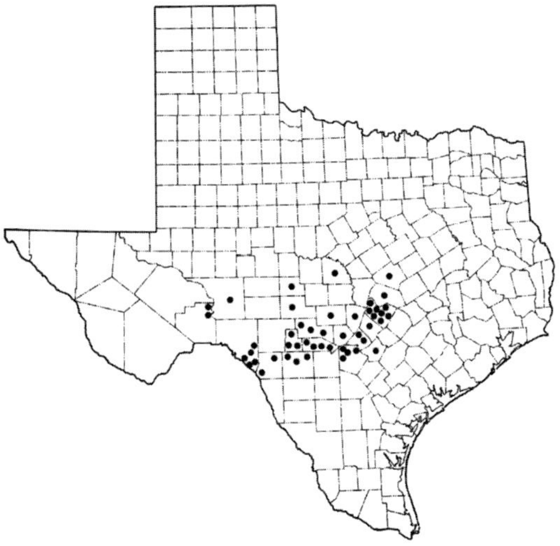
Triodanus coloradoensis

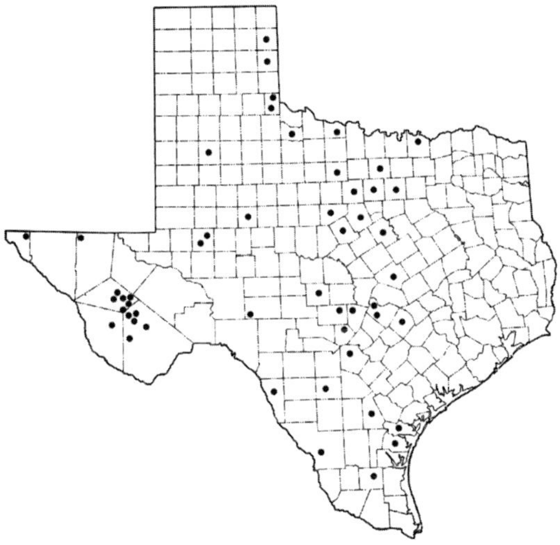
Triodanus holzingeri

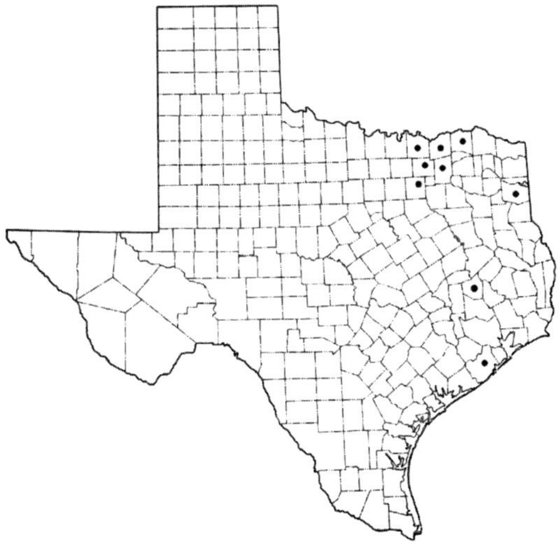
Triodanus lamprosperma

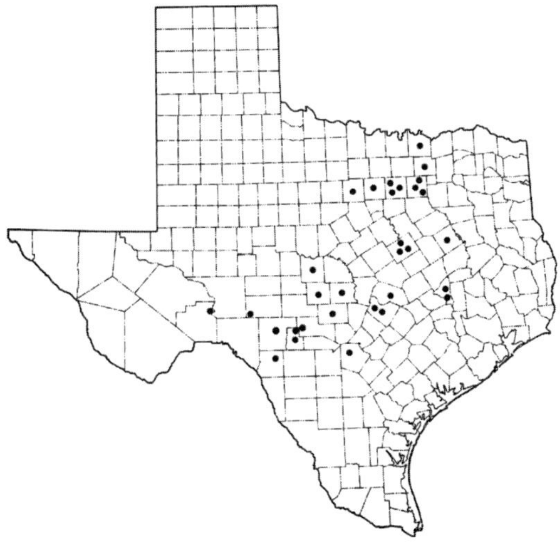
Triodanus leptocarpa

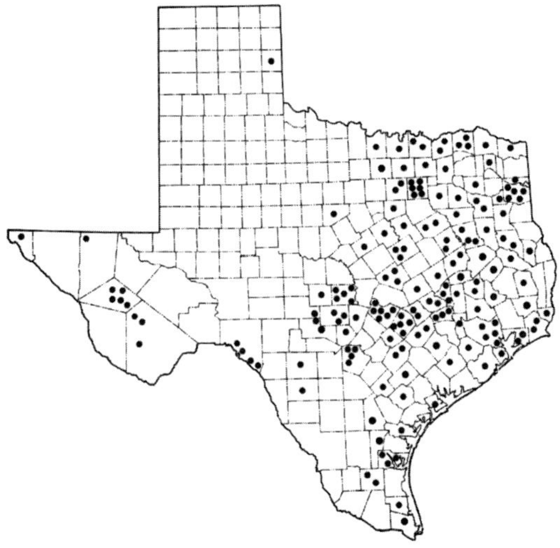
Triodanus perfoliata

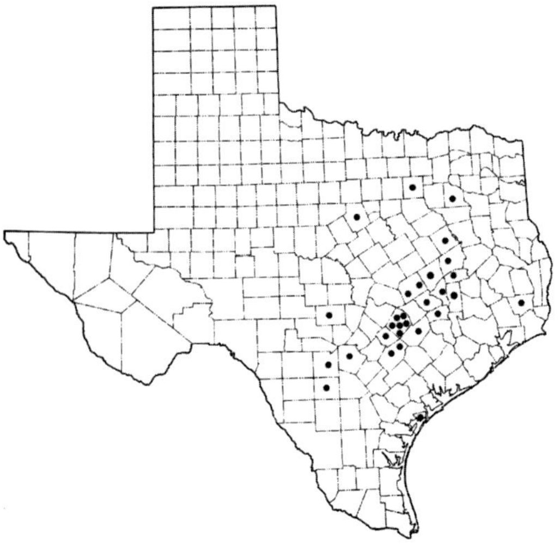

Triodanus texana

Wahlenbergia marginata

Cannabis sativa

Capparis incana

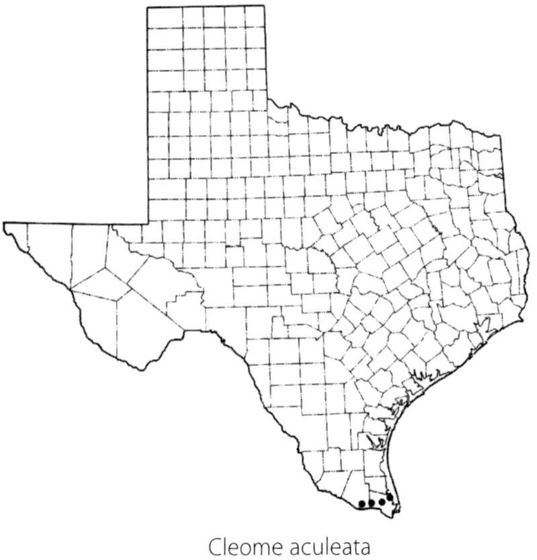

Cleome aculeata

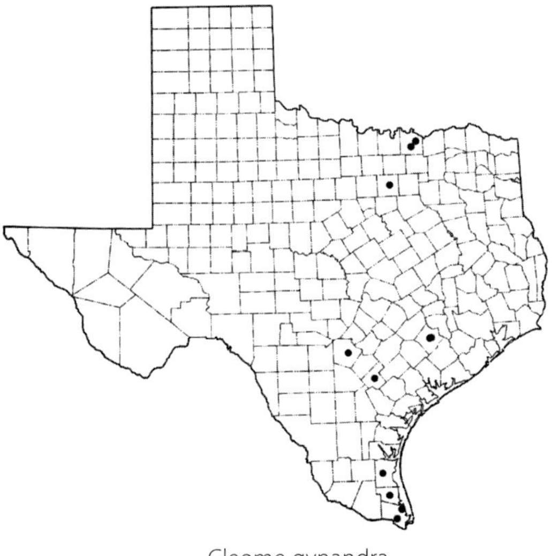

Cleome gynandra

Cleome multicaulis

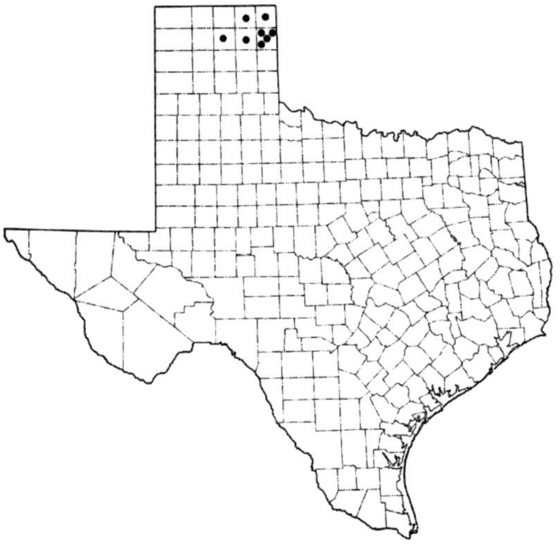

Cleome serrulata

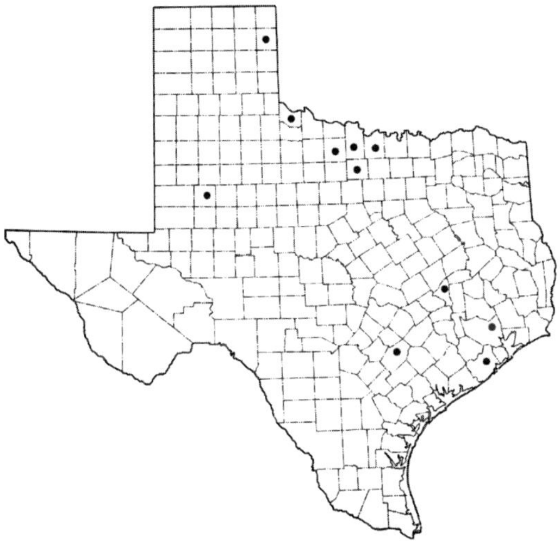

Cleomella angustifolia

Cleomella longipes

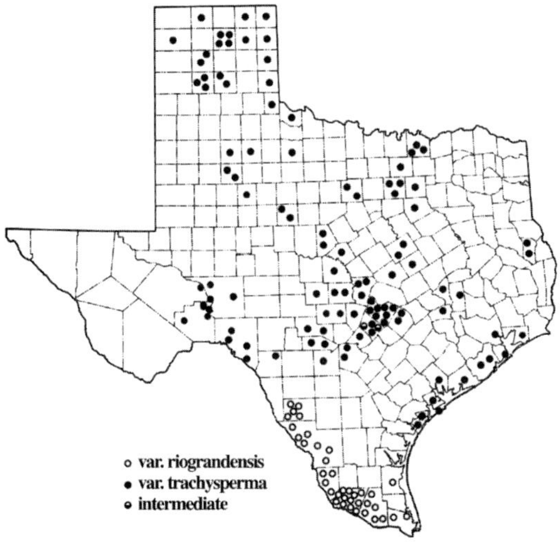
var. riograndensis
var. trachysperma
intermediate

Polanisia dodecandra

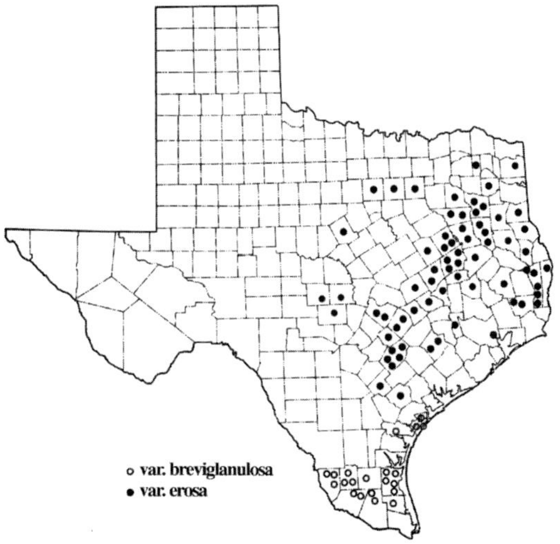
var. breviglanulosa
var. erosa

Polanisia erosa

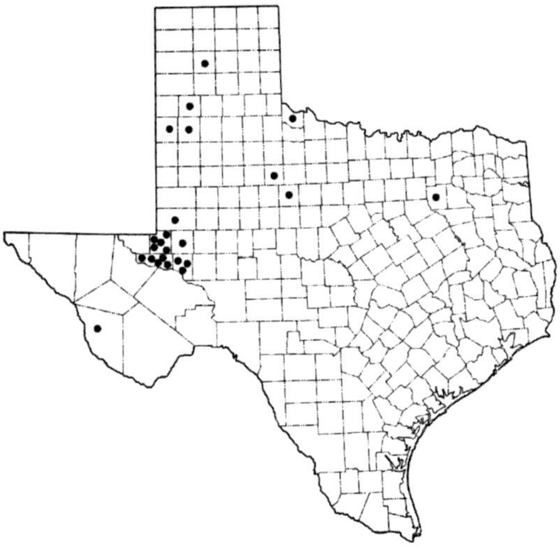

Polanisia jamesii

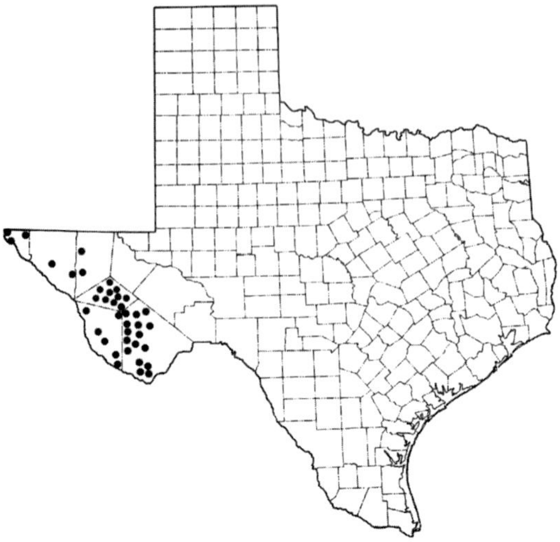

Polanisia uniglandulosa

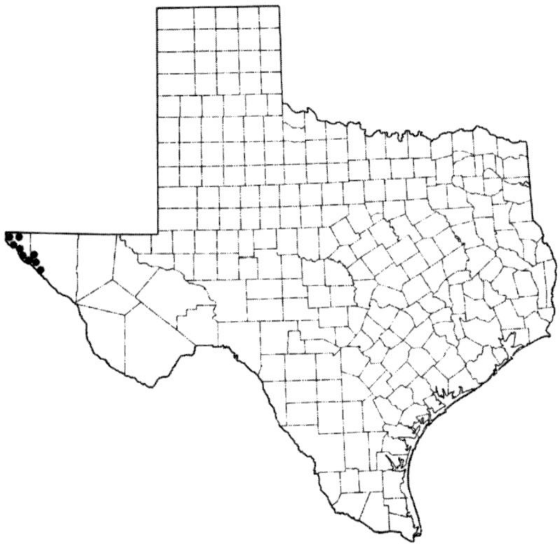

Wislizenia refracta

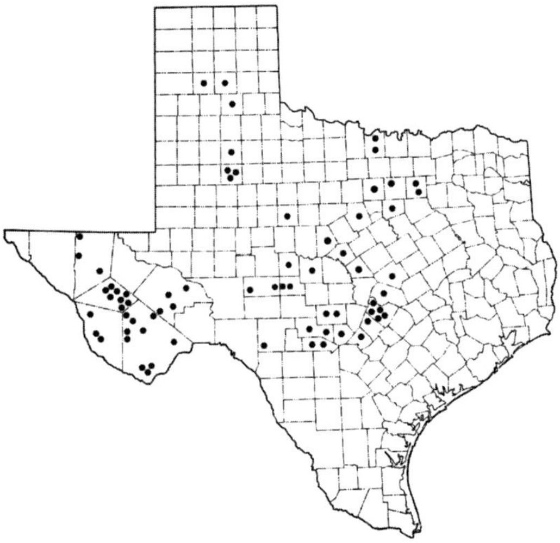

Lonicera albiflora

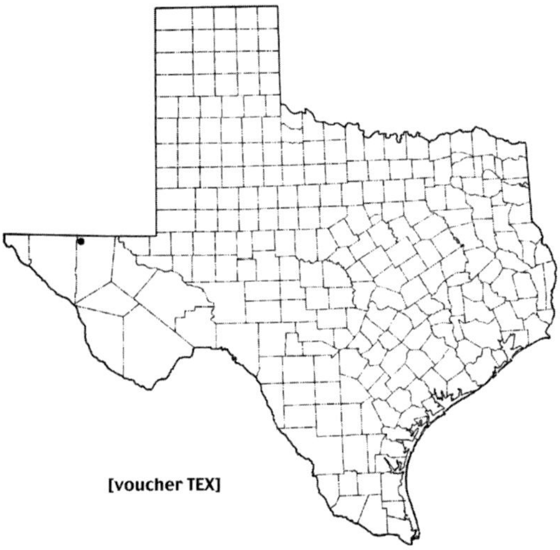
[voucher TEX]

Lonicera arizonica

Lonicera fragrantissima

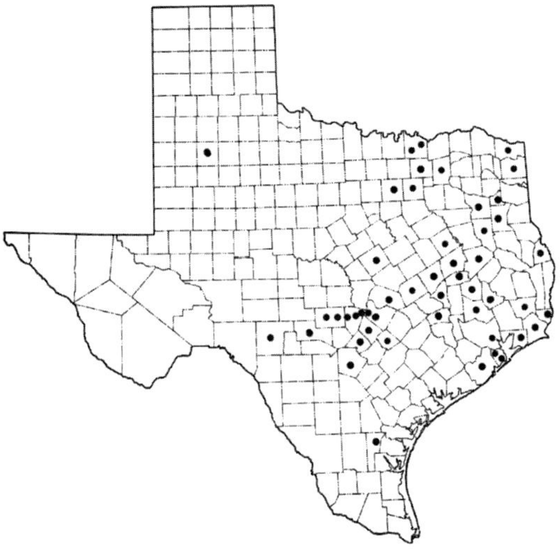

Lonicera japonica

Lonicera maackii

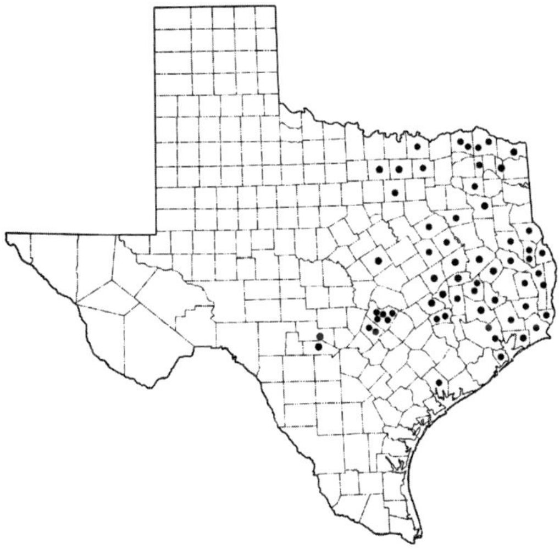

Lonicera sempervirens

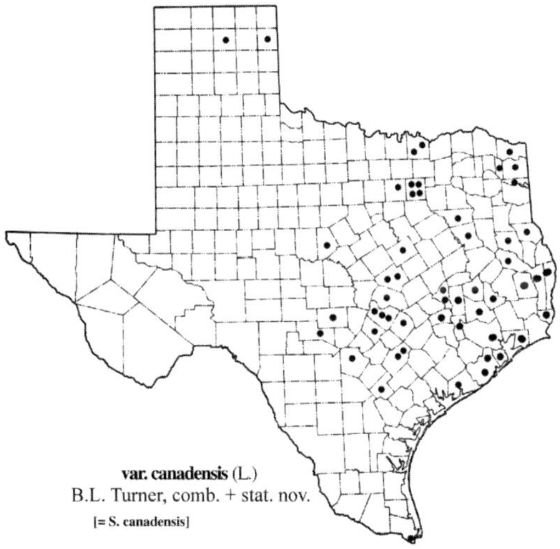
var. canadensis (L.)
B.L. Turner, comb. + stat. nov.
[= S. canadensis]

Sambucus nigra

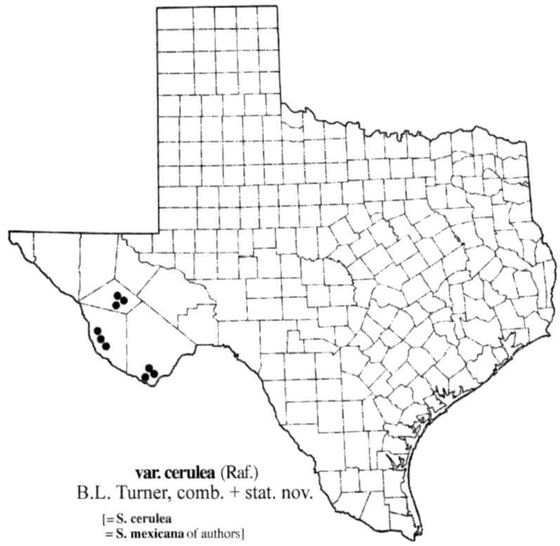
var. cerulea (Raf.)
B.L. Turner, comb. + stat. nov.
[= S. cerulea
= S. mexicana of authors]

Sambucus nigra

[includes S. guadalupensis]

Symphoricarpos longiflorus

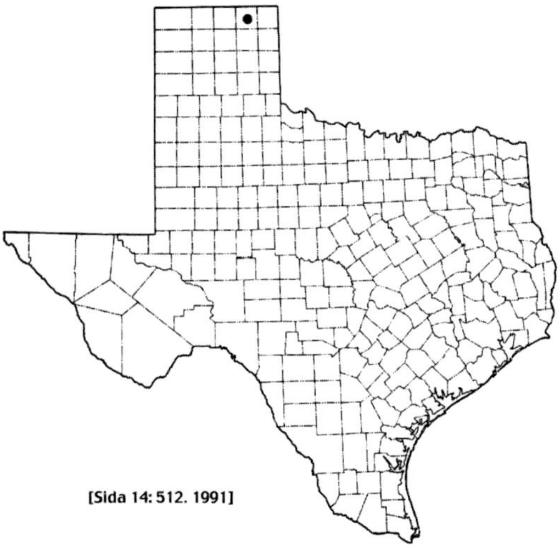
[Sida 14: 512. 1991]

Symphoricarpos occidentalis

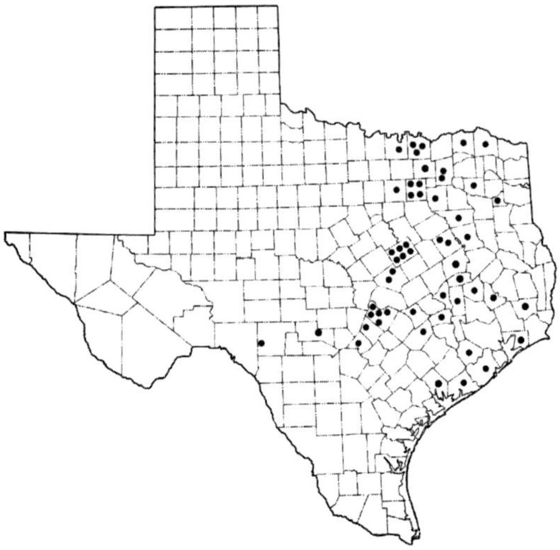

Symphoricarpos orbiculatus

Symphoricarpos oreophilus

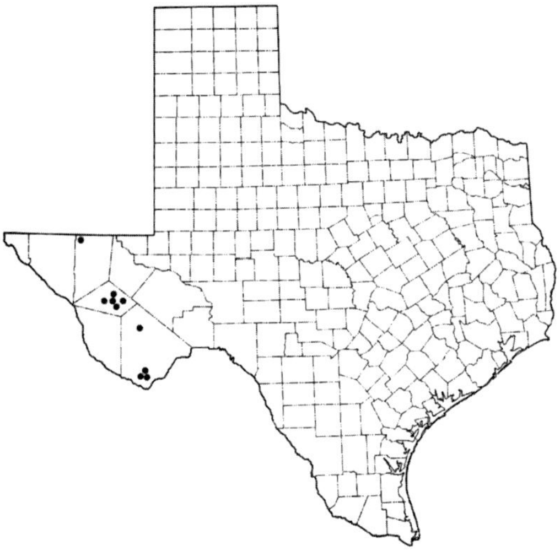

Symphoricarpos palmeri

Symphoricarpos rotundifolius

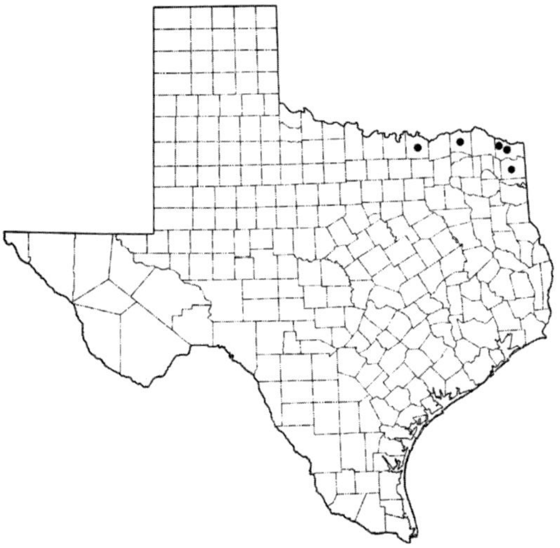

Triosteum angustifolium

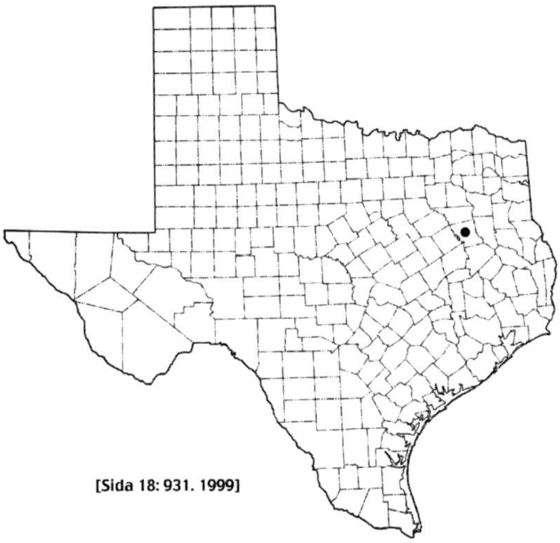
[Sida 18: 931. 1999]

Triosteum perfoliatum

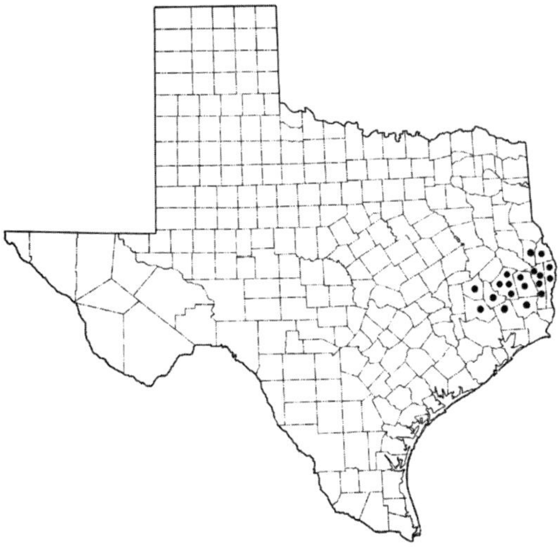

Viburnum acerifolium

Viburnum ashei

Viburnum nitidum

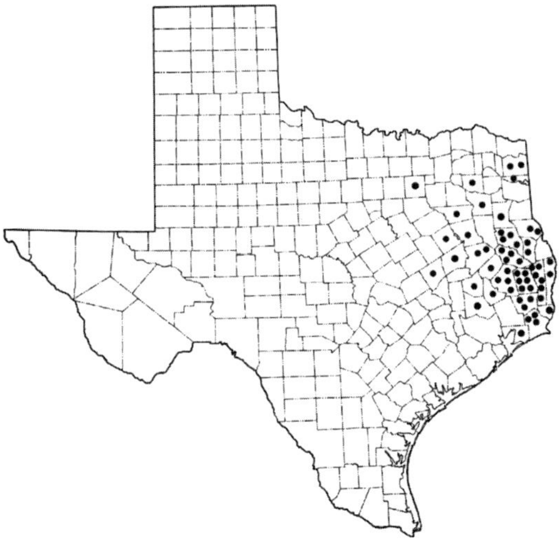

Viburnum nudum

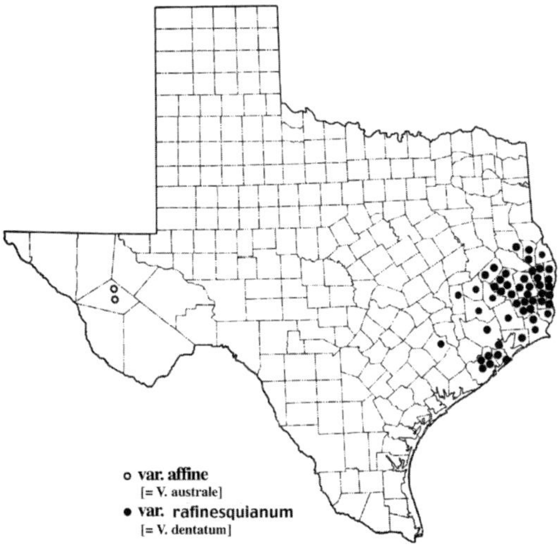
var. affine
[= V. australe]
var. rafinesquianum
[= V. dentatum]

Viburnum rafinesquianum

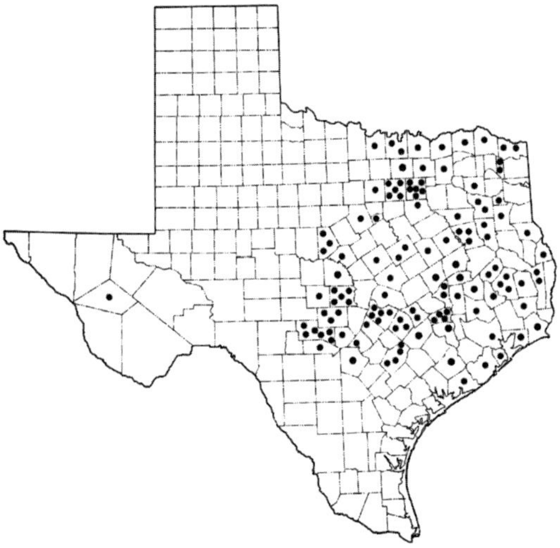

Viburnum rufidulum

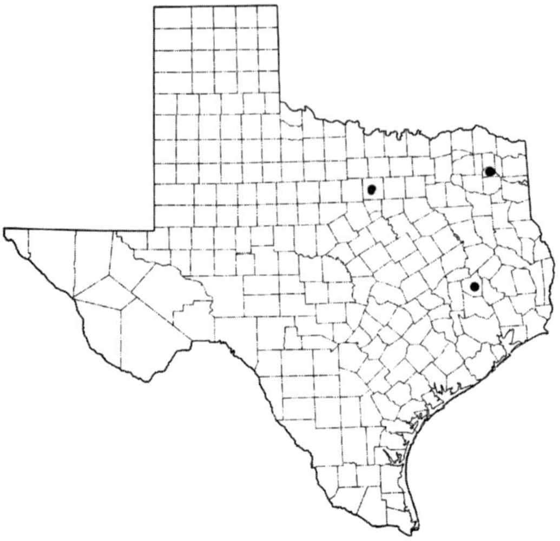

Agrostemma githago

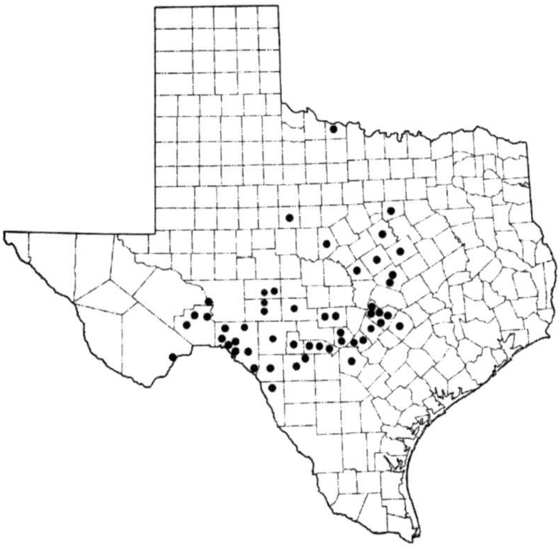

Arenaria benthamii

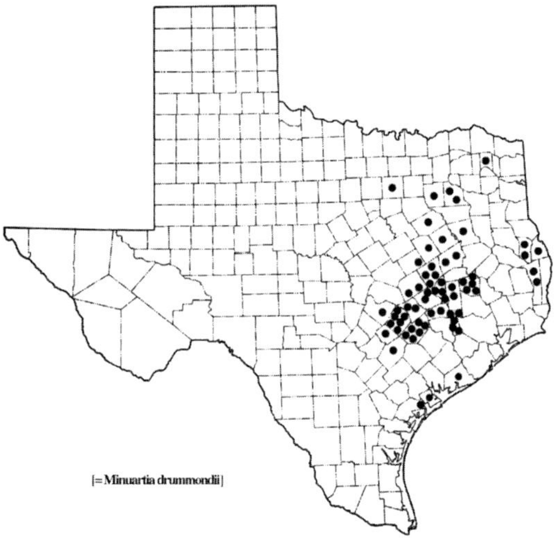
[=Minuartia drummondii]

Arenaria drummondii

Arenaria fendleri

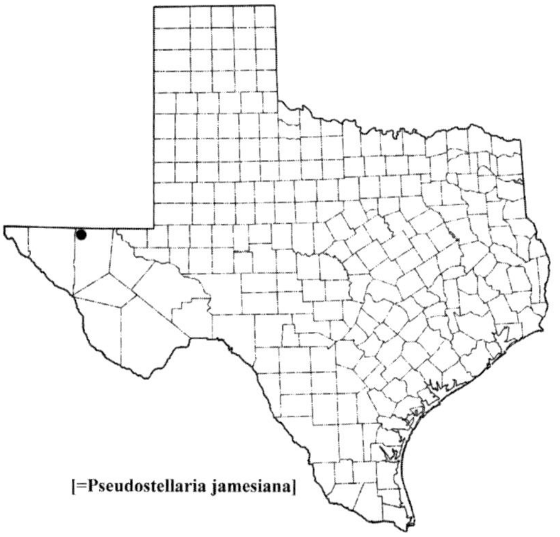
[=Pseudostellaria jamesiana]

Arenaria jamesiana

Arenaria lanuginosa

Arenaria livermorensis

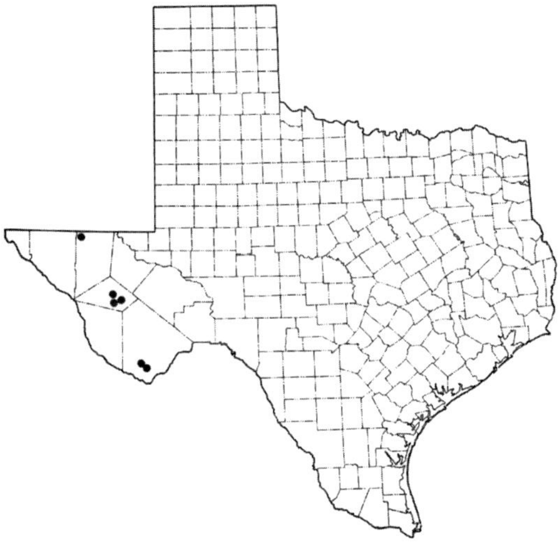

Arenaria ludens

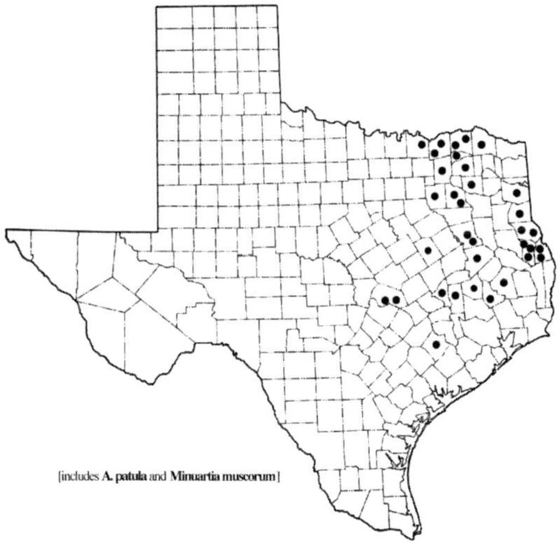
[includes A. patula and Minuartia muscorum]

Arenaria muriculata

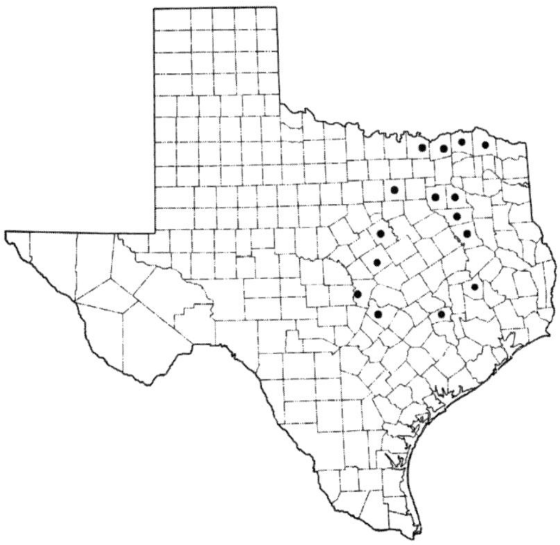

Arenaria serpyllifolia

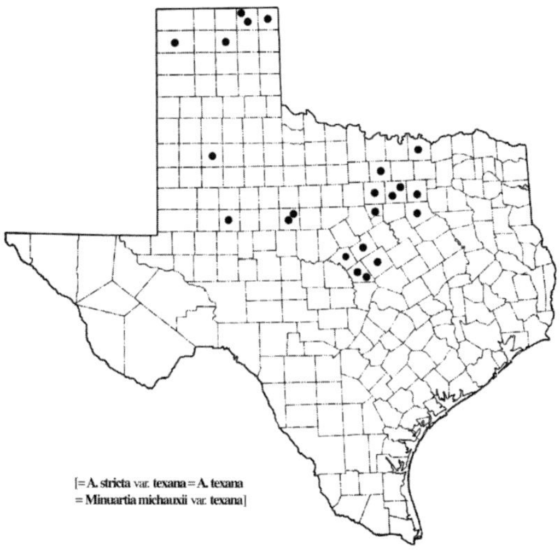
[= A. stricta var. texana = A. texana
= Minuartia michauxii var. texana]

Arenaria stricta

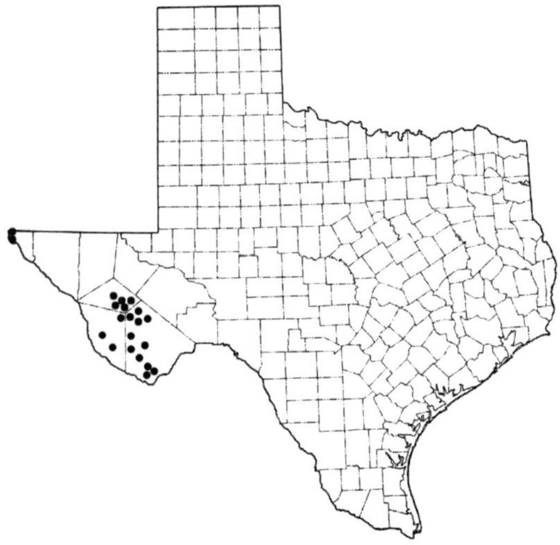

Cerastium axillare

Cerastium brachypetalum

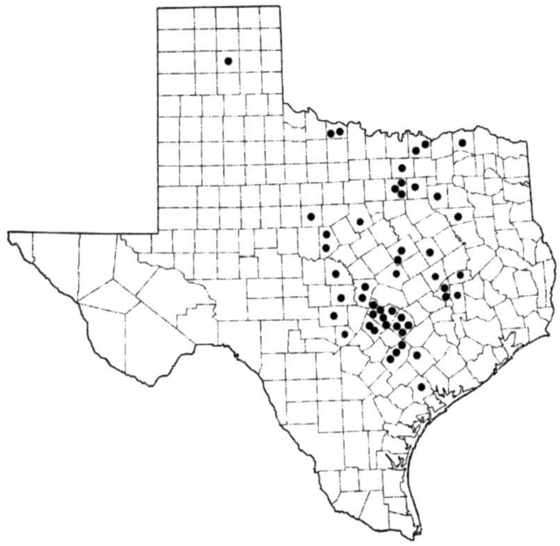

Cerastium brachypodum

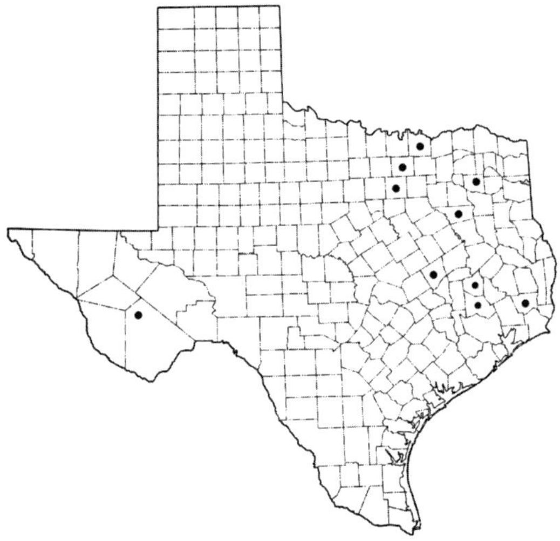

Cerastium fontanum

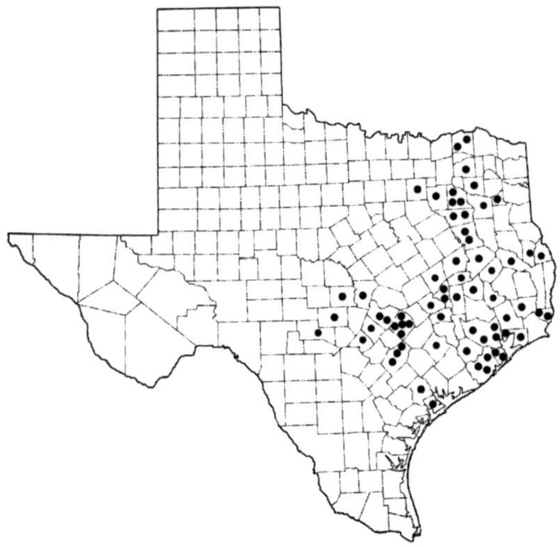

Cerastium glomeratum

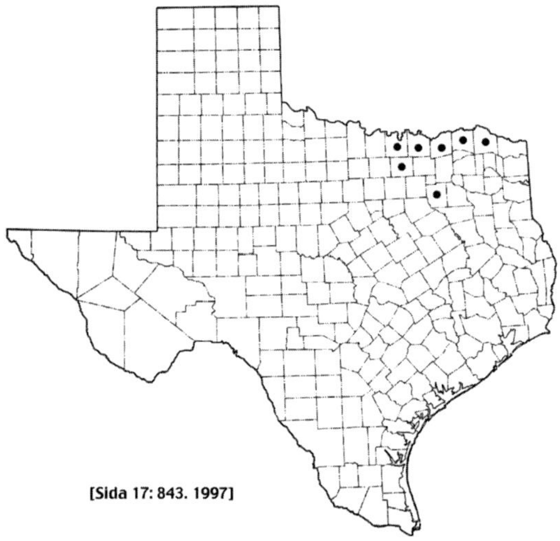
[Sida 17: 843. 1997]

Cerastium pumilum

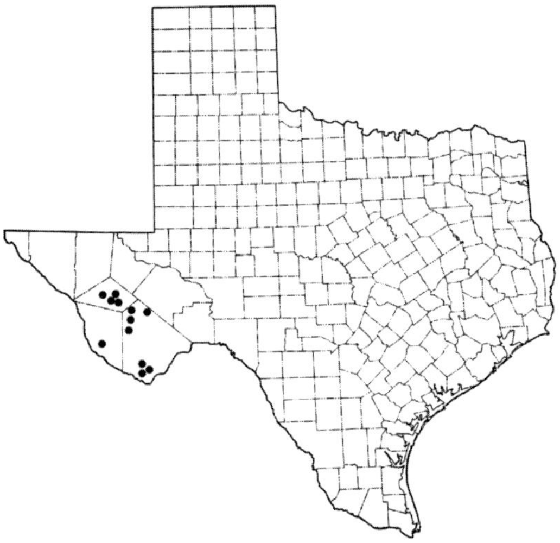

Drymaria laxiflora

Drymaria leptophylla

Drymaria molluginea

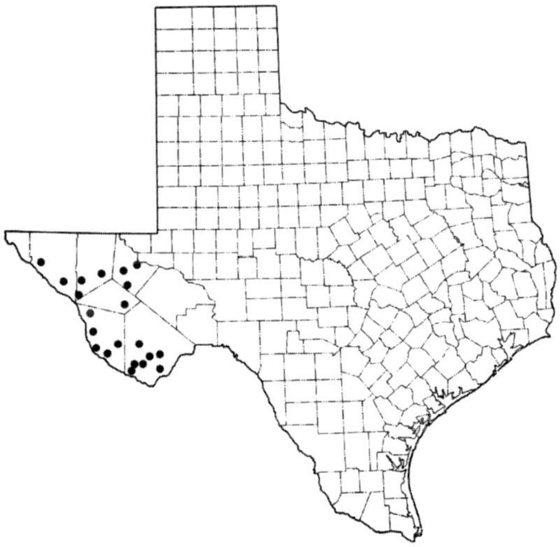

Drymaria pachyphylla

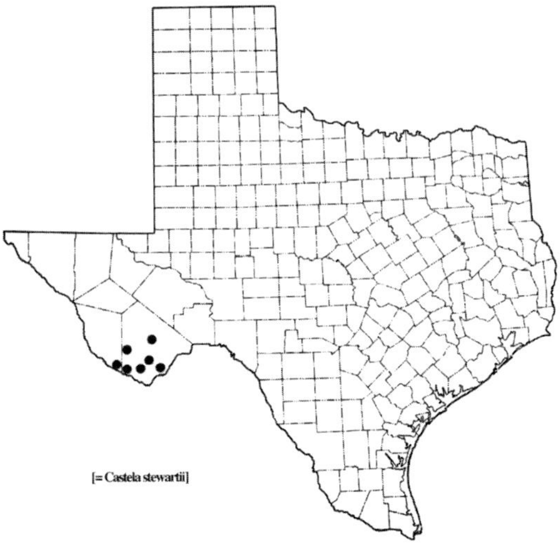
[= Castela stewartii]

Holosteum umbellatum

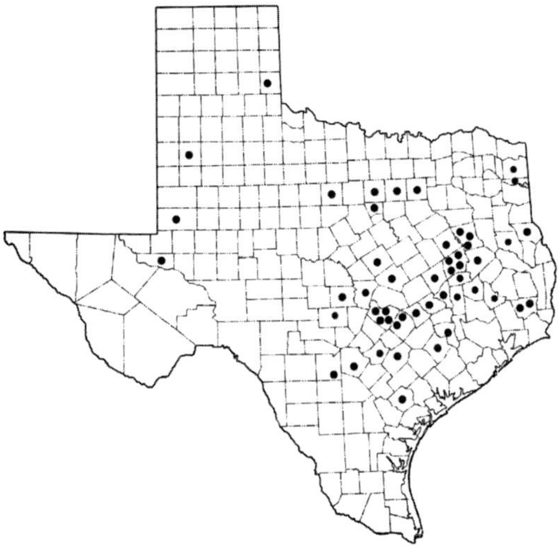

Loeflingia squarrosa

Paronychia congesta

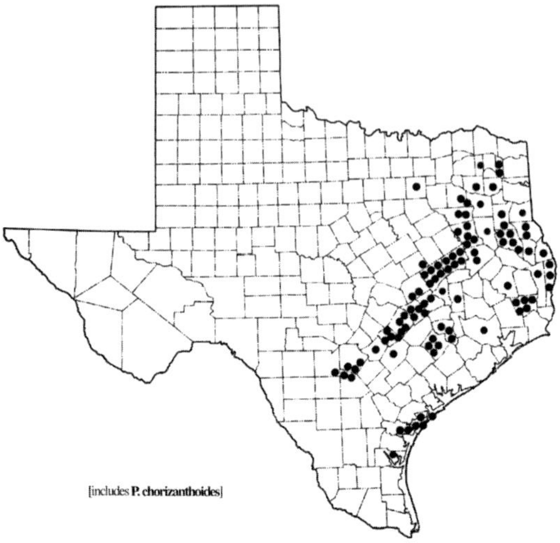
[includes P. chorizanthoides]

Paronychia drummondii

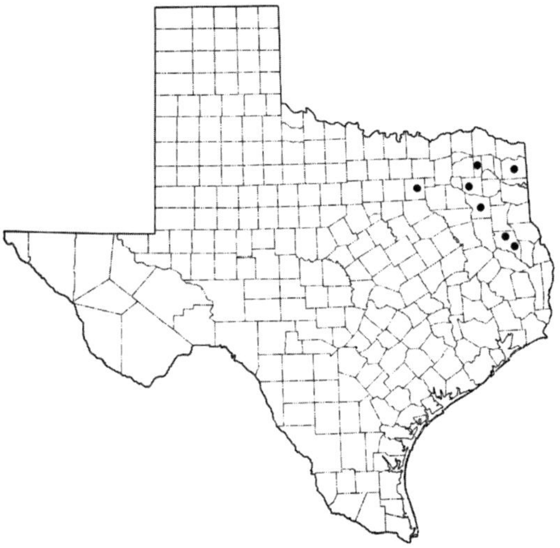

Paronychia fastigiata

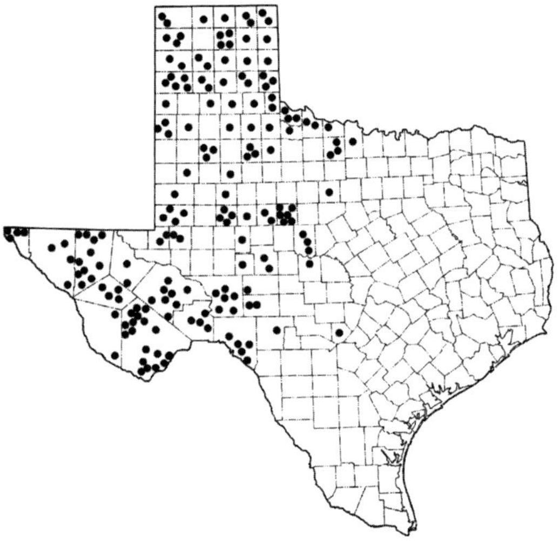

Paronychia jamesii

Paronychia jonesii

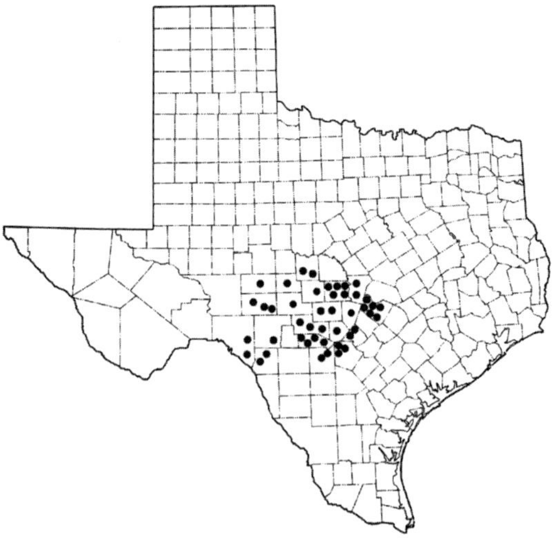

Paronychia lindheimeri

Paronychia lundelliorum

Paronychia maccartii

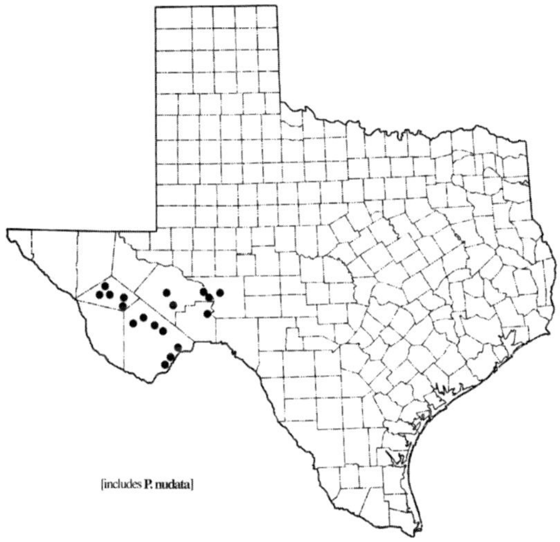
[includes P. nudata]

Paronychia monticola

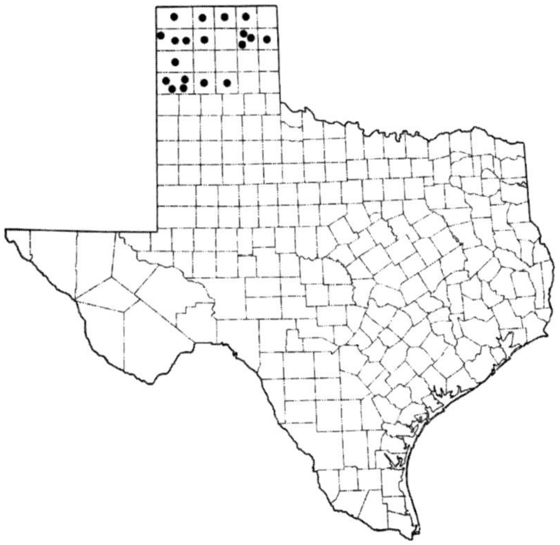

Paronychia sessiliflora

Paronychia setacea

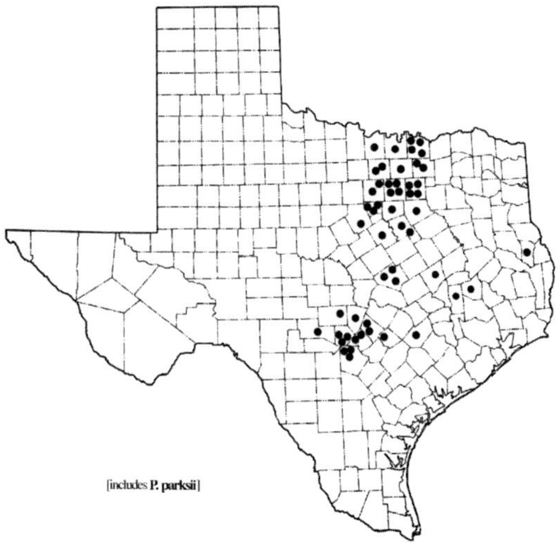
[includes P. parksii]

Paronychia virginica

Paronychia wilkinsonii

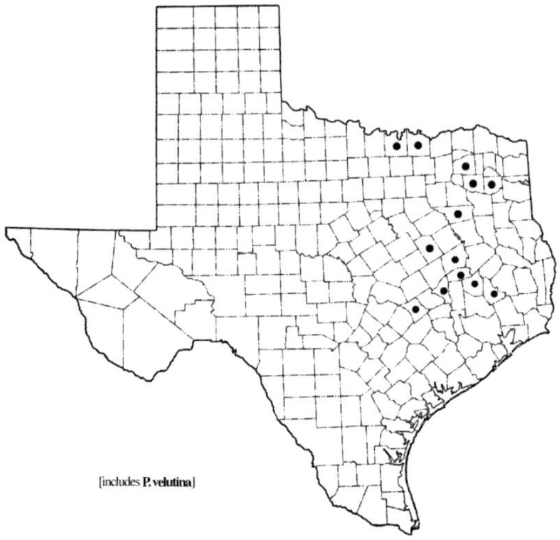
[includes **P. velutina**]

Petrorhagia dubia

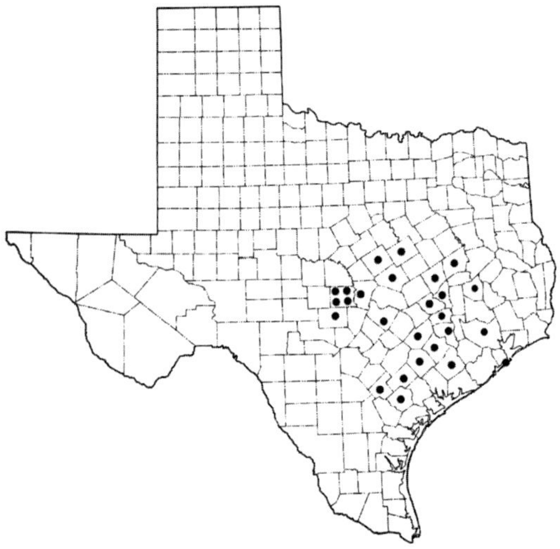

Polycarpon tetraphyllum

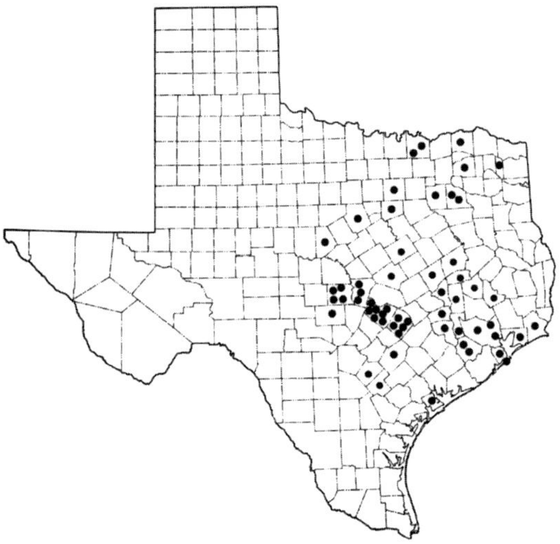

Sagina decumbens

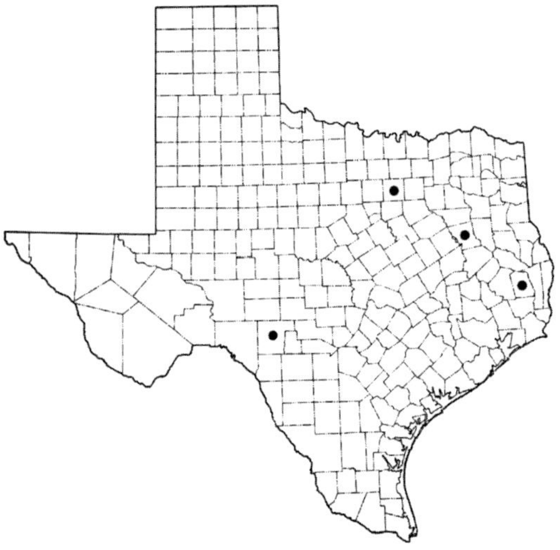

Saponaria officinalis

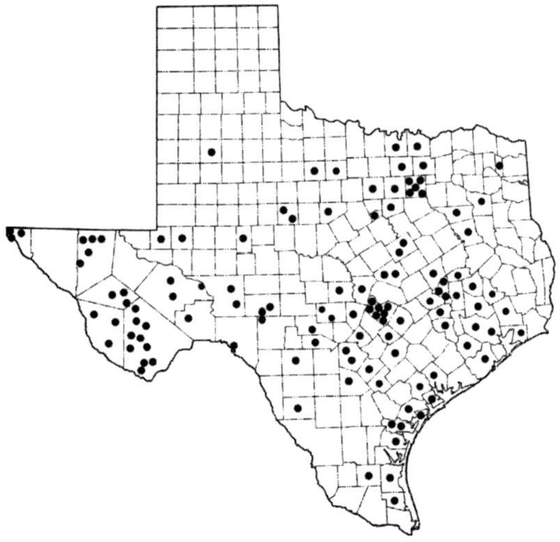

Silene antirrhina

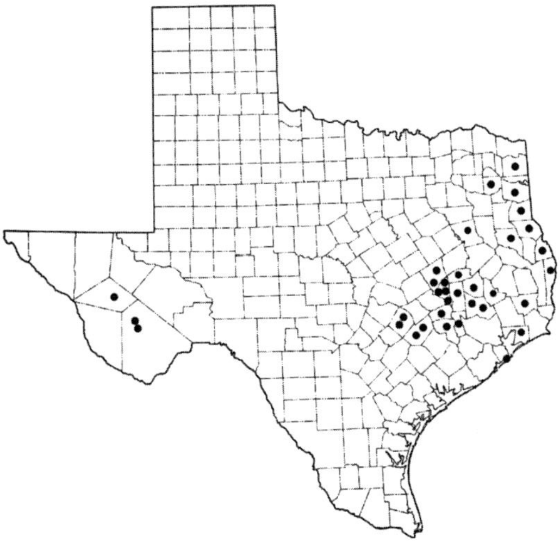

Silene gallica

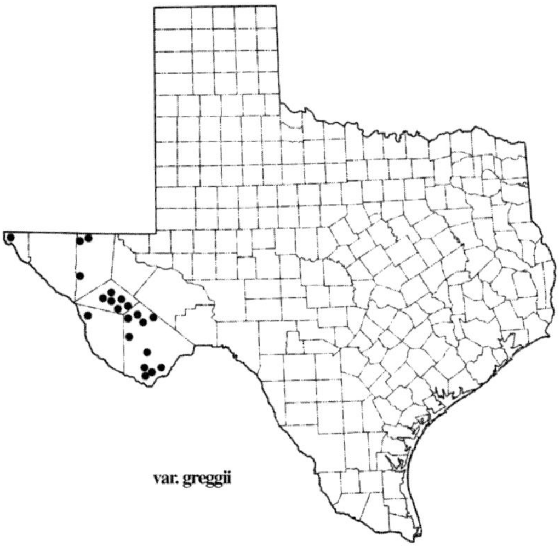
var. greggii

Silene laciniata

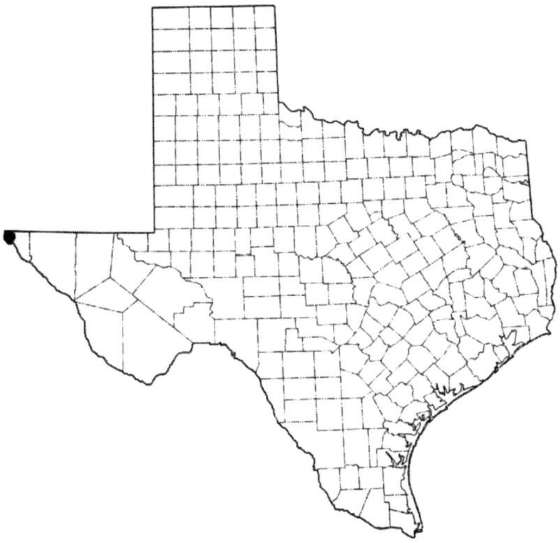

Silene plankii

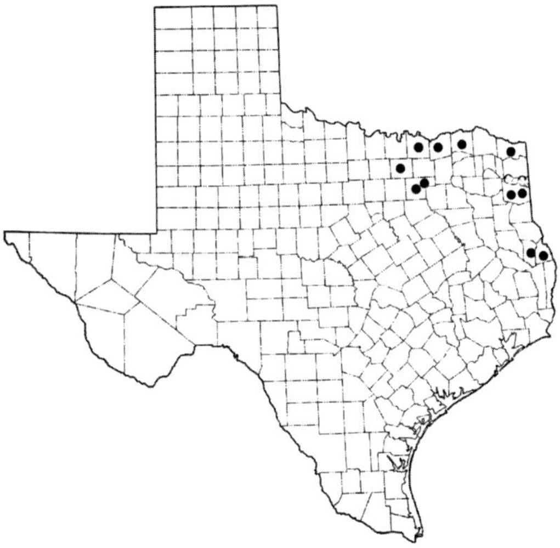

Silene stellata

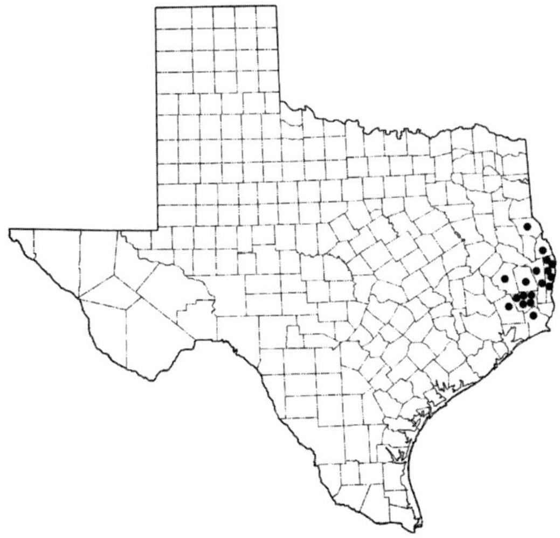

Silene subciliata

Spergula arvensis

Spergularia echinosperma

Spergularia platensis

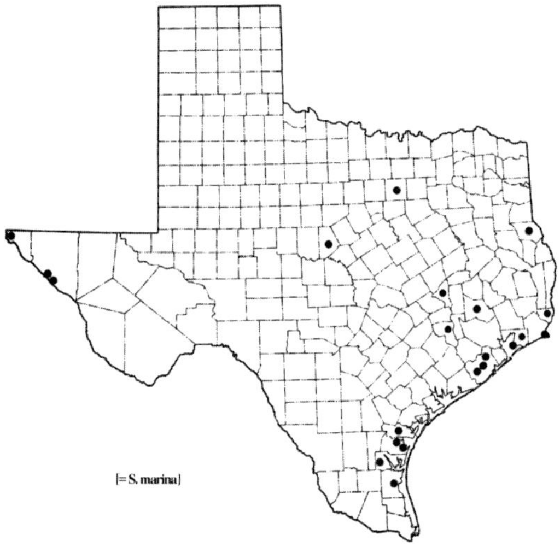
[= S. marina]

Spergularia salina

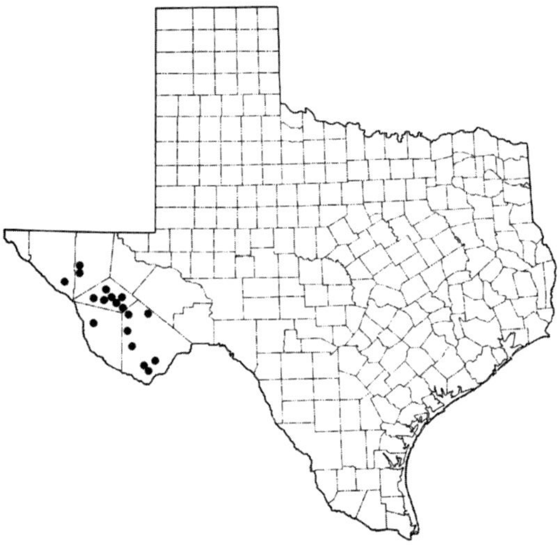

Stellaria cuspidata

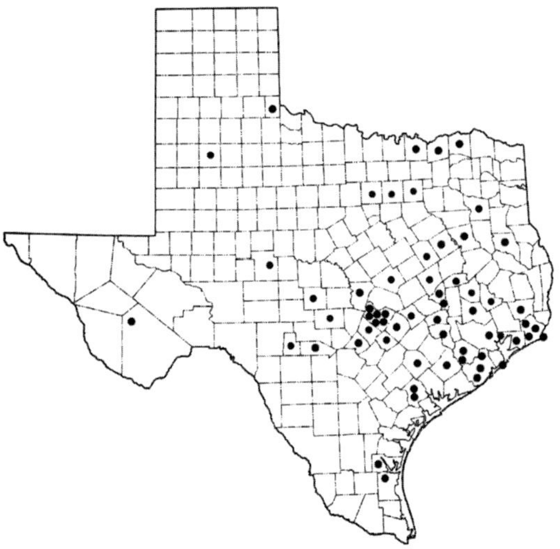

Stellaria media

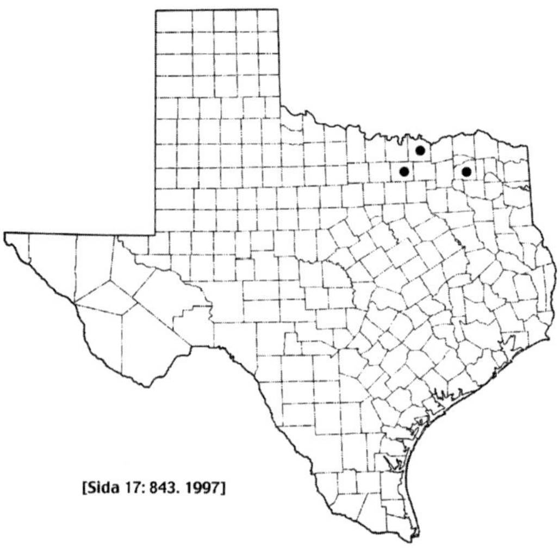
[Sida 17: 843. 1997]

Stellaria pallida

[Sida 18: 317. 1998]

Stellaria parva

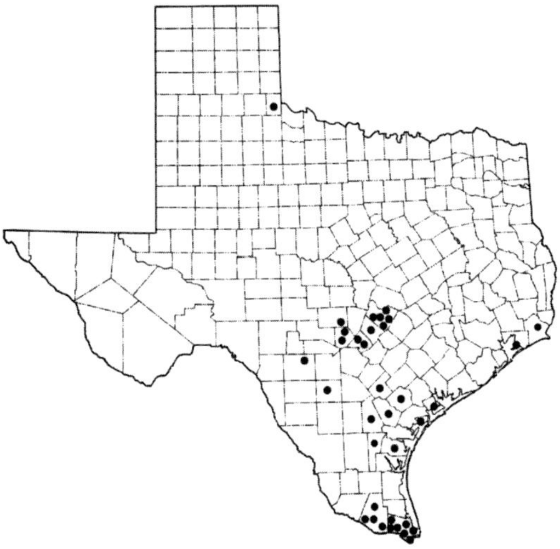

Stellaria prostrata

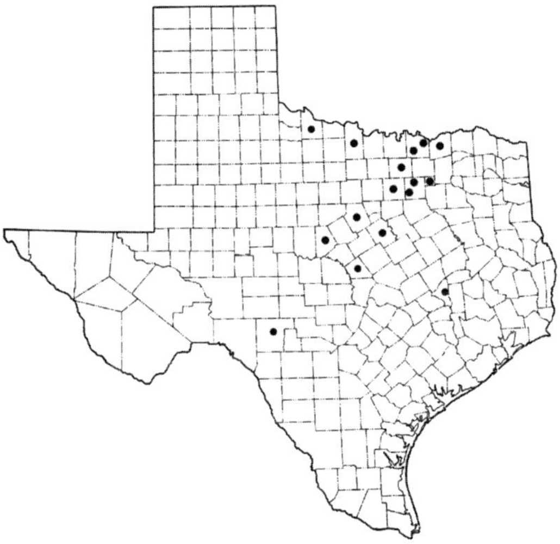

Vaccaria pyramidata

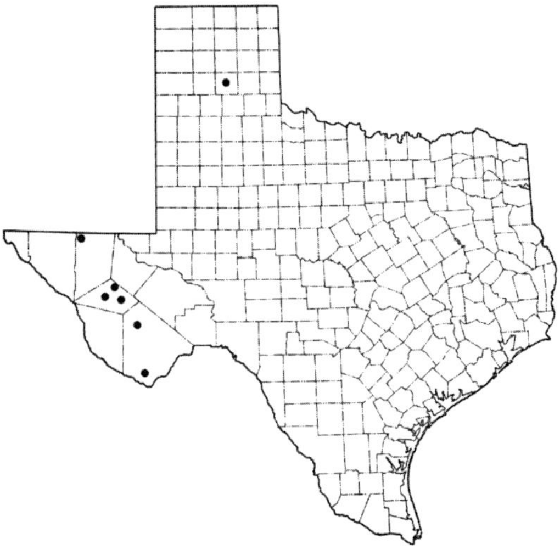

Celastrus scandens

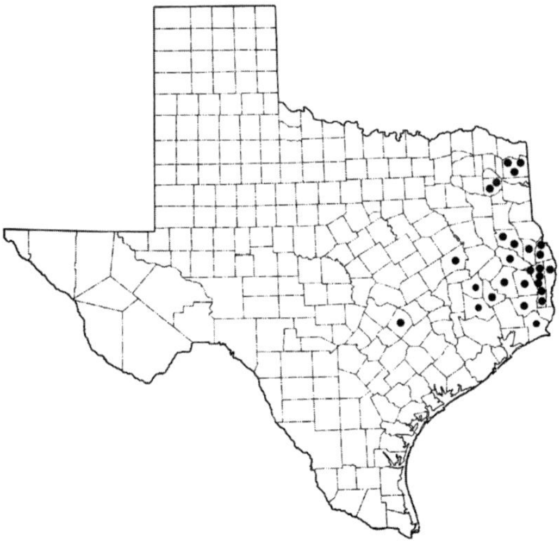

Euonymus americana

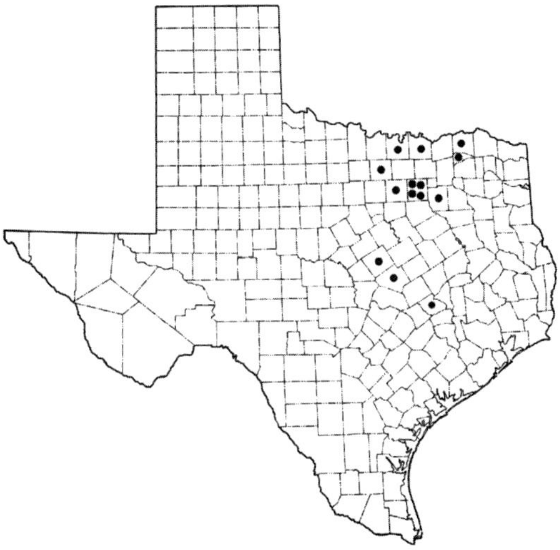

Euonymus atropurpurea

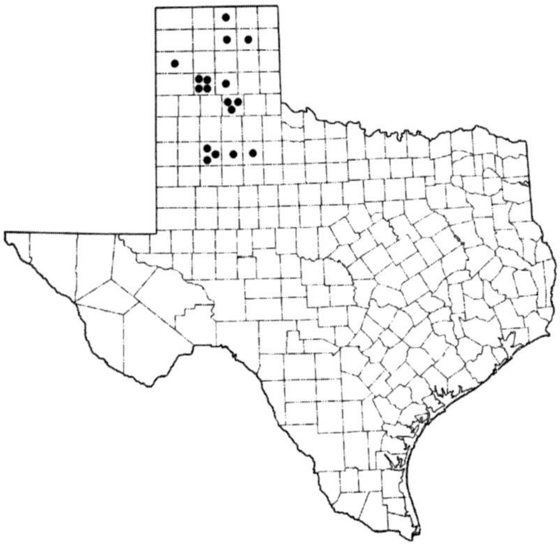

Forsellesia planitierum

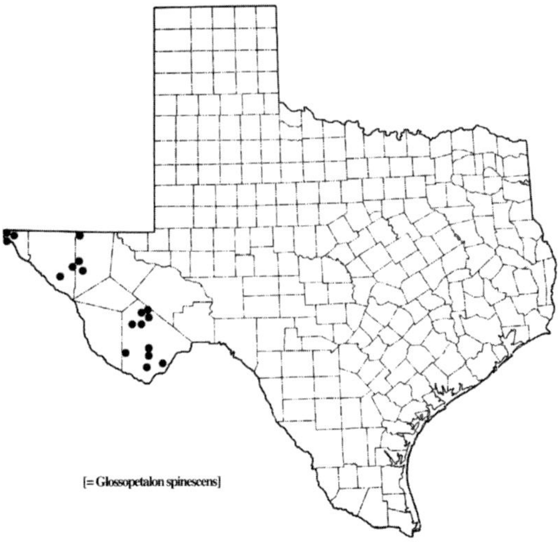
[= Glossopetalon spinescens]

Forsellesia spinescens

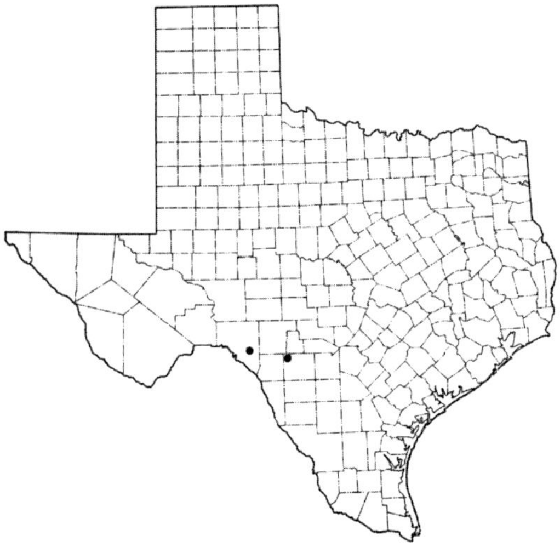

Forsellesia texensis

[= Tricerma phyllanthoides;
includes Maytenus texana]

Maytenus phyllanthoides

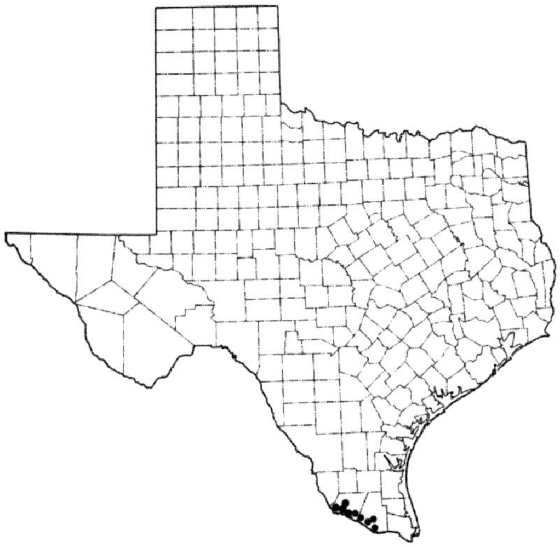

Mortonia greggii

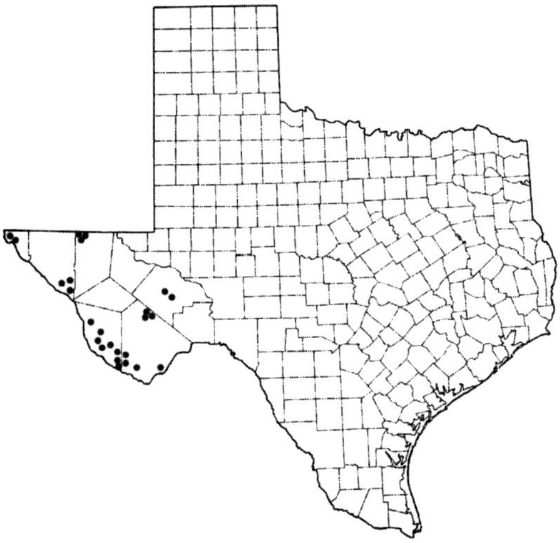

Mortonia scabrella

Mortonia sempervirens

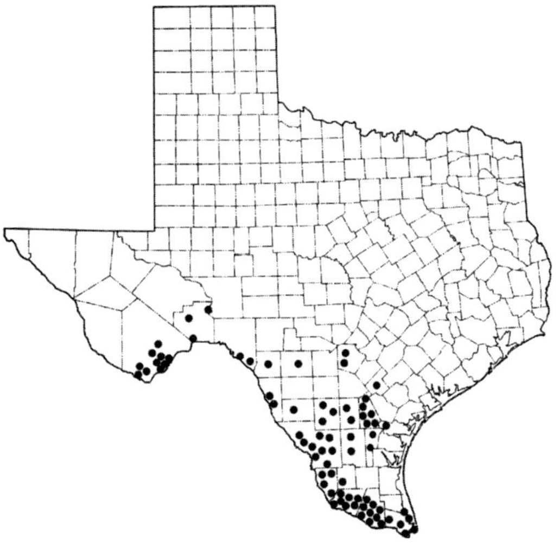

Schaefferia cuneifolia

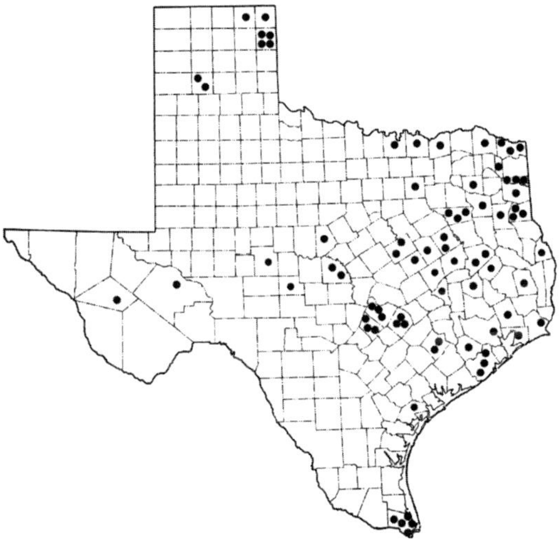

Ceratophyllum demersum

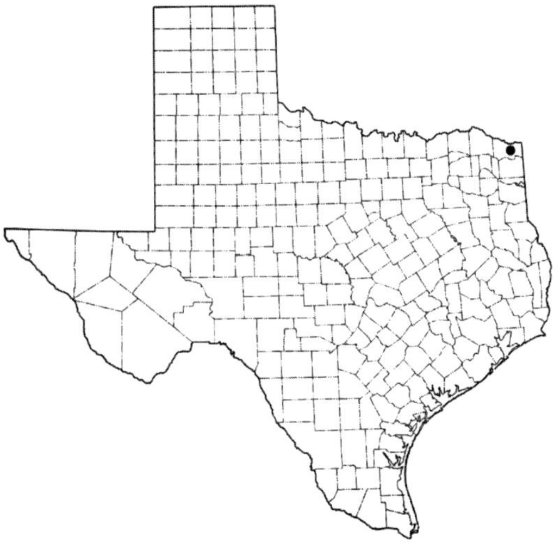

Ceratophyllum echinatum

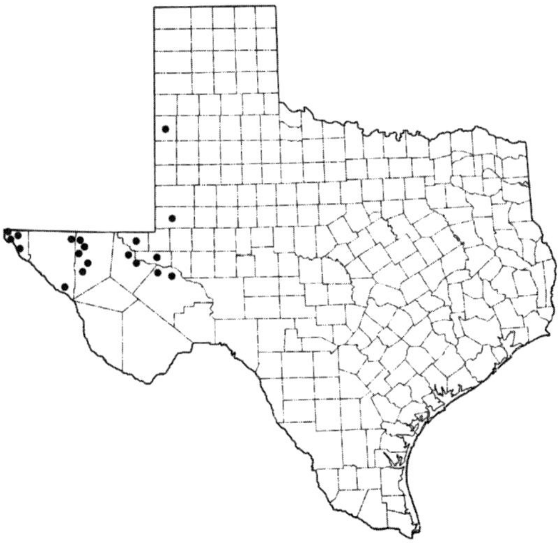

Allenrolfea occidentalis

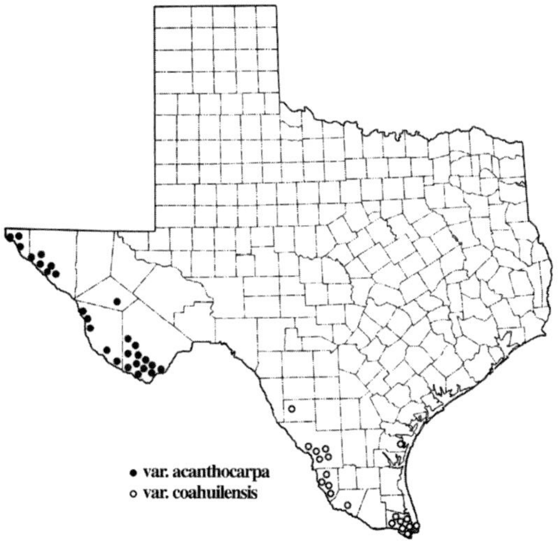
• var. acanthocarpa
○ var. coahuilensis

Atriplex acanthocarpa

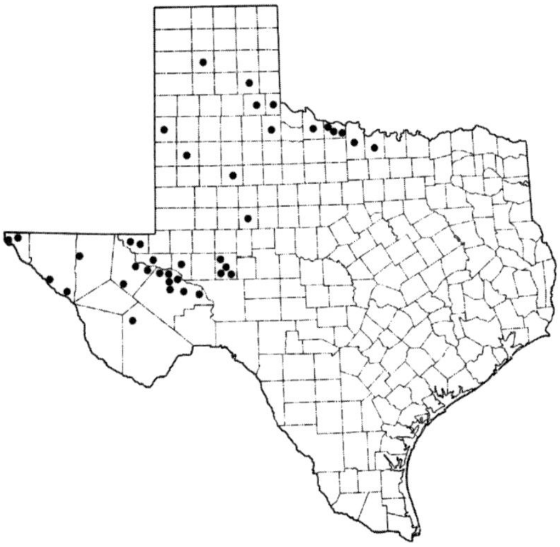

Atriplex argentea

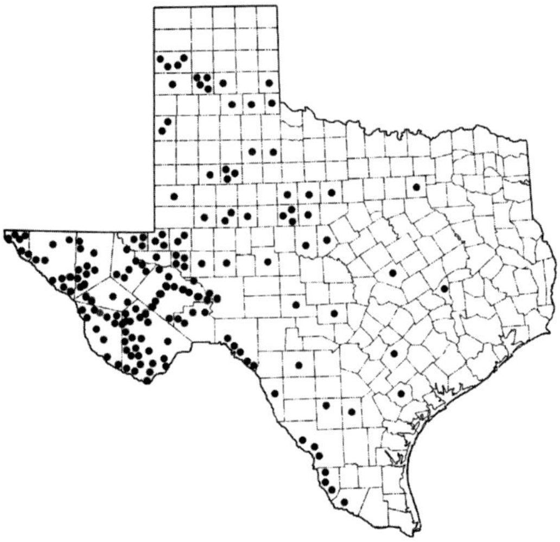

Atriplex canescens

Atriplex confertifolia

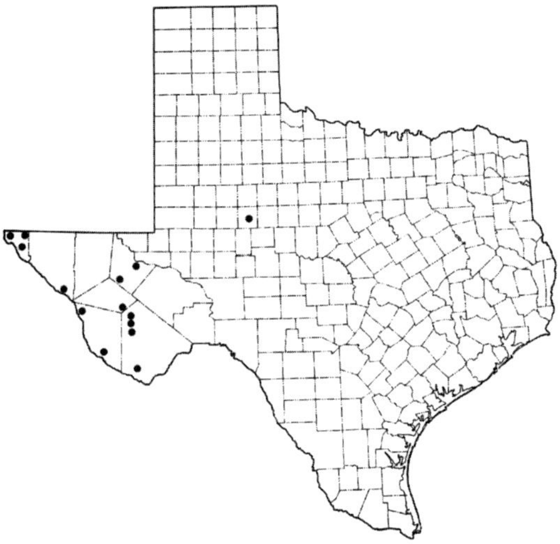

Atriplex elegans

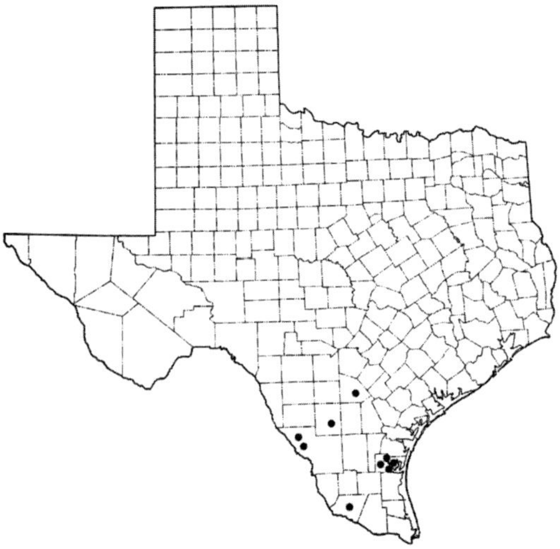

Atriplex klebergorum

Atriplex matamorensis

[=A. arenaria]

Atriplex mucronata

Atriplex muricata

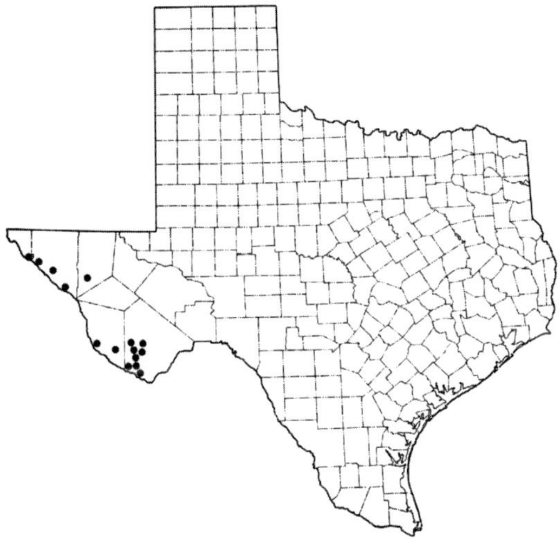

Atriplex obovata

Atriplex patula

Atriplex pentandra

Atriplex powellii

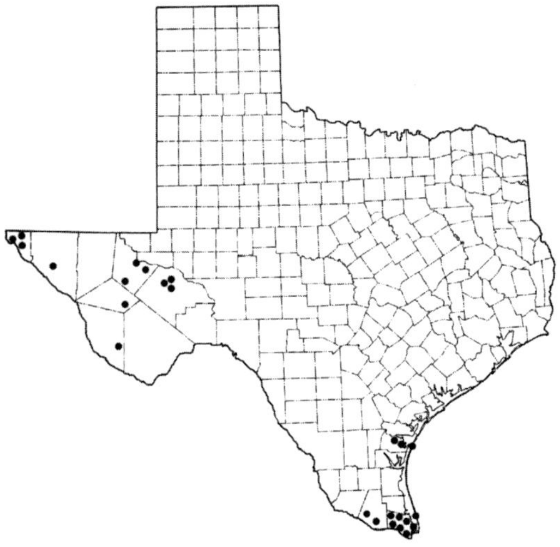
Atriplex semibaccata

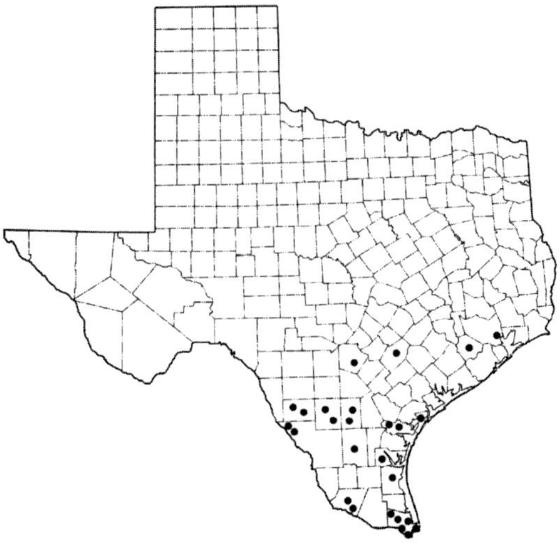
Atriplex texana

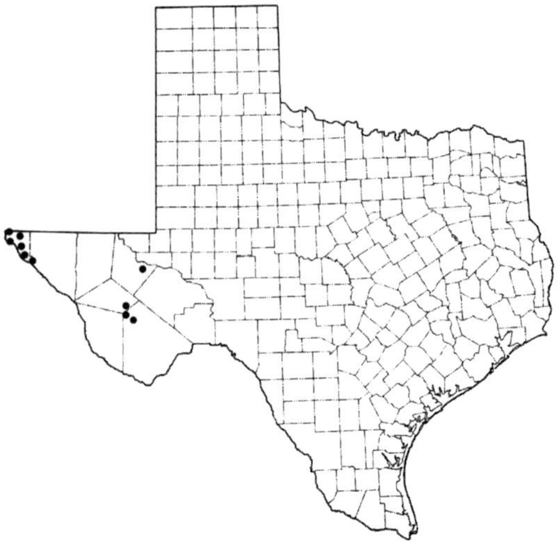
Atriplex wrightii

Bassia hyssopifolia

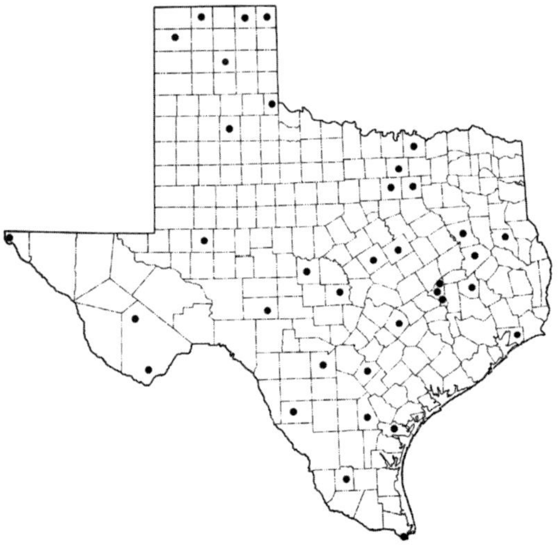
Chenopodium album

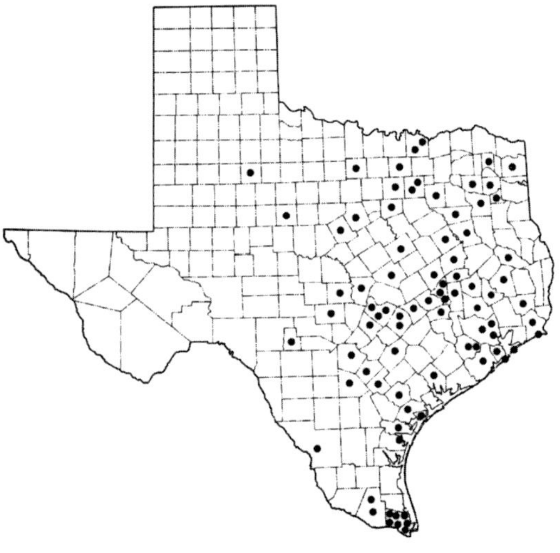
Chenopodium ambrosioides

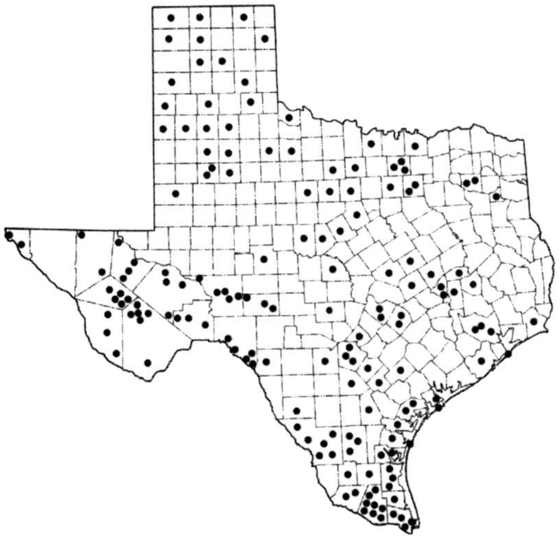

Chenopodium berlandieri

Chenopodium botrys

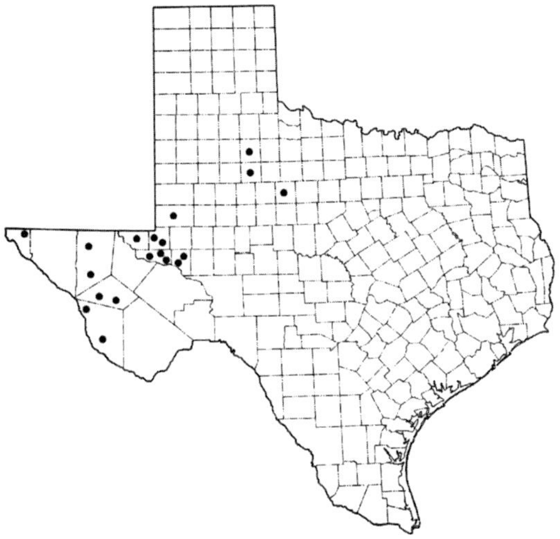

Chenopodium cycloides

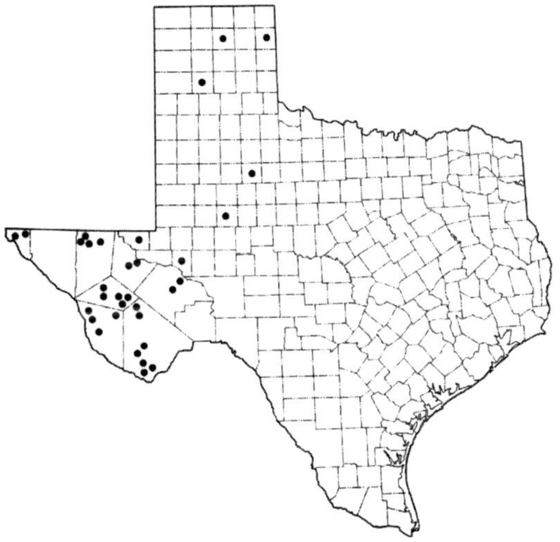

Chenopodium fremontii

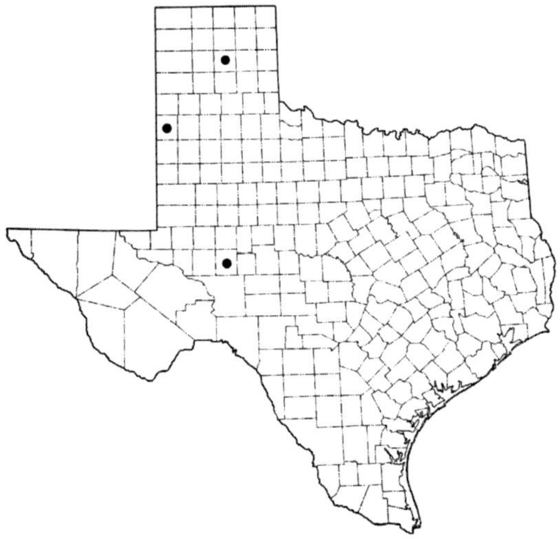

Chenopodium glaucum

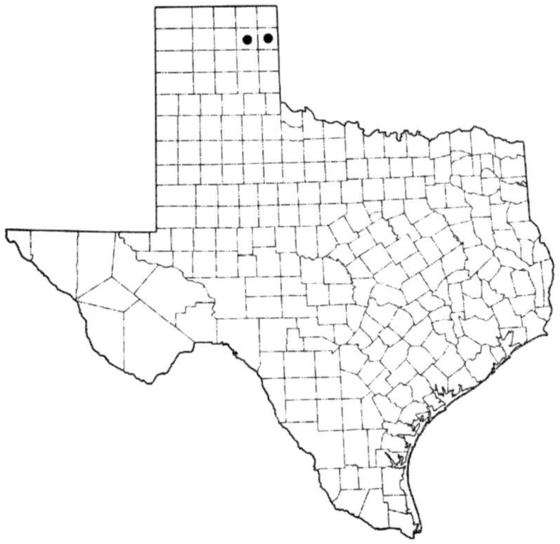

Chenopodium hybridum

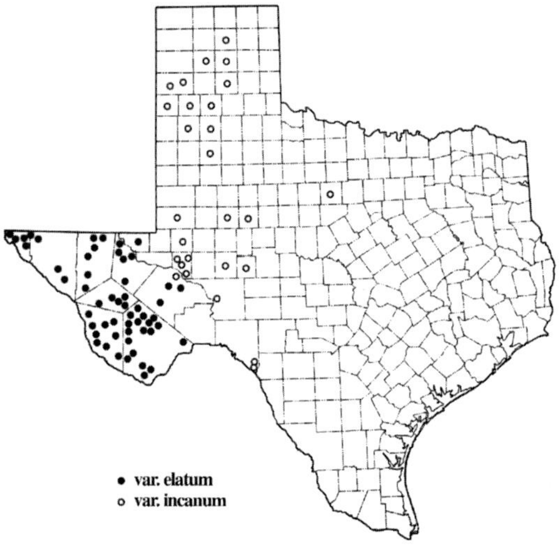
• var. elatum
○ var. incanum

Chenopodium incanum

Chenopodium incisum

Chenopodium incognitum

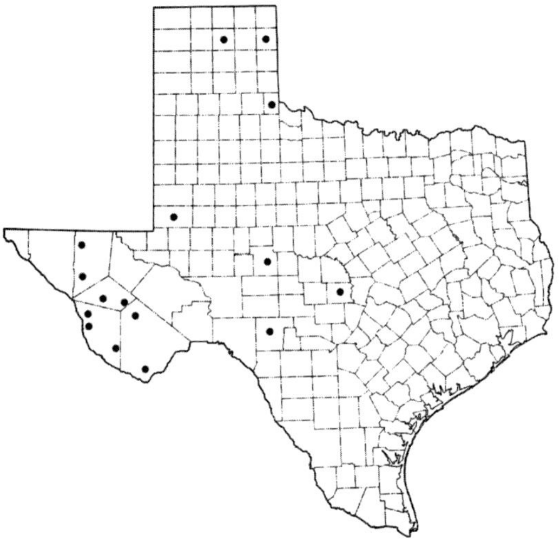

Chenopodium leptophyllum

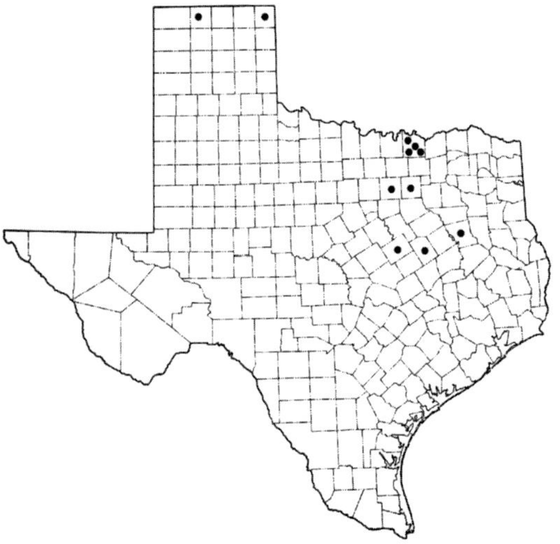

Chenopodium missouriense

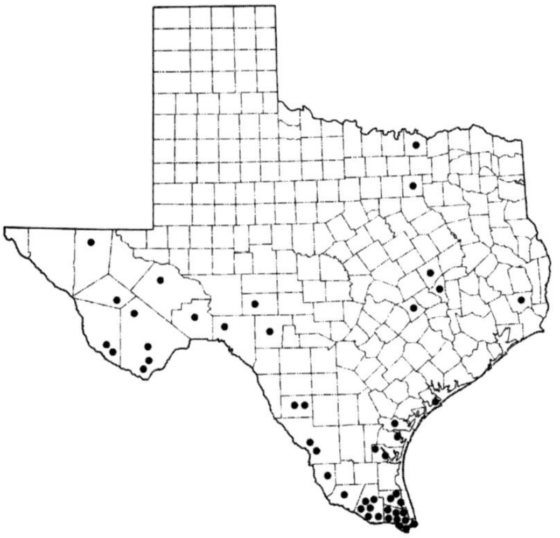

Chenopodium murale

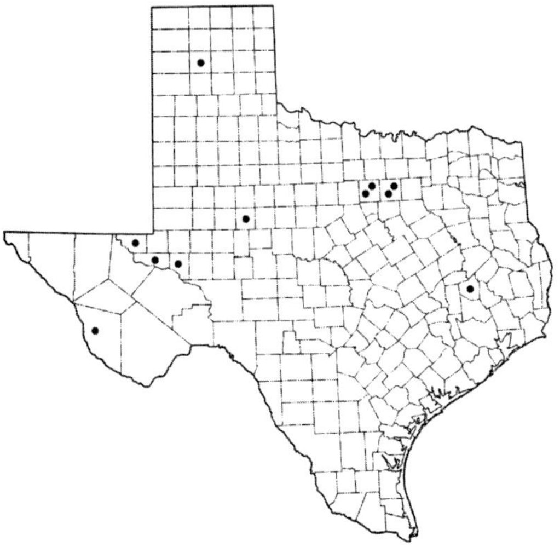

Chenopodium pallescens

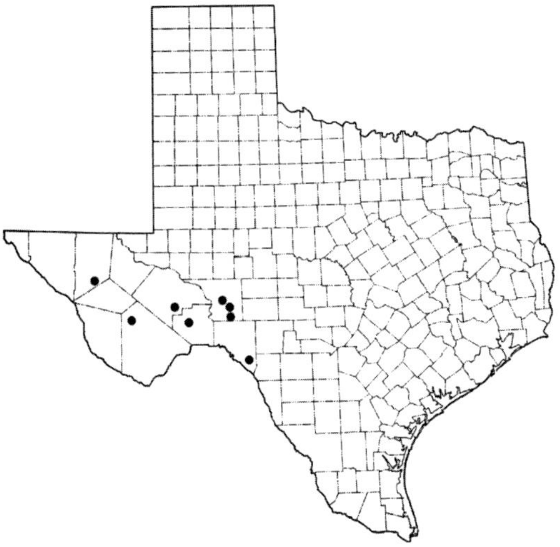

Chenopodium pratericola

Chenopodium pumilio

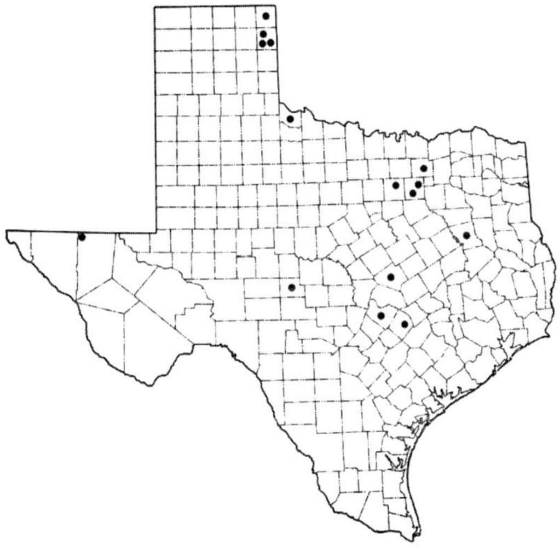

Chenopodium simplex

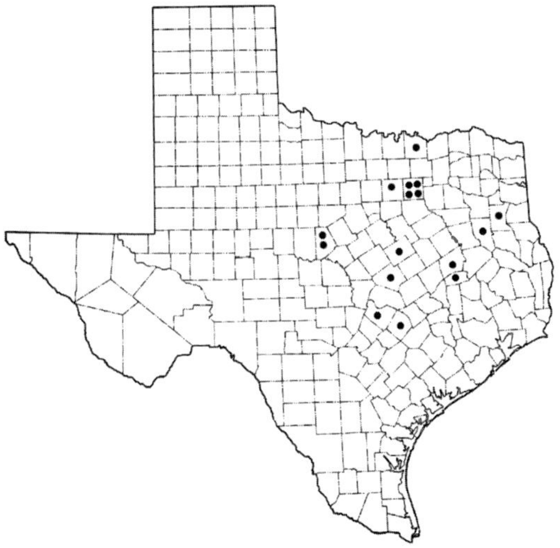

Chenopodium standleyanum

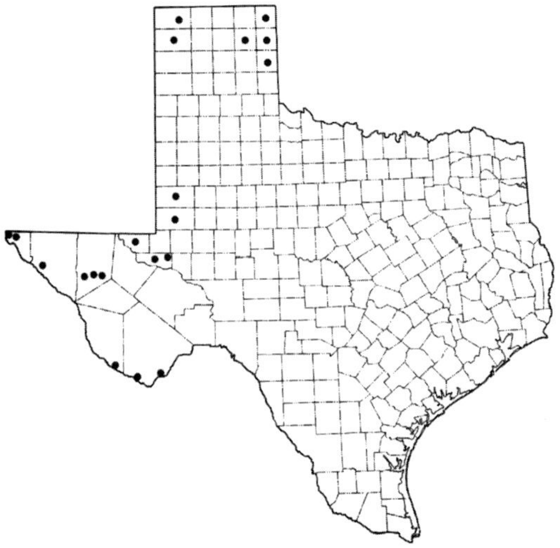

Corispermum americanum

Corispermum orientale

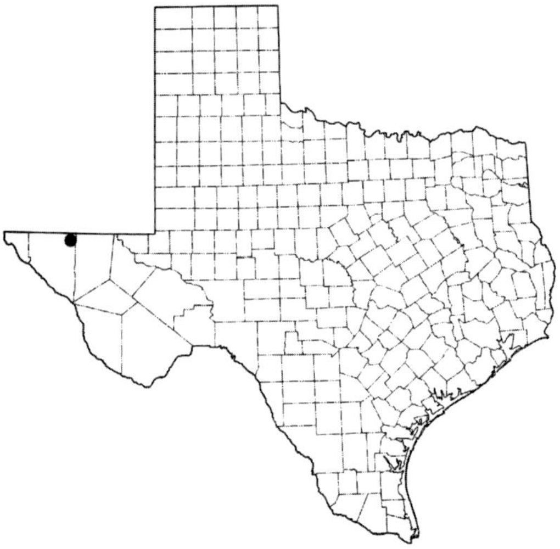

Corispermum welshii

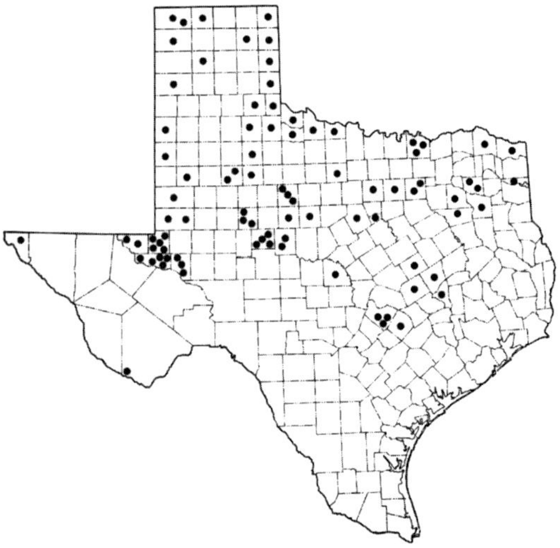

Cycloloma atriplicifolium

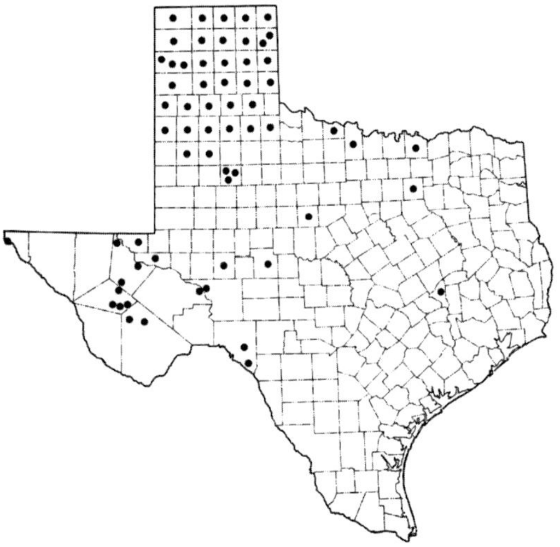

Kochia scoparia

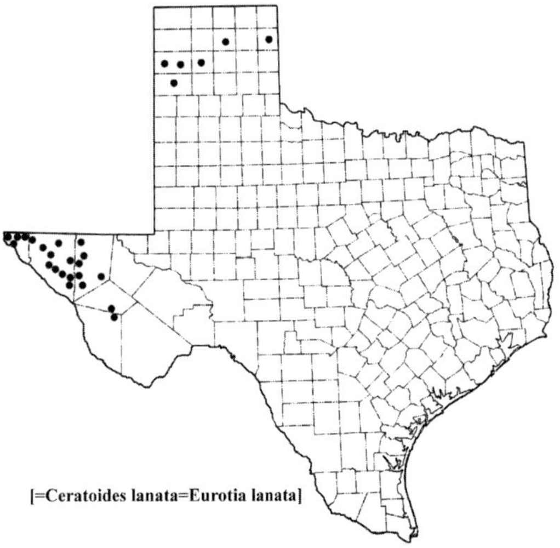
[=Ceratoides lanata=Eurotia lanata]

Krascheninnikova lanata

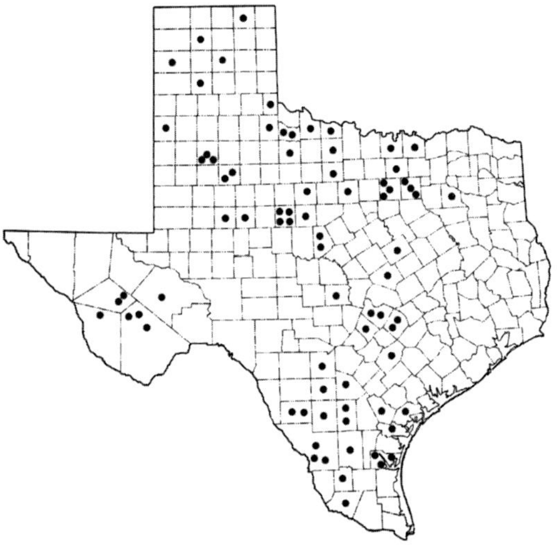

Monolepis nuttalliana

Salicornia bigelovii

Salicornia virginica

Salsola collina

Salsola paulsenii

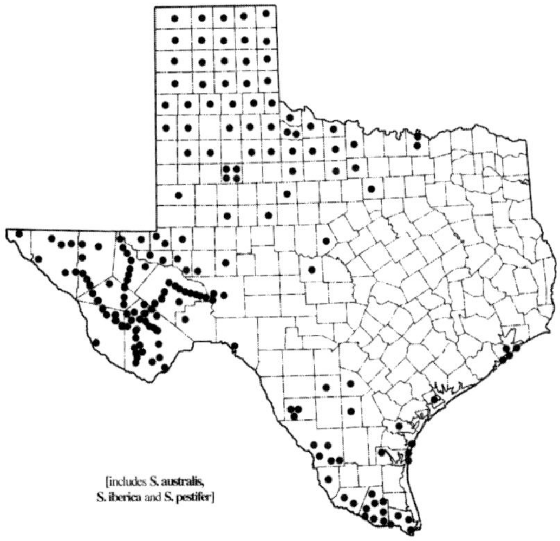
[includes **S. australis**,
S. iberica and **S. pestifer**]

Salsola tragus

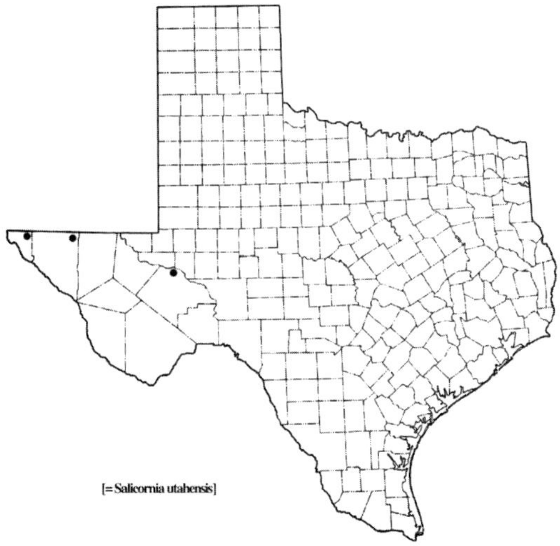
[= **Salicornia utahensis**]

Sarcocornia utahensis

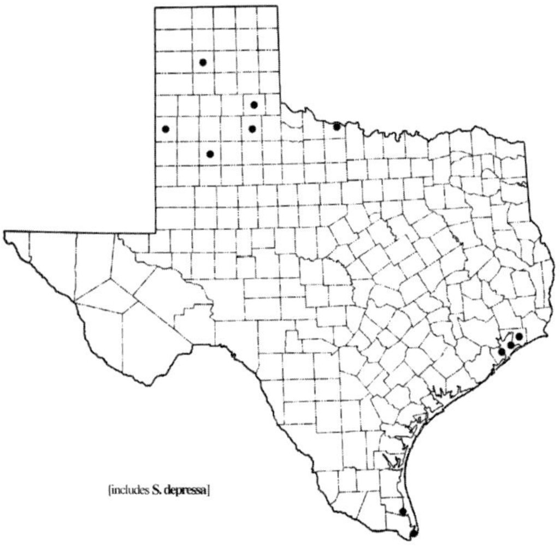
[includes S. depressa]

Suaeda calceoliformis

Suaeda conferta

Suaeda linearis

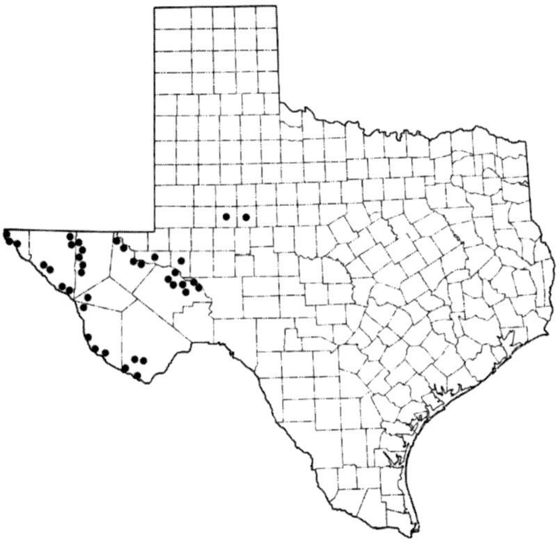

Suaeda suffrutescens

Suaeda tampicensis

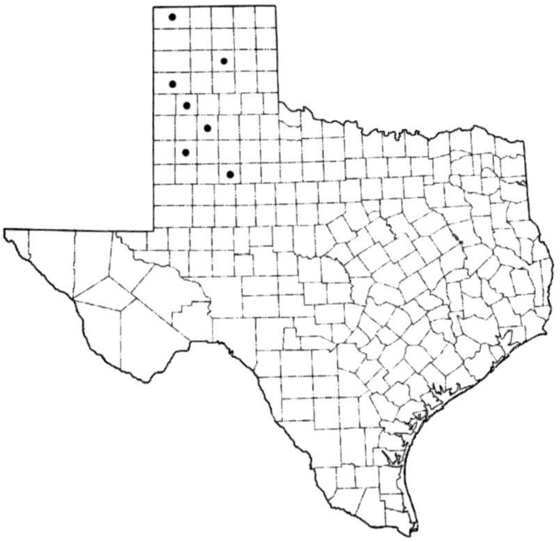

Suckleya suckleyana

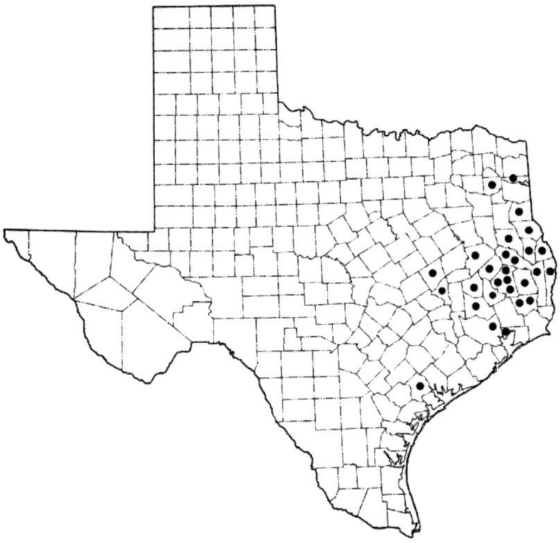

Helianthemum carolinianum

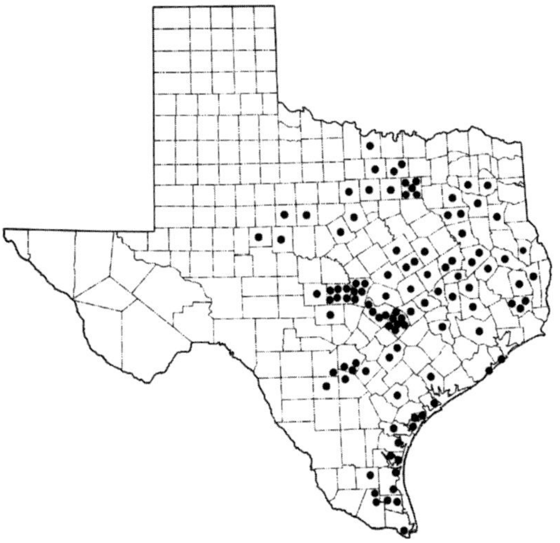

Helianthemum georgianum

Helianthemum glomeratum

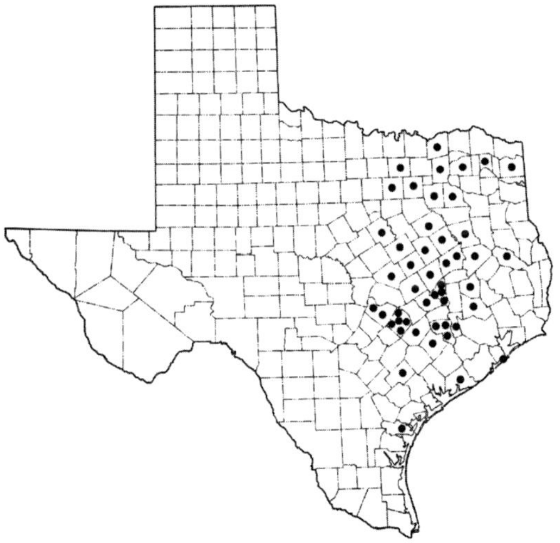

Helianthemum rosmarinifolium

Lechea mensalis

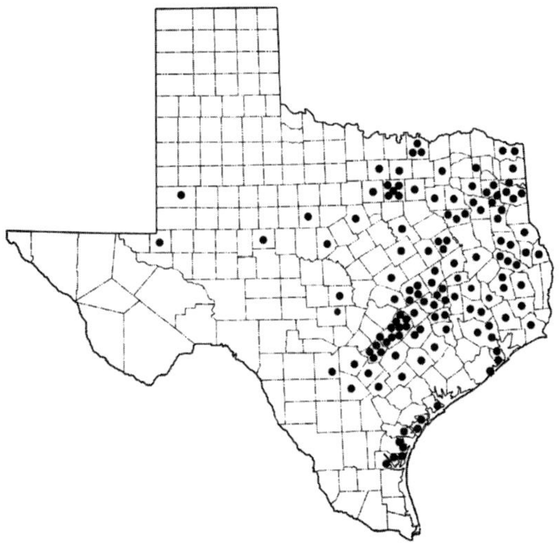

Lechea mucronata

Lechea pulchella

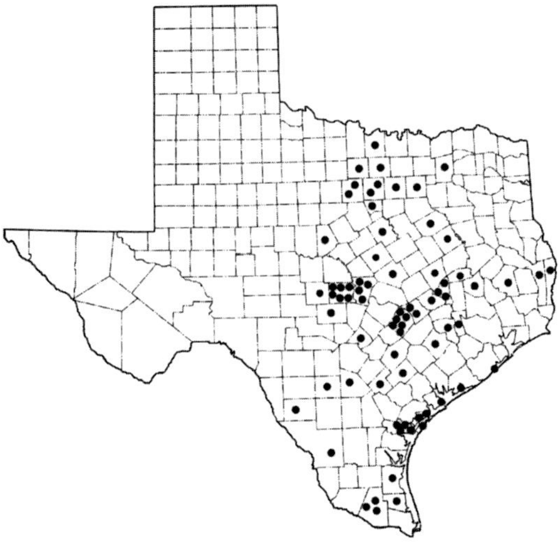

Lechea san-sabeana

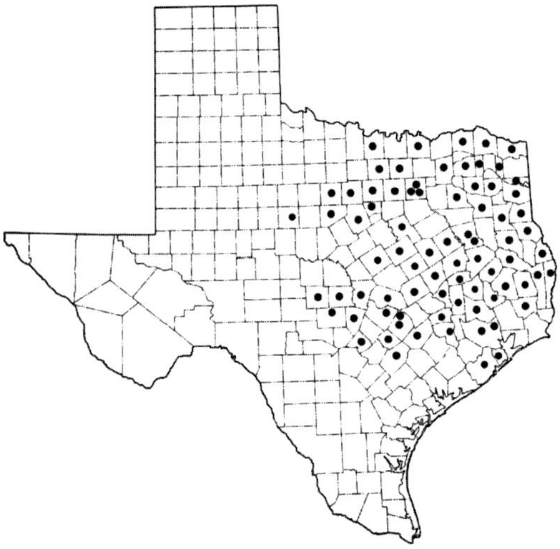

Lechea tenuifolia

Clethra alnifolia

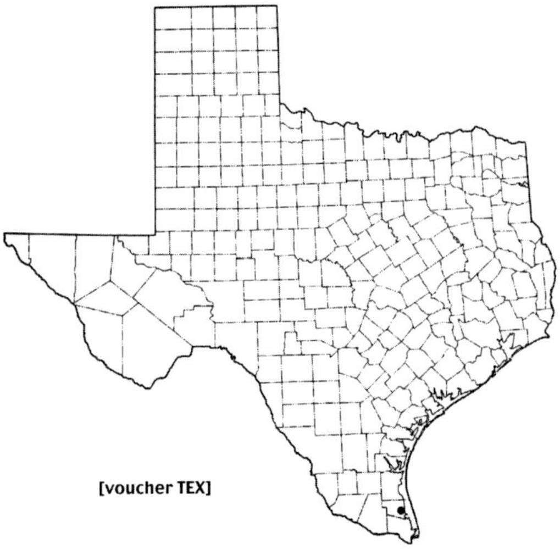
[voucher TEX]

Conocarpus erectus

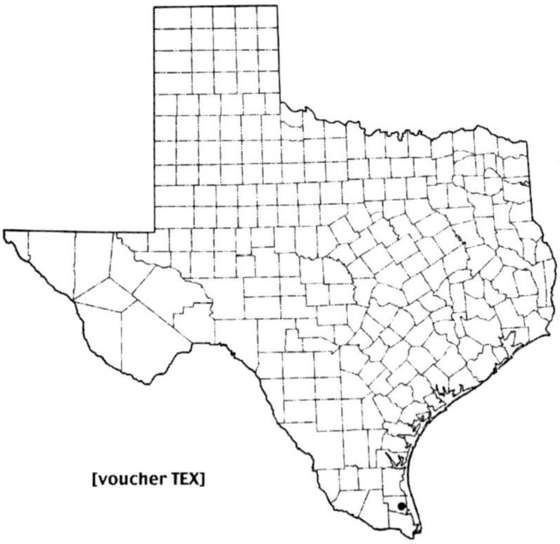
[voucher TEX]

Laguncularia racemosa

Bonamia ovalifolia

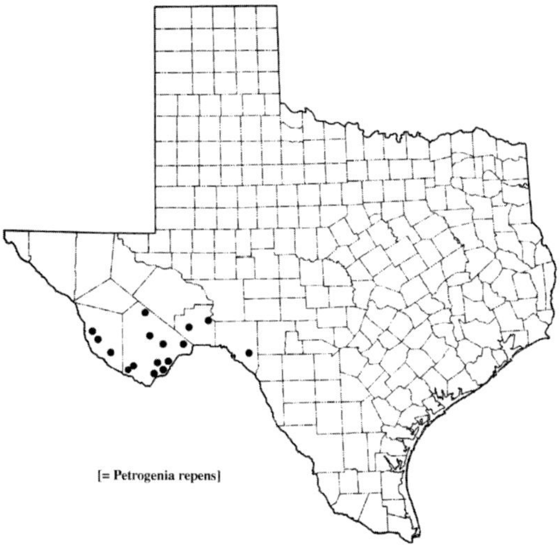
[= Petrogenia repens]

Bonamia repens

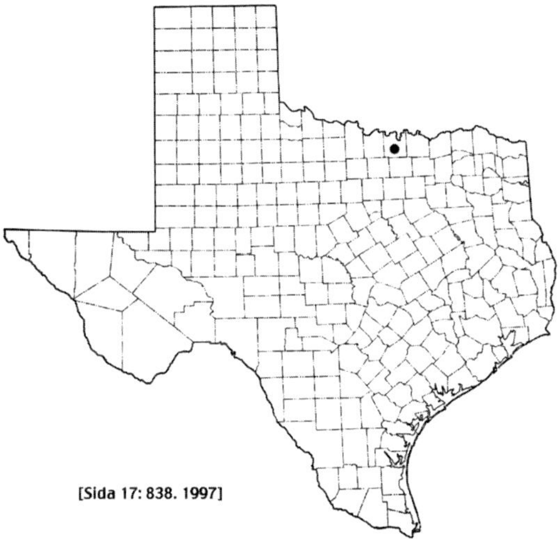
[Sida 17: 838. 1997]

Calystegia macounii

var. repens
[in Texas we recognize but
a single infraspecific taxon]

Calystegia sepium

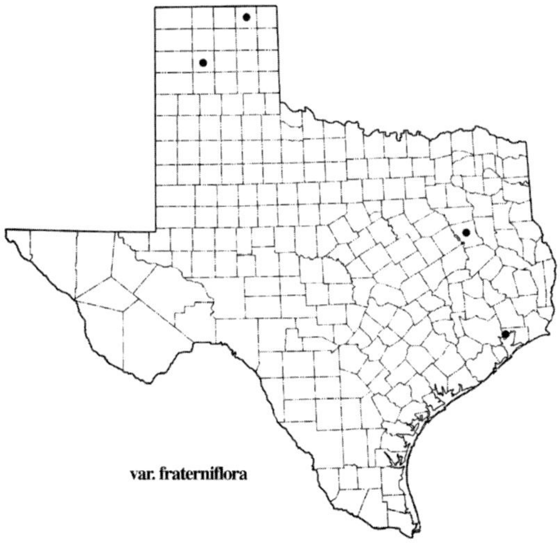
var. fraterniflora

Calystegia silvatica

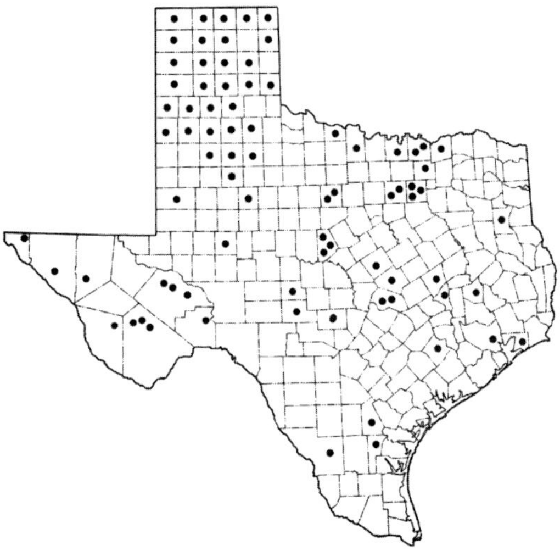

Convolvulus arvensis

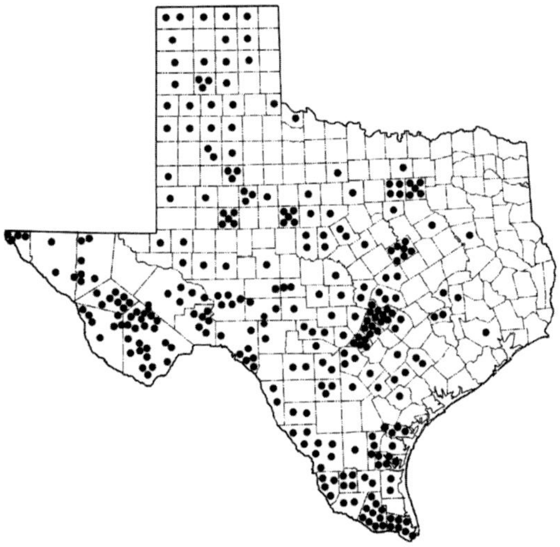

Convolvulus equitans

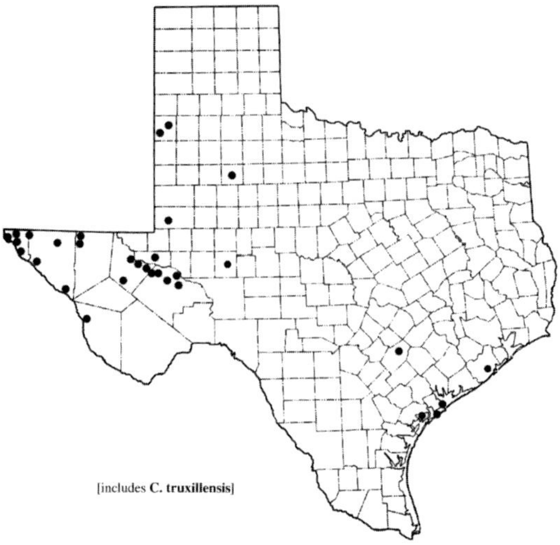
[includes C. truxillensis]

Cressa depressa

Cressa nudicaulis

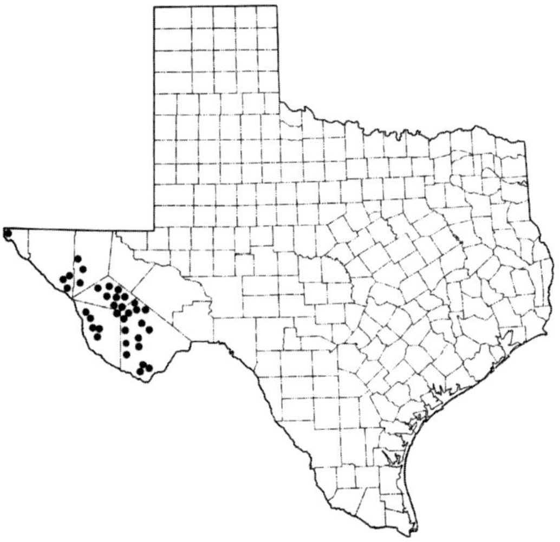

Dichondra argentea

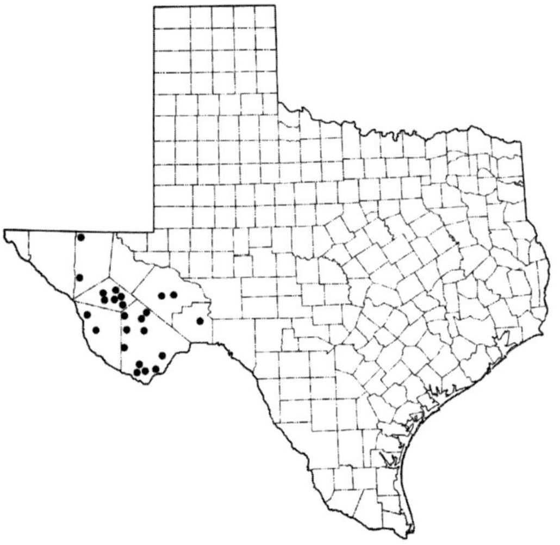

Dichondra brachypoda

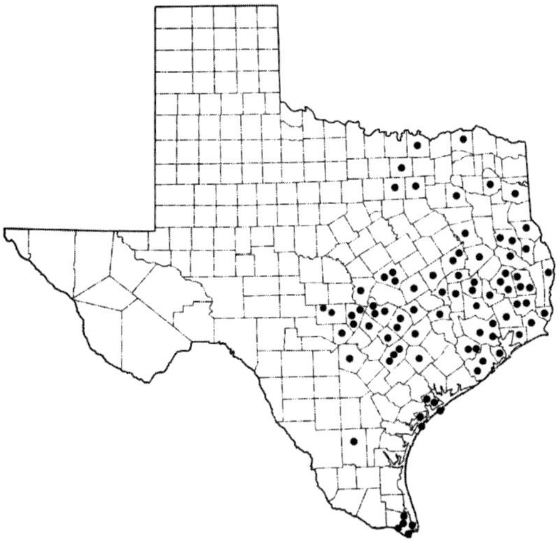

Dichondra carolinensis

Dichondra micrantha

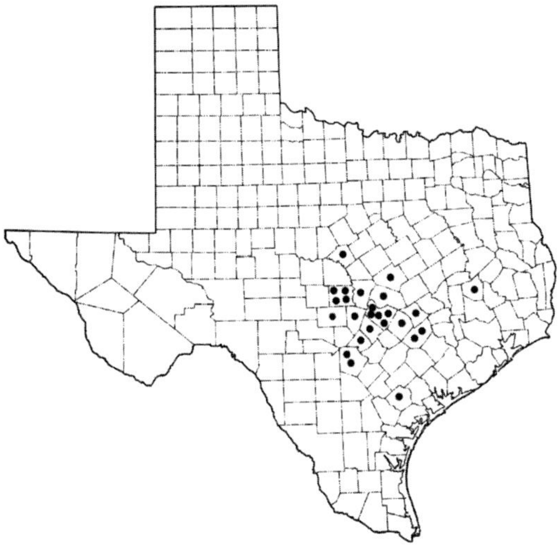
Dichondra recurvata

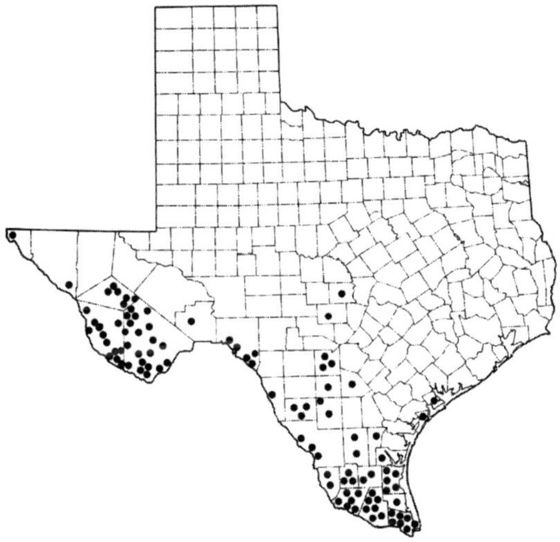
Evolvulus alsinoides

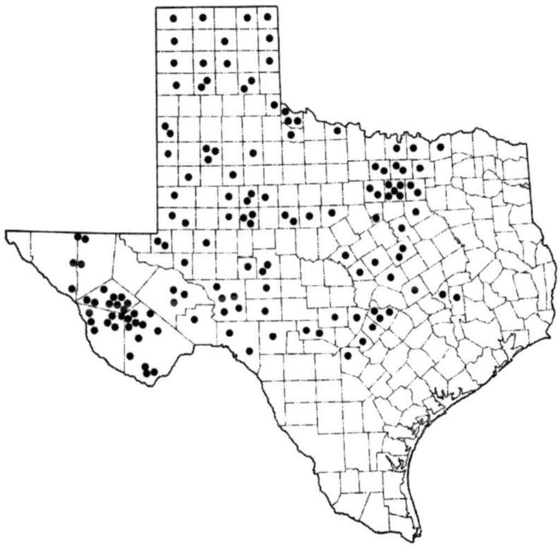
Evolvulus nuttallianus

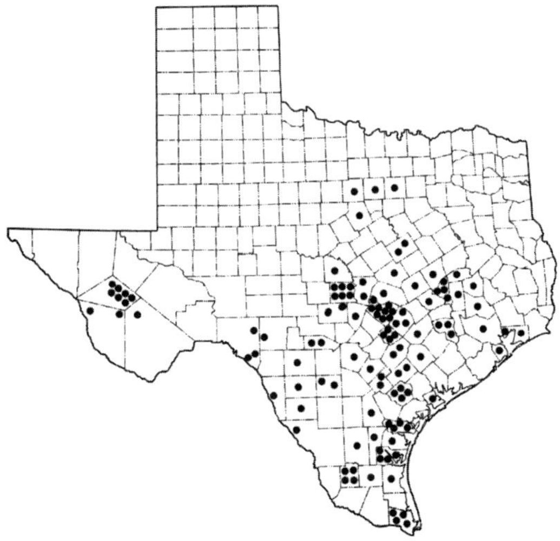
Evolvulus sericeus

Ipomoea alba

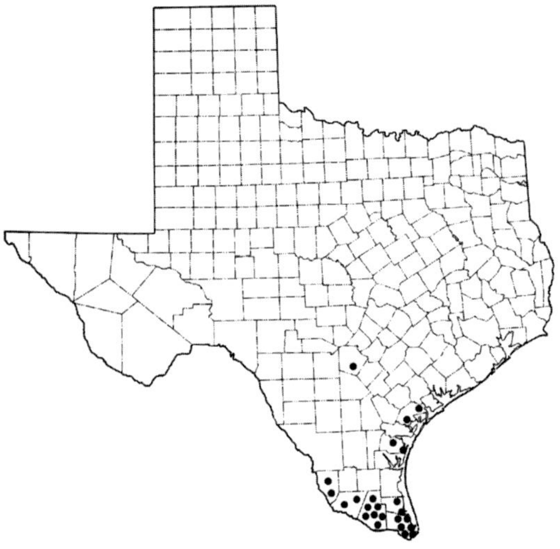

Ipomoea amnicola

Ipomoea aristolochiifolia

Ipomoea barbatisepala

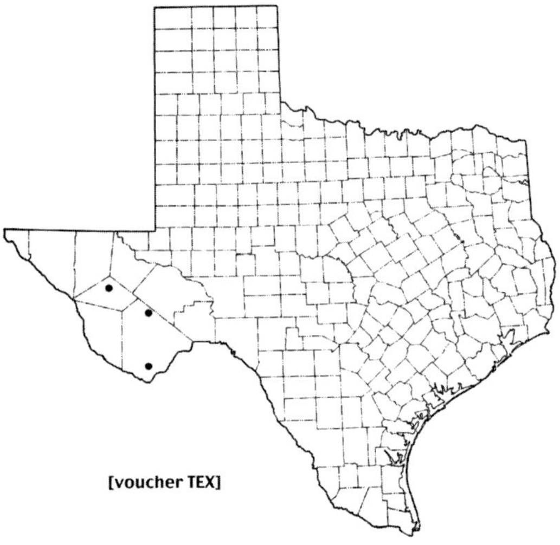
[voucher TEX]

Ipomoea capillacea

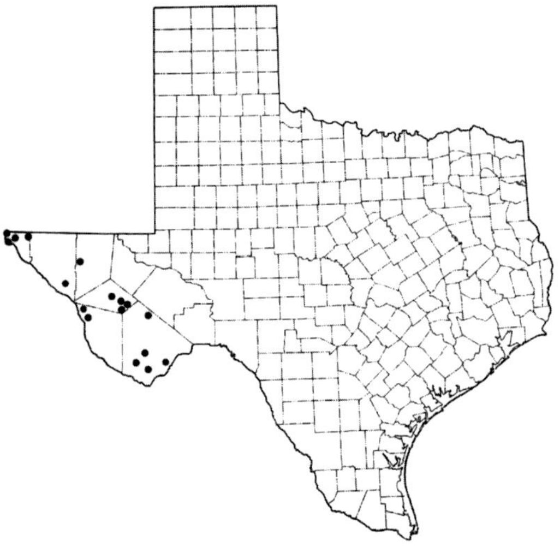

Ipomoea cardiophylla

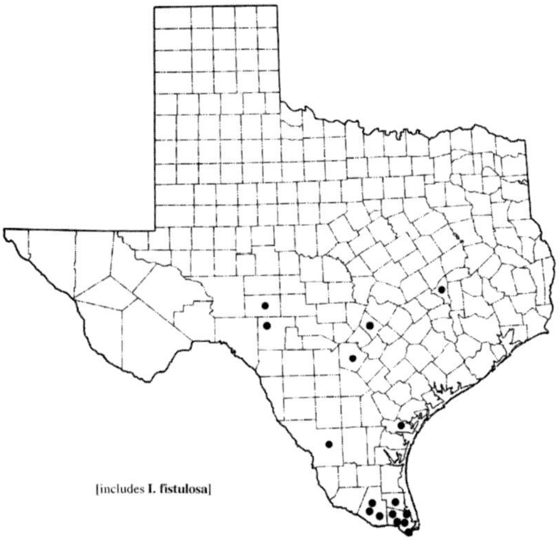
[includes I. fistulosa]

Ipomoea carnea

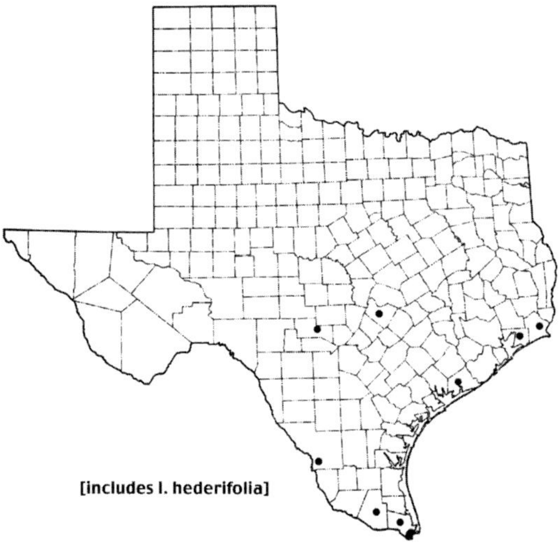
[includes I. hederifolia]

Ipomoea coccinea

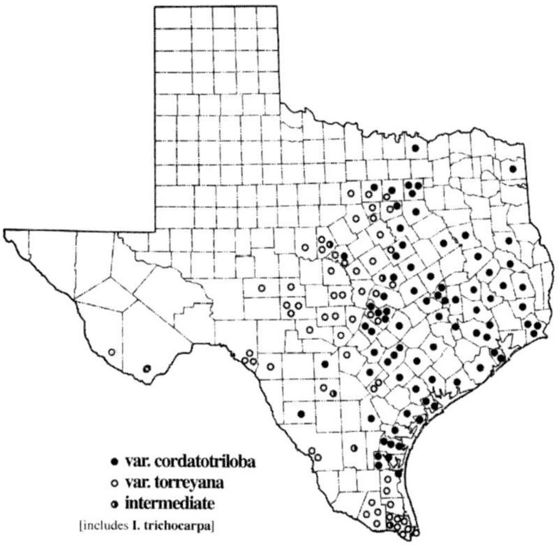
var. cordatotriloba
var. torreyana
intermediate
[includes I. trichocarpa]

Ipomoea cordatotriloba

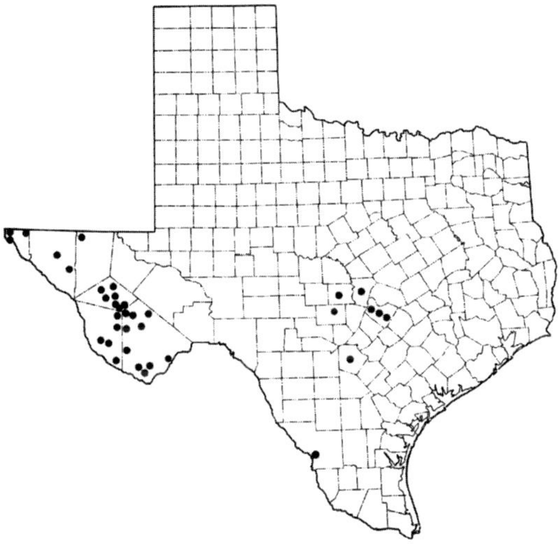

Ipomoea costellata

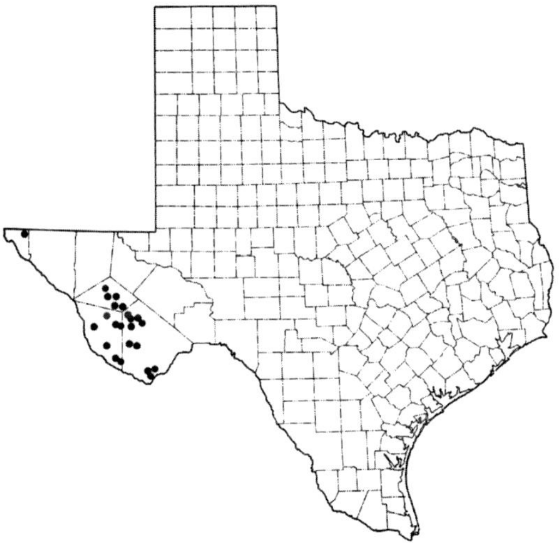

Ipomoea cristulata

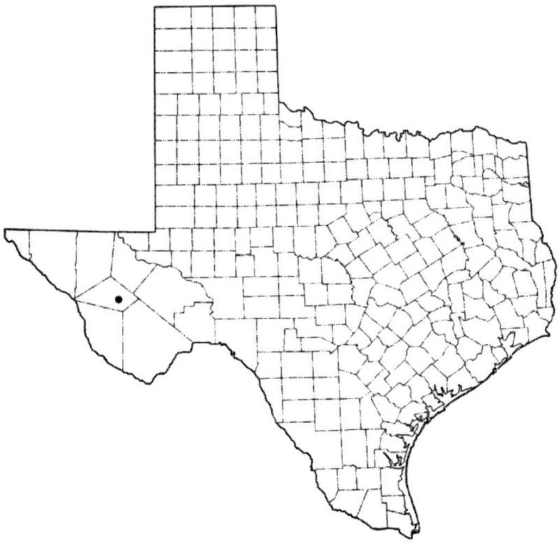

Ipomoea dumetorum

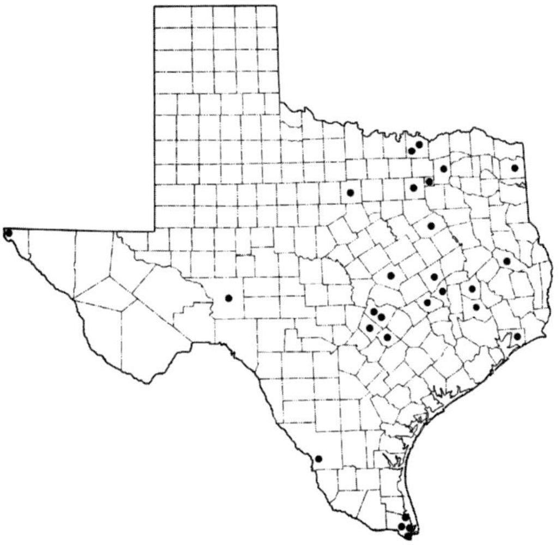

Ipomoea hederacea

Ipomoea heptaphylla

[includes I. stolonifera]

Ipomoea imperati

[= I. acuminata]

Ipomoea indica

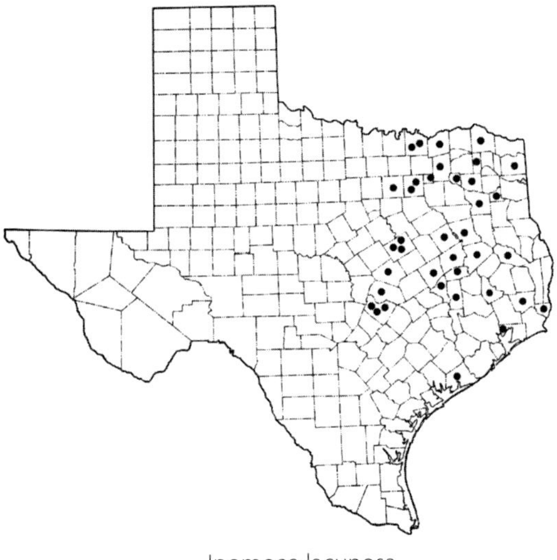

Ipomoea lacunosa

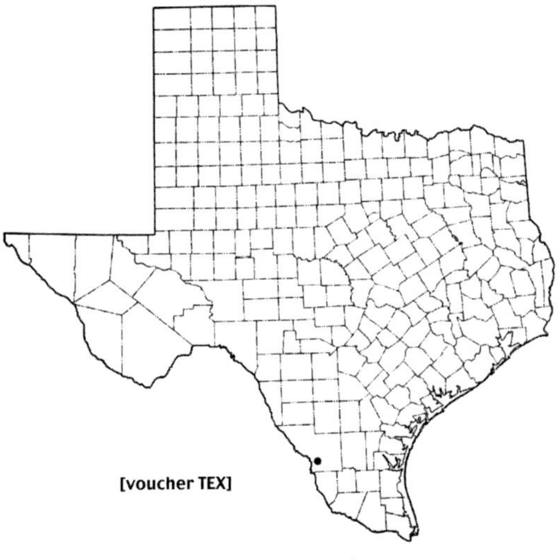
[voucher TEX]

Ipomoea leptoloma

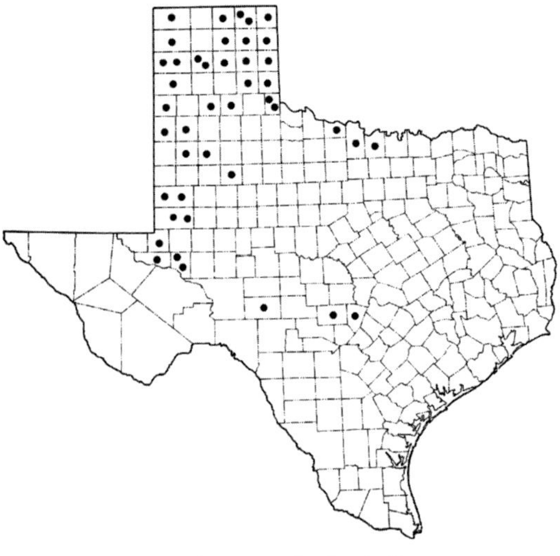

Ipomoea leptophylla

Ipomoea xleucantha

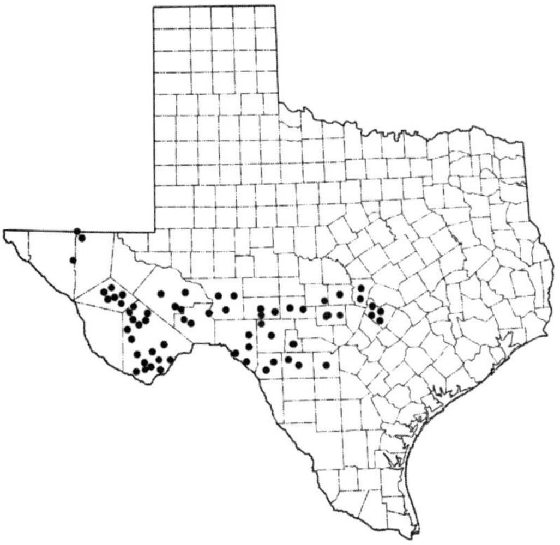

Ipomoea lindheimeri

Ipomoea nil

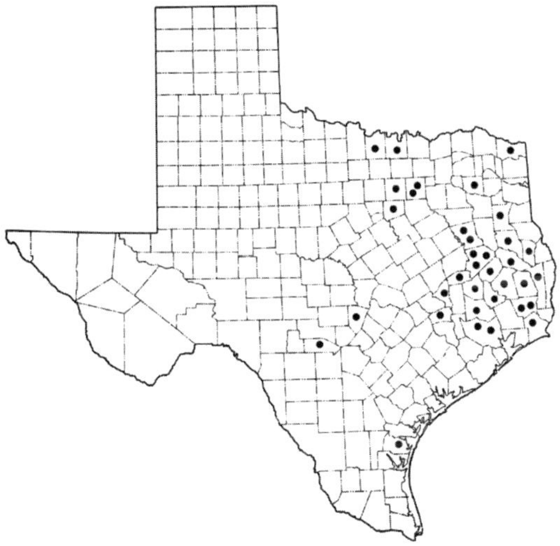

Ipomoea pandurata

Ipomoea pes-caprae

[= Operculina pinnatifida]

Ipomoea pinnatifida

Ipomoea pubescens

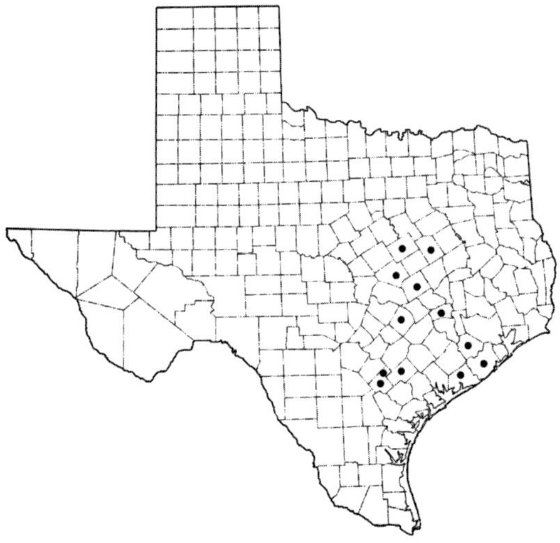

Ipomoea pulchella

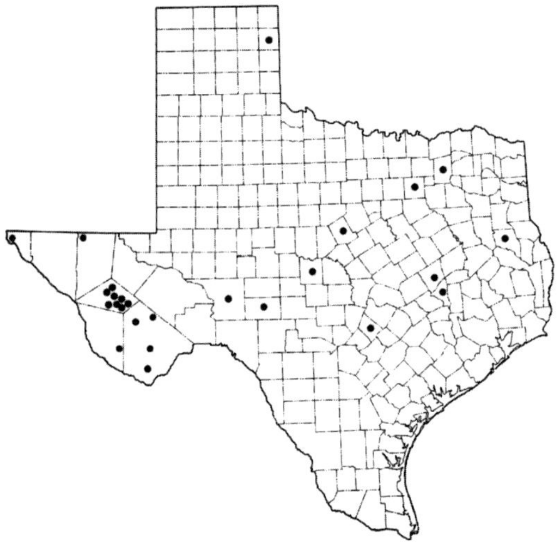

Ipomoea purpurea

Ipomoea quamoclit

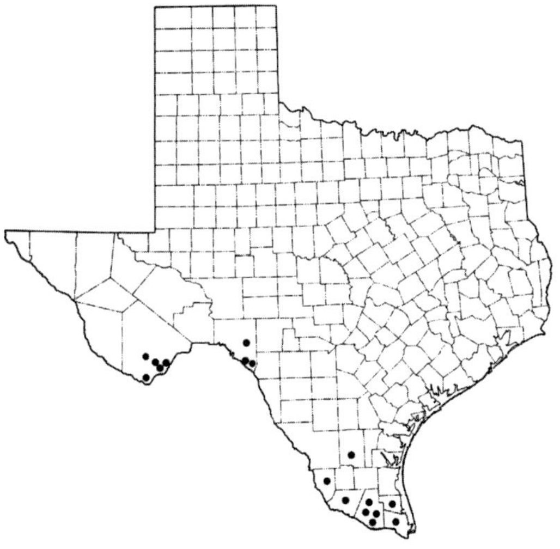

Ipomoea rupicola

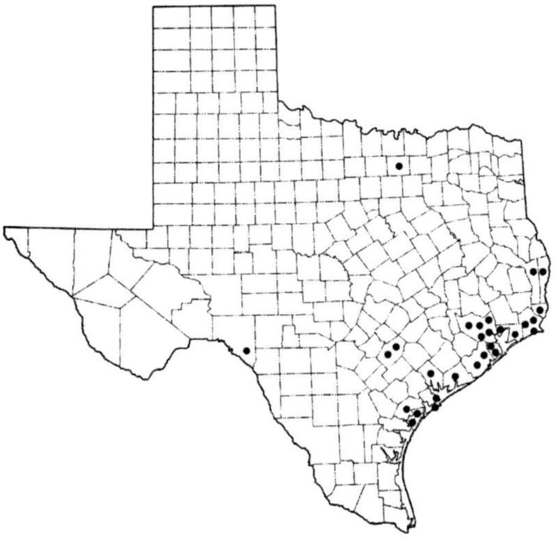

Ipomoea sagittata

Ipomoea shumardiana

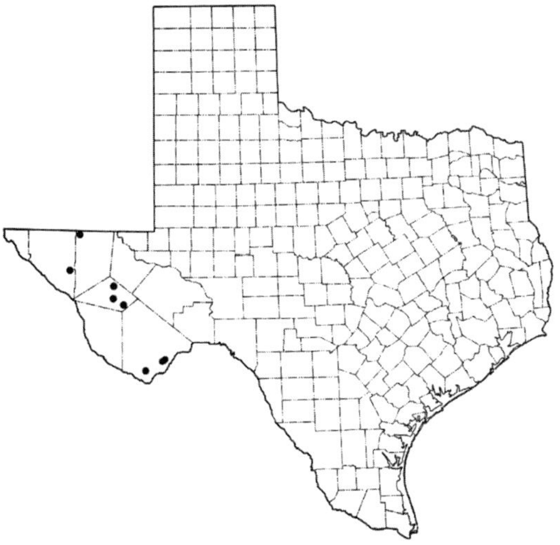

Ipomoea tenuiloba

Ipomoea tricolor

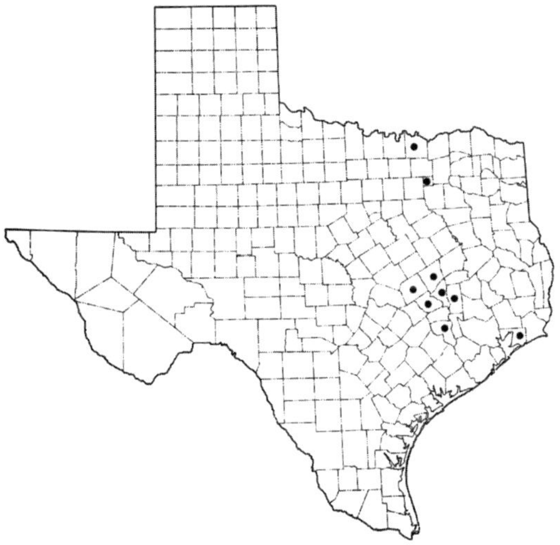

Ipomoea wrightii

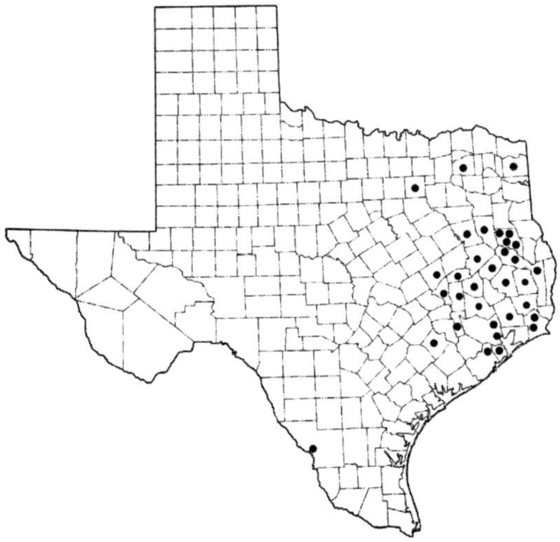

Jacquemontia tamnifolia

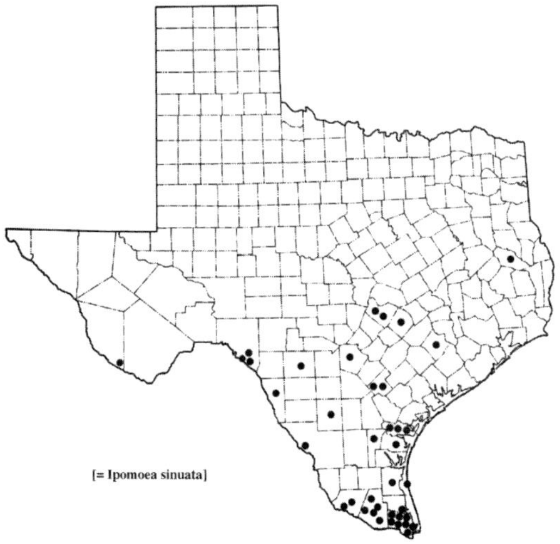
[= Ipomoea sinuata]

Merremia dissecta

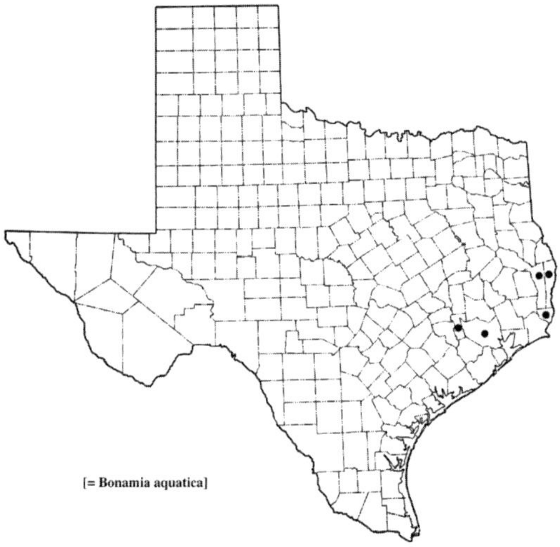
[= Bonamia aquatica]

Stylisma aquatica

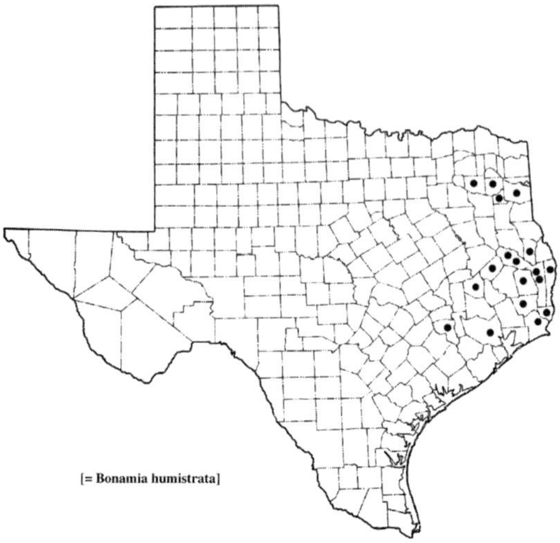
[= Bonamia humistrata]

Stylisma humistrata

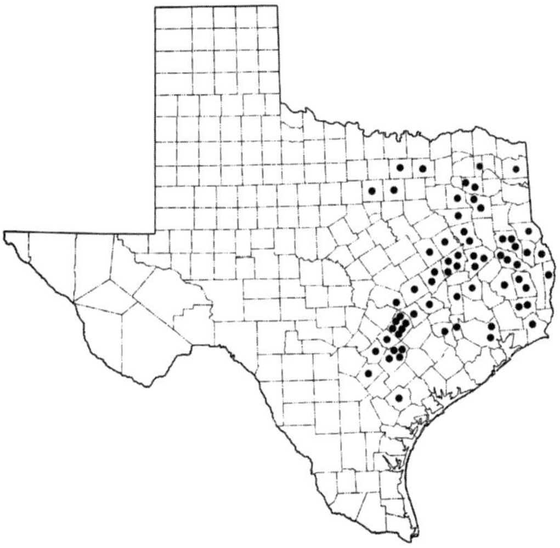

Stylisma pickeringii

[= Bonamia villosa]

Stylisma villosa

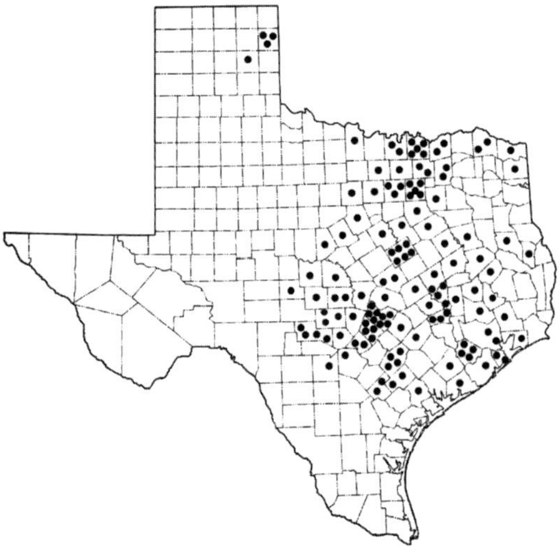

Cornus drummondii

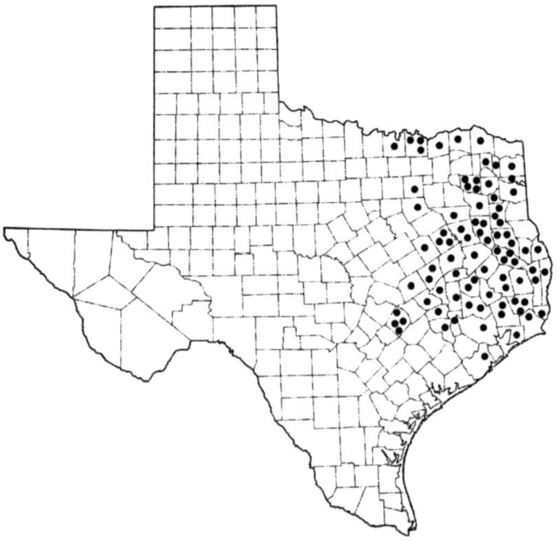

Cornus florida

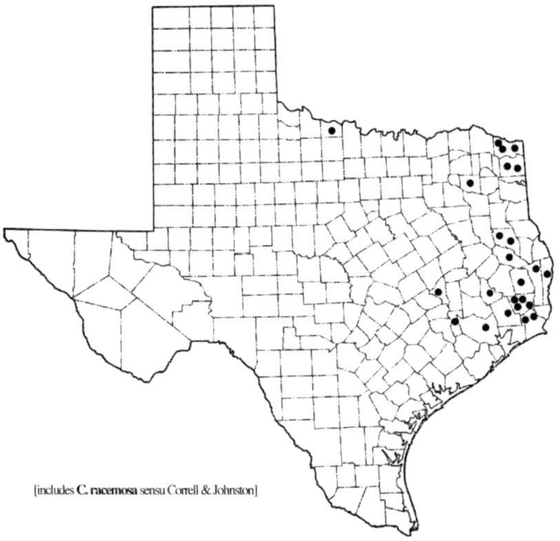
[includes **C. racemosa** sensu Correll & Johnston]

Cornus foemina

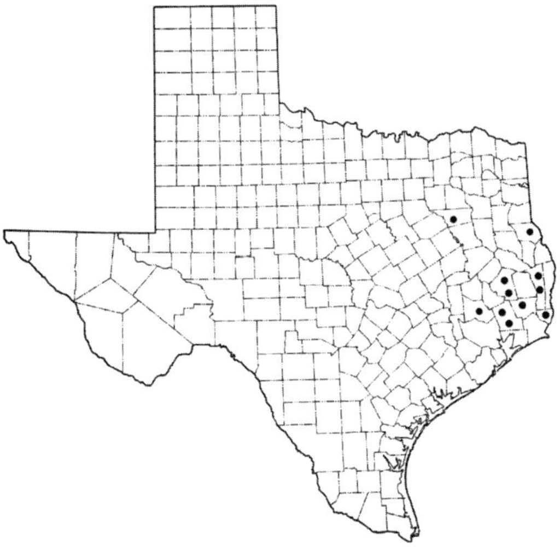

Nyssa aquatica

Nyssa biflora

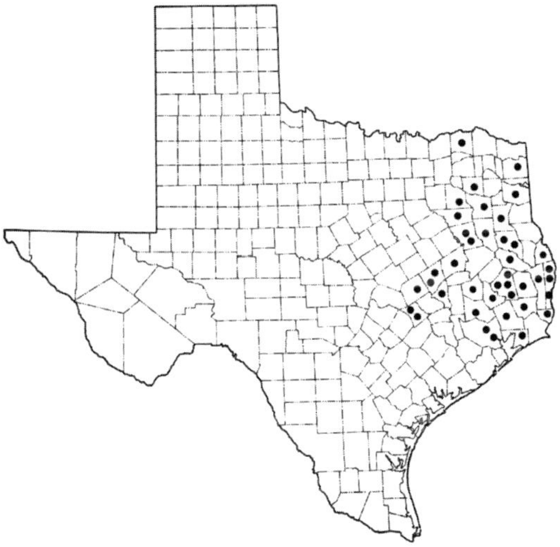

Nyssa sylvatica

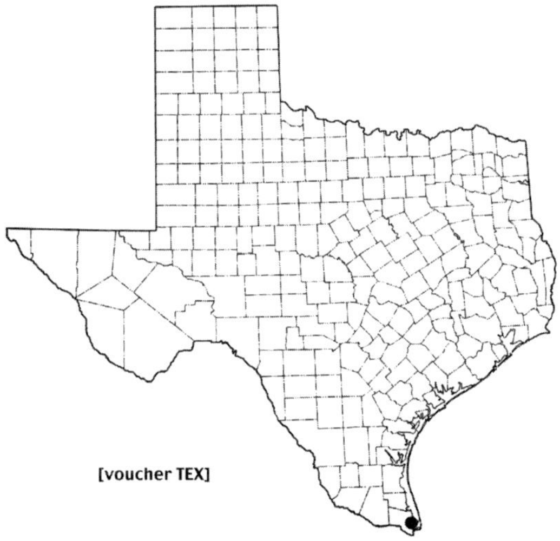
[voucher TEX]

Bryophyllum fedtschenkoi

Bryophyllum tubiflorum

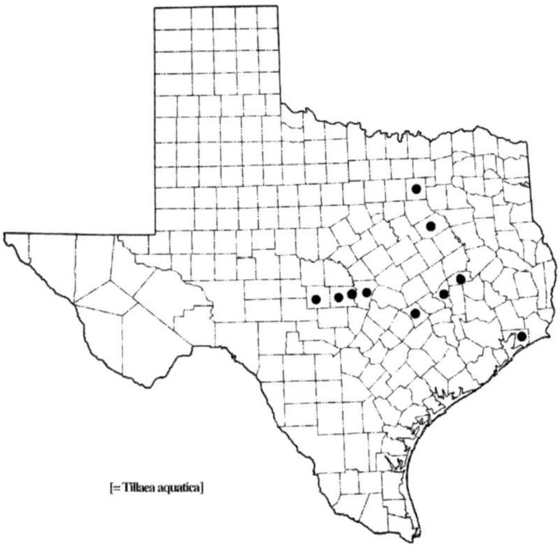
[= Tillaea aquatica]

Crassula aquatica

[voucher TEX]
[= Tillaea connata]

Crassula connata

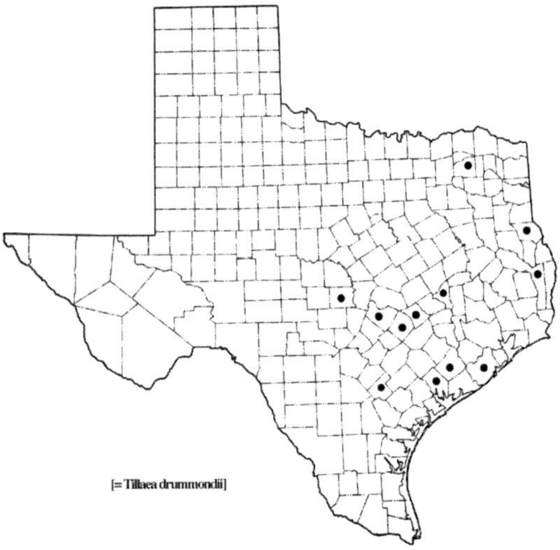
[= Tillaea drummondii]

Crassula drummondii

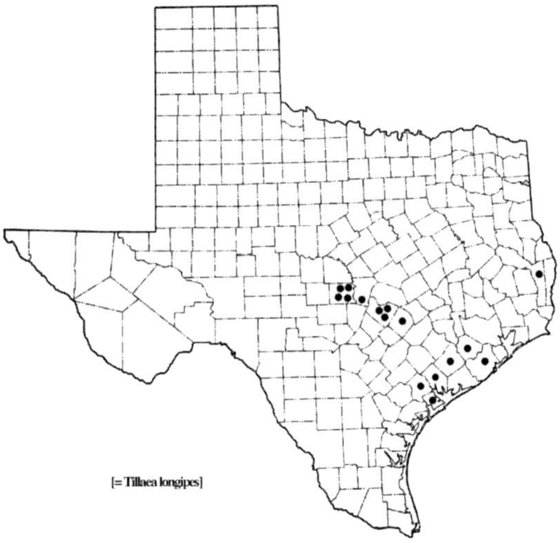
[= Tillaea longipes]

Crassula longipes

[= Tillaea solieri]

Crassula solierii

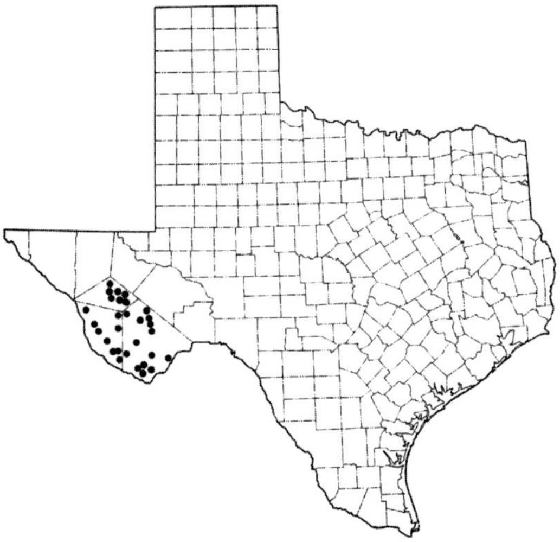

Echeveria strictiflora

[= Sedum texanum]

Lenophyllum texanum

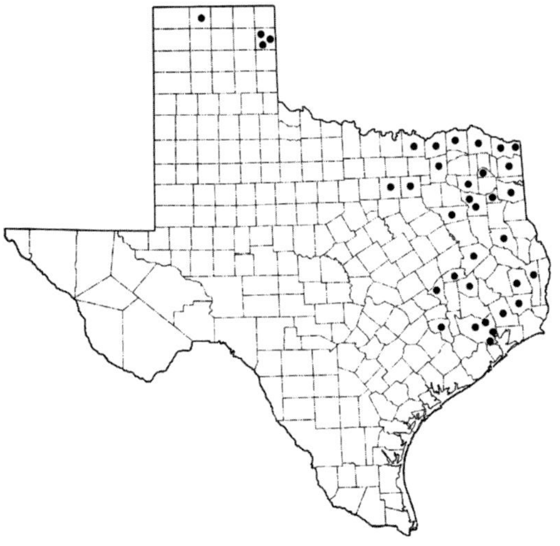

Penthorum sedoides

Sedum cockerellii

Sedum diffusum

Sedum havardii

[= S. parvum subsp. nanifolium]

Sedum nanifolium

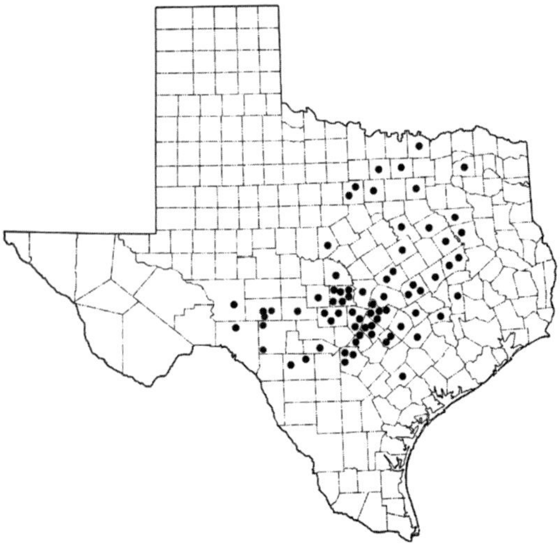

Sedum nuttallianum

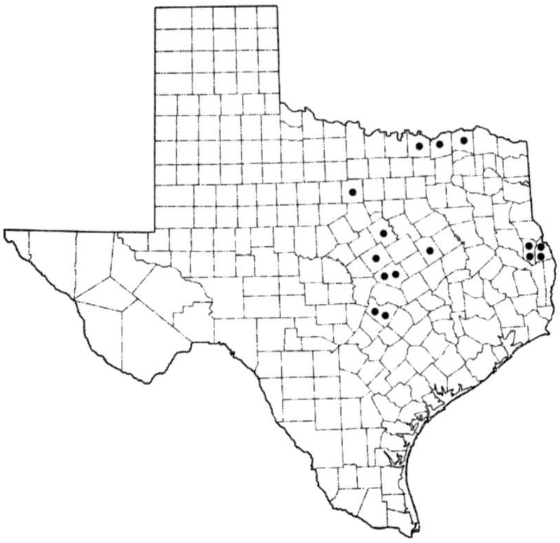

Sedum pulchellum

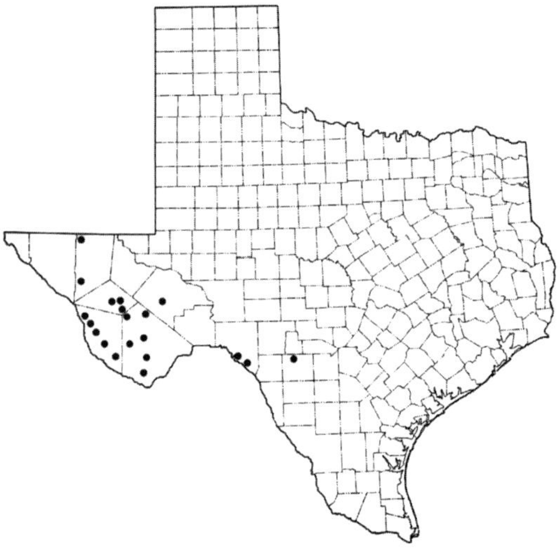

Sedum wrightii

Villadia squamulosa

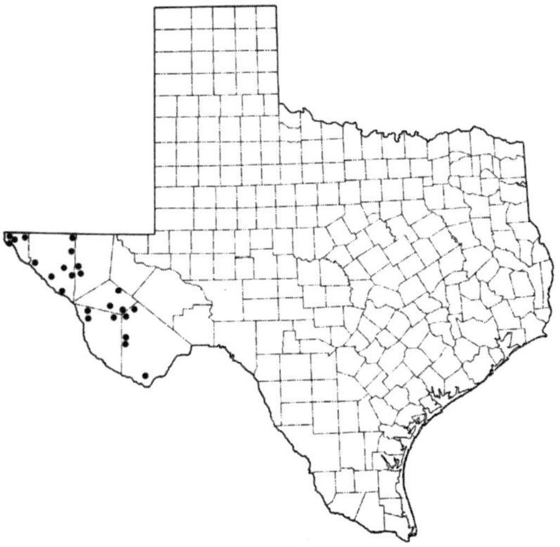

Apodanthera undulata

Cayaponia quinqueloba

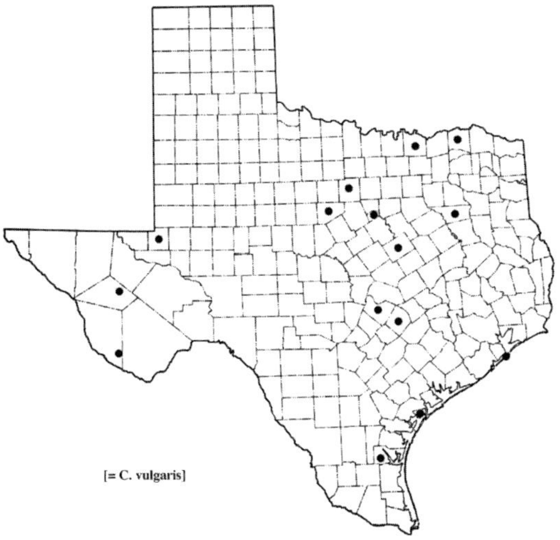
[= C. vulgaris]

Citrullus lanatus

Cucumis anguria

Cucumis dipsaceus

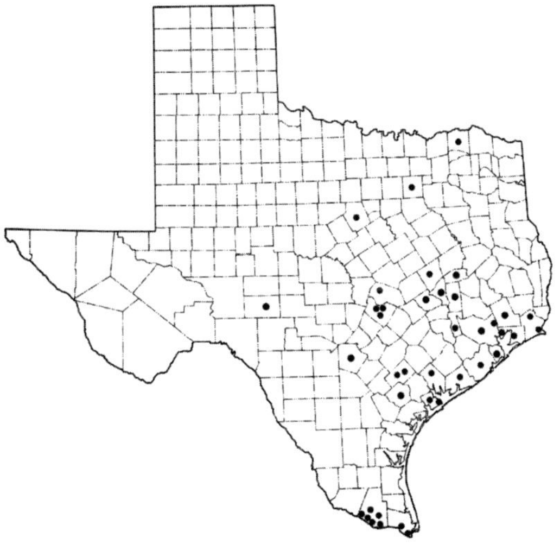

Cucumis melo

Cucurbita digitata

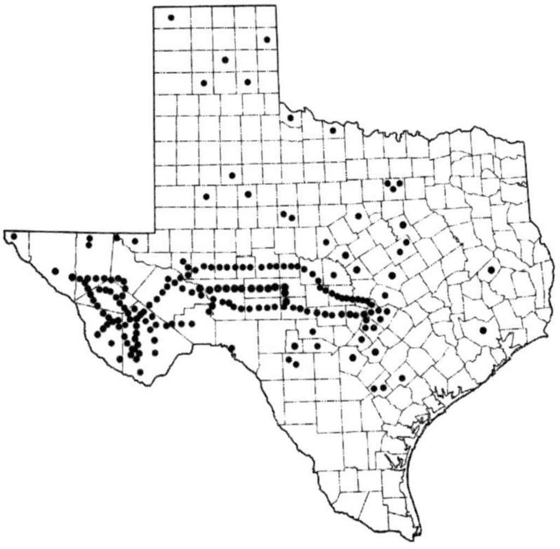

Cucurbita foetidissima

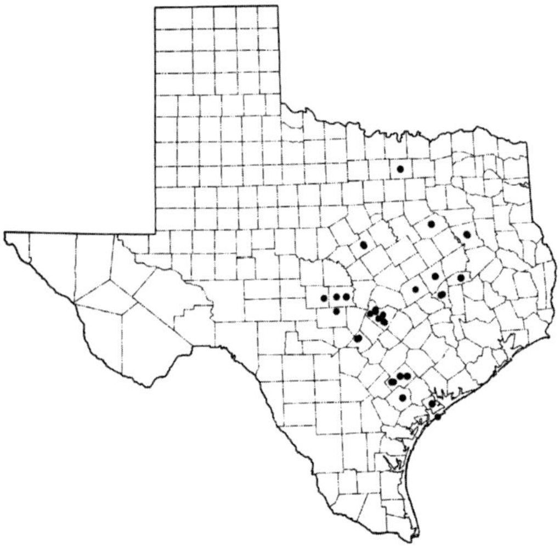

Cucurbita texana

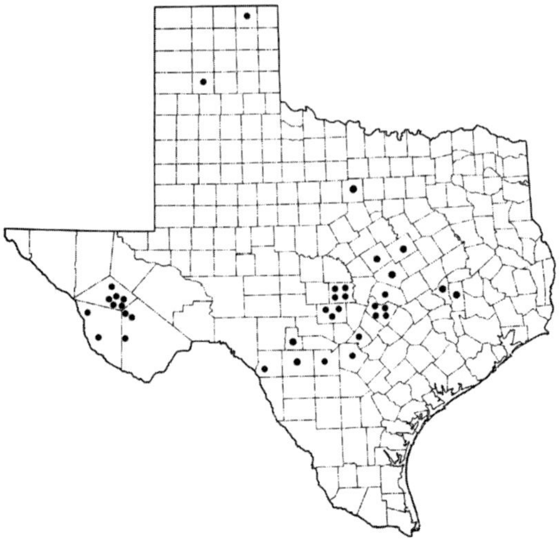

Cyclanthera dissecta

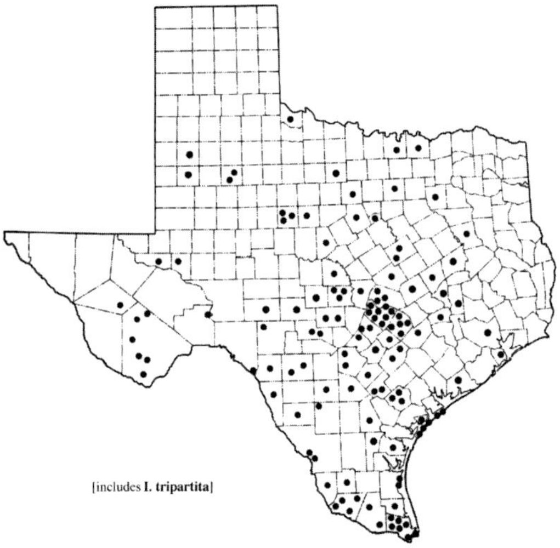
[includes **I. tripartita**]

Ibervillea lindheimeri

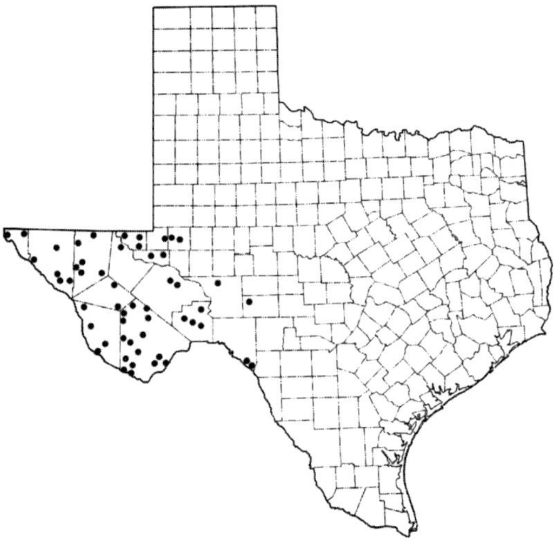

Ibervillea tenuisecta

Lagenaria siceraria

Luffa aegyptica

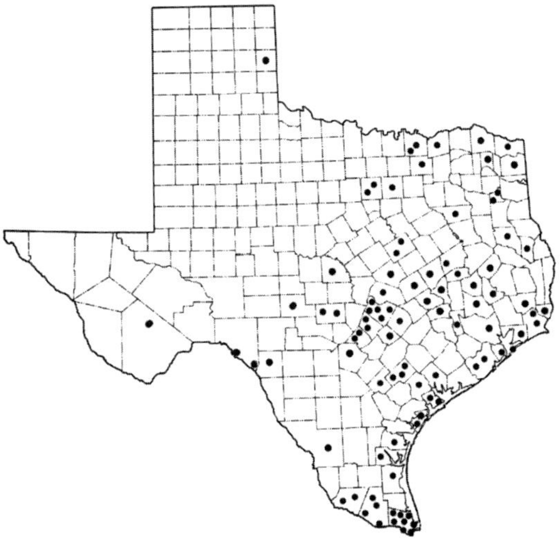
Melothria pendula

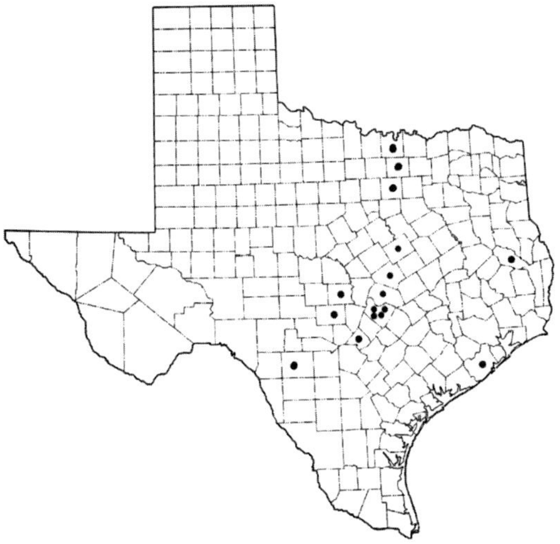
Sicyos angulatus

[includes **S. ampelophyllus**, **S. glaber**
and **S. laciniatus** sensu Correll & Johnston]
Sicyos microphyllus

Cuscuta applanata

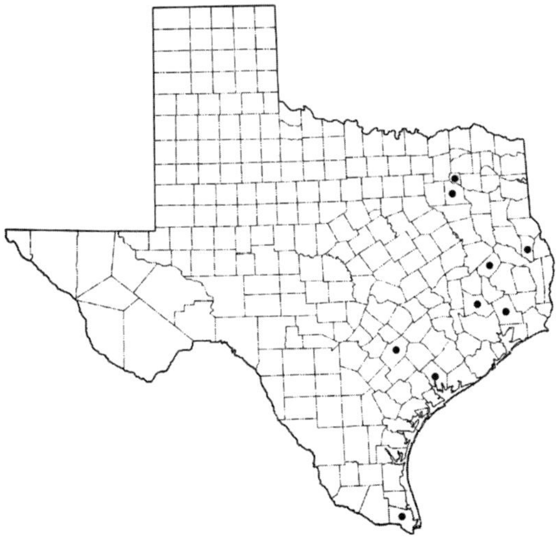
Cuscuta attenuata

Cuscuta cephalanthi

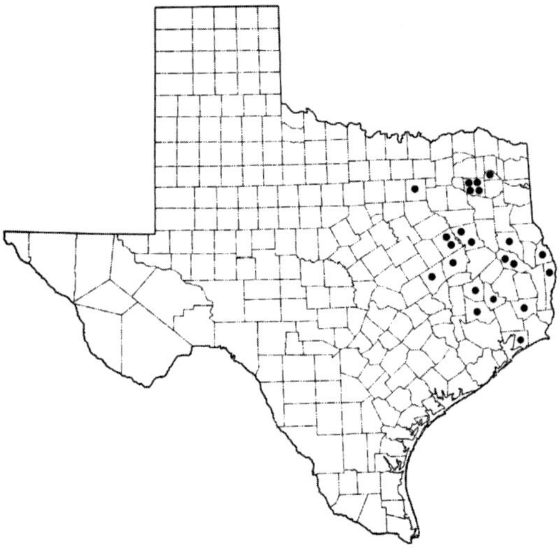

Cuscuta compacta

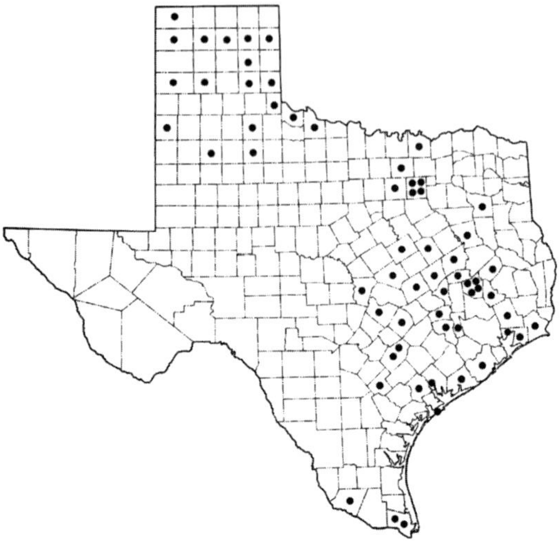

Cuscuta cuspidata

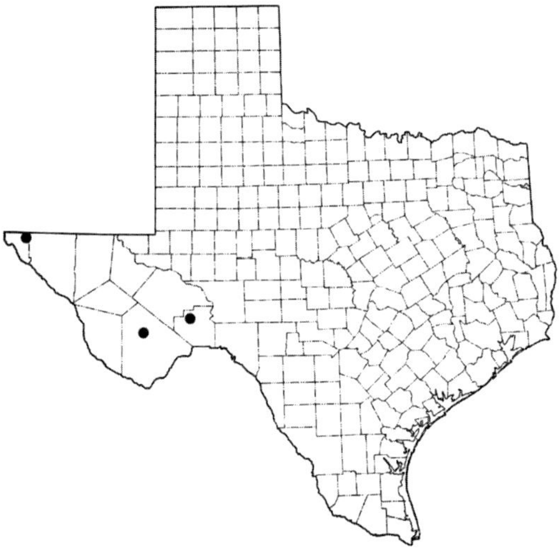

Cuscuta decipiens

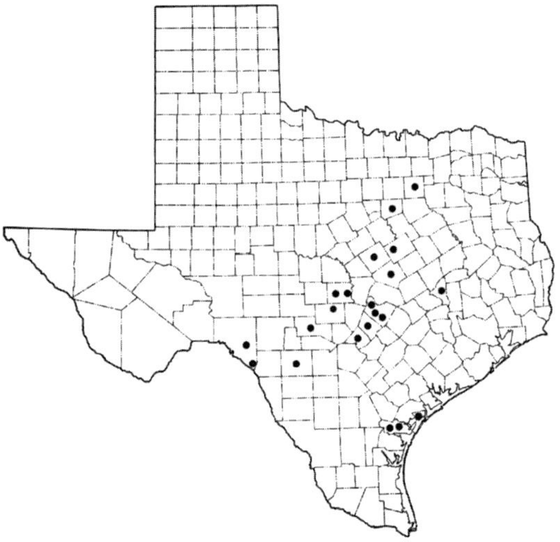

Cuscuta exaltata

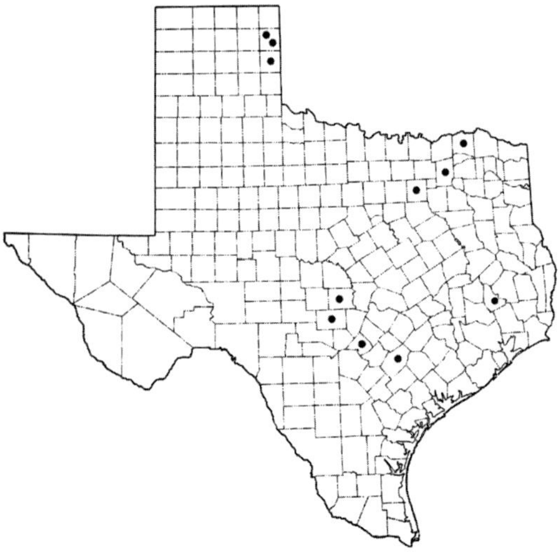

Cuscuta glomerata

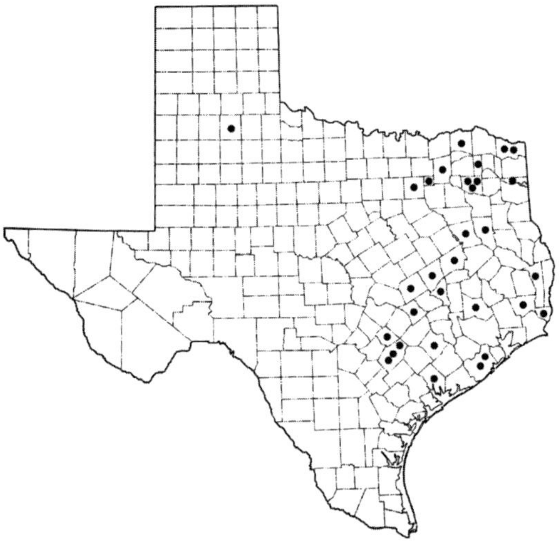

Cuscuta gronovii

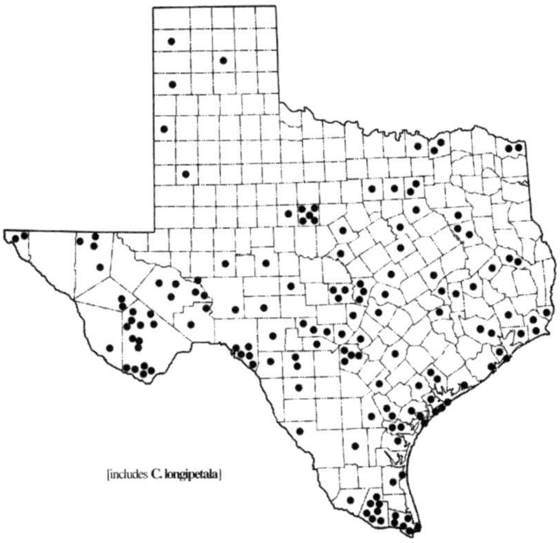
[includes **C. longipetala**]

Cuscuta indecora

Cuscuta leptantha

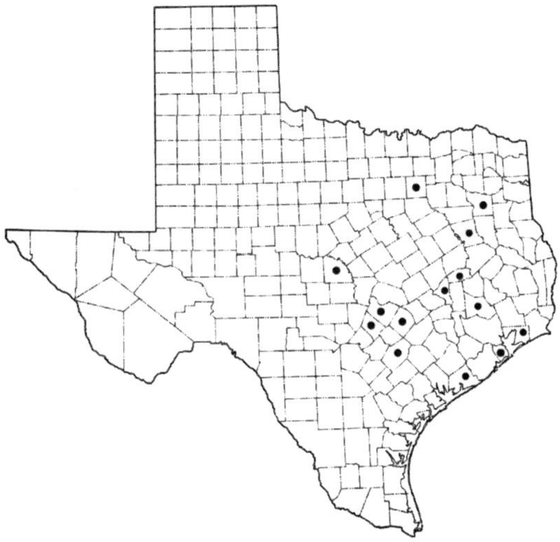

Cuscuta obtusiflora

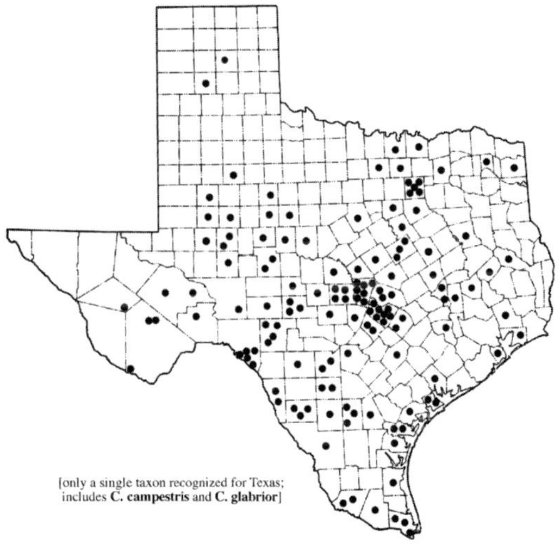
[only a single taxon recognized for Texas;
includes **C. campestris** and **C. glabrior**]

Cuscuta pentagona

[Sida 18: 317. 1998]

Cuscuta polygonorum

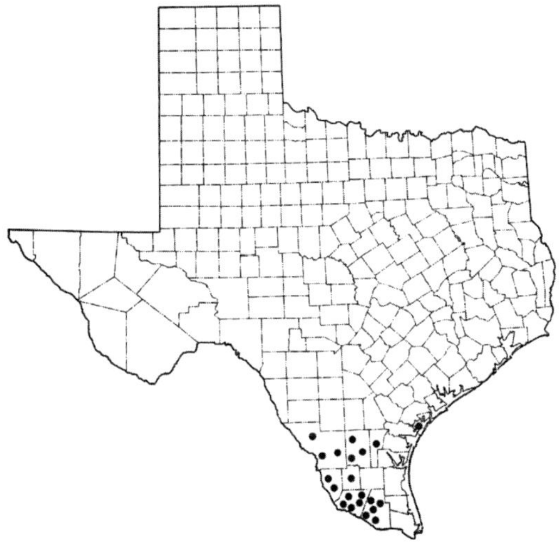

Cuscuta runyonii

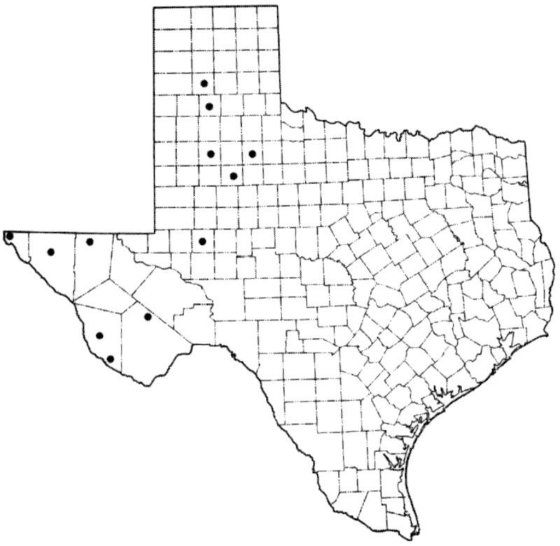

Cuscuta squamata

Cuscuta suaveolens

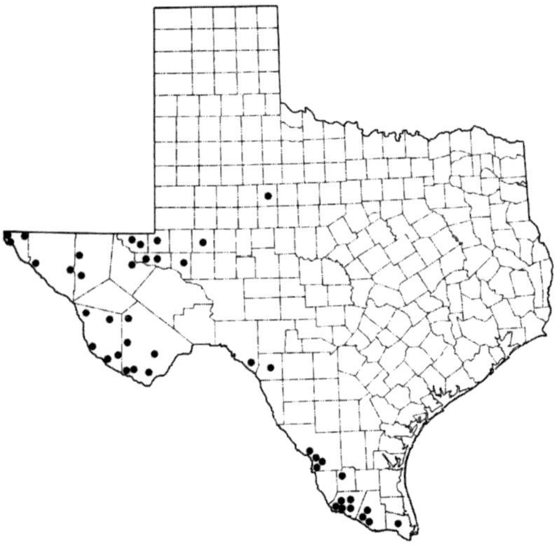

Cuscuta umbellata

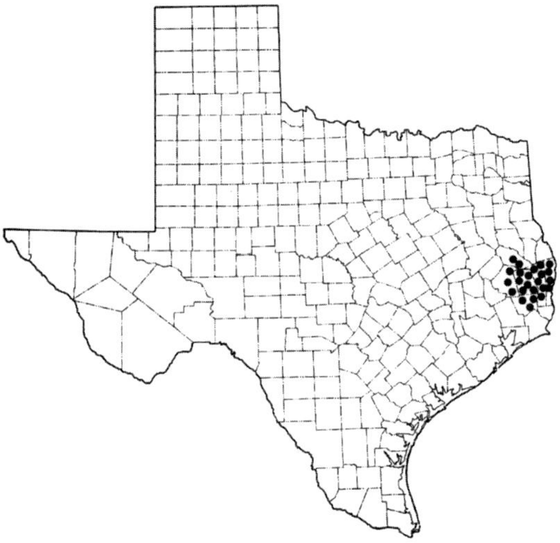

Cyrilla racemiflora

[Singhurst 9379, TEX]

Dipsacus fullonum

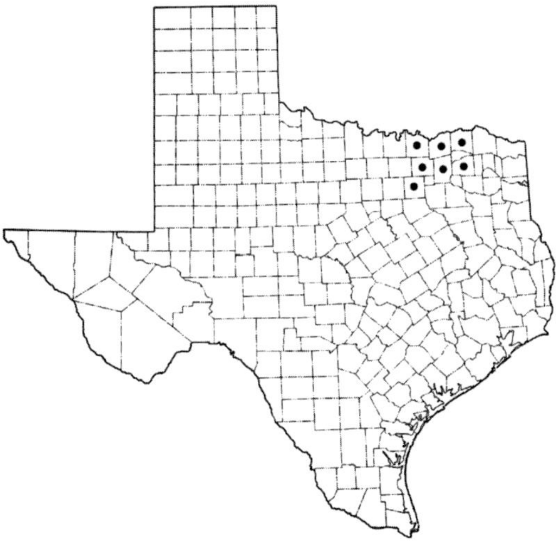

Scabiosa atropurpurea

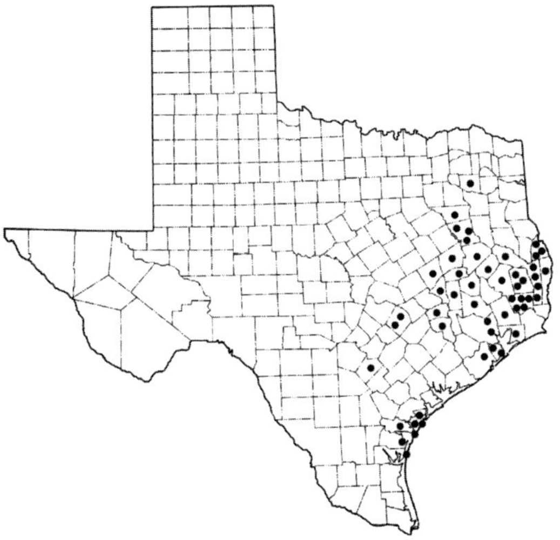

Drosera brevifolia

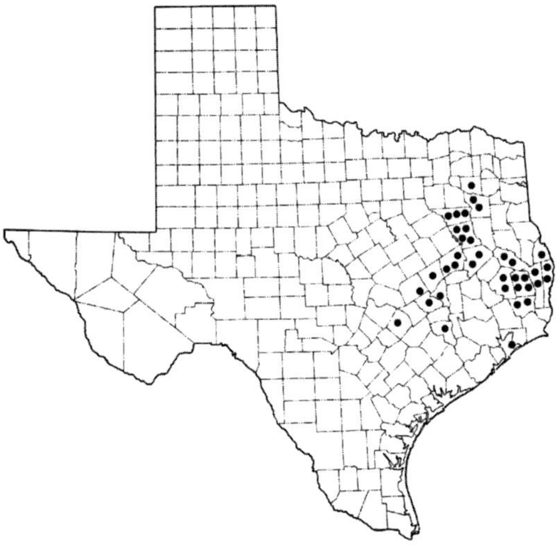

Drosera capillaris

Drosera intermedia

Drosera tracyi

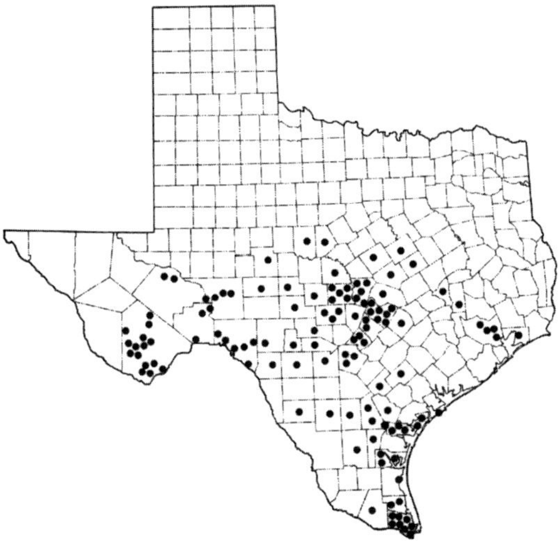

Diospyros texana

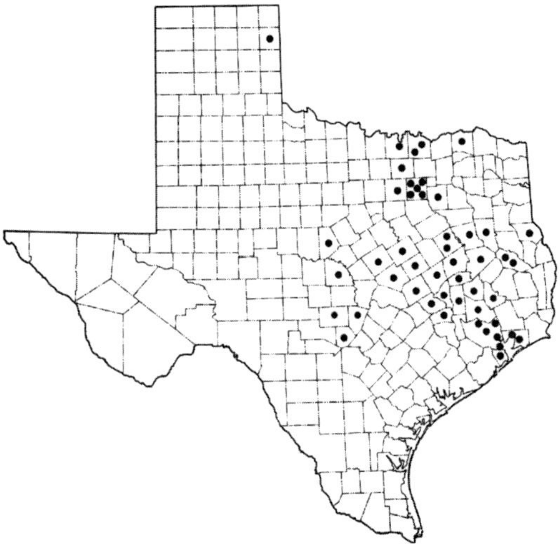

Diospyros virginiana

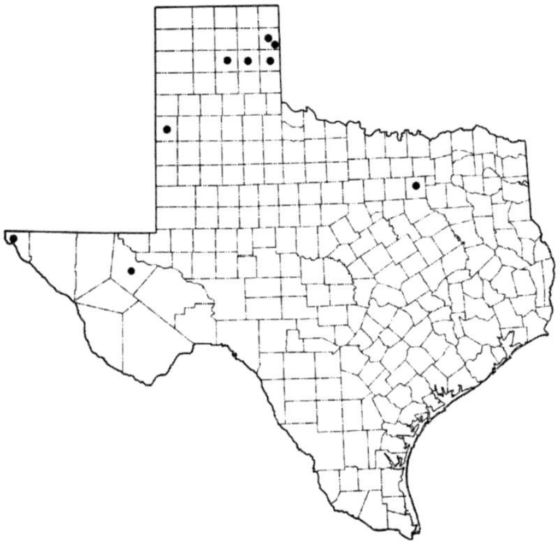

Elaeagnus angustifolia

Elaeagnus pungens

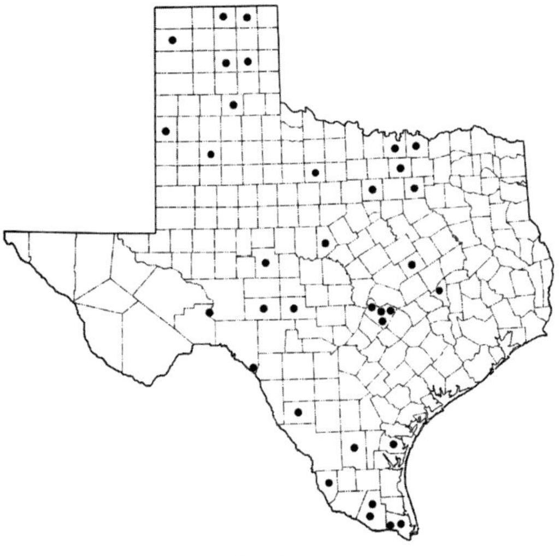

Bergia texana

Elatine brachysperma

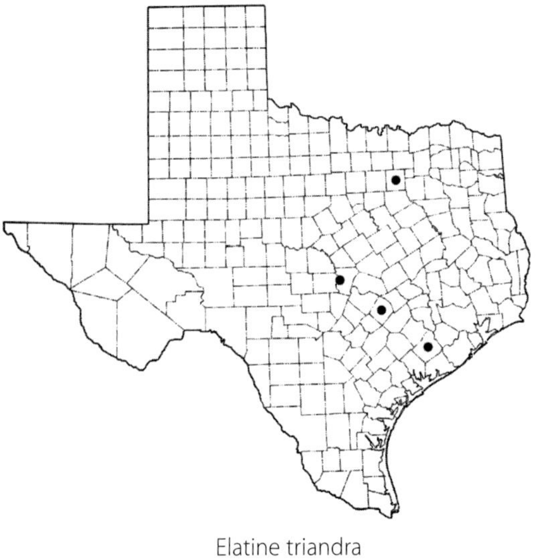

Elatine triandra

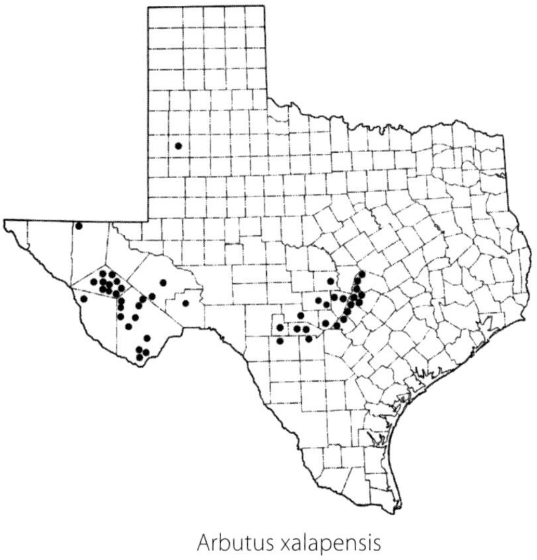

Arbutus xalapensis

Arctostaphylos pungens

Leucothoe racemosa

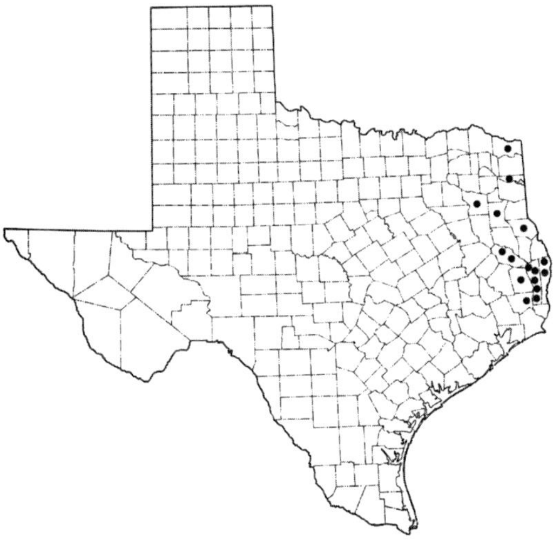

Lyonia ligustrina

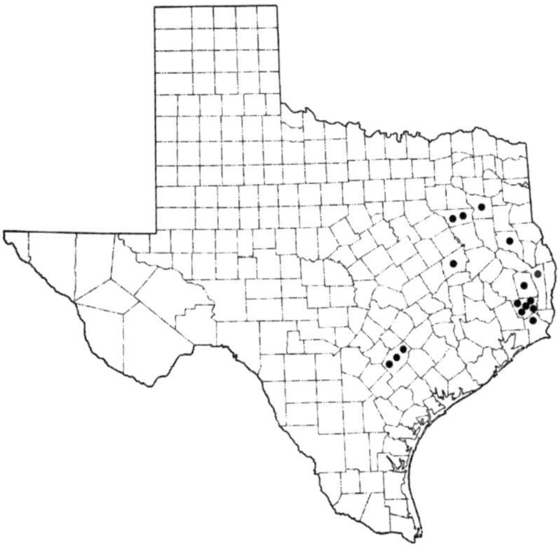

Lyonia mariana

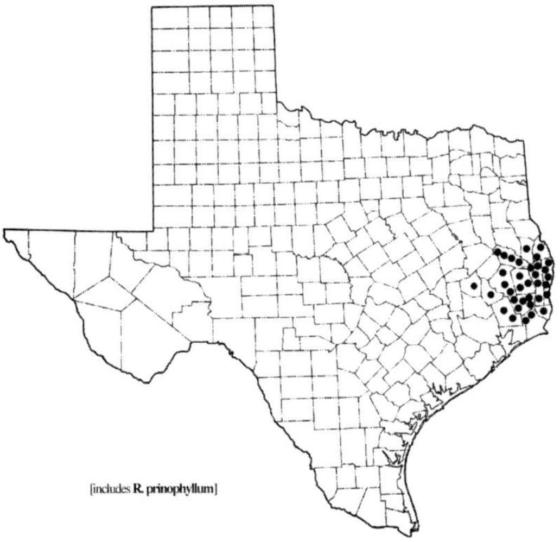
[includes **R. prinophyllum**]

Rhododendron canescens

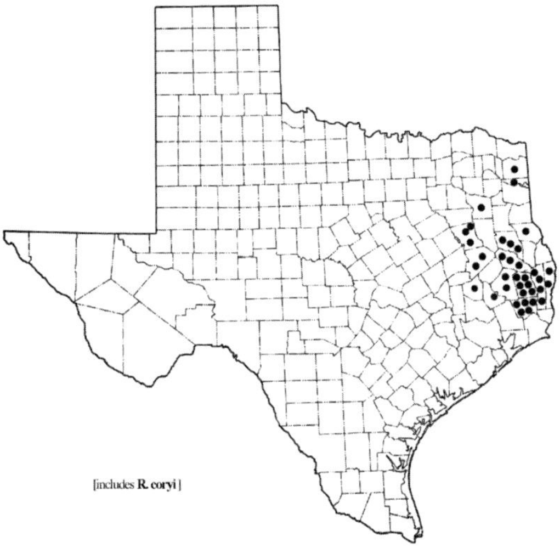
[includes **R. coryi**]

Rhododendron viscosum

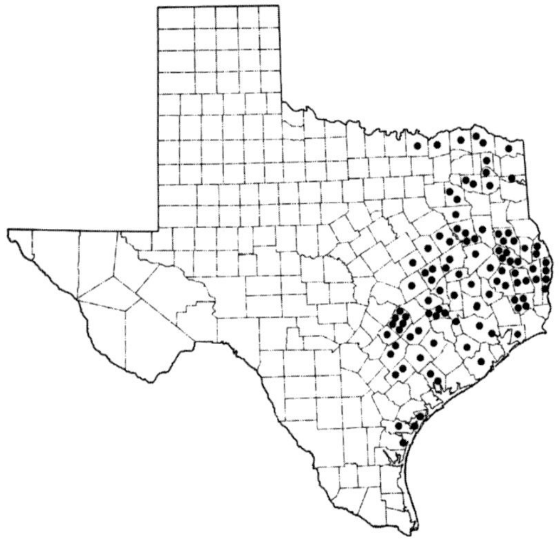

Vaccinium arboreum

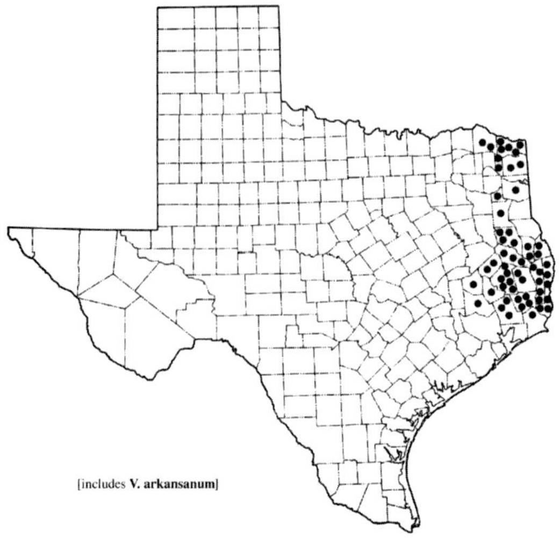
[includes V. arkansanum]

Vaccinium corymbosum

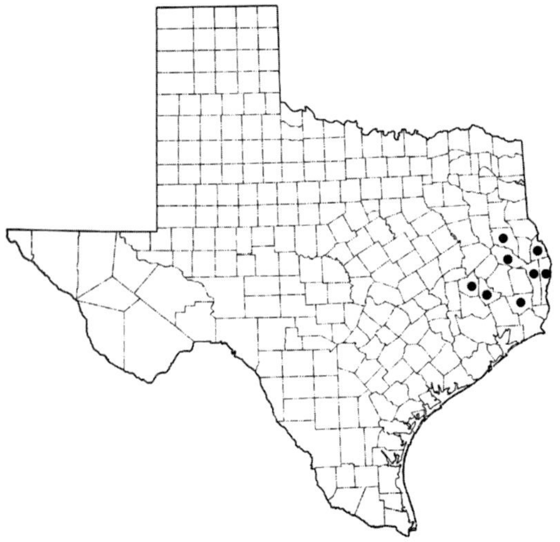

Vaccinium elliotii

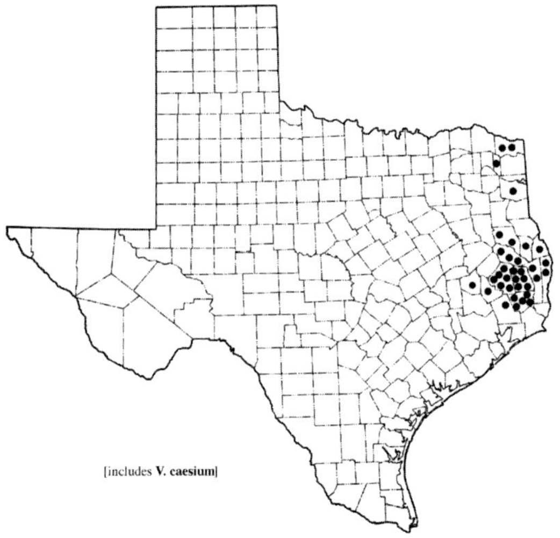
[includes V. caesium]

Vaccinium stamineum

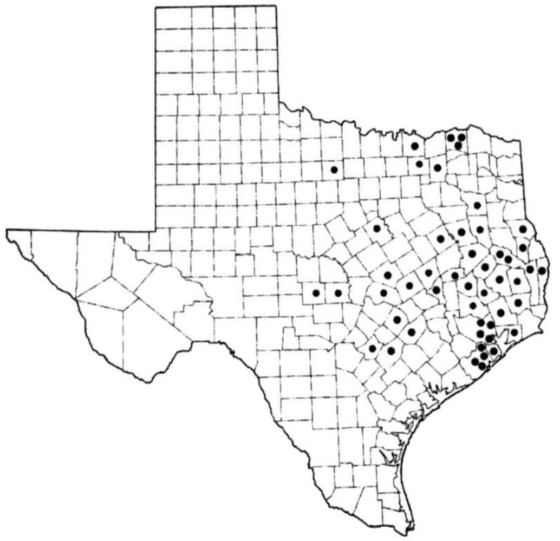

Acalypha gracilens

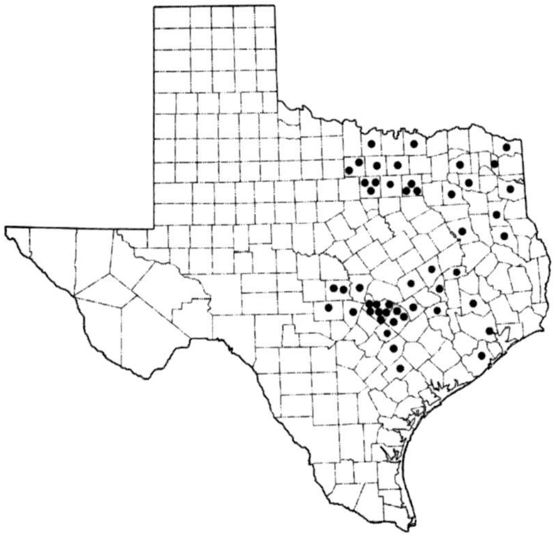

Acalypha monococca

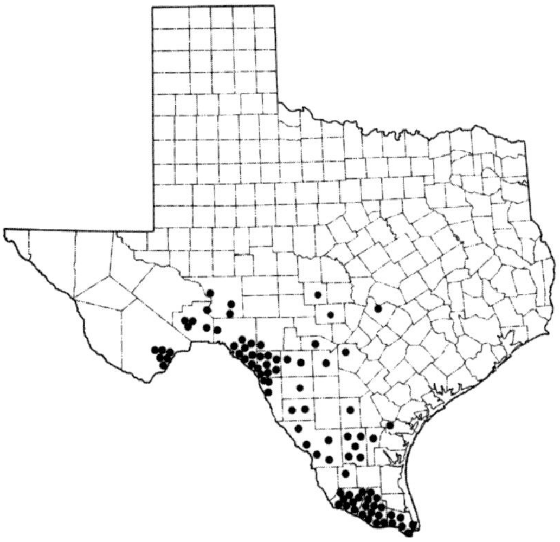

Acalypha monostachya

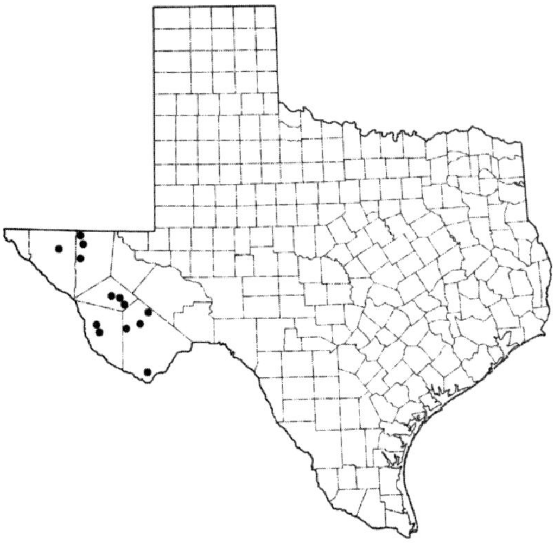

Acalypha neomexicana

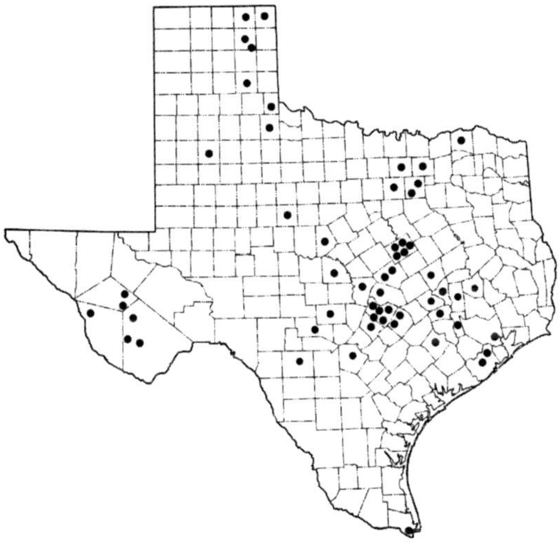

Acalypha ostryifolia

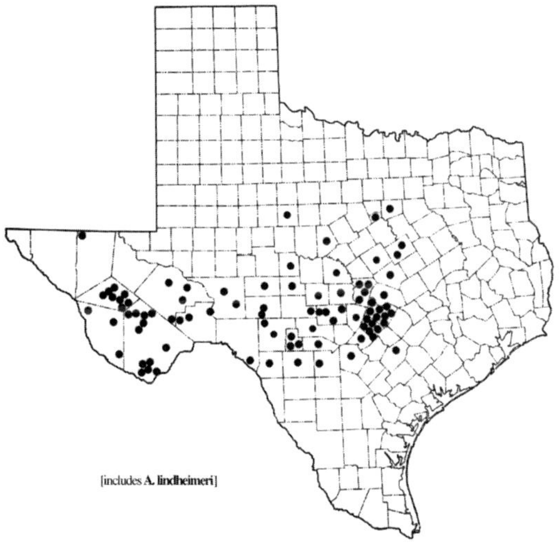
[includes **A. lindheimeri**]

Acalypha phleoides

Acalypha poiretii

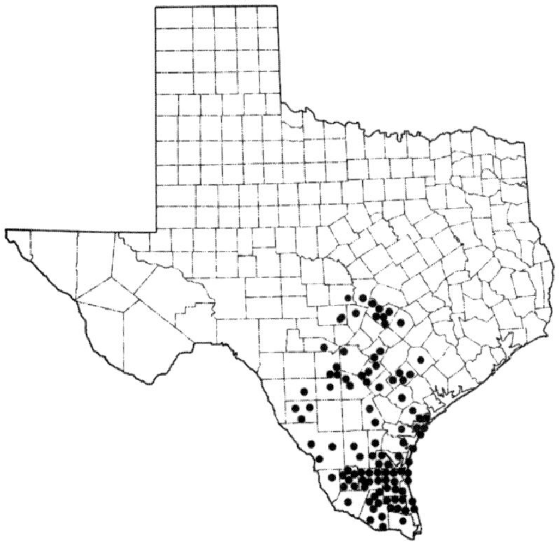

Acalypha radians

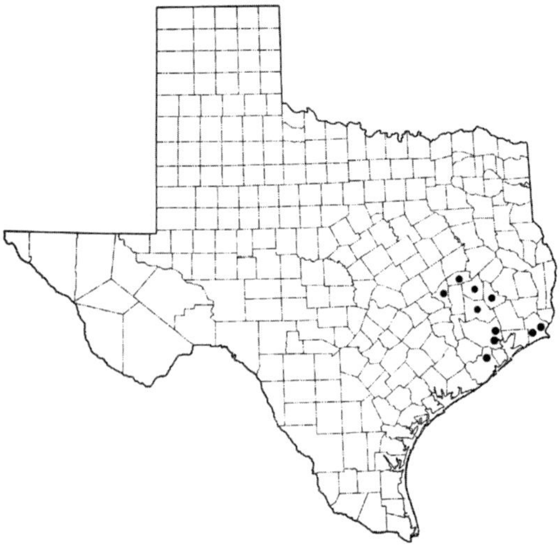

Acalypha rhomboidea

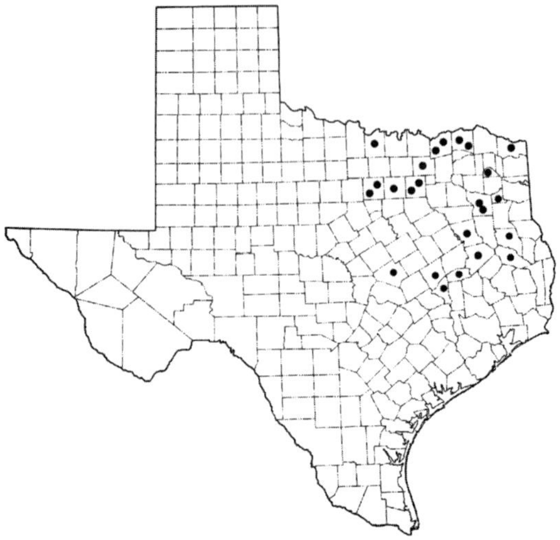

Acalypha virginica

Adelia vaseyi

Andrachne arida

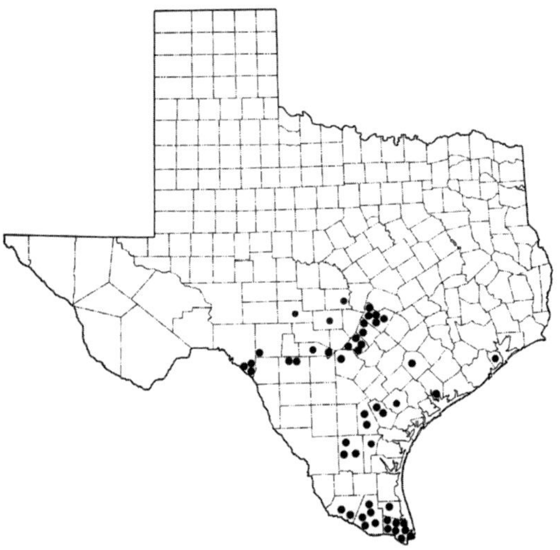

Bernardia myricifolia

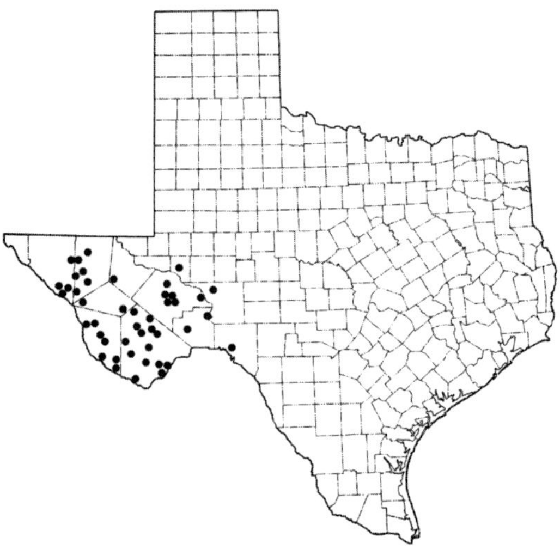

Bernardia obovata

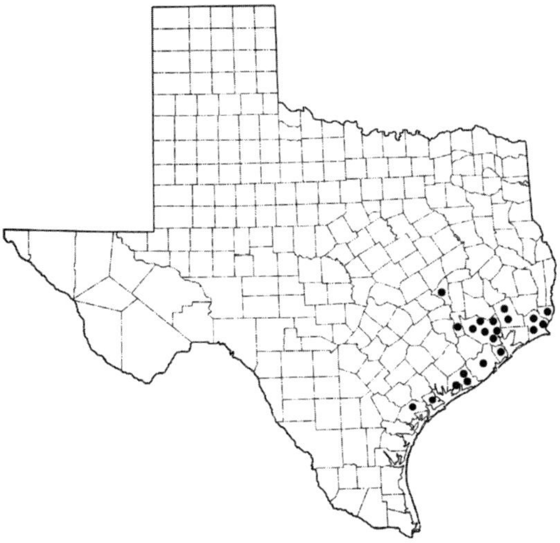
Caperonia palustris

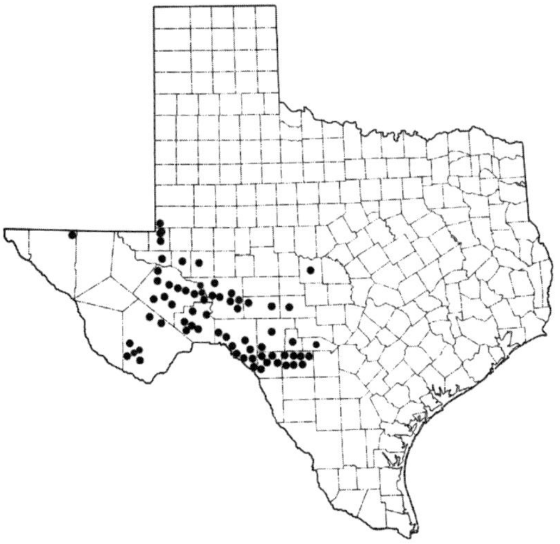
Chamaesyce acuta

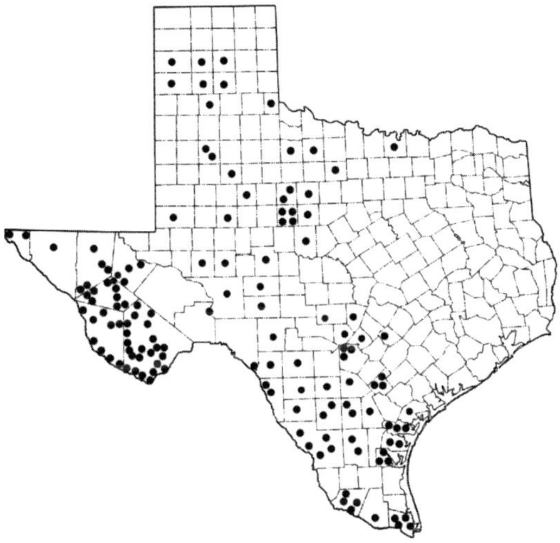
Chamaesyce albomarginata

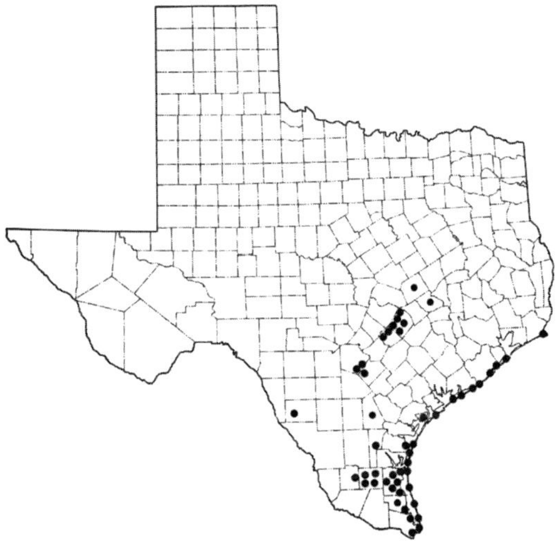
Chamaesyce ammannioides

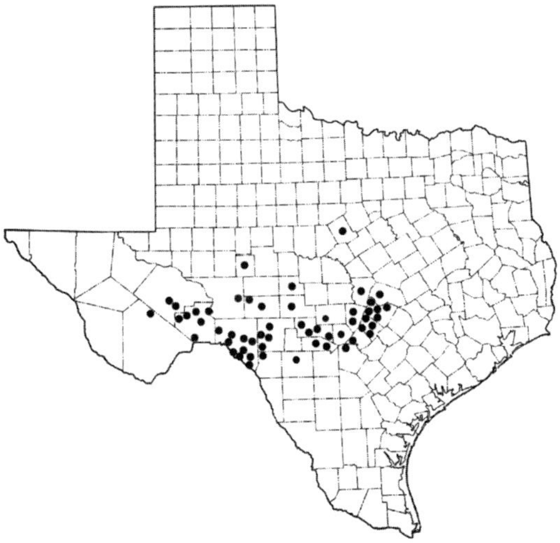
Chamaesyce angusta

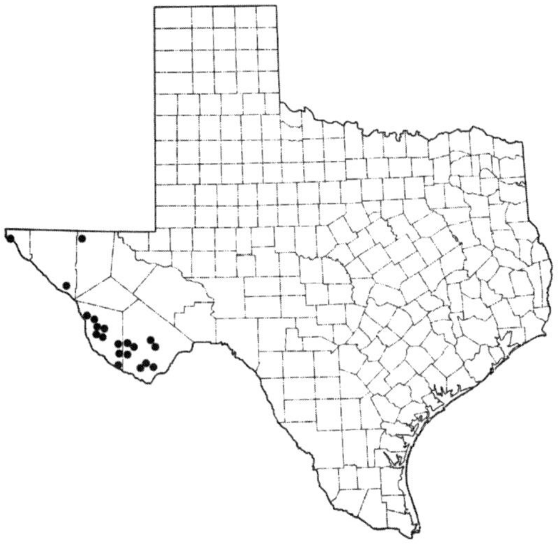
Chamaesyce arizonica

Chamaesyce astyla

[= Euphorbia pycnanthema]

Chamaesyce capitellata

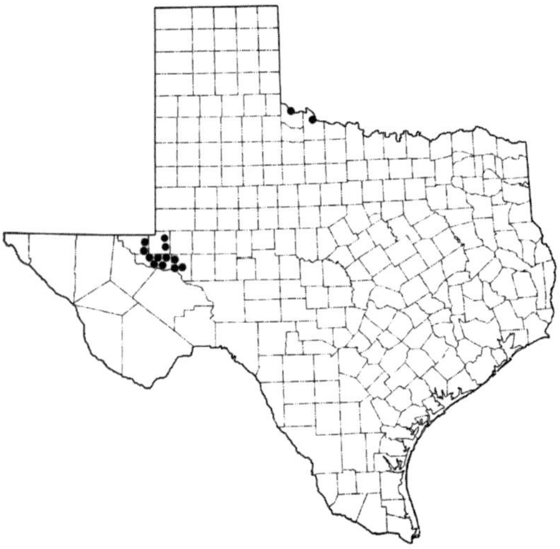

Chamaesyce carunculata

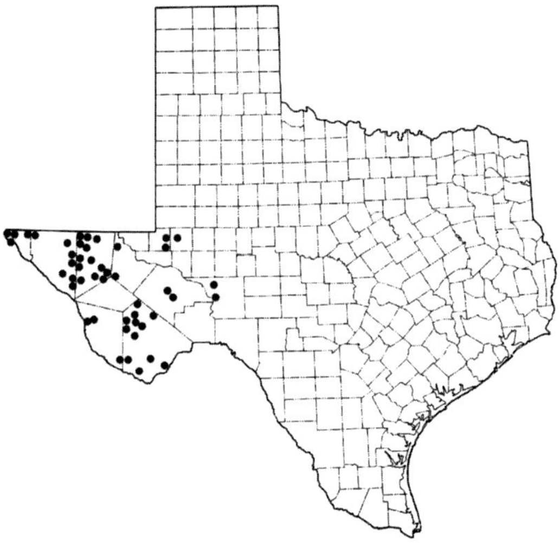

Chamaesyce chaetocalyx

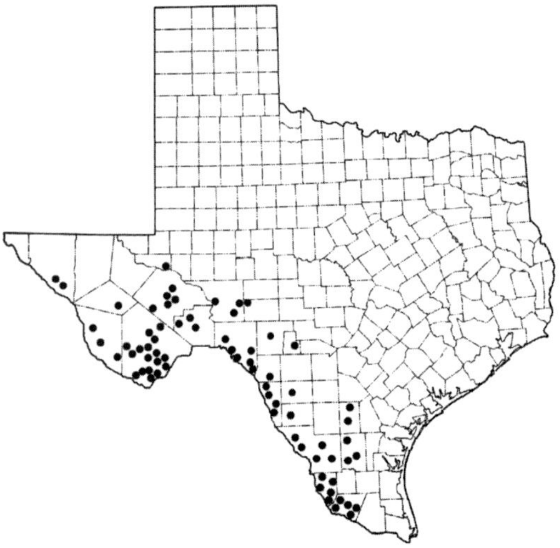

Chamaesyce cinerascens

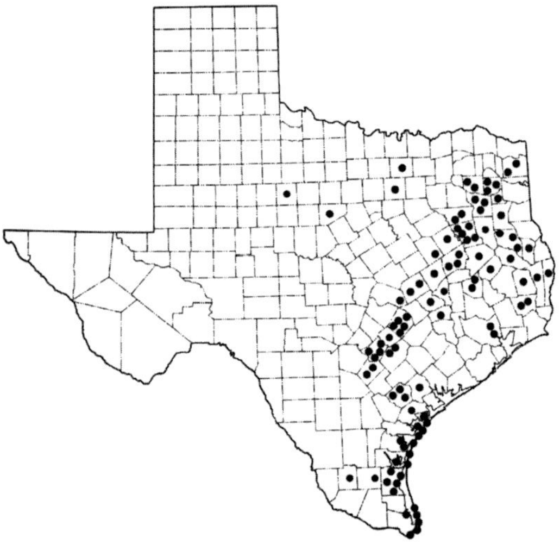

Chamaesyce cordifolia

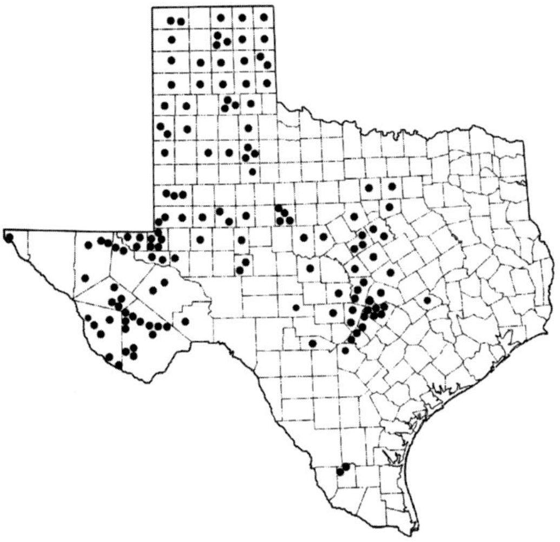

Chamaesyce fendleri

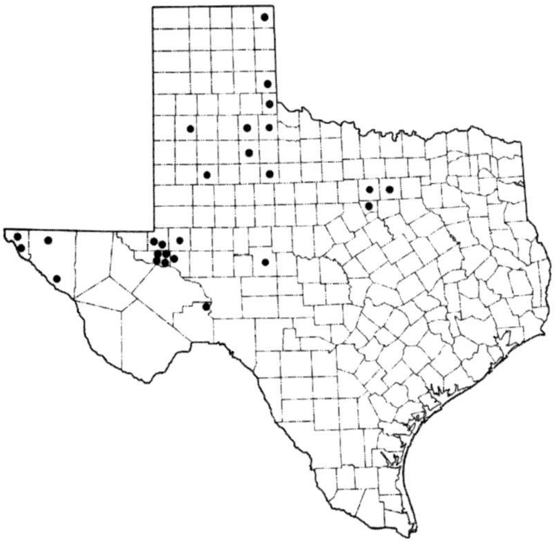

Chamaesyce geyeri

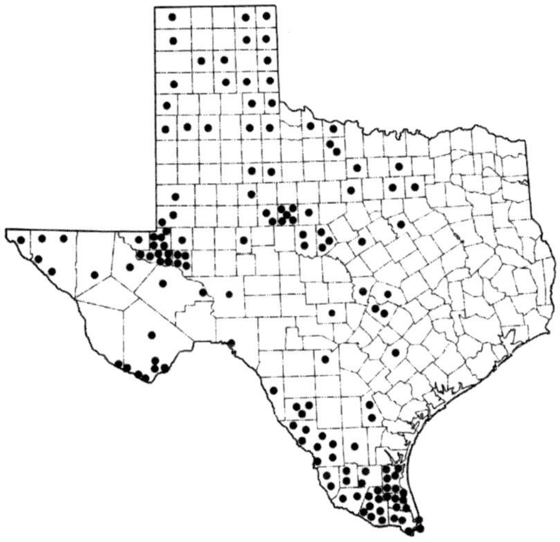

Chamaesyce glyptosperma

Chamaesyce golondrina

Chamaesyce hirta

Chamaesyce humistrata

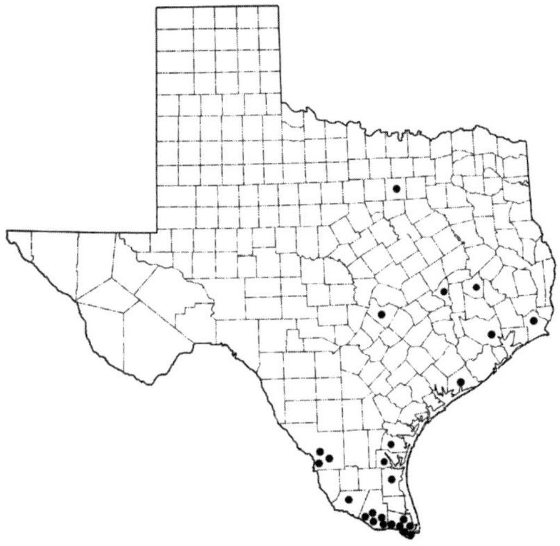

Chamaesyce hypericifolia

Chamaesyce indivisa

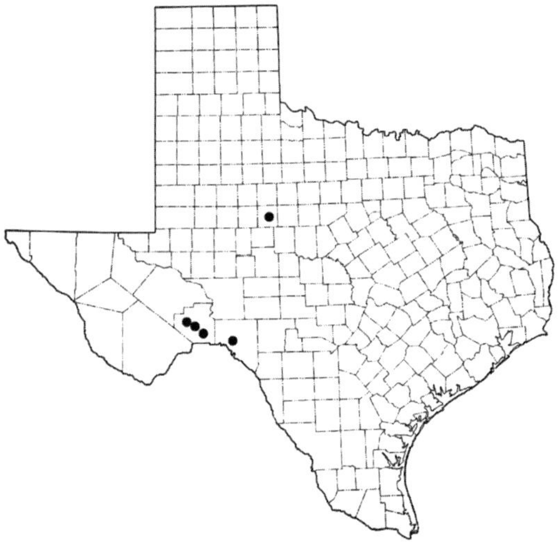

Chamaesyce jejuna

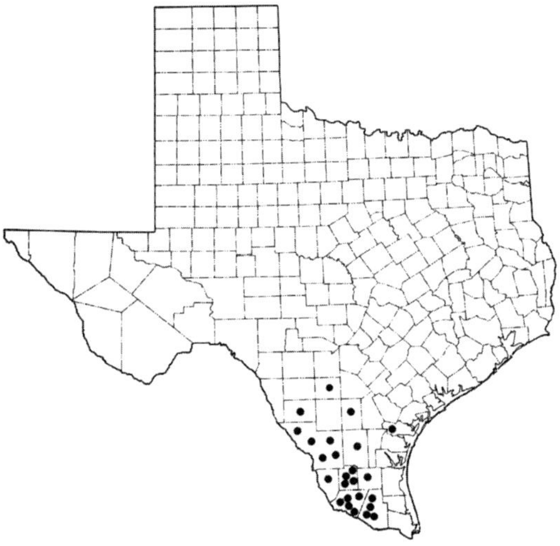

Chamaesyce laredana

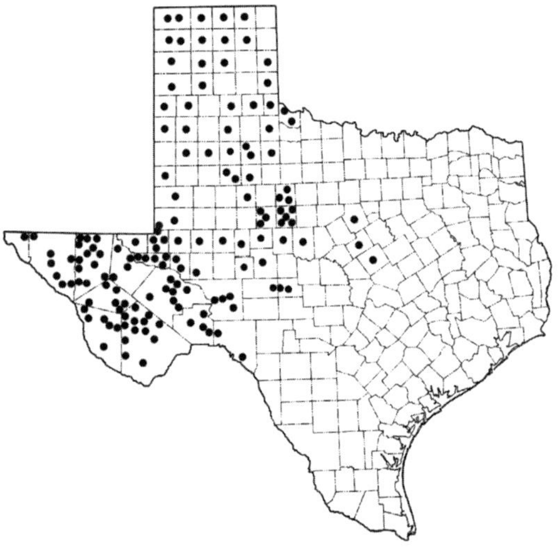

Chamaesyce lata

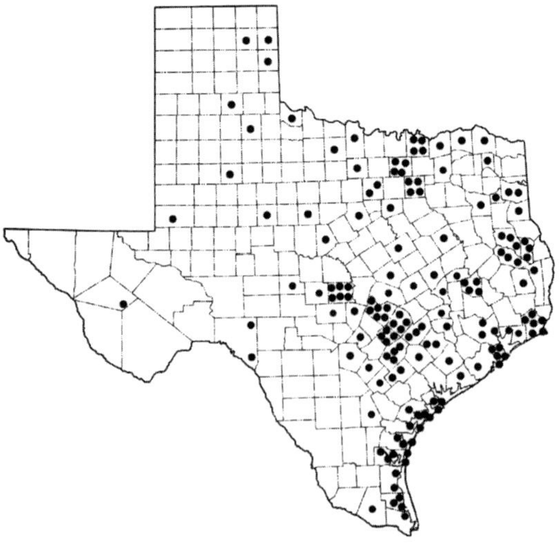

Chamaesyce maculata

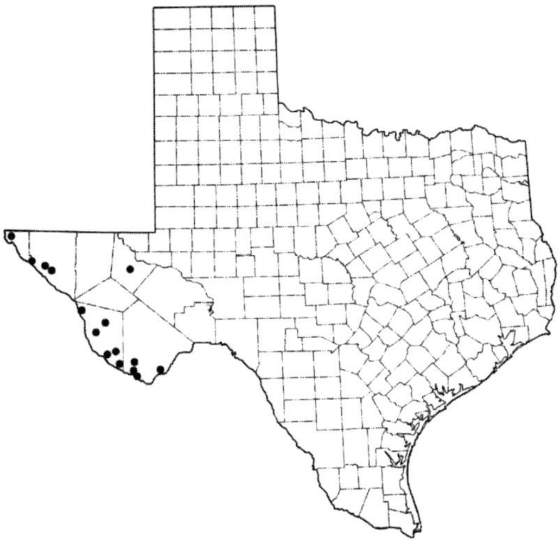

Chamaesyce micromera

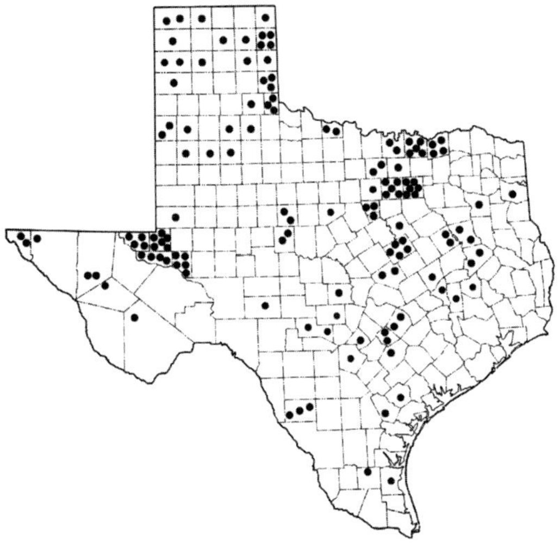

Chamaesyce missurica

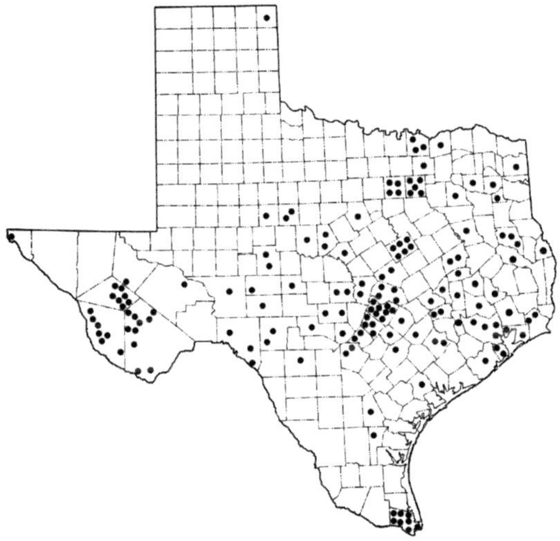

Chamaesyce nutans

Chamaesyce parryi

Chamaesyce perennans

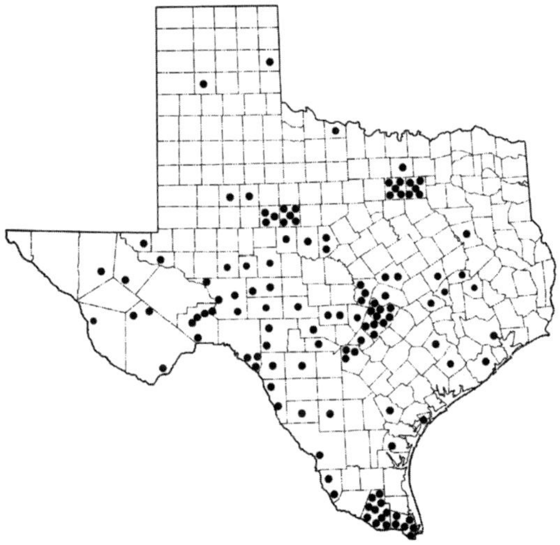

Chamaesyce prostrata

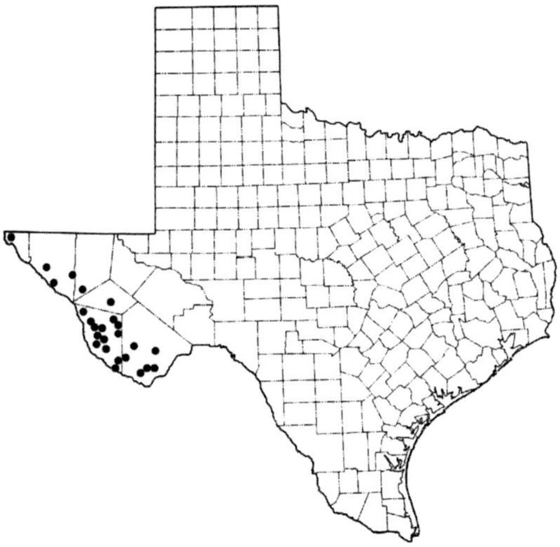

Chamaesyce revoluta

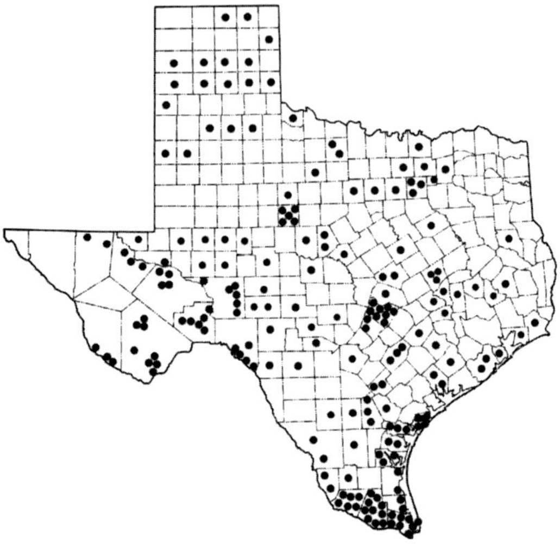

Chamaesyce serpens

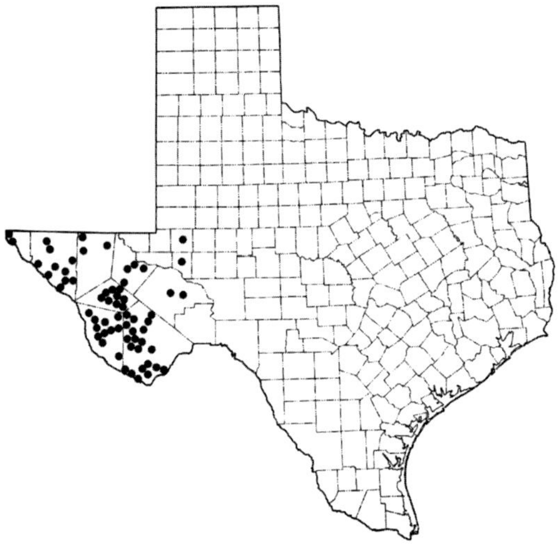

Chamaesyce serpyllifolia

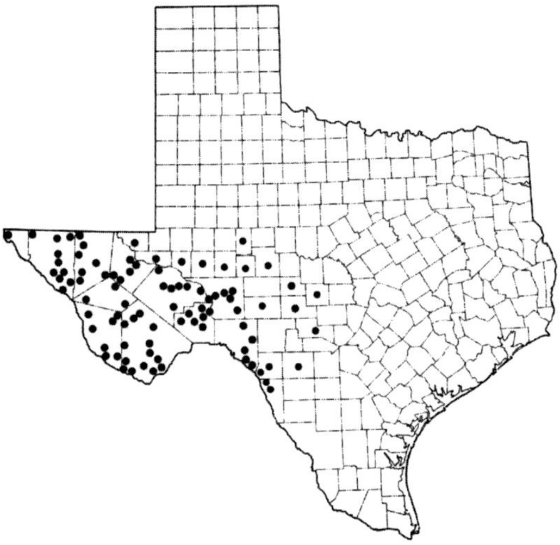

Chamaesyce serrula

Chamaesyce setiloba

Chamaesyce spurca

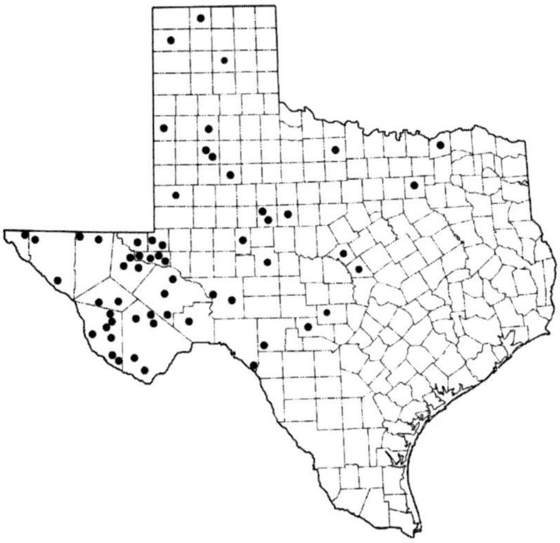

Chamaesyce stictospora

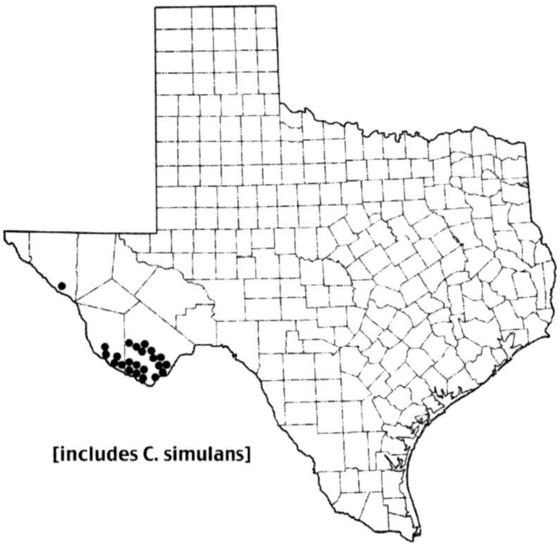
[includes C. simulans]

Chamaesyce theriaca

Chamaesyce triligulata

Chamaesyce velleriflora

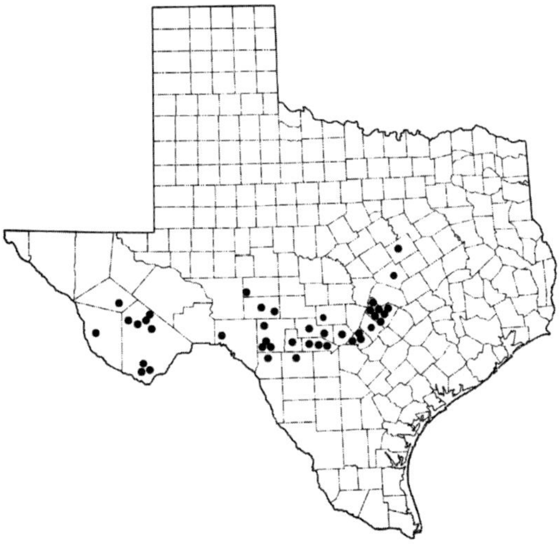

Chamaesyce villifera

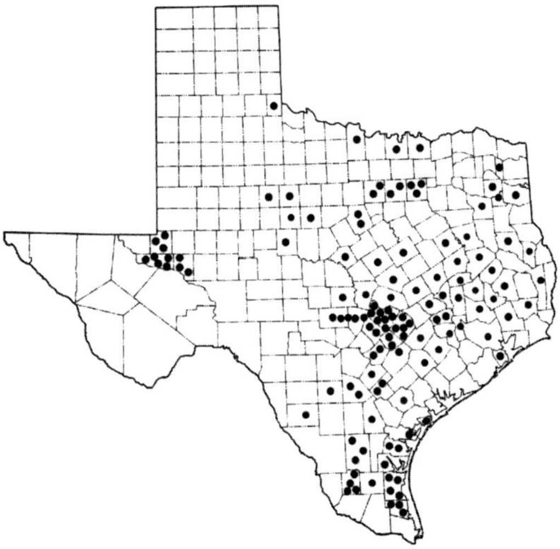

Cnidoscolus texanus

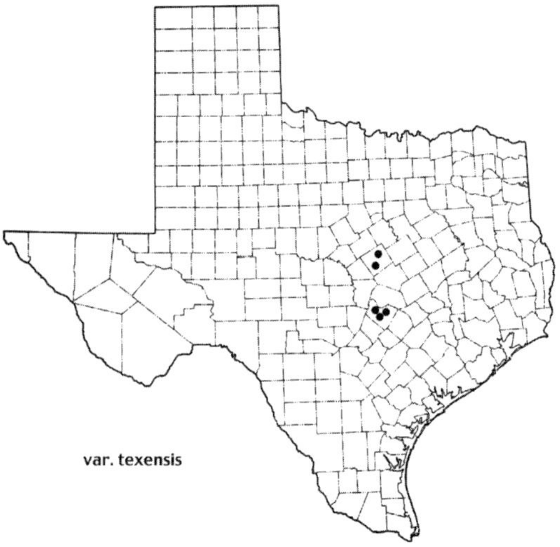
var. texensis

Croton alabamensis

[= Julocroton argenteus]

Croton argenteus

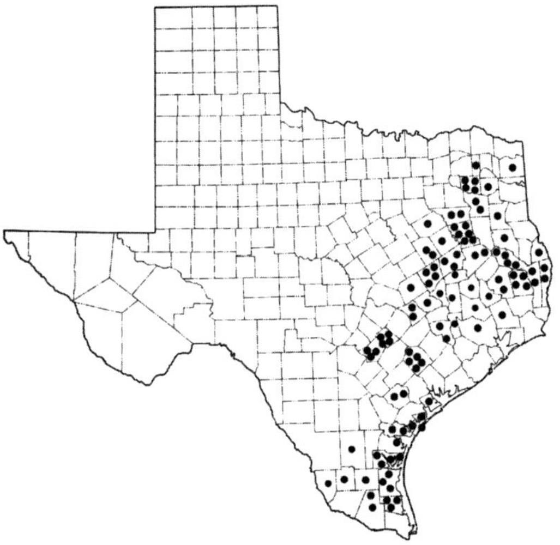

Croton argyranthemus

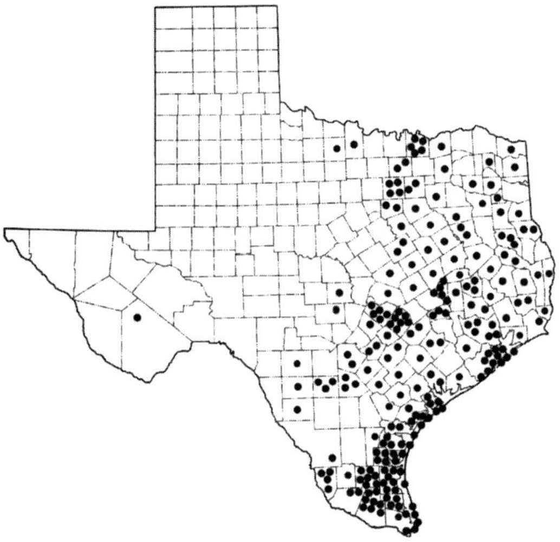

Croton capitatus

Croton ciliatoglandulifer

Croton cortesianus

Croton coryi

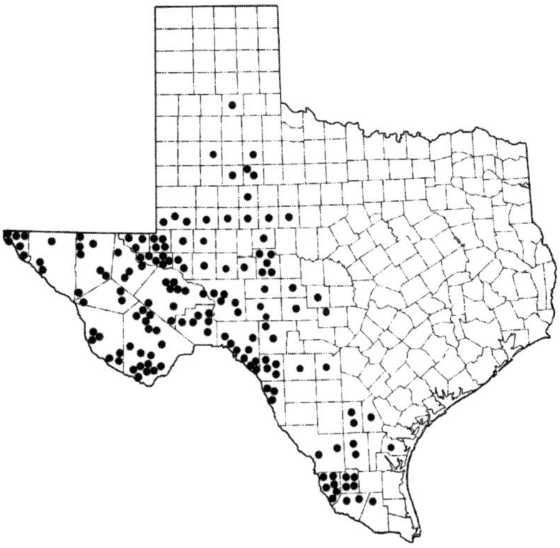

Croton dioicus

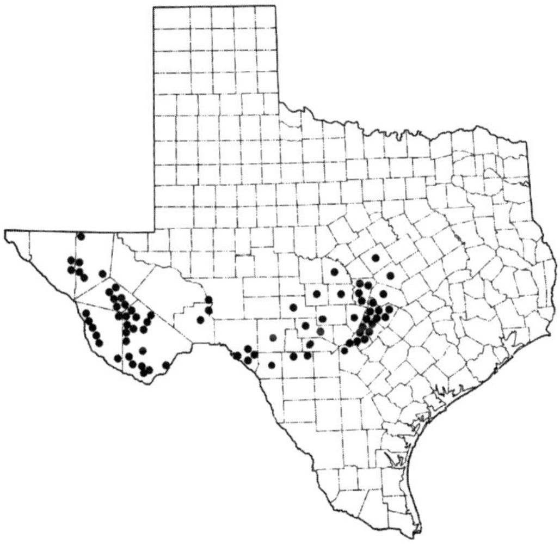

Croton fruticulosus

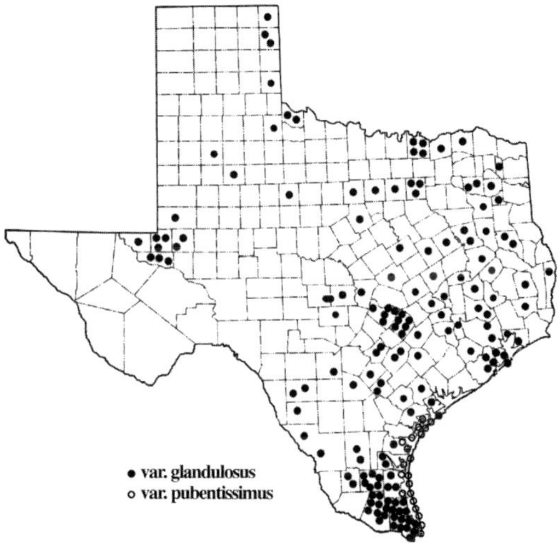
• var. glandulosus
○ var. pubentissimus

Croton glandulosus

Croton humilis

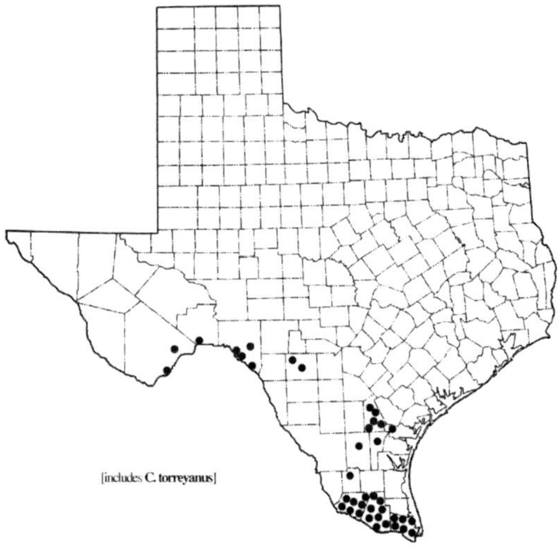
[includes C. torreyanus]

Croton incanus

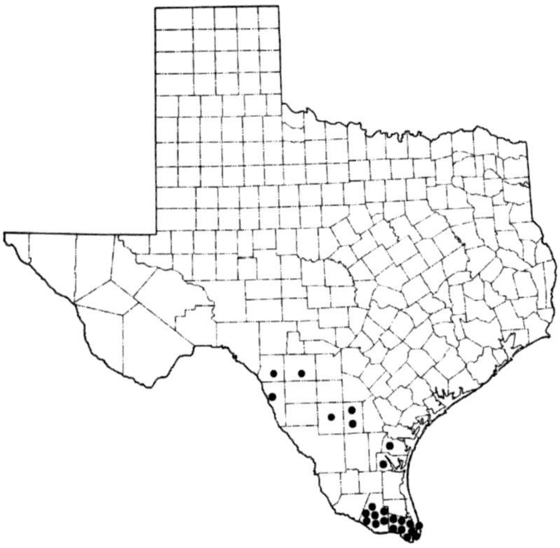

Croton leucophyllus

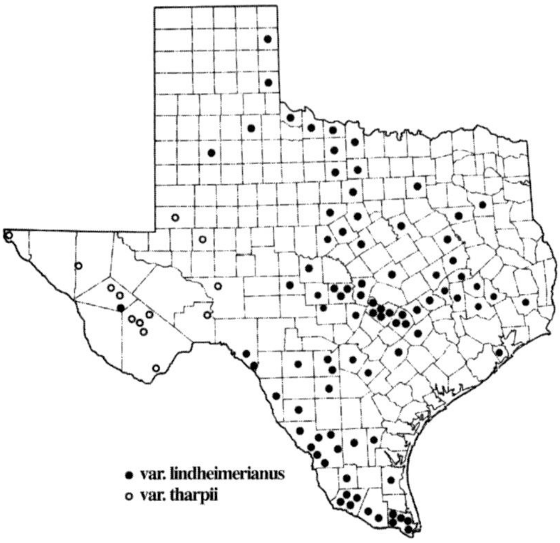
var. lindheimerianus
var. tharpii

Croton lindheimerianus

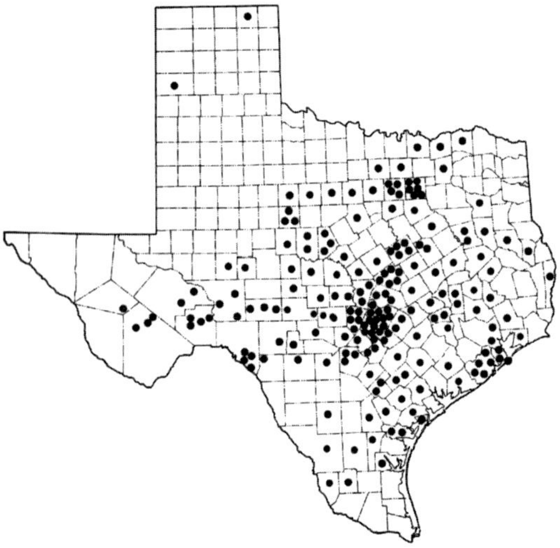

Croton monanthogynus

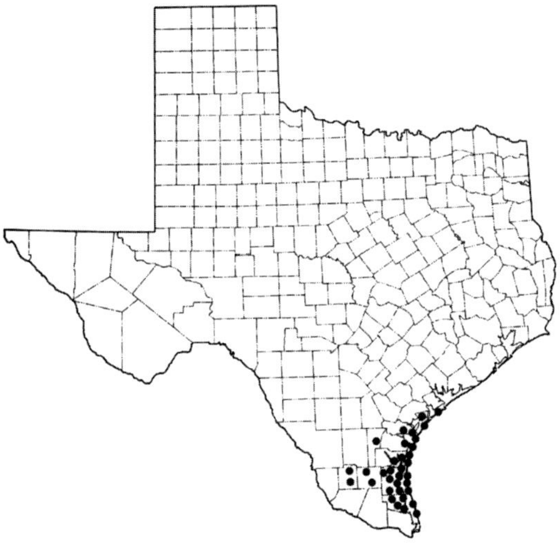

Croton parksii

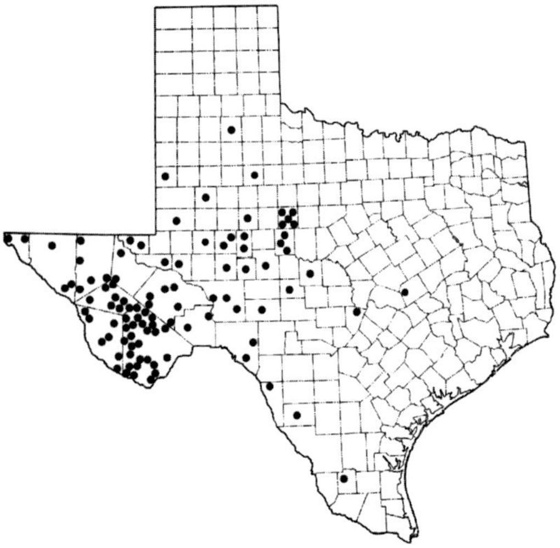

Croton pottsii

Croton punctatus

Croton sancti-lazari

Croton soliman

Croton suaveolens

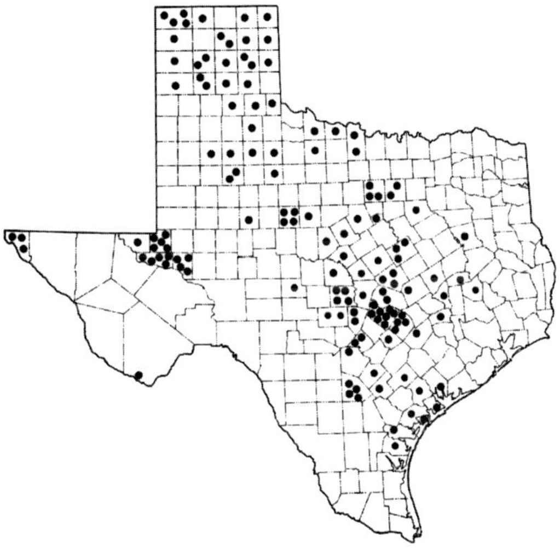

Croton texensis

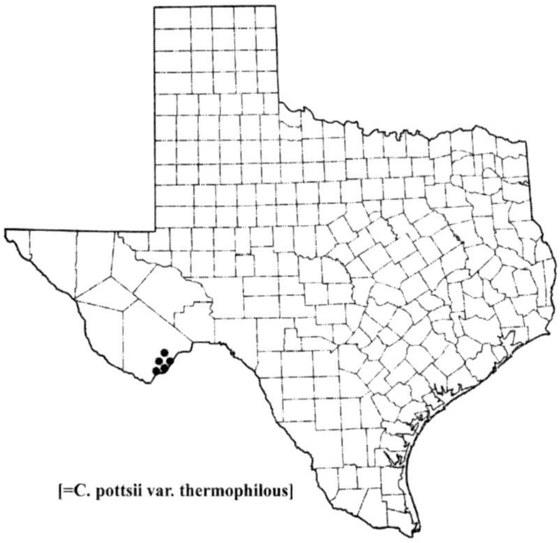
[=C. pottsii var. thermophilous]

Croton thermophilus

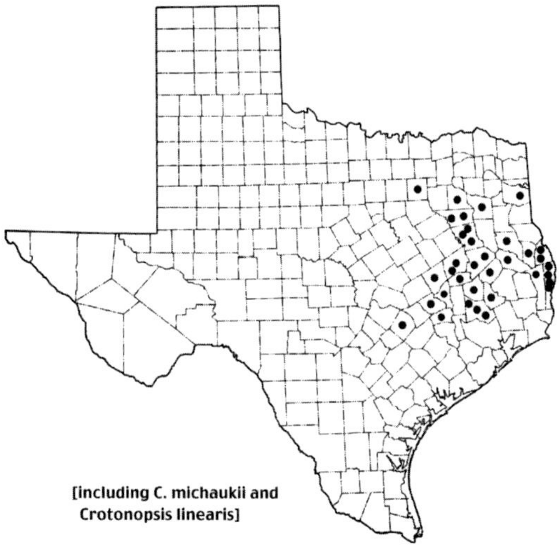
[including C. michaukii and
Crotonopsis linearis]

Croton willdenowii

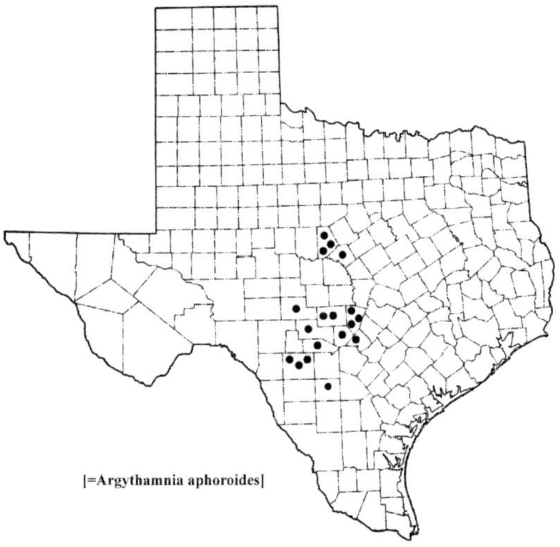
[=Argythamnia aphoroides]

Ditaxis aphoroides

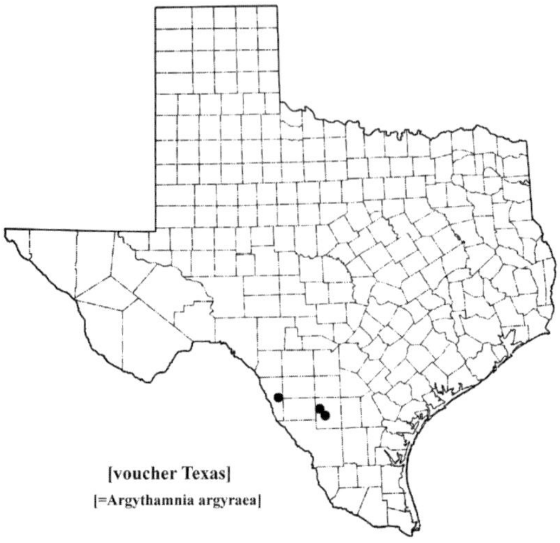
[voucher Texas]
[=Argythamnia argyraea]

Ditaxis argyraea

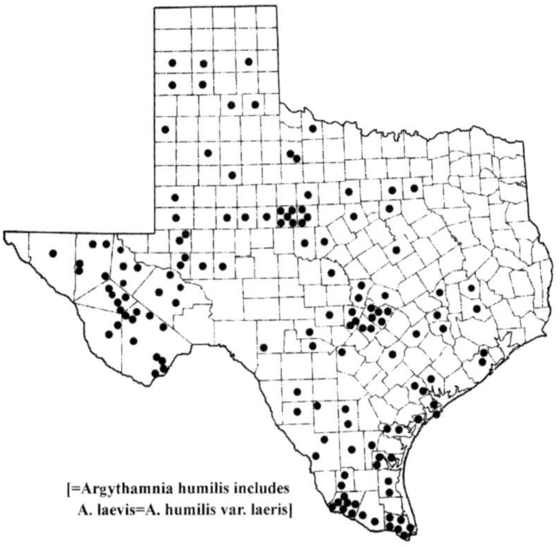
[=Argythamnia humilis includes
A. laevis=A. humilis var. laeris]

Ditaxis humilis

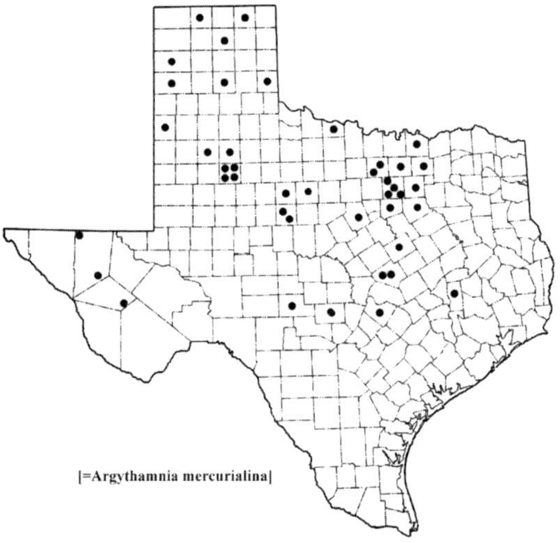
[=Argythamnia mercurialina]

Ditaxis mercurialina

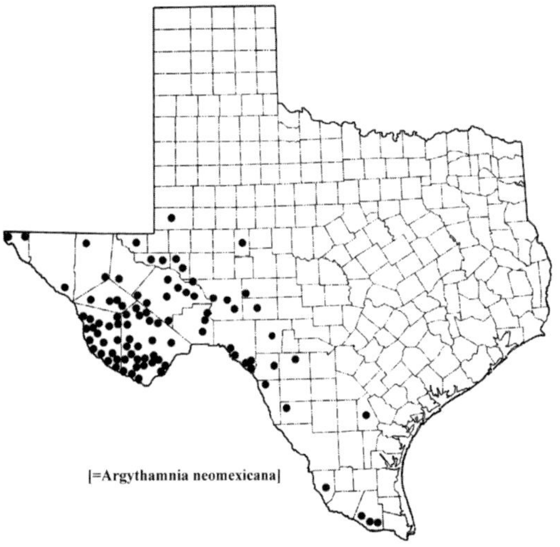
[=Argythamnia neomexicana]

Ditaxis neomexicana

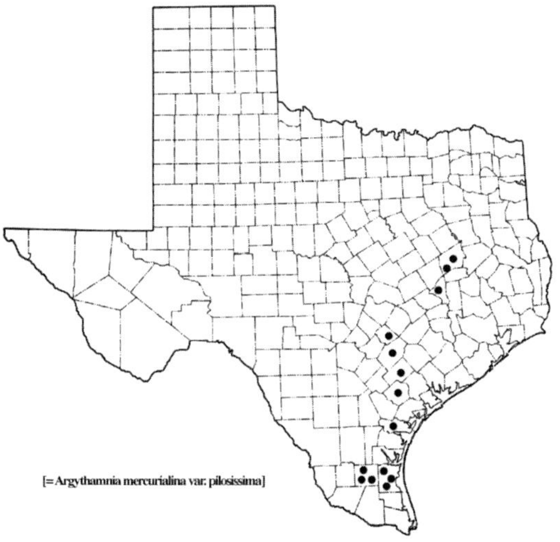
[=Argythamnia mercurialina var. pilosissima]

Ditaxis pilosissima

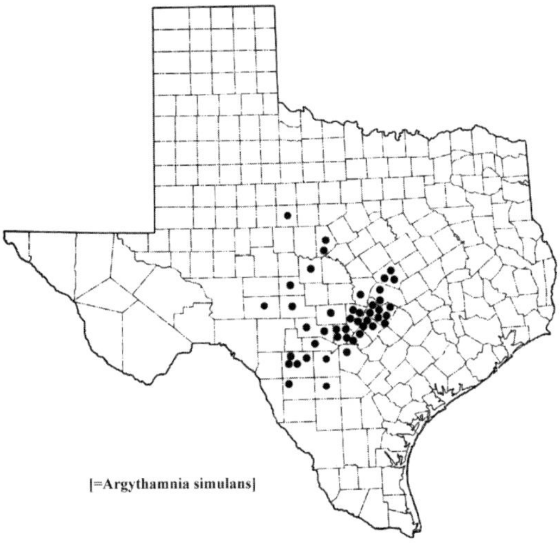
[=Argythamnia simulans]

Ditaxis simulans

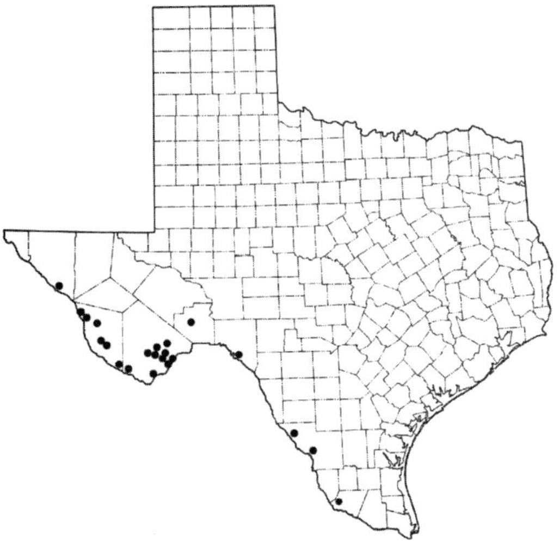

Euphorbia antisyphilitica

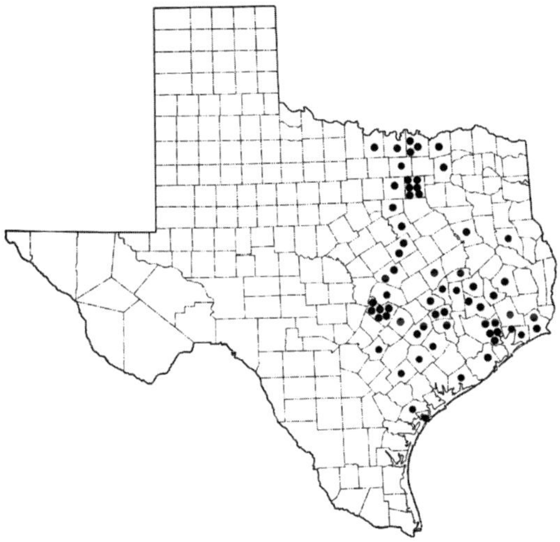

Euphorbia bicolor

Euphorbia bifurcata

Euphorbia bilobata

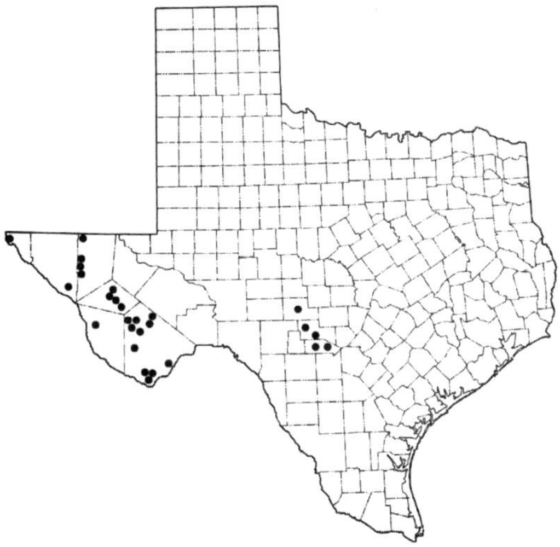

Euphorbia brachycera

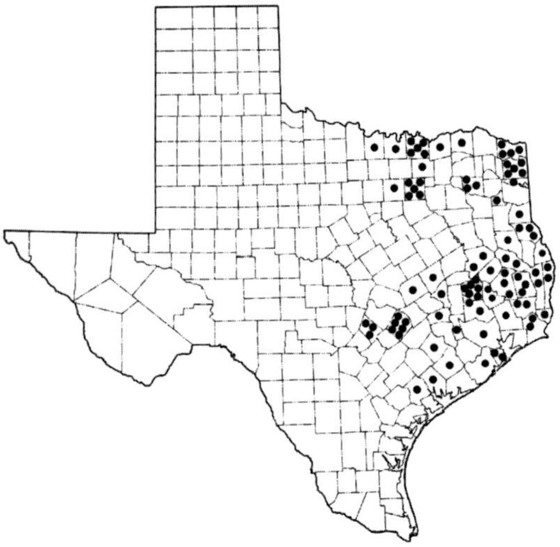

Euphorbia corollata

Euphorbia eriantha

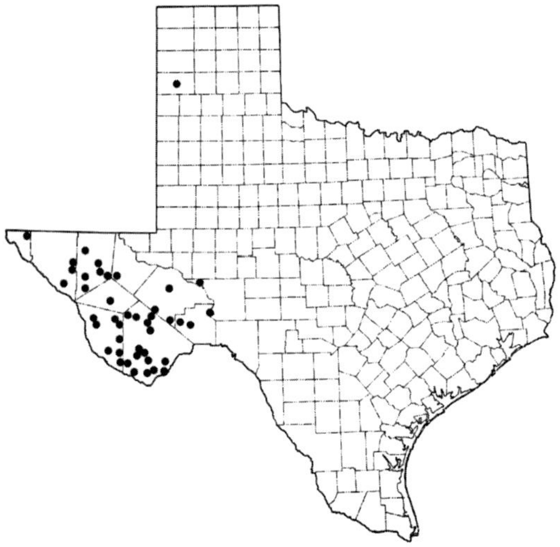

Euphorbia exstipulata

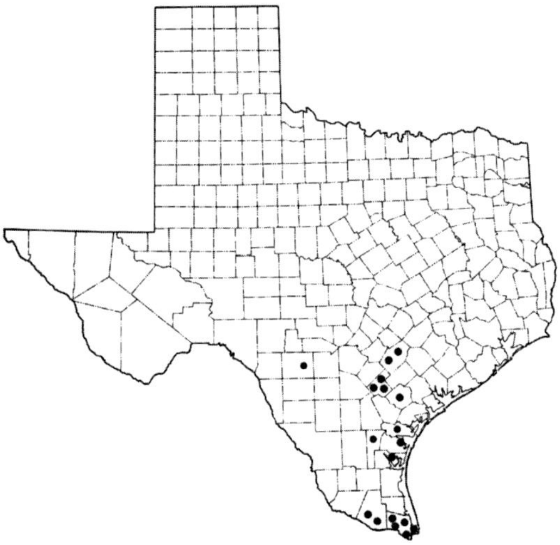

Euphorbia helleri

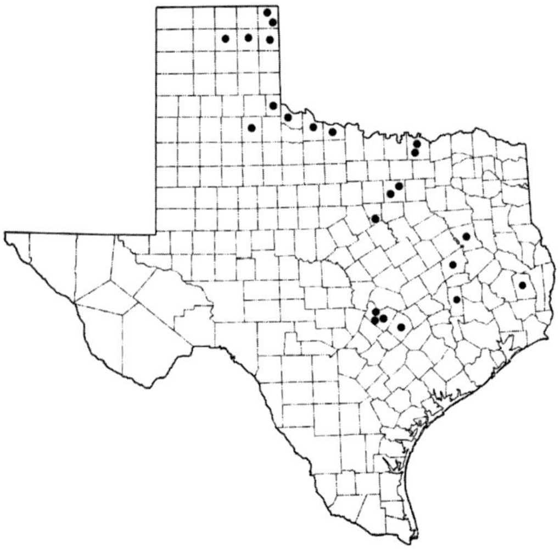

Euphorbia hexagona

Euphorbia innocua

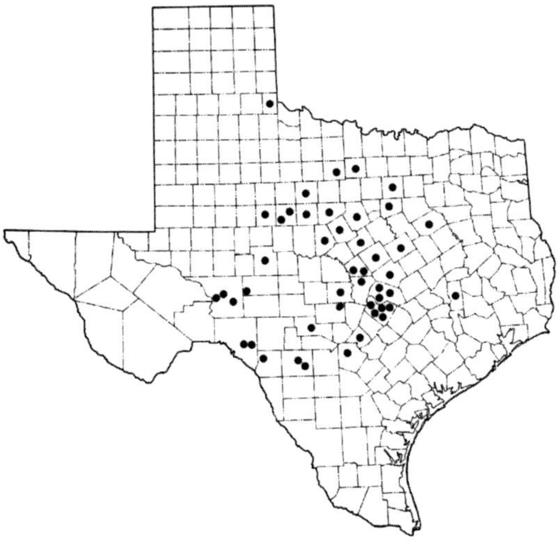
Euphorbia longicruris

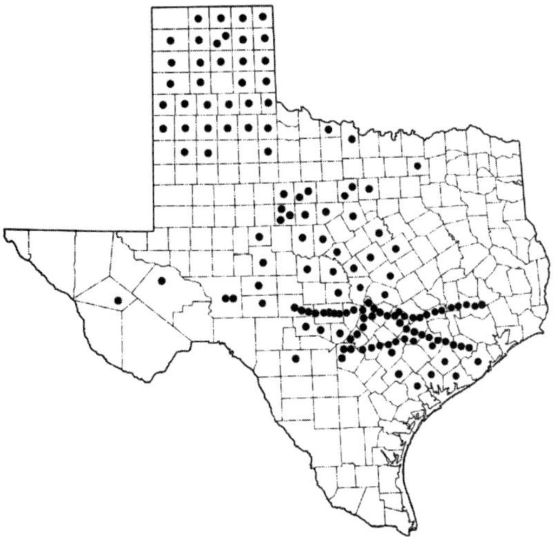
Euphorbia marginata

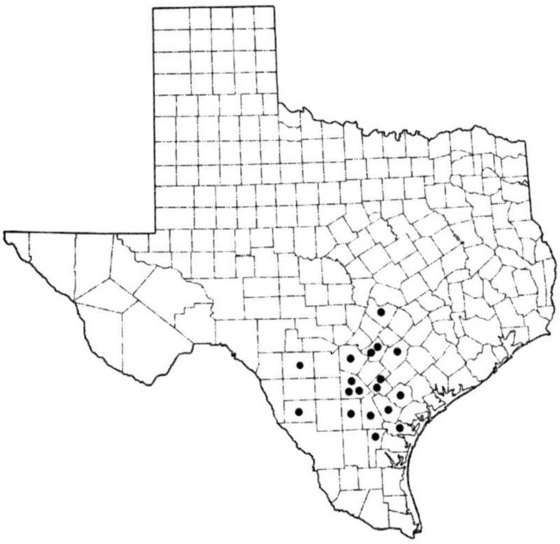
Euphorbia peplidion

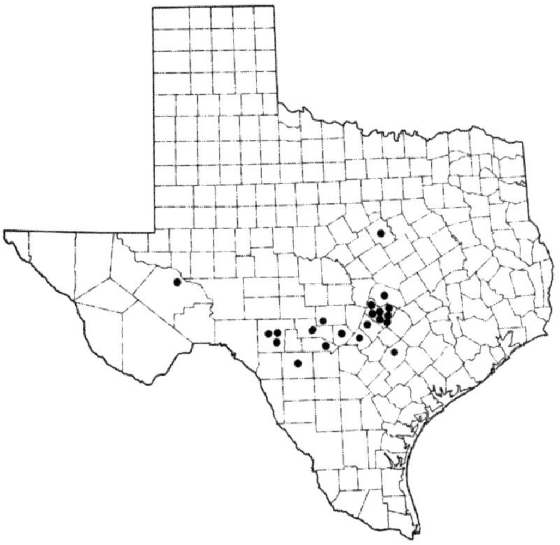
Euphorbia roemeriana

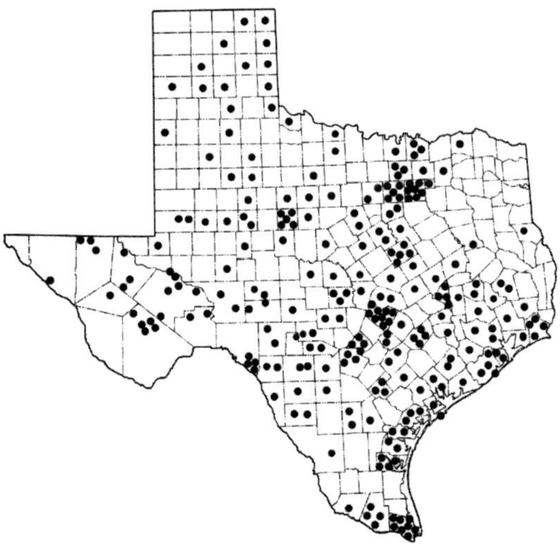
Euphorbia spathulata

Euphorbia strictior

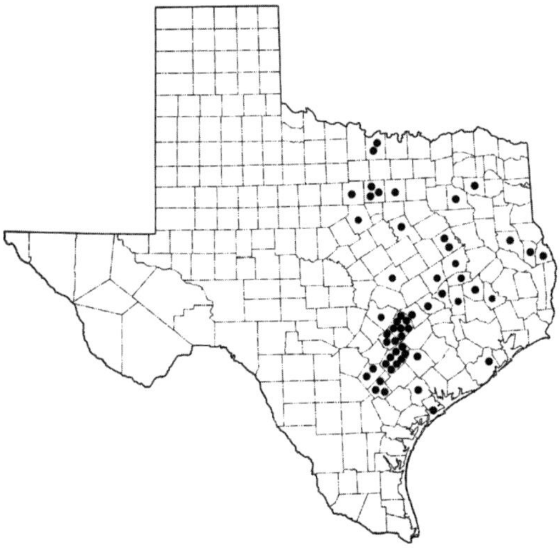

Euphorbia tetrapora

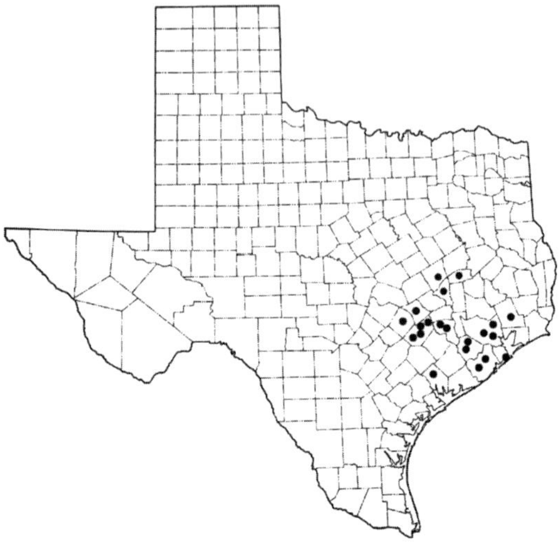

Euphorbia texana

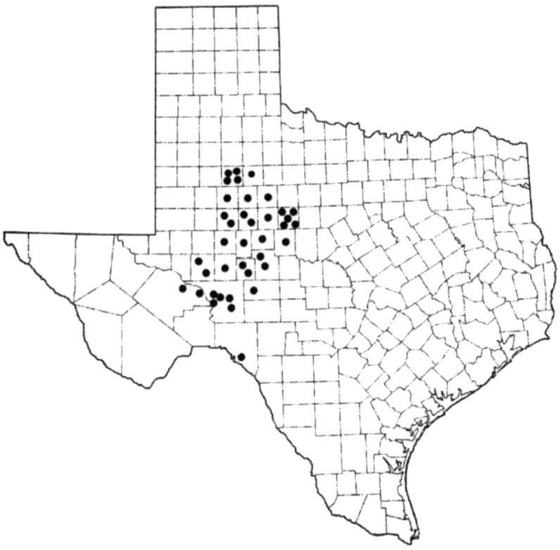

Euphorbia wrightii

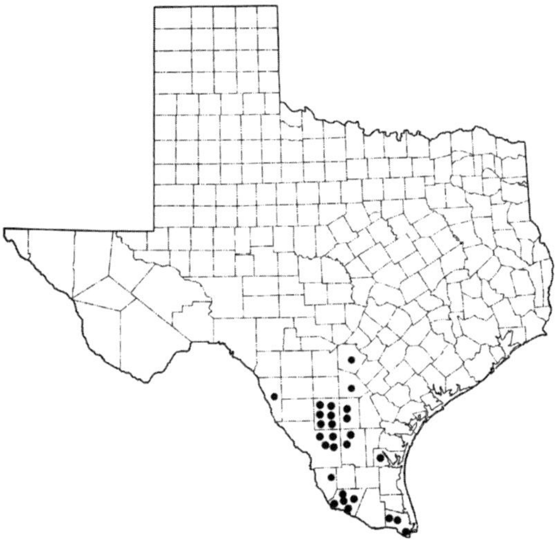

Jatropha cathartica

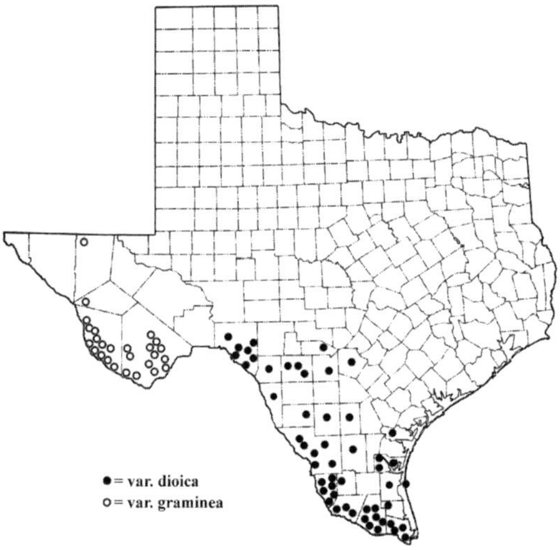
● = var. dioica
○ = var. graminea

Jatropha dioica

Jatropha macrorhiza

[= Andrachne phyllanthoides]
Leptopus phyllanthoides

Manihot esculenta

Manihot subspicata

Manihot walkerae

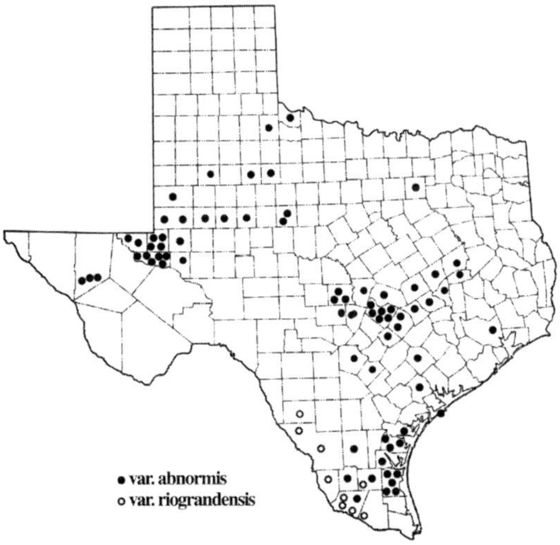
var. abnormis
var. riograndensis
Phyllanthus abnormis

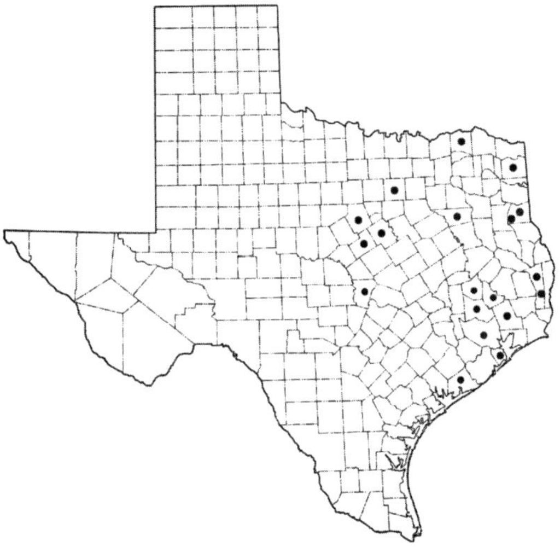
Phyllanthus caroliniensis

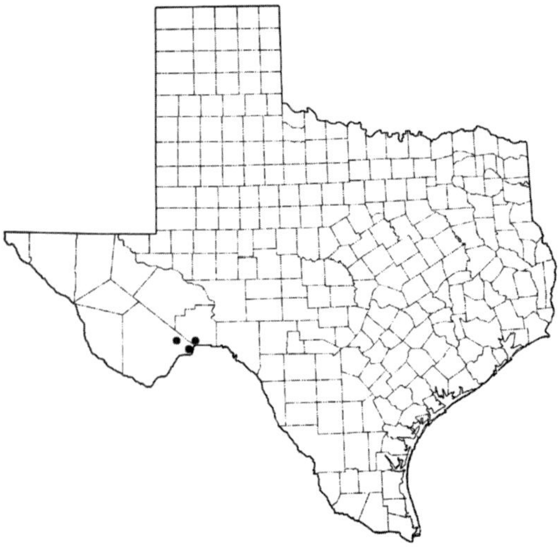

Phyllanthus ericoides

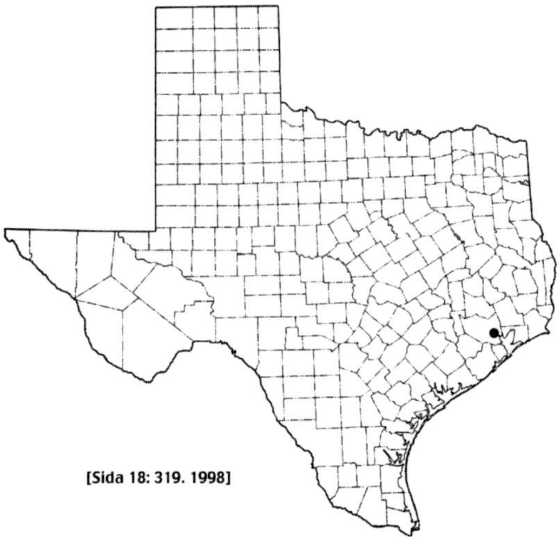
[Sida 18: 319. 1998]

Phyllanthus fraternus

Phyllanthus niruri

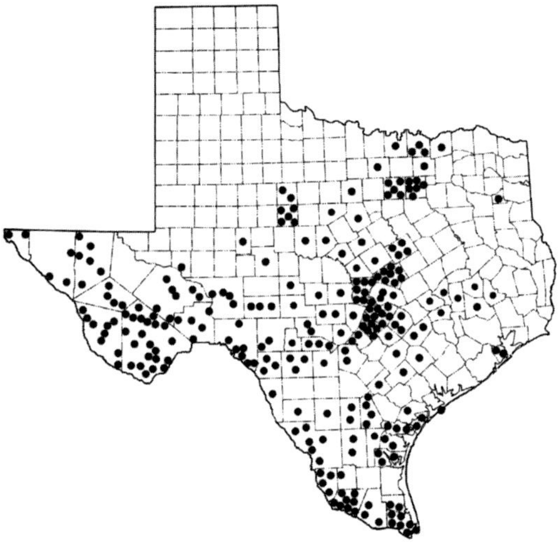

Phyllanthus polygonoides

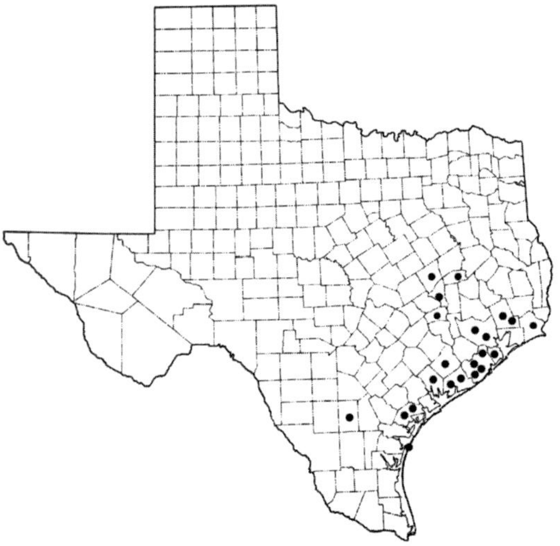

Phyllanthus pudens

Phyllanthus tenellus

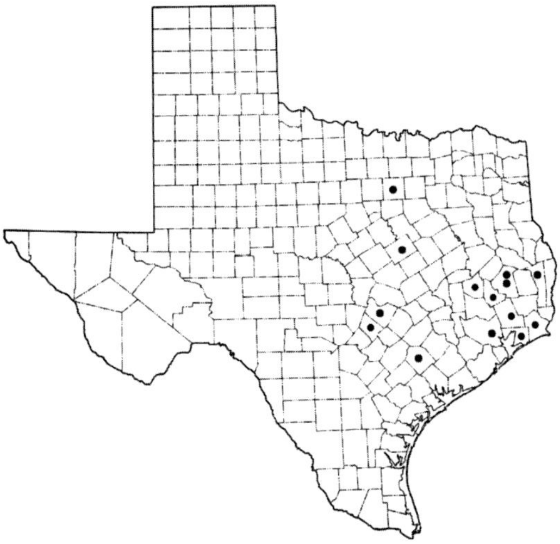
Phyllanthus urinaria

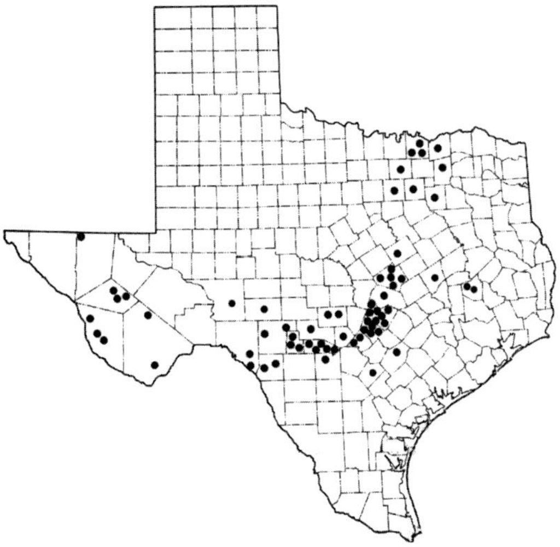
Poinsettia cyathophora

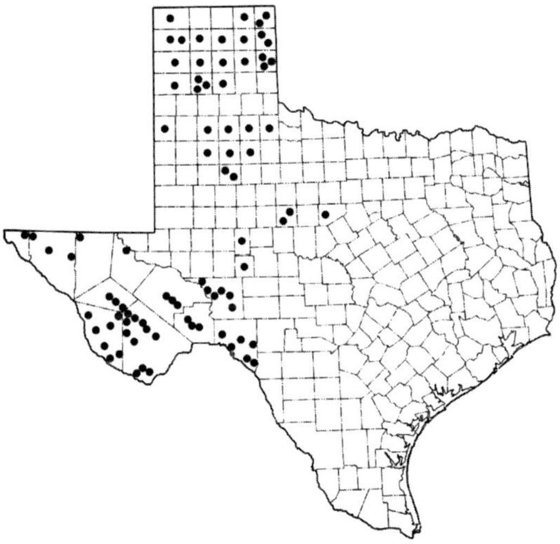
Poinsettia davidii

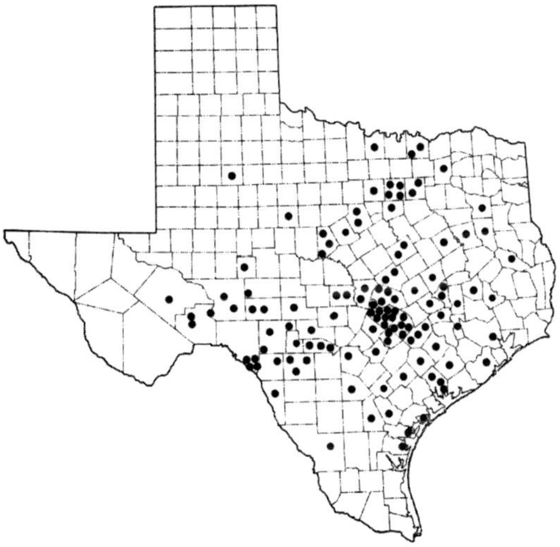
Poinsettia dentata

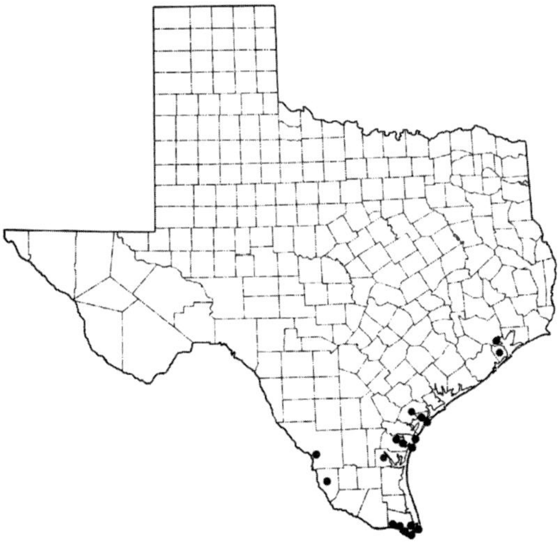
Poinsettia heterophylla

Poinsettia radians

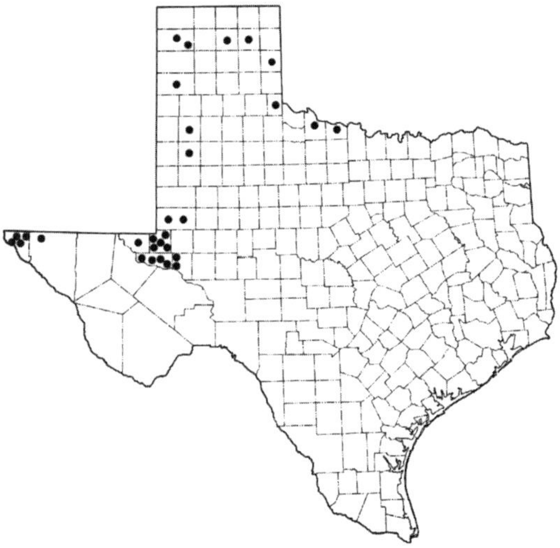
Reverchonia arenaria

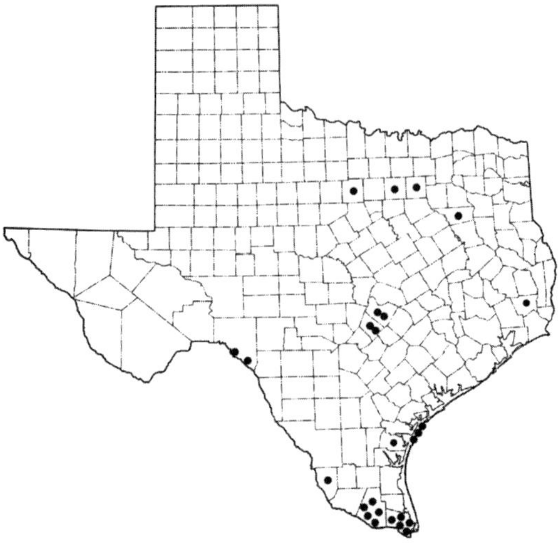
Ricinus communis

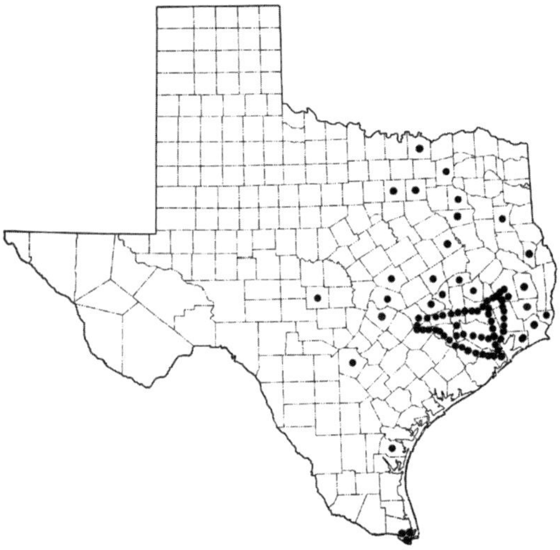
Sapium sebiferum

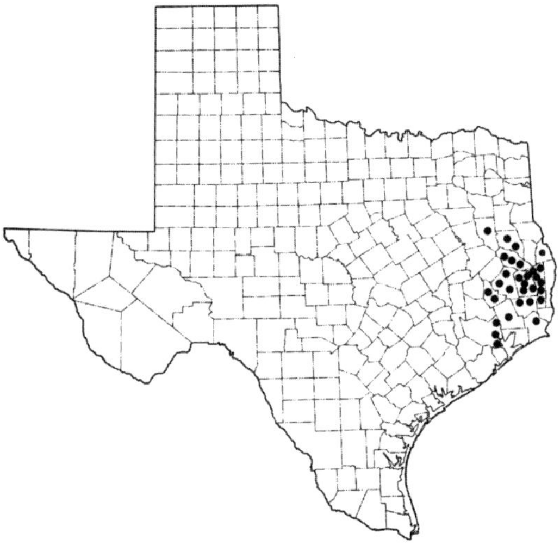
Sebastiania fruticosa

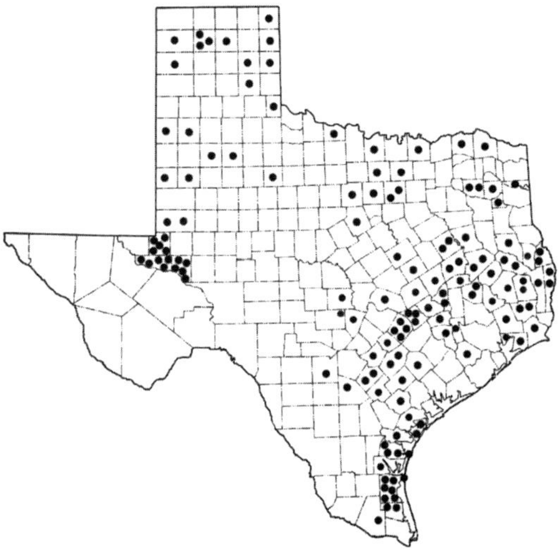
Stillingia sylvatica

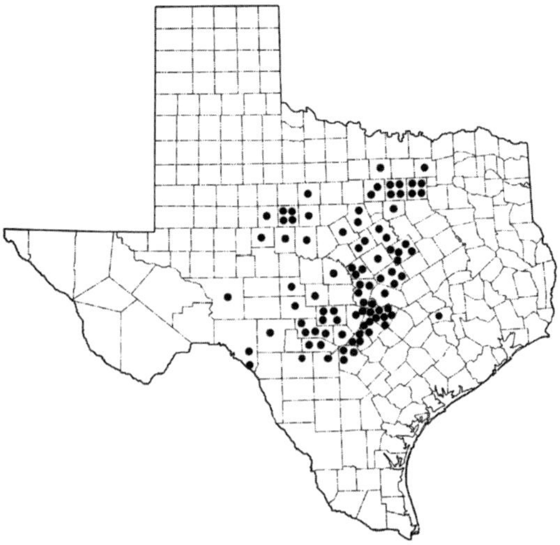
Stillingia texana

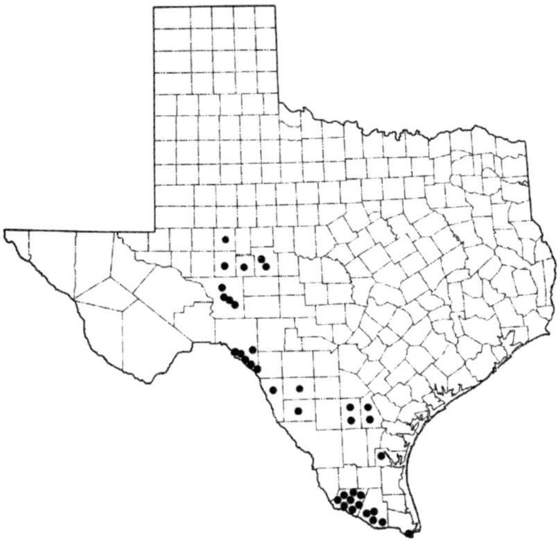

Stillingia treculiana

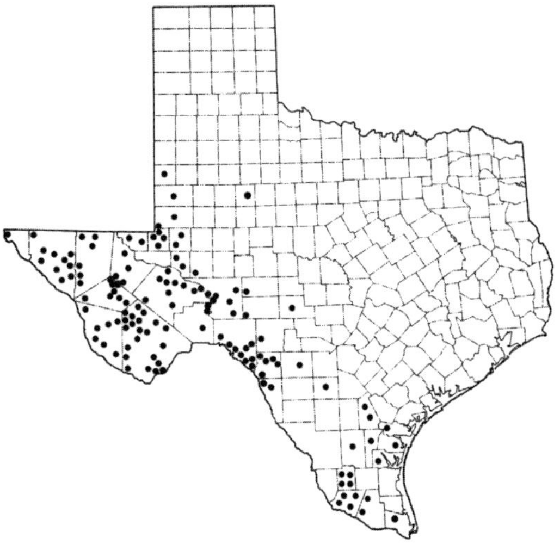

Tragia amblyodonta

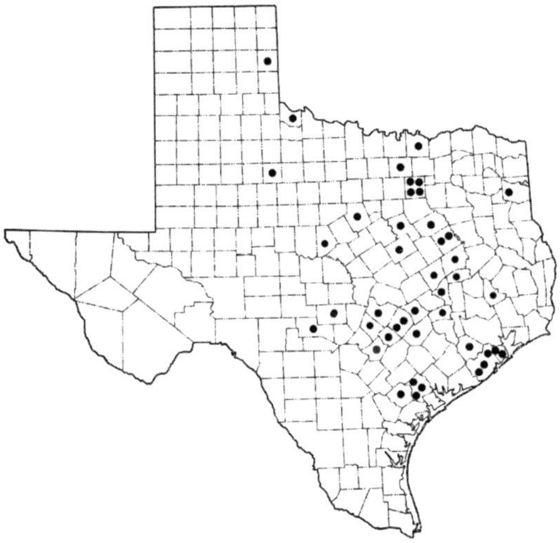

Tragia betonicifolia

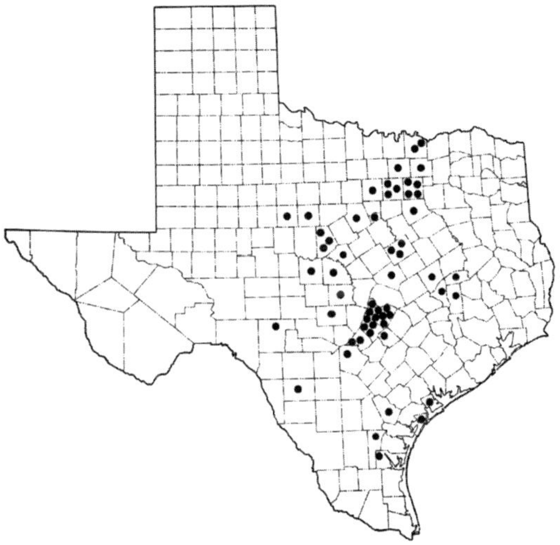

Tragia brevispica

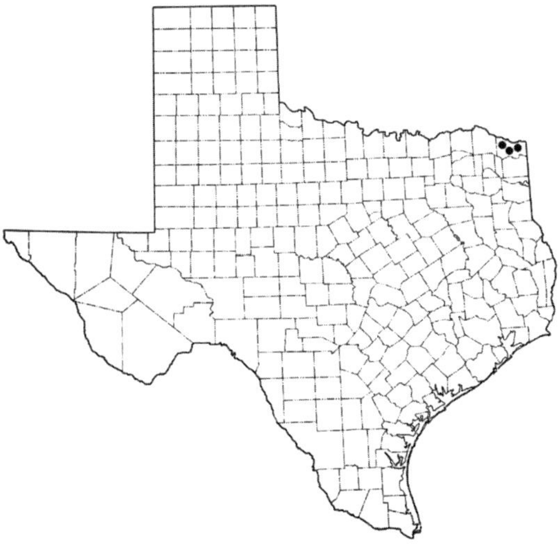

Tragia cordata

Tragia glanduligera

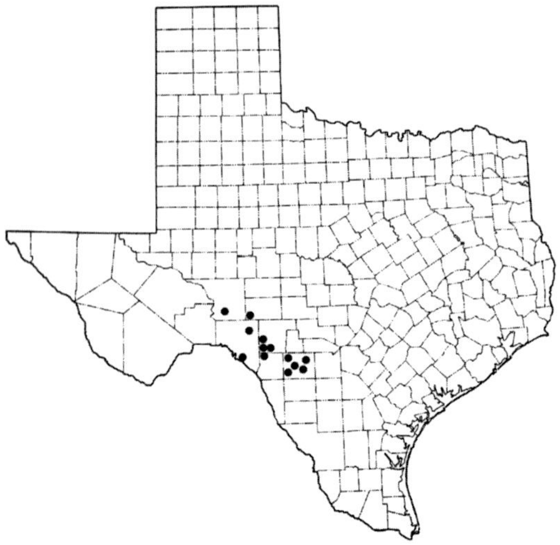

Tragia leptophylla

Tragia nigricans

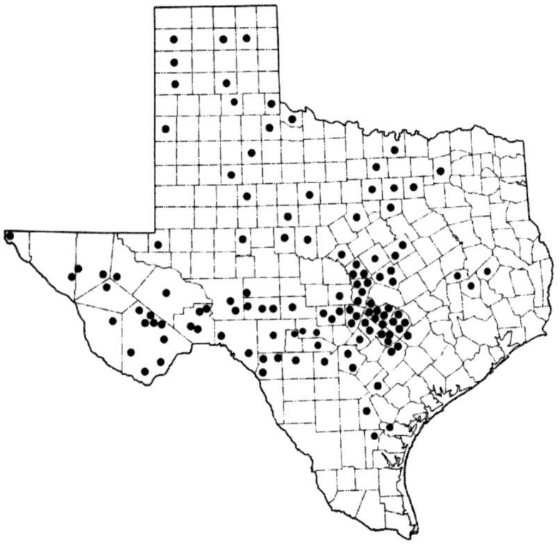

Tragia ramosa

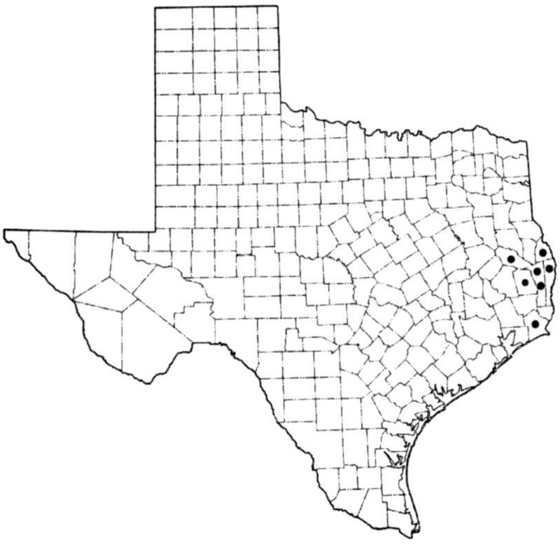

Tragia smallii

Tragia urens

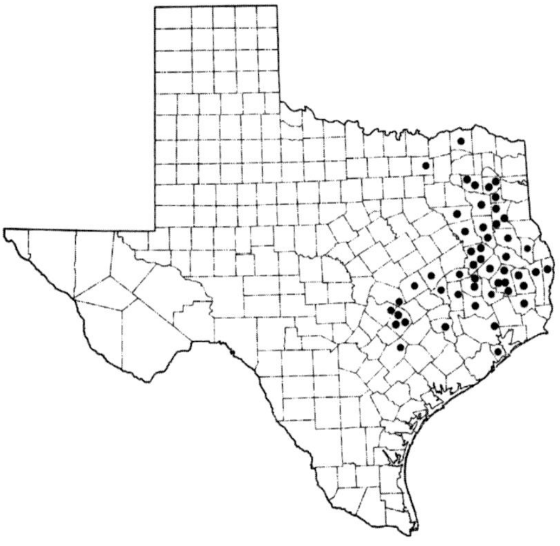

Tragia urticifolia

Vernicia fordii

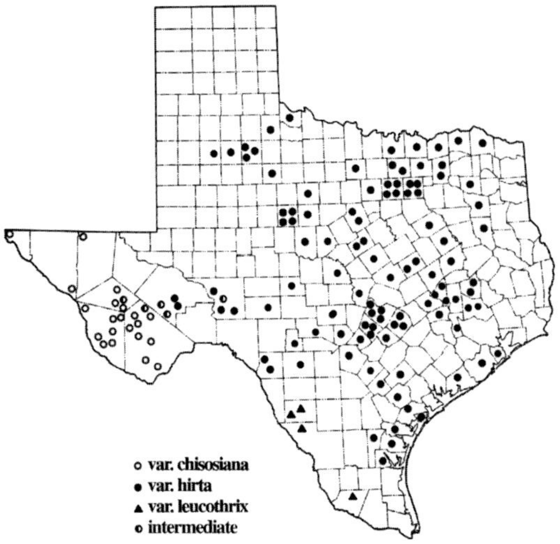
var. chisosiana
var. hirta
var. leucothrix
intermediate

Acacia angustissima

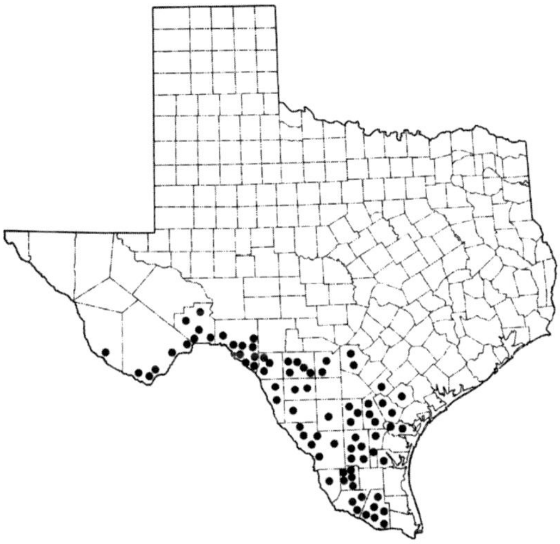

Acacia berlandieri

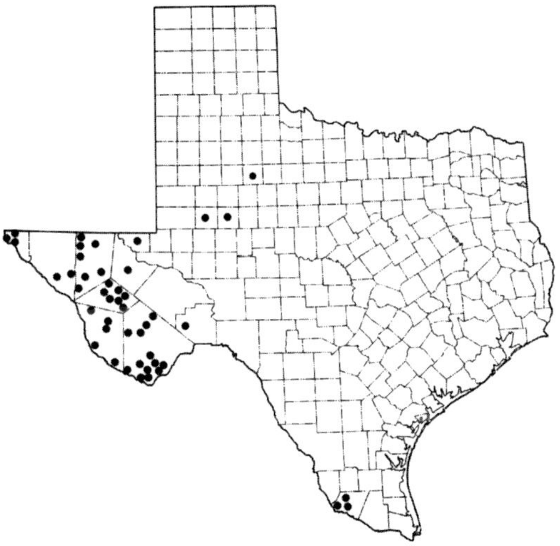

Acacia constricta

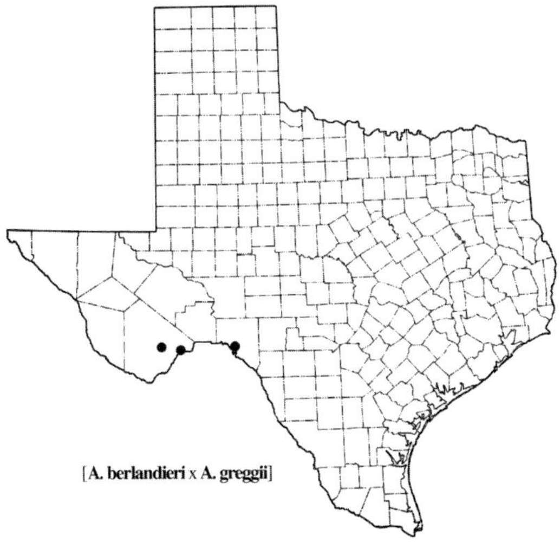
[A. berlandieri x A. greggii]

Acacia x emoryana

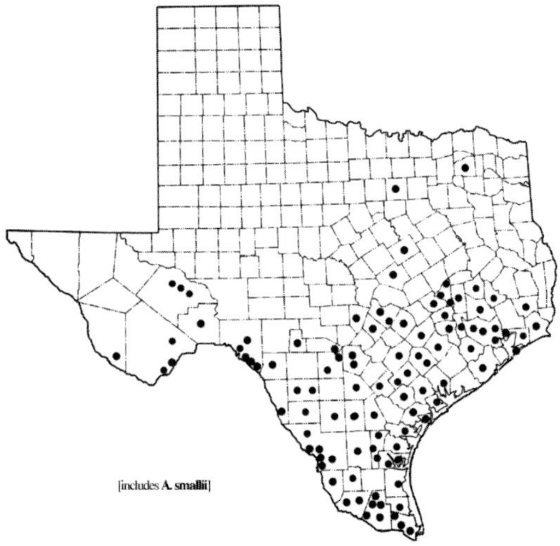
[includes A. smallii]

Acacia farnesiana

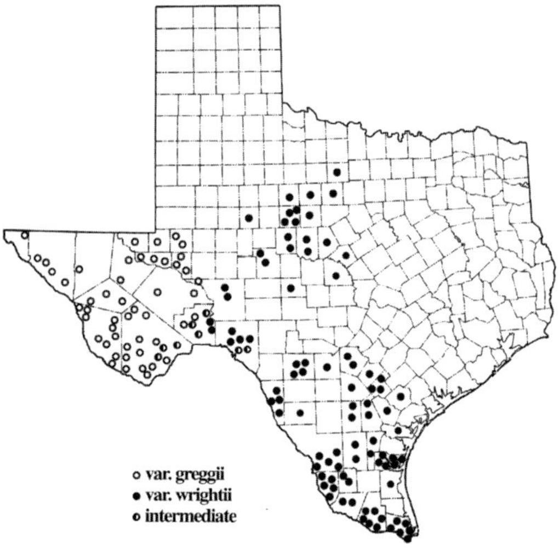
var. greggii
var. wrightii
intermediate

Acacia greggii

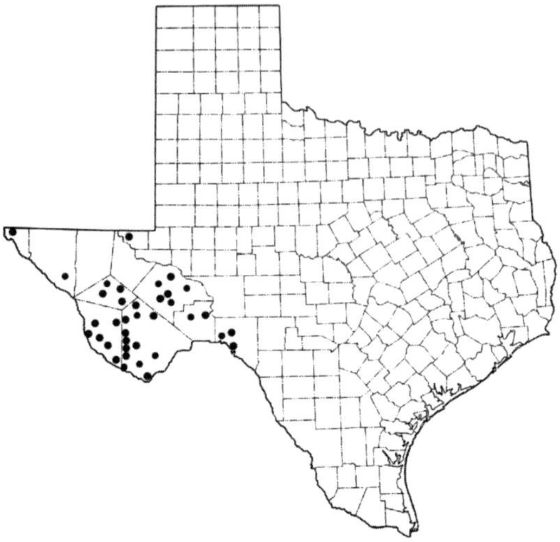

Acacia neovernicosa

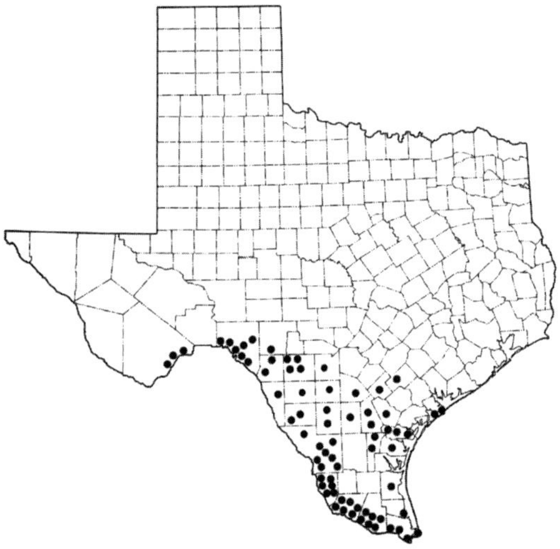

Acacia rigidula

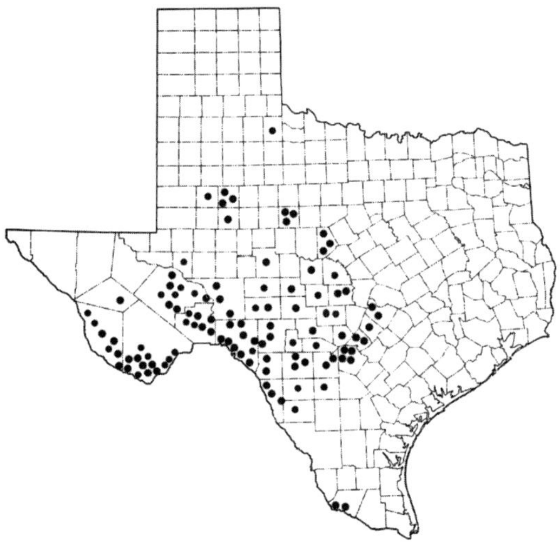

Acacia roemeriana

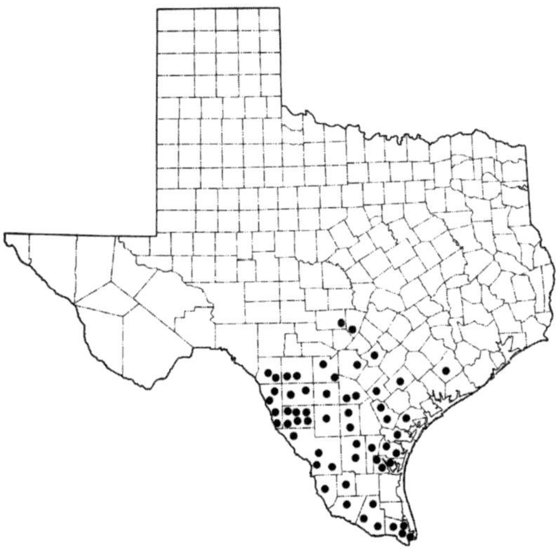

Acacia schaffneri

Acacia schottii

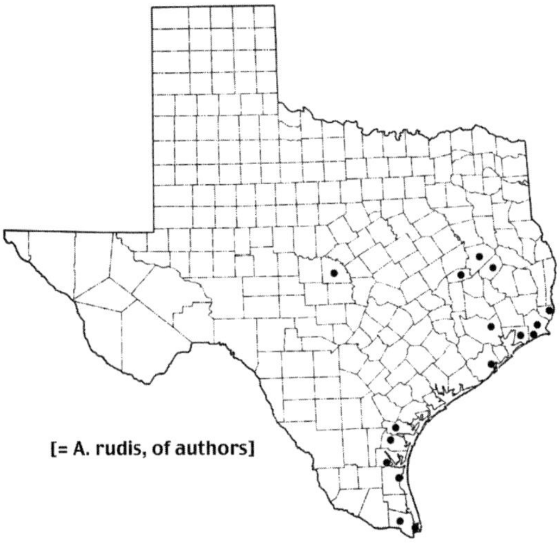
[= A. rudis, of authors]

Aeschynomene indica

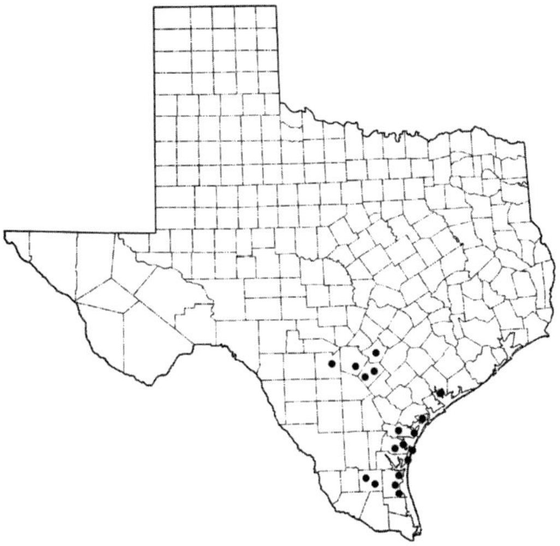

Aeschynomene viscidula

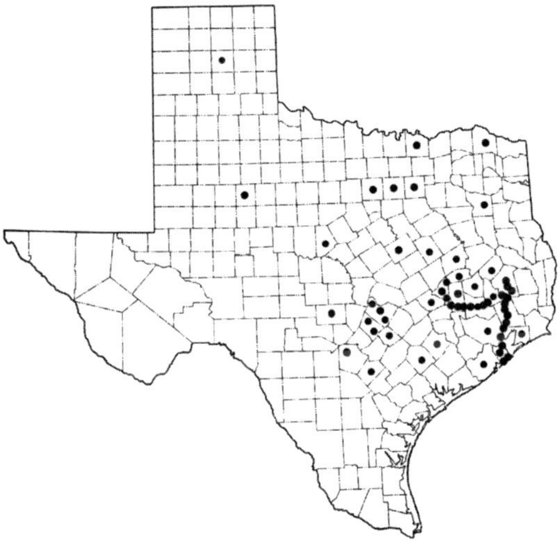

Albizia julibrissin

Albizia lebbeck

Alhagi maurorum

Alysicarpus vaginalis

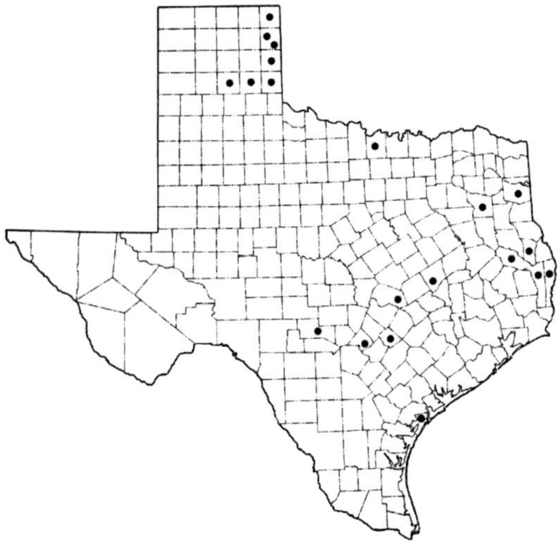

Amorpha canescens

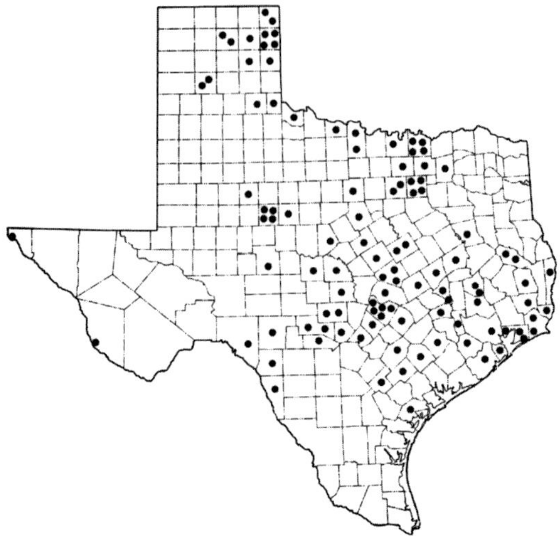

Amorpha fruticosa

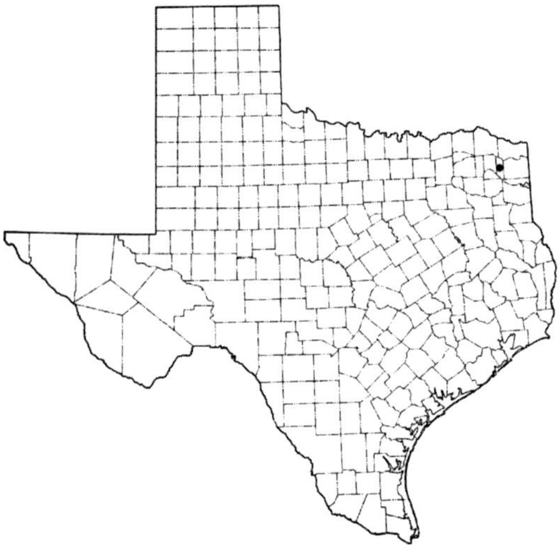

Amorpha laevigata

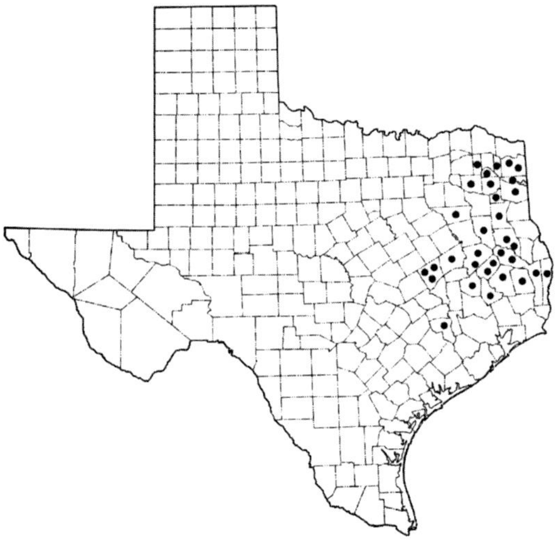

Amorpha paniculata

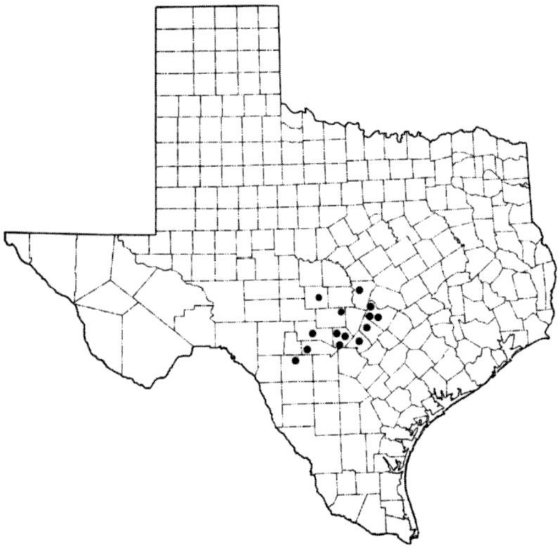

Amorpha roemeriana

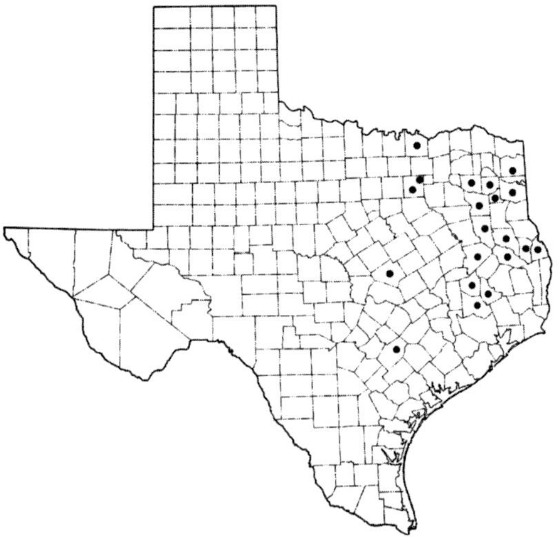

Amphicarpaea bracteata

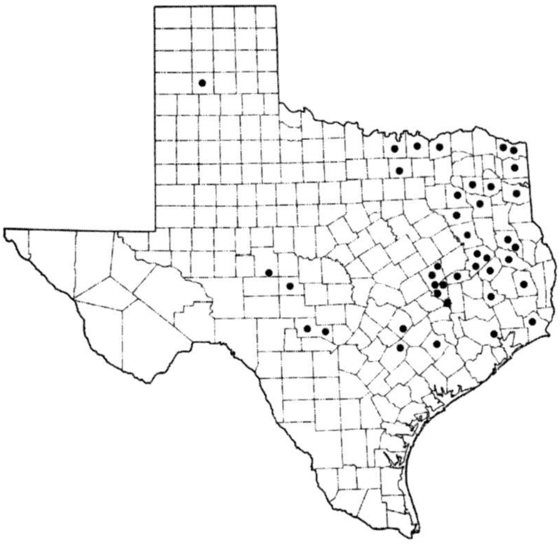

Apios americana

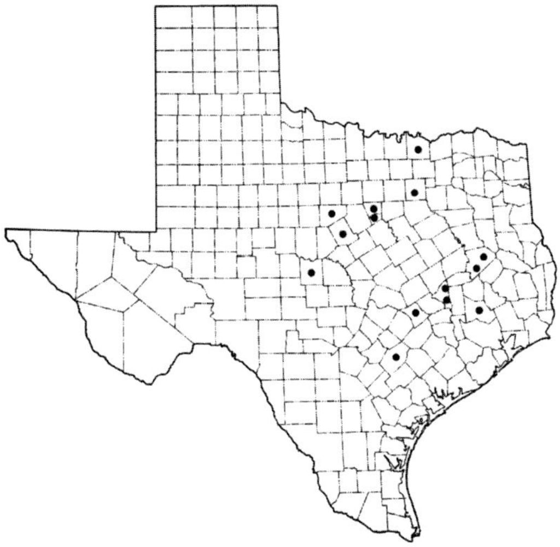

Arachis hypogaea

Astragalus albulus

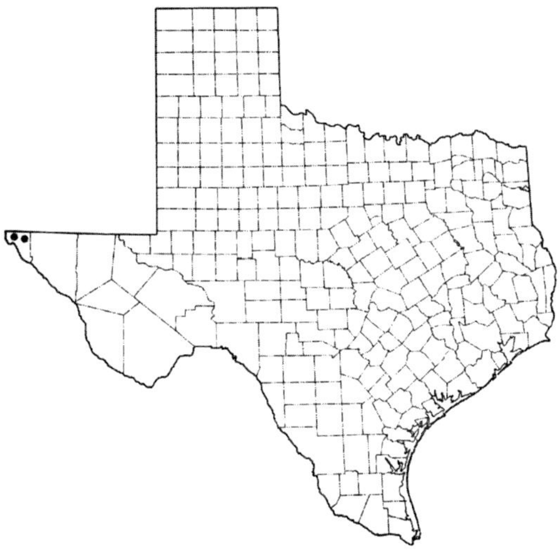

Astragalus allochrous

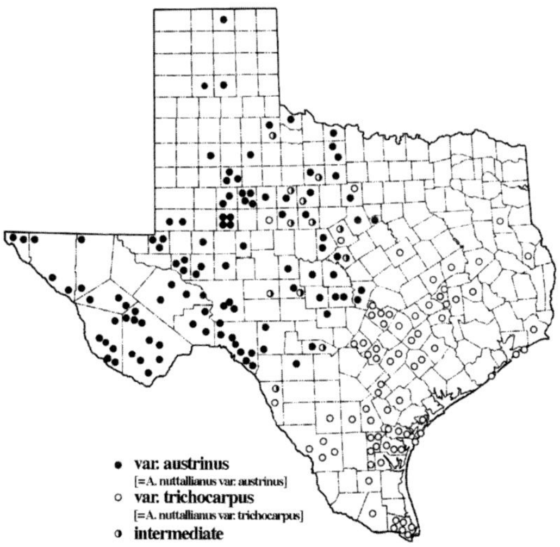
var. austrinus
[=A. nuttallianus var. austrinus]
var. trichocarpus
[=A. nuttallianus var. trichocarpus]
intermediate

Astragalus austrinus

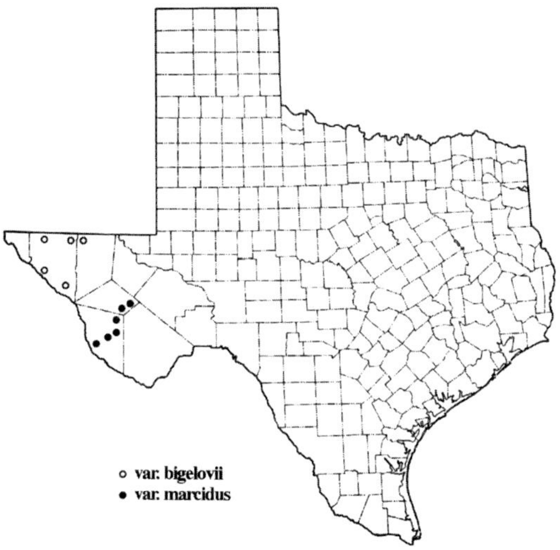
var. bigelovii
var. marcidus

Astragalus bigelovii

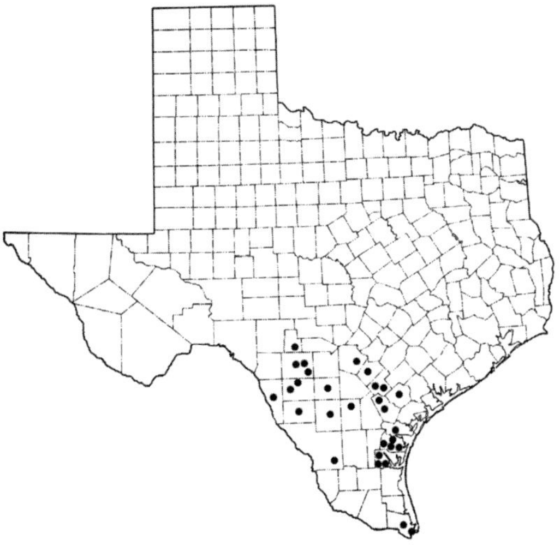

Astragalus brazoensis

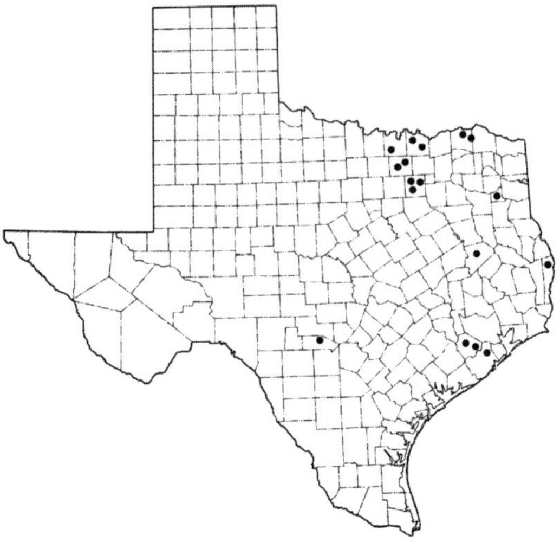

Astragalus canadensis

var. filifolius

Astragalus ceramicus

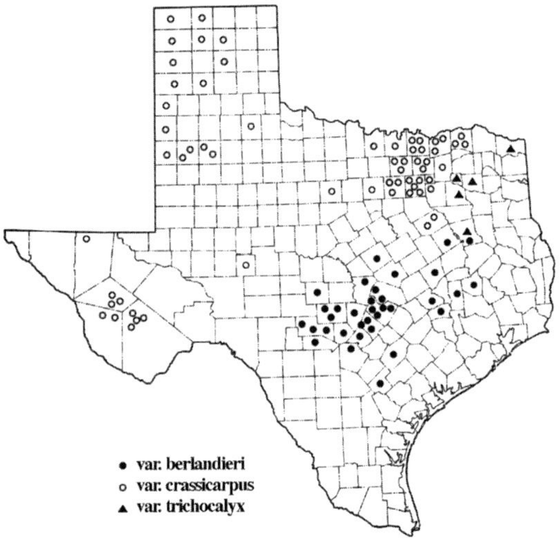
var. berlandieri
var. crassicarpus
var. trichocalyx

Astragalus crassicarpus

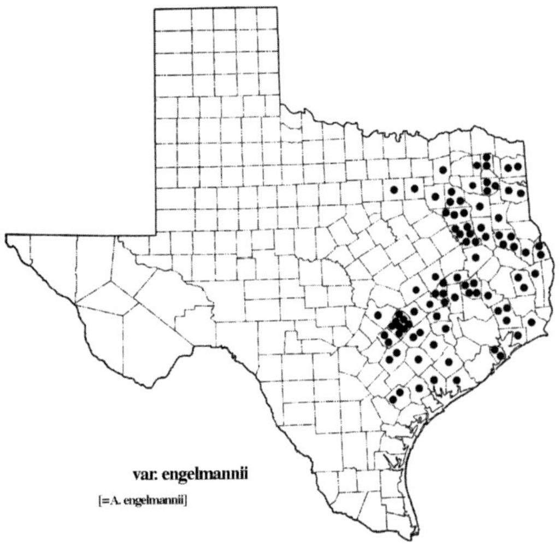
var. engelmannii
[=A. engelmannii]

Astragalus distortus

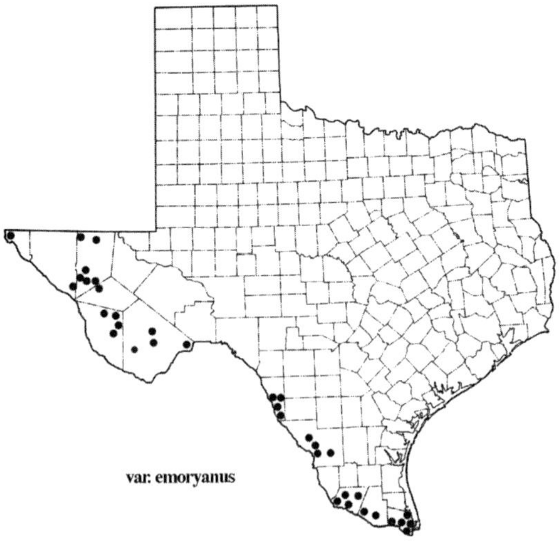
var. emoryanus

Astragalus emoryanus

Astragalus giganteus

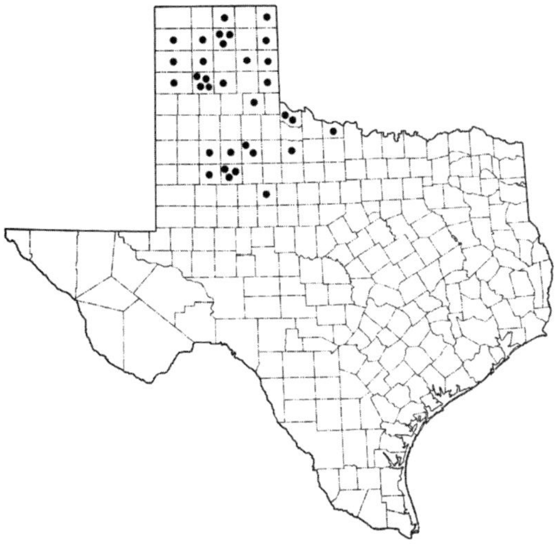

Astragalus gracilis

Astragalus gypsodes

Astragalus humistratus

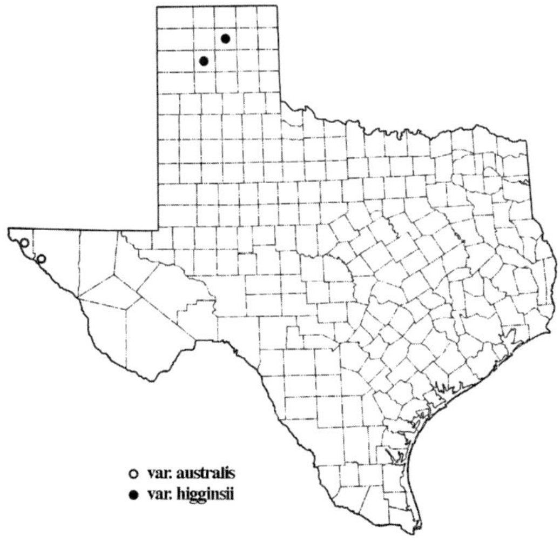
○ var. australis
● var. higginsii

Astragalus lentiginosus

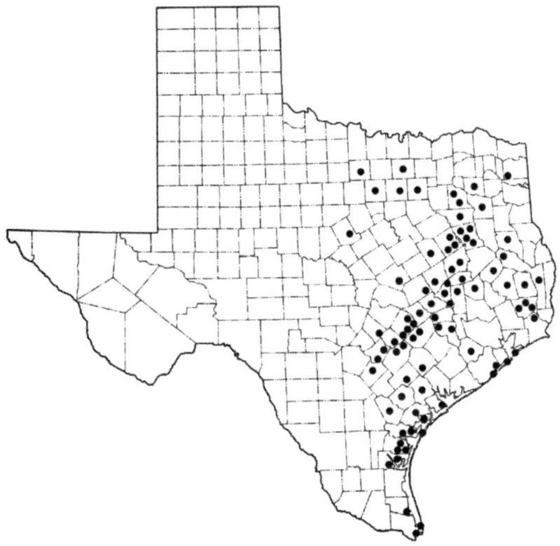

Astragalus leptocarpus

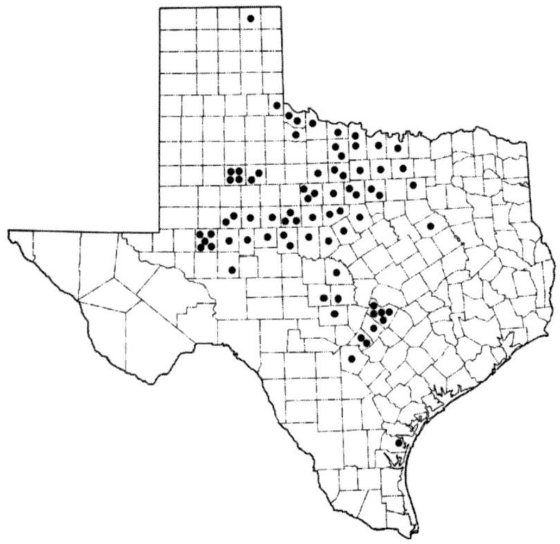

Astragalus lindheimeri

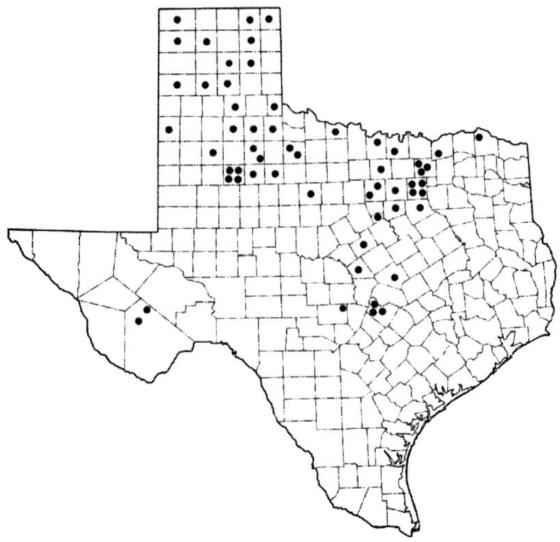

Astragalus lotiflorus

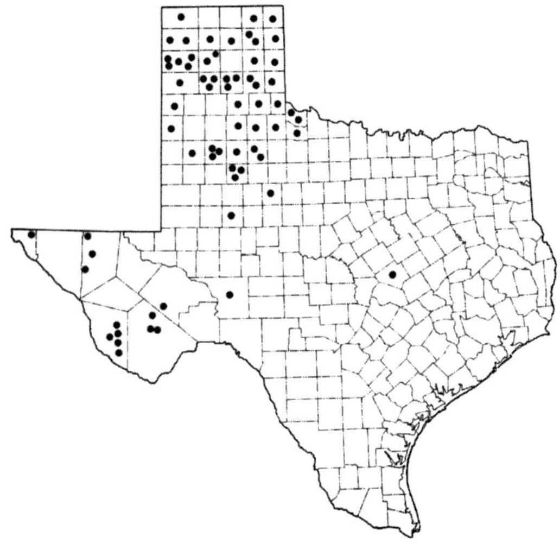

Astragalus missouriensis

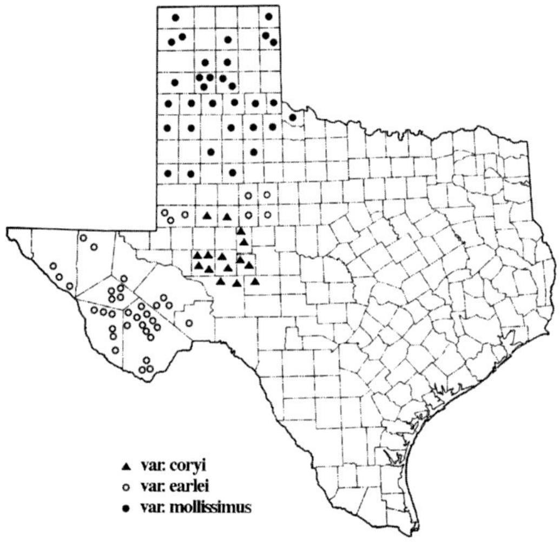
▲ var. coryi
○ var. earlei
● var. mollissimus

Astragalus mollissimus

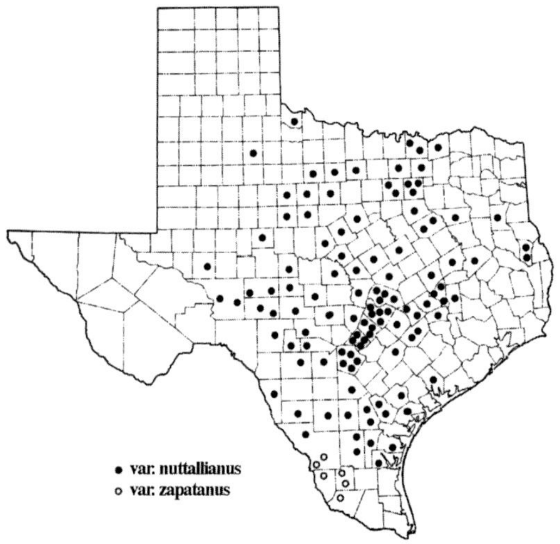
● var. nuttallianus
○ var. zapatanus

Astragalus nuttallianus

Astragalus pictiformis

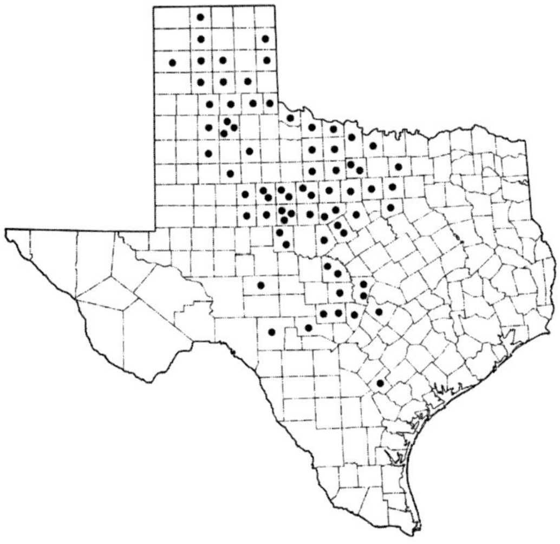

Astragalus plattensis

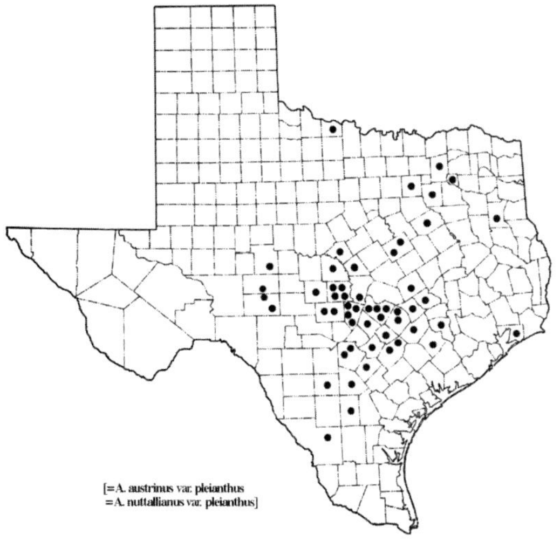
[=A. austrinus var. pleianthus
=A. nuttallianus var. pleianthus]

Astragalus pleianthus

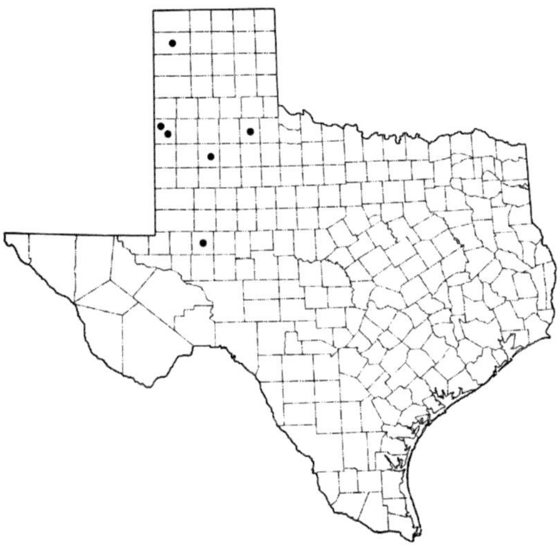

Astragalus praelongus

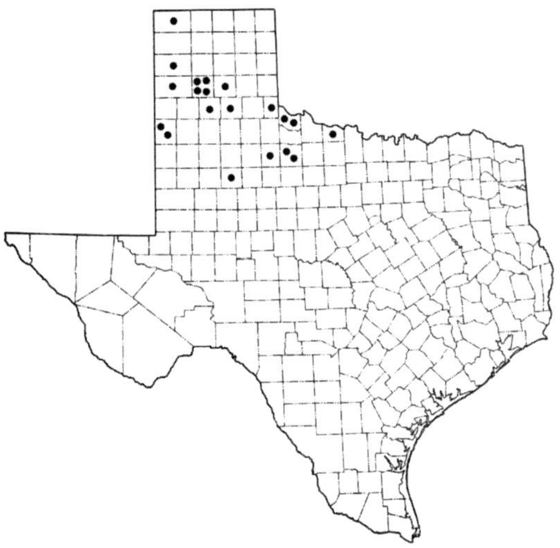

Astragalus racemosus

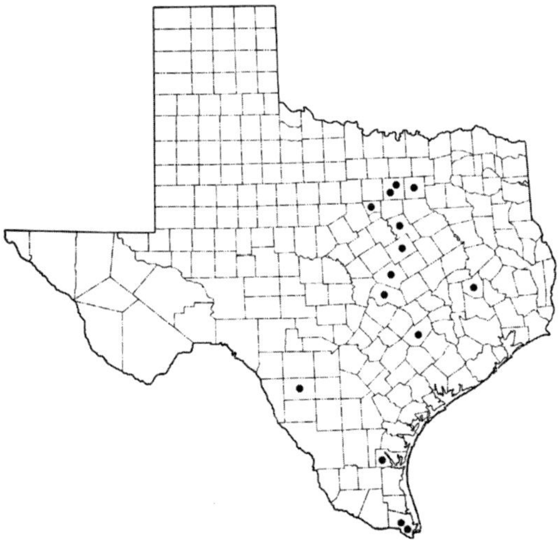

Astragalus reflexus

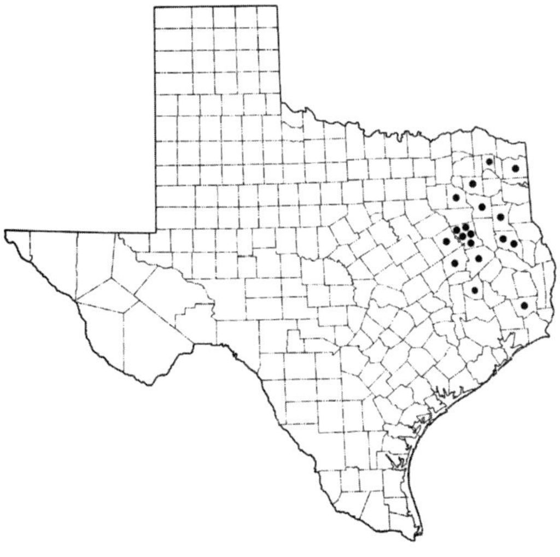

Astragalus soxmaniorum

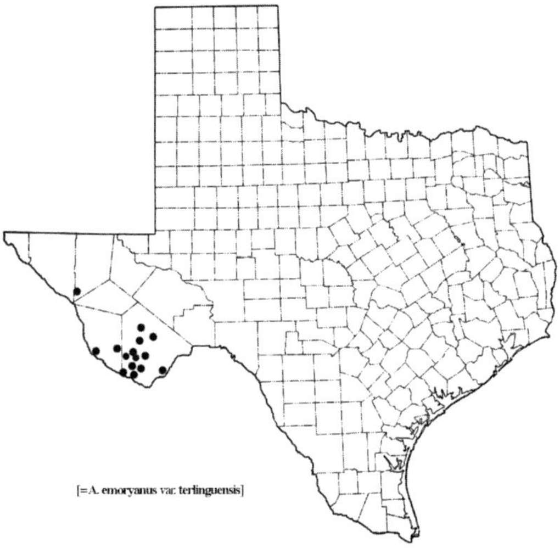
[=A. emoryanus var. terlinguensis]

Astragalus terlinguensis

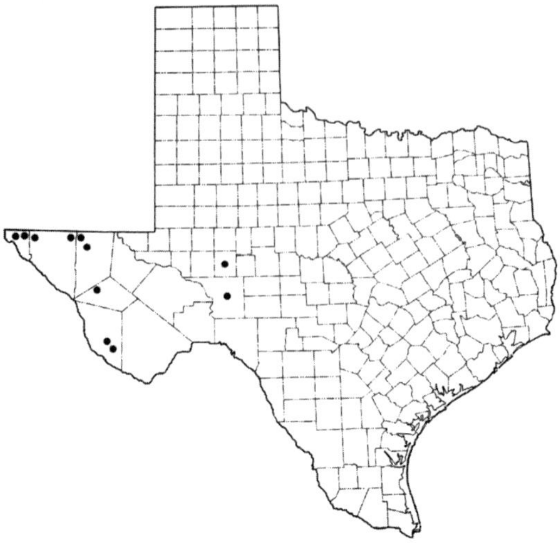

Astragalus waterfallii

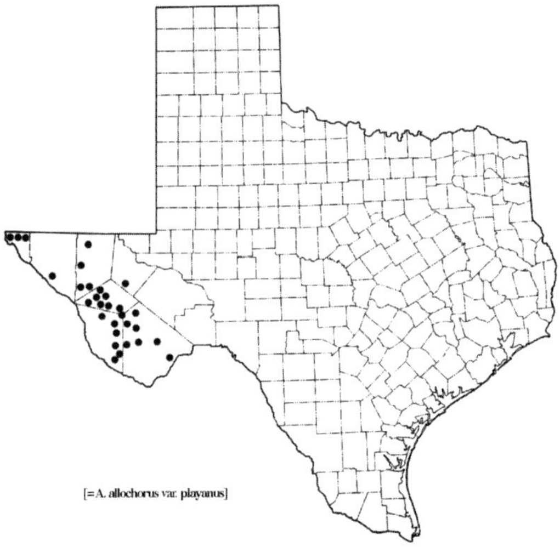
[=A. allochorus var. playanus]

Astragalus wootonii

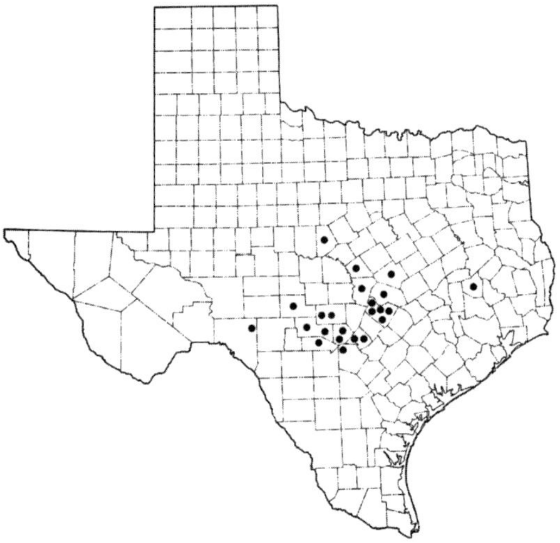

Astragalus wrightii

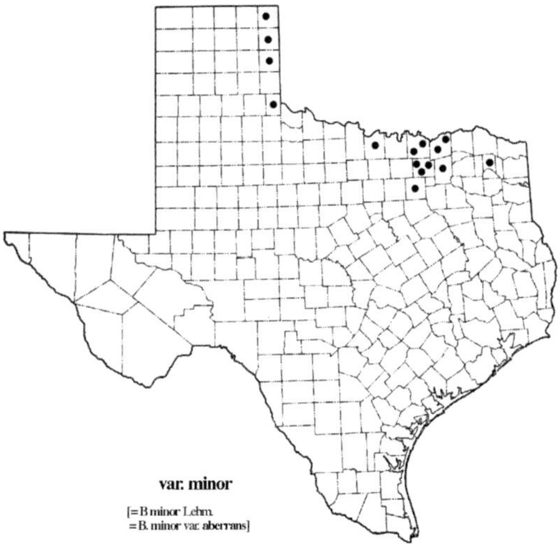
var. minor
[=B minor Lehm.
=B. minor var. aberrans]

Baptisia australis

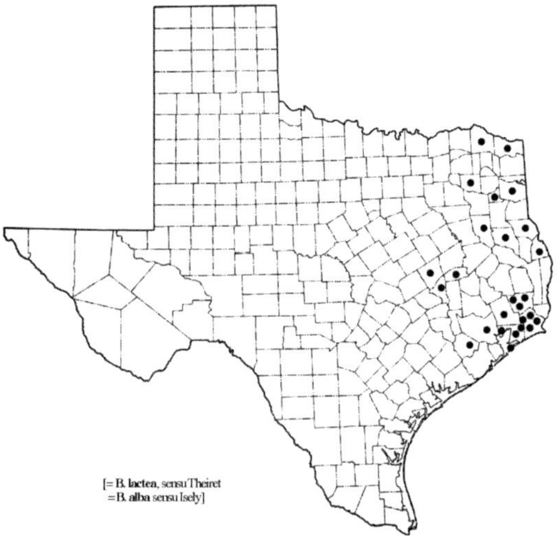
[= B. lactea, sensu Theiret
=B. alba sensu Isely]

Baptisia leucantha

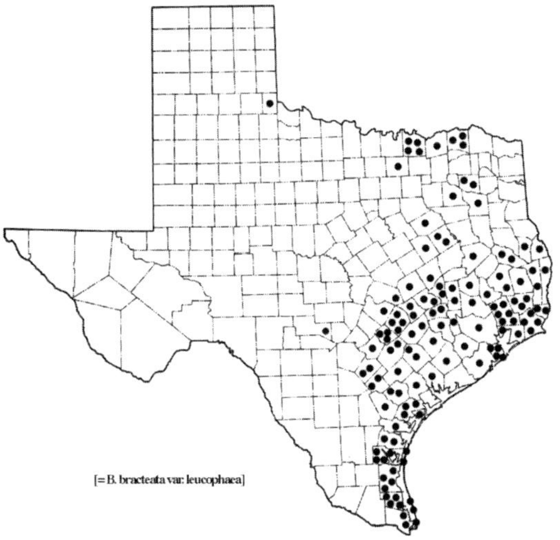
[= B. bracteata var. leucophaea]

Baptisia leucophaea

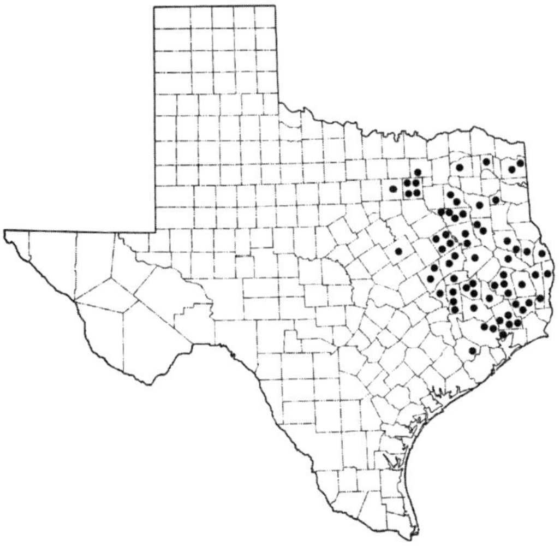

Baptisia nuttalliana

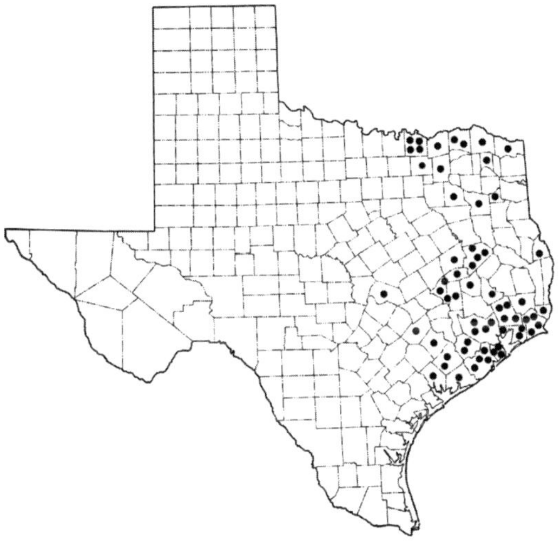

Baptisia sphaerocarpa

Bauhinia lunarioides

Brongniartia minutifolia

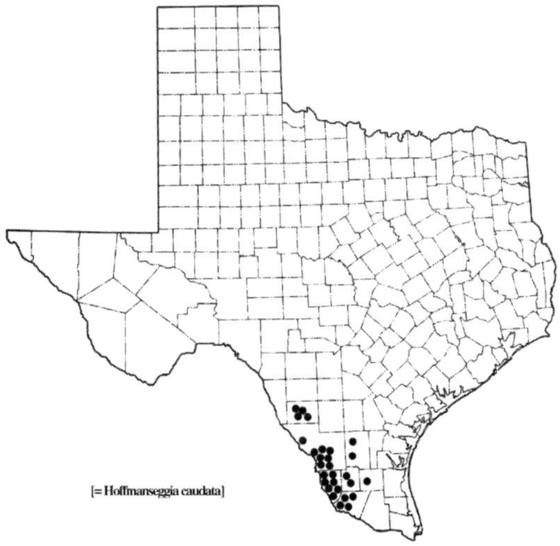
[= Hoffmanseggia caudata]

Caesalpinia caudata

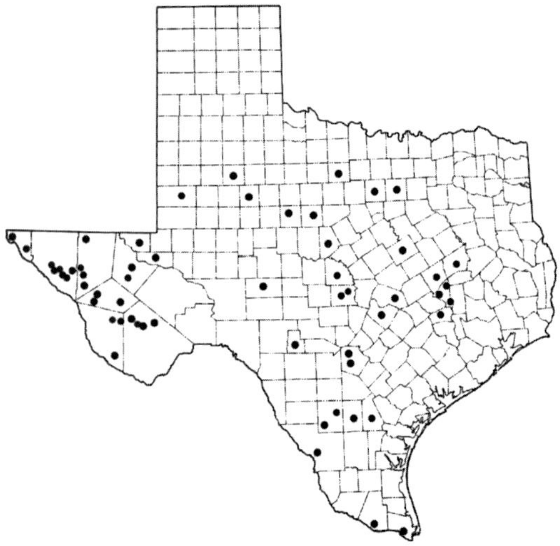

Caesalpinia gilliesii

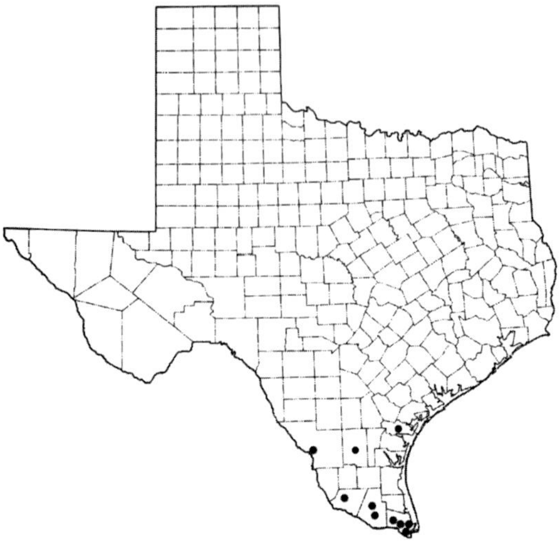

Caesalpinia mexicana

Caesalpinia phyllanthoides

Caesalpinia pulcherrima

Calliandra biflora

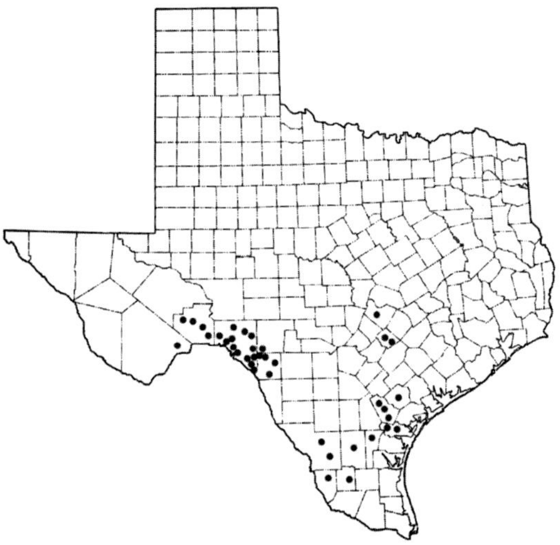

Calliandra conferta

var. eriophylla

Calliandra eriophylla

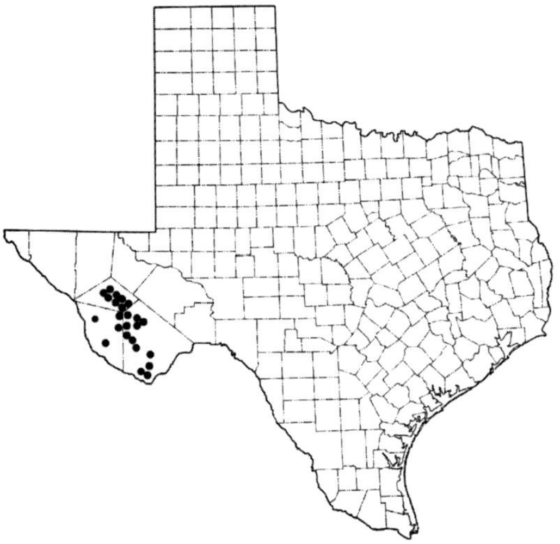

Calliandra humilis

Calliandra isleyi

Canavalia rosea

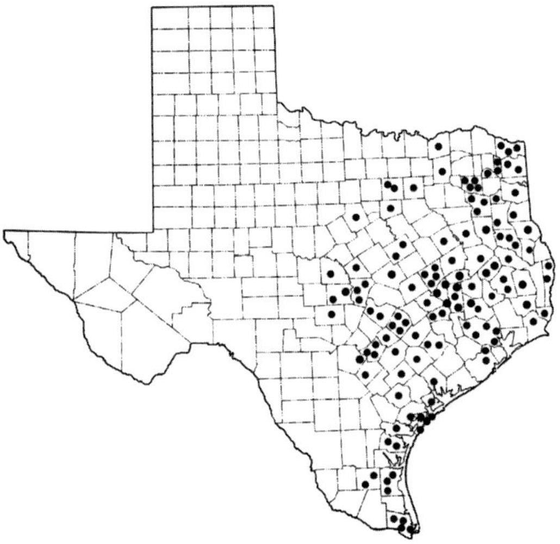

Centrosema virginianum

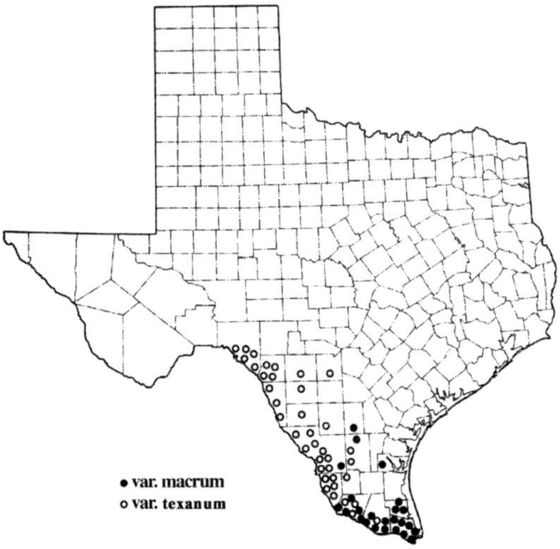
• var. macrum
○ var. texanum

Cercidium texanum

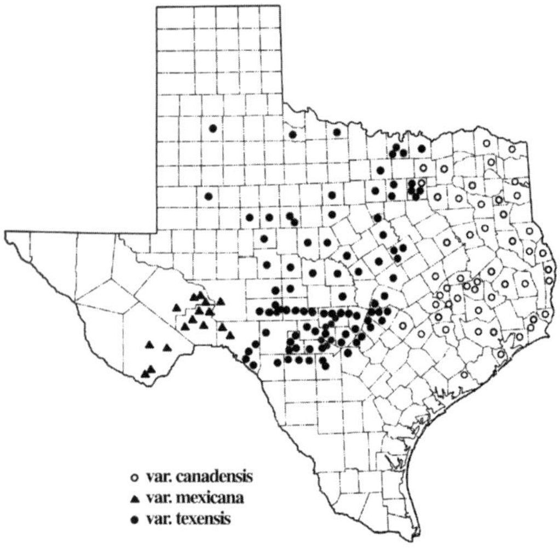
var. canadensis
var. mexicana
var. texensis

Cercis canadensis

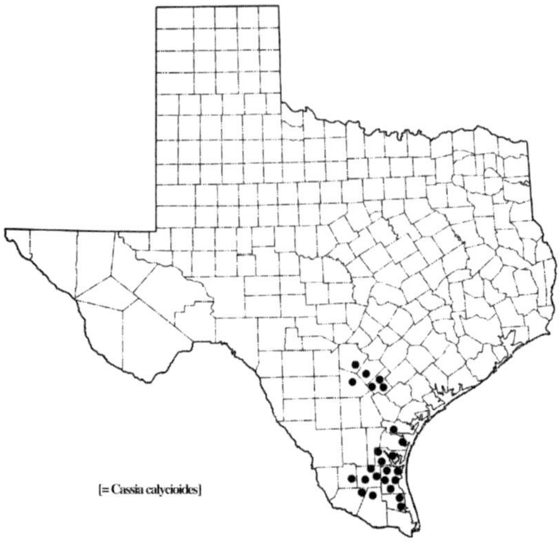
[= Cassia calycioides]

Chamaecrista calycioides

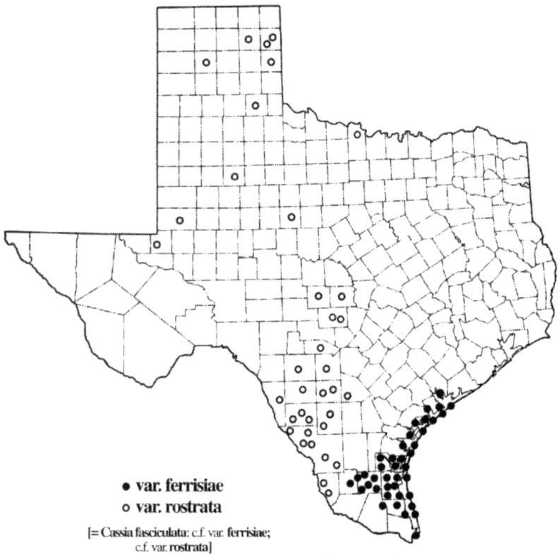
var. ferrisiae
var. rostrata
[= Cassia fasciculata: c.f. var. ferrisiae; c.f. var. rostrata]

Chamaecrista fasciculata

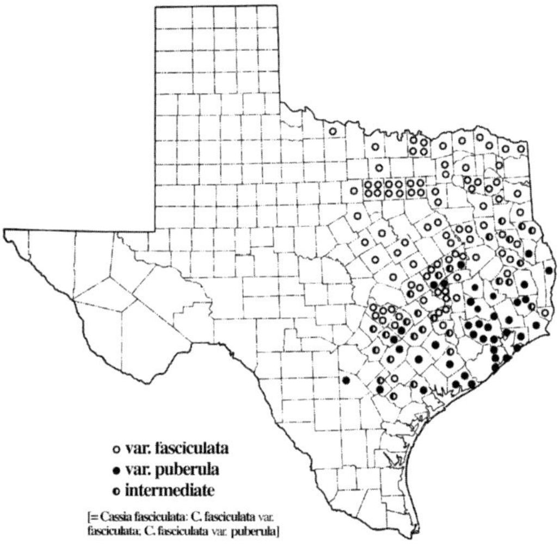
var. fasciculata
var. puberula
intermediate
[= Cassia fasciculata: C. fasciculata var. fasciculata; C. fasciculata var. puberula]

Chamaecrista fasciculata

[= Cassia greggii]

Chamaecrista greggii

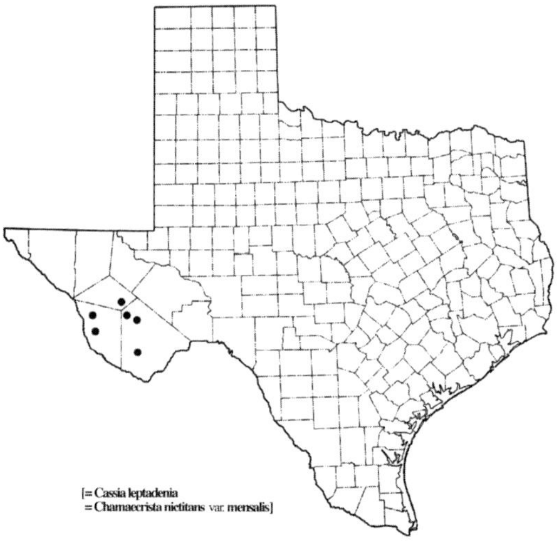
[= Cassia leptadenia
= Chamaecrista nictitans var. mensalis]

Chamaecrista leptadenia

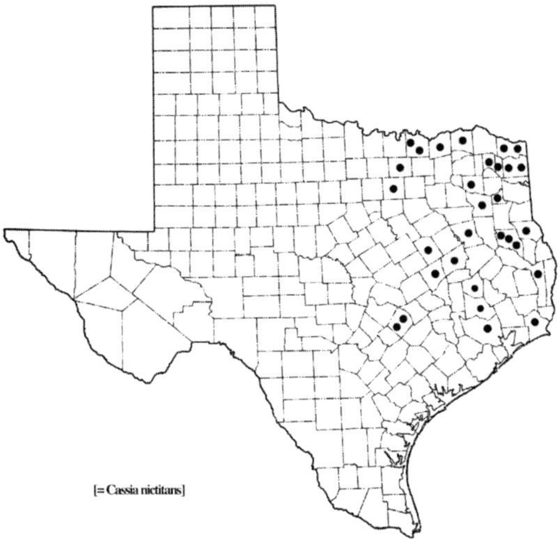
[= Cassia nictitans]

Chamaecrista nictitans

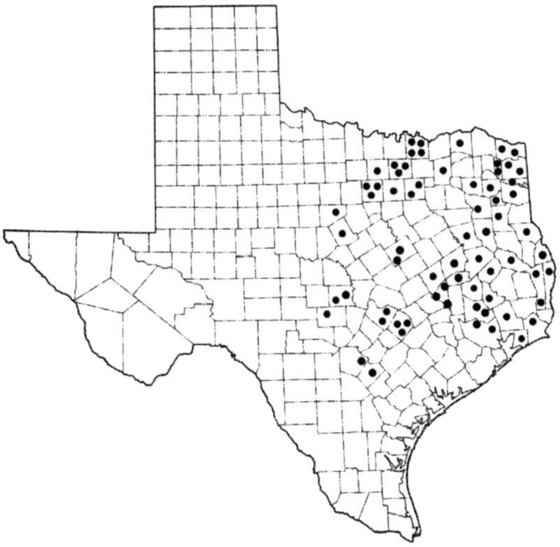

Clitoria mariana

Clitoria ternatea

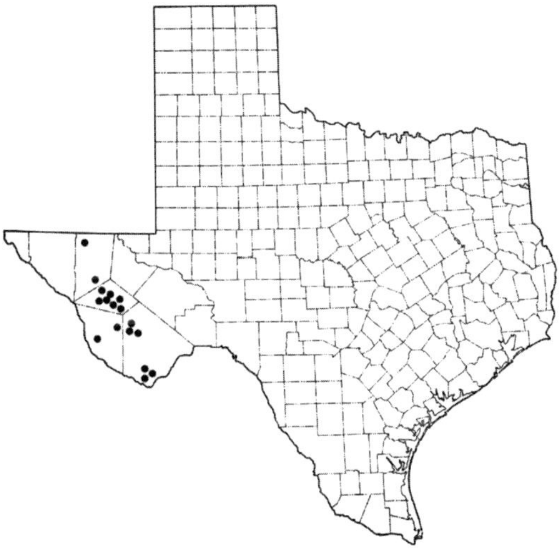

Cologania angustifolia

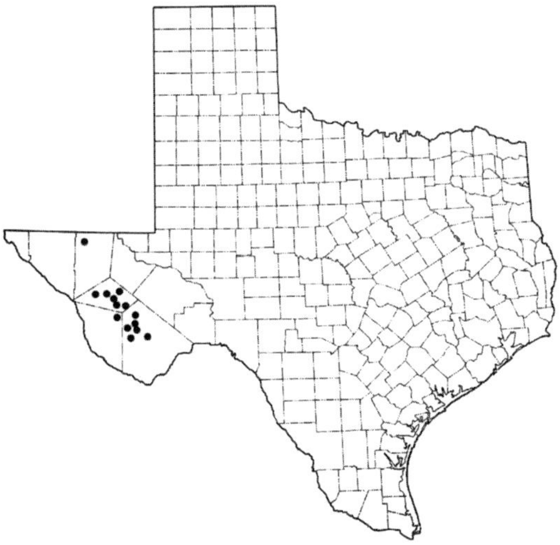

Cologania pallida

Coursetia axillaris

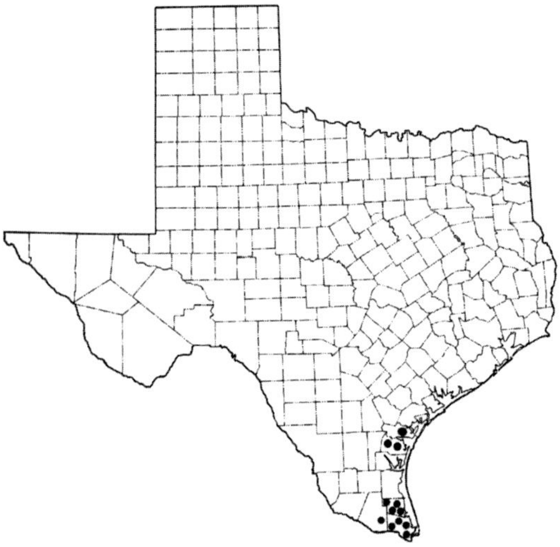
Crotalaria incana

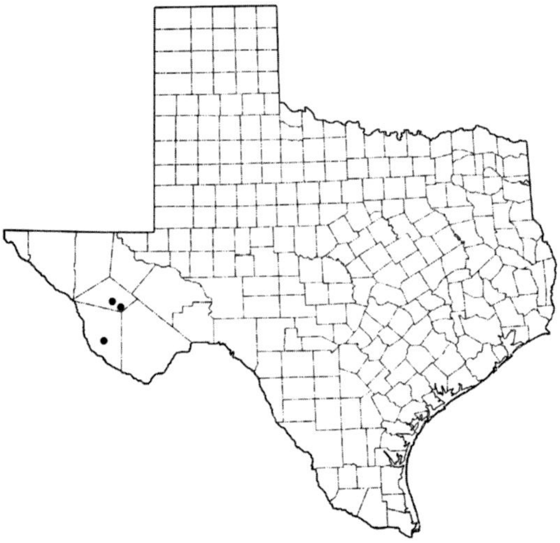
Crotalaria pumila

Crotalaria purshii

Crotalaria retusa

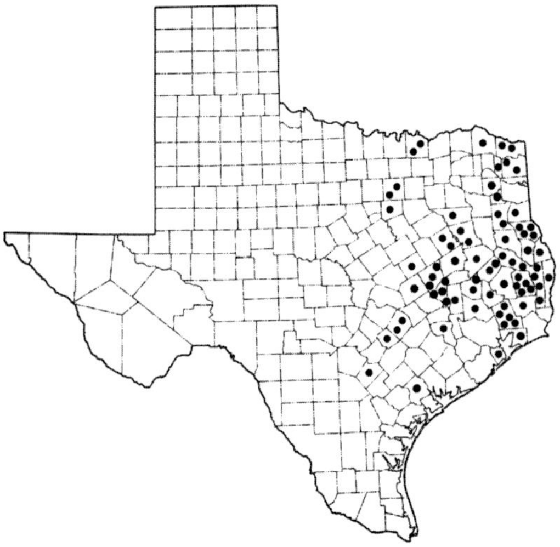
Crotalaria sagittalis

Crotalaria spectabilis

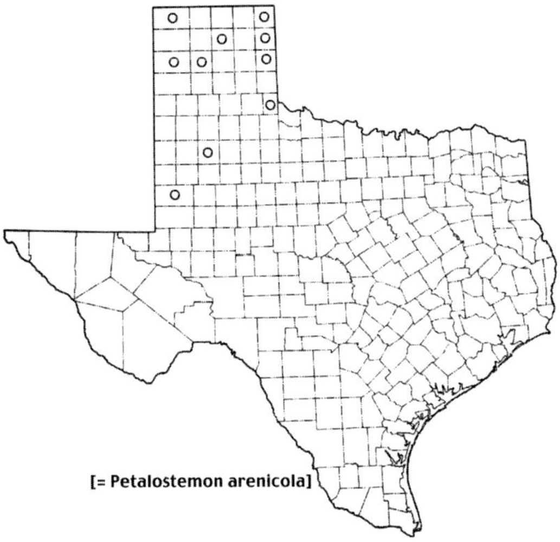
[= Petalostemon arenicola]

Dalea arenicola

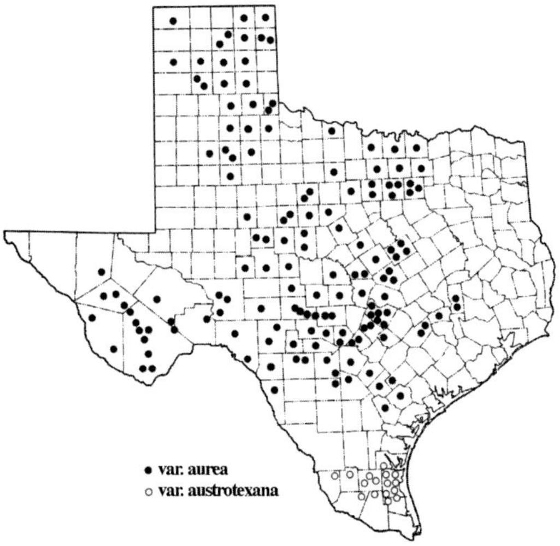
var. aurea
var. austrotexana

Dalea aurea

Dalea bartonii

var. argyrea
[= D. argyrea]

Dalea bicolor

Dalea brachystachya

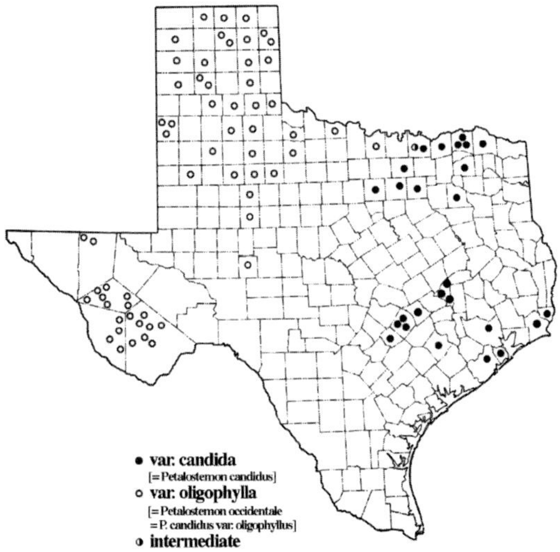
var. candida
[= Petalostemon candidus]
var. oligophylla
[= Petalostemon occidentale
= P. candidus var. oligophyllus]
intermediate

Dalea candida

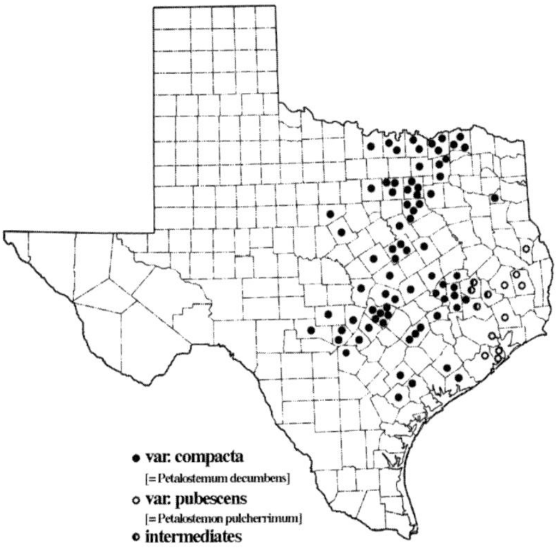
var. compacta
[=Petalostemum decumbens]
var. pubescens
[=Petalostemon pulcherrimum]
intermediates
Dalea compacta

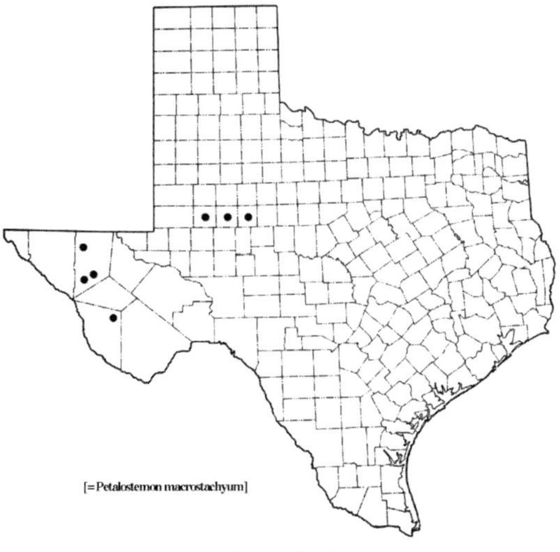
[=Petalostemon macrostachyum]
Dalea cylindriceps

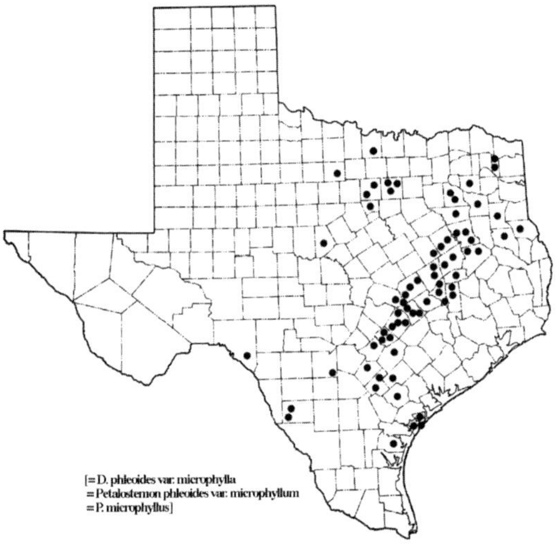
[=D. phleoides var. microphylla
=Petalostemon phleoides var. microphyllum
=P. microphyllus]
Dalea drummondiana

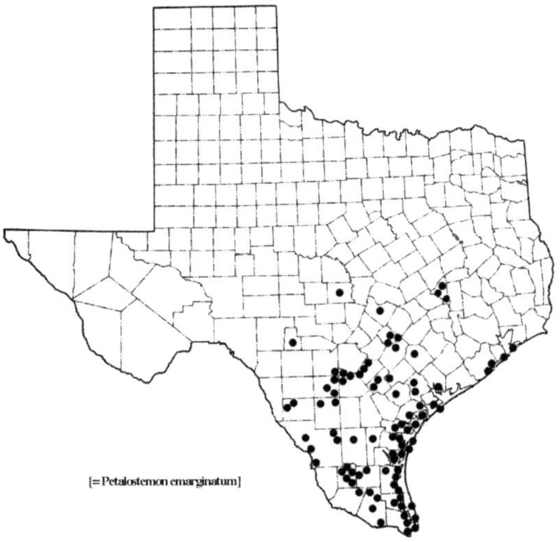
[=Petalostemon emarginatum]
Dalea emarginata

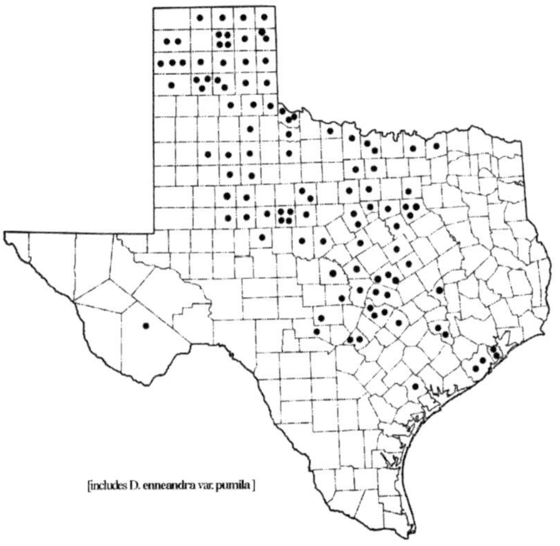
[includes D. enneandra var. pumila]
Dalea enneandra

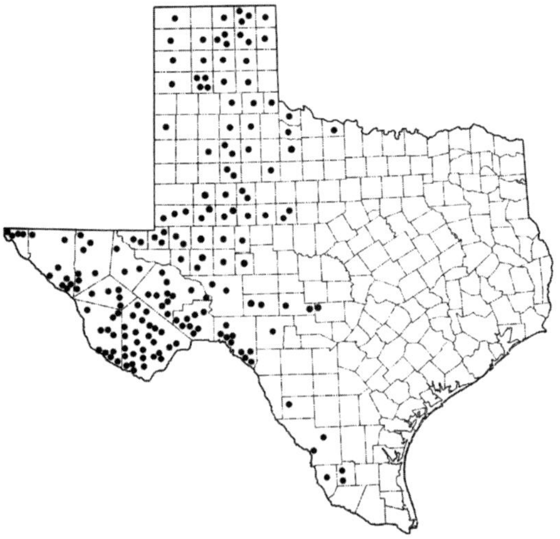
Dalea formosa

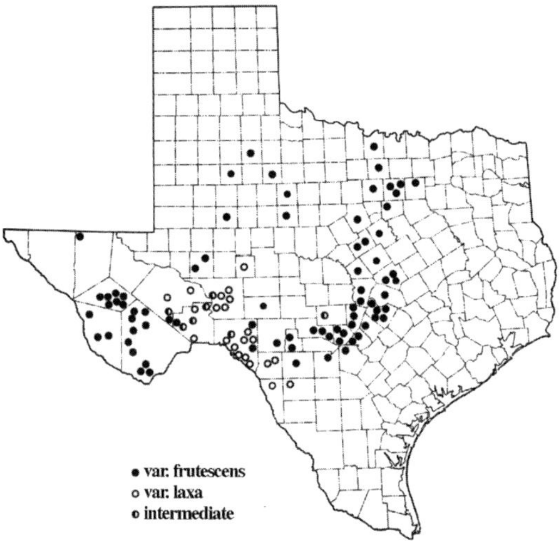
• var. frutescens
○ var. laxa
◦ intermediate

Dalea frutescens

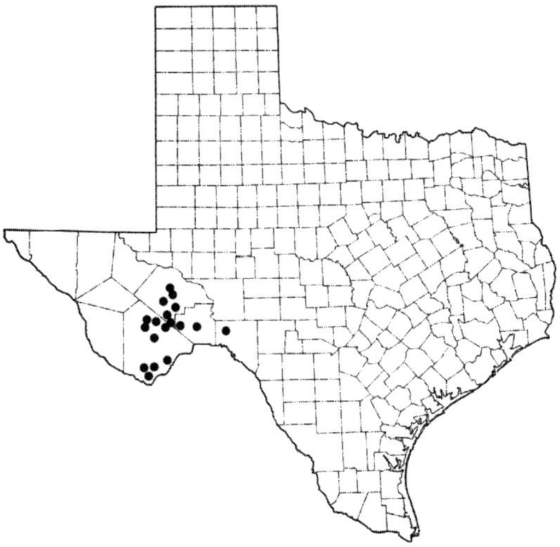

Dalea greggii

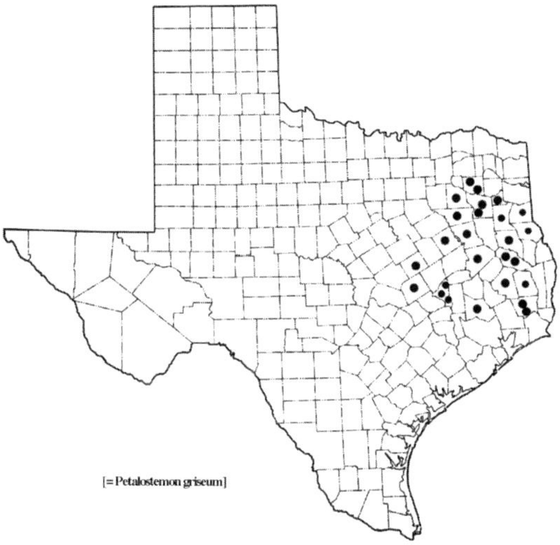
[= Petalostemon griseum]

Dalea grisea

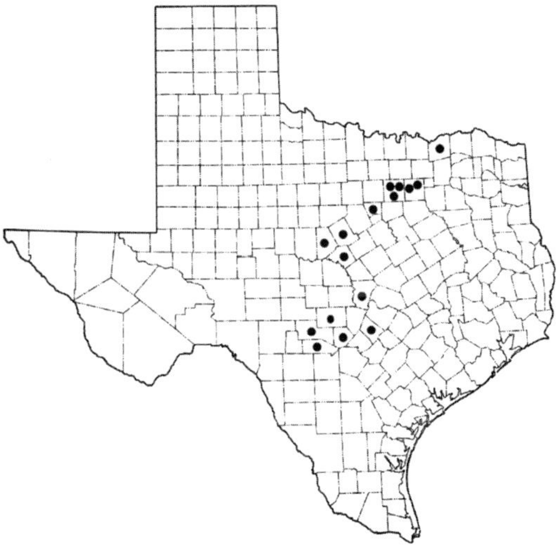

Dalea hallii

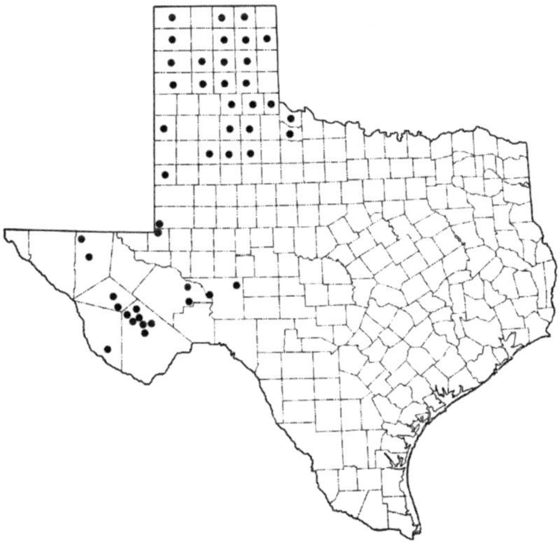

Dalea jamesii

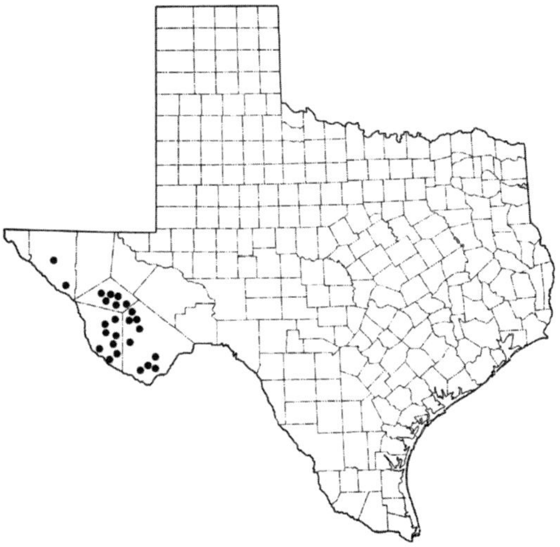

Dalea lachnostachys

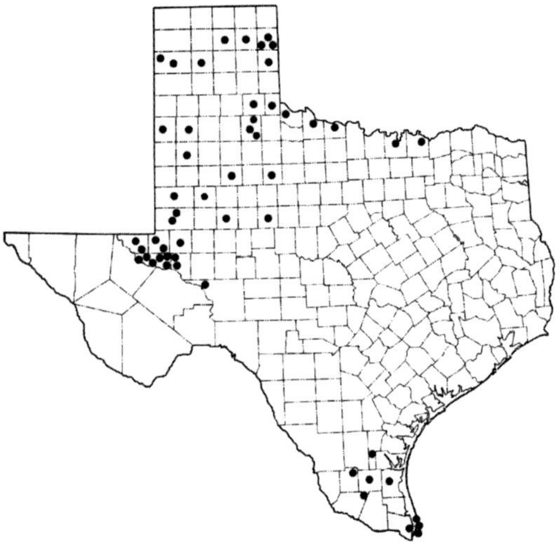

Dalea lanata

Dalea laniceps

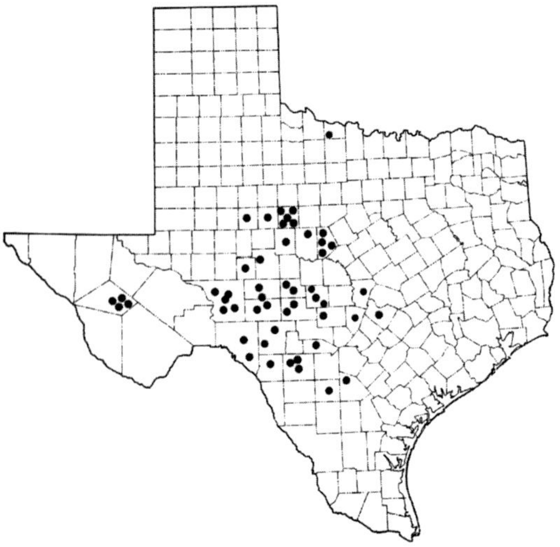

Dalea lasiathera

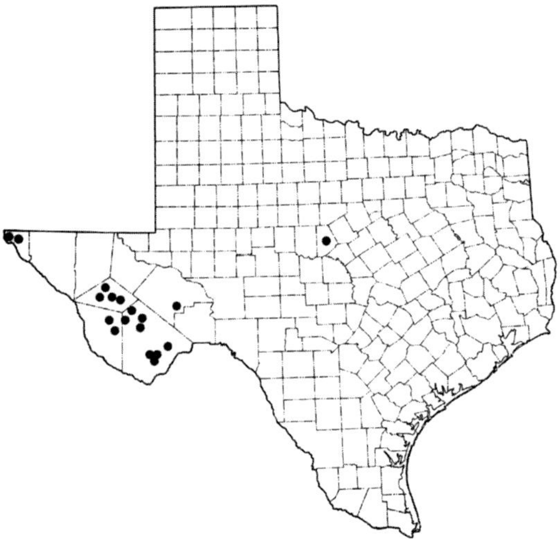

Dalea leporina

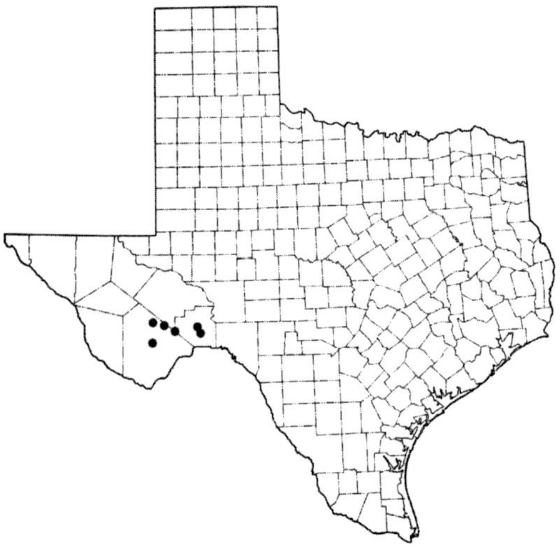

Dalea longipila

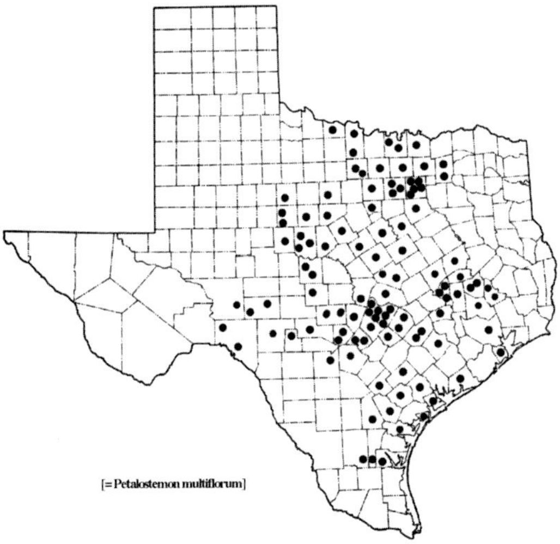
[= Petalostemon multiflorum]

Dalea multiflora

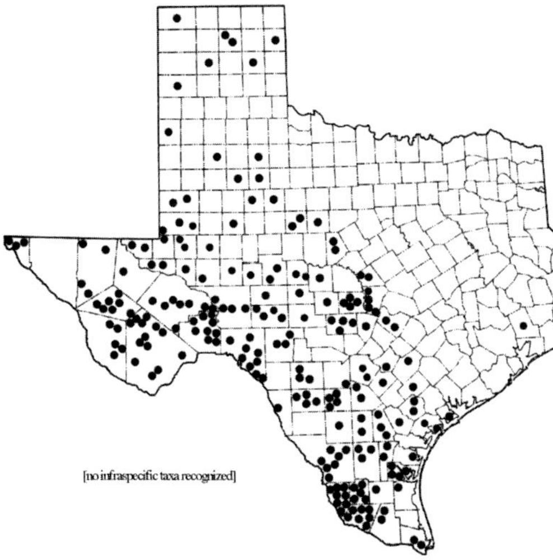
[no infraspecific taxa recognized]

Dalea nana

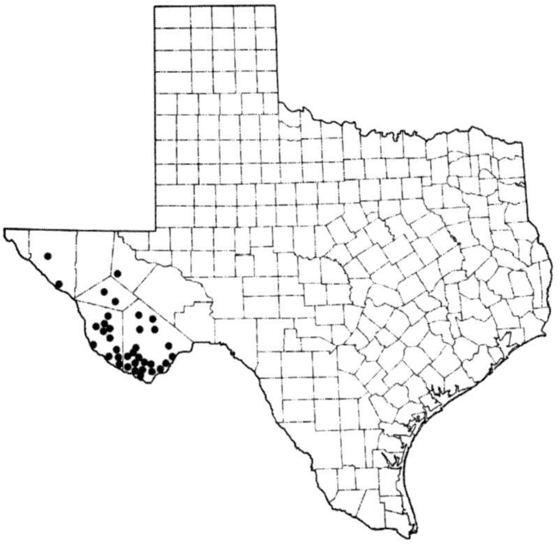

Dalea neomexicana

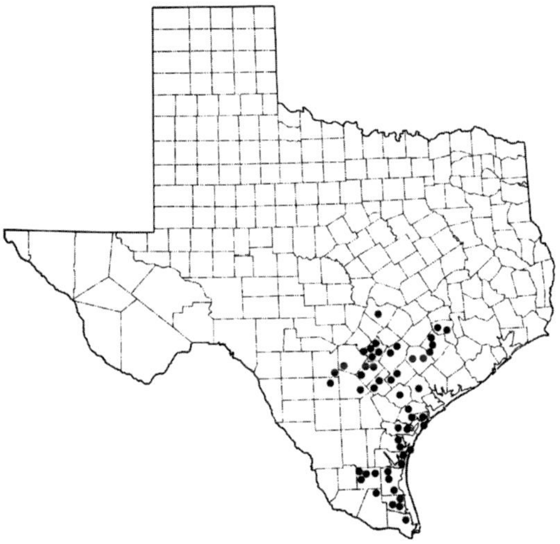

Dalea obovata

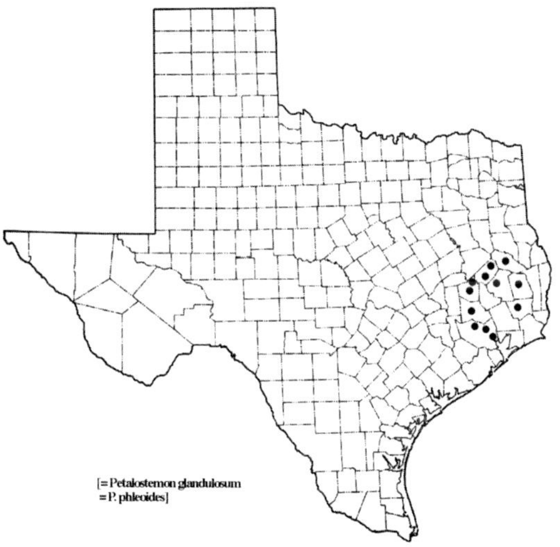
[= Petalostemon glandulosum
= P. phleoides]

Dalea phleoides

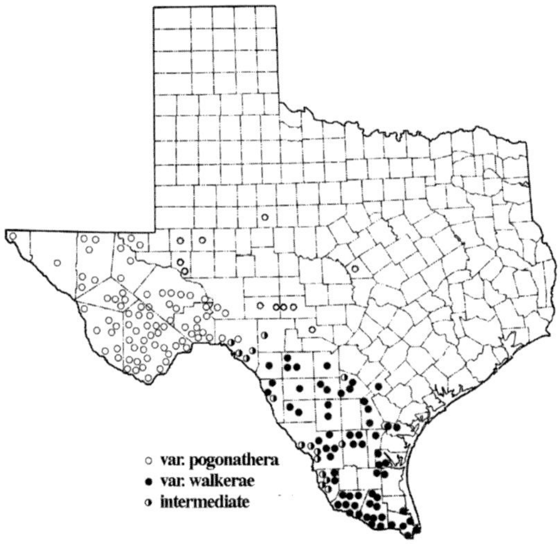
var. pogonathera
var. walkerae
intermediate

Dalea pogonathera

Dalea polygonoides

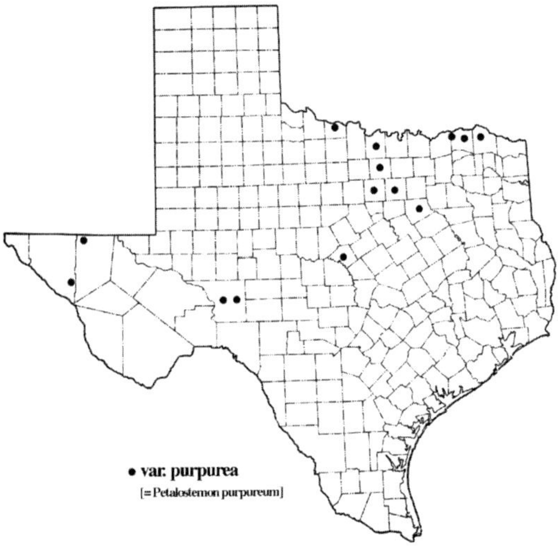
• var. purpurea
[= Petalostemon purpureum]

Dalea purpurea

[= Petalostemon reverchonii]

Dalea reverchonii

[= Petalostemon sabinale]

Dalea sabinalis

var. pauciflora

Dalea scandens

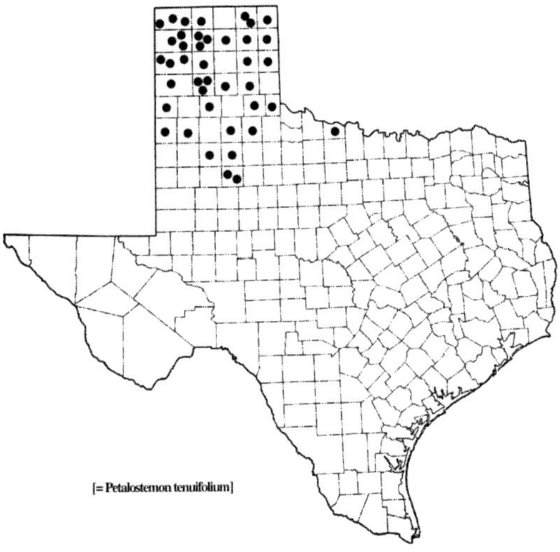
[= Petalostemon tenuifolium]

Dalea tenuifolia

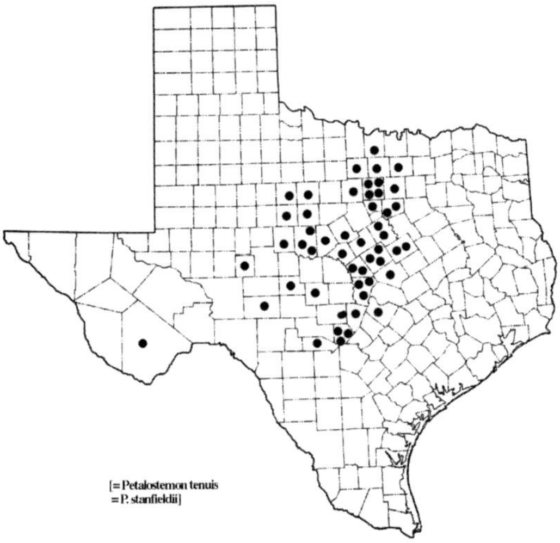
[= Petalostemon tenuis
= P. stanfieldii]

Dalea tenuis

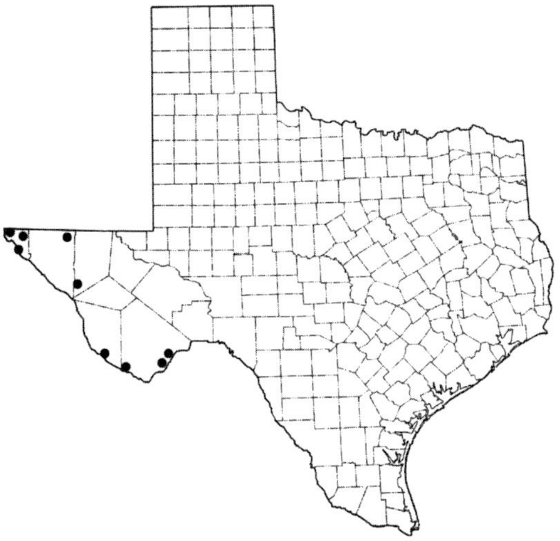

Dalea terminalis

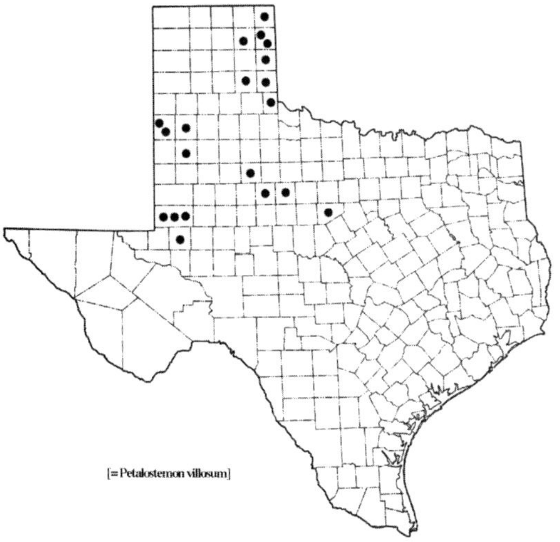
[= Petalostemon villosum]

Dalea villosa

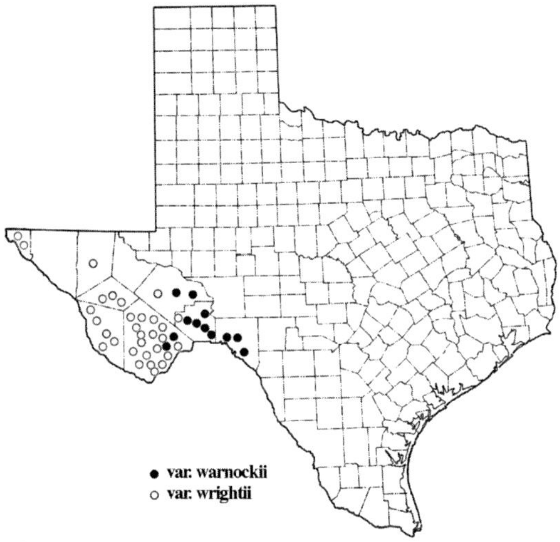
var. warnockii
var. wrightii

Dalea wrightii

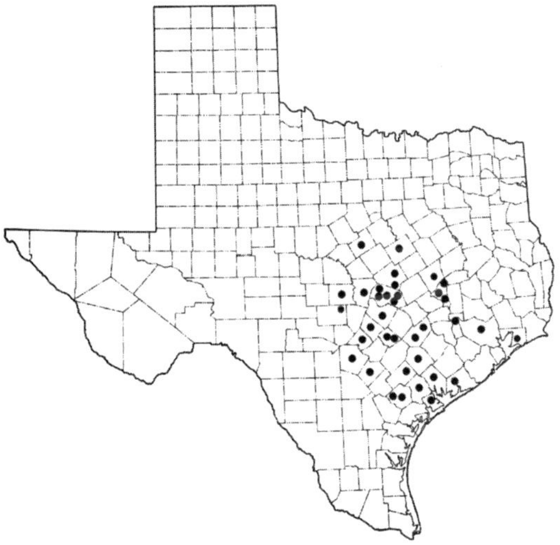

Desmanthus acuminatus

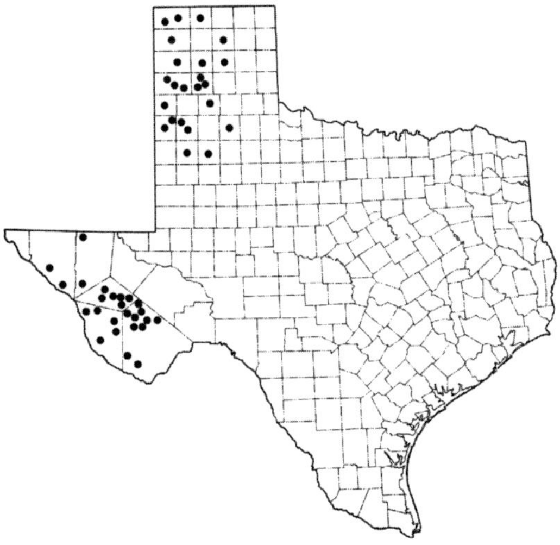

Desmanthus cooleyi

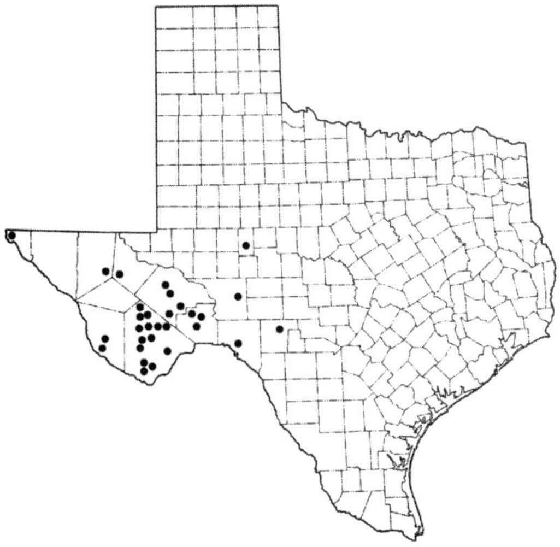

Desmanthus glandulosus

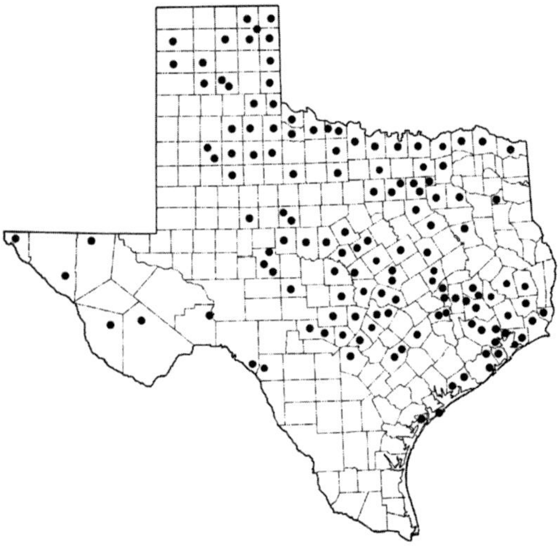

Desmanthus illinoensis

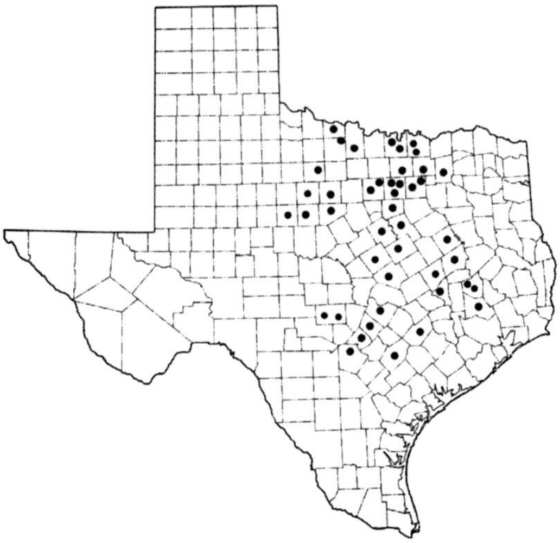

Desmanthus leptolobus

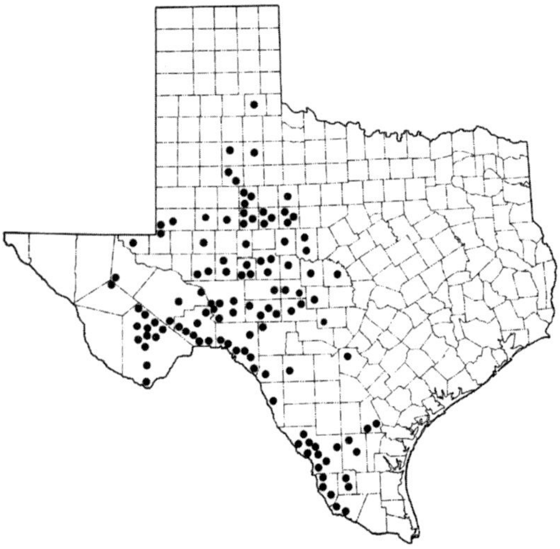

Desmanthus obtusus

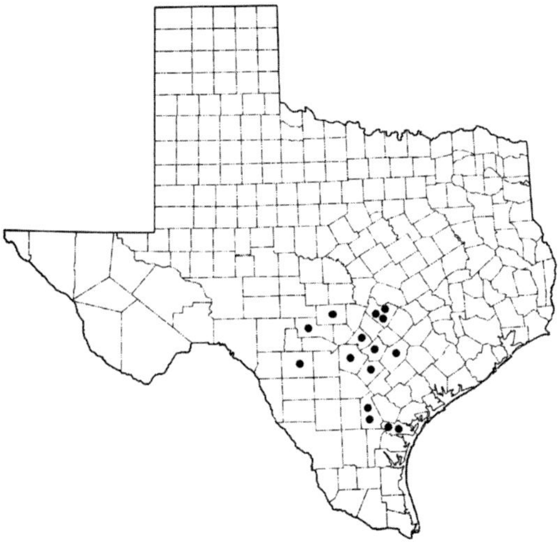

Desmanthus reticulatus

var. brevipes

Desmanthus tatuhyensis

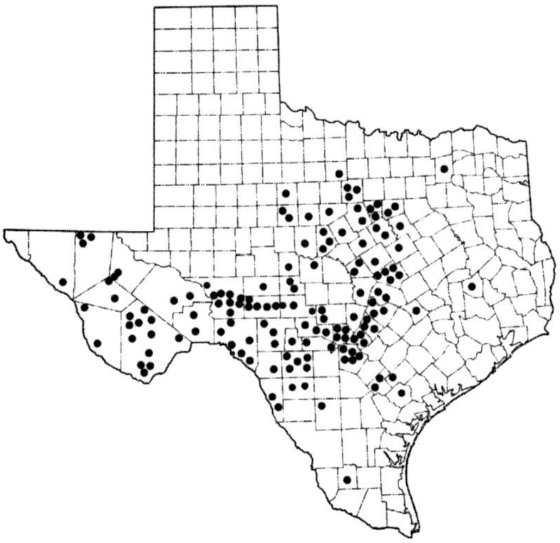

Desmanthus velutinus

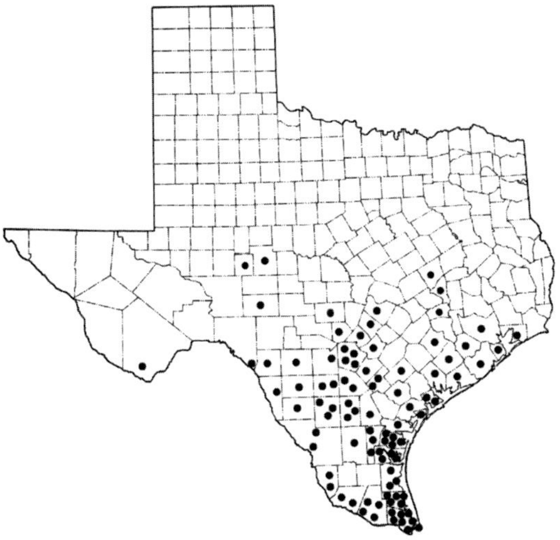

Desmanthus virgatus

Desmodium canadense

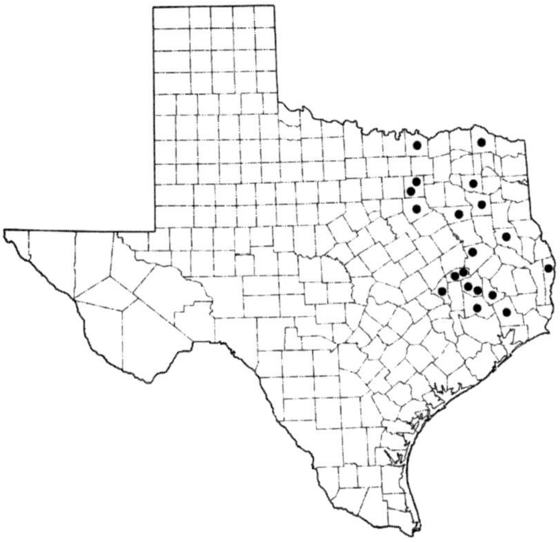

Desmodium canescens

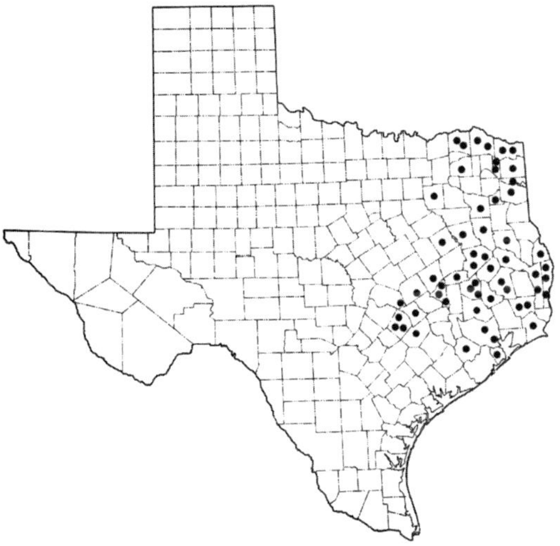

Desmodium ciliare

Desmodium cuspidatum

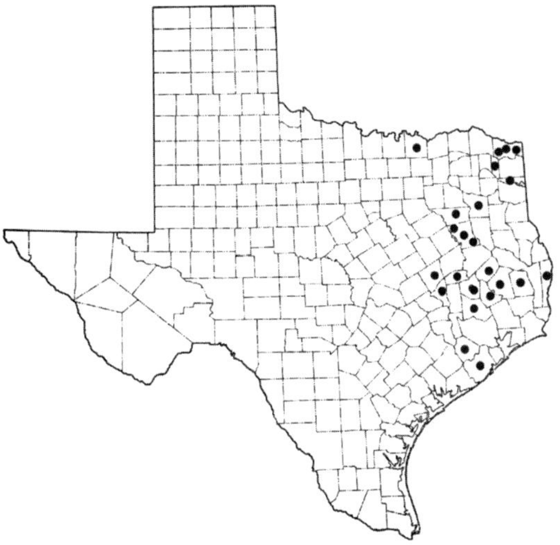

Desmodium glabellum

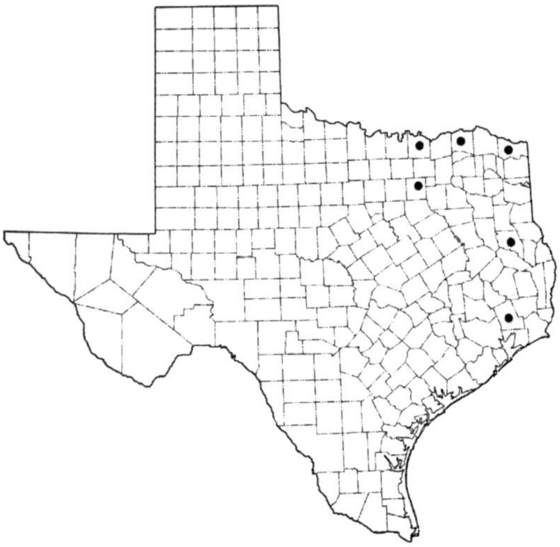

Desmodium glutinosum

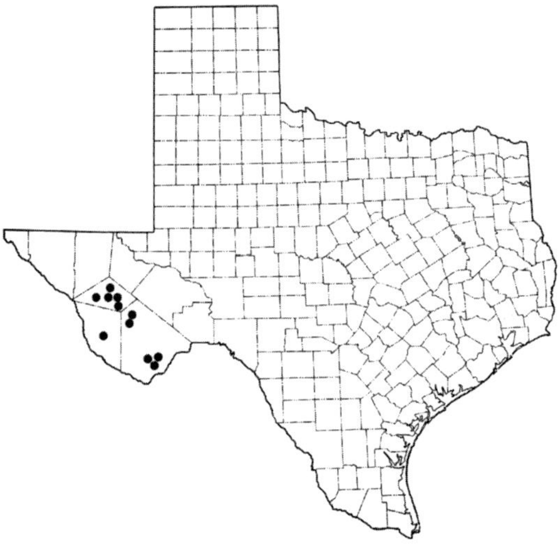

Desmodium grahamii

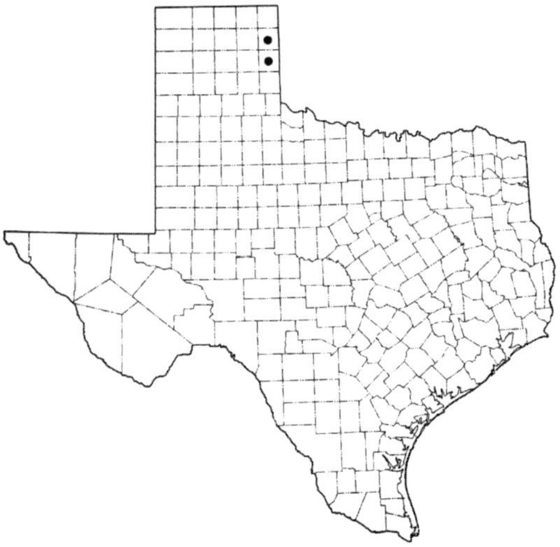

Desmodium illinoense

Desmodium incanum

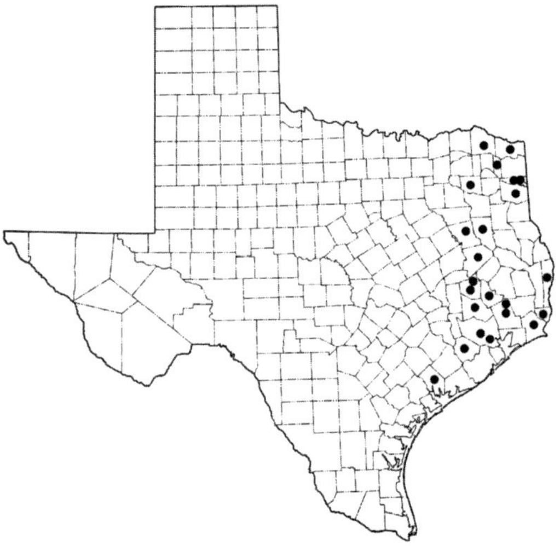

Desmodium laevigatum

Desmodium lindheimeri

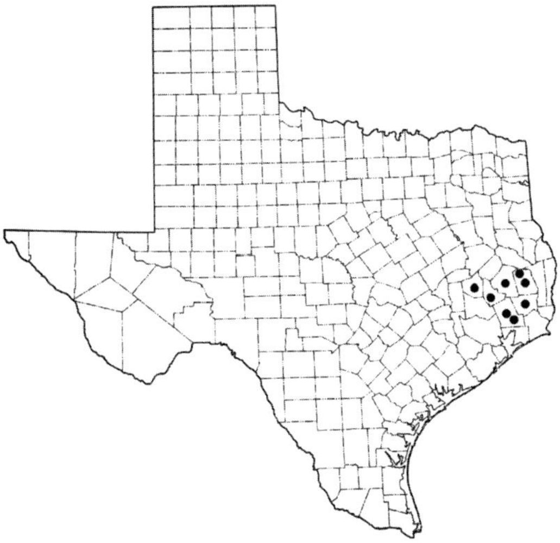

Desmodium lineatum

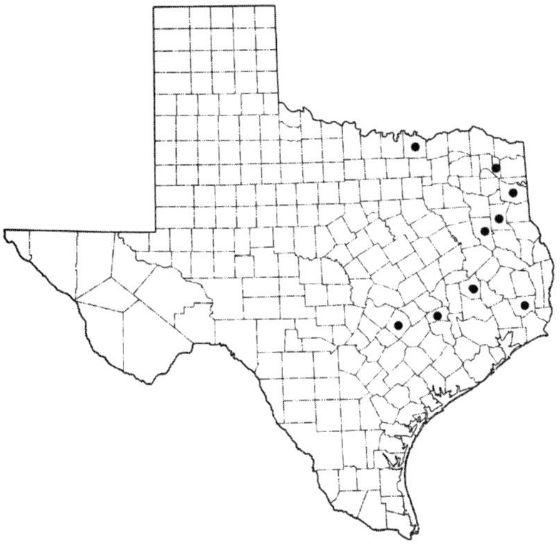

Desmodium marilandicum

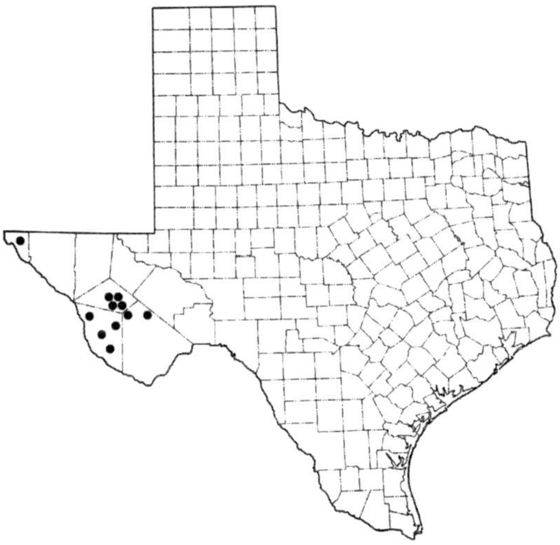

Desmodium neomexicanum

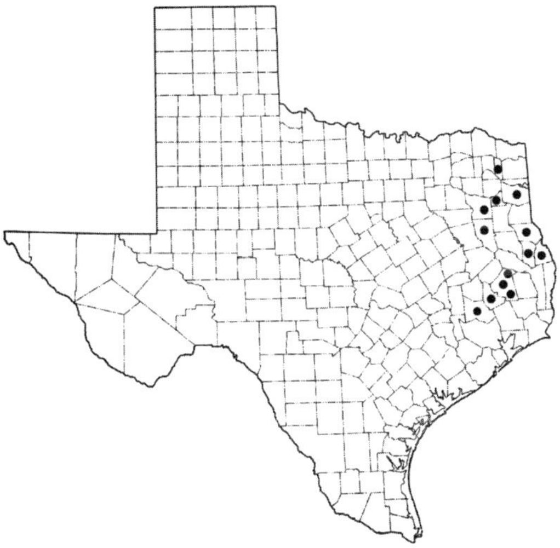

Desmodium nudiflorum

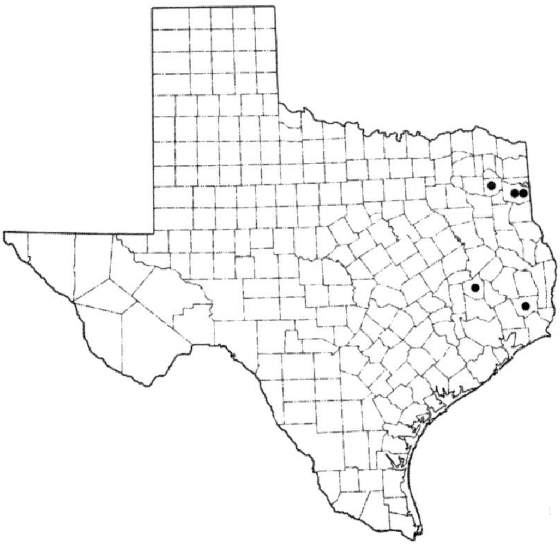

Desmodium nuttallii

Desmodium obtusum

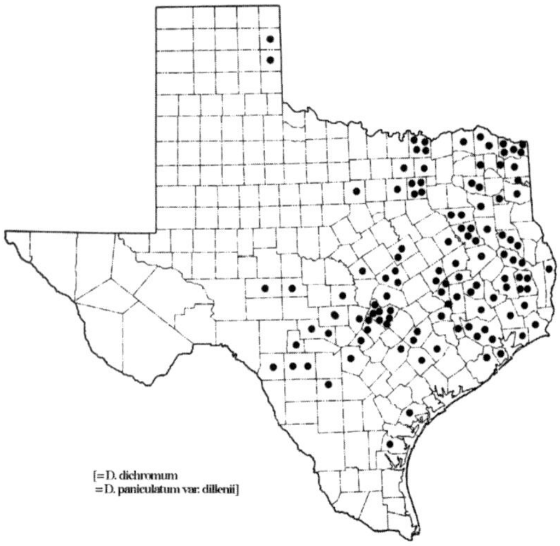
[=D. dichromum
=D. paniculatum var. dillenii]

Desmodium paniculatum

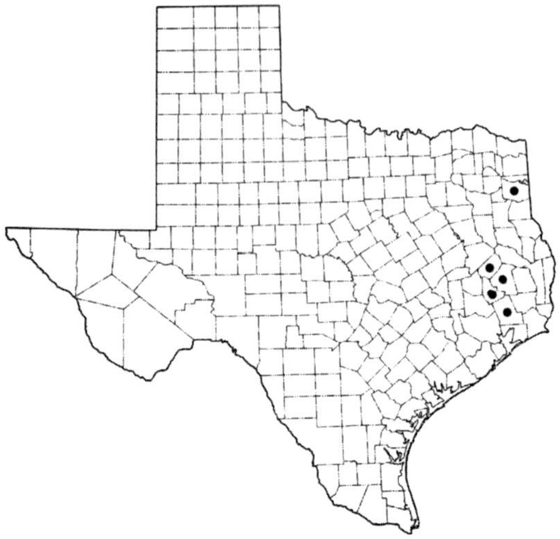

Desmodium pauciflorum

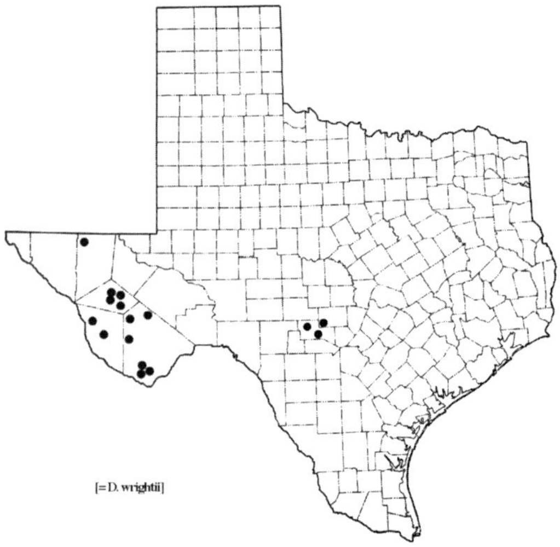
[=D. wrightii]

Desmodium psilophyllum

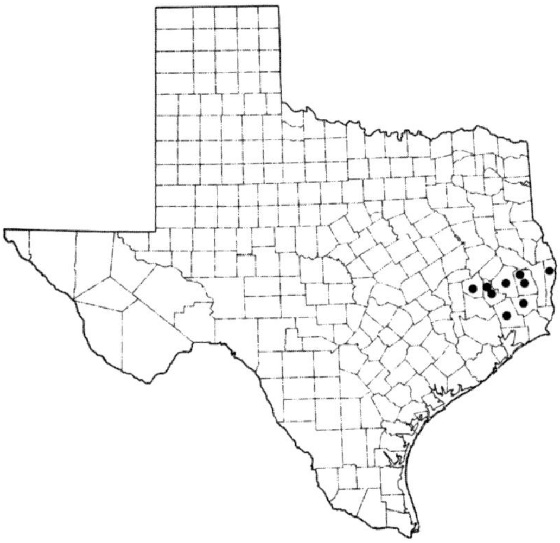

Desmodium rotundifolium

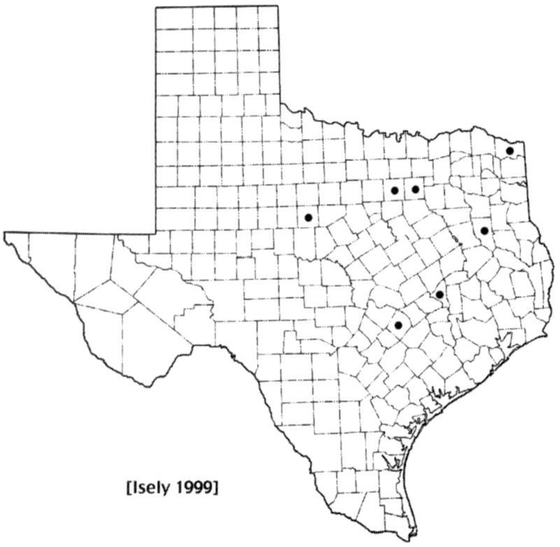
[Isely 1999]

Desmodium scorpiurus

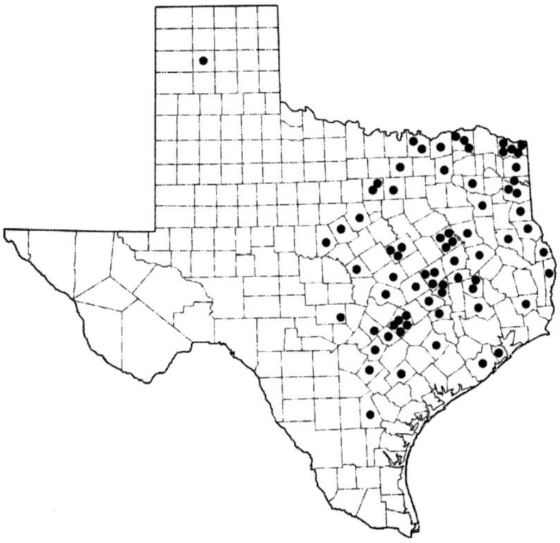

Desmodium sessilifolium

Desmodium tortuosum

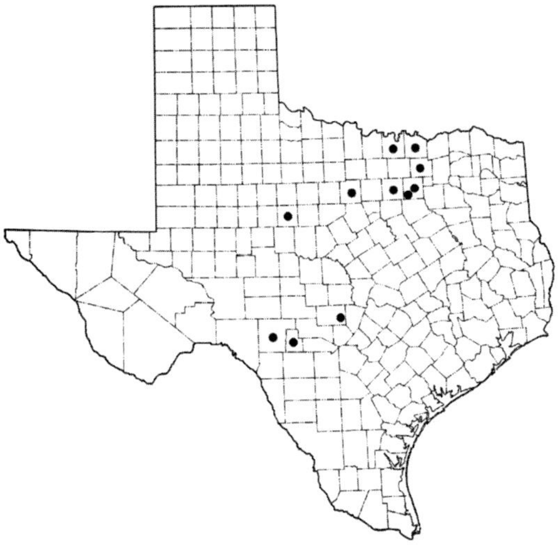

Desmodium tweedyi

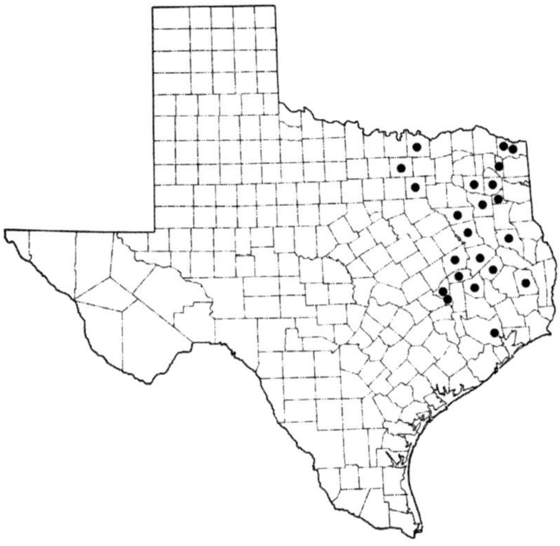

Desmodium viridiflorum

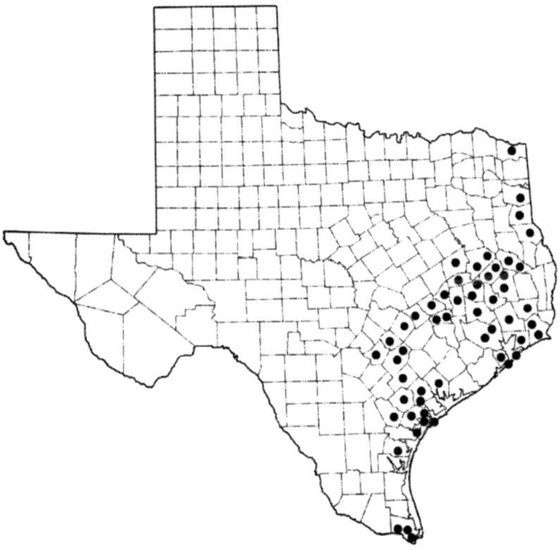

Erythrina herbacea

Eysenhardtia spinosa

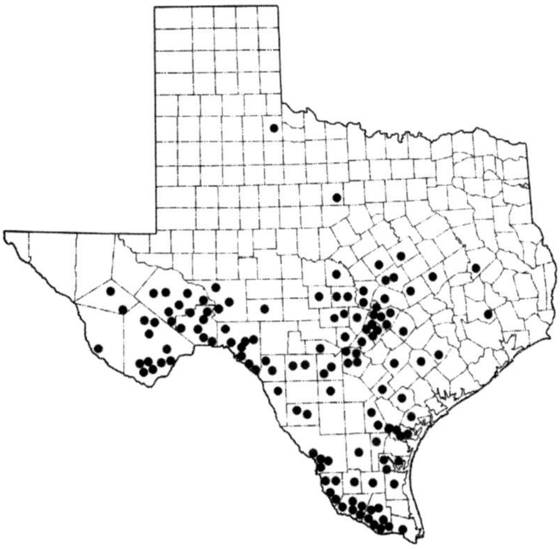

Eysenhardtia texana

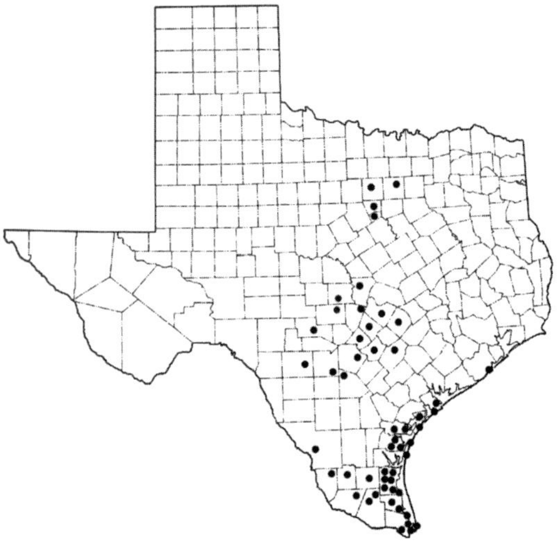

Galactia canescens

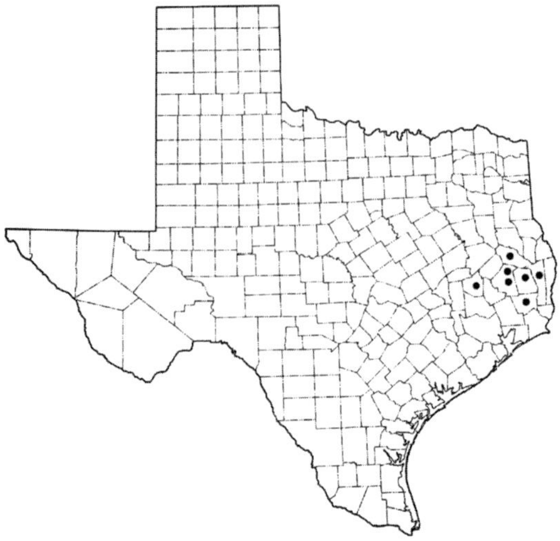

Galactia erecta

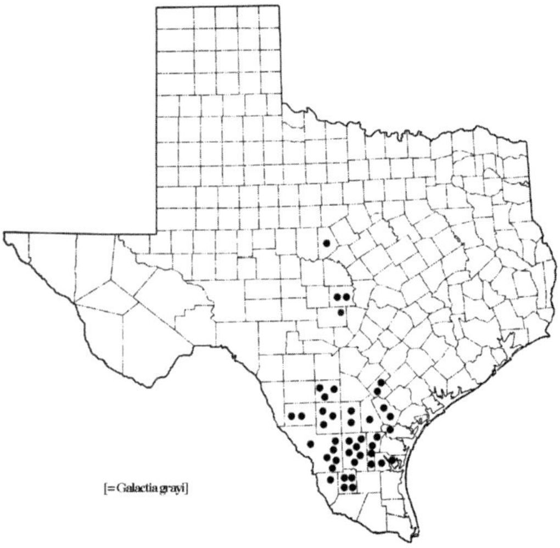
[= Galactia grayi]

Galactia heterophylla

Galactia longifolia

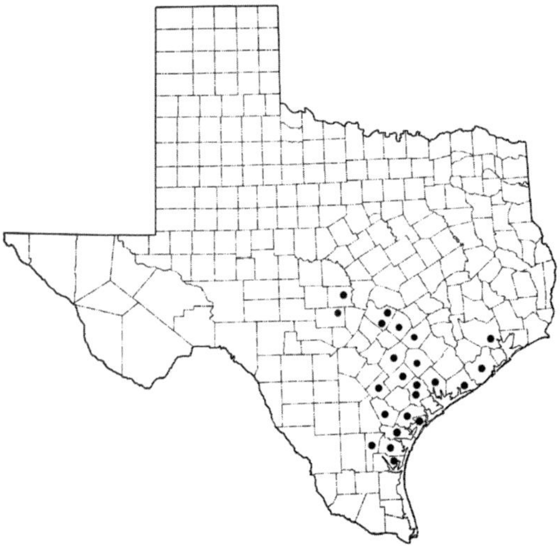

Galactia marginalis

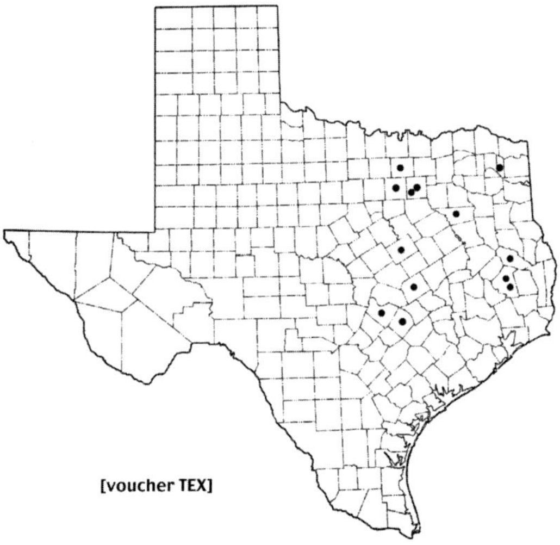
[voucher TEX]

Galactia striata

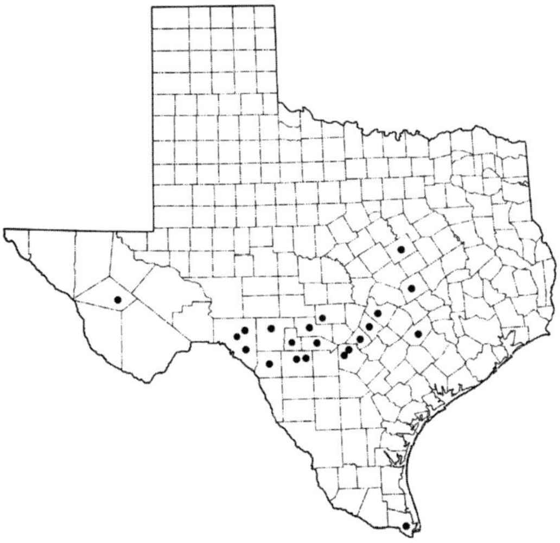

Galactia texana

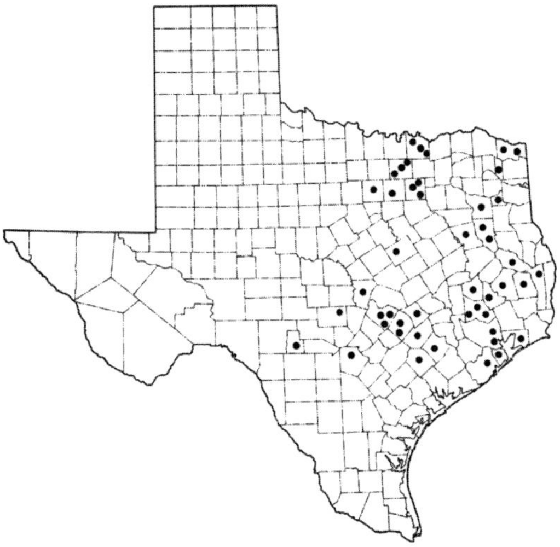

Galactia volubilis

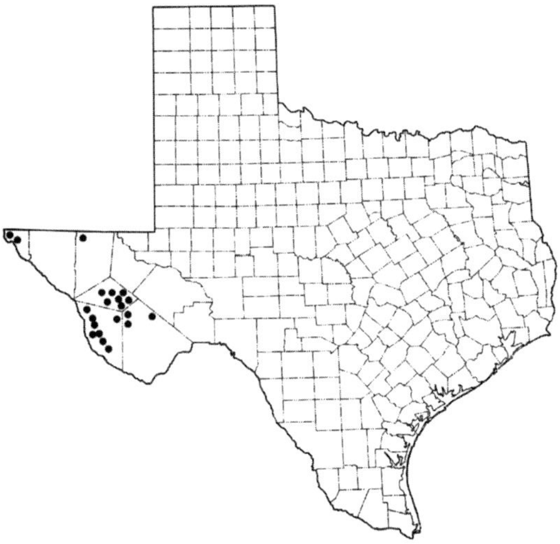

Galactia wrightii

Genistidium dumosum

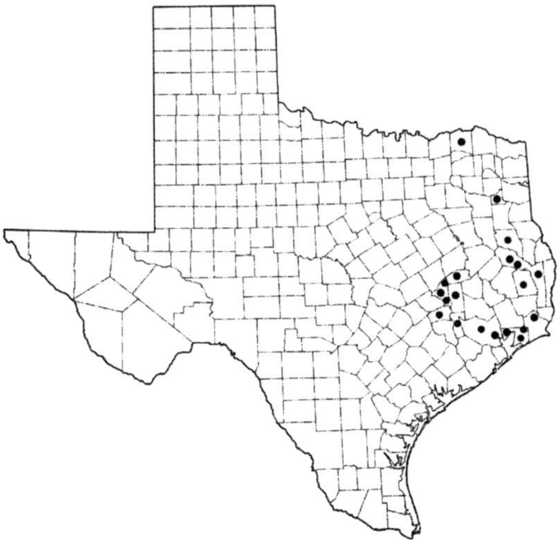

Gleditsia aquatica

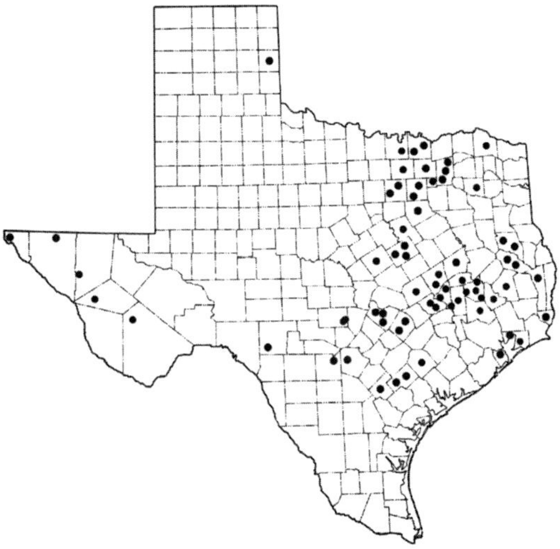

Gleditsia triacanthos

Gleditsia x texana

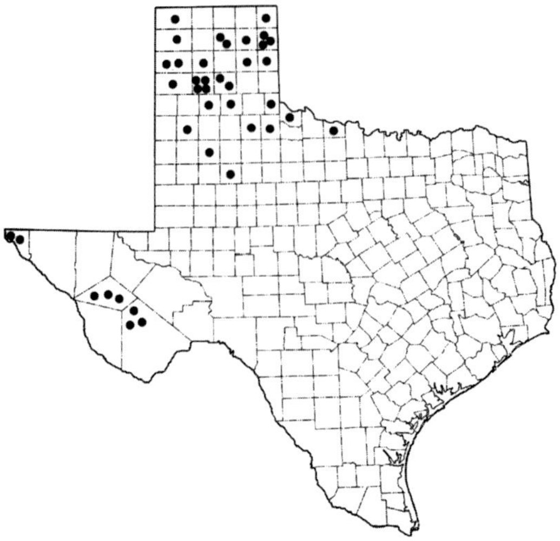

Glychrrhiza lepidota

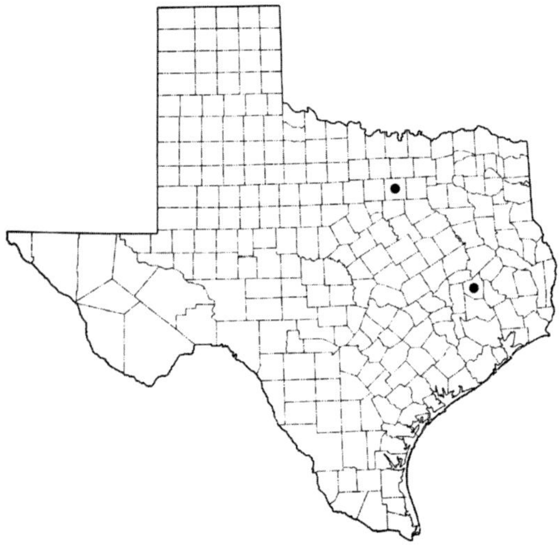

Gymnocladus dioicus

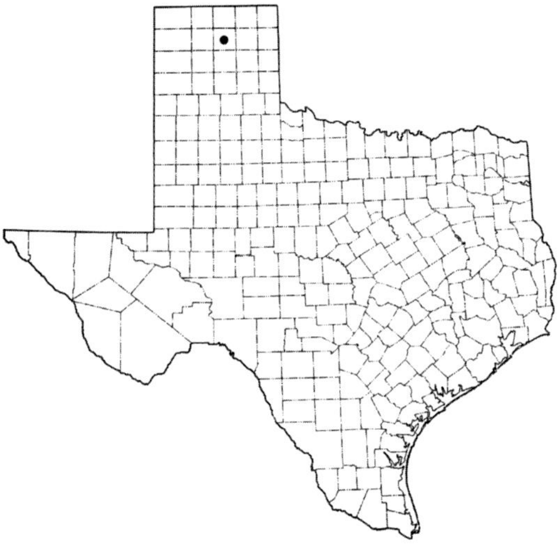

Hedysarum boreale

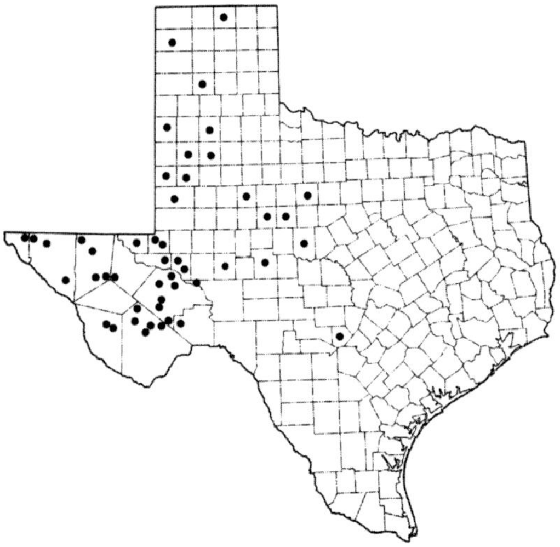

Hoffmannseggia drepanocarpa

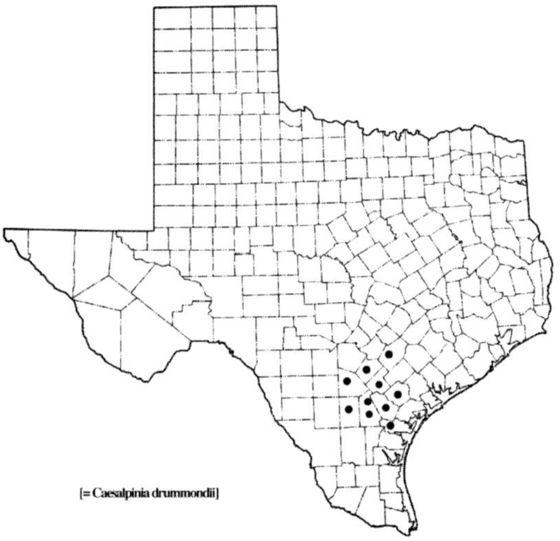
[= Caesalpinia drummondii]

Hoffmannseggia drummondii

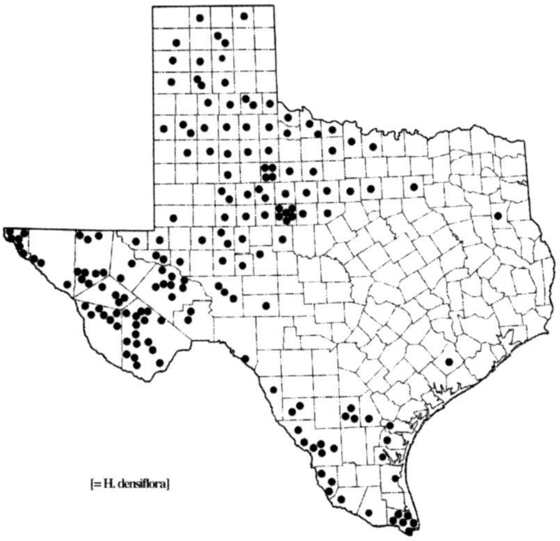
[= H. densiflora]

Hoffmannseggia glauca

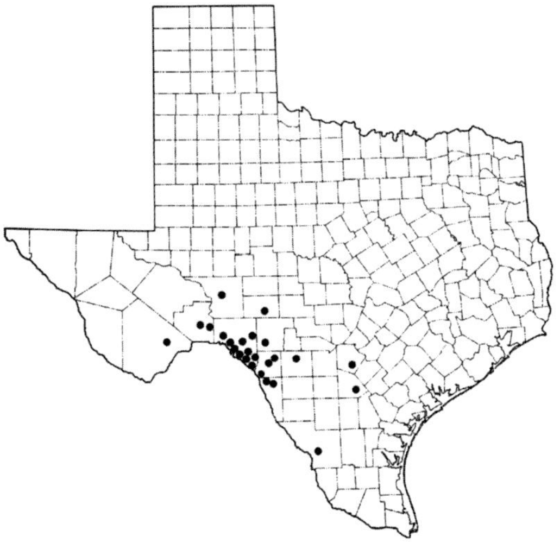

Hoffmannseggia oxycarpa

Hoffmannseggia tenella

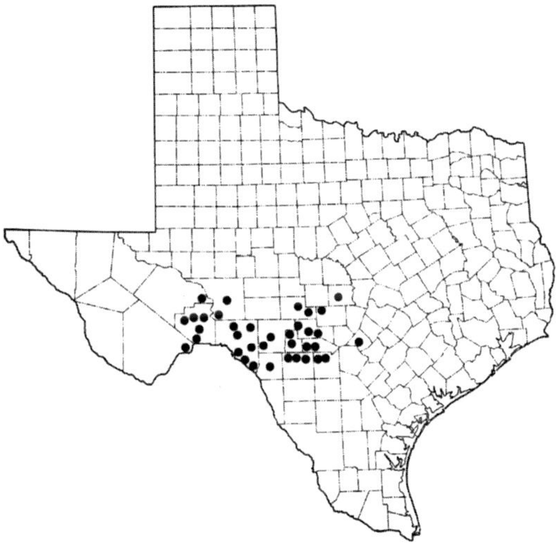

Indigofera lindheimeriana

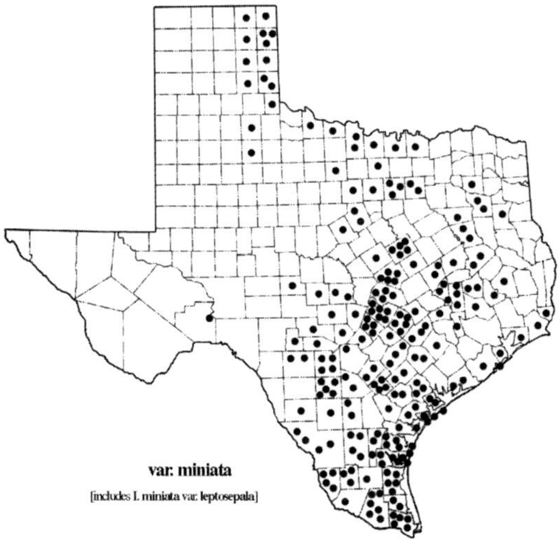
var. miniata
[includes I. miniata var. leptosepala]

Indigofera miniata

var. texana

Indigofera miniata

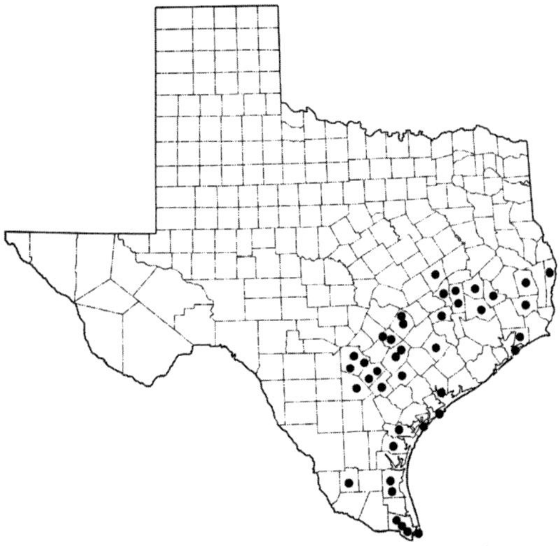

Indigofera suffruticosa

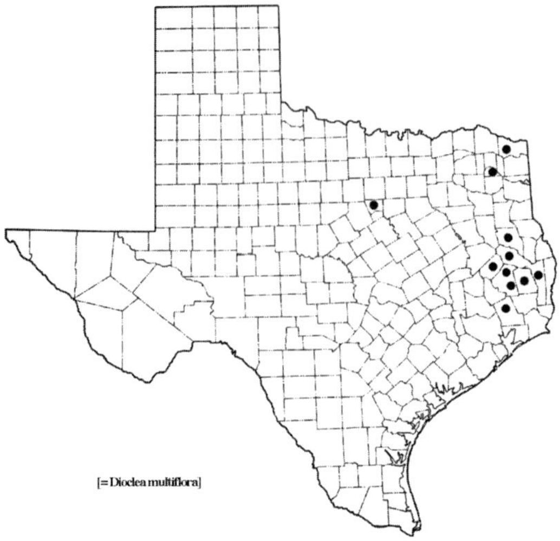
[=Dioclea multiflora]

Lackeya multiflora

Lathyrus aphaca

Lathyrus graminifolius

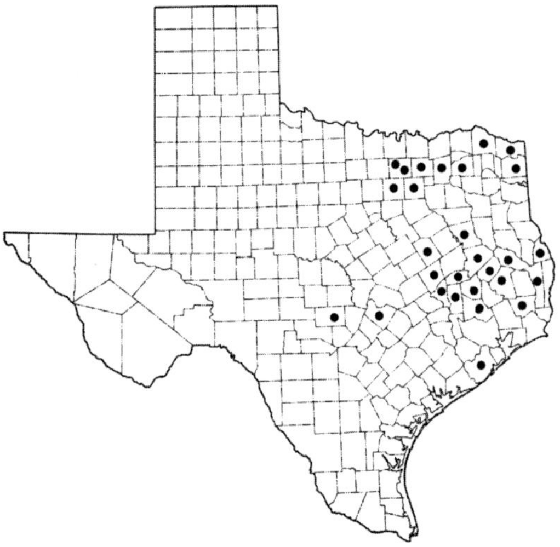

Lathyrus hirsutus

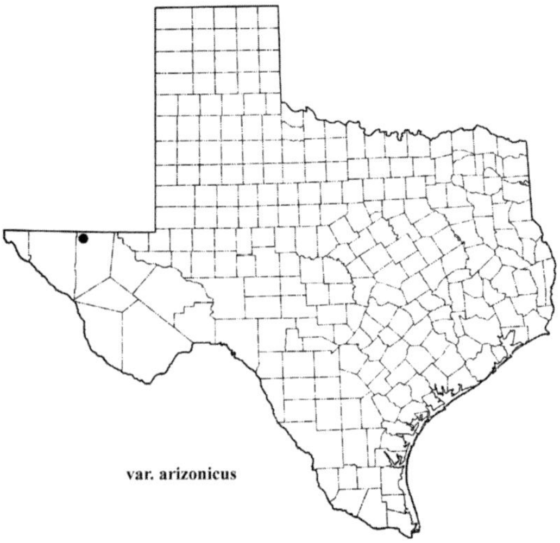
var. arizonicus

Lathyrus lanszwertii

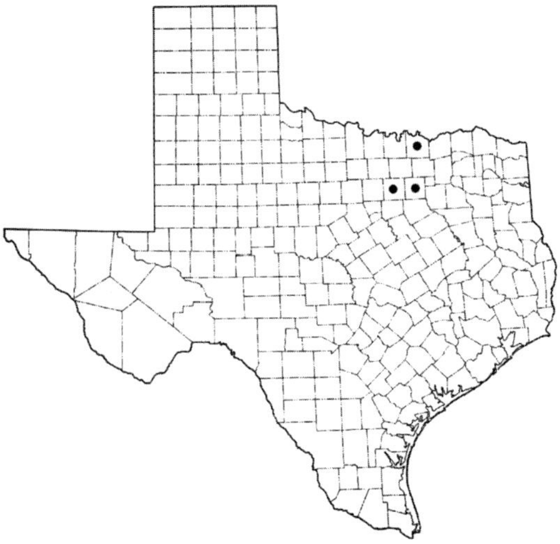
Lathyrus latifolius

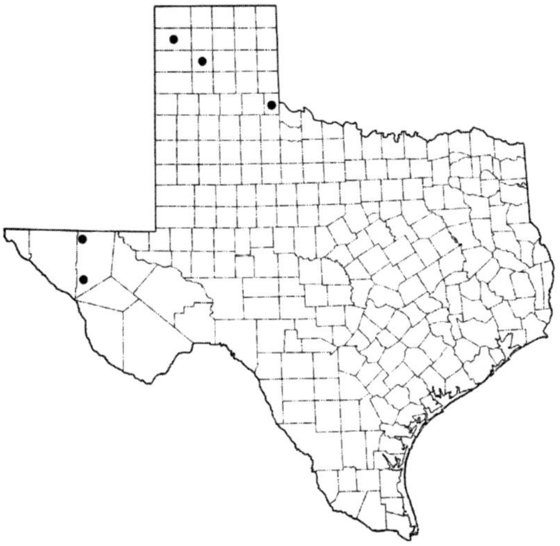
Lathyrus polymorphus

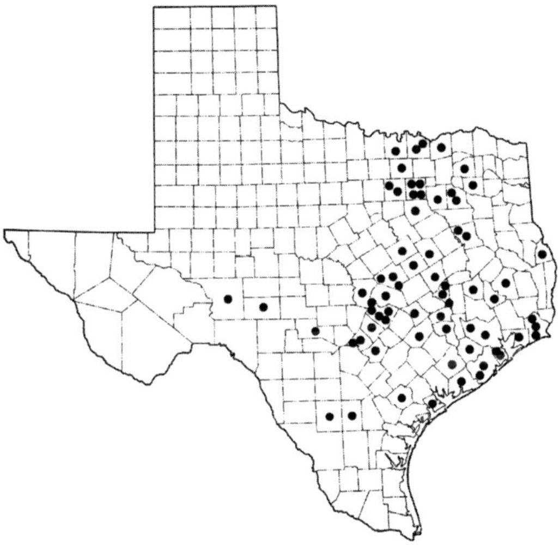
Lathyrus pusillus

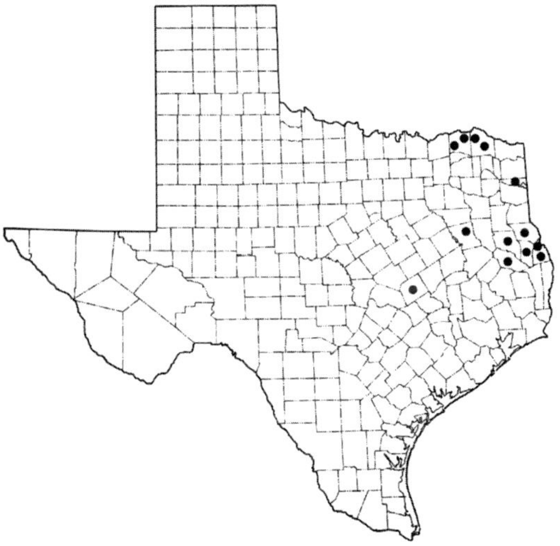
Lathyrus venosus

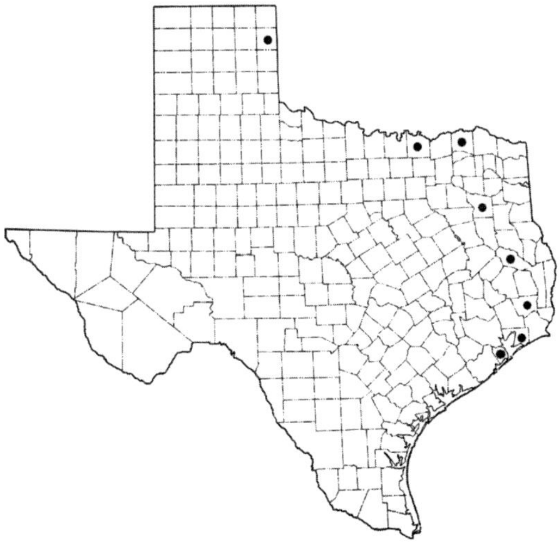
Lespedeza capitata

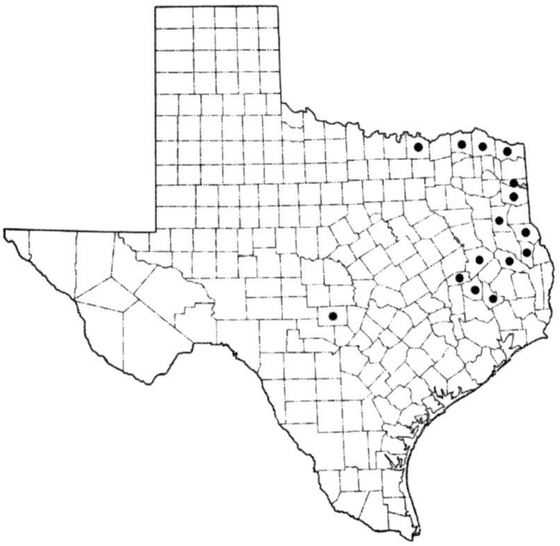
Lespedeza cuneata

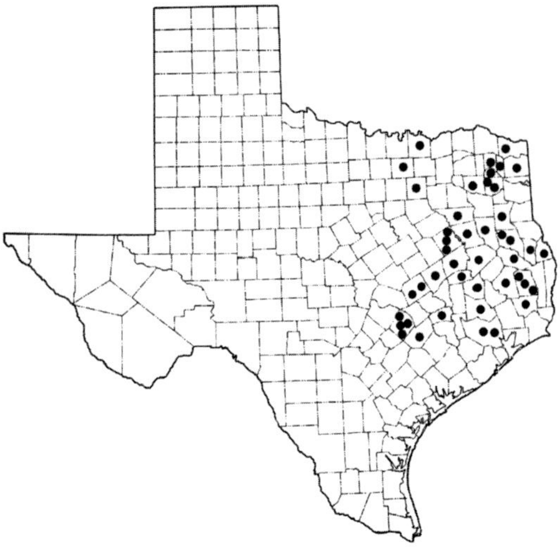

Lespedeza hirta

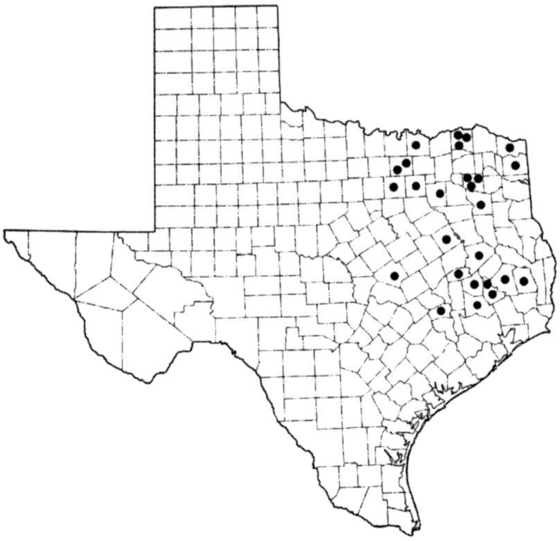

Lespedeza procumbens

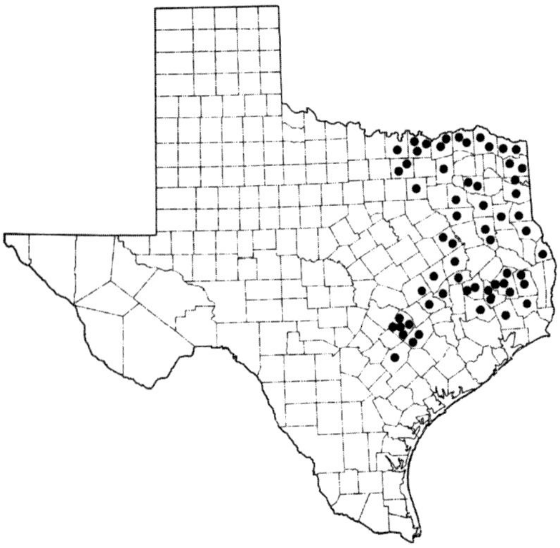

Lespedeza repens

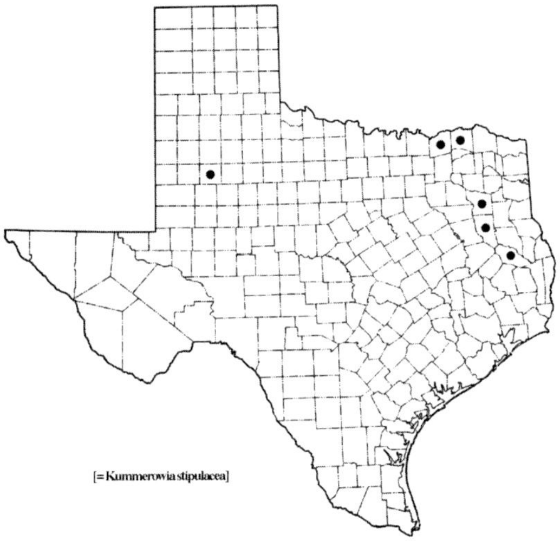
[= Kummerowia stipulacea]

Lespedeza stipulacea

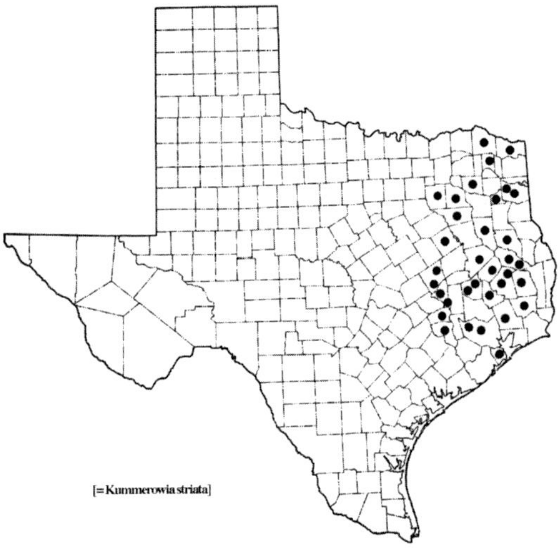
[= Kummerowia striata]

Lespedeza striata

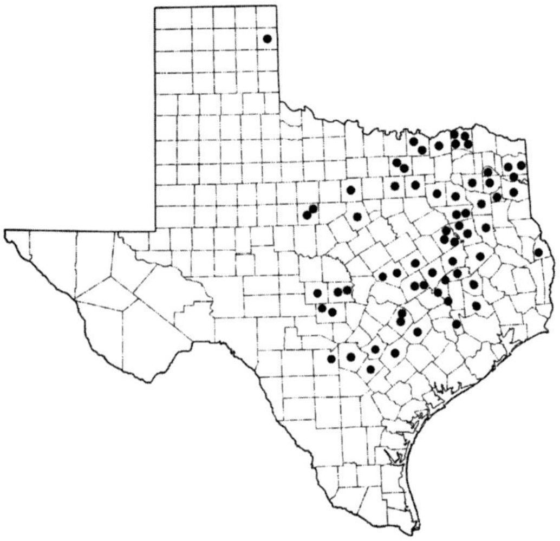

Lespedeza stuevei

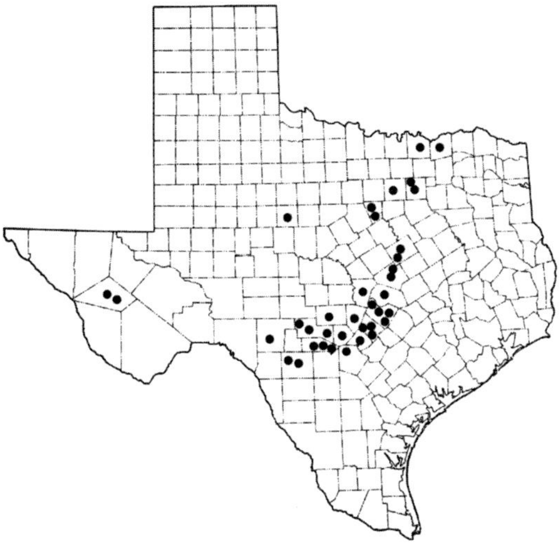

Lespedeza texana

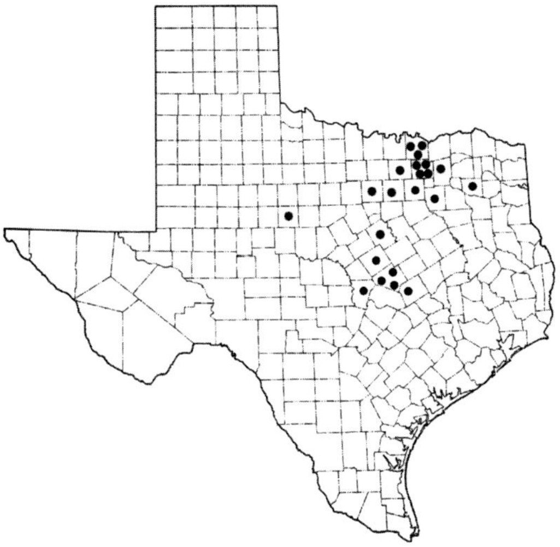

Lespedeza violacea

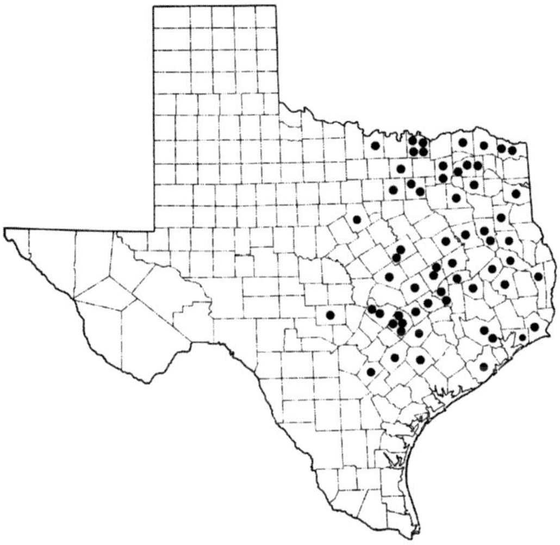

Lespedeza virginica

Leucaena leucocephala

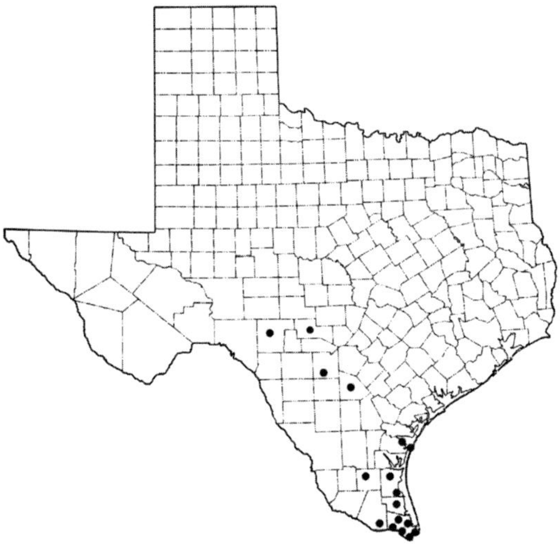

Leucaena pulverulenta

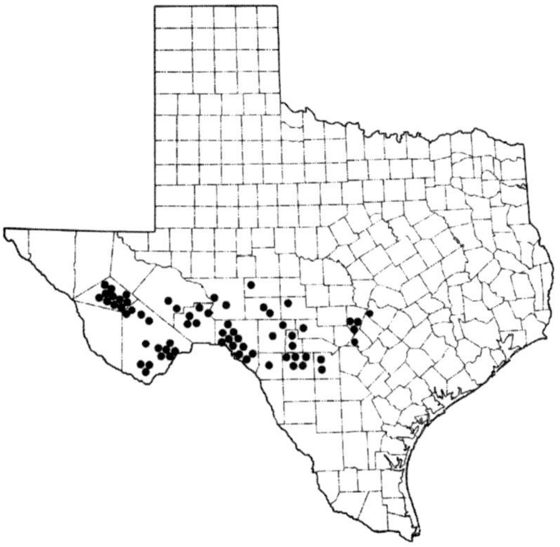

Leucaena retusa

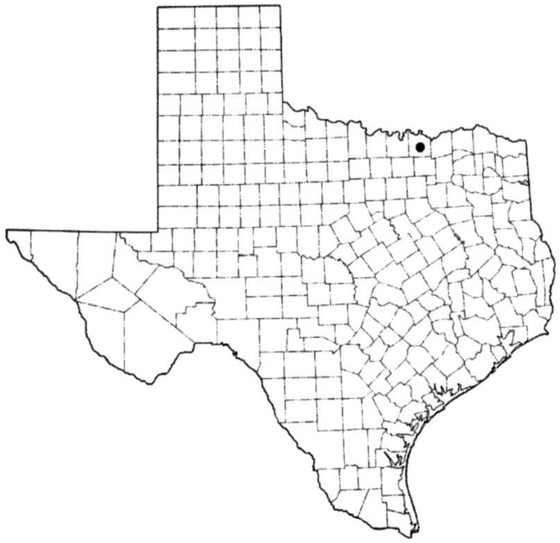

Lotus corniculatus

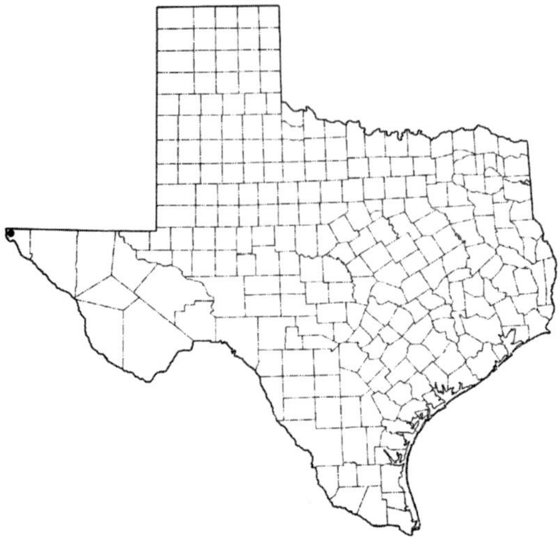

Lotus humistratus

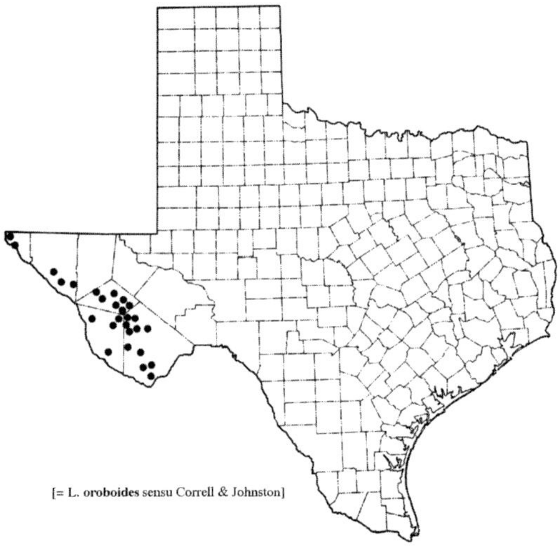
[= L. **oroboides** sensu Correll & Johnston]

Lotus plebeius

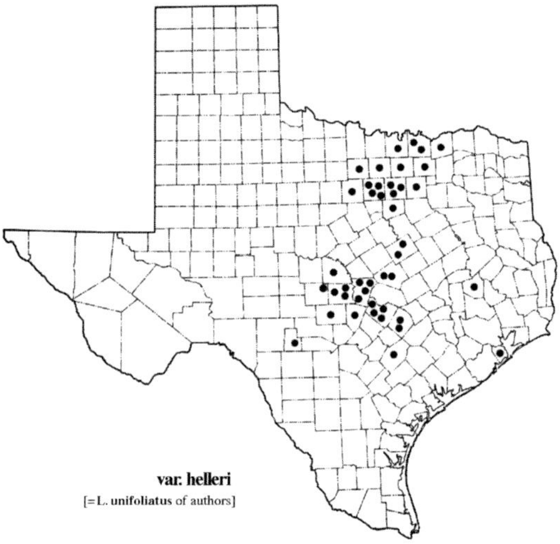
var. helleri
[=L. **unifoliatus** of authors]

Lotus purshianus

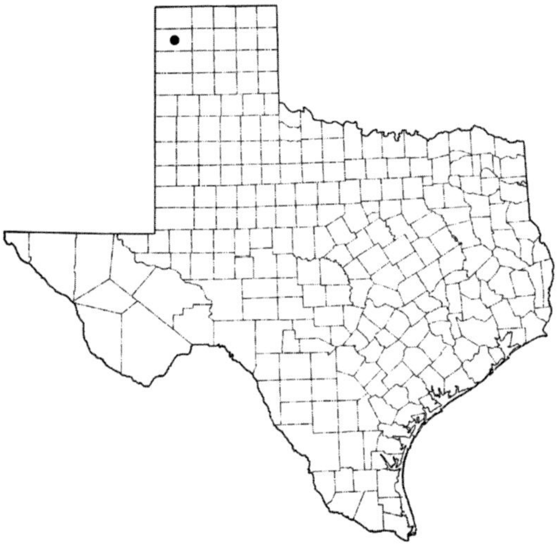

Lupinus argenteus

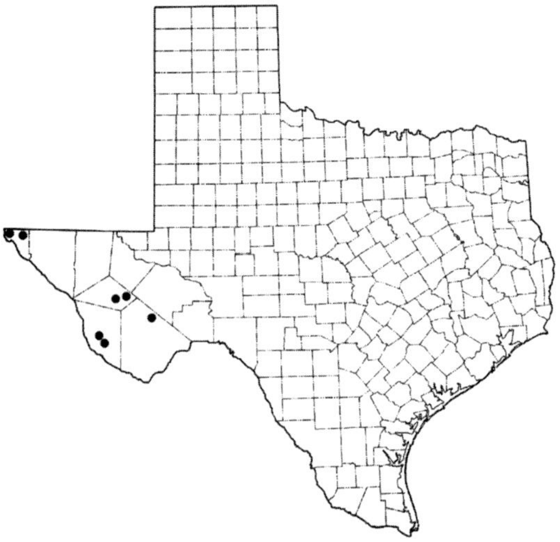

Lupinus concinnus

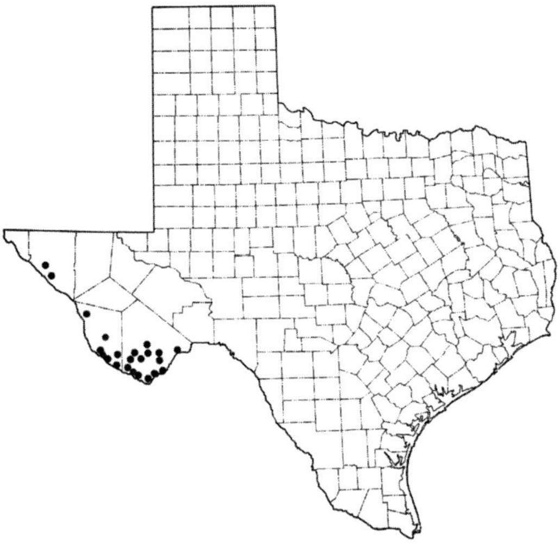

Lupinus havardii

Lupinus perennis

Lupinus plattensis

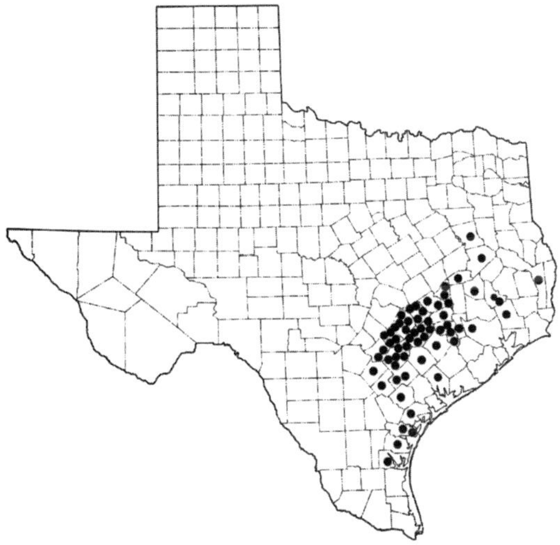

Lupinus subcarnosus

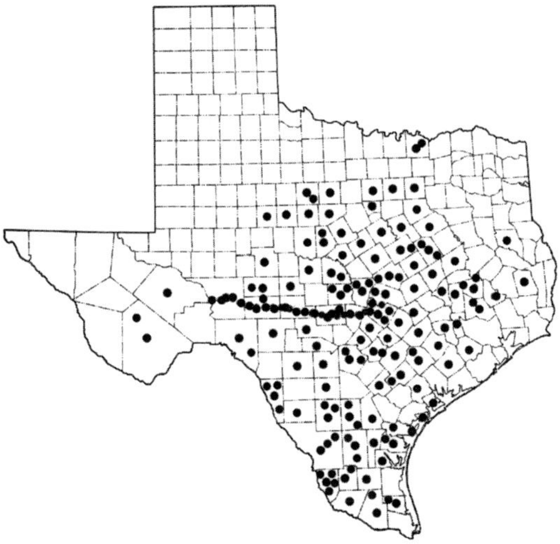

Lupinus texensis

Macroptilium atropurpureum

Macroptilium gibbosifolium

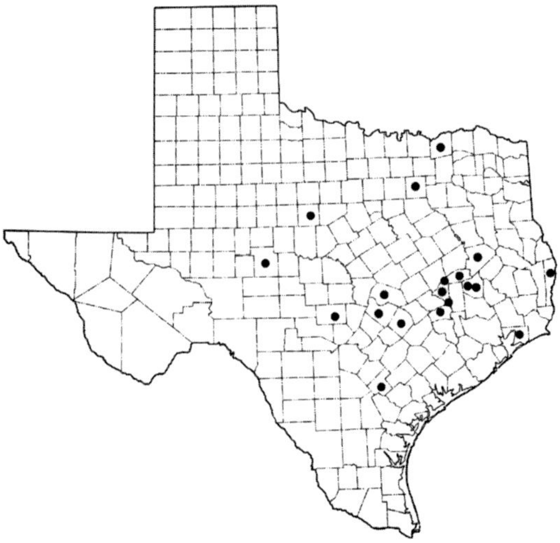

Medicago arabica

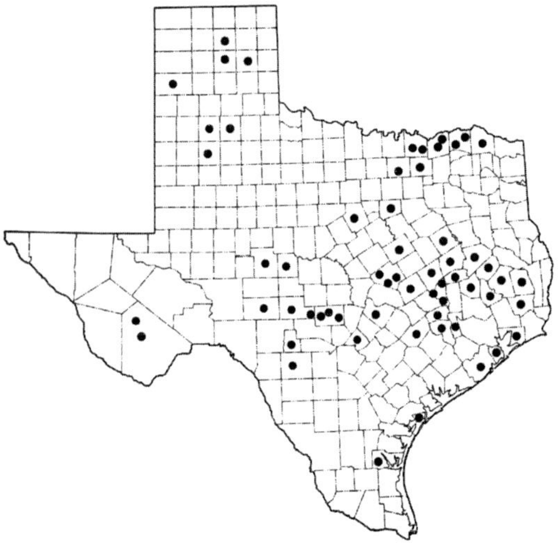

Medicago lupulina

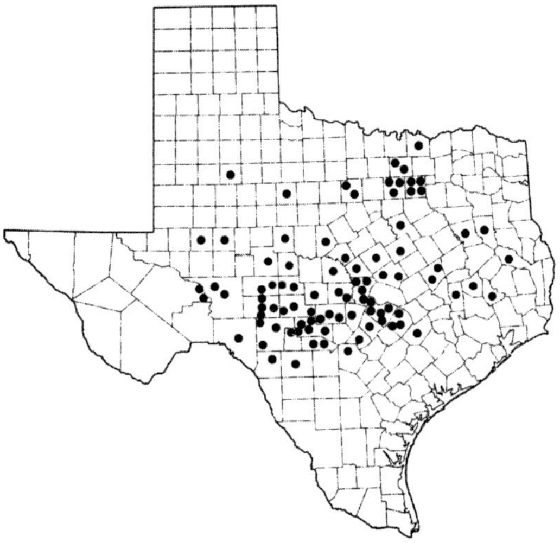

Medicago minima

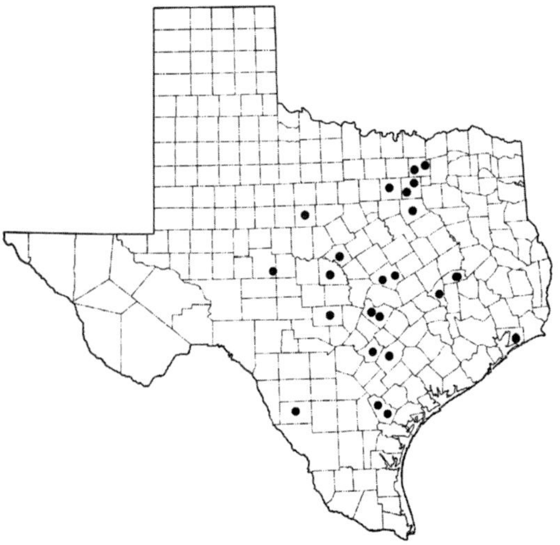

Medicago orbicularis

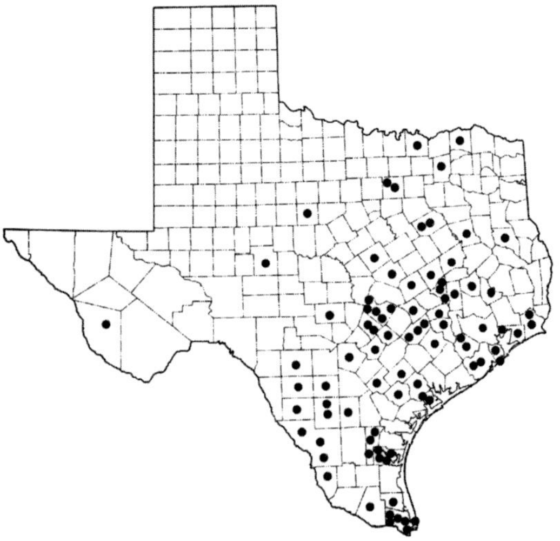

Medicago polymorpha

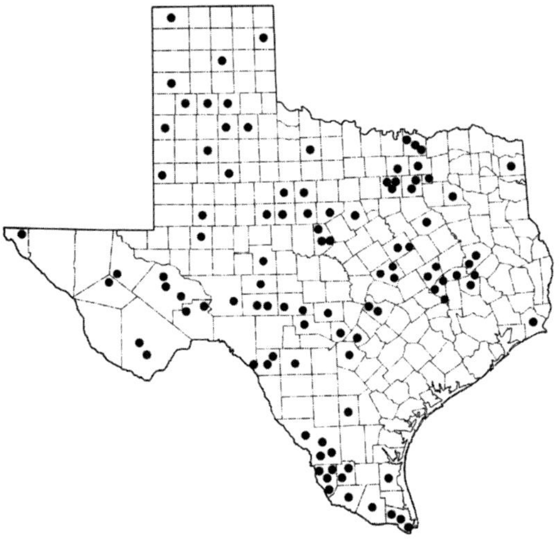

Medicago sativa

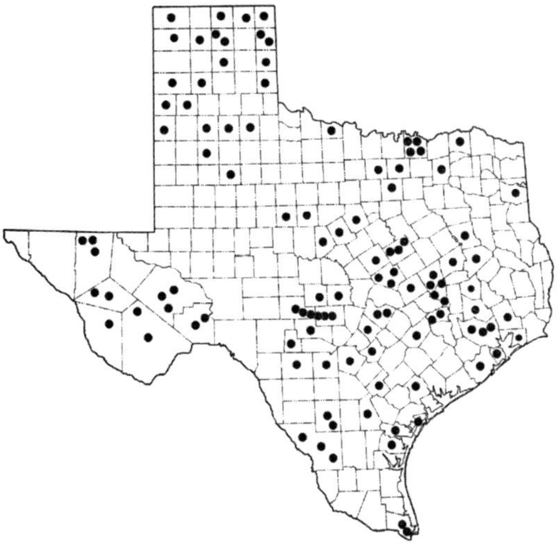

Melilotus albus

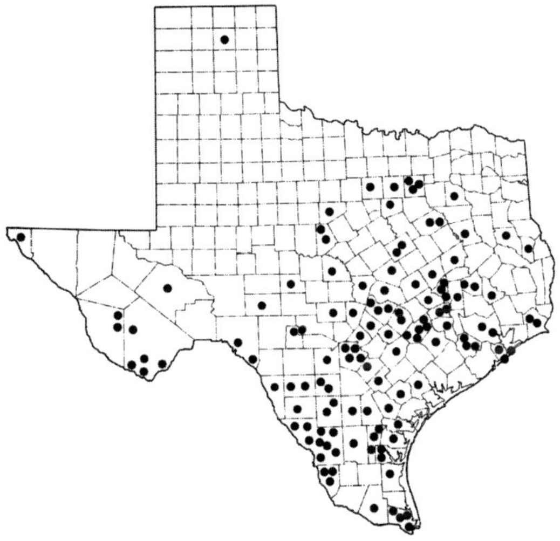

Melilotus indicus

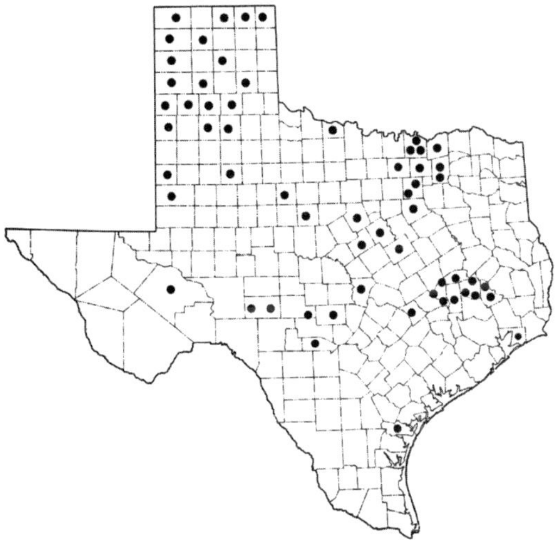

Melilotus officinalis

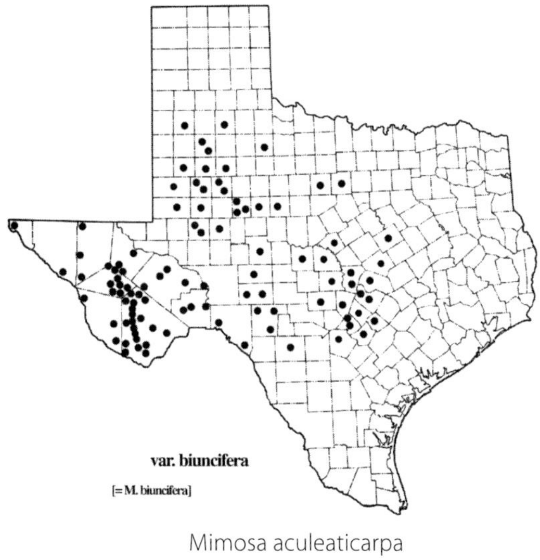
var. biuncifera
[= M. biuncifera]

Mimosa aculeaticarpa

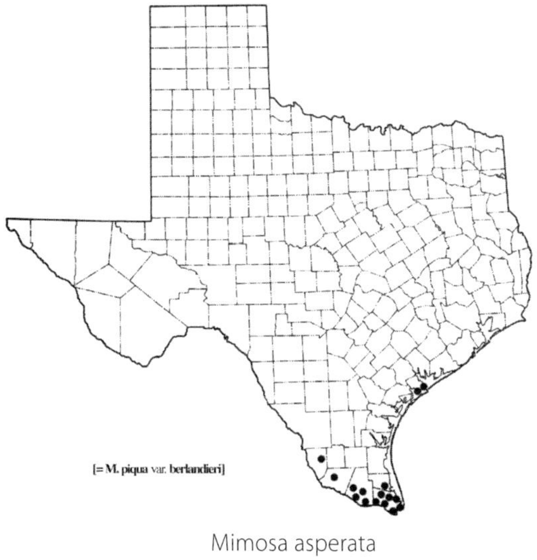
[= M. piqua var. berlandieri]

Mimosa asperata

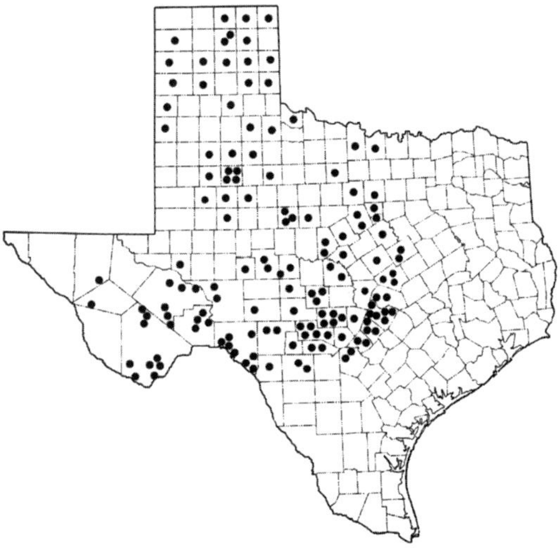

Mimosa borealis

Mimosa dysocarpa

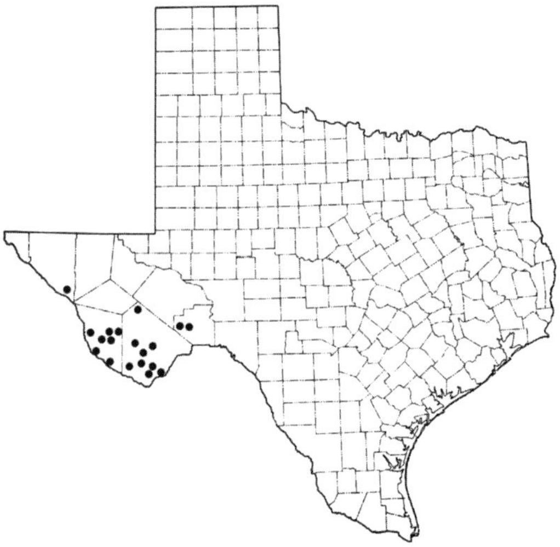

Mimosa emoryana

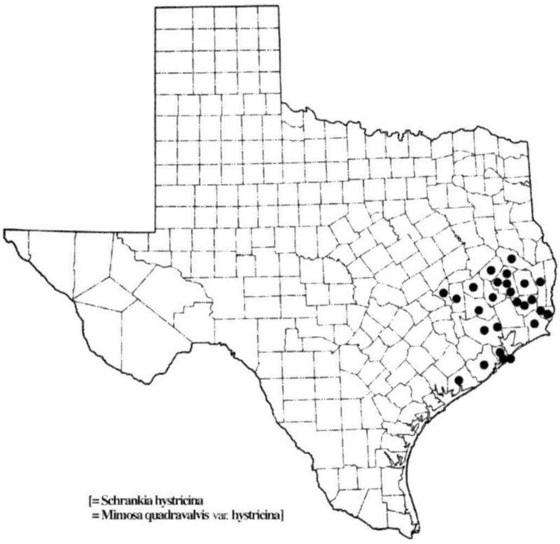
[= Schrankia hystricina
= Mimosa quadravalvis var. hystricina]

Mimosa hystricina

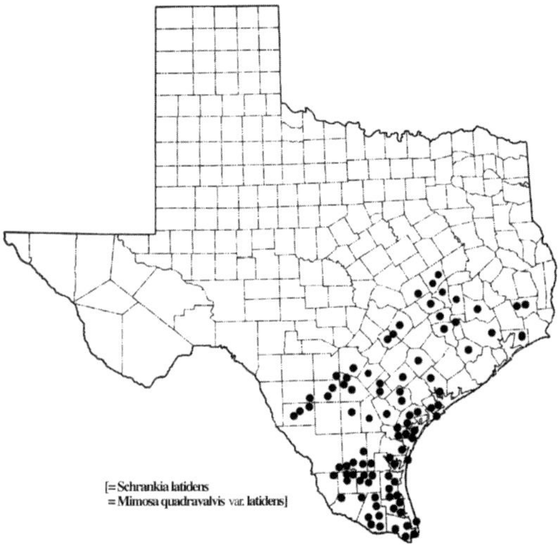
[= Schrankia latidens
= Mimosa quadravalvis var. latidens]

Mimosa latidens

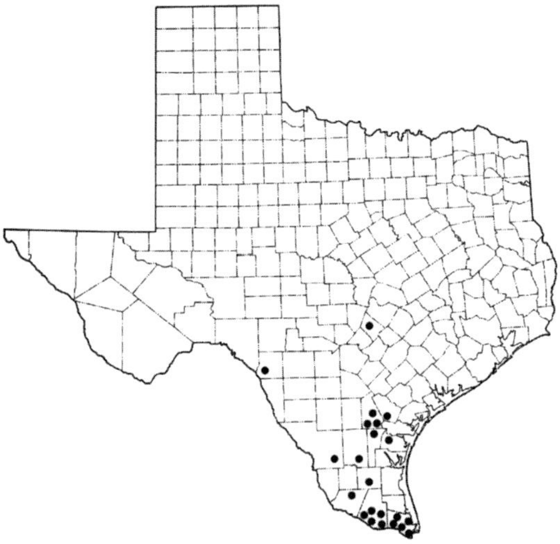

Mimosa malacophylla

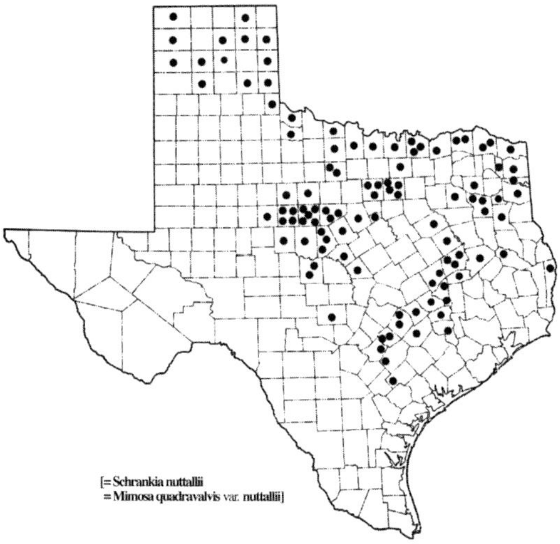
[= Schrankia nuttallii
= Mimosa quadravalvis var. nuttallii]

Mimosa nuttallii

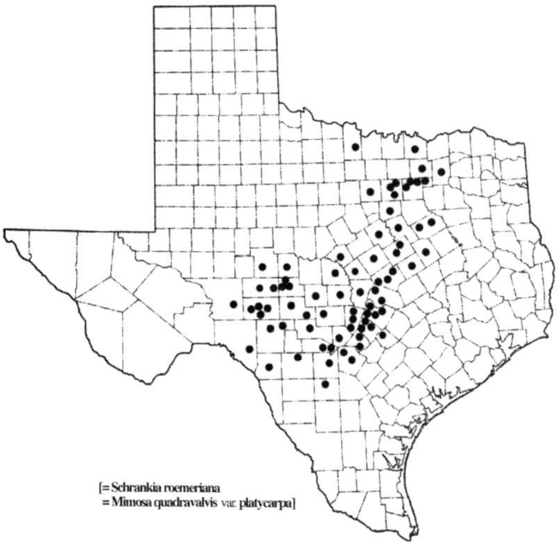
[= Schrankia roemeriana
= Mimosa quadravalvis var. platycarpa]

Mimosa roemeriana

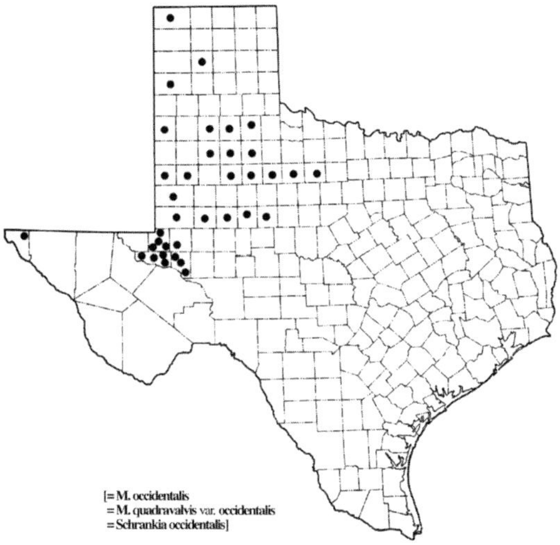
[= M. occidentalis
= M. quadravalvis var. occidentalis
= Schrankia occidentalis]

Mimosa rupertiana

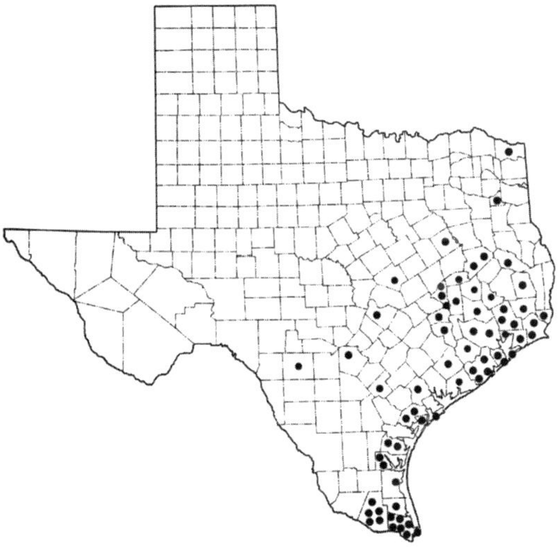

Mimosa strigillosa

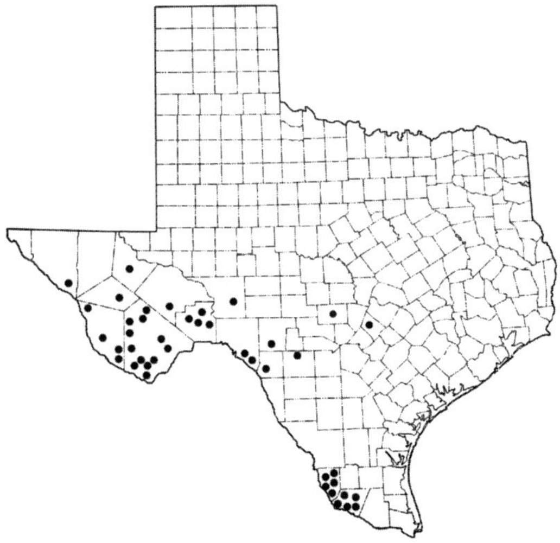

Mimosa texana

Mimosa turneri

Mimosa warnockii

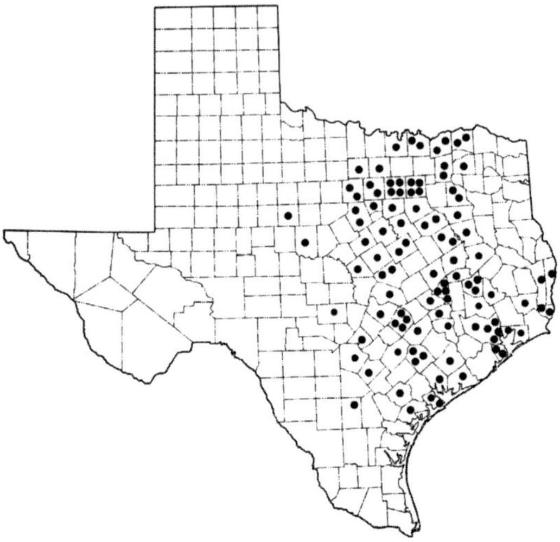

Neptunia lutea

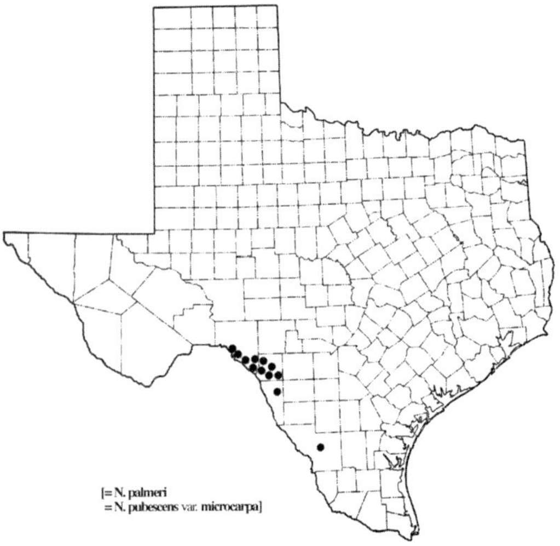
[= N. palmeri
= N. pubescens var. microcarpa]

Neptunia microcarpa

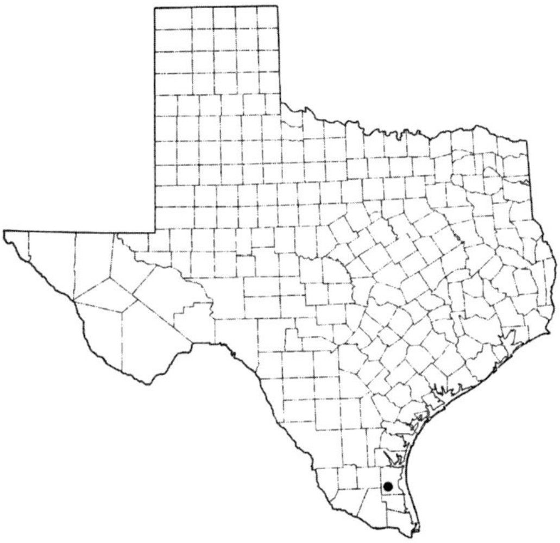

Neptunia plena

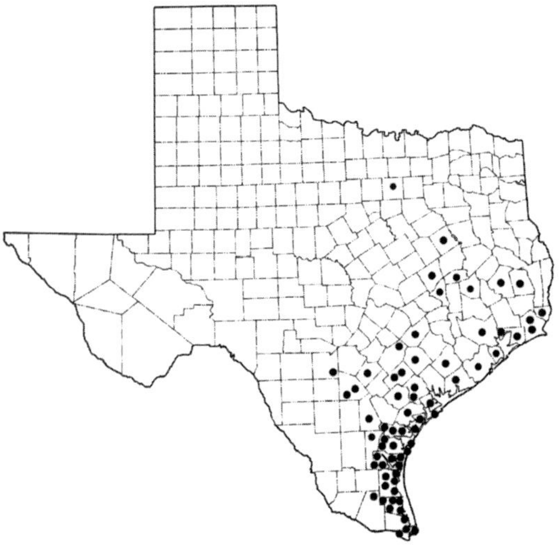

Neptunia pubescens

Nissolia platycalyx

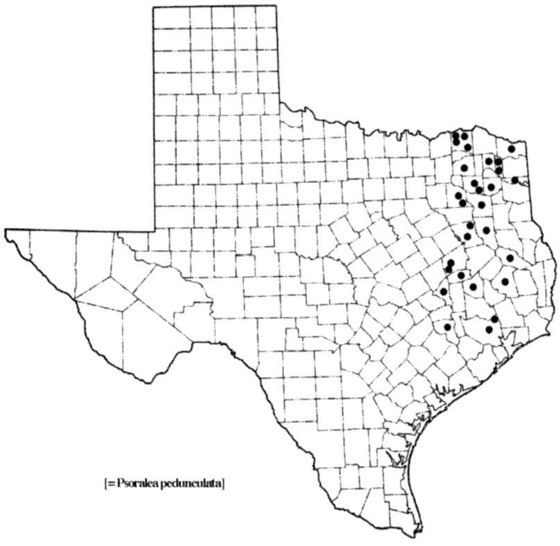
[= Psoralea pedunculata]

Orbexilum pedunculatum

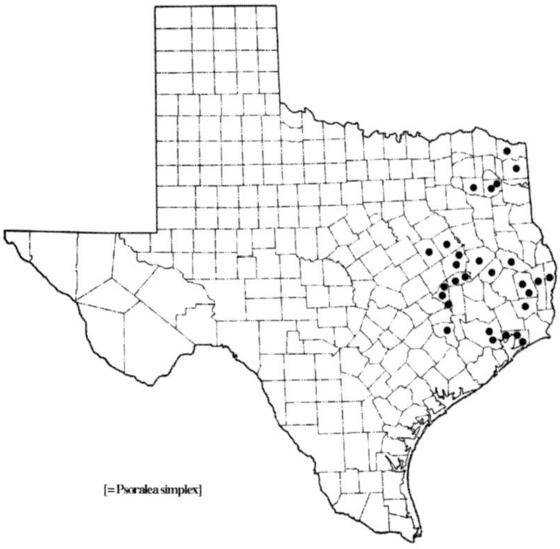
[= Psoralea simplex]

Orbexilum simplex

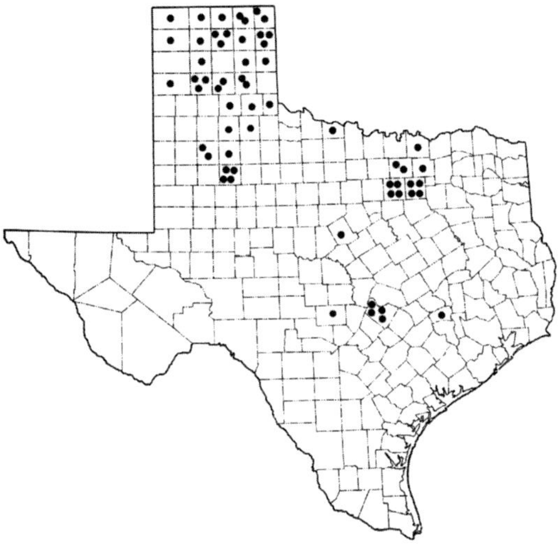

Oxytropis lambertii

Oxytropis sericea

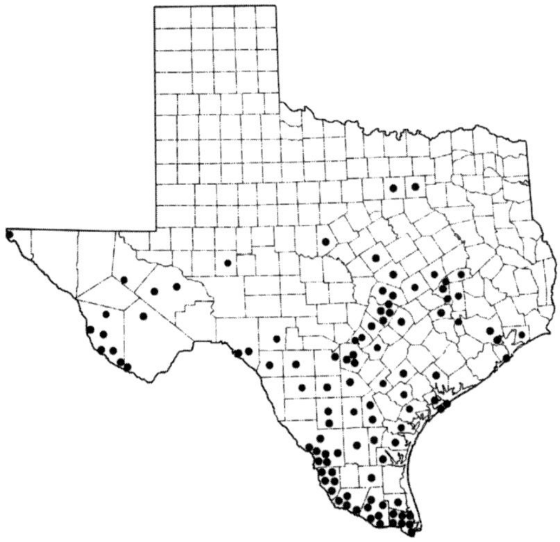

Parkinsonia aculeata

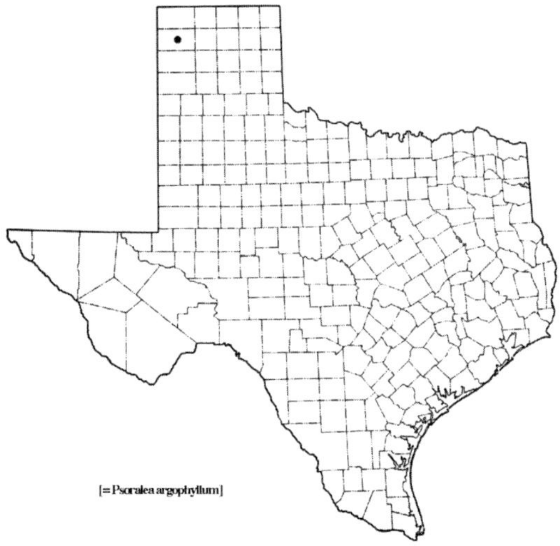
[= Psoralea argophyllum]

Pediomelum argophyllum

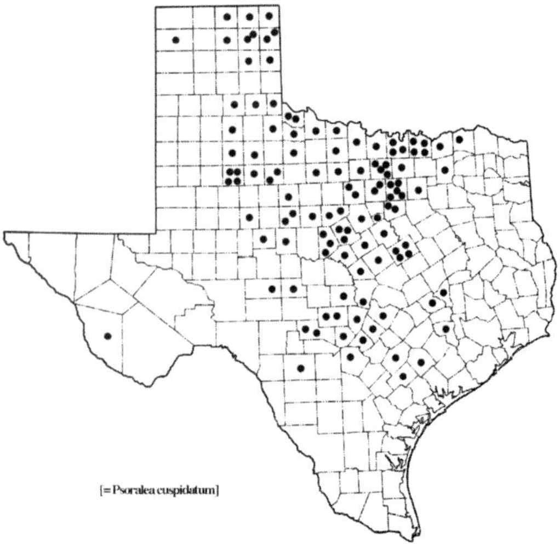
[= Psoralea cuspidatum]
Pediomelum cuspidatum

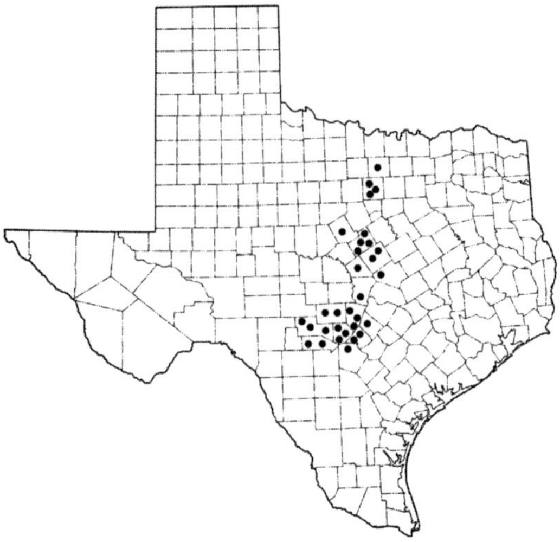
Pediomelum cyphocalyx

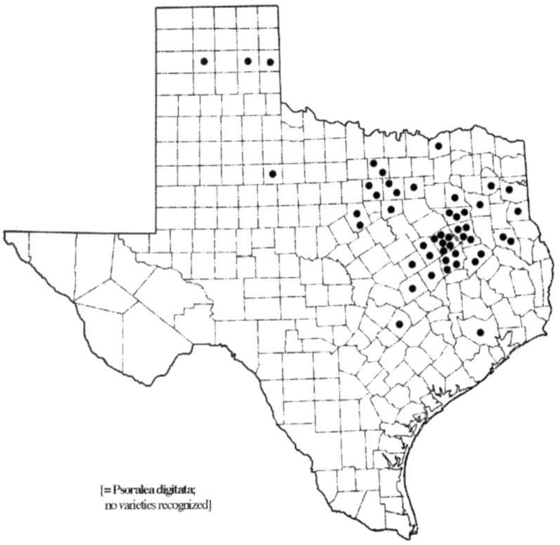
[= Psoralea digitata;
no varieties recognized]
Pediomelum digitatum

[= Psoralea humilis]
Pediomelum humile

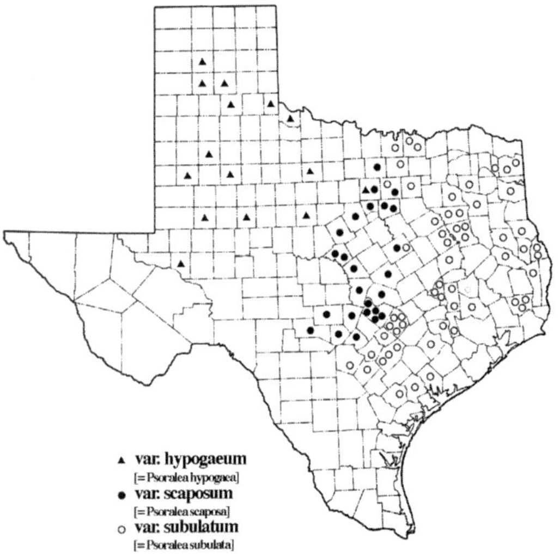
▲ var. hypogaeum
[= Psoralea hypogaea]
● var. scaposum
[= Psoralea scaposa]
○ var. subulatum
[= Psoralea subulata]
Pediomelum hypogaeum

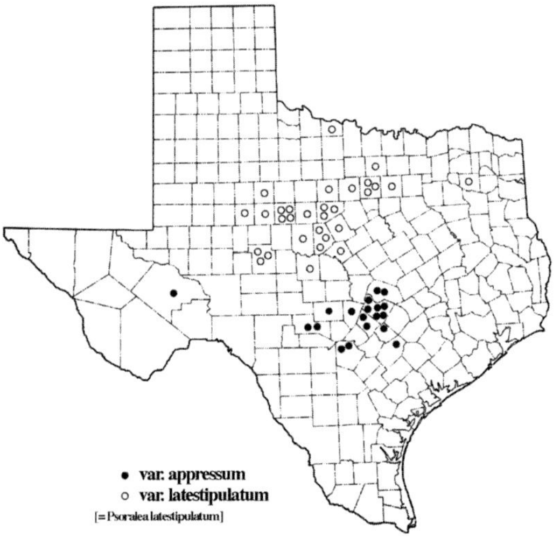
● var. appressum
○ var. latestipulatum
[= Psoralea latestipulatum]
Pediomelum latestipulatum

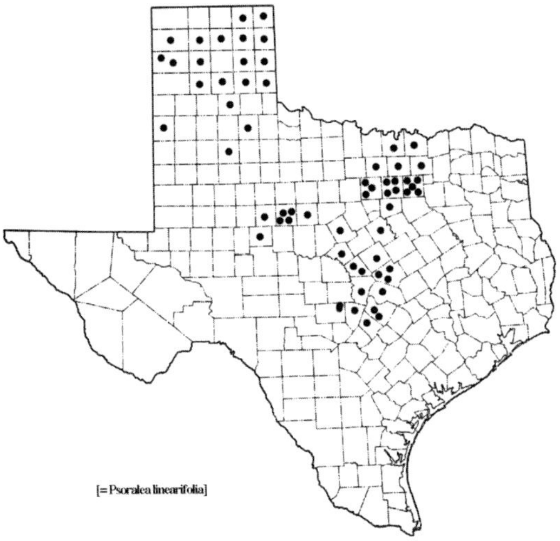
[= Psoralea linearifolia]

Pediomelum linearifolium

Pediomelum pentaphyllum

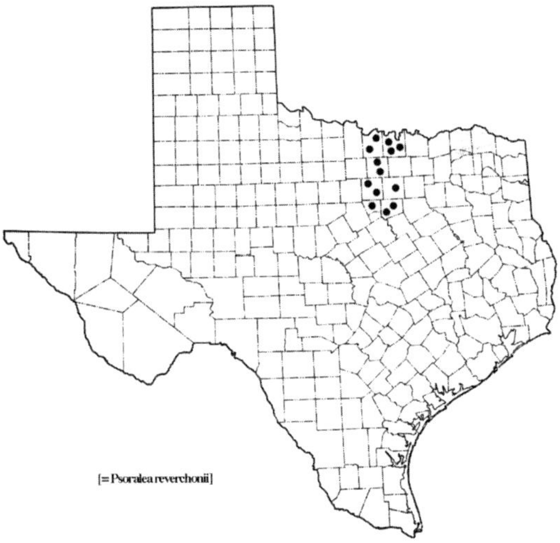
[= Psoralea reverchonii]

Pediomelum reverchonii

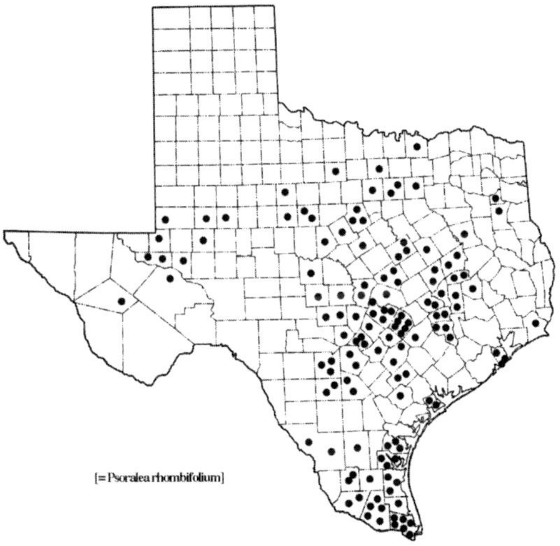
[= Psoralea rhombifolium]

Pediomelum rhombifolium

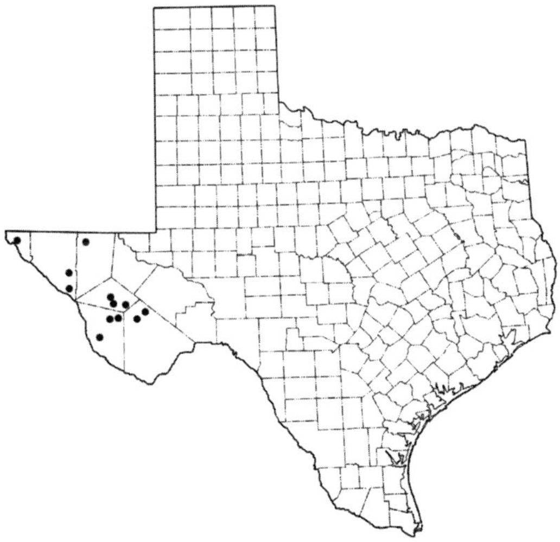

Peteria scoparia

Phaseolus acutifolius

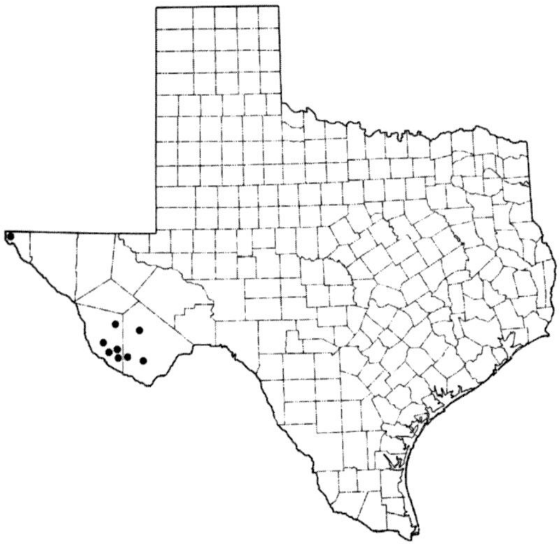

Phaseolus angustissimus

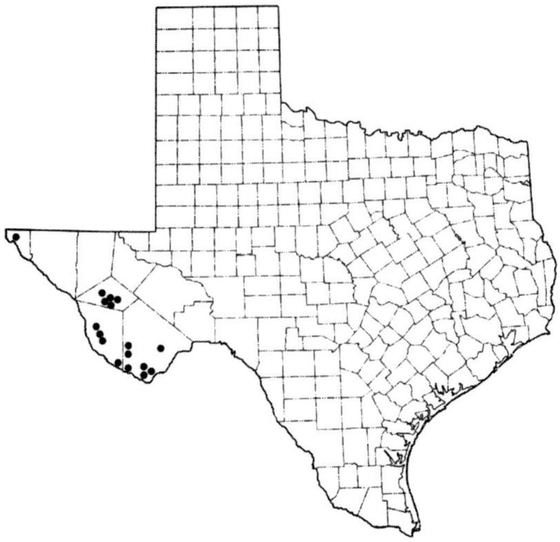

Phaseolus filiformis

Phaseolus maculatus

Phaseolus pedicellatus

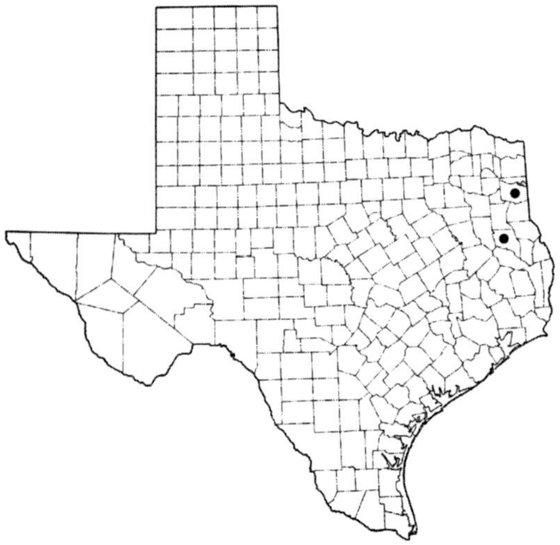

Phaseolus polystachios

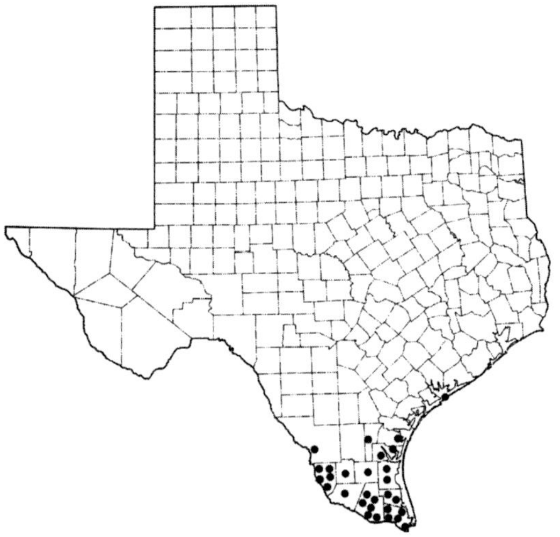

Pithecellobium ebano

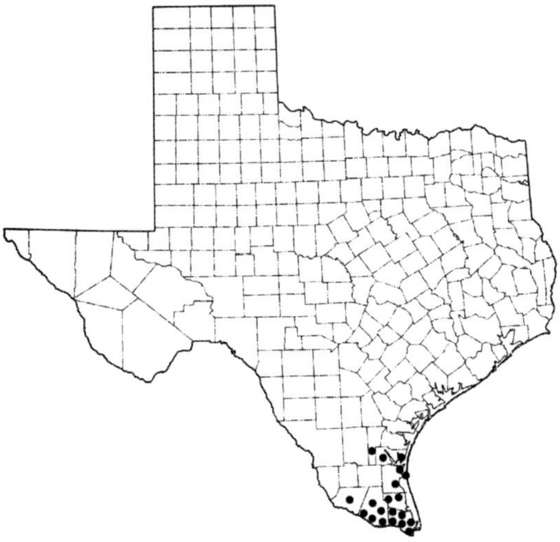

Pithecellobium pallens

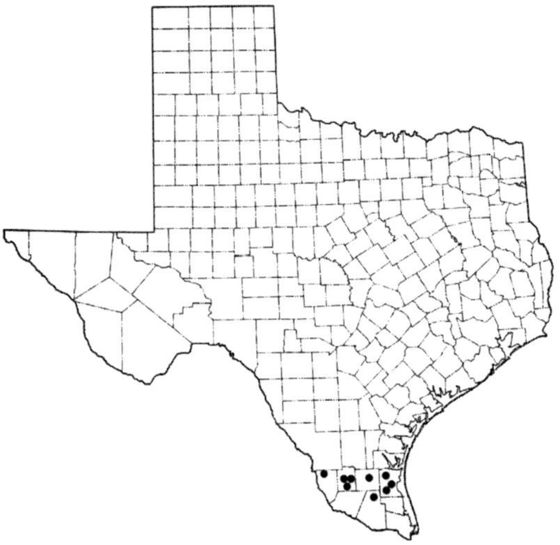

Pomaria austrotexana

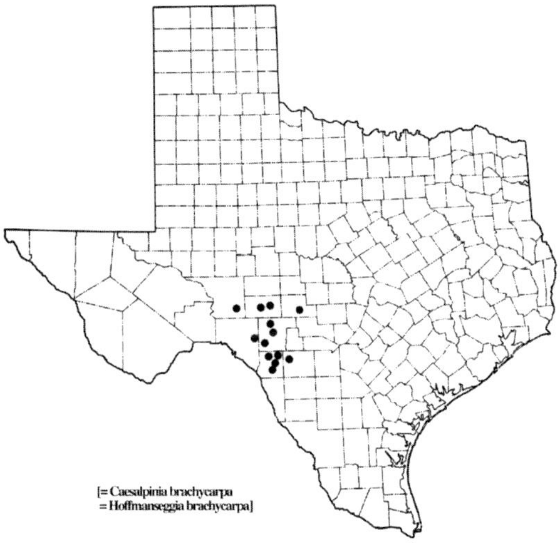
[= Caesalpinia brachycarpa
= Hoffmanseggia brachycarpa]

Pomaria brachycarpa

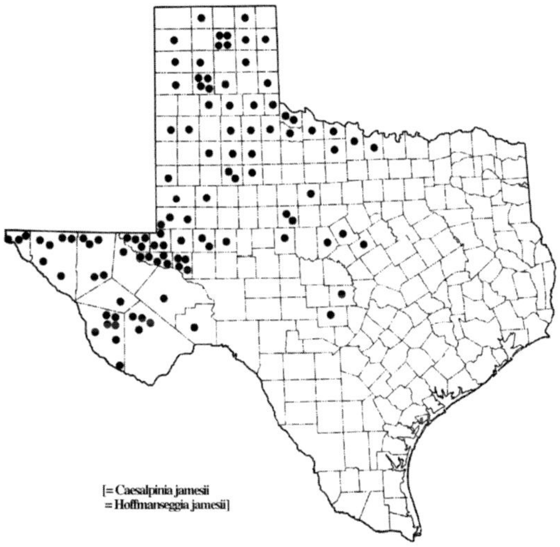
[= Caesalpinia jamesii
= Hoffmanseggia jamesii]

Pomaria jamesii

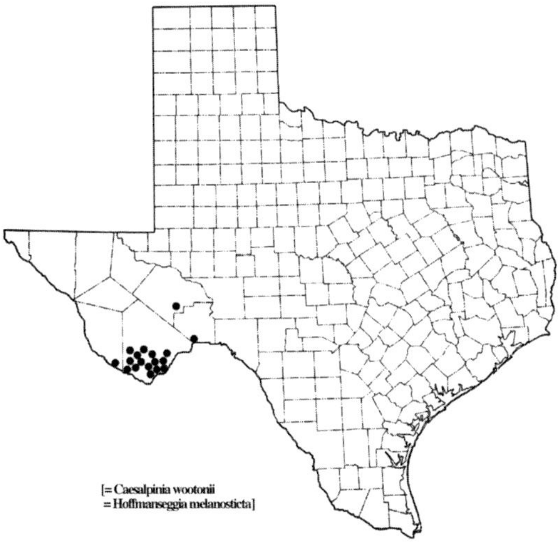
[= Caesalpinia wootonii
= Hoffmanseggia melanosticta]

Pomaria melanosticta

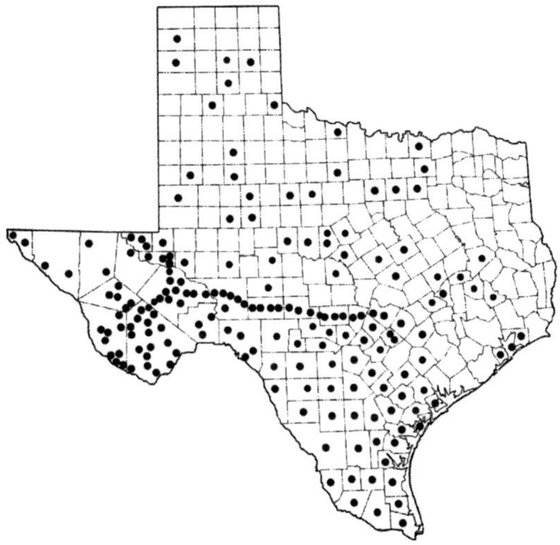

Prosopis glandulosa

Prosopis laevigata

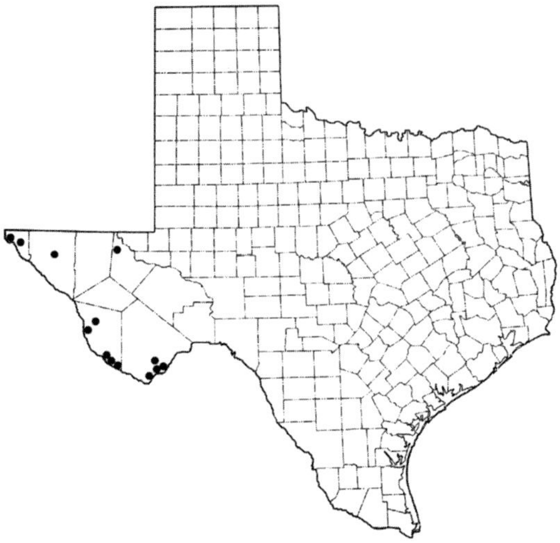

Prosopis pubescens

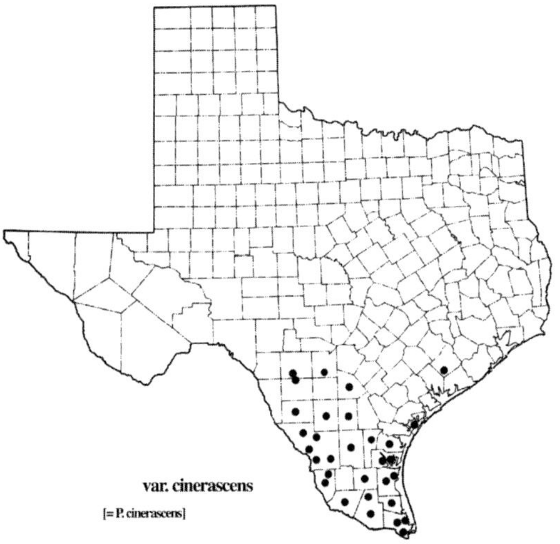
var. cinerascens
[= P. cinerascens]

Prosopis reptans

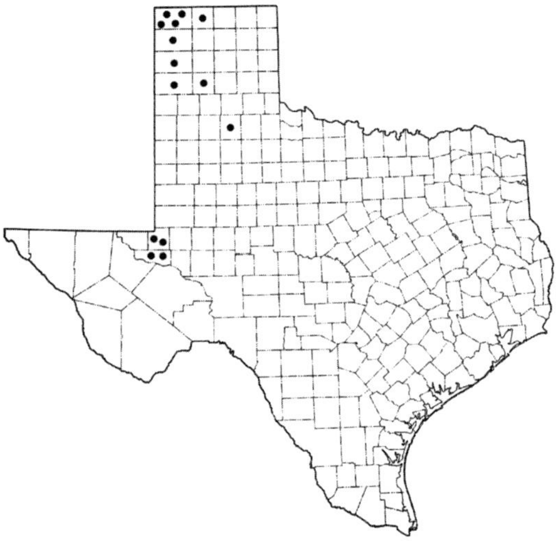

Psoralidium lanceolatum

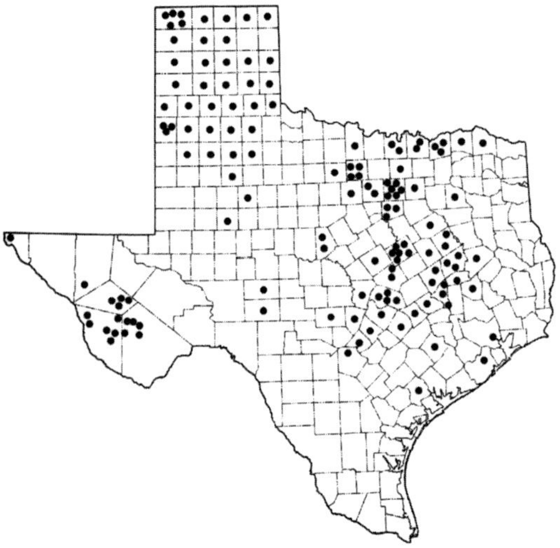

Psoralidium tenuiflorum

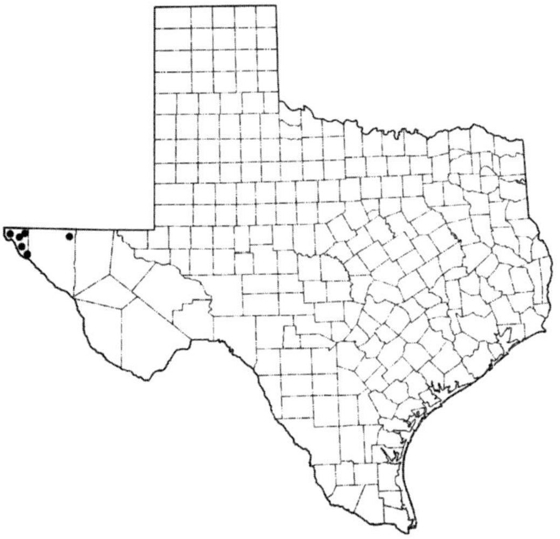

Psorothamnus scoparius

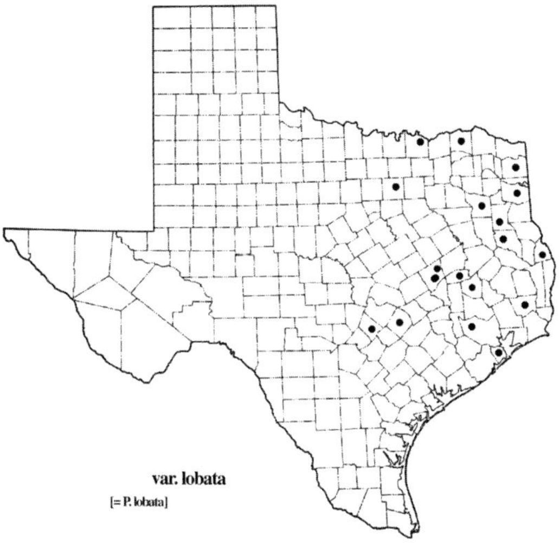
var. lobata
[= P. lobata]

Pueraria montana

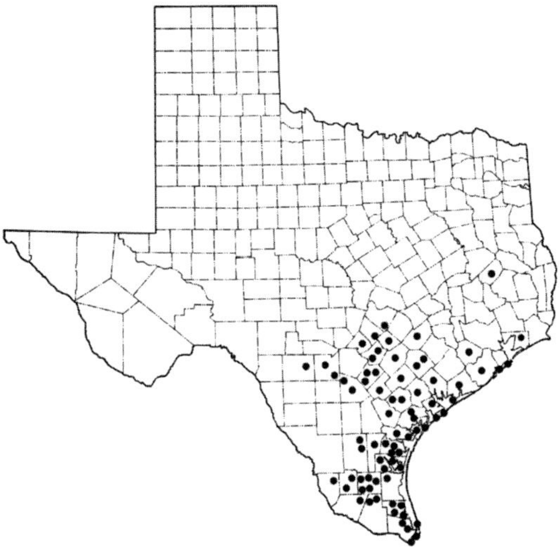

Rhynchosia americana

Rhynchosia difformis

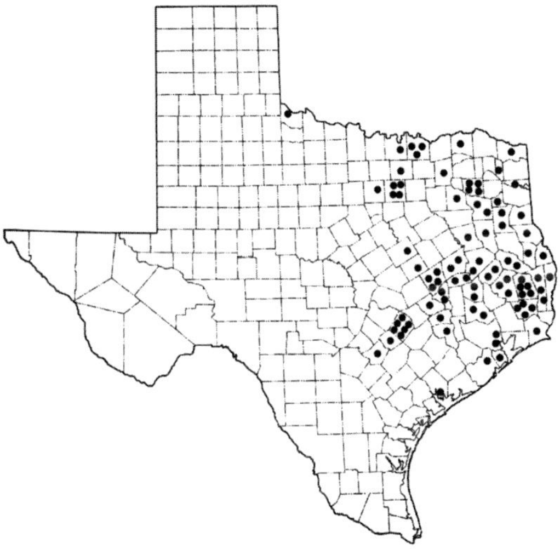

Rhynchosia latifolia

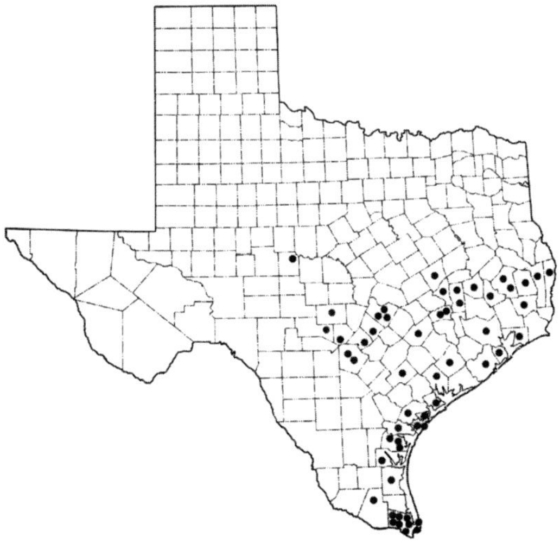

Rhynchosia minima

Rhynchosia reniformis

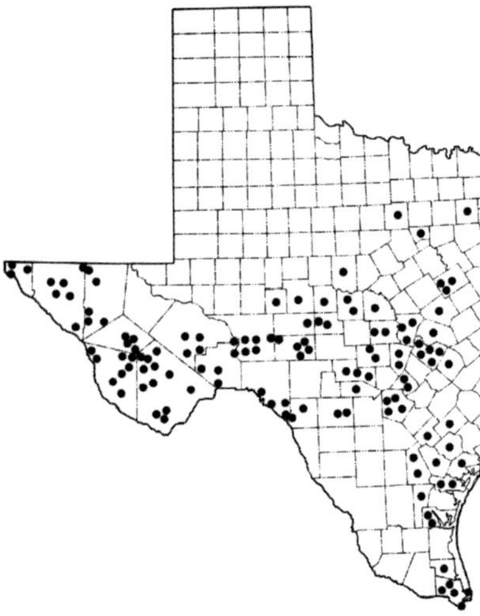

Rhynchosia senna

Robinia hispida

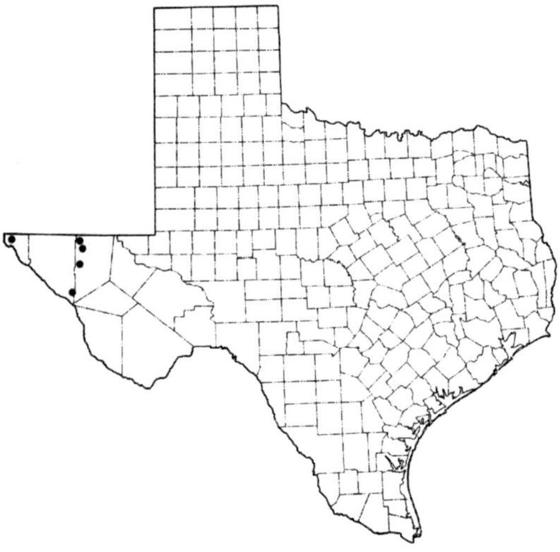

Robinia neomexicana

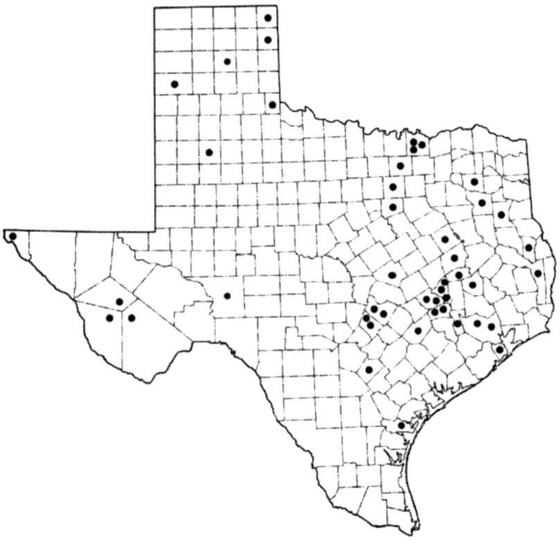

Robinia pseudoacacia

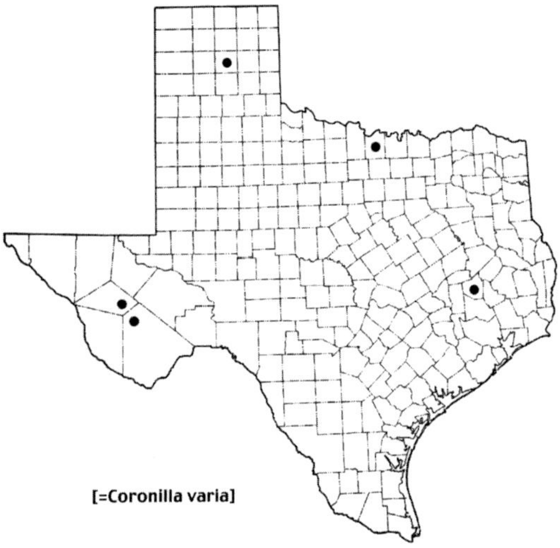
[=Coronilla varia]

Securigera varia

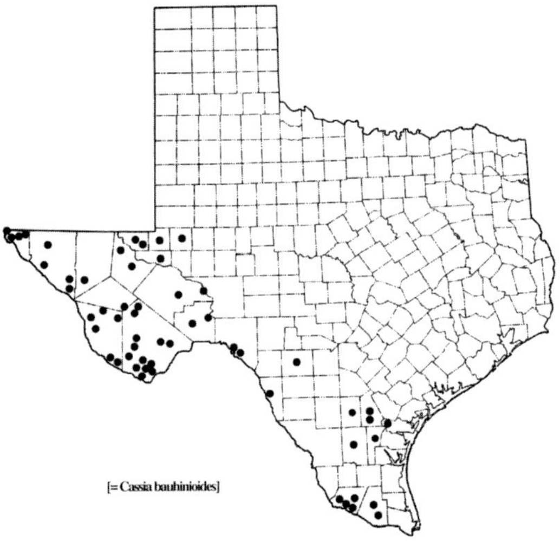
[= Cassia bauhinioides]

Senna bauhinioides

Senna corymbosa

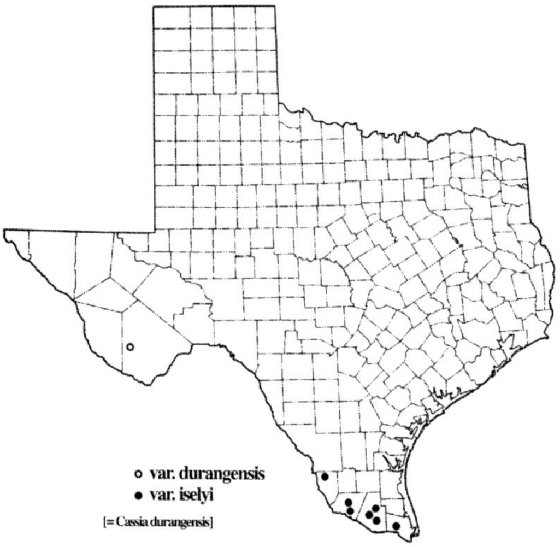
var. durangensis
var. iselyi
[= Cassia durangensis]

Senna durangensis

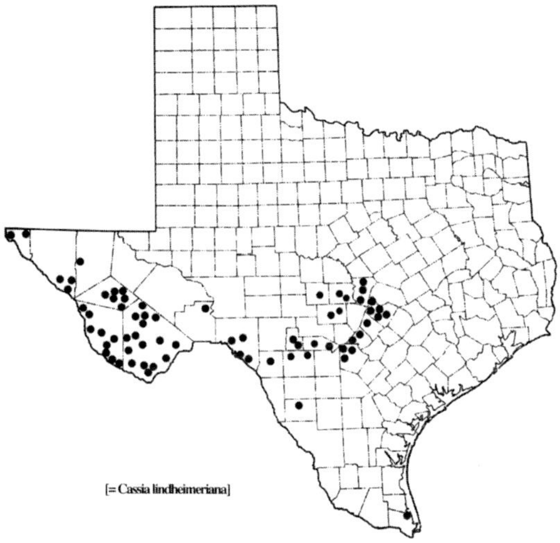
[= Cassia lindheimeriana]

Senna lindheimeriana

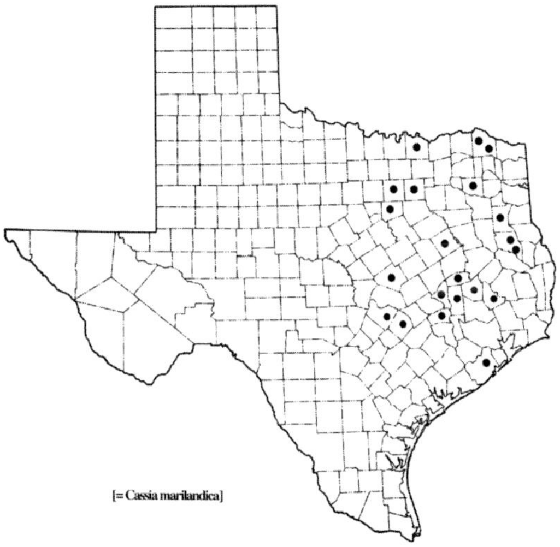
[= Cassia marilandica]

Senna marilandica

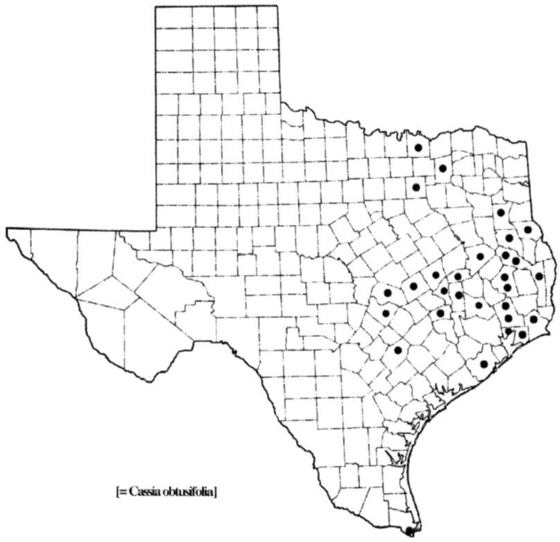
[= Cassia obtusifolia]

Senna obtusifolia

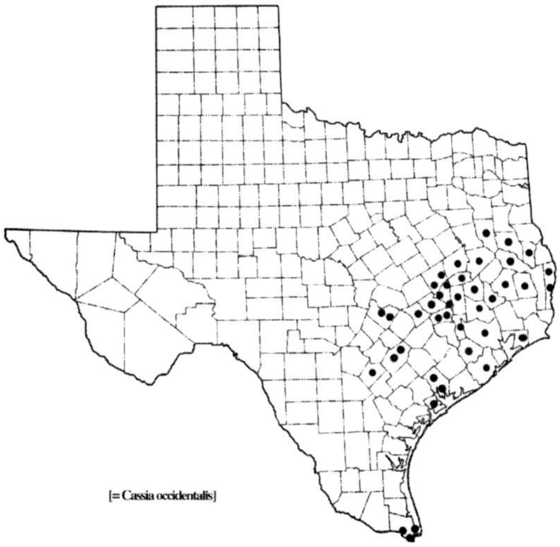
[= Cassia occidentalis]

Senna occidentalis

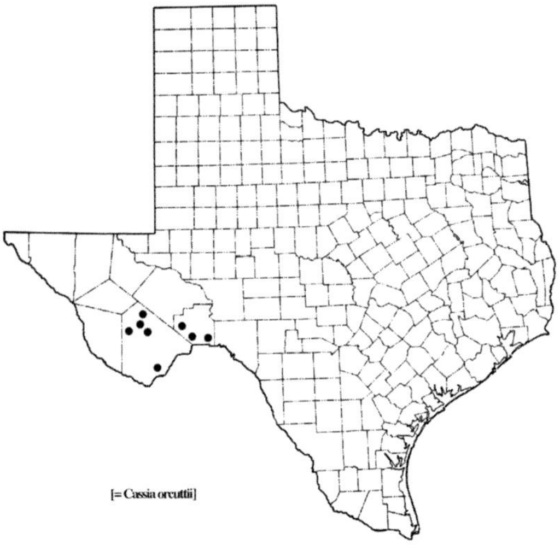
[= Cassia orcuttii]

Senna orcuttii

Senna pendula

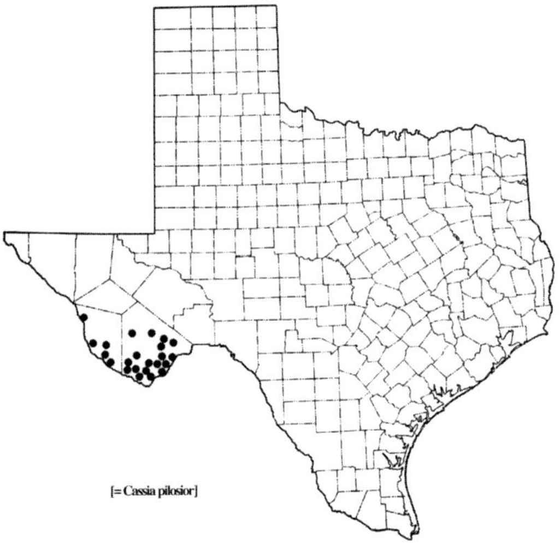
[= Cassia pilosior]

Senna pilosior

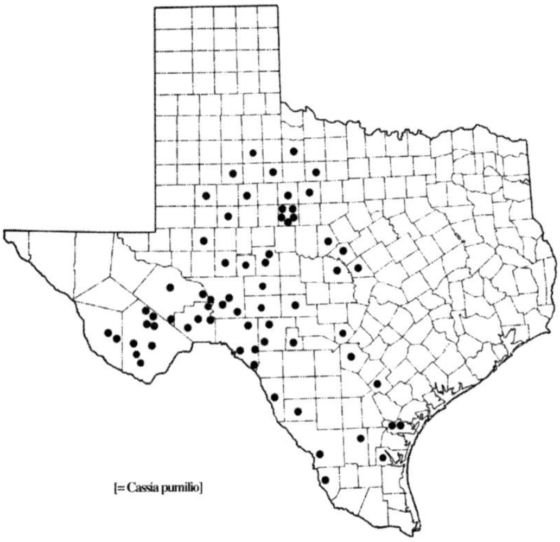
[= Cassia pumilio]

Senna pumilio

[= Cassia ripleyana]

Senna ripleyana

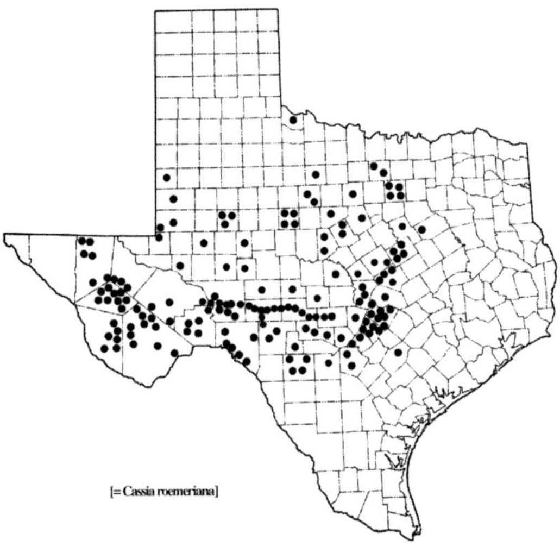
[= Cassia roemeriana]

Senna roemeriana

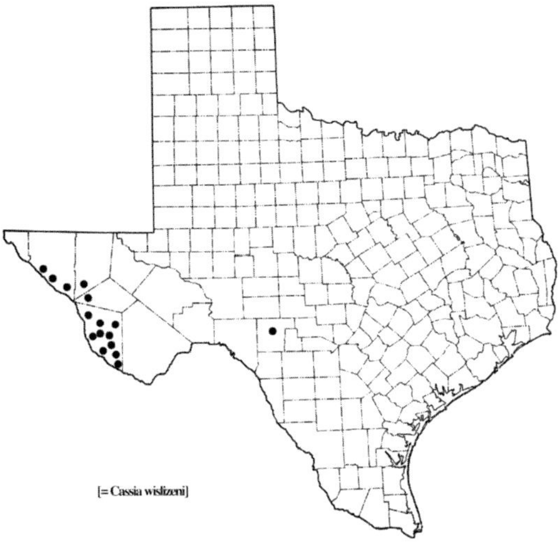
[= Cassia wislizeni]

Senna wislizeni

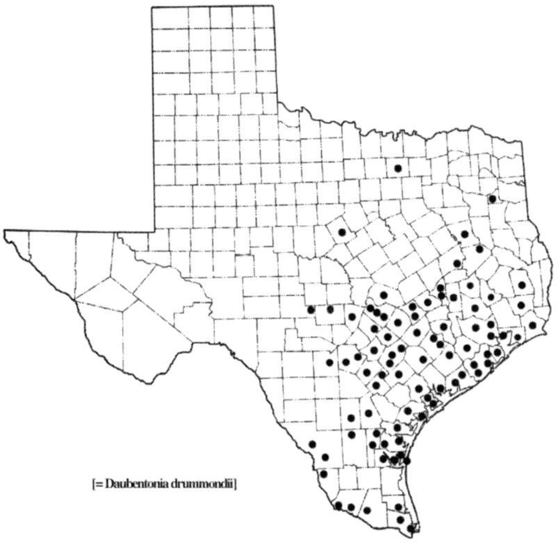
[= Daubentonia drummondii]

Sesbania drummondii

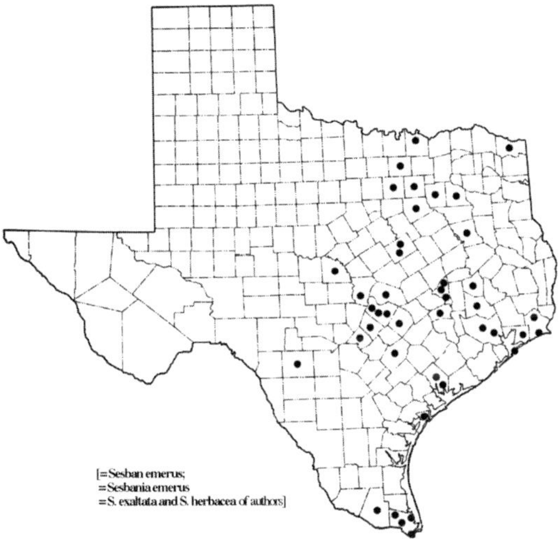
[=Sesban emerus;
=Sesbania emerus
=S. exaltata and S. herbacea of authors]

Sesbania emerus

[=Sesbania exaltatus]

Sesbania macrocarpa

Sesbania punicea

Sesbania sericea

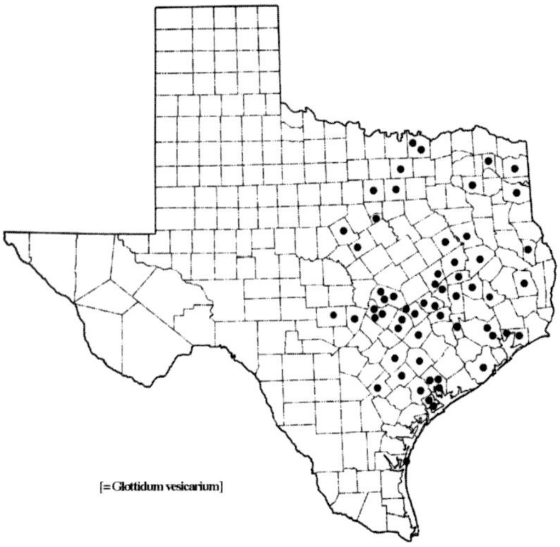
[=Glottidum vesicarium]

Sesbania vesicaria

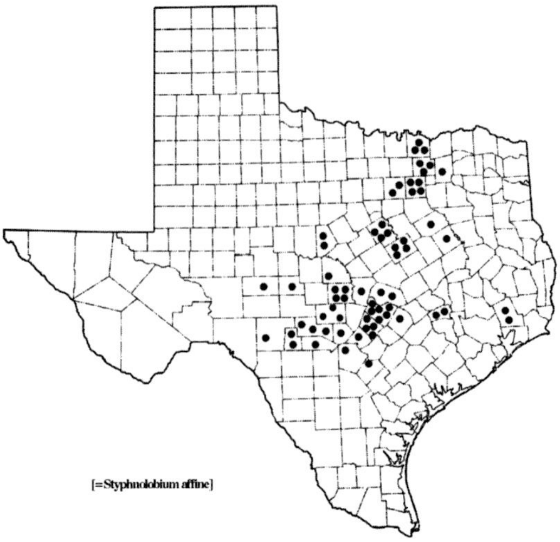
[=Styphnolobium affine]

Sophora affinis

var. guadalupensis

Sophora gypsophila

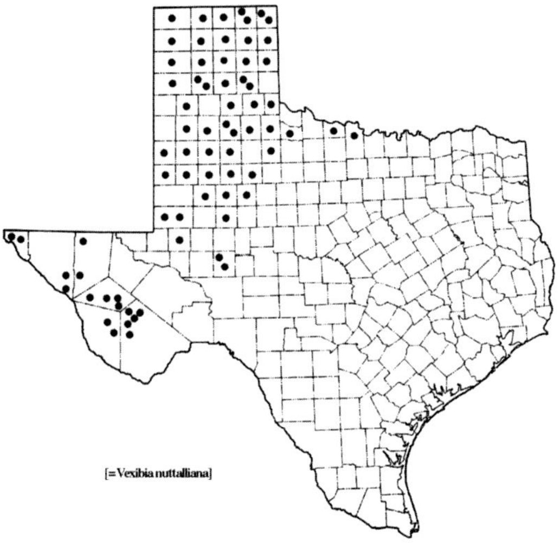
[=Vexibia nuttalliana]

Sophora nuttalliana

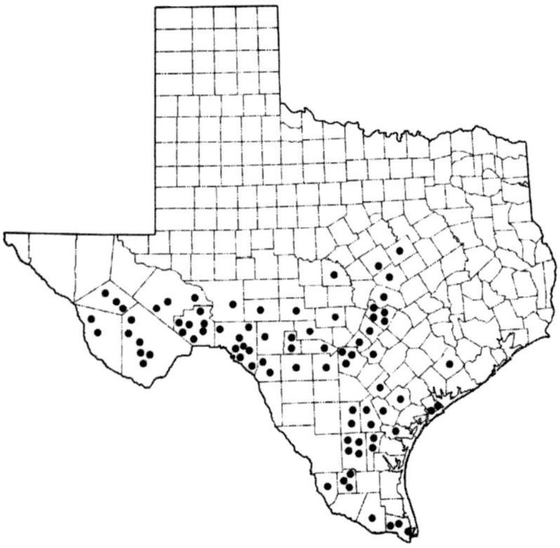

Sophora secundiflora

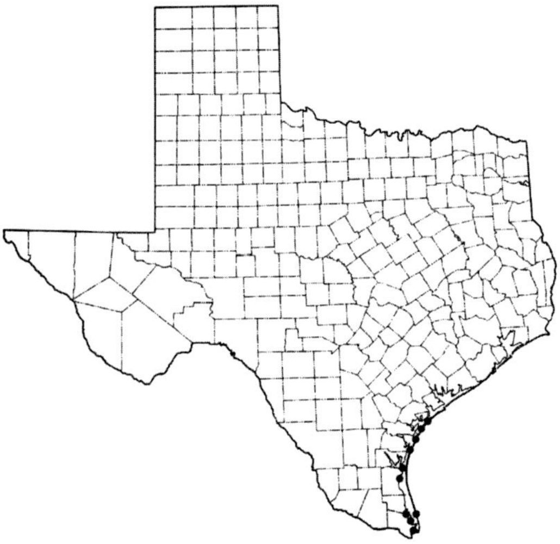

Sophora tomentosa

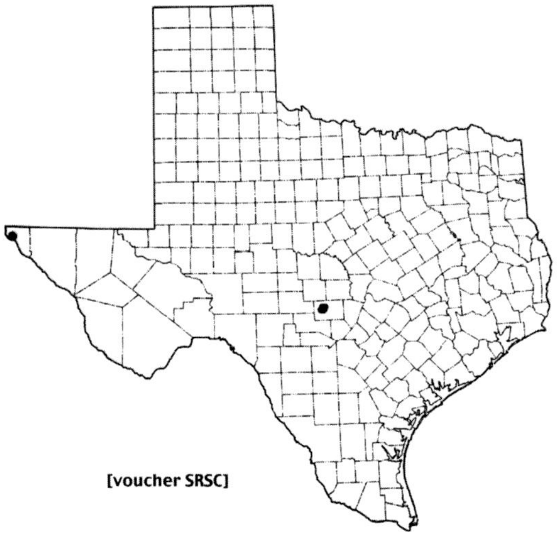
[voucher SRSC]

Spartium junceum

Sphaerophysa salsula

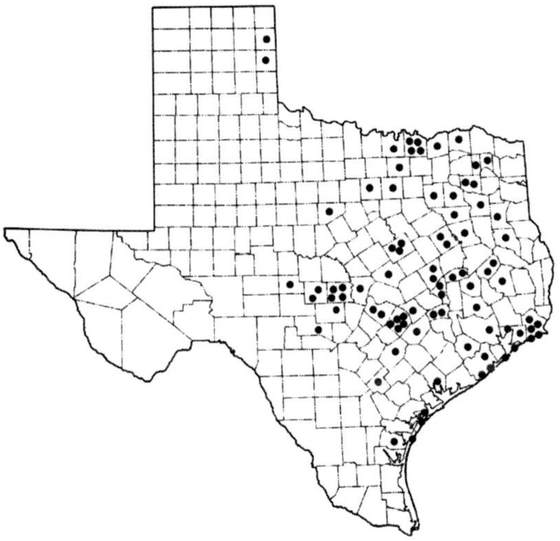

Strophostyles helvula

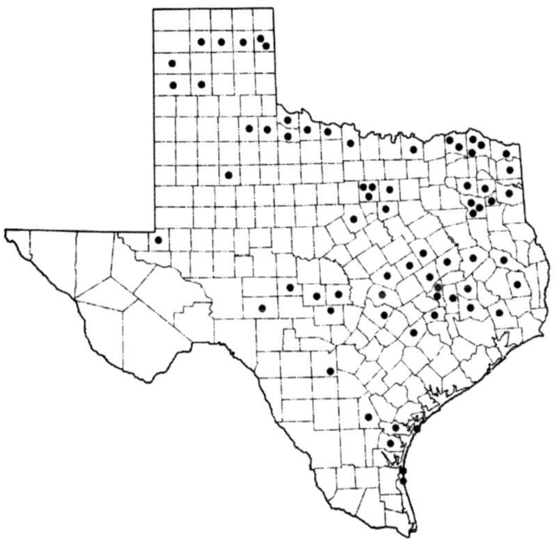

Strophostyles leiosperma

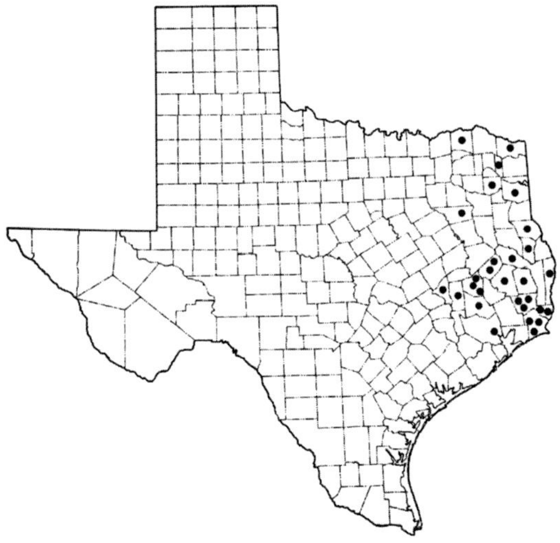

Strophostyles umbellata

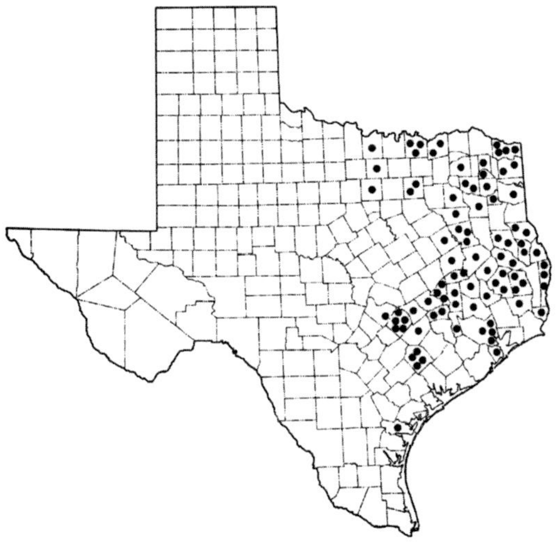

Stylosanthes biflora

Stylosanthes viscosa

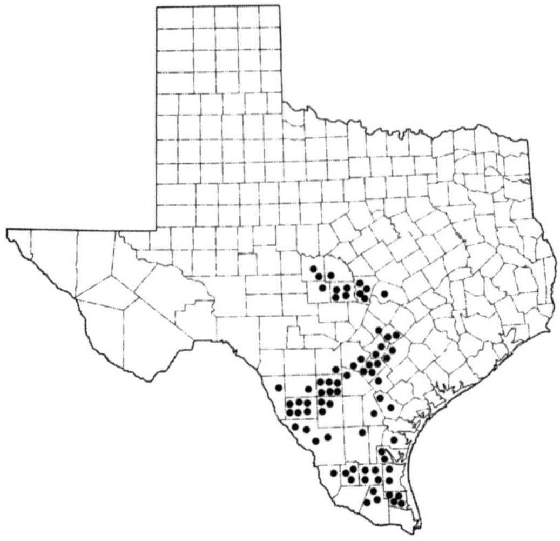

Tephrosia lindheimeri

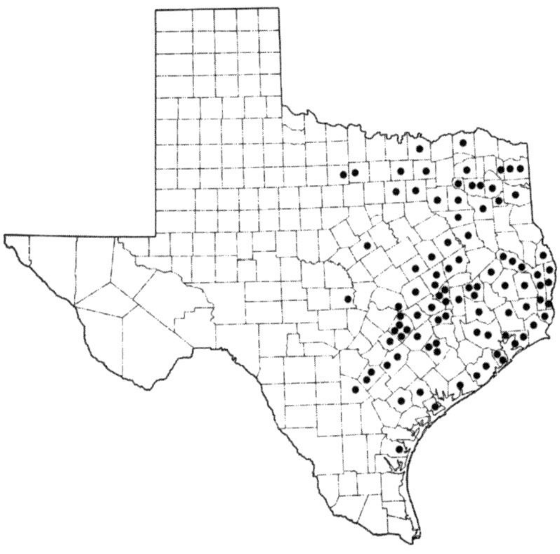

Tephrosia onobrychoides

Tephrosia potosina

Tephrosia tenella

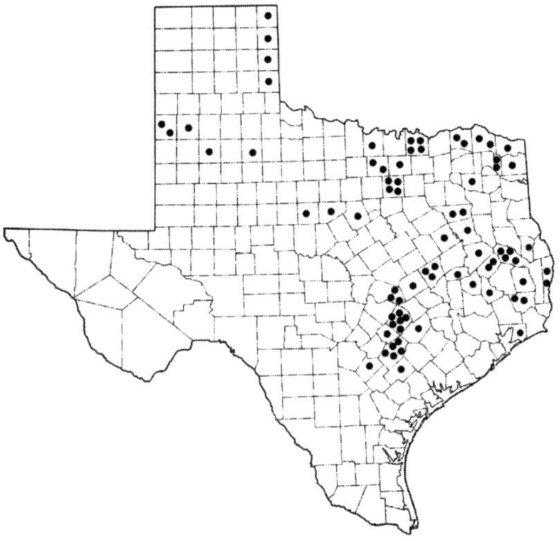

Tephrosia virginiana

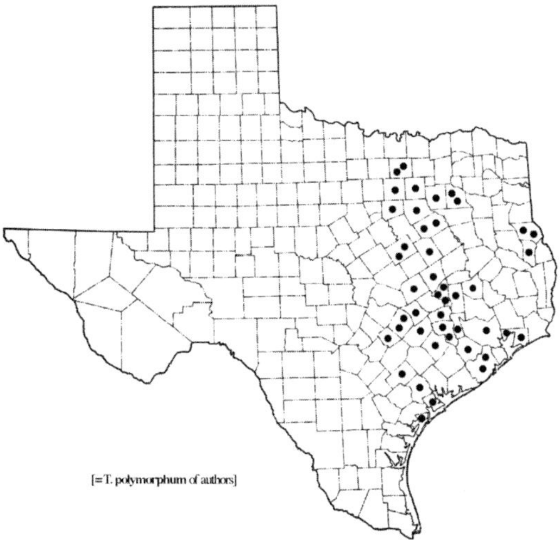
[= T. polymorphum of authors]

Trifolium amphianthum

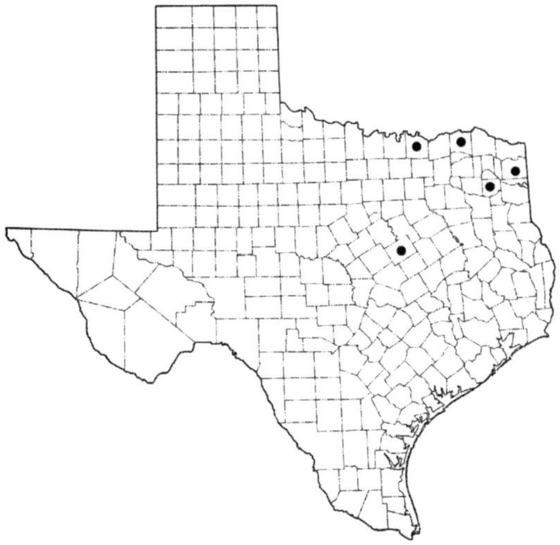

Trifolium arvense

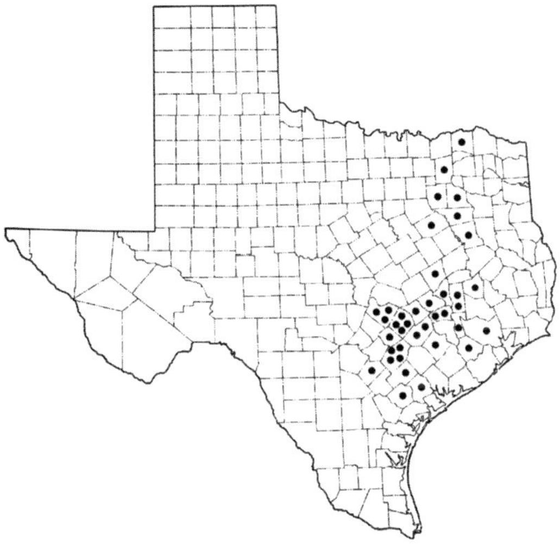

Trifolium bejariense

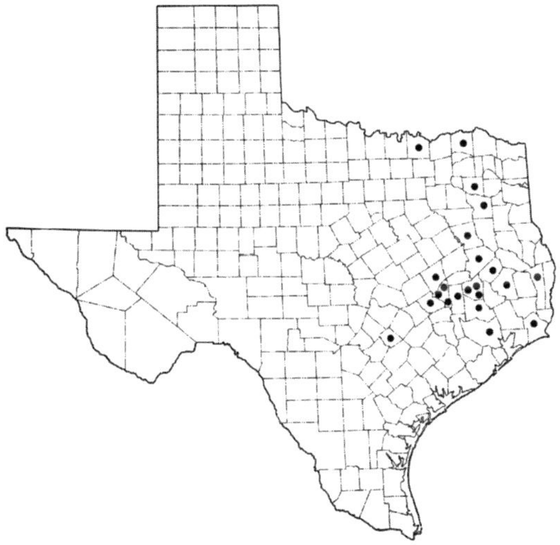

Trifolium campestre

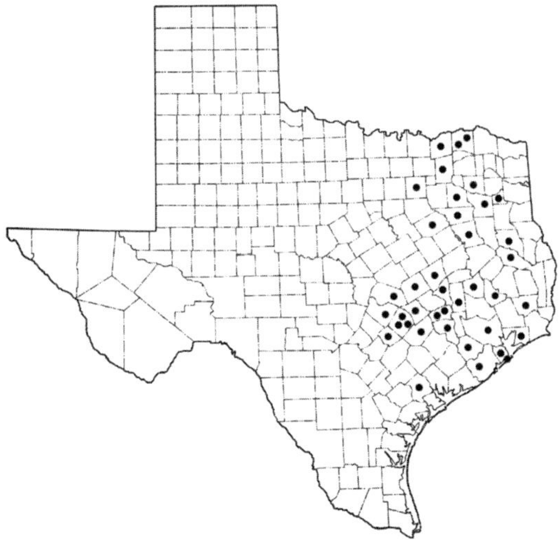

Trifolium carolinianum

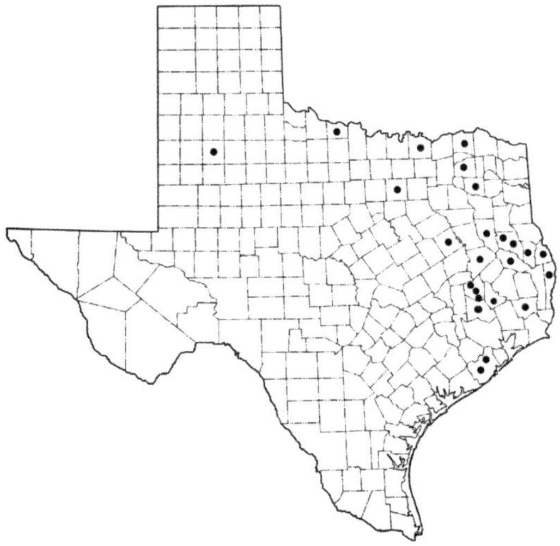

Trifolium dubium

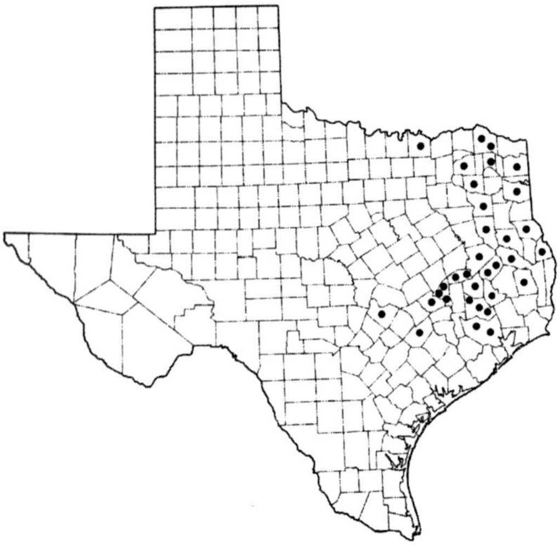

Trifolium incarnatum

Trifolium lappaceum

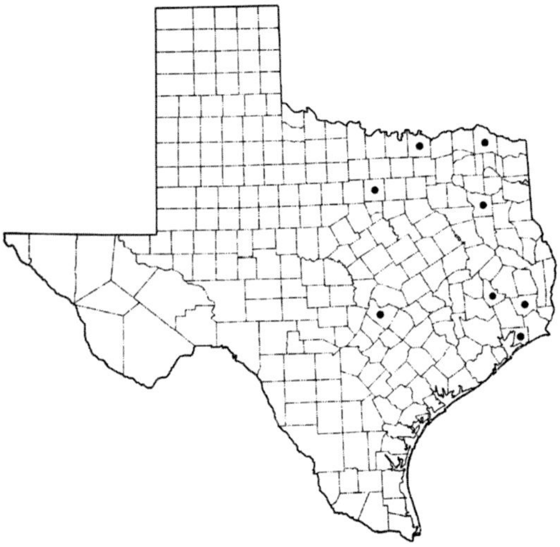

Trifolium pratense

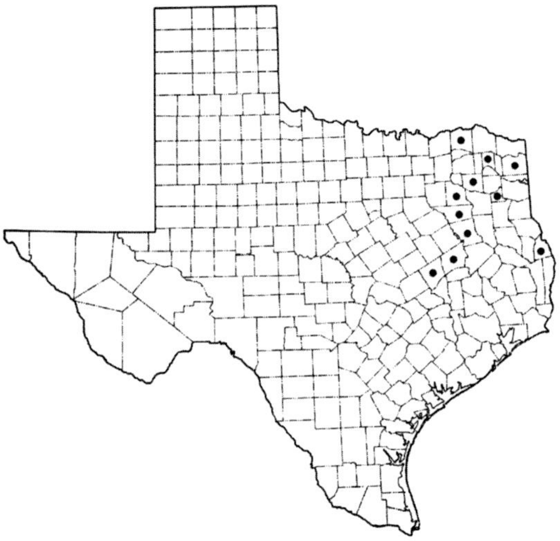

Trifolium reflexum

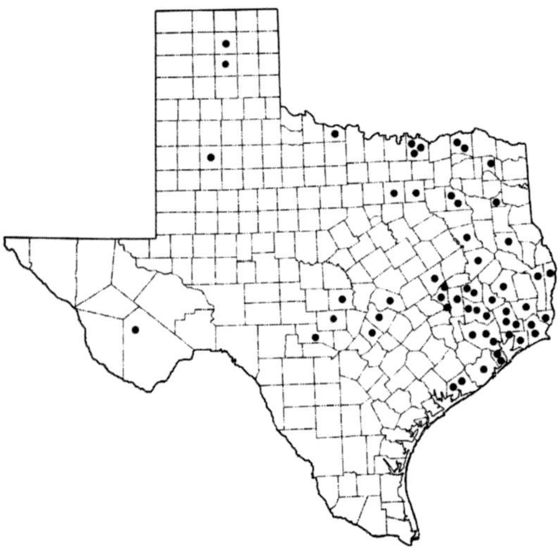

Trifolium repens

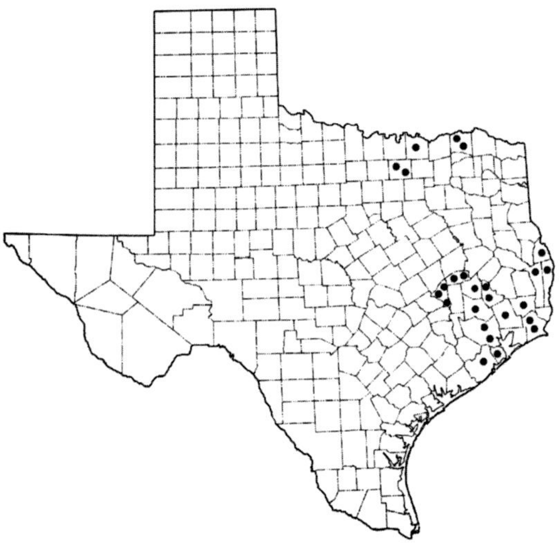

Trifolium resupinatum

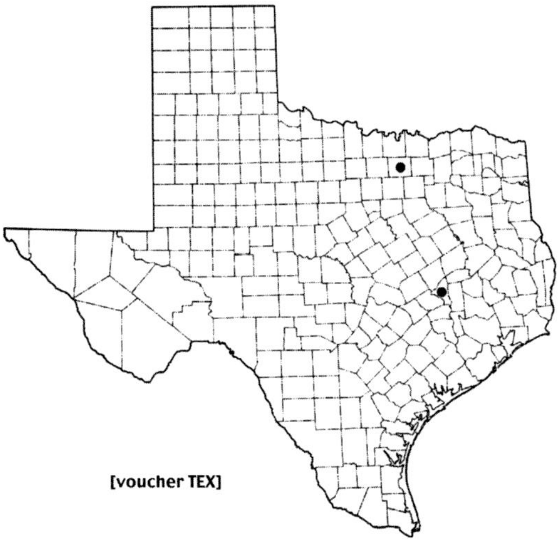
[voucher TEX]

Trifolium subterraneum

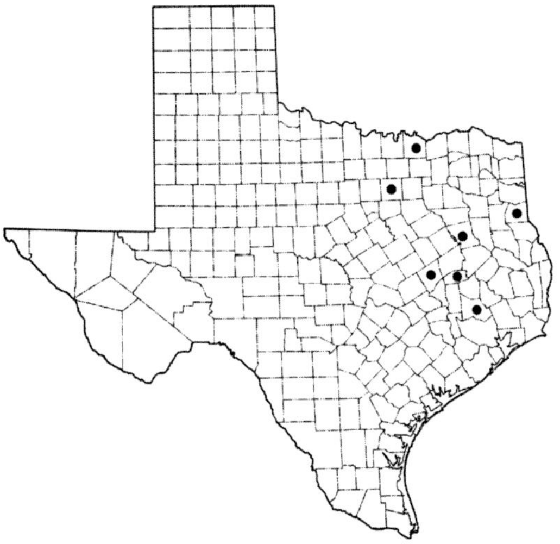

Trifolium vesiculosum

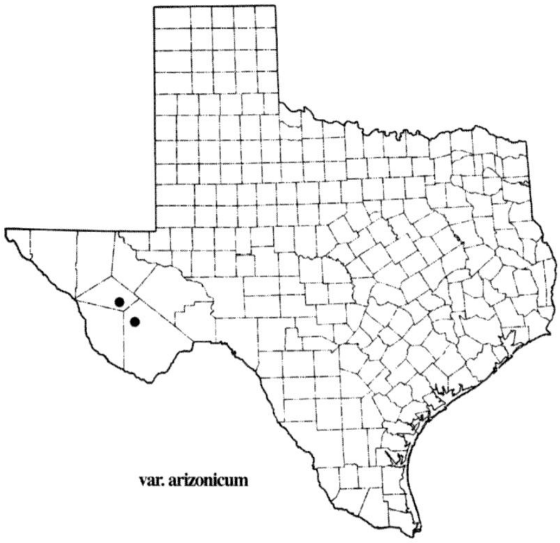
var. arizonicum

Trifolium wormskioldii

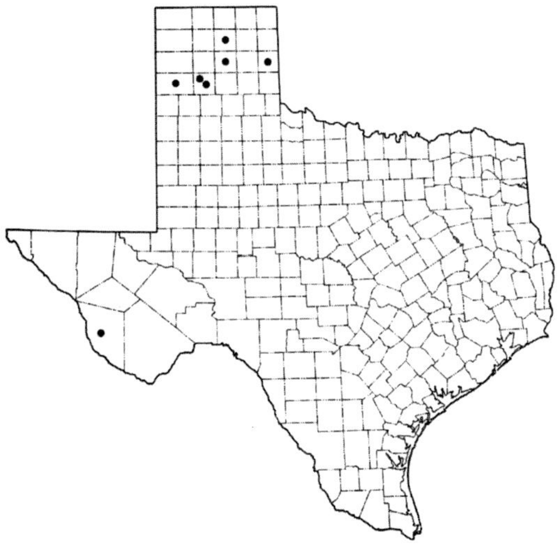

Vicia americana

Vicia angustifolia

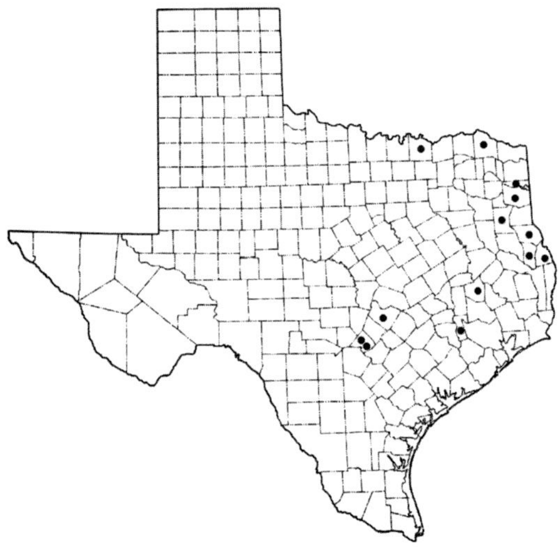

Vicia caroliniana

[Sida 6: 114. 1975]

Vicia hirsuta

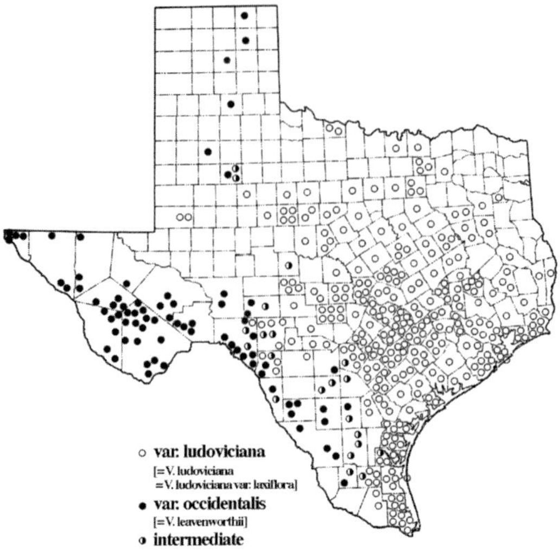
var. ludoviciana
[=V. ludoviciana
=V. ludoviciana var. laxiflora]
var. occidentalis
[=V. leavenworthii]
intermediate

Vicia ludoviciana

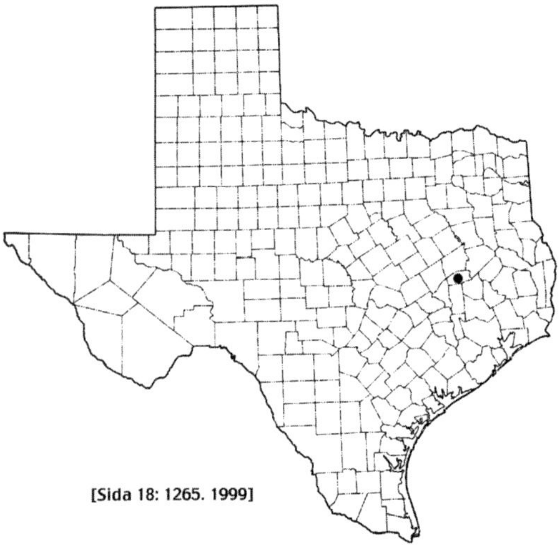
[Sida 18: 1265. 1999]

Vicia lutea

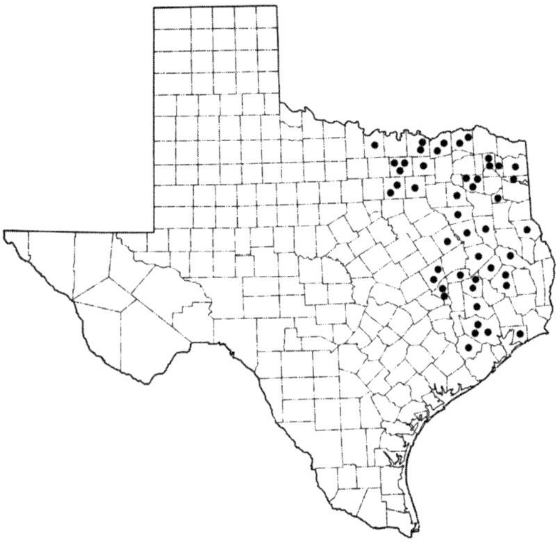

Vicia minutiflora

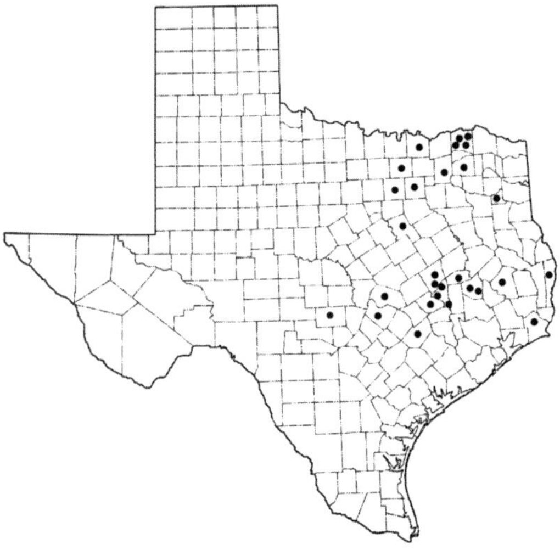

Vicia sativa

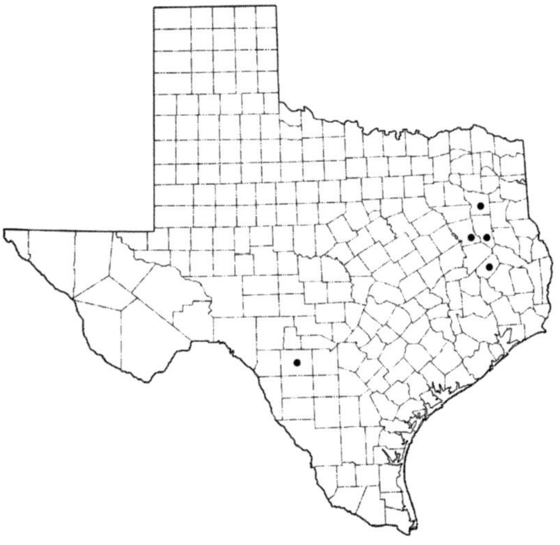

Vicia tetrasperma

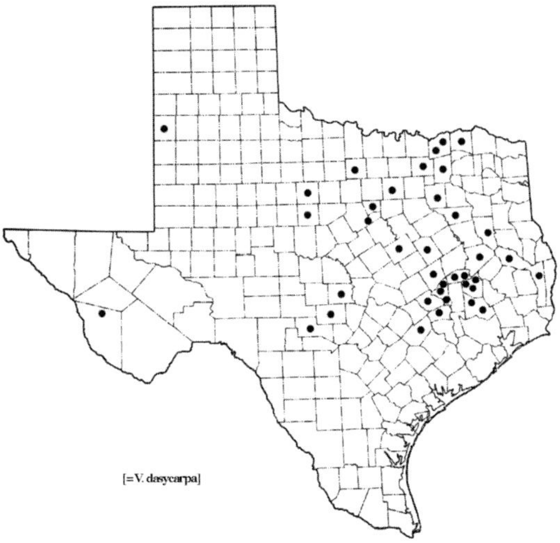
[=V. dasycarpa]

Vicia villosa

Vigna luteola

Vigna unguiculata

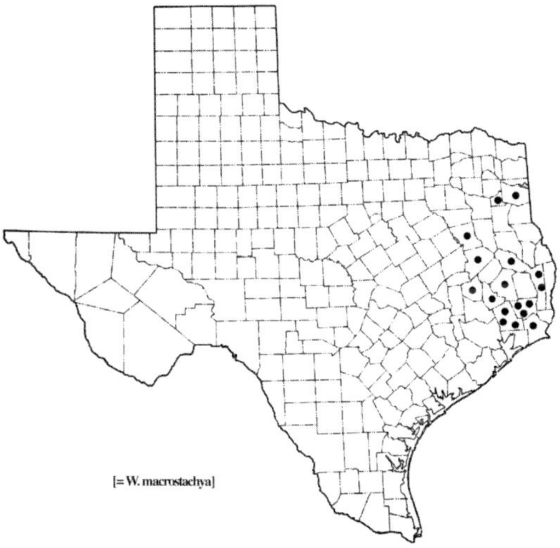
[= W. macrostachya]

Wisteria frutescens

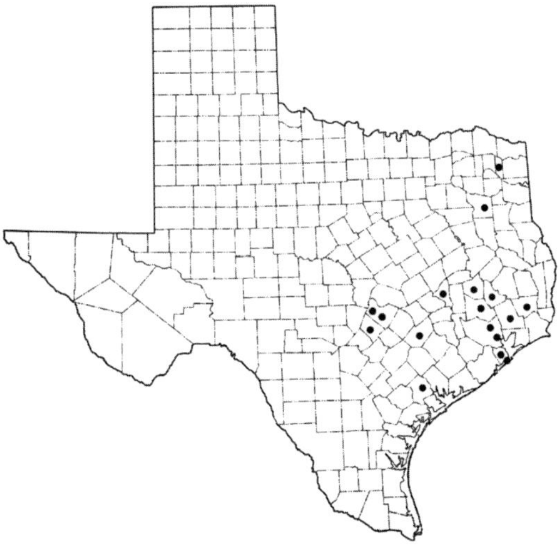

Wisteria sinensis

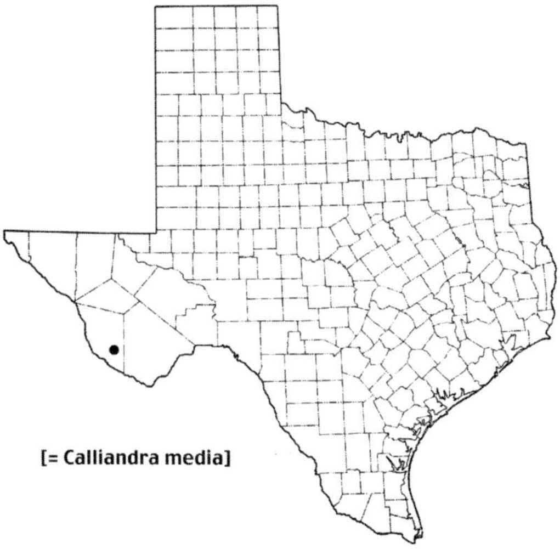
[= Calliandra media]

Zapoteca coulteri

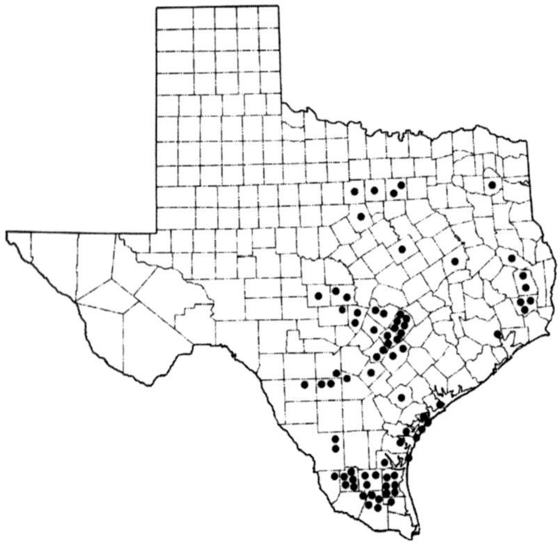

Zornia bracteata

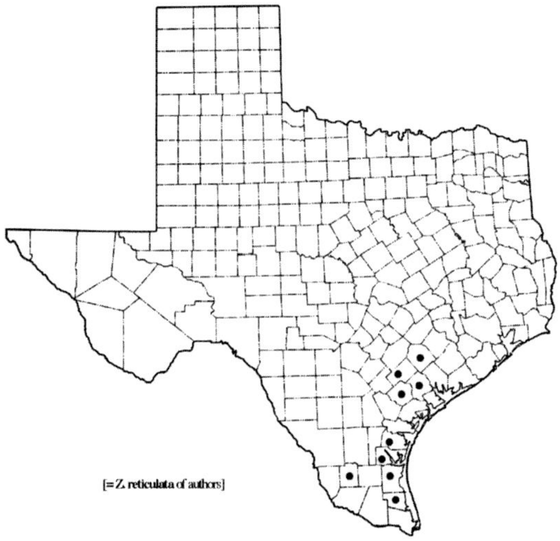
[= Z. reticulata of authors]

Zornia gemella

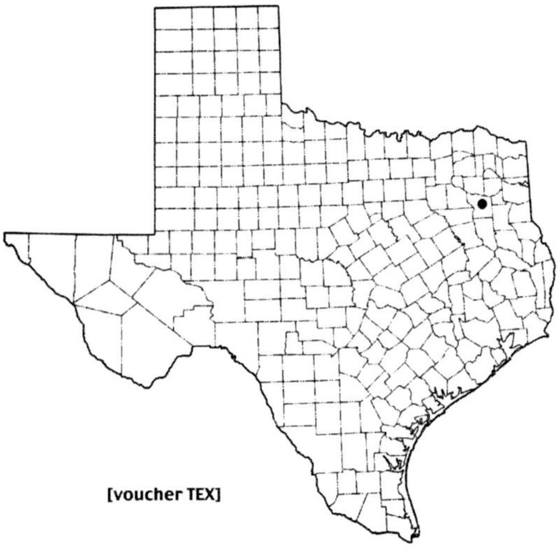
[voucher TEX]

Castanea mollissima

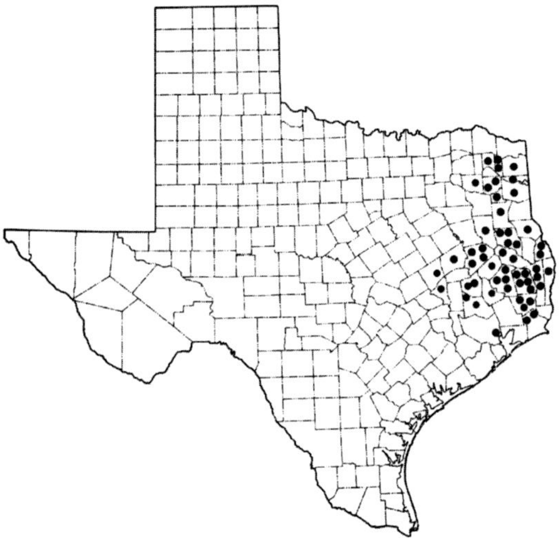

Castanea pumila

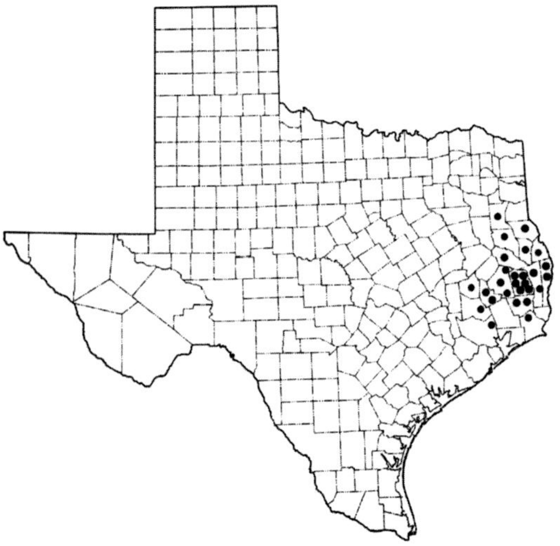

Fagus grandifolia

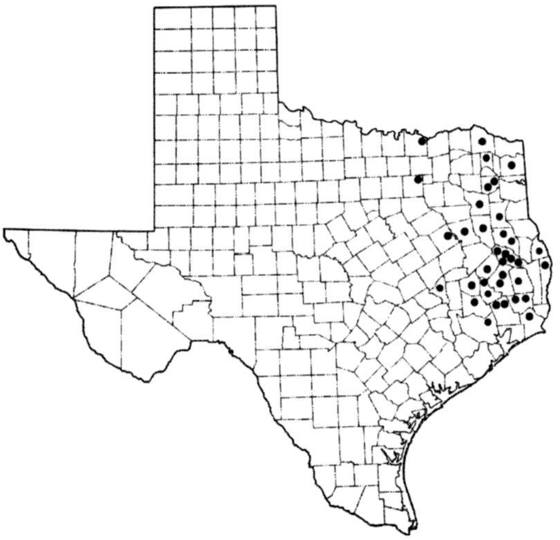

Quercus alba

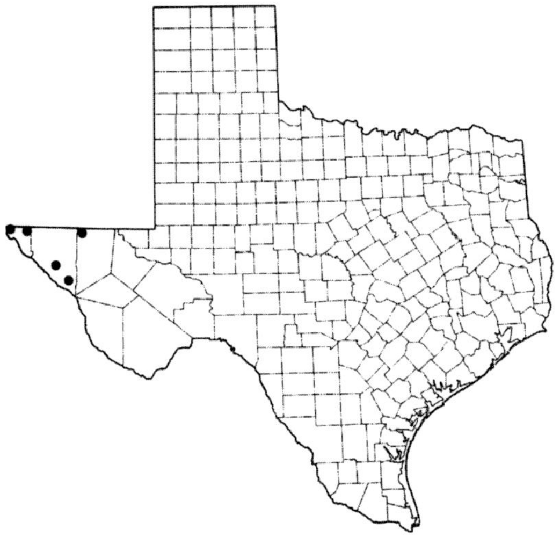

Quercus arizonica

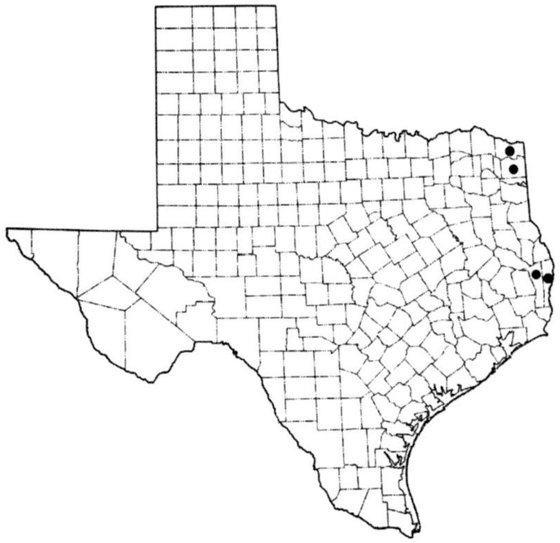

Quercus arkansana

Quercus boyntonii

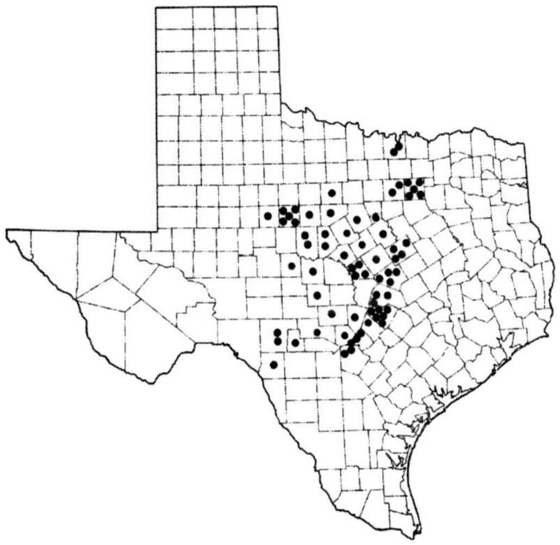

Quercus buckleyi

Quercus carmenensis

Quercus chihuahuensis

Quercus depressipes

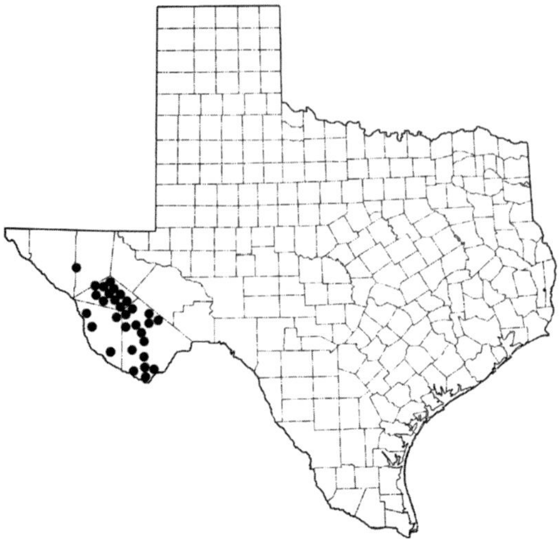

Quercus emoryi

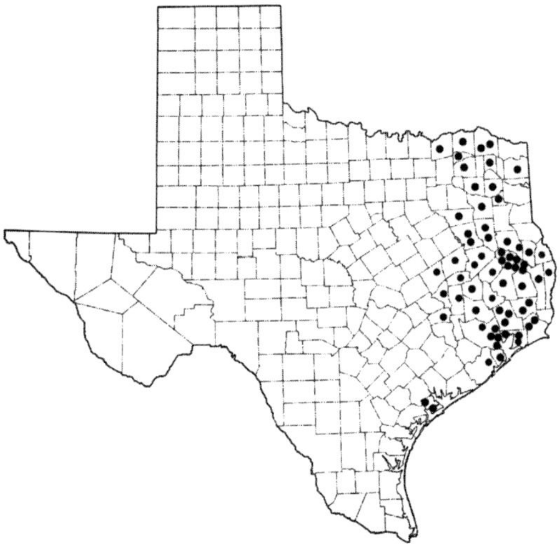

Quercus falcata

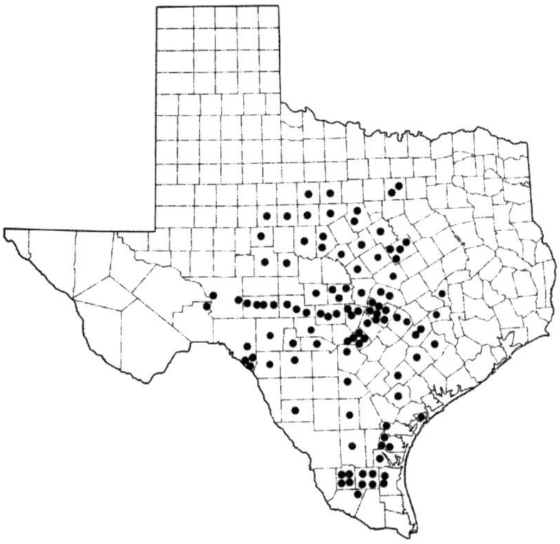

Quercus fusiformis

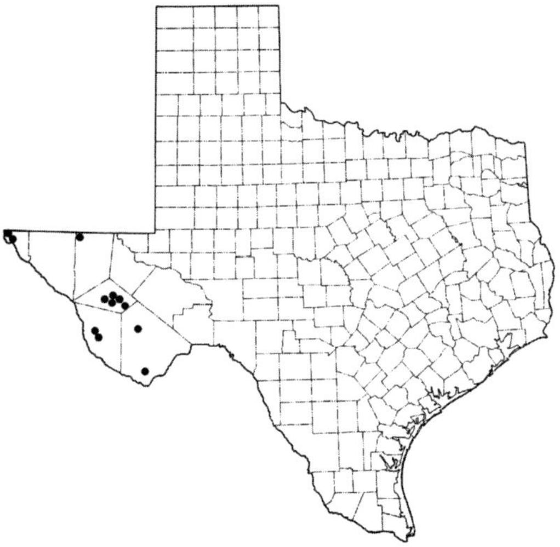

Quercus gambelii

Quercus graciliformis

Quercus gravesii

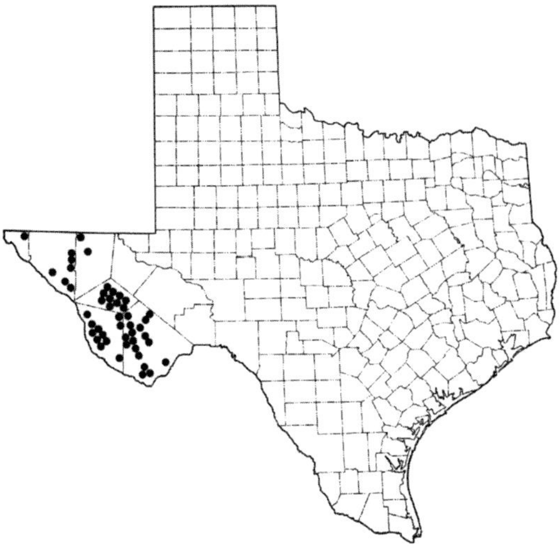

Quercus grisea

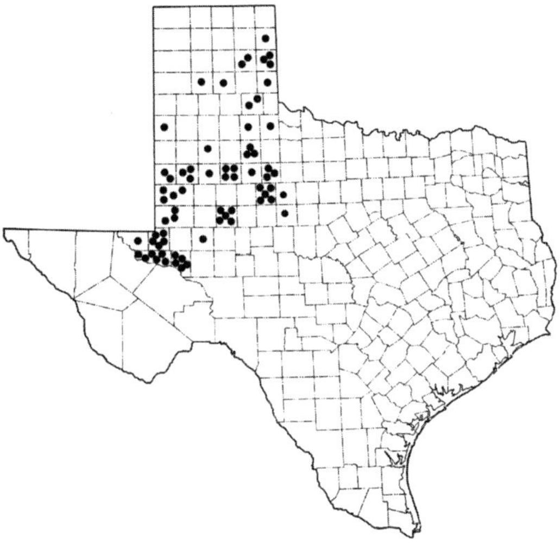

Quercus havardii

Quercus hemisphaerica

Quercus hinckleyi

Quercus hypoleucoides

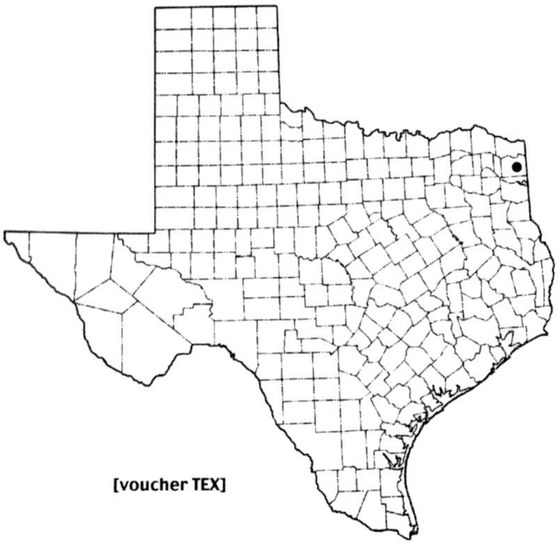
[voucher TEX]
Quercus imbricaria

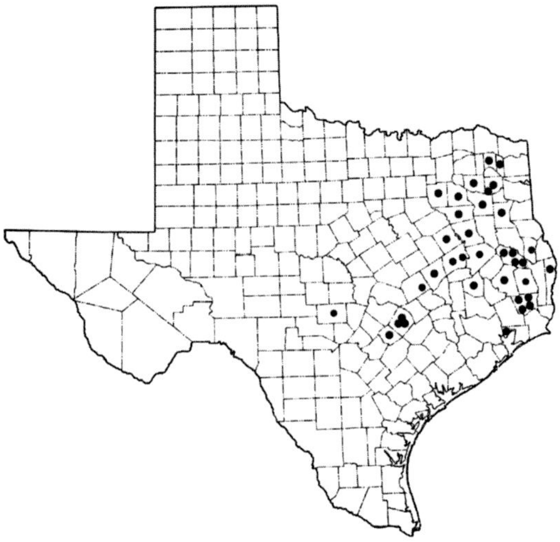
Quercus incana

Quercus intricata

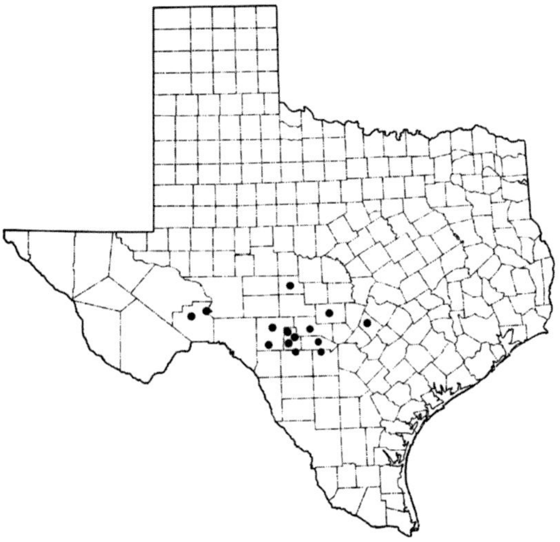
Quercus laceyi

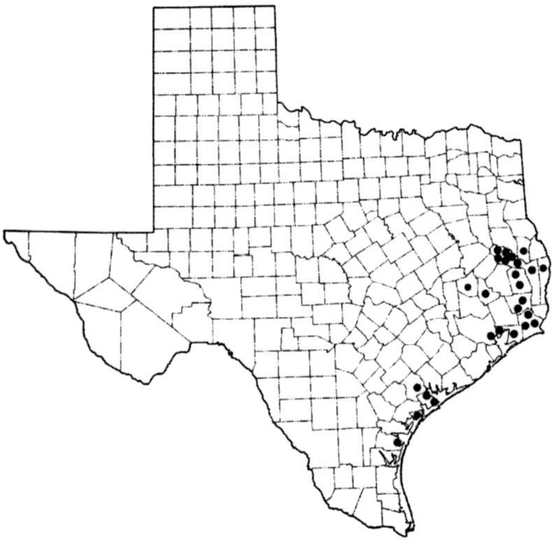
Quercus laurifolia

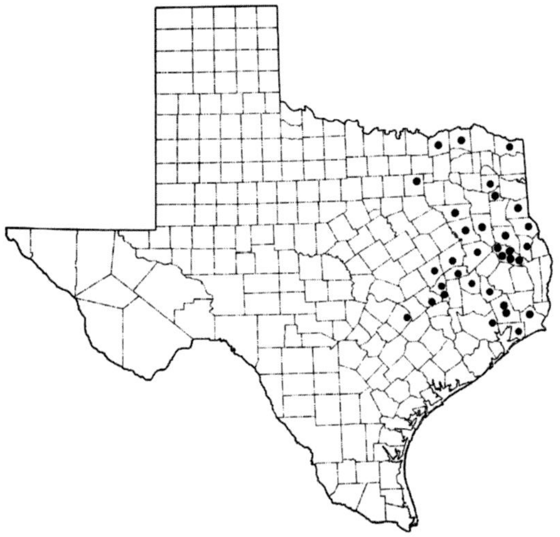
Quercus lyrata

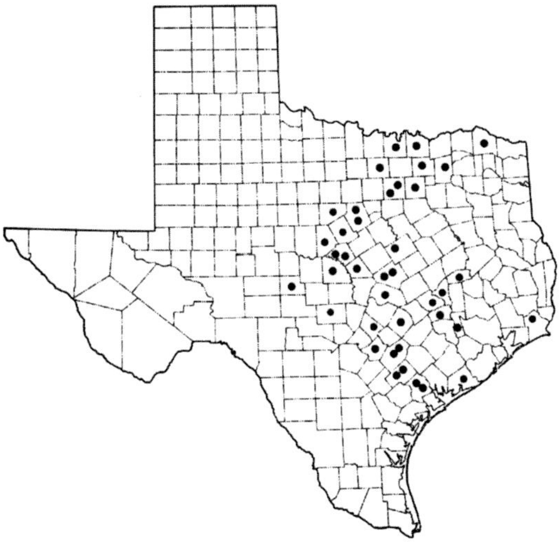

Quercus macrocarpa

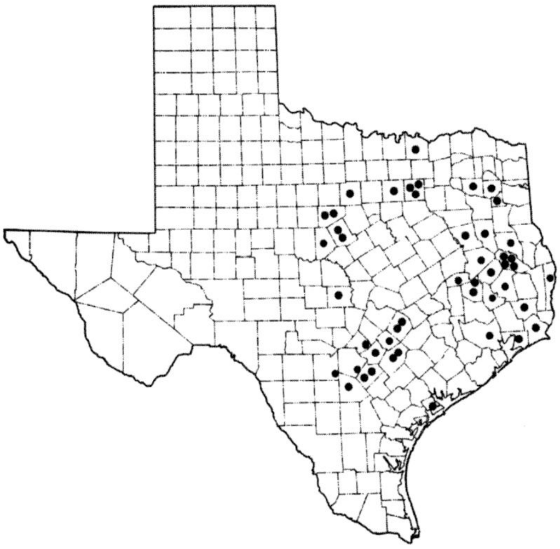

Quercus margarettiae

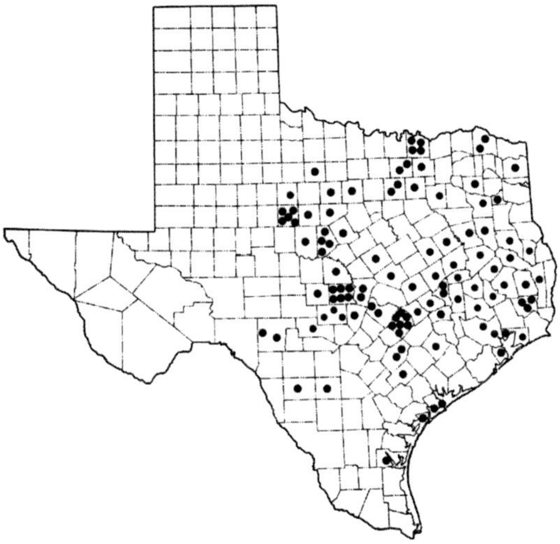

Quercus marilandica

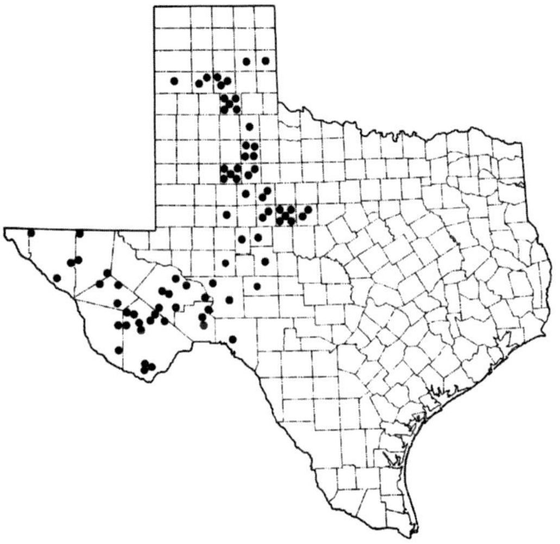

Quercus mohriana

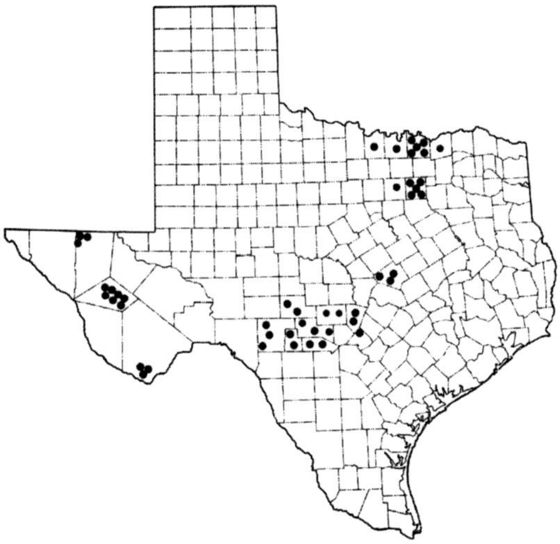

Quercus muhlenbergii

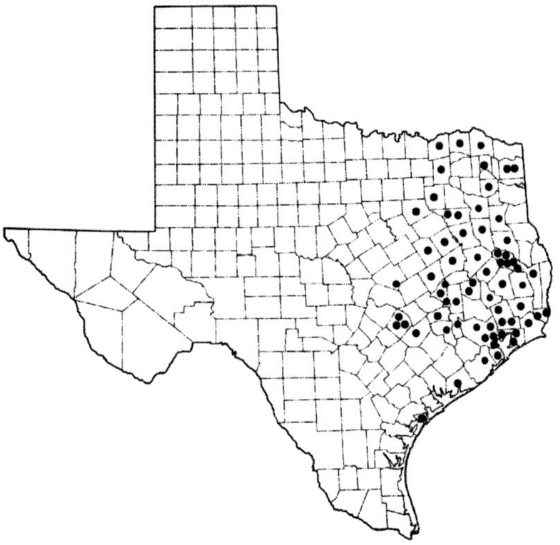

Quercus nigra

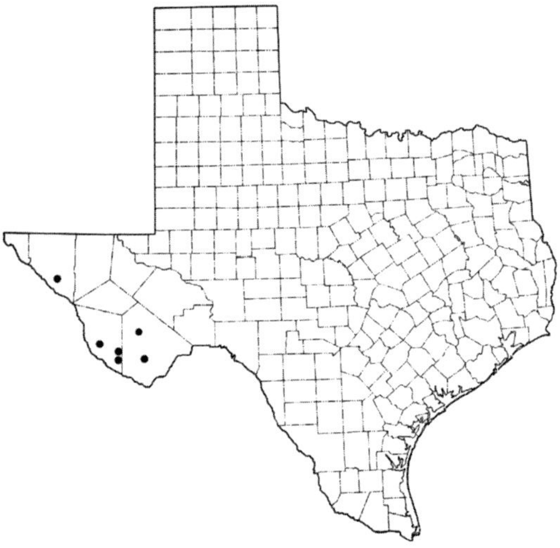

Quercus oblongifolia

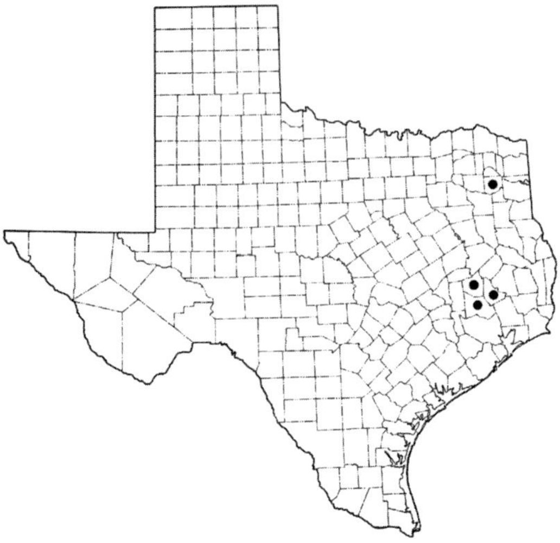

Quercus pagoda

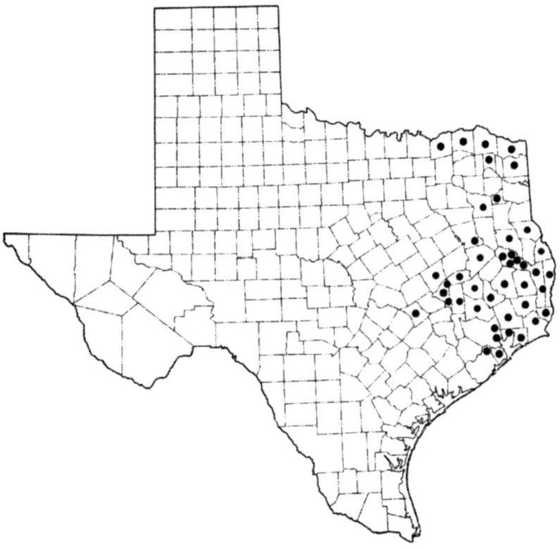

Quercus phellos

Quercus polymorpha

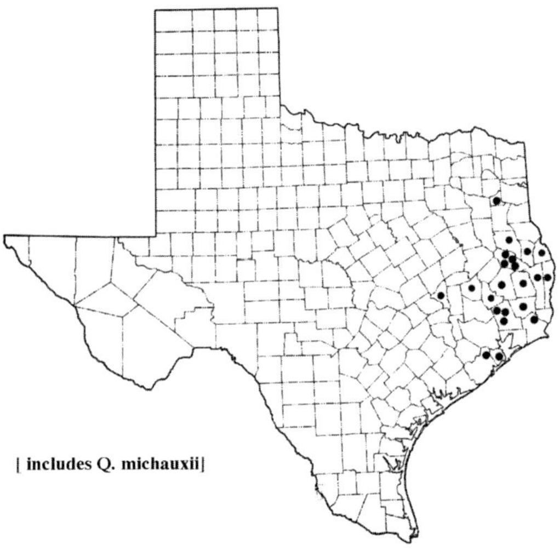
[includes Q. michauxii]

Quercus prinus

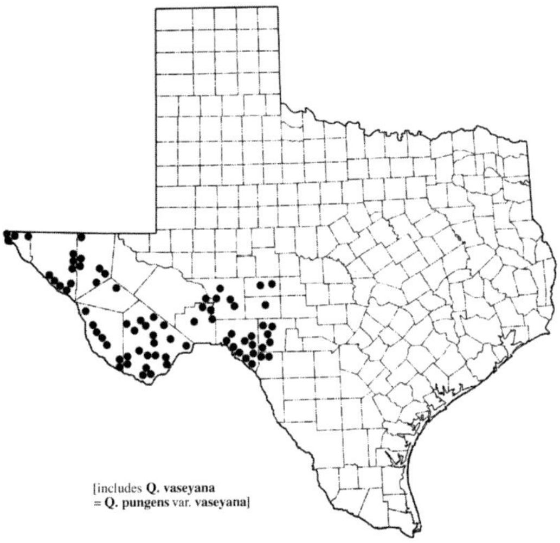
[includes Q. vaseyana
= Q. pungens var. vaseyana]

Quercus pungens

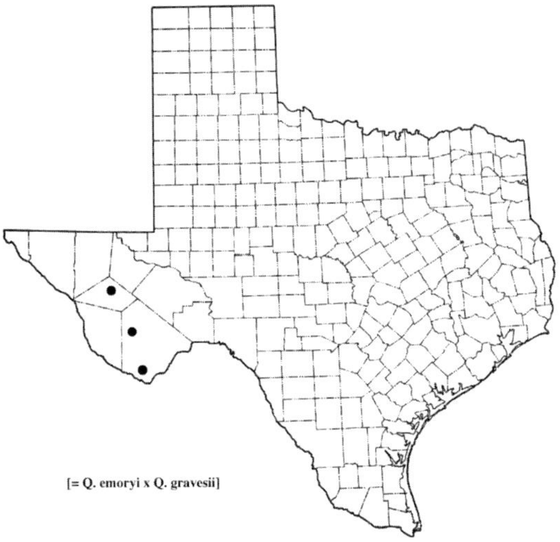
[= Q. emoryi x Q. gravesii]

Quercus robusta

Quercus rugosa

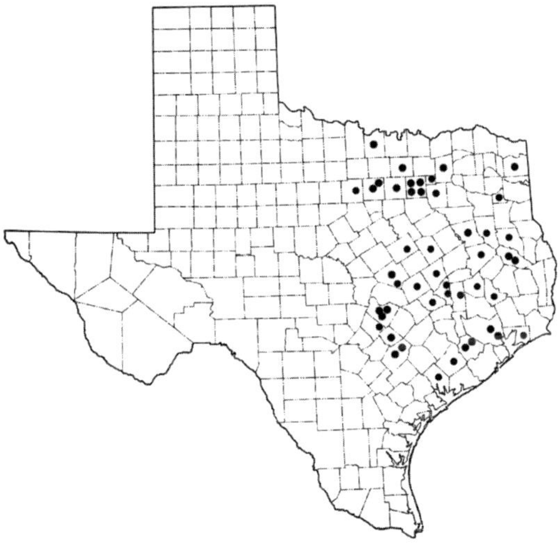

Quercus shumardii

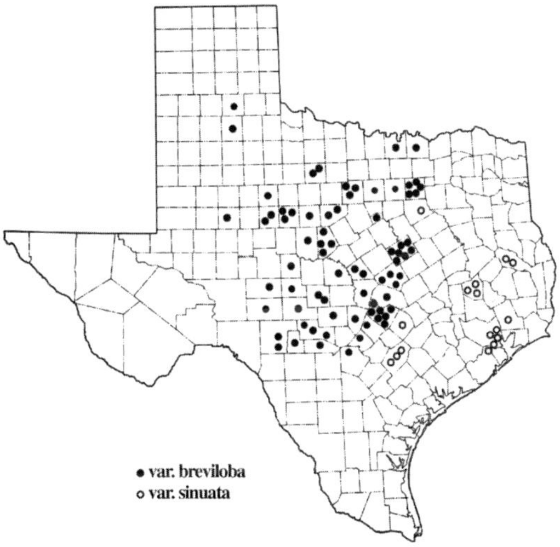
• var. breviloba
○ var. sinuata

Quercus sinuata

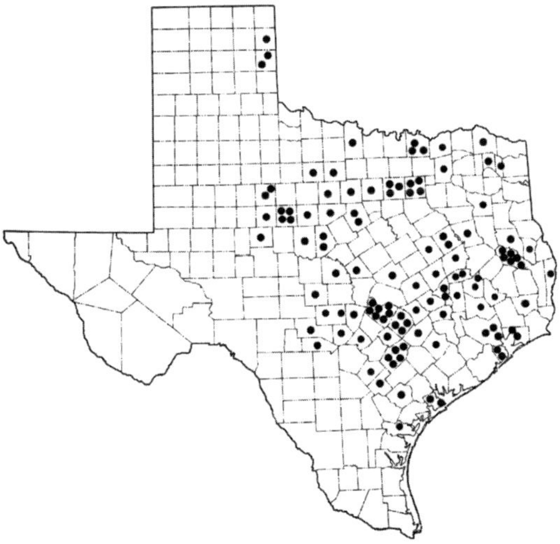

Quercus stellata

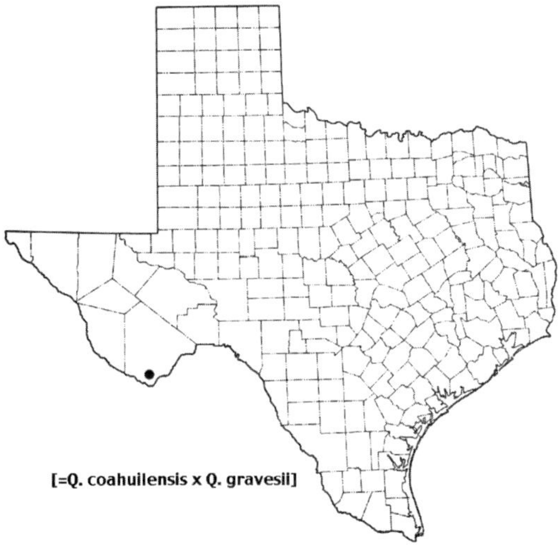
[=Q. coahuilensis x Q. gravesii]

Quercus tardifolia

Quercus texana

Quercus toumeyi

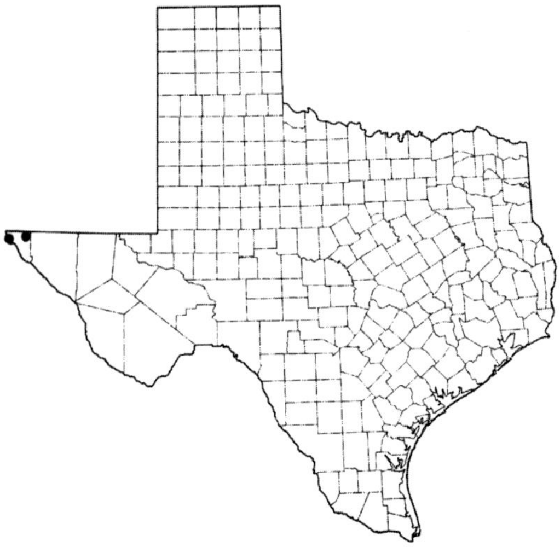

Quercus turbinella

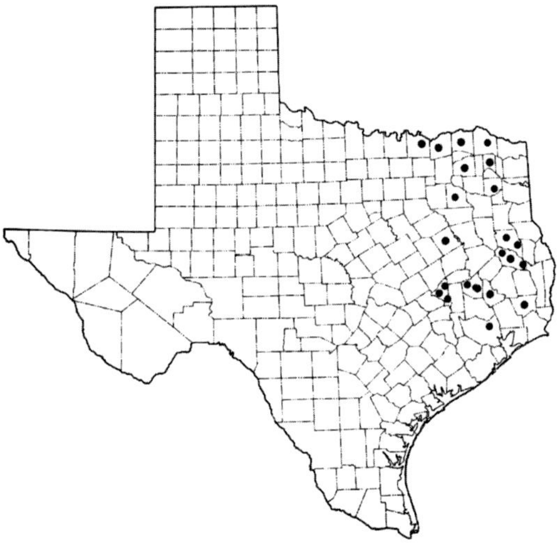

Quercus velutina

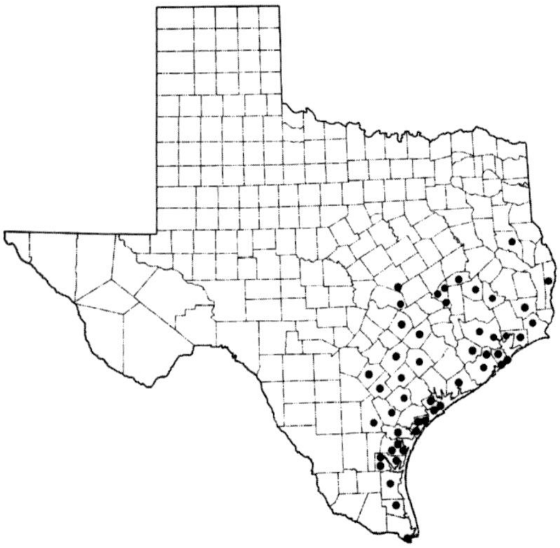

Quercus virginiana

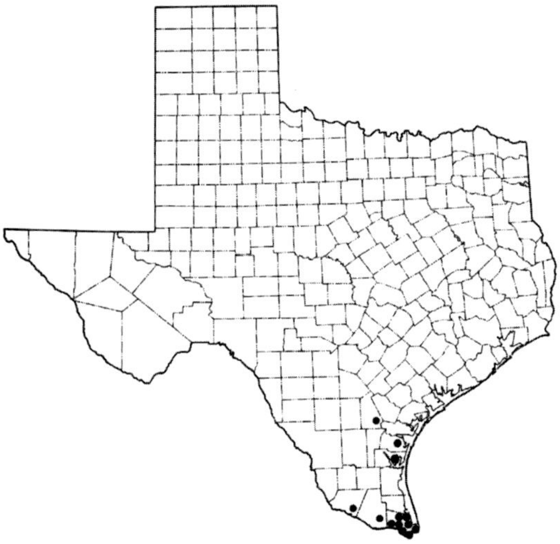

Xylosma flexuosa

Fouquiera splendens

Frankenia jamesii

Frankenia johnstonii

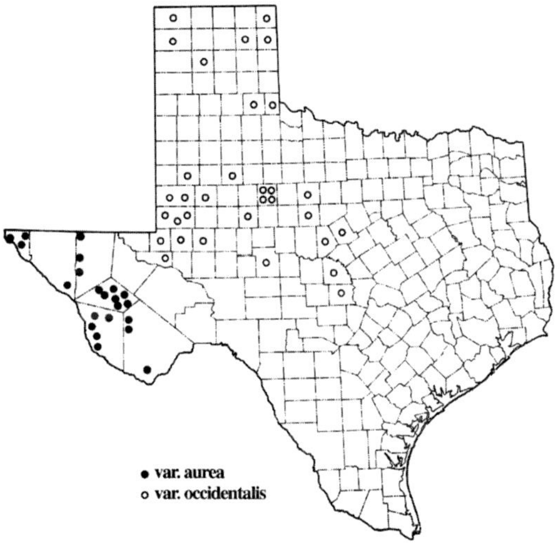
• var. aurea
○ var. occidentalis

Corydalis aurea

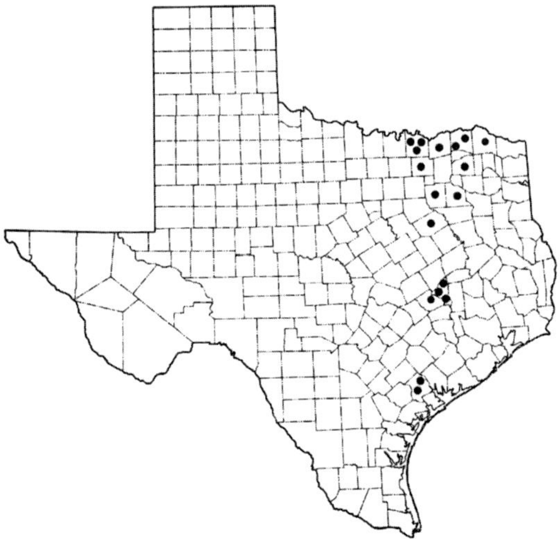

Corydalis crystallina

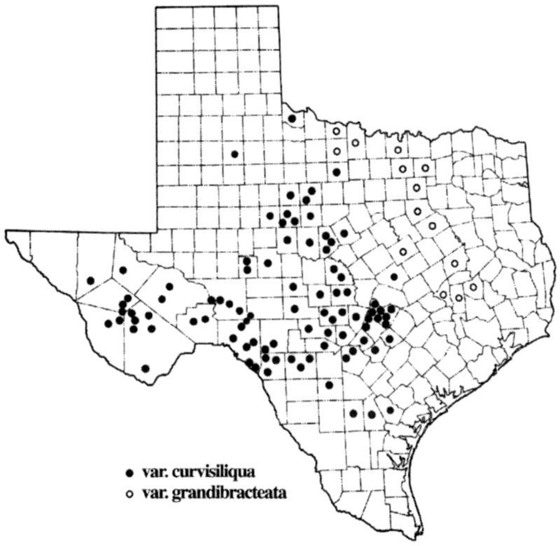
• var. curvisiliqua
○ var. grandibracteata

Corydalis curvisiliqua

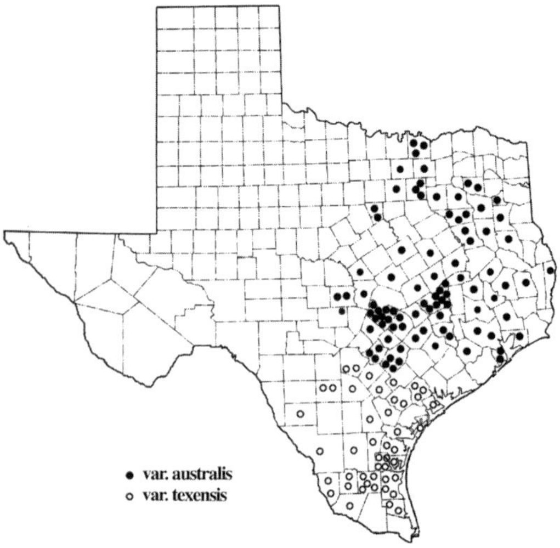
• var. australis
○ var. texensis

Corydalis micrantha

Fumaria officinalis

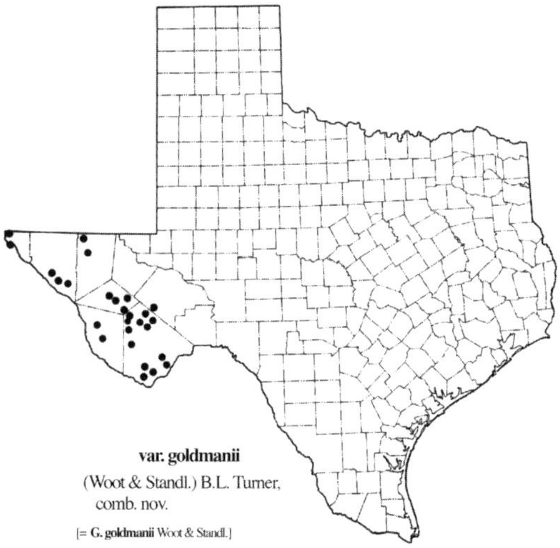
var. goldmanii
(Woot & Standl.) B.L. Turner, comb. nov.
[= G. goldmanii Woot & Standl.]

Garrya ovata

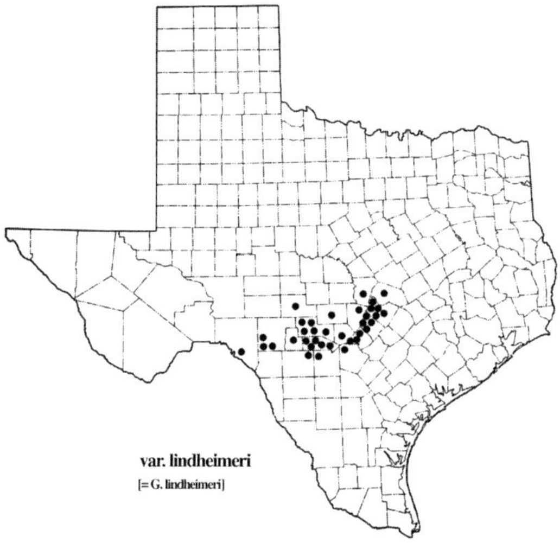
var. lindheimeri
[= G. lindheimeri]

Garrya ovata

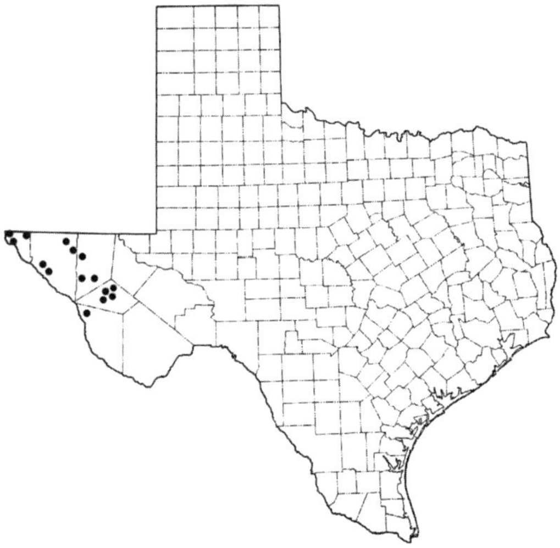

Garrya wrightii

Bartonia paniculata

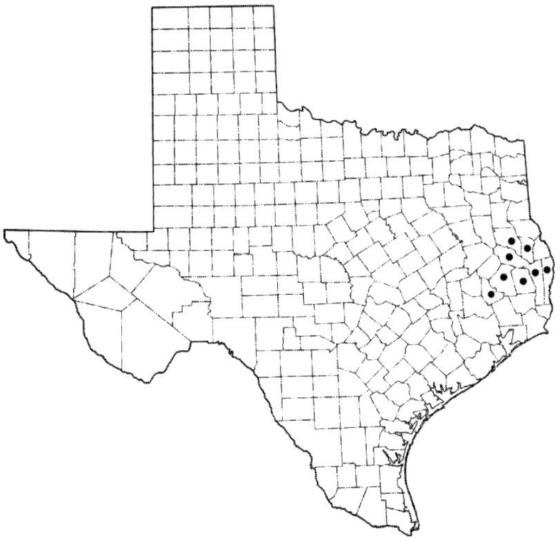

Bartonia texana

Bartonia verna

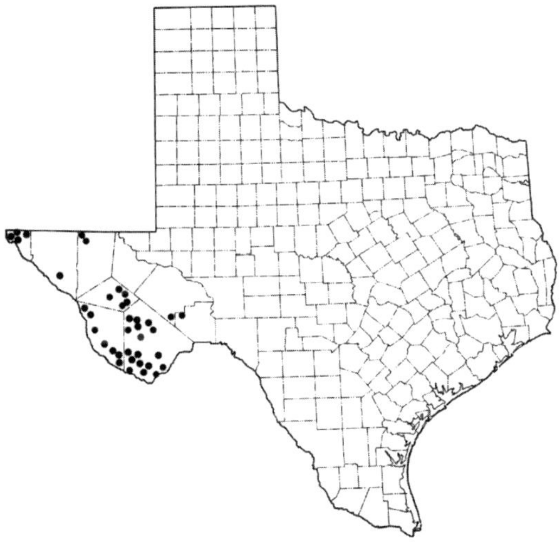

Centaurium arizonicum

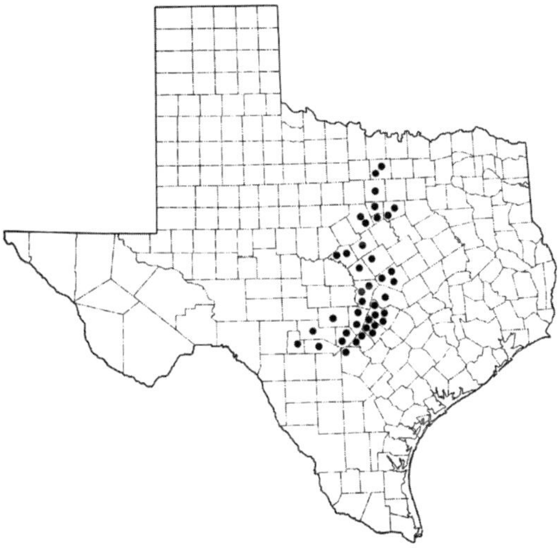

Centaurium beyrichii

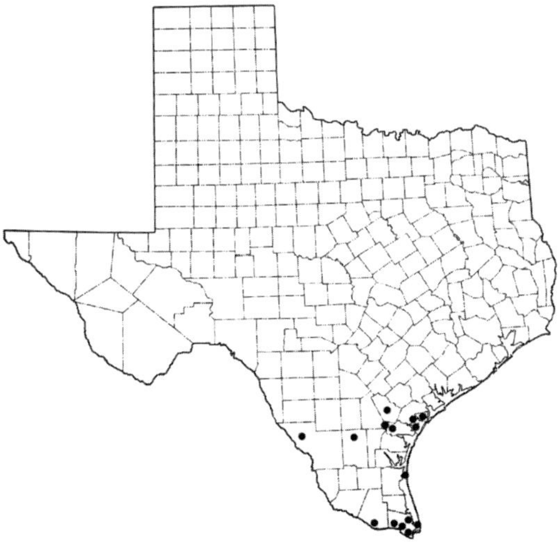

Centaurium breviflorum

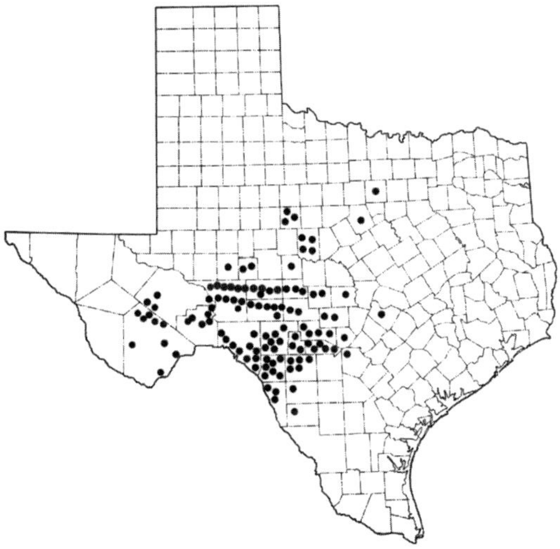

Centaurium calycosum

Centaurium glanduliferum

Centaurium maryannum

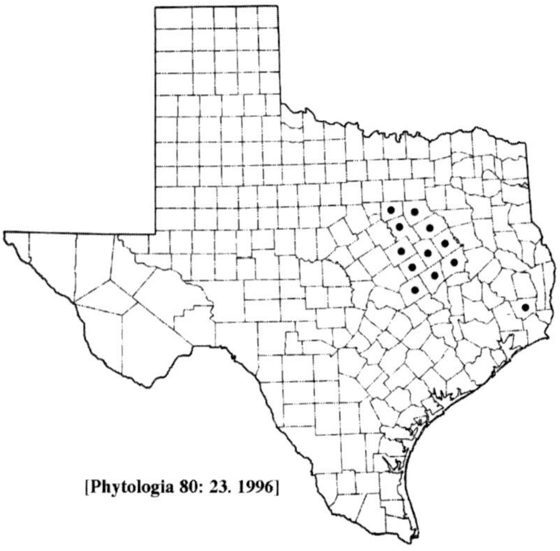
[Phytologia 80: 23. 1996]

Centaurium muhlenbergii

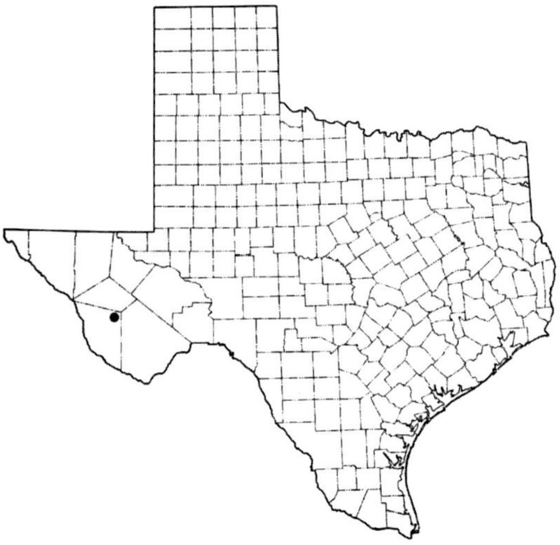

Centaurium multicaule

Centaurium nudicaule

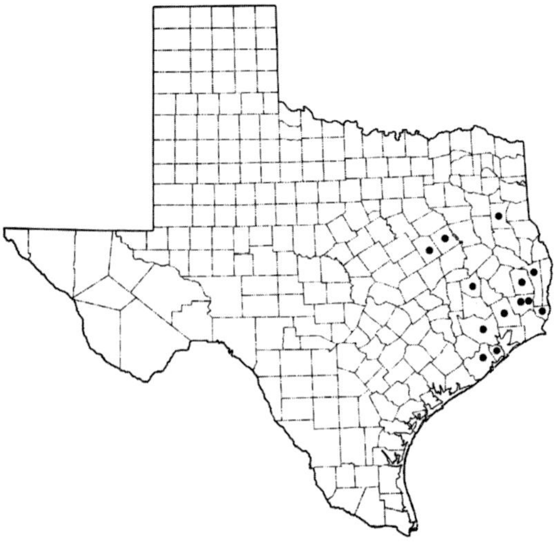

Centaurium pulchellum

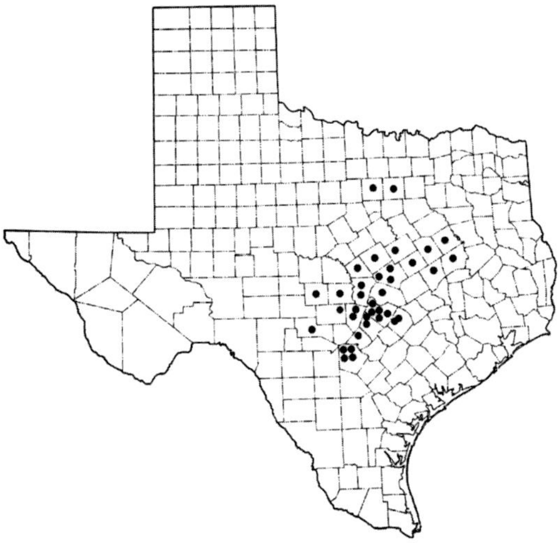

Centaurium texense

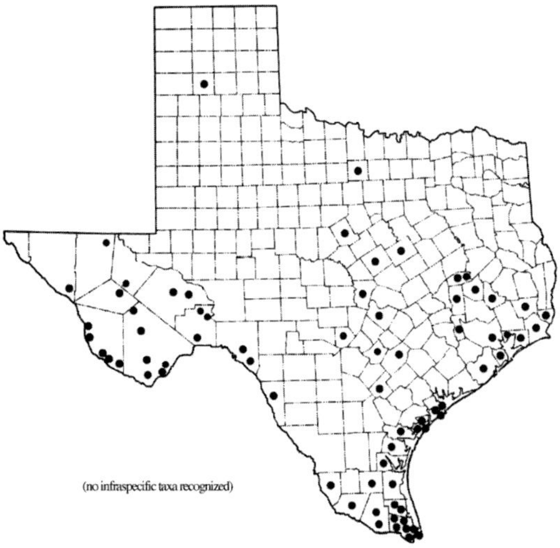
(no infraspecific taxa recognized)

Eustoma exaltatum

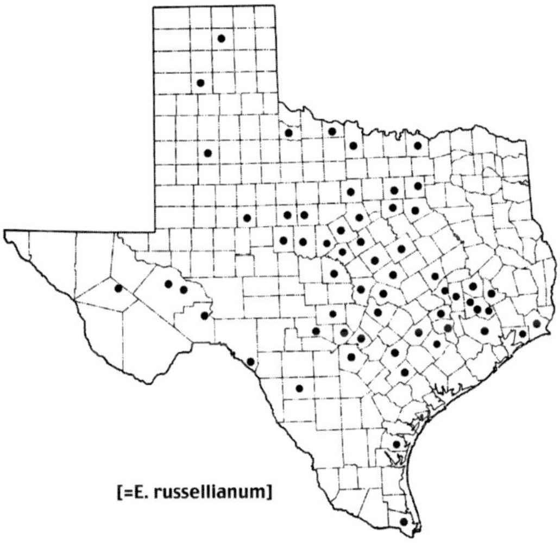
[=E. russellianum]

Eustoma grandiflorum

Gentiana affinis

Gentiana saponaria

Obolaria virginica

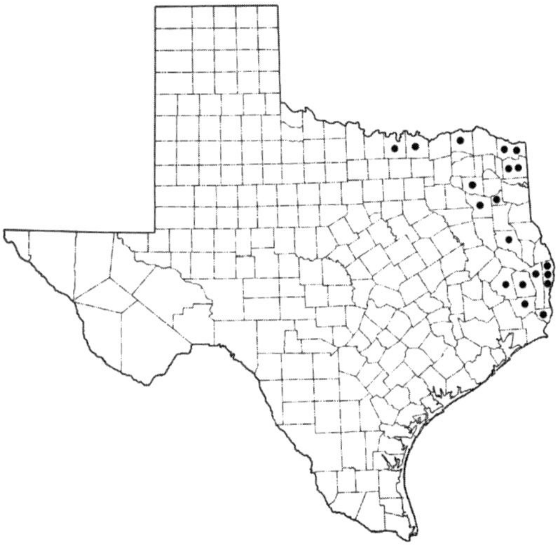

Sabatia angularis

Sabatia arenicola

[voucher TEX]

Sabatia brachiata

Sabatia calycina

Sabatia campanulata

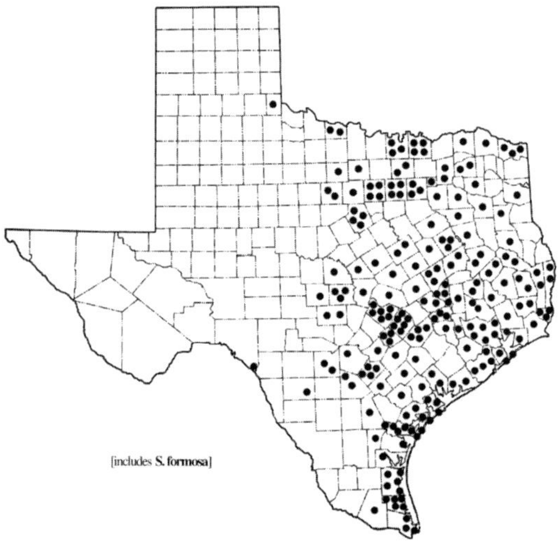
[includes S. formosa]

Sabatia campestris

Sabatia dodecandra

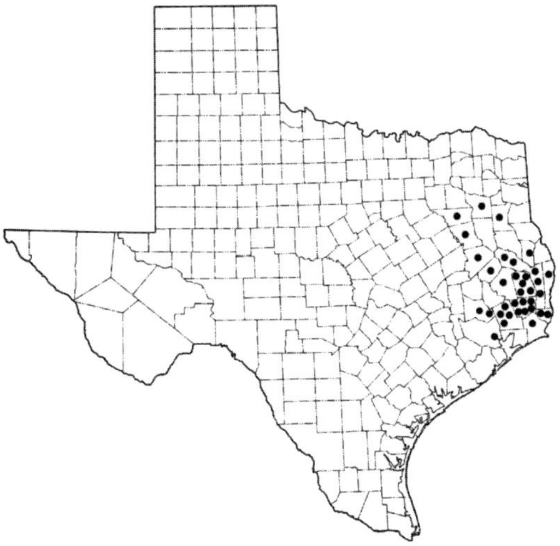

Sabatia gentianoides

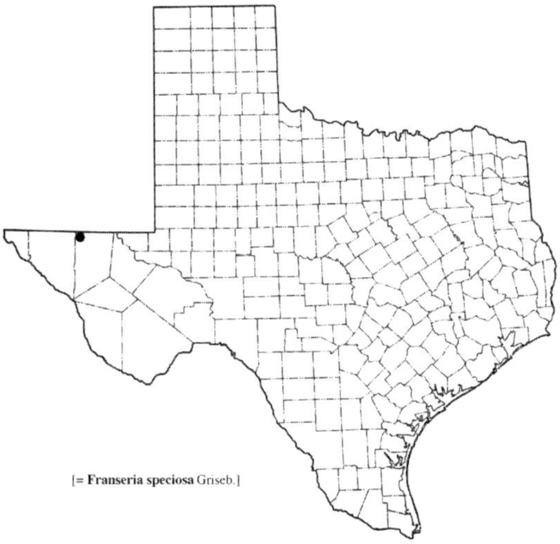
[= Franseria speciosa Griseb.]

Swertia speciosa

Erodium botrys

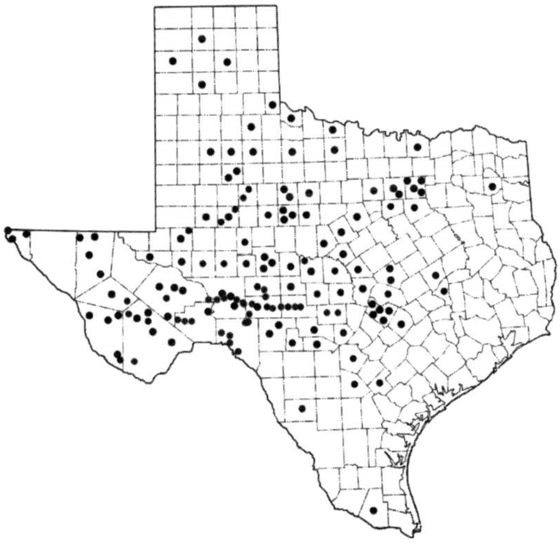

Erodium cicutarium

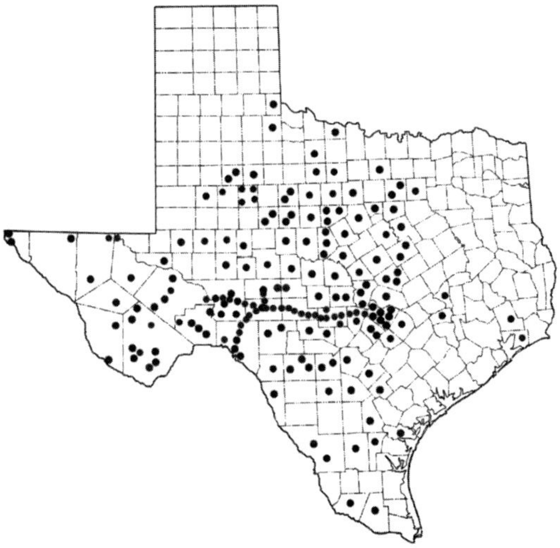

Erodium texanum

Geranium caespitosum

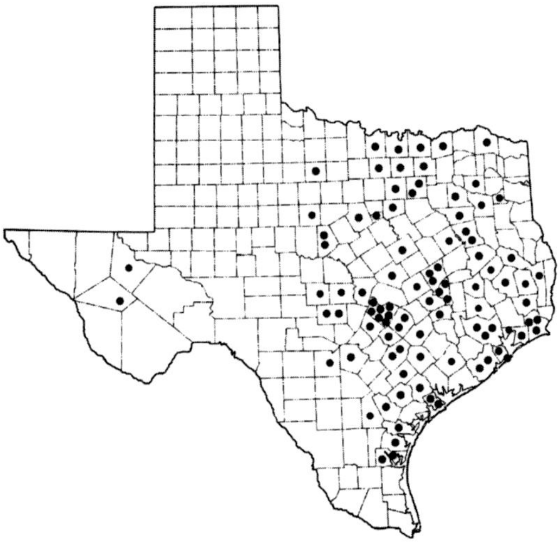

Geranium carolinianum

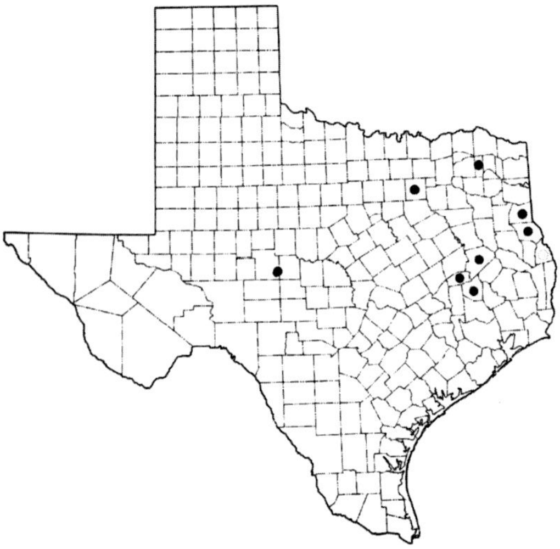

Geranium dissectum

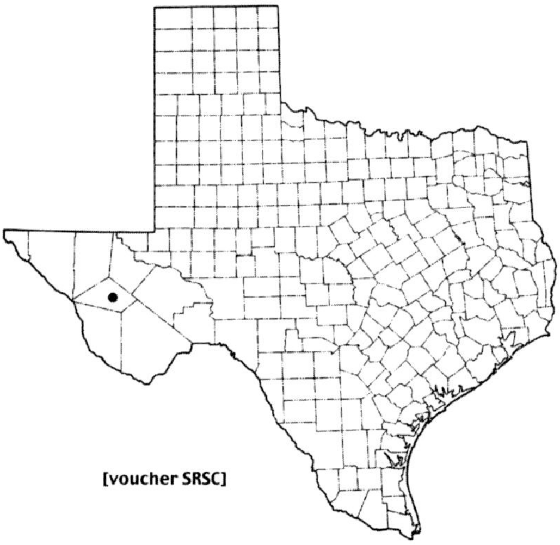
[voucher SRSC]

Geranium seemannii

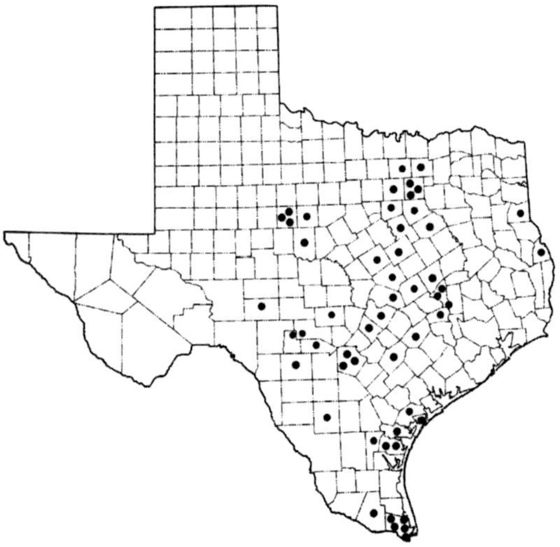

Geranium texanum

Scaevola plumieri

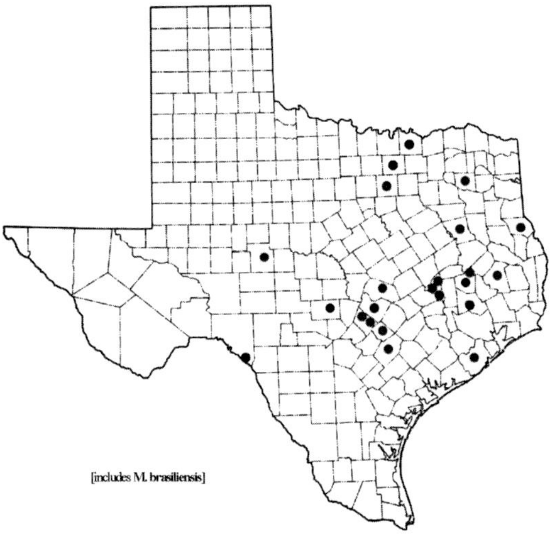
[includes M. brasiliensis]

Myriophyllum aquaticum

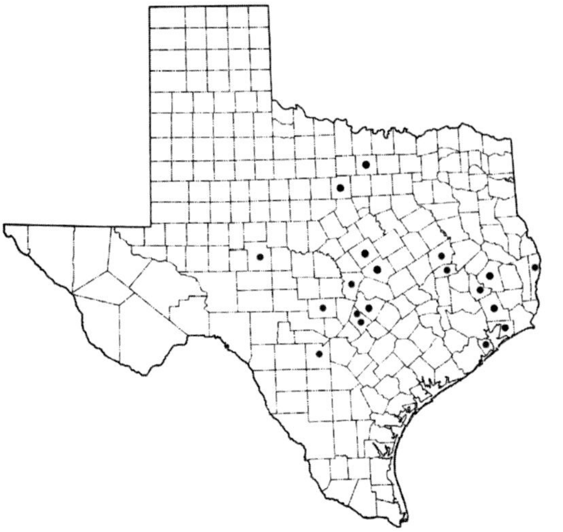

Myriophyllum heterophyllum

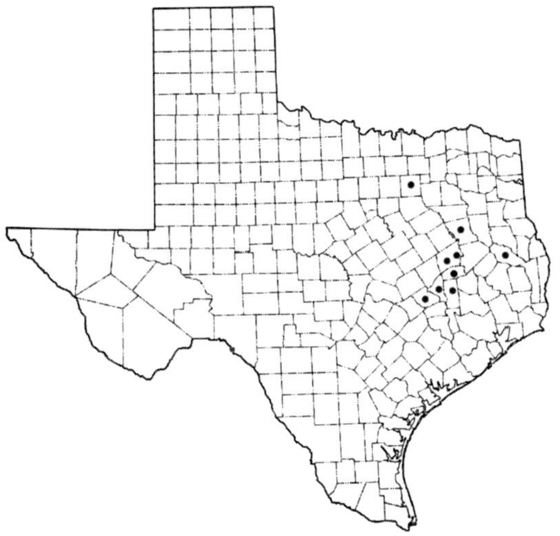

Myriophyllum pinnatum

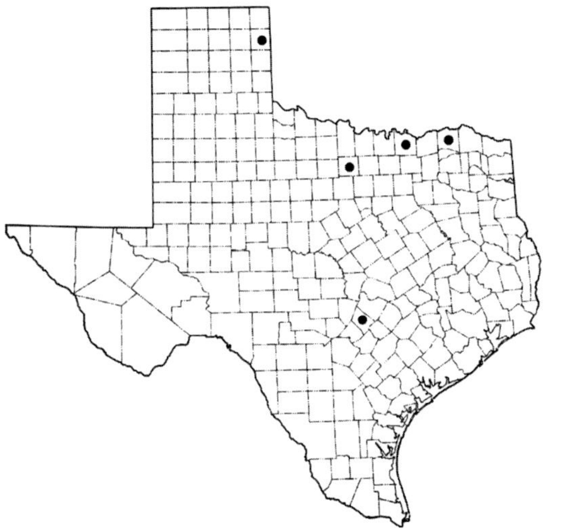

Myriophyllum spicatum

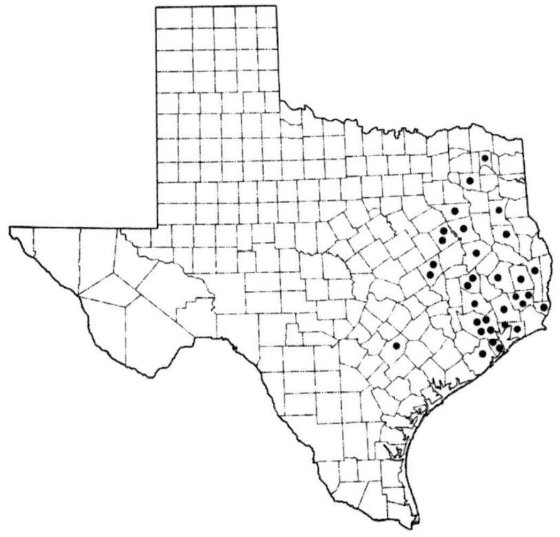

Proserpinaca palustris

Proserpinaca pectinata

Hamamelis vernalis

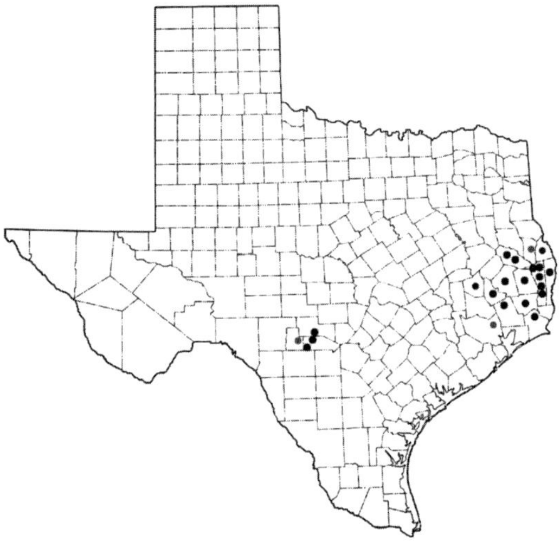

Hamamelis virginiana

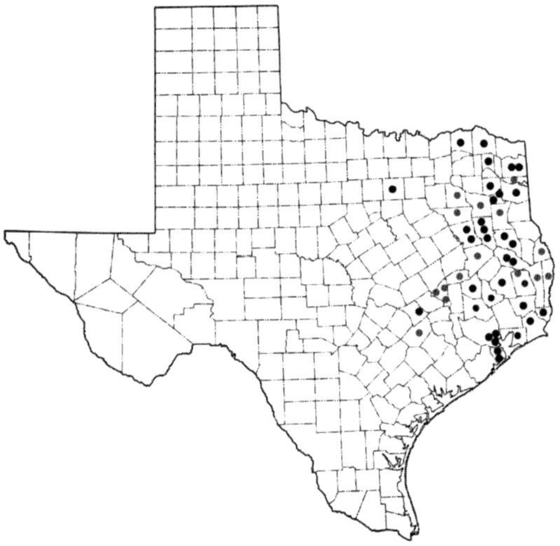

Liquidambar styraciflua

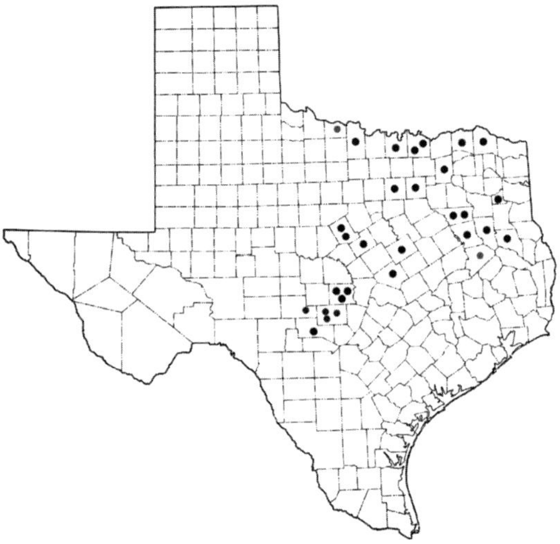

Aesculus arguta

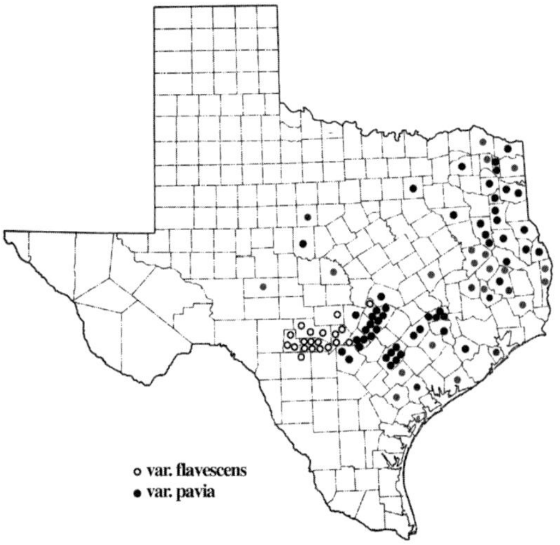
○ var. flavescens
● var. pavia

Aesculus pavia

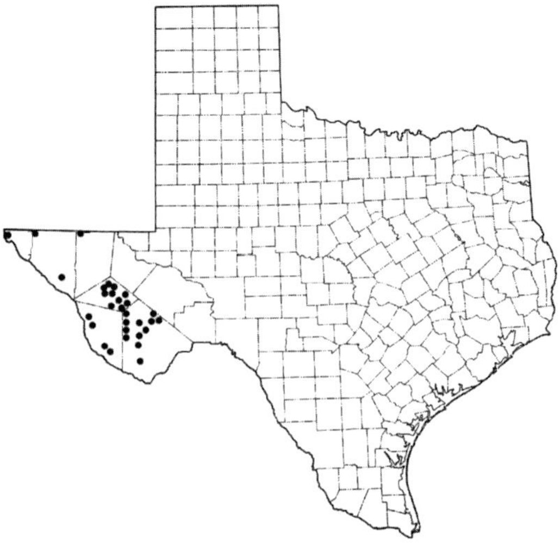

Fendlera falcata

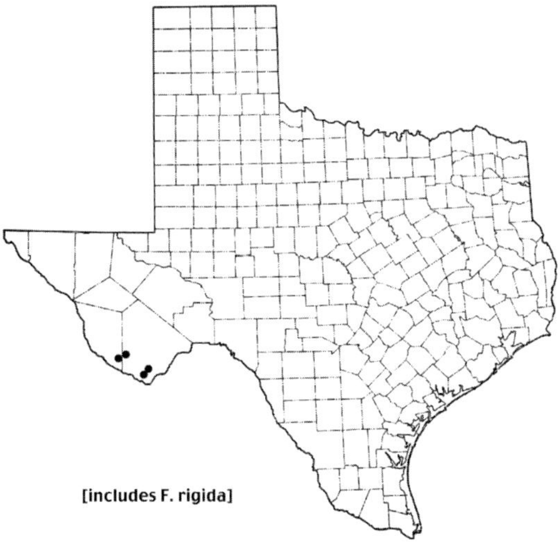
[includes F. rigida]

Fendlera linearis

Fendlera rupicola

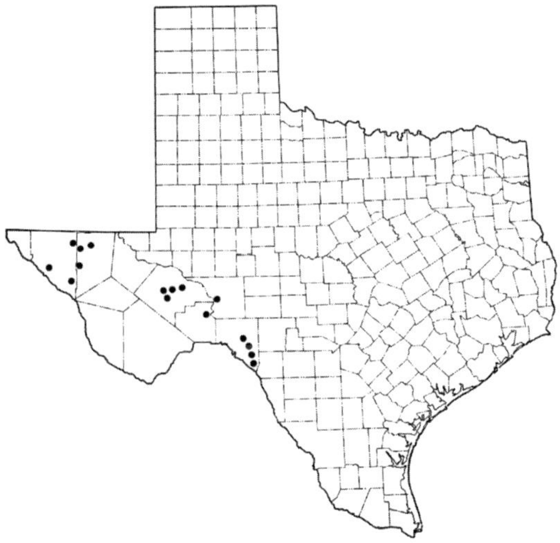

Fendlera wrightii

Fendlerella utahensis

[includes P. crinitus]

Philadelphus argenteus

Philadelphus ernestii

[includes P. hitchcockianus]

Philadelphus mearnsii

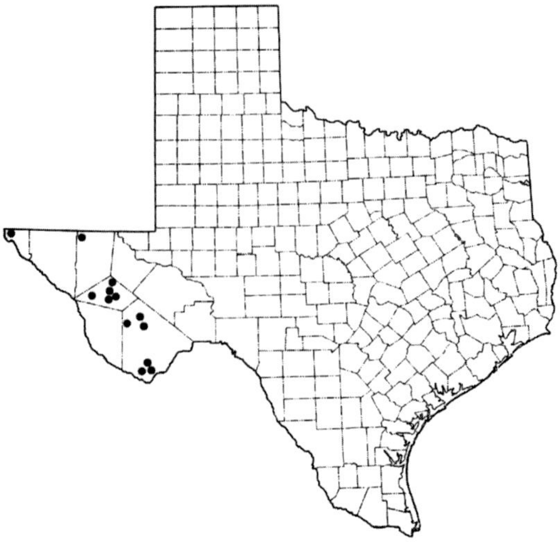

Philadelphus microphyllus

Philadelphus occidentalis

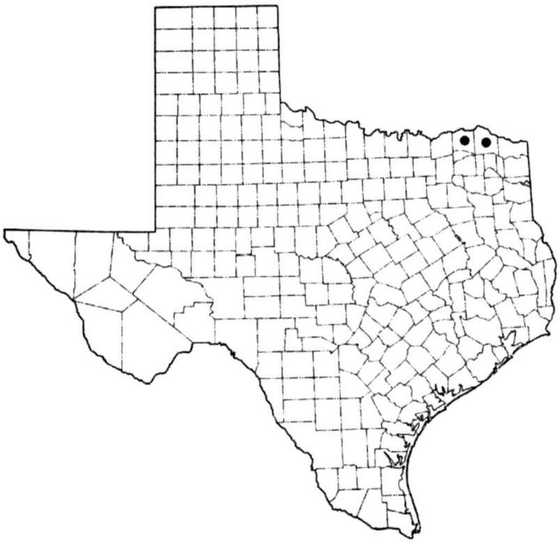

Philadelphus pubescens

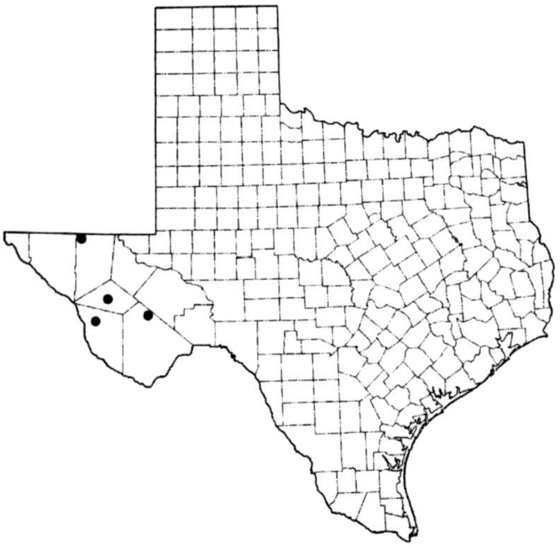

Philadelphus serpyllifolius

Philadelphus texanus

Ellisia nyctelea

Eucrypta micrantha

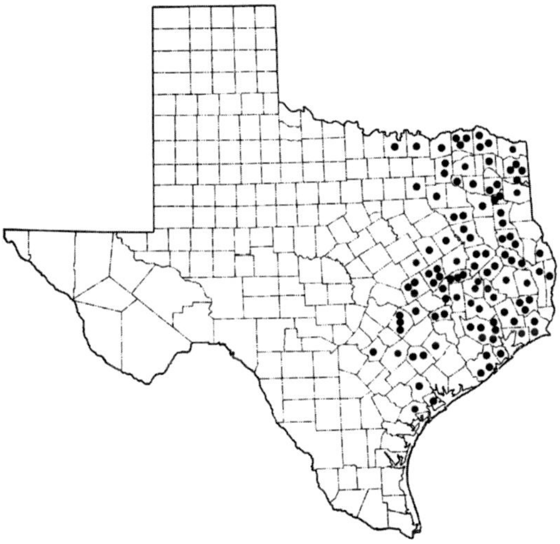

Hydrolea ovata

Hydrolea spinosa

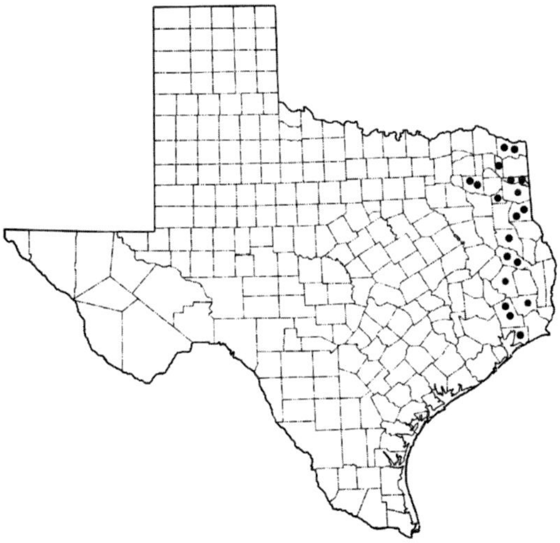

Hydrolea uniflora

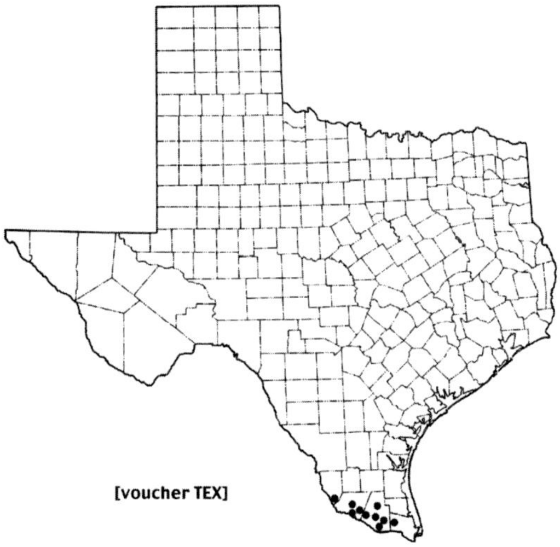
[voucher TEX]

Nama berlandieri

Nama carnosum

Nama dichotomum

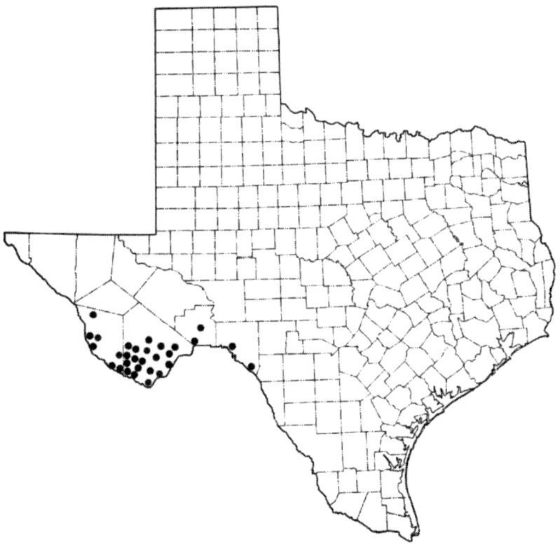

Nama havardii

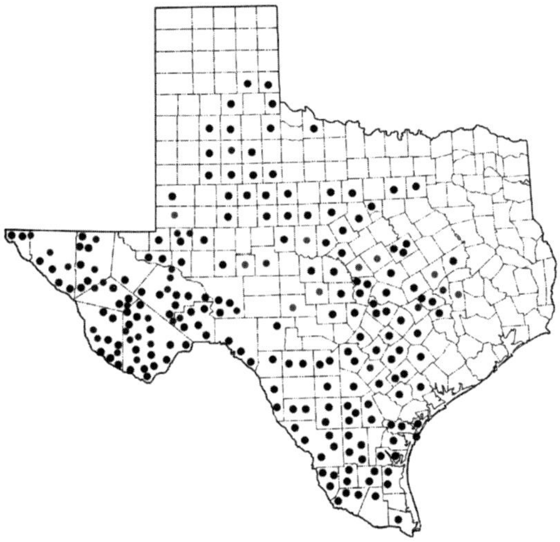

Nama hispida

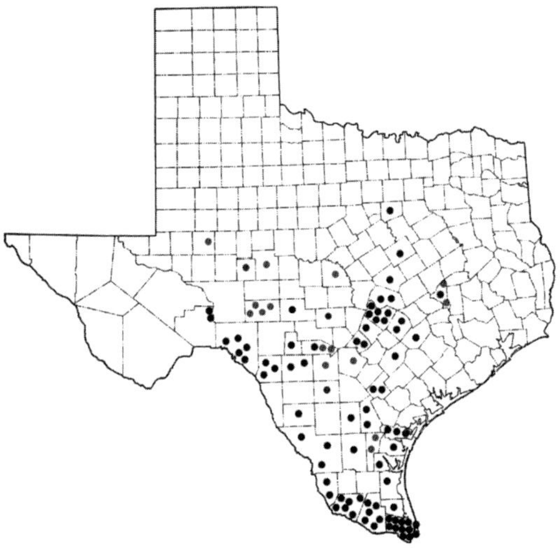

Nama jamaicense

Nama parvifolium

Nama stenocarpum

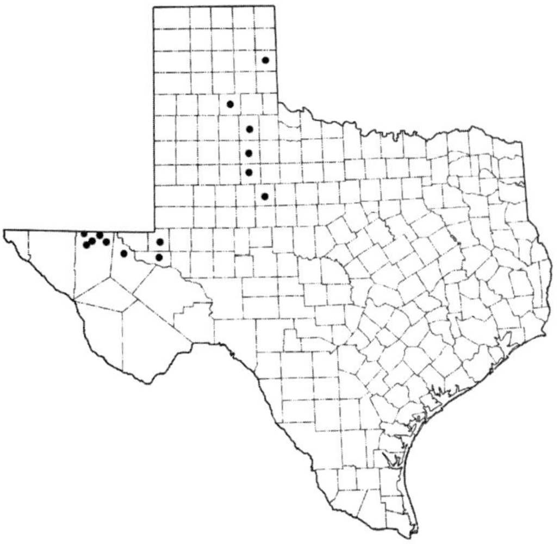

Nama stevensii

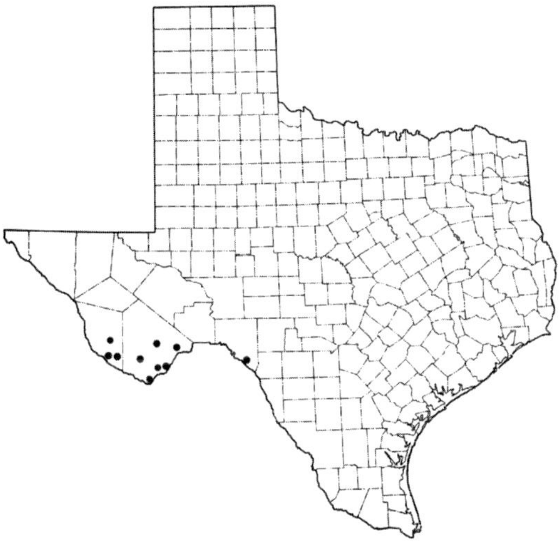

Nama torynophyllum

Nama undulatum

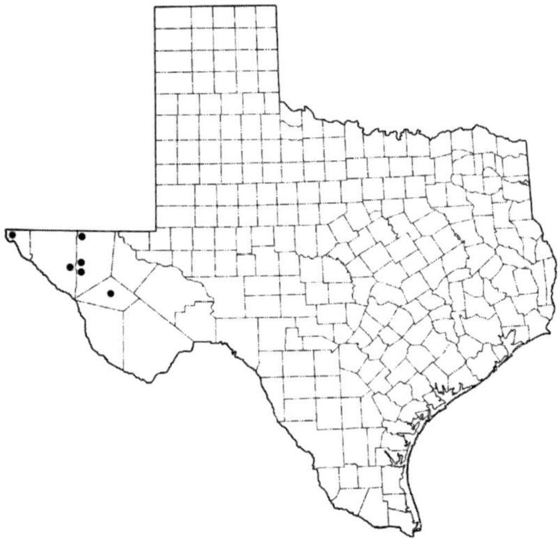

Nama xylopodum

[= N. triloba]

Nemophila aphylla

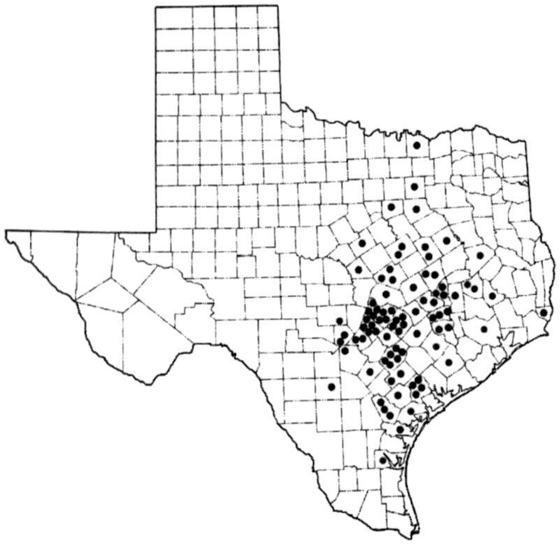

Nemophila phacelioides

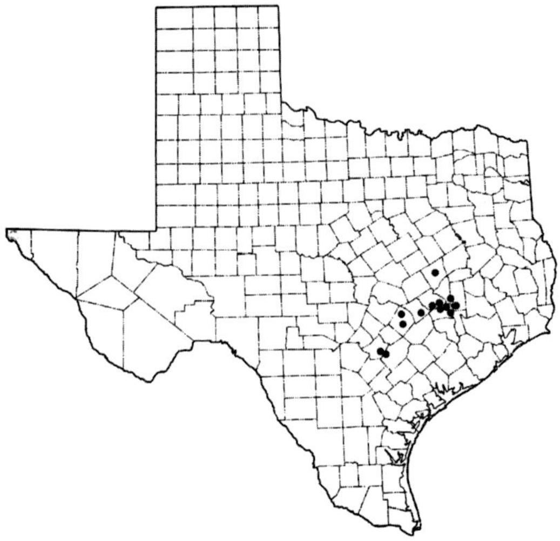

Nemophila sayersensis

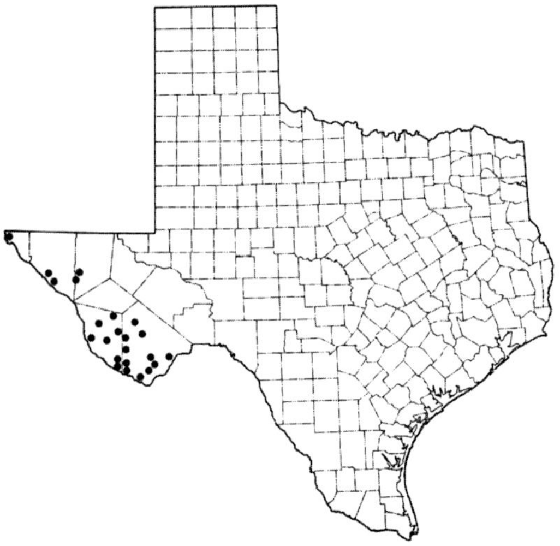

Phacelia coerulea

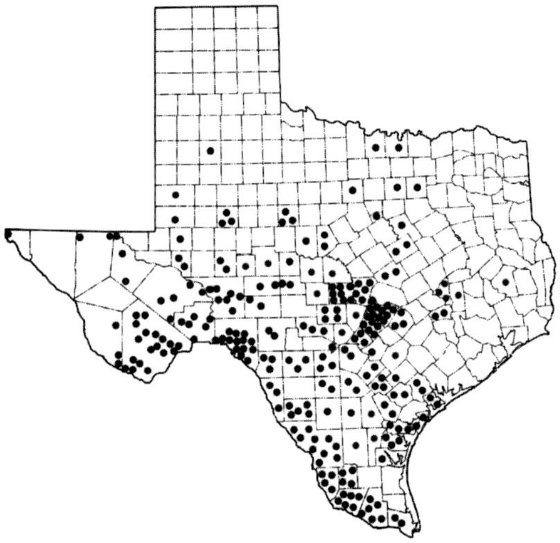

Phacelia congesta

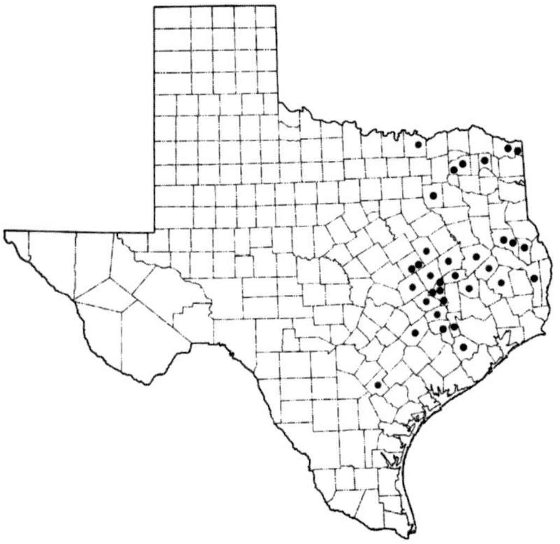

Phacelia glabra

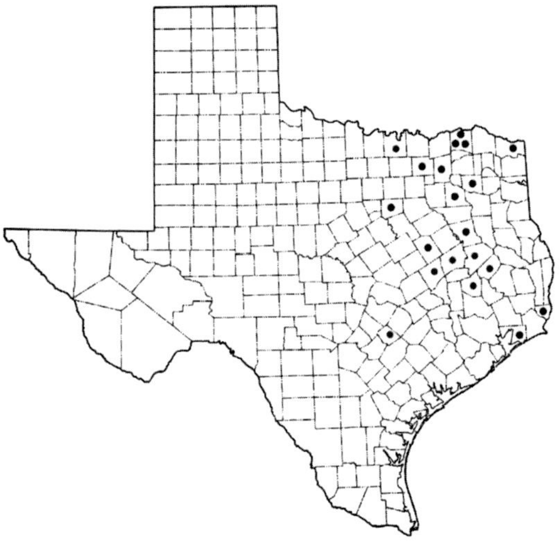

Phacelia hirsuta

Phacelia infundibuliformis

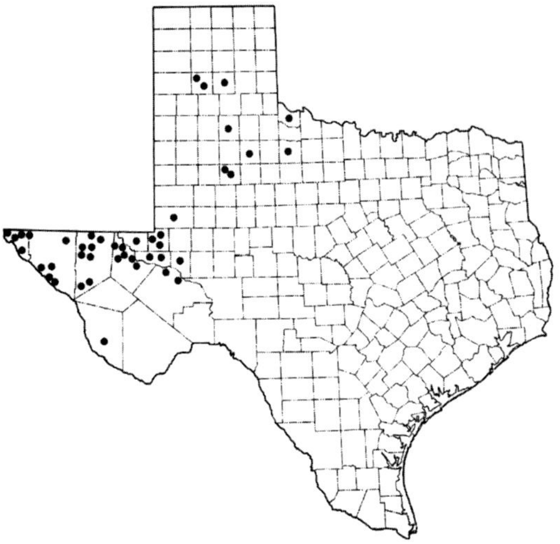

Phacelia integrifolia

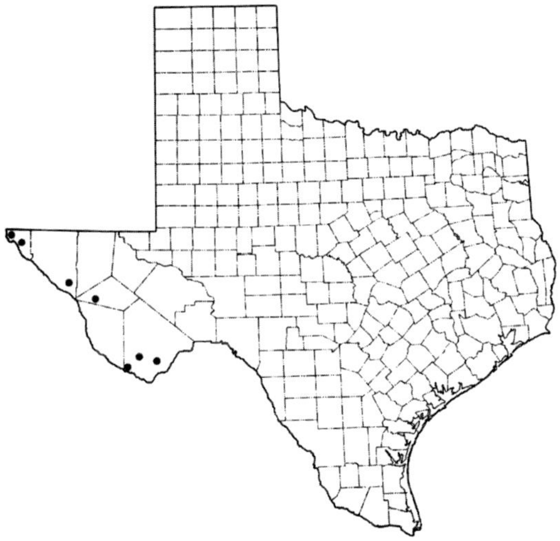

Phacelia intermedia

Phacelia laxa

Phacelia pallida

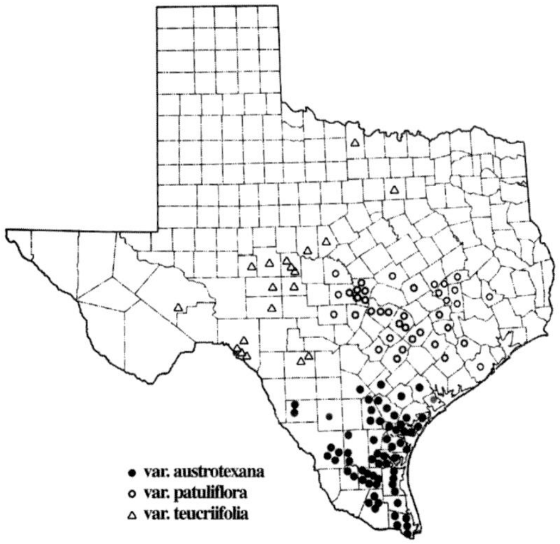
• var. austrotexana
○ var. patuliflora
△ var. teucriifolia

Phacelia patuliflora

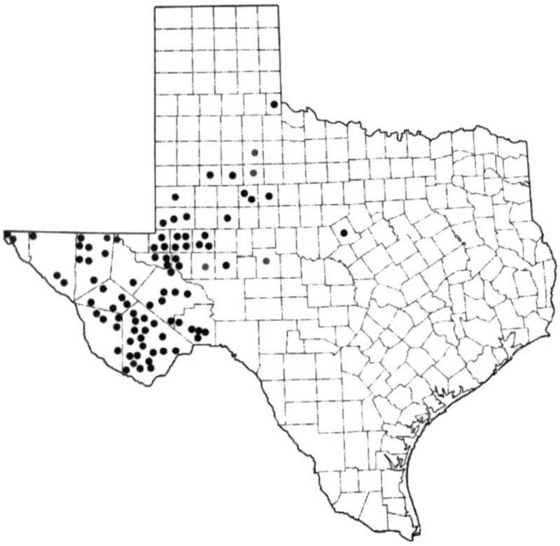

Phacelia popei

Phacelia robusta

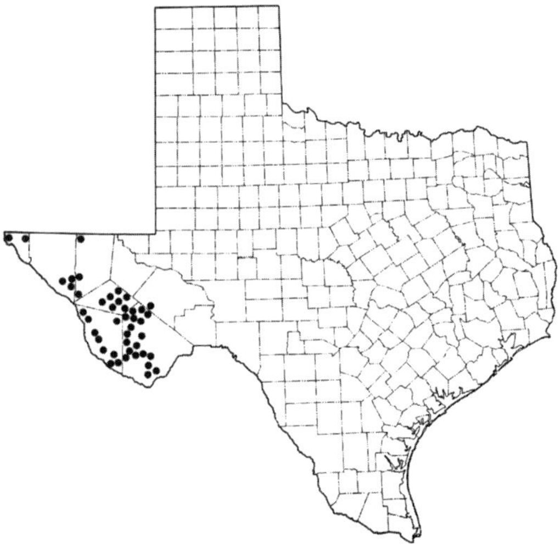

Phacelia rupestris

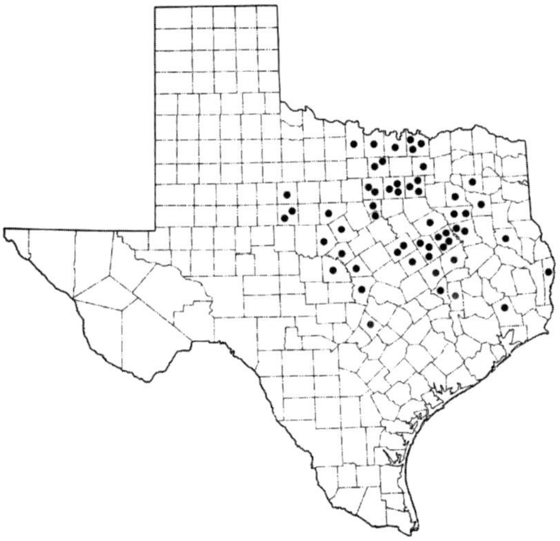

Phacelia strictiflora

Hypericum cistifolium

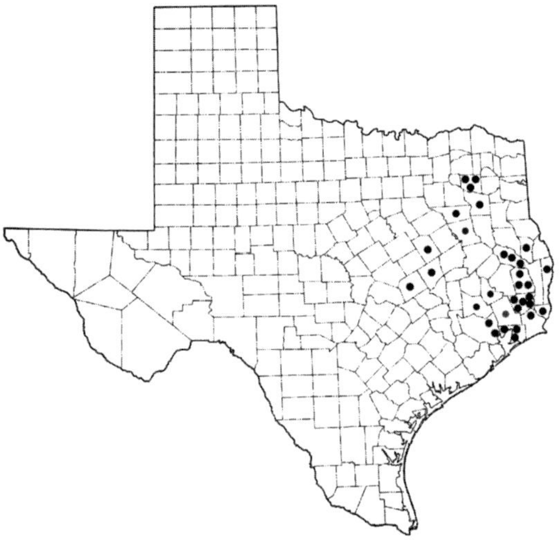

Hypericum crux-andreae

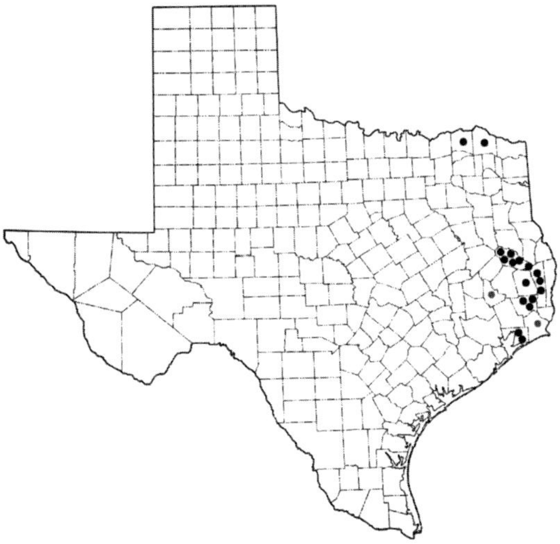

Hypericum densiflorum

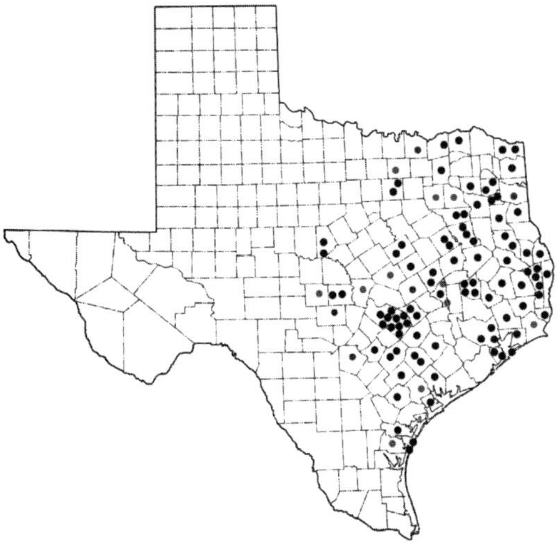

Hypericum drummondii

Hypericum frondosum

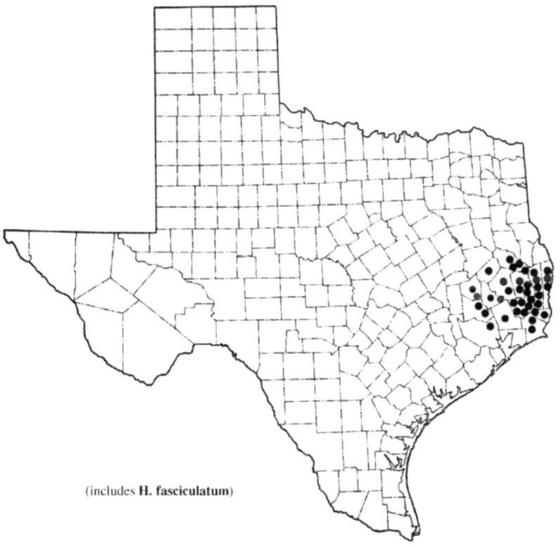
(includes **H. fasciculatum**)

Hypericum galioides

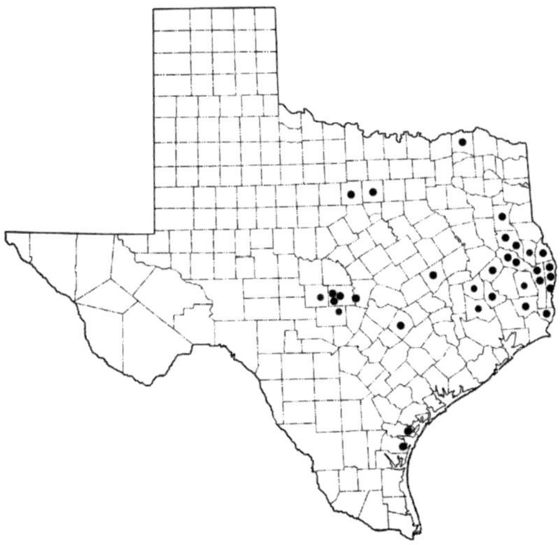

Hypericum gentianoides

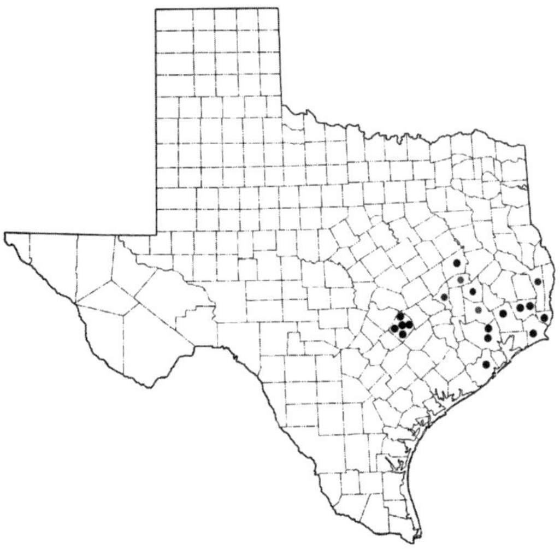

Hypericum gymnanthum

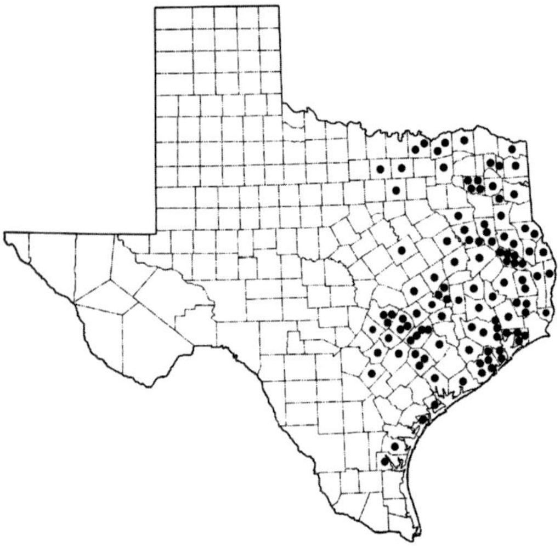

Hypericum hypericoides

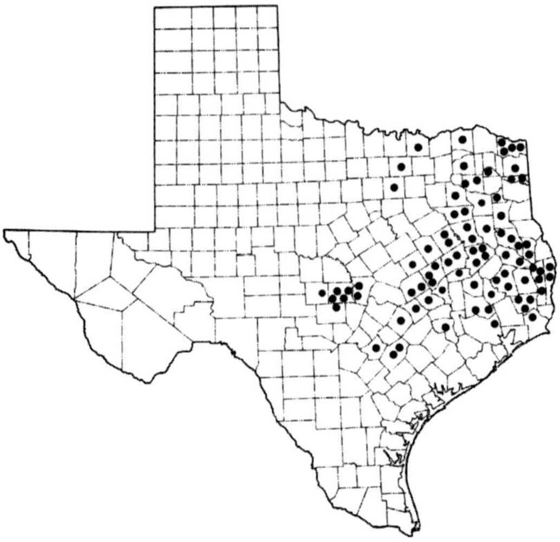

Hypericum mutilum

Hypericum nudiflorum

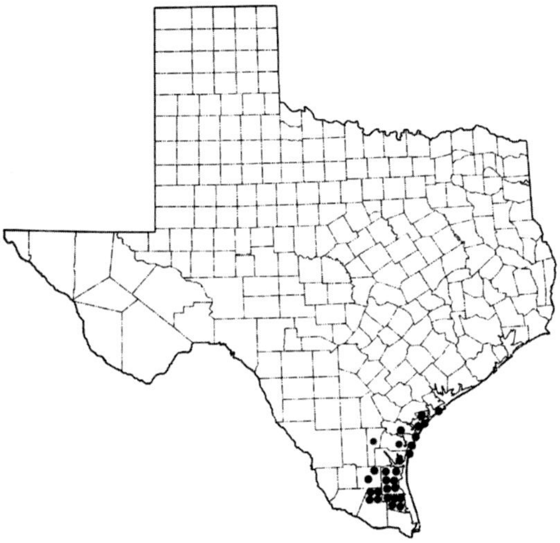

Hypericum pauciflorum

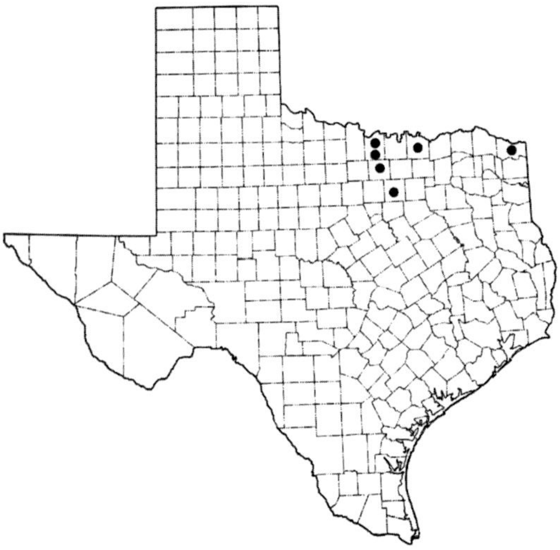

Hypericum perforatum

Hypericum prolificum

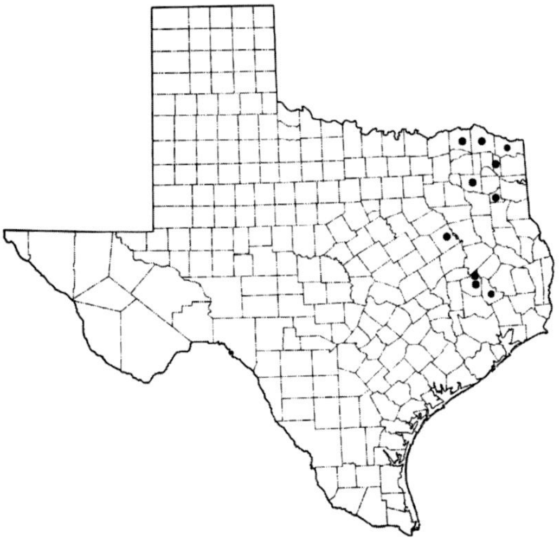
Hypericum pseudomaculatum

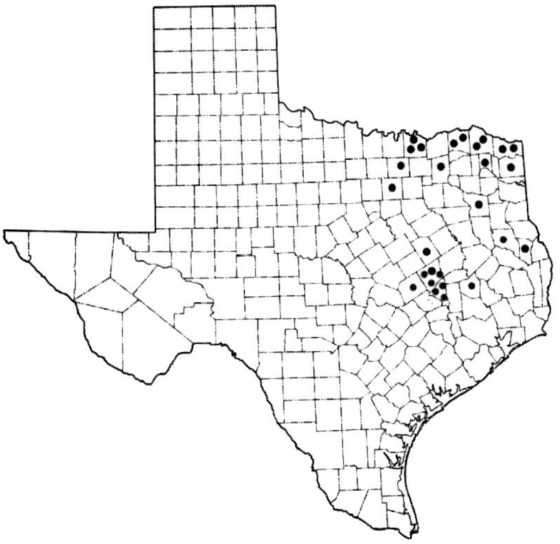
Hypericum punctatum

Hypericum setosum

Hypericum sphaerocarpum

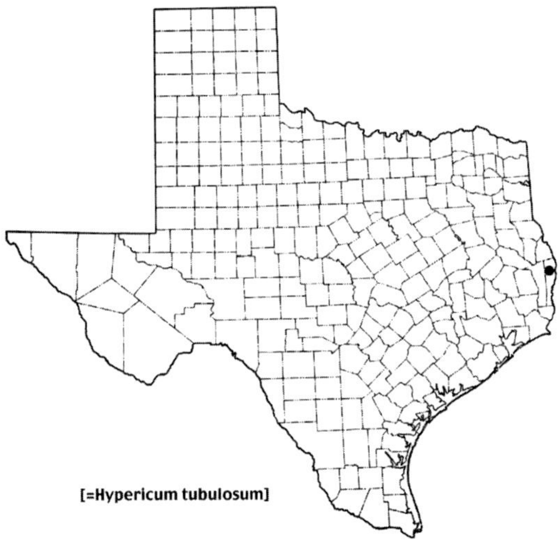
[=Hypericum tubulosum]

Triadenum tubulosum

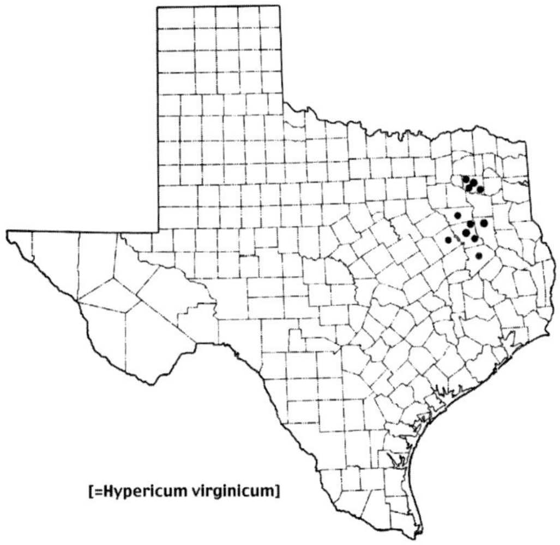
[=Hypericum virginicum]

Triadenum virginicum

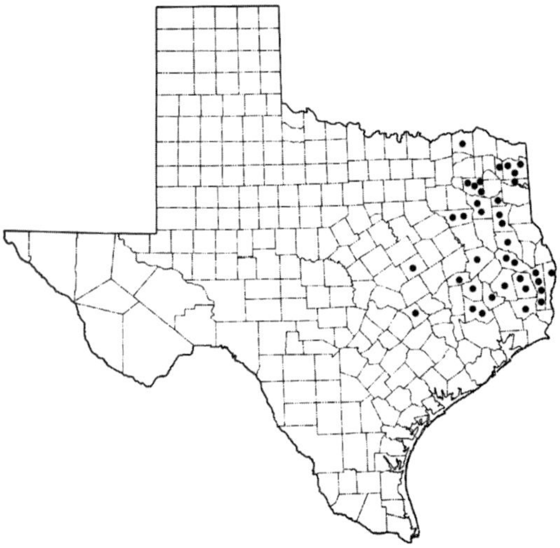

Triadenum walteri

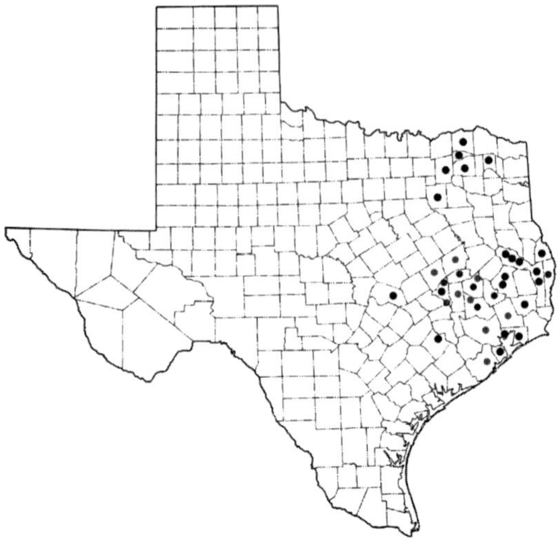

Carya aquatica

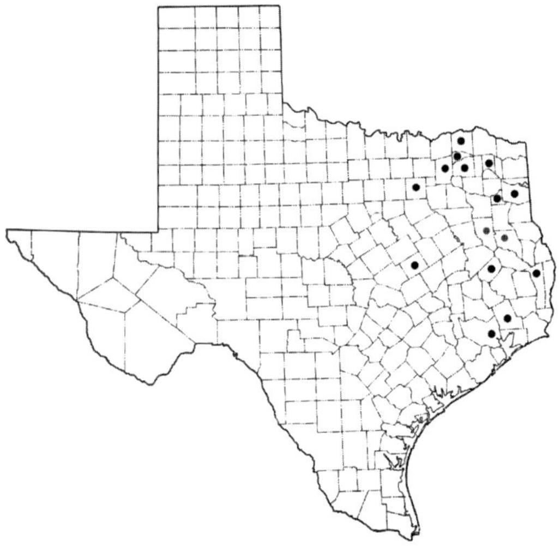

Carya cordiformis

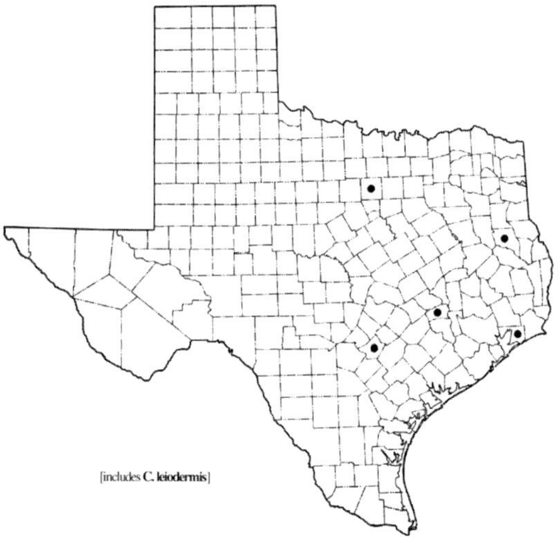
[includes C. leiodermis]

Carya glabra

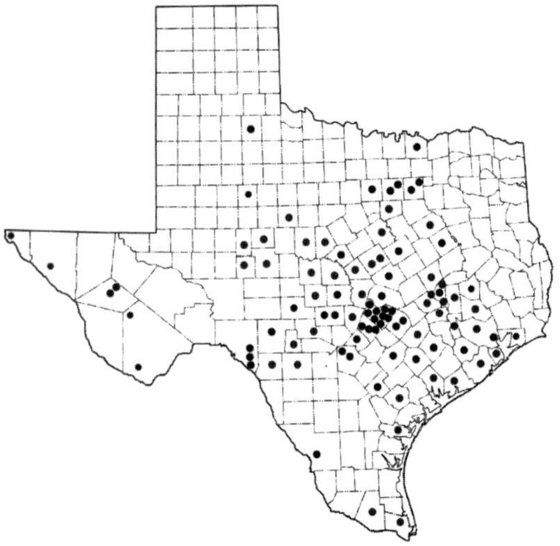

Carya illinoinensis

Carya laciniosa

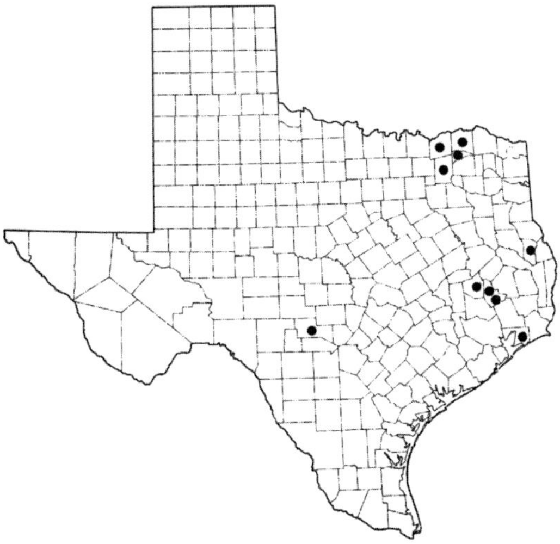

Carya myristiciformis

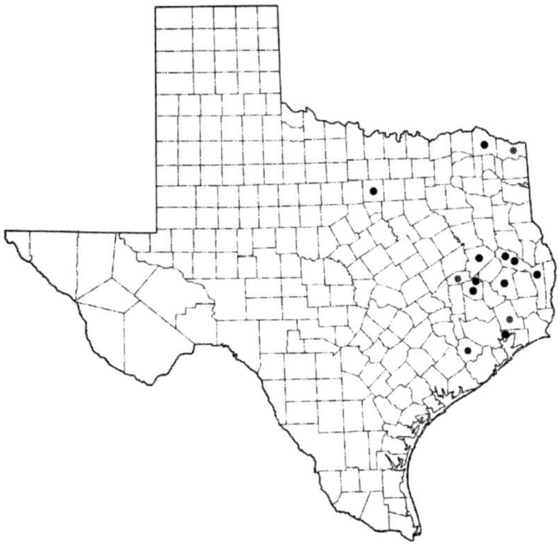

Carya ovata

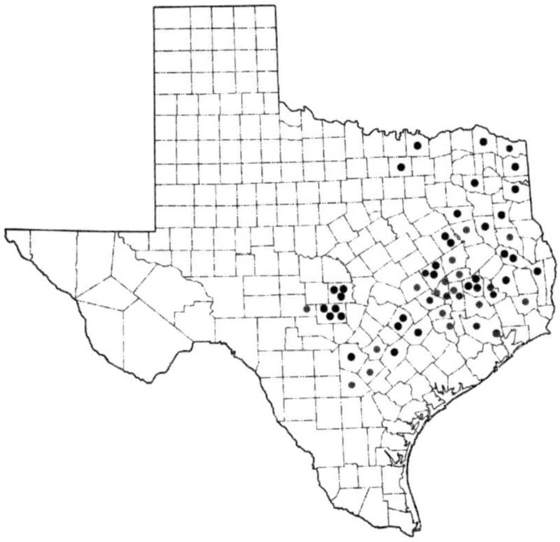

Carya texana

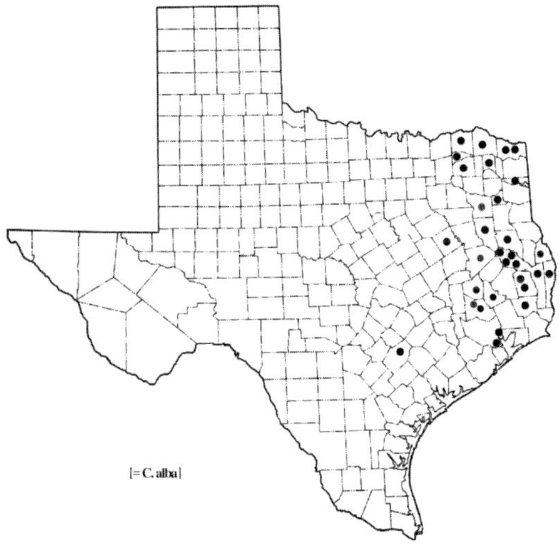
[= C. alba]

Carya tomentosa

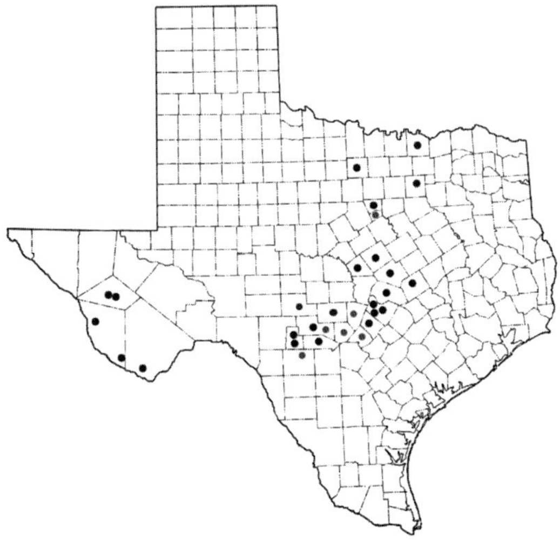

Juglans major

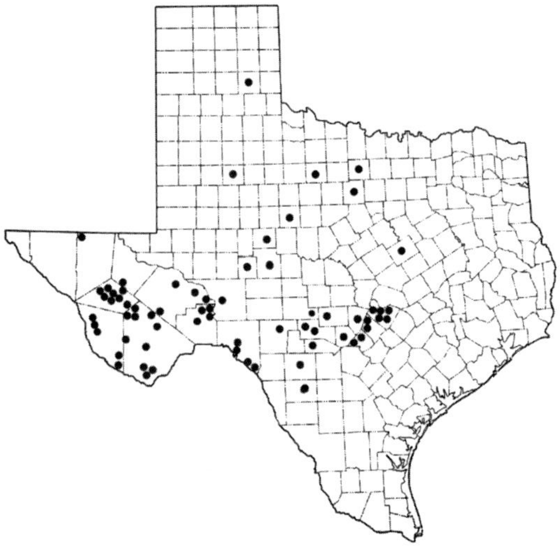

Juglans microcarpa

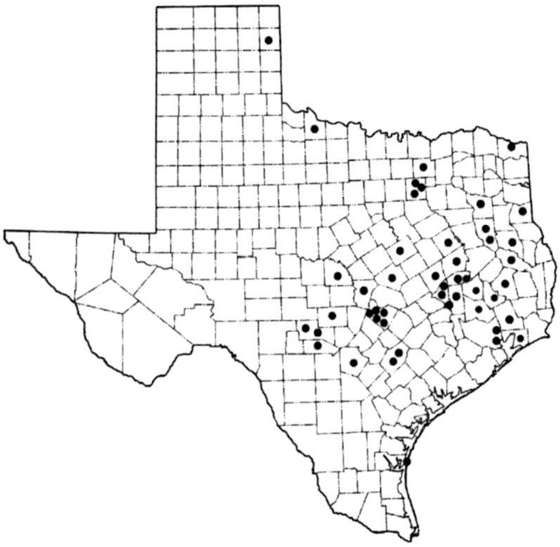
Juglans nigra

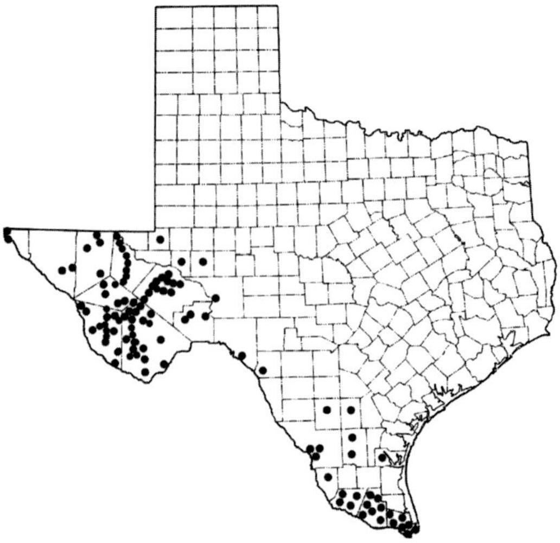
Koeberlinia spinosa

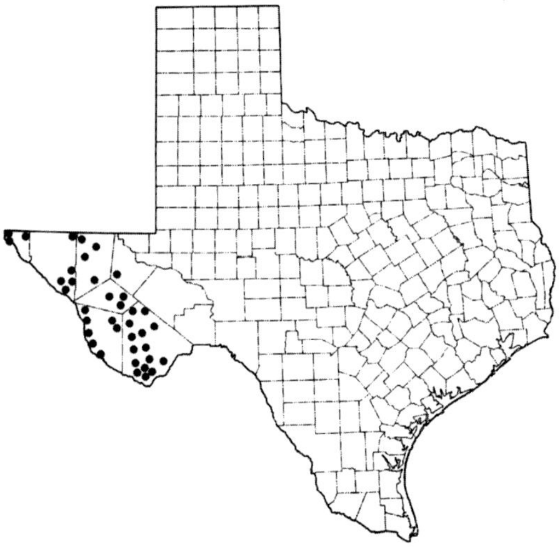
Krameria erecta

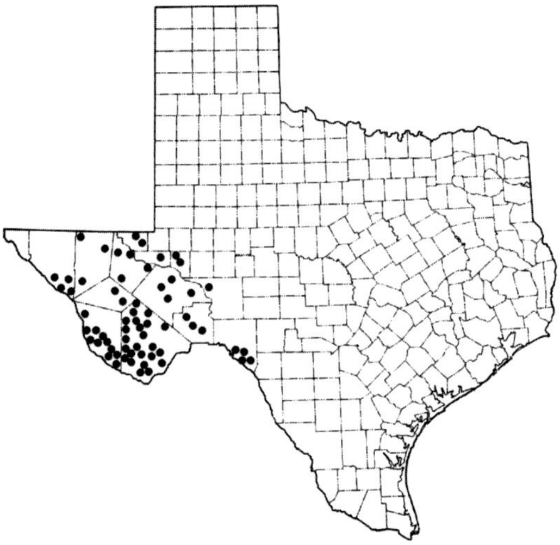
Krameria grayi

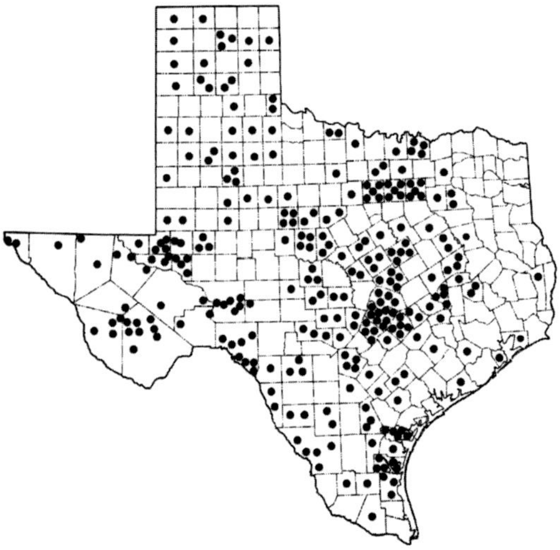
Krameria lanceolata

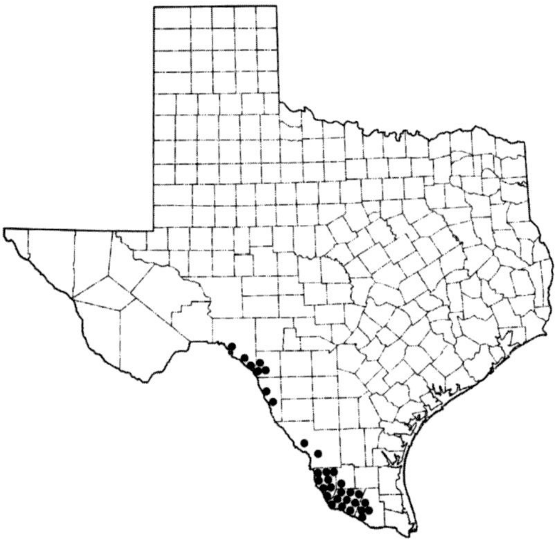
Krameria ramosissima

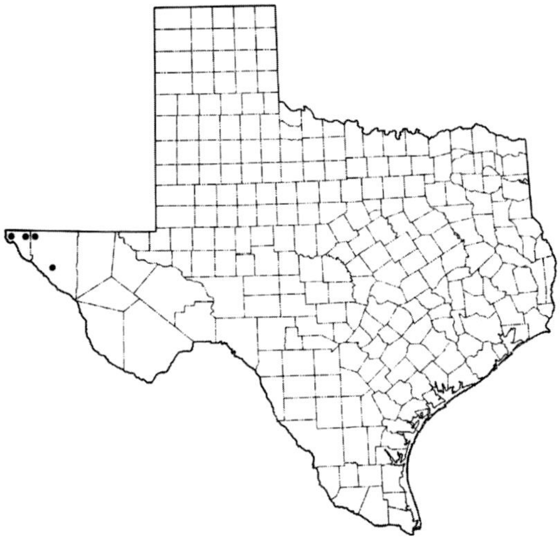

Agastache cana

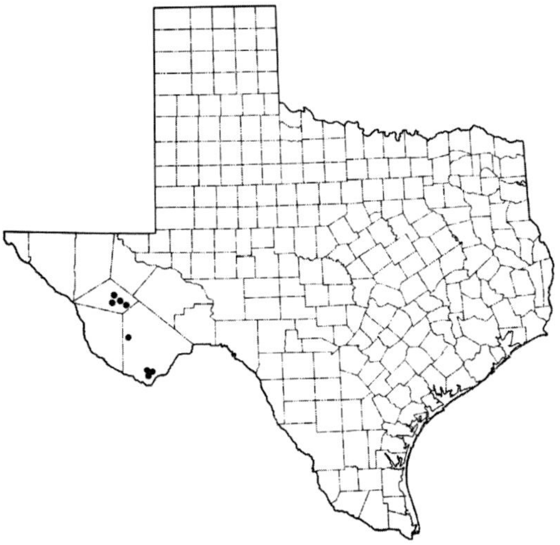

Agastache micrantha

Agastache pallidiflora

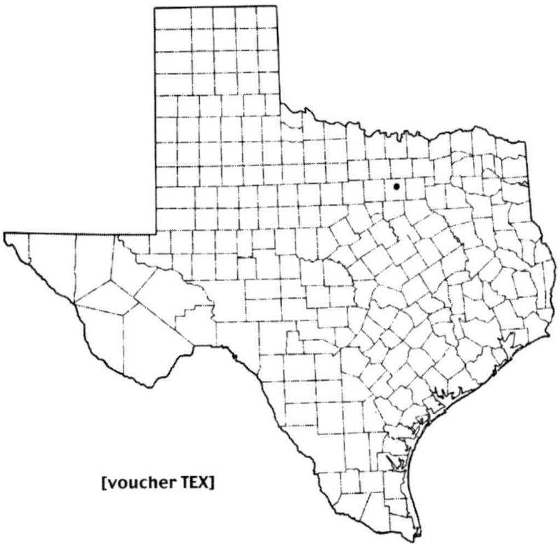
[voucher TEX]

Agastache scrophulariifolia

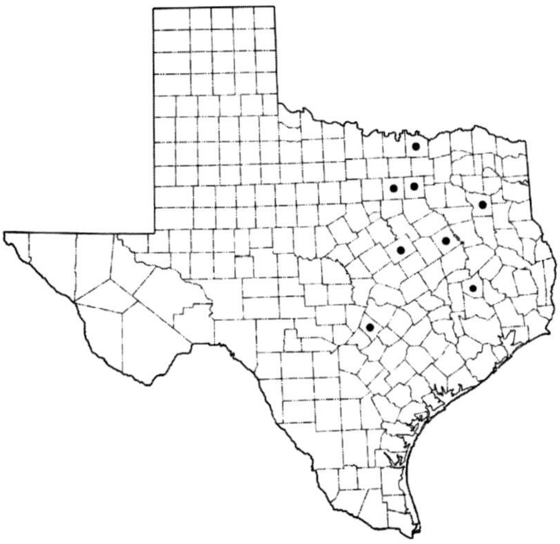

Ajuga reptans

Brazoria arenaria

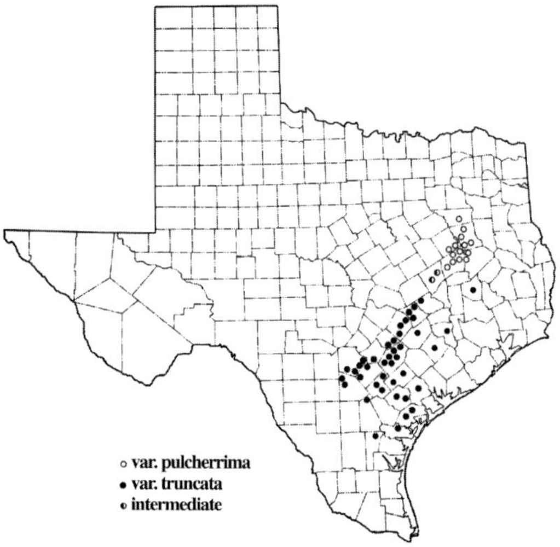
var. pulcherrima
var. truncata
intermediate

Brazoria truncata

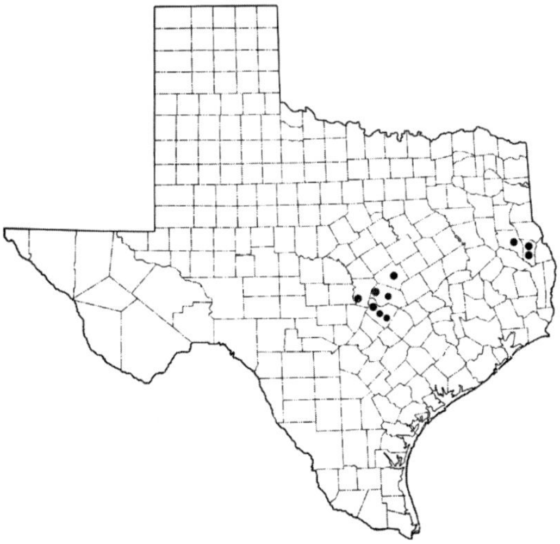

Calamintha arkansana

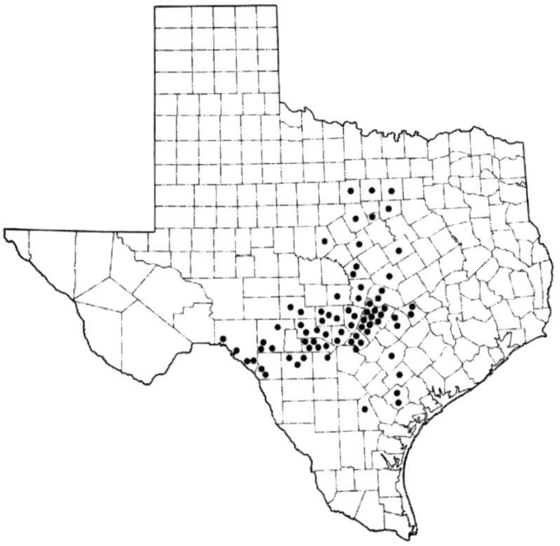

Hedeoma acinoides

Hedeoma apiculata

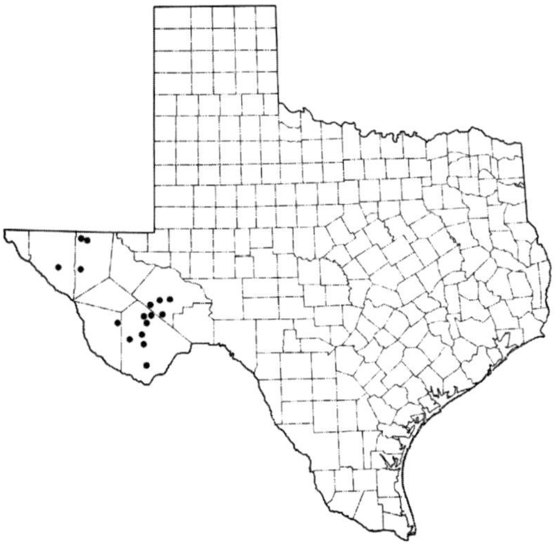

Hedeoma costata

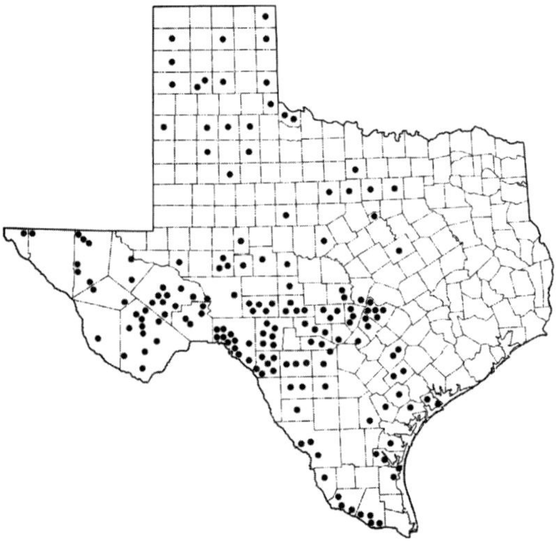

Hedeoma drummondii

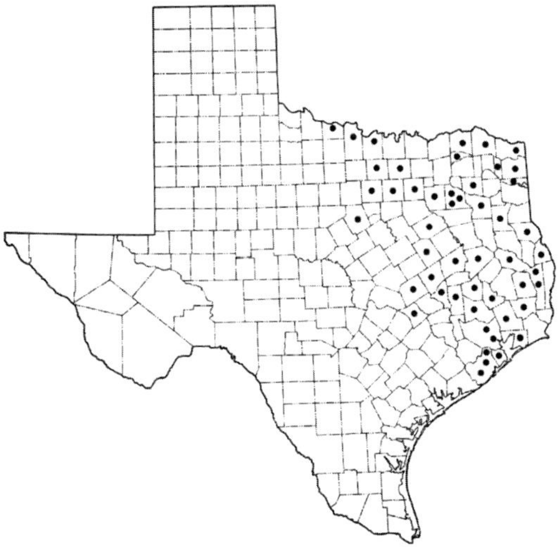

Hedeoma hispida

Hedeoma mollis

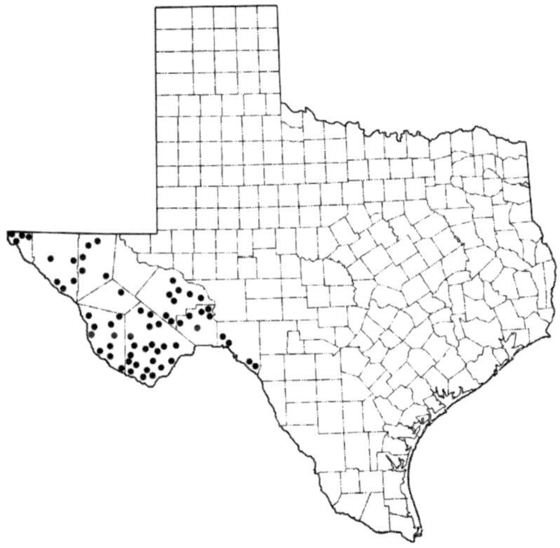

Hedeoma nana

Hedeoma pilosa

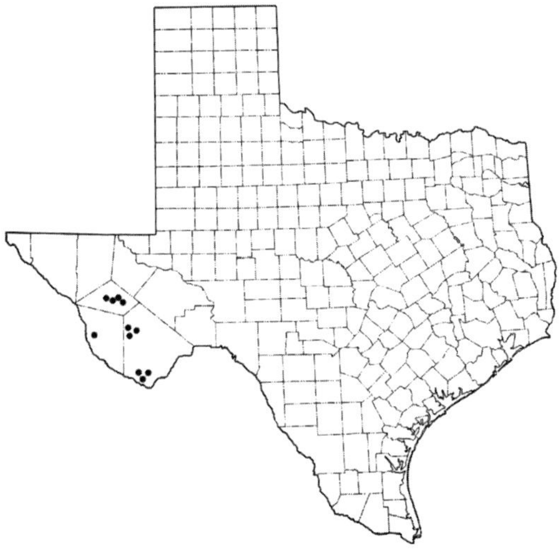

Hedeoma plicata

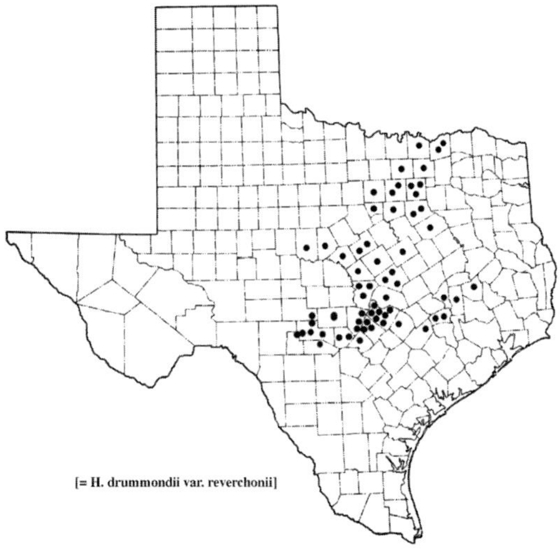
[= H. drummondii var. reverchonii]

Hedeoma reverchonii

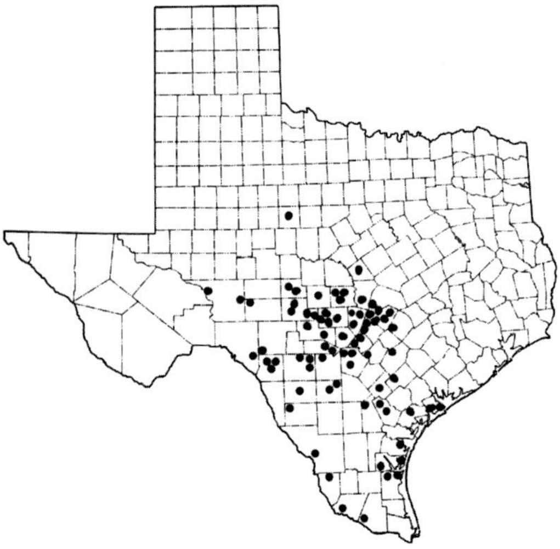

Hedeoma serpyllifolia

Hyptis alata

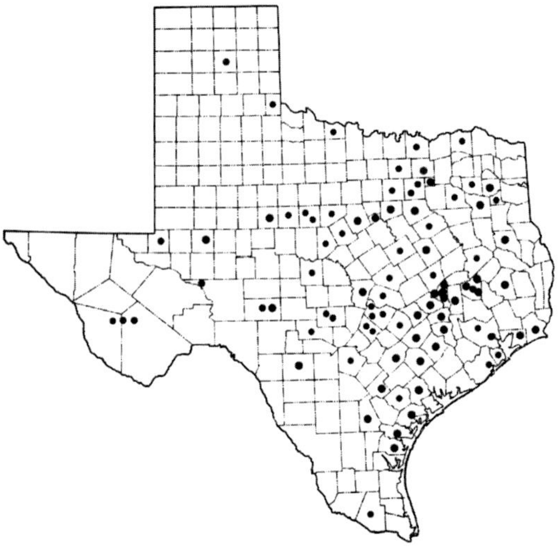

Lamium amplexicaule

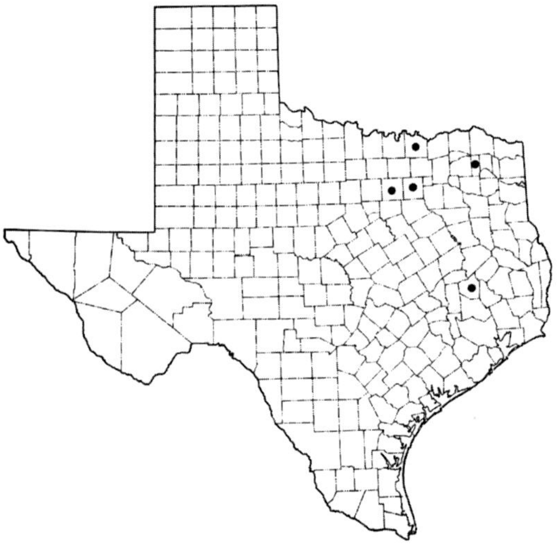

Lamium purpureum

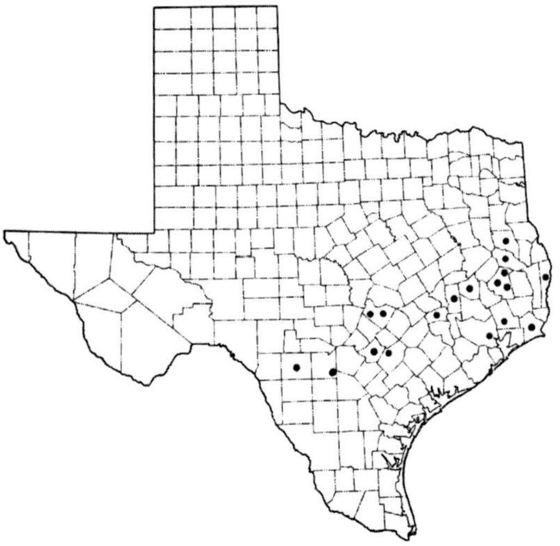

Leonotis nepetifolia

Leonurus cardiaca

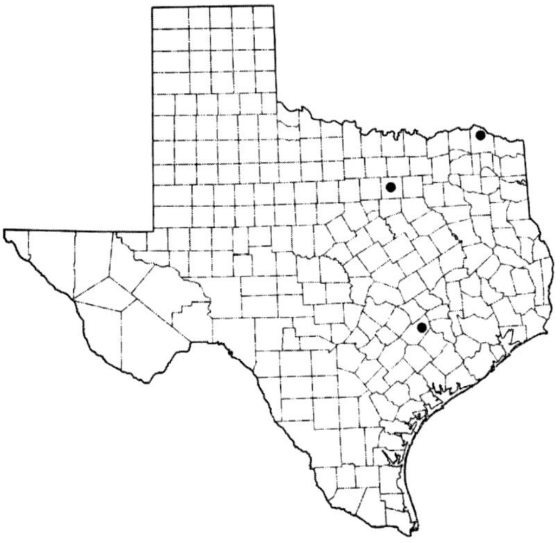

Leonurus sibiricus

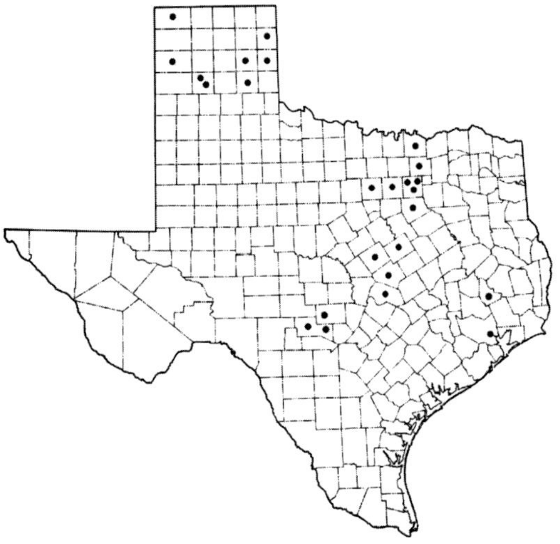

Lycopus americanus

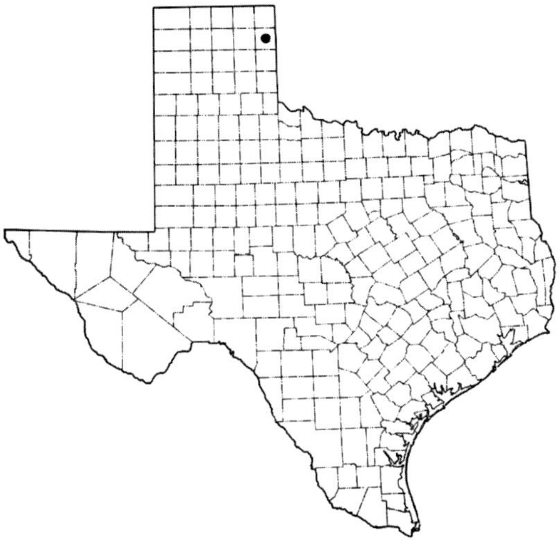

Lycopus asper

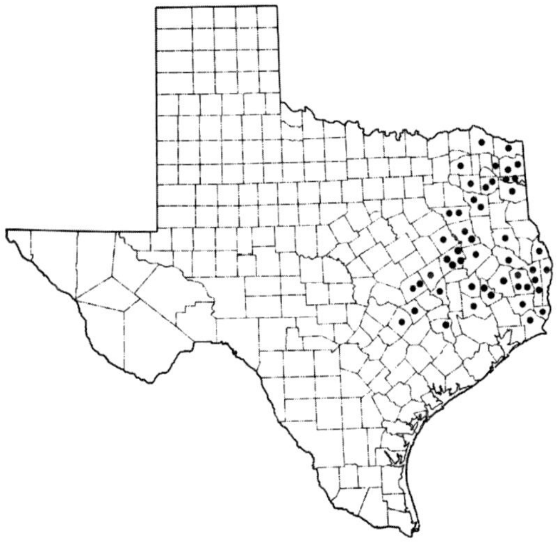

Lycopus rubellus

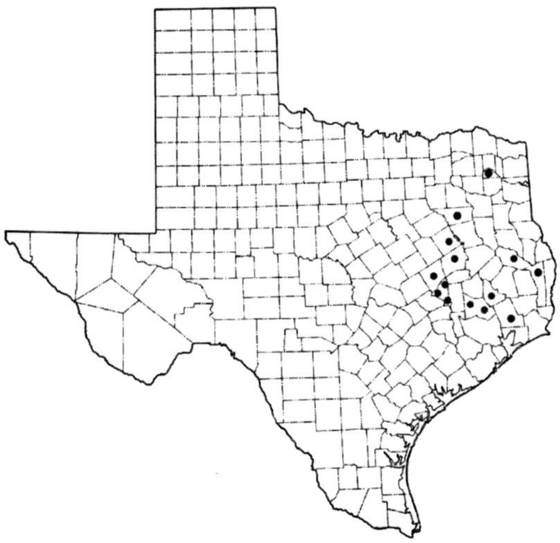

Lycopus virginicus

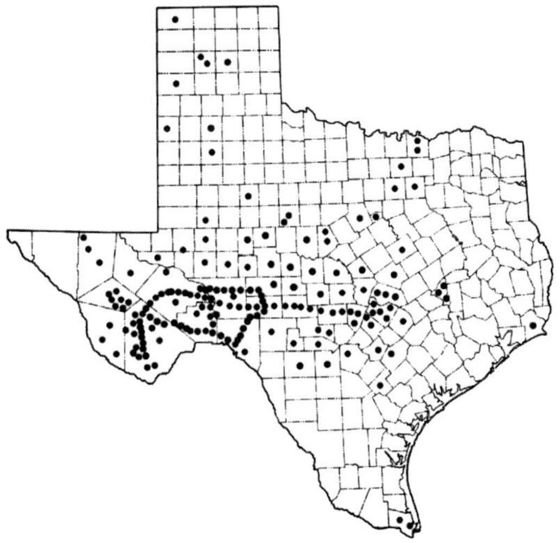

Marrubium vulgare

Mentha arvensis

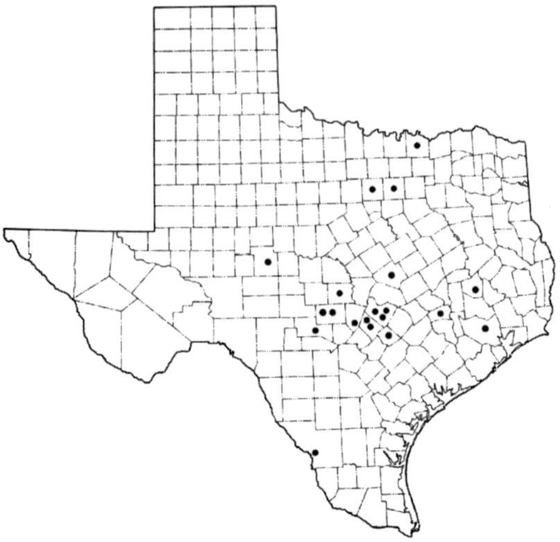

Mentha piperita

Mentha rotundifolia

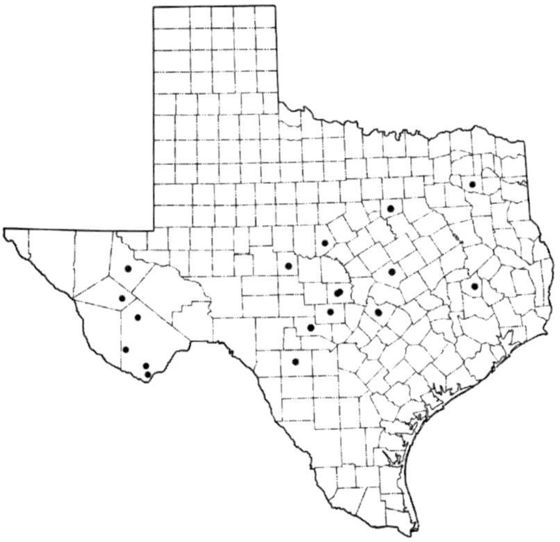

Mentha spicata

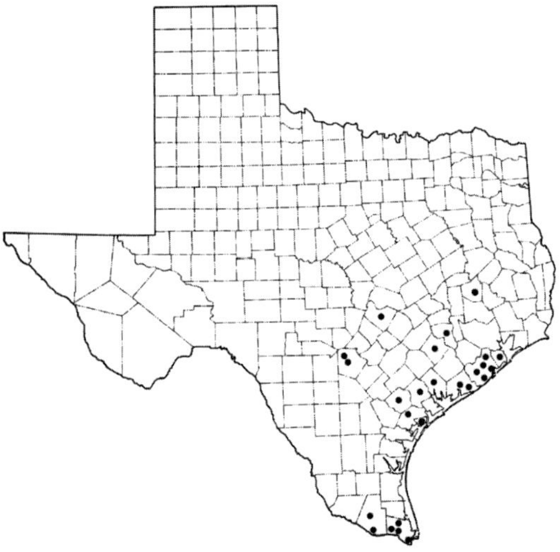

Micromeria brownei

Moluccella laevis

var. parva
Monarda citriodora

var. parva
Monarda citriodora

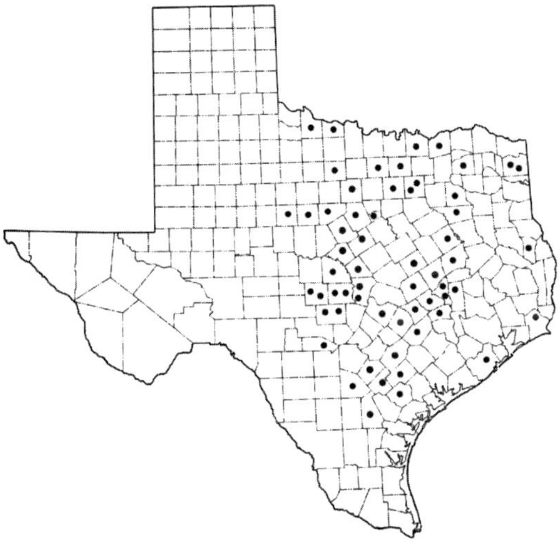
Monarda clinopodioides

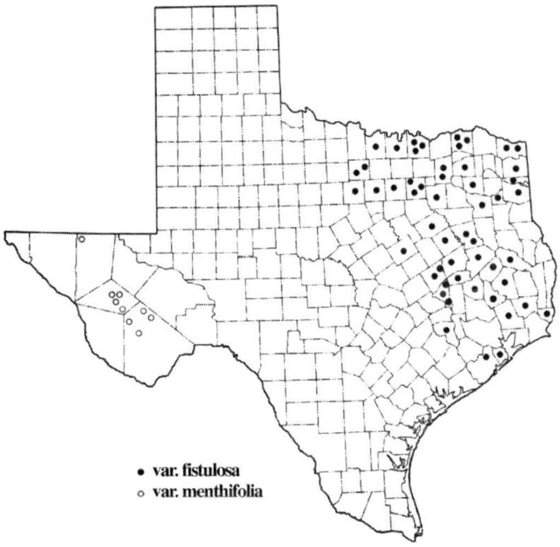
• var. fistulosa
○ var. menthifolia
Monarda fistulosa

Monarda fruticulosa

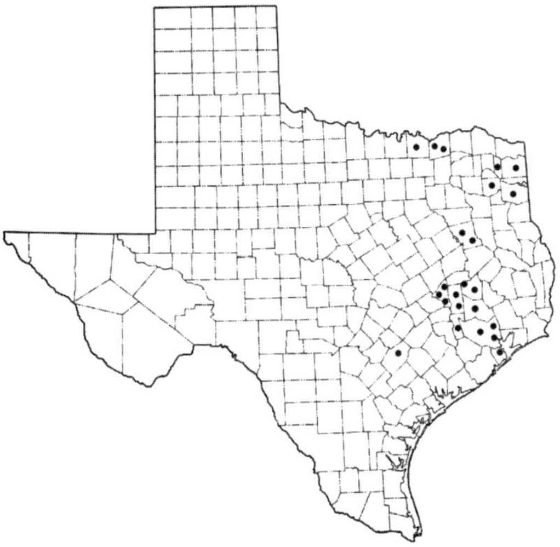
Monarda lindheimeri

Monarda maritima

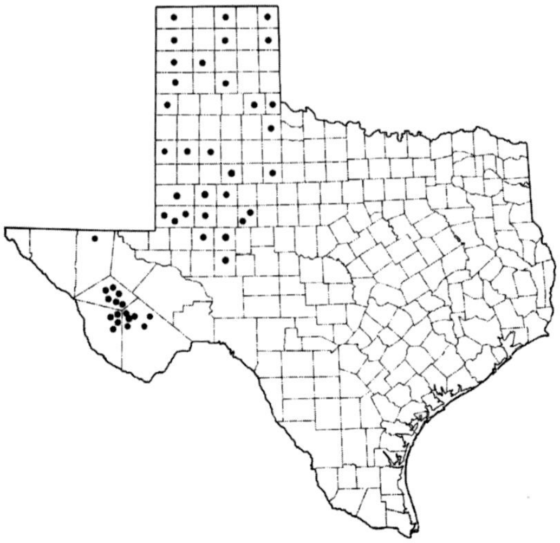

Monarda pectinata

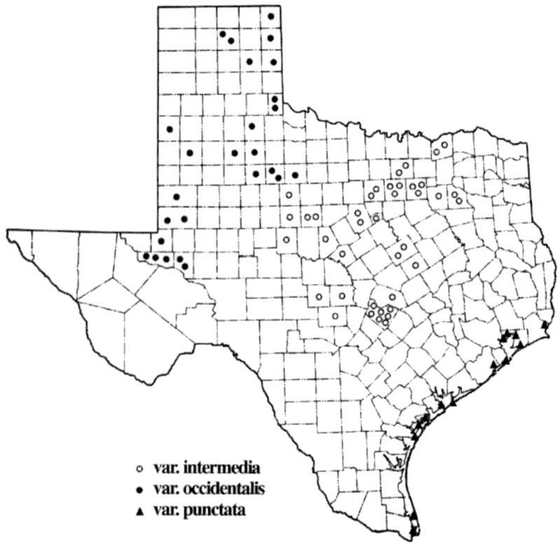
var. intermedia
var. occidentalis
var. punctata

Monarda punctata

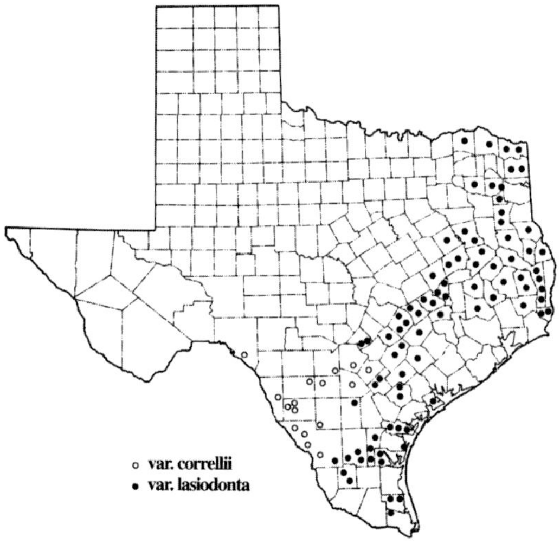
var. correllii
var. lasiodonta

Monarda punctata

Monarda russeliana

Monarda stanfieldii

Monarda viridissima

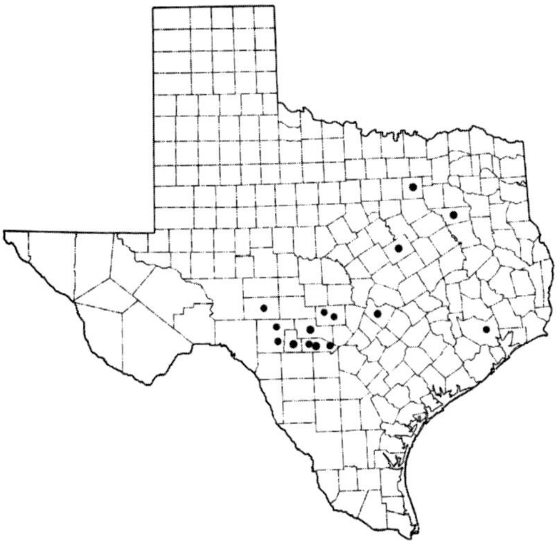

Nepeta cataria

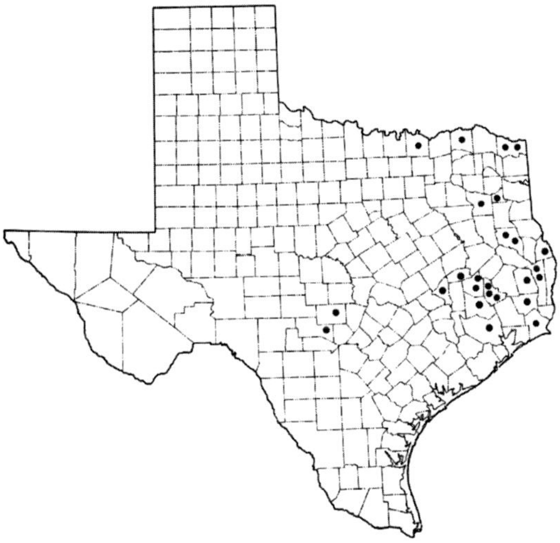

Perilla frutescens

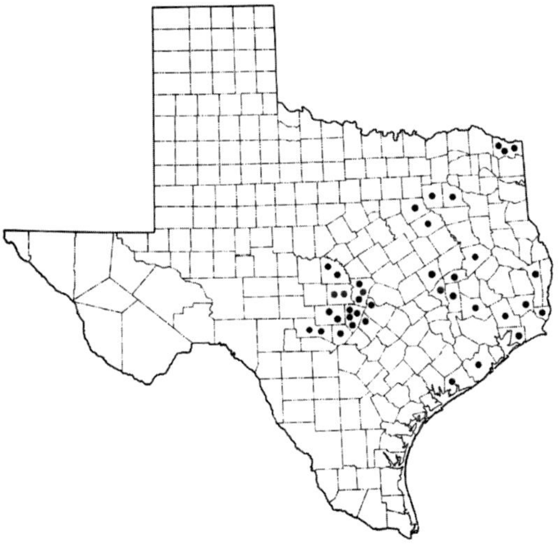

Physostegia angustifolia

Physostegia correllii

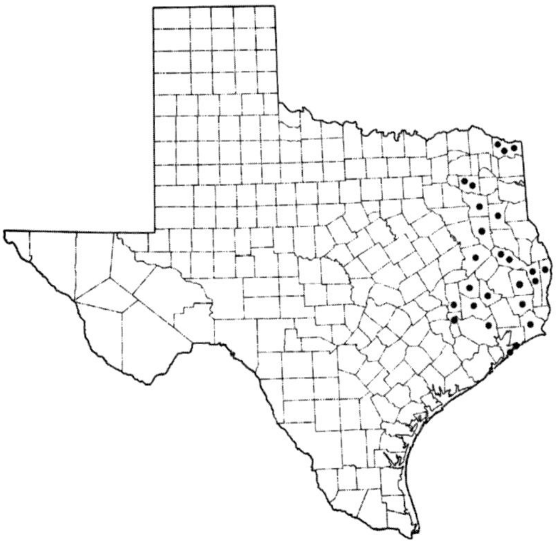

Physostegia digitalis

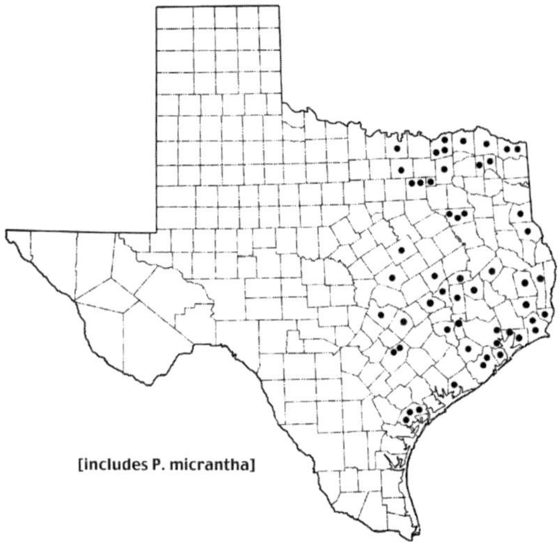
[includes P. micrantha]

Physostegia intermedia

Physostegia longisepala

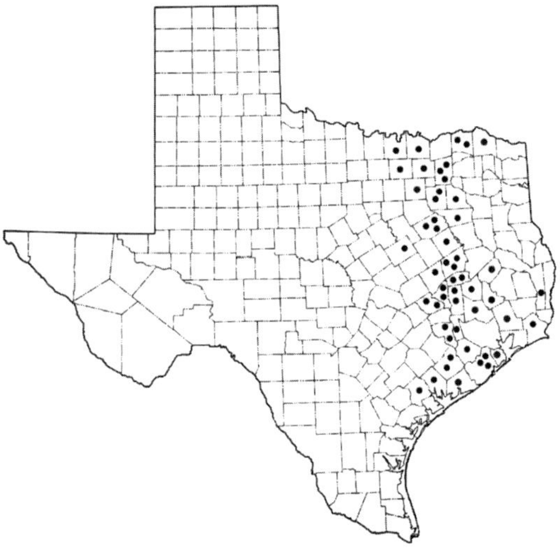

Physostegia pulchella

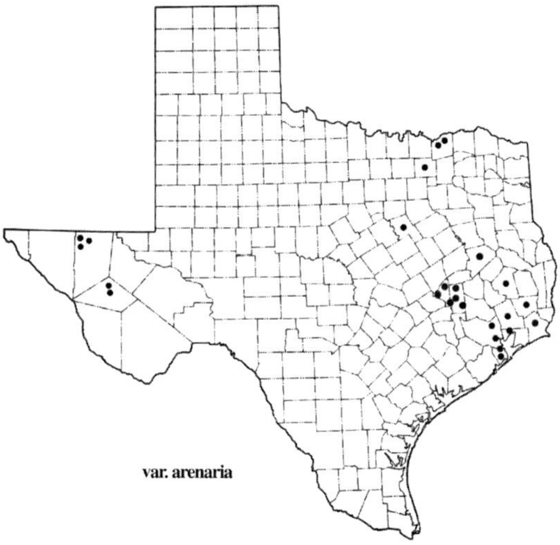
var. arenaria

Physostegia virginiana

Poliomintha glabrescens

Poliomintha incana

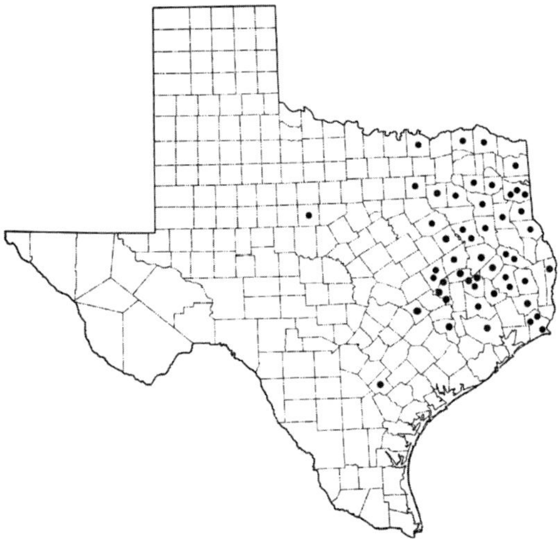

Prunella vulgaris

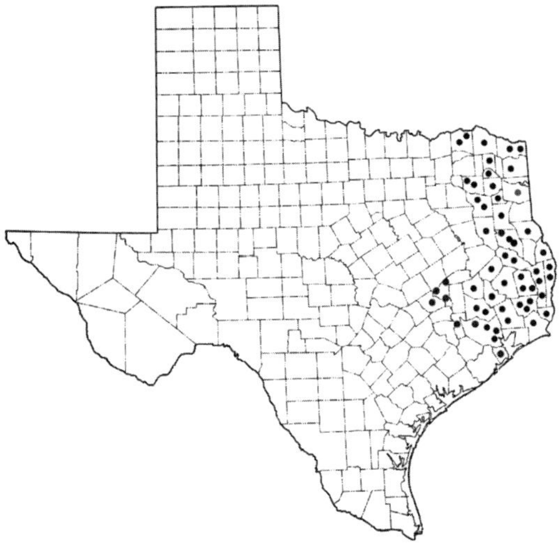

Pycnanthemum albescens

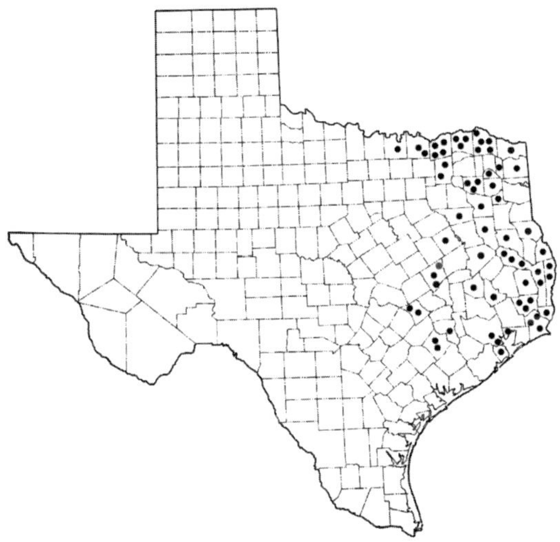

Pycnanthemum tenuifolium

Rhododon angulatus

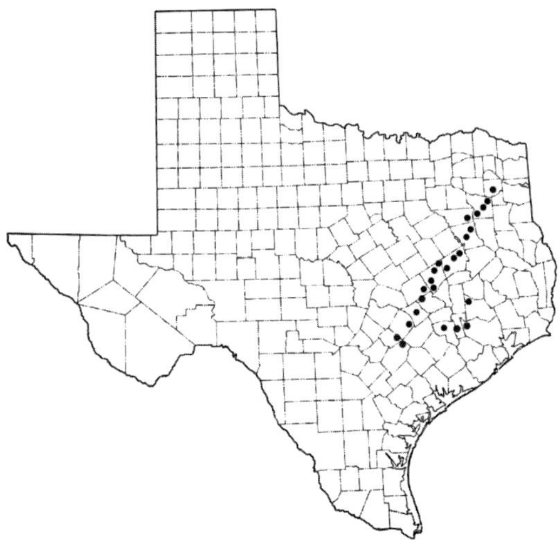

Rhododon ciliatus

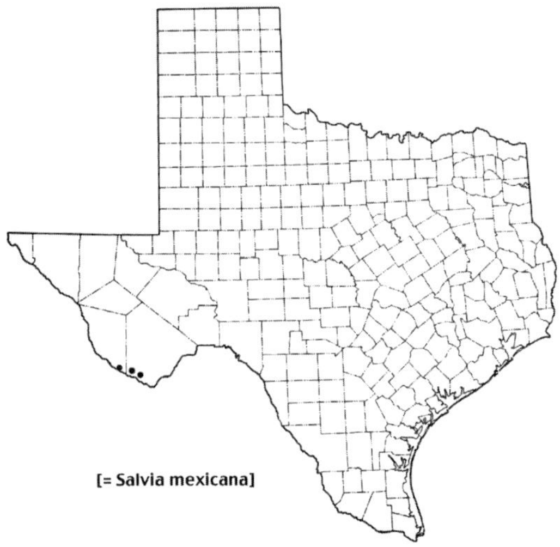
[= Salvia mexicana]

Salazaria mexicana

Salvia arizonica

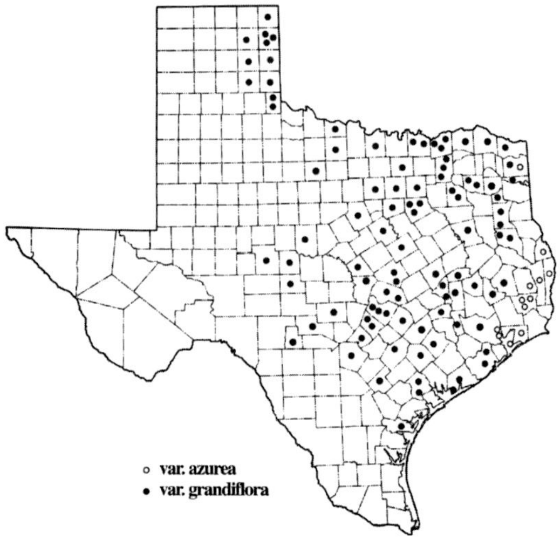
var. azurea
var. grandiflora

Salvia azurea

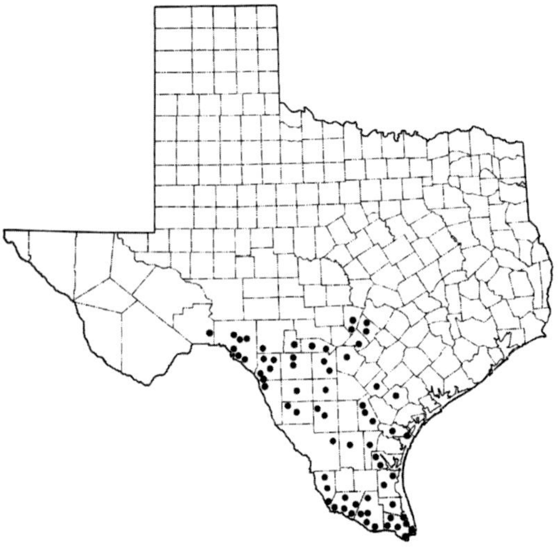

Salvia ballotiflora

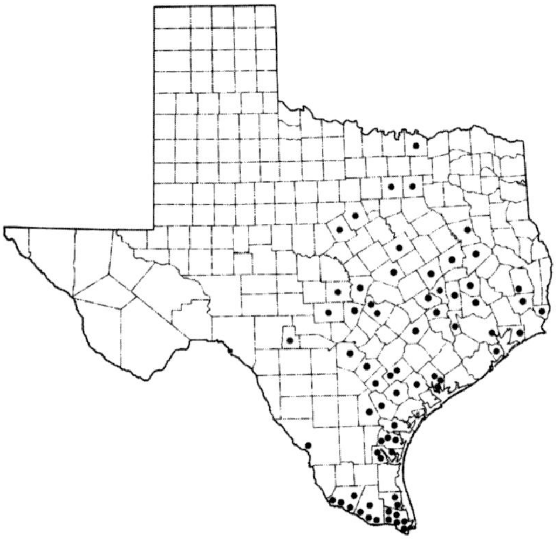

Salvia coccinea

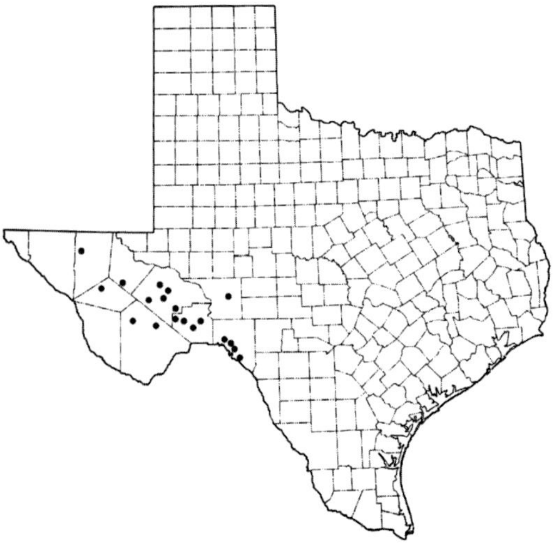

Salvia dolichantha

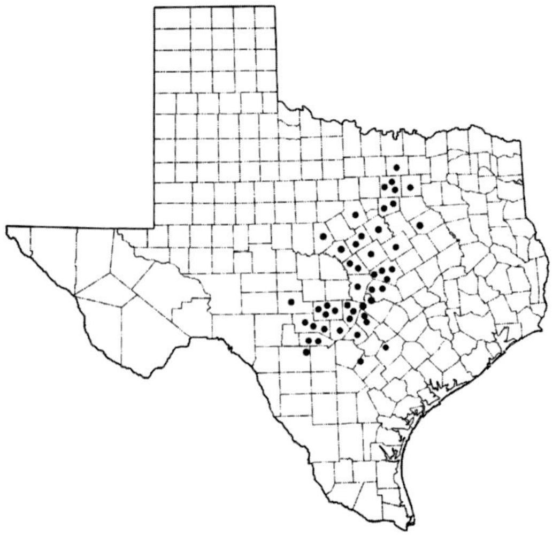

Salvia engelmannii

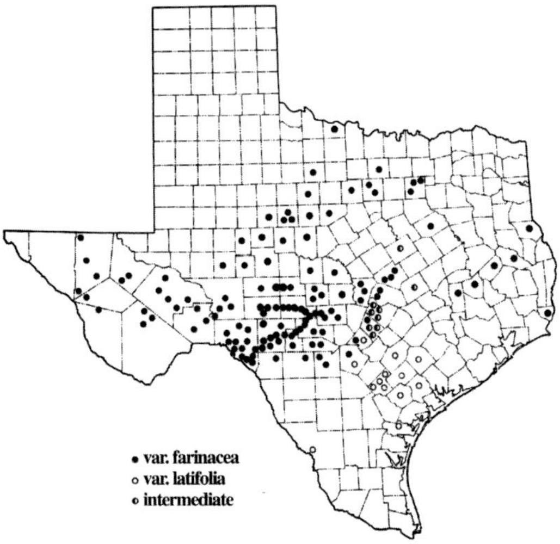
• var. farinacea
○ var. latifolia
⊙ intermediate

Salvia farinacea

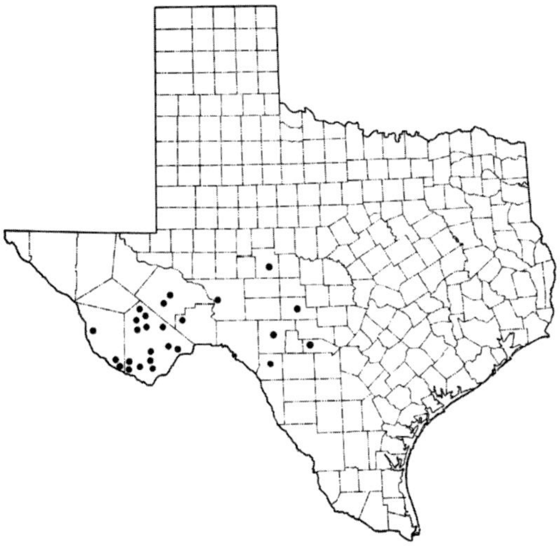

Salvia greggii

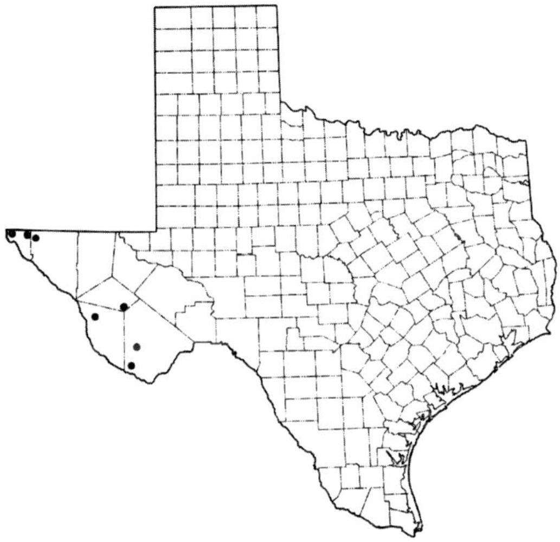

Salvia henryi

Salvia lycioides

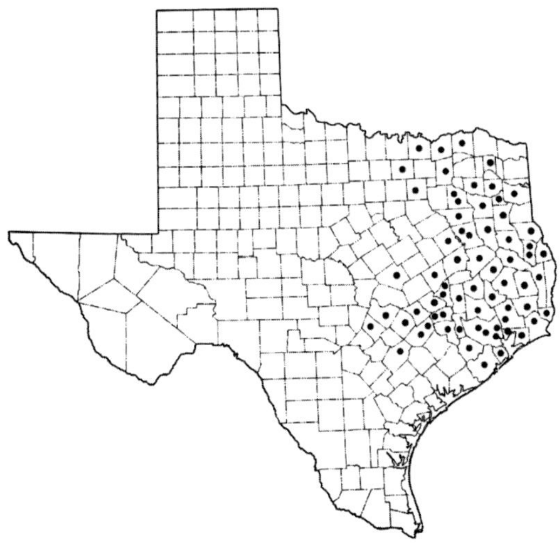

Salvia lyrata

Salvia penstemonoides

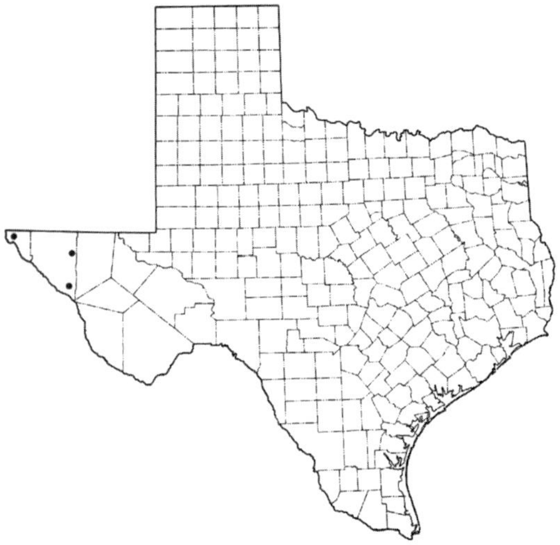

Salvia pinguifolia

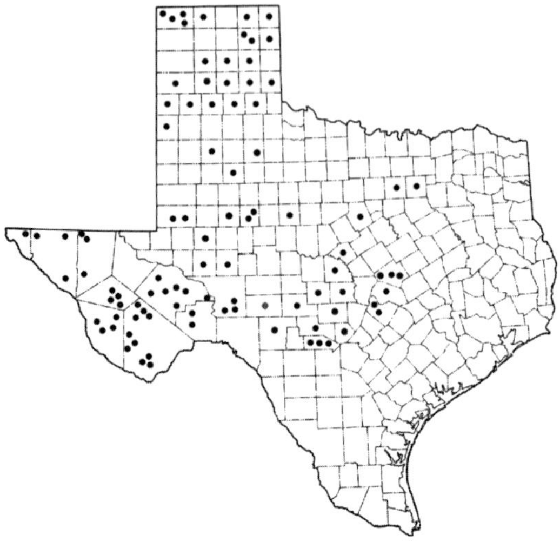

Salvia reflexa

Salvia regla

var. glabra

Salvia reptans

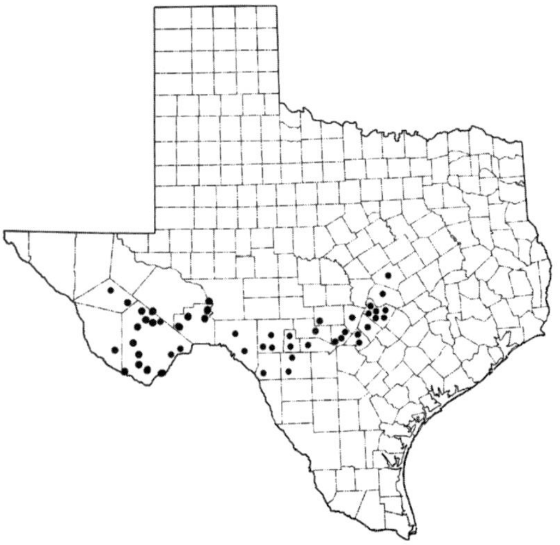

Salvia roemeriana

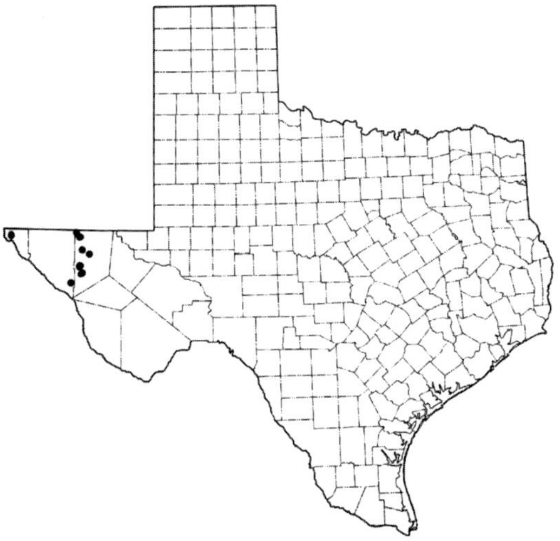

Salvia summa

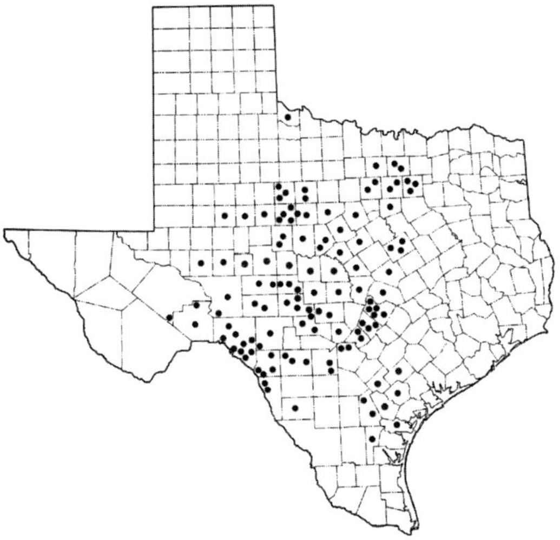

Salvia texana

Salvia tiliifolia

Salvia vinacea

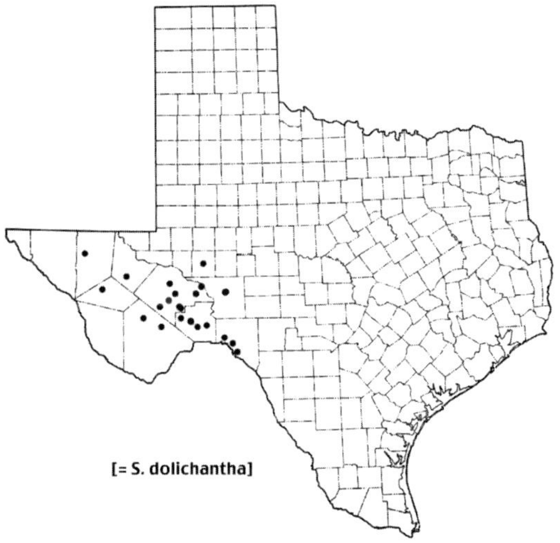
[= S. dolichantha]

Salvia whitehouseae

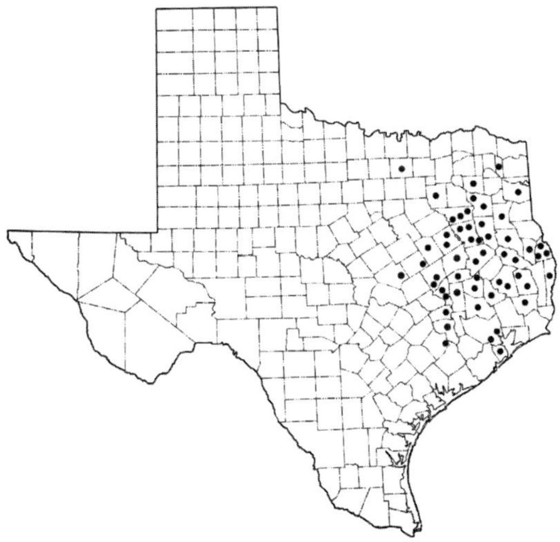

Scutellaria cardiophylla

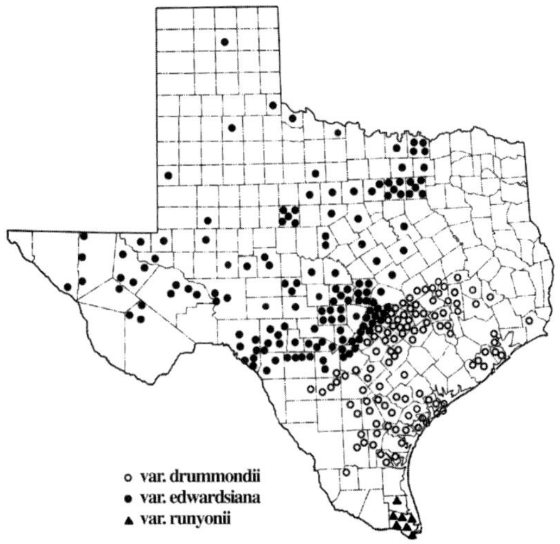
var. drummondii
var. edwardsiana
var. runyonii

Scutellaria drummondii

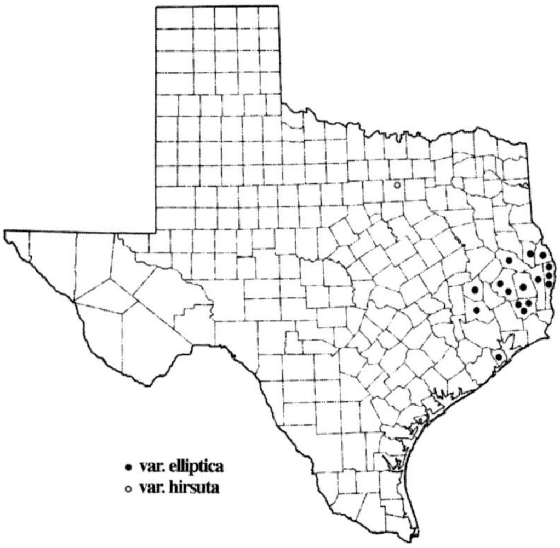
• var. elliptica
∘ var. hirsuta
Scutellaria elliptica

Scutellaria galericulata

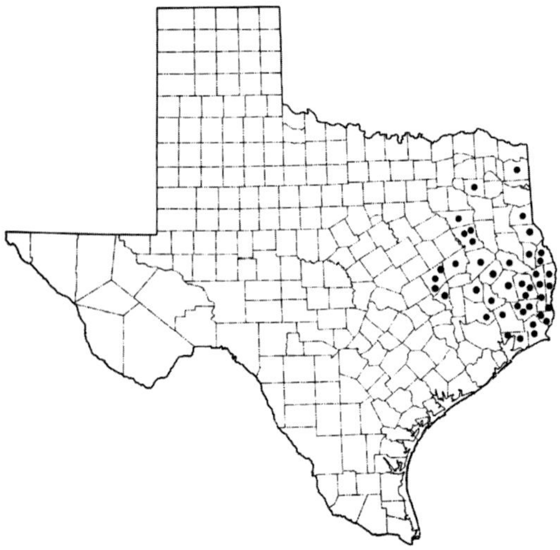
Scutellaria integrifolia

Scutellaria laevis

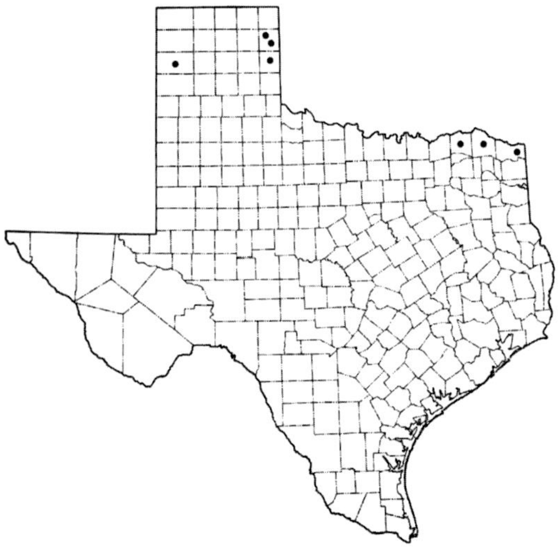
Scutellaria lateriflora

Scutellaria minor

Scutellaria muriculata

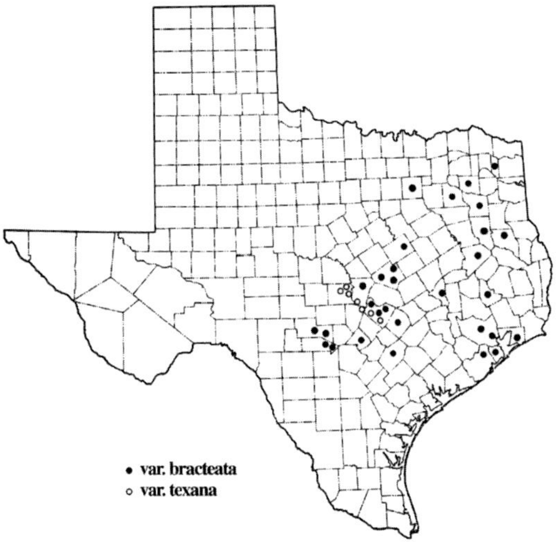
• var. bracteata
○ var. texana

Scutellaria ovata

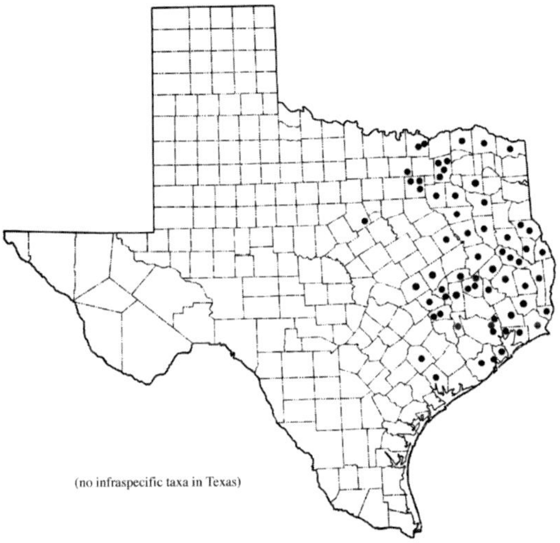
(no infraspecific taxa in Texas)

Scutellaria parvula

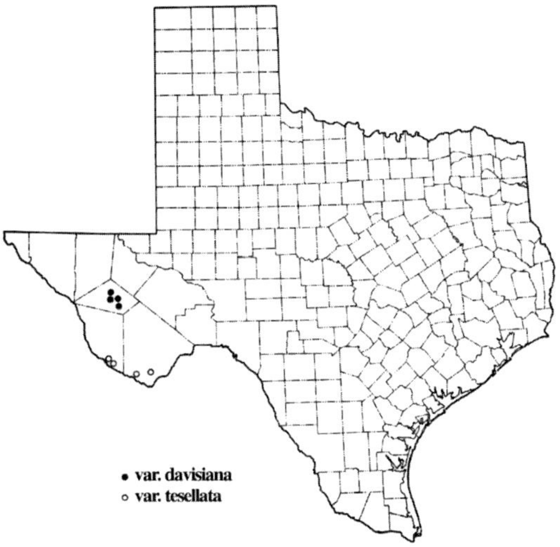
• var. davisiana
○ var. tesellata

Scutellaria potosina

Scutellaria racemosa

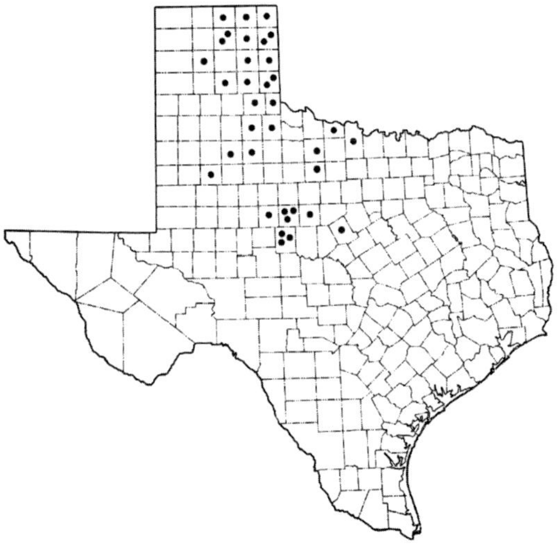

Scutellaria resinosa

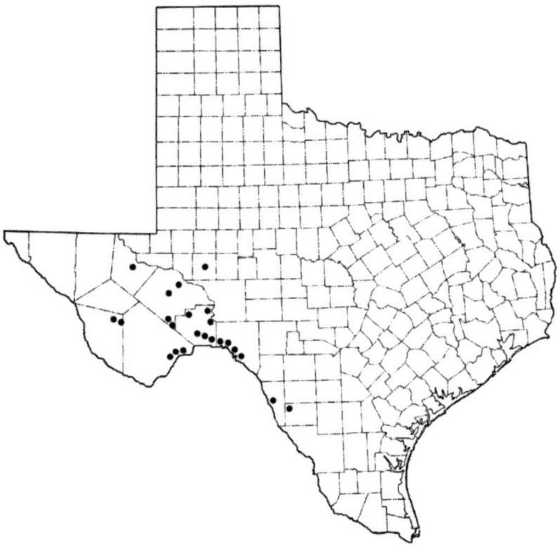

Scutellaria texana

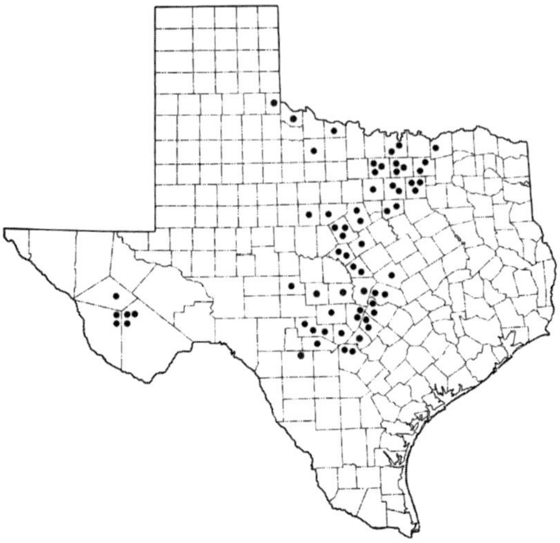

Scutellaria wrightii

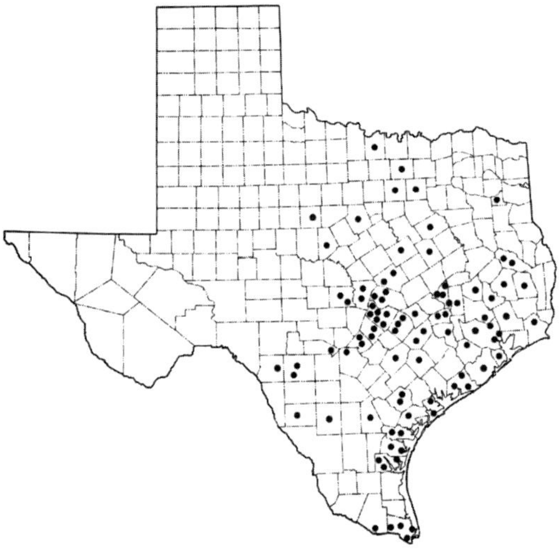

Stachys agraria

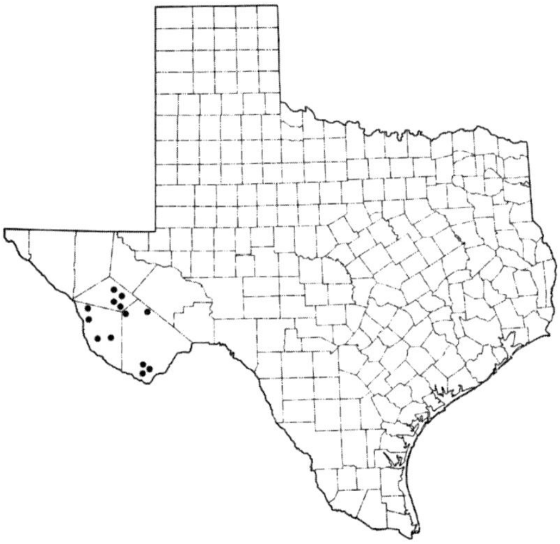

Stachys bigelovii

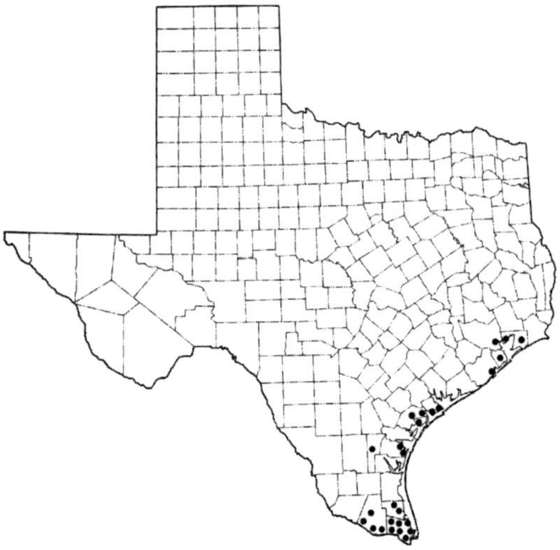

Stachys drummondii

Stachys floridana

Stachys grahamii

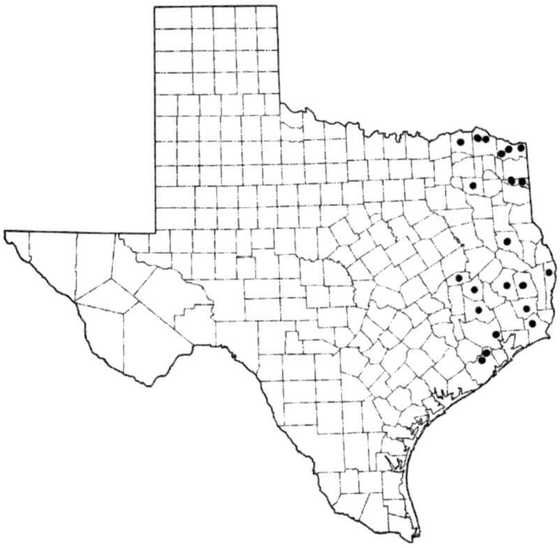

Stachys tenuifolia

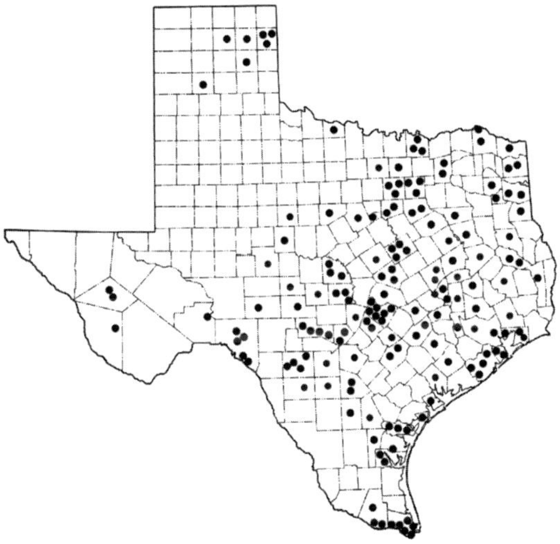

Teucrium canadense

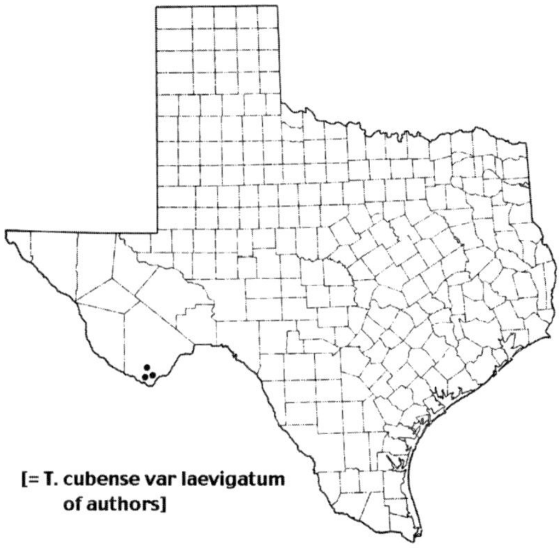
[= T. cubense var laevigatum
of authors]

Teucrium coahuilanum

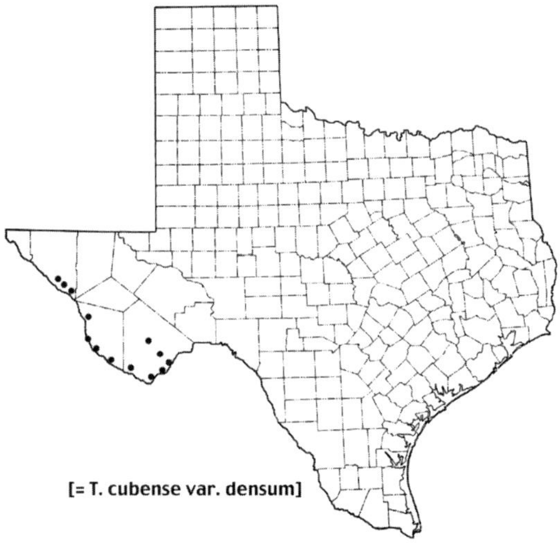
[= T. cubense var. densum]

Teucrium depressum

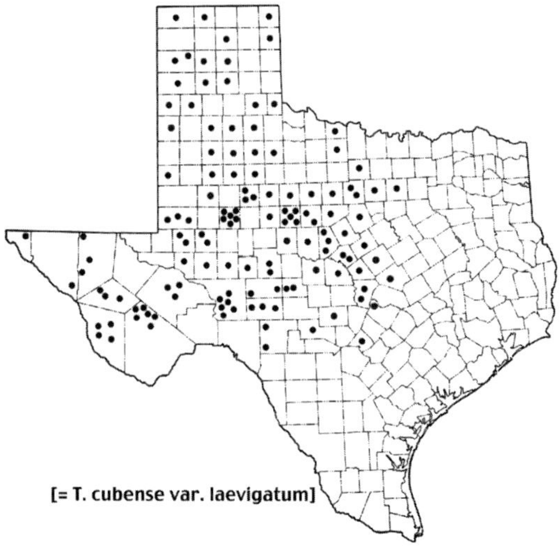
[= T. cubense var. laevigatum]

Teucrium laciniatum

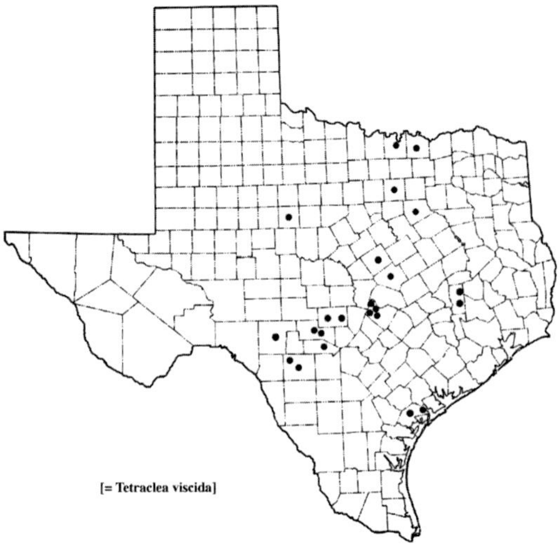
[= Tetraclea viscida]

Trichostema brachiatum

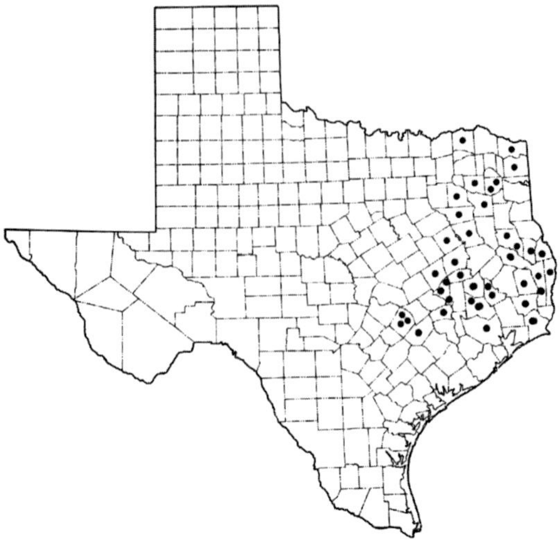

Trichostema dichotomum

Trichostema setaceum

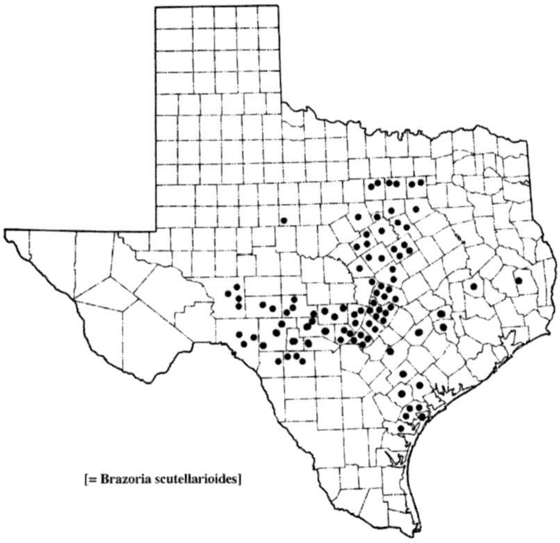
[= Brazoria scutellarioides]

Warnockia scutellarioides

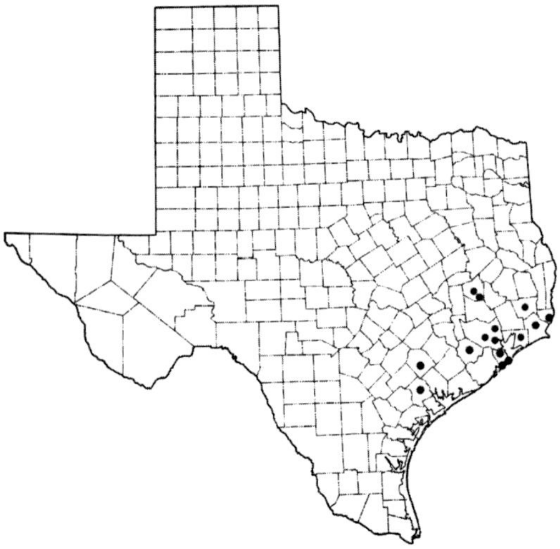

Cinnamomum camphora

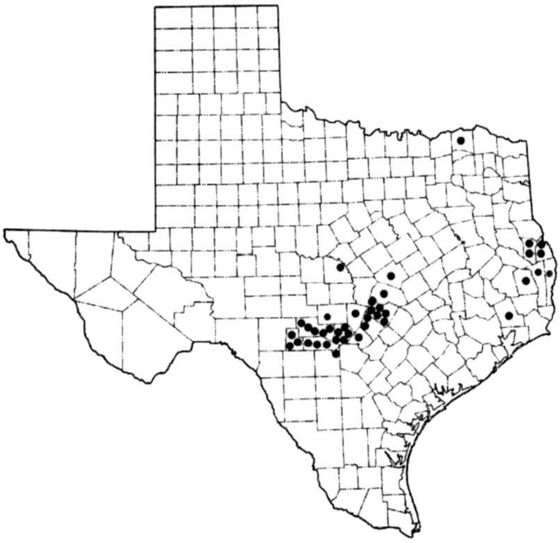

Lindera benzoin

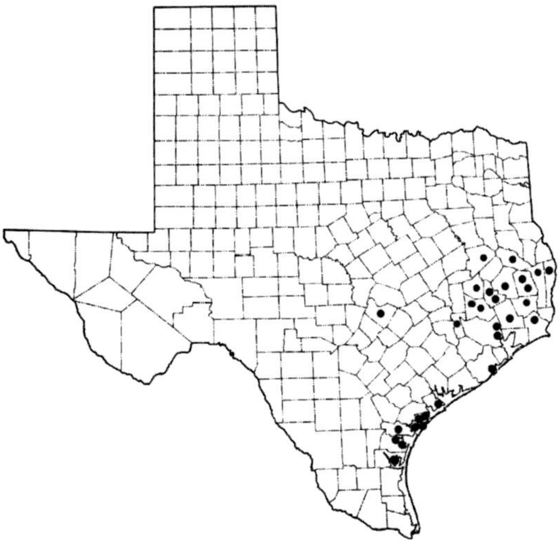

Persea borbonia

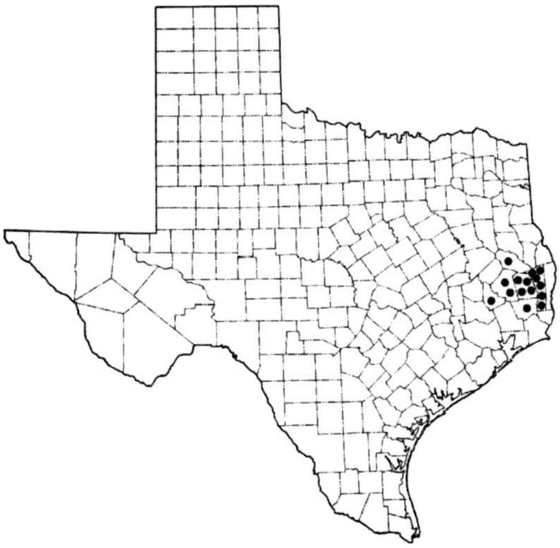

Persea palustris

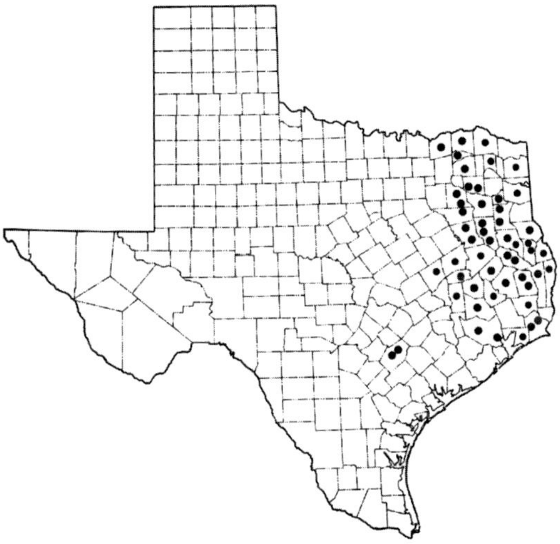

Sassafras albidum

Leitneria floridana

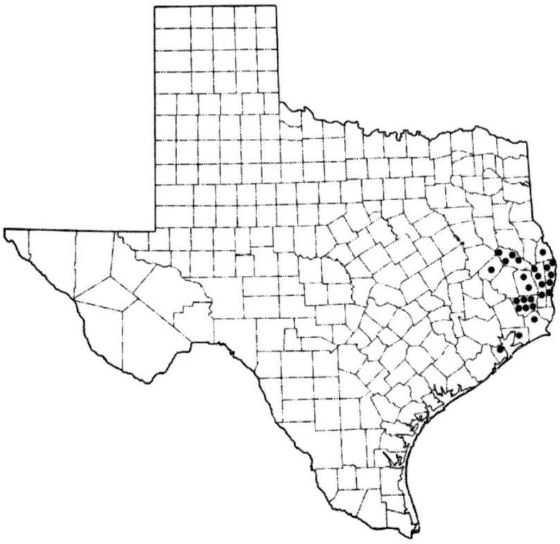

Pinguicula pumila

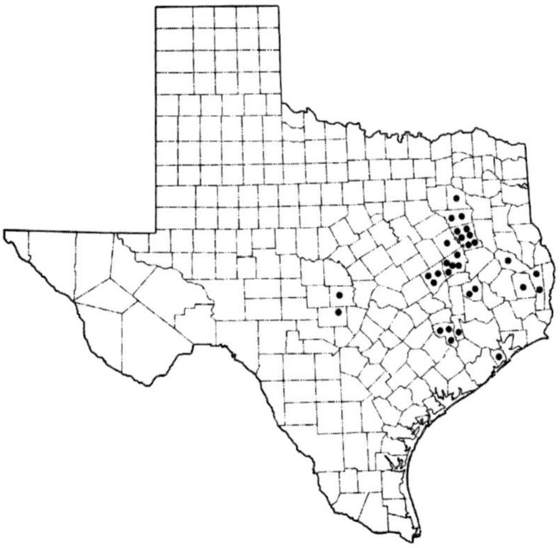

Utricularia cornuta

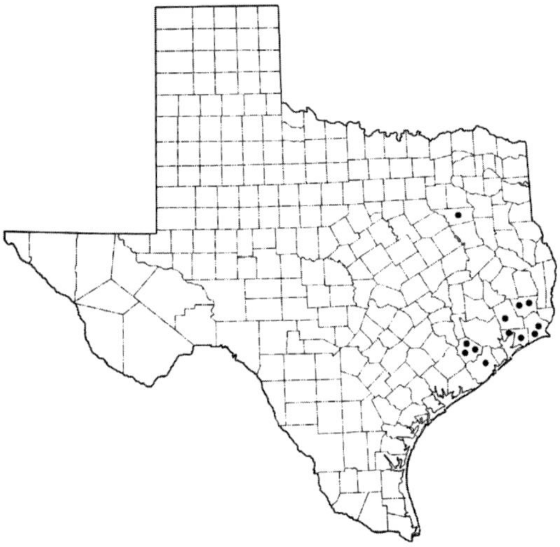

Utricularia foliosa

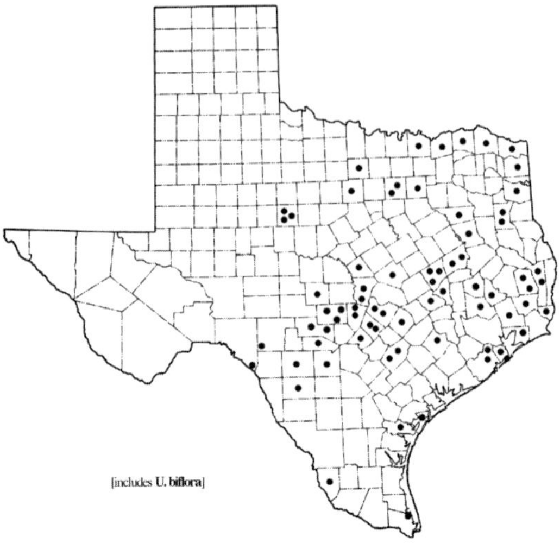
[includes U. biflora]

Utricularia gibba

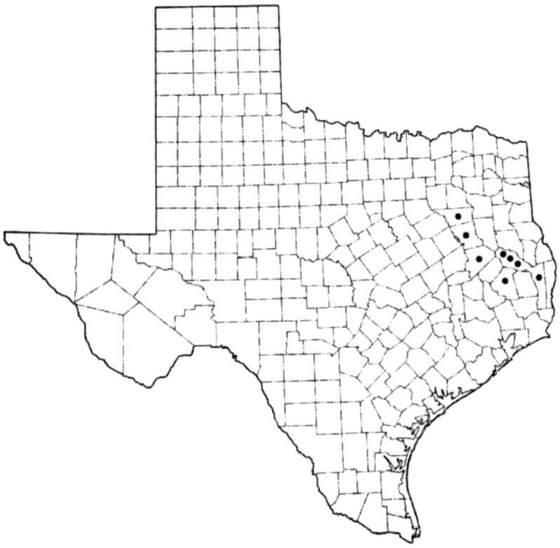

Utricularia juncea

[includes U. vulgaris]

Utricularia macrorhiza

Utricularia purpurea

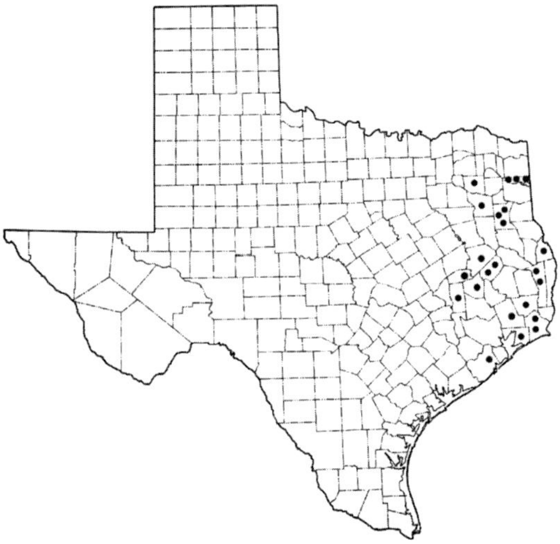

Utricularia radiata

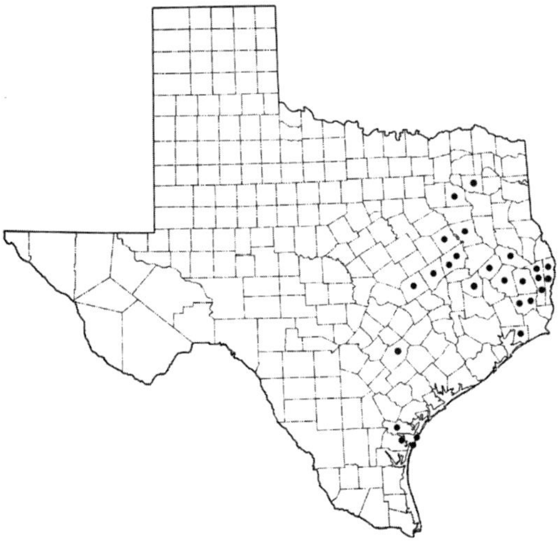

Utricularia subulata

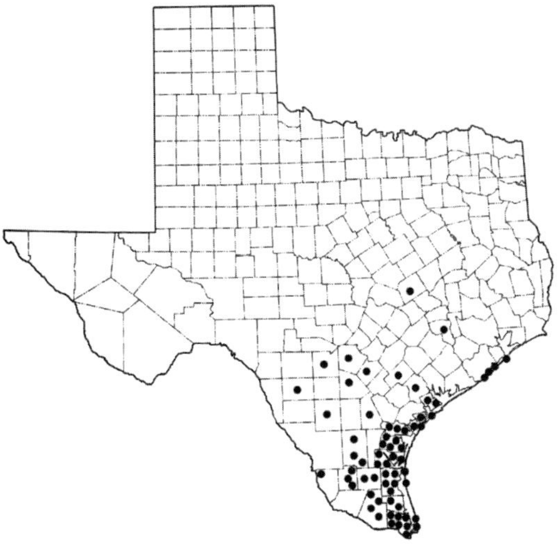

Linum alatum

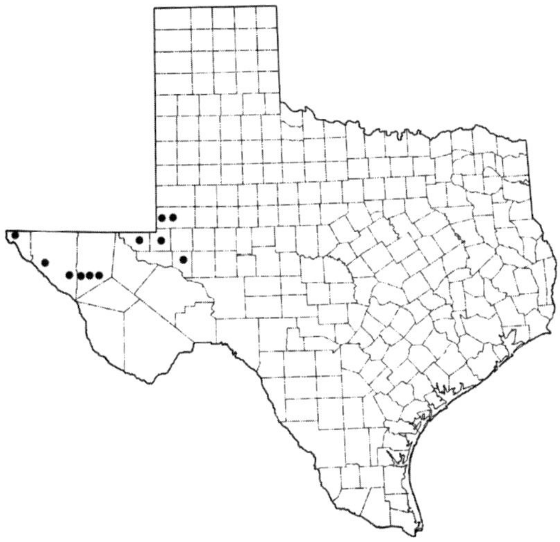

Linum aristatum

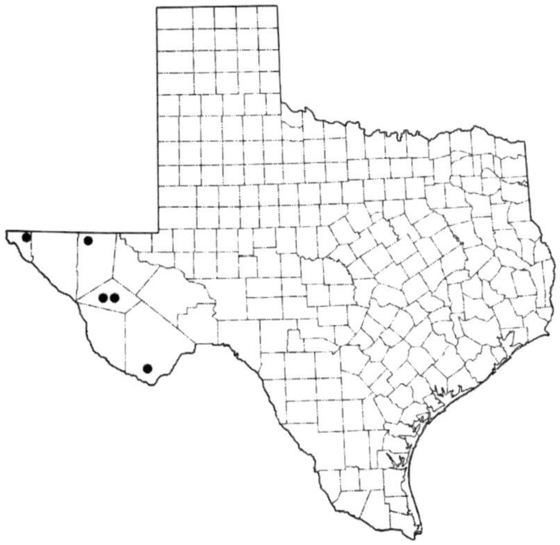

Linum australe

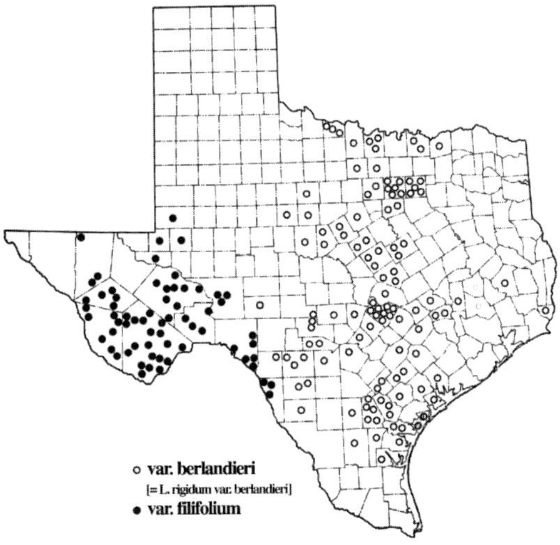
○ var. berlandieri
[= L. rigidum var. berlandieri]
● var. filifolium

Linum berlandieri

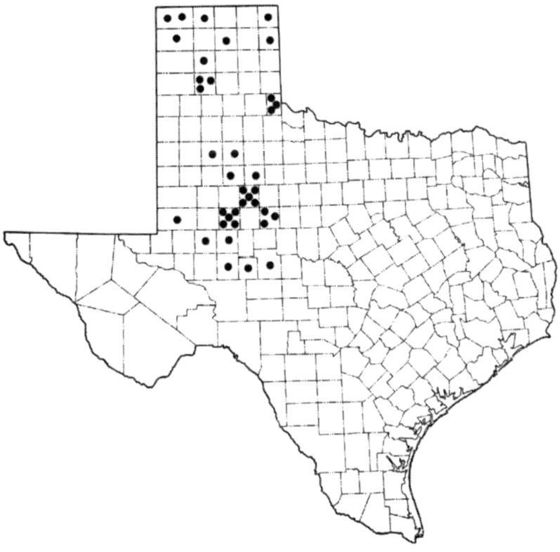

Linum compactum

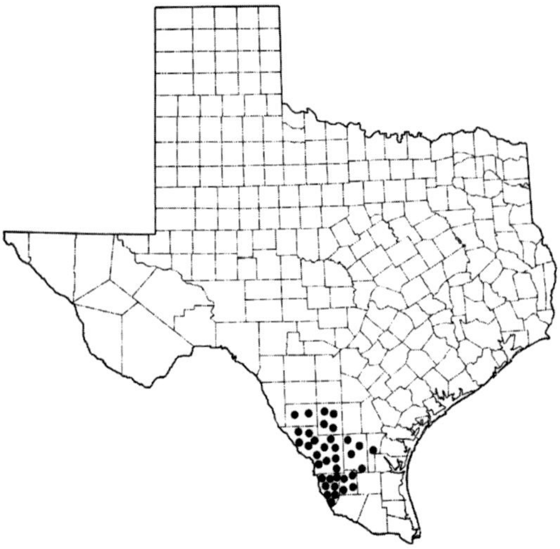

Linum elongatum

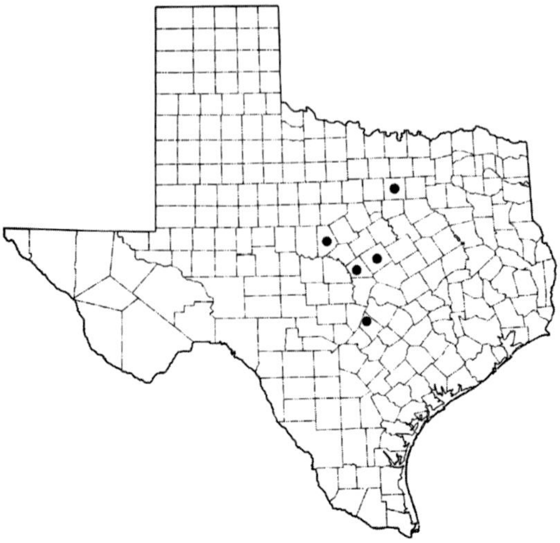

Linum grandiflorum

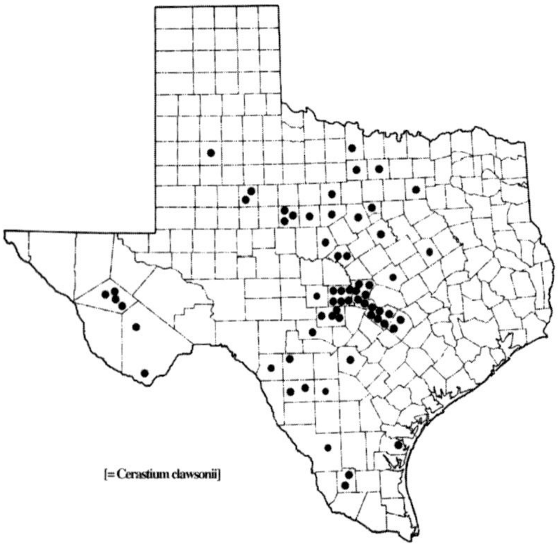
[= Cerastium clawsonii]

Linum hudsonioides

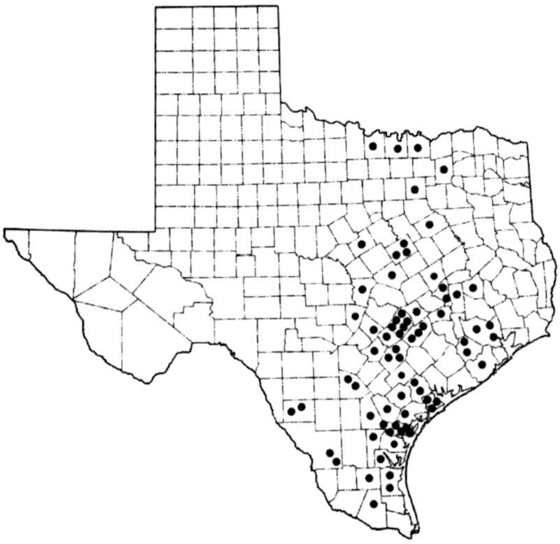

Linum imbricatum

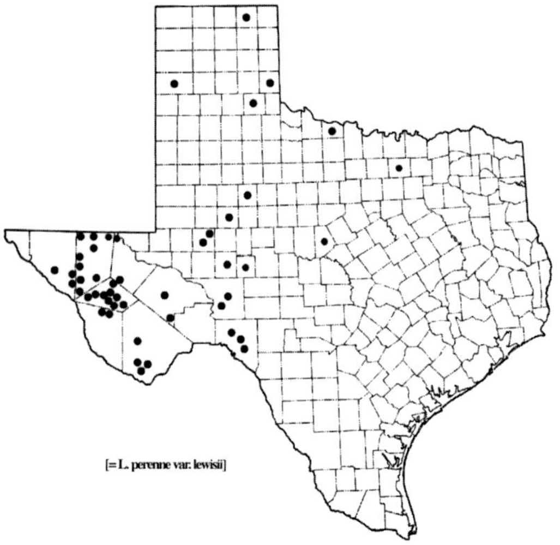
[= L. perenne var. lewisii]

Linum lewisii

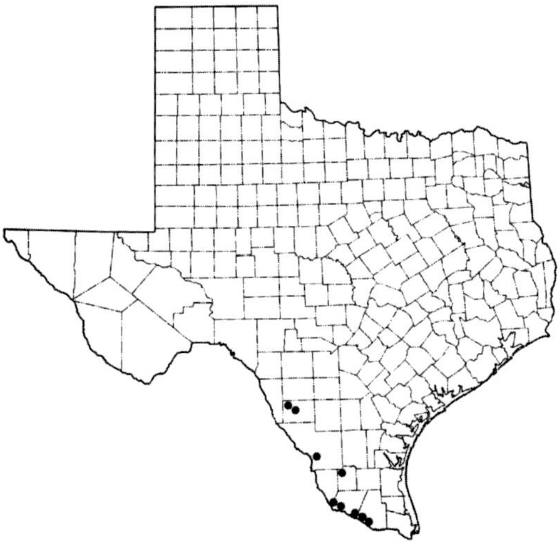

Linum lundellii

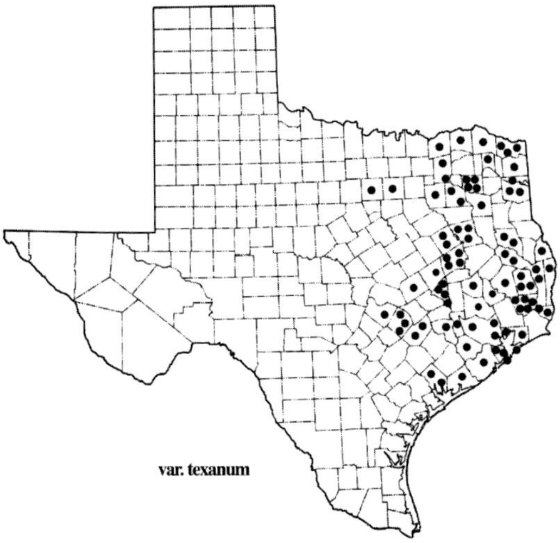
var. texanum

Linum medium

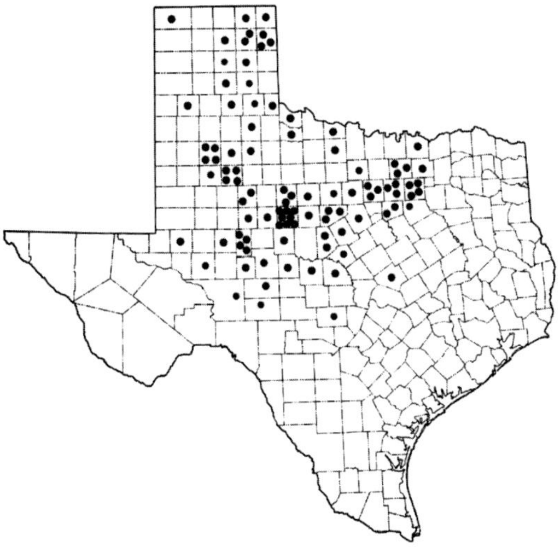

Linum pratense

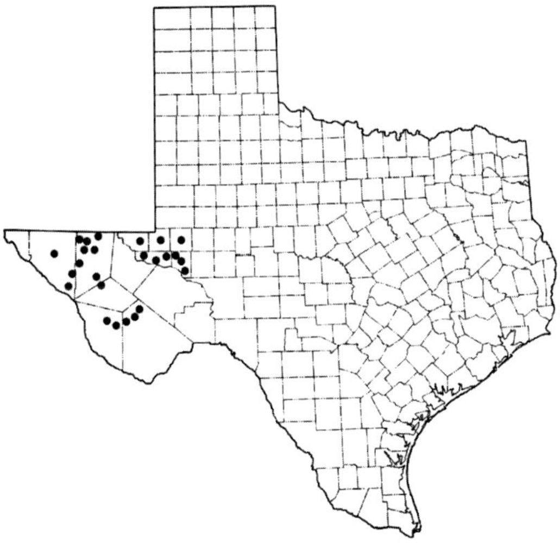

Linum puberulum

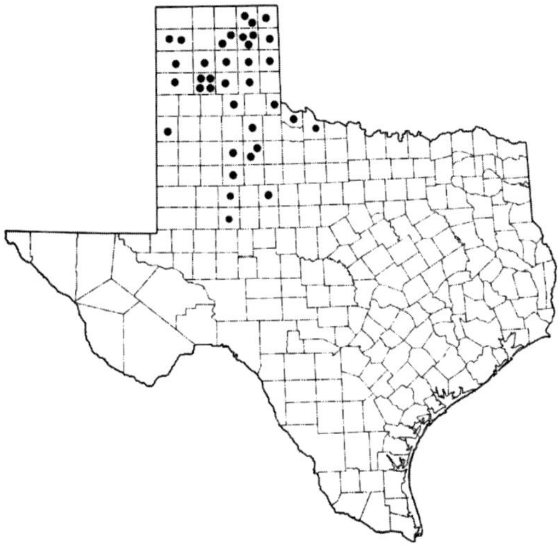

Linum rigidum

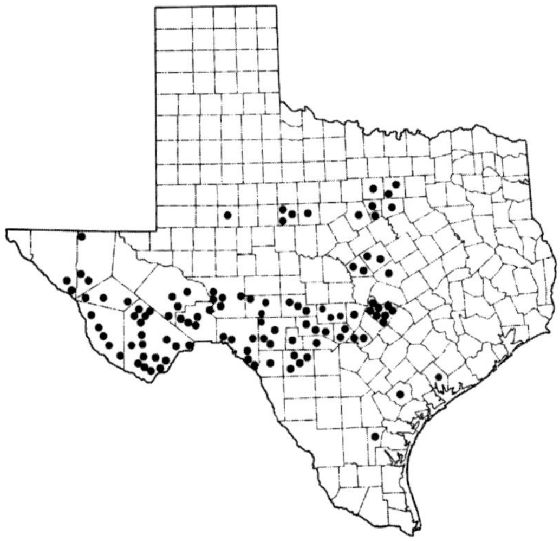

Linum rupestre

Linum schiedeanum

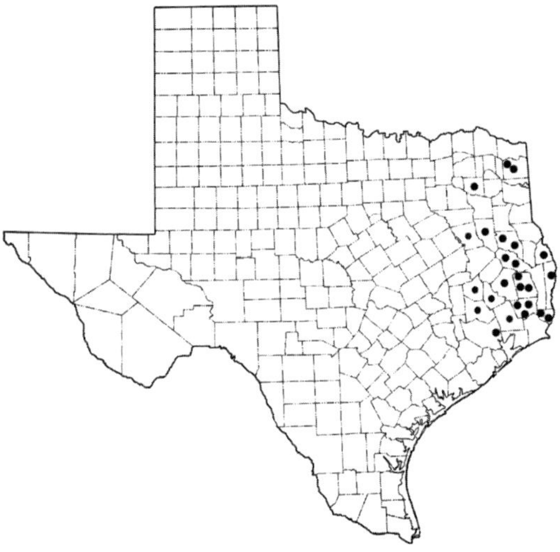

Linum striatum

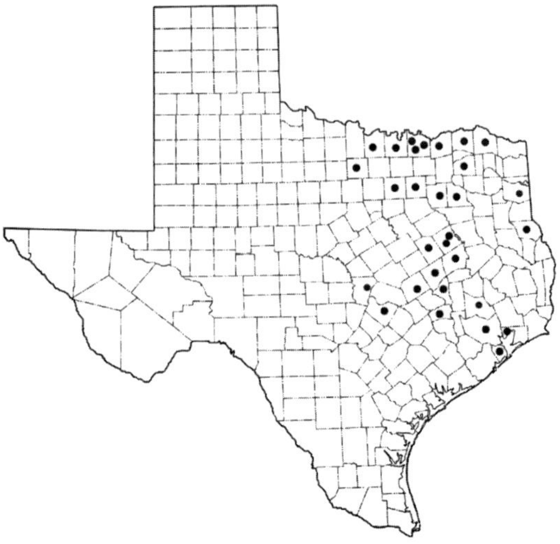

Linum sulcatum

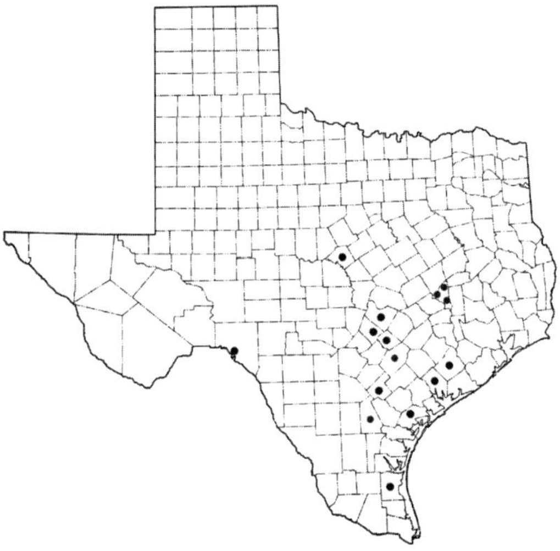

Linum usitatissimum

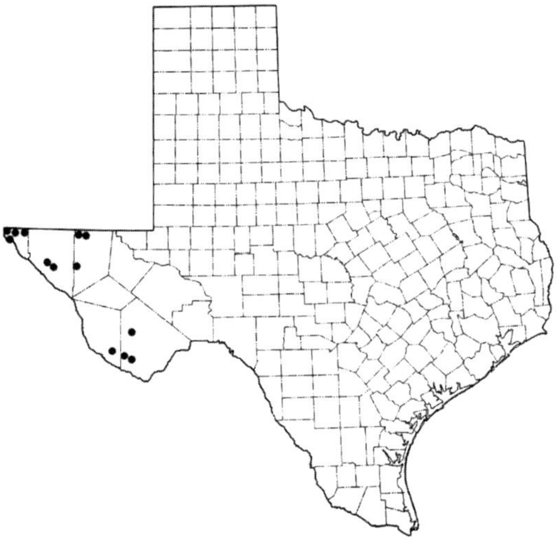

Linum vernale

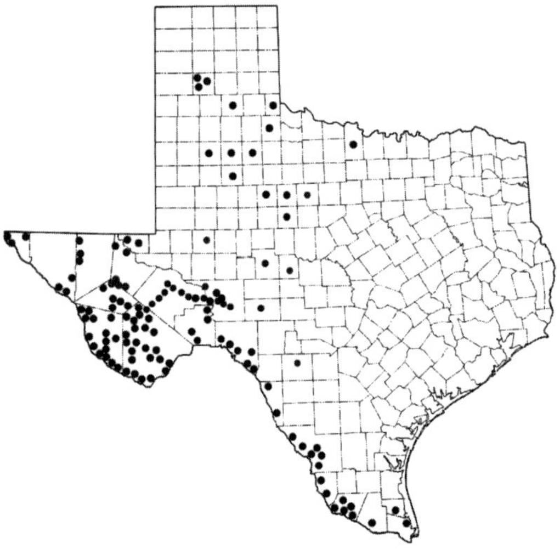

Cevallia sinuata

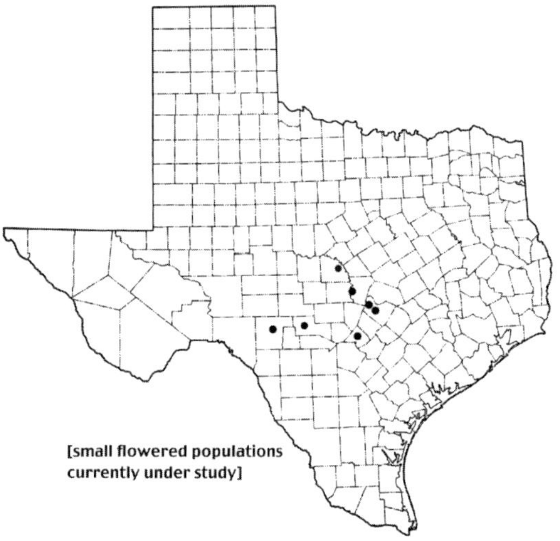
[small flowered populations
currently under study]

Eucnide bartonioides

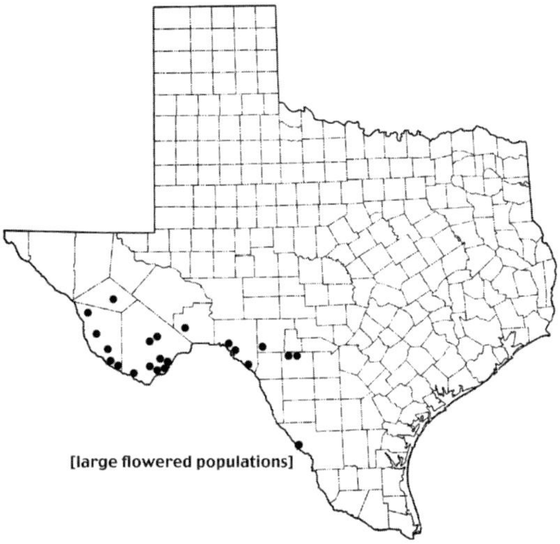
[large flowered populations]

Eucnide bartonioides

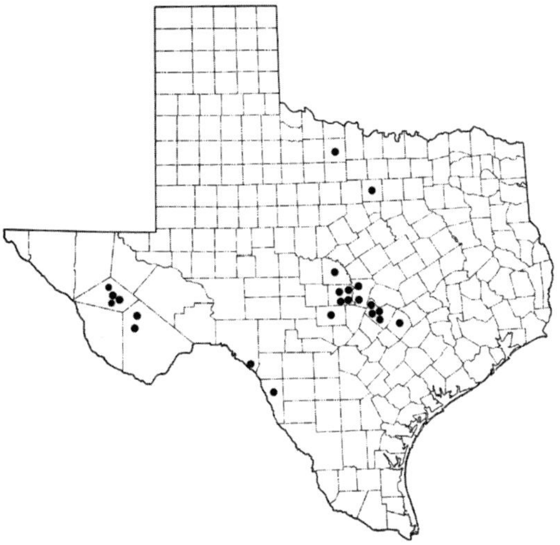

Mentzelia albescens

Mentzelia albicaulis

Mentzelia asperula

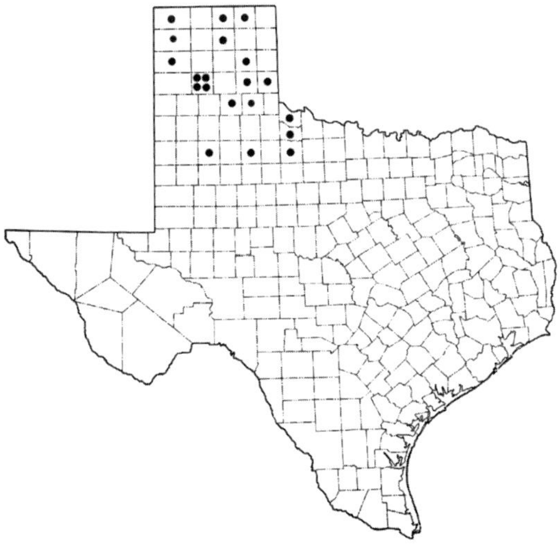

Mentzelia decapetala

Mentzelia hispida

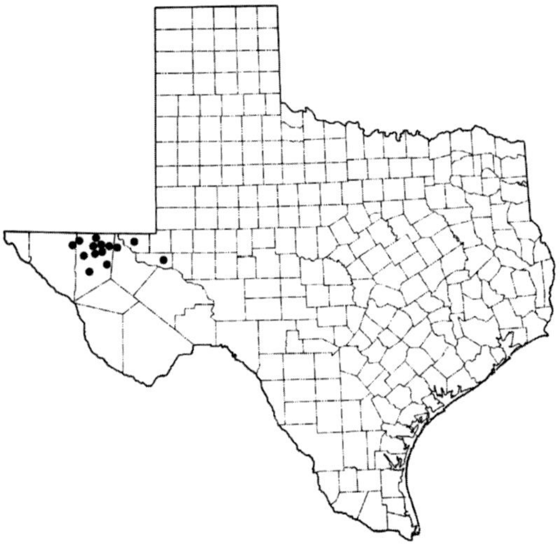

Mentzelia humilis

Mentzelia incisa

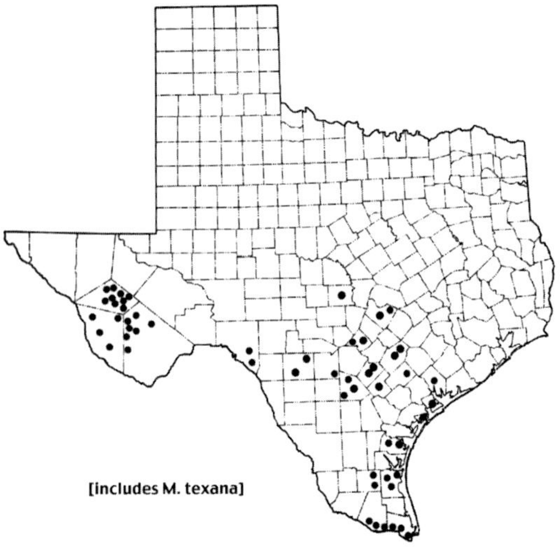
[includes M. texana]

Mentzelia lindheimeri

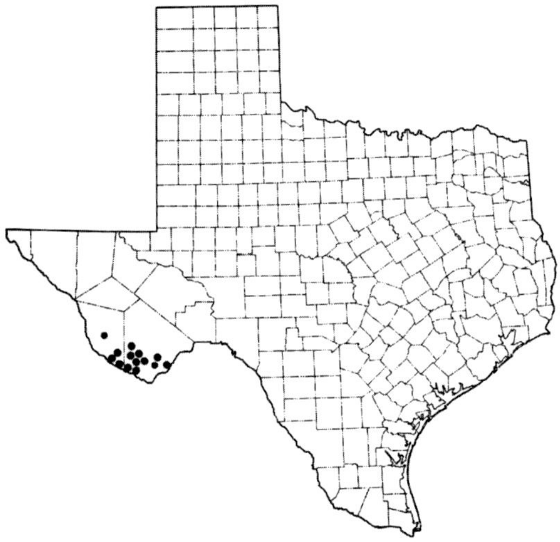

Mentzelia mexicana

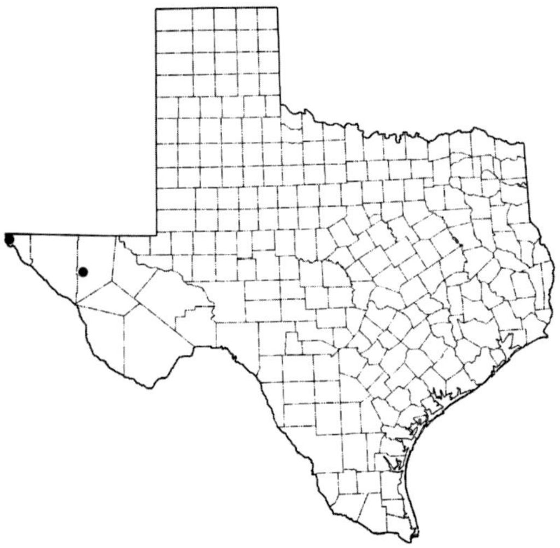

Mentzelia montana

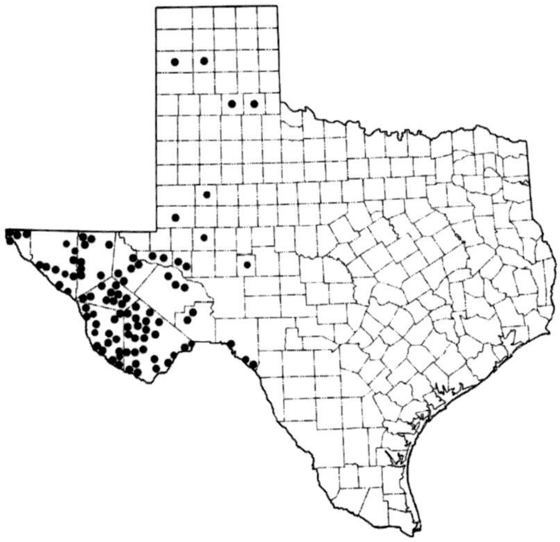

Mentzelia multiflora

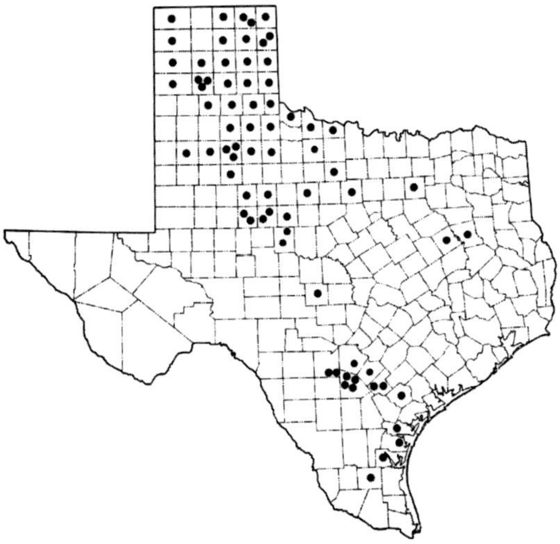

Mentzelia nuda

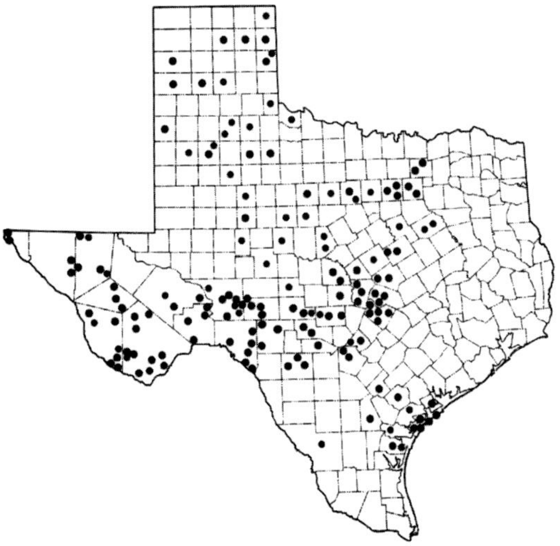

Mentzelia oligosperma

Mentzelia pachyrhiza

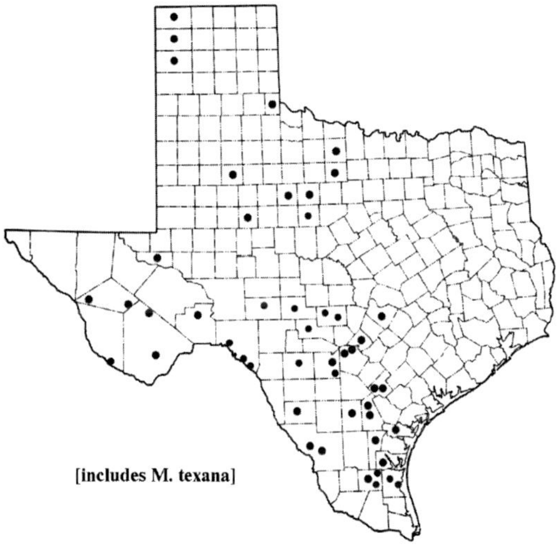
[includes M. texana]

Mentzelia reverchonii

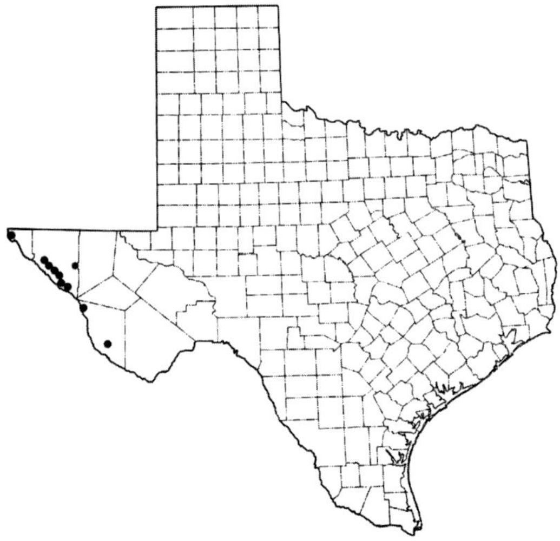

Mentzelia saxicola

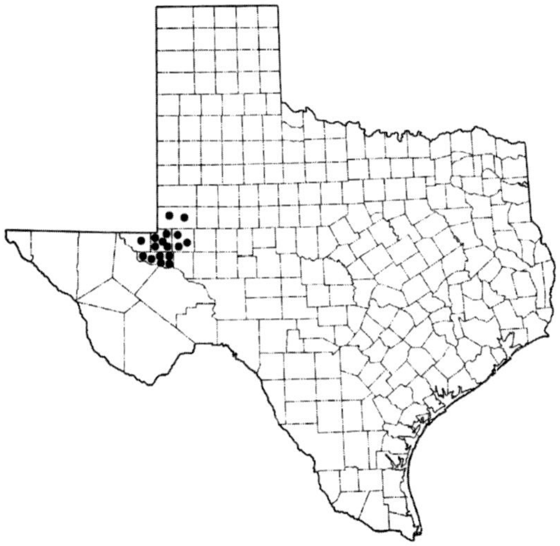

Mentzelia strictissima

Buddleja marrubiifolia

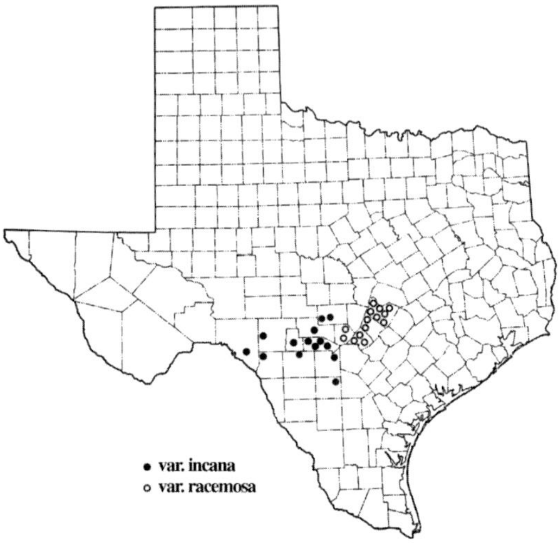
• var. incana
○ var. racemosa

Buddleja racemosa

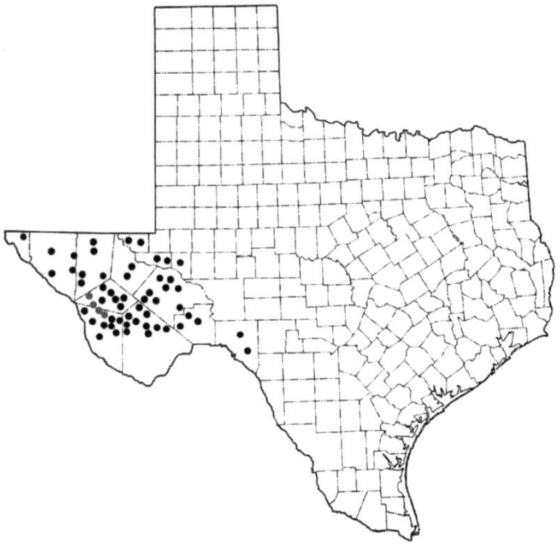

Buddleja scordioides

Buddleja sessiliflora

Emorya suaveolens

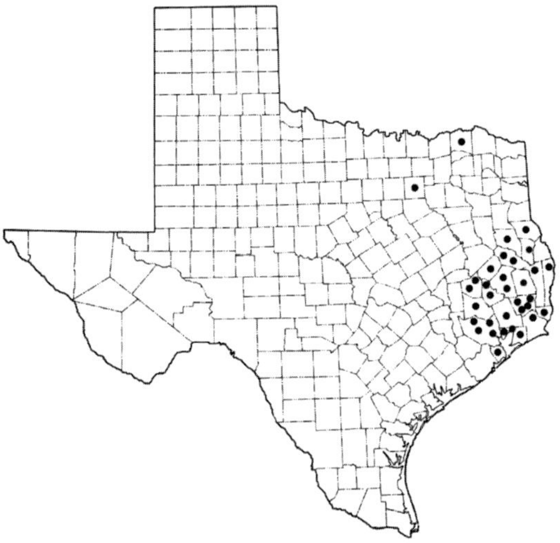

Gelsemium sempervirens

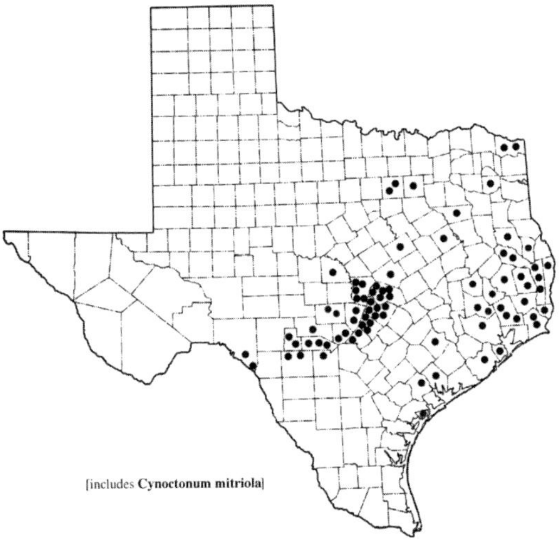
[includes Cynoctonum mitriola]

Mitreola petiolata

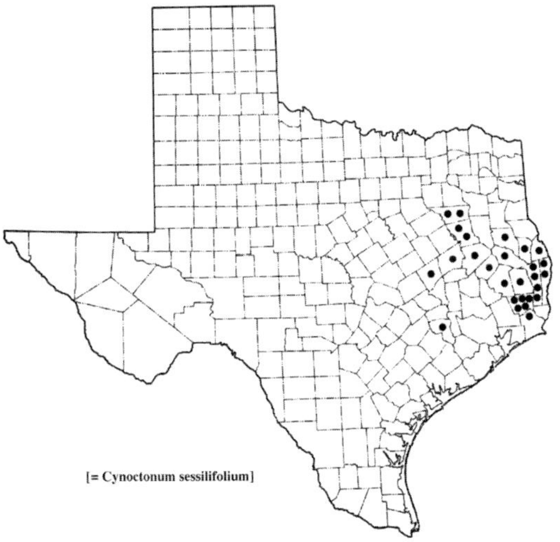
[= Cynoctonum sessilifolium]

Mitreola sessilifolia

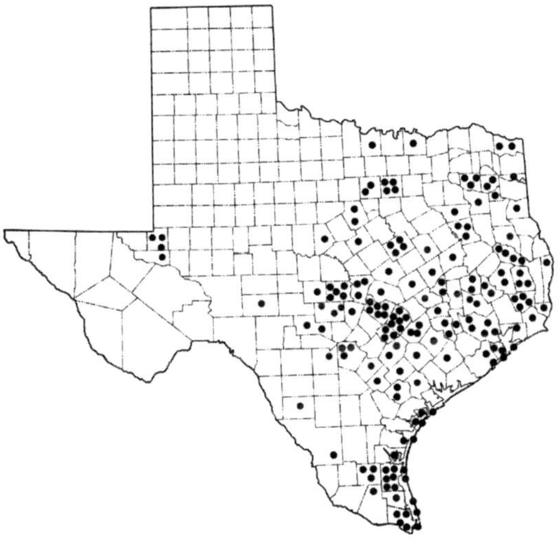

Polypremum procumbens

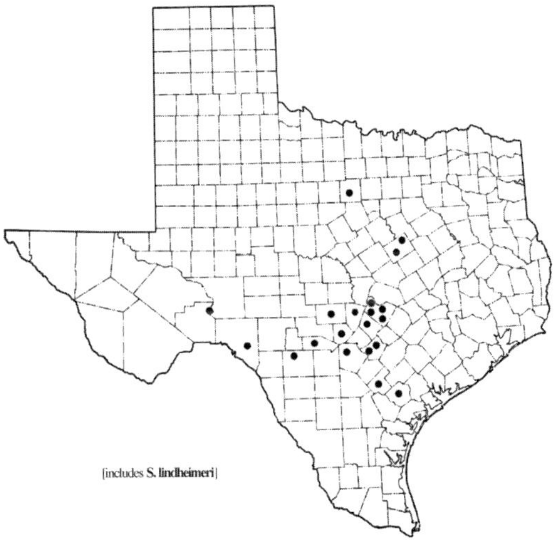
[includes S. lindheimeri]

Spigelia hedyotidea

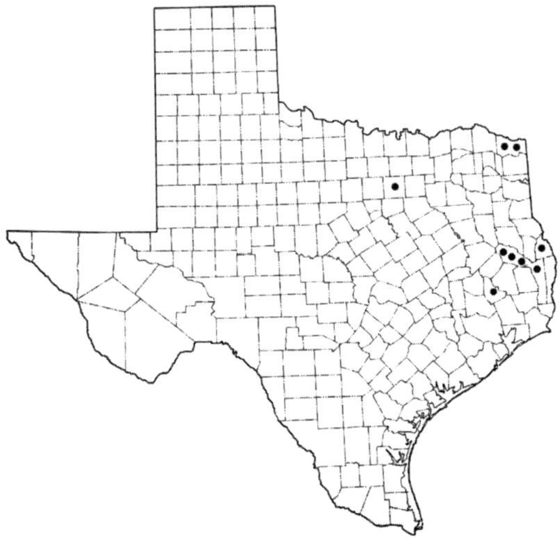

Spigelia marilandica

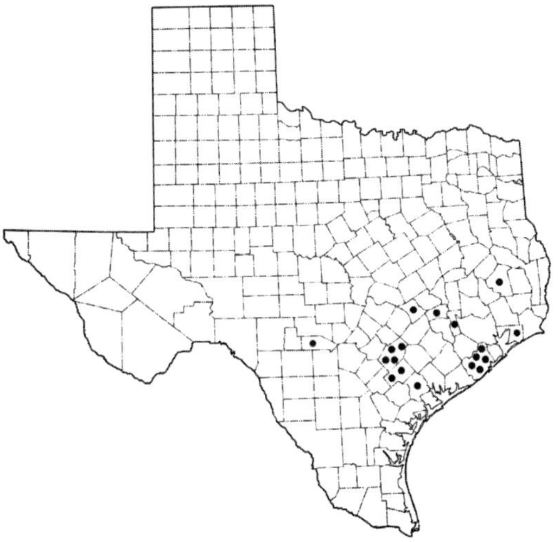

Spigelia texana

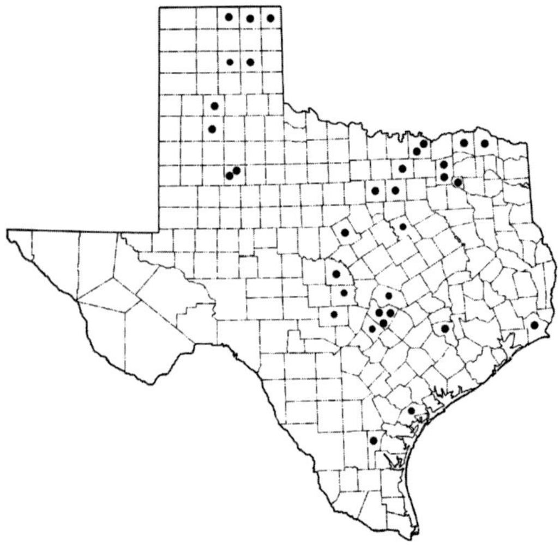

Ammannia auriculata

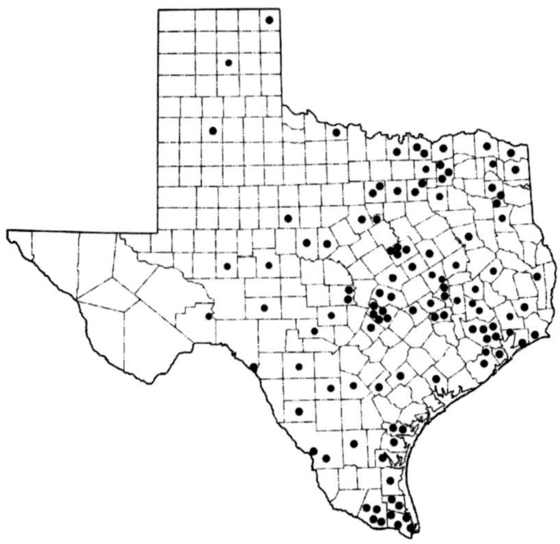

Ammannia coccinea

[includes A. teres]

Ammannia latifolia

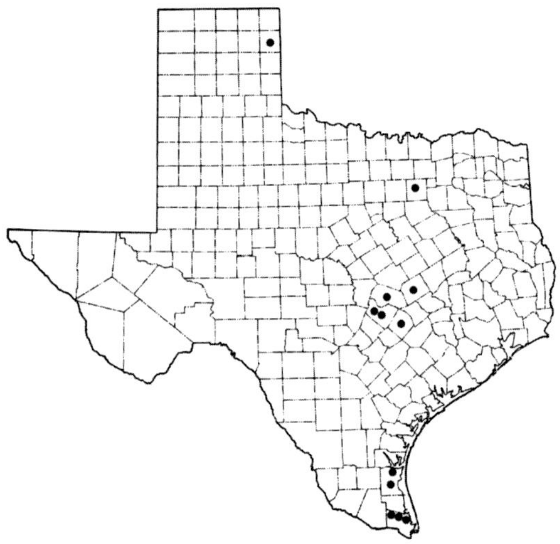

Ammannia robusta

Cuphea carthagenensis

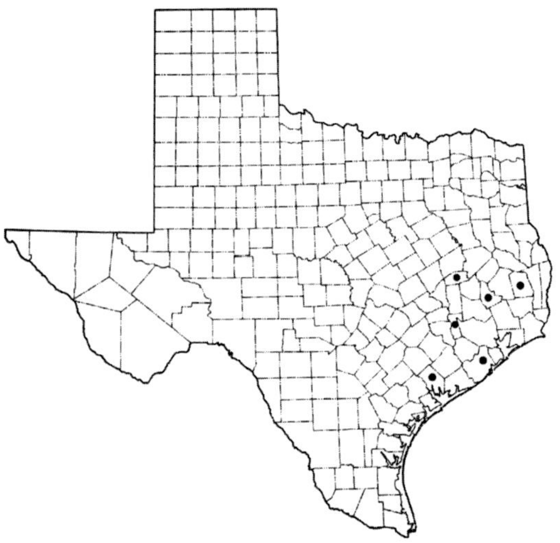

Cuphea glutinosa

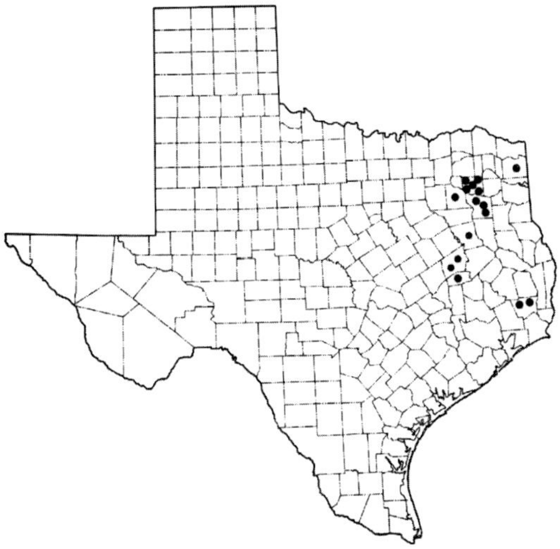

Decodon verticillatus

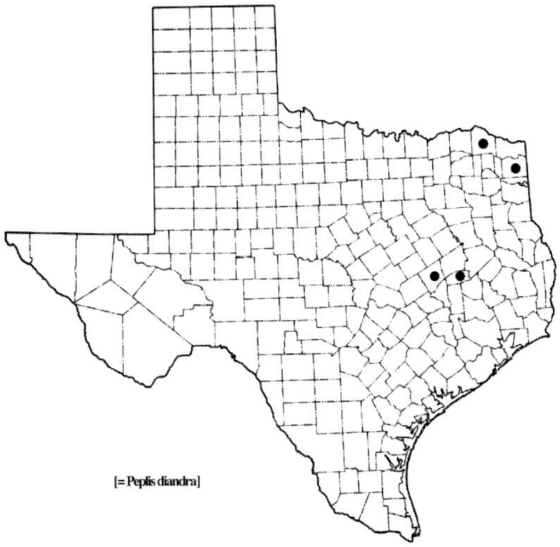
[= Peplis diandra]

Didiplis diandra

Heimia salicifolia

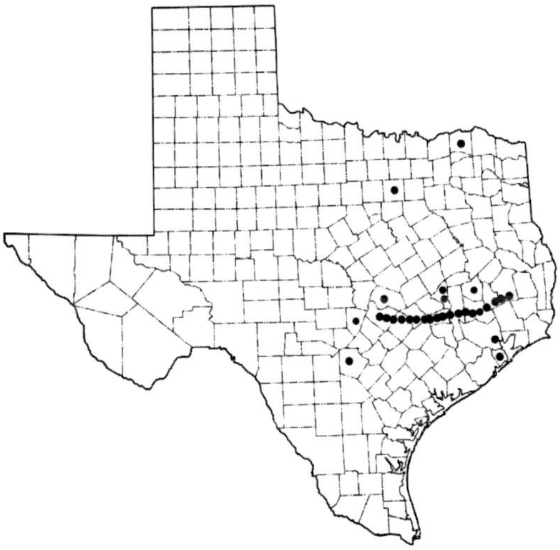

Lagerstroemia indica

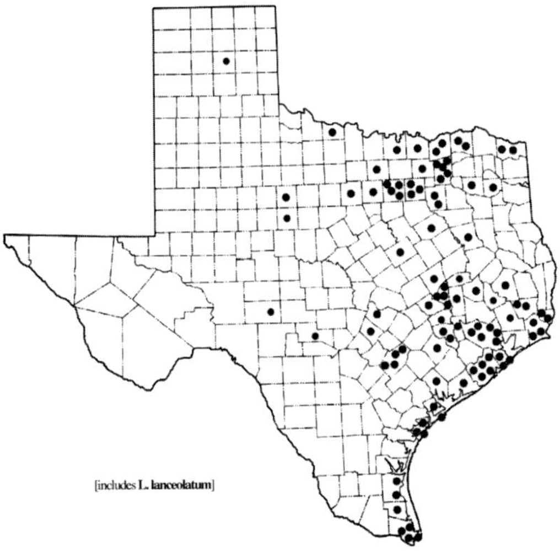
[includes L. lanceolatum]

Lythrum alatum

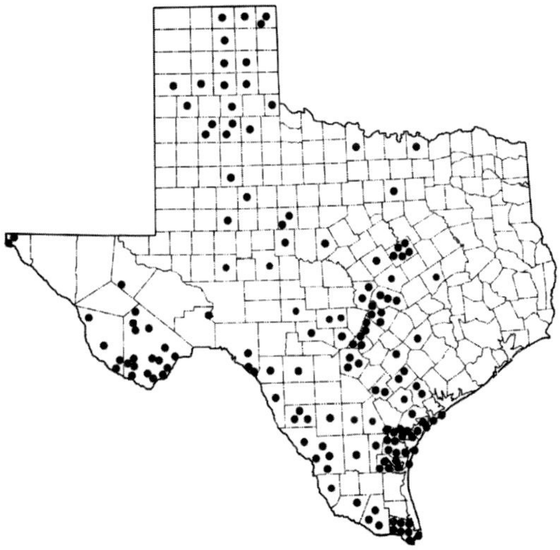

Lythrum californicum

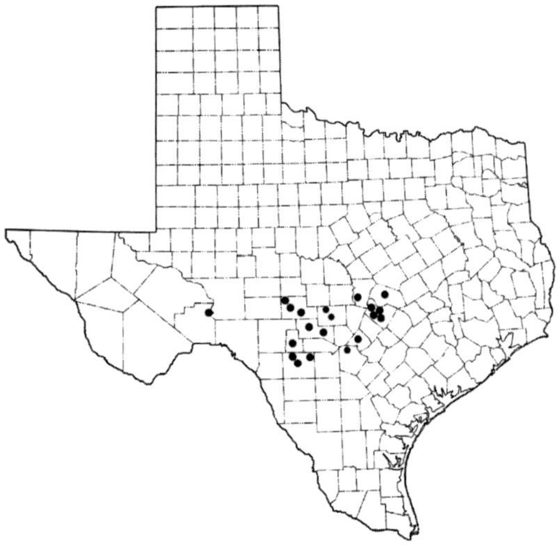

Lythrum ovalifolium

Lythrum salicaria

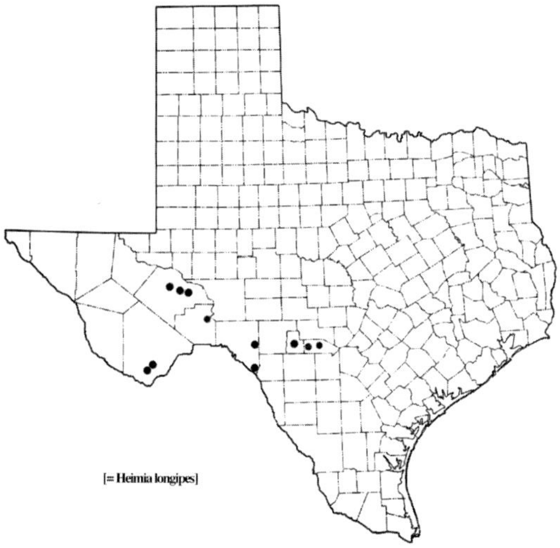
[= Heimia longipes]

Nesaea longipes

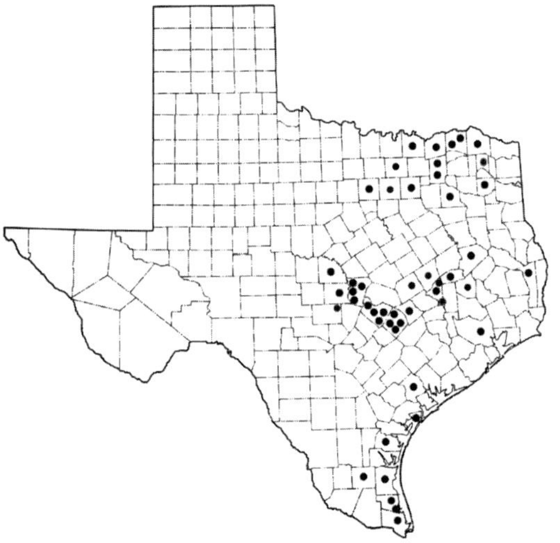

Rotala ramosior

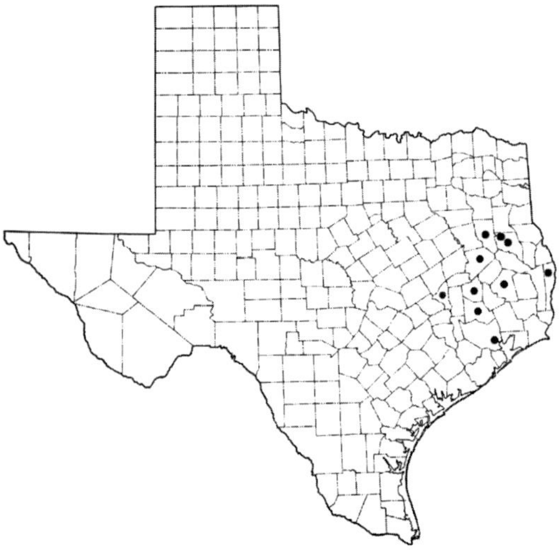

Liriodendron tulipifera

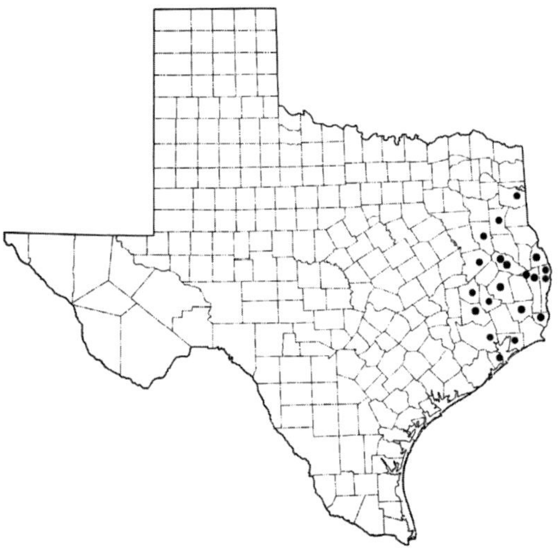

Magnolia grandiflora

Magnolia pyramidata

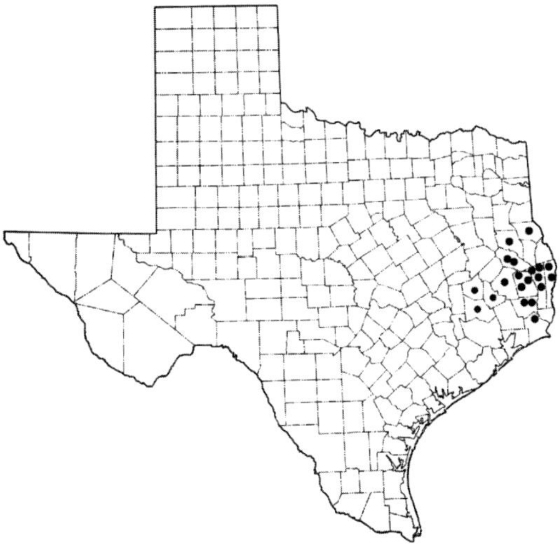

Magnolia virginiana

Aspicarpa hirtella

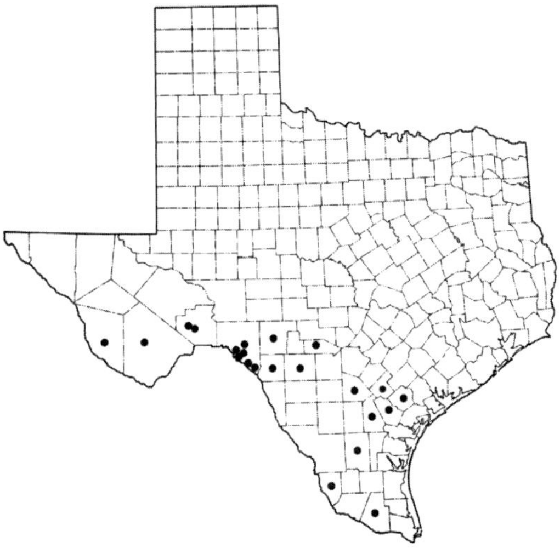

Aspicarpa hyssopifolia

Callaeum septentrionale

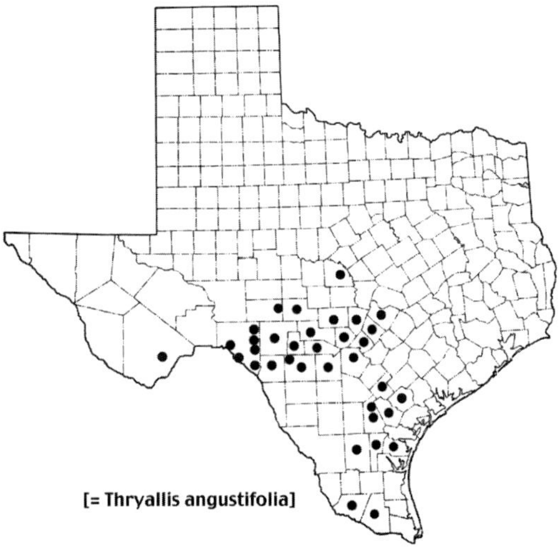
[= Thryallis angustifolia]

Galphimia angustifolia

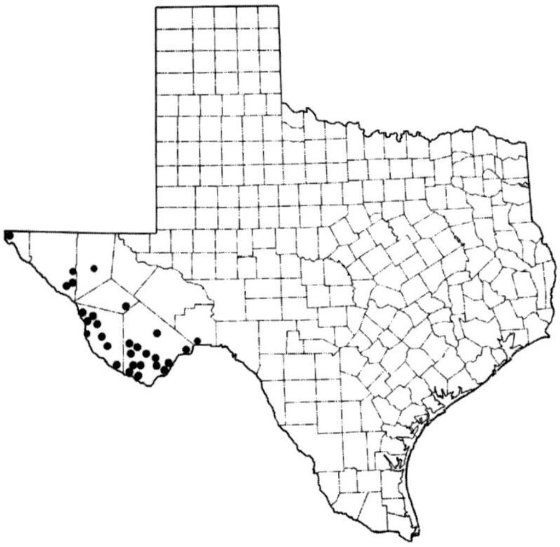

Janusia gracilis

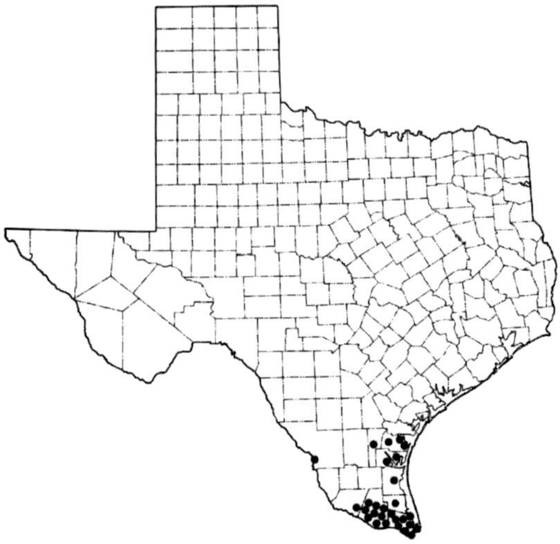

Malpighia glabra

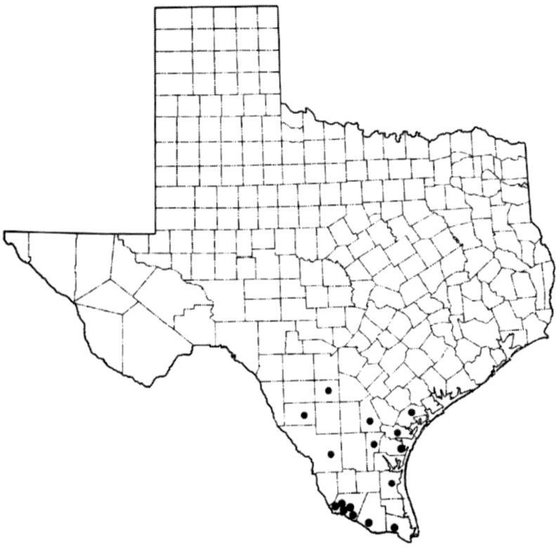

Abutilon abutiloides

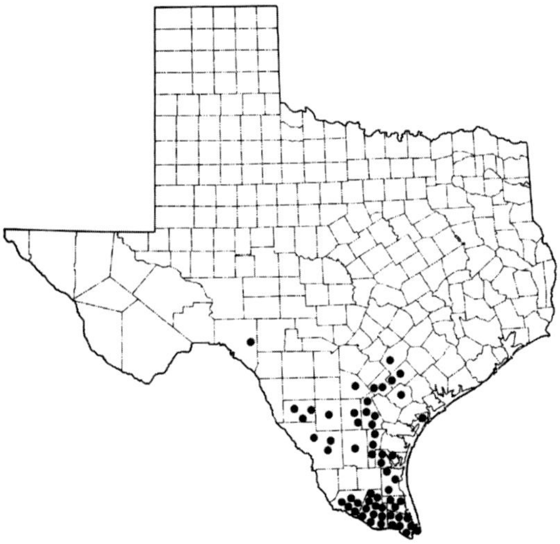

Abutilon berlandieri

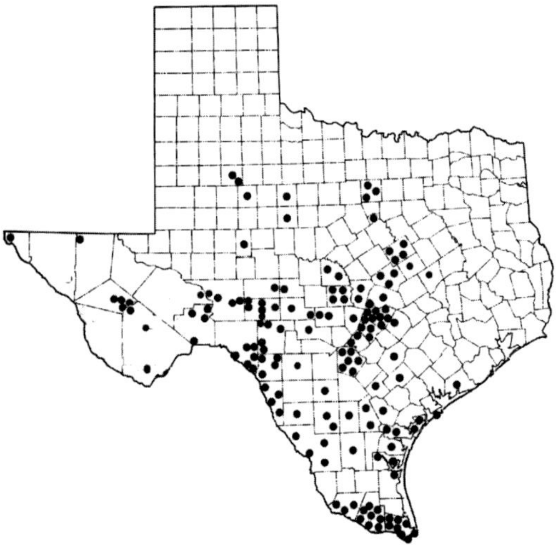

Abutilon fruticosum

Abutilon hulseanum

Abutilon hypoleucum

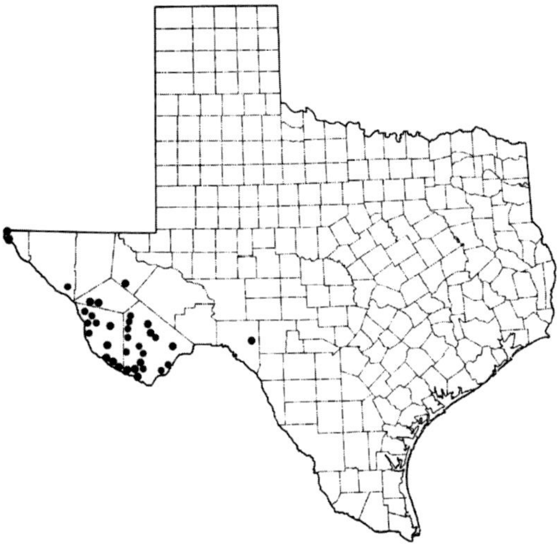

Abutilon malacum

Abutilon mollicomum

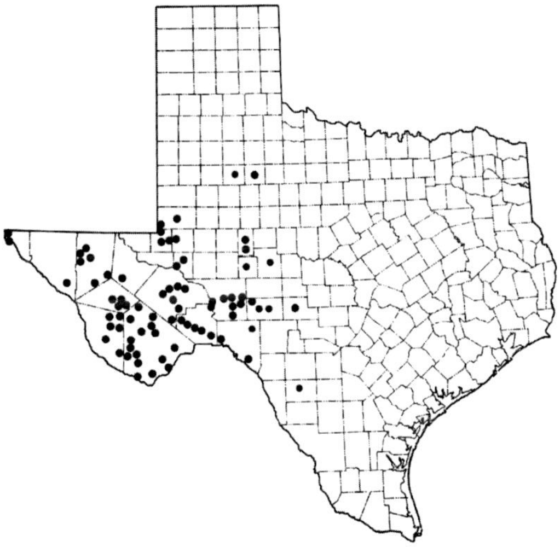

Abutilon parvulum

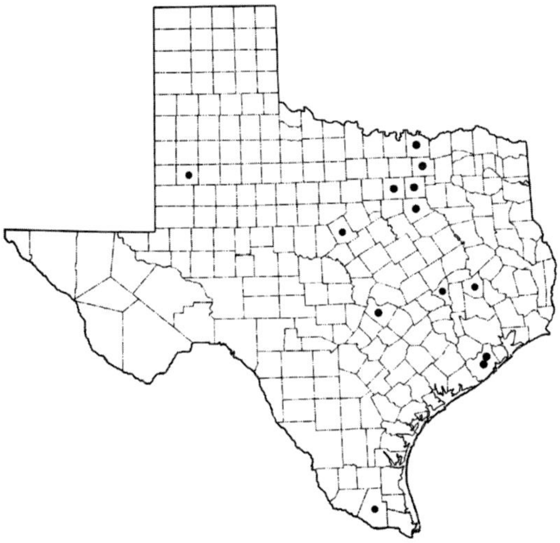

Abutilon theophrasti

Abutilon trisulcatum

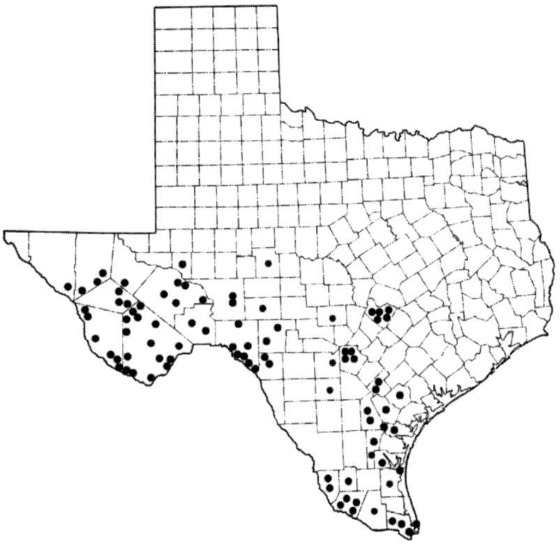

Abutilon wrightii

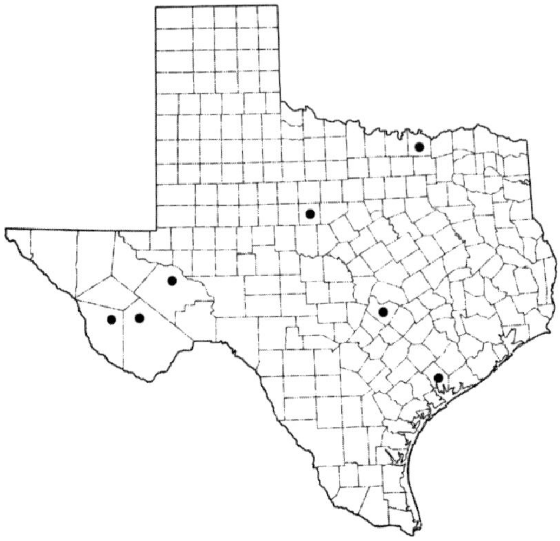

Alcea rosea

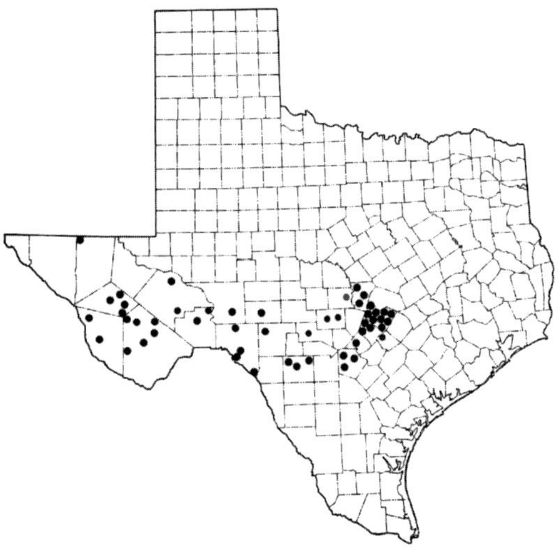

Allowissadula holosericea

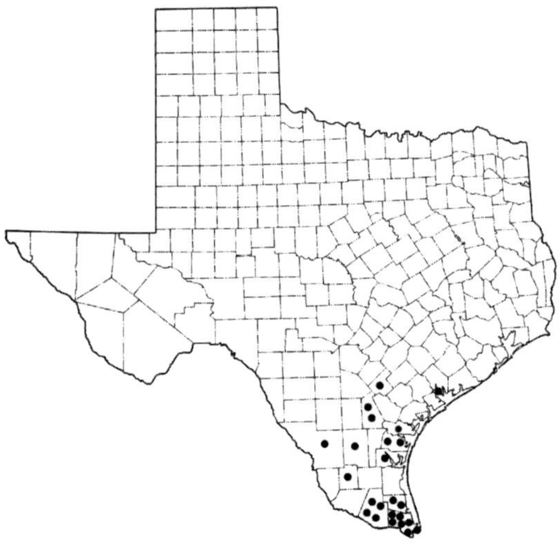

Allowissadula lozanii

Anoda crenatiflora

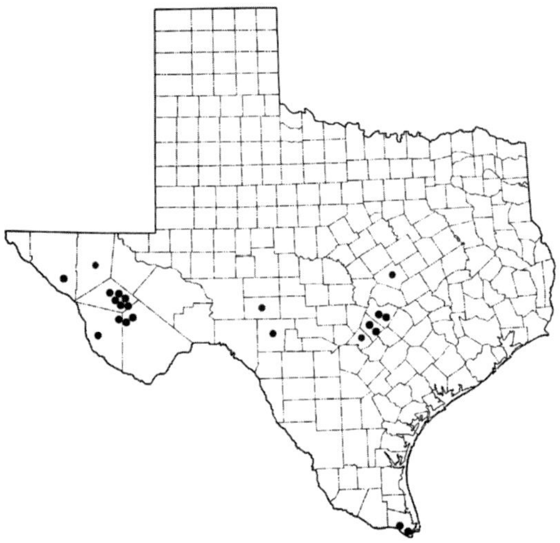

Anoda cristata

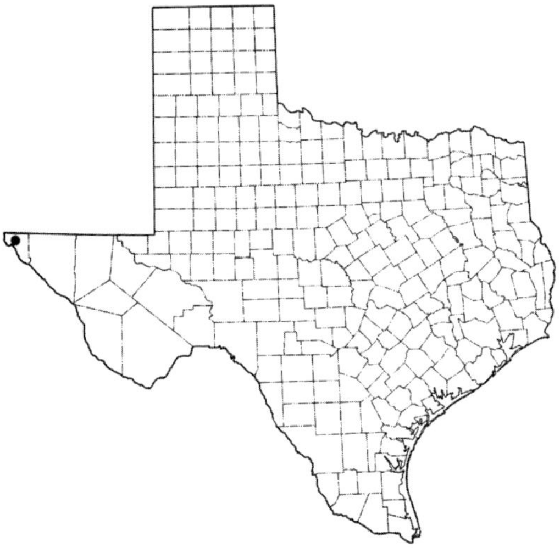

Anoda lanceolata

Anoda pentaschista

Bastardia viscosa

Batesimalva violacea

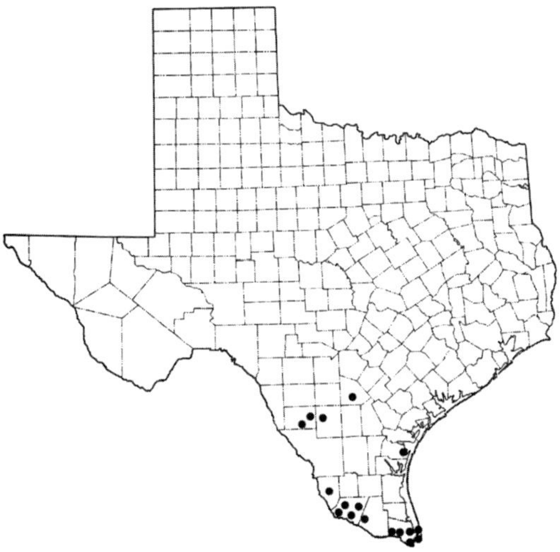

Billieturnera helleri

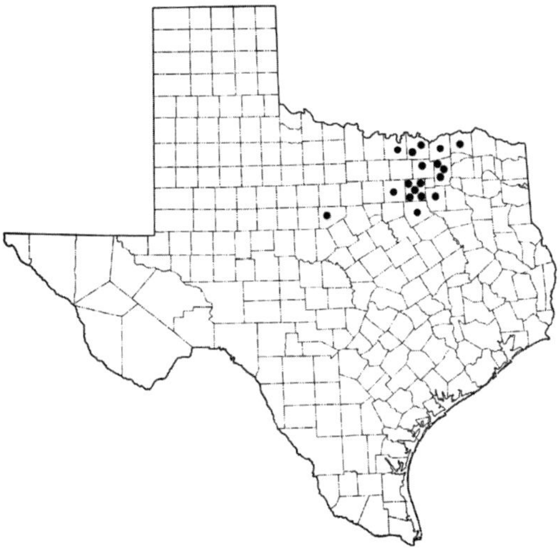

Callirhoe alcaeoides

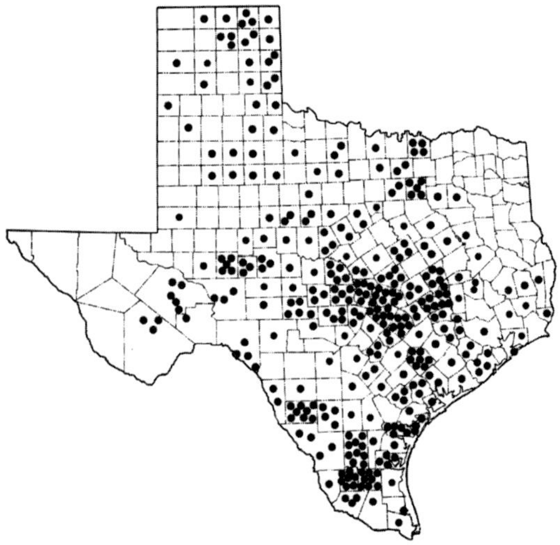

Callirhoe involucrata

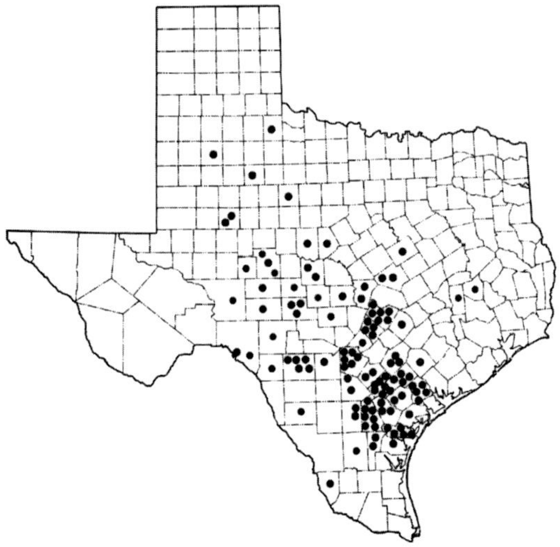

Callirhoe leiocarpa

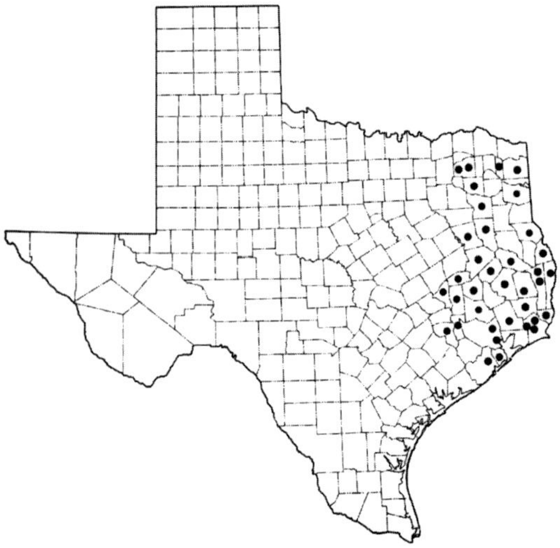

Callirhoe papaver

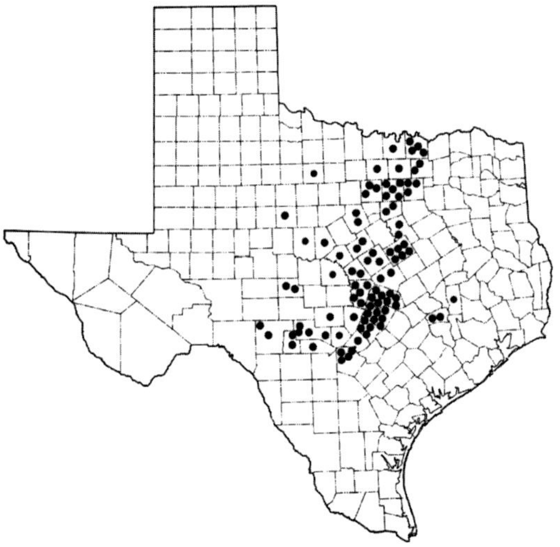

Callirhoe pedata

Callirhoe scabriuscula

Cienfuegosia drummondii

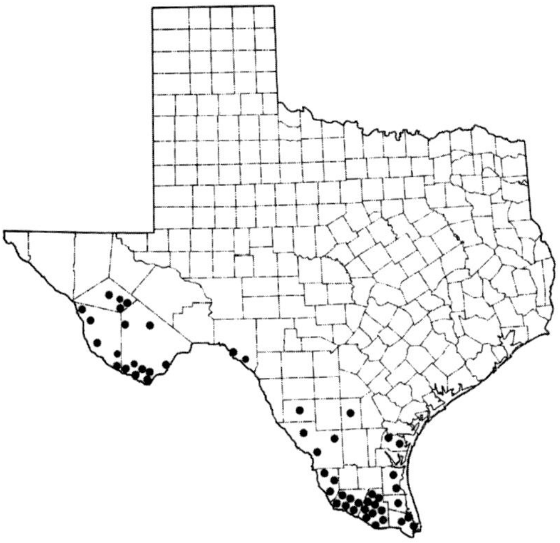

Herissantia crispa

Hibiscus aculeatus

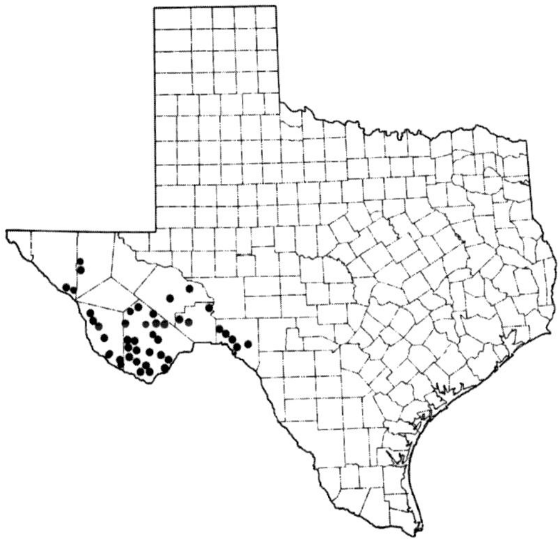

Hibiscus coulteri

Hibiscus dasycalyx

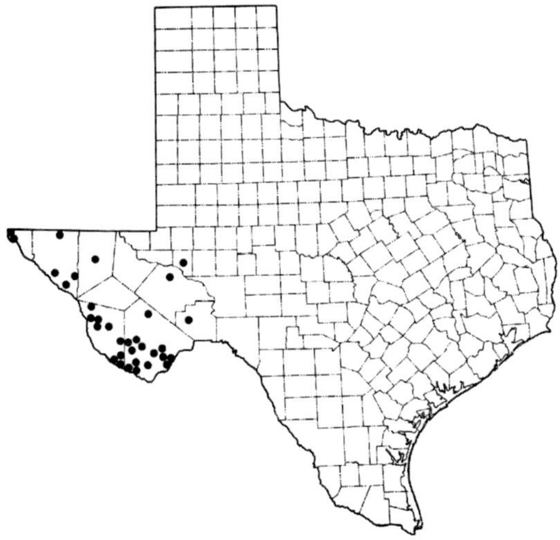

Hibiscus denudatus

Hibiscus esculentus

Hibiscus grandiflorus

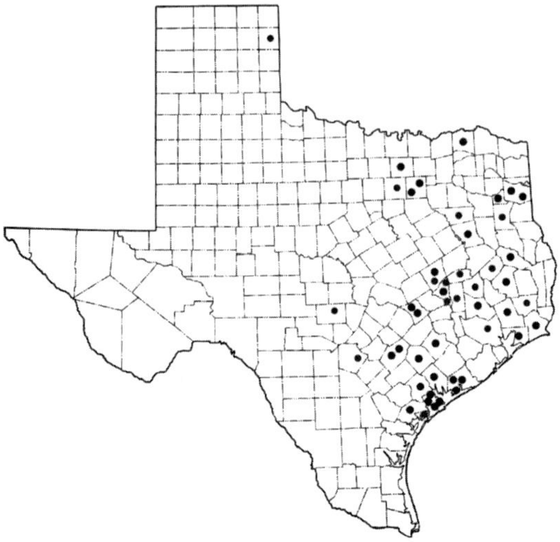

Hibiscus laevis

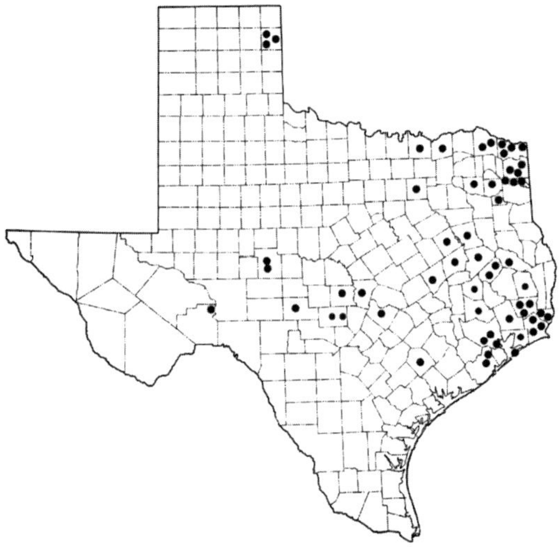

Hibiscus lasiocarpos

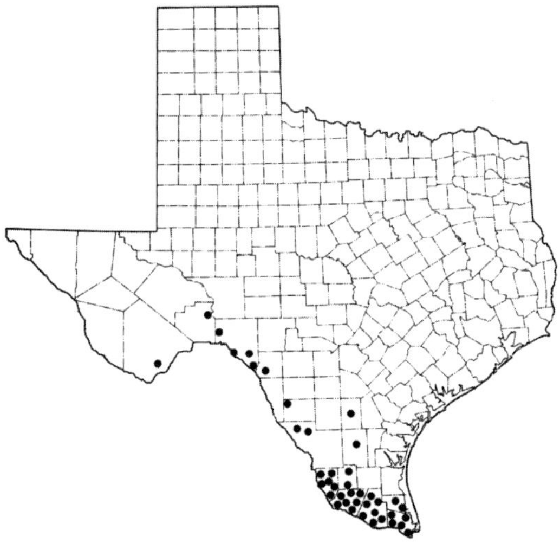

Hibiscus martianus

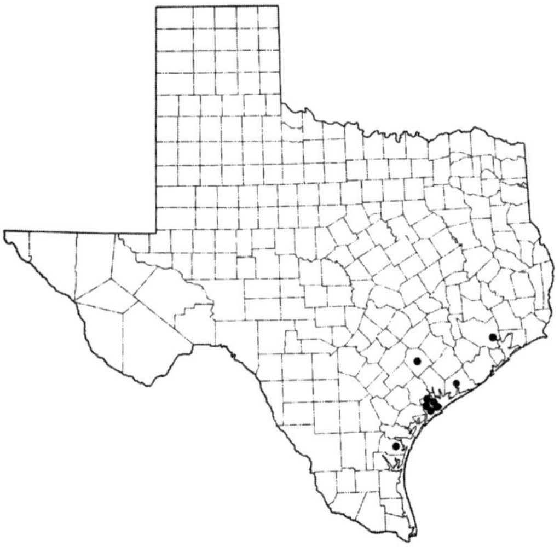

Hibiscus striatus

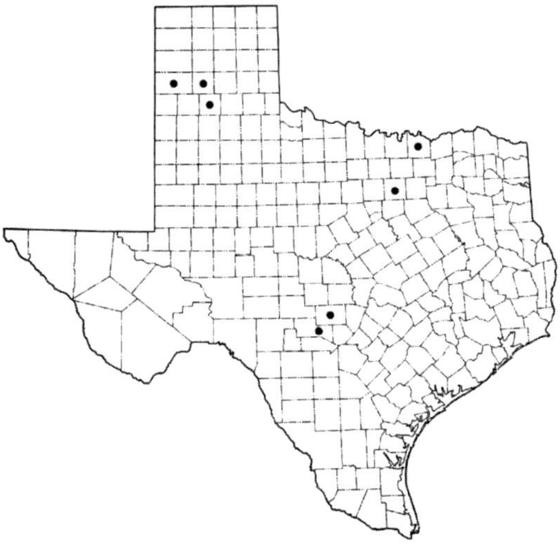

Hibiscus trionum

Kosteletzkya depressa

Kosteletzkya virginica

Krapovickasia physaloides

Lavatera trimestris

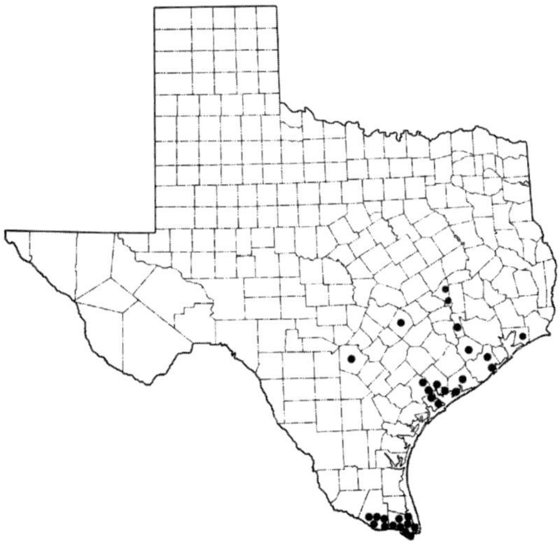

Malachra capitata

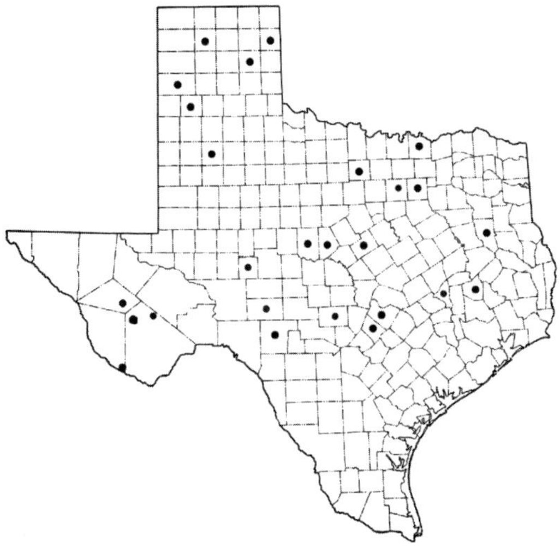

Malva neglecta

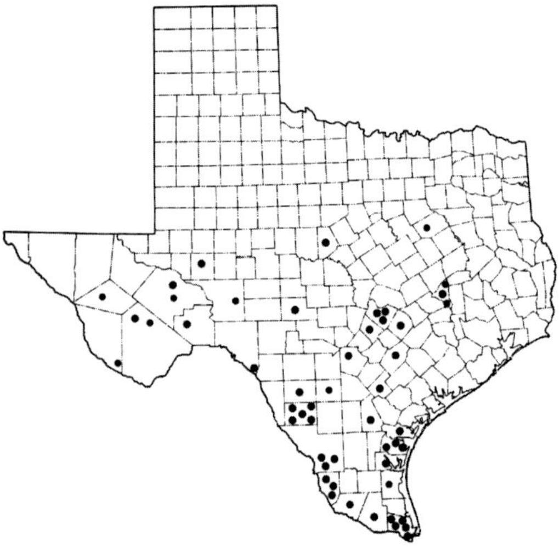

Malva parviflora

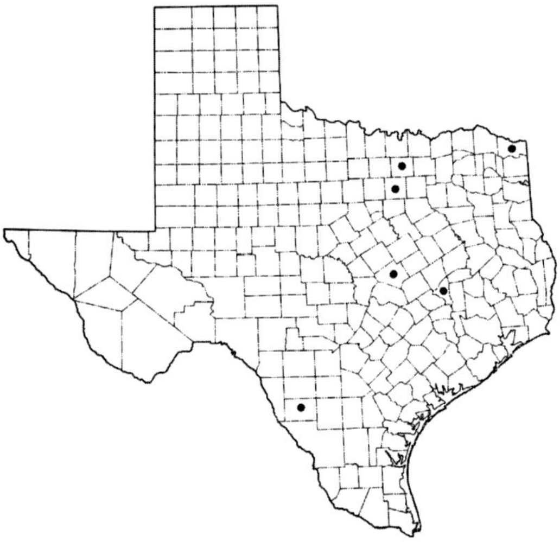

Malva rotundifolia

Malva sylvestris

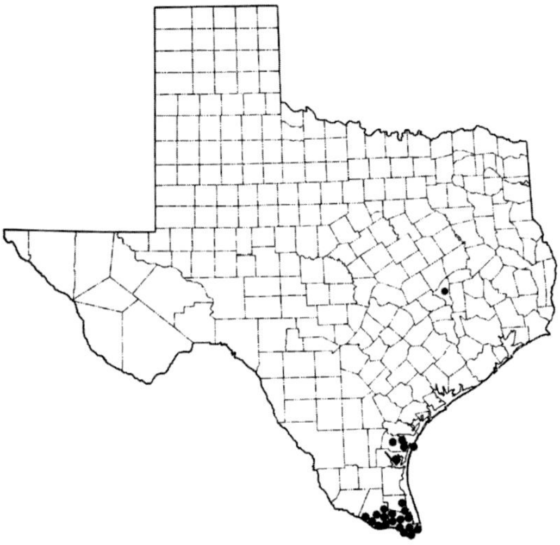

Malvastrum americanum

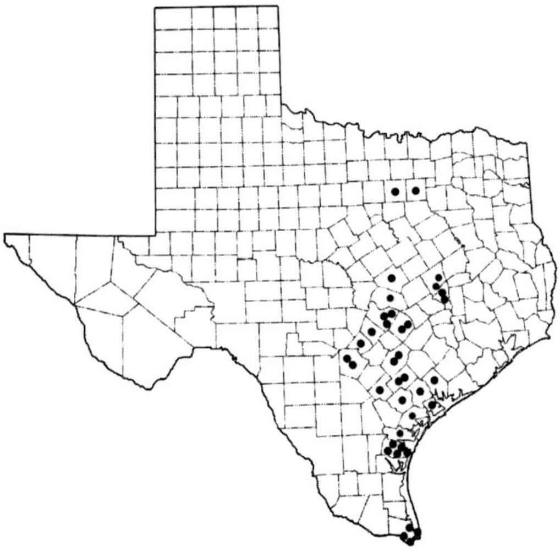

Malvastrum aurantiacum

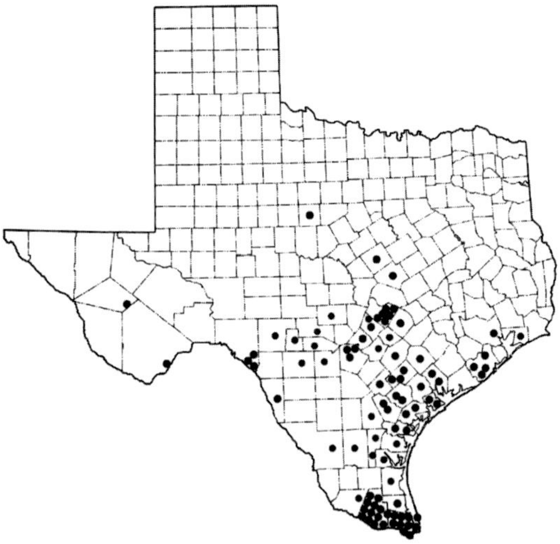

Malvastrum coromandelianum

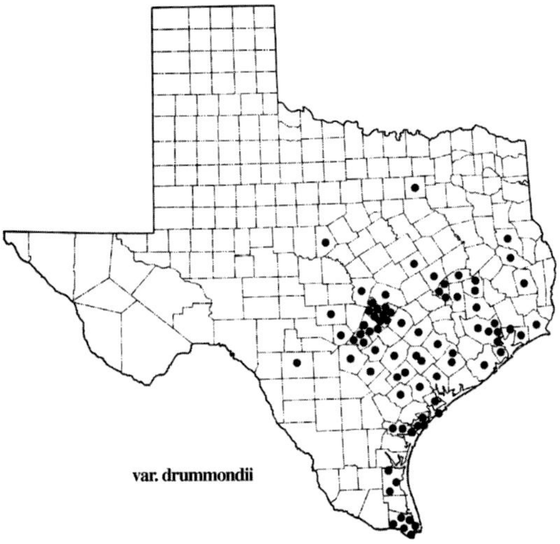
var. drummondii

Malvaviscus arboreus

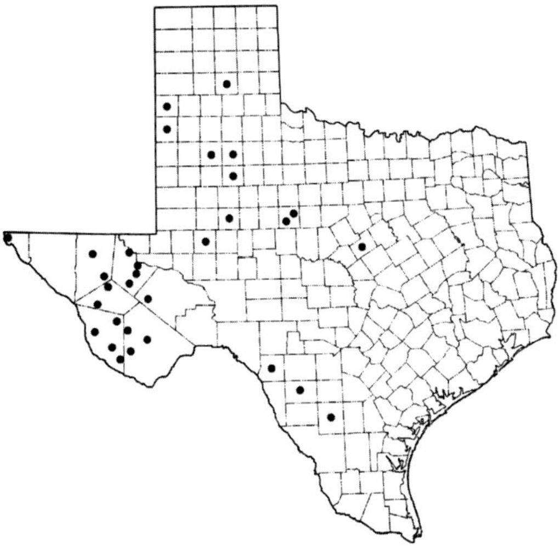

Malvella lepidota

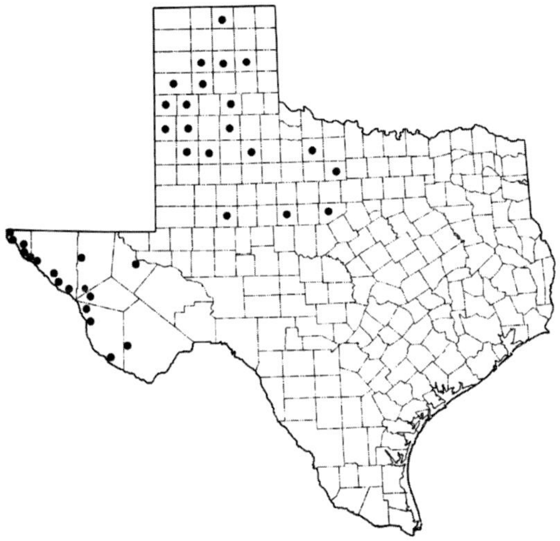

Malvella leprosa

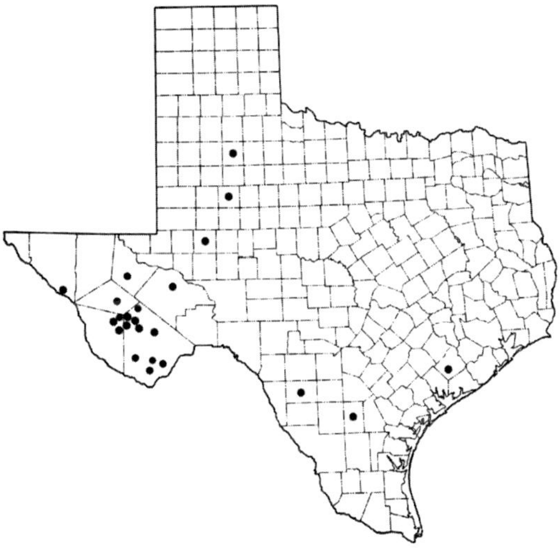

Malvella sagittifolia

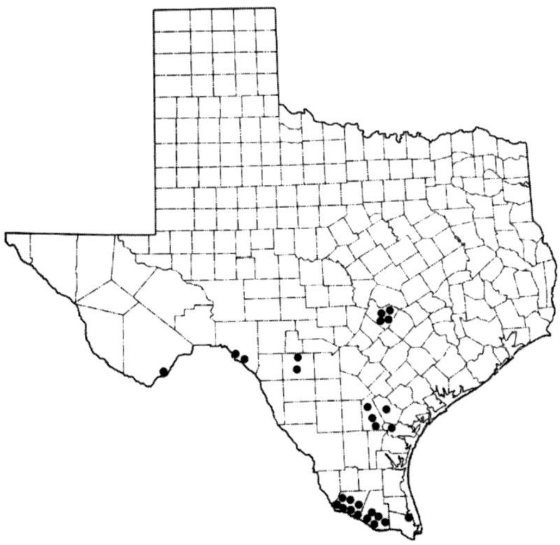

Meximalva filipes

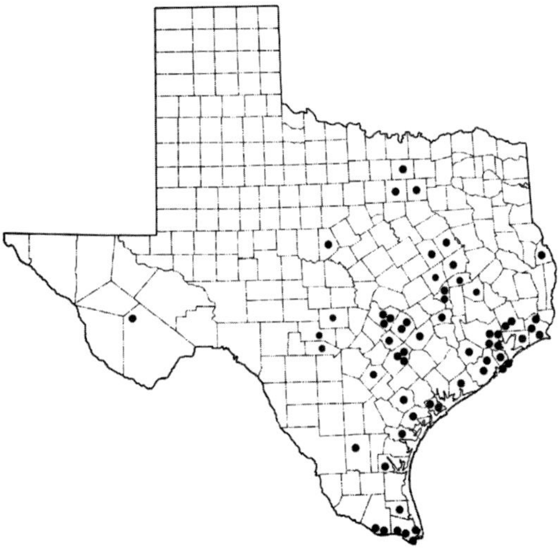

Modiola caroliniana

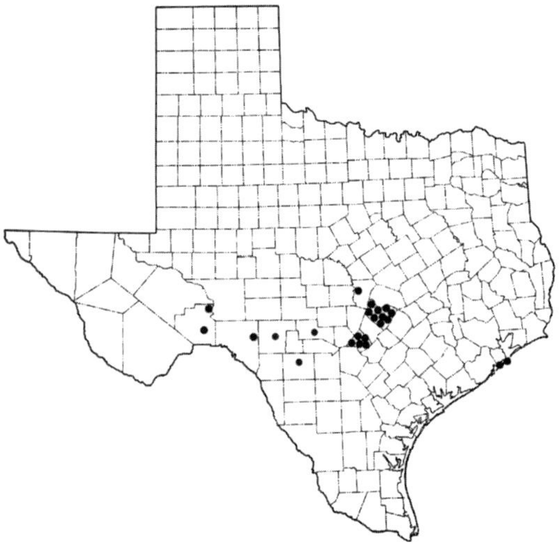

Pavonia lasiopetala

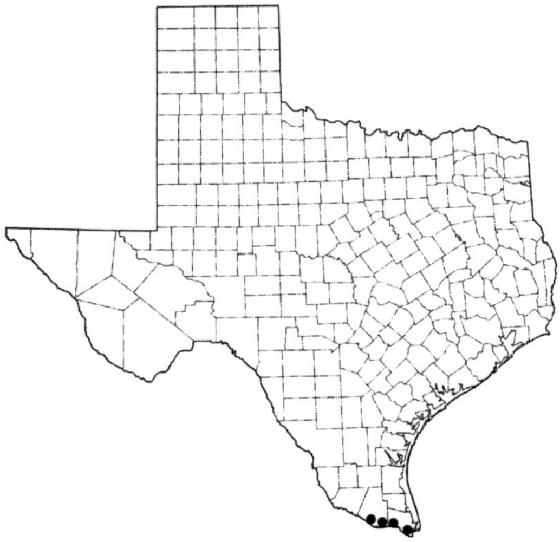

Pseudabutilon umbellatum

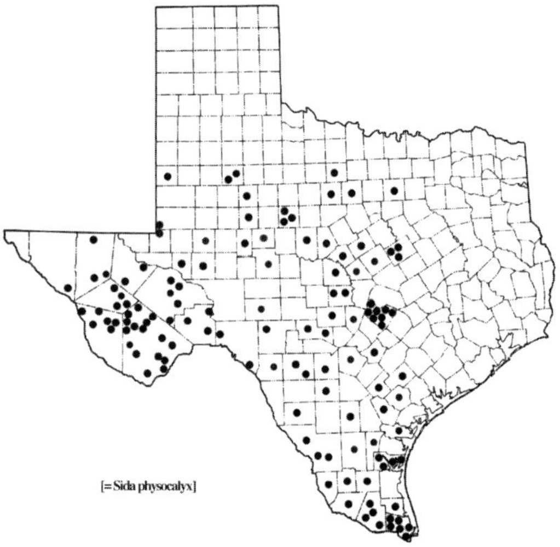
[= Sida physocalyx]

Rhynchosida physocalyx

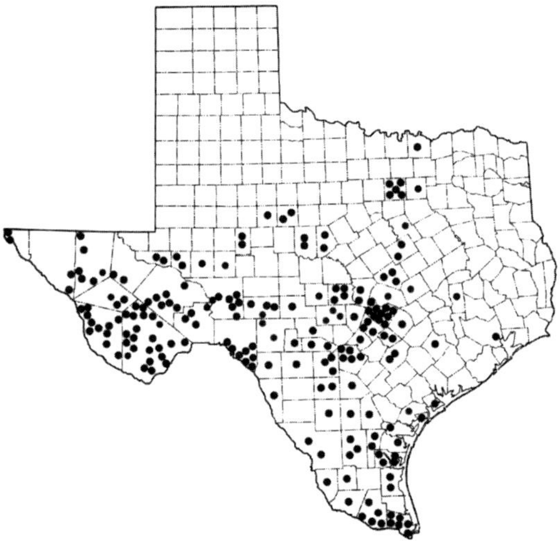

Sida abutifolia

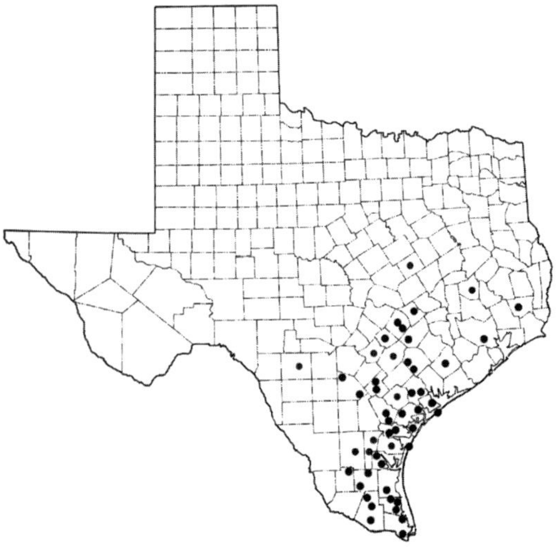

Sida ciliaris

Sida cordifolia

Sida elliotii

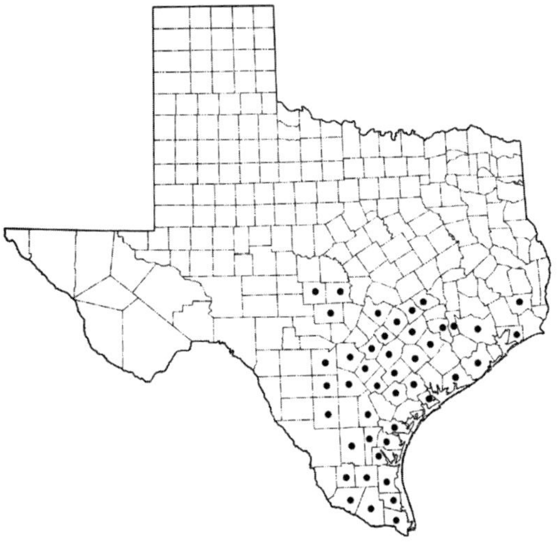

Sida lindheimeri

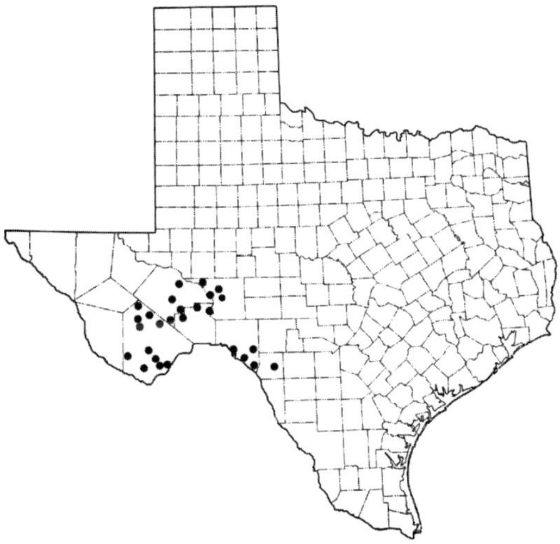

Sida longipes

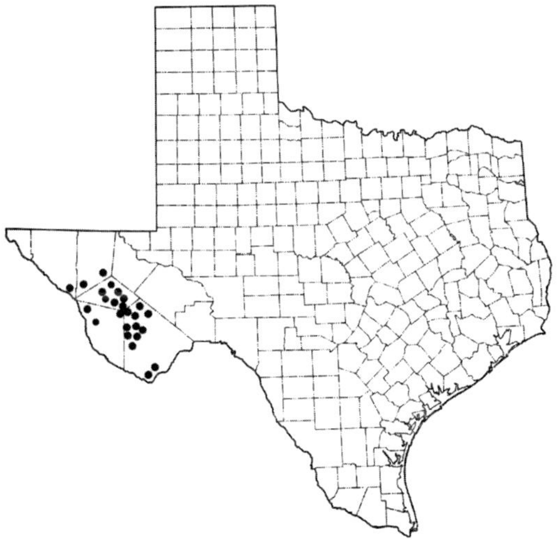

Sida neomexicana

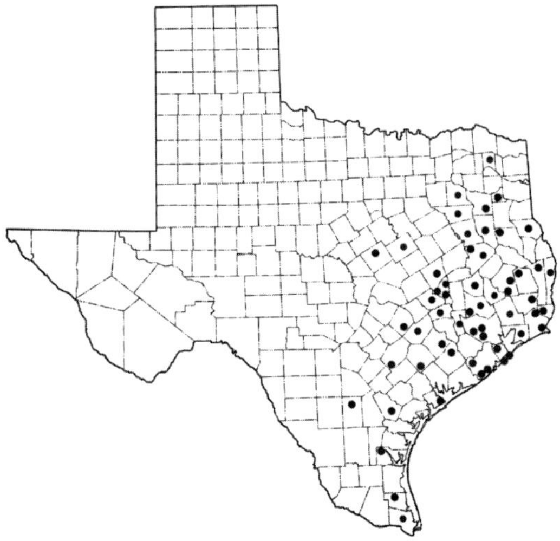

Sida rhombifolia

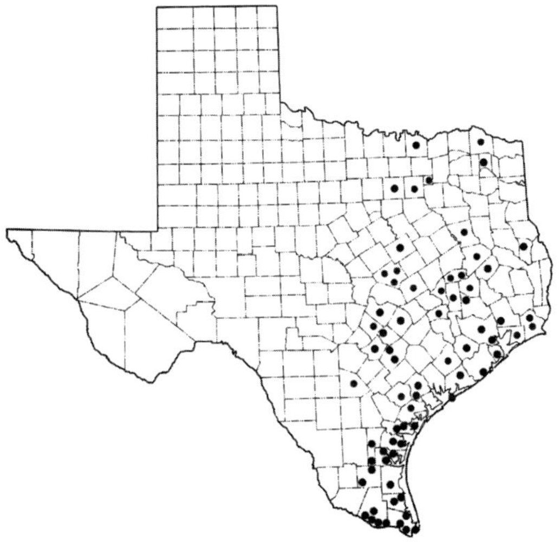

Sida spinosa

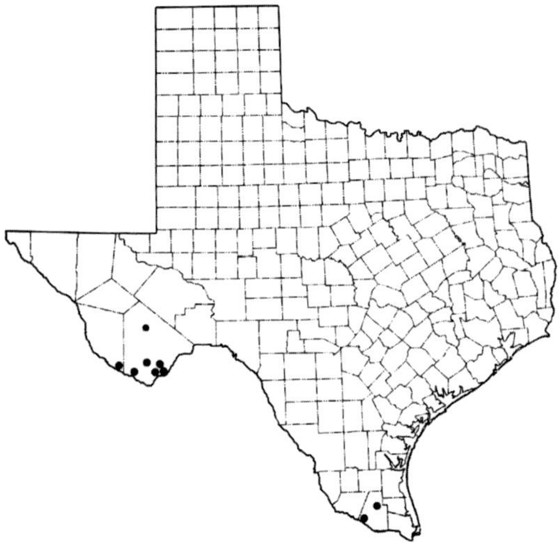

Sida tragiifolia

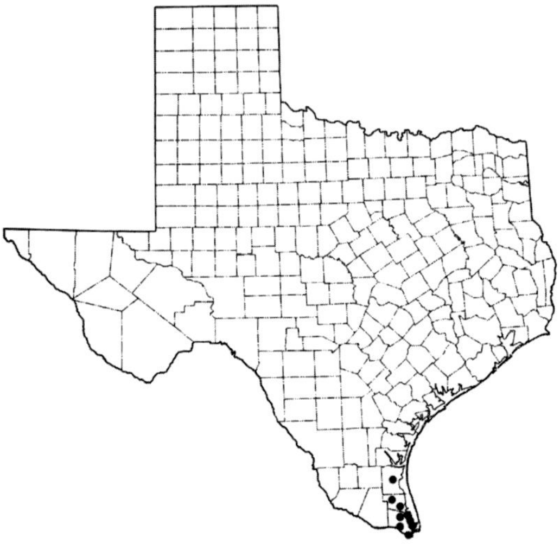

Sidastrum paniculatum

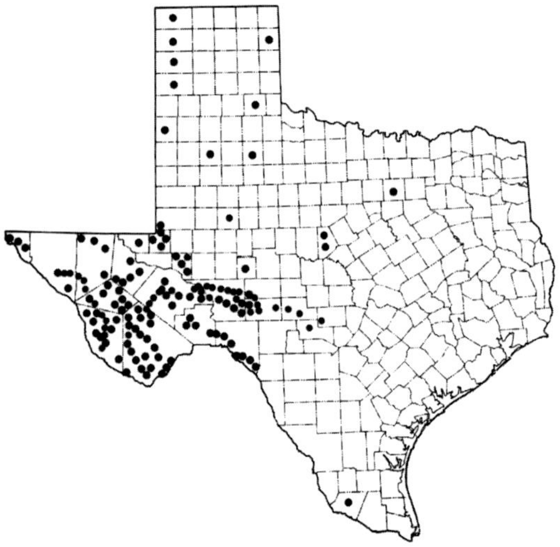

Sphaeralcea angustifolia

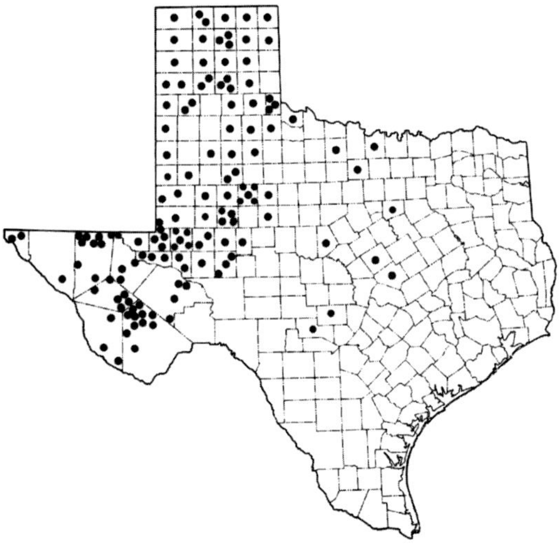

Sphaeralcea coccinea

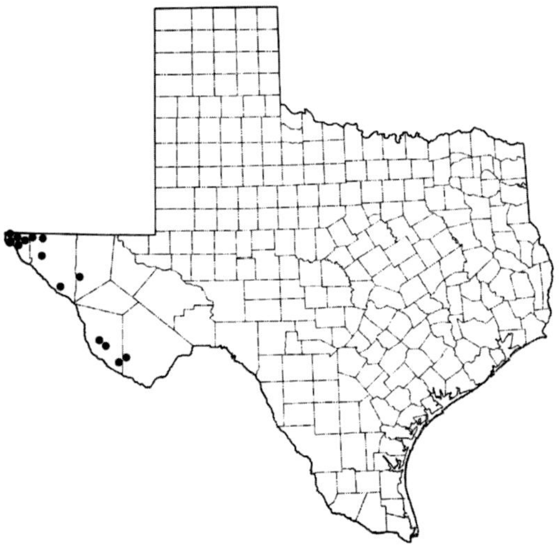

Sphaeralcea digitata

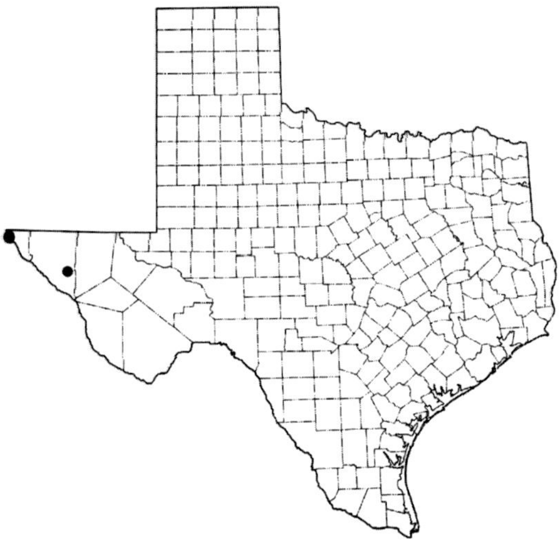

Sphaeralcea fendleri

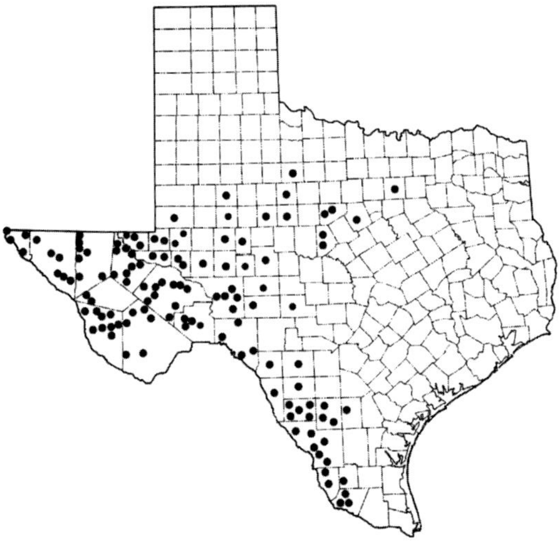

Sphaeralcea hastulata

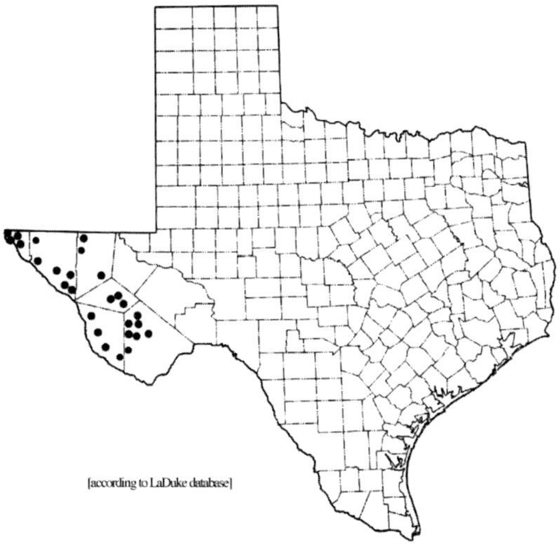
[according to LaDuke database]

Sphaeralcea incana

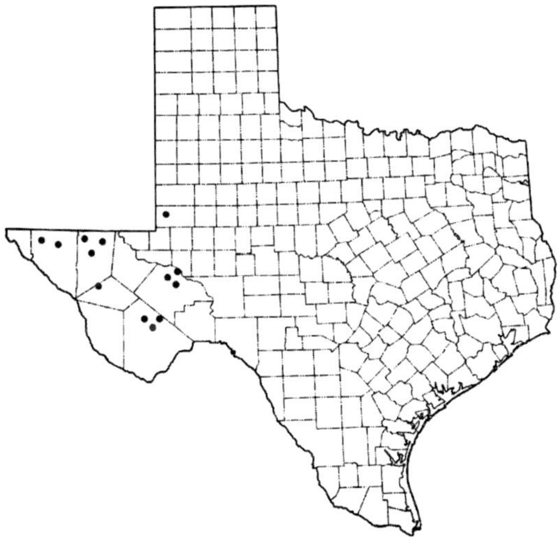

Sphaeralcea leptophylla

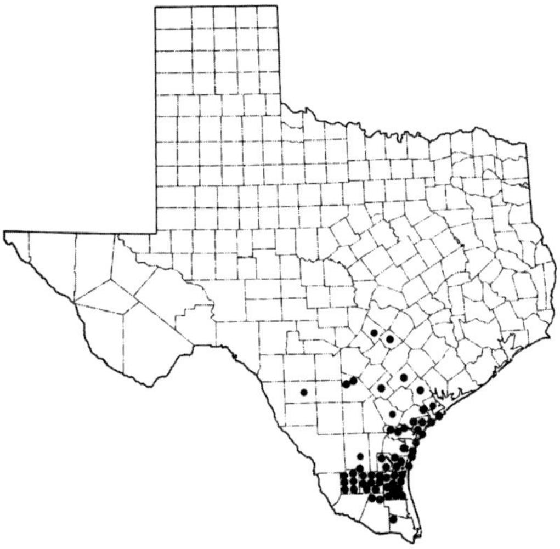

Sphaeralcea lindheimeri

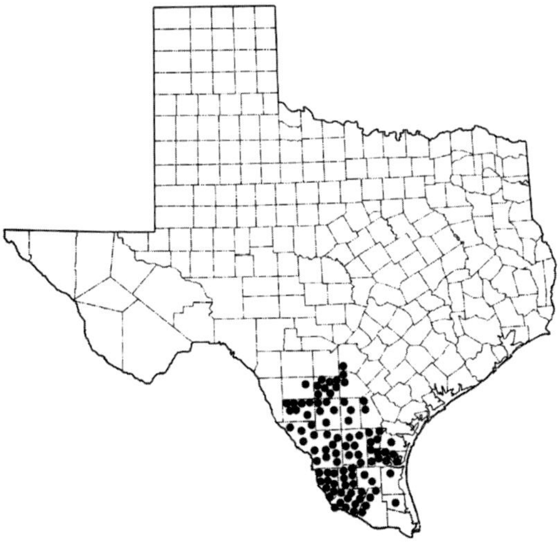

Sphaeralcea pedatifida

[voucher TEX]

Wissadula hernandioides

Wissadula periplocifolia

Rhexia alifanus

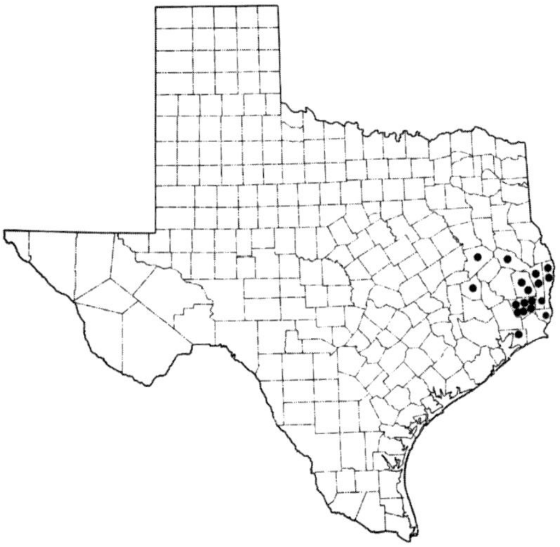

Rhexia lutea

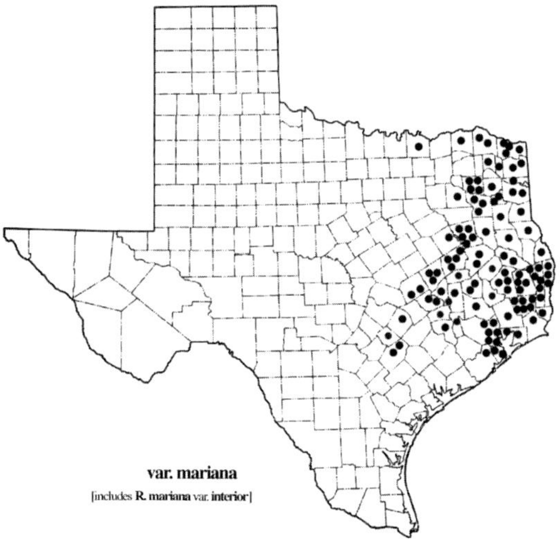
var. mariana
[includes R. mariana var. interior]

Rhexia mariana

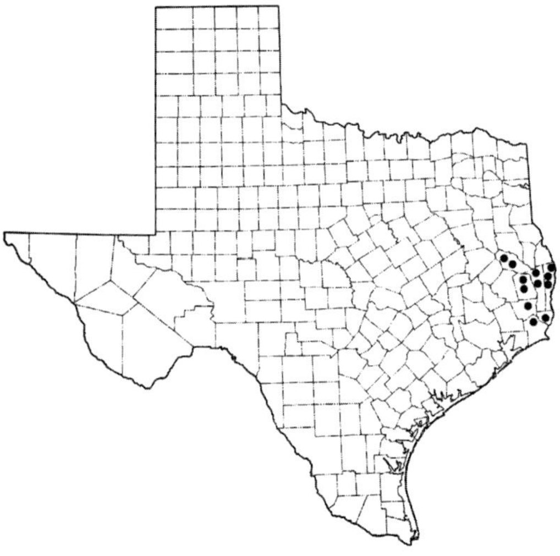

Rhexia petiolata

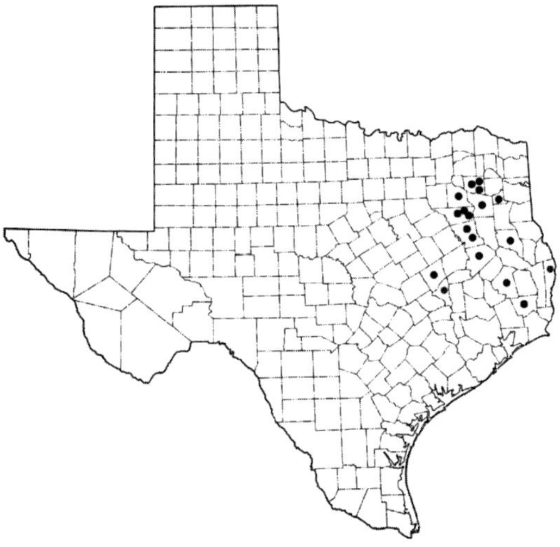

Rhexia virginica

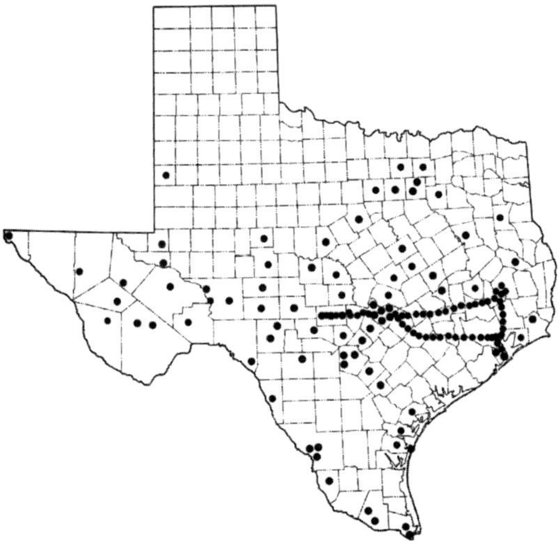

Melia azedarach

Calycocarpum lyonii

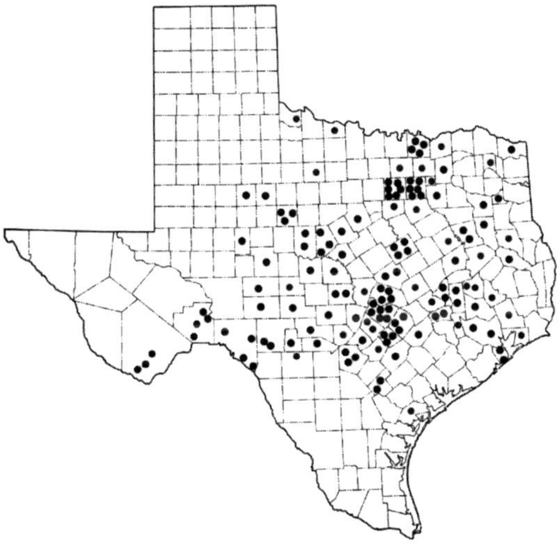

Cocculus carolinus

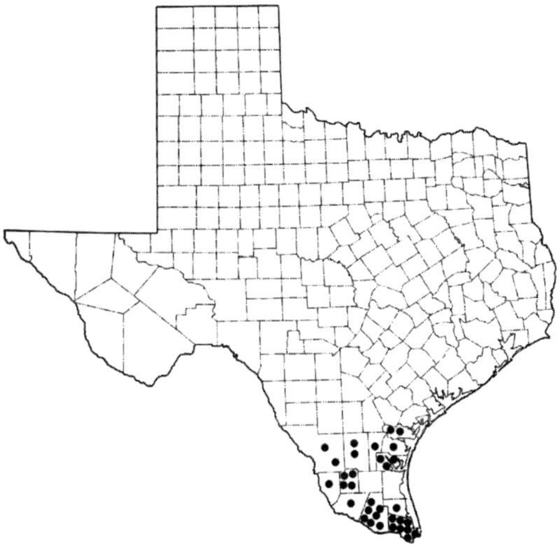

Cocculus diversifolius

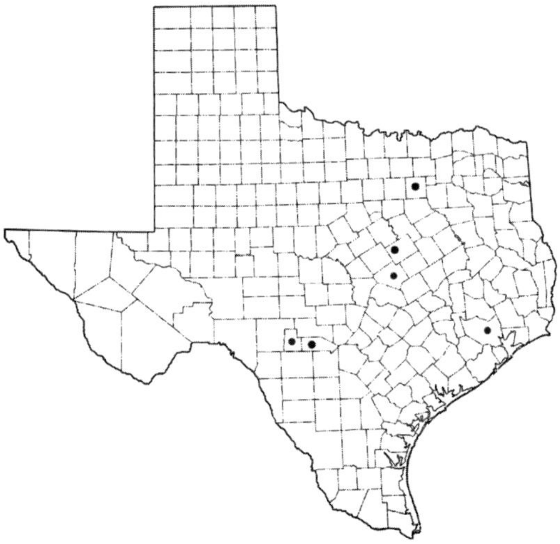

Menispermum canadense

Nymphoides aquatica

Nymphoides peltata

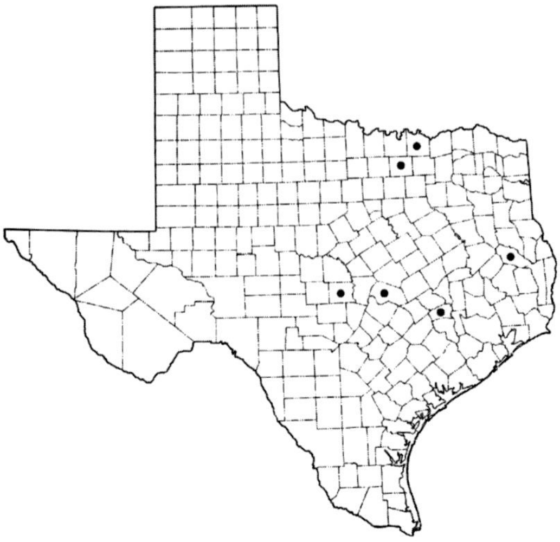

Glinus lotoides

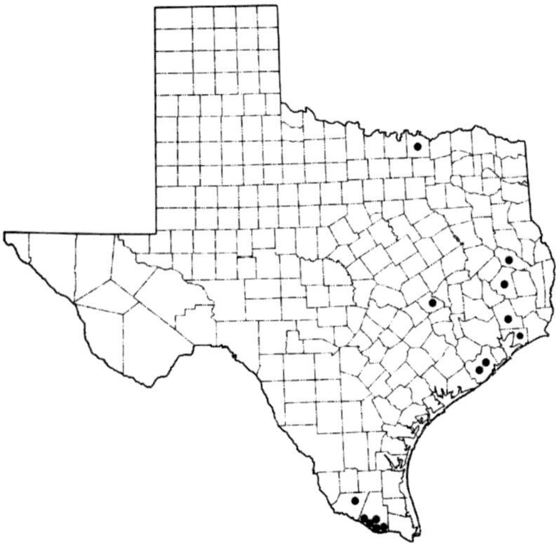

Glinus radiatus

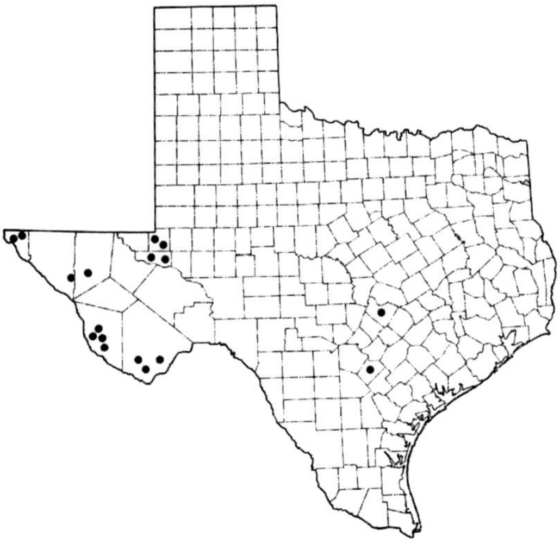

Mollugo cerviana

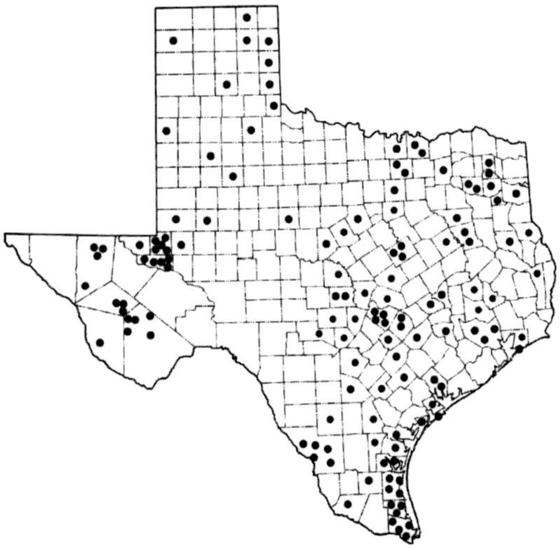

Mollugo verticillata

Monotropa hypopithys

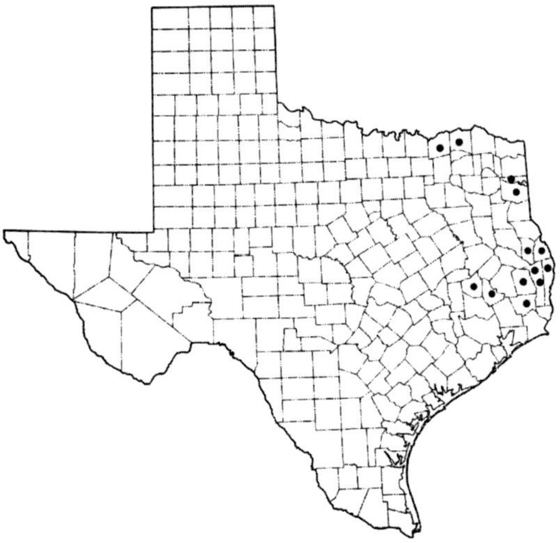

Monotropa uniflora

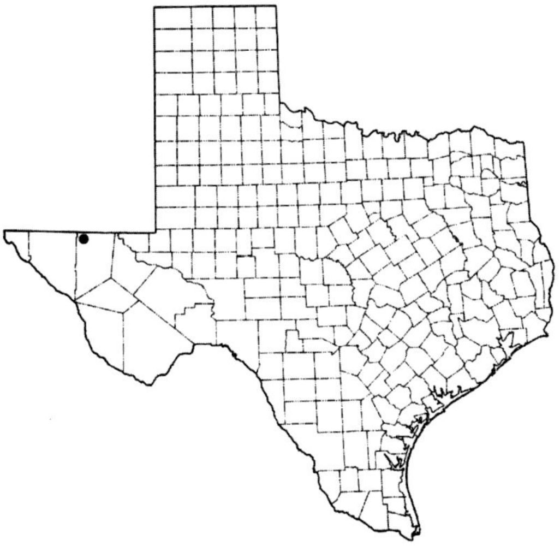

Pterospora andromedea

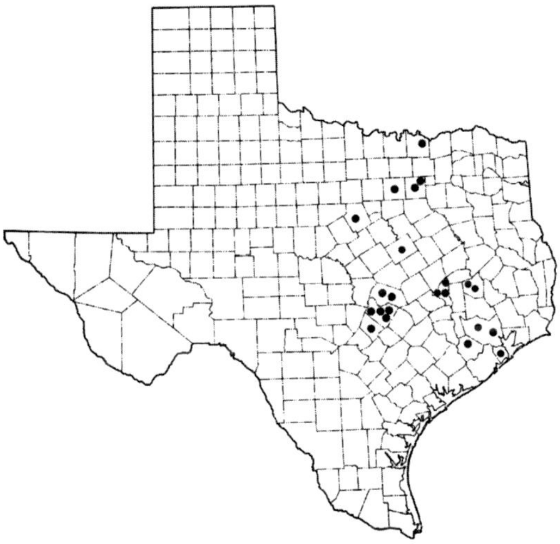

Broussonetia papyrifera

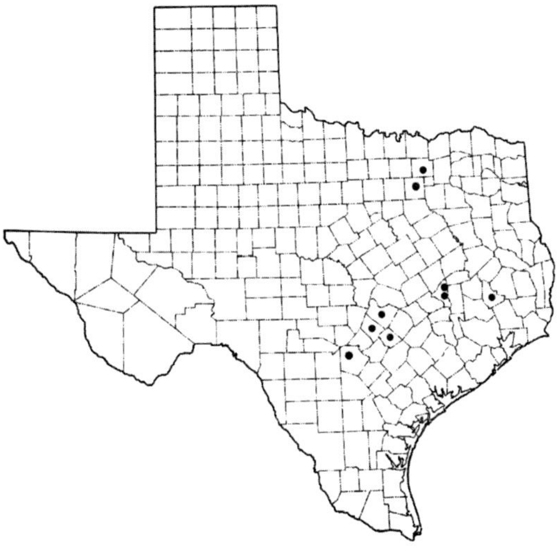

Fatoua villosa

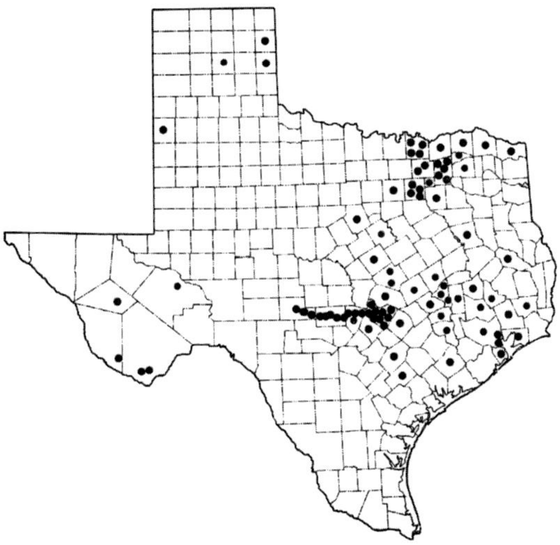

Maclura pomifera

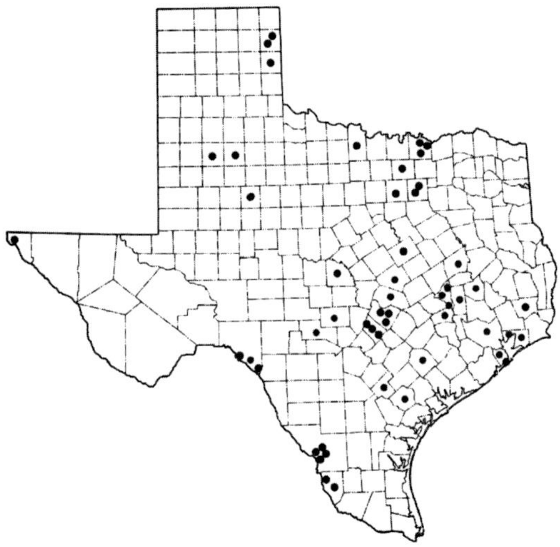

Morus alba

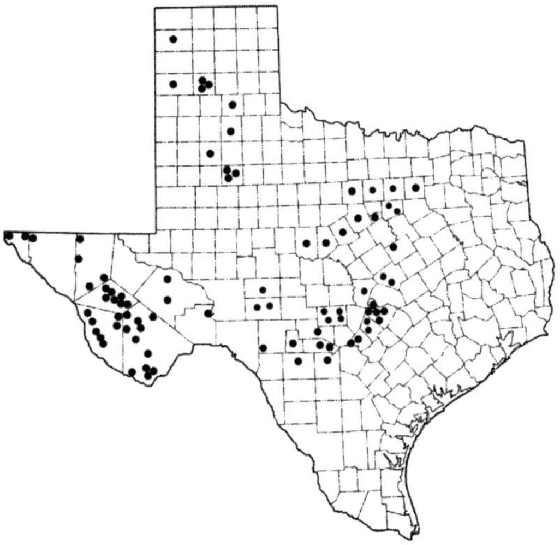

Morus microphylla

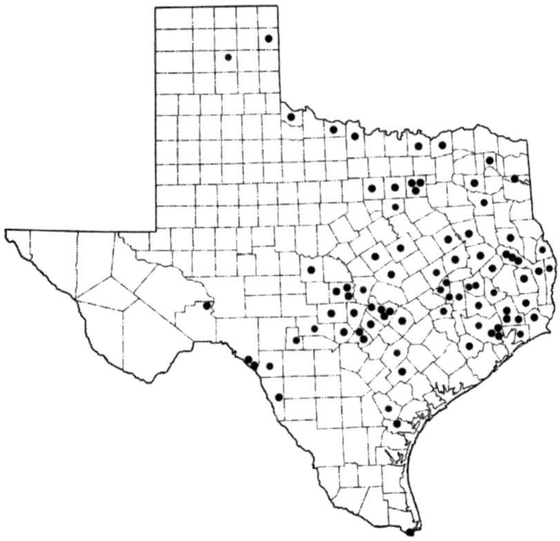

Morus rubra

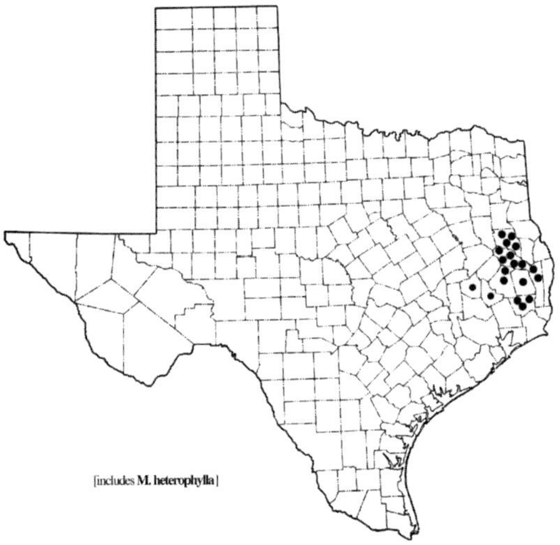
[includes **M. heterophylla**]

Morella caroliniensis

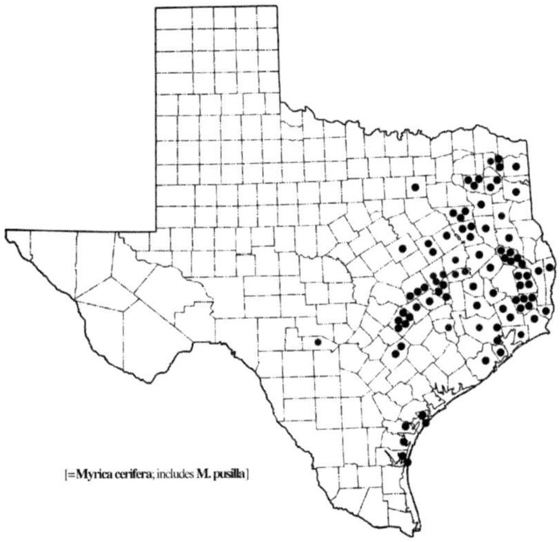
[=**Myrica cerifera**; includes **M. pusilla**]

Morella cerifera

Ardisia crenata

Abronia ameliae

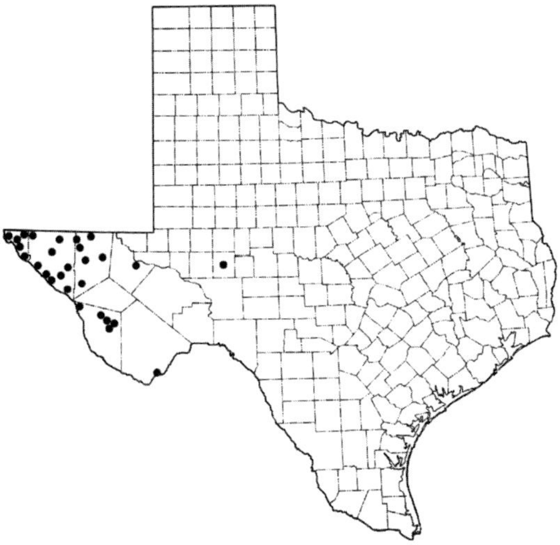

Abronia angustifolia

Abronia carletonii

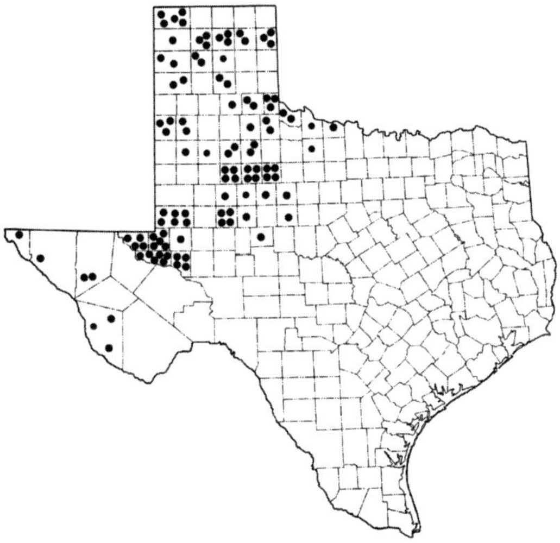

Abronia fragrans

Abronia macrocarpa

Acleisanthes acutifolia

Acleisanthes anisophylla

Acleisanthes crassifolia

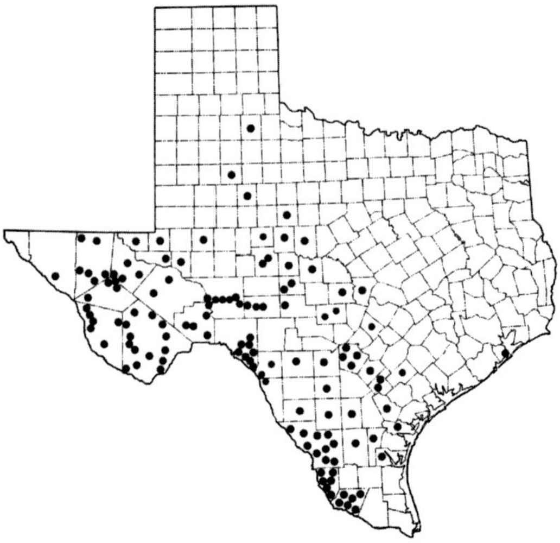
Acleisanthes longiflora

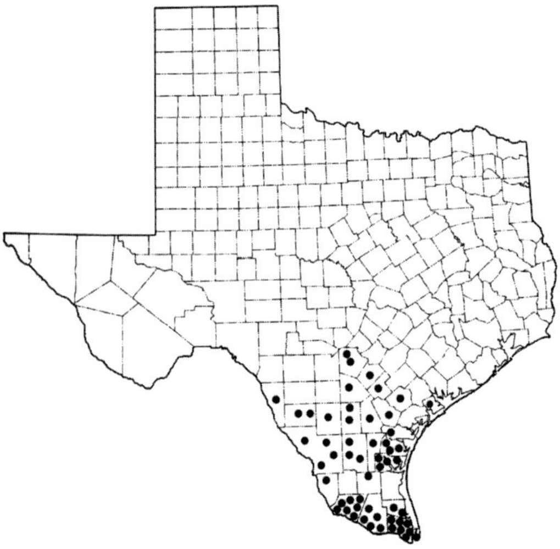
Acleisanthes obtusa

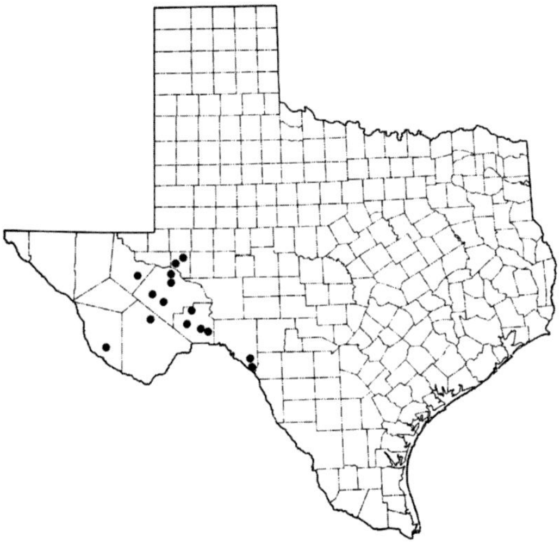
Acleisanthes wrightii

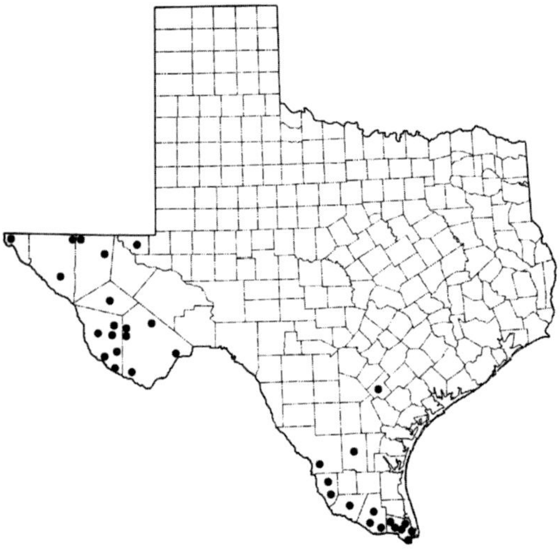
Allionia choisyi

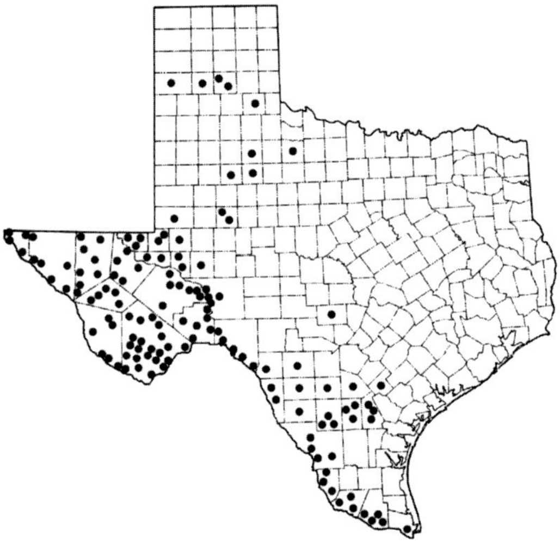
Allionia incarnata

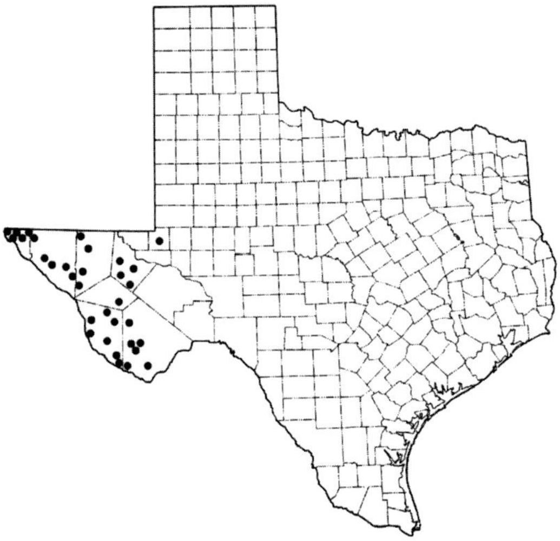

Ammocodon chenopodioides

Anulocaulis eriosolenus

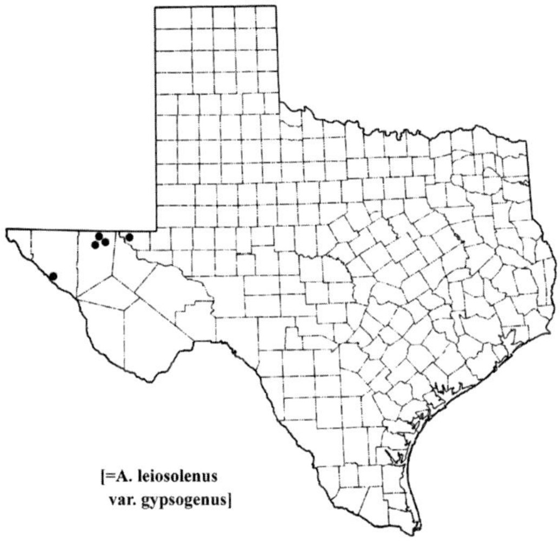
[=A. leiosolenus
var. gypsogenus]

Anulocaulis gypsogenus

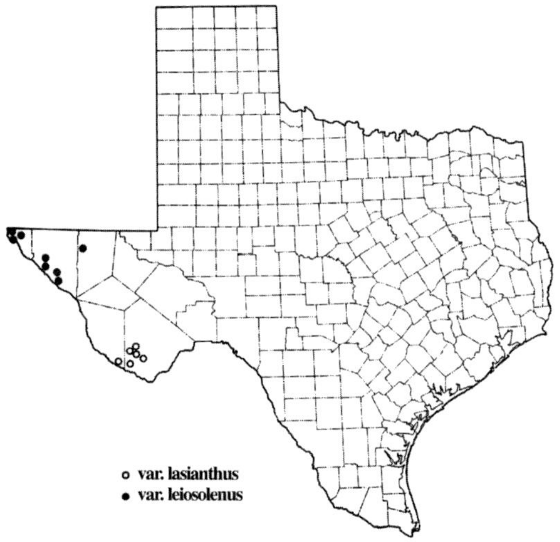
○ var. lasianthus
● var. leiosolenus

Anulocaulis leiosolenus

Anulocaulis reflexus

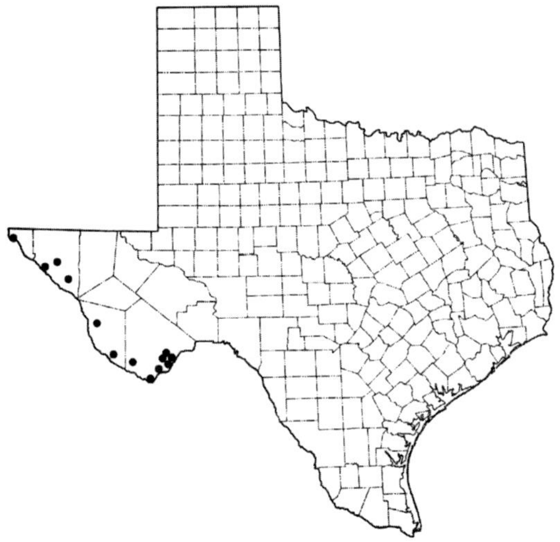

Boerhavia anisophylla

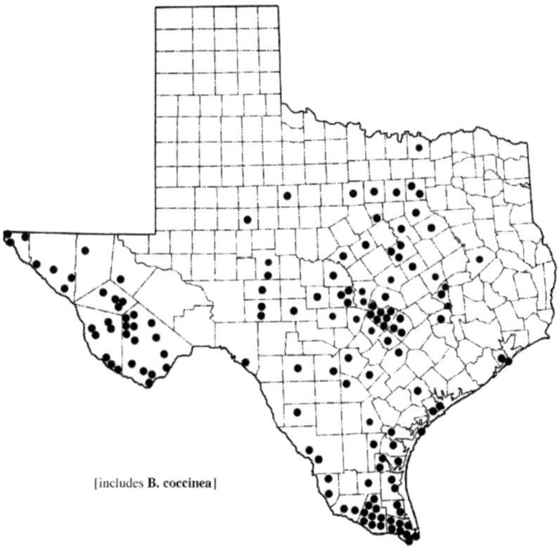
[includes B. coccinea]

Boerhavia diffusa

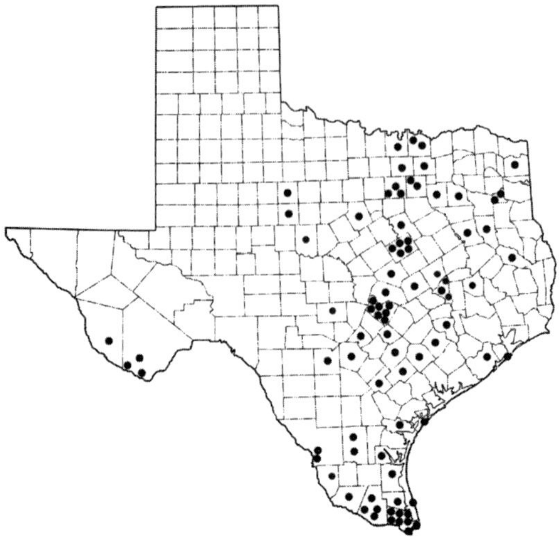

Boerhavia erecta

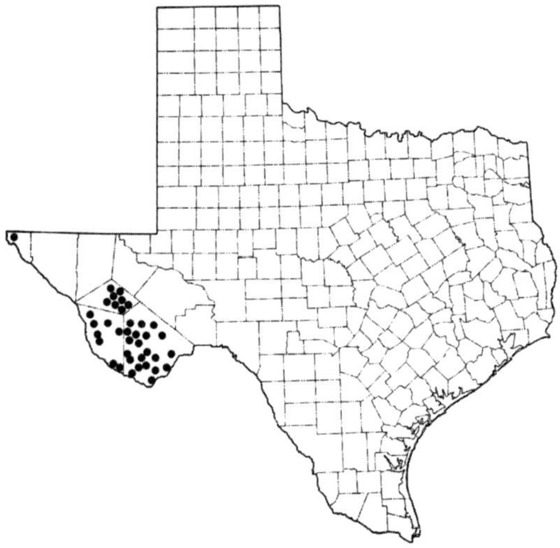

Boerhavia gracillima

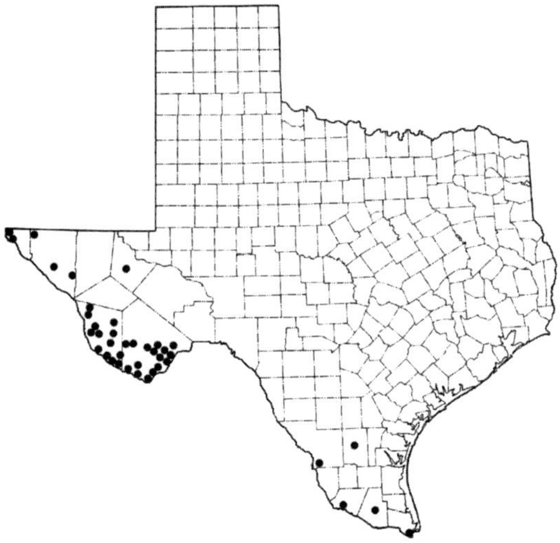

Boerhavia intermedia

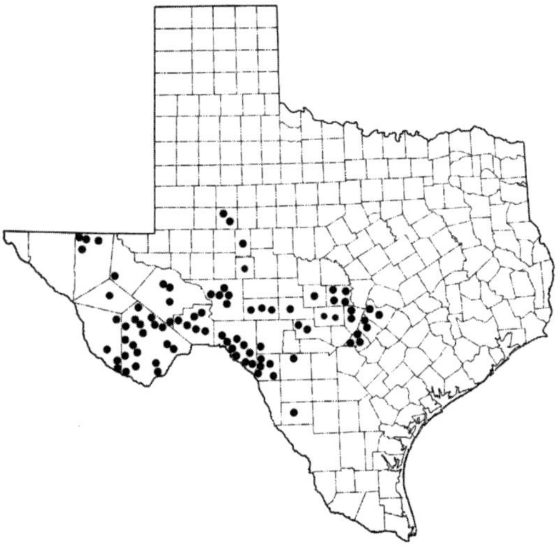

Boerhavia linearifolia

Boerhavia mathisiana

Boerhavia purpurascens

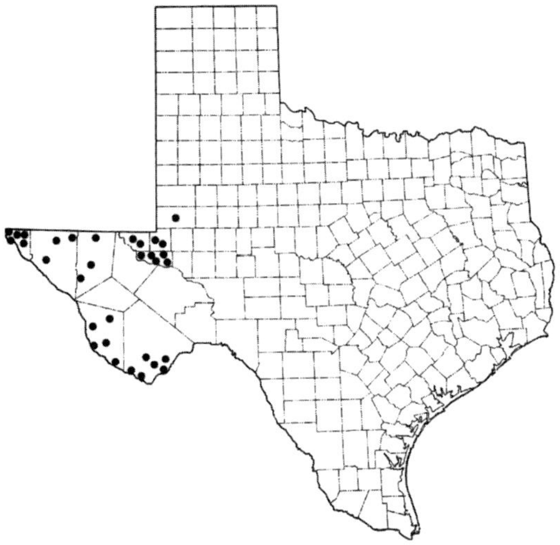

Boerhavia spicata

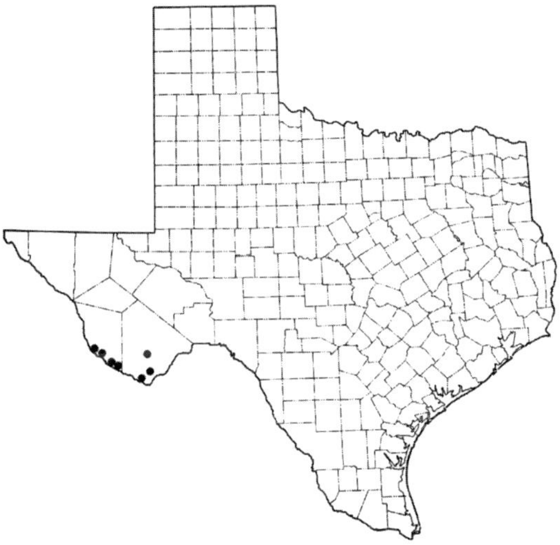

Boerhavia wrightii

Commicarpus scandens

Cyphomeris crassifolia

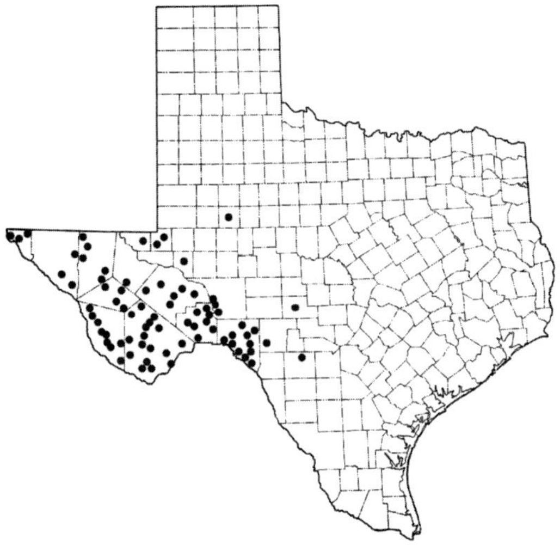

Cyphomeris gypsophiloides

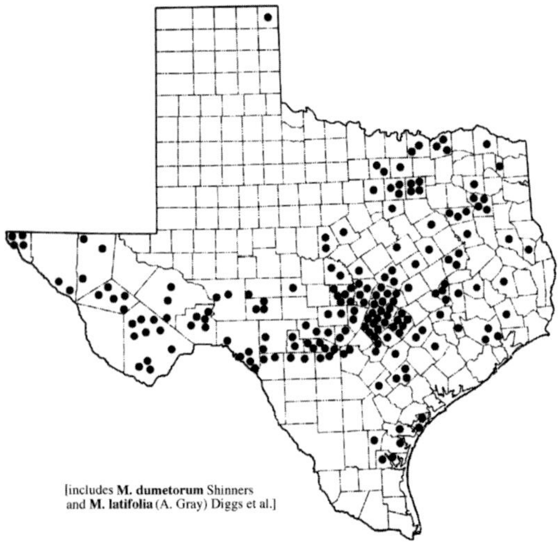
[includes M. dumetorum Shinners and M. latifolia (A. Gray) Diggs et al.]

Mirabilis albida

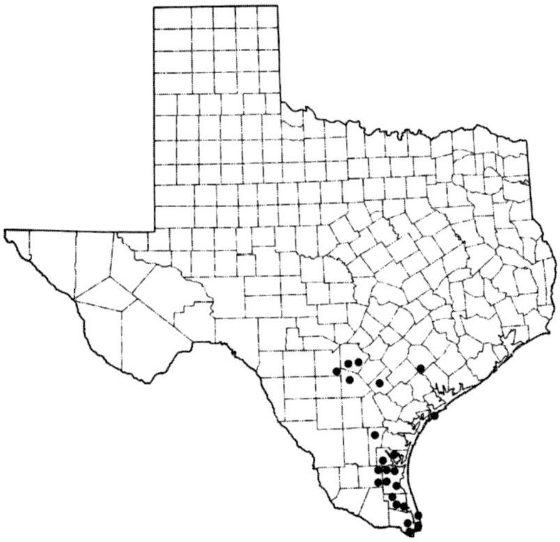

Mirabilis austrotexana

Mirabilis comata

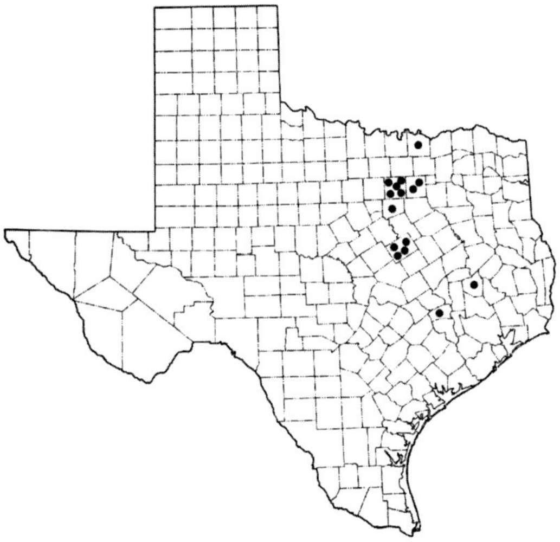

Mirabilis gigantea

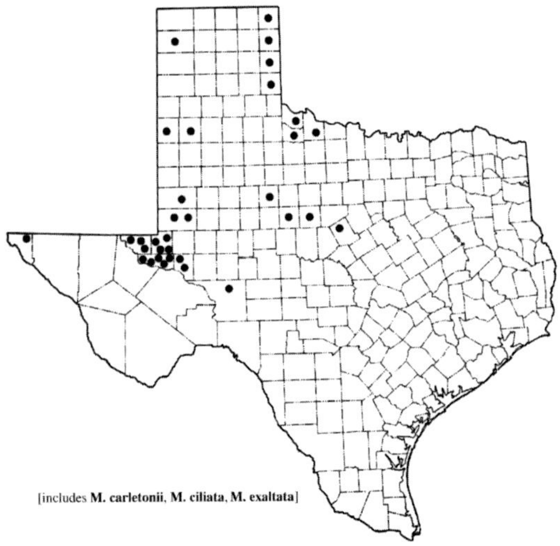
[includes M. carletonii, M. ciliata, M. exaltata]

Mirabilis glabra

Mirabilis hirsuta

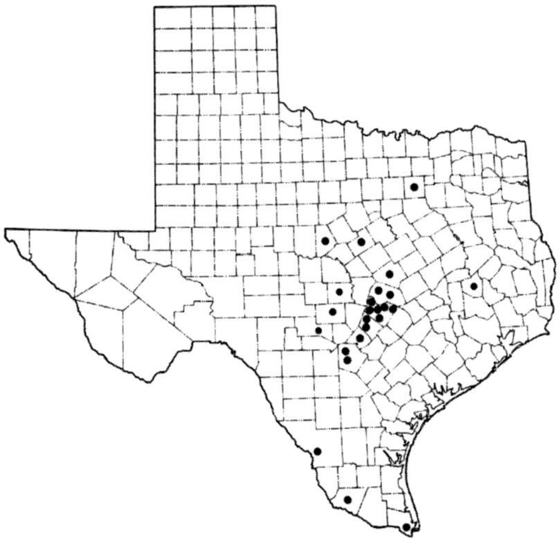

Mirabilis jalapa

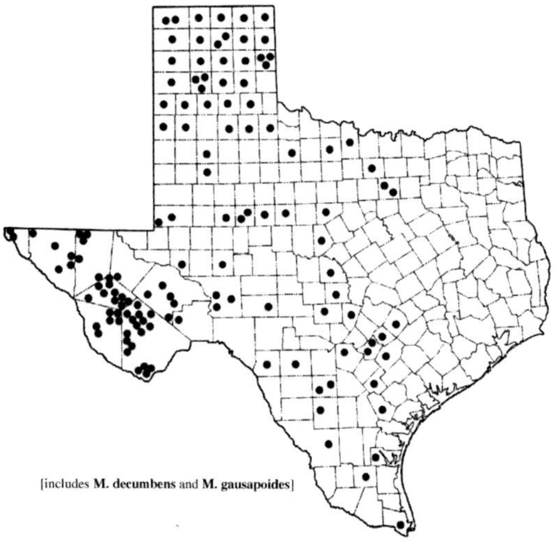
[includes M. decumbens and M. gausapoides]

Mirabilis linearis

Mirabilis longiflora

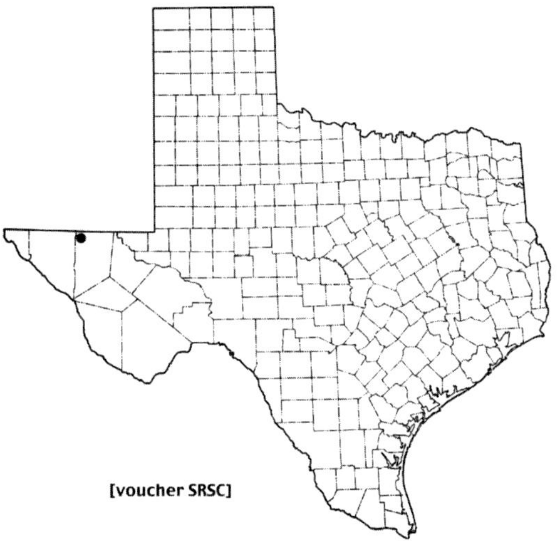
[voucher SRSC]

Mirabilis melanotricha

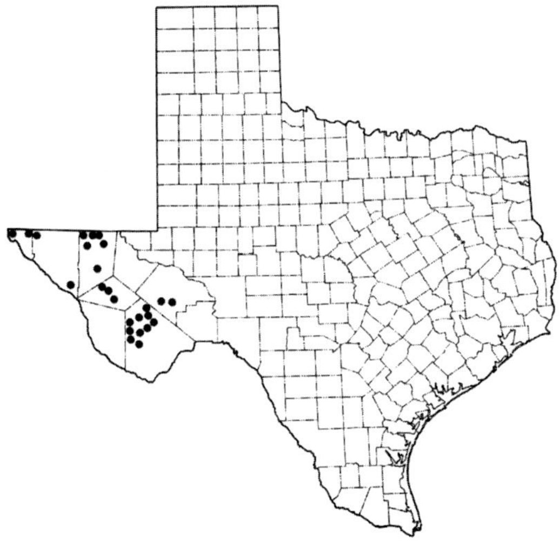

Mirabilis multiflora

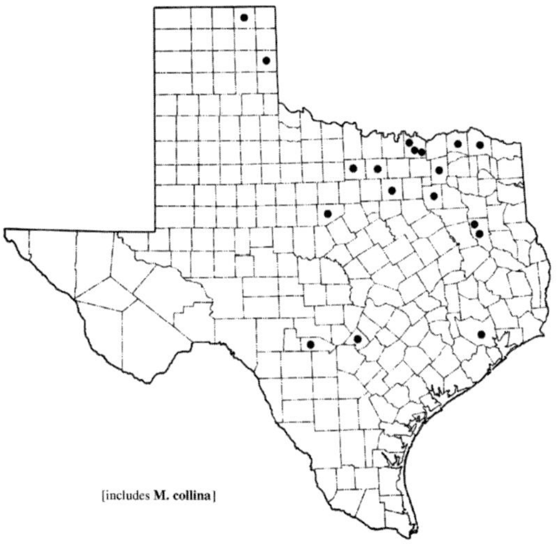
[includes M. collina]

Mirabilis nyctaginea

Mirabilis oxybaphoides

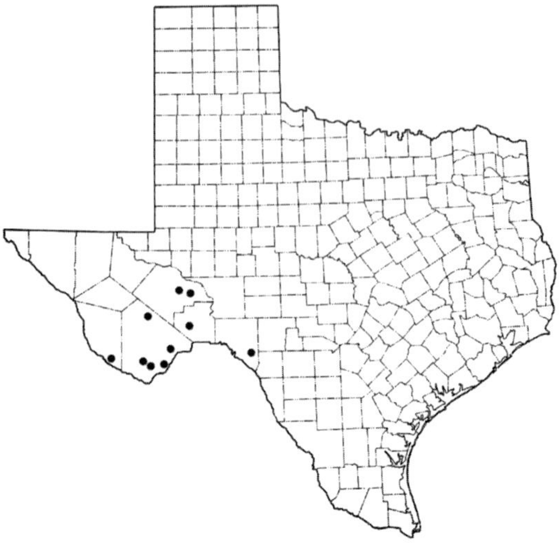

Mirabilis texensis

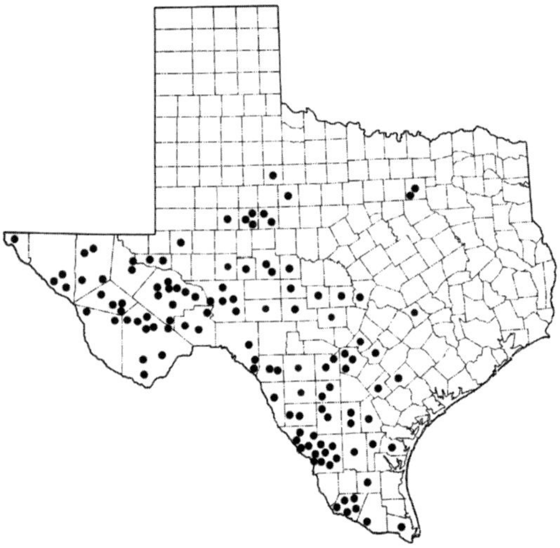

Nyctaginia capitata

Pisonia aculeata

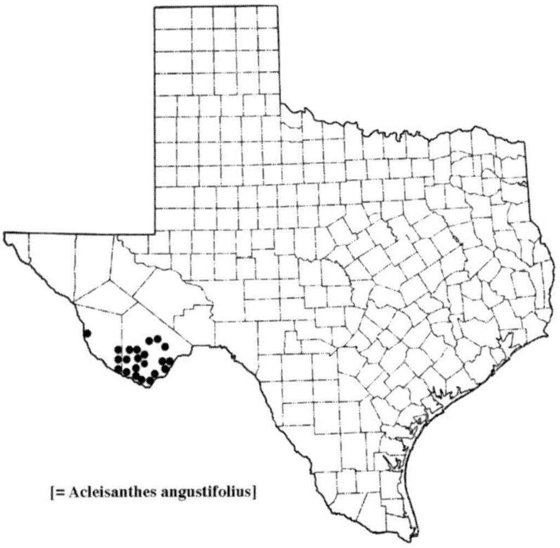
[= Acleisanthes angustifolius]

Selinocarpus angustifolius

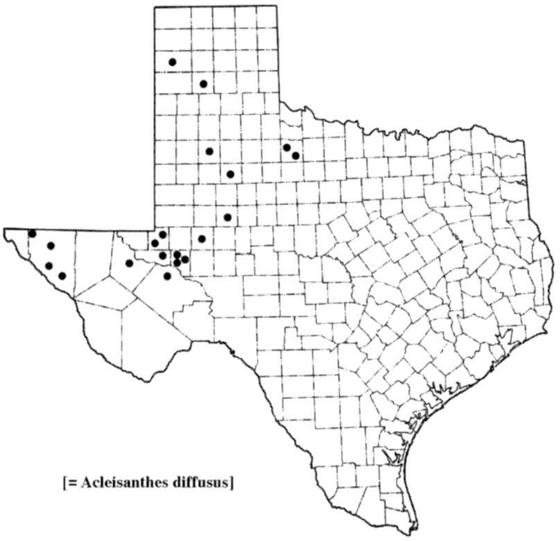
[= Acleisanthes diffusus]

Selinocarpus diffusus

[= Acleisanthes lanceolatus
var. lanceolatus]

Selinocarpus lanceolatus

[= Acleisanthes lanceolatus
var. maloneanus]

Selinocarpus maloneanus

Selinocarpus parvifolius

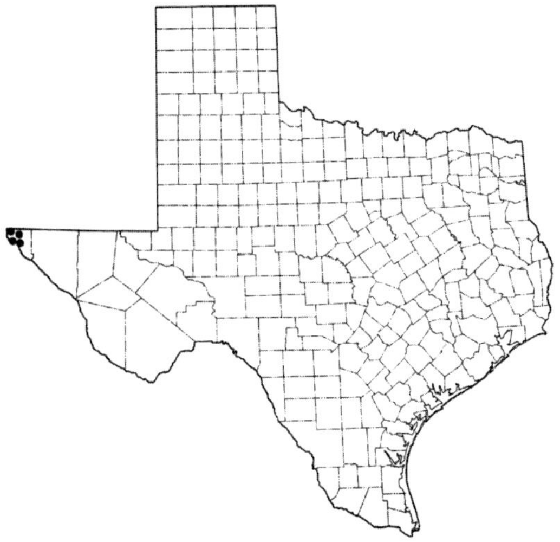

Tripterocalyx carnea

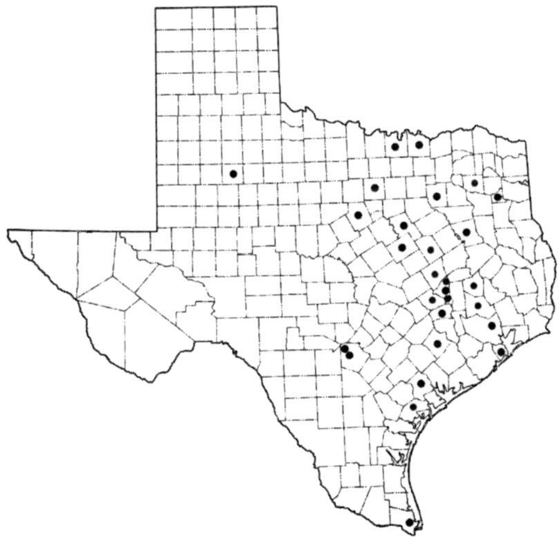

Nelumbo lutea

[= N. lutea var. advena]

Nuphar advena

Nymphaea ampla

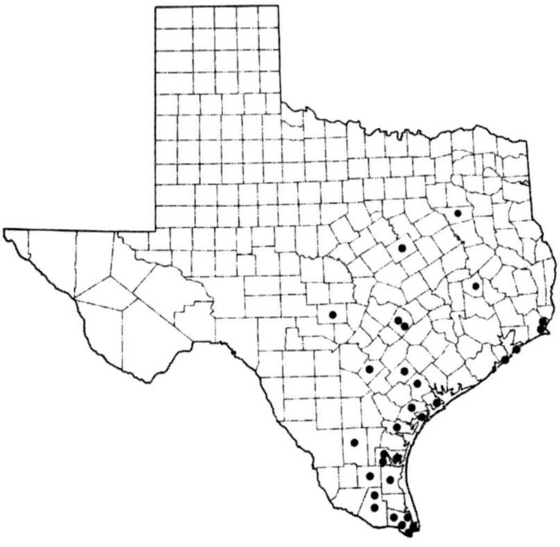

Nymphaea elegans

Nymphaea mexicana

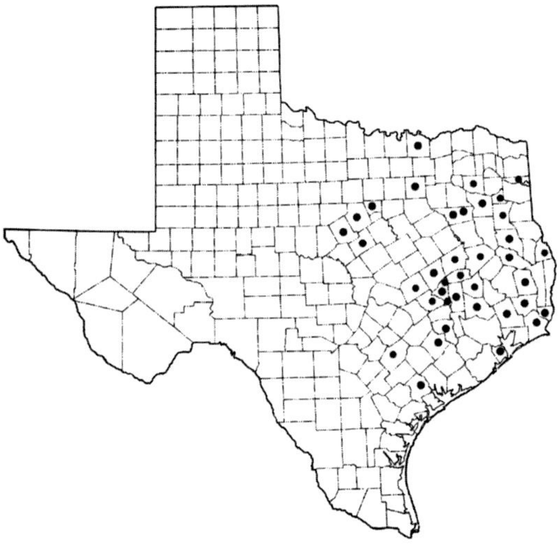

Nymphaea odorata

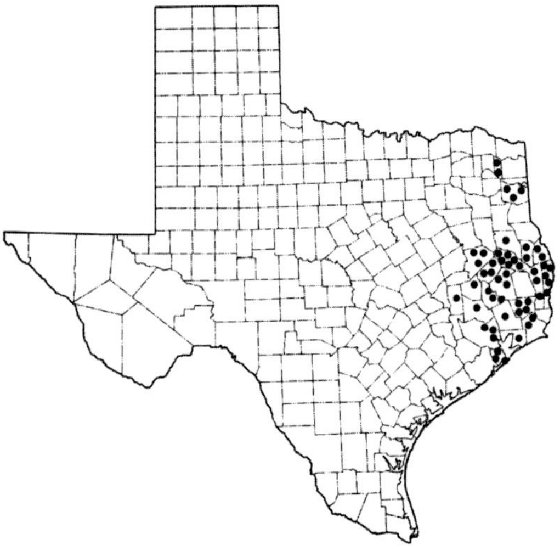

Chionanthus virginicus

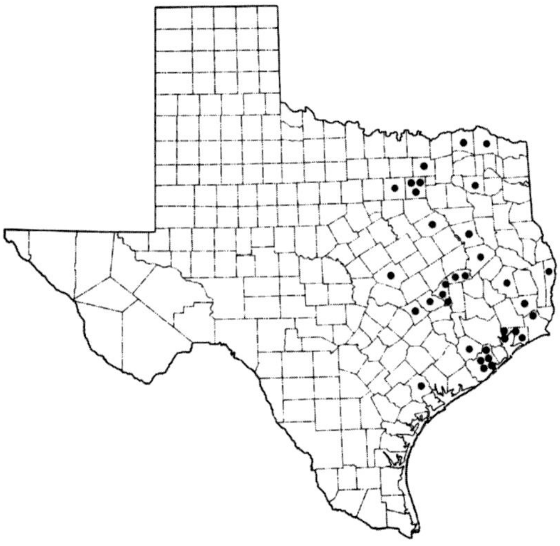

Forestiera acuminata

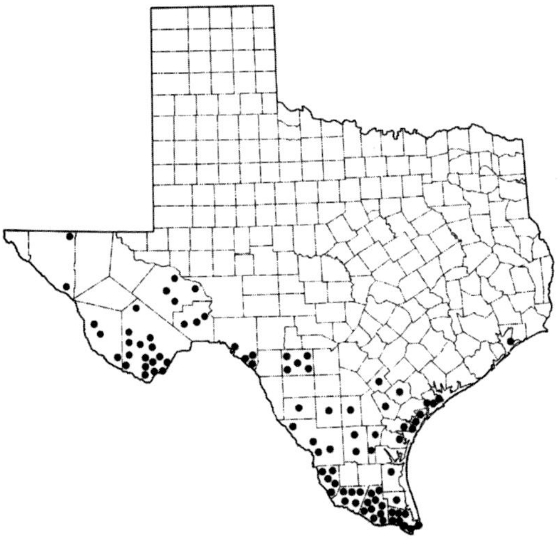

Forestiera angustifolia

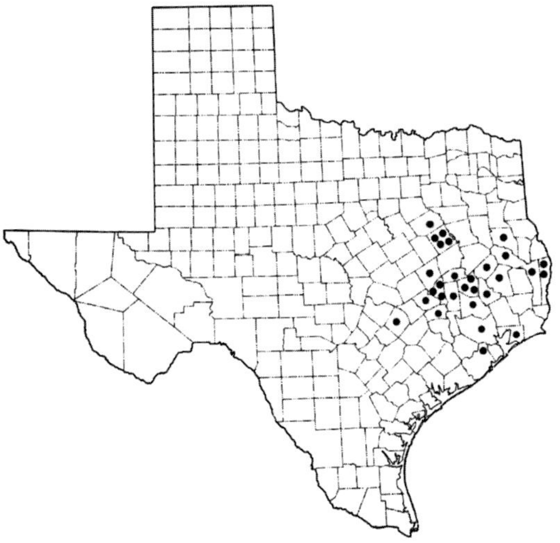

Forestiera ligustrina

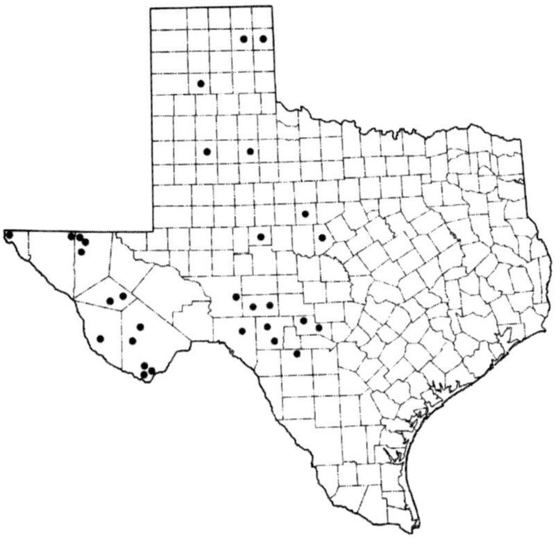

Forestiera neomexicana

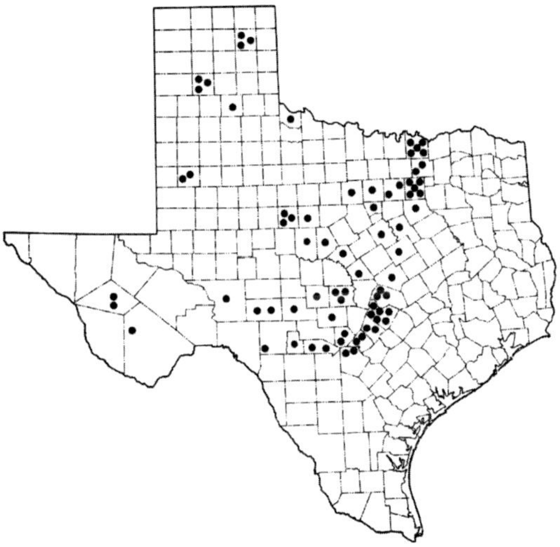

Forestiera pubescens

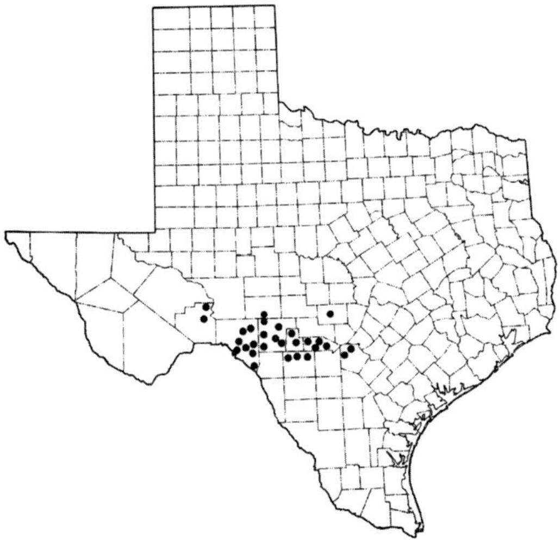

Forestiera reticulata

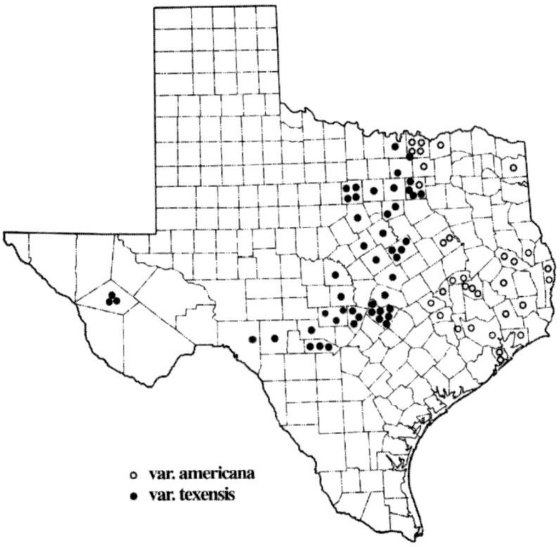
var. americana
var. texensis

Fraxinus americana

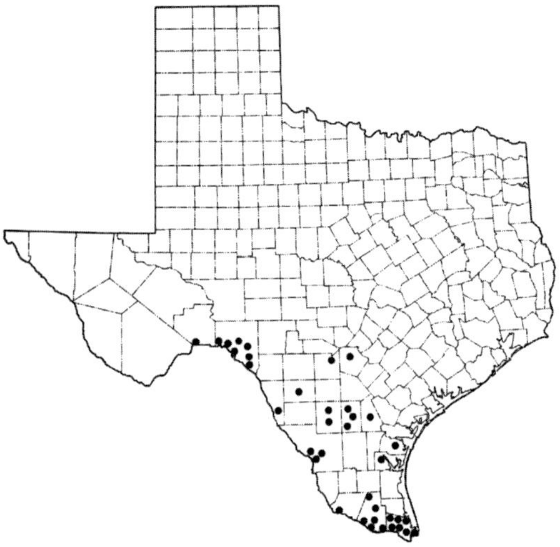

Fraxinus berlandieriana

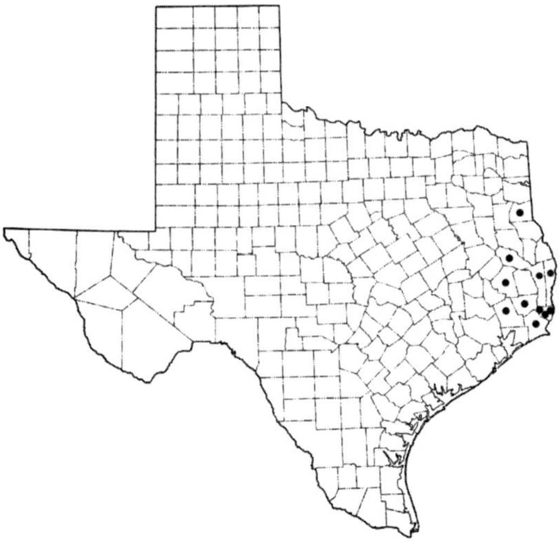

Fraxinus caroliniana

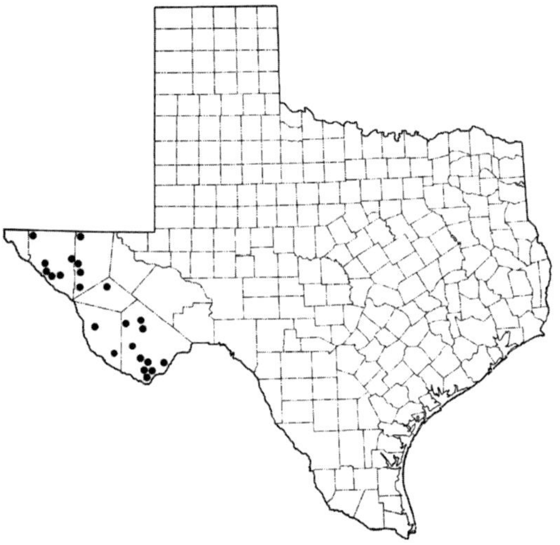

Fraxinus cuspidata

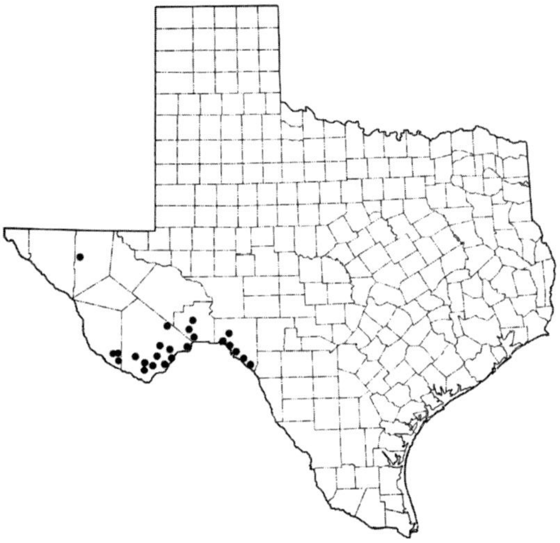

Fraxinus greggii

Fraxinus papillosa

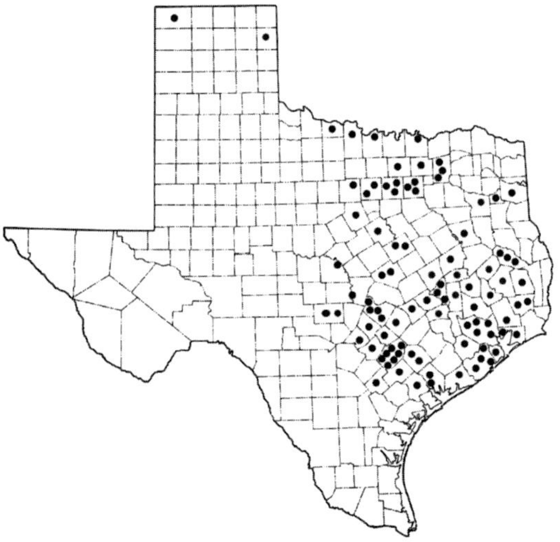

Fraxinus pennsylvanica

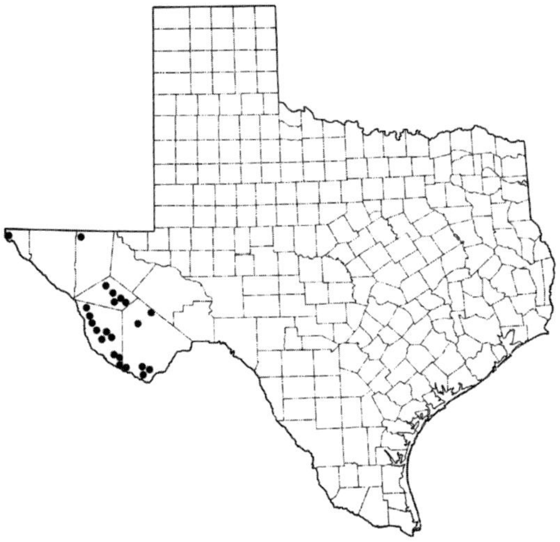

Fraxinus velutina

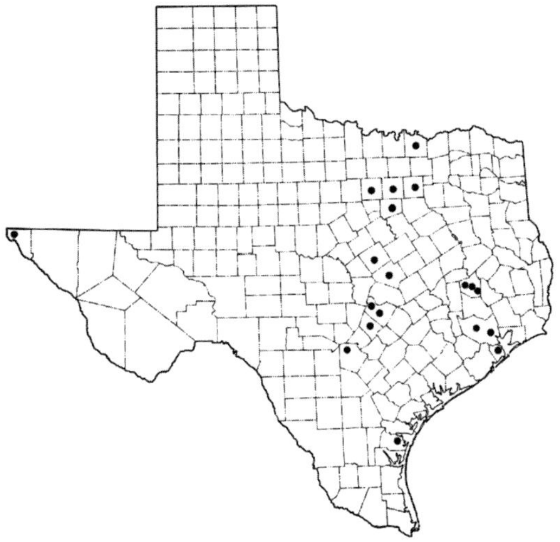

Ligustrum japonicum

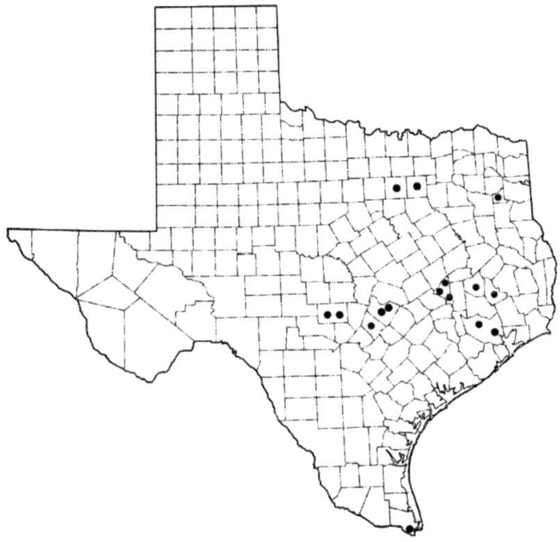

Ligustrum lucidum

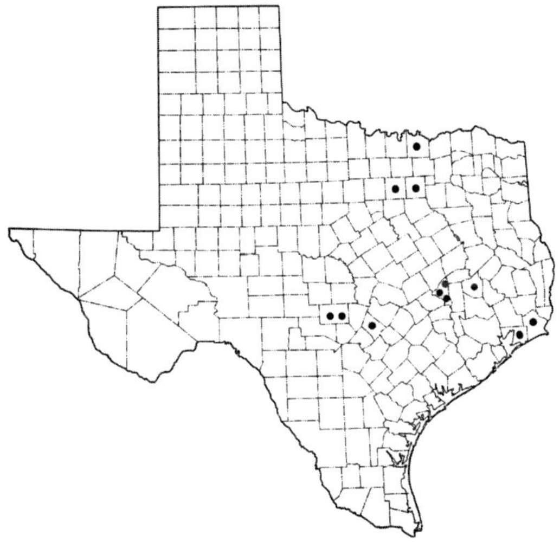

Ligustrum quihoui

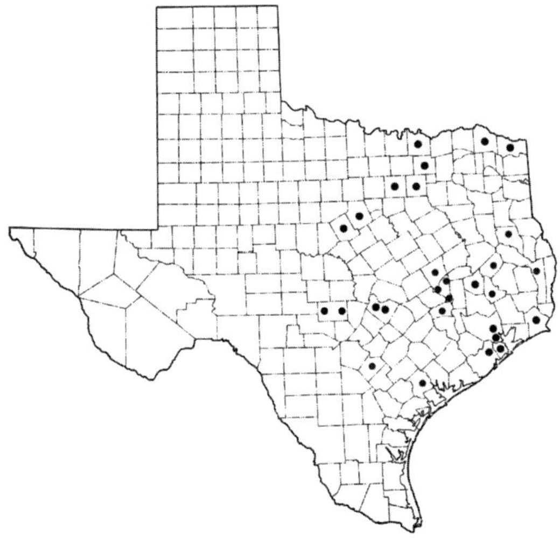

Ligustrum sinense

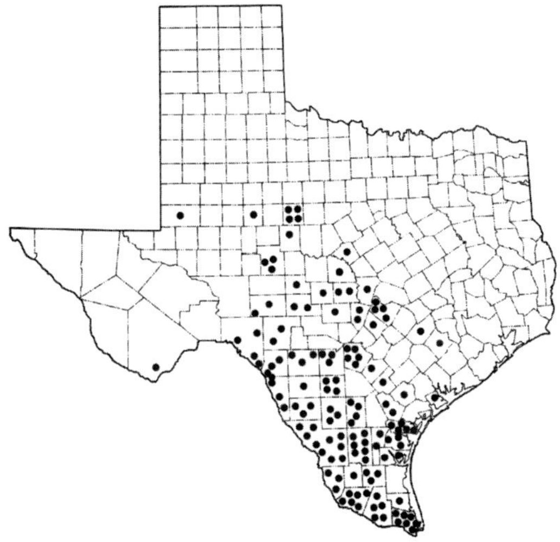

Menodora heterophylla

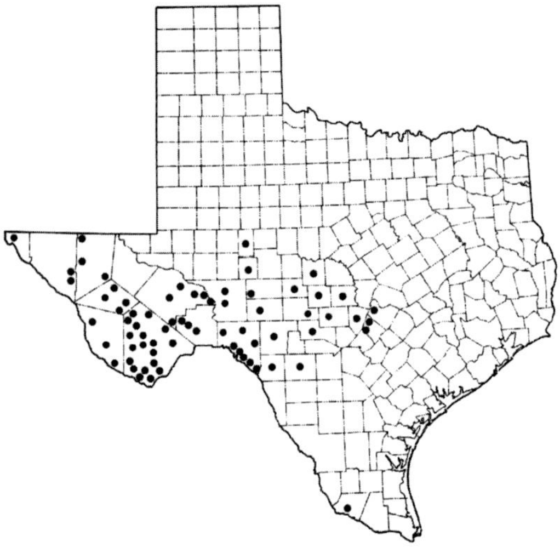

Menodora longiflora

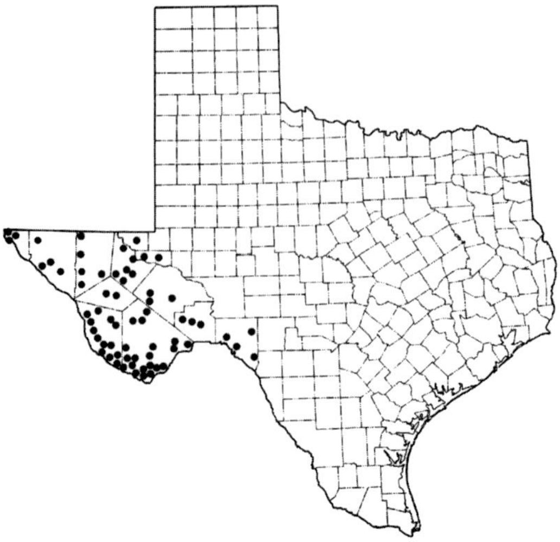

Menodora scabra

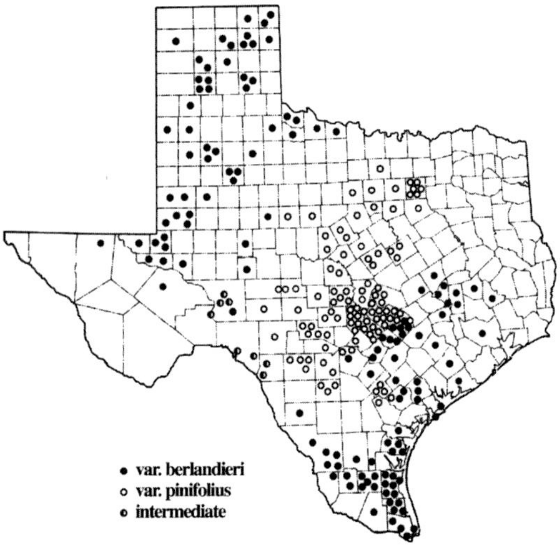
• var. berlandieri
○ var. pinifolius
◐ intermediate

Calylophus berlandieri

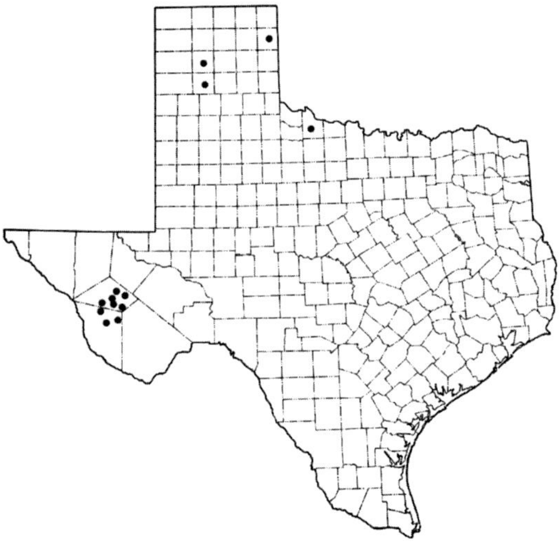

Calylophus fendleri

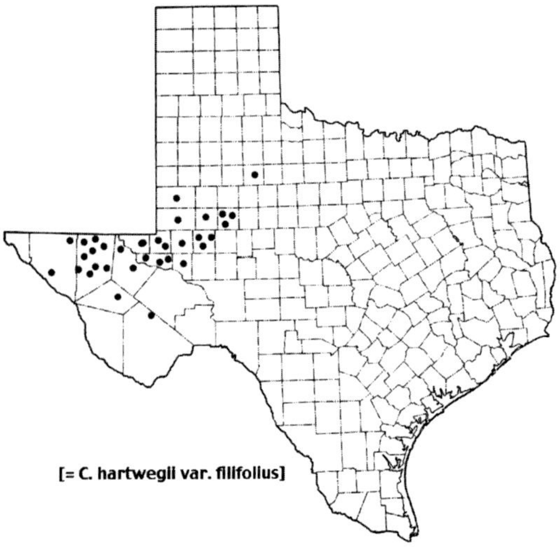
[= C. hartwegii var. filifolius]

Calylophus filifolius

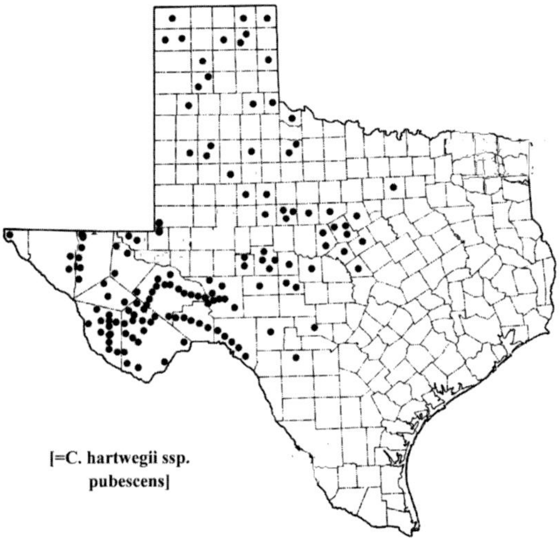
[=C. hartwegii ssp. pubescens]

Calylophus greggii

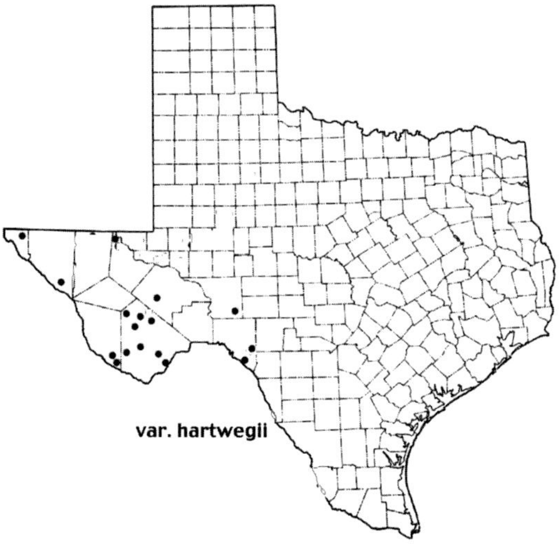
var. hartwegii

Calylophus hartwegii

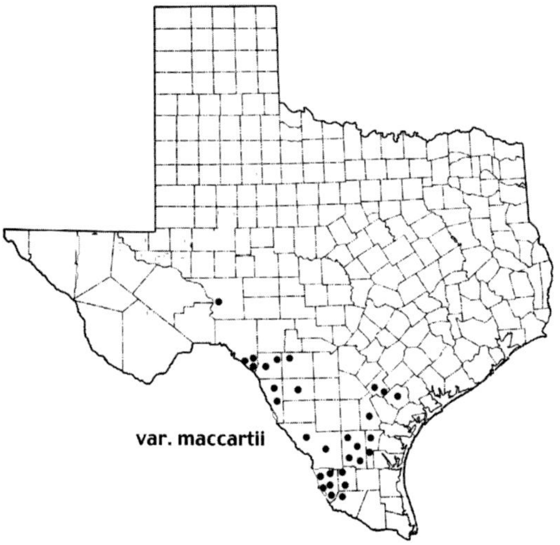
var. maccartii

Calylophus hartwegii

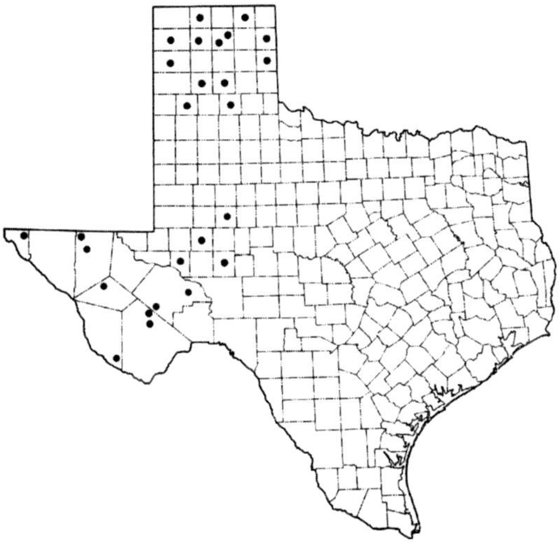

Calylophus lavandulifolius

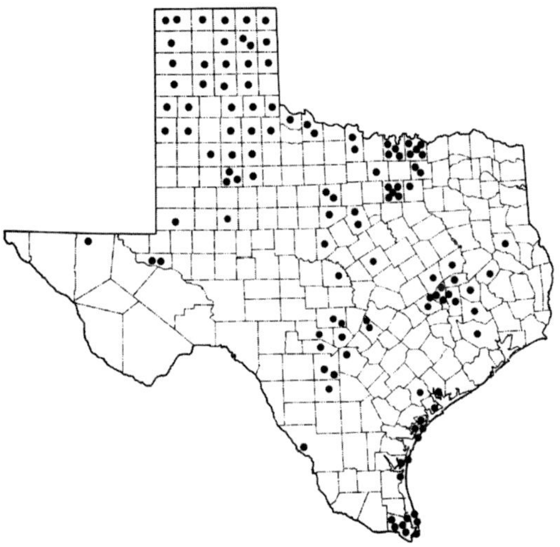

Calylophus serrulatus

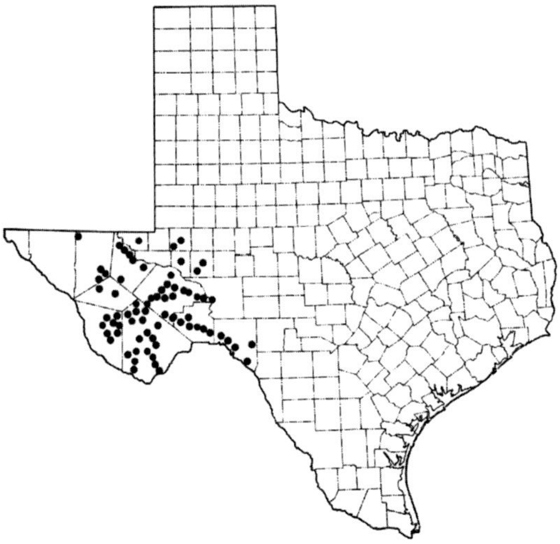

Calylophus tubicula

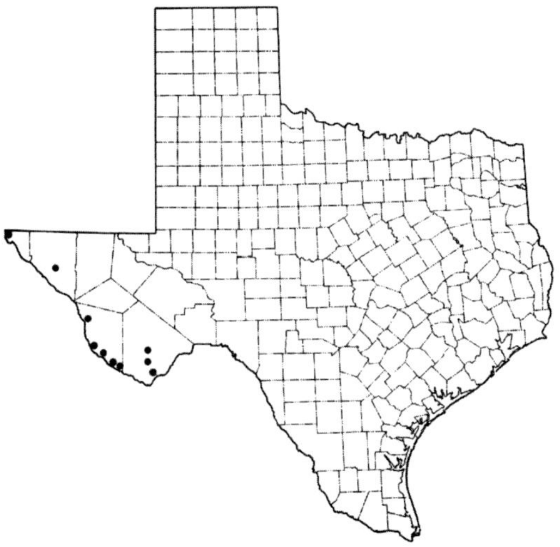

Camissonia chamaenerioides

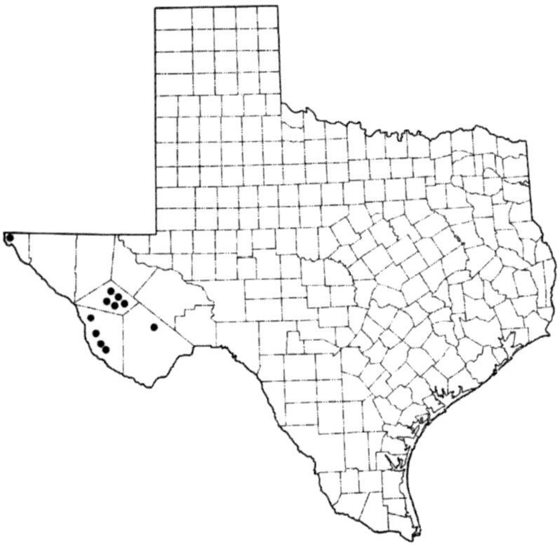

Epilobium ciliatum

Epilobium coloratum

Gaura boquillensis

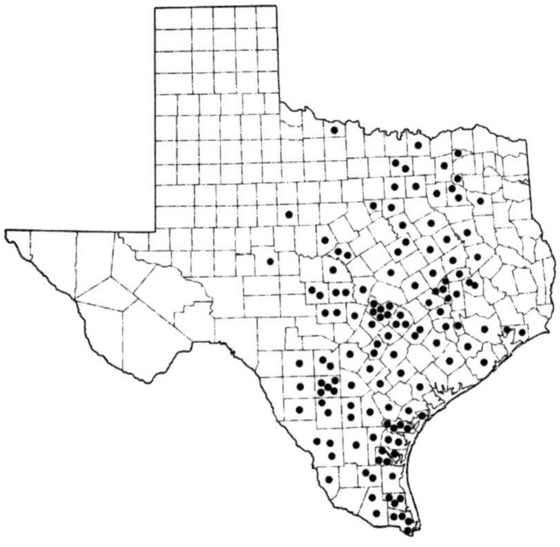

Gaura brachycarpa

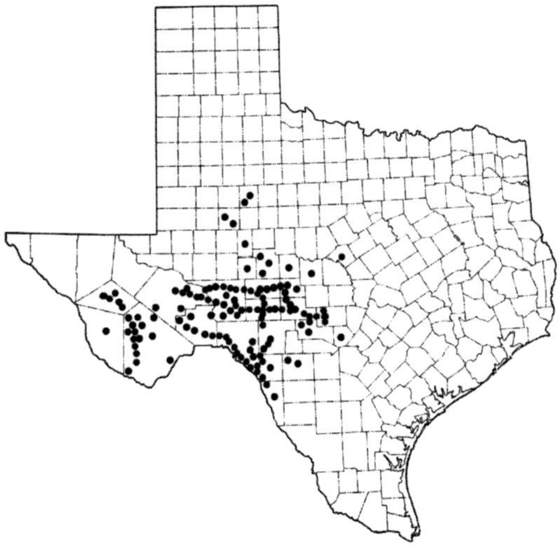

Gaura calcicola

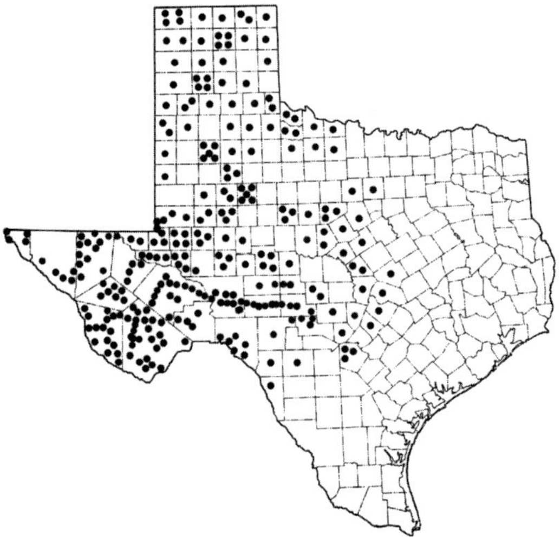

Gaura coccinea

Gaura demareei

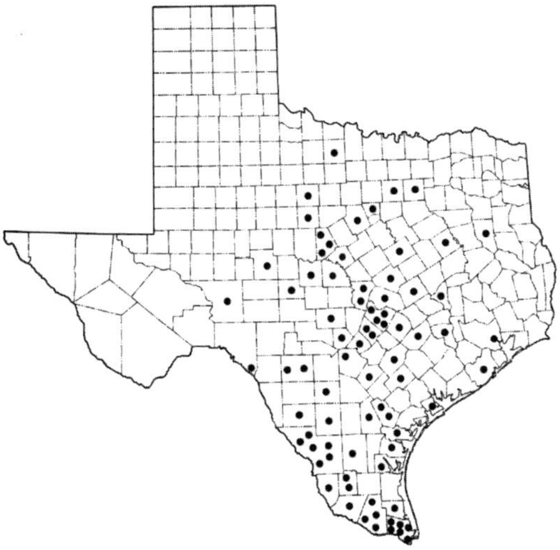

Gaura drummondii

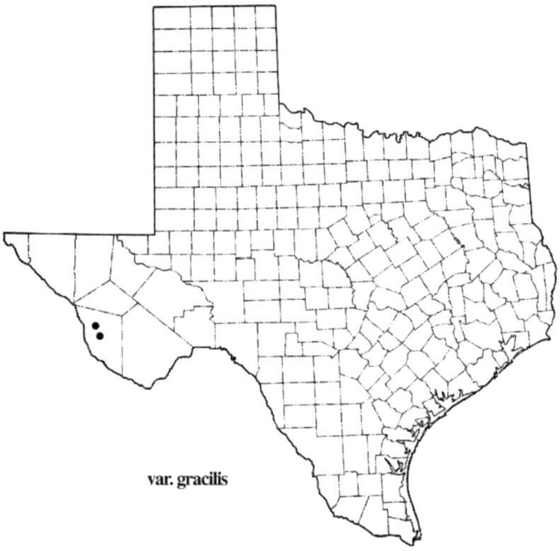
var. gracilis

Gaura hexandra

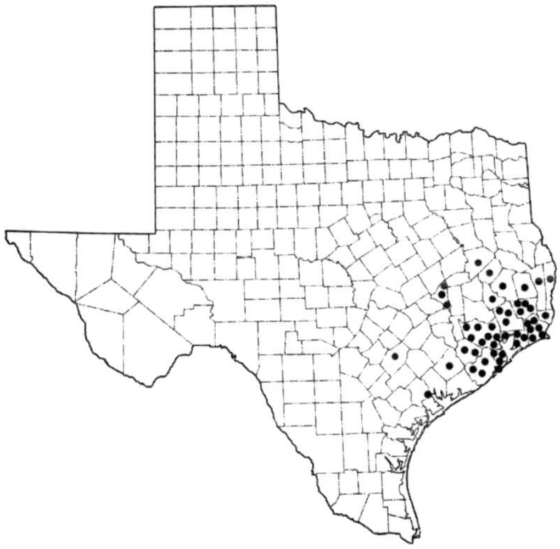

Gaura lindheimeri

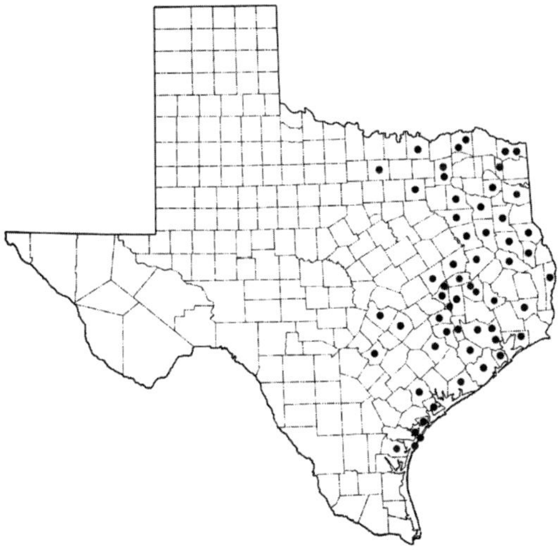

Gaura longiflora

Gaura macrocarpa

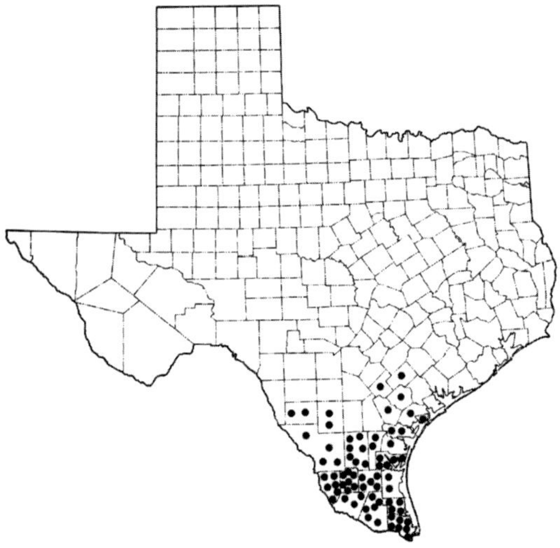

Gaura mckelveyae

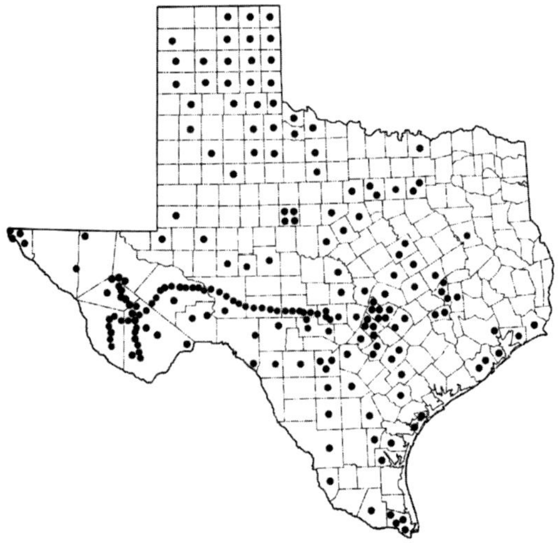

Gaura parviflora

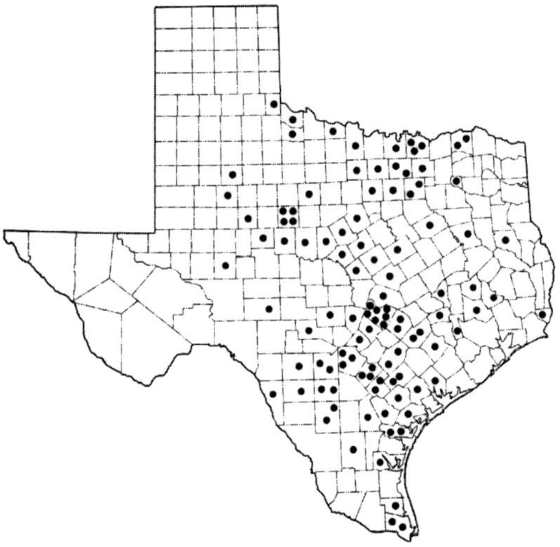

Gaura sinuata

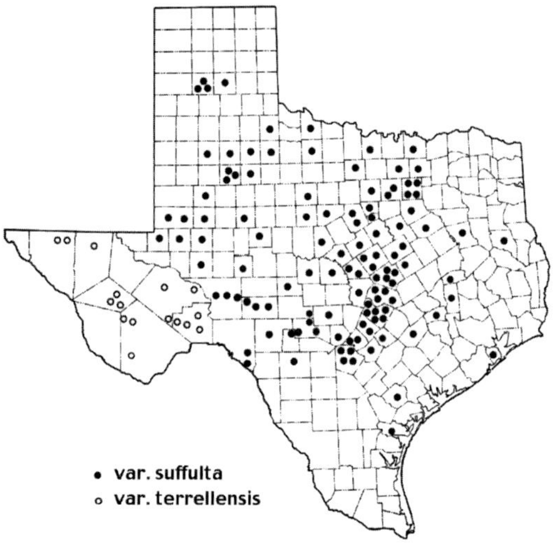
• var. suffulta
o var. terrellensis

Gaura suffulta

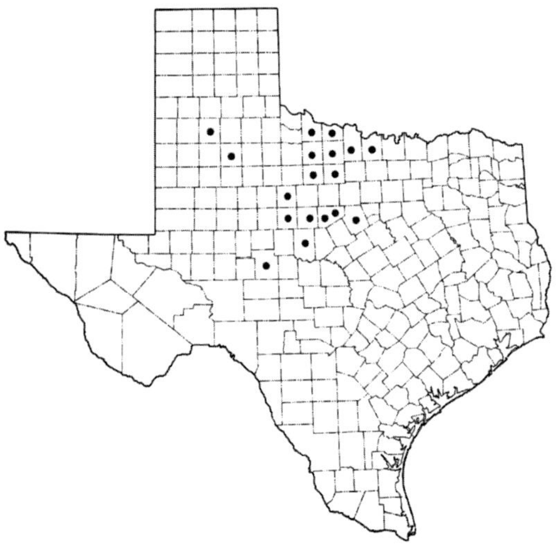

Gaura triangulata

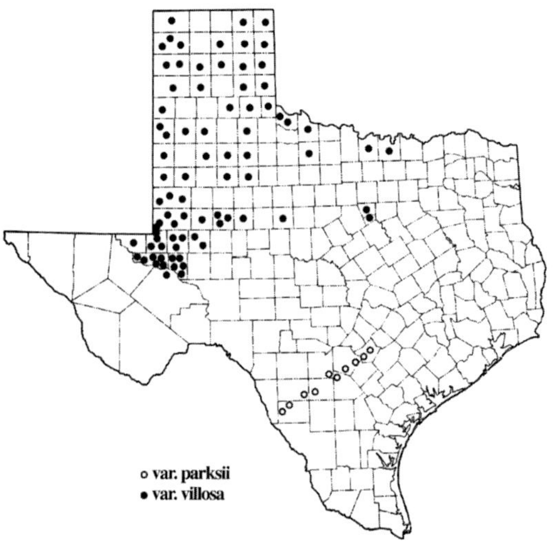
o var. parksii
• var. villosa

Gaura villosa

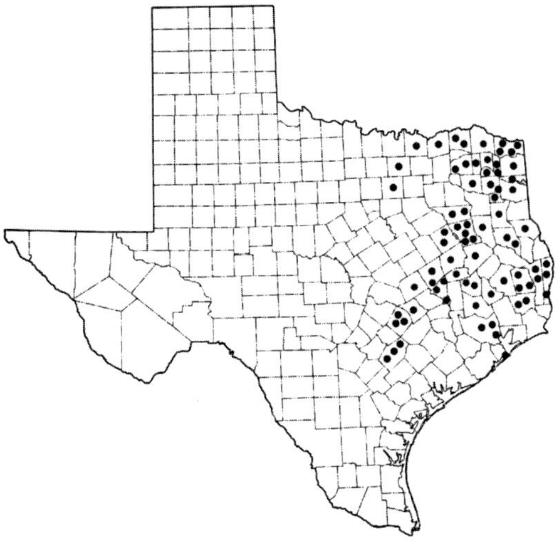

Ludwigia alternifolia

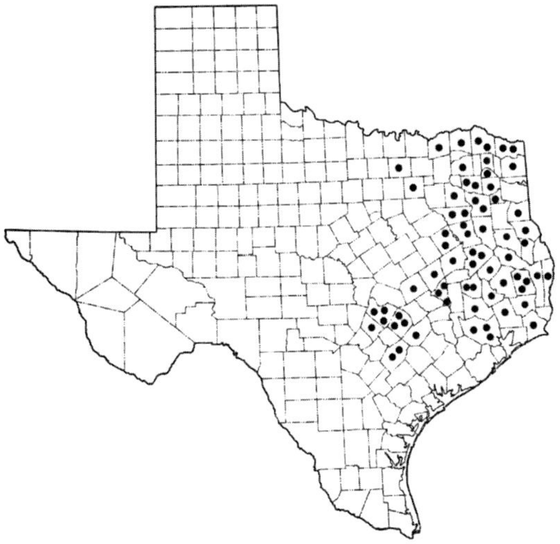

Ludwigia decurrens

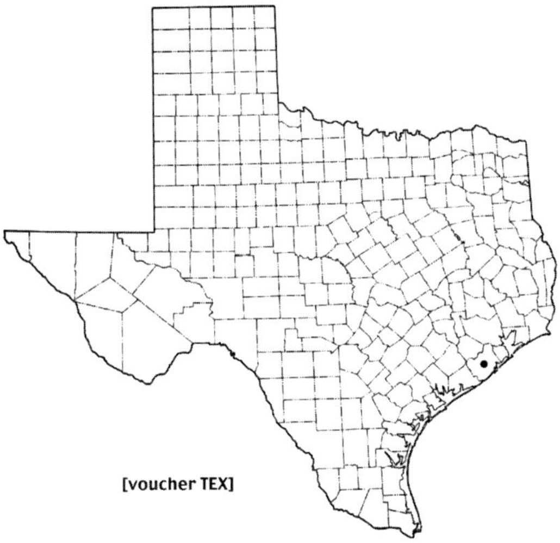
[voucher TEX]

Ludwigia erecta

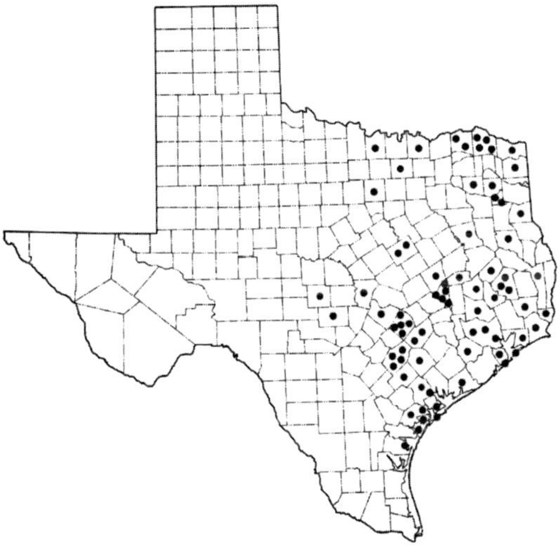

Ludwigia glandulosa

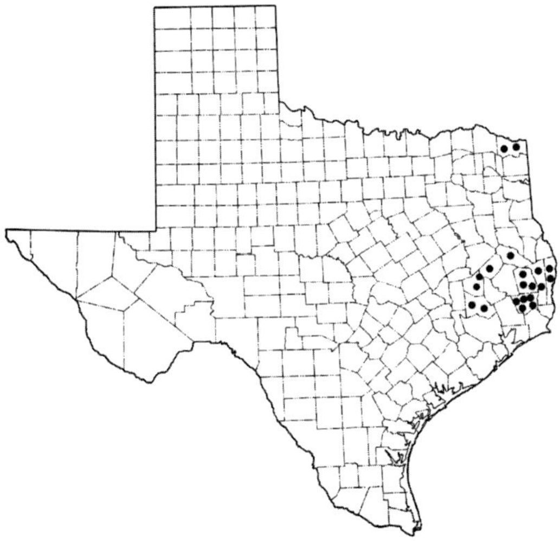

Ludwigia hirtella

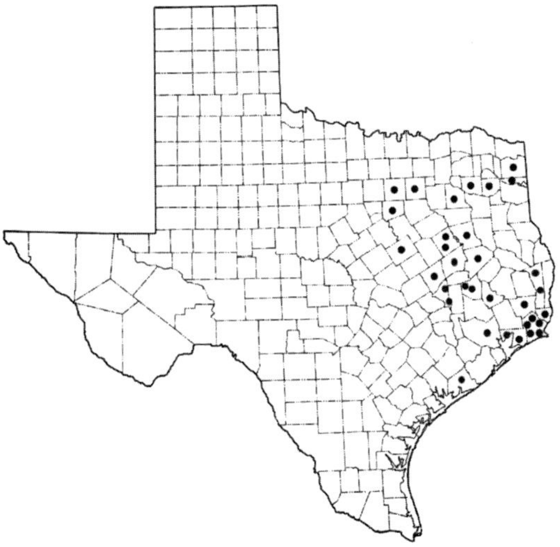

Ludwigia leptocarpa

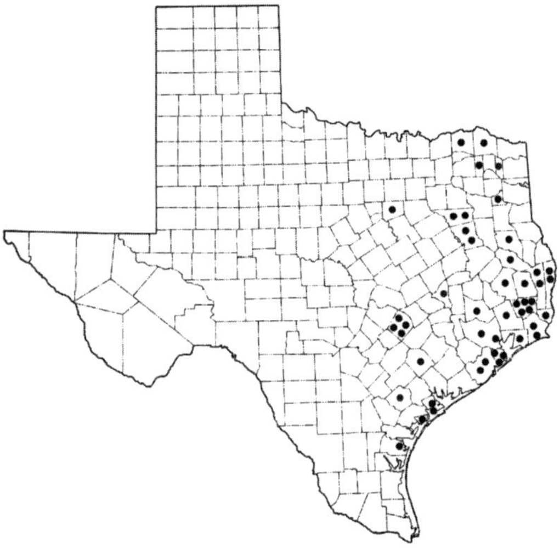

Ludwigia linearis

Ludwigia microcarpa

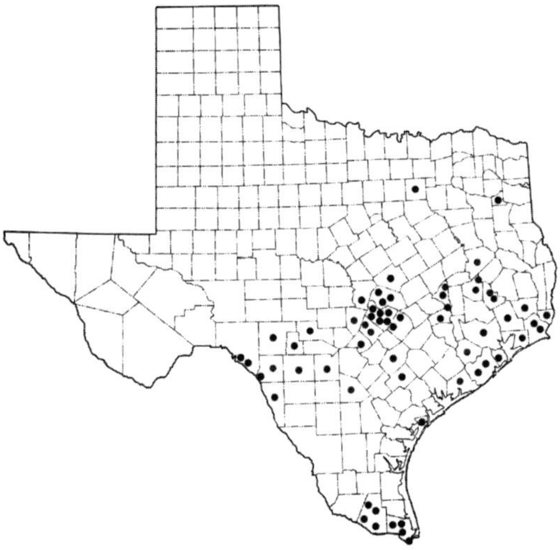

Ludwigia octovalvis

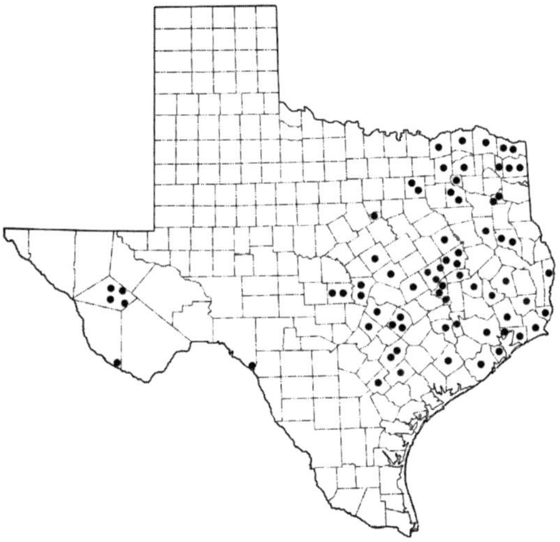

Ludwigia palustris

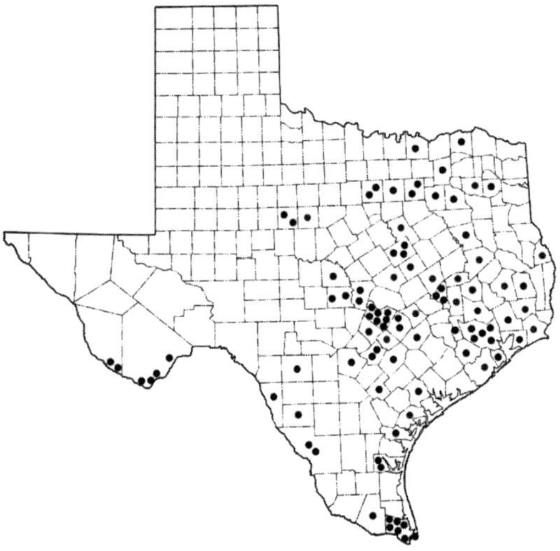

Ludwigia peploides

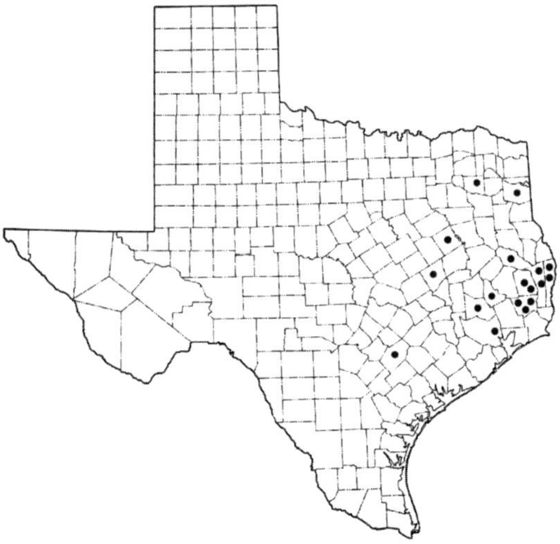

Ludwigia pilosa

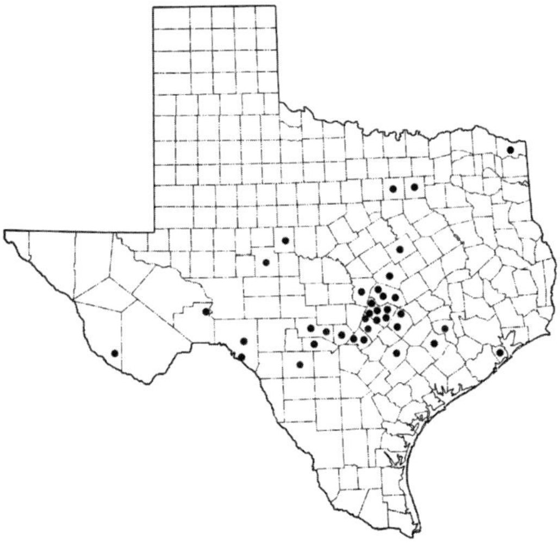

Ludwigia repens

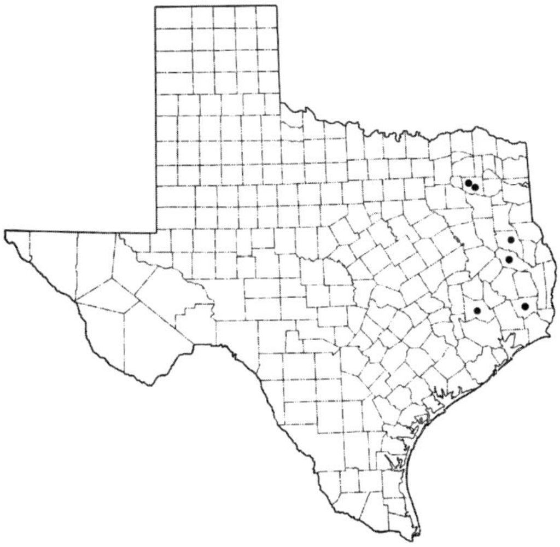

Ludwigia sphaerocarpa

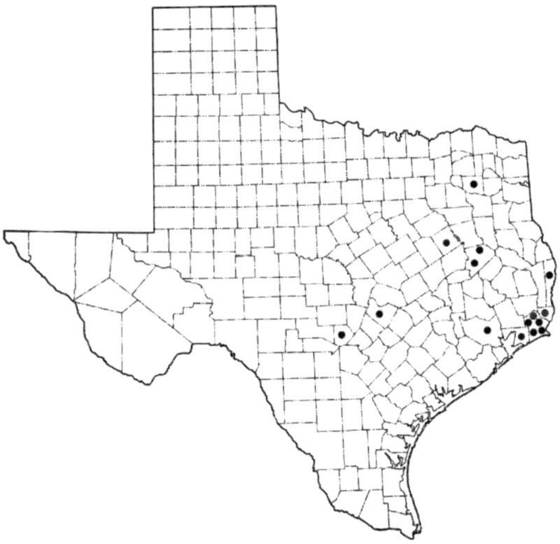

Ludwigia uruguayensis

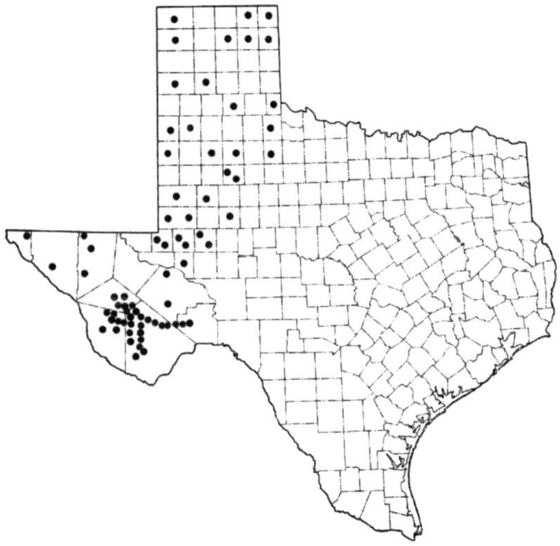

Oenothera albicaulis

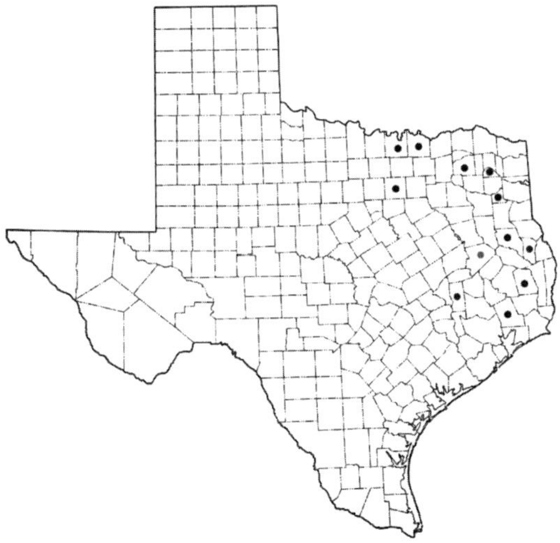

Oenothera biennis

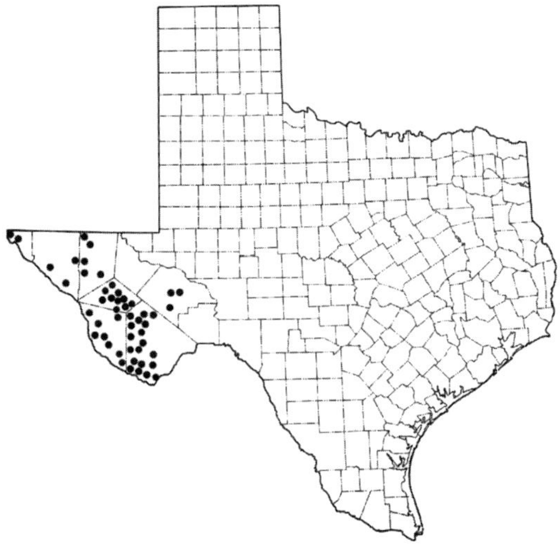

Oenothera brachycarpa

Oenothera caespitosa

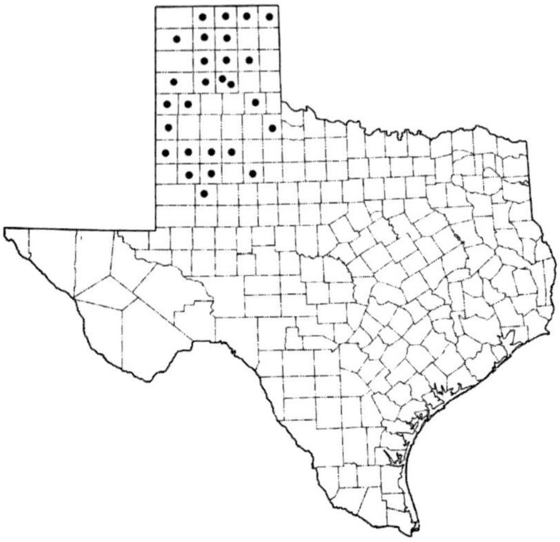

Oenothera canescens

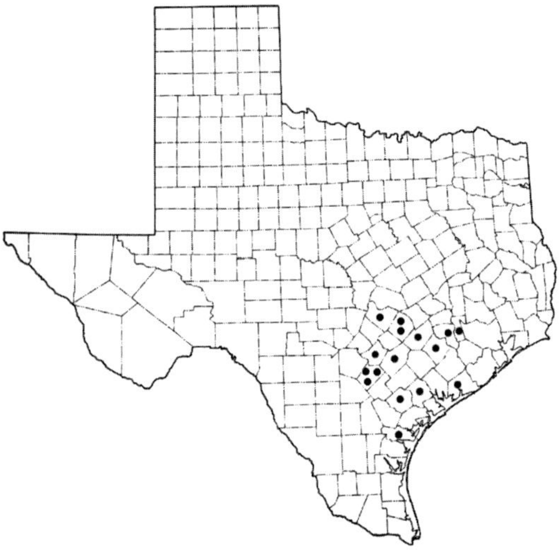

Oenothera cordata

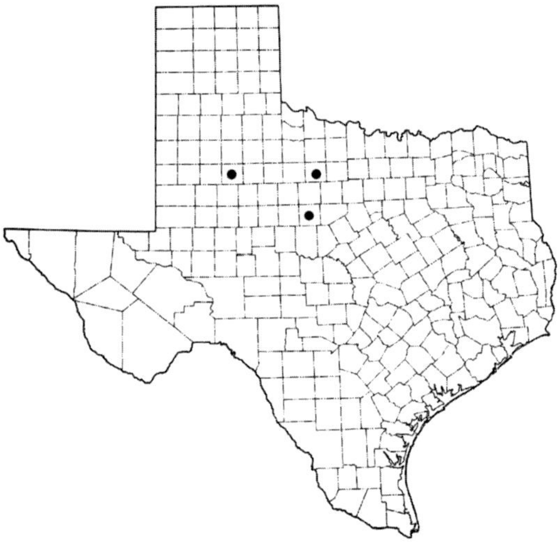

Oenothera coryi

Oenothera drummondii

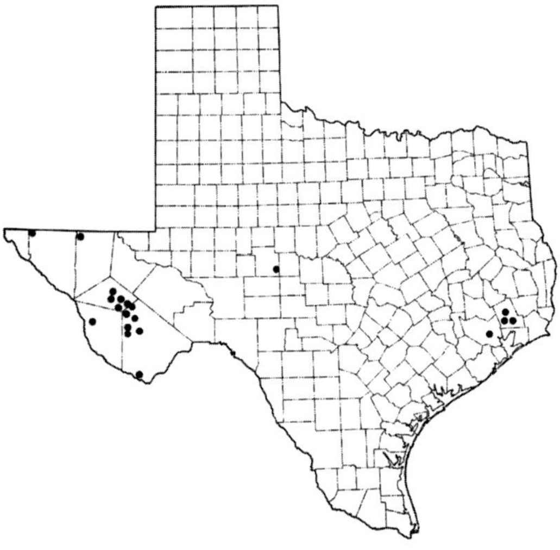

Oenothera elata

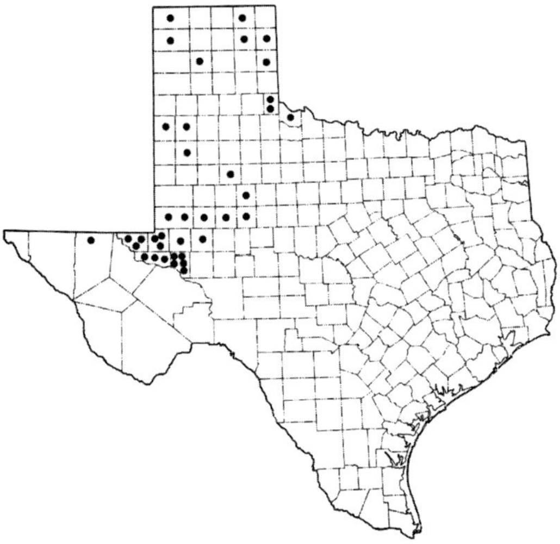
Oenothera engelmannii

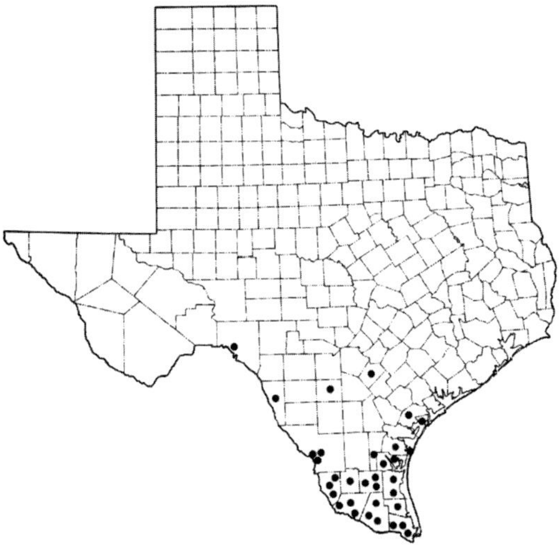
Oenothera falfurriae

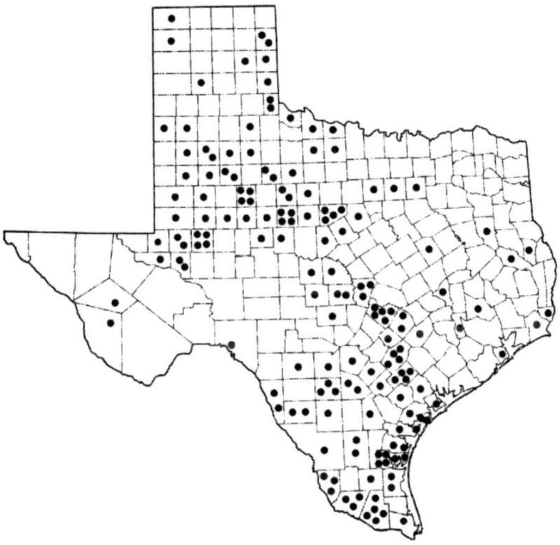
Oenothera grandis

Oenothera havardii

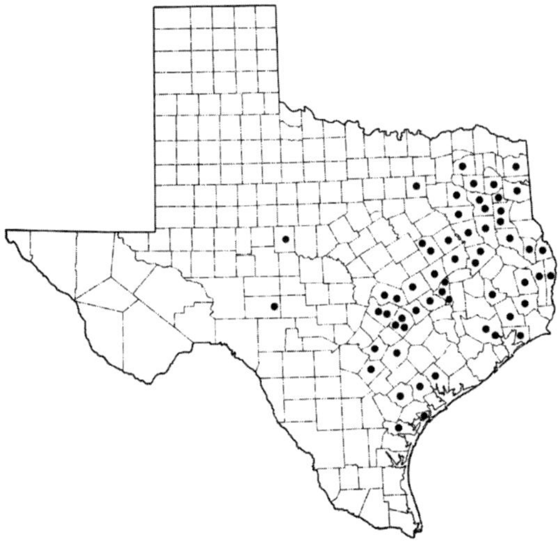
Oenothera heterophylla

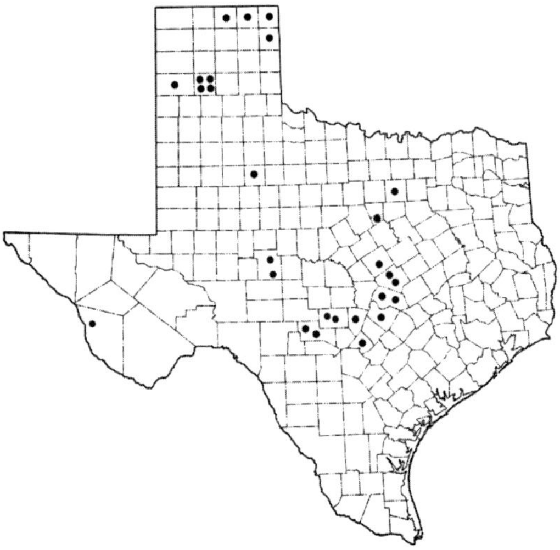
Oenothera jamesii

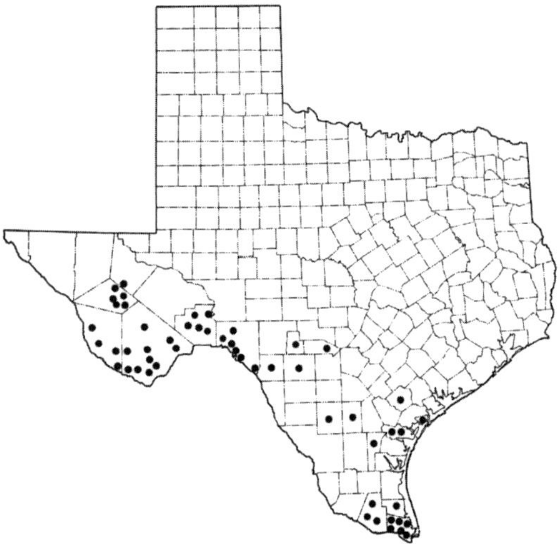

Oenothera kunthiana

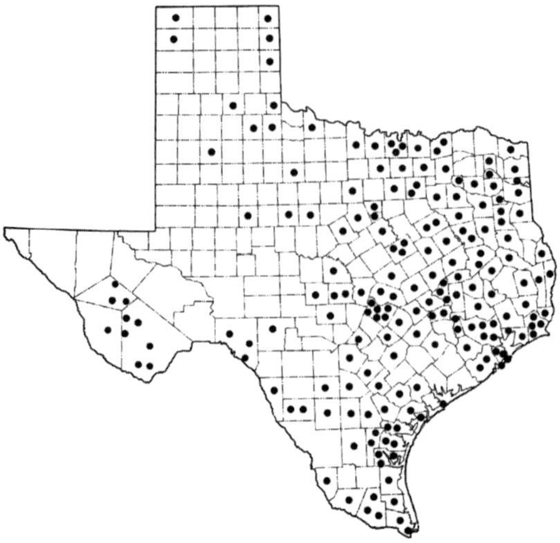

Oenothera laciniata

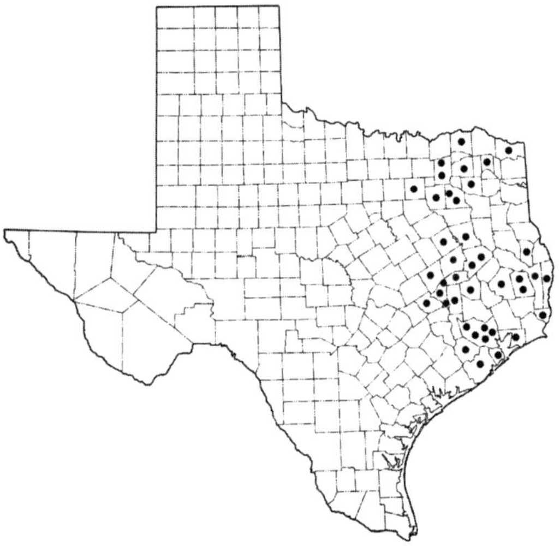

Oenothera linifolia

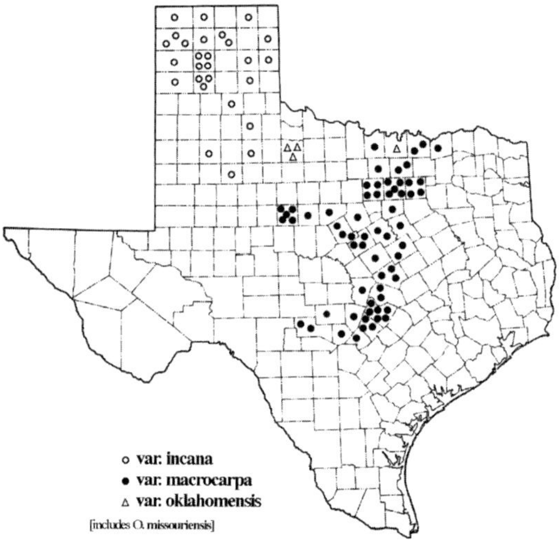
○ var. incana
● var. macrocarpa
△ var. oklahomensis
[includes O. missouriensis]

Oenothera macrocarpa

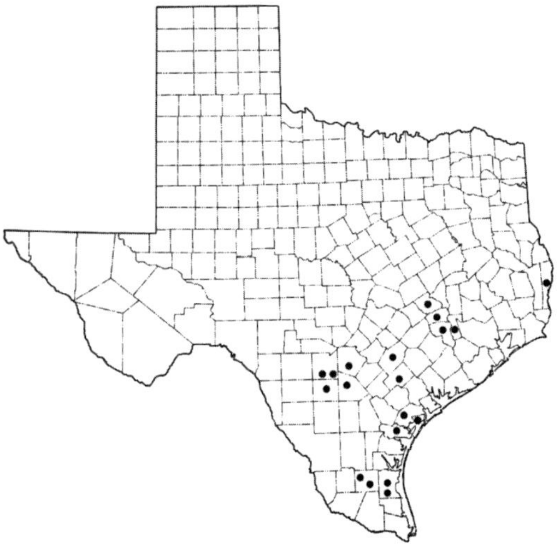

Oenothera mexicana

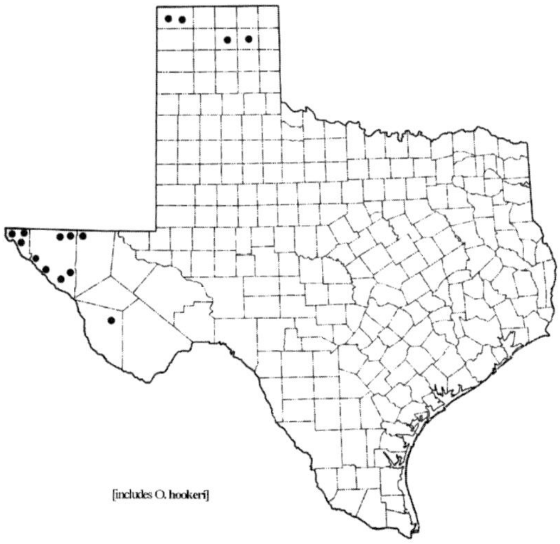
[includes O. hookeri]

Oenothera pallida

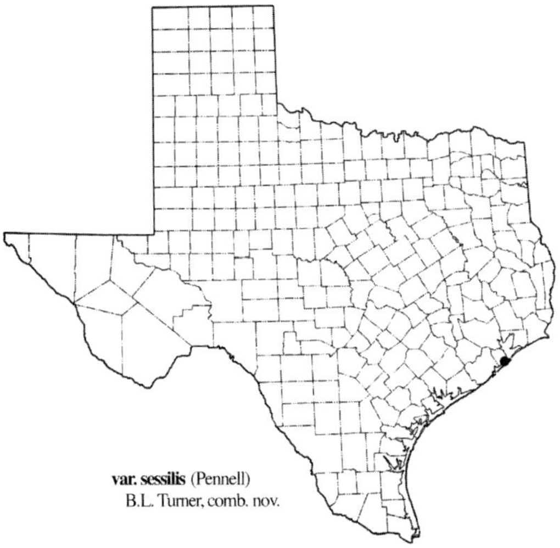
var. sessilis (Pennell)
B.L. Turner, comb. nov.
Oenothera pilosella

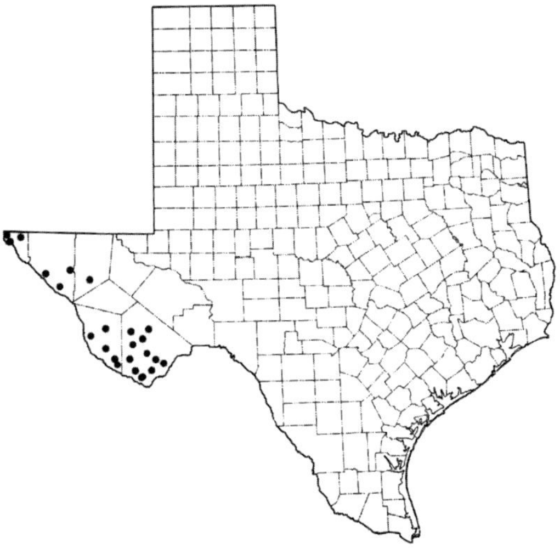
Oenothera primiveris

Oenothera pubescens

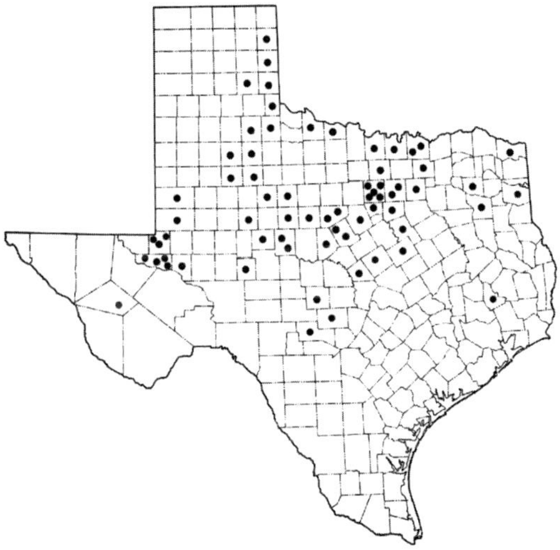
Oenothera rhombipetala

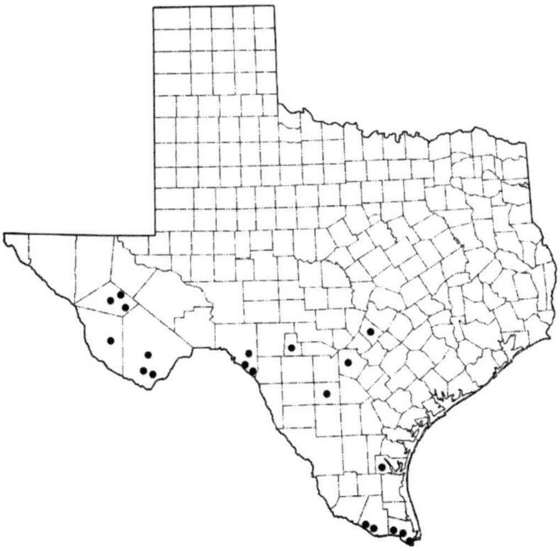
Oenothera rosea

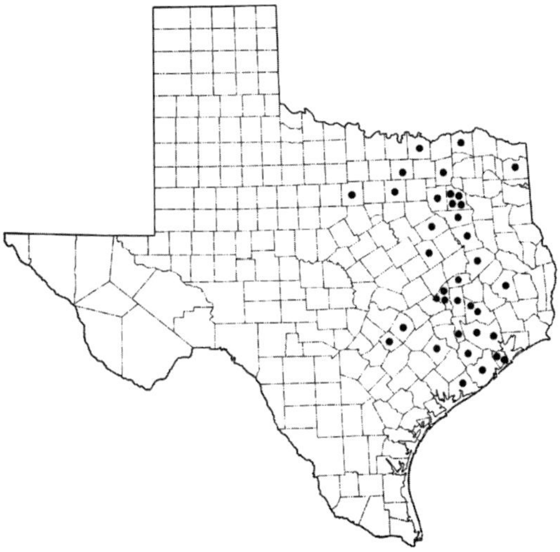
Oenothera spachiana

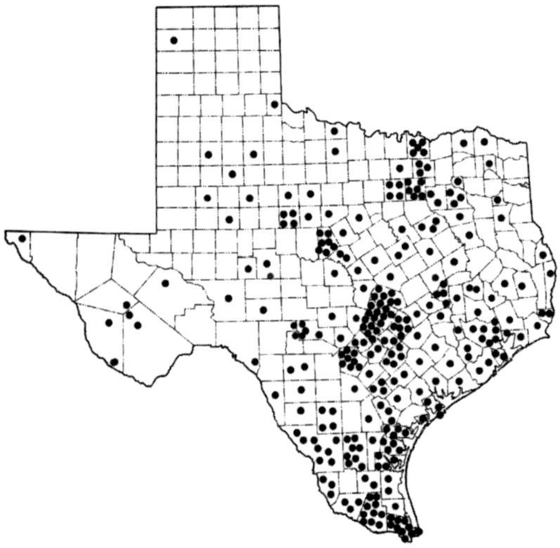

Oenothera speciosa

Oenothera tetraptera

Oenothera texensis

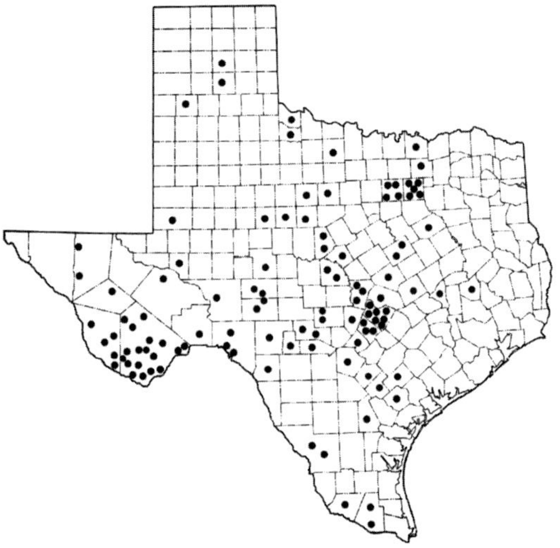

Oenothera triloba

Oenothera villosa

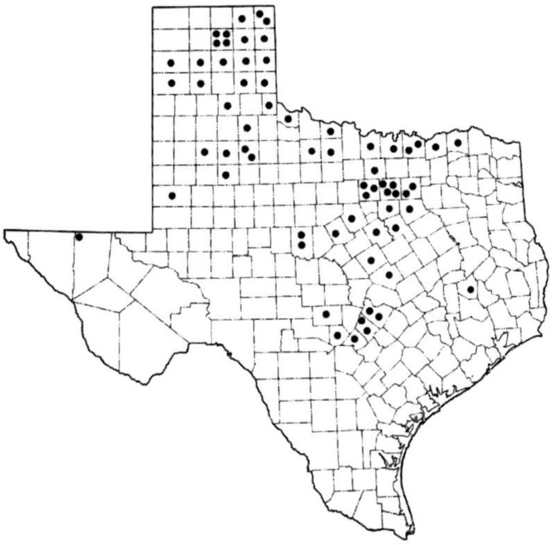

Stenosiphon linifolius

Conopholis alpina

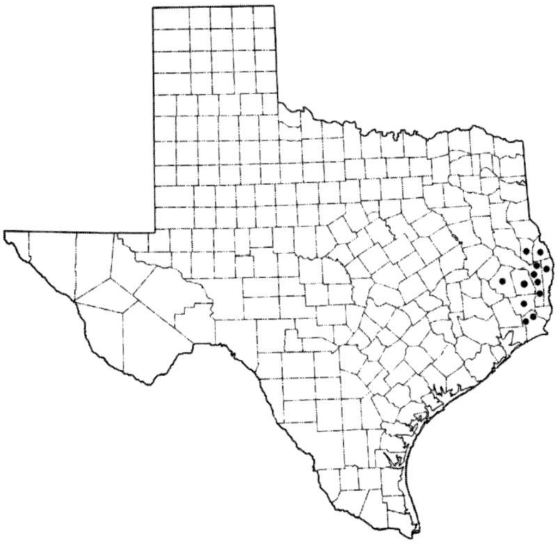

Epifagus virginiana

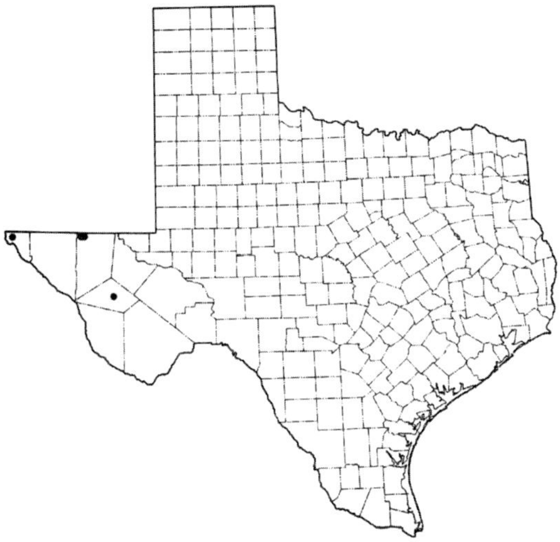

Orobanche fasciculata

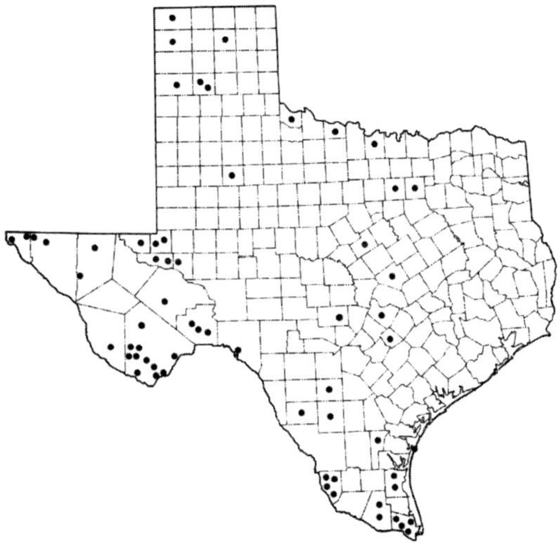

Orobanche ludoviciana

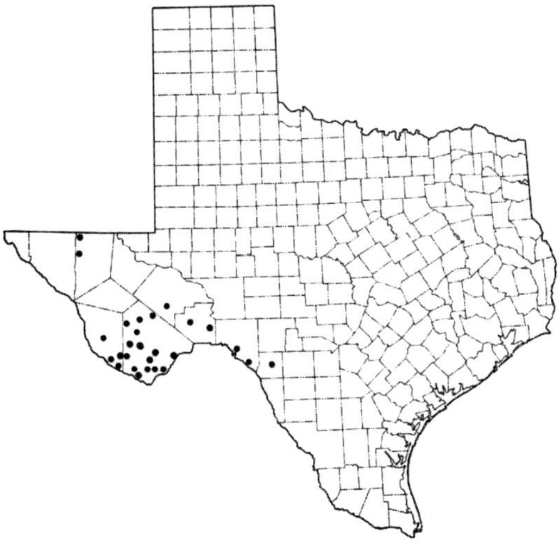

Orobanche multicaulis

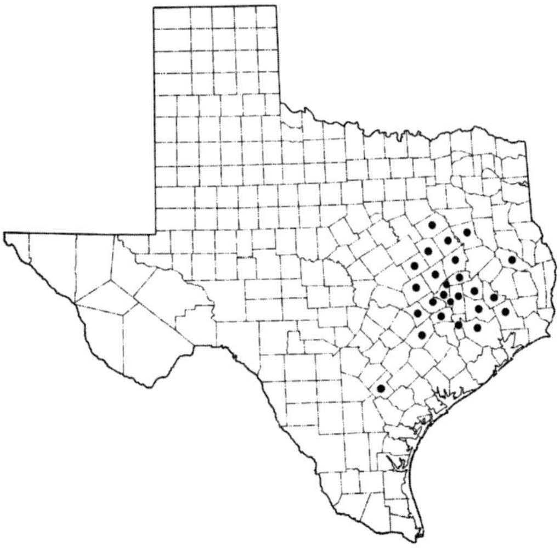

Orobanche ramosa

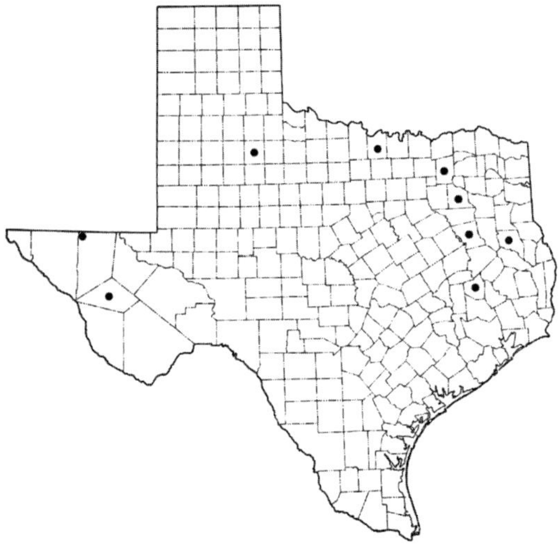

Orobanche uniflora

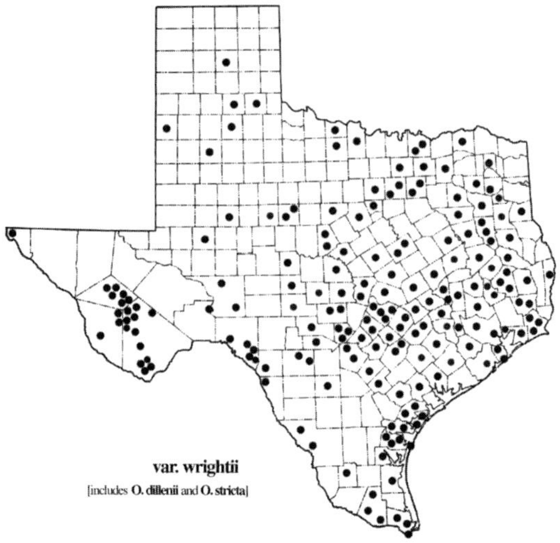
var. wrightii
[includes O. dillenii and O. stricta]

Oxalis corniculata

Oxalis corymbosa

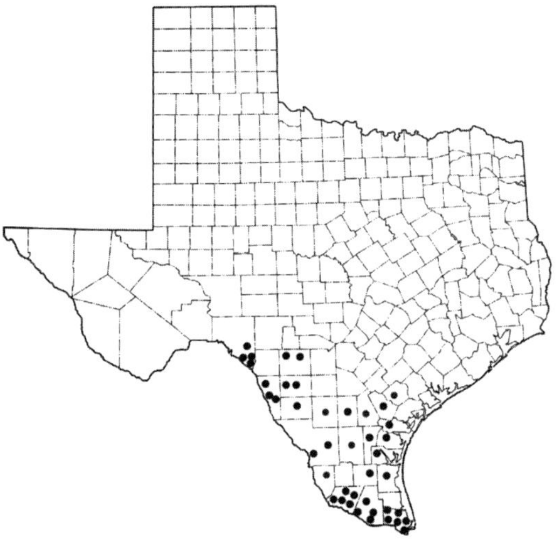

Oxalis dichondrifolia

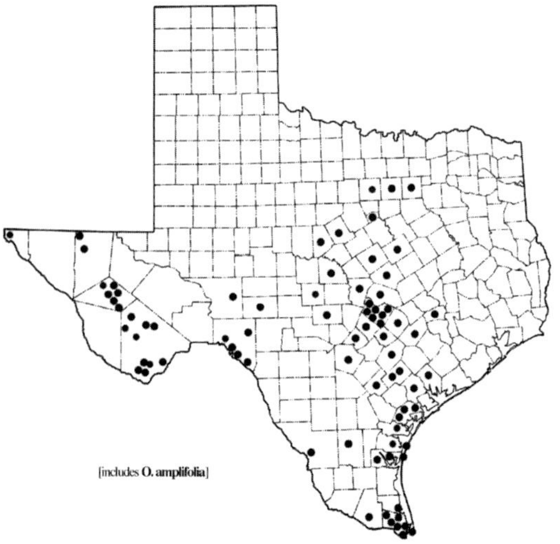
[includes O. amplifolia]

Oxalis drummondii

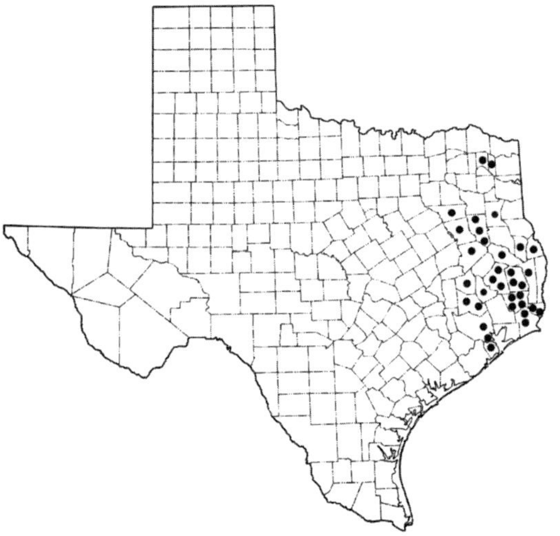

Oxalis lyonii

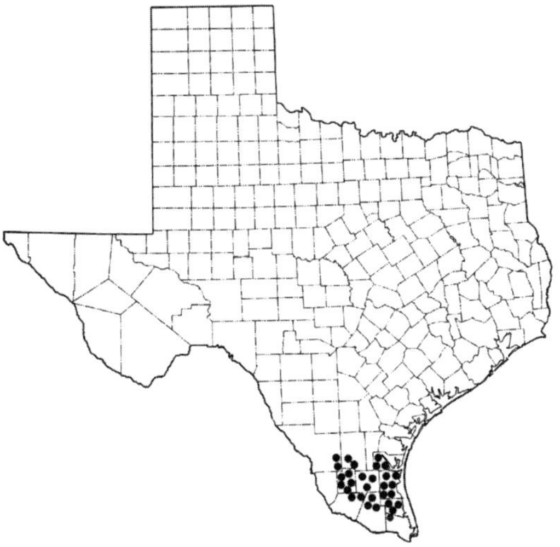

Oxalis neaei

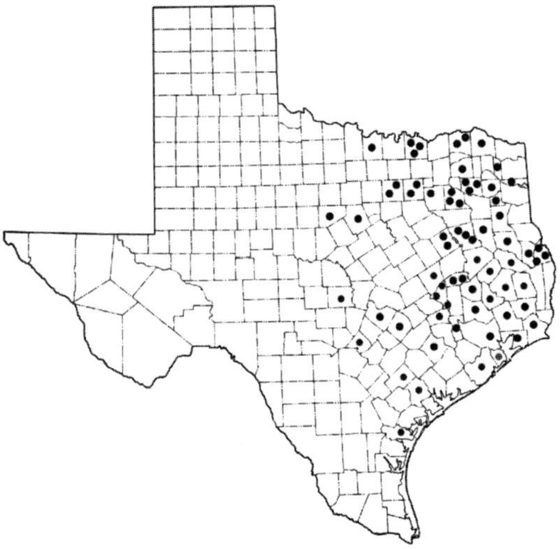

Oxalis violacea

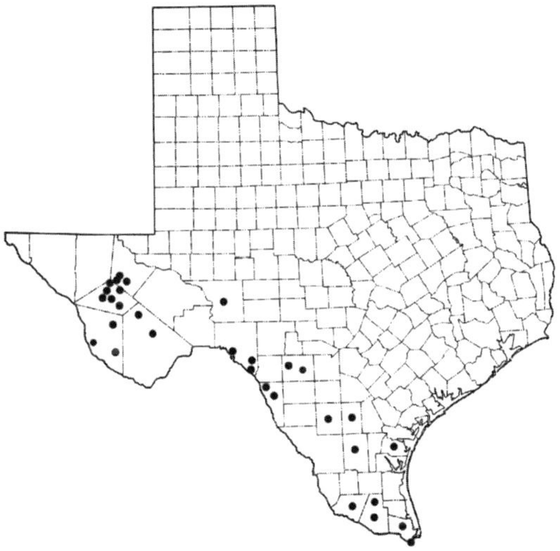

Argemone aenea

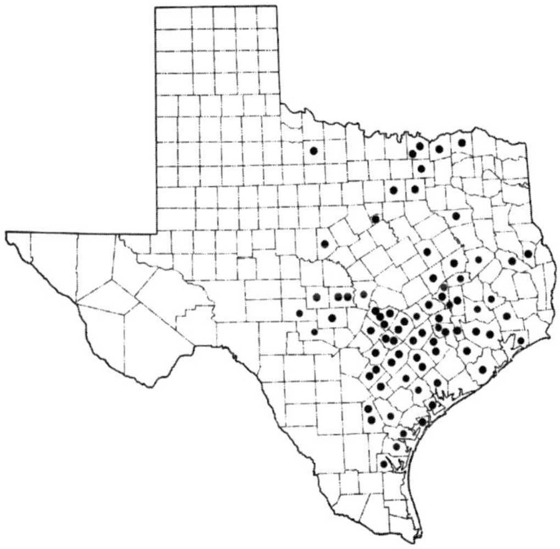

Argemone albiflora

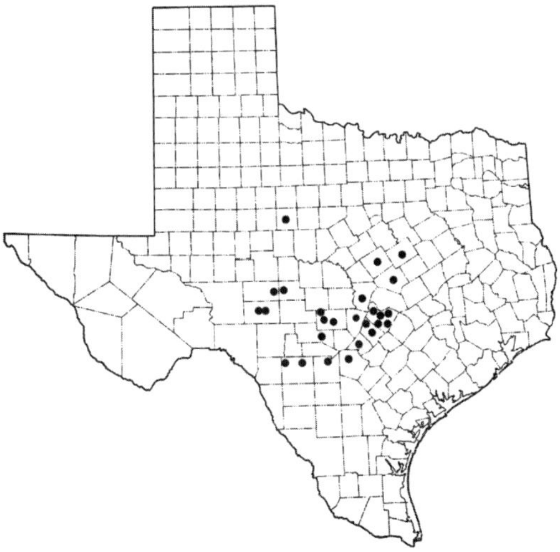

Argemone aurantiaca

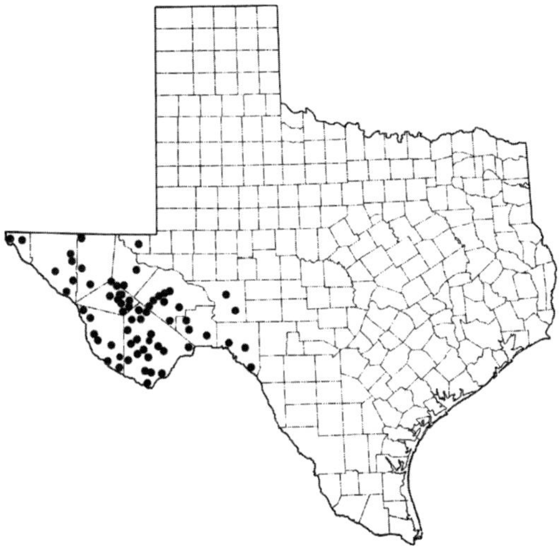

Argemone chisosensis

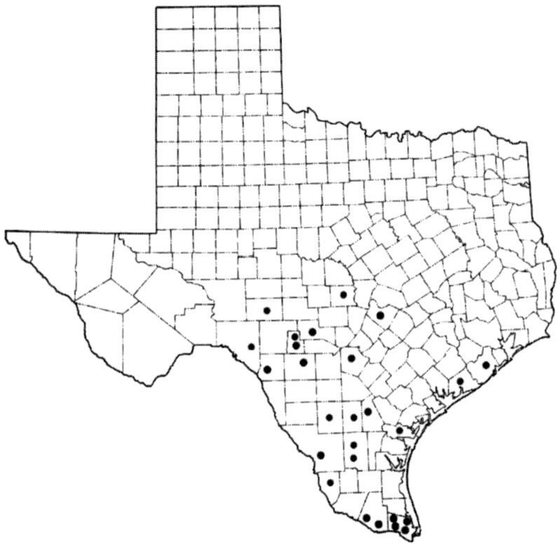

Argemone mexicana

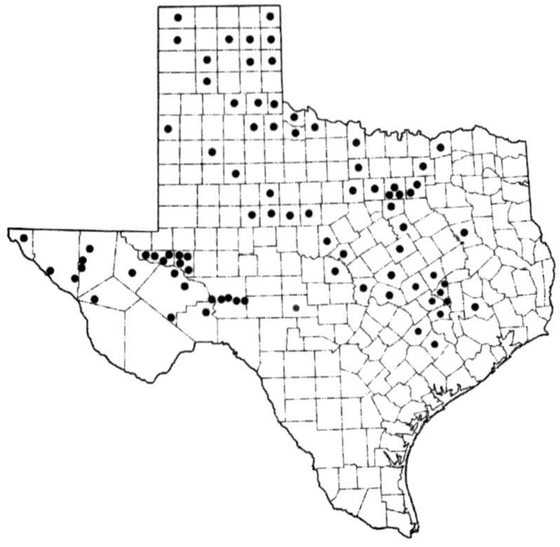

Argemone polyanthemos

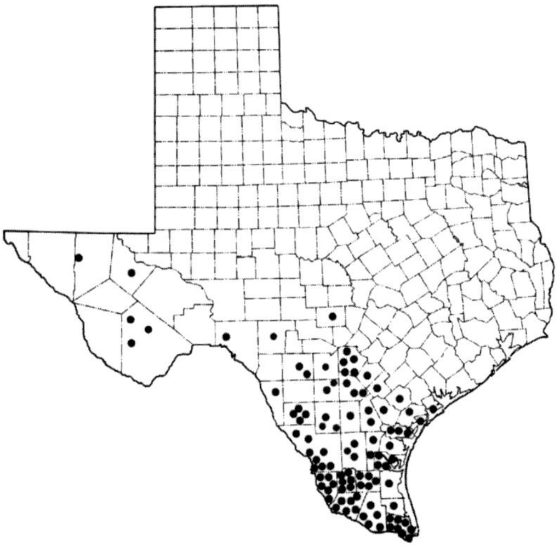

Argemone sanguinea

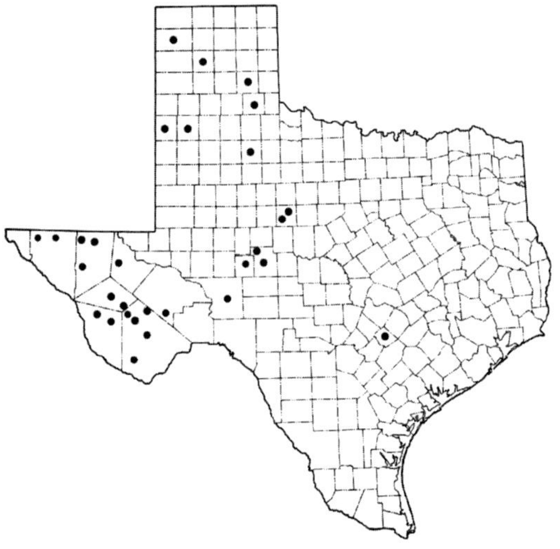

Argemone squarrosa

Eschscholtzia californica

[= E. californica
var. mexicana]

Eschscholtzia mexicana

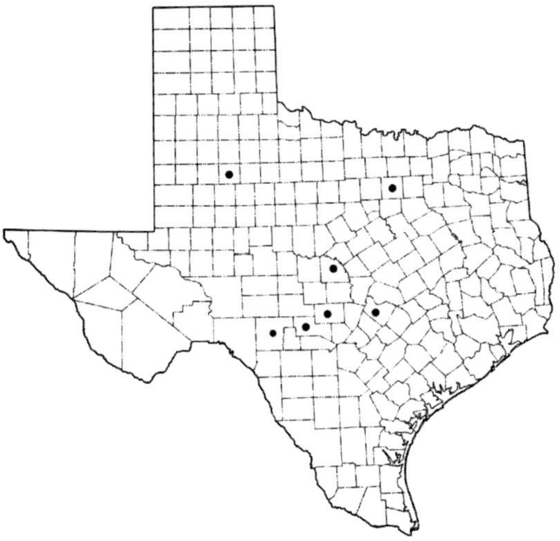

Glaucium corniculatum

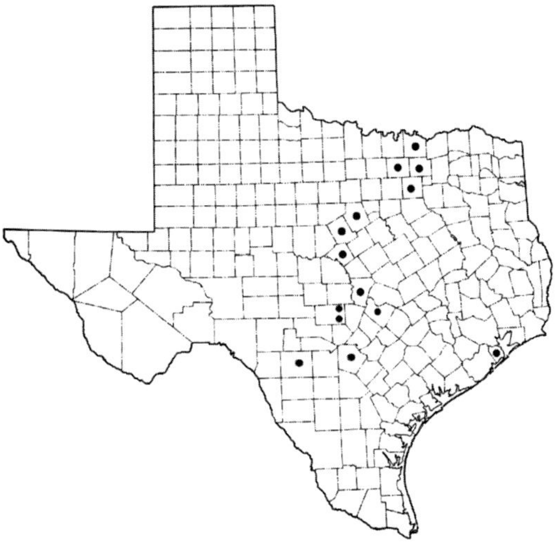

Papaver rhoeas

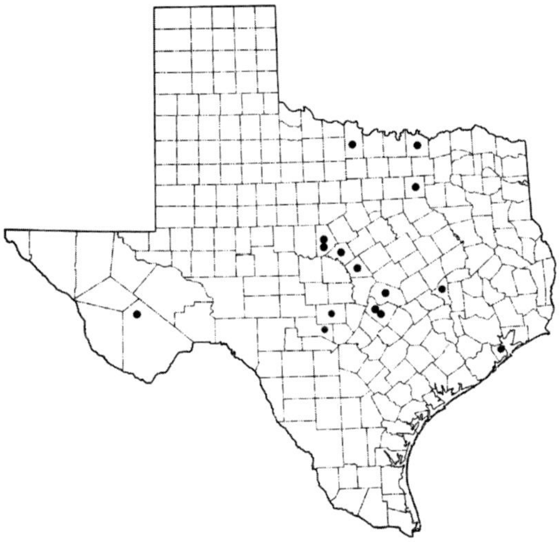

Papaver somniferum

Sanguinaria canadensis

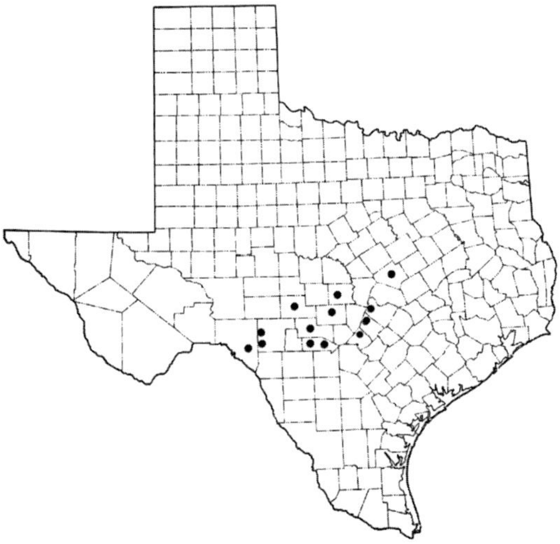

Passiflora affinis

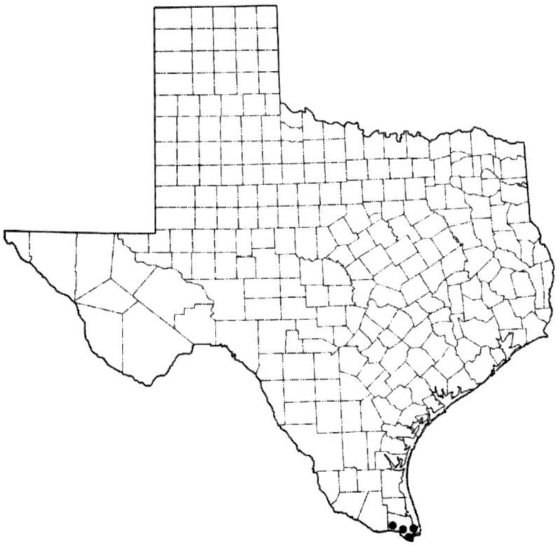

Passiflora filipes

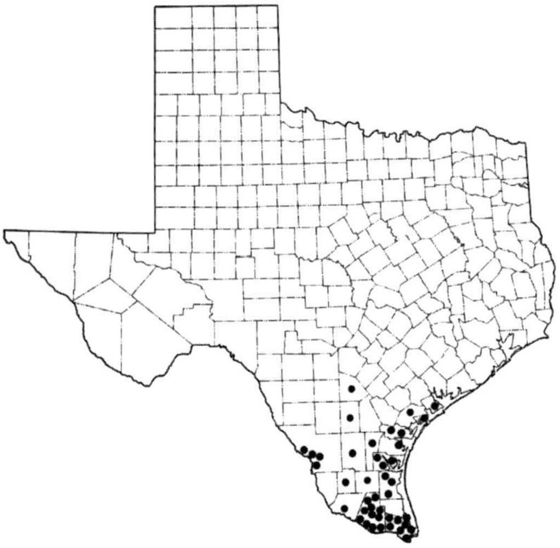

Passiflora foetida

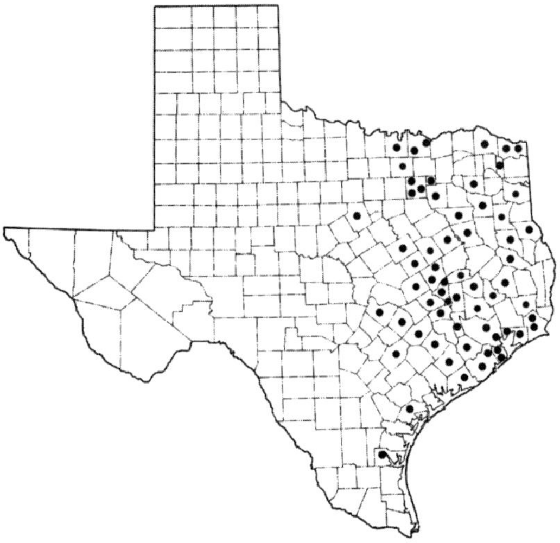

Passiflora incarnata

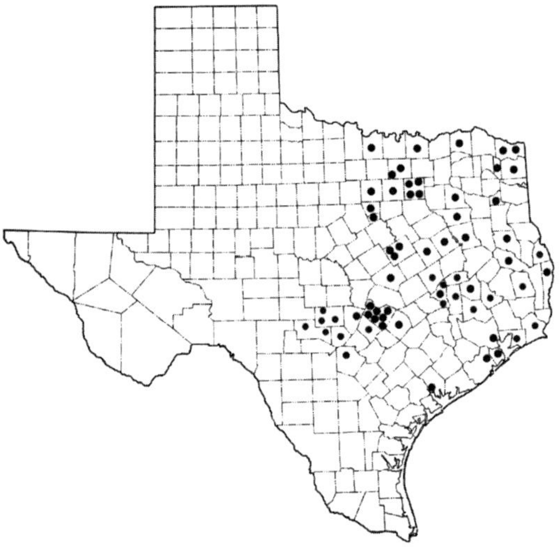

Passiflora lutea

Passiflora suberosa

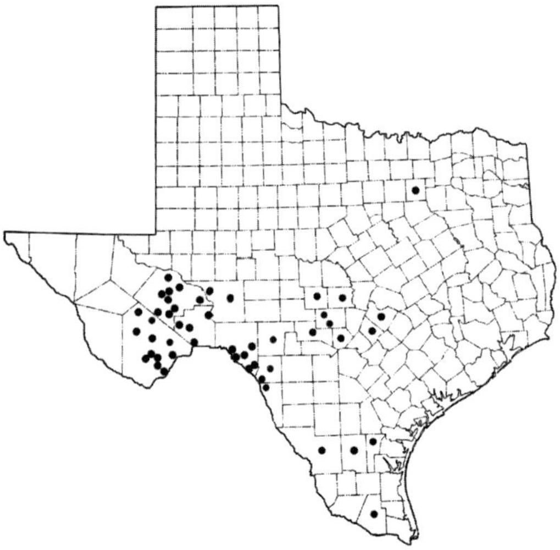

Passiflora tenuiloba

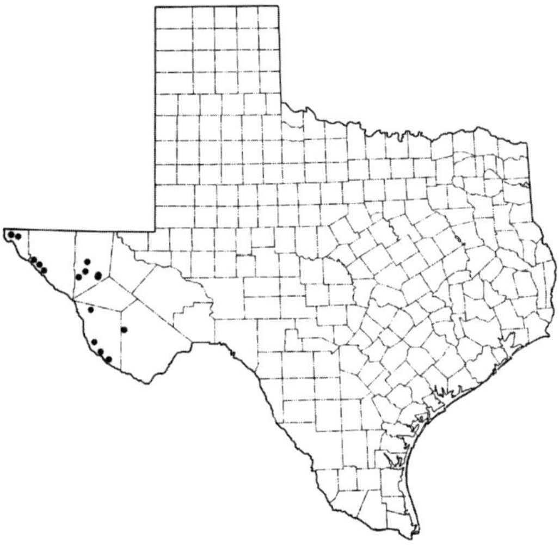

Proboscidea althaeifolia

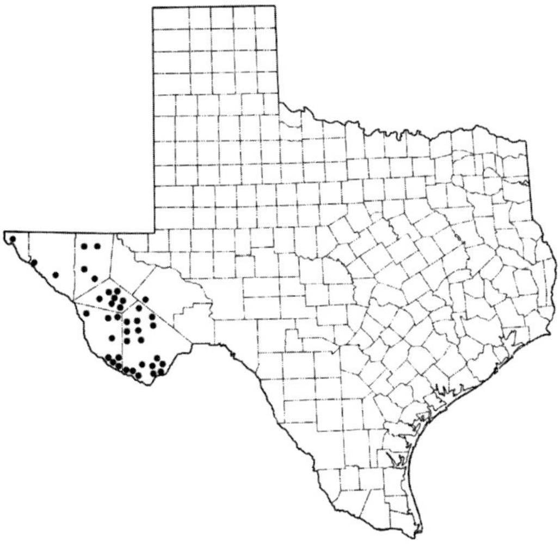

Proboscidea formosa

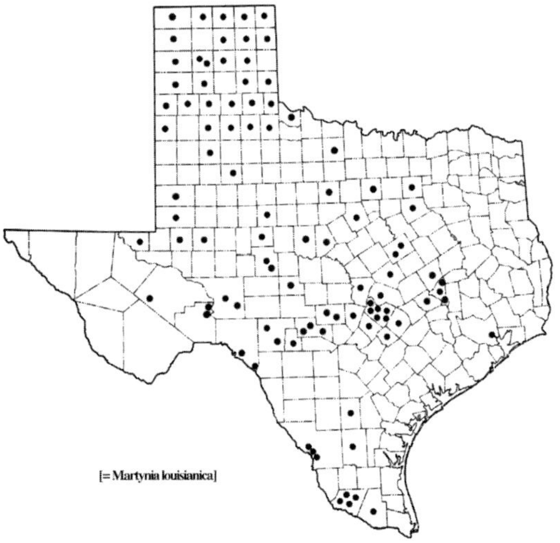
[= Martynia louisianica]

Proboscidea louisianica

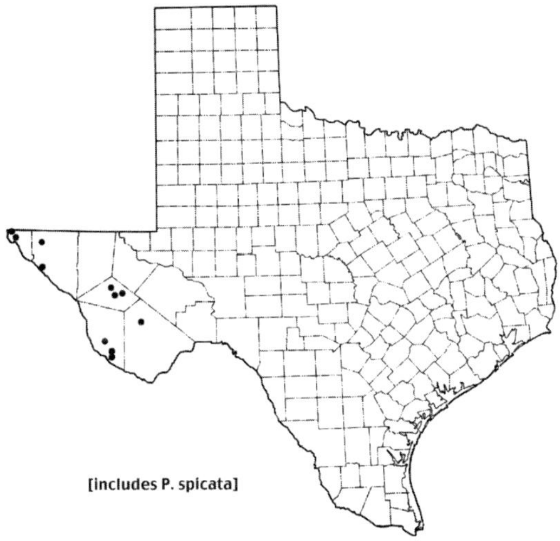
[includes P. spicata]

Proboscidea parviflora

Proboscidea sabulosa

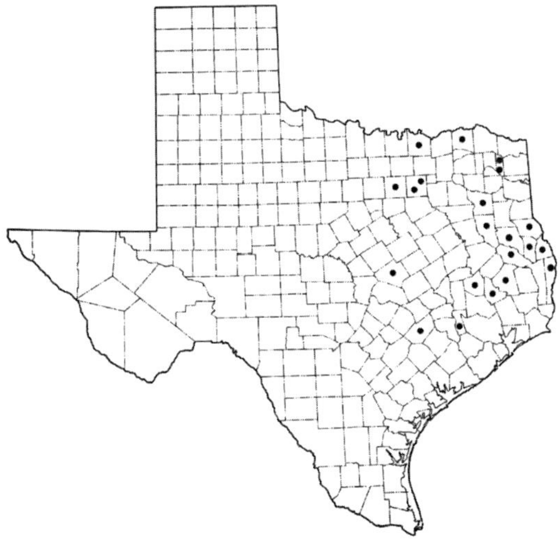

Phryma leptostachya

Petiveria alliacea

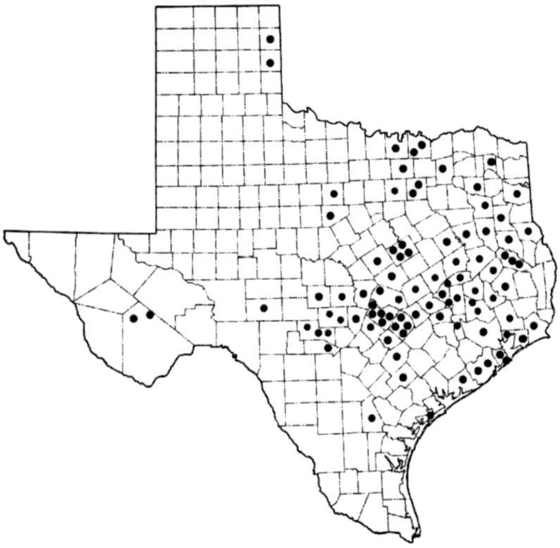

Phytolacca americana

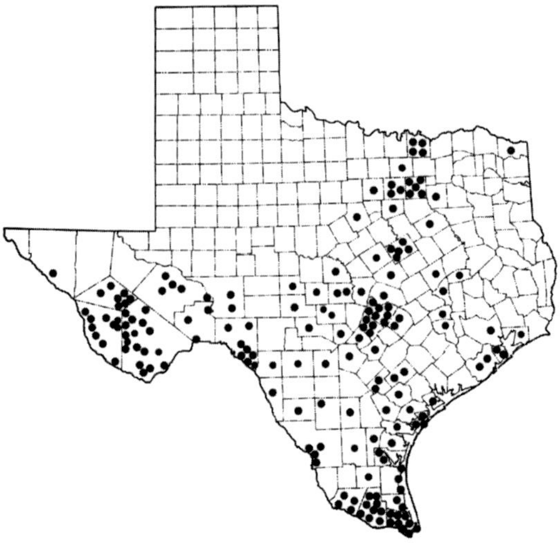

Rivina humilis

1973 Harris Co.*

Peperomia pellucida

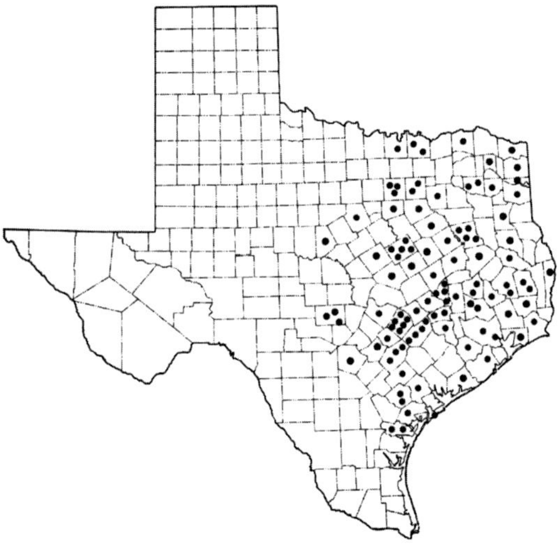

Plantago aristata

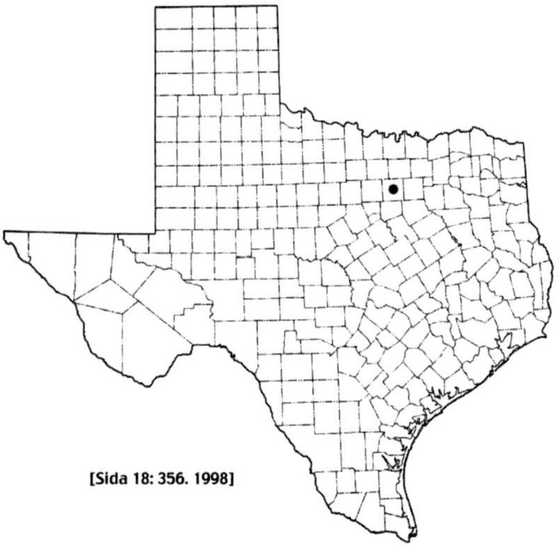
[Sida 18: 356. 1998]

Plantago coronopus

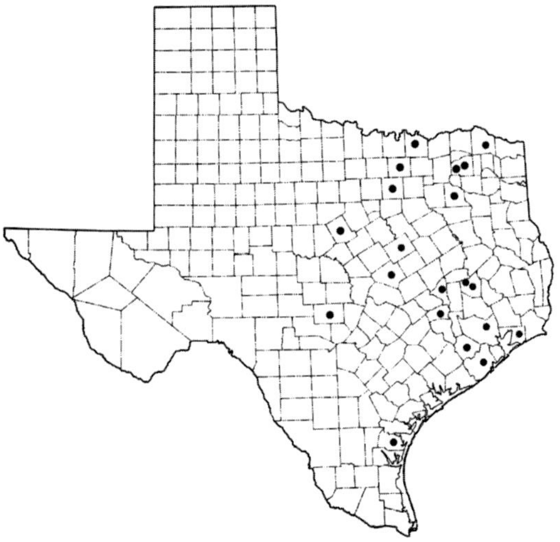

Plantago elongata

Plantago fastigiata

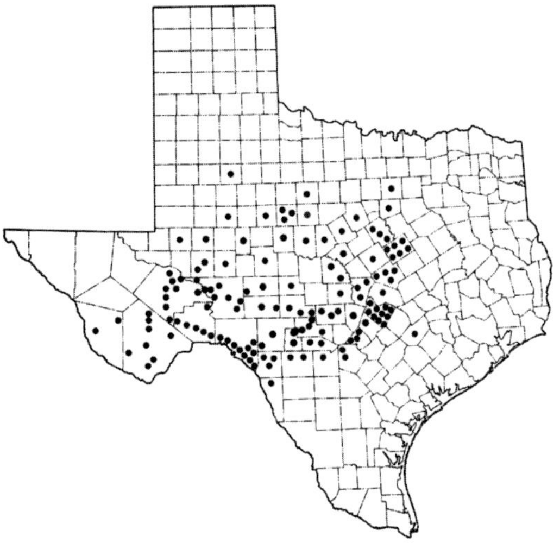

Plantago helleri

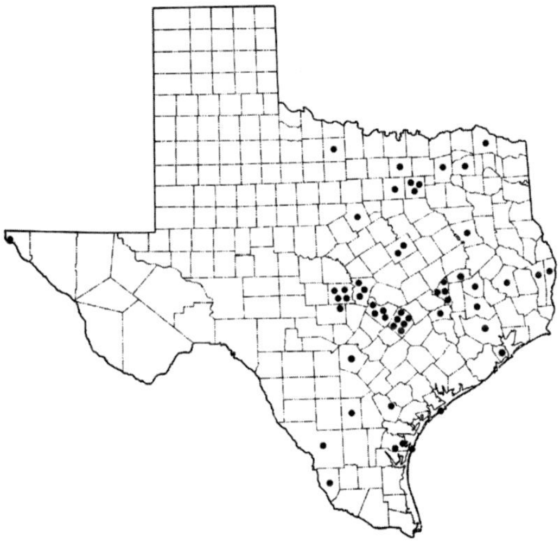

Plantago heterophylla

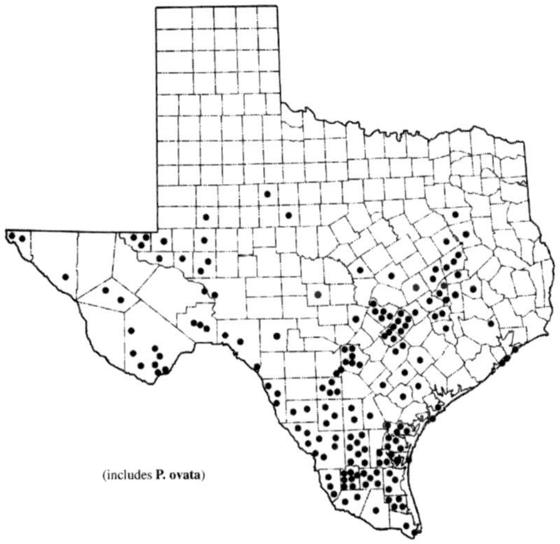
(includes **P. ovata**)

Plantago hookeriana

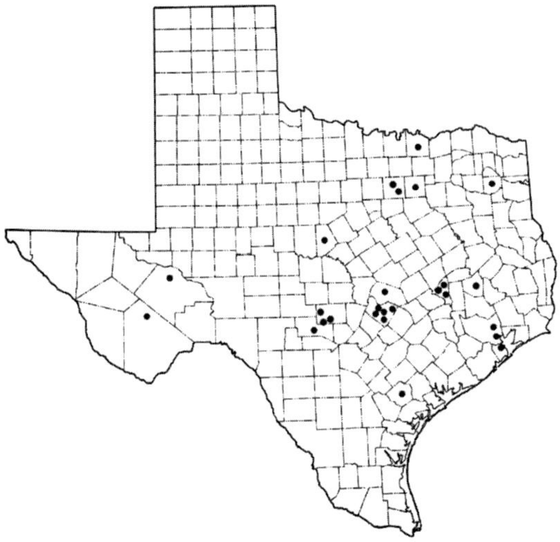

Plantago lanceolata

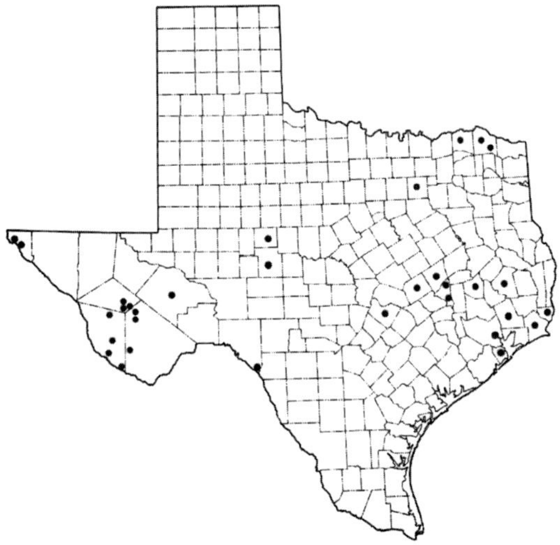

Plantago major

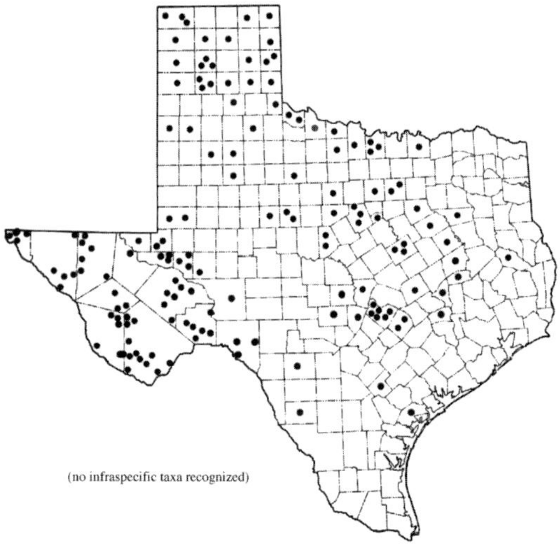
(no infraspecific taxa recognized)

Plantago patagonica

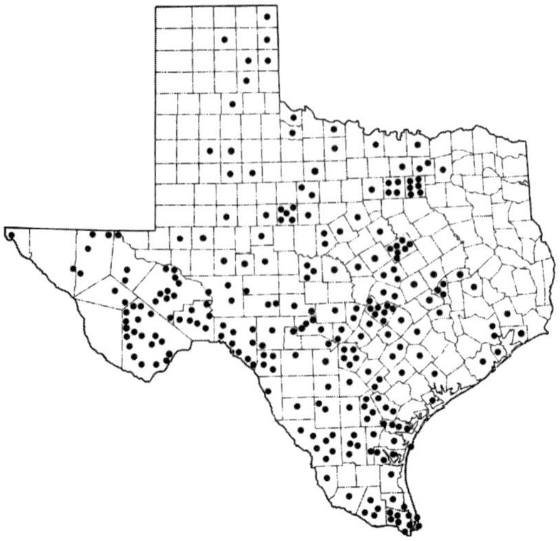

Plantago rhodosperma

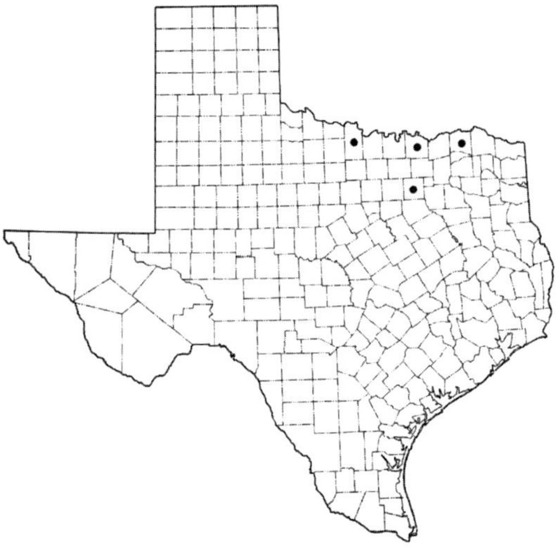

Plantago rugelii

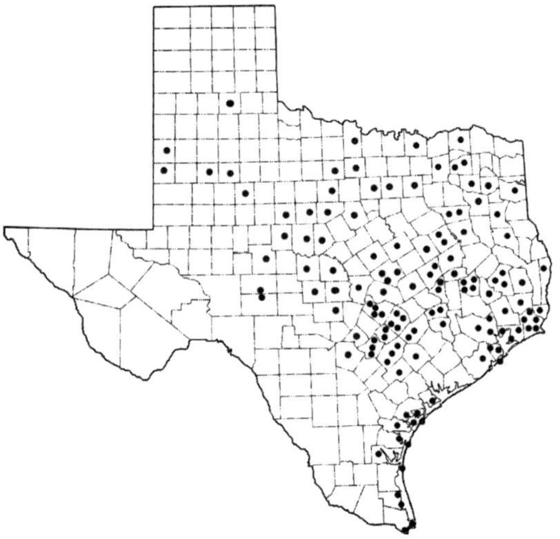

Plantago virginica

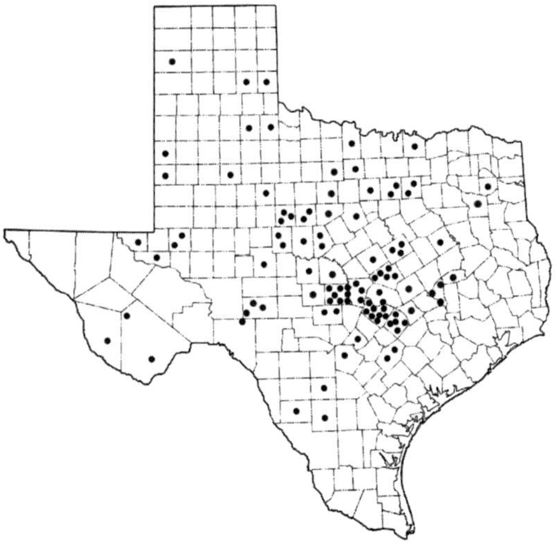

Plantago wrightiana

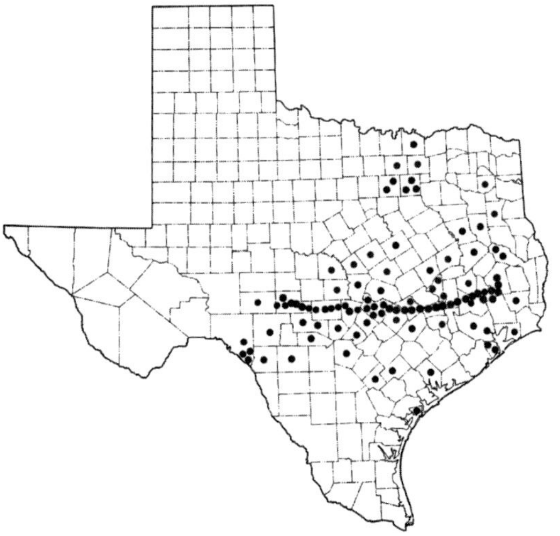

Platanus occidentalis

Limonium carolinianum

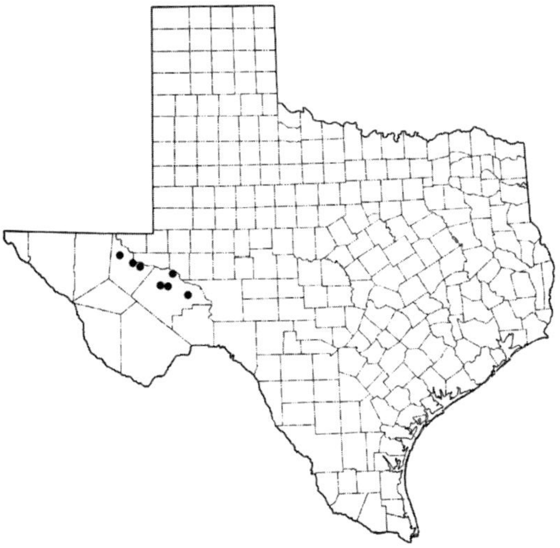

Limonium limbatum

Plumbago scandens

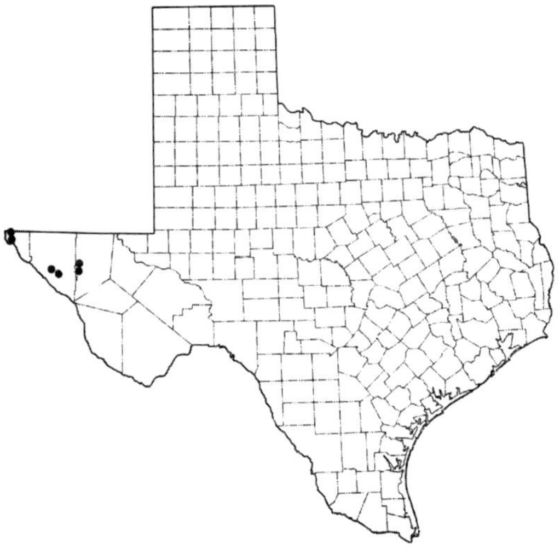

Eriastrum diffusum

Gilia flavocincta

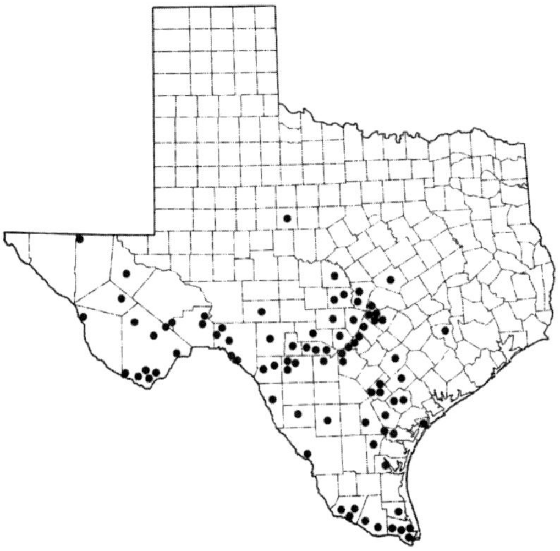

Gilia incisa

Gilia insignis

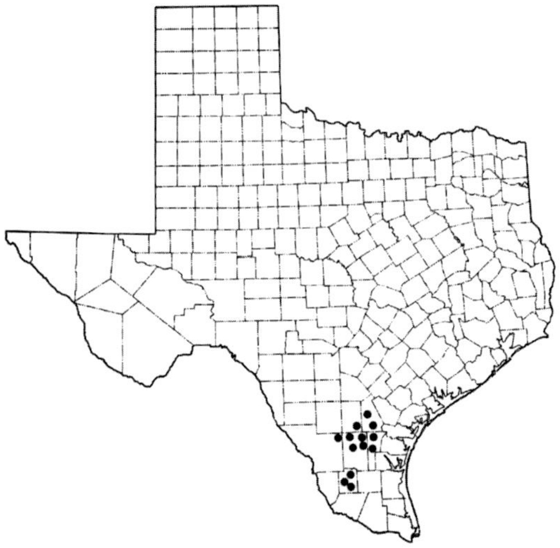

Gilia ludens

Gilia mexicana

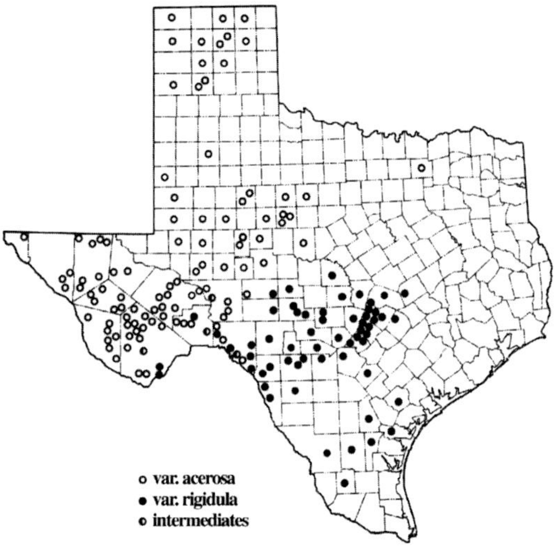
var. acerosa
var. rigidula
intermediates

Gilia rigidula

Gilia stewartii

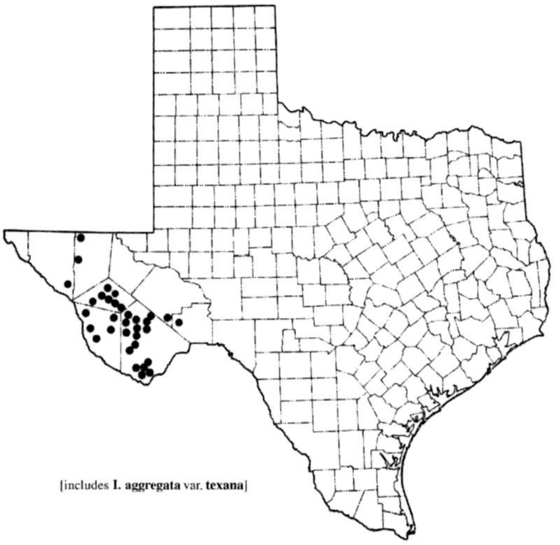
[includes I. aggregata var. texana]

Ipomopsis aggregata

Ipomopsis havardii

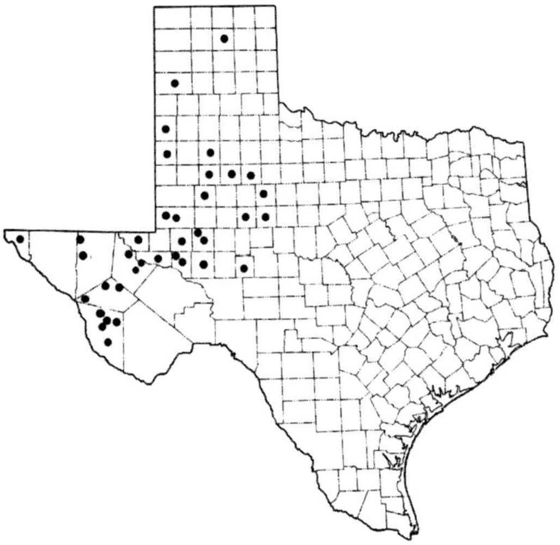

Ipomopsis laxiflora

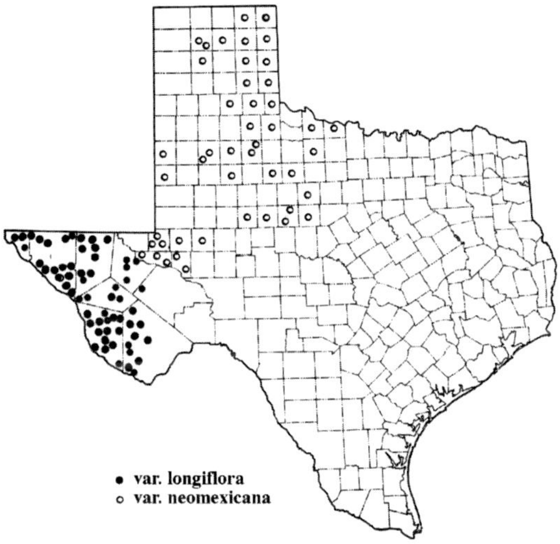
• var. longiflora
○ var. neomexicana

Ipomopsis longiflora

Ipomopsis polycladon

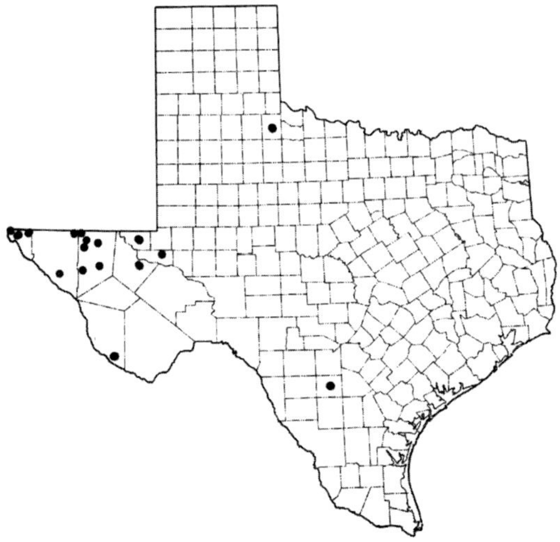

Ipomopsis pumila

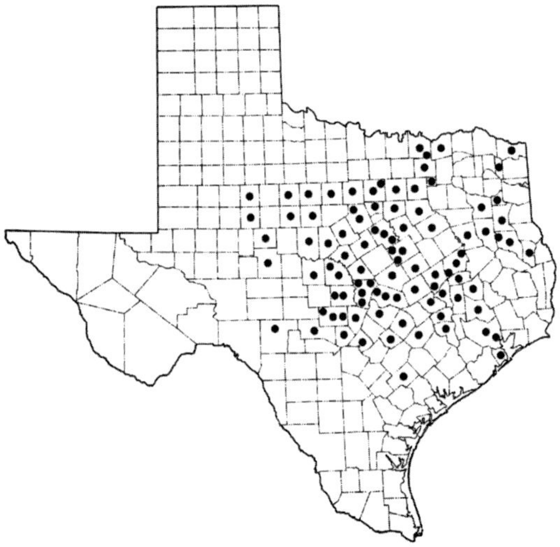

Ipomopsis rubra

Ipomopsis wrightii

Linanthus bigelovii

Loeselia greggii

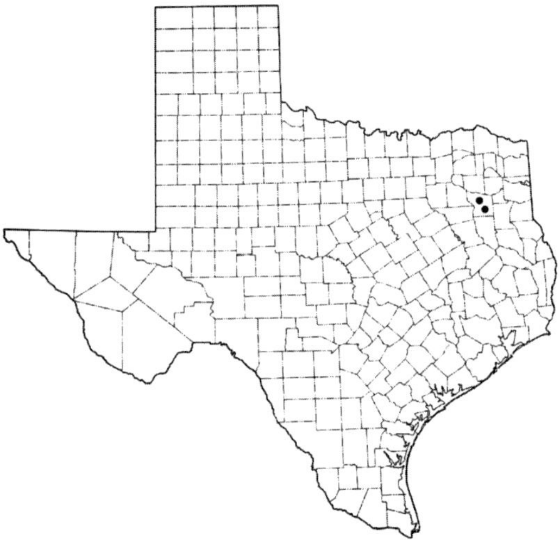

Phlox carolina

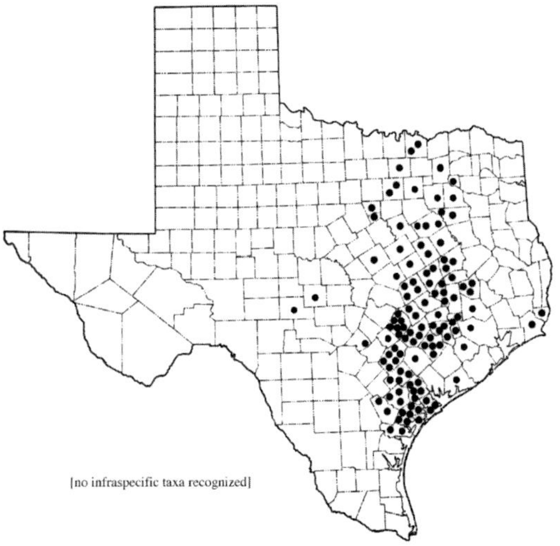
[no infraspecific taxa recognized]

Phlox cuspidata

Phlox divaricata

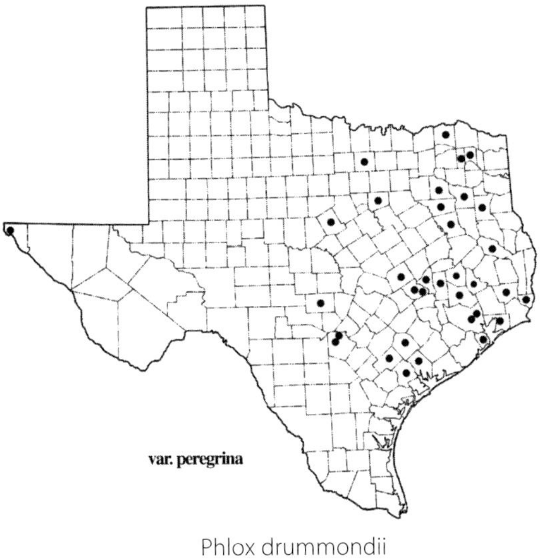
var. peregrina

Phlox drummondii

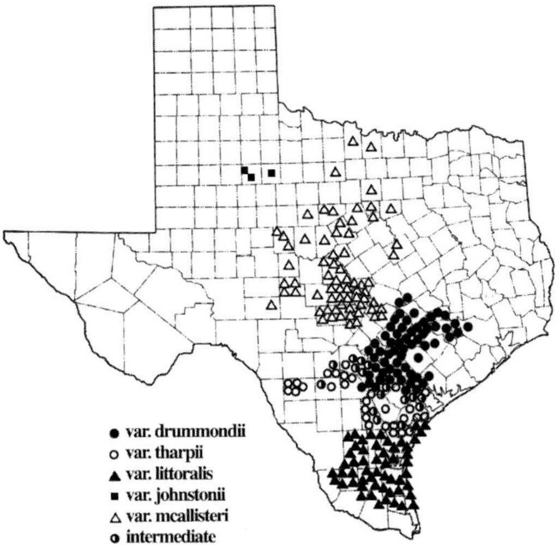
var. drummondii
var. tharpii
var. littoralis
var. johnstonii
var. mcallisteri
intermediate

Phlox drummondii

[includes P. stansburyi]

Phlox longifolia

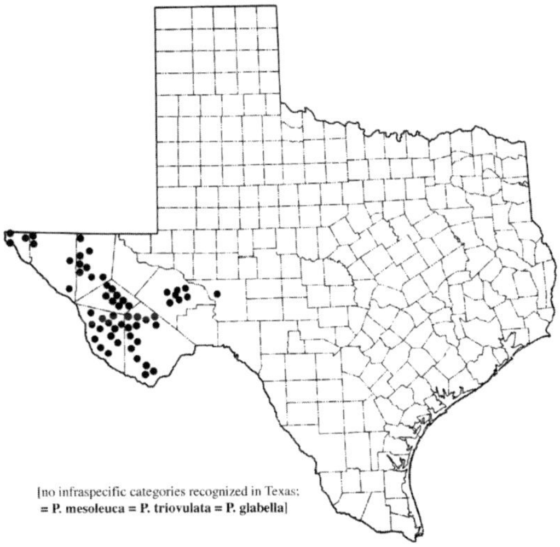
[no infraspecific categories recognized in Texas:
= P. mesoleuca = P. triovulata = P. glabella]

Phlox nana

Phlox nivalis

Phlox oklahomensis

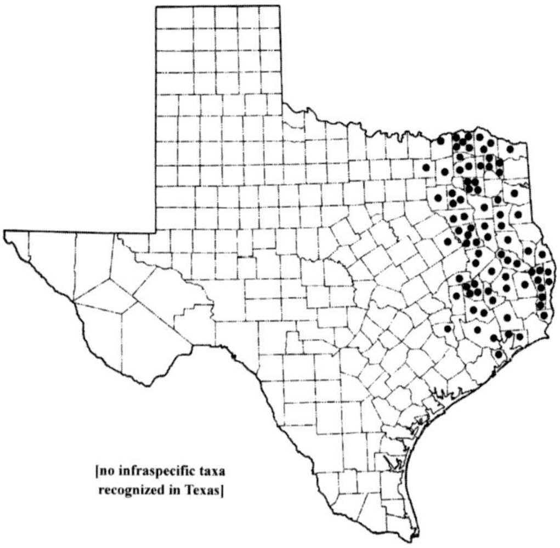
[no infraspecific taxa
recognized in Texas]

Phlox pilosa

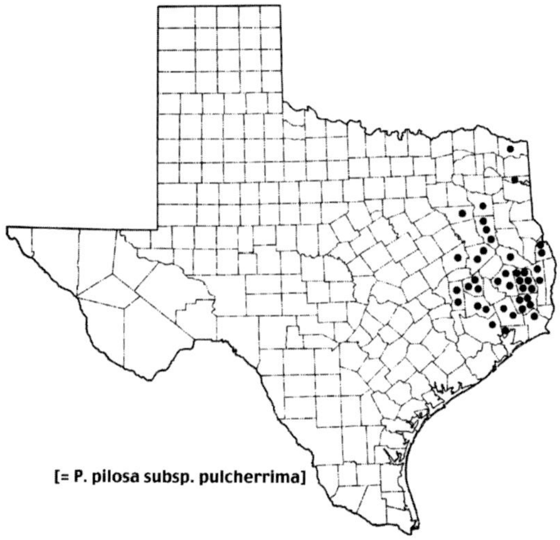
[= P. pilosa subsp. pulcherrima]

Phlox pulcherrima

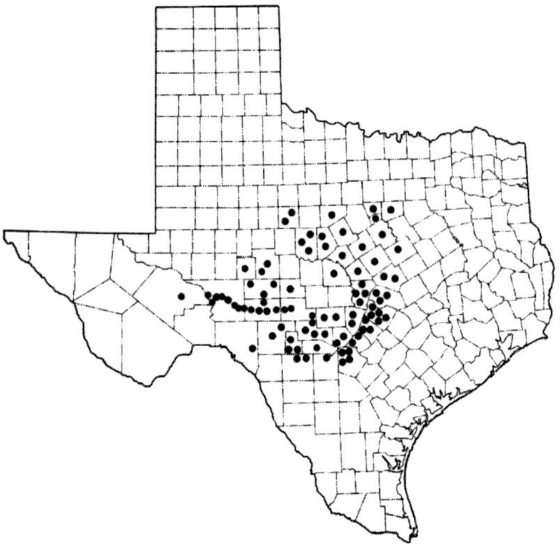

Phlox roemeriana

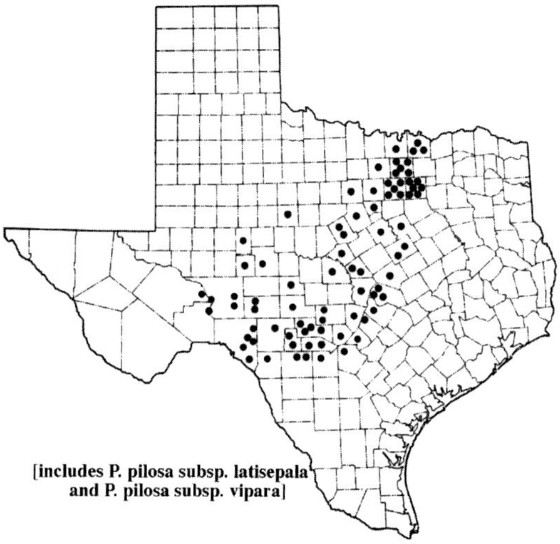
[includes P. pilosa subsp. latisepala
and P. pilosa subsp. vipara]

Phlox villosissima

Polemonium pauciflorum

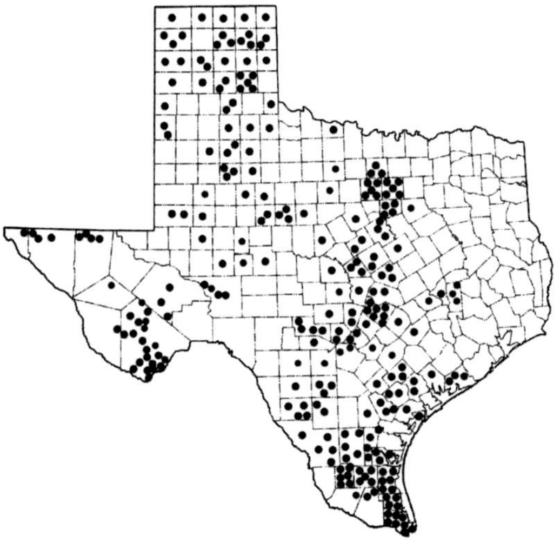

Polygala alba

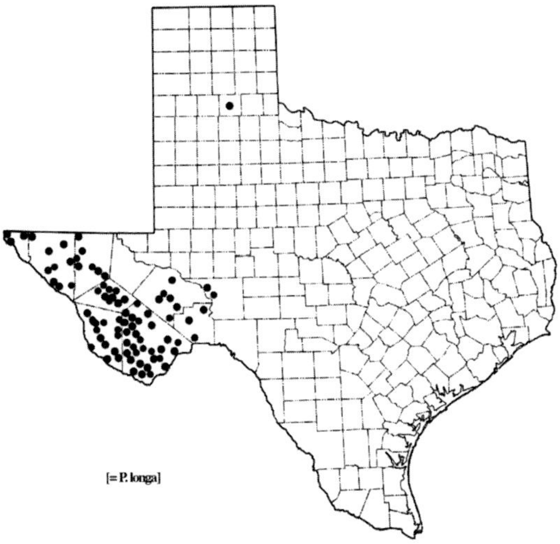
[= P. longa]
Polygala barbeyana

Polygala crenata

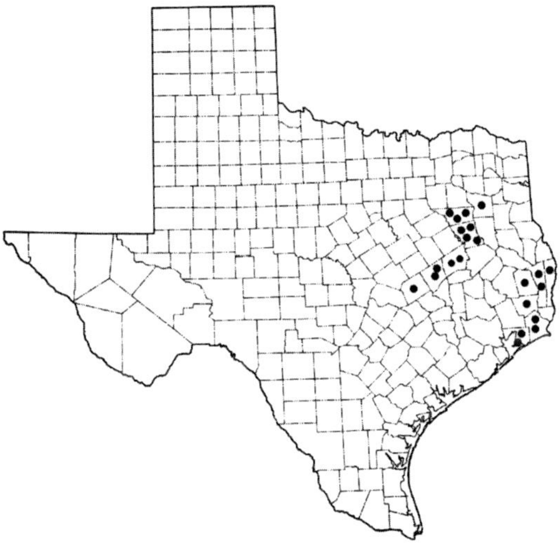
Polygala cruciata

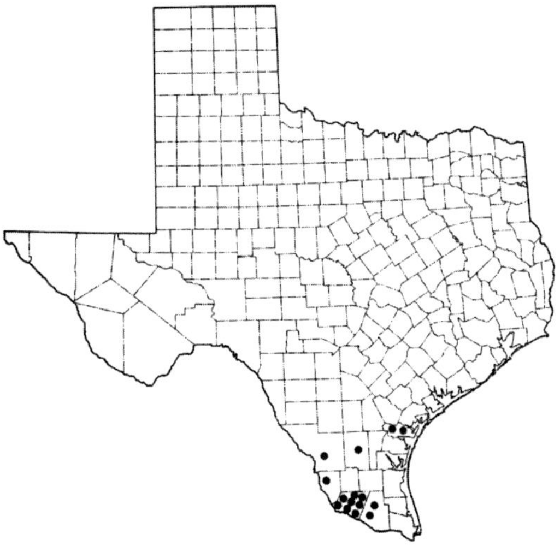
Polygala glandulosa

Polygala hemipterocarpa

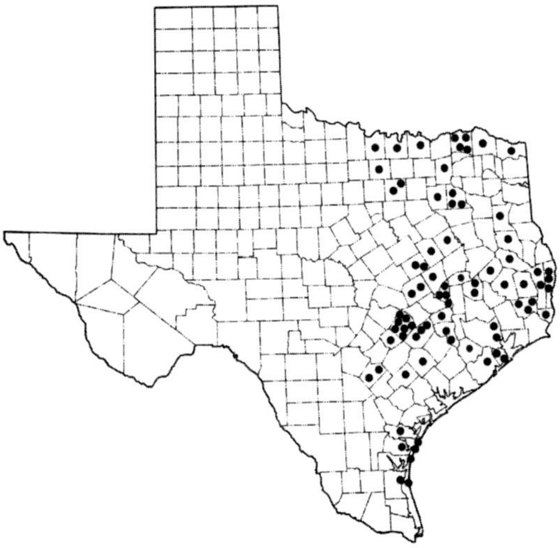
Polygala incarnata

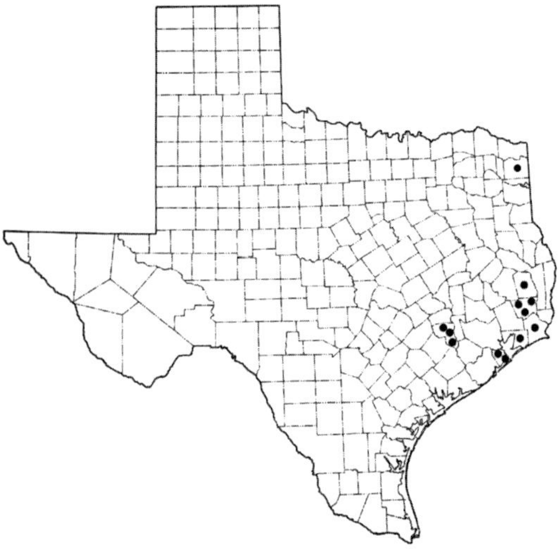

Polygala leptocaulis

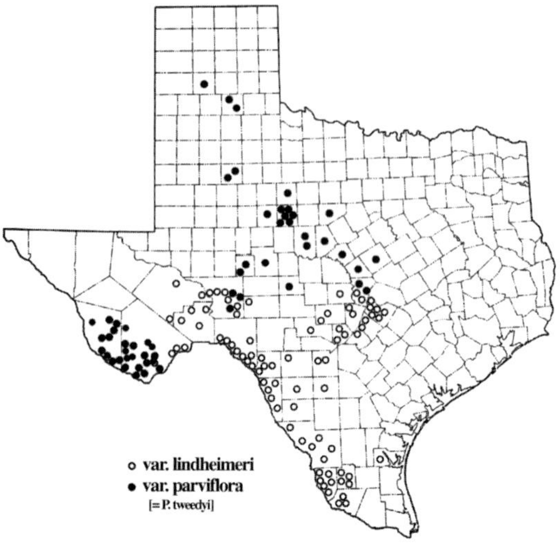
var. lindheimeri
var. parviflora
[= P. tweedyi]

Polygala lindheimeri

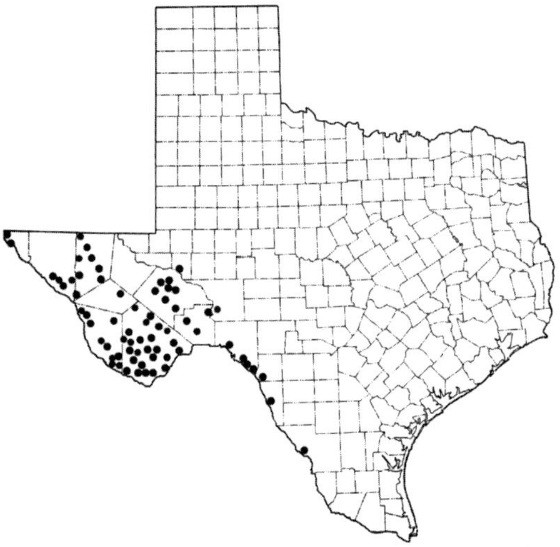

Polygala macradenia

Polygala maravillasensis

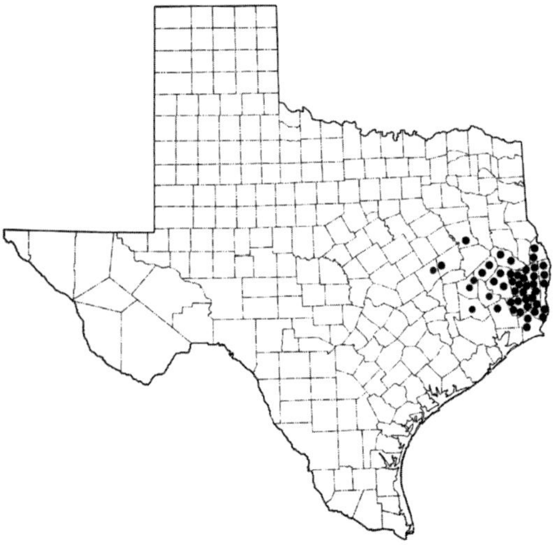

Polygala mariana

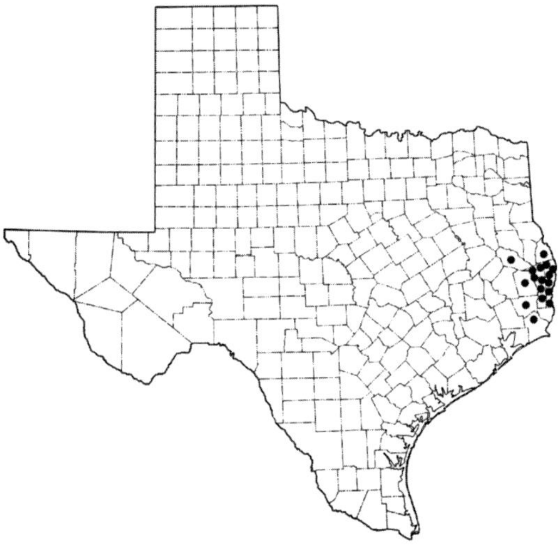

Polygala nana

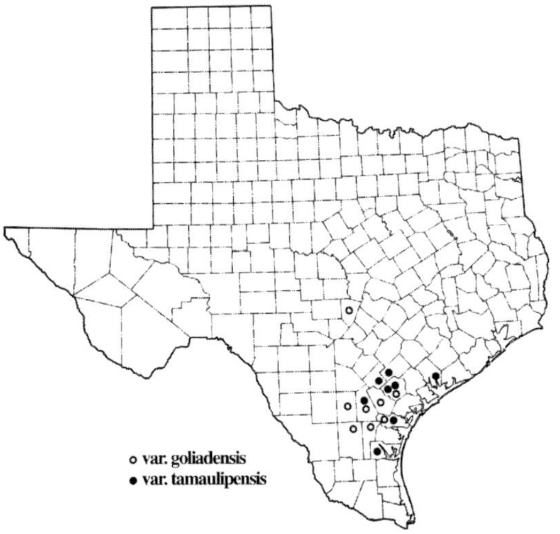
var. goliadensis
var. tamaulipensis

Polygala nitida

Polygala nudata

Polygala obscura

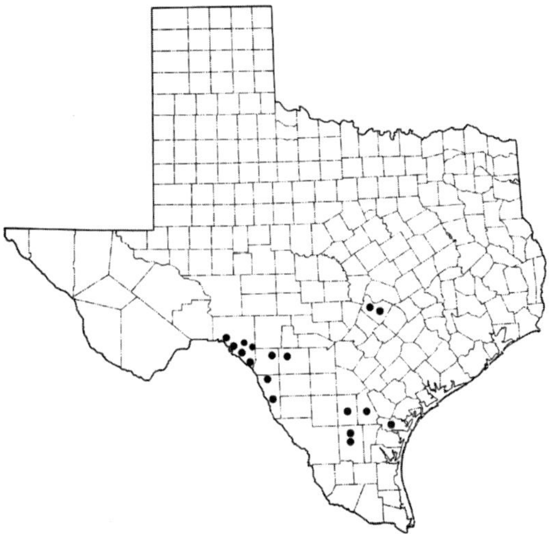

Polygala ovatifolia

Polygala palmeri

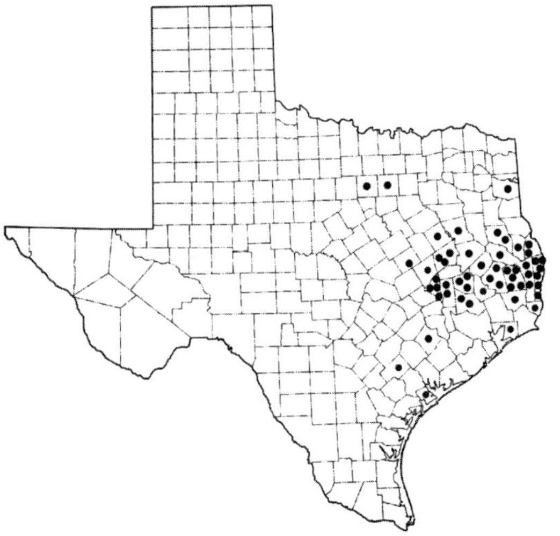

Polygala polygama

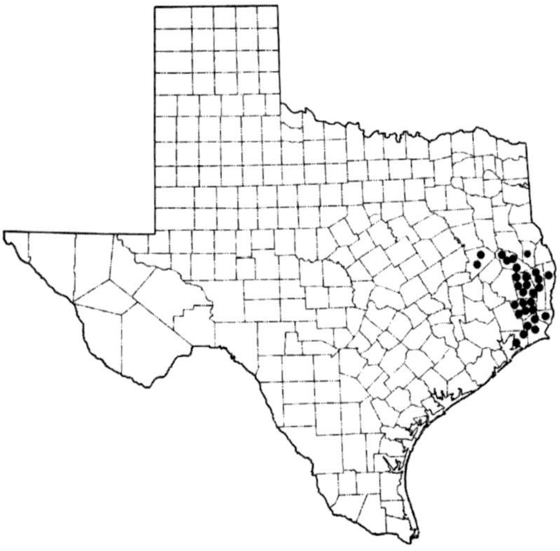

Polygala ramosa

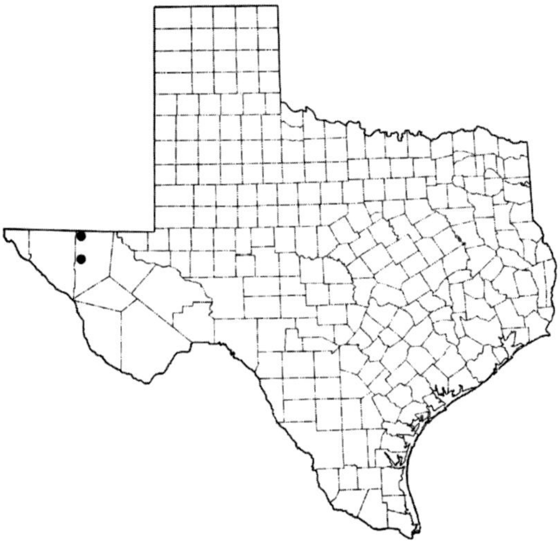

Polygala rimulicola

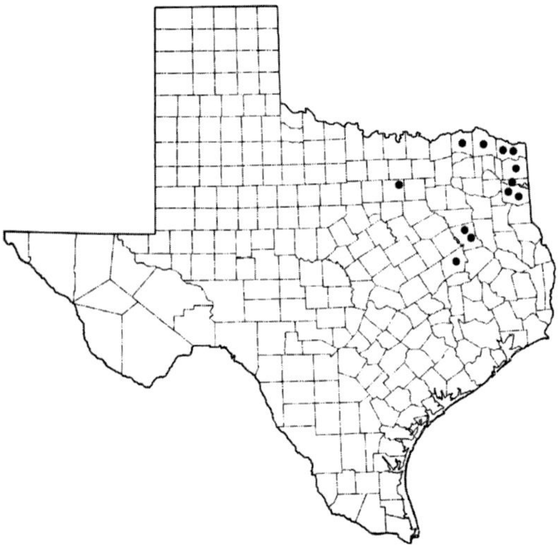

Polygala sanguinea

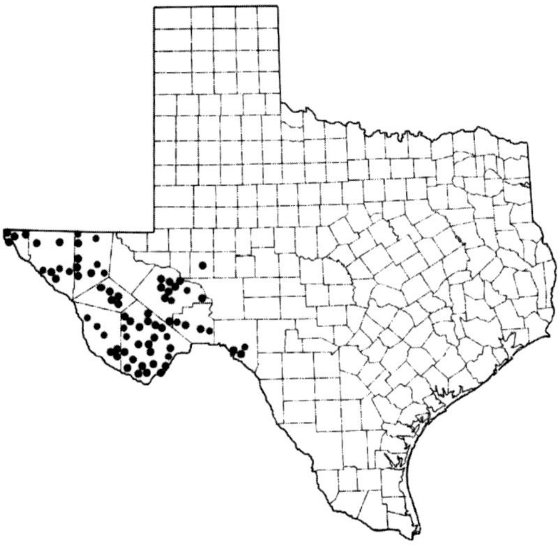

Polygala scoparioides

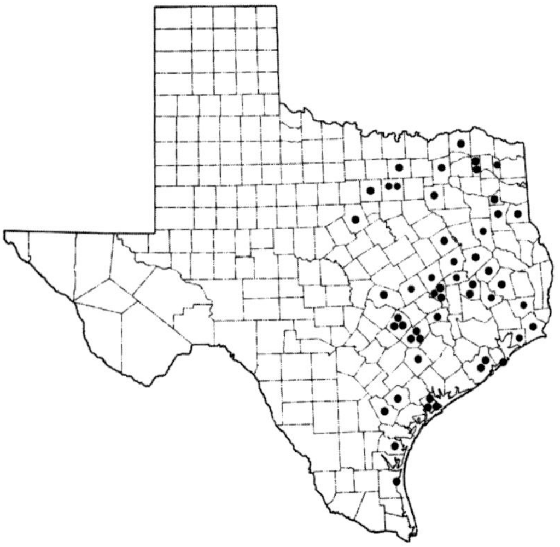

Polygala verticillata

Polygala watsonii

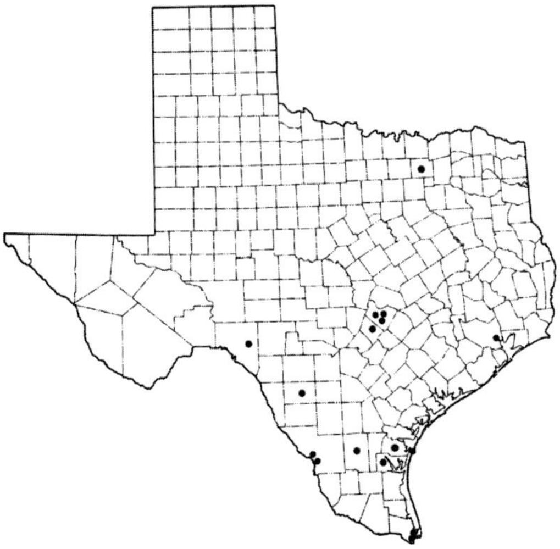

Antigonon leptopus

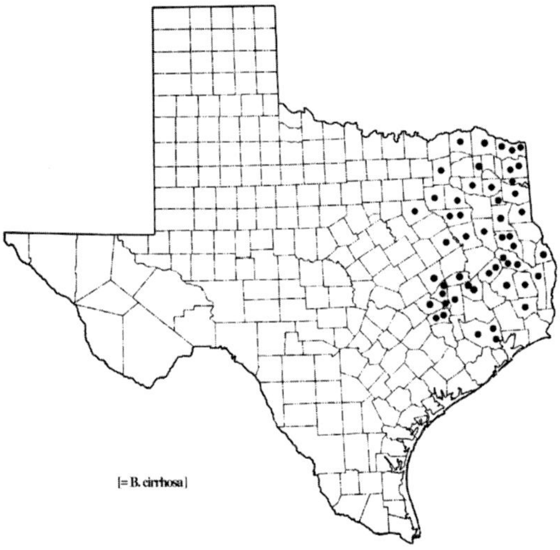
[= B. cirrhosa]

Brunnichia ovata

Emex spinosa

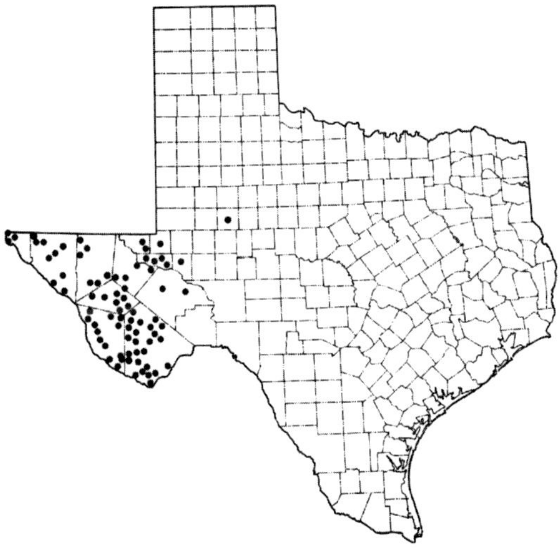

Eriogonum abertianum

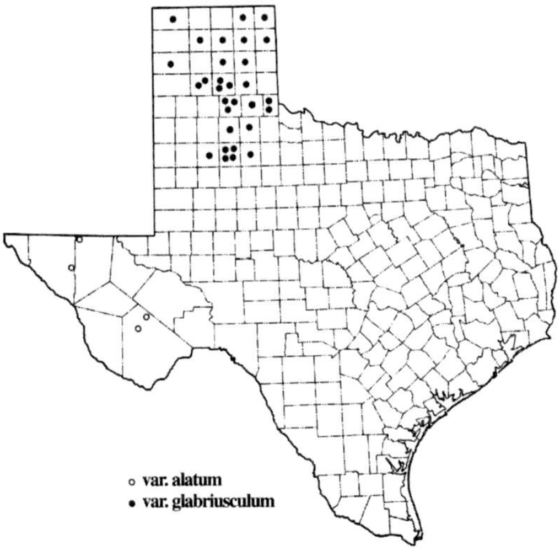
○ var. alatum
● var. glabriusculum

Eriogonum alatum

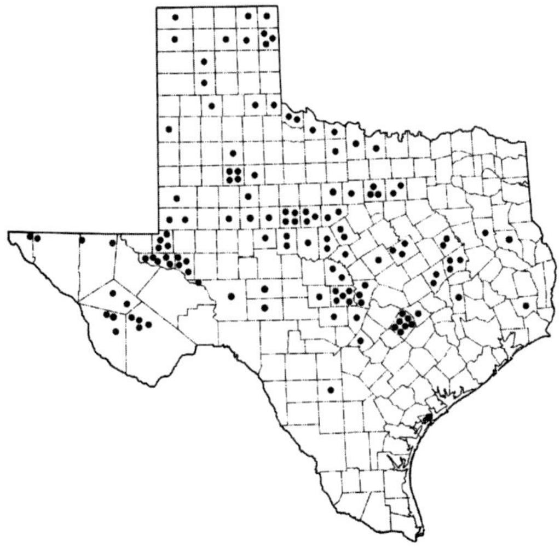

Eriogonum annuum

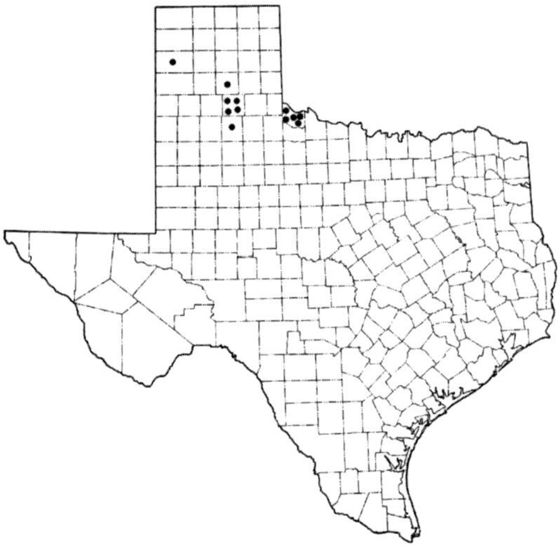

Eriogonum correllii

Eriogonum greggii

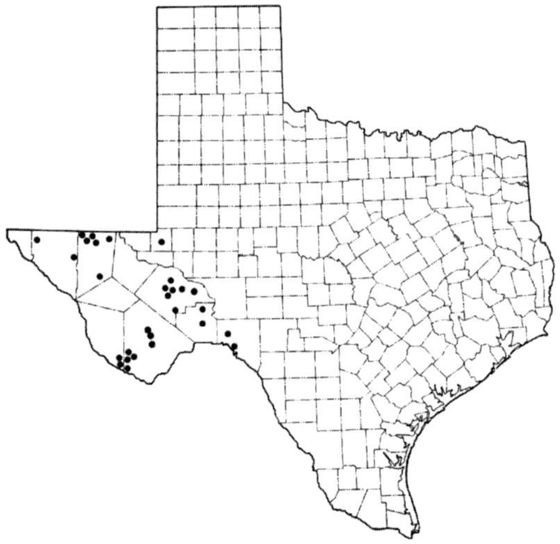

Eriogonum havardii

Eriogonum hemipterum

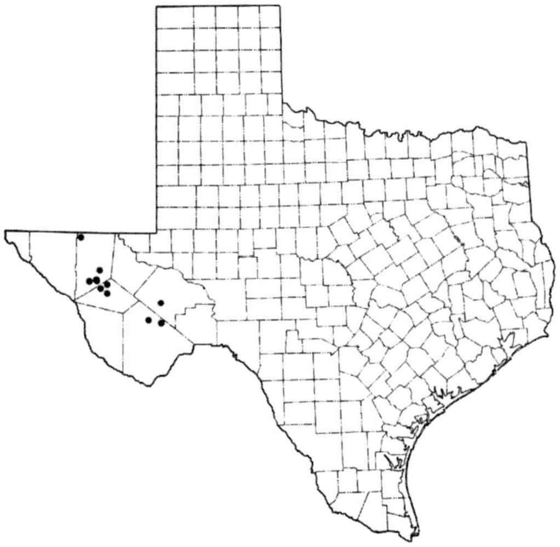

Eriogonum hieraciifolium

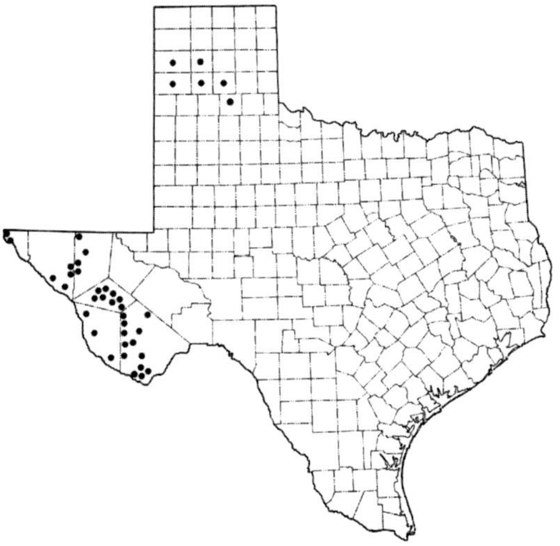

Eriogonum jamesii

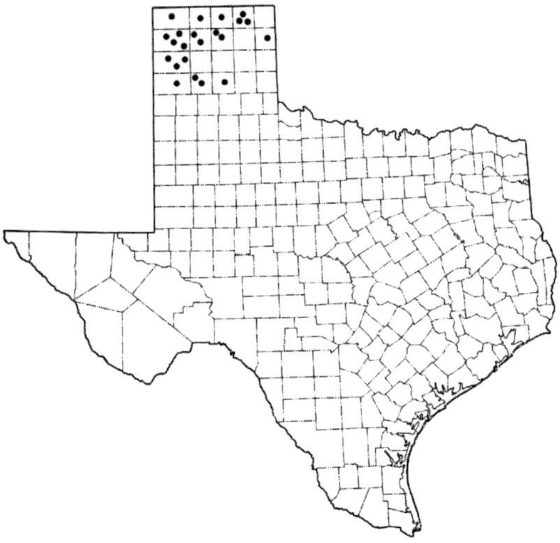

Eriogonum lachnogynum

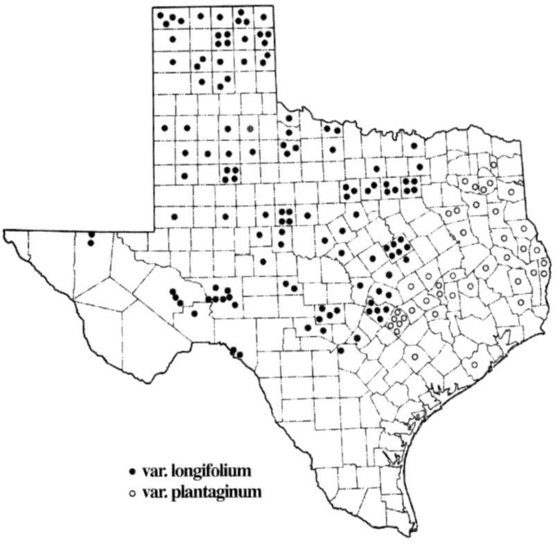
• var. longifolium
○ var. plantaginum

Eriogonum longifolium

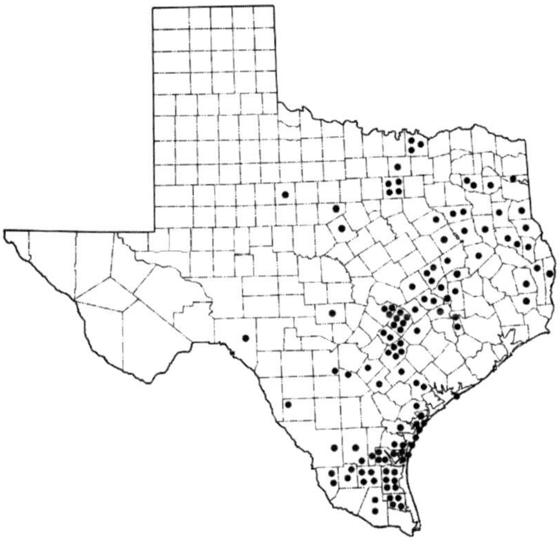

Eriogonum multiflorum

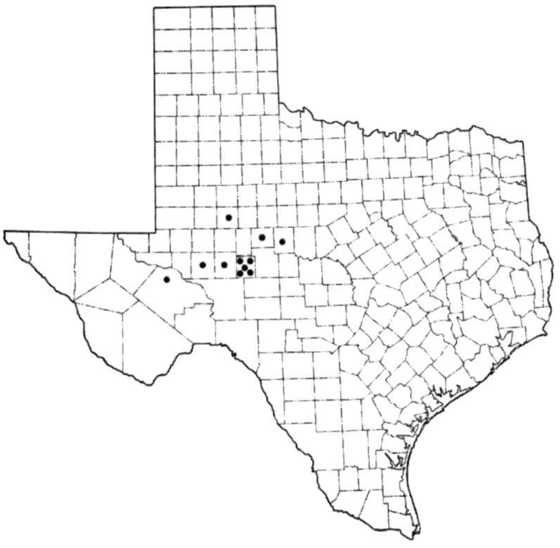

Eriogonum nealleyi

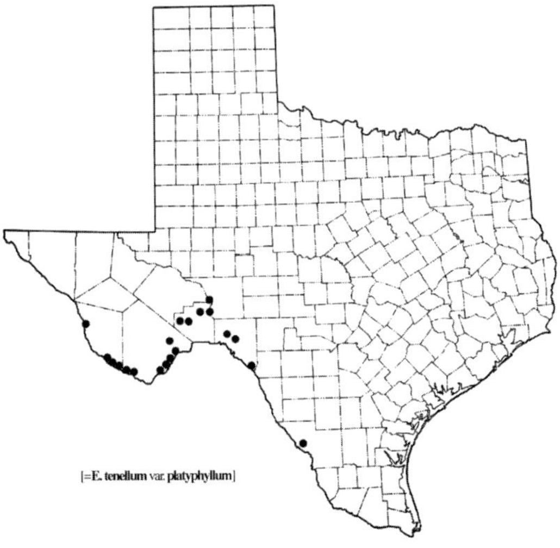
[=E. tenellum var. platyphyllum]

Eriogonum platyphyllum

Eriogonum polycladon

[=E. tenellum var. ramosissimum]

Eriogonum ramosissimum

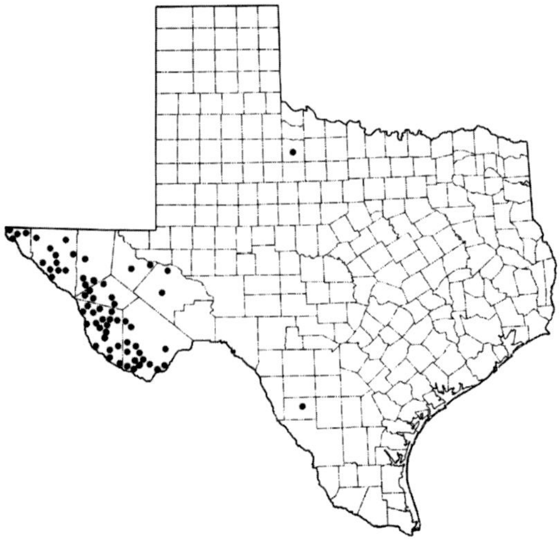

Eriogonum rotundifolium

Eriogonum suffruticosum

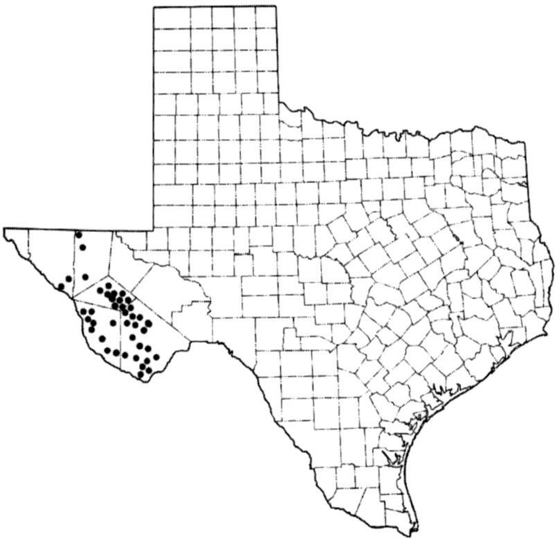

Eriogonum tenellum

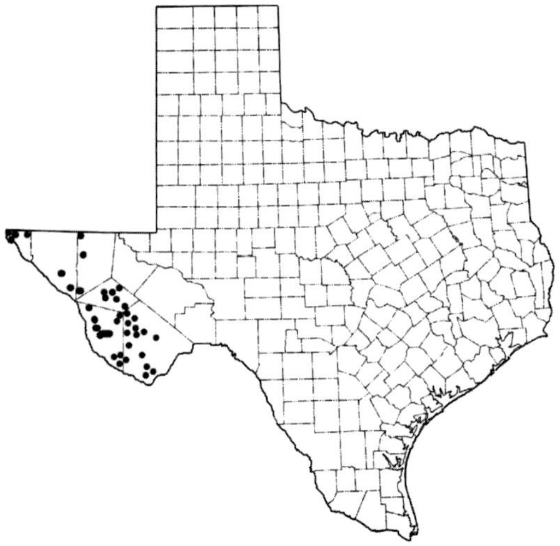

Eriogonum wrightii

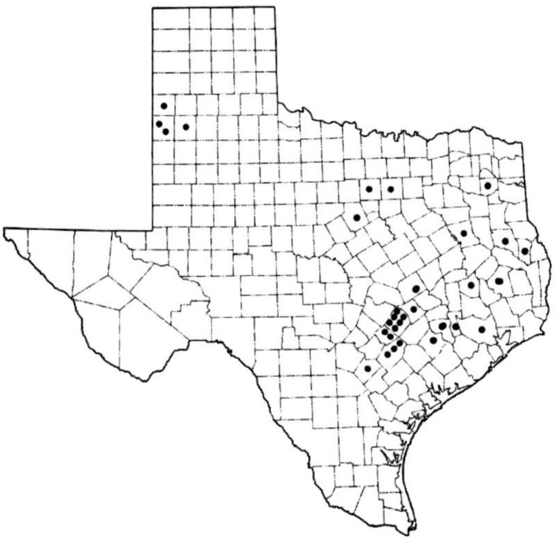

Polygonella americana

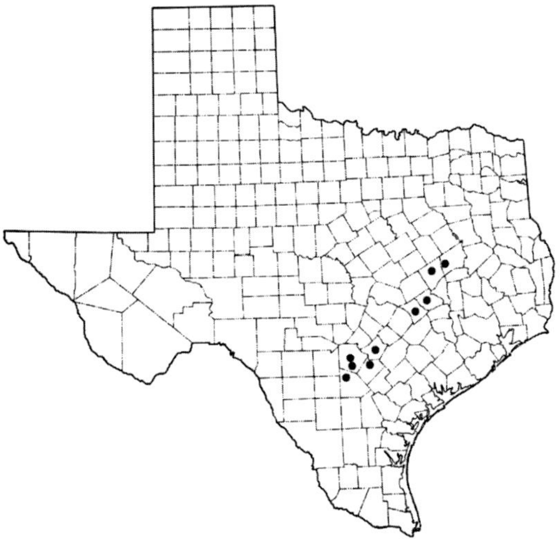
Polygonella parksii

Polygonella polygama

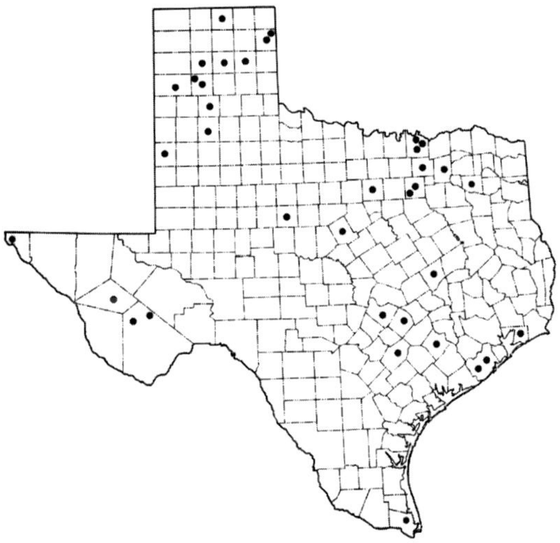
Polygonum amphibium

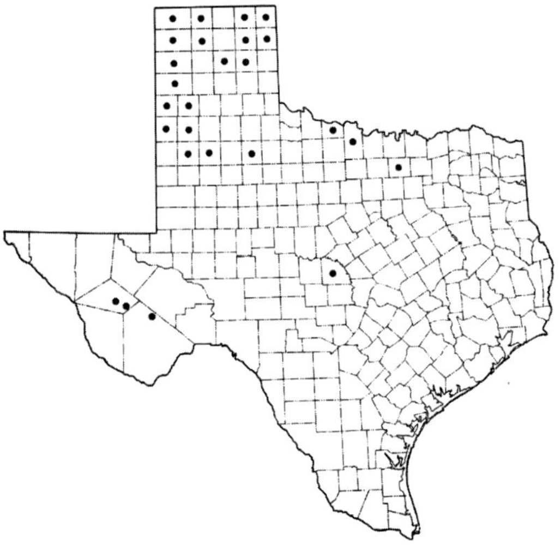
Polygonum arenastrum

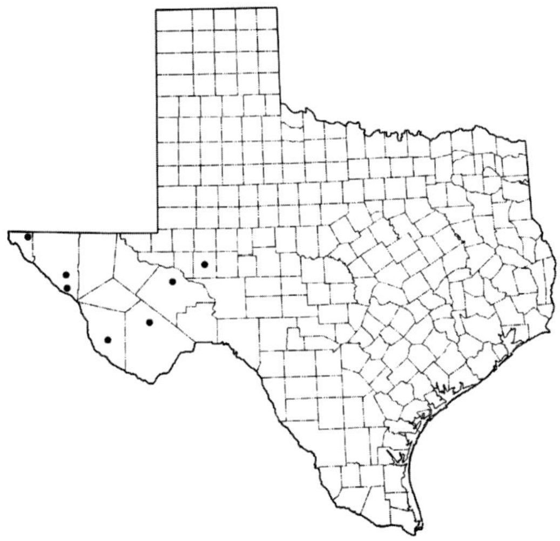
Polygonum argyrocoleon

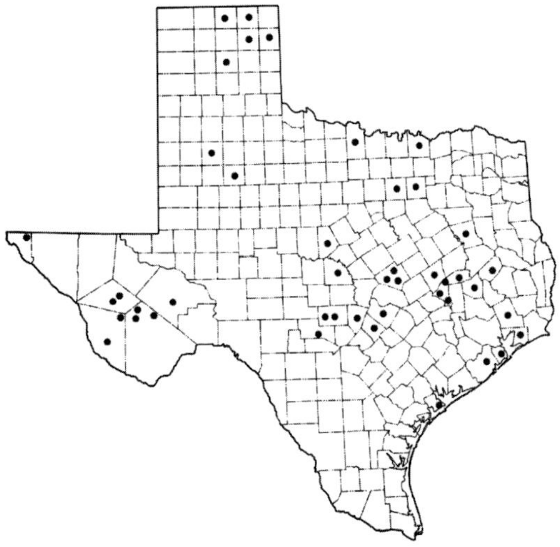
Polygonum aviculare

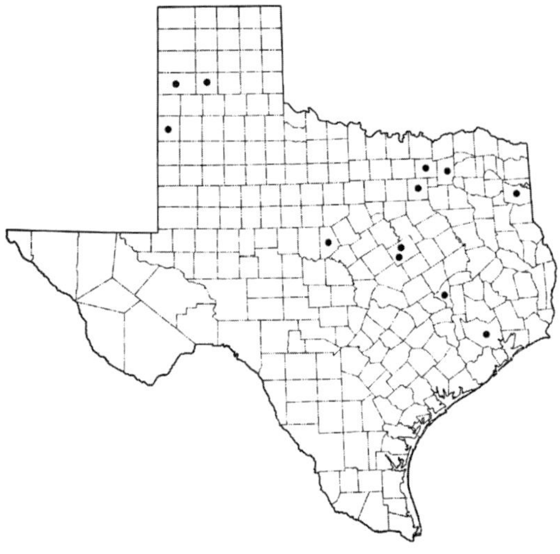

Polygonum buxiforme

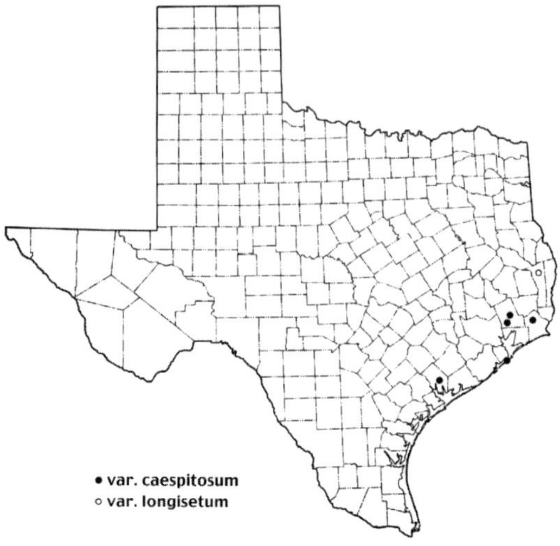
• var. caespitosum
○ var. longisetum

Polygonum caespitosum

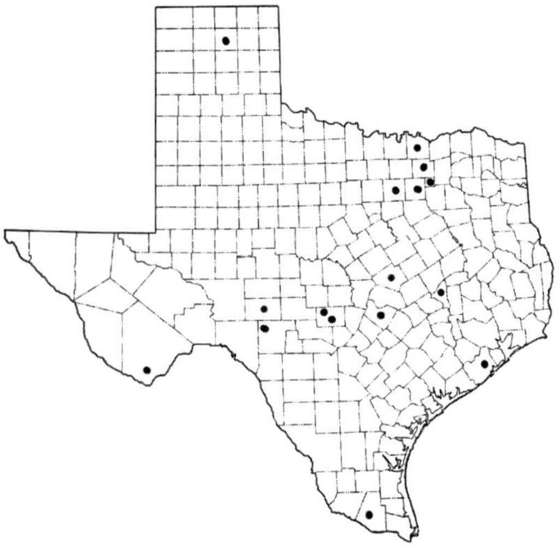

Polygonum convolvulus

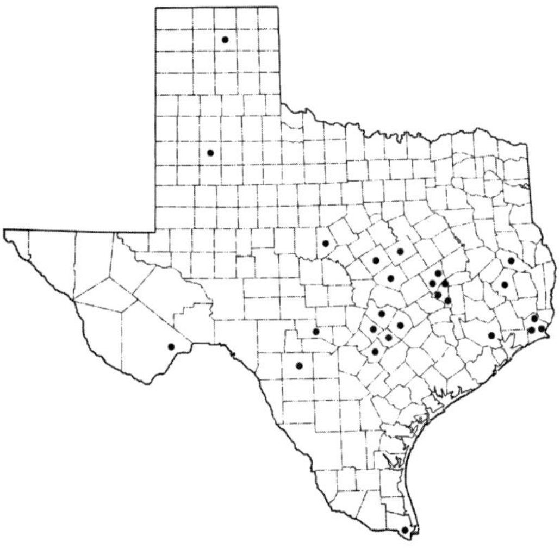

Polygonum densiflorum

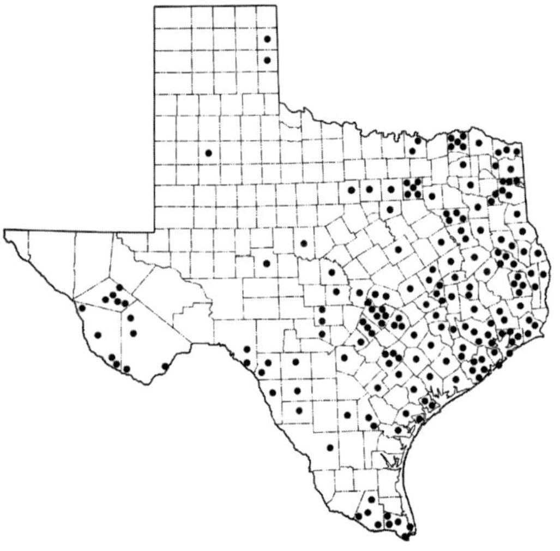

Polygonum hydropiperoides

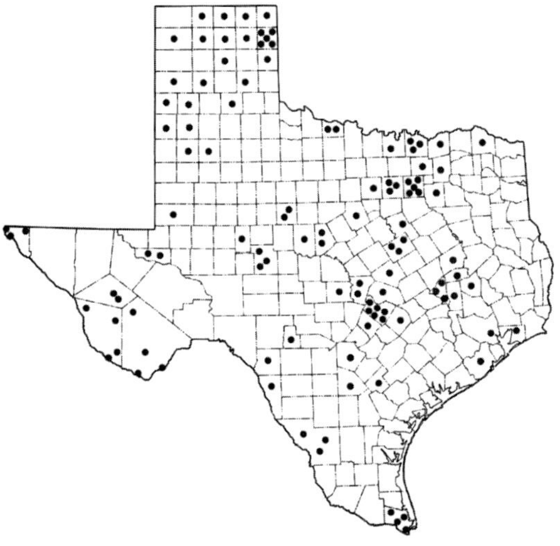

Polygonum lapanthifolium

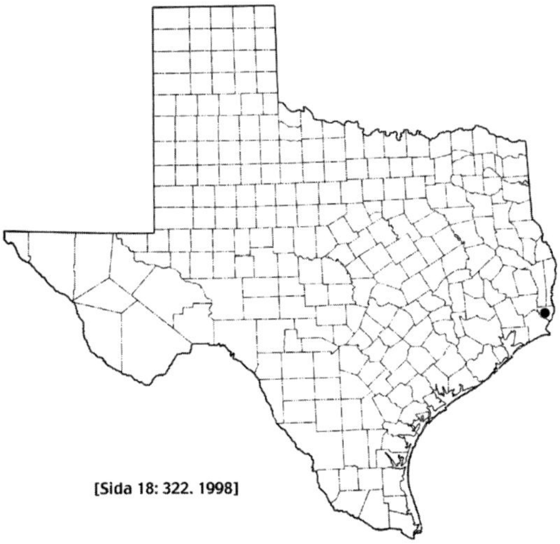
[Sida 18: 322. 1998]

Polygonum meisnerianum

Polygonum orientale

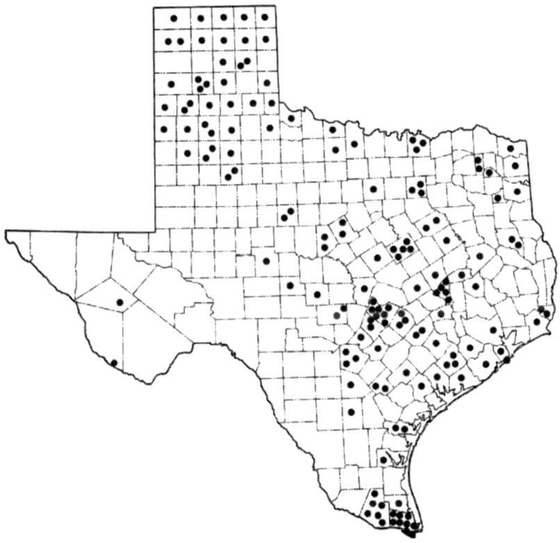

Polygonum pensylvanicum

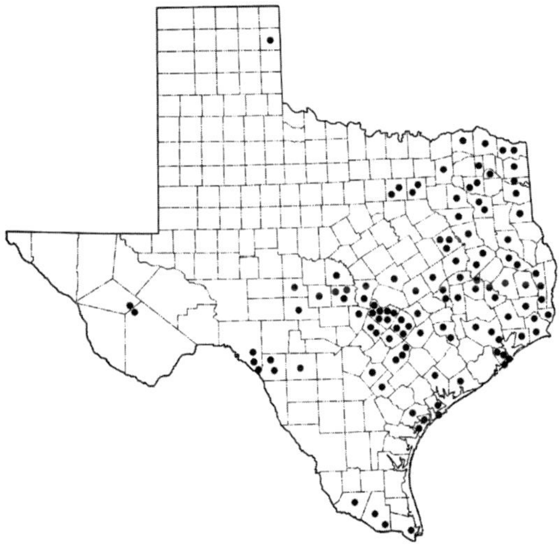

Polygonum punctatum

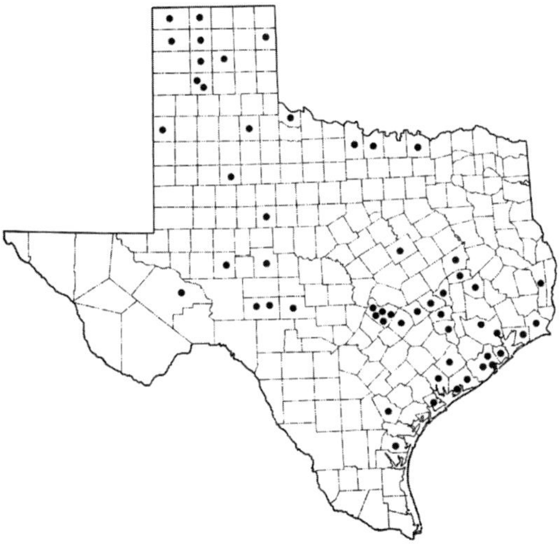

Polygonum ramosissimum

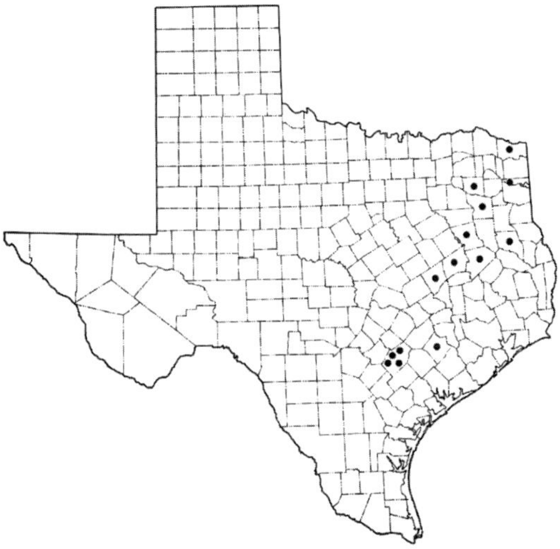

Polygonum sagittatum

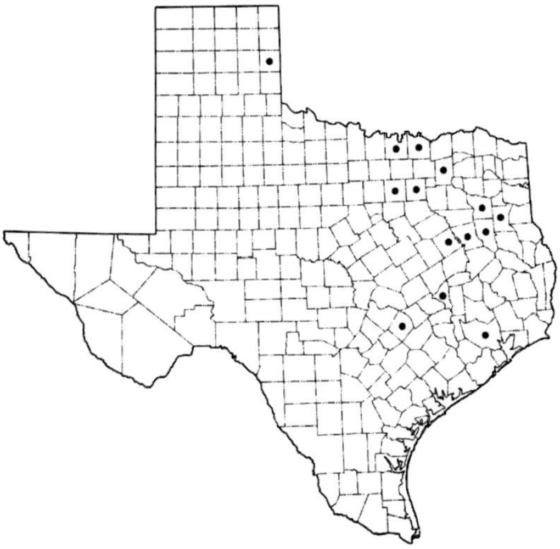

Polygonum scandens

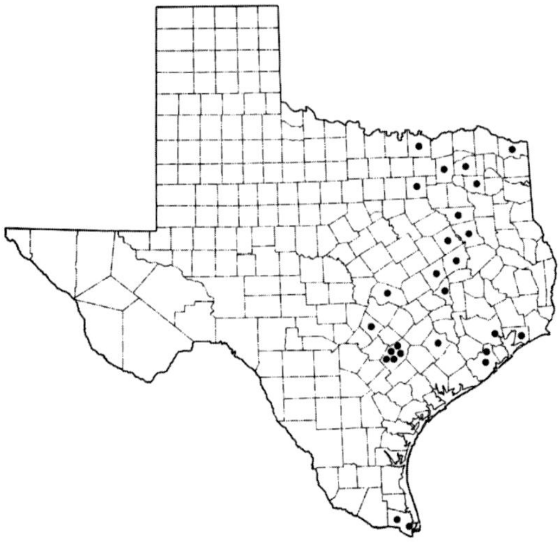

Polygonum setaceum

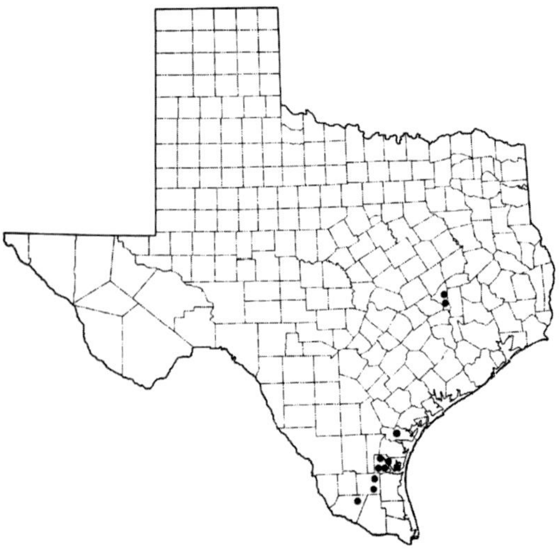

Polygonum striatulum

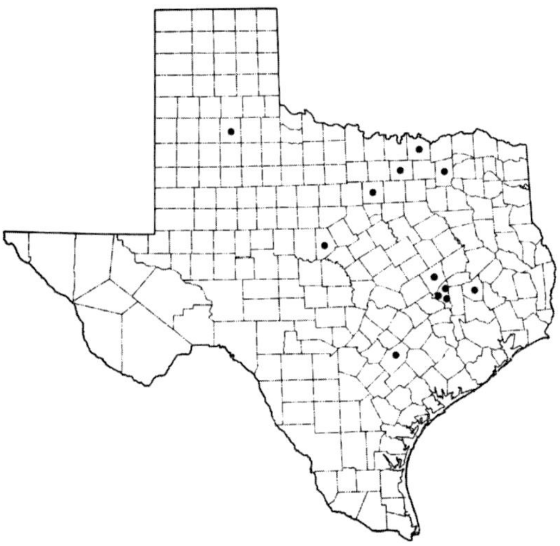

Polygonum tenue

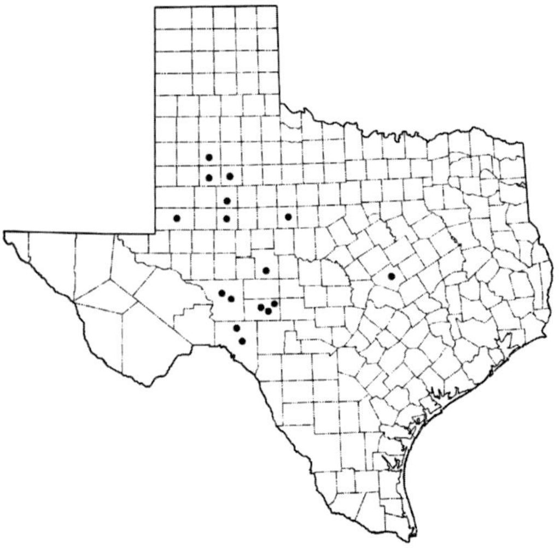

Polygonum texense

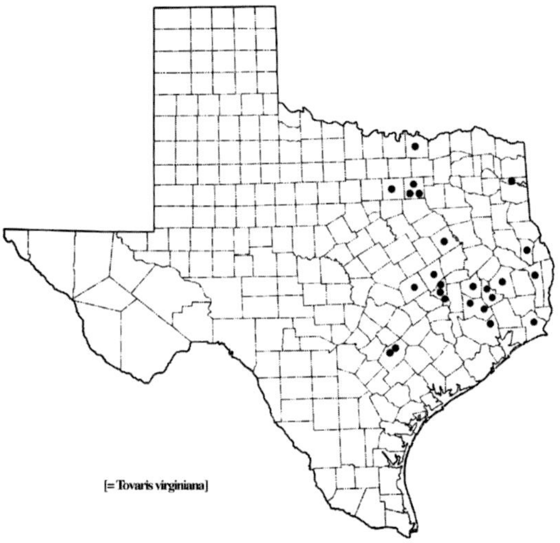
[= Tovaris virginiana]

Polygonum virginianum

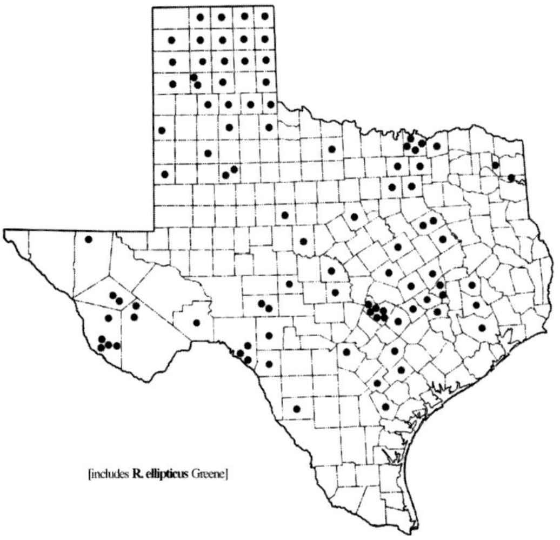
[includes R. ellipticus Greene]

Rumex altissimus

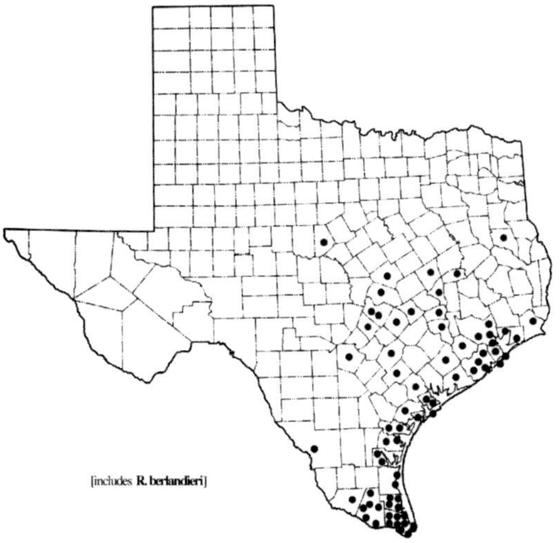
[includes R. berlandieri]

Rumex chrysocarpus

Rumex conglomeratus

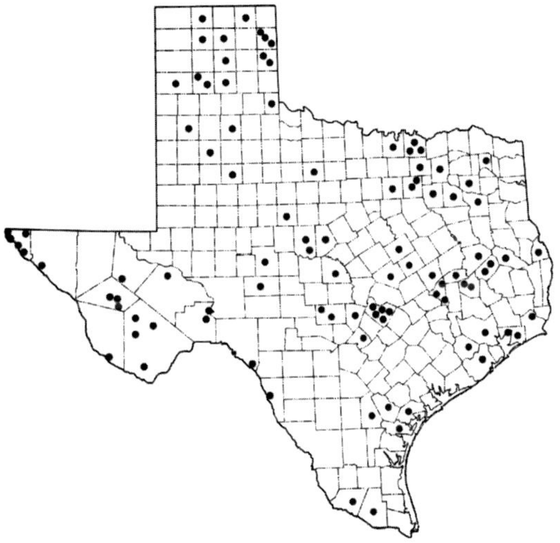

Rumex crispus

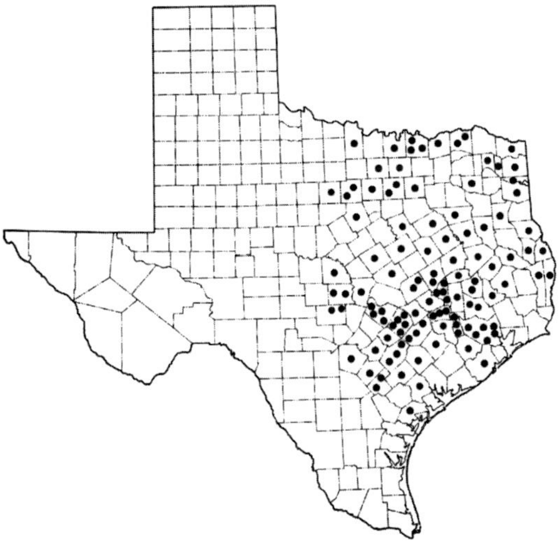

Rumex hastatulus

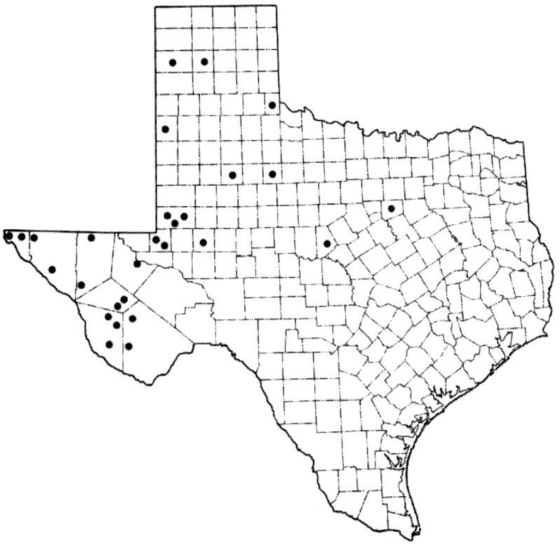

Rumex hymenosepalus

Rumex maritimus

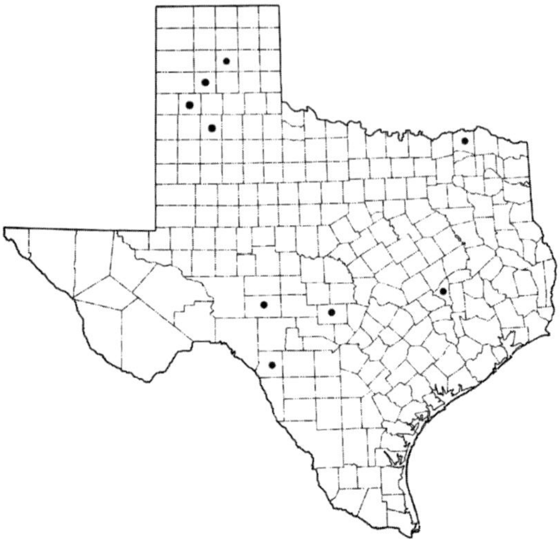

Rumex obtusifolius

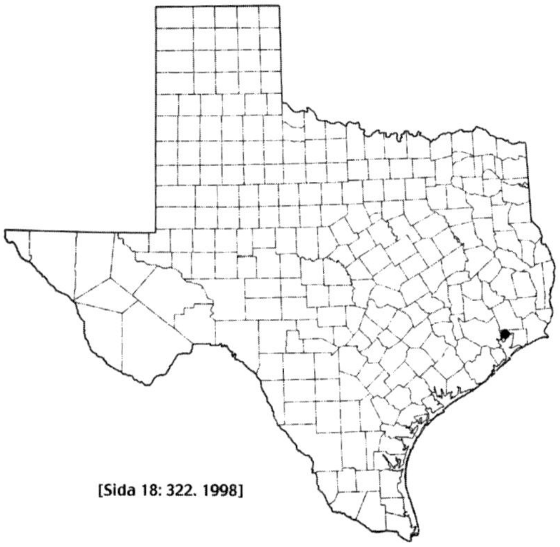
[Sida 18: 322. 1998]

Rumex paraguayensis

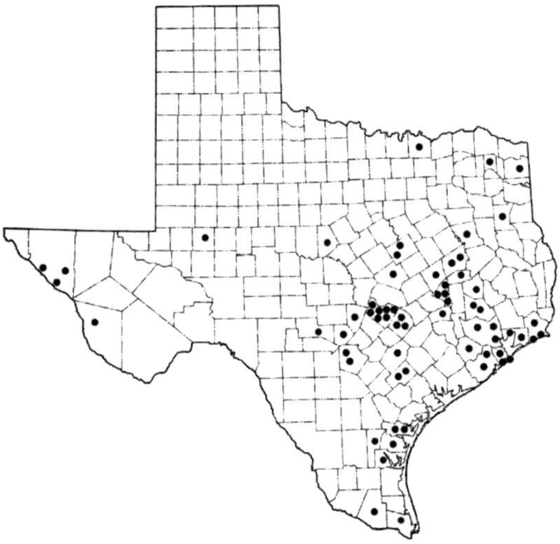

Rumex pulcher

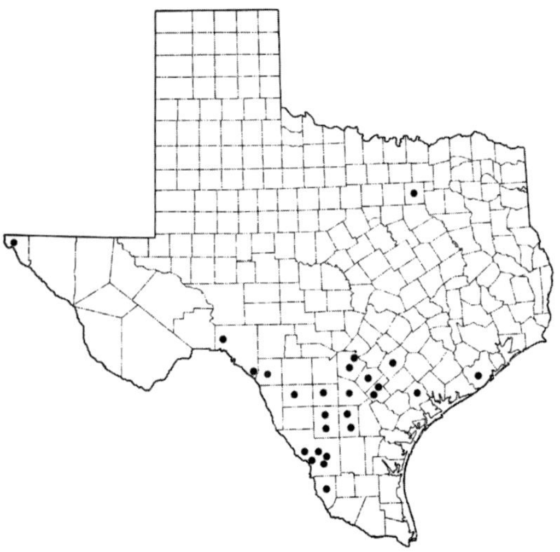

Rumex spiralis

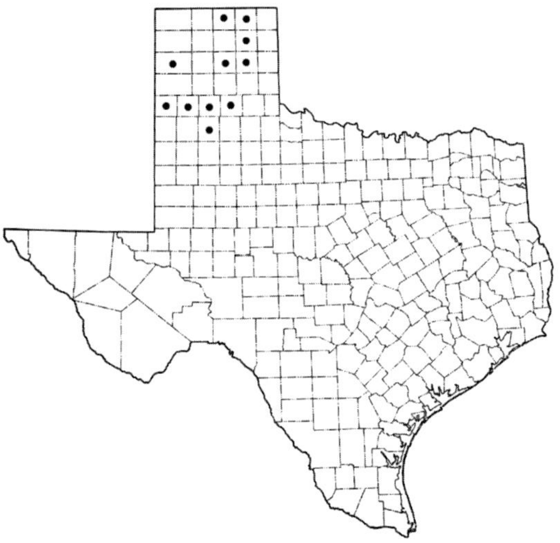

Rumex stenophyllus

Rumex venosus

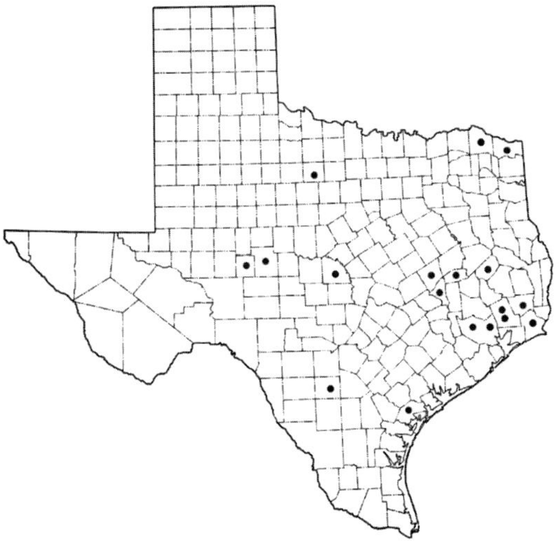

Rumex verticillatus

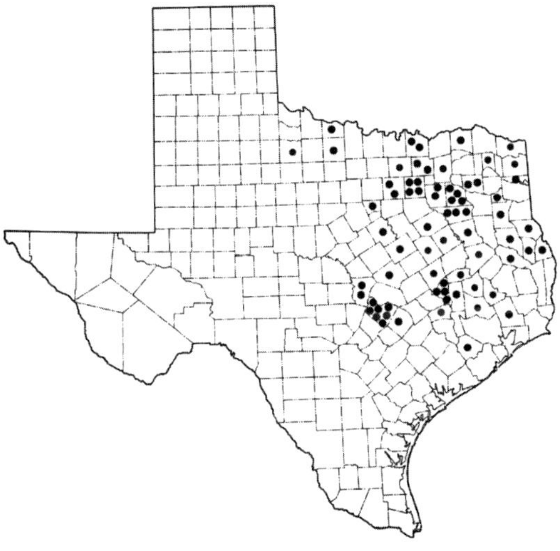

Claytonia virginica

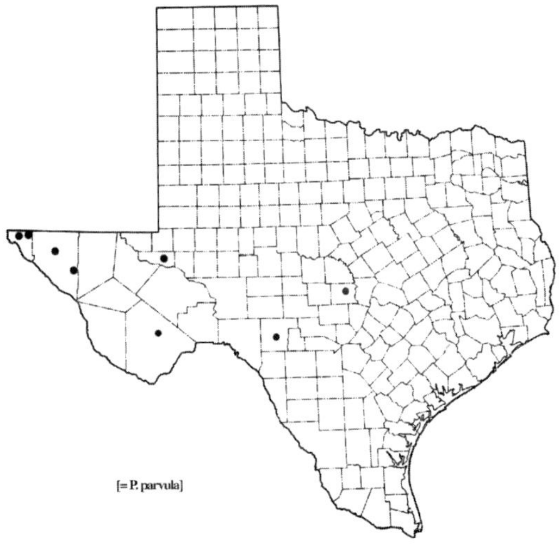
[= P. parvula]

Portulaca halimoides

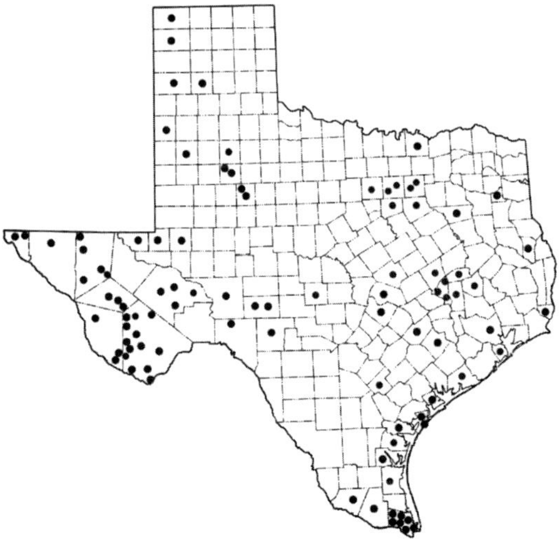

Portulaca oleracea

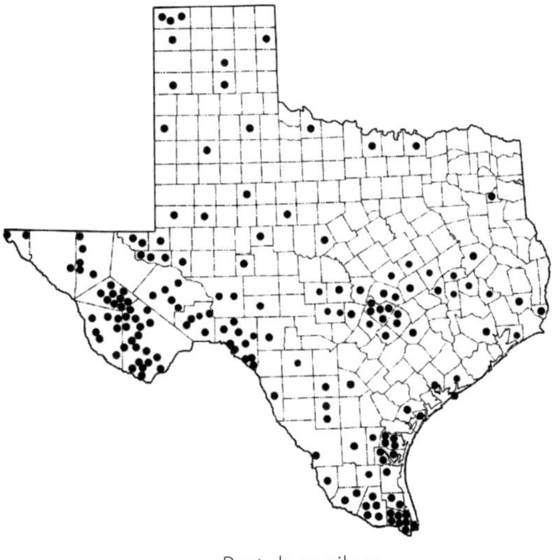

Portulaca pilosa

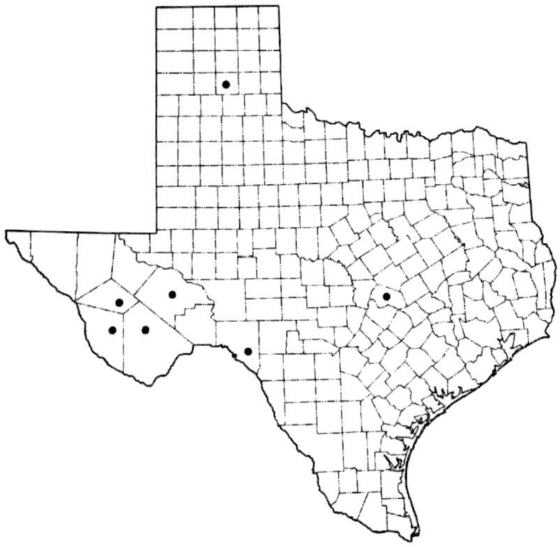

Portulaca retusa

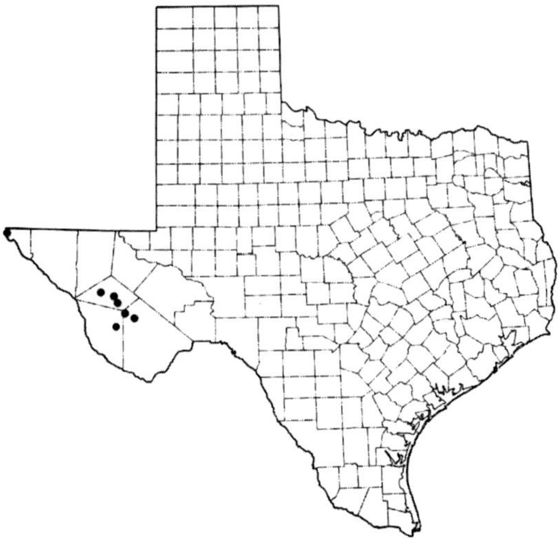

Portulaca suffrutescens

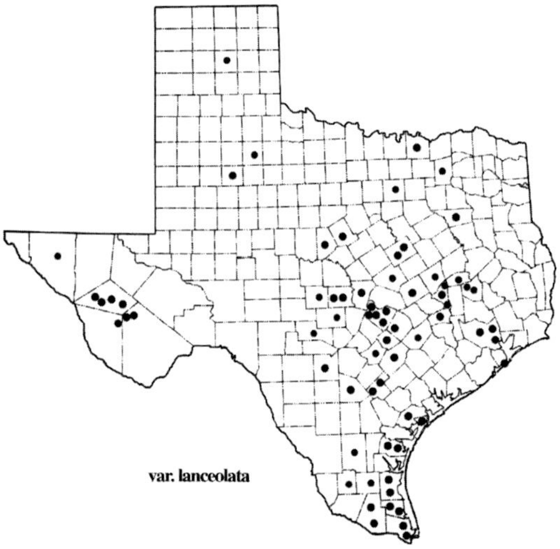
var. lanceolata

Portulaca umbraticola

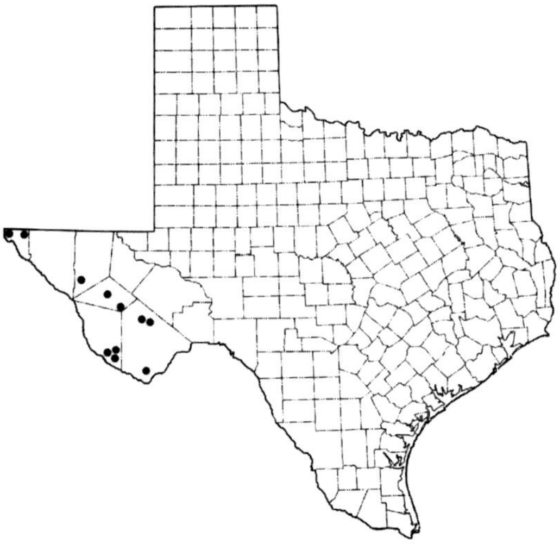

Talinopsis frutescens

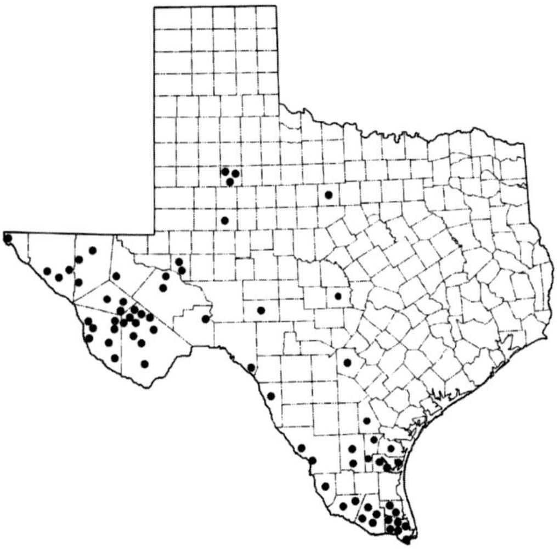

Talinum angustissimum

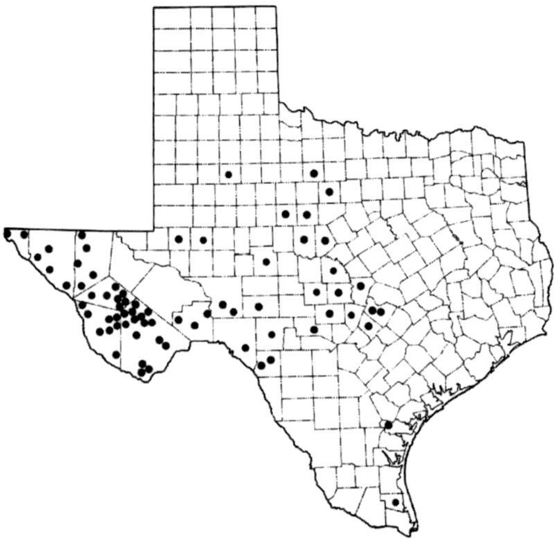

Talinum aurantiacum

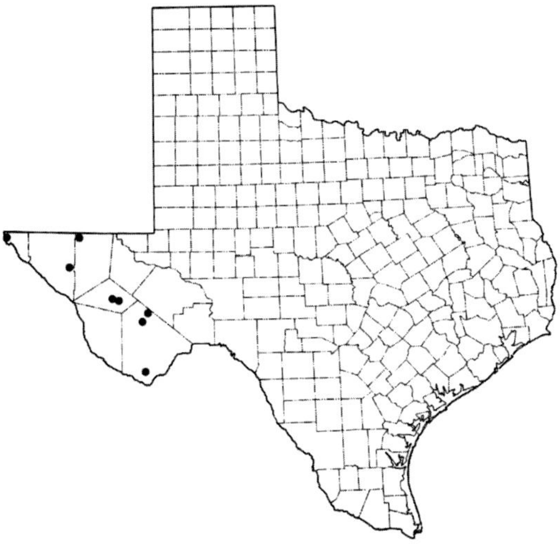

Talinum brevicaule

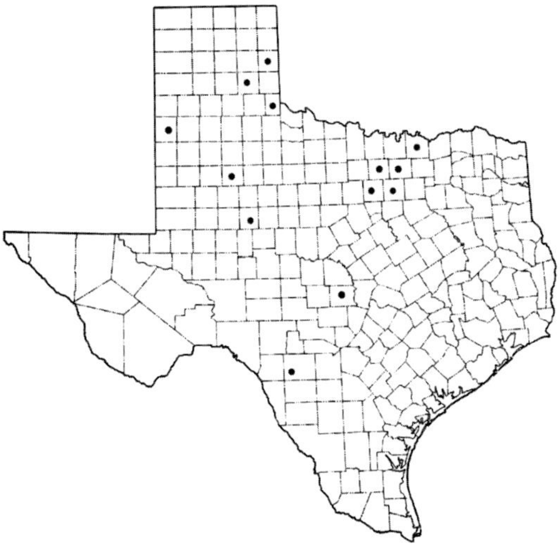

Talinum calycinum

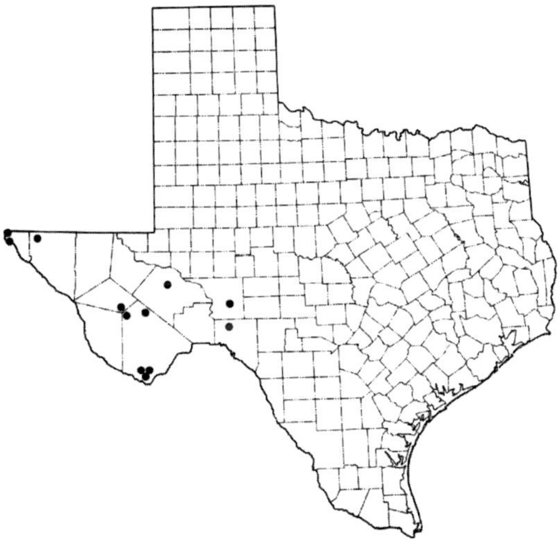

Talinum longipes

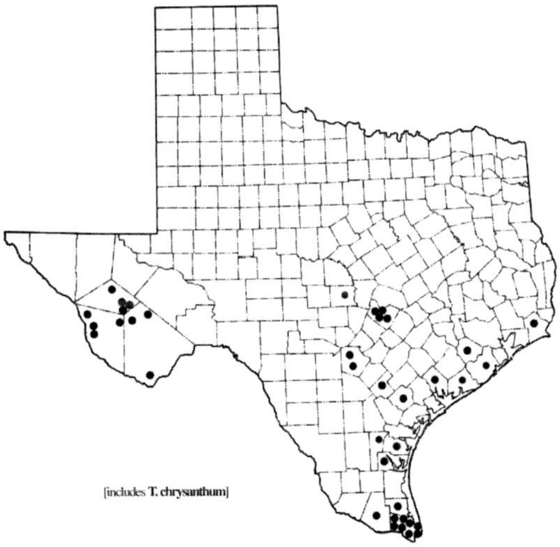
[includes T. chrysanthum]

Talinum paniculatum

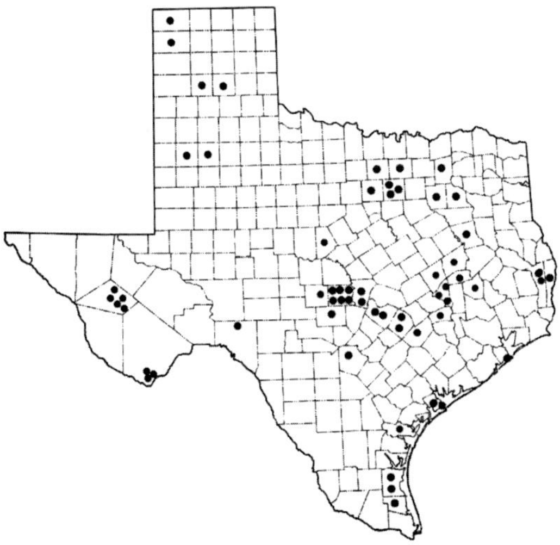

Talinum parviflorum

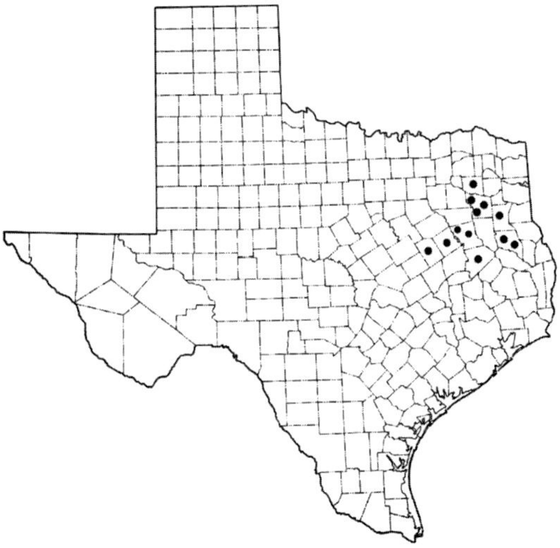

Talinum rugospermum

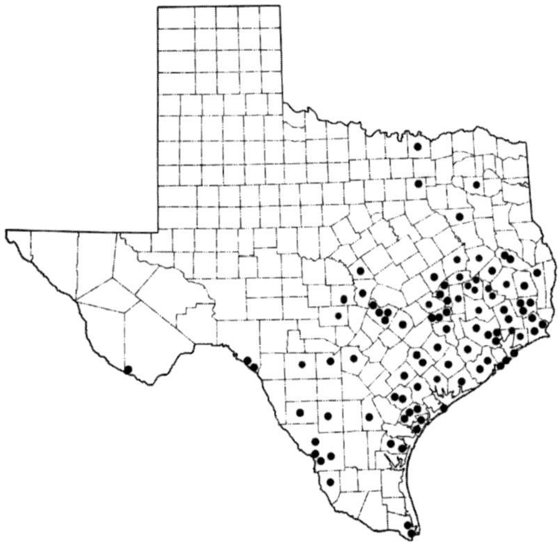

Anagallis arvensis

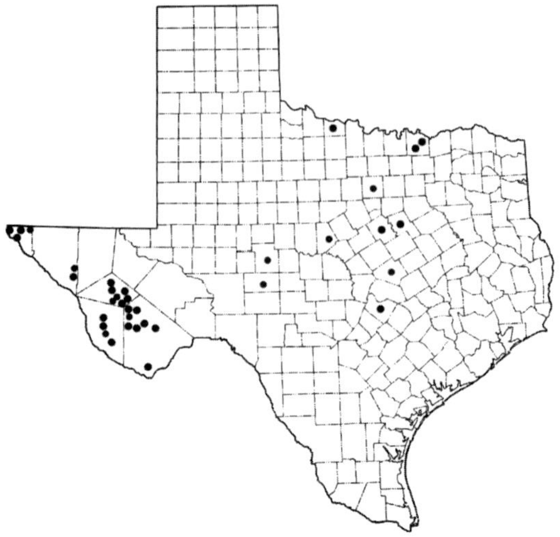

Androsace occidentalis

Androsace septentrionalis

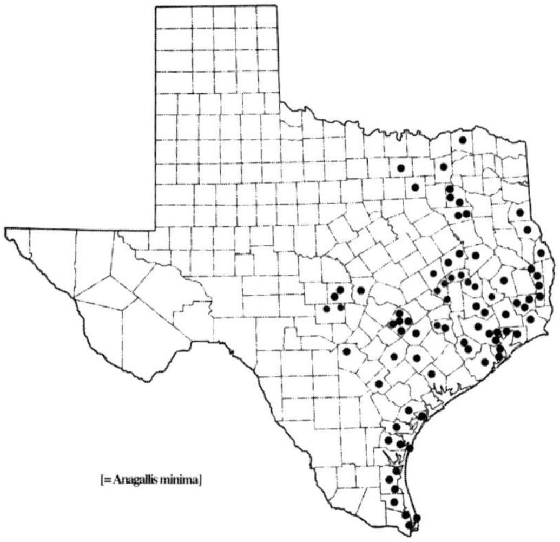
[= Anagallis minima]

Centunculus minimus

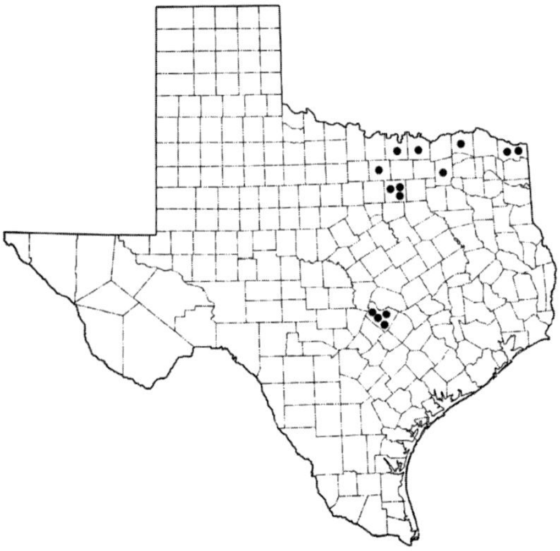

Dodecatheon meadia

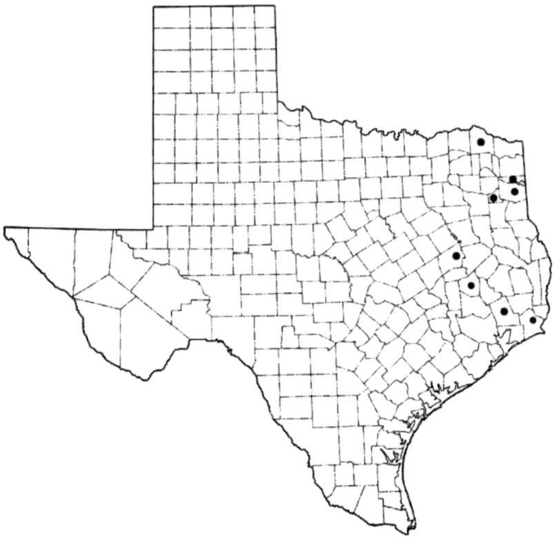

Hottonia inflata

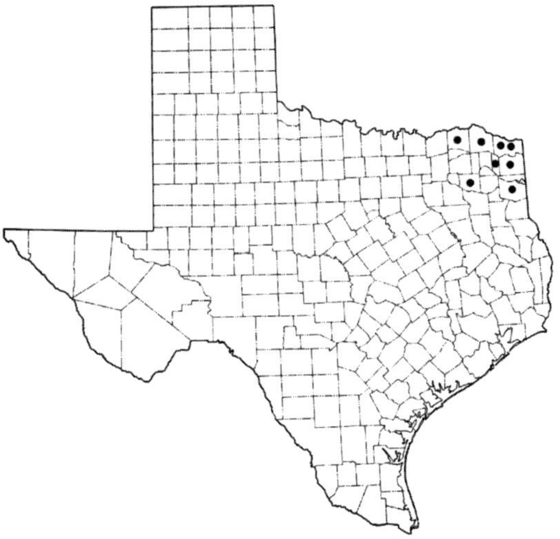

Lysimachia lanceolata

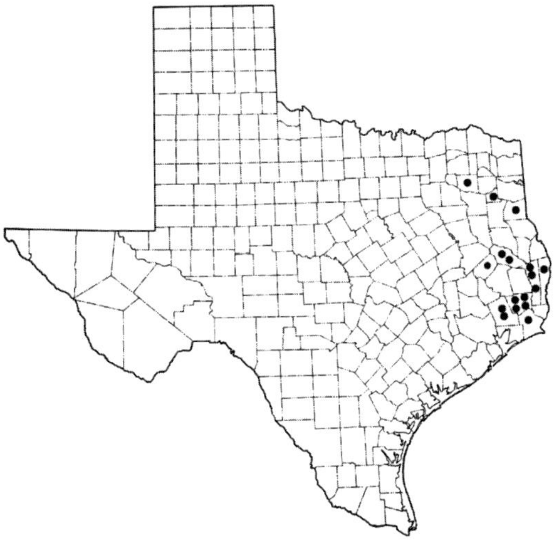

Lysimachia radicans

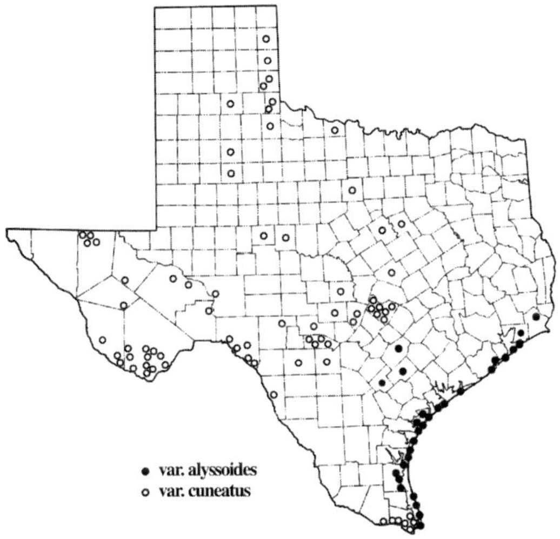
• var. alyssoides
○ var. cuneatus

Samolus ebracteatus

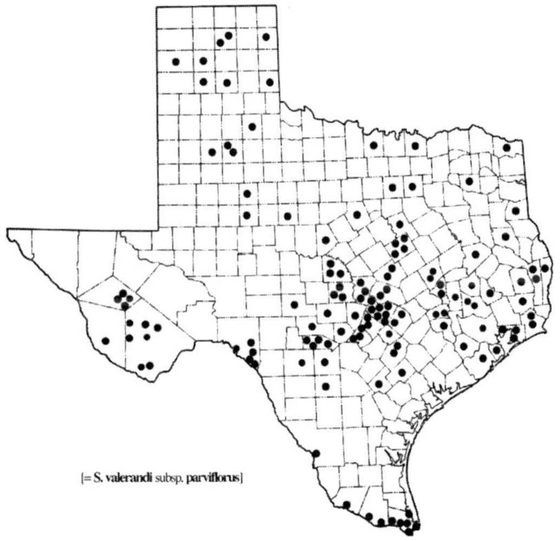
[= S. valerandi subsp. parviflorus]

Samolus parviflorus

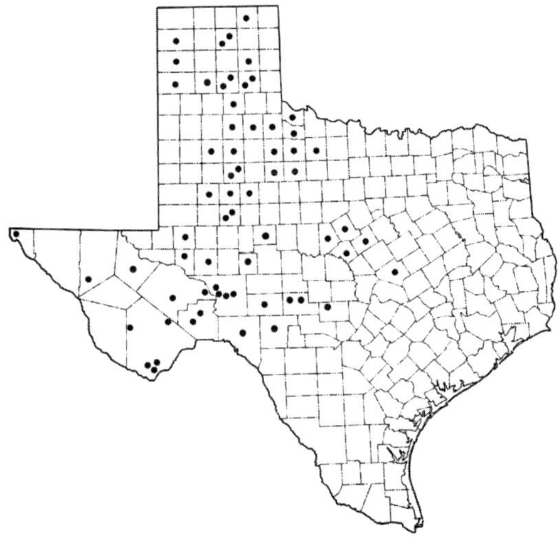

Pilostyles thurberi

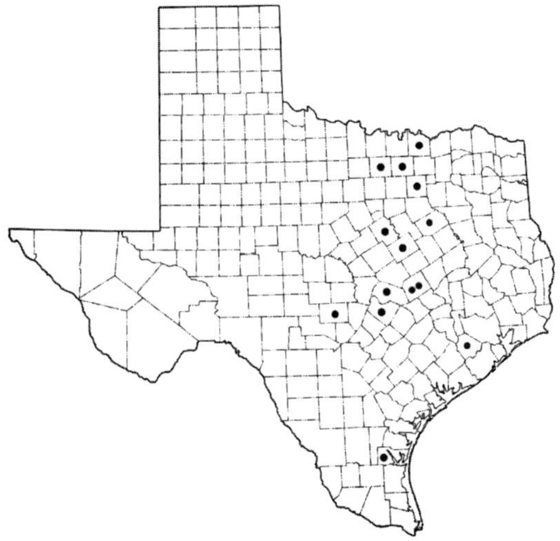

Adonis annua

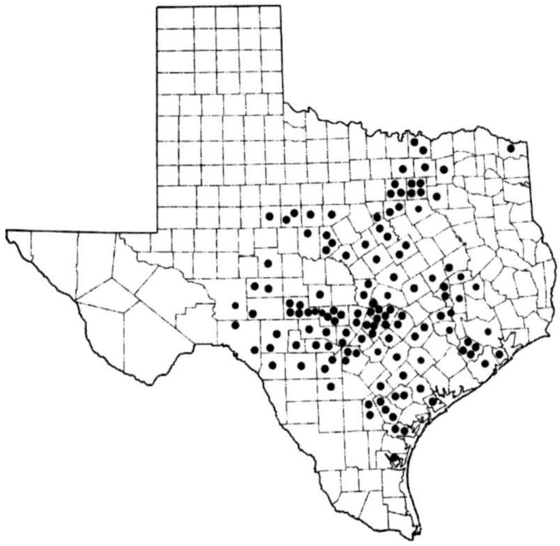

Anemone berlandieri

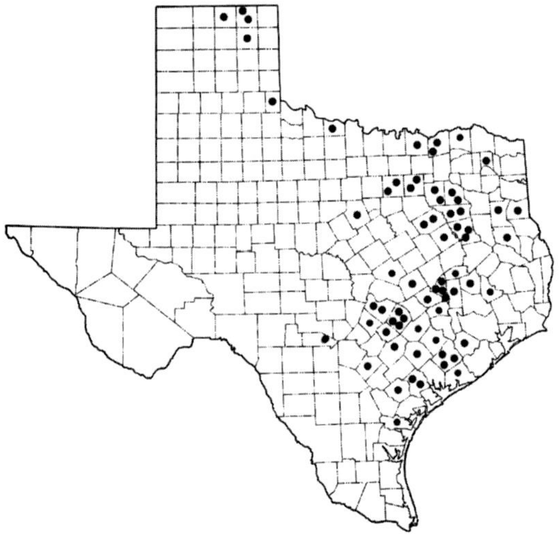

Anemone caroliniana

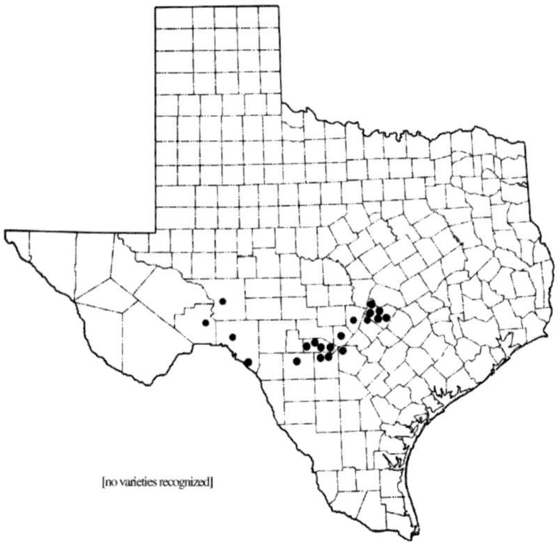
[no varieties recognized]

Anemone edwardsiana

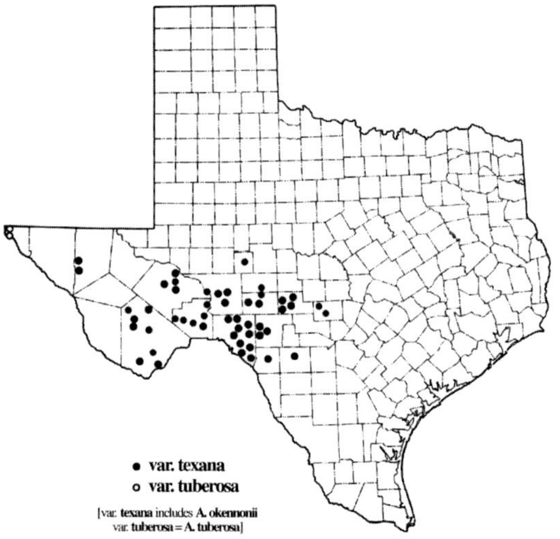
● var. texana
○ var. tuberosa
[var. texana includes A. okennonii
var. tuberosa = A. tuberosa]

Anemone tuberosa

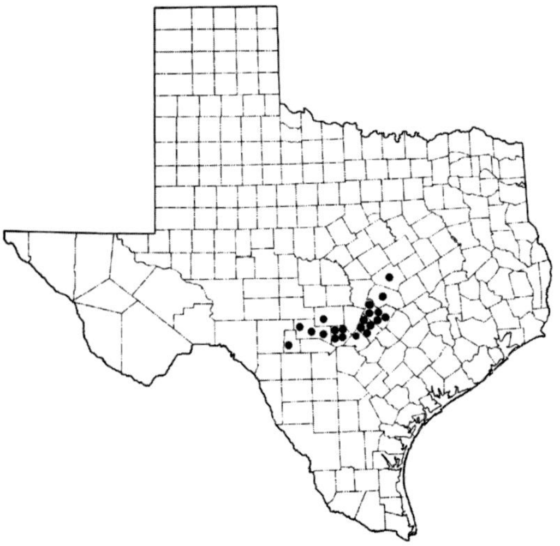

Aquilegia canadensis

[= A. chrysantha var. chaplinei]

Aquilegia chaplinei

Aquilegia chrysantha

[= A. chrysantha var. hinckleyana]

Aquilegia hinckleyana

Aquilegia longissima

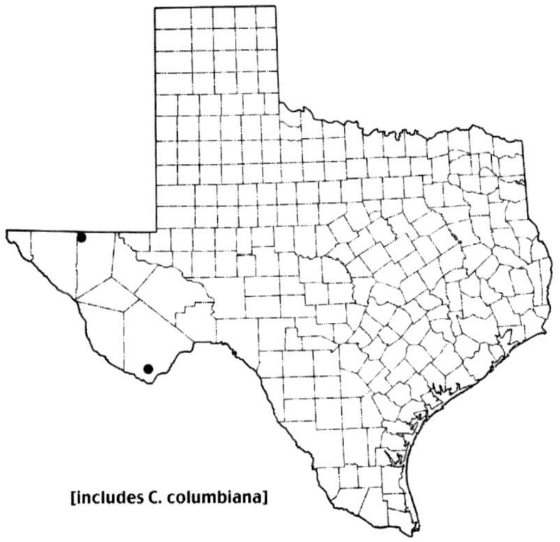
[includes C. columbiana]

Clematis alpina

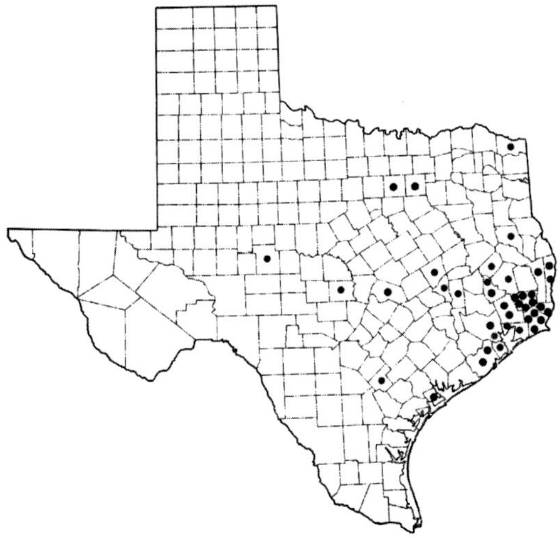

Clematis crispa

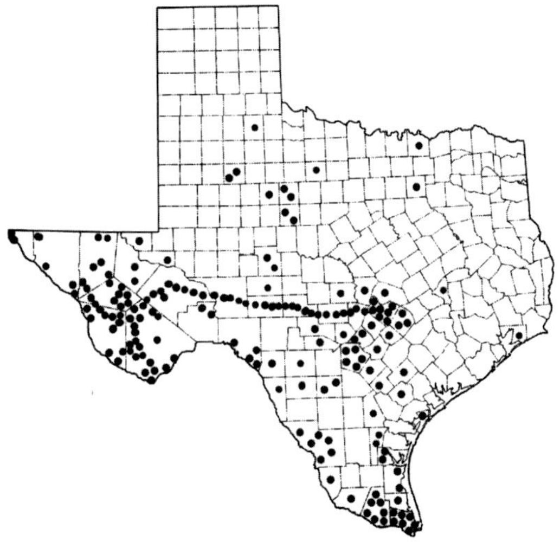

Clematis drummondii

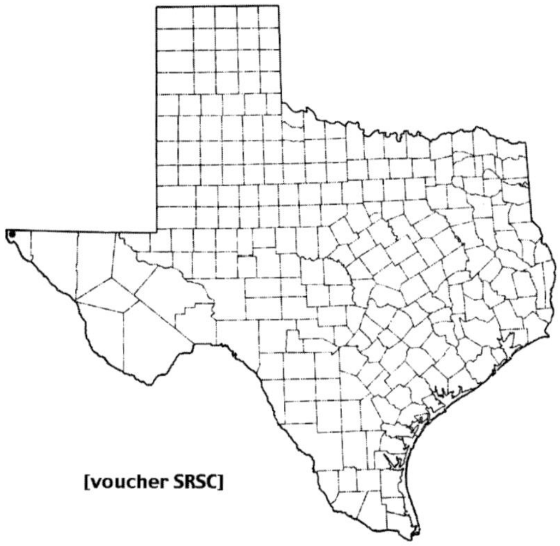
[voucher SRSC]

Clematis ligusticifolia

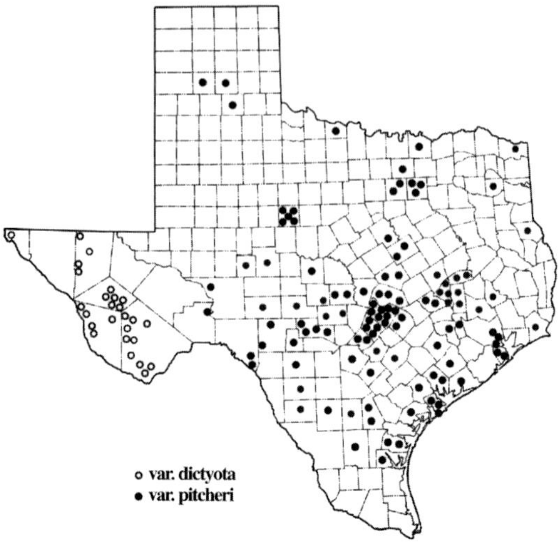
○ var. dictyota
● var. pitcheri

Clematis pitcheri

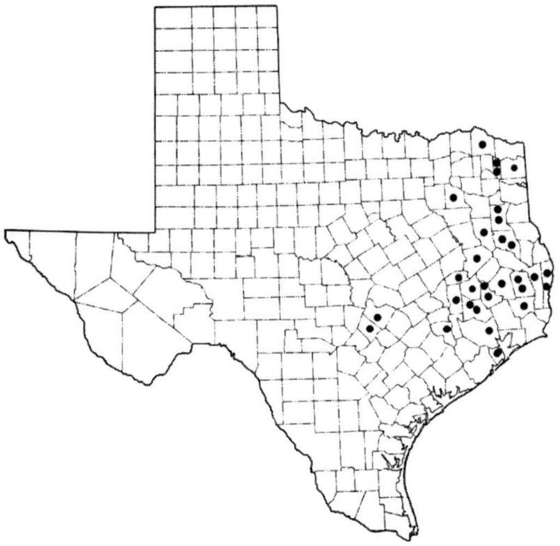

Clematis reticulata

Clematis ternifolia

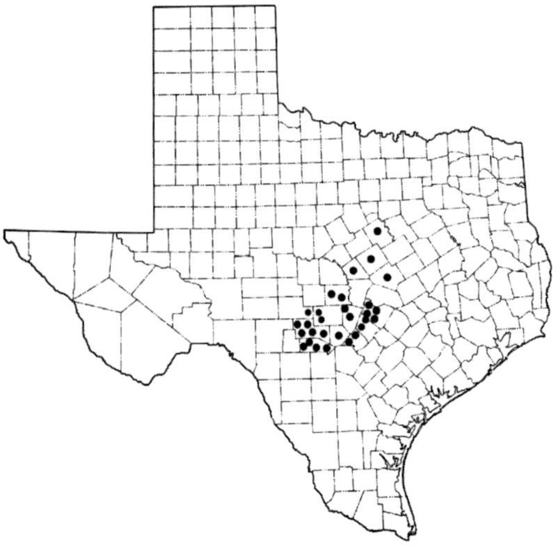

Clematis texensis

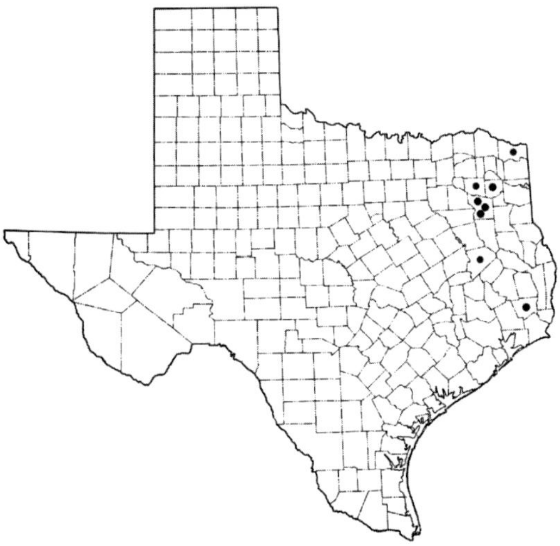

Clematis versicolor

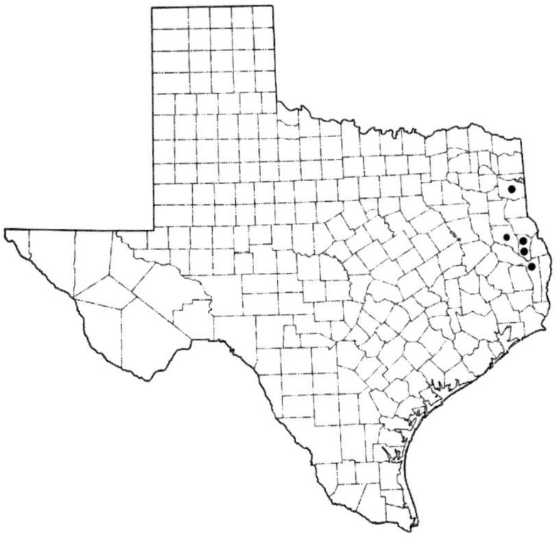

Clematis virginiana

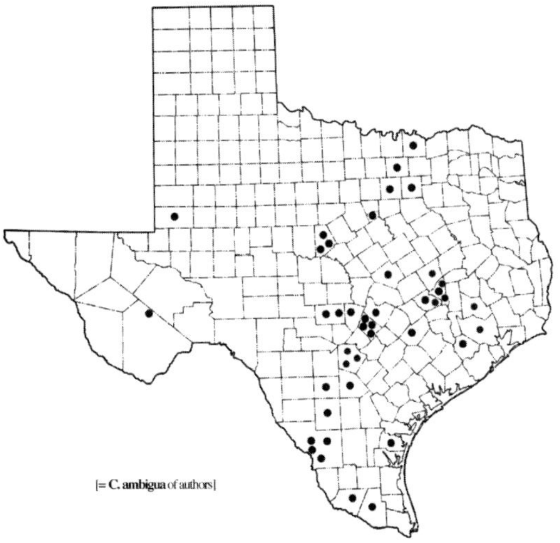
[= C. ambigua of authors]

Consolida ajacis

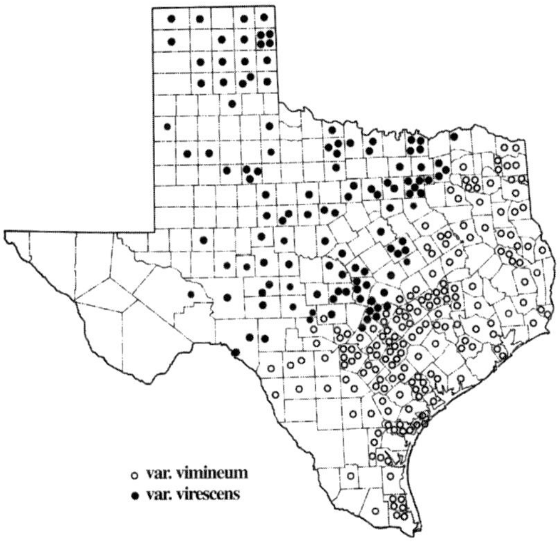
o var. vimineum
• var. virescens

Delphinium carolinianum

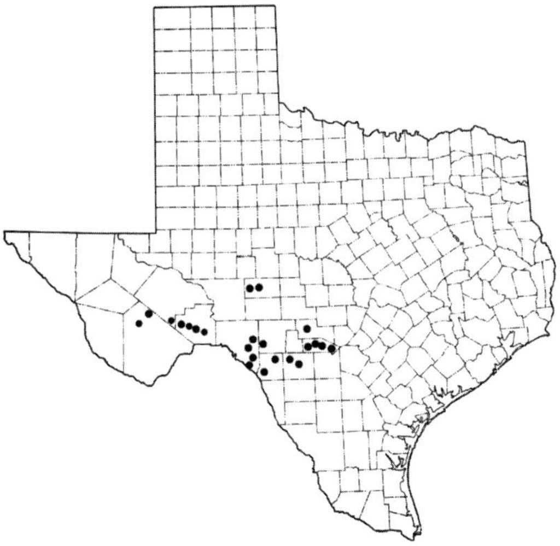

Delphinium madrense

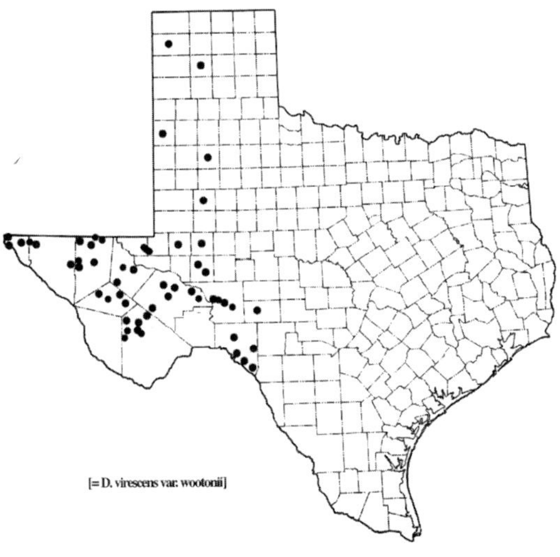
[= D. virescens var. wootonii]

Delphinium wootonii

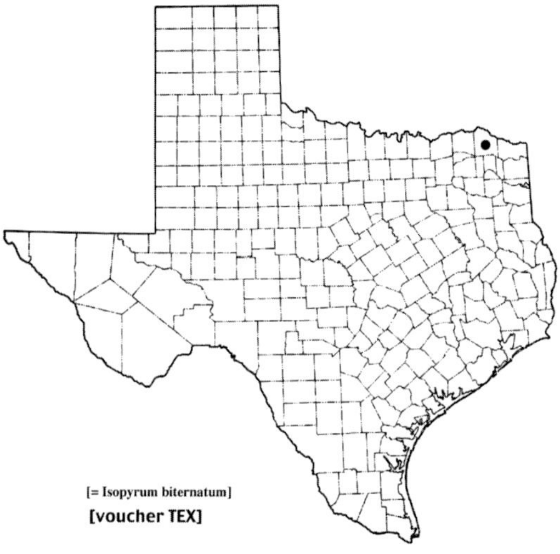
[= Isopyrum biternatum]
[voucher TEX]

Enemion biternatum

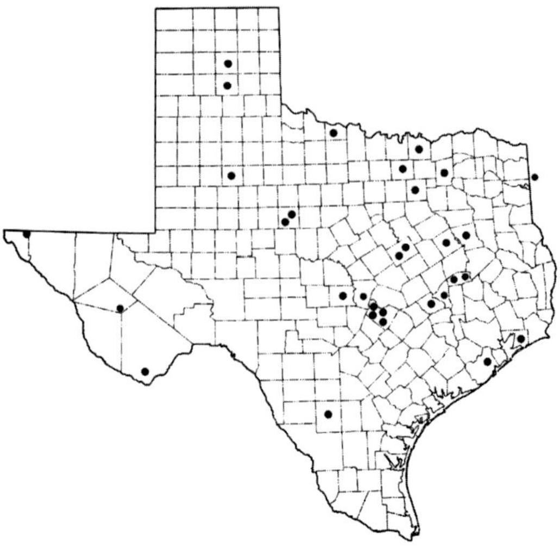

Myosurus minimus

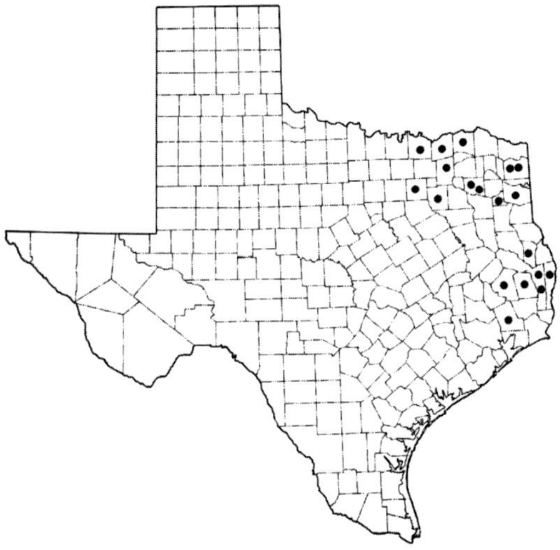

Ranunculus abortivus

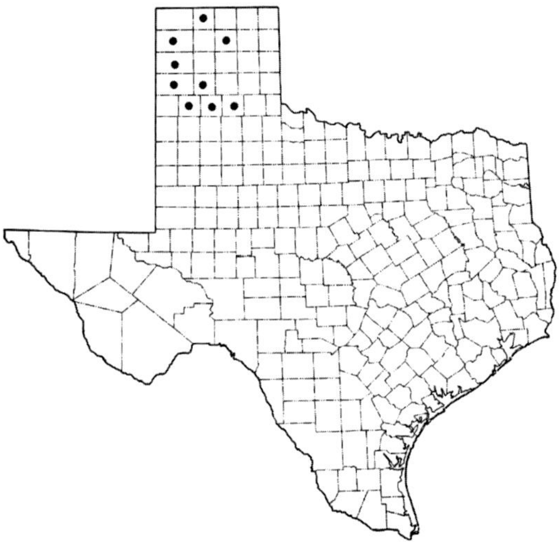

Ranunculus cymbalaria

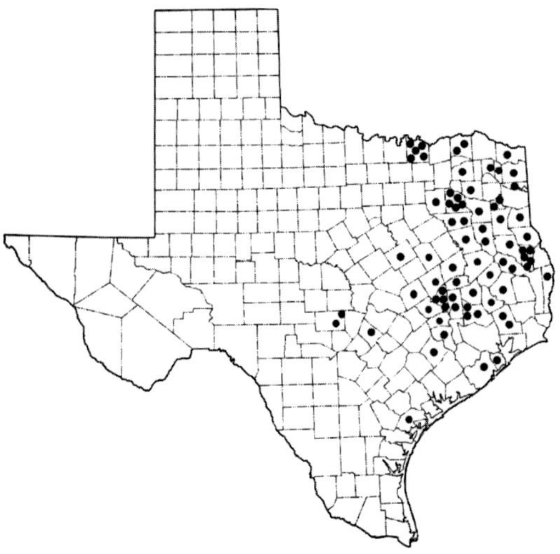

Ranunculus fascicularis

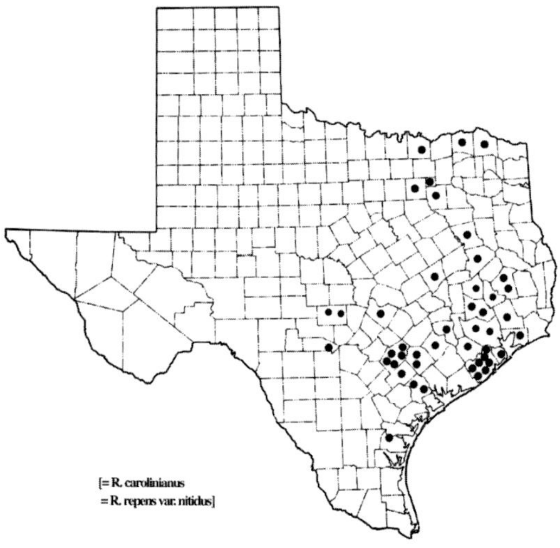
[= R. carolinianus
= R. repens var. nitidus]

Ranunculus hispidus

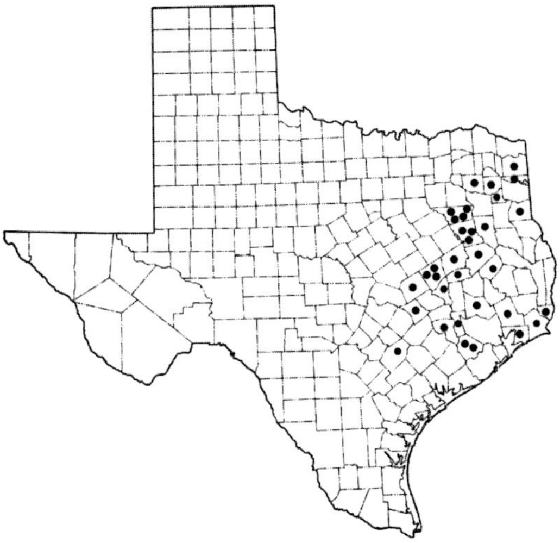

Ranunculus laxicaulis

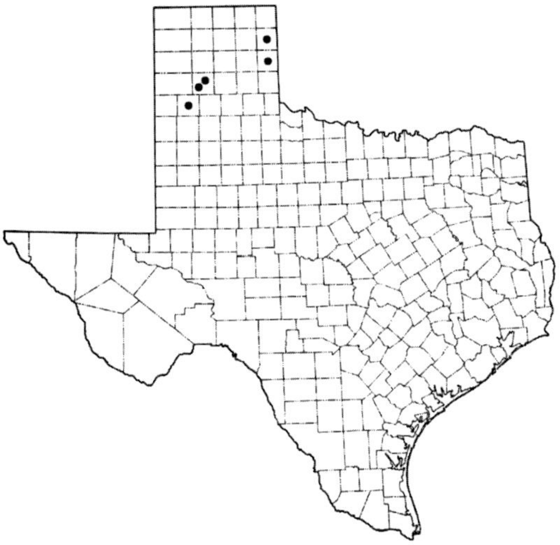

Ranunculus longirostris

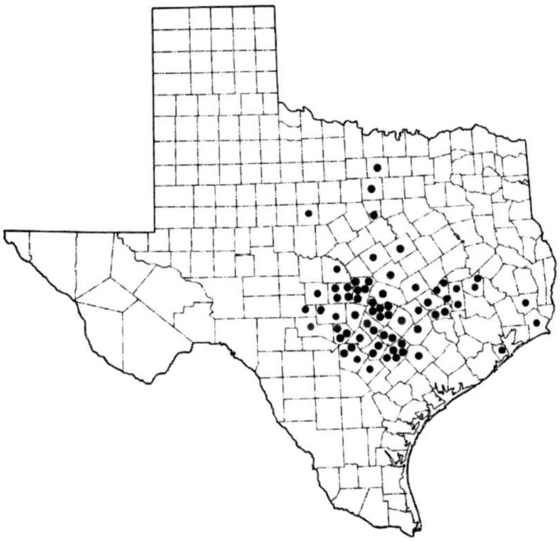

Ranunculus macranthus

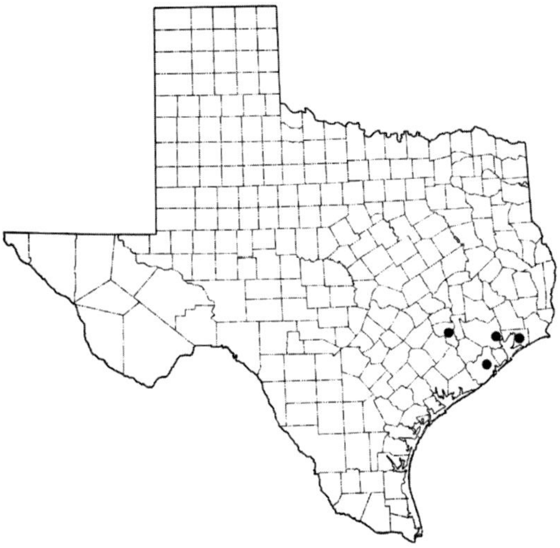

Ranunculus marginatus

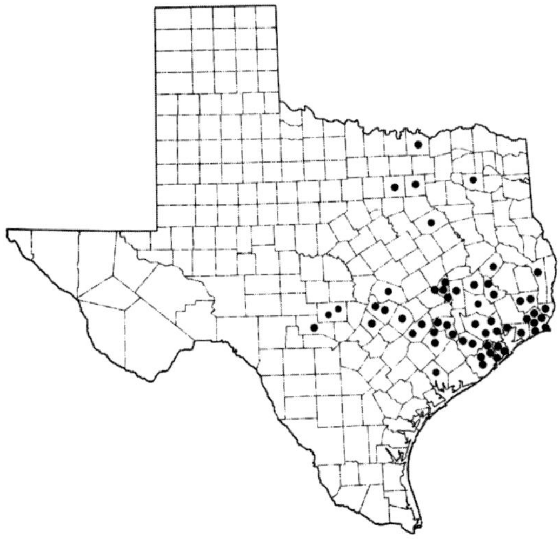

Ranunculus muricatus

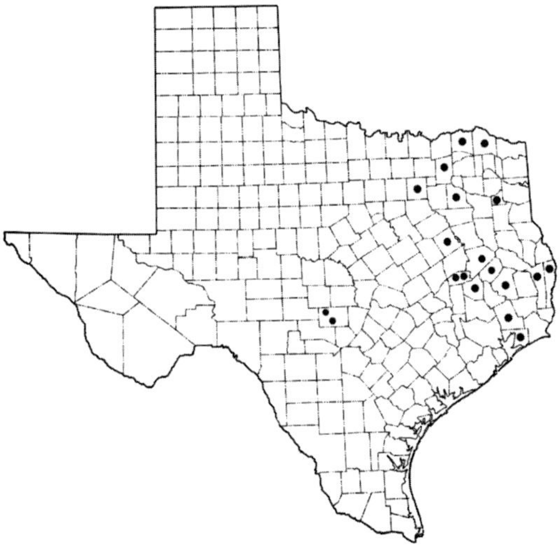

Ranunculus parviflorus

Ranunculus petiolaris

Ranunculus platensis

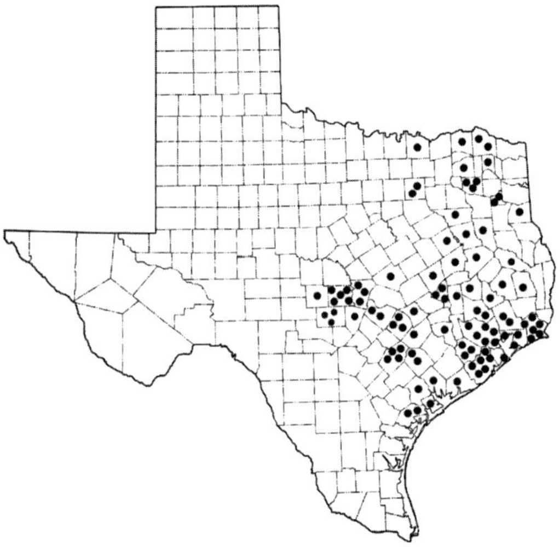

Ranunculus pusillus

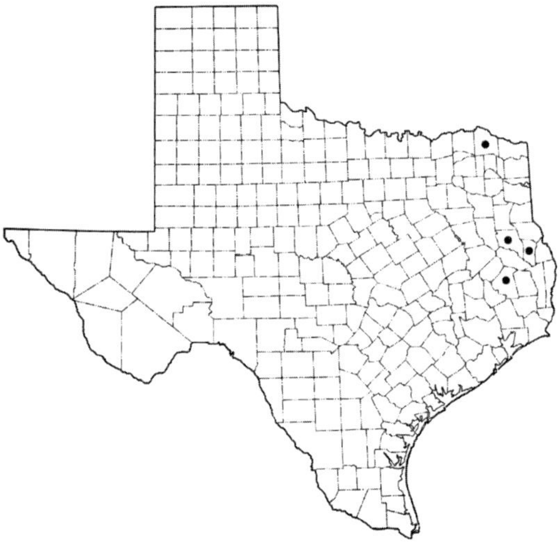

Ranunculus recurvatus

Ranunculus repens

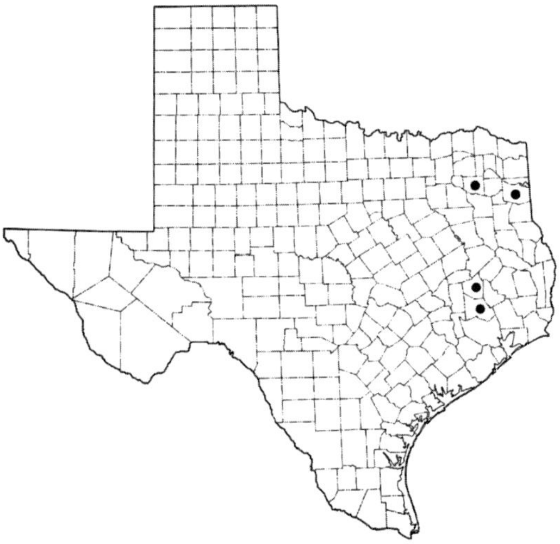

Ranunculus sardous

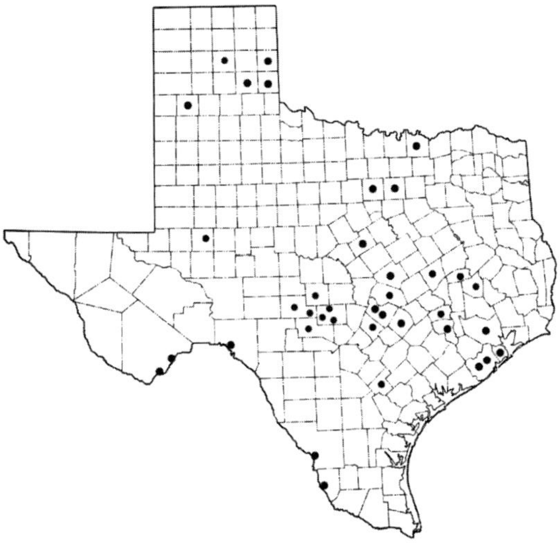

Ranunculus scleratus

Ranunculus trilobus

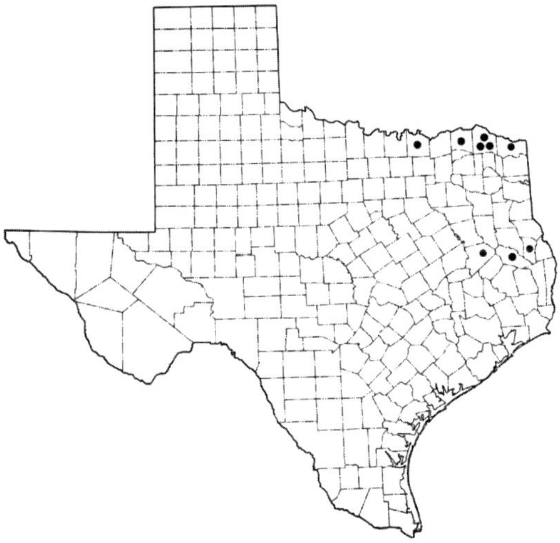

Thalictrum arkansanum

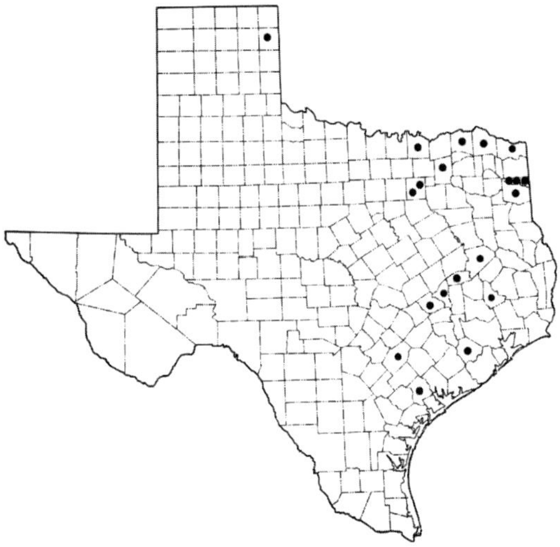

Thalictrum dasycarpum

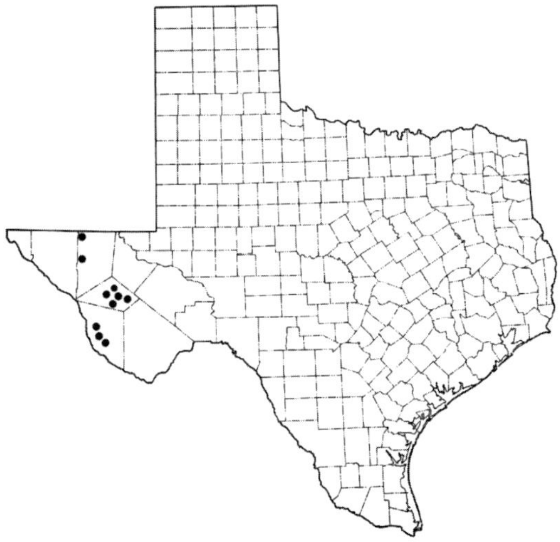

Thalictrum fendleri

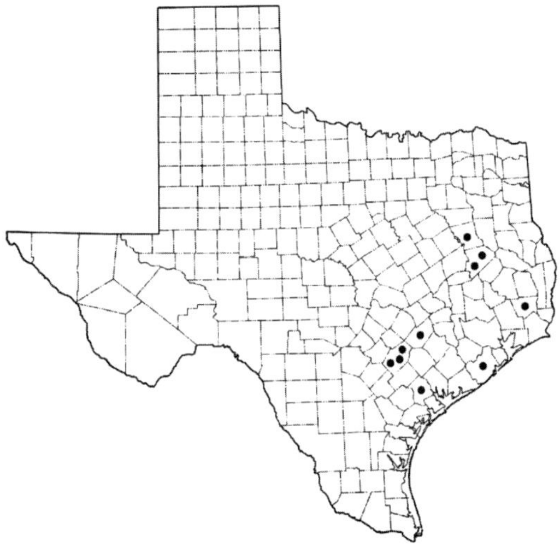

Thalictrum revolutum

Thalictrum texanum

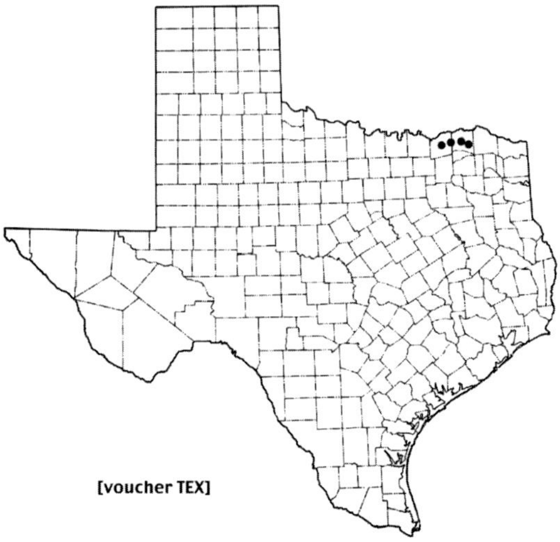
[voucher TEX]

Thalictrum thalictroides

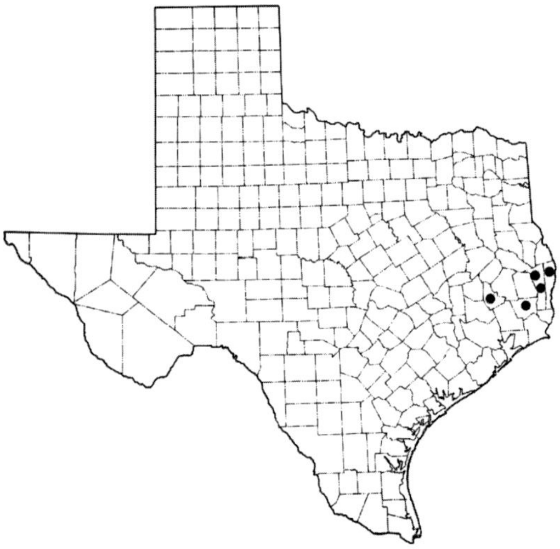

Xanthorhiza simplicissima

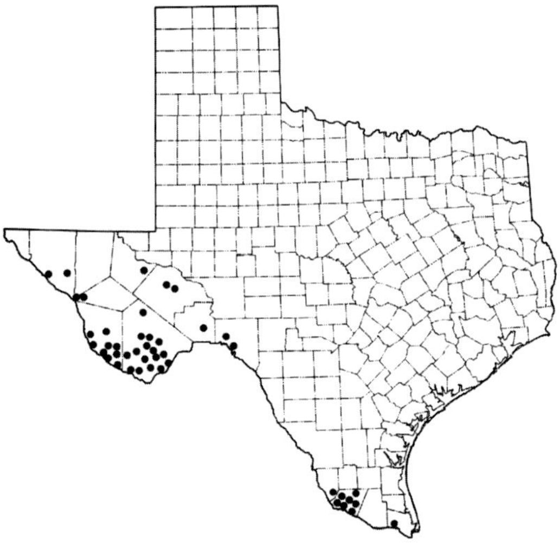

Oligomeris linifolia

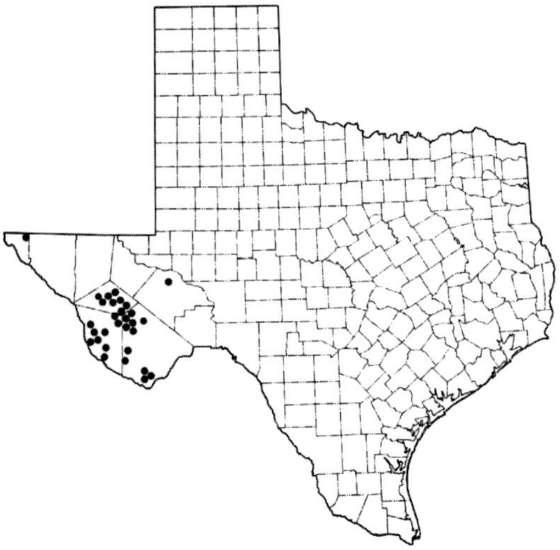

Adolphia infesta

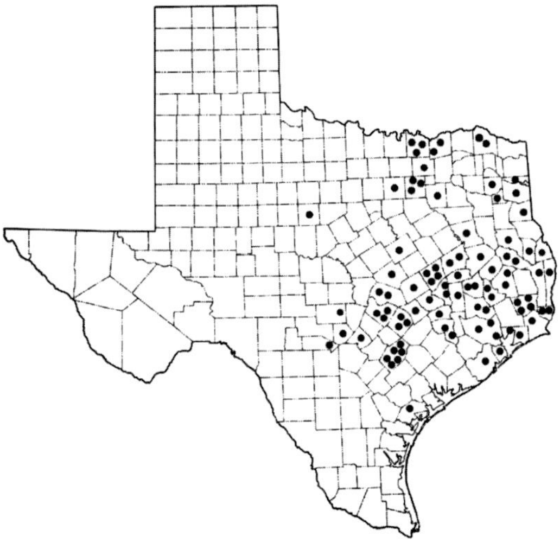

Berchemia scandens

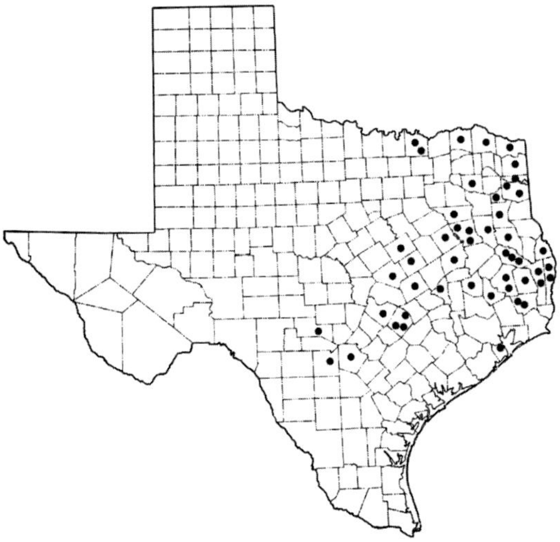

Ceanothus americanus

Ceanothus fendleri

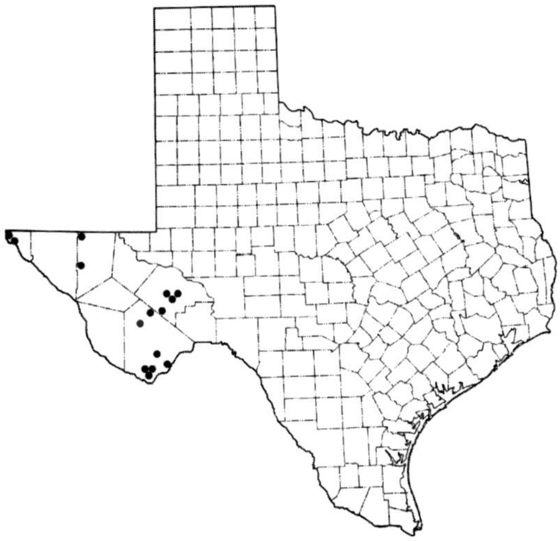

Ceanothus greggii

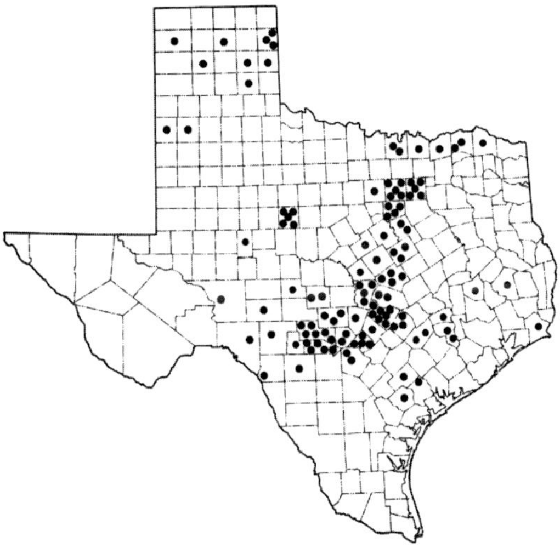

Ceanothus herbaceus

Colubrina greggii

Colubrina stricta

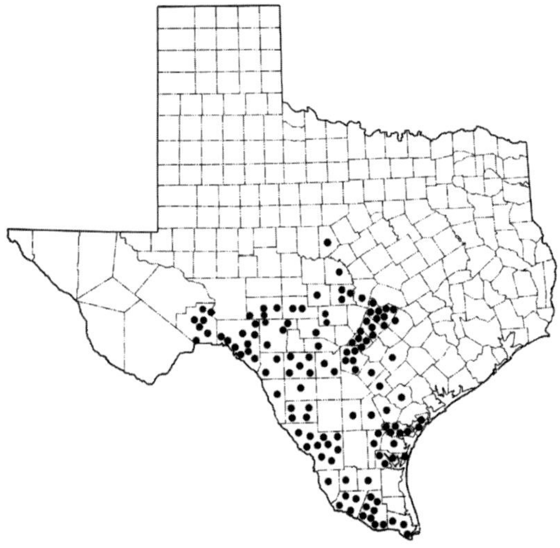

Colubrina texensis

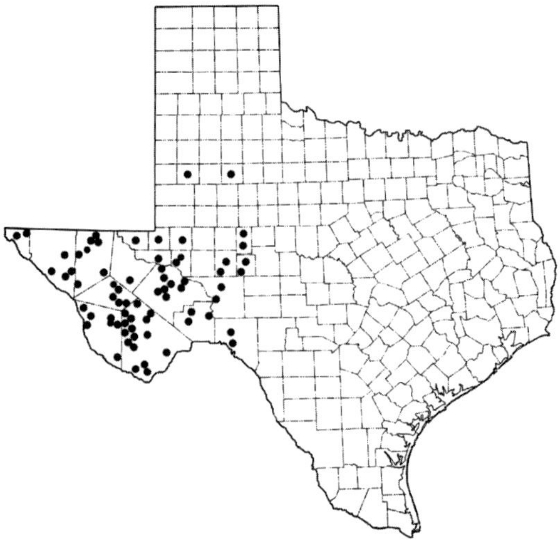

Condalia ericoides

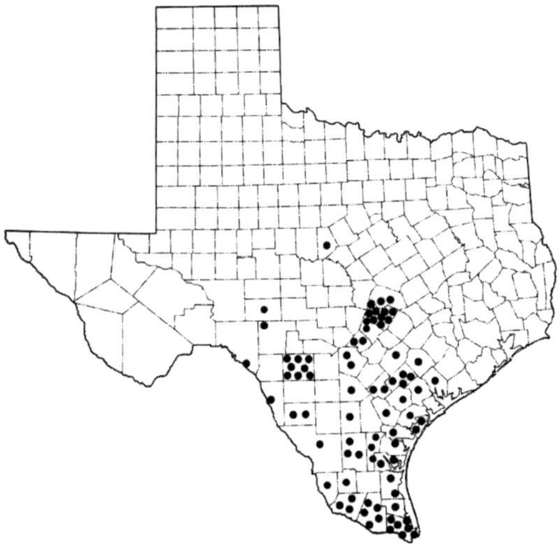

Condalia hookeri

Condalia spathulata

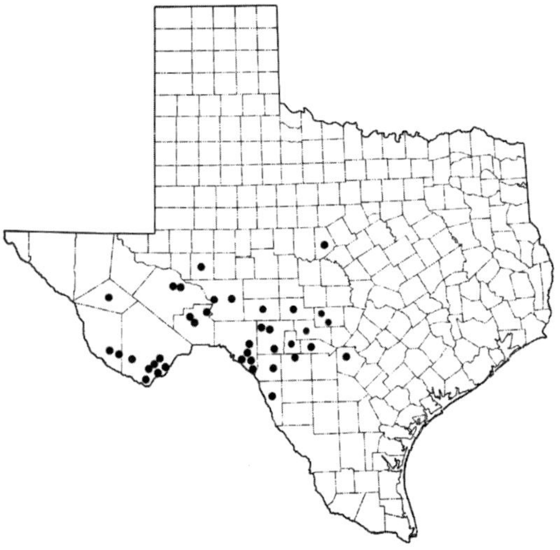

Condalia viridis

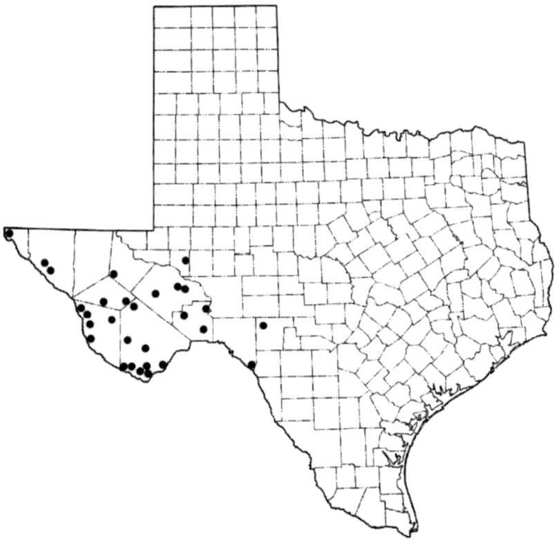

Condalia warnockii

[Sida 18: 322. 1998]

Hovenia dulcis

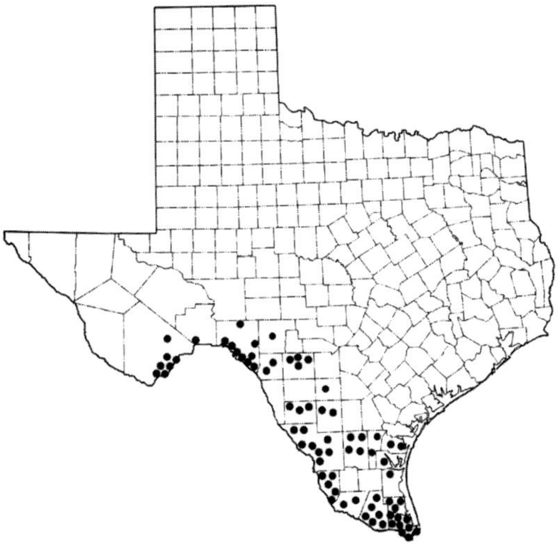

Karwinskia humboldtiana

Rhamnus betulifolia

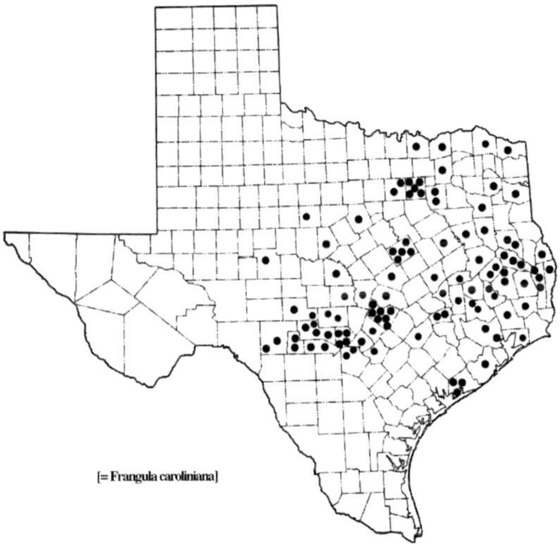
[= Frangula caroliniana]

Rhamnus caroliniana

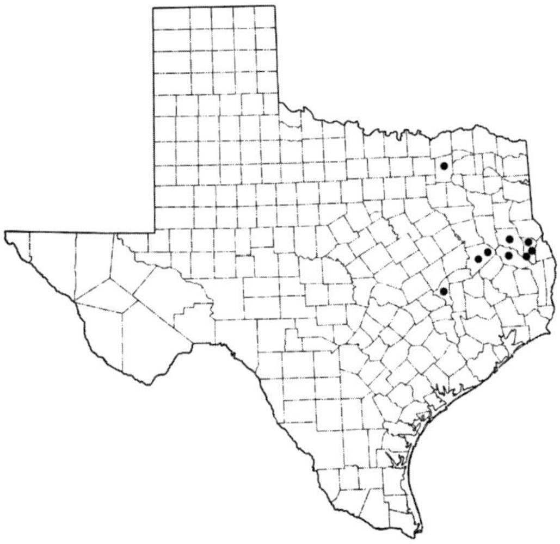

Rhamnus lanceolata

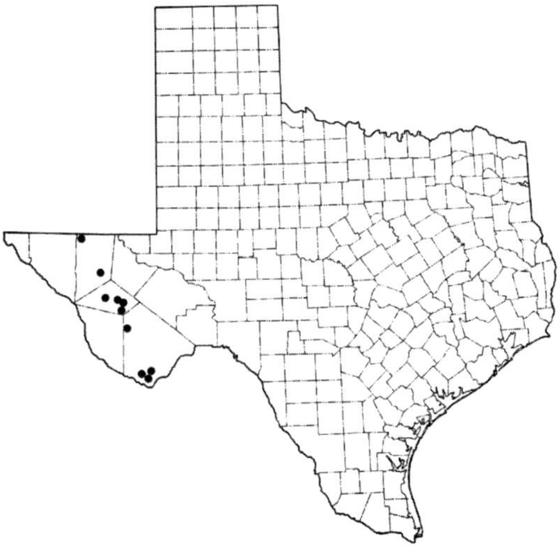

Rhamnus serrata

Sageretia thea

Sageretia wrightii

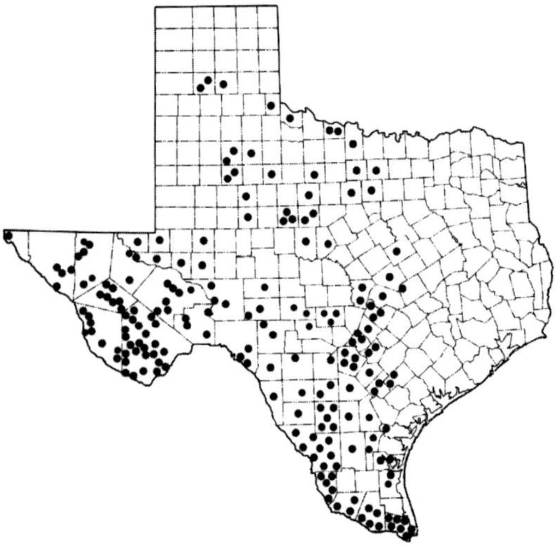

Ziziphus obtusifolia

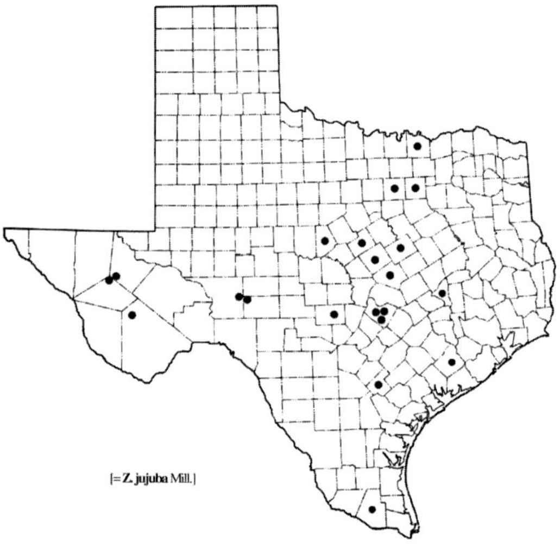
[= Z. jujuba Mill.]

Ziziphus zizyphus

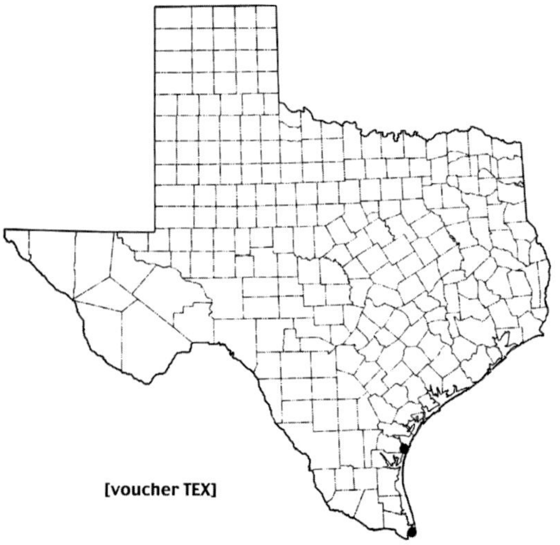
[voucher TEX]

Rhizophora mangle

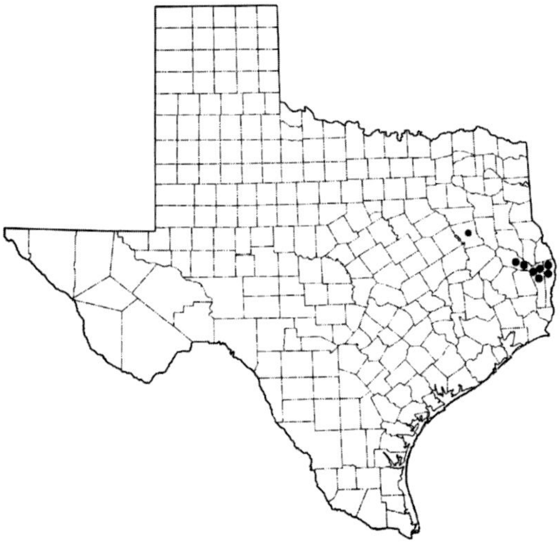

Agrimonia incisa

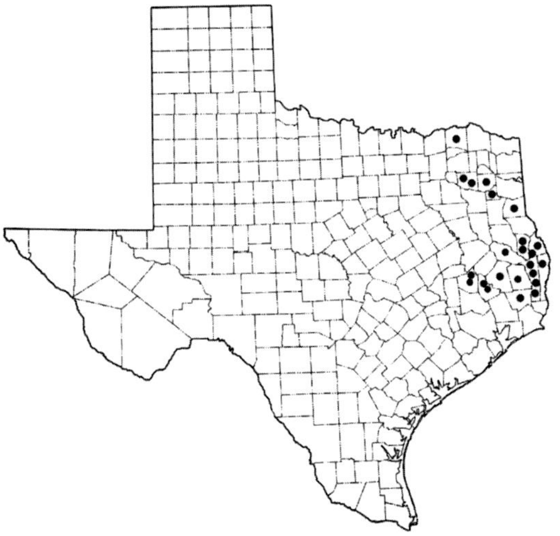

Agrimonia microcarpa

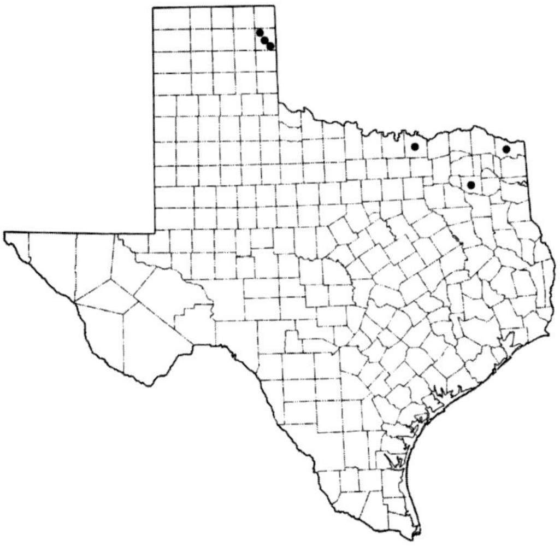

Agrimonia parviflora

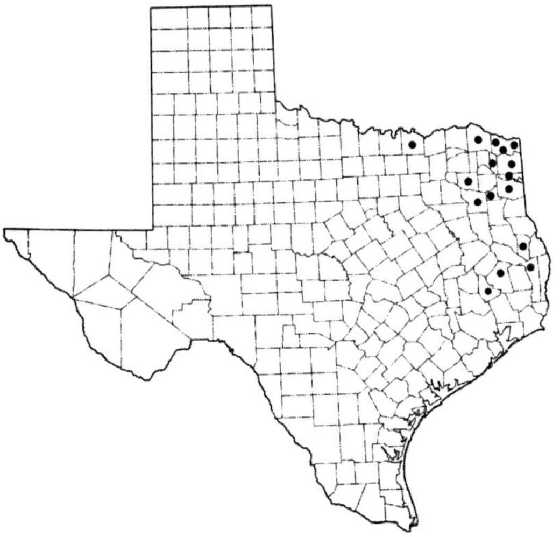

Agrimonia rostellata

Amelanchier utahensis

[= Alchemilla microcarpa]

Aphanes microcarpa

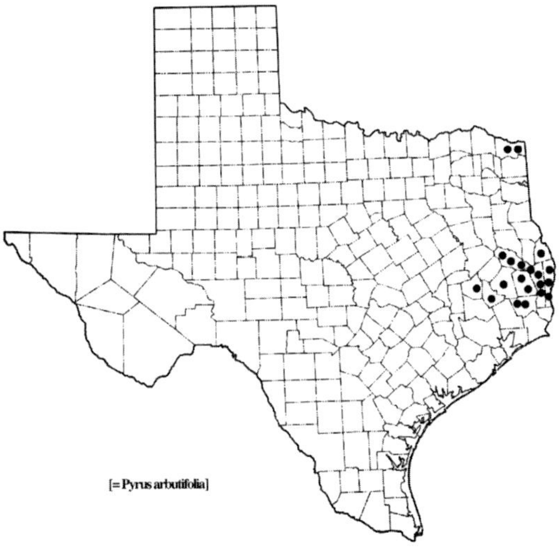
[= Pyrus arbutifolia]

Aronia abutifolia

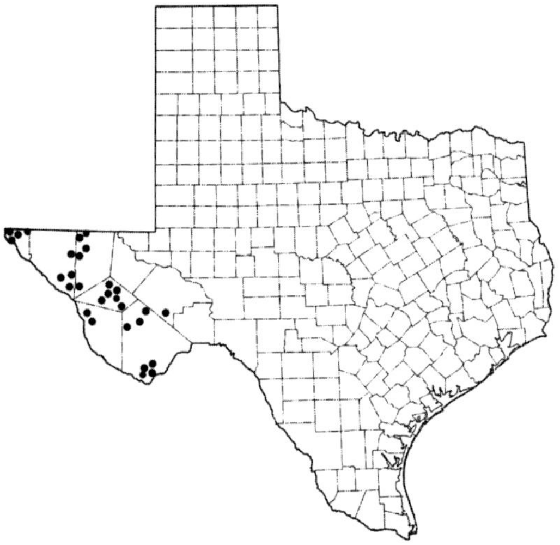

Cercocarpus breviflorus

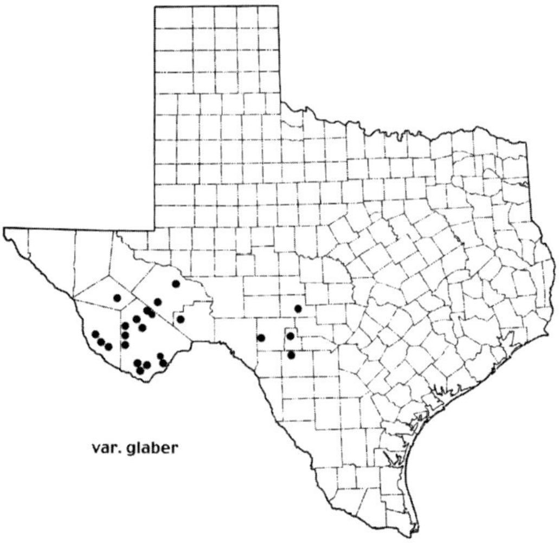
var. glaber

Cercocarpus glaberoides

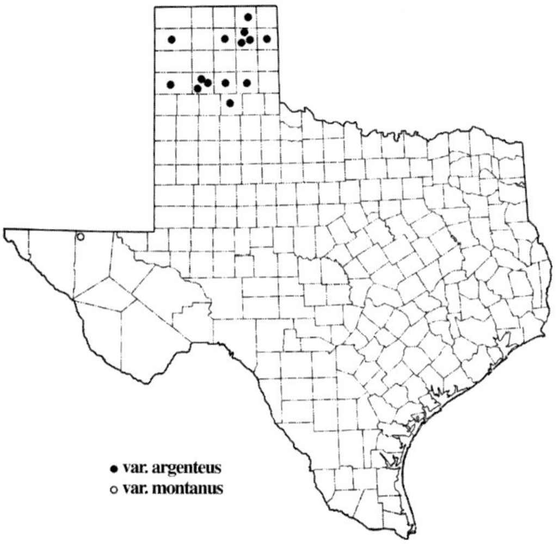
• var. argenteus
○ var. montanus

Cercocarpus montanus

Crataegus anamesa

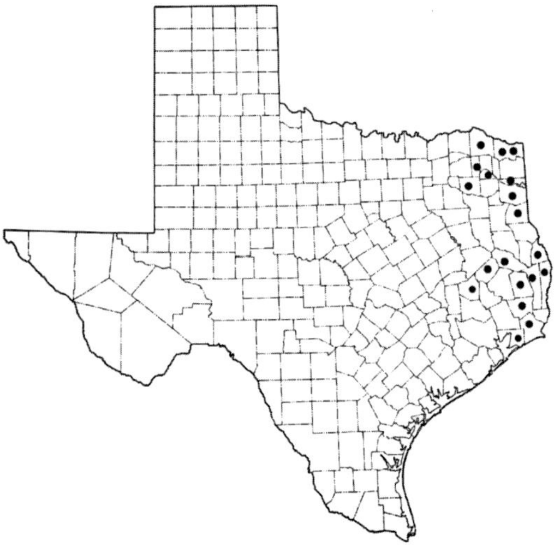

Crataegus brachyacantha

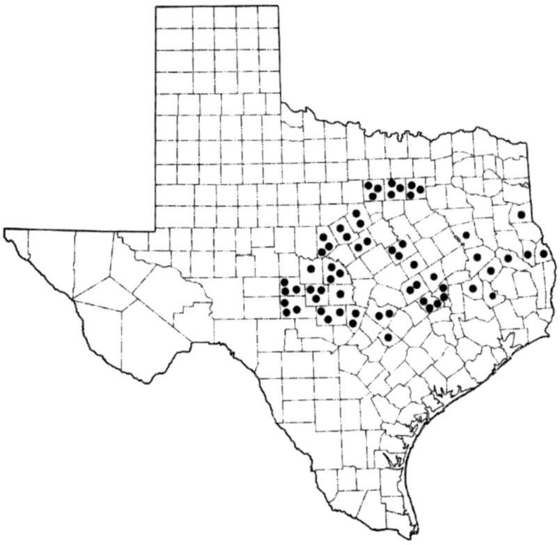

Crataegus crus-galli

Crataegus dallasiana

Crataegus engelmannii

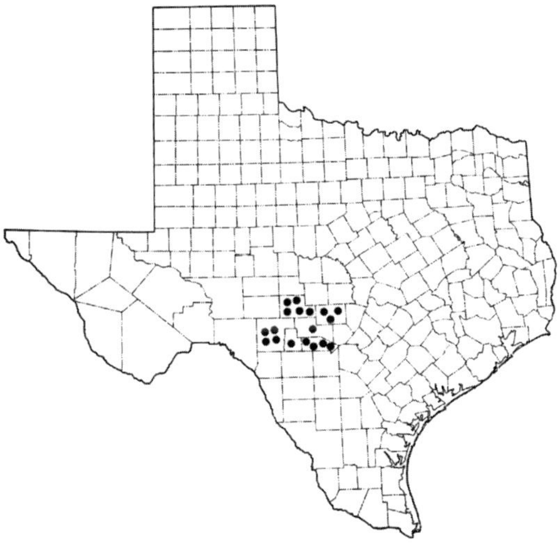
Crataegus greggiana

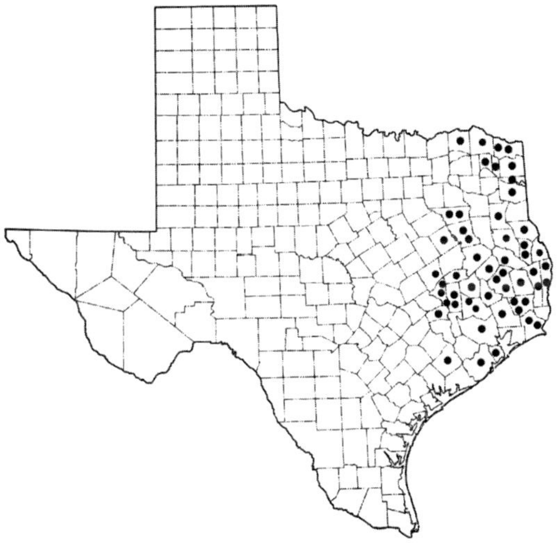
Crataegus marshallii

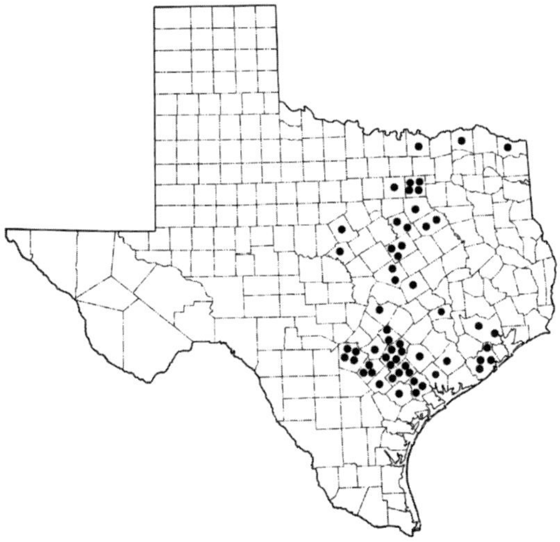
Crataegus mollis

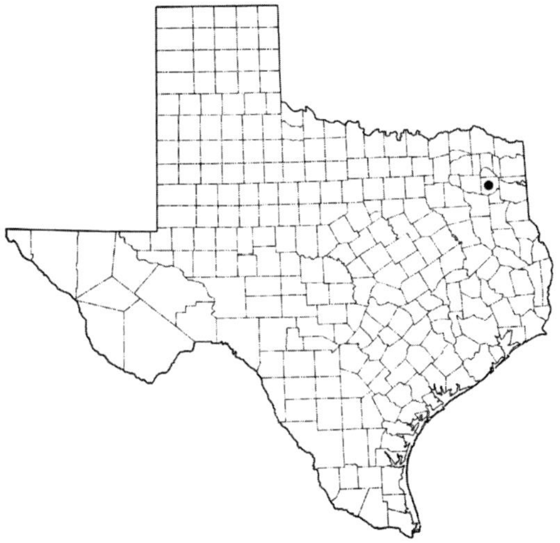
Crataegus nananixonii

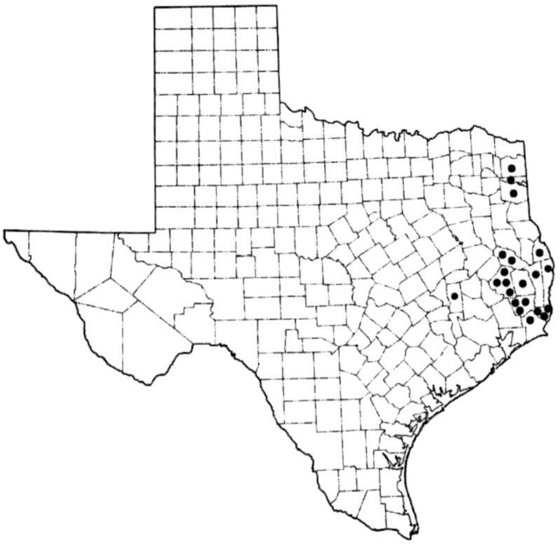

Crataegus opaca

Crataegus pearsonii

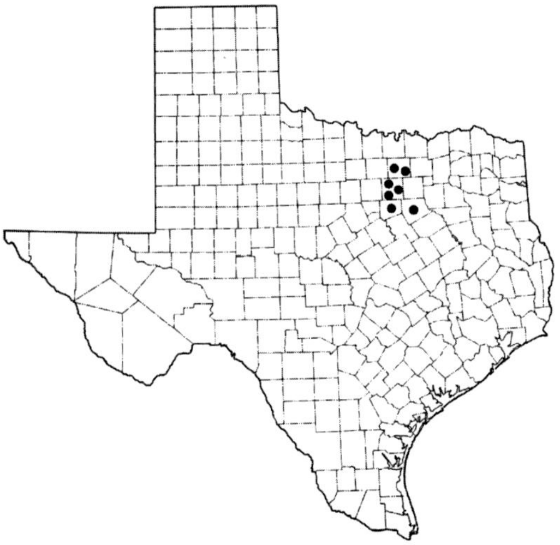

Crataegus reverchonii

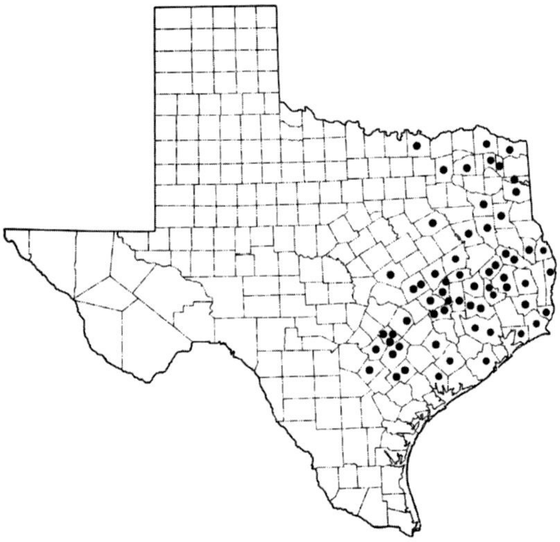

Crataegus spathulata

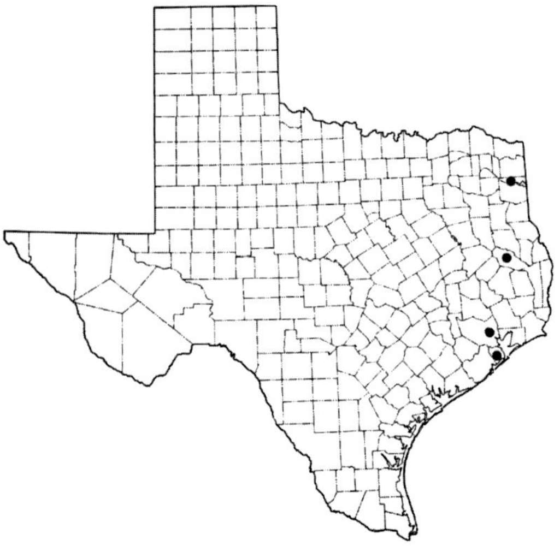

Crataegus stenosepala

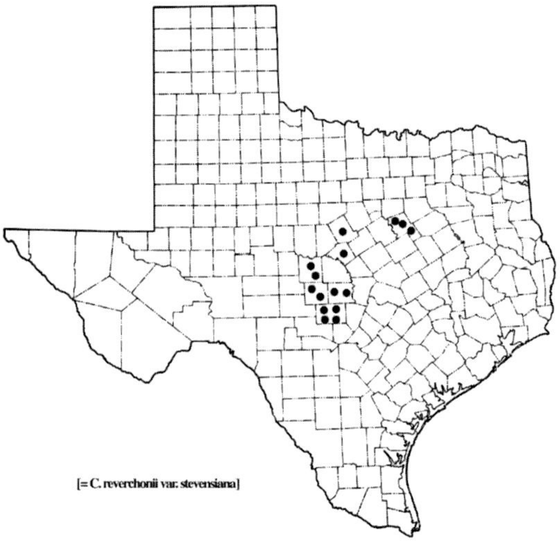
[= C. reverchonii var. stevensiana]

Crataegus stevensiana

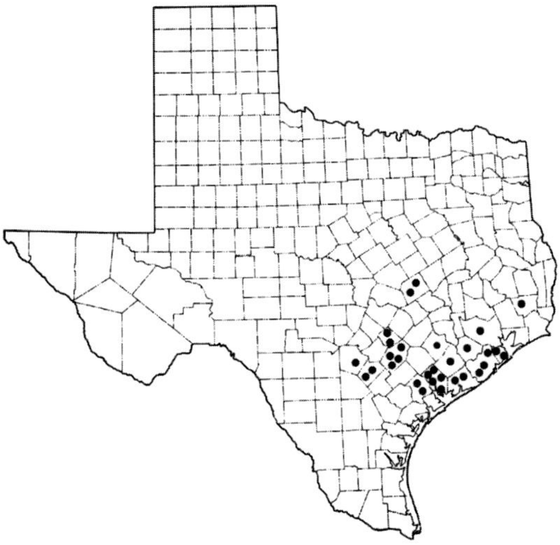

Crataegus texana

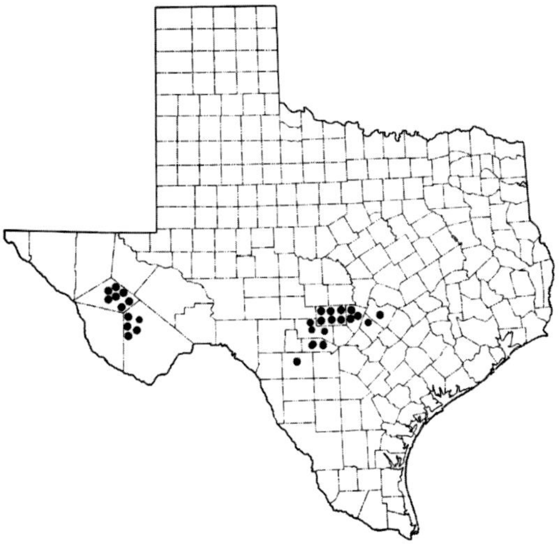

Crataegus tracyi

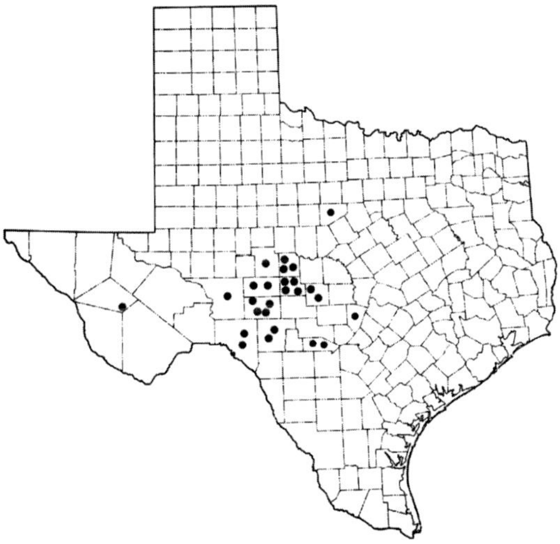

Crataegus turnerorum

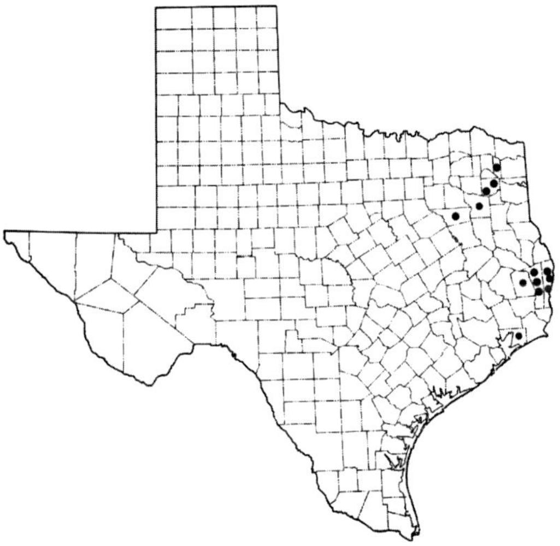

Crataegus uniflora

Crataegus viburnifolia

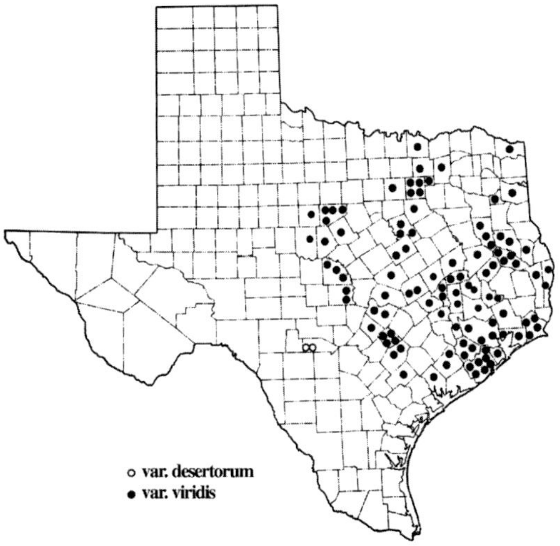
○ var. desertorum
● var. viridis

Crataegus viridis

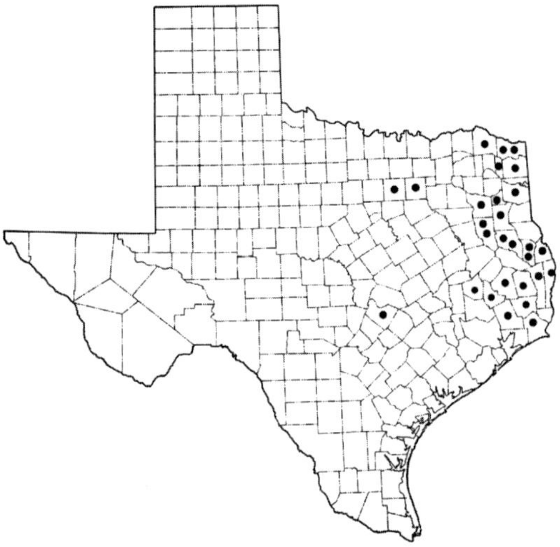

Duchesnea indica

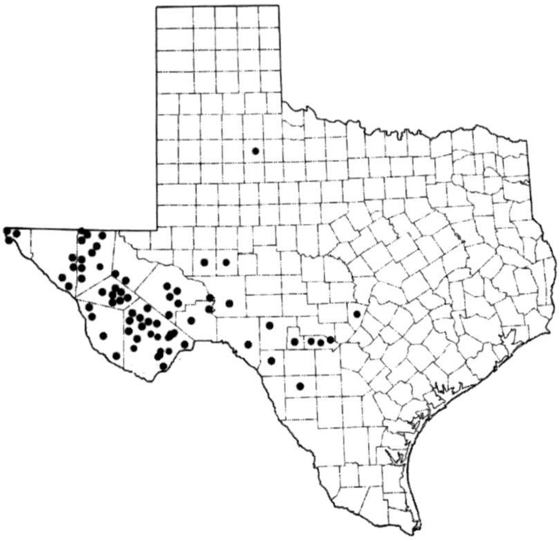

Fallugia paradoxa

Fragaria chiloensis

Fragaria vesca

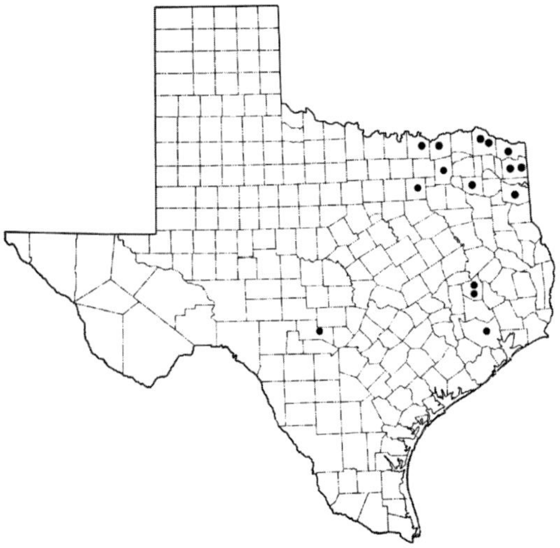

Fragaria virginiana

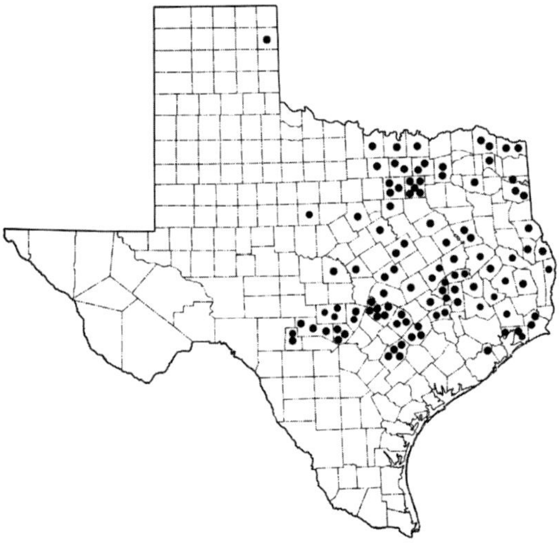

Geum canadense

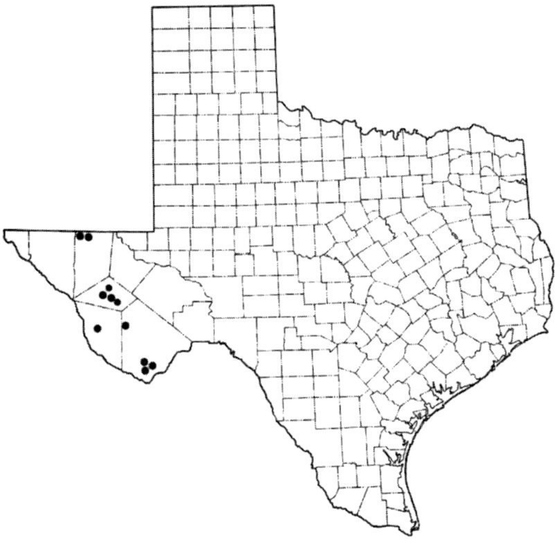

Holodiscus discolor

[= Amelanchier denticulata]

Malacomeles denticulata

[= Pyrus ioensis]

Malus ioensis

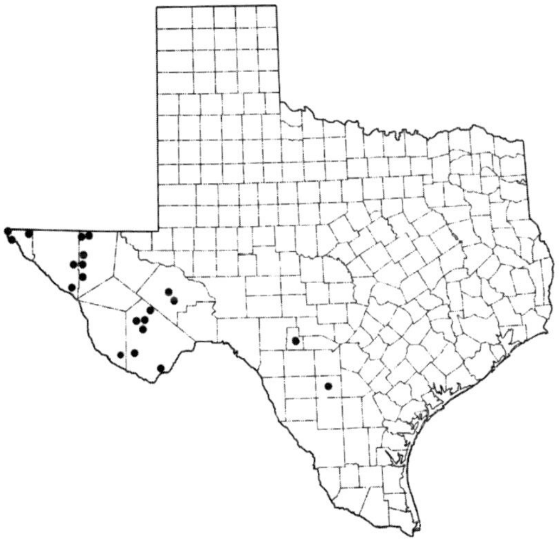

Petrophyton caespitosum

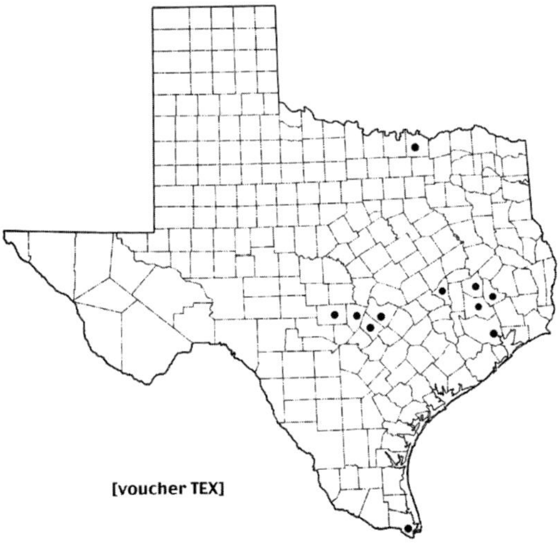
[voucher TEX]

Photinia serratifolia

Physocarpus monogynus

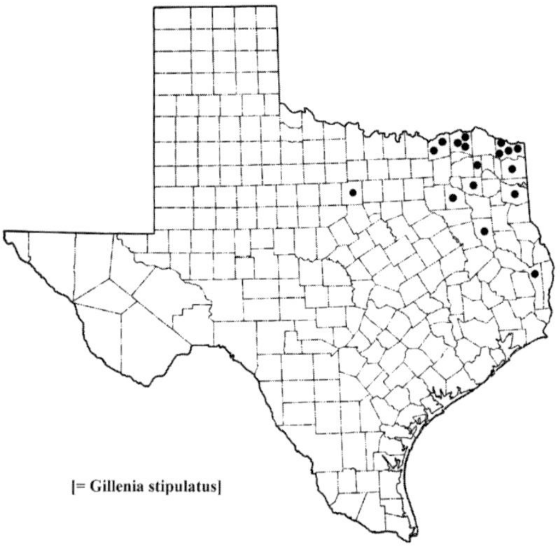
[= Gillenia stipulatus]

Porteranthus stipulatus

Potentilla norvegica

Potentilla paradoxa

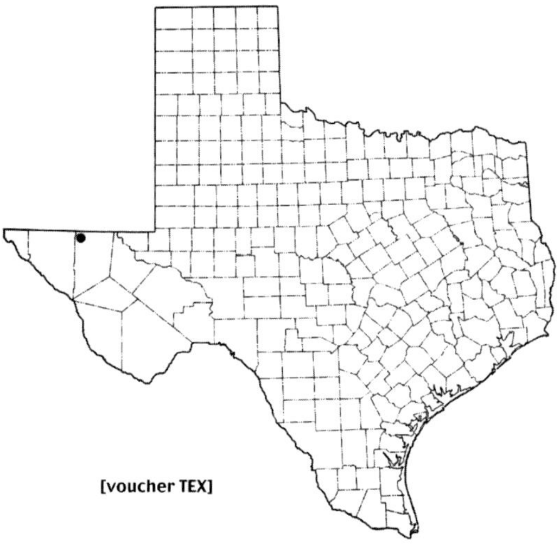
[voucher TEX]

Potentilla pensylvanica

Potentilla recta

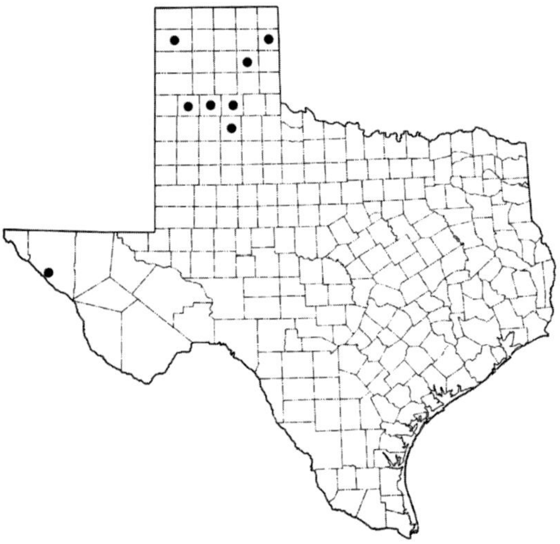

Potentilla rivalis

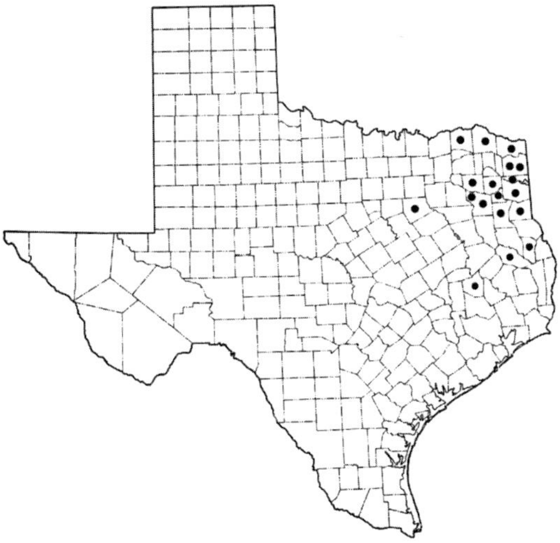

Potentilla simplex

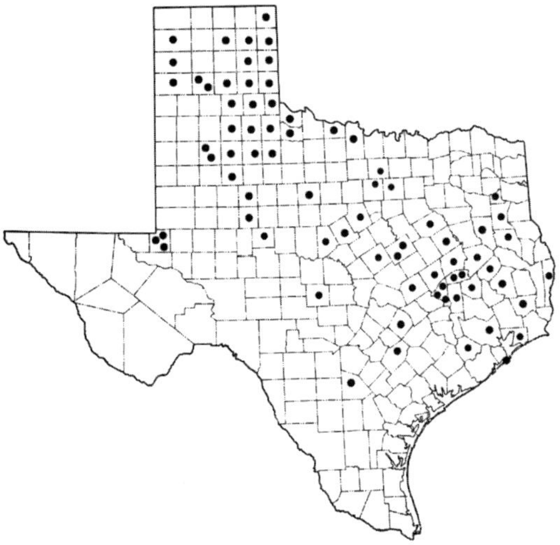

Prunus angustifolia

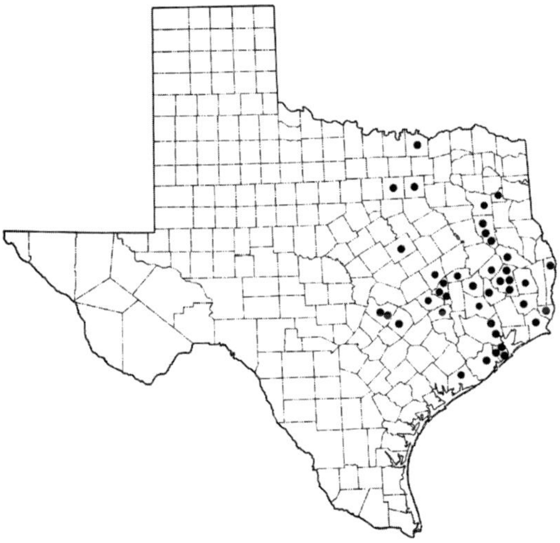

Prunus caroliniana

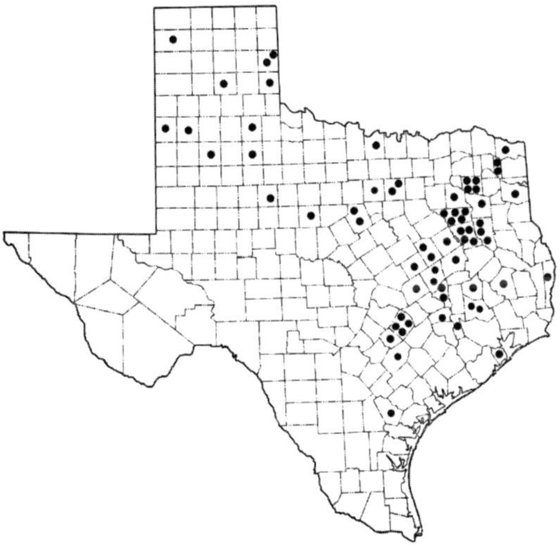

Prunus gracilis

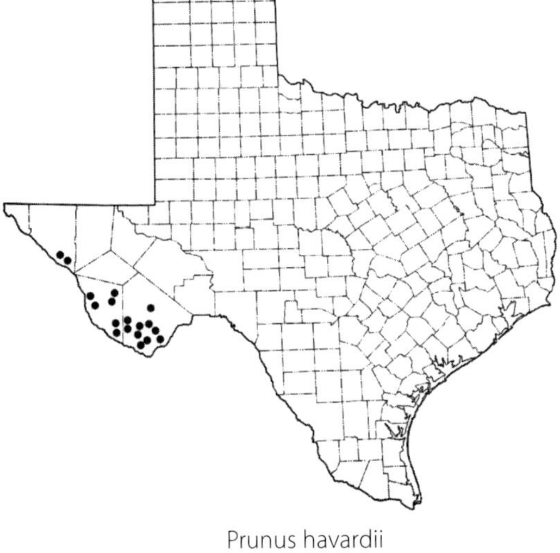

Prunus havardii

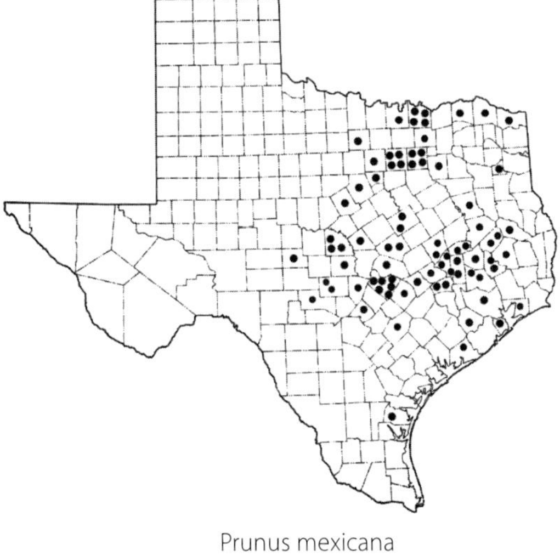

Prunus mexicana

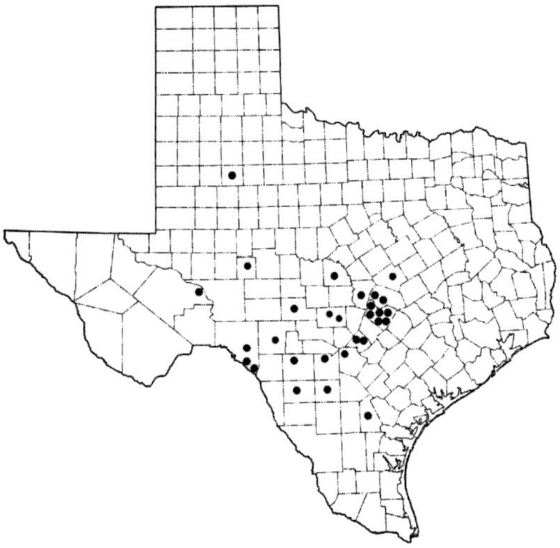

Prunus minutiflora

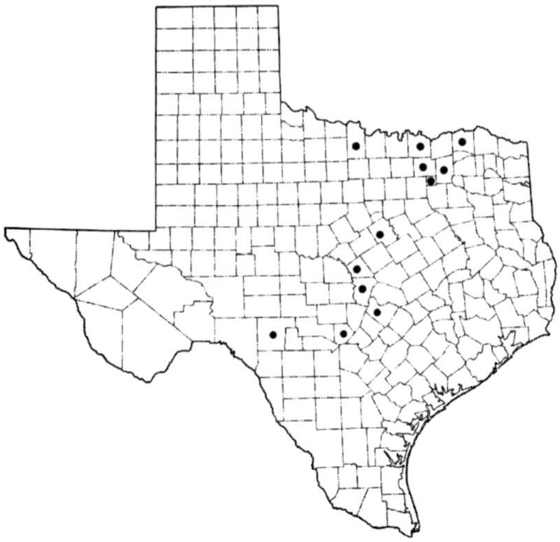

Prunus munsoniana

Prunus murrayana

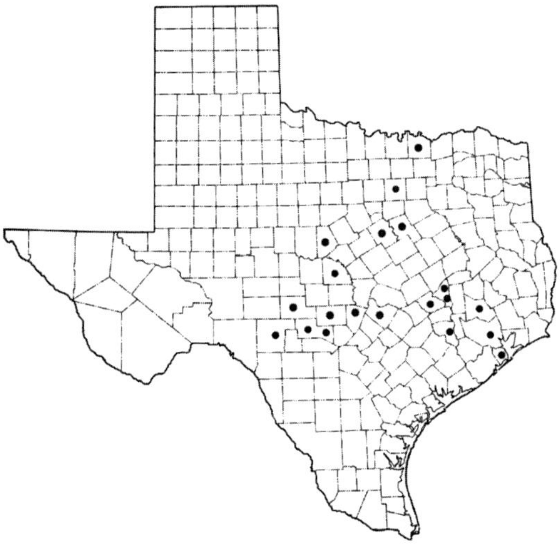

Prunus persica

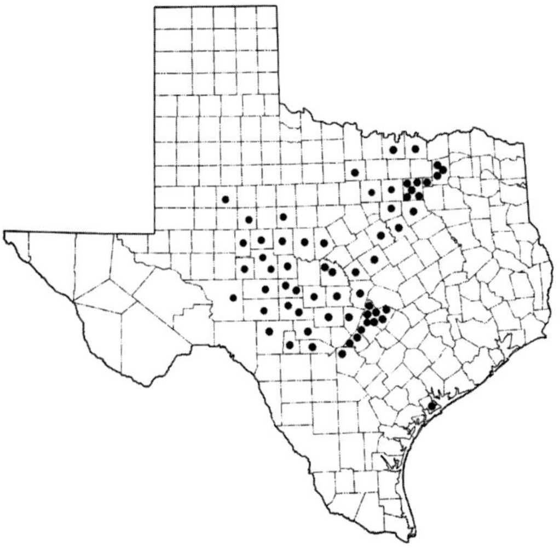

Prunus rivularis

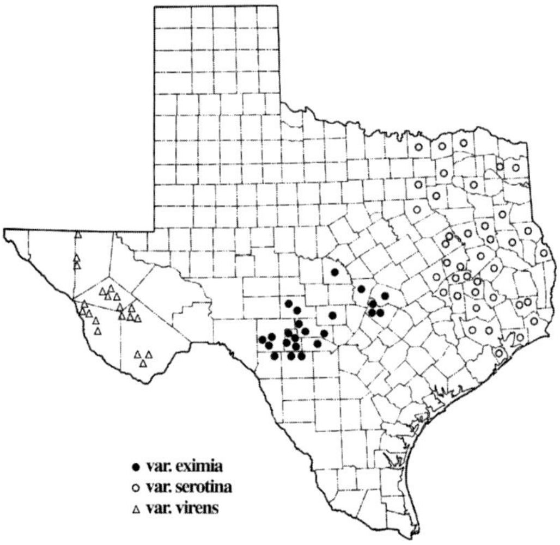
● var. eximia
○ var. serotina
△ var. virens

Prunus serotina

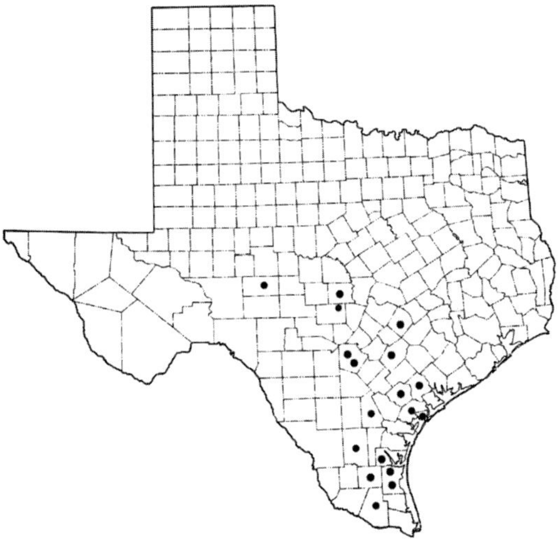

Prunus texana

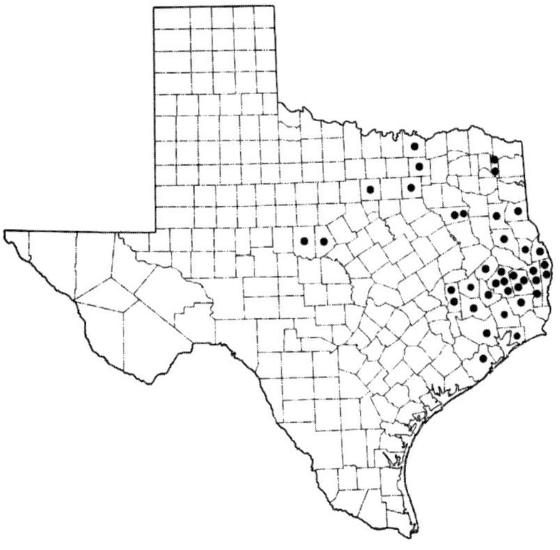

Prunus umbellata

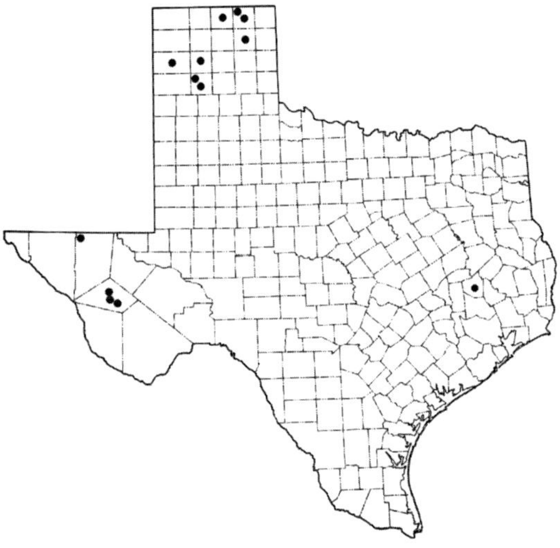

Prunus virginiana

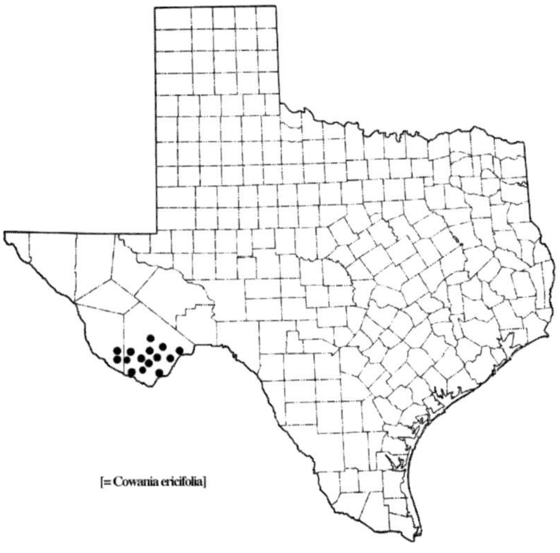
[= Cowania ericifolia]

Purshia ericifolia

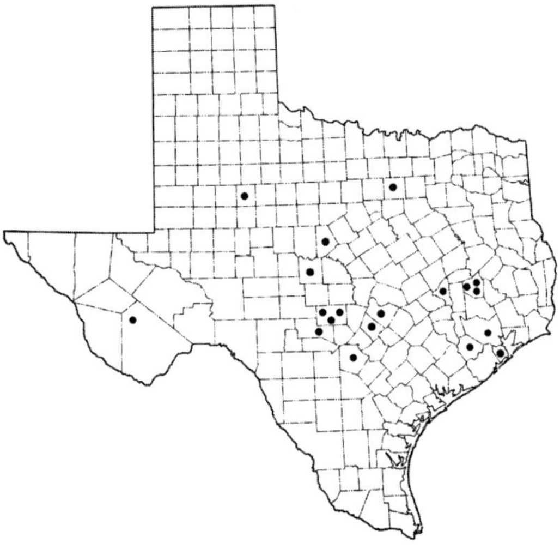

Pyracantha coccinea

Pyrus angustifolia

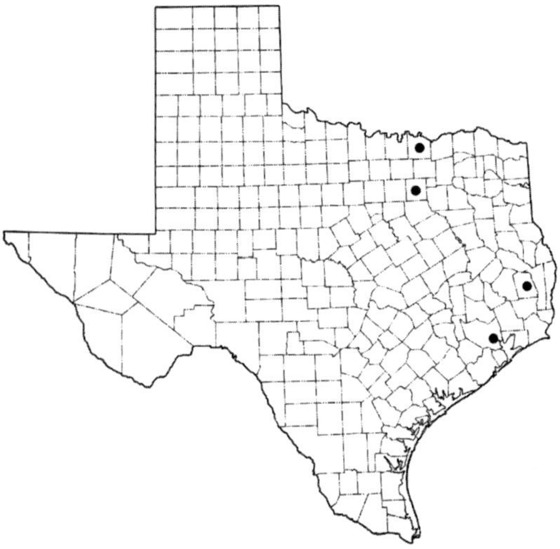

Pyrus calleryana

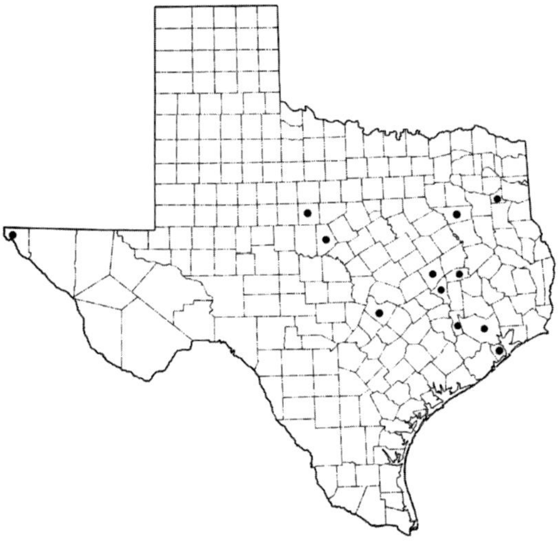

Pyrus communis

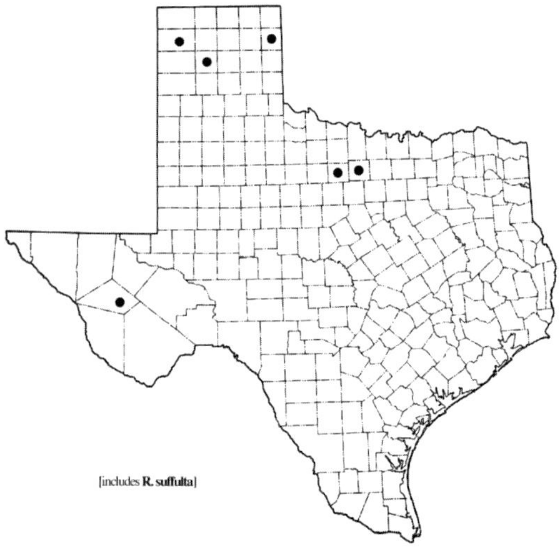
[includes R. suffulta]

Rosa arkansana

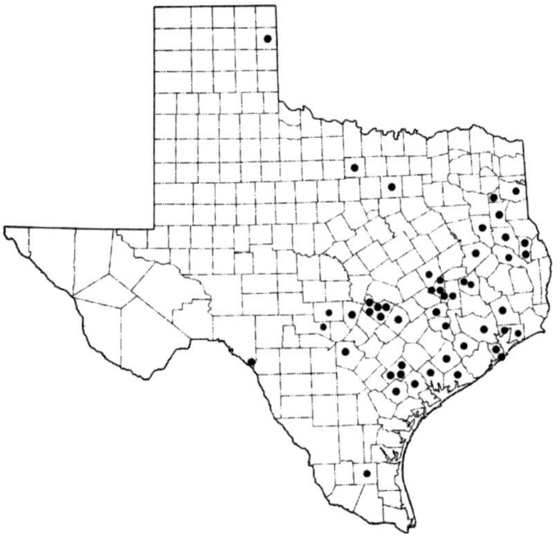

Rosa bracteata

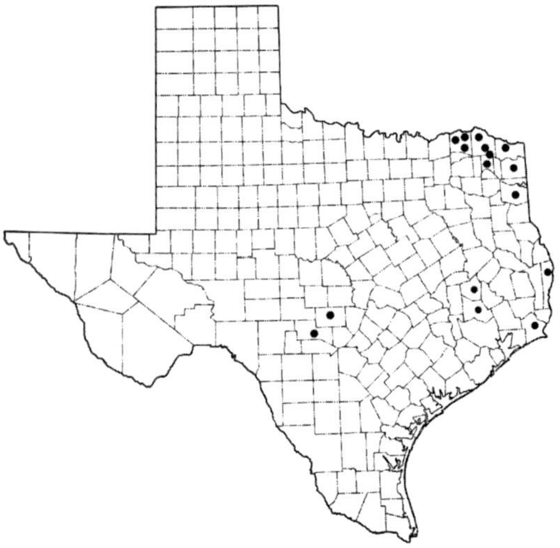

Rosa carolina

Rosa eglanteria

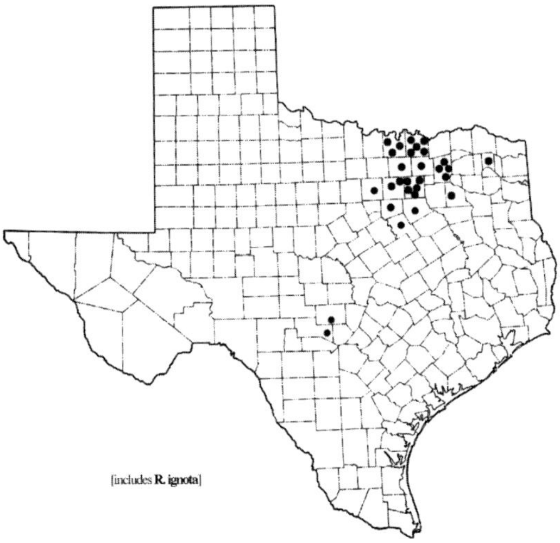
[includes R. ignota]

Rosa foliolosa

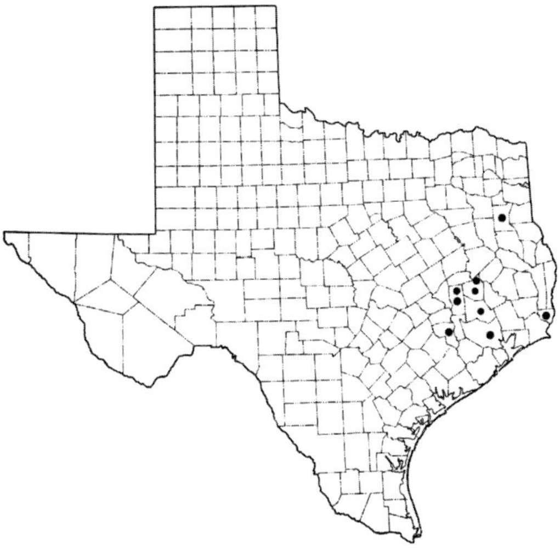

Rosa laevigata

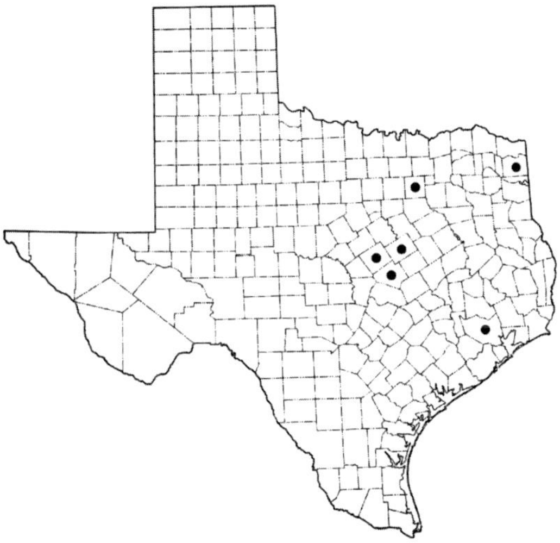

Rosa multiflora

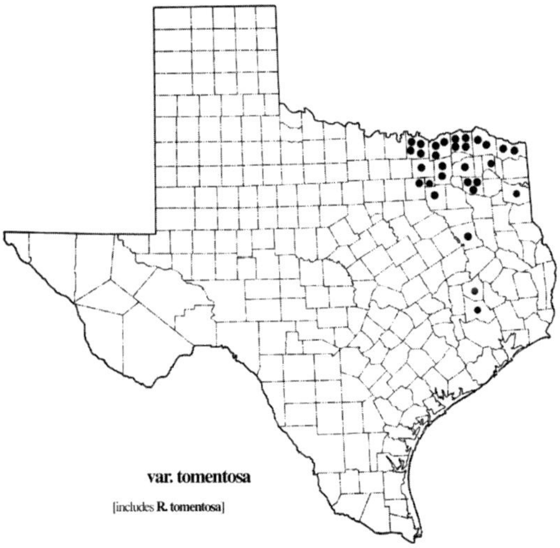
var. tomentosa
[includes R. tomentosa]

Rosa setigera

Rosa stellata

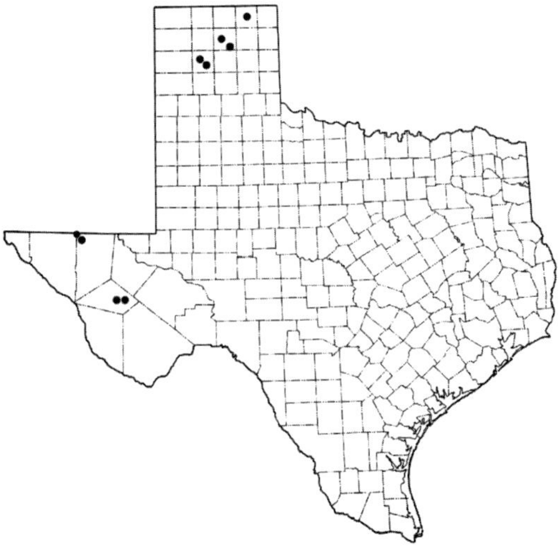

Rosa woodsii

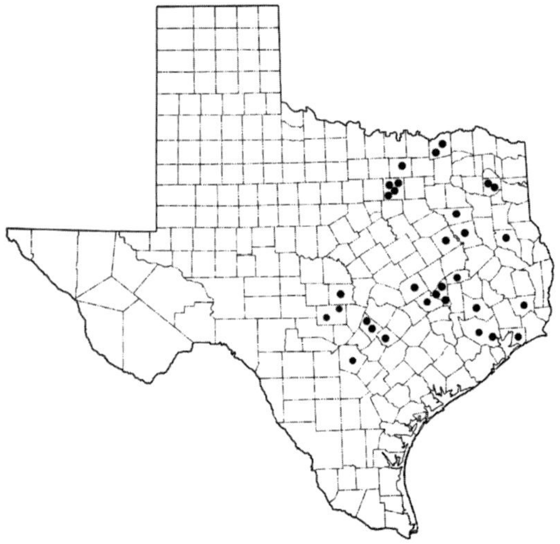
Rubus aboriginum

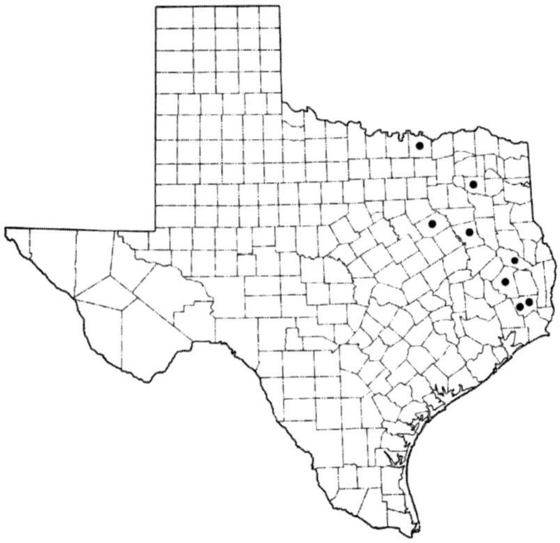
Rubus apogaeus

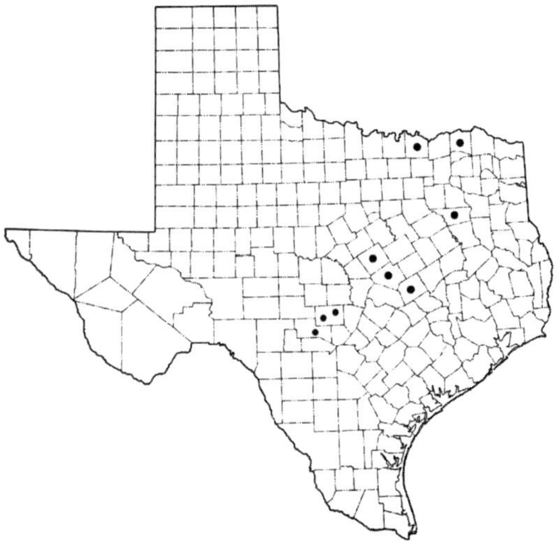
Rubus bifrons

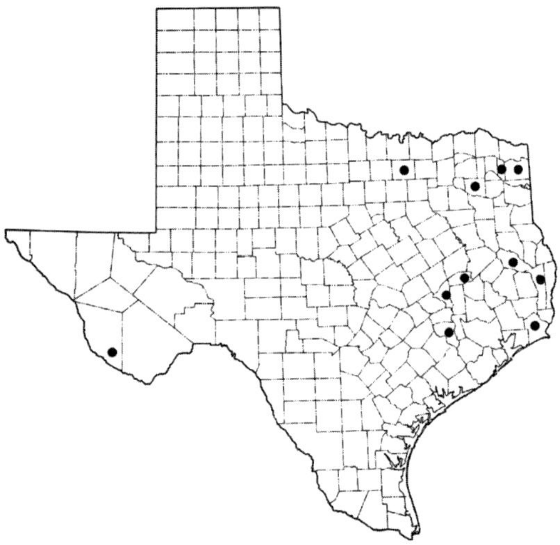
Rubus flagellaris

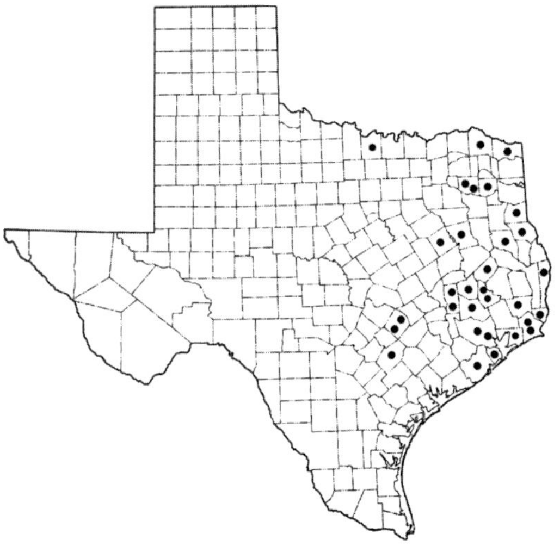
Rubus louisianus

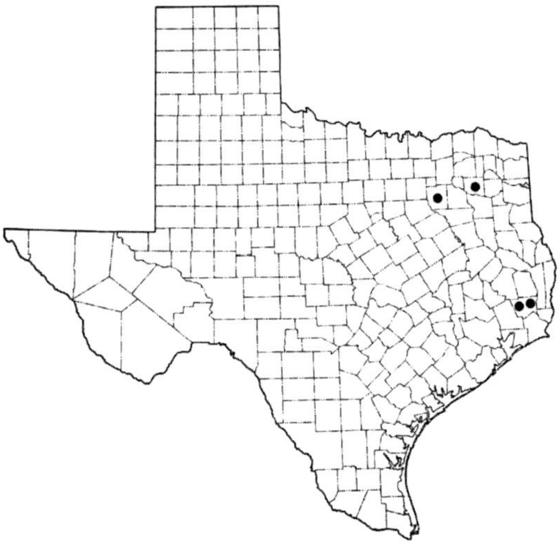
Rubus lucidus

Rubus oklahomus

Rubus persistens

Rubus saepescandens

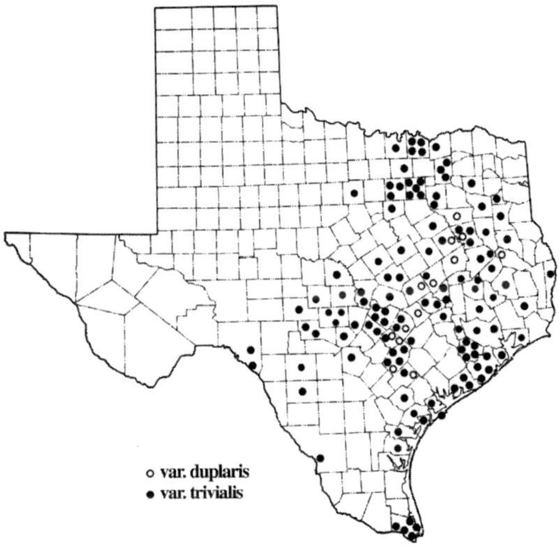
o var. duplaris
• var. trivialis

Rubus trivialis

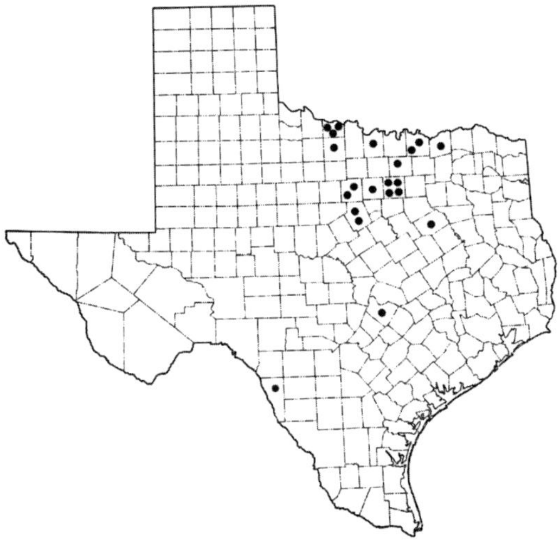

Sanguisorba annua

var. angustifolia
[= V. angustifolia]

Vauquelinia corymbosa

Asperula arvensis

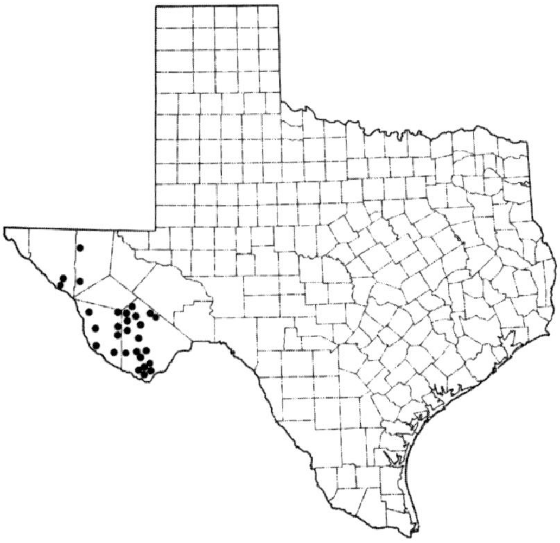

Bouvardia ternifolia

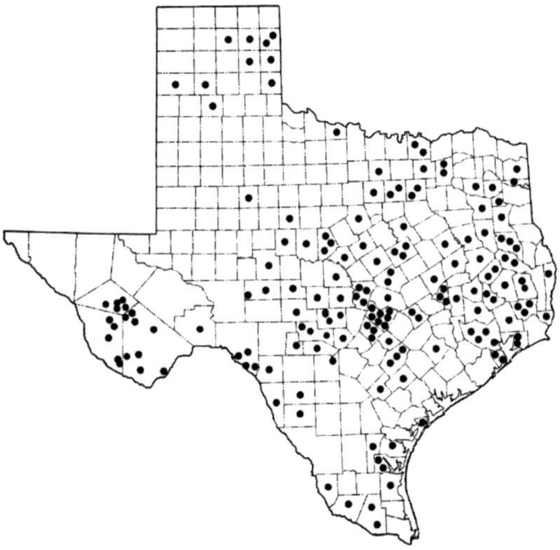

Cephalanthus occidentalis

Cephalanthus salicifolius

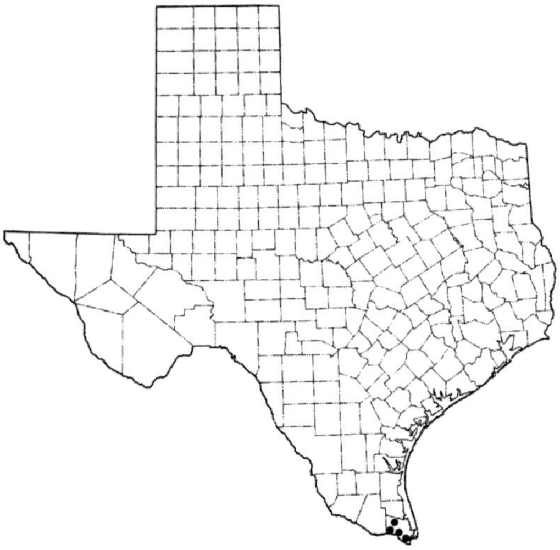

Chiococca alba

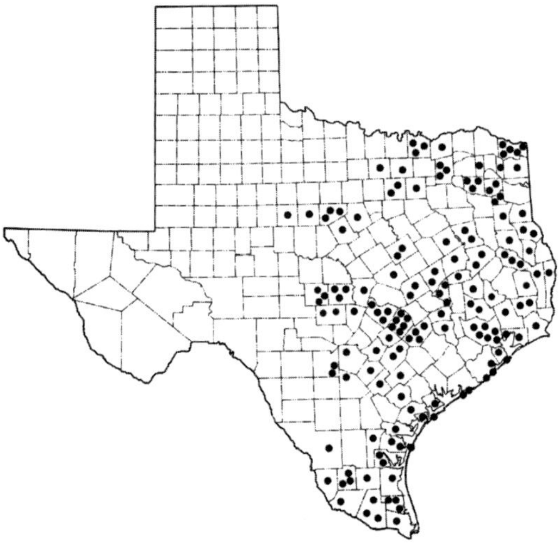

Diodia teres

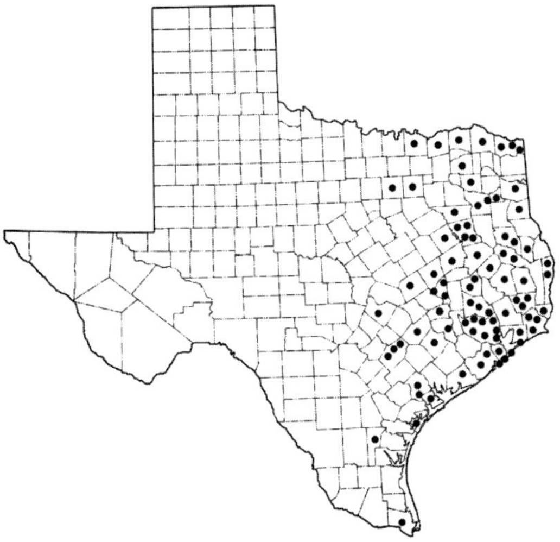

Diodia virginiana

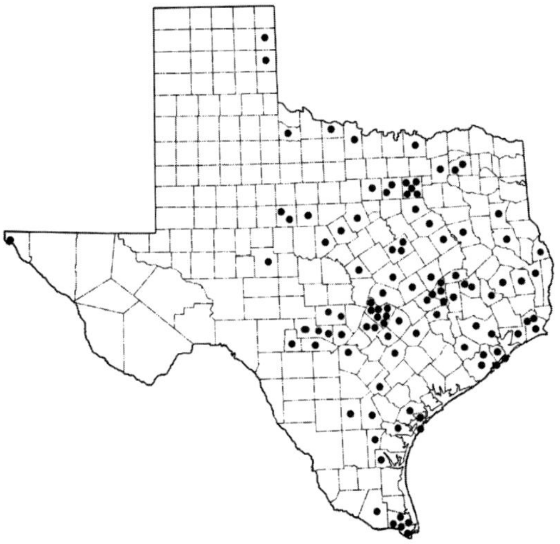

Galium aparine

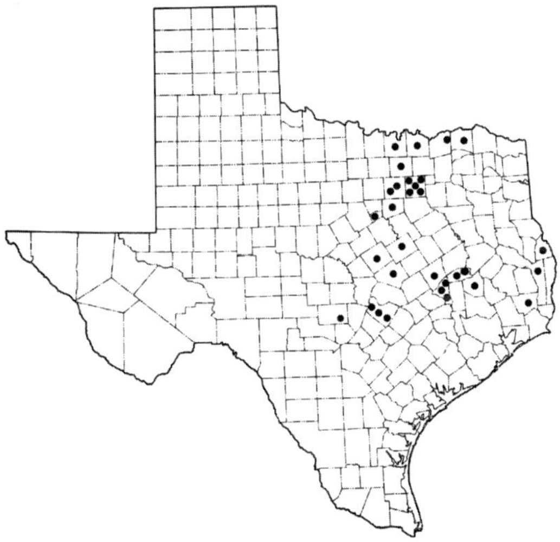

Galium circaezans

Galium frankliniense

Galium hispidulum

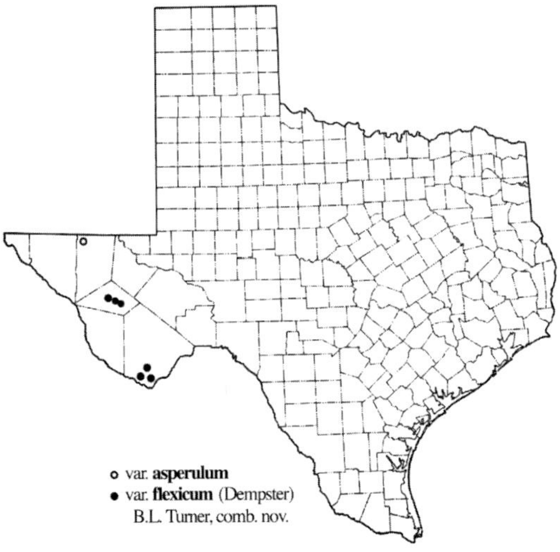
○ var. **asperulum**
● var. **flexicum** (Dempster) B.L. Turner, comb. nov.

Galium mexicanum

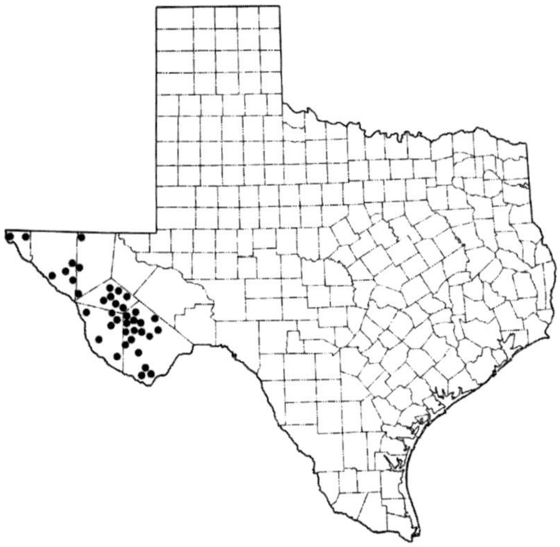
Galium microphyllum

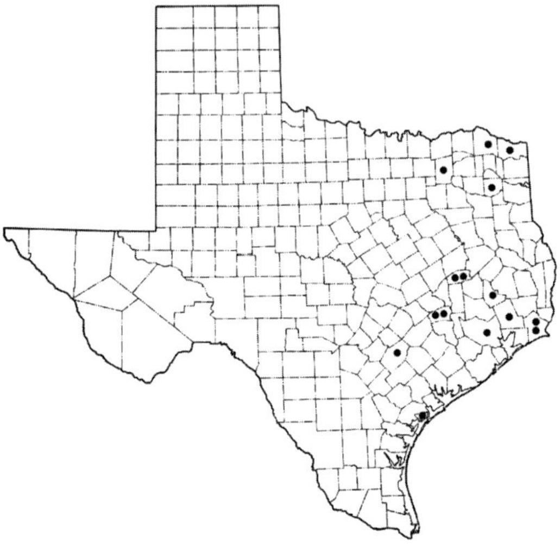
Galium obtusum

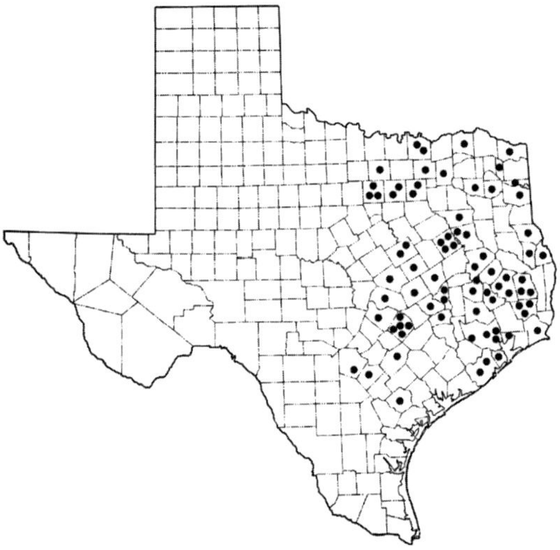
Galium pilosum

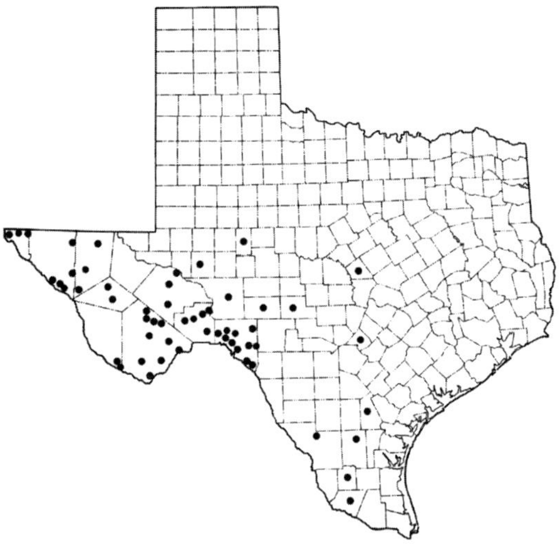
Galium proliferum

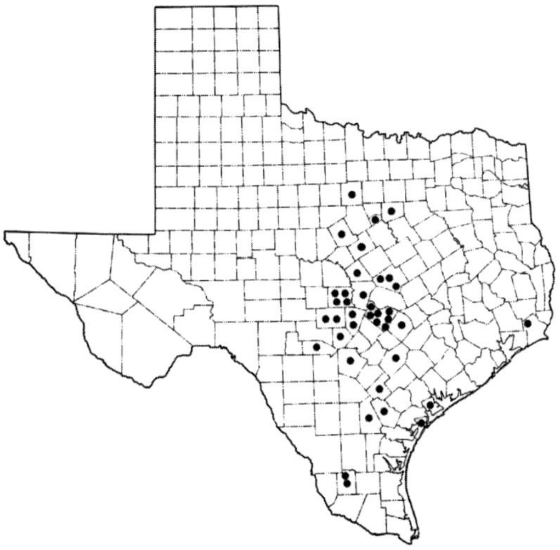
Galium texense

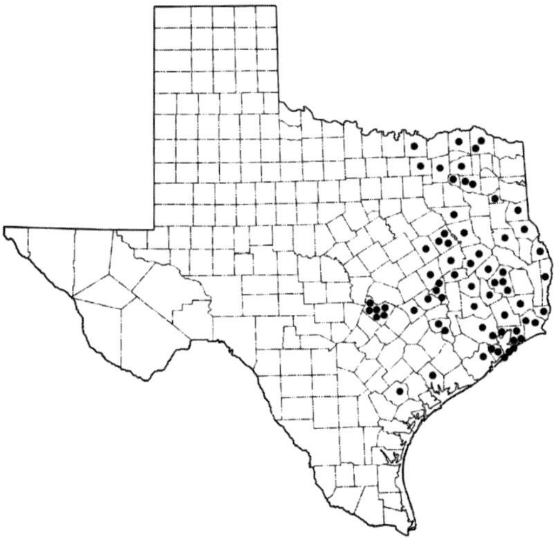
Galium tinctorium

Galium trifidum

Galium triflorum

Galium uncinulatum

Galium uniflorum

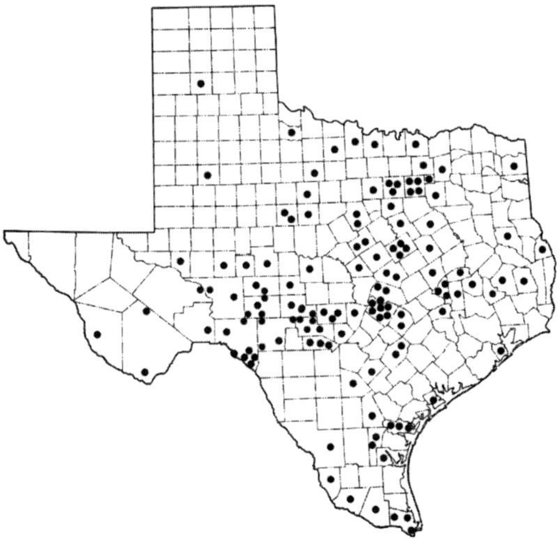

Galium virgatum

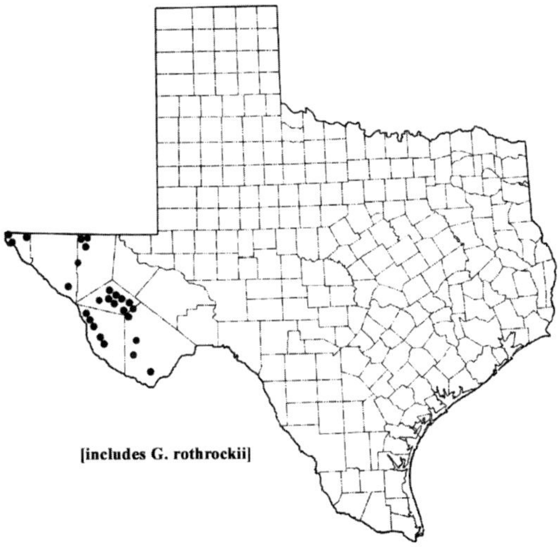
[includes G. rothrockii]

Galium wrightii

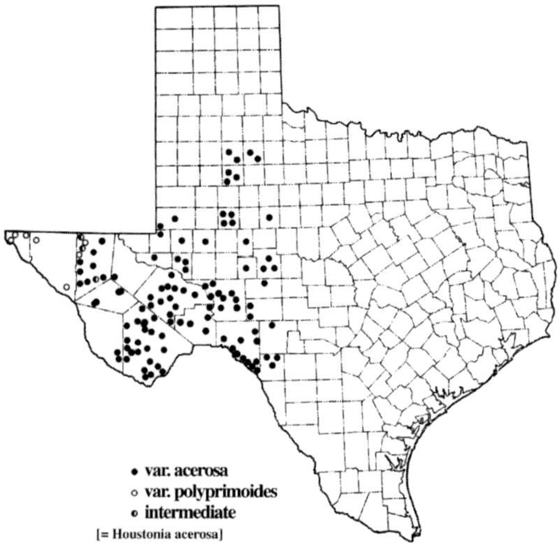
• var. acerosa
○ var. polyprimoides
◦ intermediate
[= Houstonia acerosa]

Hedyotis acerosa

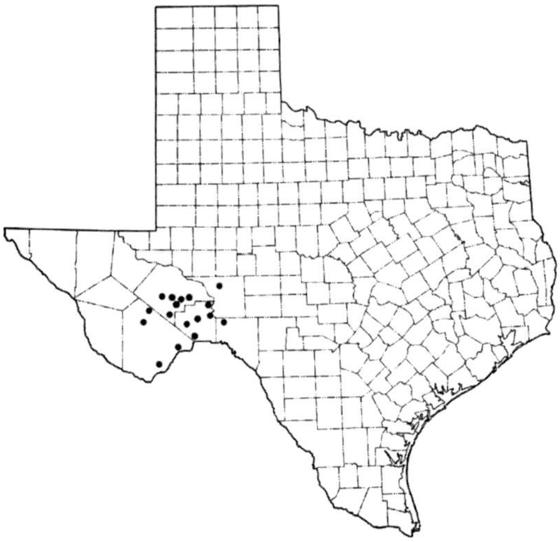

Hedyotis angulata

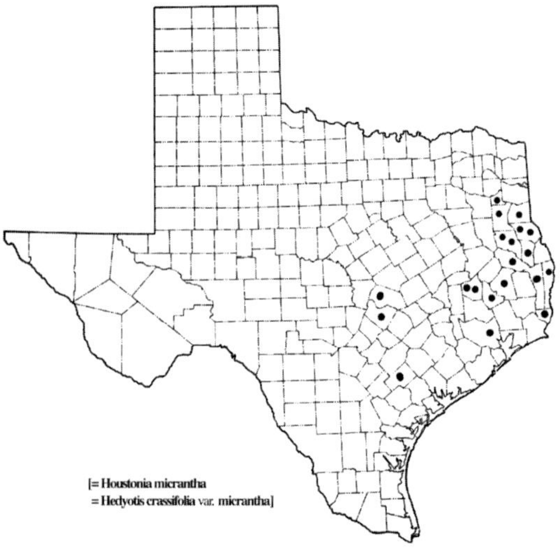
[= Houstonia micrantha
= Hedyotis crassifolia var. micrantha]

Hedyotis australis

[= Houstonia butterwickae]

Hedyotis butterwickiae

[= Houstonia correllii]

Hedyotis correllii

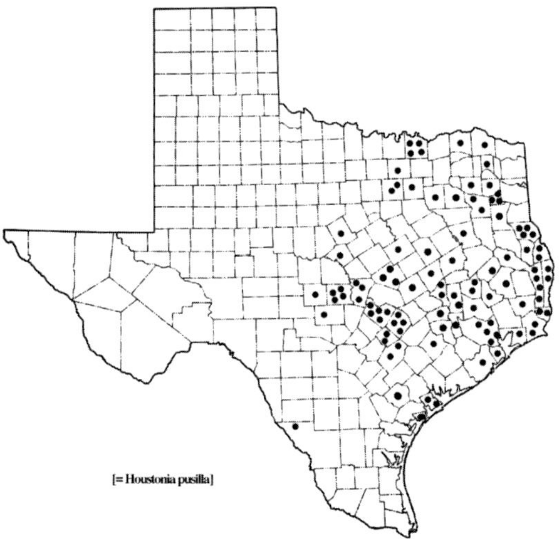
[= Houstonia pusilla]

Hedyotis crassifolia

Hedyotis croftiae

[= Houstonia parviflora]

Hedyotis greenmanii

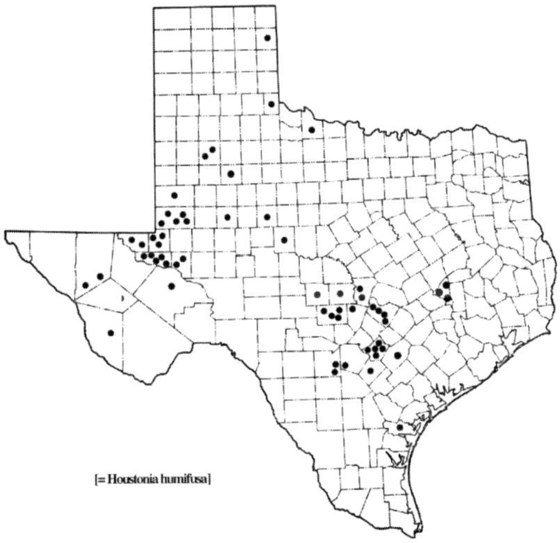
[= Houstonia humifusa]

Hedyotis humifusa

[= Houstonia intricata]

Hedyotis intricata

[= Houstonia longifolia]

Hedyotis longifolia

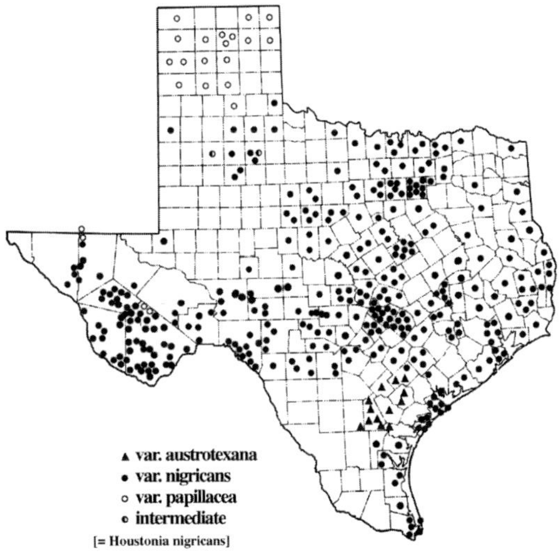
var. austrotexana
var. nigricans
var. papillacea
intermediate
[= Houstonia nigricans]

Hedyotis nigricans

Hedyotis pooleana

[= Houstonia purpurea]

Hedyotis purpurea

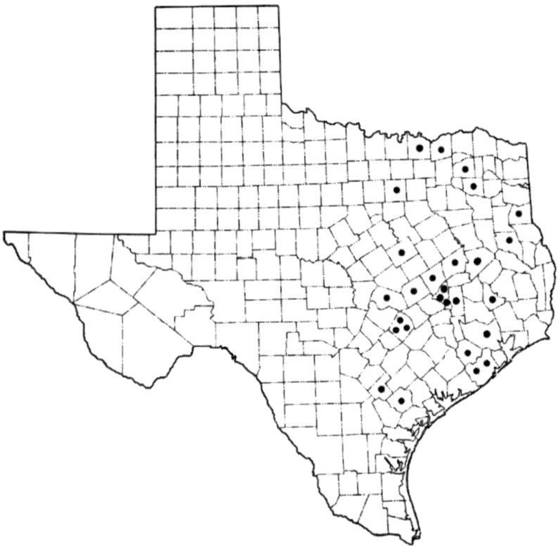

Hedyotis rosea

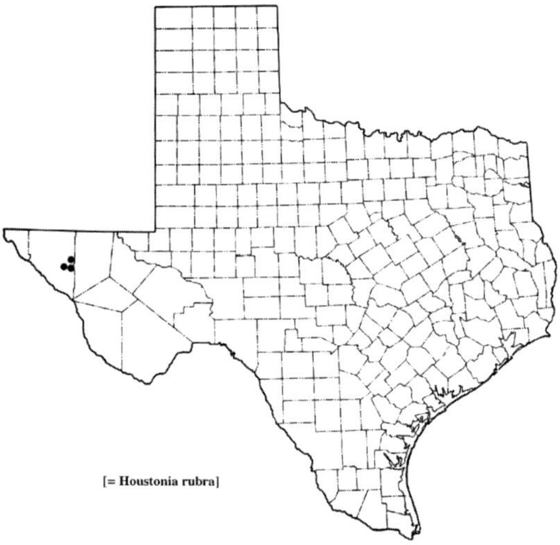
[= Houstonia rubra]

Hedyotis rubra

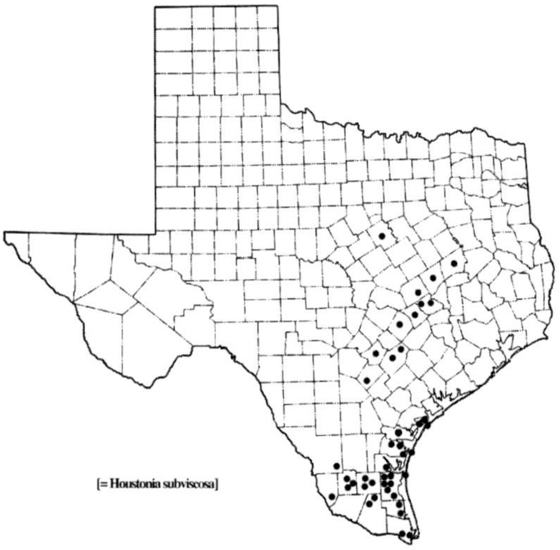
[= Houstonia subviscosa]

Hedyotis subviscosa

Hedyotis wrightii

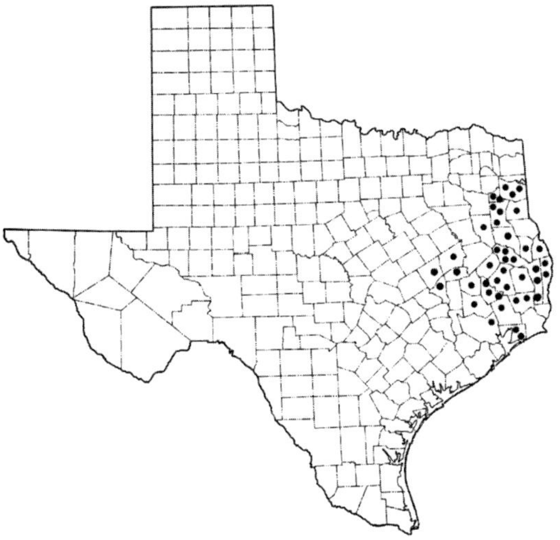

Mitchella repens

[= M. villosus;
includes M. breviflorus]

Mitracarpus hirtus

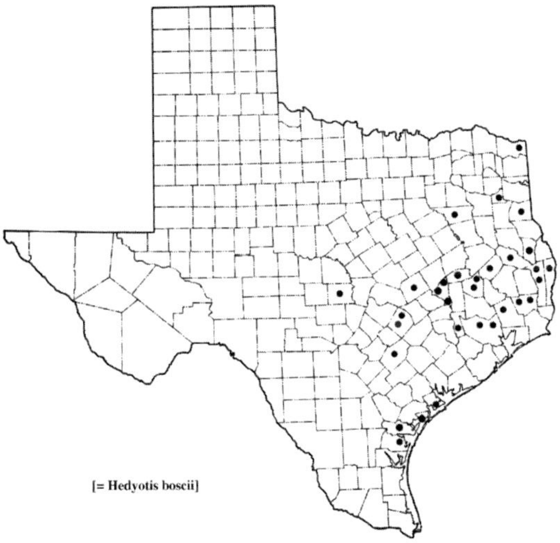
[= Hedyotis boscii]

Oldenlandia boscii

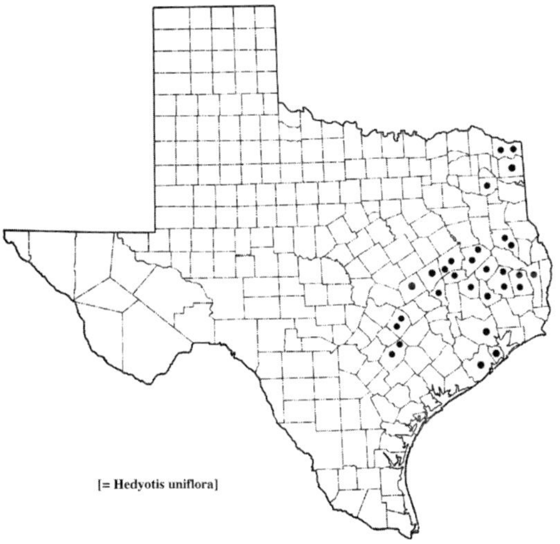
[= Hedyotis uniflora]

Oldenlandia uniflora

Paederia foetida

Pentodon pentandrus

Randia rhagocarpa

Richardia brasiliensis

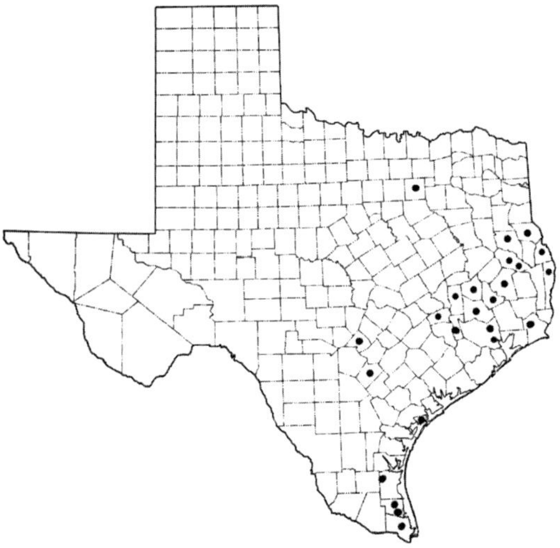

Richardia scabra

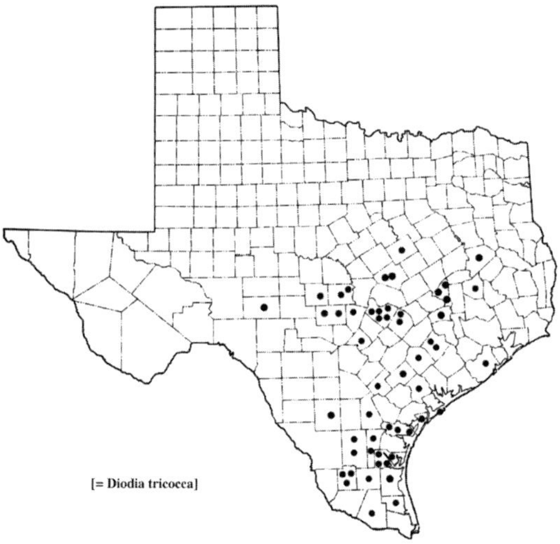
[= Diodia tricocca]

Richardia tricocca

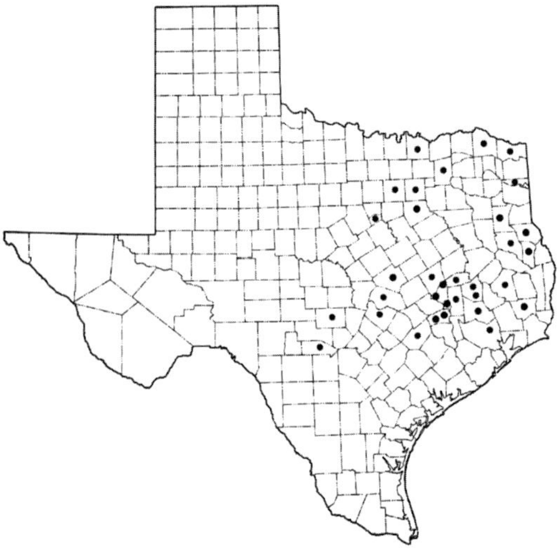

Sherardia arvensis

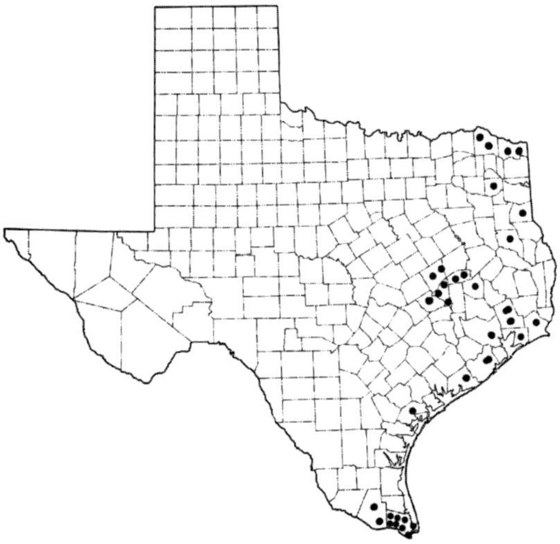

Spermacoce glabra

[includes S. floridana
sensu Correll & Johnston]

Spermacoce tenuior

Amyris madrensis

Amyris texana

Choisya dumosa

Citrus aurantium

Esenbeckia berlandieri

Helietta parvifolia

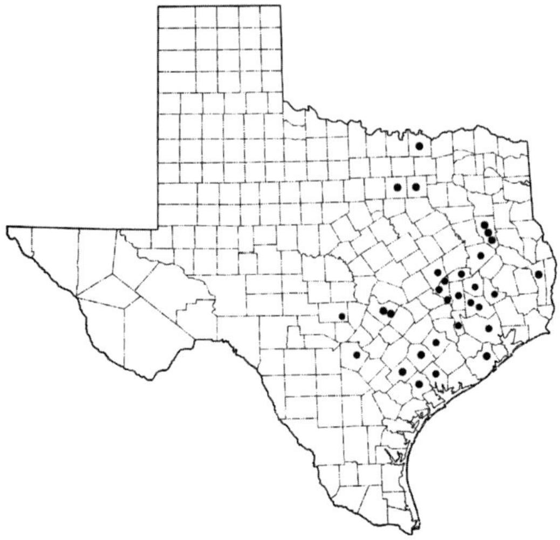

Poncirus trifoliata

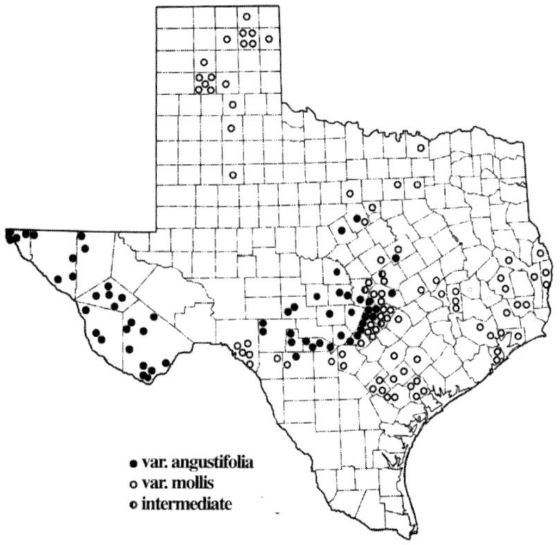
• var. angustifolia
○ var. mollis
o intermediate

Ptelea trifoliata

Ruta chalapensis

Ruta graveolens

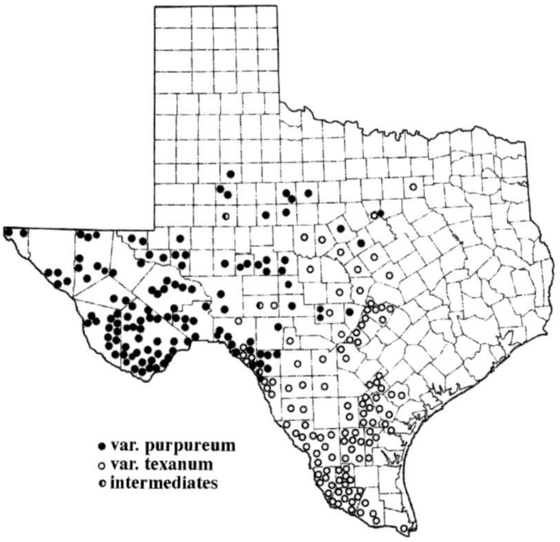
• var. purpureum
○ var. texanum
o intermediates

Thamnosma texanum

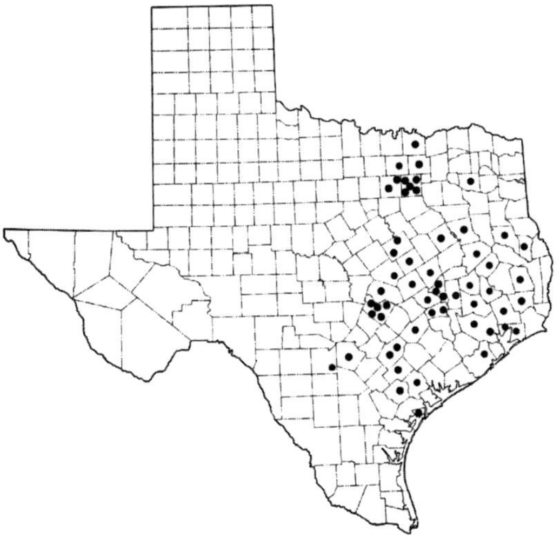

Zanthoxylum clava-herculis

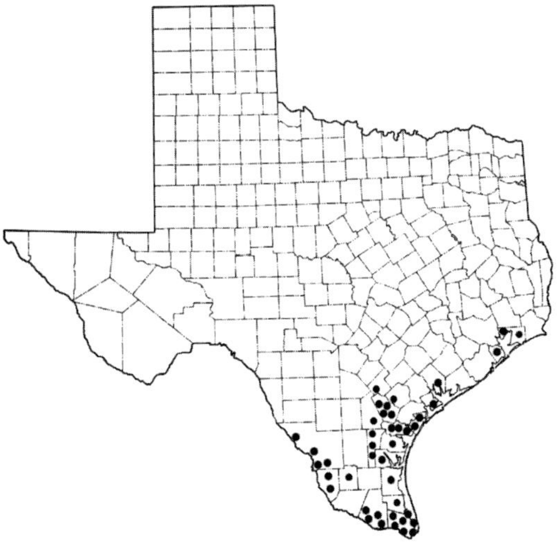

Zanthoxylum fagara

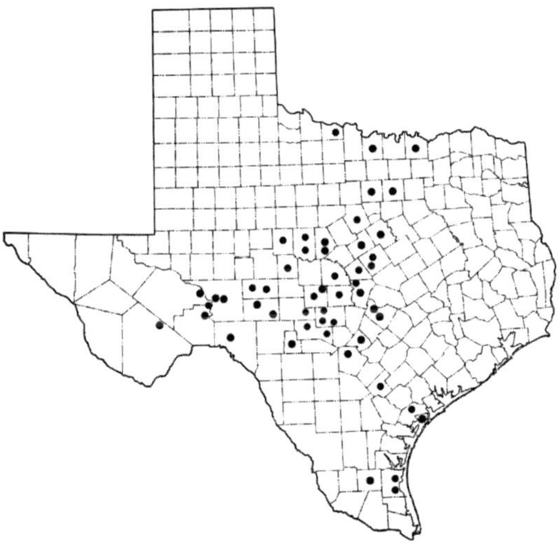

Zanthoxylum hirsutum

Zanthoxylum parvum

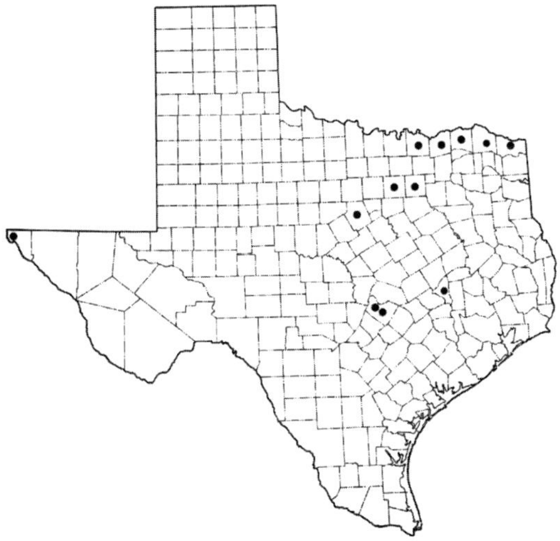

Populus alba

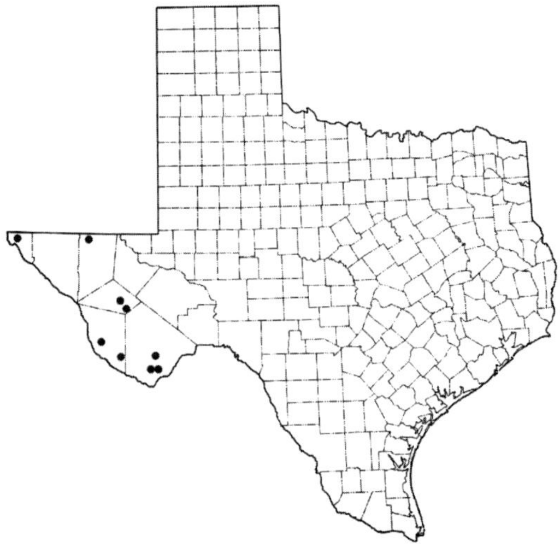

Populus angustifolia

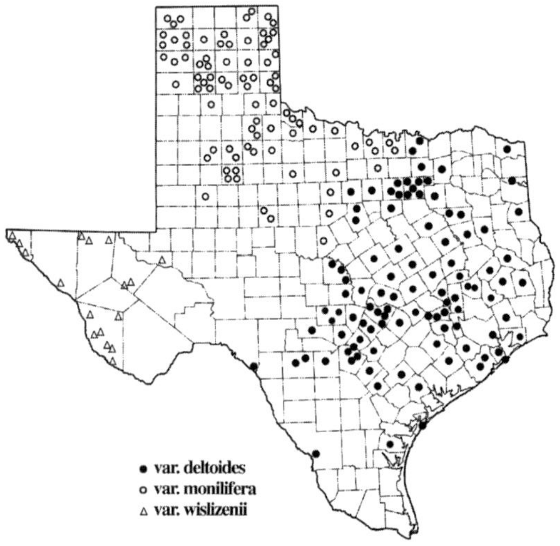
var. deltoides
var. monilifera
var. wislizenii

Populus deltoides

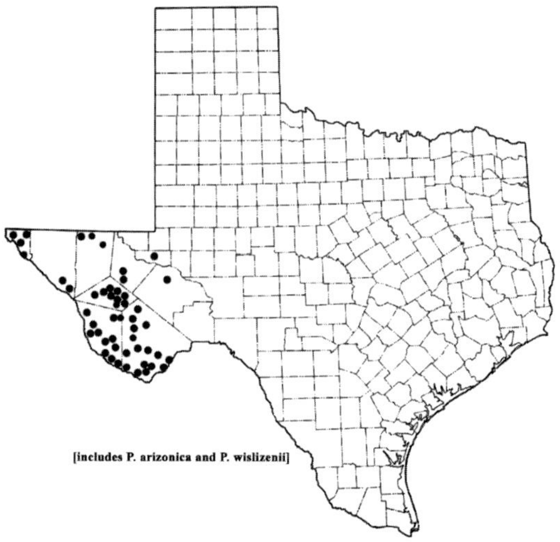
[includes P. arizonica and P. wislizenii]

Populus fremontii

[P. angustifolia x P. fremontii;
= P. hinckleyana]

Populus × hinkleyana

Populus nigra

Populus tremuloides

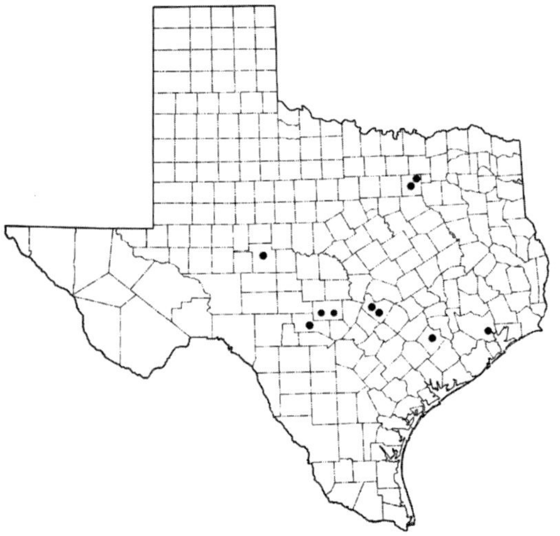

Salix babylonica

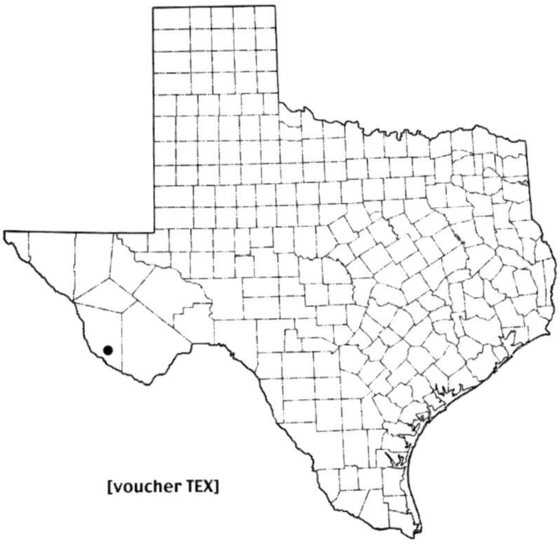
[voucher TEX]

Salix bonplandiana

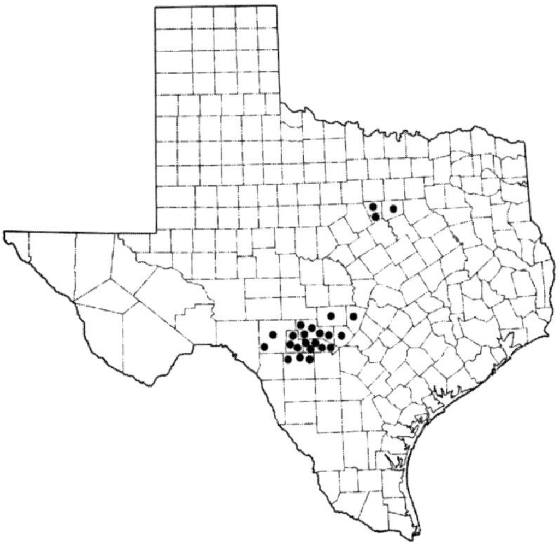

Salix caroliniana

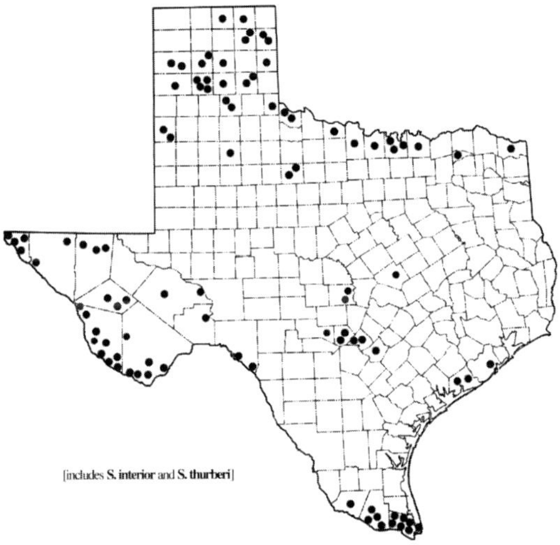
[includes S. interior and S. thurberi]

Salix exigua

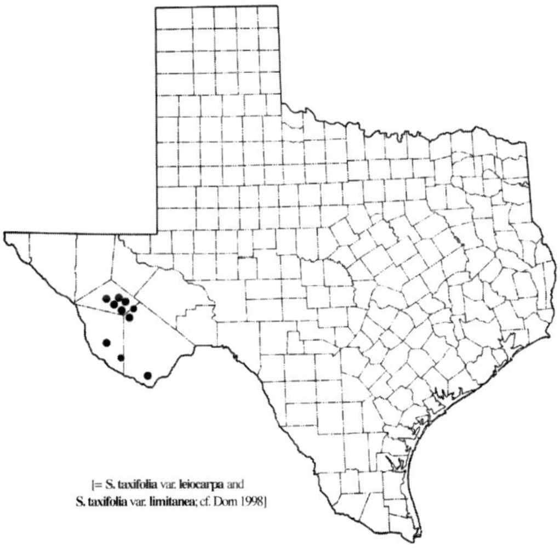
[= S. taxifolia var. leiocarpa and
S. taxifolia var. limitanea; cf. Dorn 1998]

Salix exilifolia

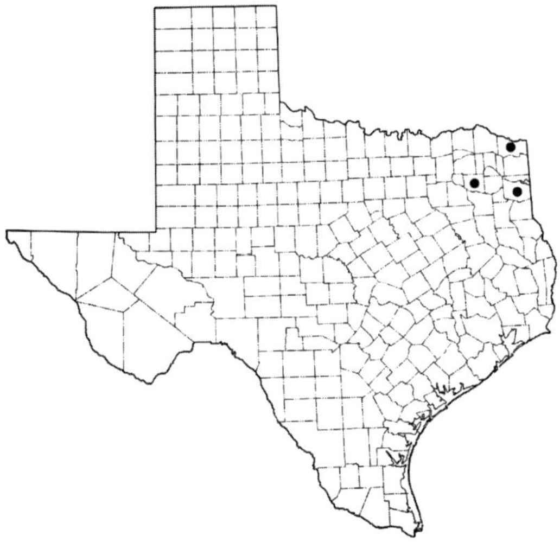

Salix humilis

Salix lasiolepis

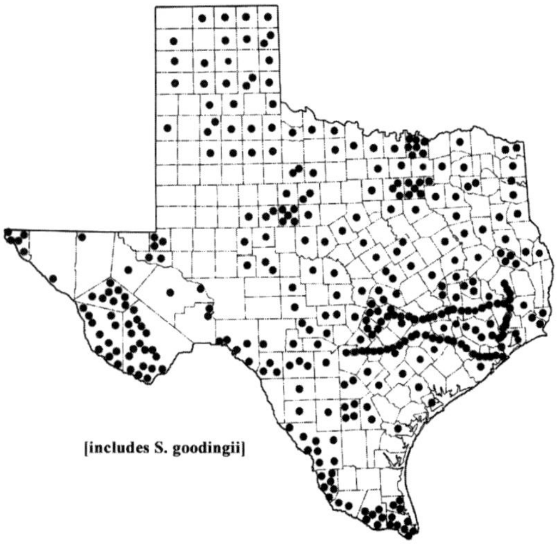
[includes S. goodingii]

Salix nigra

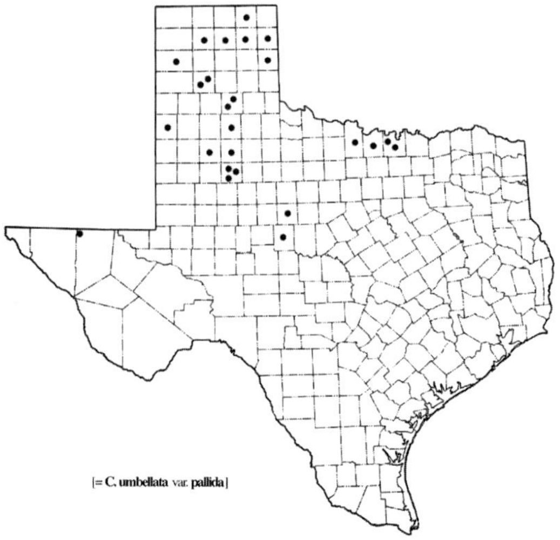
[= C. umbellata var. pallida]

Comandra pallida

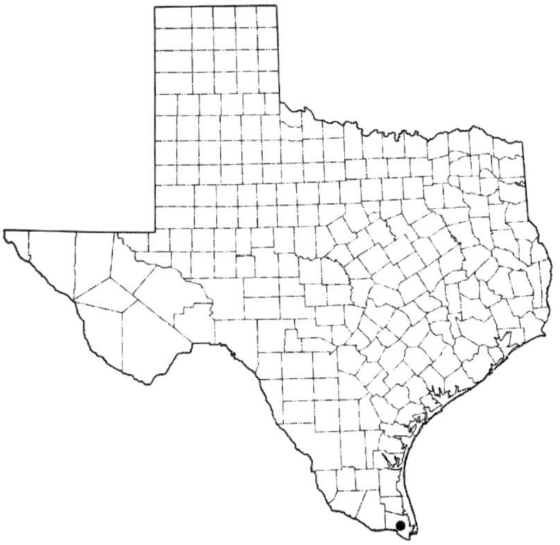

Cardiospermum corindum

Cardiospermum dissectum

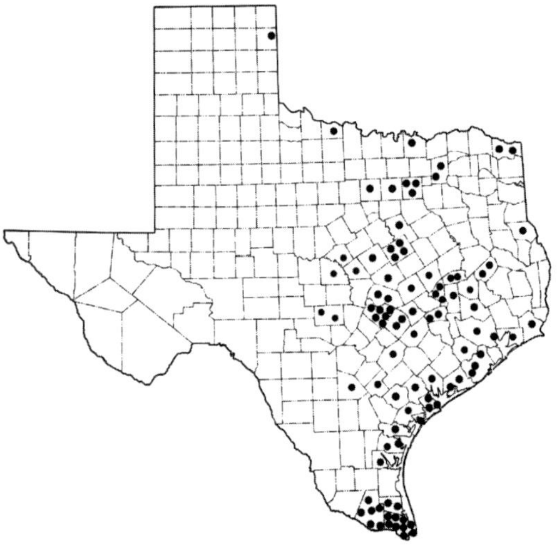

Cardiospermum halicacabum

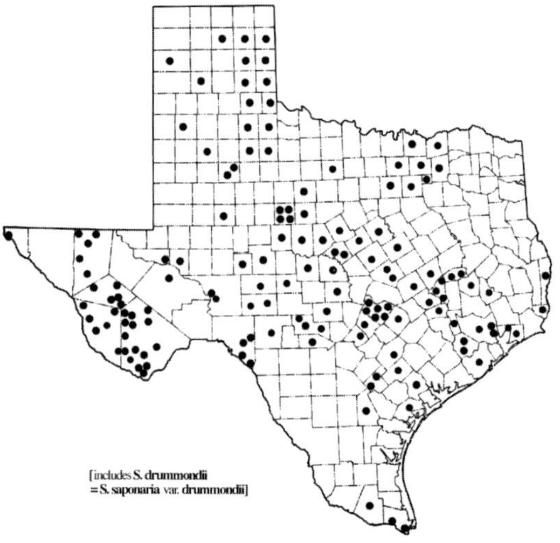
[includes S. drummondii
= S. saponaria var. drummondii]

Sapindus saponaria

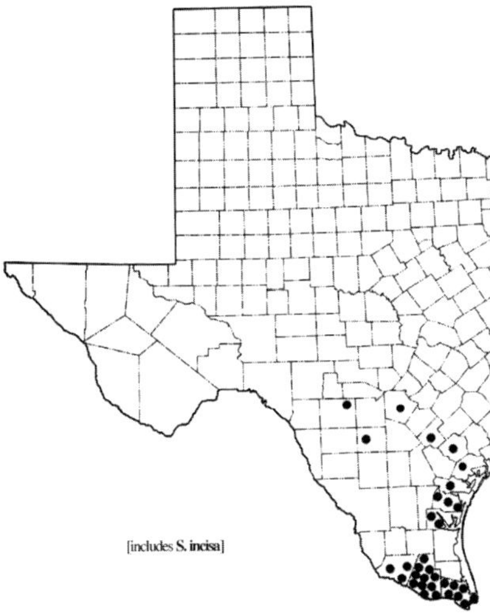
[includes S. incisa]

Serjania brachycarpa

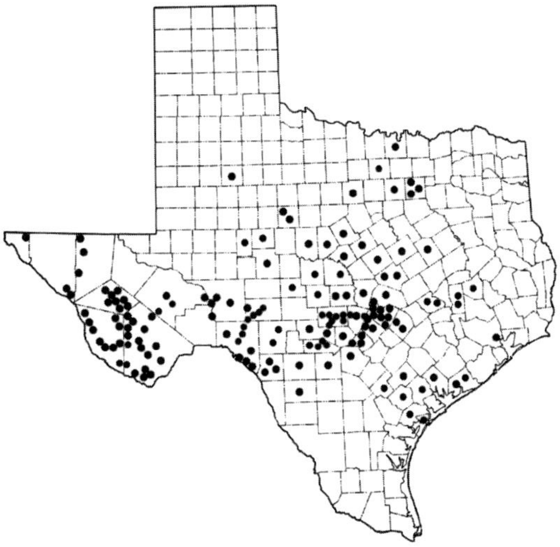

Ungnadia speciosa

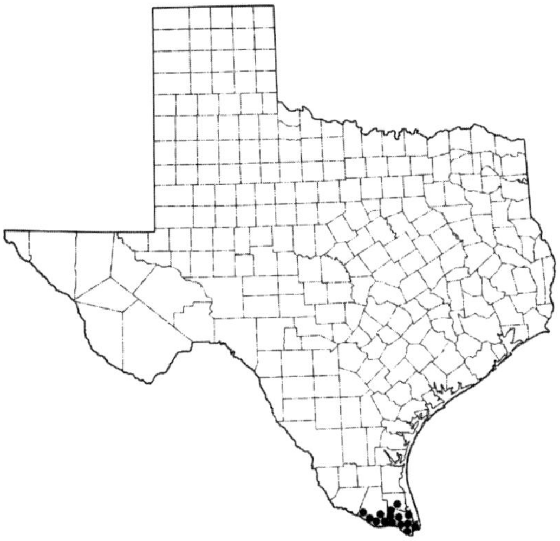

Urvillea ulmacea

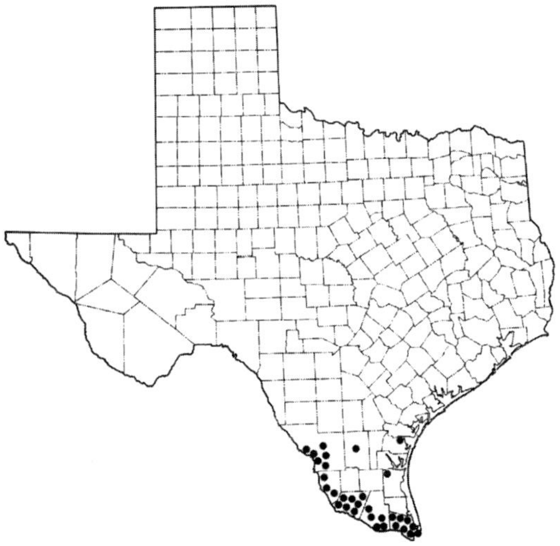

Sideroxylon celastrinum

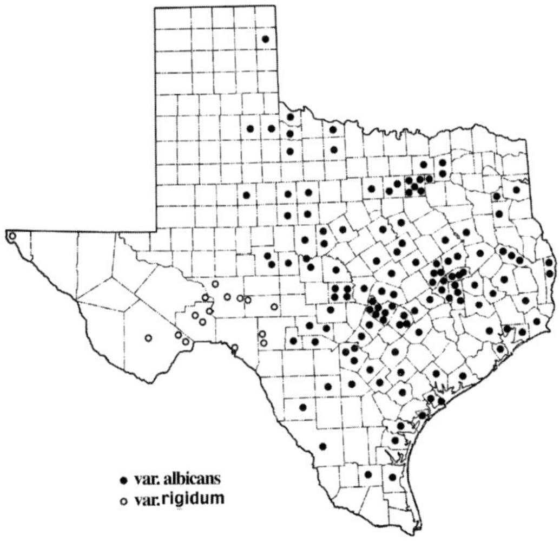
• var. albicans
○ var. rigidum

Sideroxylon lanuginosum

Sideroxylon lycioides

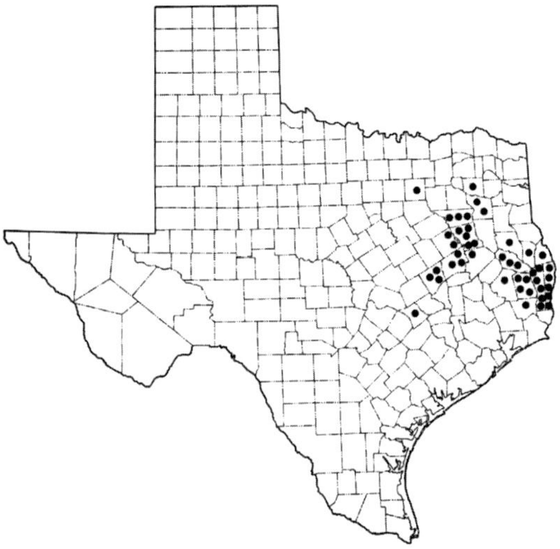

Sarracenia alata

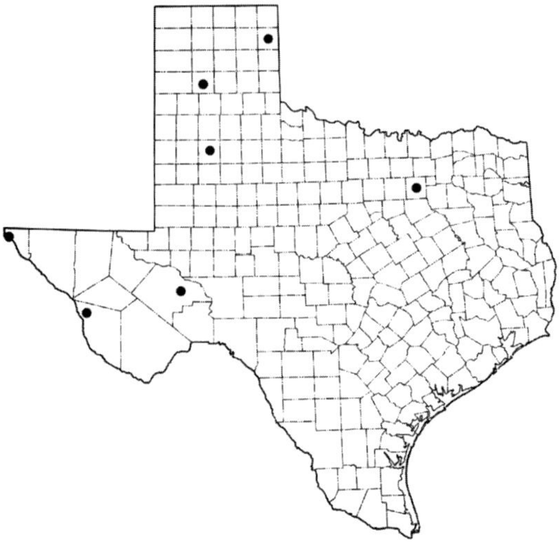

Anemopsis californica

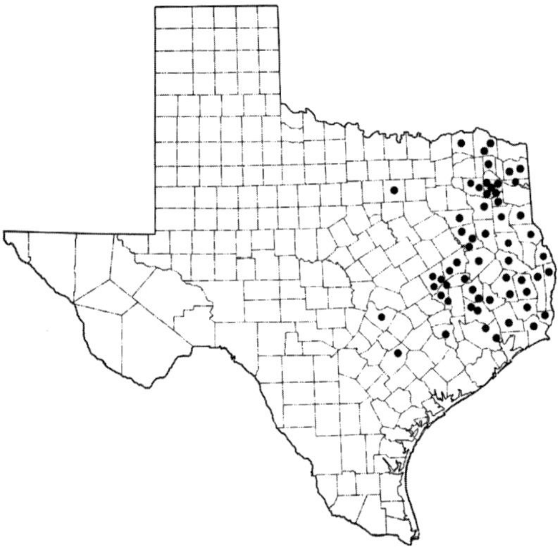

Saururus cernuus

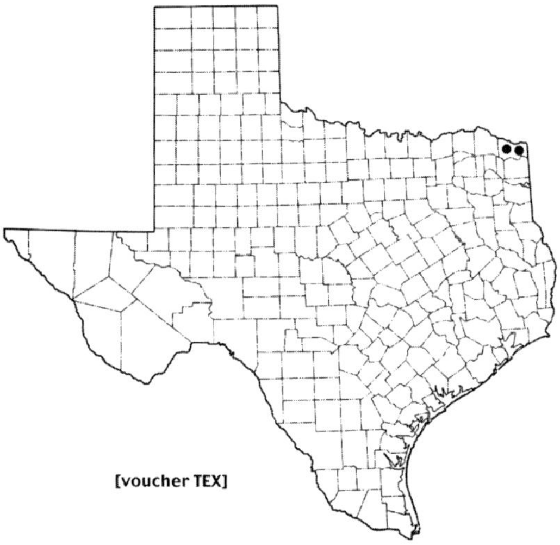
[voucher TEX]

Heuchera americana

[includes H. leptomeria]

Heuchera rubescens

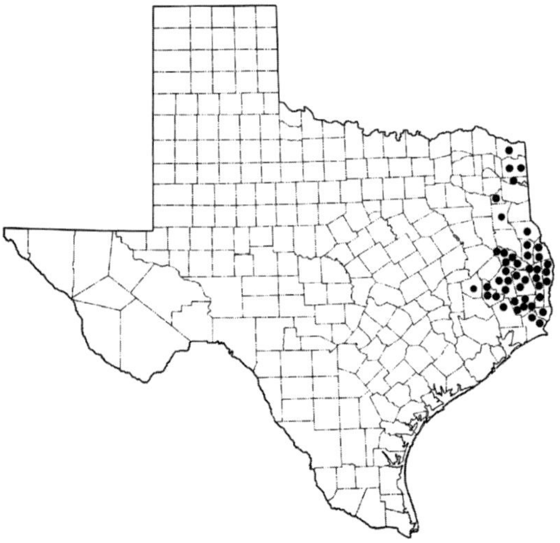

Itea virginica

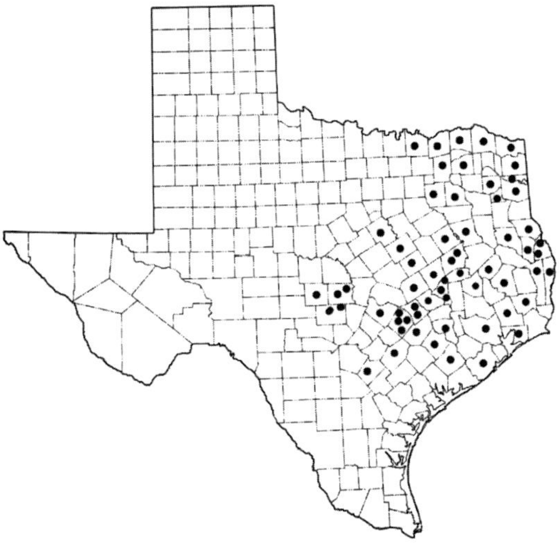

Lepuropetalon spathulatum

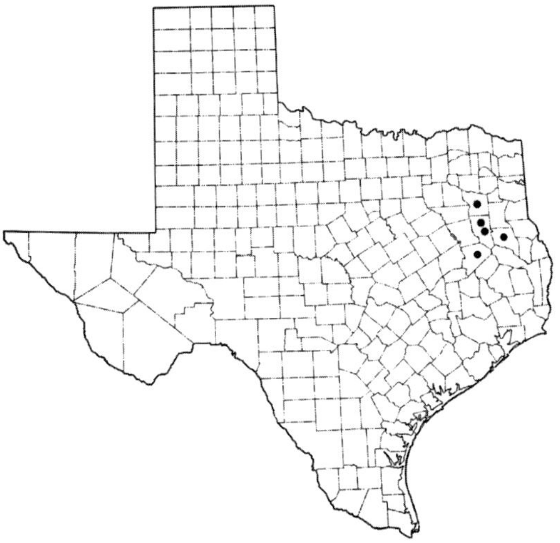

Parnassia asarifolia

Parnassia grandiflora

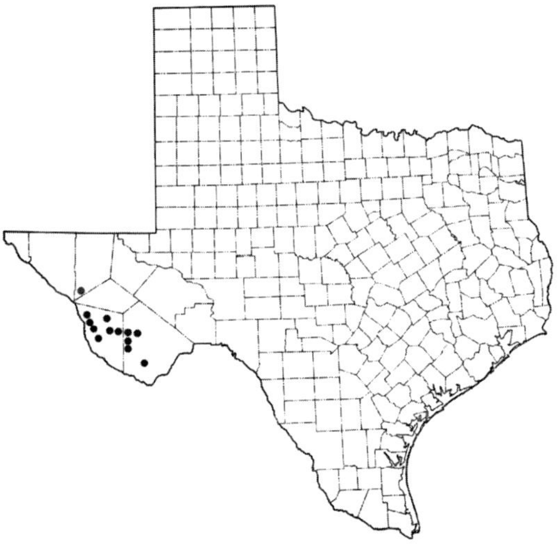

Ribes aureum

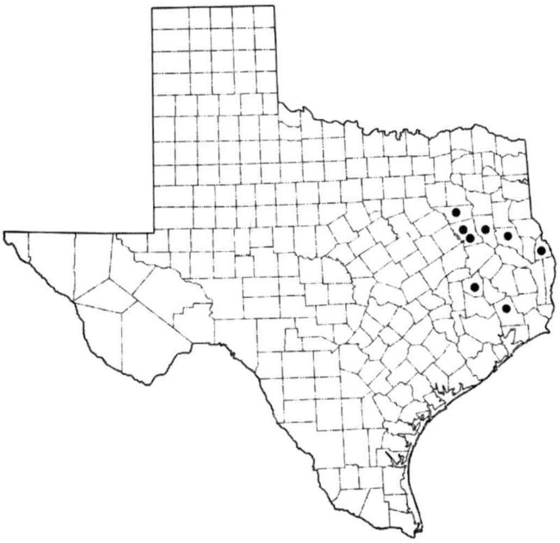

Ribes curvatum

Ribes leptanthum

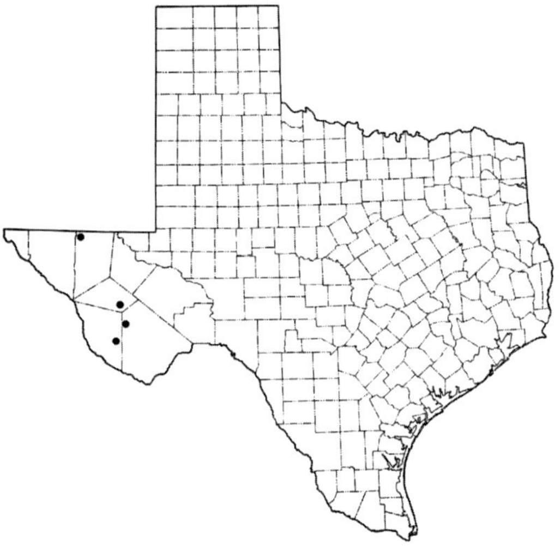
Ribes mescalerum

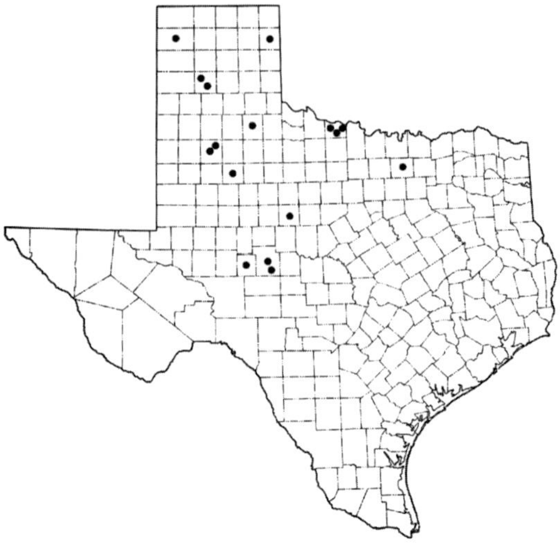
Ribes odoratum

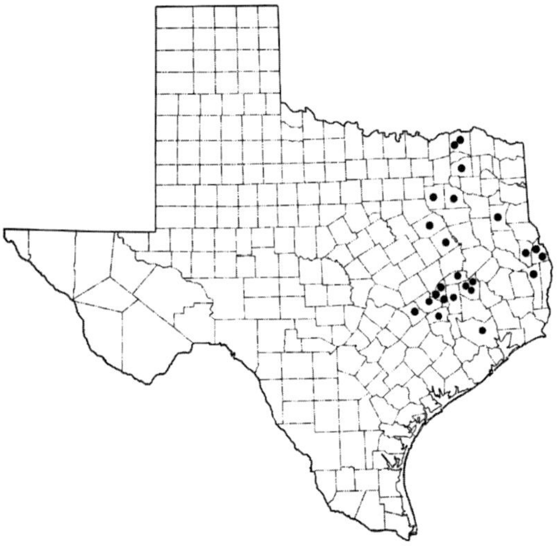
Saxifraga texana

Agalinis calycina

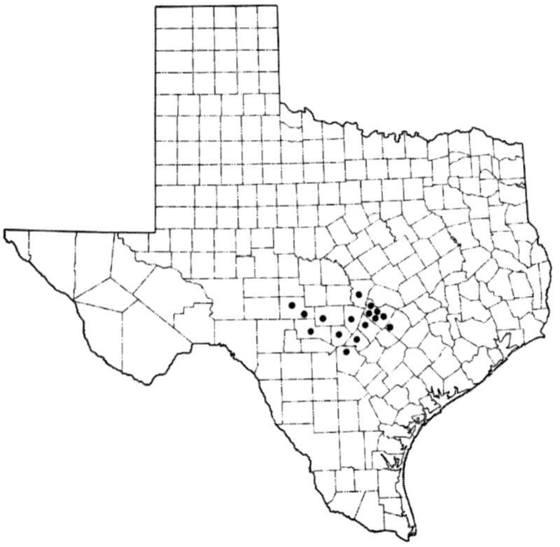
Agalinis edwardsiana

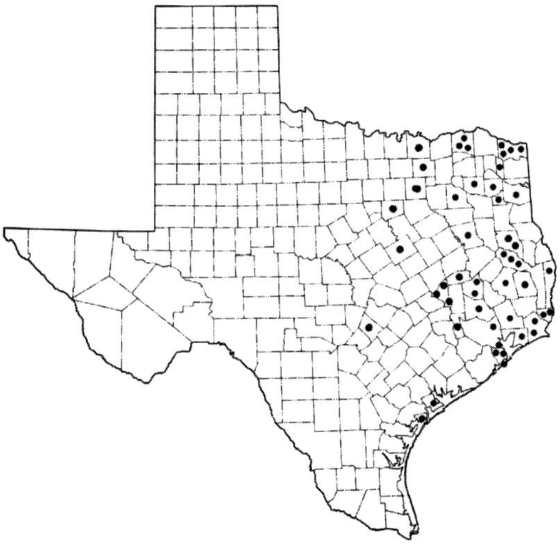
Agalinis fasciculata

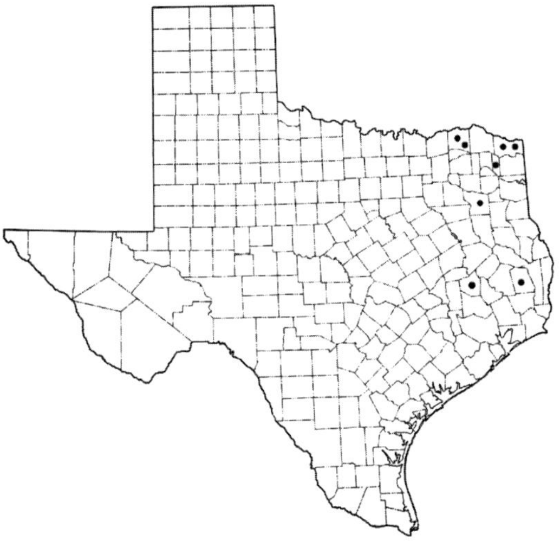

Agalinis gattingeri

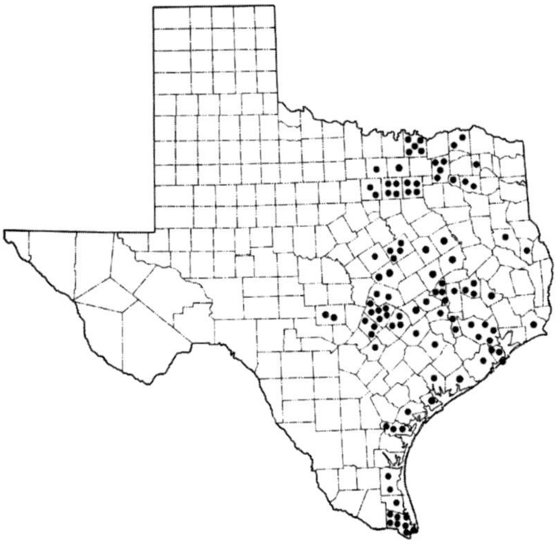

Agalinis heterophylla

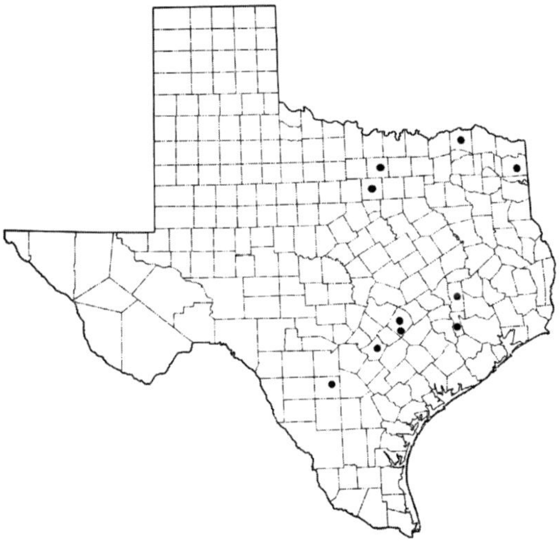

Agalinis homalantha

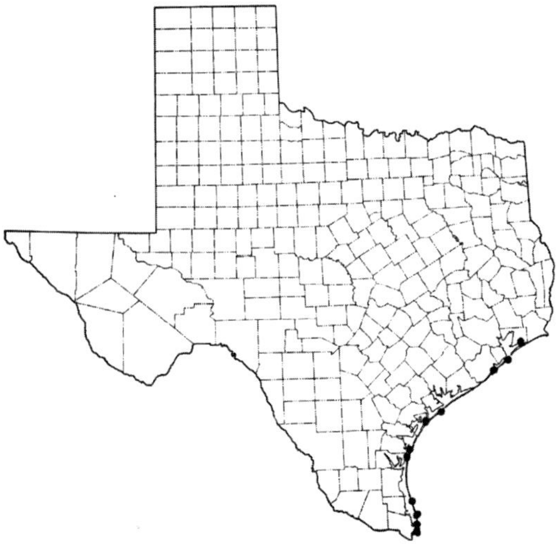

Agalinis maritima

Agalinis navasotensis

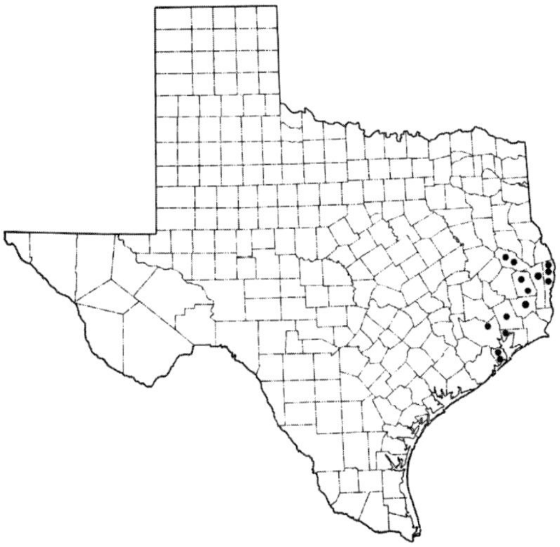

Agalinis oligophylla

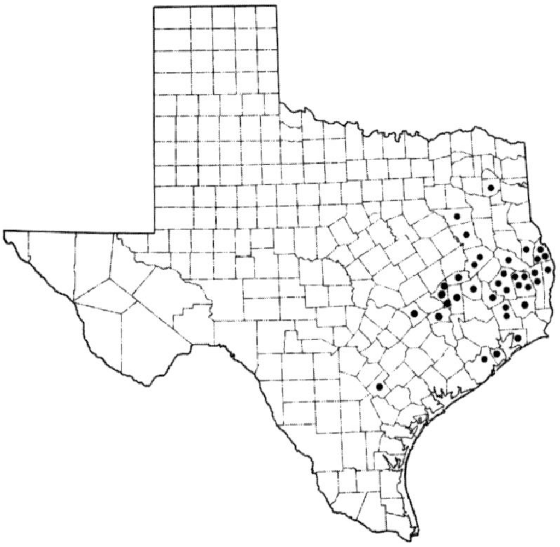

Agalinis purpurea

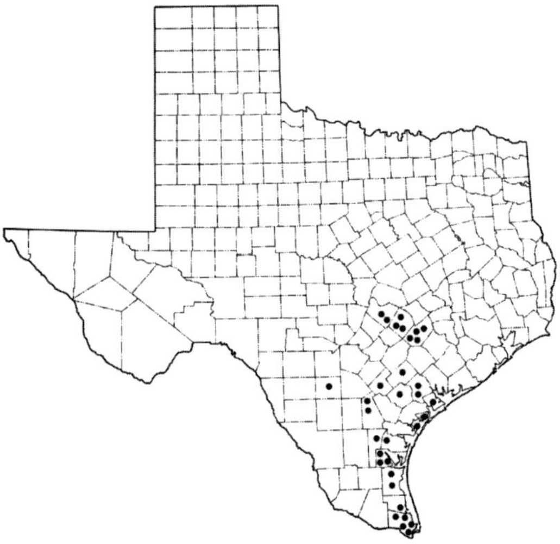

Agalinis strictifolia

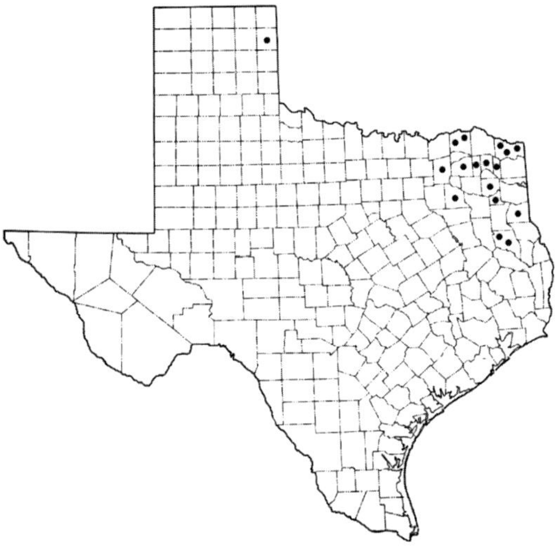

Agalinis tenuifolia

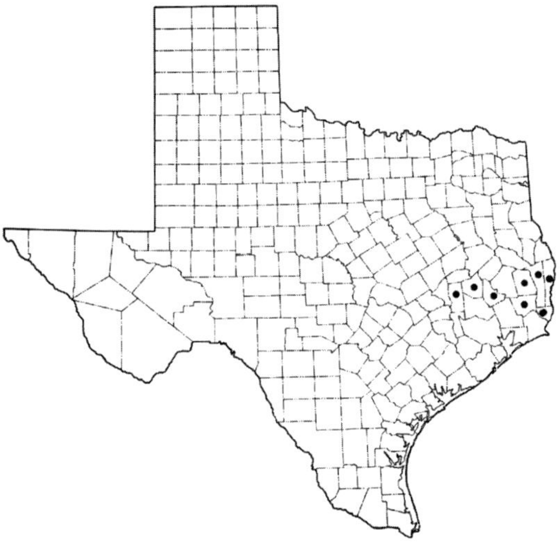

Agalinis viridis

Aureolaria flava

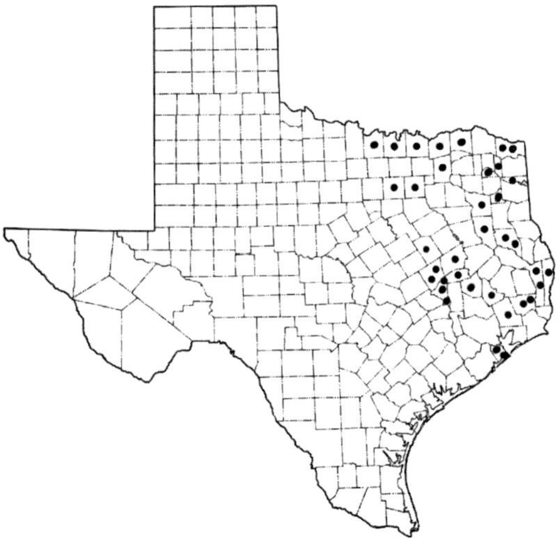

Aureolaria grandiflora

Aureolaria pectinata

Bacopa caroliniana

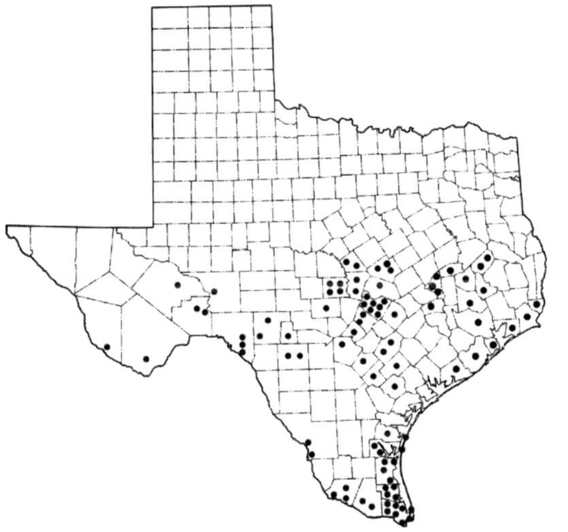

Bacopa monnieri

introduced 1986

Bacopa repens

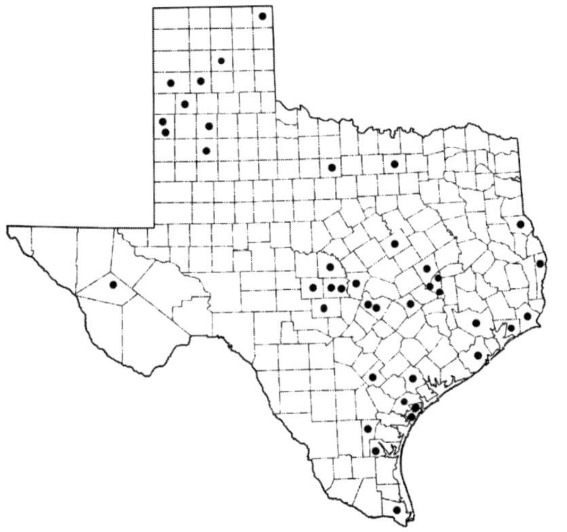

Bacopa rotundifolia

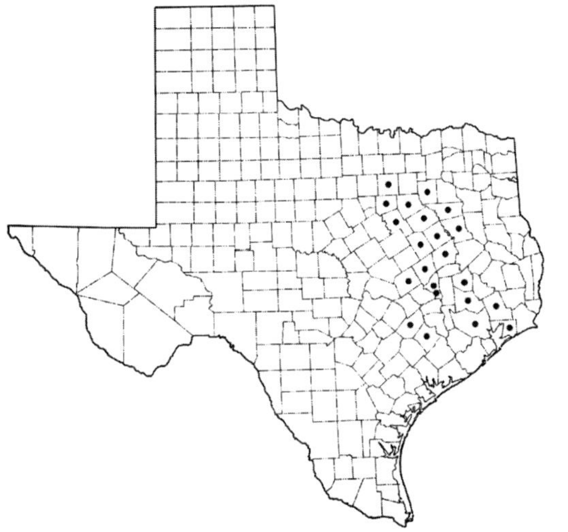

Bellardia trixago

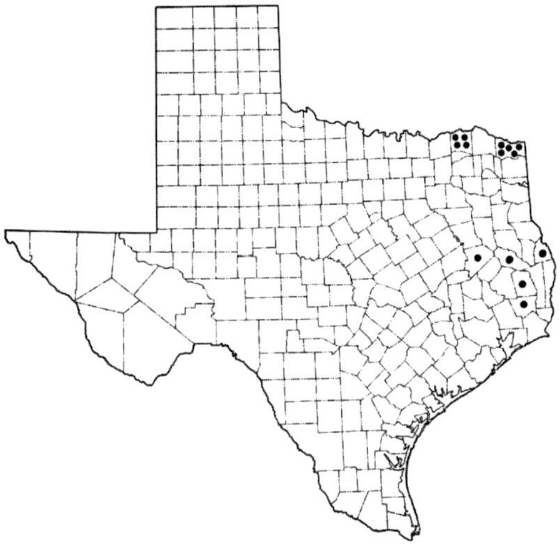

Buchnera americana

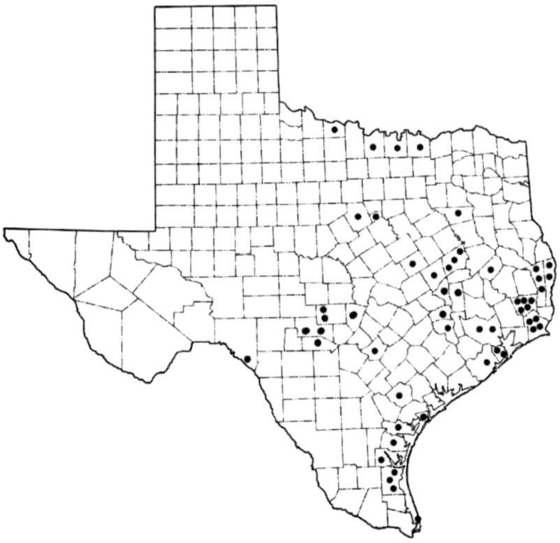

Buchnera floridana

[Sida 17: 523. 1996]

Capraria mexicana

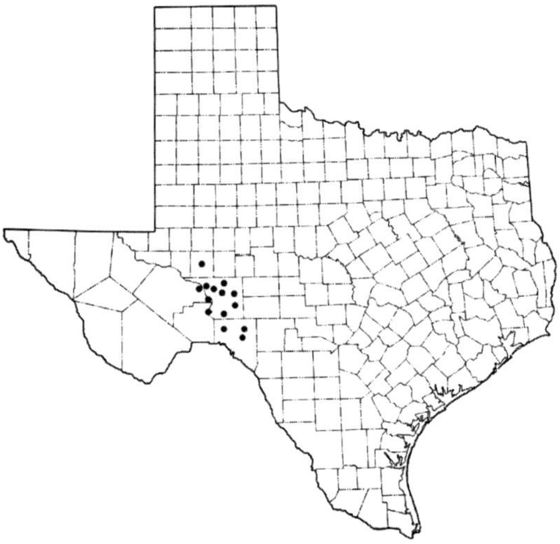

Castilleja genevievana

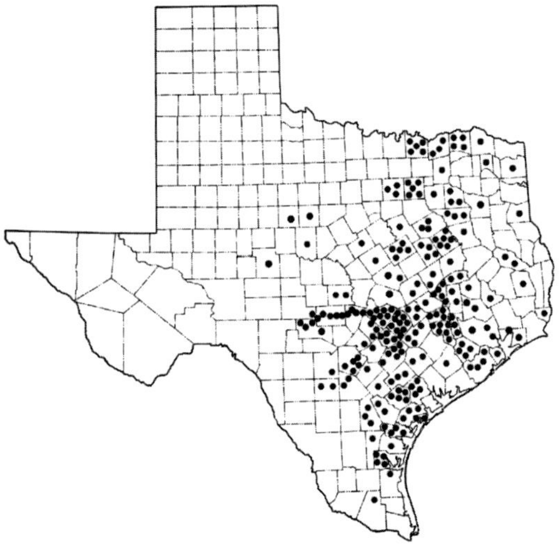

Castilleja indivisa

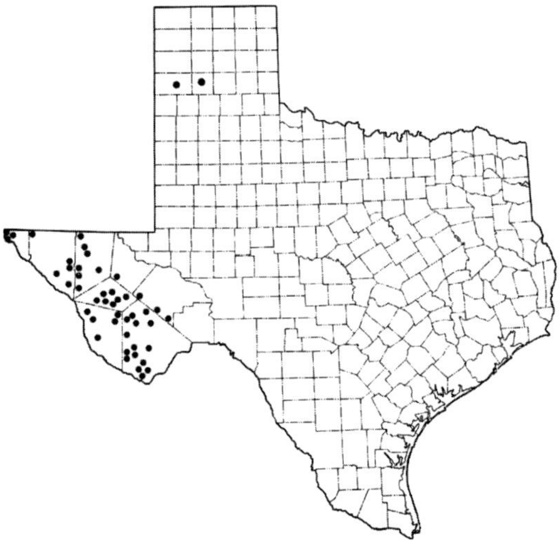

Castilleja integra

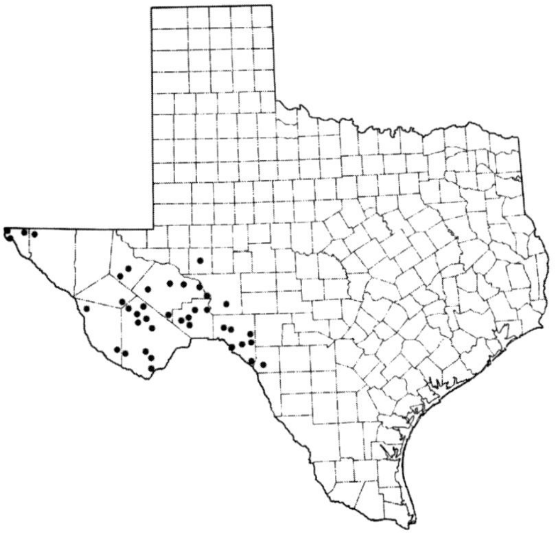

Castilleja lanata

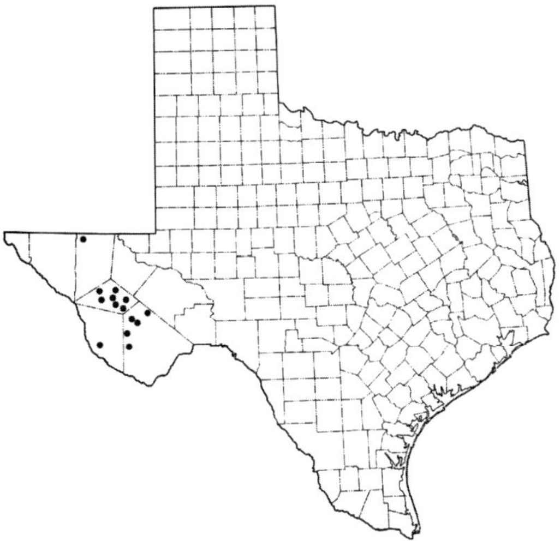

Castilleja mexicana

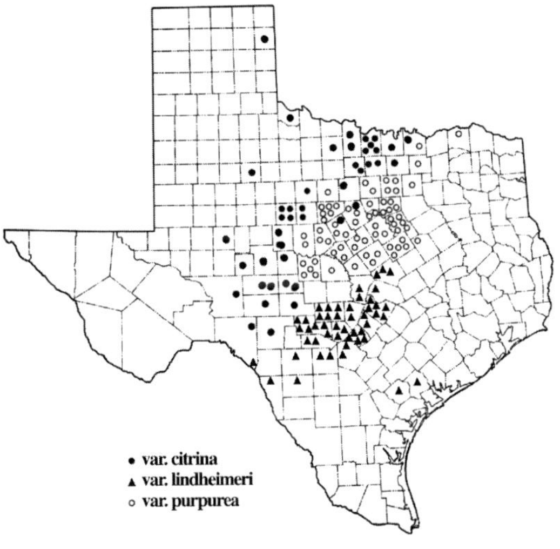
• var. citrina
▲ var. lindheimeri
○ var. purpurea

Castilleja purpurea

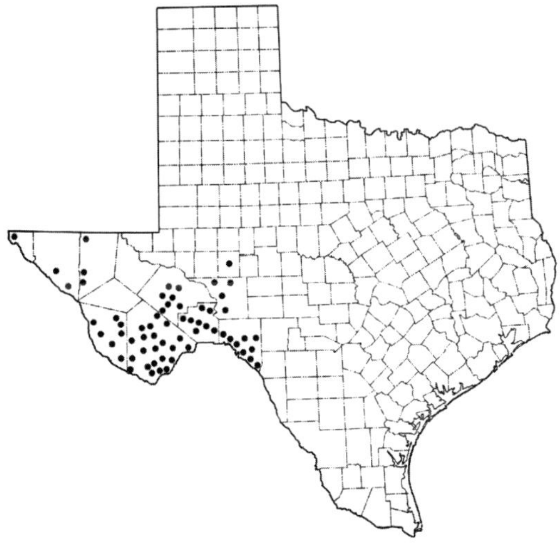

Castilleja rigida

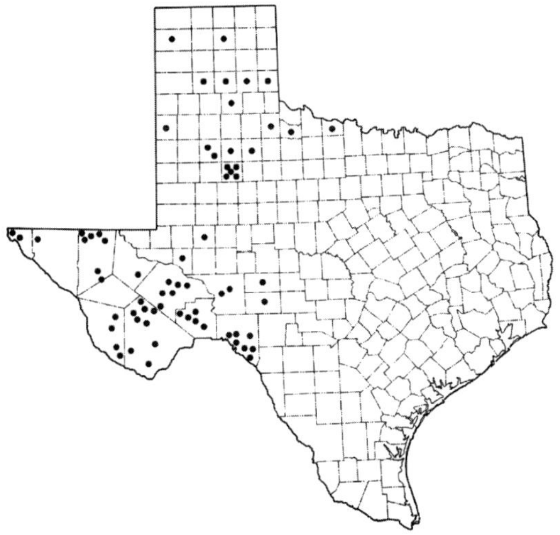

Castilleja sessiliflora

[includes C. ciliata and
C. livermorensis]

Castilleja wootonii

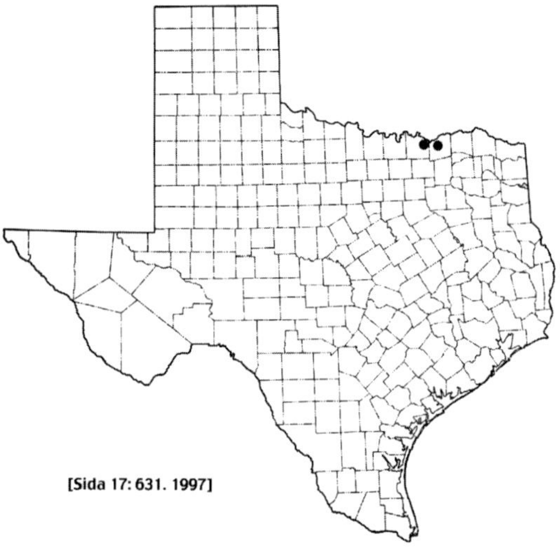
[Sida 17: 631. 1997]
Chaenorhinum minus

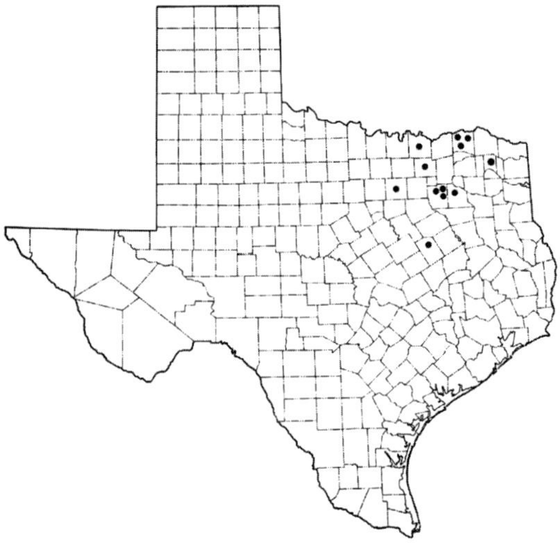
Collinsia violacea

Cordylanthus wrightii

Dasistoma macrophylla

Elatine brachysperma

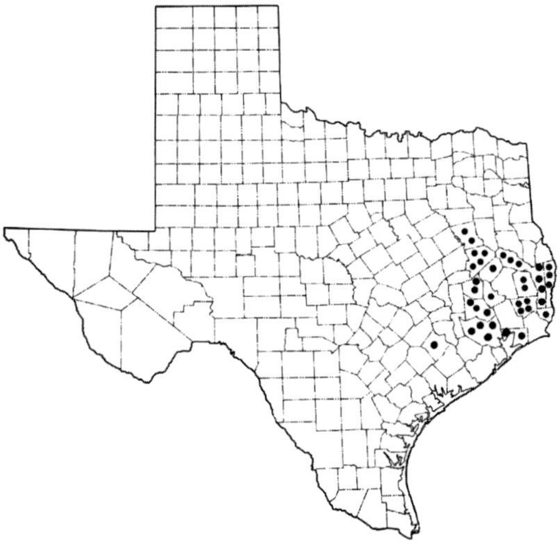
Gratiola brevifolia

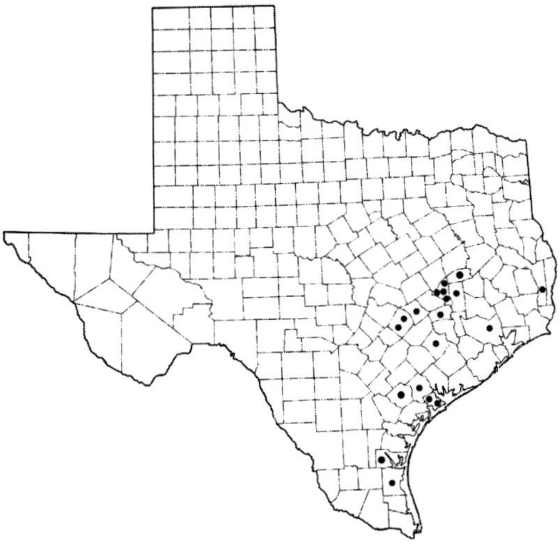

Gratiola flava

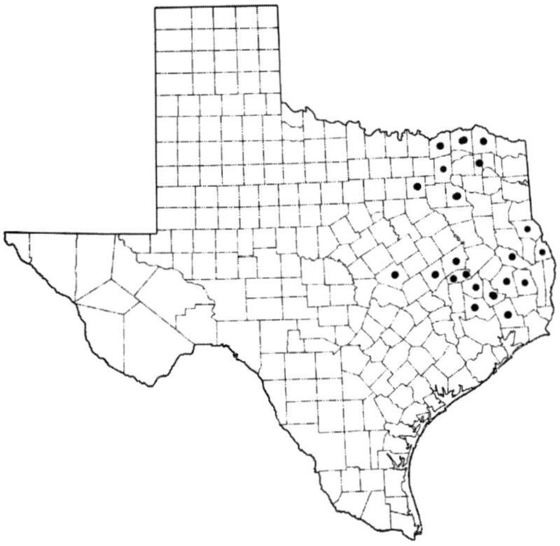

Gratiola neglecta

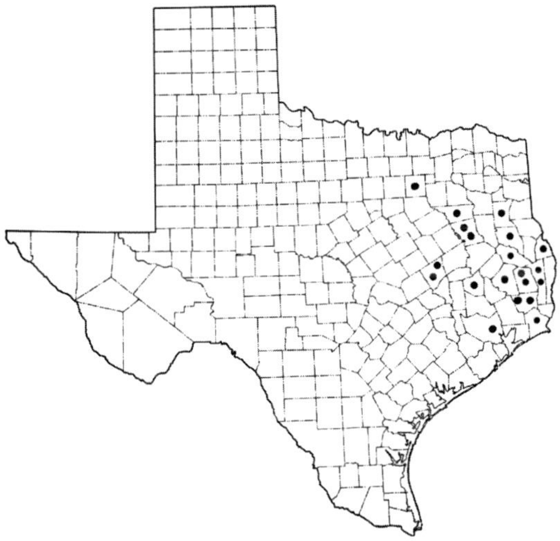

Gratiola pilosa

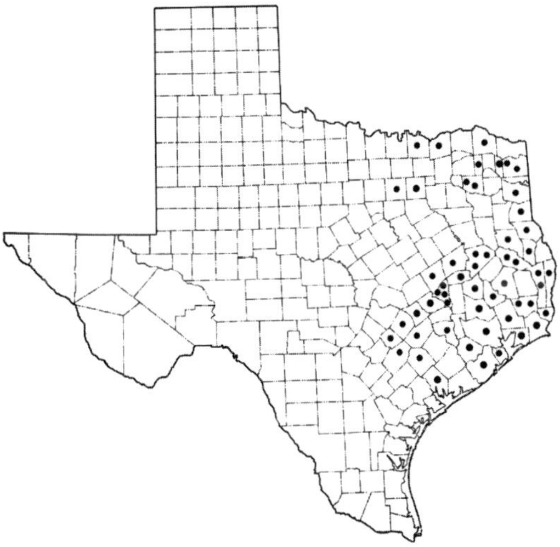

Gratiola virginiana

Kickxia elatine

Leucophyllum candidum

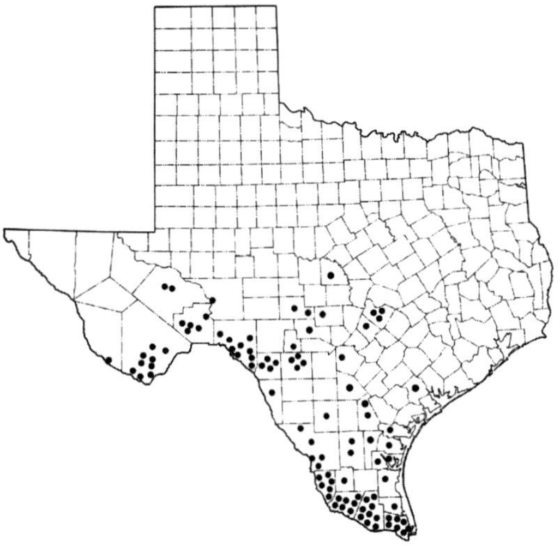

Leucophyllum frutescens

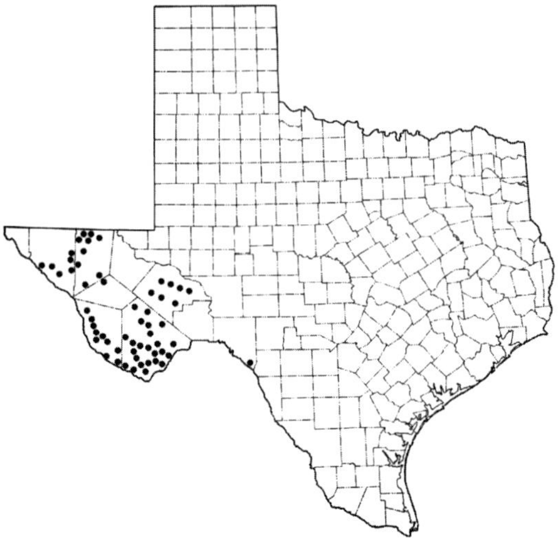

Leucophyllum minus

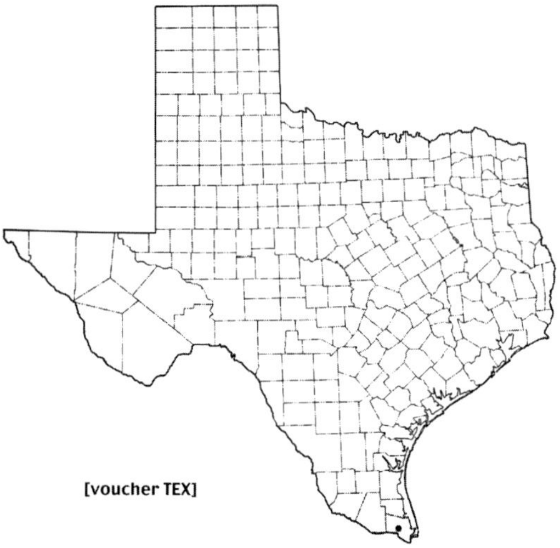
[voucher TEX]

Leucophyllum zygophyllum

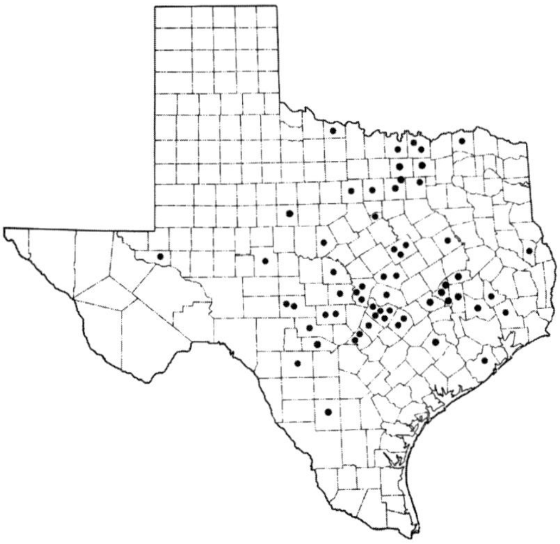

Leucospora multifida

Limnophila sessiliflora

Linaria anagallidea

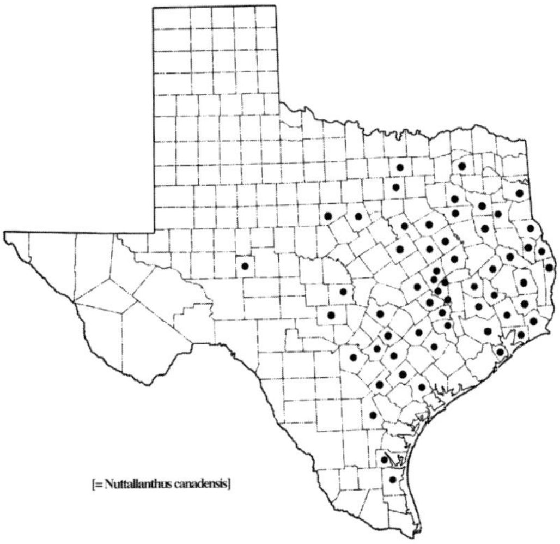
[= Nuttallanthus canadensis]

Linaria canadensis

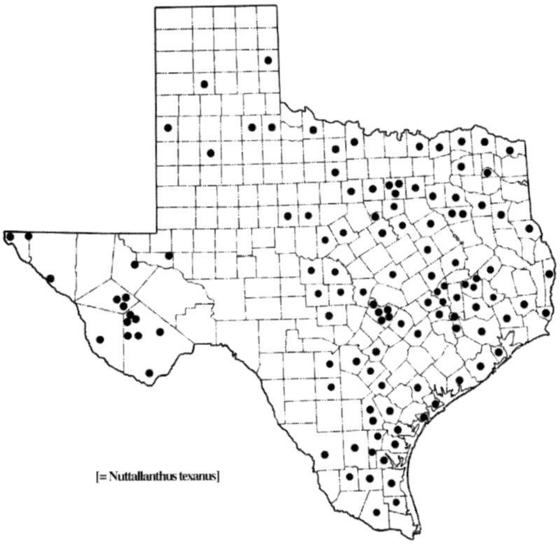
[= Nuttallanthus texanus]

Linaria texana

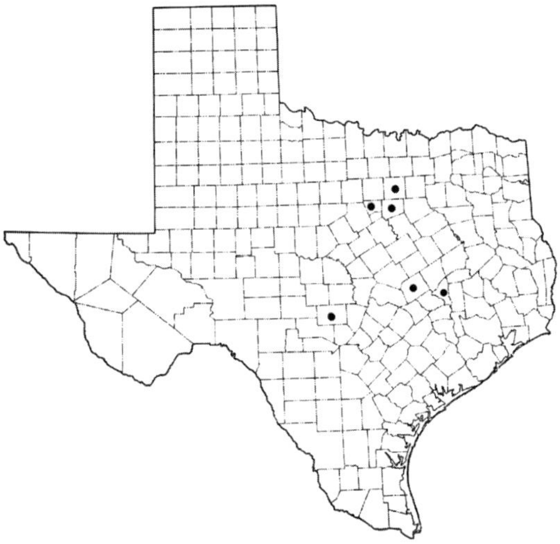

Linaria vulgaris

[Sida 18: 322. 1998]

Lindernia crustacea

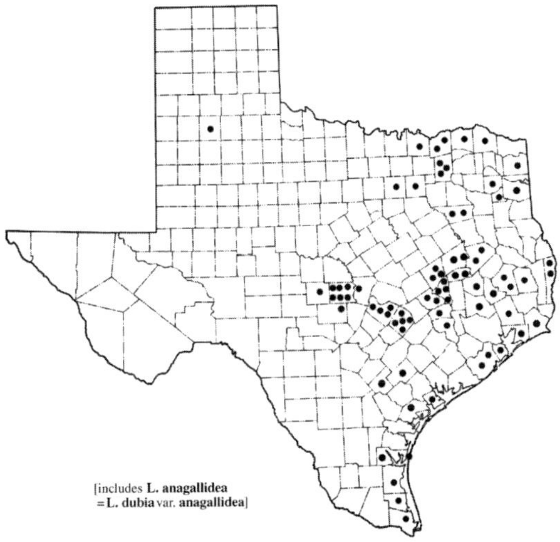
[includes L. anagallidea
= L. dubia var. anagallidea]

Lindernia dubia

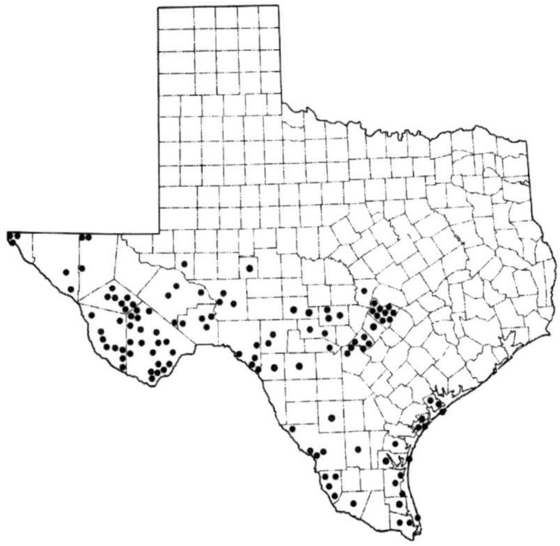

Maurandya antirrhiniflora

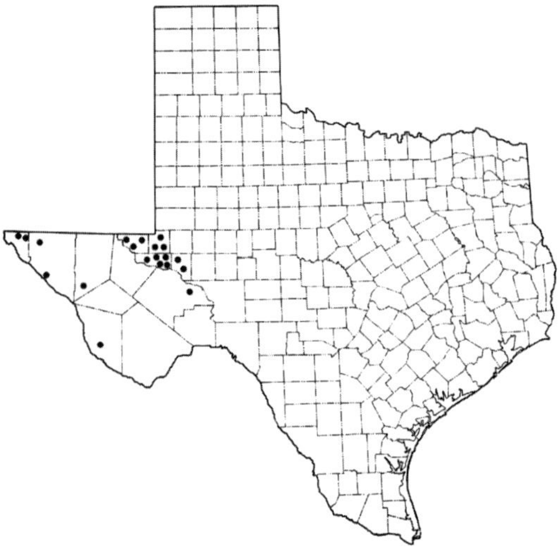

Maurandya wislizeni

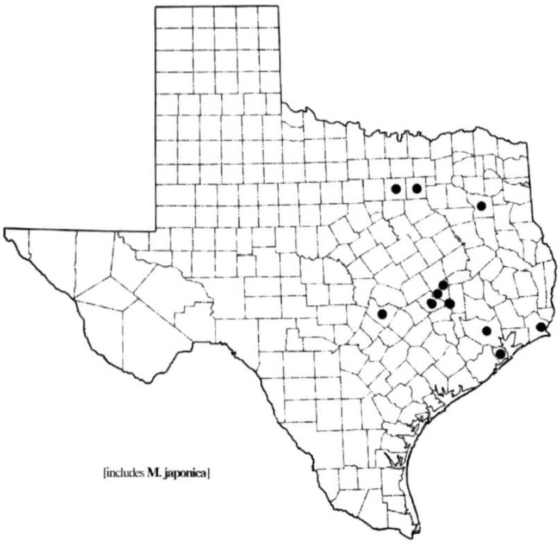
[includes M. japonica]

Mazus pumilus

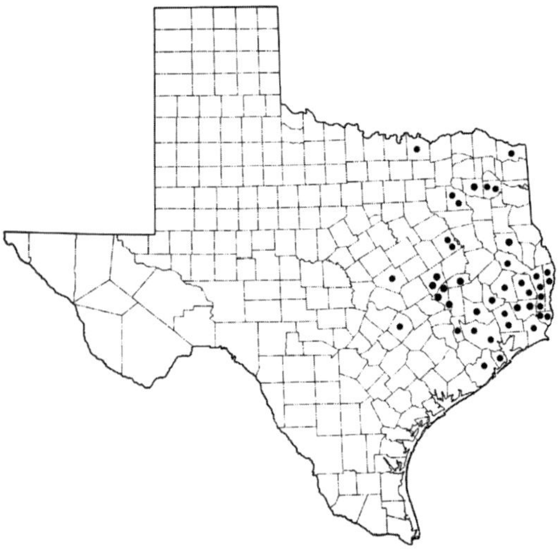

Mecardonia acuminata

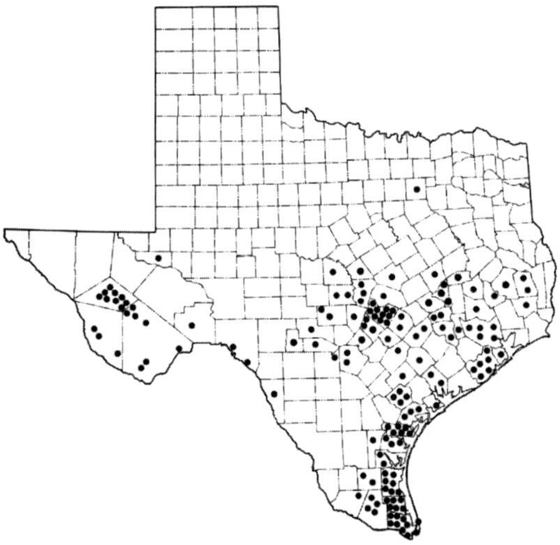

Mecardonia procumbens

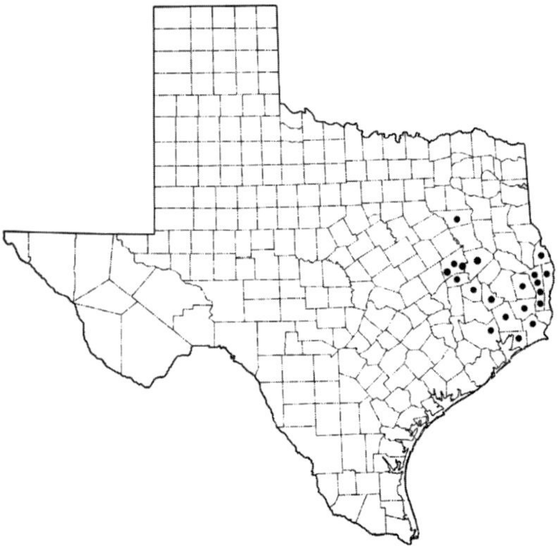

Micranthemum umbrosum

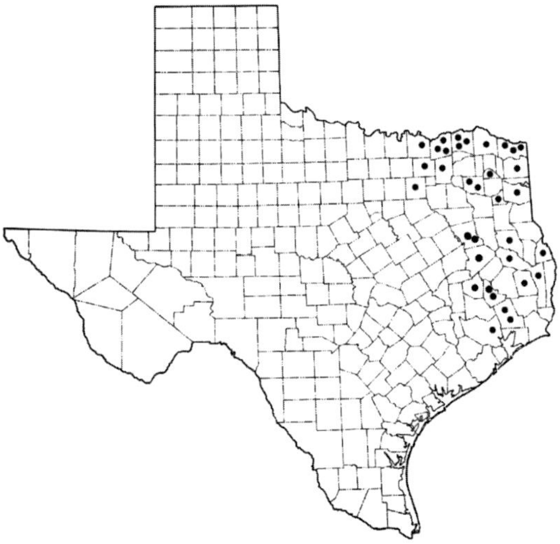

Mimulus alatus

Mimulus dentilobus

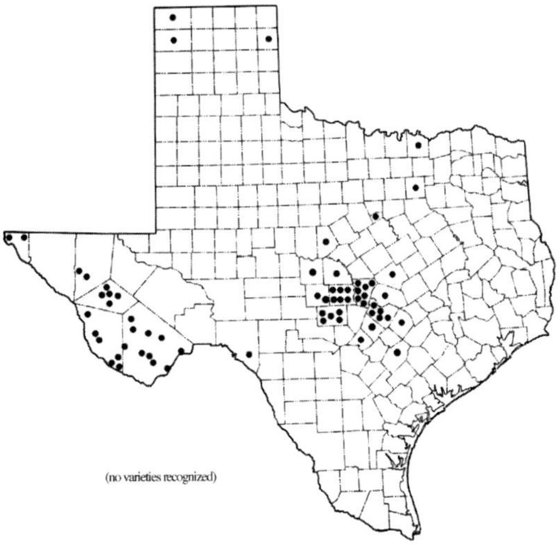
(no varieties recognized)

Mimulus glabratus

Mimulus rubellus

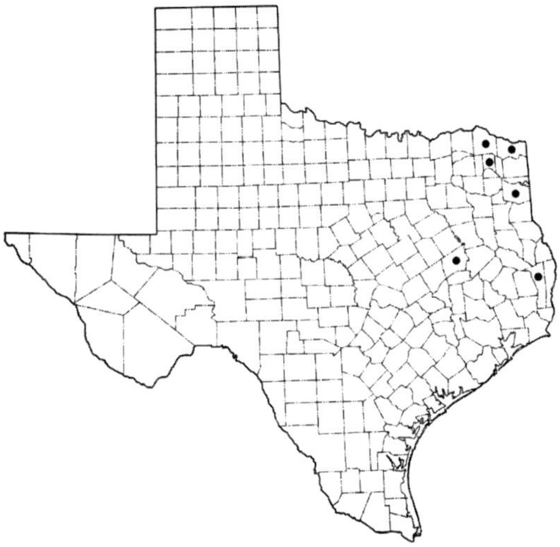

Parentucellia viscosa

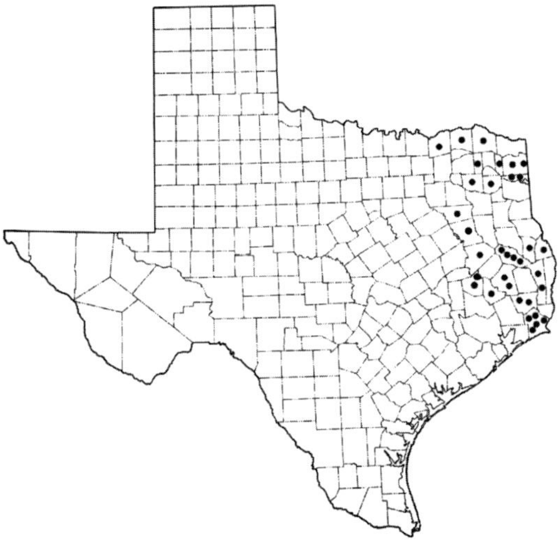

Pedicularis canadensis

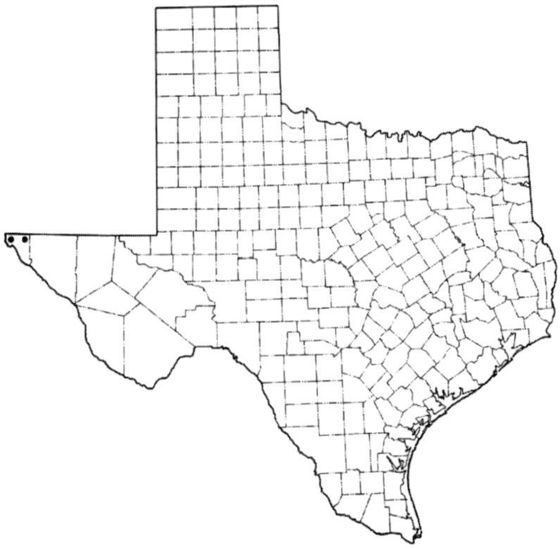

Penstemon alamosensis

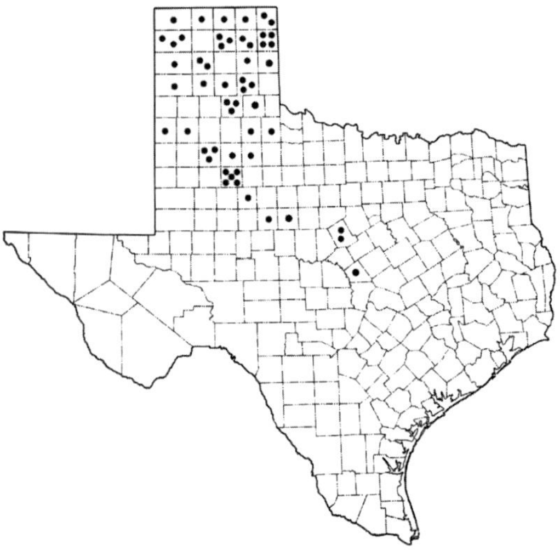

Penstemon albidus

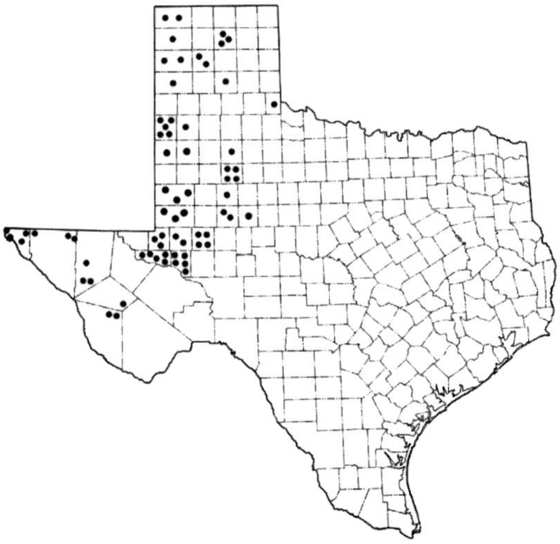

Penstemon ambiguus

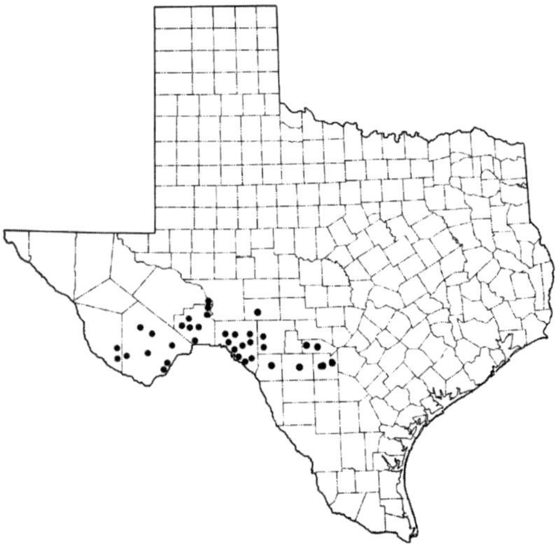

Penstemon baccharifolius

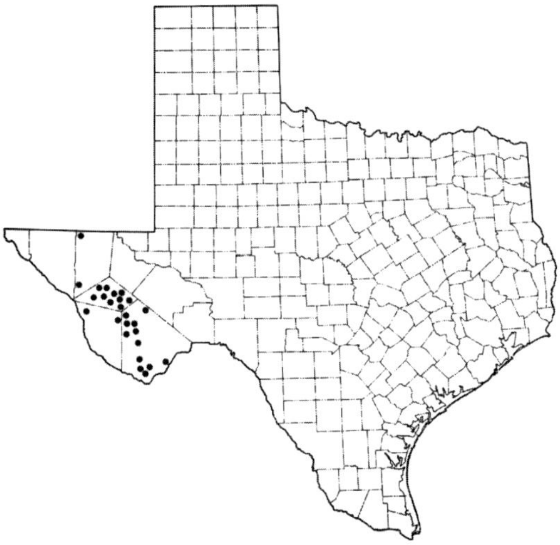

Penstemon barbatus

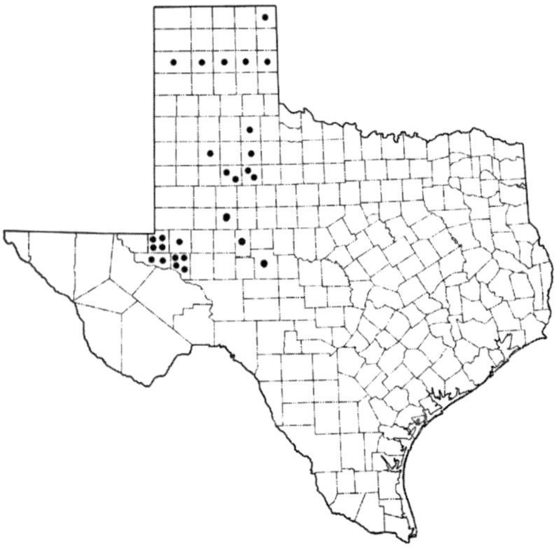

Penstemon buckleyi

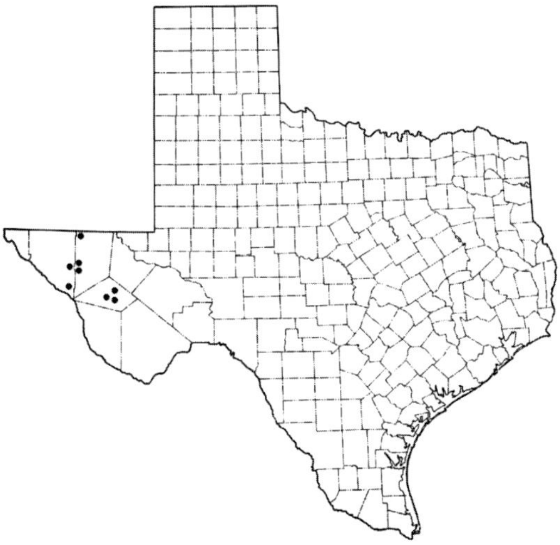

Penstemon cardinalis

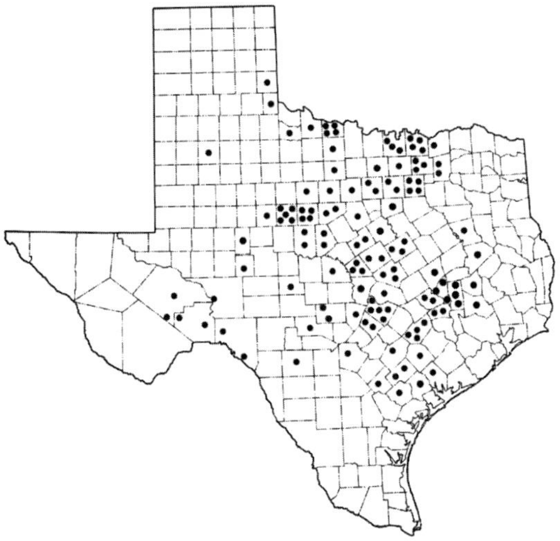

Penstemon cobaea

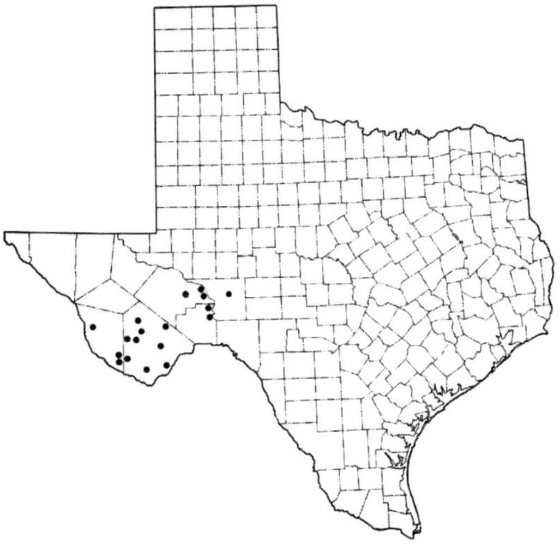

Penstemon dasyphyllus

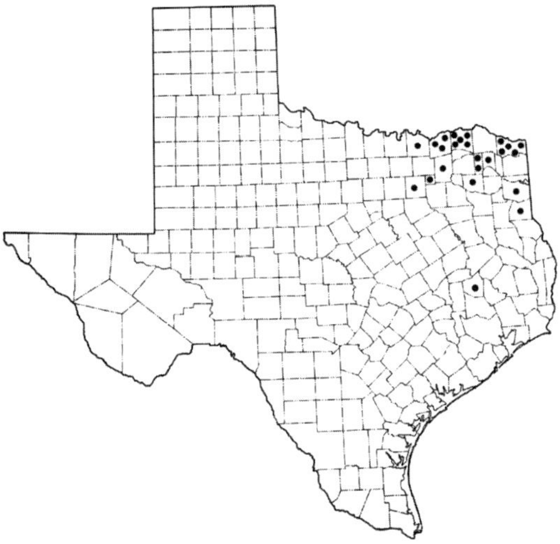

Penstemon digitalis

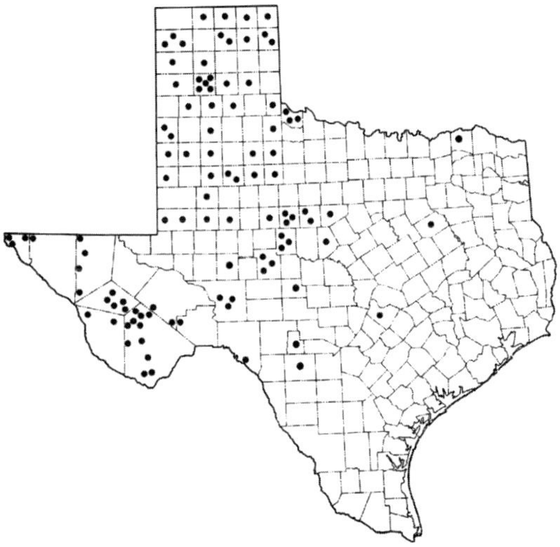

Penstemon fendleri

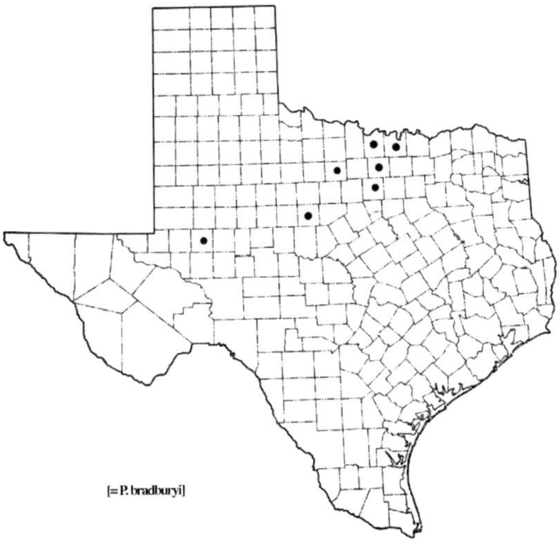
[= P. bradburyi]

Penstemon grandiflorus

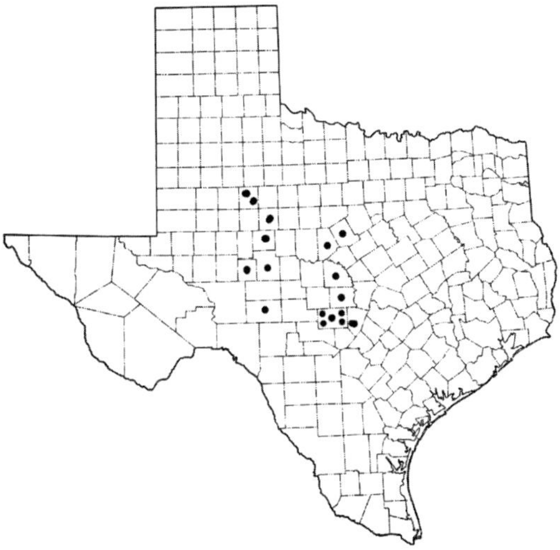

Penstemon guadalupensis

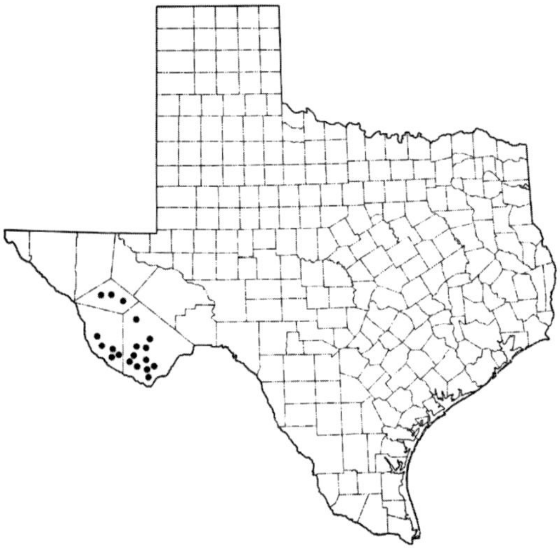

Penstemon havardii

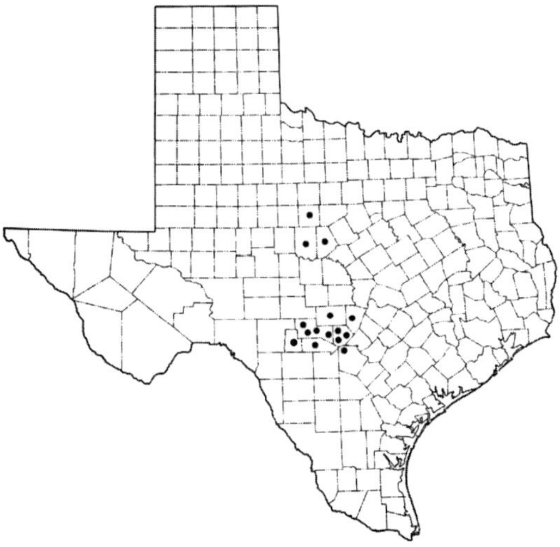

Penstemon helleri

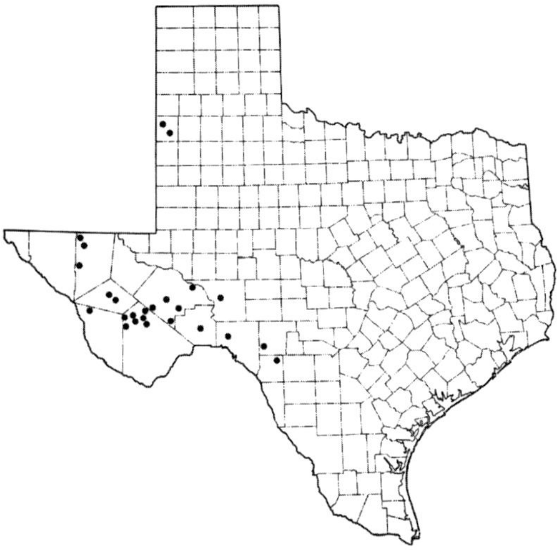

Penstemon jamesii

Penstemon lanceolatus

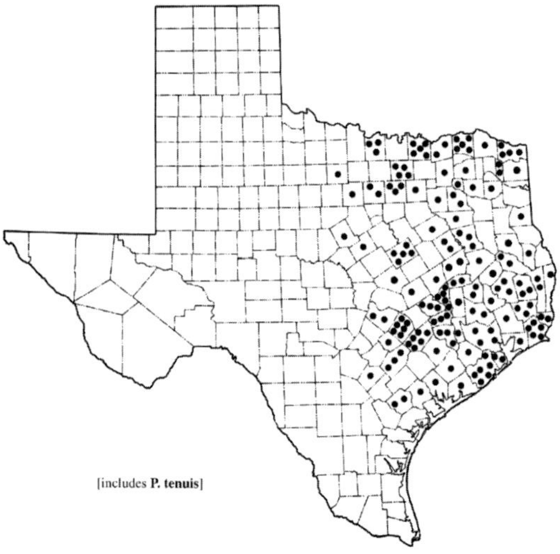
[includes P. tenuis]

Penstemon laxiflorus

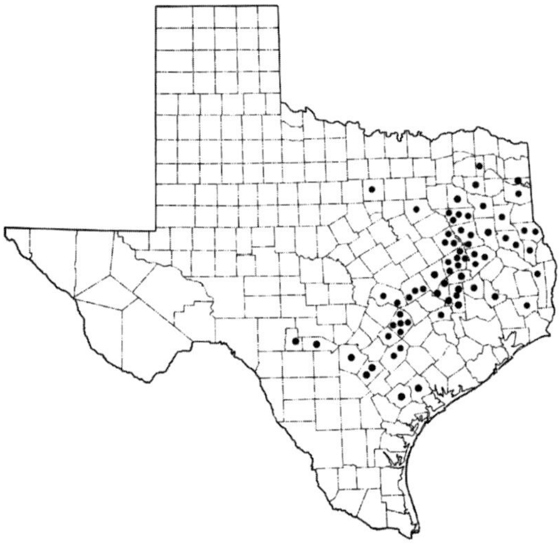

Penstemon murrayanus

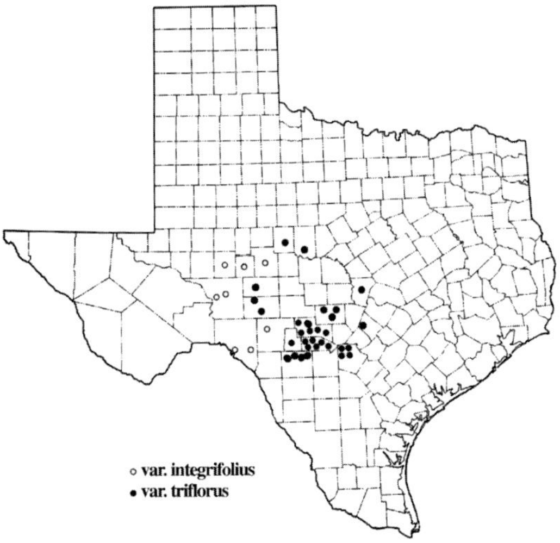
var. integrifolius
var. triflorus

Penstemon triflorus

Penstemon tubiflorus

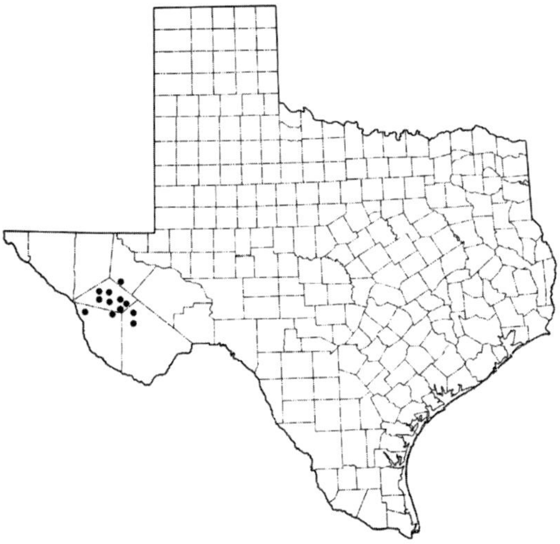

Penstemon wrightii

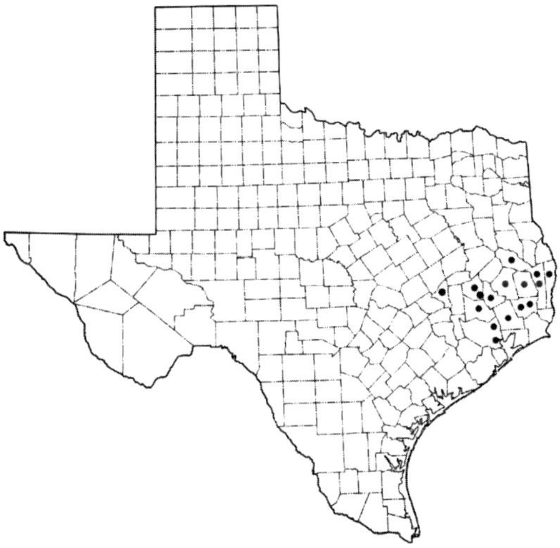

Scoparia dulcis

Scrophularia marilandica

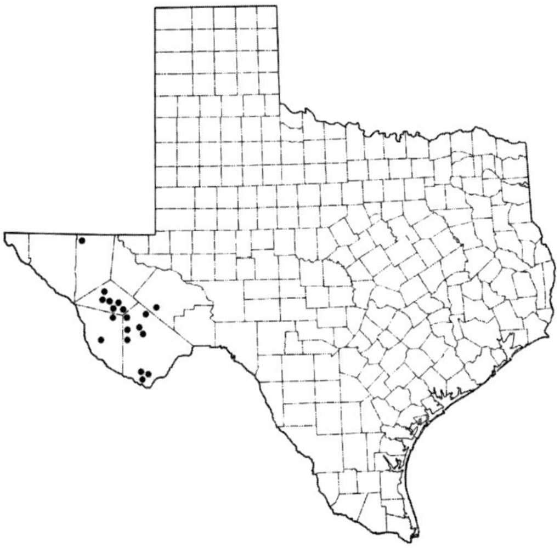

Seymeria scabra

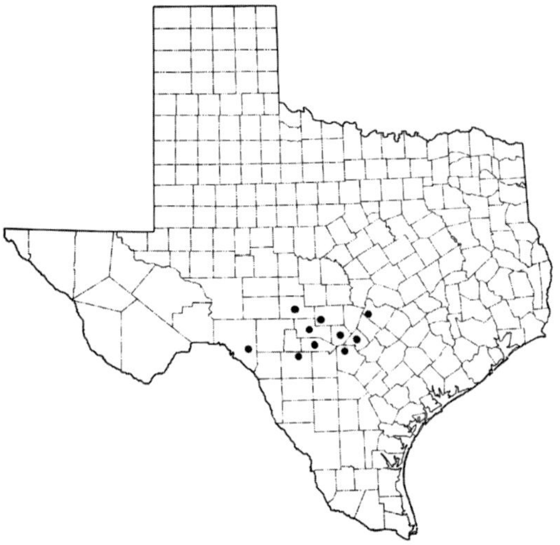

Seymeria texana

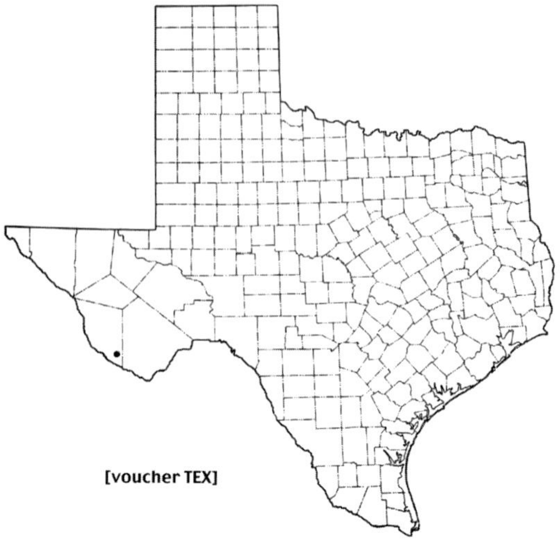
[voucher TEX]

Stemodia multifida

Stemodia schottii

Stemodia tomentosa

(includes Agalinis auriculata as
maintained by Diggs et al. 1999)

Tomanthera auriculata

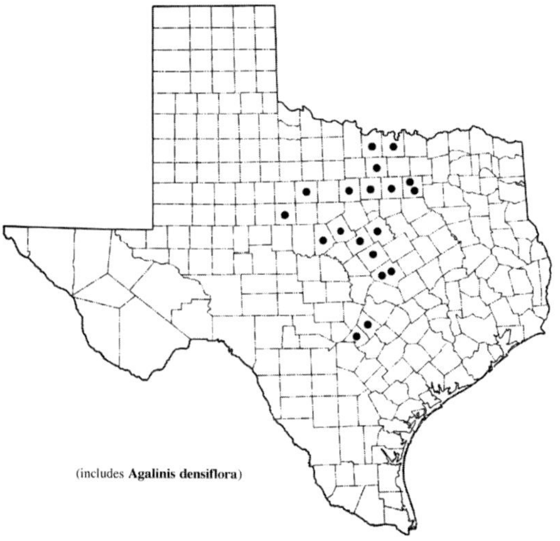
(includes Agalinis densiflora)

Tomanthera densiflora

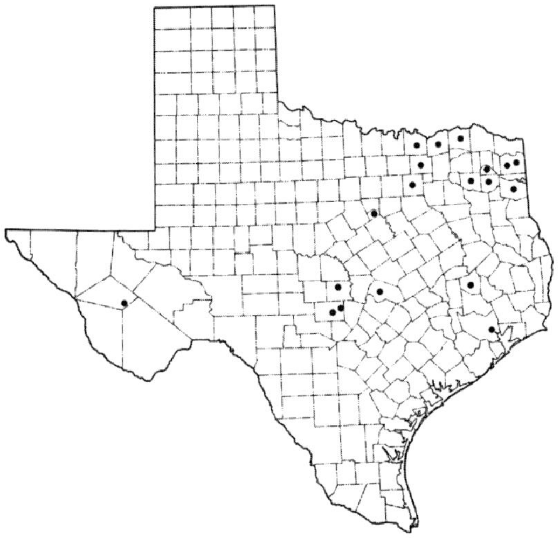

Verbascum blattaria

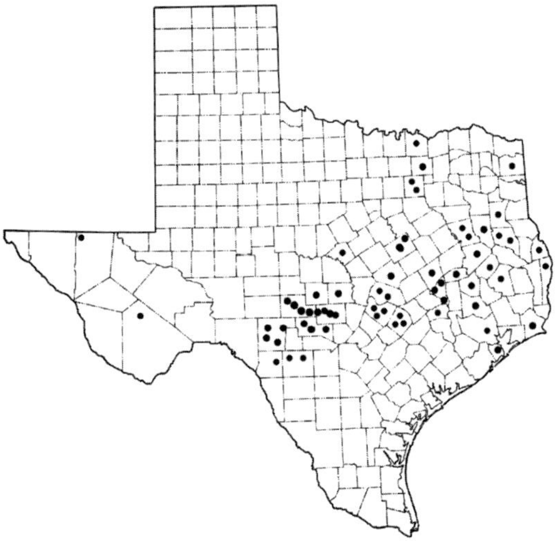

Verbascum thapsus

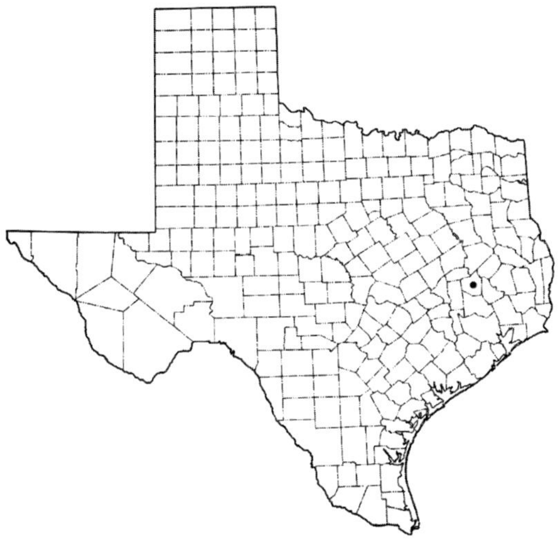

Verbascum virgatum

Veronica americana

Veronica anagallis-aquatica

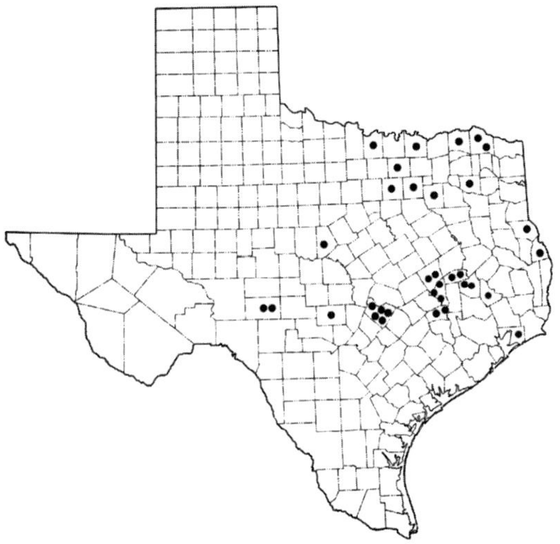

Veronica arvensis

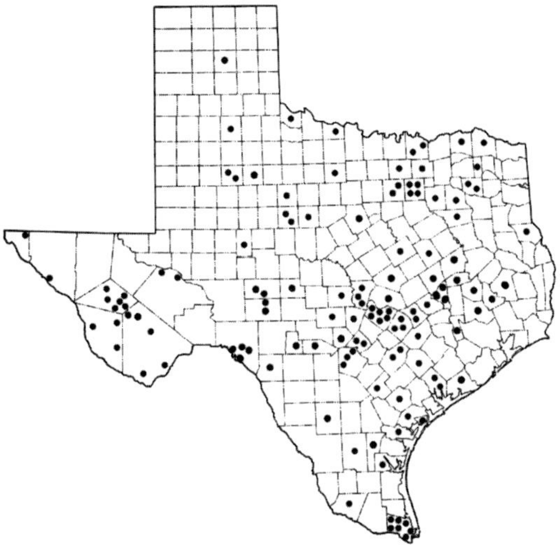

Veronica peregrina

Veronica persica

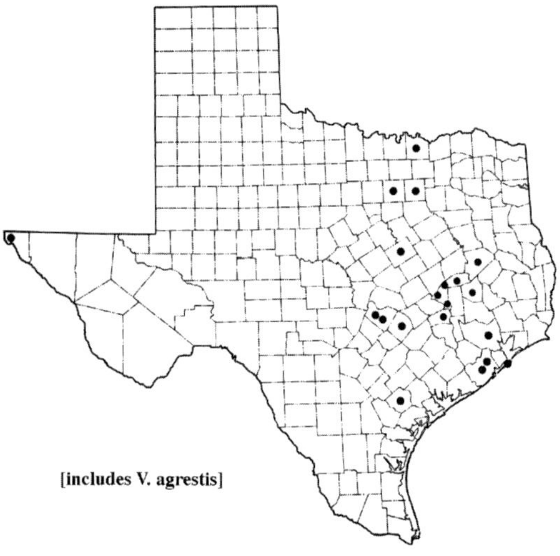
[includes V. agrestis]

Veronica polita

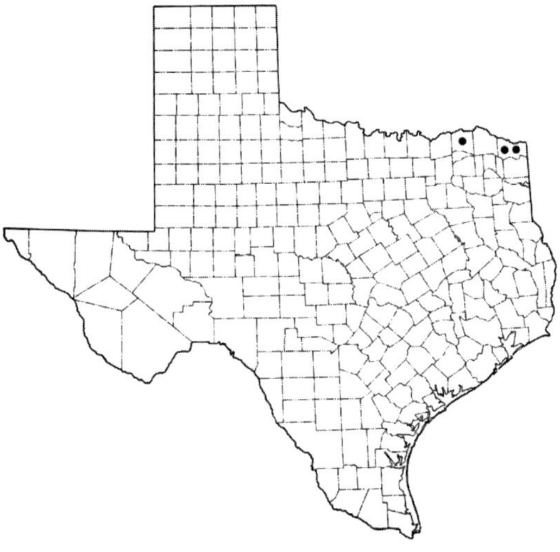

Veronicastrum virginicum

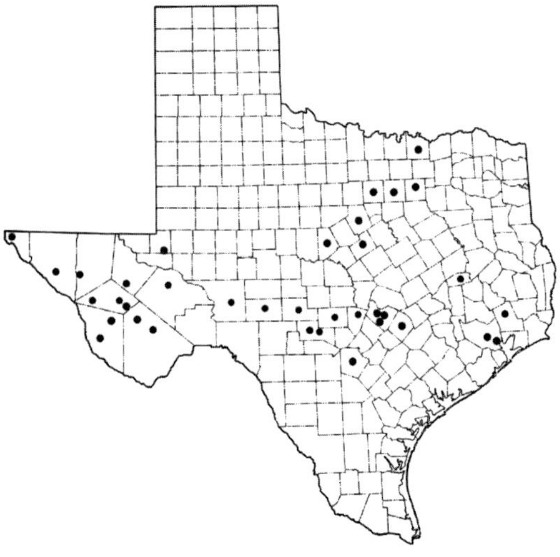

Ailanthus altissima

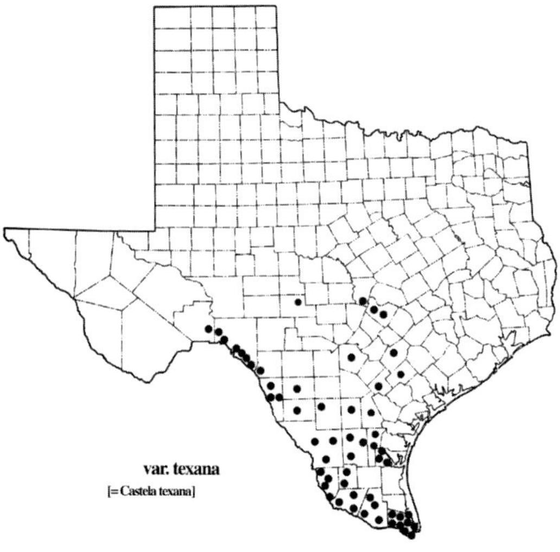
var. texana
[= Castela texana]

Castela erecta

[= Castela stewartii]

Holocantha stewartii

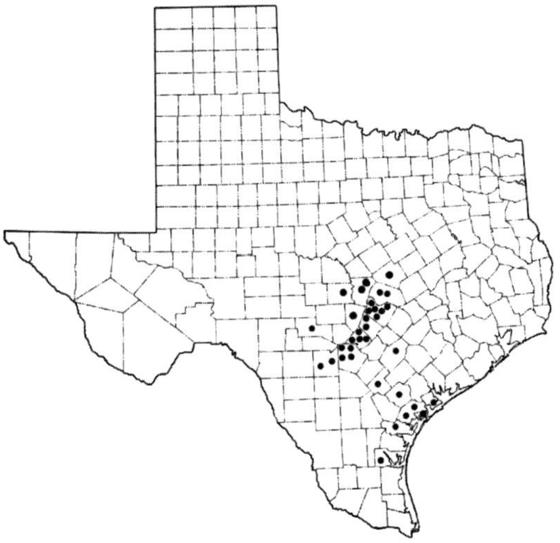

Bouchetia erecta

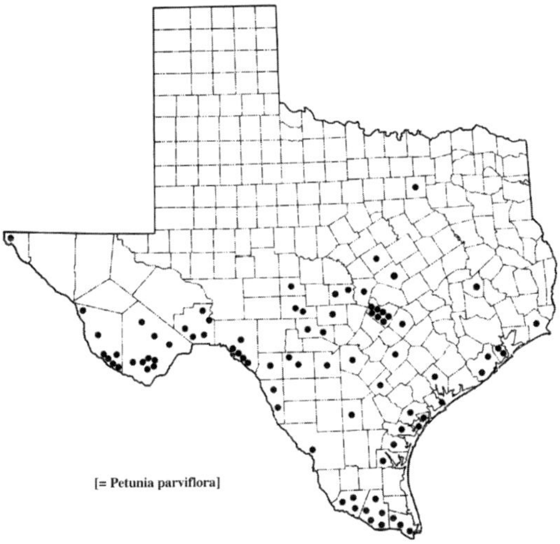
[= Petunia parviflora]

Calibrachoa parviflora

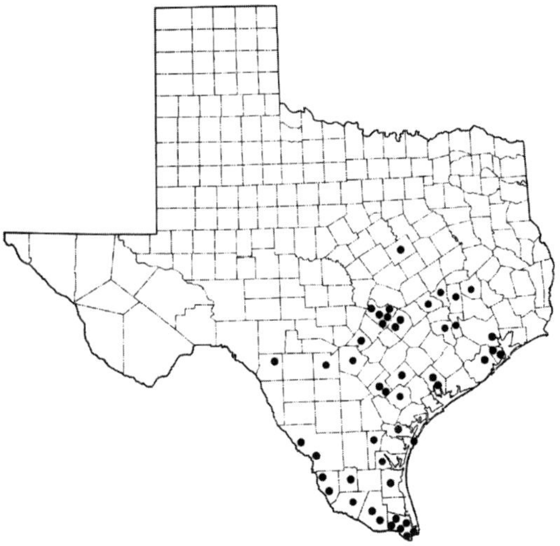

Capsicum annuum

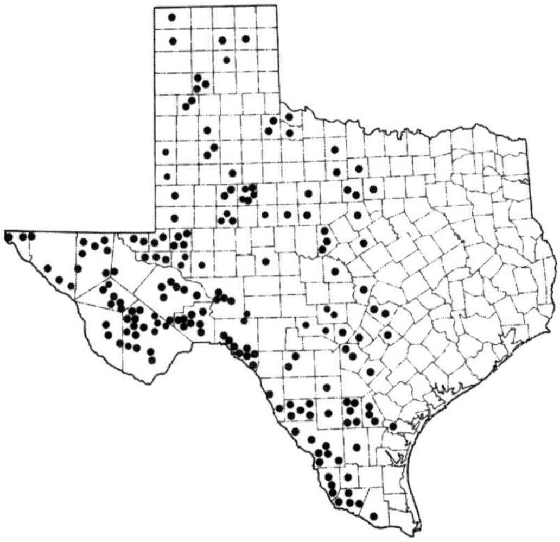

Chamaesaracha coniodes

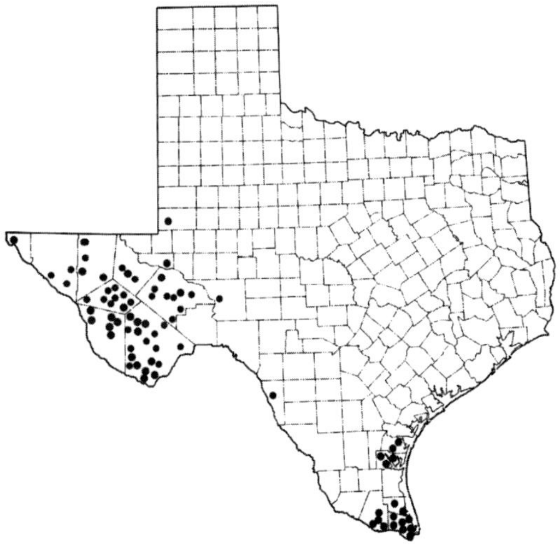

Chamaesaracha coronopus

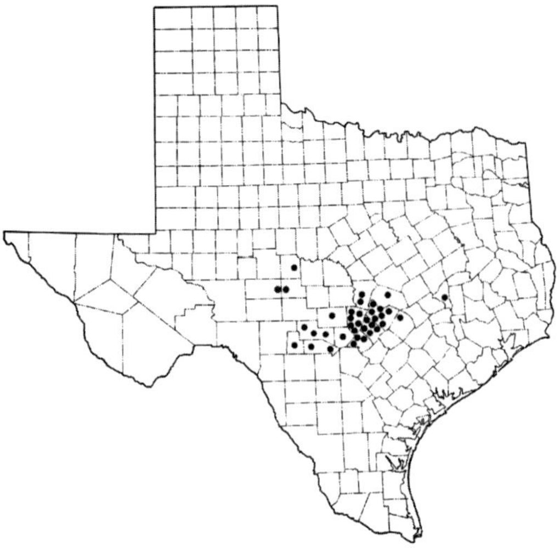

Chamaesaracha edwardsiana

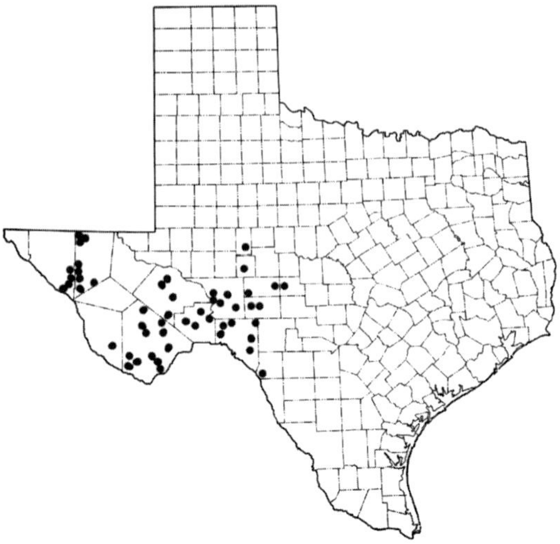

Chamaesaracha pallida

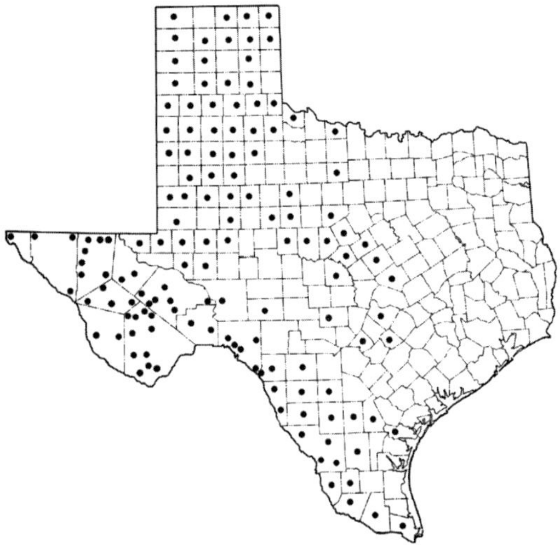

Chamaesaracha sordida

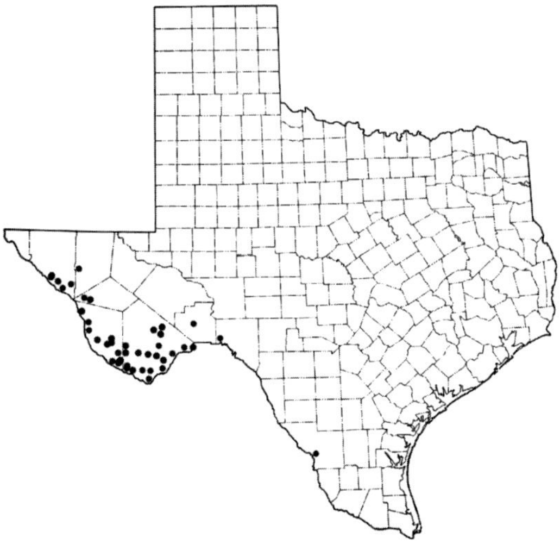

Chamaesaracha villosa

Datura inoxia

Datura metel

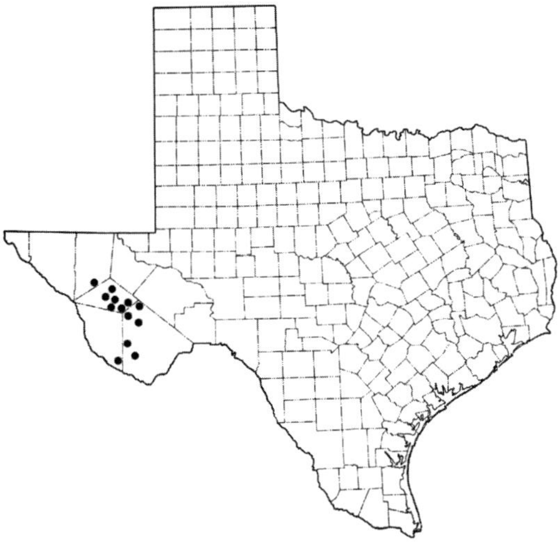

Datura quercifolia

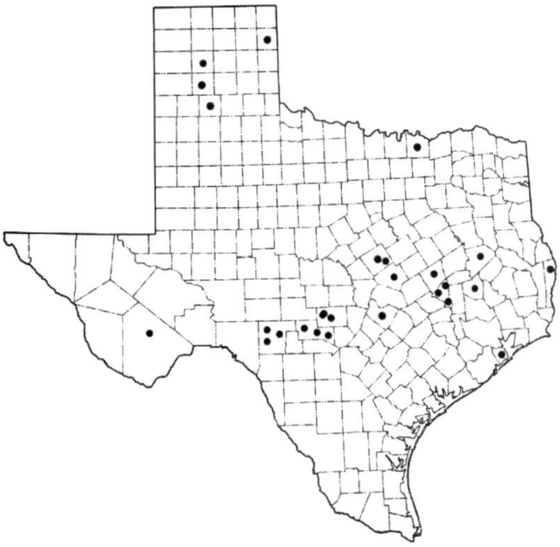

Datura stramonium

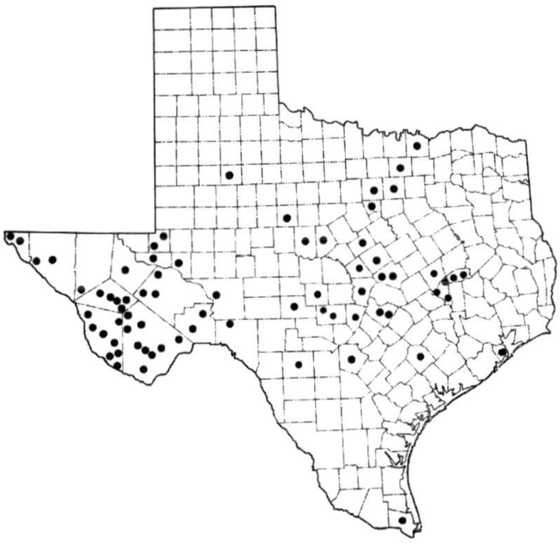

Datura wrightii

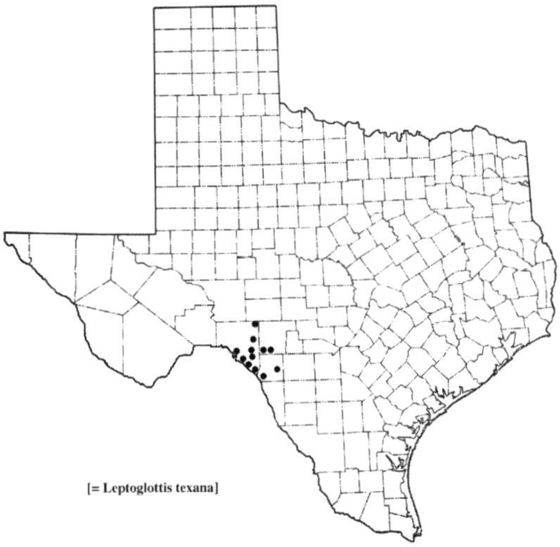
[= Leptoglottis texana]

Hunzikeria texana

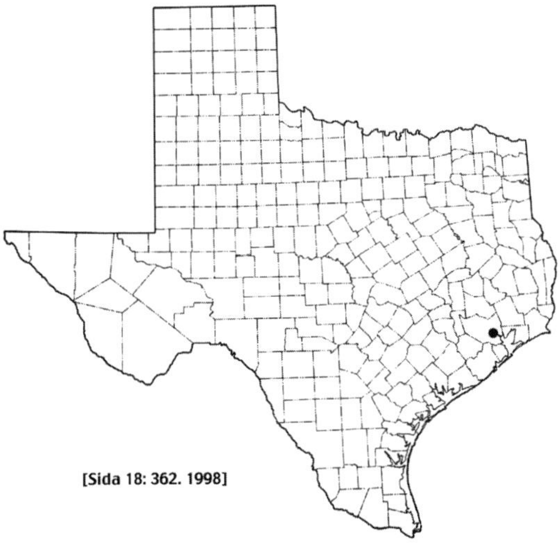
[Sida 18: 362. 1998]

Lycianthes asarifolia

Lycium barbarum

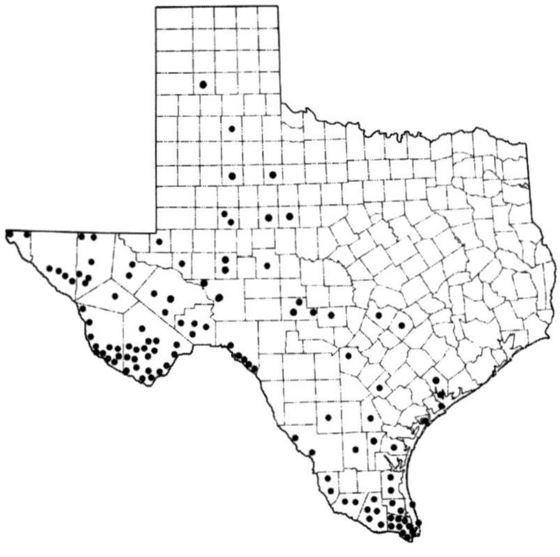

Lycium berlandieri

Lycium carolinianum

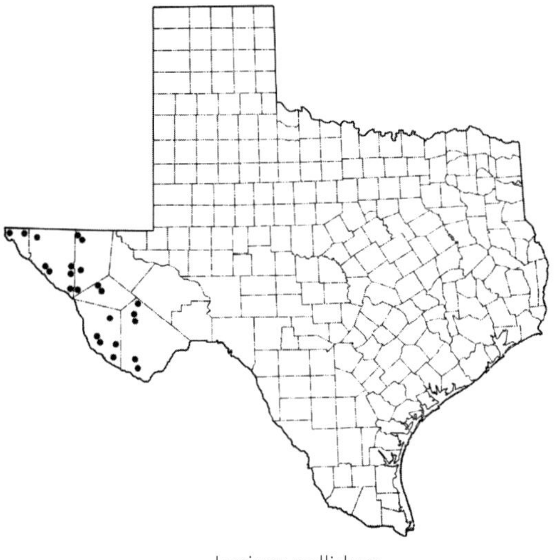

Lycium pallidum

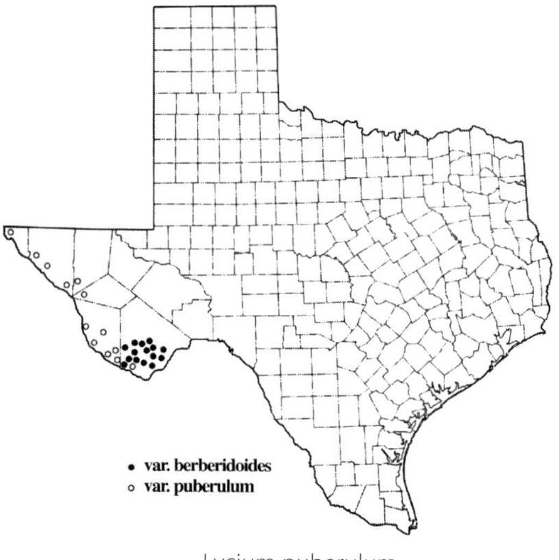
• var. berberidoides
○ var. puberulum

Lycium puberulum

Lycium texanum

Lycium torreyi

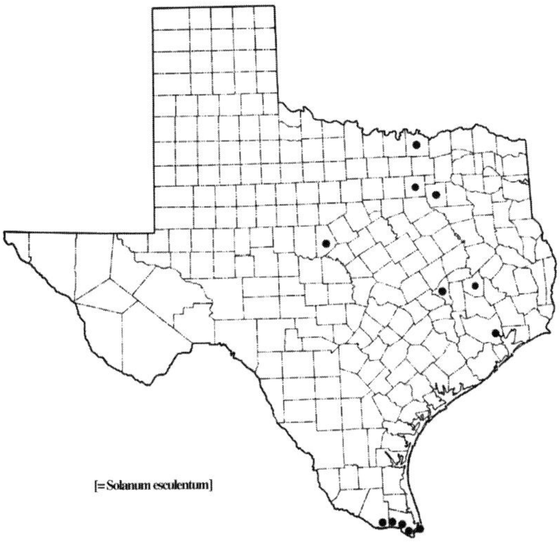
[=Solanum esculentum]

Lycopersicon esculentum

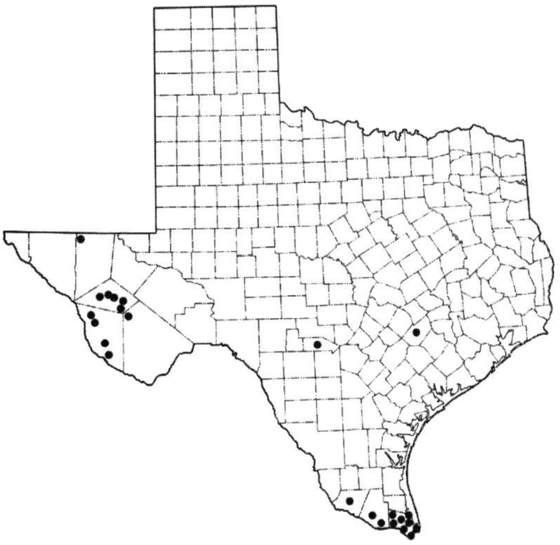

Margaranthus solanaceus

Nectouxia formosa

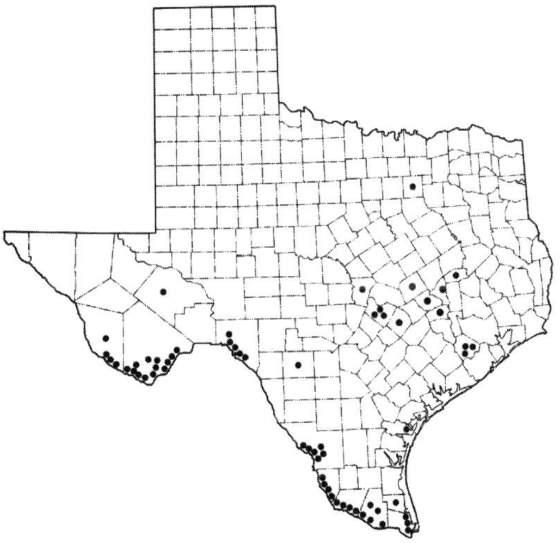

Nicotiana glauca

Nicotiana plumbaginifolia

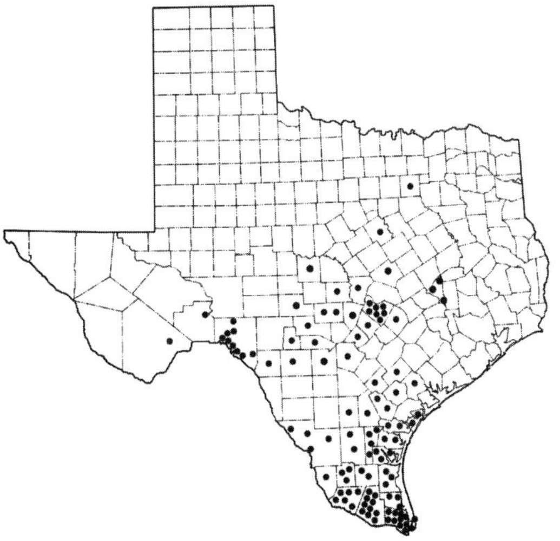

Nicotiana repanda

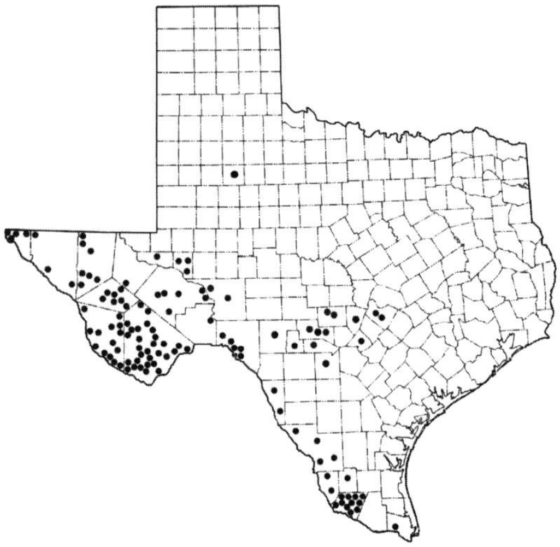

Nicotiana trigonophylla

Nierembergia hippomanica

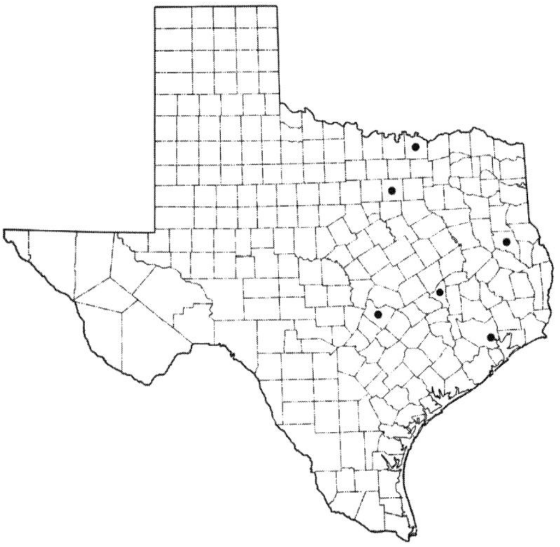

Petunia axillaris

Physalis acutifolia

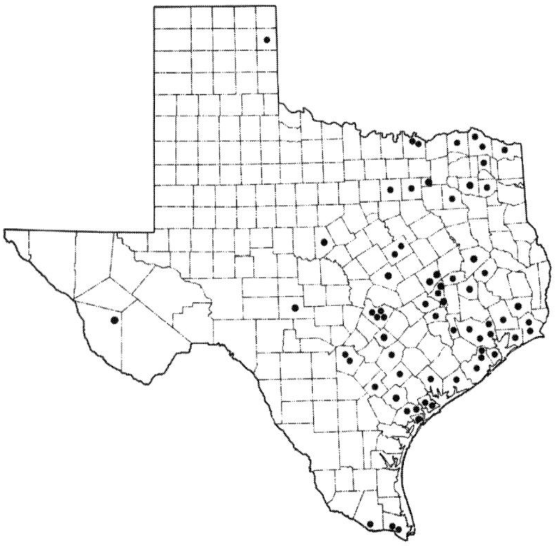

Physalis angulata

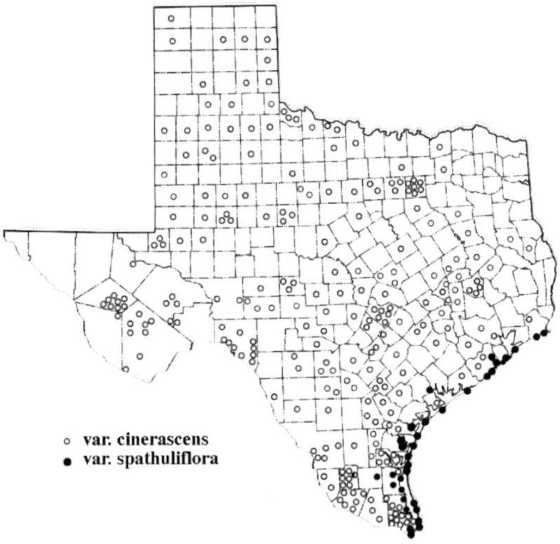
var. cinerascens
var. spathuliflora

Physalis cinerascens

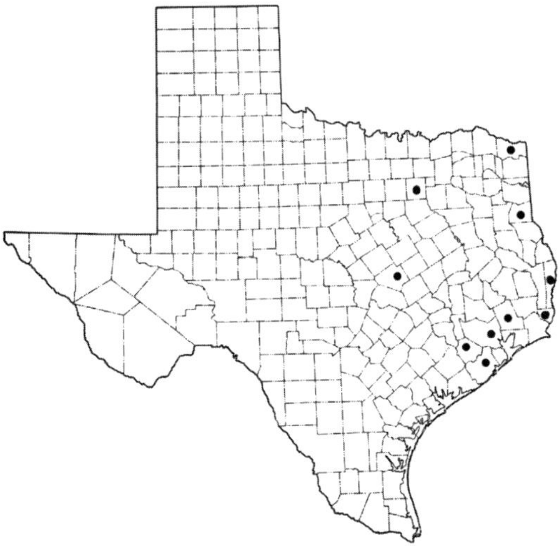

Physalis cordata

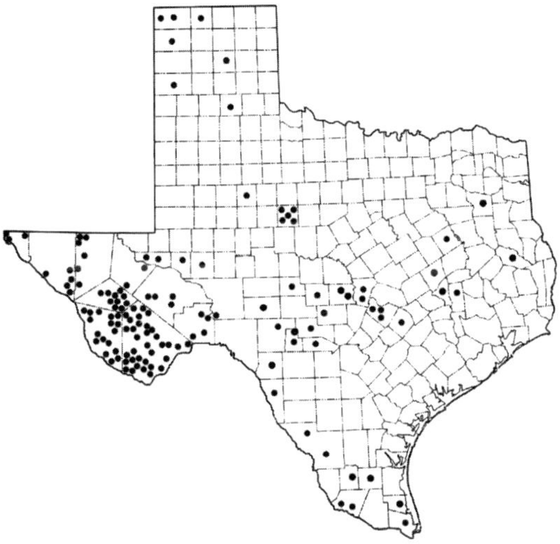

Physalis hederifolia

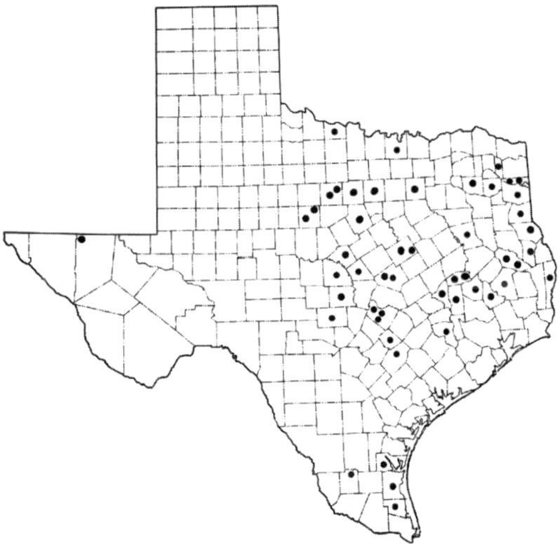

Physalis heterophylla

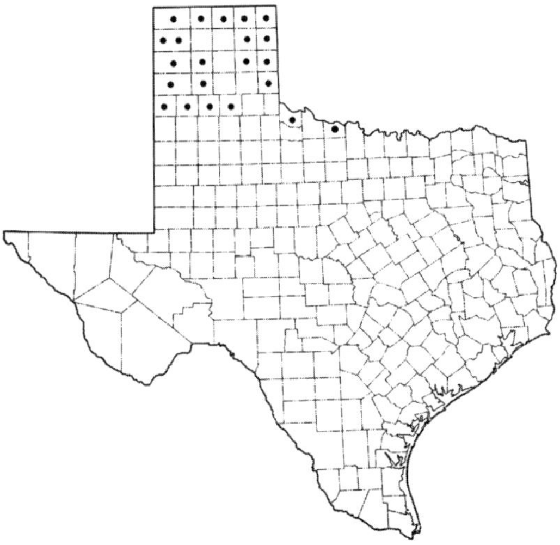

Physalis hispida

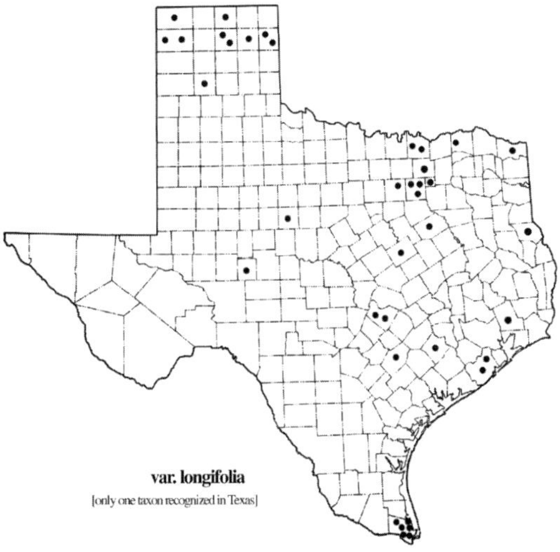
var. longifolia
[only one taxon recognized in Texas]

Physalis longifolia

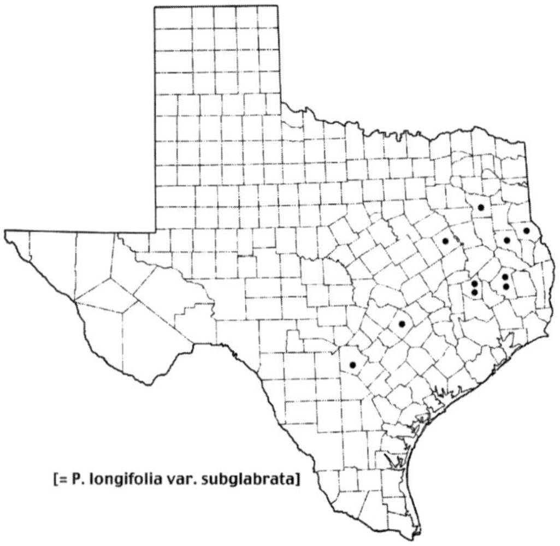
[= P. longifolia var. subglabrata]

Physalis macrophysa

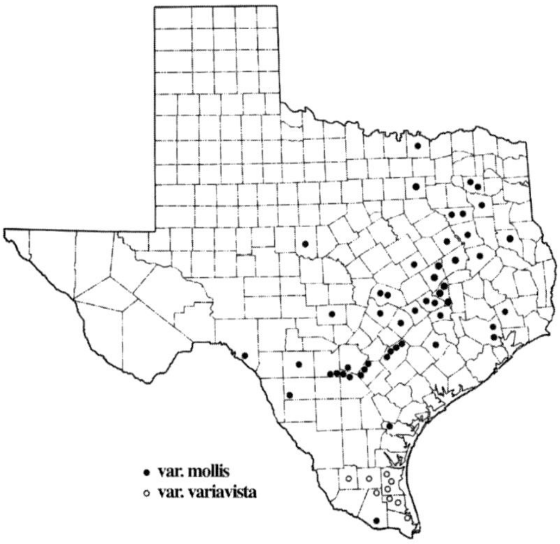
• var. mollis
○ var. variavista

Physalis mollis

Physalis neomexicana

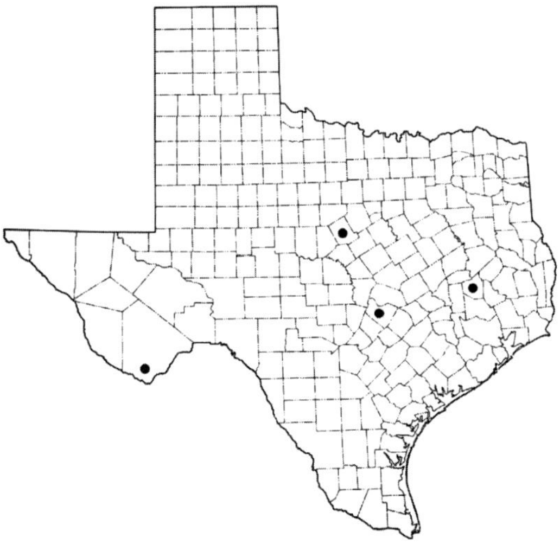

Physalis philadelphica

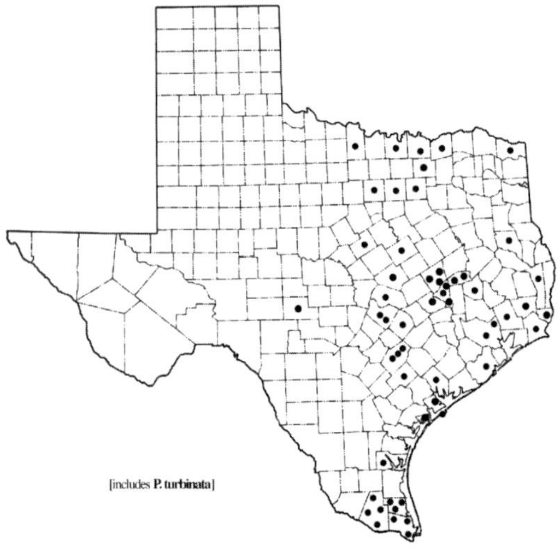
[includes P. turbinata]

Physalis pubescens

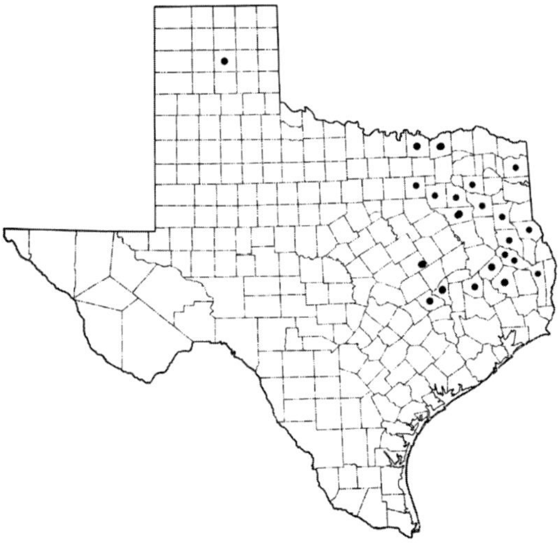

Physalis pumila

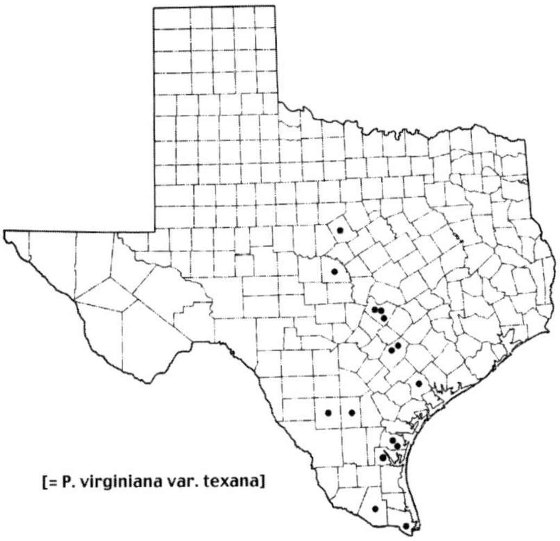
[= P. virginiana var. texana]

Physalis texana

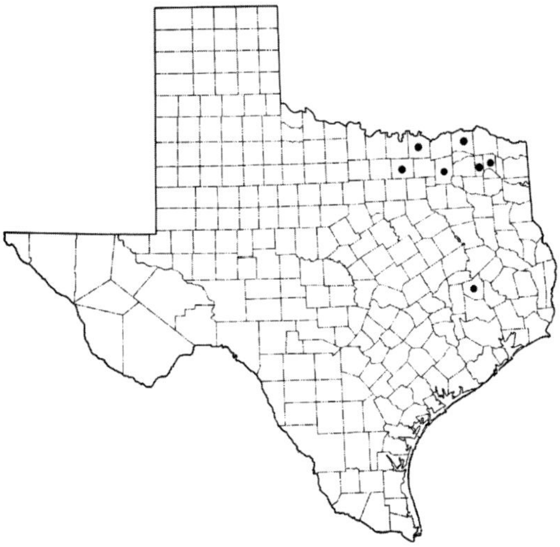

Physalis virginiana

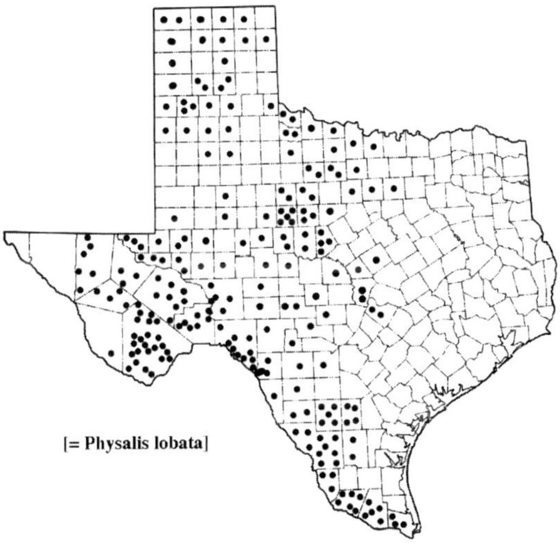
[= Physalis lobata]

Quincula lobata

Salpichroa origanifolia

Solanum campechiense

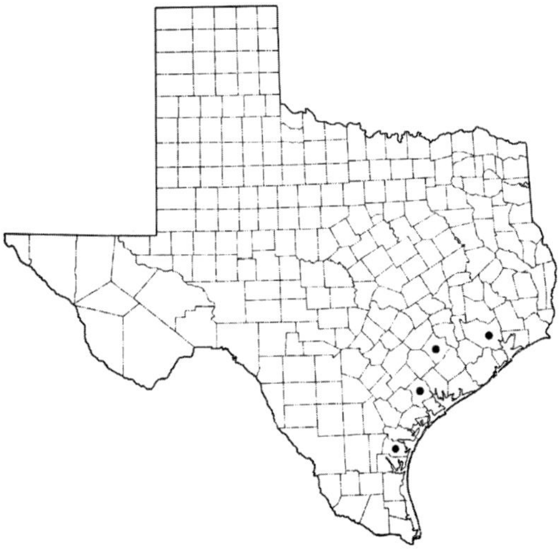

Solanum capsicastrum

Solanum capsicoides

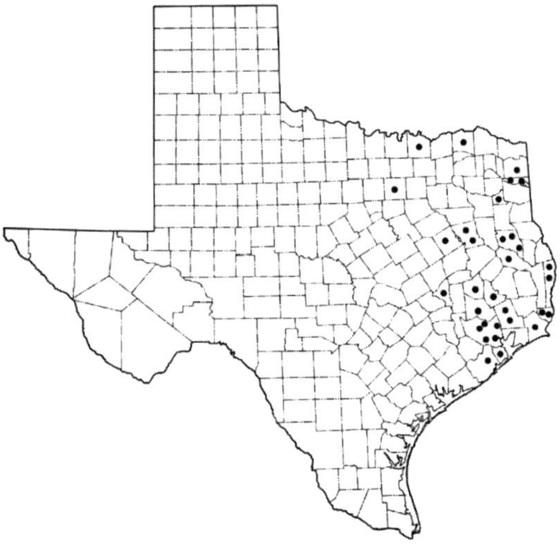

Solanum carolinense

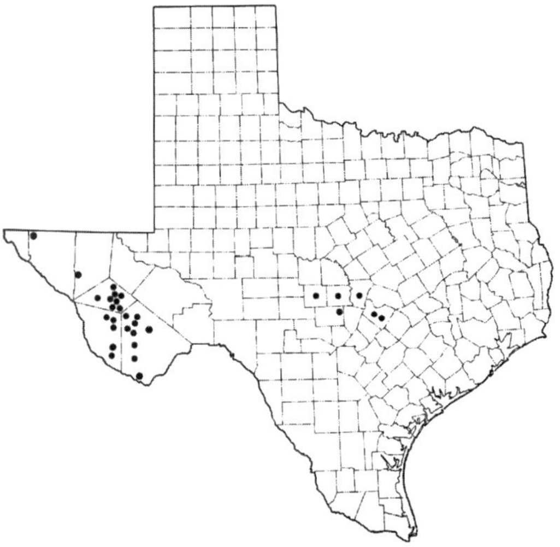

Solanum citrullifolium

Solanum davisense

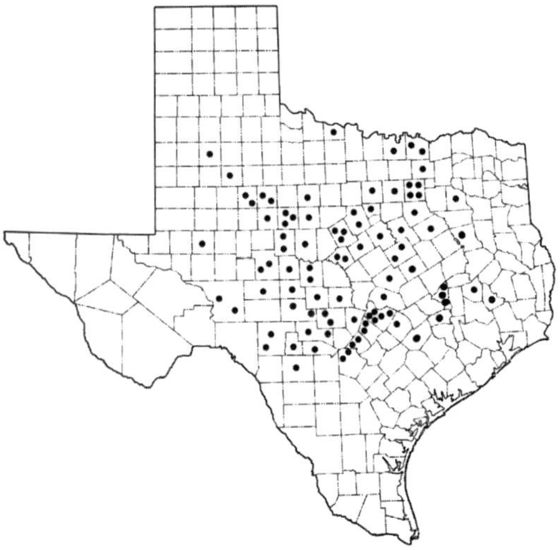

Solanum dimidiatum

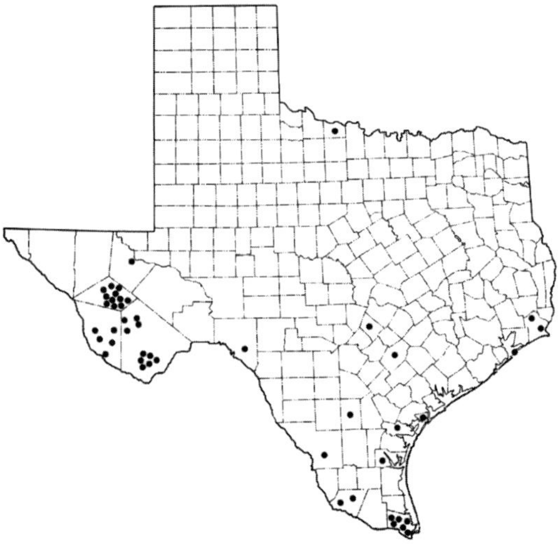

Solanum douglasii

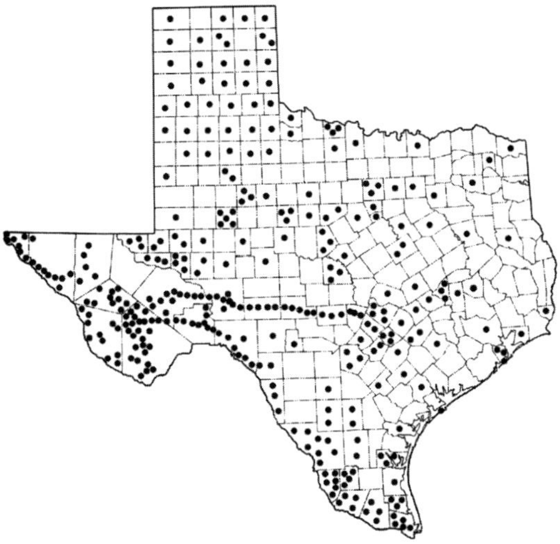

Solanum elaeagnifolium

Solanum erianthum

Solanum fendleri

Solanum heterodoxum

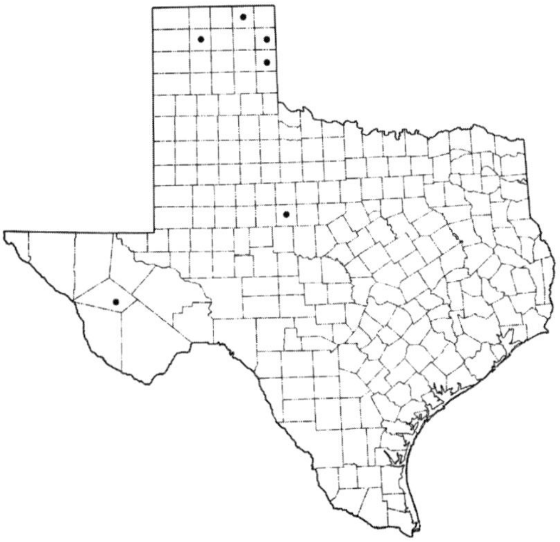

Solanum interius

Solanum jamesii

Solanum leptosepalum

Solanum pseudocapsicum

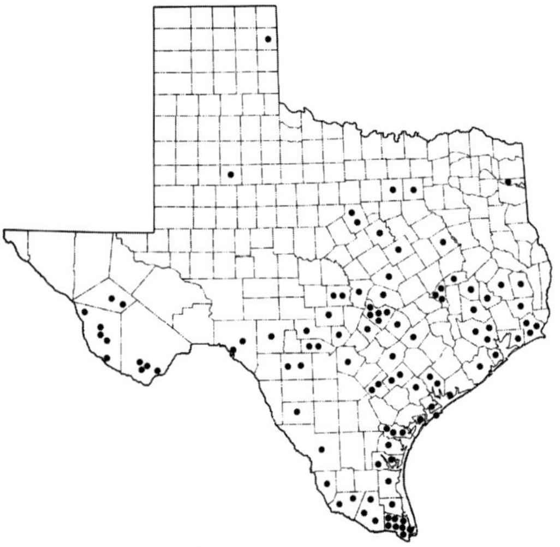

Solanum ptycanthum

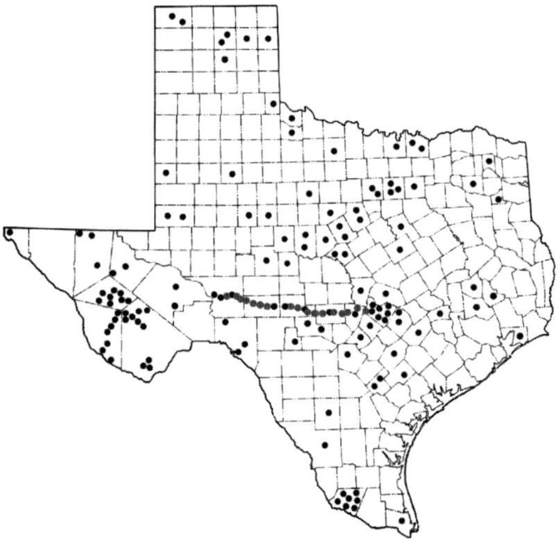

Solanum rostratum

Solanum sarrachoides

Solanum sisymbriifolium

var. latisectum
var. tenuipes

Solanum tenuipes

Solanum triflorum

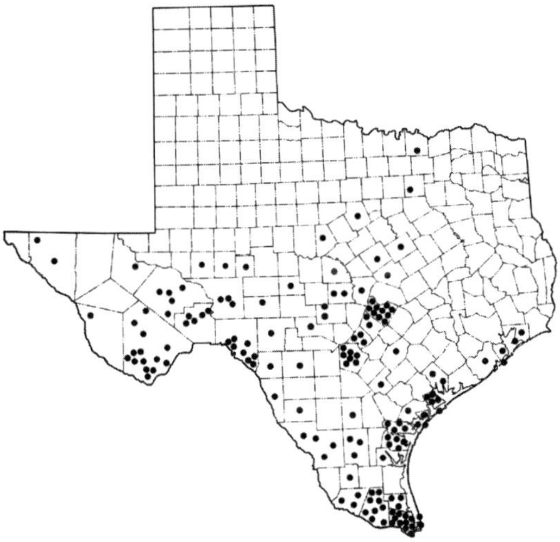

Solanum triquetrum

Sphenoclea zeylanica

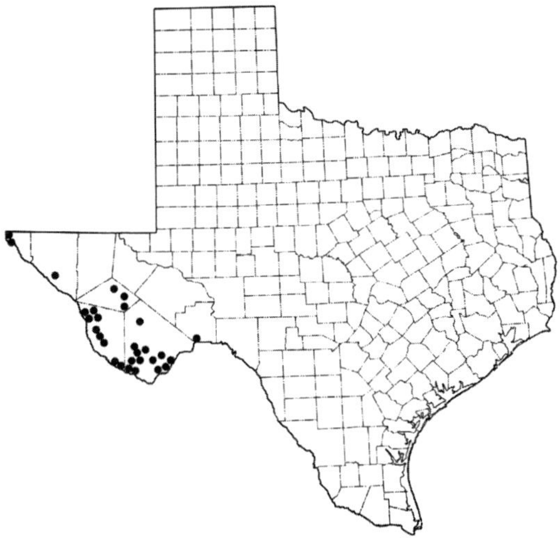

Ayenia filiformis

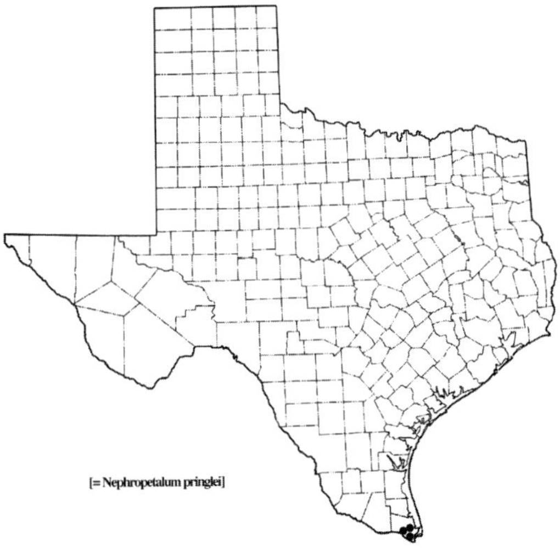
[= Nephropetalum pringlei]

Ayenia limitaris

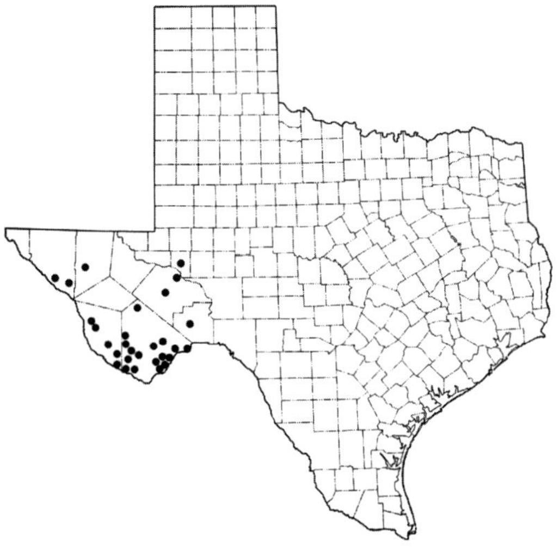

Ayenia microphylla

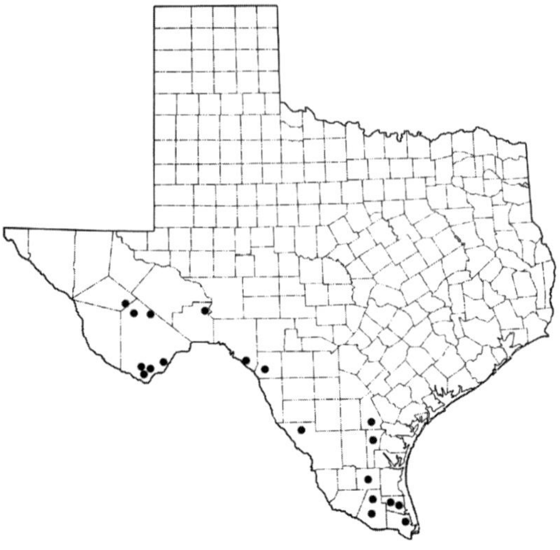

Ayenia pilosa

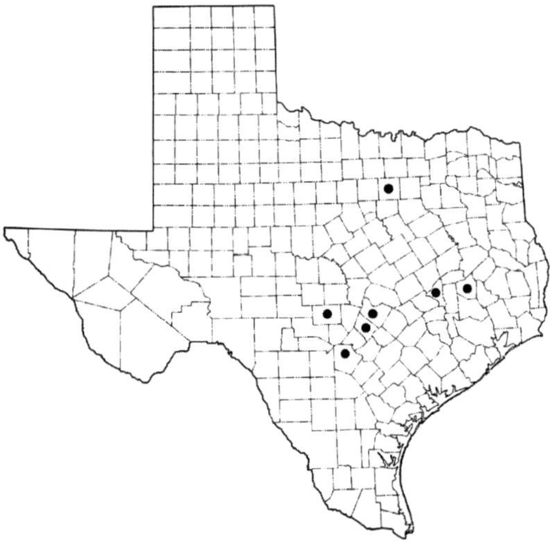

Firmiana simplex

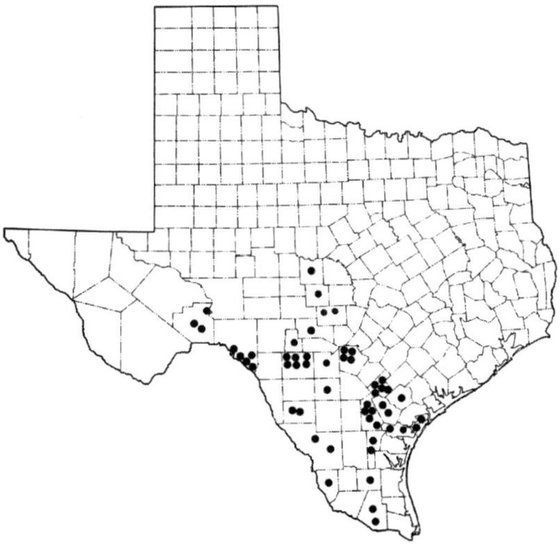

Hermannia texana

Melochia corchorifolia

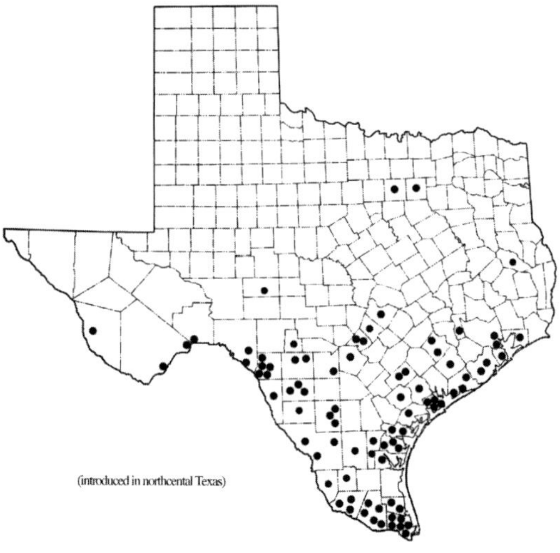
(introduced in northcental Texas)

Melochia pyramidata

Melochia tomentosa

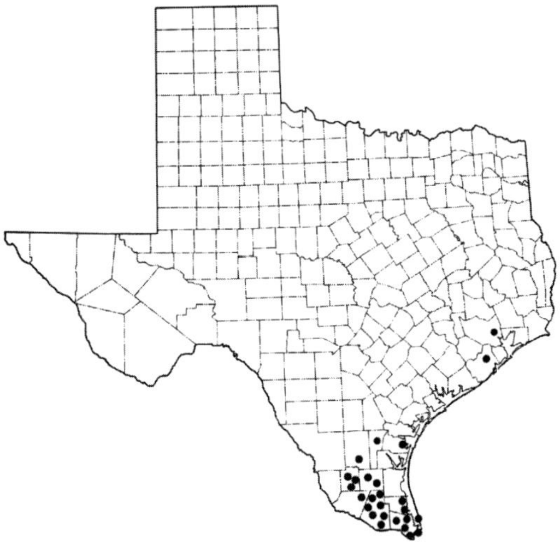

Waltheria indica

Halesia diptera

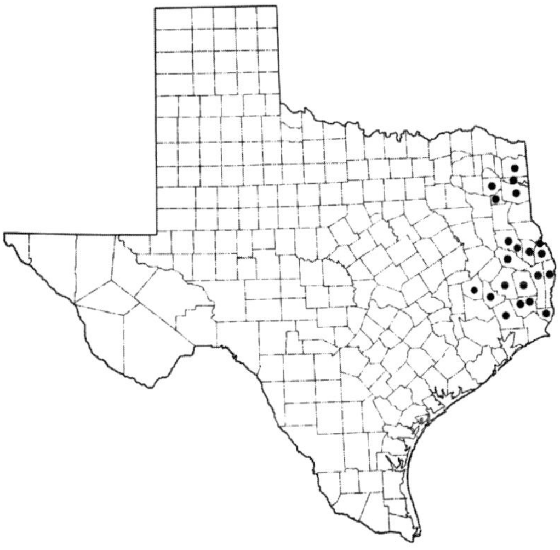

Styrax americanus

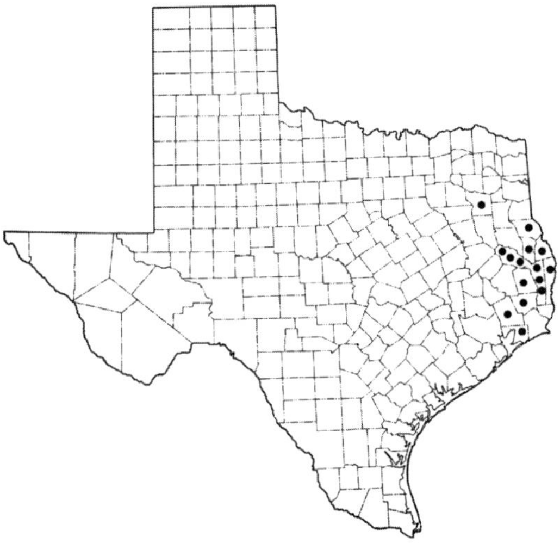

Styrax grandifolius

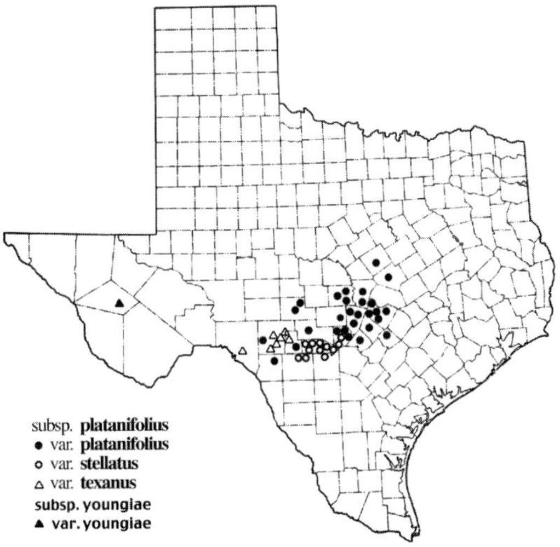
subsp. platanifolius
var. platanifolius
var. stellatus
var. texanus
subsp. youngiae
var. youngiae

Styrax platanifolius

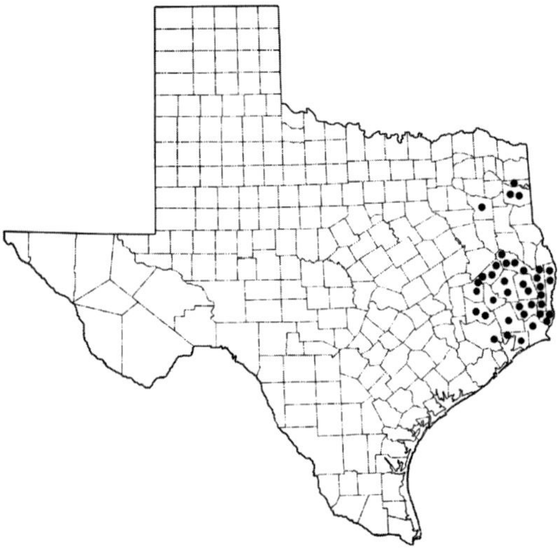

Symplocos tinctoria

Tamarix aphylla

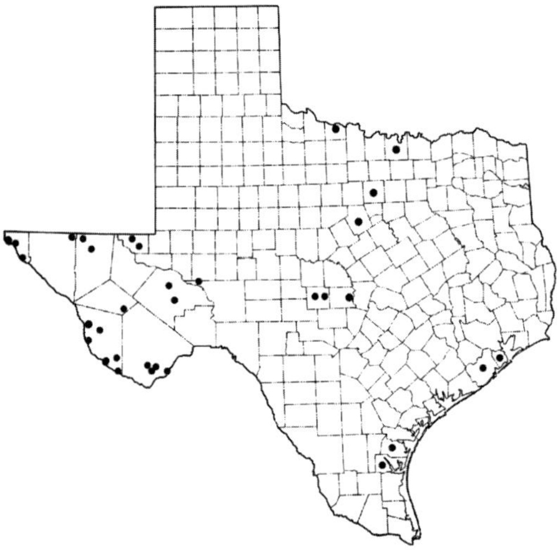

Tamarix chinensis

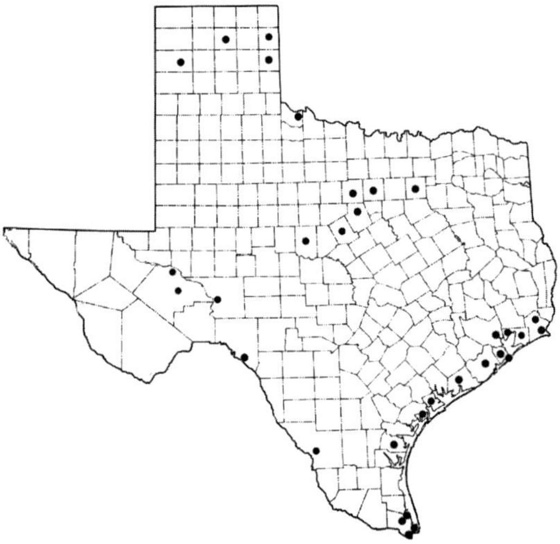

Tamarix gallica

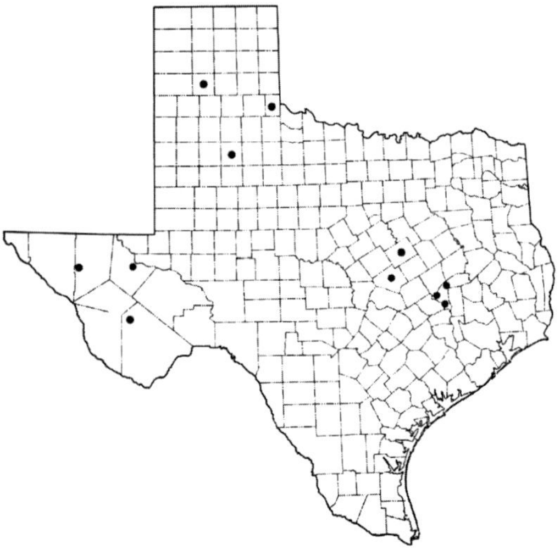

Tamarix parviflora

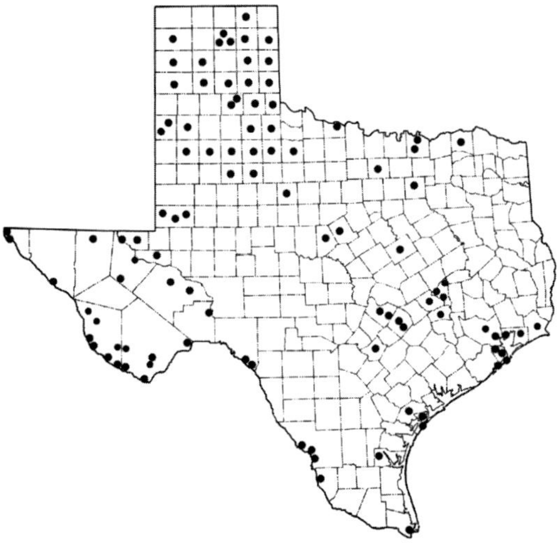

Tamarix ramosissima

Stewartia malacodendron

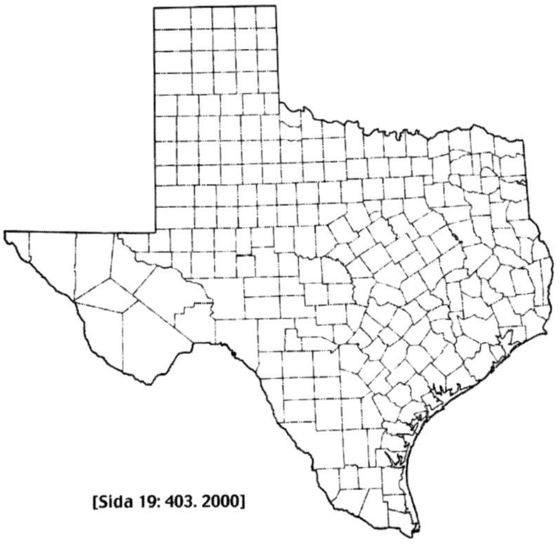
[Sida 19: 403. 2000]

Thymelaea passerina

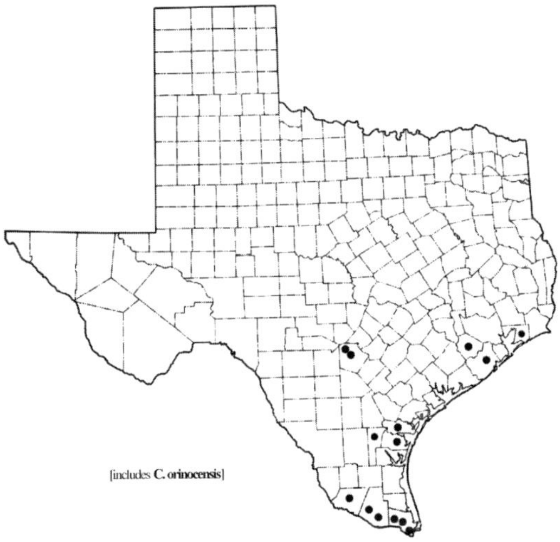
[includes C. orinocensis]

Corchorus hirtus

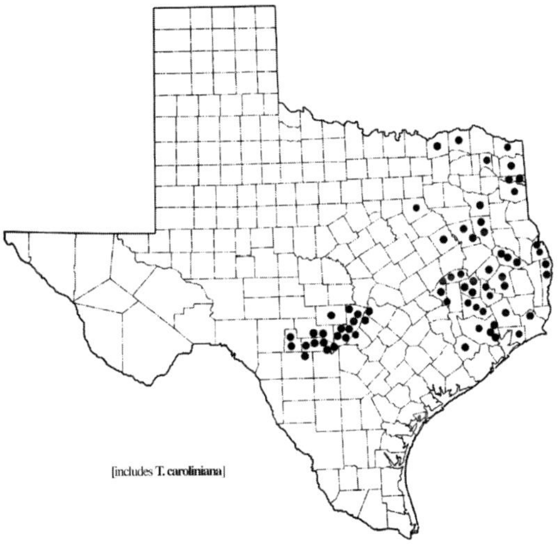
[includes T. caroliniana]

Tilia americana

Turnera diffusa

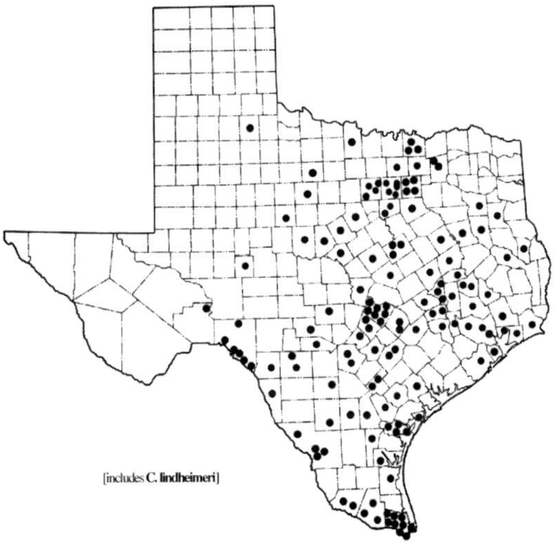
[includes C. lindheimeri]

Celtis laevigata

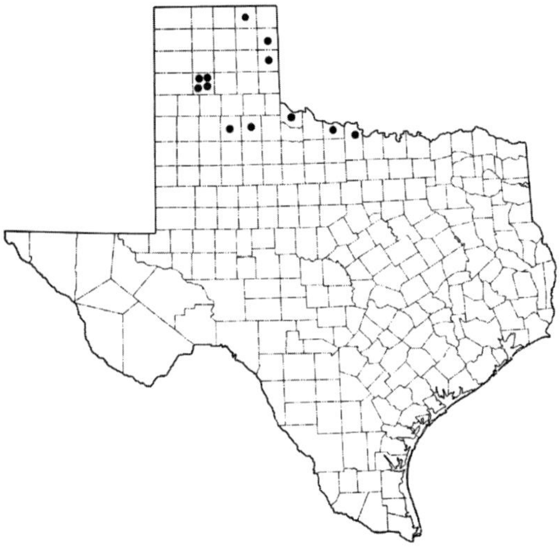

Celtis occidentalis

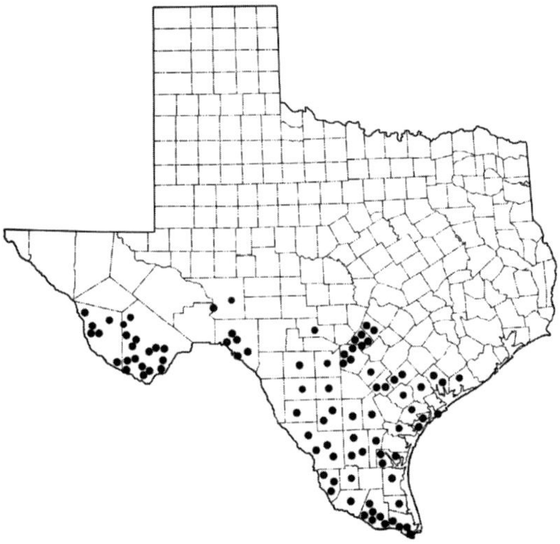

Celtis pallida

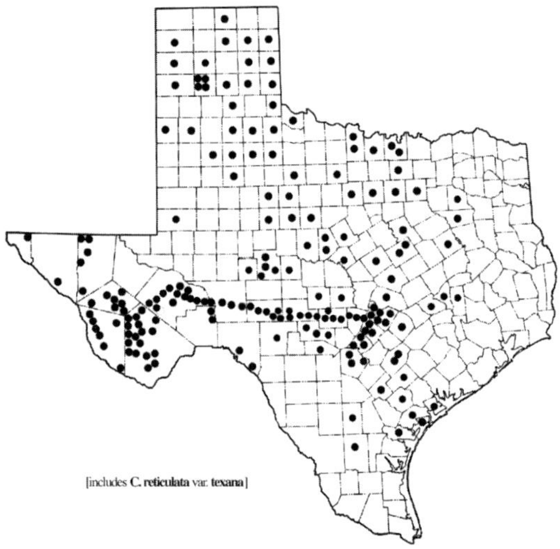
[includes C. reticulata var. texana]

Celtis reticulata

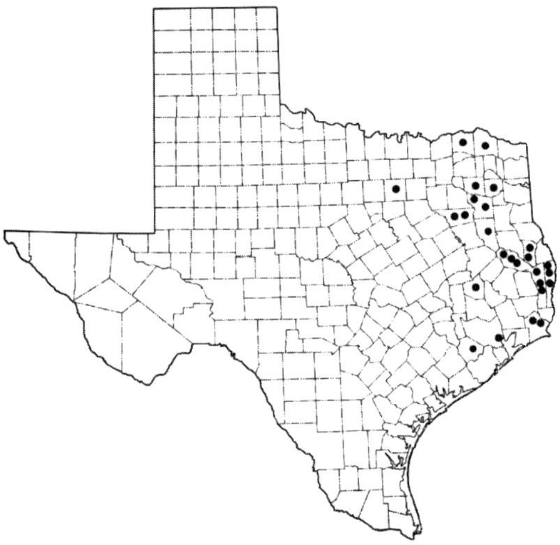

Celtis tenuifolia

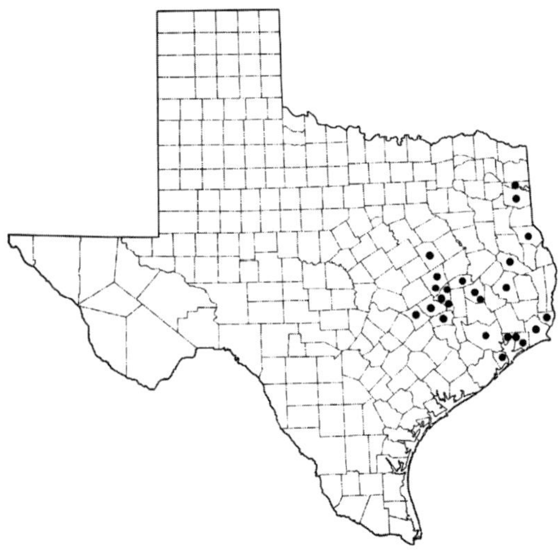

Planera aquatica

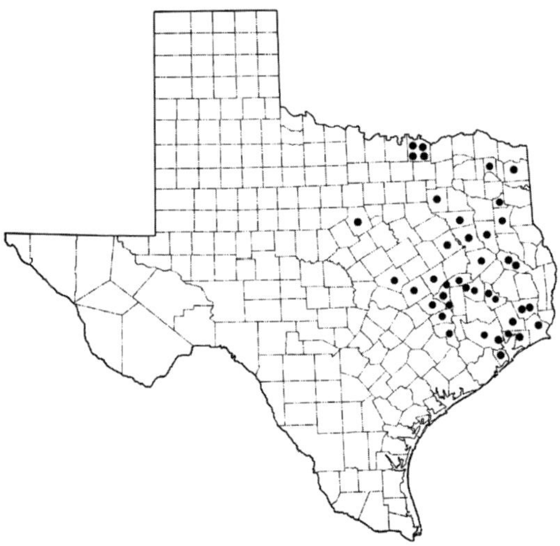

Ulmus alata

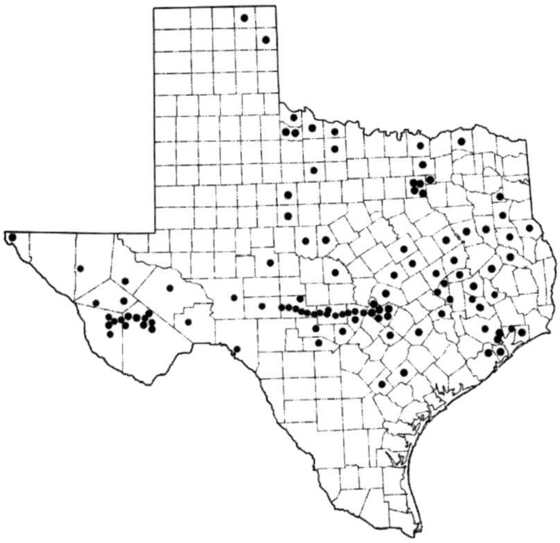

Ulmus americana

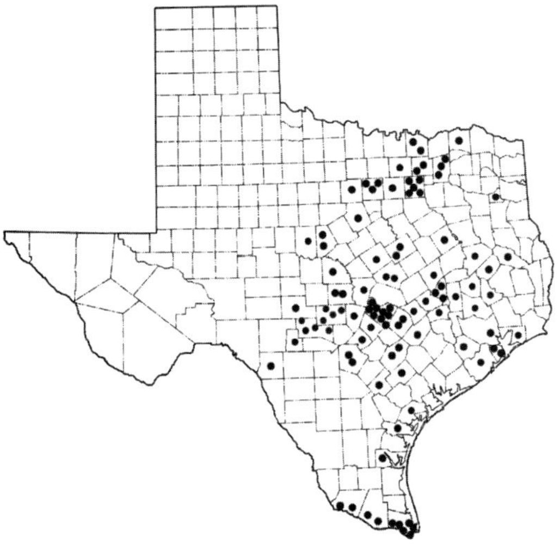

Ulmus crassifolia

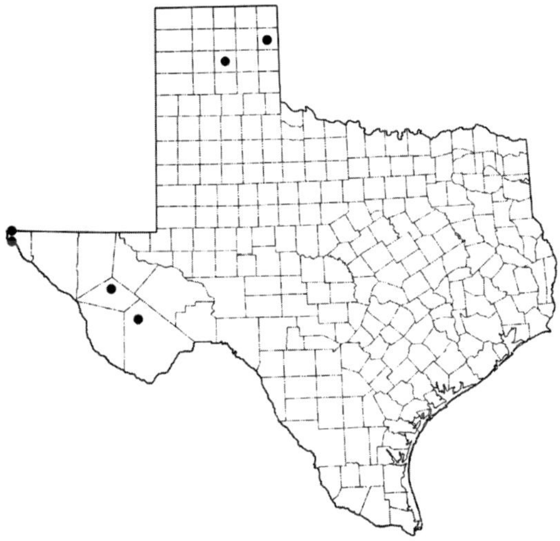

Ulmus pumila

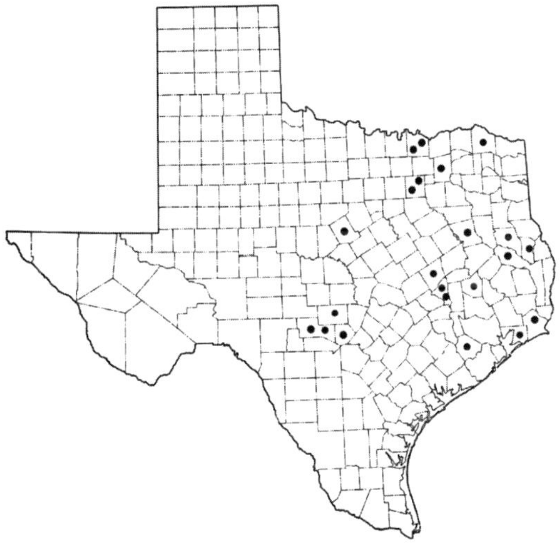

Ulmus rubra

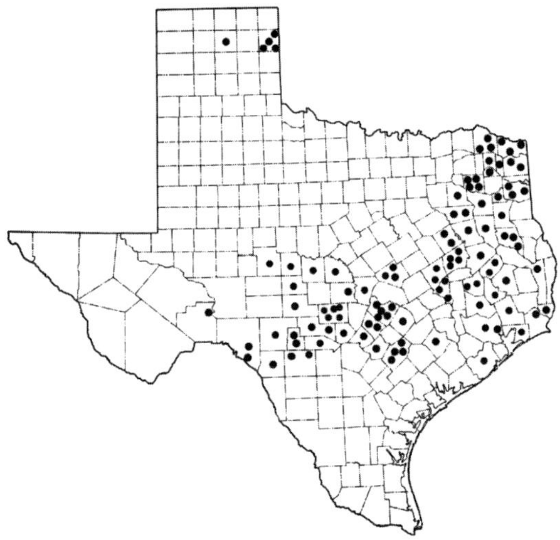

Boehmeria cylindrica

Parietaria floridana

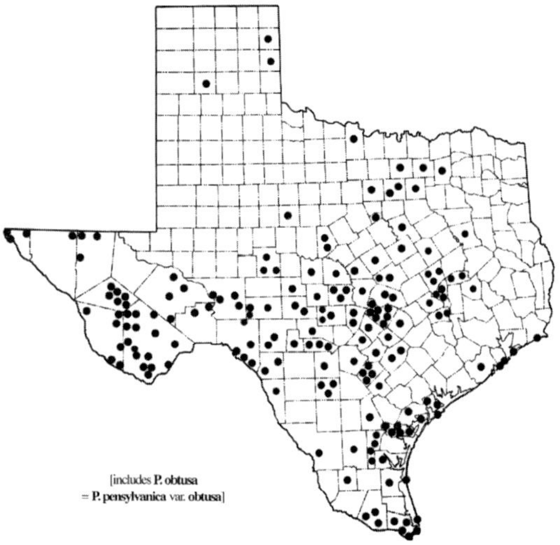
[includes **P. obtusa**
= **P. pensylvanica** var. **obtusa**]

Parietaria pensylvanica

*1988 Harris Co.: Lawn weed, Houston.

Pilea microphylla

Pilea pumila

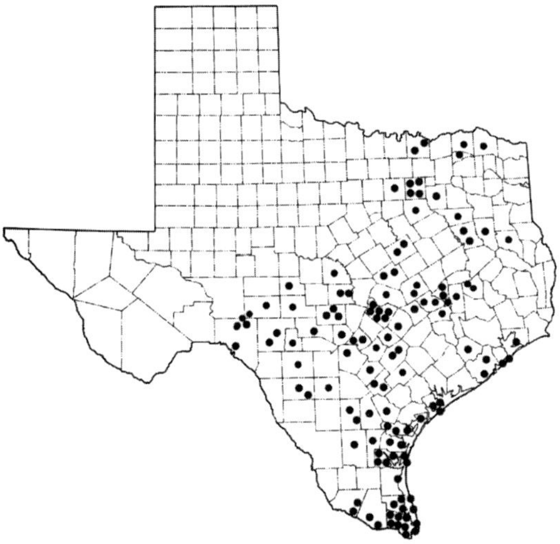

Urtica chamaedryoides

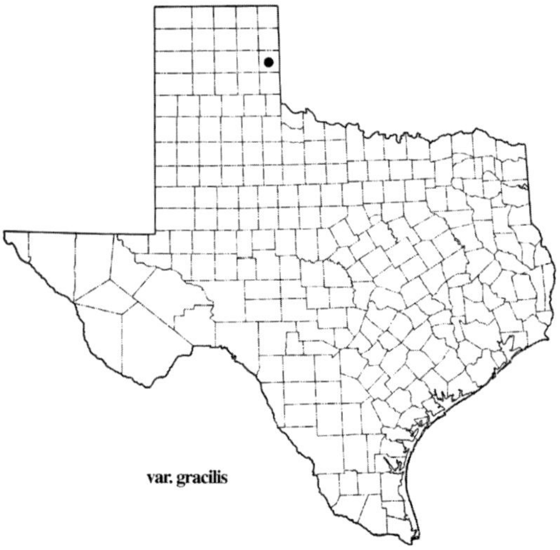
var. gracilis

Urtica dioica

Urtica gracilenta

Urtica urens

Valeriana arizonica

Valeriana texana

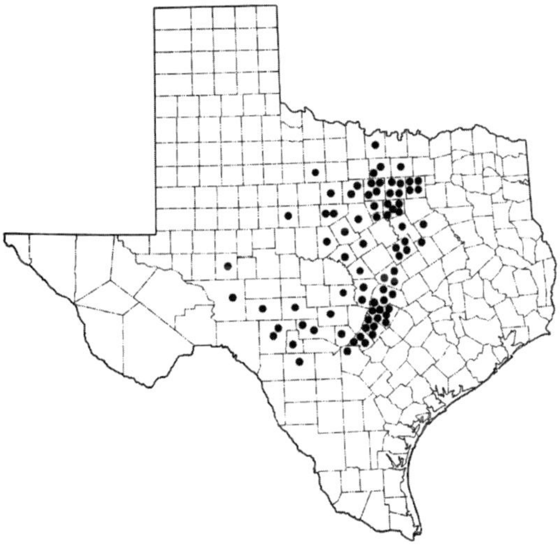

Valerianella amarella

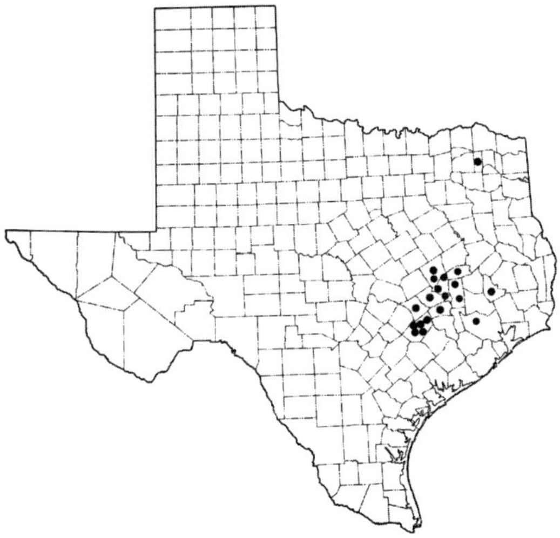

Valerianella florifera

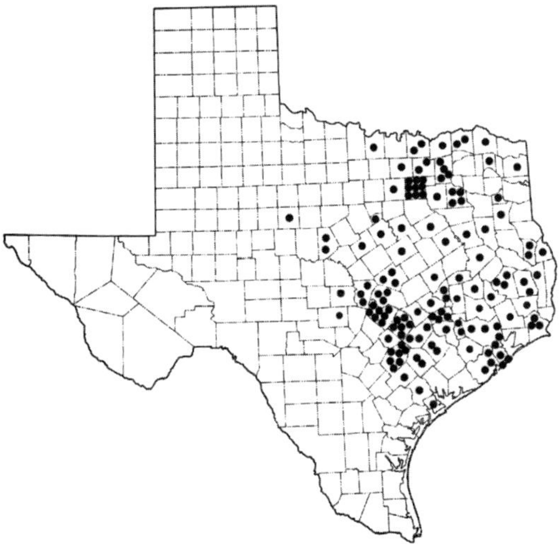

Valerianella radiata

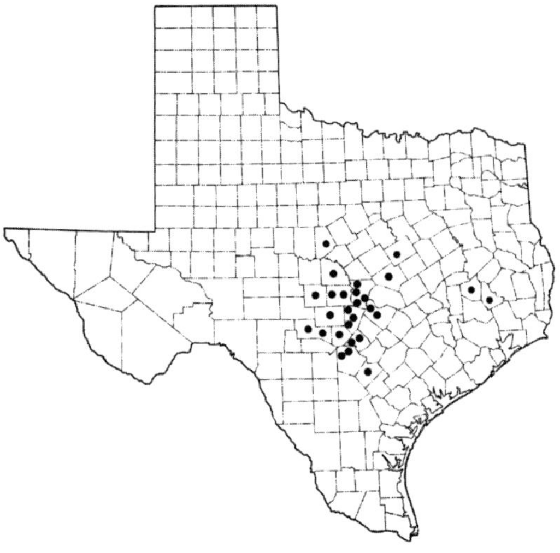
Valerianella stenocarpa

Valerianella texana

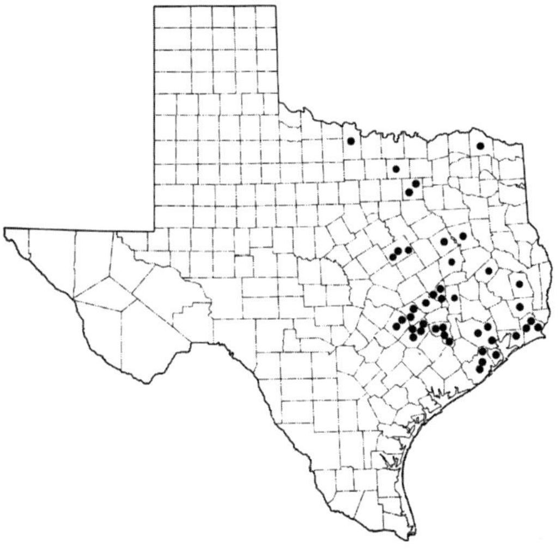
Valerianella woodsiana

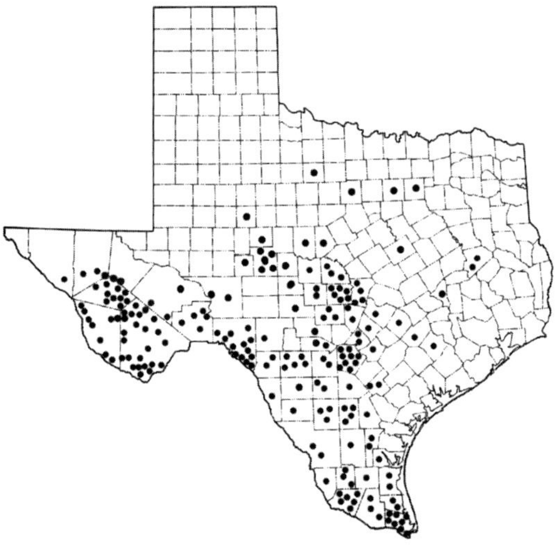
Aloysia gratissima

Aloysia macrostachya

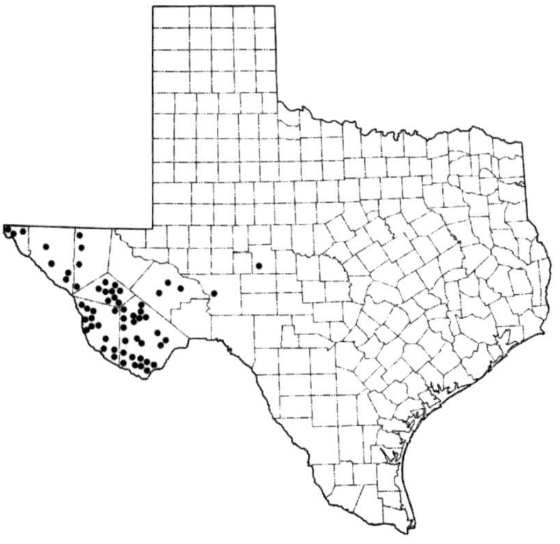
Aloysia wrightii

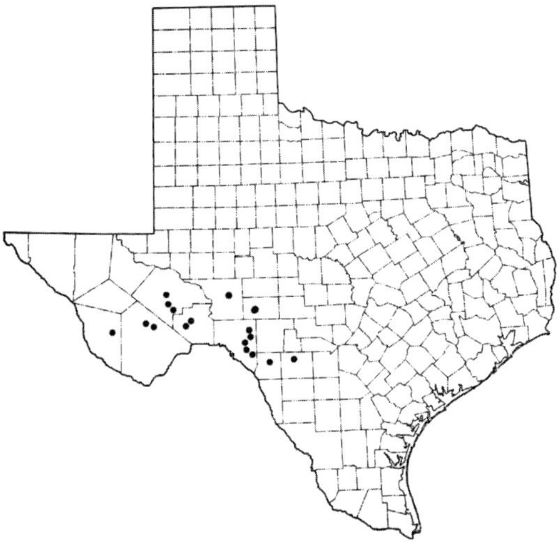

Bouchea linifolia

Bouchea spathulata

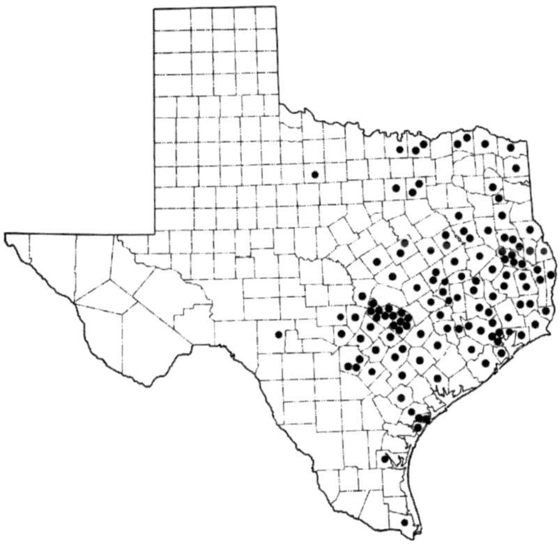

Callicarpa americana

Citharexylum berlandieri

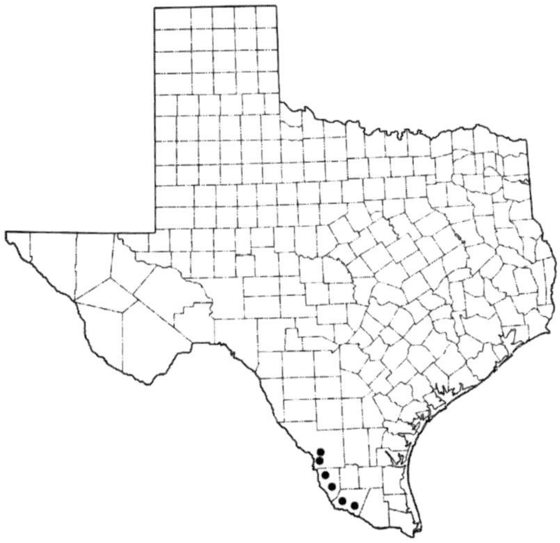

Citharexylum brachyanthum

Citharexylum spathulatum

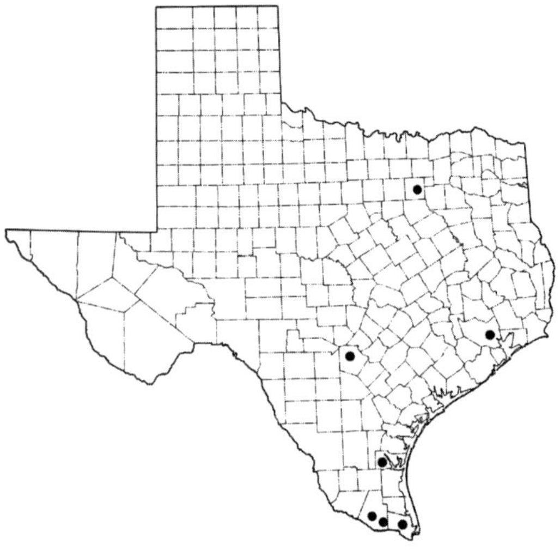

Duranta erecta

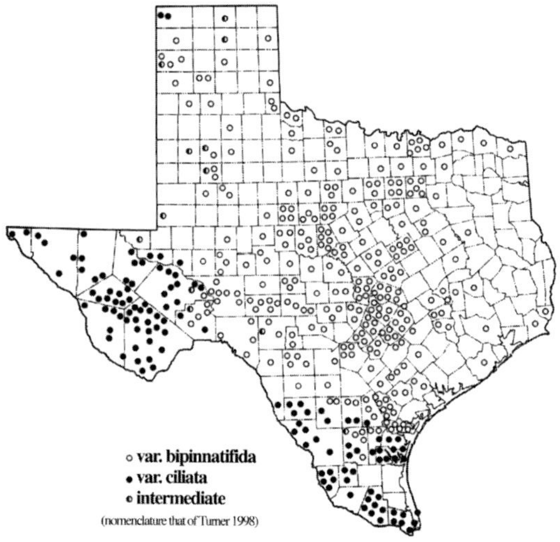
var. bipinnatifida
var. ciliata
intermediate
(nomenclature that of Turner 1998)

Glandularia bipinnatifida

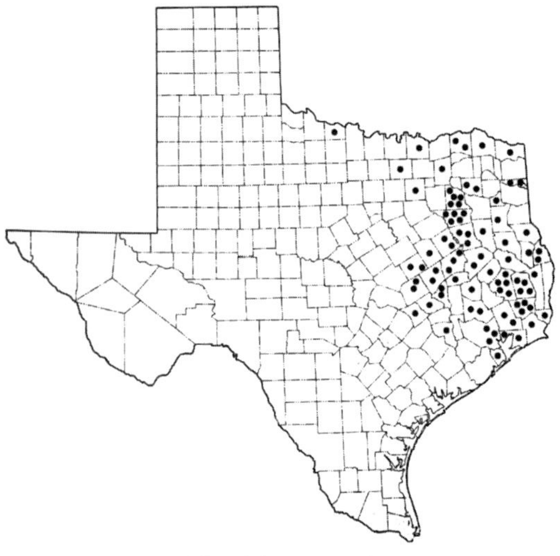

Glandularia canadensis

Glandularia delticola

Glandularia polyantha

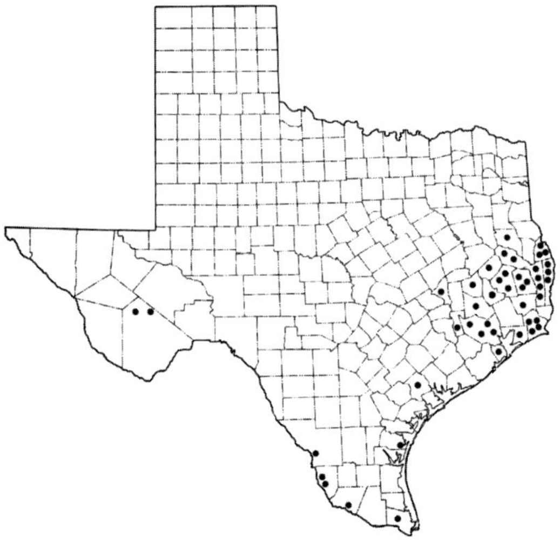

Glandularia pulchella

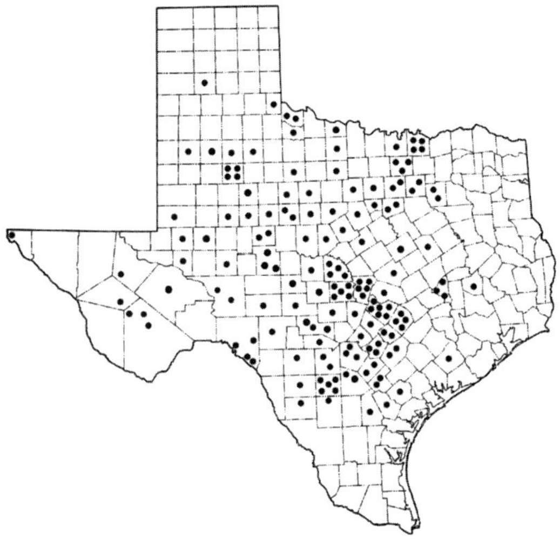

Glandularia pumila

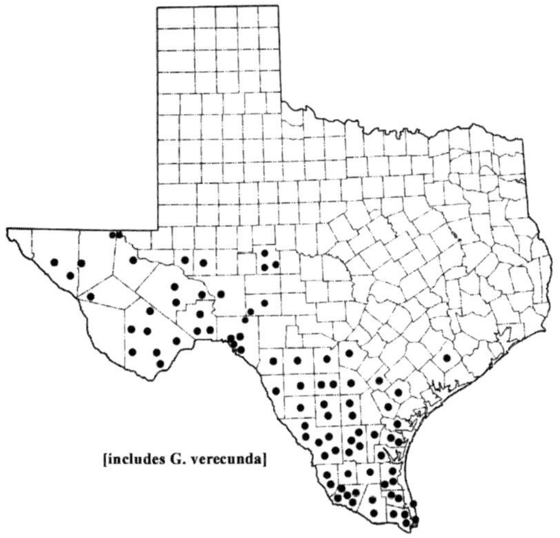
[includes G. verecunda]

Glandularia quandrangulata

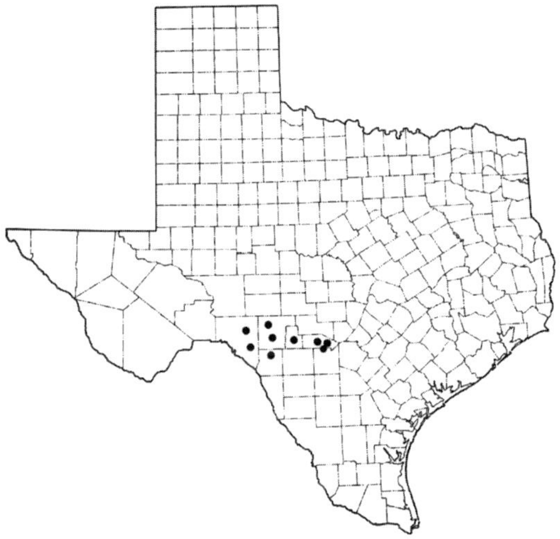

Glandularia tumidula

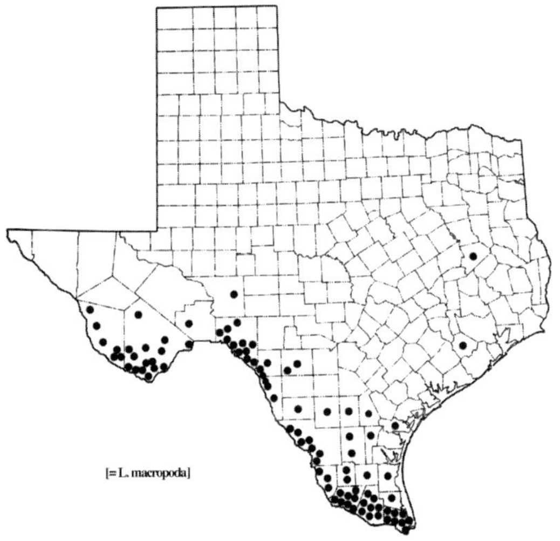
[= L. macropoda]

Lantana achyranthifolia

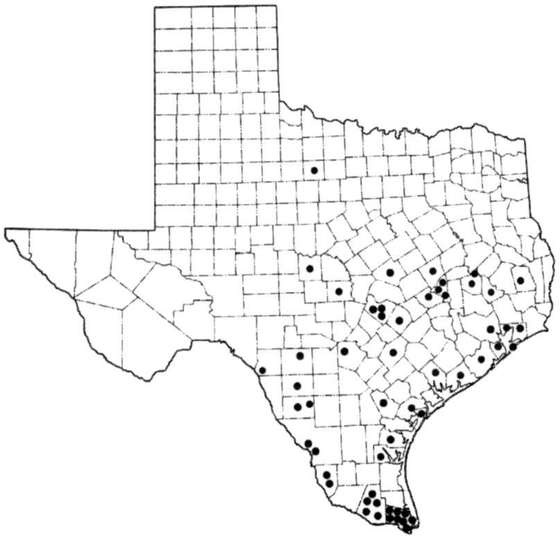

Lantana camara

Lantana canescens

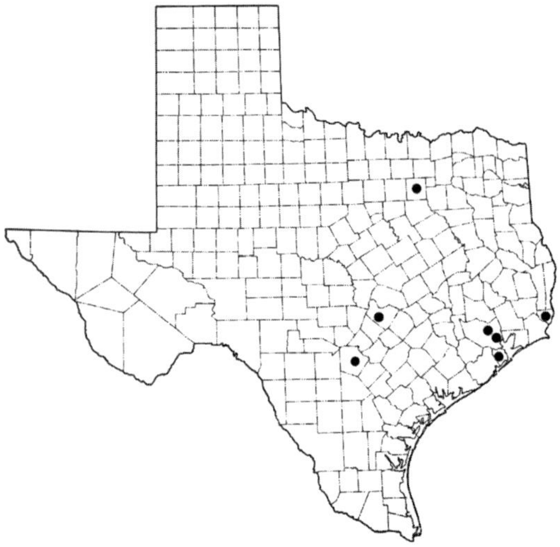

Lantana montevidensis

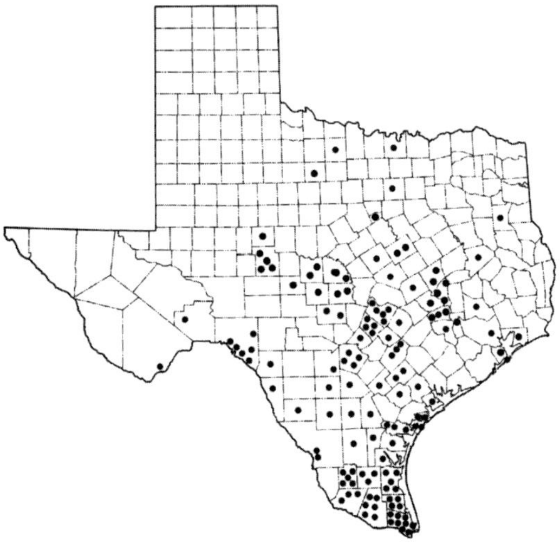

Lantana urticoides

Lippia alba

Lippia graveolens

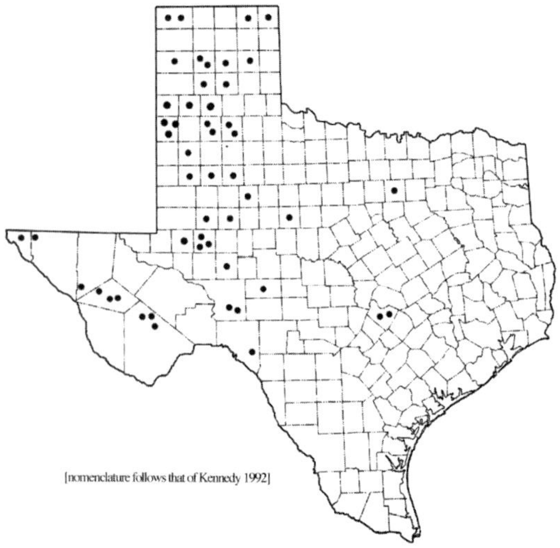
[nomenclature follows that of Kennedy 1992]

Phyla cuneifolia

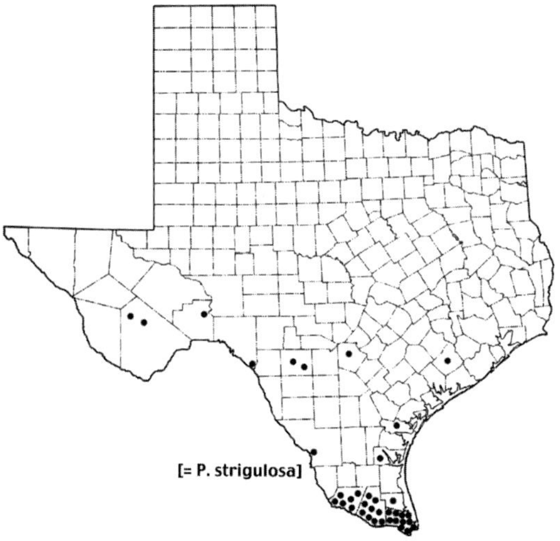
[= P. strigulosa]

Phyla fruticosa

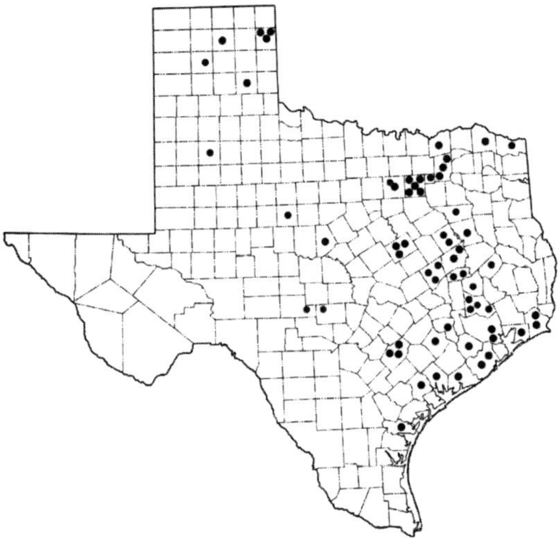

Phyla lanceolata

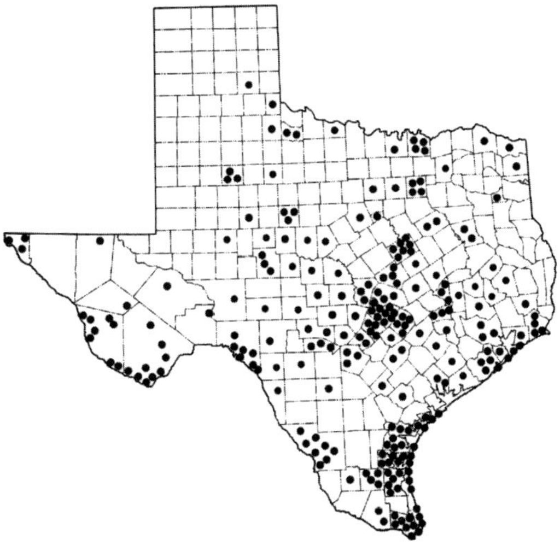

Phyla nodiflora

Priva lappulacea

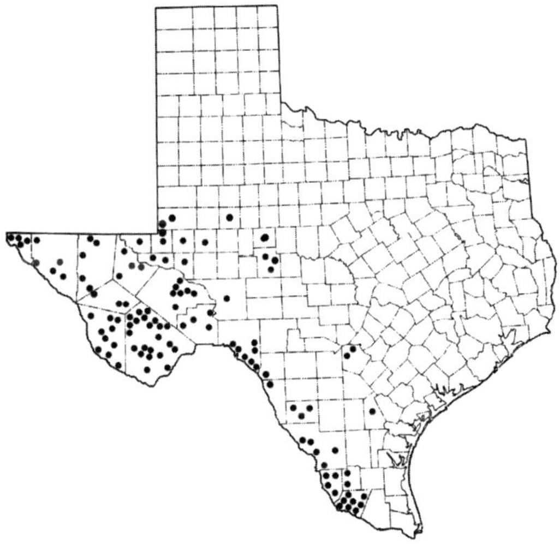

Tetraclea coulteri

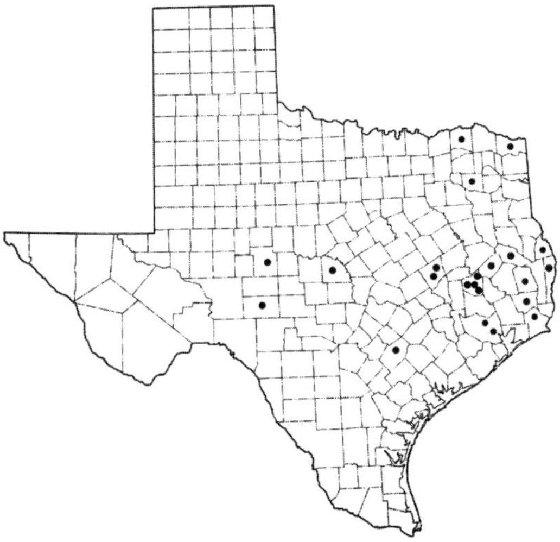

Verbena bonariensis

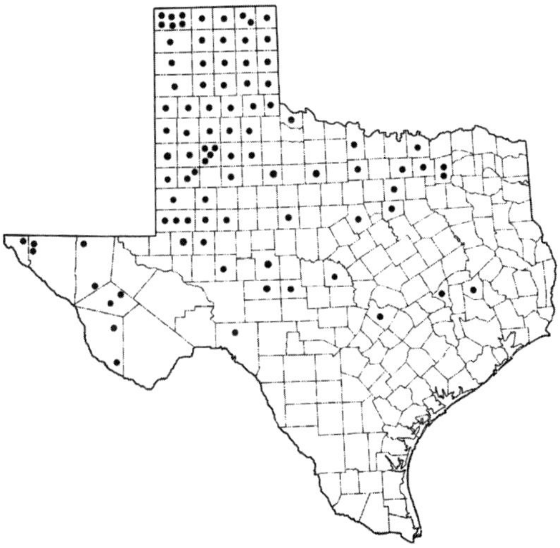

Verbena bracteata

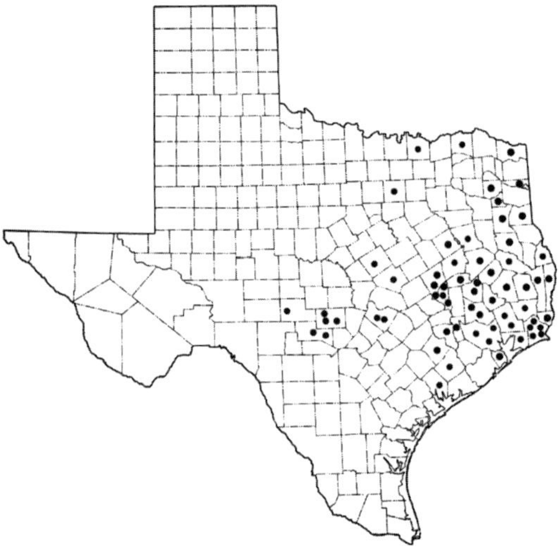

Verbena brasiliensis

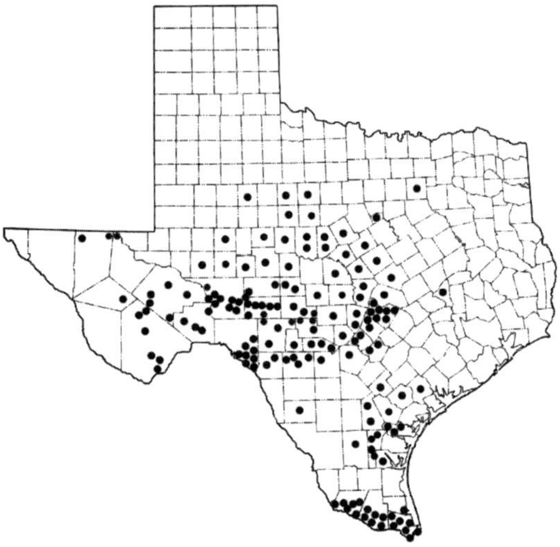

Verbena canescens

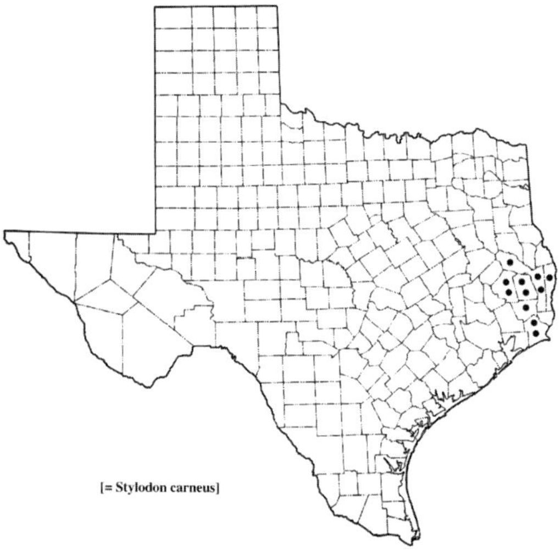
[= Stylodon carneus]

Verbena carnea

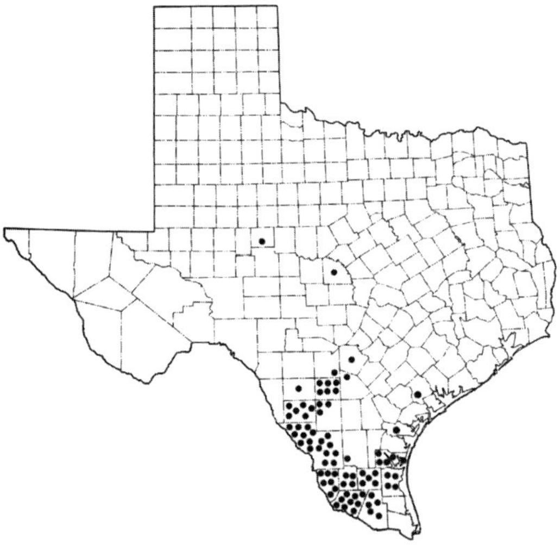

Verbena cloverae

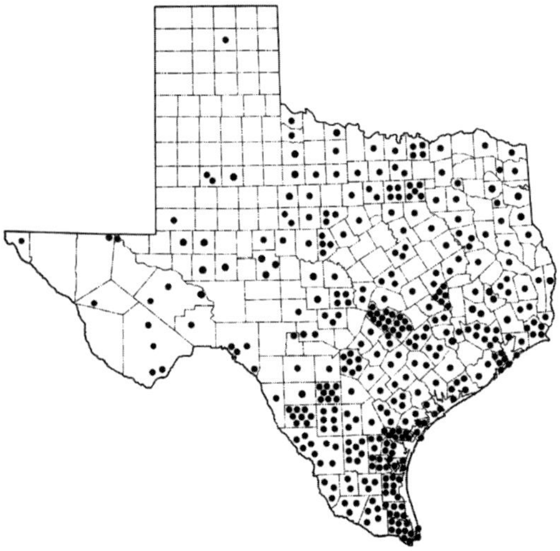

Verbena halei

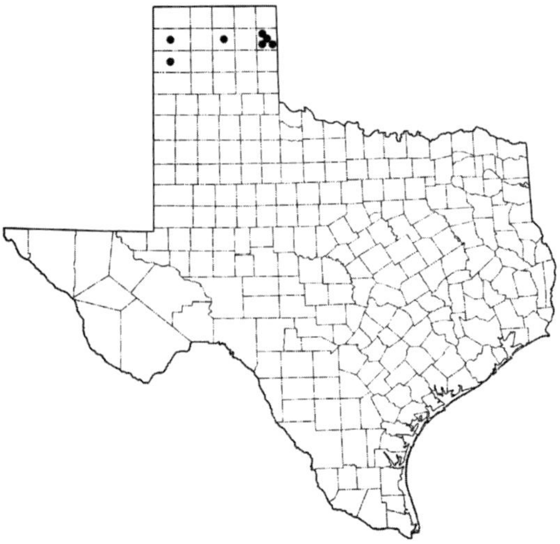

Verbena hastata

Verbena litoralis

Verbena livermorensis

Verbena macdougalii

Verbena neomexicana

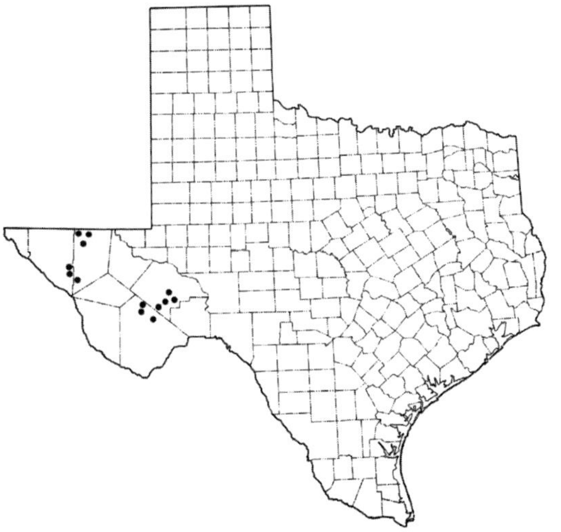

Verbena perennis

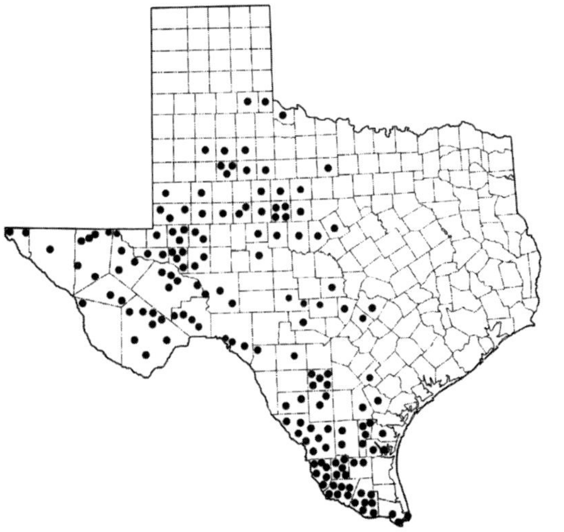

Verbena plicata

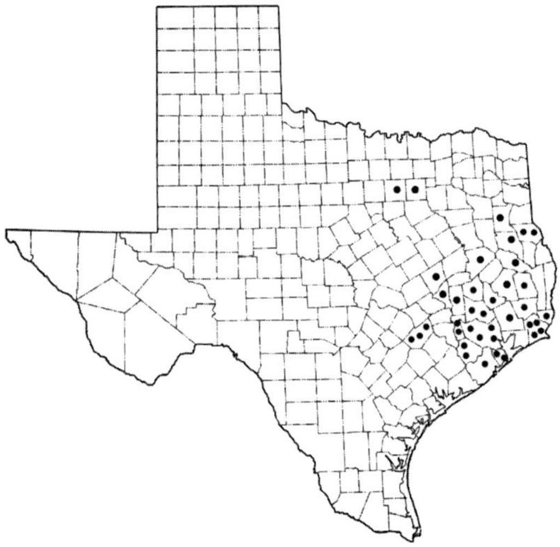

Verbena rigida

Verbena runyonii

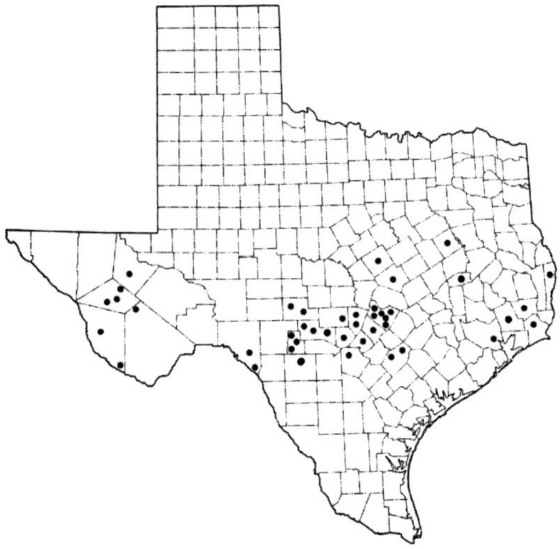

Verbena scabra

Verbena simplex

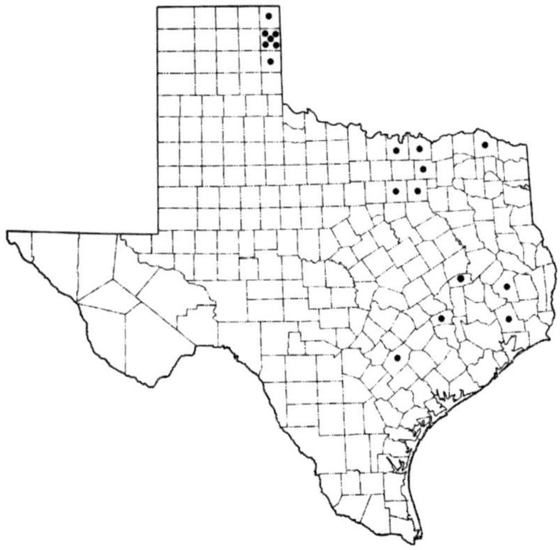

Verbena stricta

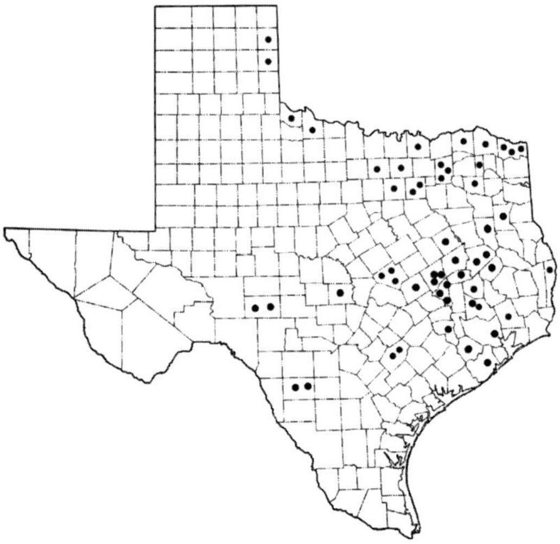

Verbena urticifolia

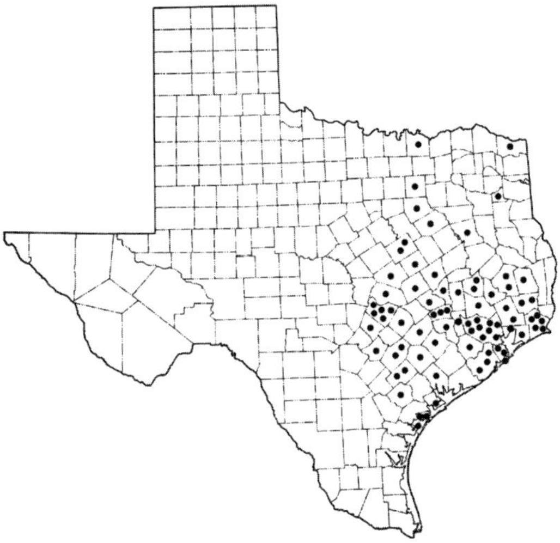

Verbena xutha

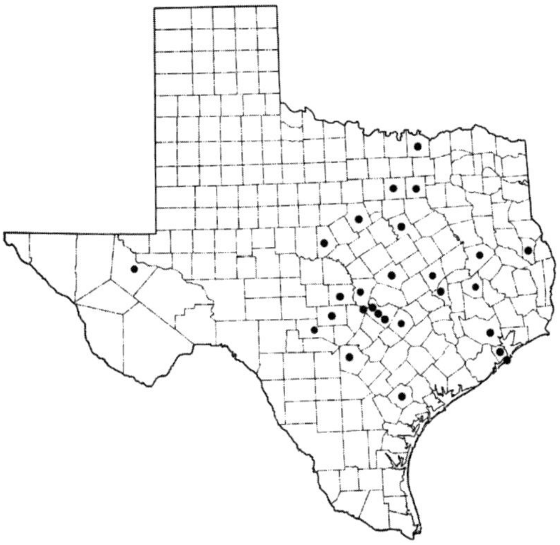

Vitex agnus-castus

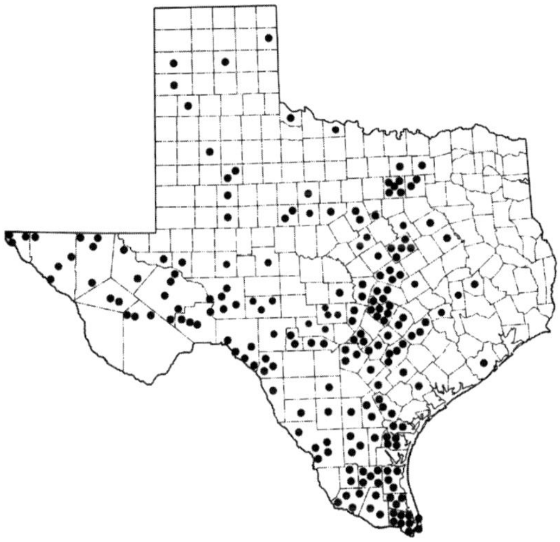

Hybanthus verticillatus

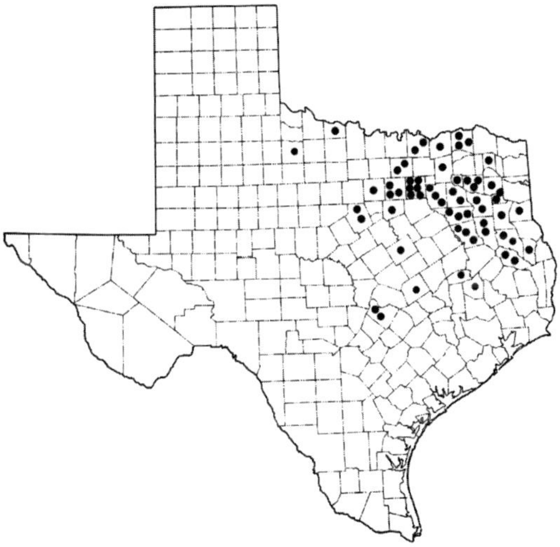

Viola bicolor

Viola guadalupensis

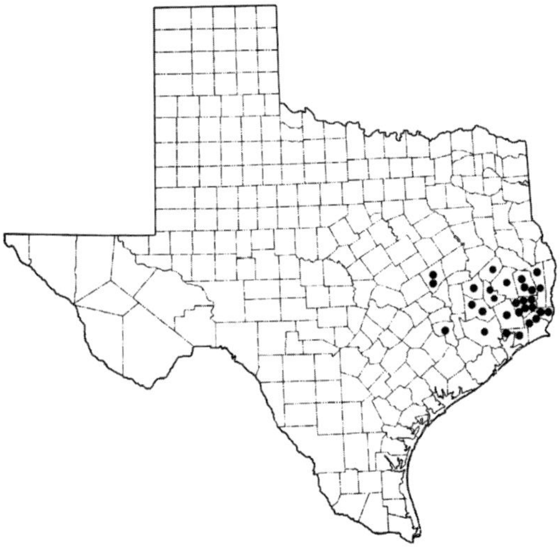

Viola lanceolata

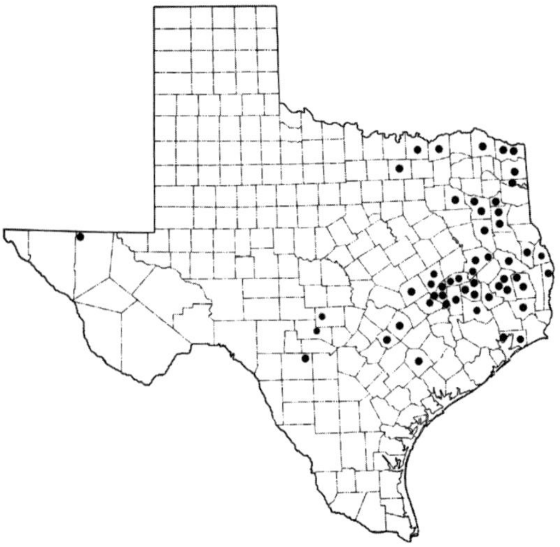

Viola palmata

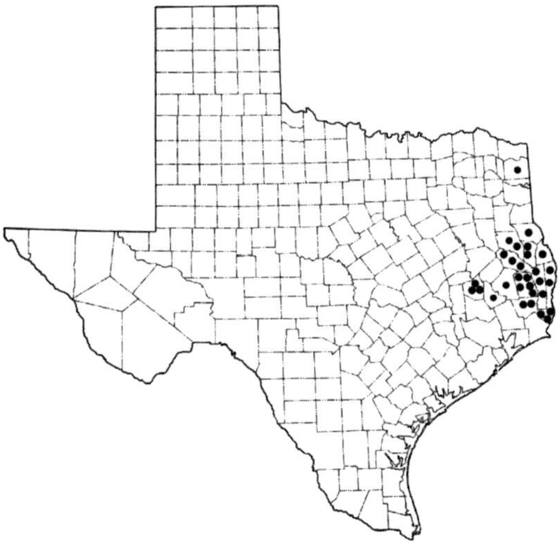

Viola pedata

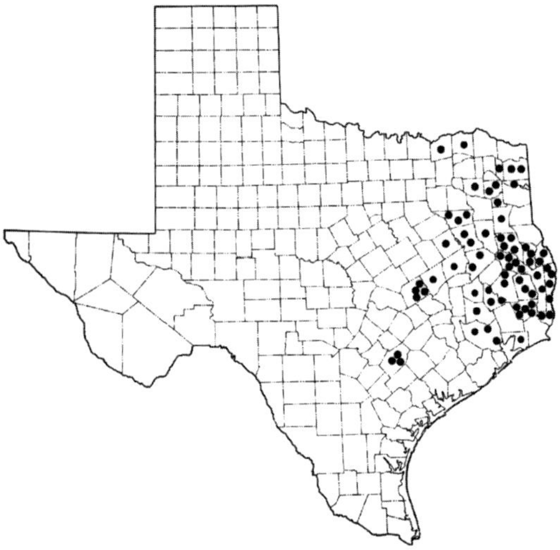

Viola primulifolia

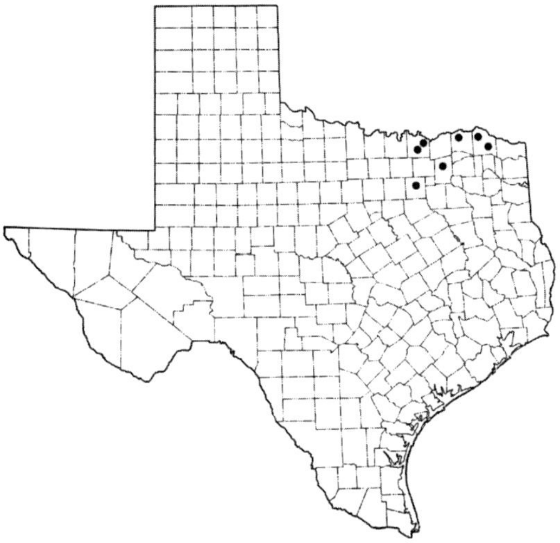

Viola pubescens

Viola purpurea

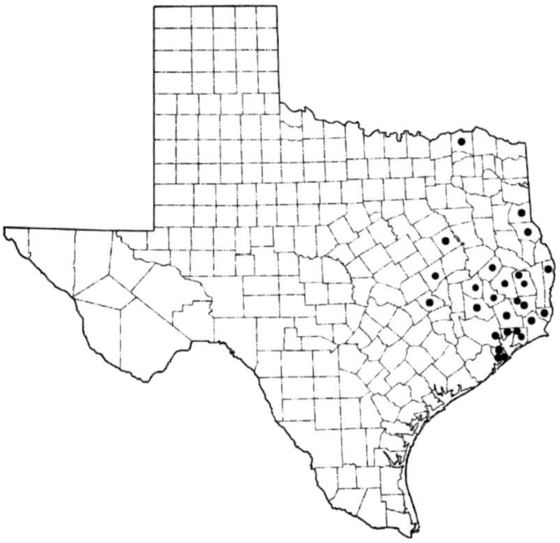

Viola sagittata

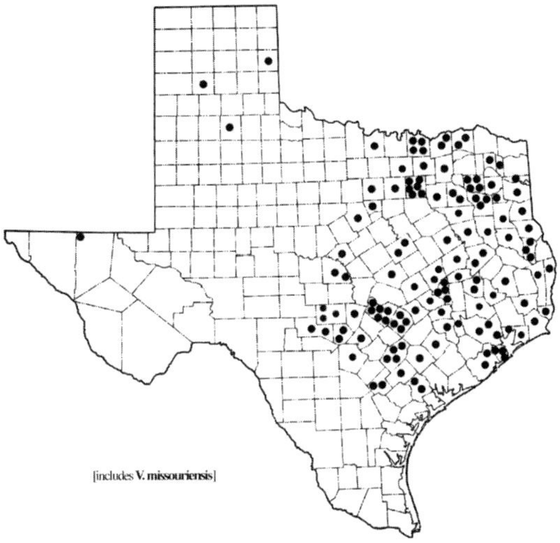
[includes **V. missouriensis**]

Viola sororia

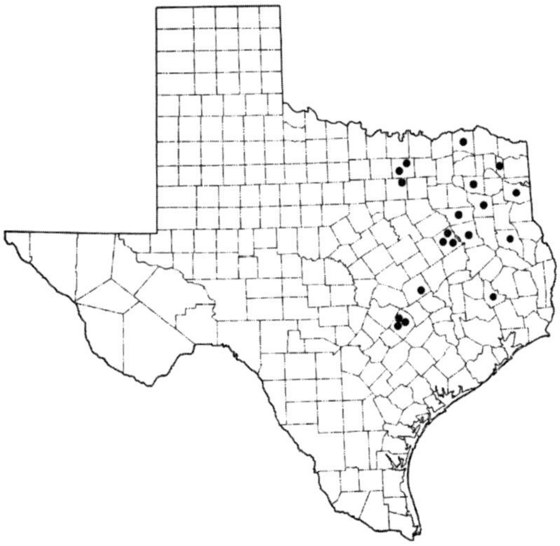

Viola villosa

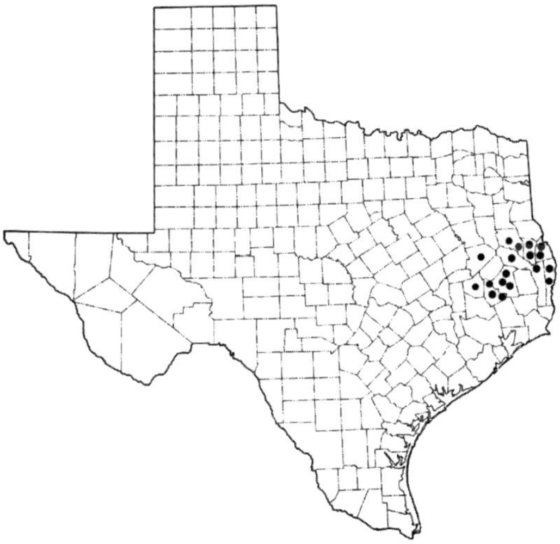

Viola walteri

Arceuthobium divaricatum

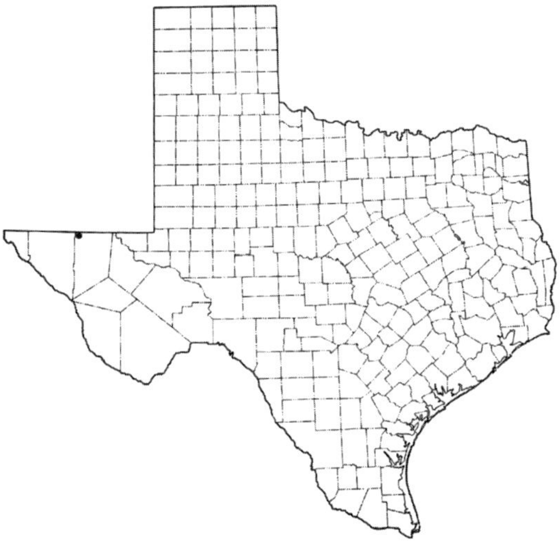

Arceuthobium douglasii

var. cryptopodum
B. L. Turner, var. nov.

Arceuthobium vaginatum

[fide Wiens, 1964]

Phoradendron bolleanum

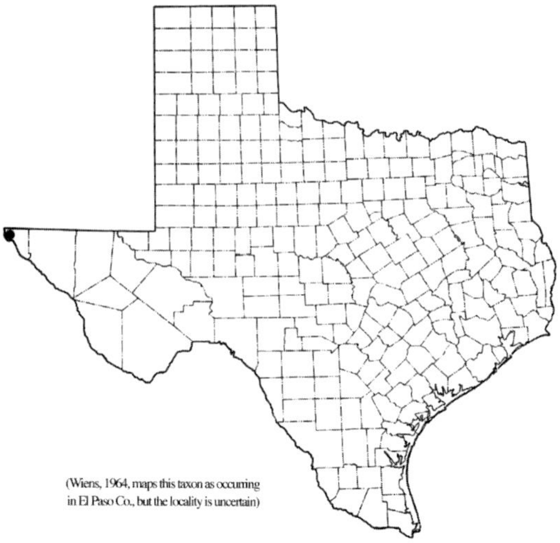
(Wiens, 1964, maps this taxon as occurring
in El Paso Co., but the locality is uncertain)

Phoradendron capitellatum

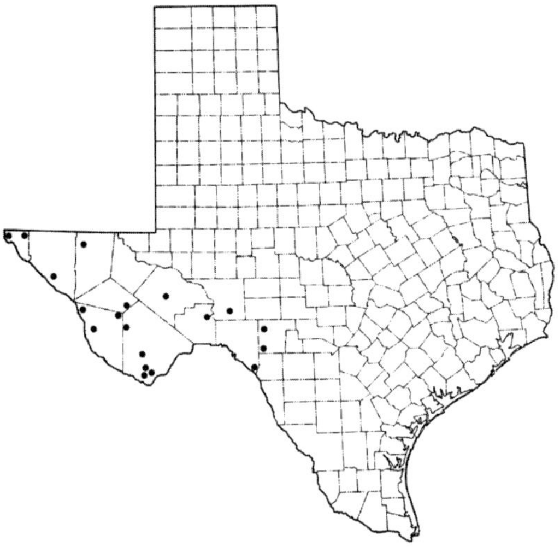

Phoradendron hawksworthii

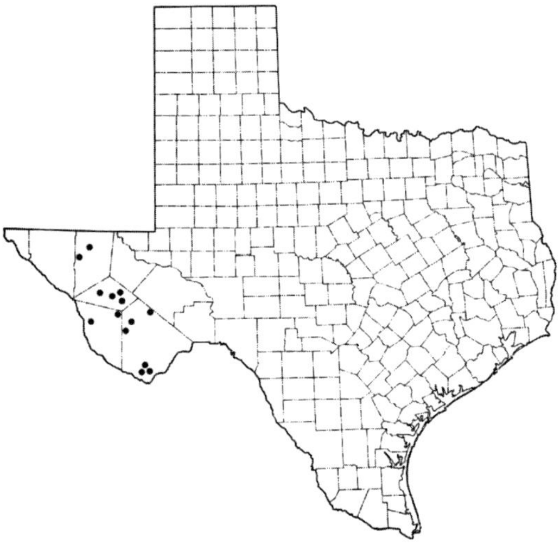

Phoradendron juniperinum

Phoradendron macrophyllum

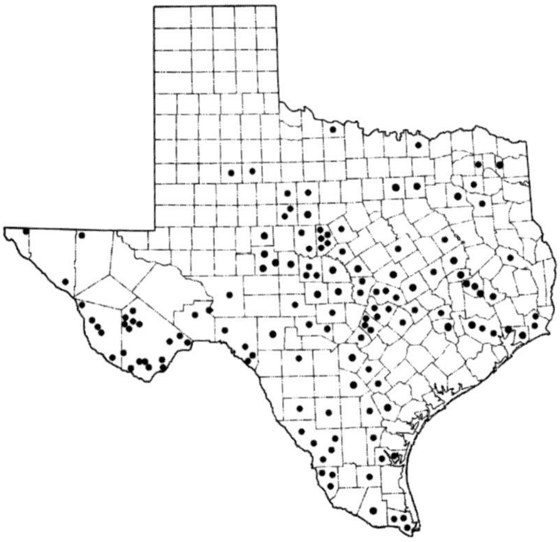

Phoradendron tomentosum

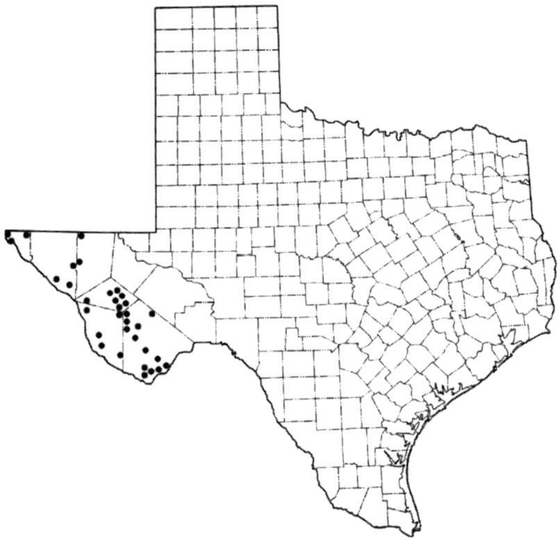

Phoradendron villosum

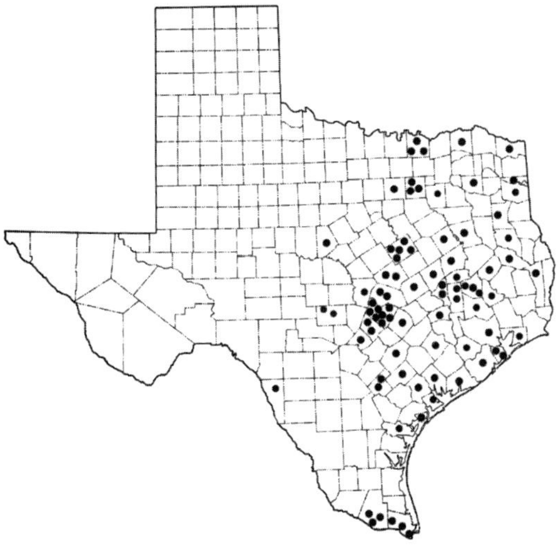

Ampelopsis arborea

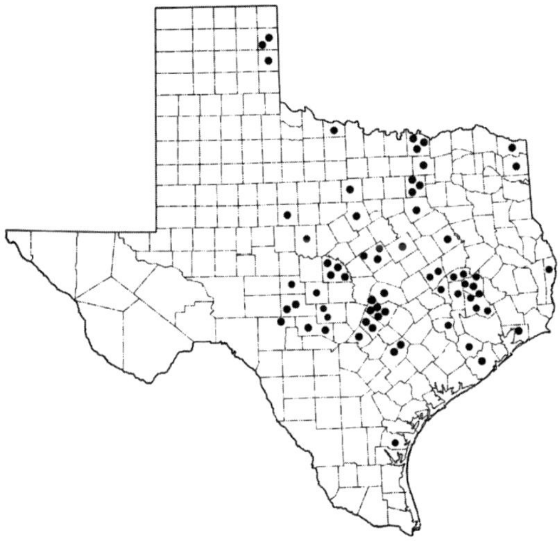

Ampelopsis cordata

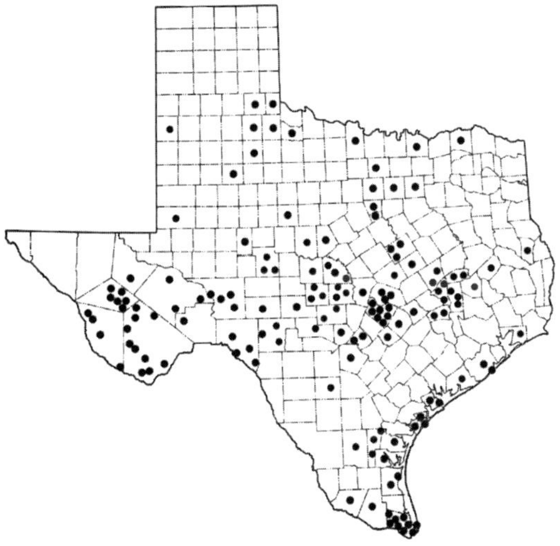

Cissus incisa

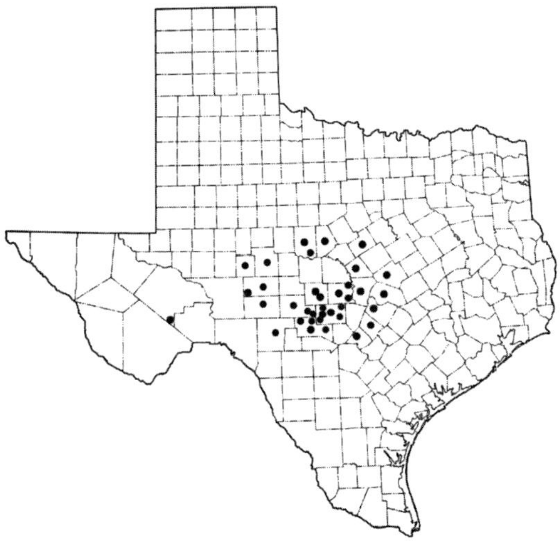

Parthenocissus heptaphylla

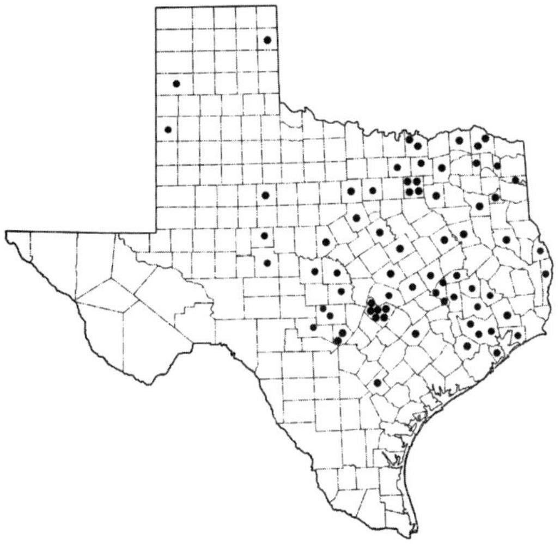

Parthenocissus quinquefolia

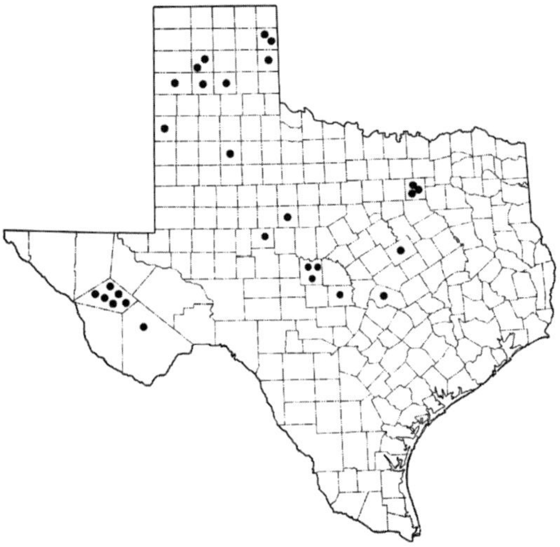

Parthenocissus vitacea

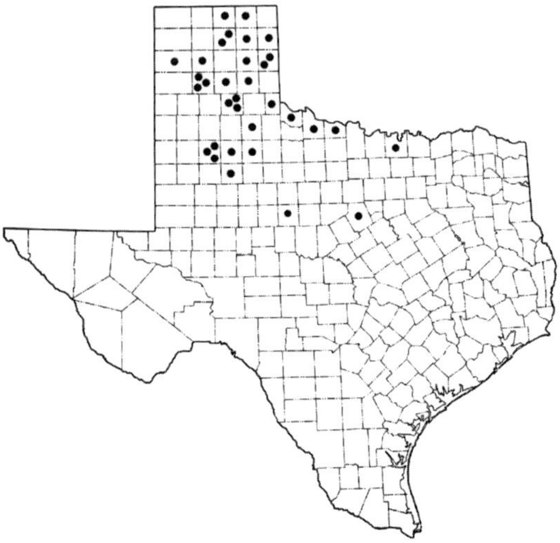

Vitis acerifolia

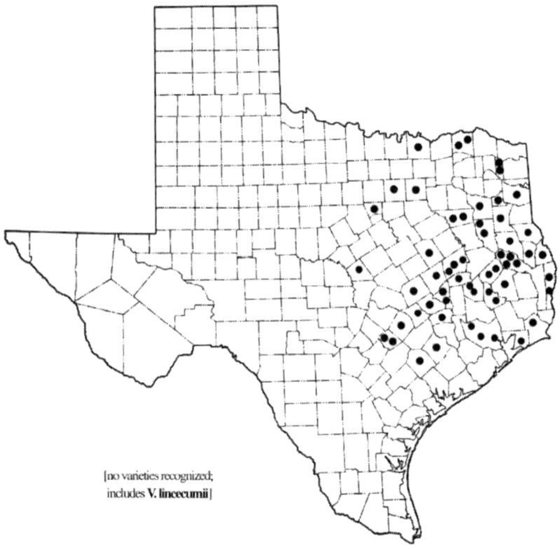
[no varieties recognized;
includes **V. lincecumii**]

Vitis aestivalis

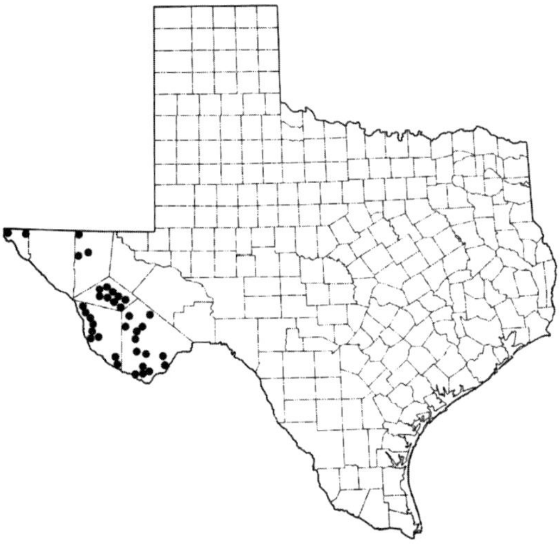

Vitis arizonica

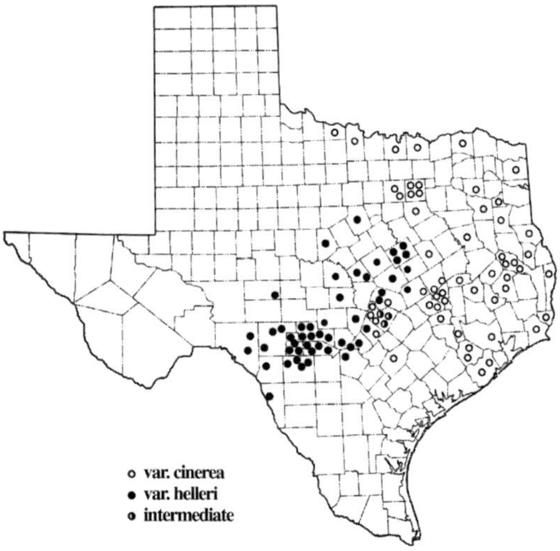
o **var. cinerea**
• **var. helleri**
o **intermediate**

Vitis cinerea

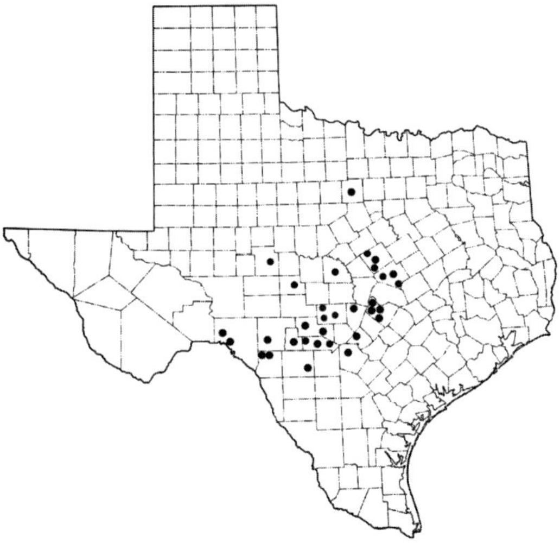

Vitis monticola

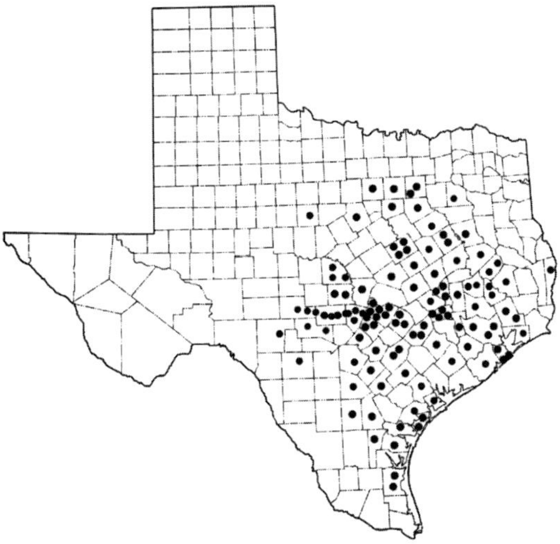

Vitis mustangensis

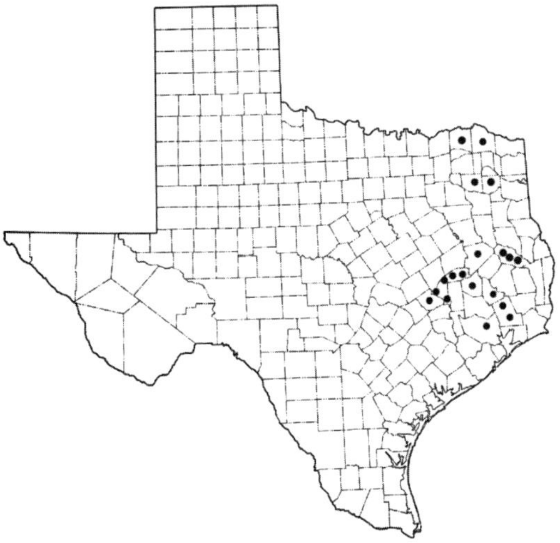

Vitis palmata

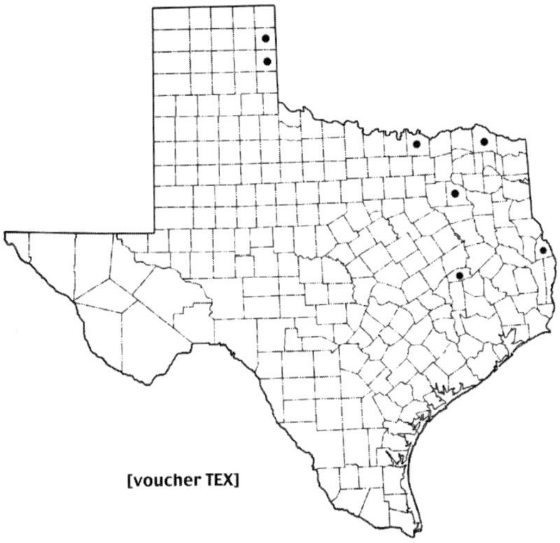
[voucher TEX]

Vitis riparia

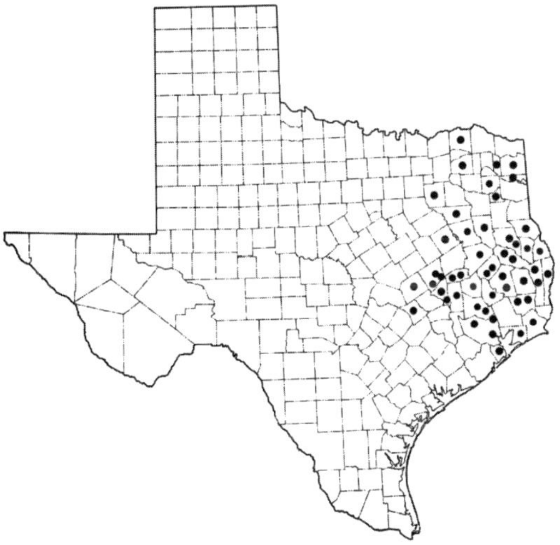

Vitis rotundifolia

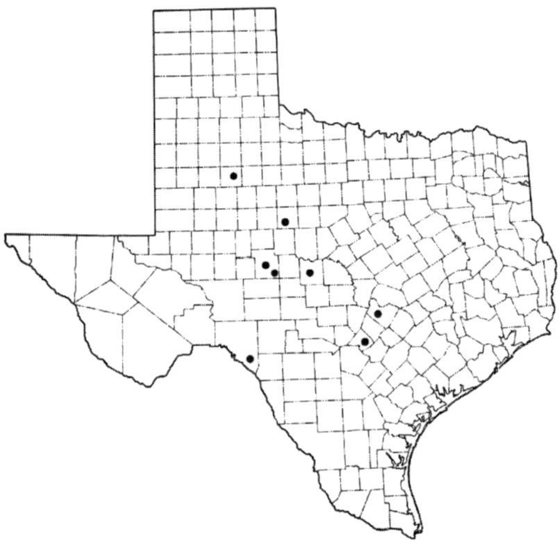

Vitis rupestris

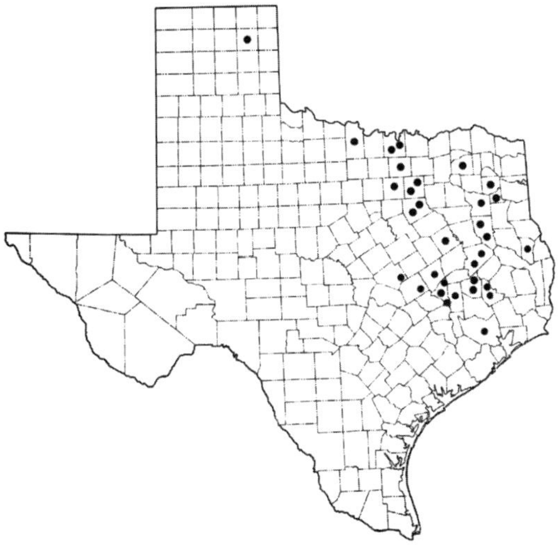

Vitis vulpina

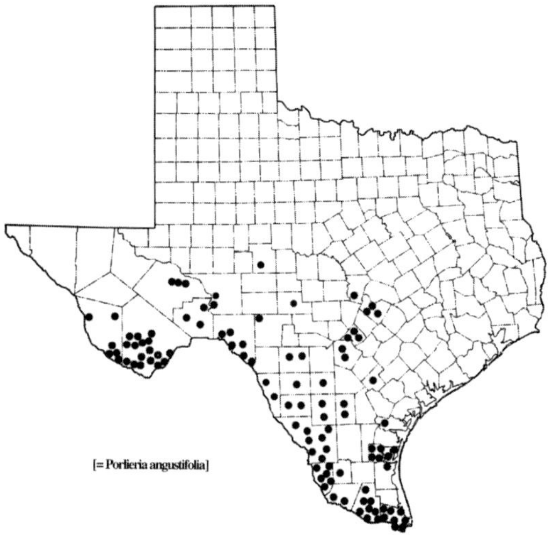
[= Porlieria angustifolia]

Guajacum angustifolium

Kallstroemia californica

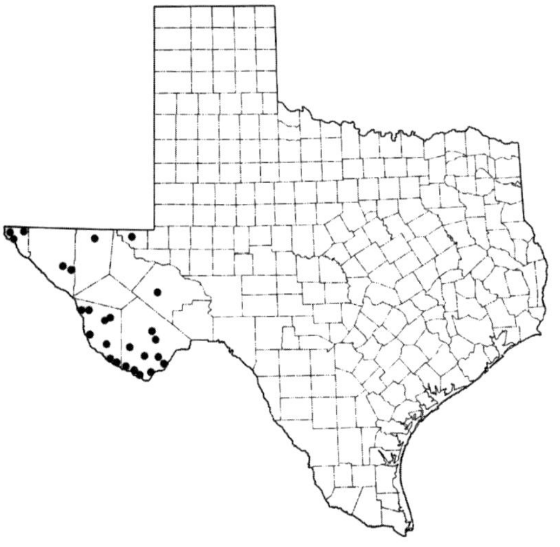

Kallstroemia grandiflora

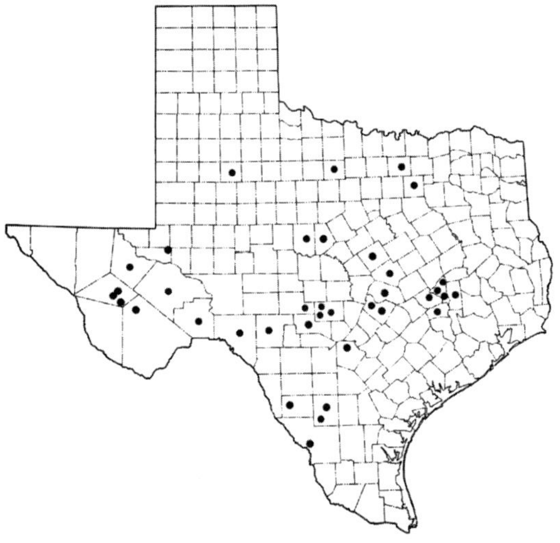

Kallstroemia hirsutissima

Kallstroemia maxima

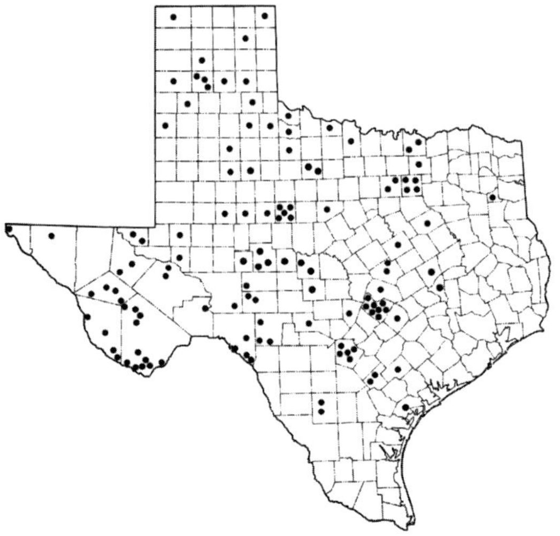

Kallstroemia parviflora

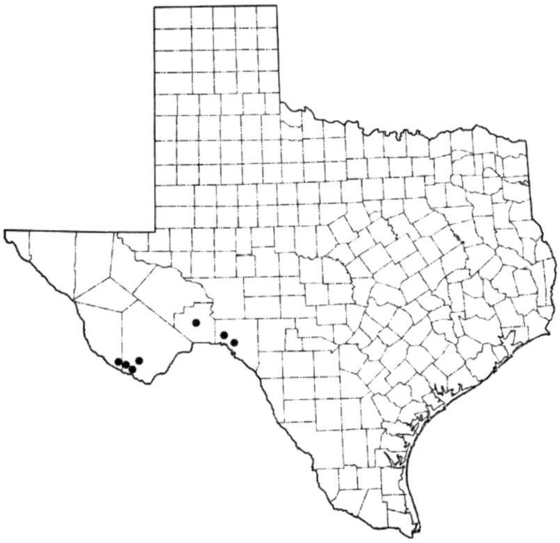

Kallstroemia perennans

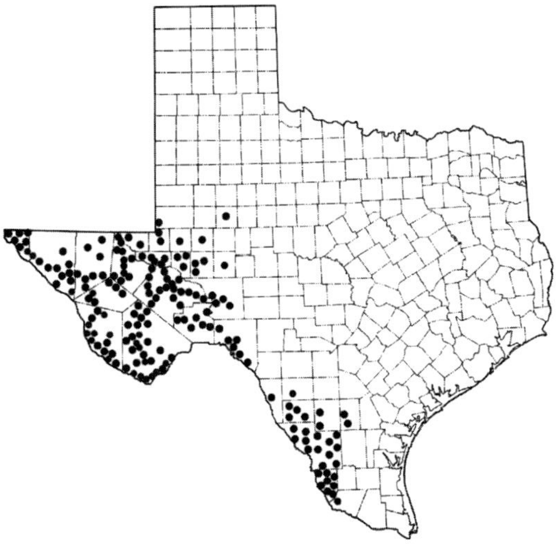

Larrea tridentata

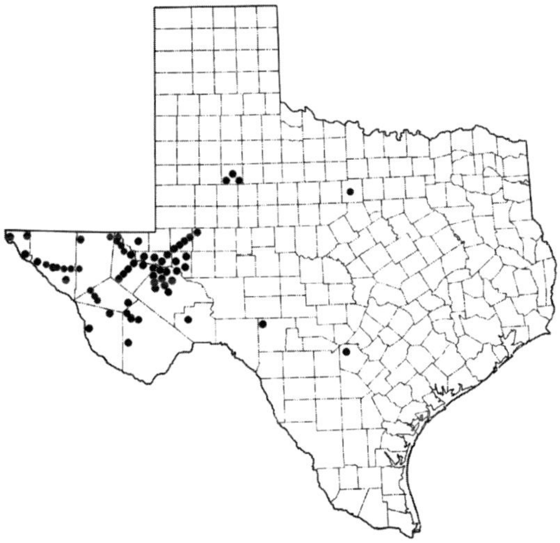

Peganum harmala

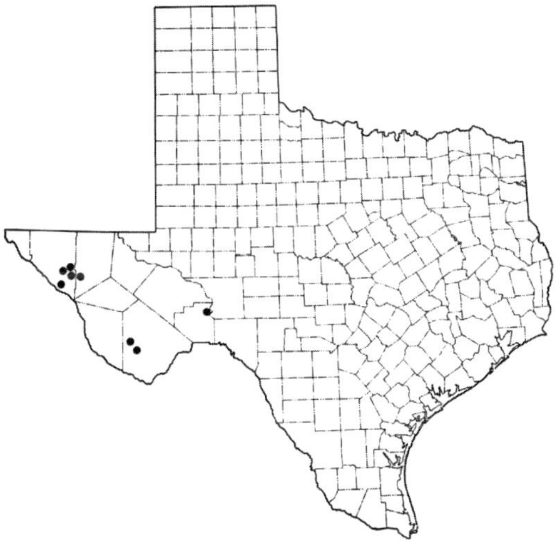

Peganum mexicanum

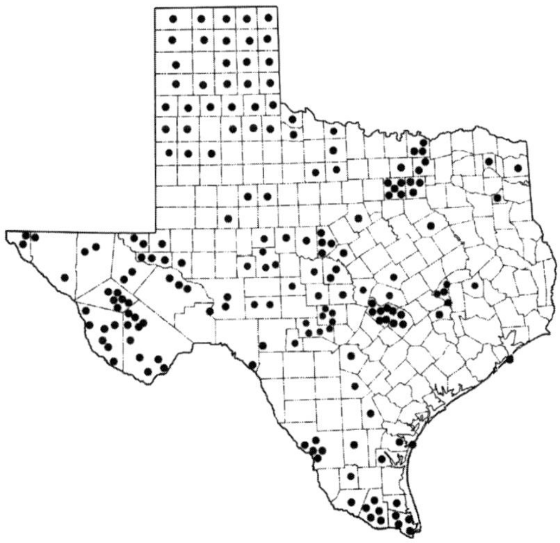

Tribulus terrestris

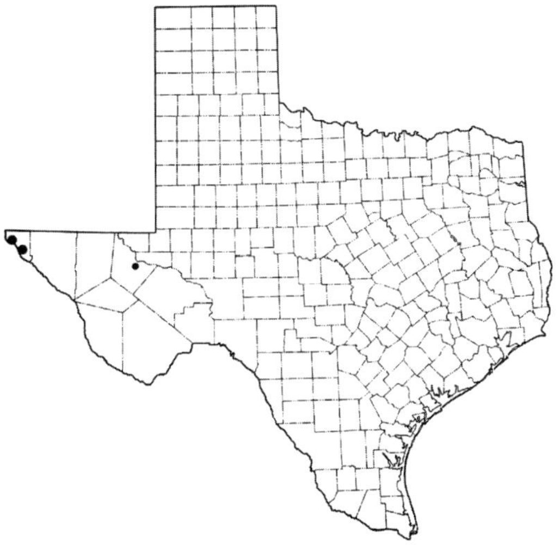

Zygophyllum fabago

Amaranthus viridis

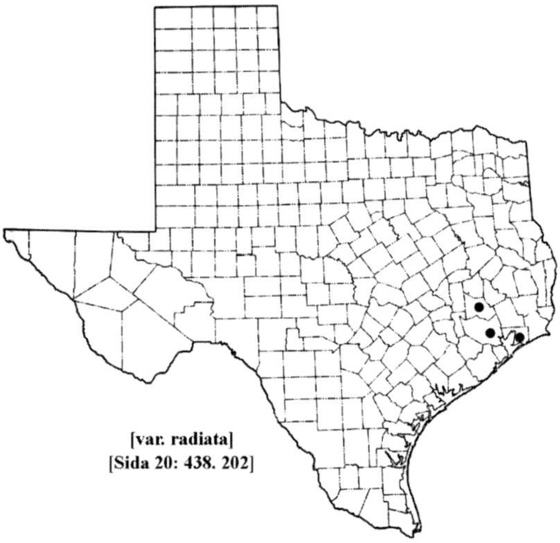
[var. radiata]
[Sida 20: 438. 202]

Bidens alba

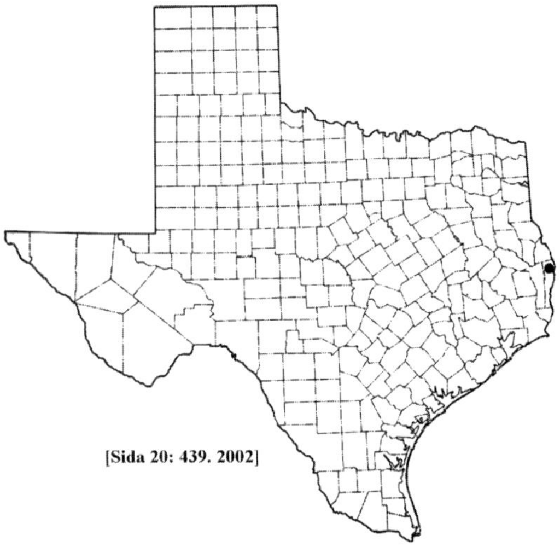
[Sida 20: 439. 2002]

Calycanthus floridus

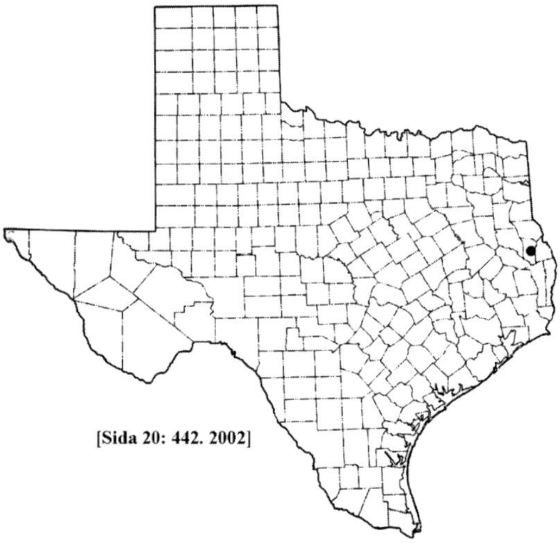
[Sida 20: 442. 2002]

Cuphea viscosissima

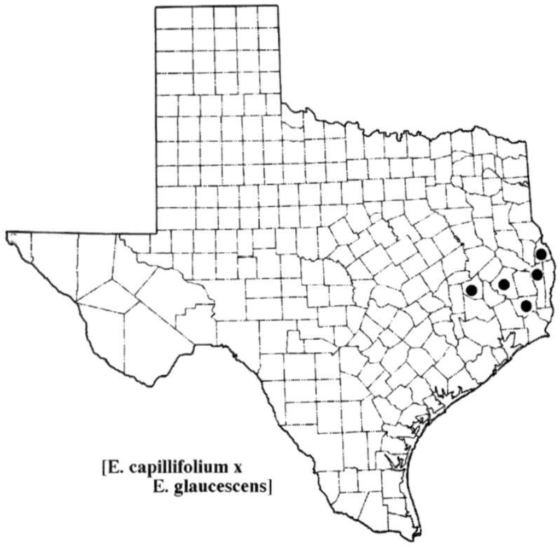
[E. capillifolium x
E. glaucescens]

Eupatorium pinnatifidum

Glechoma hederacea

[= H. annuus x H. petiolaris]

Helianthus paradoxus

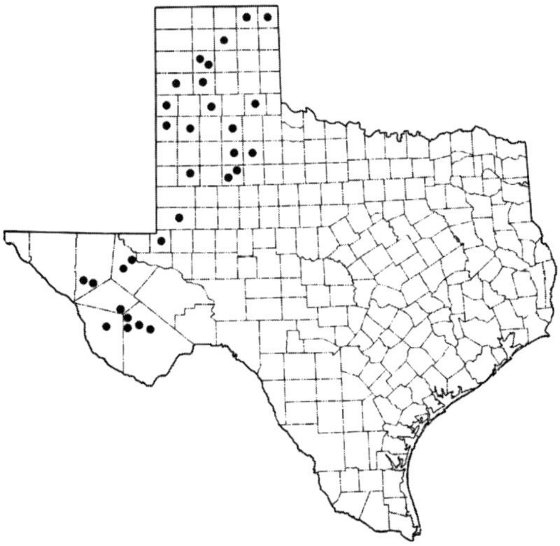

Heterotheca villosa

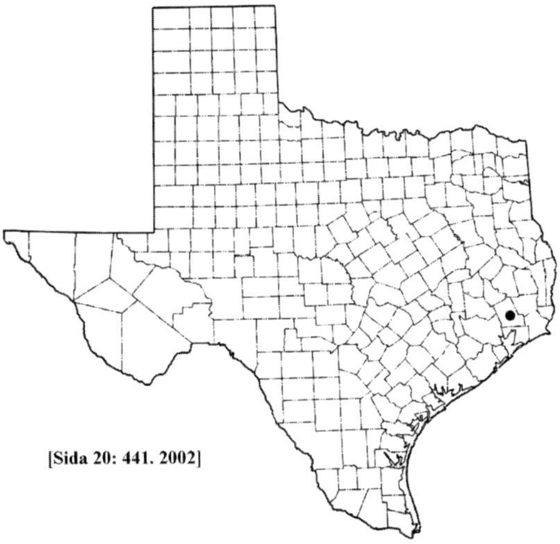
[Sida 20: 441. 2002]

Hyptis mutabilis

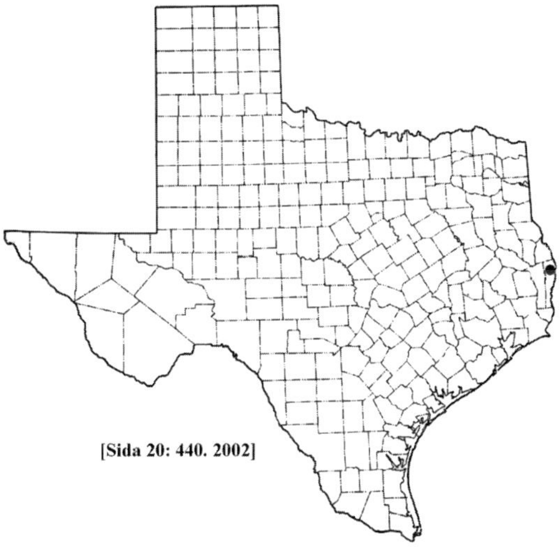
[Sida 20: 440. 2002]

Lyonia lucida

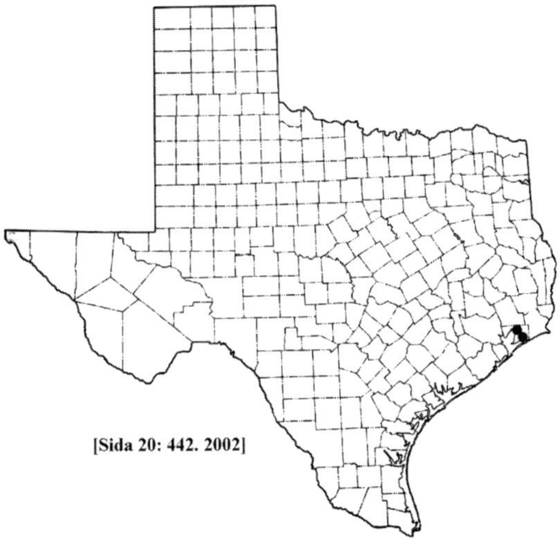
[Sida 20: 442. 2002]

Lythrum lineare

[voucher TEX]

Machaeranthera gypsitherma

[Sida 20: 442. 2002]
Myrtus communis

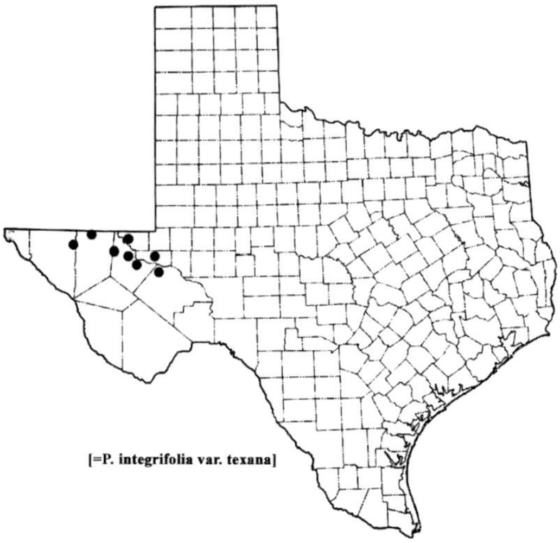
[=P. integrifolia var. texana]
Phacelia texana

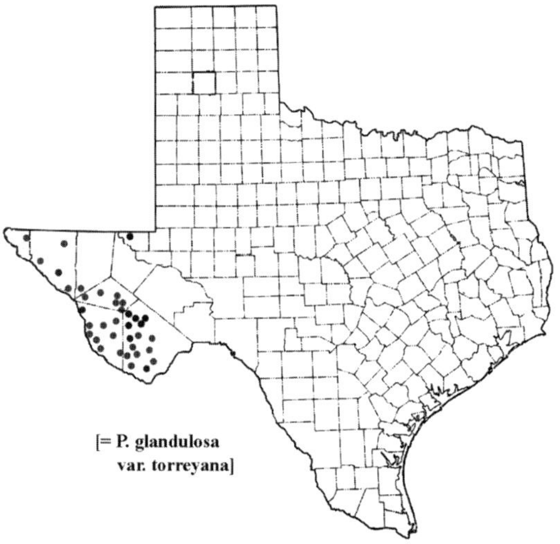
[= P. glandulosa
var. torreyana]
Prosopis odorata

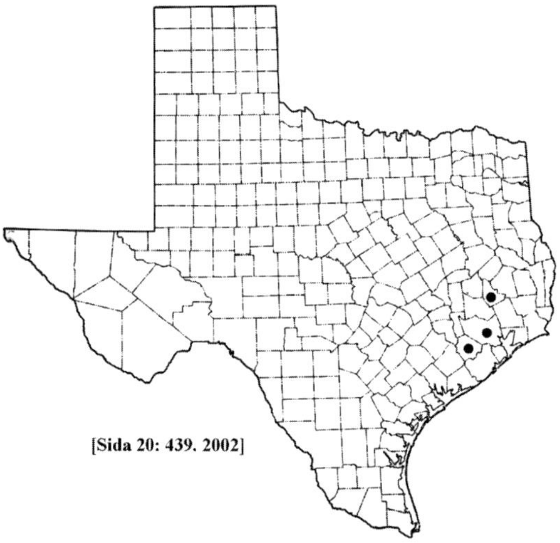
[Sida 20: 439. 2002]
Raphanus raphanistrum

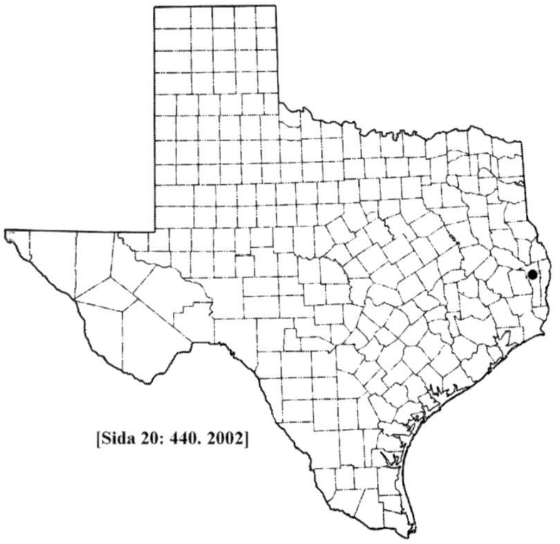
[Sida 20: 440. 2002]
Rhynchosia tomentosa

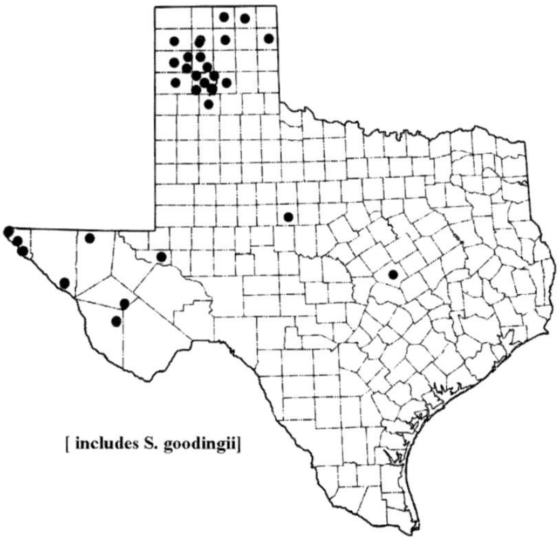
[includes S. goodingii]
Salix amygdaloides

[= S. angustifolia
var. oblongifolia]
Sphaeralcea lobata

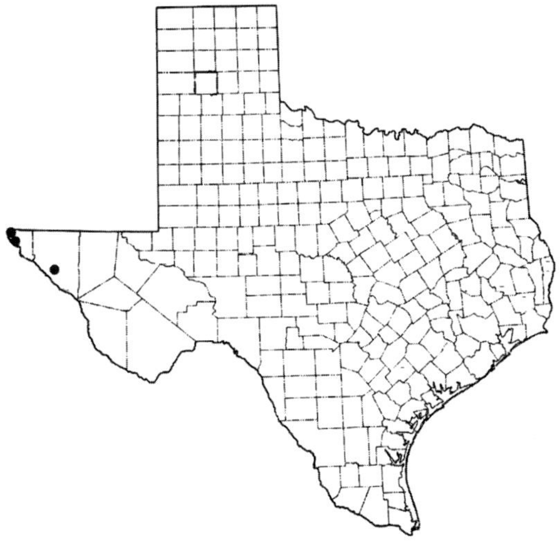
Sphaeralcea polychroma

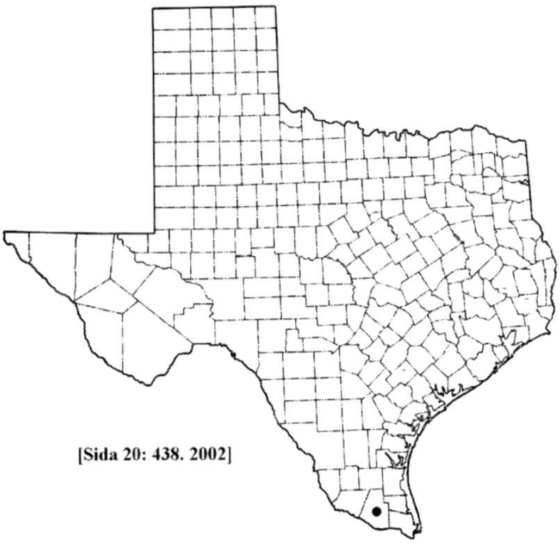
[Sida 20: 438. 2002]
Tridax procumbens

Galium correllii

ADVENTIVES IN TEXAS (VOLUME 1, DICOTS)

APIACEAE
Torilis arvensis
Torilis nodosa
Ammi visnaga
Apium graveolens var. *dulce*
Cyclospermum leptophyllum
Conium maculatum
Coriandrum sativum
Cuminum cyminum
Pastinaca ativa

APOCYNACEAE
Catharanthus roseus
Vinca major
Periploca graeca

ASTERACEAE
Acanthospermum australe
Anthemis cotula
Bidens odorata
Bidens pilosa
Carduus acanthoides
Carduus nutans
Carduus tenuiflorus
Carthamus lanatus
Carthamus tinctorius
Centaurea cyanus
Centaurea repens
Centaurea solstitialis
Chamomilla recutita
Cichorium intybus
Conyza bonariensis
Cosmos sulphureus
Cotula australis
Crepis capillaris
Crepis pulchra
Crepis setosa
Crepis zacintha
Emilia fosbergii
Facelis retusa
Hedypnois cretica
Hypochaeris glabra
Hypochaeris microcephala var. *albiflora*
Hypochaeris radicata
Lactuca intybacea
Lactuca saligna
Lactuca serriola
Lapsana communis
Leontodon hispidus
Leontodon taraxacoides
Leucanthemum vulgare
Madia elegans
Matricaria matricarioides (=Matricaria discoidea)
Onopordum acanthium
Pseudogynoxys chenopodioides
Senecio vulgaris
Silybum marianum
Soliva mutisii
Soliva sessilis
Soliva anthemifolia
Sonchus asper
Sonchus oleraceus
Taraxacum laevigatum
Taraxacum officinale
Tragopogon dubius
Tragopogon porrifolius
Viguiera phenax
Xanthium spinosum
Xanthium strumarium
Youngia japonica

BASELLACEAE
Anredera cordiolia

BERBERIDACEAE
Nandina domestica

BORAGINACEAE
Anchusa azrea
Cynoglossum zeylanicum
Echium vulgare
Heliotropium europaeum

BRASSICACEAE
Armoracia lacustris
Brassica nigra
Brassica rapa
Brassica tournefortii
Capsella bursa-pastoris
Cardamine debilis
Cardaria draba
Chorispora tenella
Conringia orientalis
Coronopus didymus
Descurainia sophia
Diplotaxis muralis
Diplotaxis tenuifolia
Eruca vesicaria
Erucastrum gallicum
Erysimum repandum
Lepidium ruderale
Lobularia maritima
Matthiola longipetala
Myagrum perfoliatum
Raphanus raphanistrum
Raphanus sativus
Rapistrum rugosum
Sinapis arvensis
Sisymbrium altissimum
Sisymbrium auriculatum
Sisymbrium irio
Sisymbrium officinale

CACTACEAE
Opuntia ficus-indica
Pereskia aculeata

CAMPANULACEAE
Campanula rapunculoides

CANNABINACEAE
Cannabis sativa

CAPRIFOLIACEAE
Lonicera fragrantissima
Lonicera japonica
Lonicera maackii

CARYOPHYLLACEAE
Agrostemma githago
Cerastium brachypetalum
Cerastium glomeratum
Cerastium pumilum
Holosteum umbellatum
Petrorhagia dubia
Polycarpon tetraphyllum
Saponaria officinalis
Spergula arvensis
Spergularia platensis
Spergularia salina
Stellaria media
Stellaria pallida
Vaccaria pyramidata

CHENOPODIACEAE
Atriplex mucronata
Atriplex muricata
Atriplex patula
Atriplex semibaccata
Bassia hyssopifolia
Chenopodium album
Chenopodium ambrosioides
Chenopodium botrys
Chenopodium murale
Chenopodium pumilio
Salsola paulsenii
Salsola tragus

CONVOLVULACEAE
Convolvulus arvensis
Ipomoea heptaphylla

Ipomoea indica
Ipomoea pubescens
Ipomoea quamoclit
Ipomoea wrightii

CRASSULACEAE
Bryophyllum fedtschenkoi
Bryophyllum tubiflorum
Crassula connata

CUCURBITACEAE
Citrullus lanatus
Cucumis anguria
Cucumis melo
Lagenaria siceraria
Luffa aegyptiaca

ELAEAGNACEAE
Elaeagnus angustifolia
Elaeagnus pungens

EUPHORBIACEAE
Chamaesyce hirta
Manihot esculenta
Manihot subspicata
Phyllanthus fraternus
Phyllanthus urinaria
Ricinus communis
Vernicia fordii

FABACEAE
Albizia julibrissin
Albizia lebbeck
Alhagi maurorum
Arachis hypogaea
Caesalpinia gilliesii
Crotalaria retusa
Crotalaria spectabilis
Desmodium scoparium
Lathyrus aphaca
Lathyrus hirsutus
Lespedeza cuneata
Lespedeza stipulacea (=Kummerowia stipulacea)
Lespedeza striata (=Kummerowia striata)
Leucaena leucocephala
Lotus corniculatus
Medicago arabica
Medicago lupulina
Medicago minima
Medicago orbicularis
Medicago polymorpha
Medicago sativa
Melilotus albus
Melilotus indicus
Melilotus officinalis
Pueraria montana
Securigera varia (=Coronilla varia)
Senna corymbosa
Sesbania sericea
Spartium junceum
Sphaerophysa salsula
Trifolium arvense
Trifolium campestre
Trifolium dubium
Trifolium incarnatum
Trifolium lappaceum
Trifolium pratense
Trifolium repens
Trifolium resupinatum
Trifolium subterraneum
Trifolium vesiculosum
Vicia hirsuta
Vicia lutea
Vicia sativa
Vicia villosa
Vigna unguiculata
Wisteria sinensis

FAGACEAE
Castanea mollissima

FUMARIACEAE
Fumria officinalis

GERANIACEAE
Erodium botrys
Erodium cicutarium
Geranium dissectum

HALORAGACEAE
Myriophyllum aquaticum
Myriophyllum spicatum

LAMIACEAE
Agastache scrophulariifolia
Ajuga reptans
Glechoma hederacea
Lamium amplexicaule
Lamium pupureum
Leootis nepetifolia
Leonurus cardiaca
Leonurus sibiricus
Marrubium vulgare
Mentha ×piperita
Mentha ×rotundifolia
Mentha spicata
Moluccella laevis
Nepeta cataria
Perilla frutescens
Stachys grahamii

LINACEAE
Linum usitatissimum

LYTHRACEAE
Lagerstroemia indica
Lythrum salicaria

MALPIGHIACEAE
Callaeum septentrionalis

MALVACEAE
Abutilon theophrasti
Alcea rosea
Hibiscus esculentus
Hibiscus trionum
Krapovickasia physaloides
Malva neglecta
Malva parviflora
Malva rotundifolia
Malva sylvestris

MELIACEAE
Melia azedarach

MENYANTHACEAE
Nymphoides peltata

MOLLUGINACEAE
Glinus lotoides
Glinus radiatus
Mollugo cerviana
Mollugo verticillata

MORACEAE
Broussonetia papyrifera
Fatoua villosa
Morus alba

MYRSINACEAE
Ardisia crenata

MYRTACEAE
Myrtus communis

NYCTAGINACEAE
Mirabilis jalapa

OLEACEAE
Ligustrum japonicum
Ligustrum lucidum
Ligustrum quihoui
Ligustrum sinense

ONAGRACEAE
Ludwigia erecta
Ludwigia uruguayensis

PAPAVERACEAE
Eschscholzia californica
Glaucium corniculatum
Papaver rhoeas
Papaver somniferum

PIPERACEAE
Peperomia pellucida

PLANTAGINACEAE
Plantago coronopus
Plantago lanceolata
Plantago major

POLYGONACEAE
Antigonon leptopus
Emex spinosa
Polygonum aviculare
Polygonum argyrocoleon
Polygonum convolvulus
Polygonum orientale
Rumex crispus
Rumex paraguayensis
Rumex pulcher

PORTULACACEAE
Portulaca oleracea
Portulaca retusa

PRIMULACEAE
Anagallis arvensis

RANUNCULACEAE
Clematis terniflora
Consolida ajacis
Enemion biternatum
Ranunculus muricatus
Ranunculus parviflorus
Ranunculus platensis

RHAMNACEAE
Hovenia dulcis
Sageretia thea
Ziziphus zizyphus

ROSACEAE
Duchesnea indica
Fragaria chiloensis
Photinia serratifolia
Potentilla norvegica
Potentilla recta
Prunus persica
Pyracantha coccinea
Pyrus calleryana
Pyrus communis
Rosa bracteata
Rosa eglanteria
Rosa laevigata
Rosa multiflora
Rubus bifrons

RUBIACEAE
Asperula arvensis
Paederia foetida
Sherardia arvensis

RUTACEAE
Citrus aurantium
Poncirus trifoliata
Ruta chalepensis
Ruta graveolens

SALICACEAE
Populus alba
Populus nigra
Salix babylonica

SCROPHULARIACEAE
Bacopa repens
Bellardia trixago
Chaenorhinum minus
Linaria vulgaris
Lindernia crustacea
Mazus pumilus
Verbascum blattaria
Verbascum thapsus
Verbascum virgatum
Veronica arvensis
Veronica persica
Veronica polita

SIMAROUBACEAE
Ailanthus altissima

SOLANACEAE
Datura stramonium
Lycianthes asarifolia
Lycium barbarum
Lycopersicon esculentum
Nicotiana glauca
Nierembergia hippomanica
Petunia axillaris
Salpichroa origanifolia
Solanum capsicastrum
Solanum capsicoides

SPHENOCLEACEAE
Sphenoclea zeylanica

STERCULIACEAE
Firmiana simplex

TAMARICACEAE
Tamarix aphylla
Tamarix chinensis
Tamarix gallica
Tamarix parviflora
Tamarix ramosissima

THYMELIACEAE
Thymelaea passerina

ULMACEAE
Ulmus pumila

URTICACEAE
Pilea microphylla

VERBENACEAE
Lantana montevidensis
Verbena bonariensis
Verbena brasiliensis
Verbena rigida
Vitex agnus-castus

ZYGOPHYLLACEAE
Kallstroemia maxima
Peganum harmala
Tribulus terrestris
Zygophyllum fabago

VOLUME 1 INDEX (DICOTS)

The accepted scientific names of native or naturalized members of the Texas flora are given in [Roman type]. In addition, accepted generic and new combinations of taxa of plants in the flora are in [**bold**]. Taxonomic synonyms are in [*italic*]. Family names are in [**bold Cap/Small Cap**].

Abronia 452
ameliae 452
angustifolia 453
carletonii 453
fragrans 453
macrocarpa 453
Abutilon 434
abutiloides 434
berlandieri 434
fruticosum 434
hulseanum 434
hypoleucum 435
malacum 435
mollicomum 435
parvulum 435
theophrasti 435
trisulcatum 435
wrightii 436
Acacia 303
angustissima 303
var. chisosiana 303
var. hirta 303
var. leucothrix 303
berlandieri 303
× *greggii* 304
constricta 303
farnesiana 303
greggii 303
var. greggii 303
var. wrightii 303
neovernicosa 304
rigidula 304
roemeriana 304
schaffneri 304
schottii 304
smallii 303
× emoryana 304
Acalypha 278
gracilens 278
lindheimeri 279
monococca 278
monostachya 279
neomexicana 279
ostryifolia 279
phleoides 279
poiretii 279
radians 279
rhomboidea 280
virginica 280
Acanthaceae 11
Acanthocereus 206
pentagonus 206
Acanthochiton 21
wrightii 21
Acanthospermum 60
australe 60
Aceraceae 18
Acer 18
grandidentatum 18
negundo 19
rubrum 19
saccharinum 19
saccharum 19
Achillea 60
lanulosa 60
millefolium 60
var. *lanulosa* 60
Achyranthes 20
aspera 20
Acleisanthes 453
acutifolia 453
anisophylla 453
crassifolia 454
diffusa 7
lanceolata 7
var. **maloneana 7**
longiflora 454
megaphylla 7
nevadensis 8
obtusa 454
purpusiana 8
var. **marshii 8**
wrightii 454
Acmella 60
oppositifolia 60
Acnida 22
cuspidata 22
tamariscina 24
Acourtia 60
nana 60
runcinata 60
wrightii 60
Adelia 280
vaseyi 280
Adolphia 522
infesta 522
Adonis 513
annua 513
Aeschynomene 305
indica 305
rudis 305
viscidula 305
Aesculus 384
arguta 384
pavia 384
var. flavescens 384
var. pavia 384
Agalinis 576
auriculata 576
calycina 560
densiflora 576
edwardsiana 560
fasciculata 560
gattingeri 561
heterophylla 561
homalantha 561
maritima 561
navasotensis 561
oligophylla 561
purpurea 562
strictifolia 562
tenuifolia 562
viridis 562
Agastache 399
cana 399
micrantha 399
pallidiflora 399
scrophulariifolia 399
Ageratina 61
altissima 61
havanensis 61
herbacea 61
rothrockii 61
wrightii 61
Agrimonia 526
incisa 526
microcarpa 527
parviflora 527
rostellata 527
Agrostemma 232
githago 232
Ailanthus 578
altissima 578
Aizoaceae 19
Ajuga 399
reptans 399
Albizia 305
julibrissin 305
lebbeck 305
Alcea 436
rosea 436
Alchemilla 527
microcarpa 527
Aletes 6
acaulis 6
filifolius 6, 35
montanus 35
Alhagi 305
maurorum 305
Allenrolfea 243
occidentalis 243
Allionia 454
choisyi 454
incarnata 454
Allowissadula 436
holosericea 436
lozanii 436
Alnus 171
serrulata 171
Aloysia 600
gratissima 600
macrostachya 600
wrightii 600
Alternanthera 20
caracasana 20
paronichyoides 21
philoxeroides 21
pungens 20
sessilis 21
Alysicarpus 305
vaginalis 305
Amaranthaceae 20
Amaranthus 21
acanthochiton 21
albus 21
arenicola 21
australis 22
berlandieri 24
blitoides 22
blitum 22
californicus 22
crassipes 22
var. crassipes 22
var. warnockii 22
cruentus 22
fimbriatus 23
greggii 23
hybridus 23
hypochondriacus 23
lividus 23
palmeri 23
polygonoides 24
powellii 24
retroflexus 24
rudis 24
scleropoides 24
spinosus 24
tamaulipensis 25
torreyi 25
× texensis 25
viridis 618
Amblyolepis 61
setigera 61
Ambrosia 62
acanthicarpa 62
artemisiifolia 62
bidentata 62
cheiranthifolia 62
confertiflora 62
grayi 62
psilostachya 63
trifida 63
var. *texana* 63
Amelanchier 527
denticulata 533
utahensis 527
Ammannia 430
auriculata 430

coccinea 430
latifolia 430
robusta 430
teres 430
Ammi 32
majus 32
visnaga 32
Ammocodon 7
chenopodioides 455
Ammoselinum 32
butleri 32
popei 32
Amoreuxia 173
wrightii 173
Amorpha 306
canescens 306
fruticosa 306
laevigata 306
paniculata 306
roemeriana 306
Ampelopsis 613
arborea 613
cordata 613
Amphiachyris 63
amoena 63
dracunculoides 63
Amphicarpaea 306
bracteata 306
Amsinckia 173
intermedia 173
menziesii 173
micrantha 173
Amsonia 44
arenaria 44
ciliata 44
longiflora 44
var. longiflora 44
var. salpignantha 44
palmeri 45
tabernaemontana 45
var. repens 45
var. tabernaemontana 45
tharpii 45
Amyris 551
madrensis 551
texana 551
ANACARDIACEAE 29
Anagallis 512
arvensis 512
minima 512
Anaphalis 63
margaritacea 63
Anchusa 173
azurea 173
Ancistrocactus 206
brevihamatus 206
var. brevihamatus 206
var. pallidus 206
scheeri 206
tobuschii 207
uncinatus 214
Andrachne 280
arida 280
phyllanthoides 297
Androsace 512
occidentalis 512
septentrionalis 512
Anemone 514
berlandieri 514
caroliniana 514
edwardsiana 514
okennonii 514
var. *tuberosa* 514
tuberosa 514
var. texana 514
var. tuberosa 514
Anemopsis 558
californica 558
Anethum 32
graveolens 32
Anisacanthus 11
linearis 11
puberulus 11
quadrifidus 11
var. wrightii 11
Anoda 436
crenatiflora 436
cristata 436
lanceolata 437
pentaschista 437
ANNONACEAE 31
Anredera 169
cordifolia 169
vesicaria 169
Antennaria 6
fallax 6
marginata 63
parlinii 64
var. **fallax** **6**, 64
parvifolia 64
plantaginifolia 64
Anthemis 64
cotula 64
Antigonon 499
leptopus 499
Antiphytum 173
floribundum 173
heliotropoides 174
Anulocaulis 455
eriosolenus 455
gypsogenus 455
leiosolenus 455
var. lasianthus 455
var. leiosolenus 455
var. *gypsogenus* 455
reflexus 455
Aphanes 527
microcarpa 527
Aphanostephus 64
pilosus 64
ramosissimus 64
var. humilis 64
var. ramosissimus 64
riddellii 65
skirrhobasis 65
var. kidderi 65
var. skirrhobasis 65
var. thalassius 65
APIACEAE 32
Apios 307
americana 307
Apium 33
graveolens 33
leptophyllum 35
APOCYNACEAE 45
Apocynum 45
androsaemifolium 45
cannabinum 45
sibiricum 45
Apodanthera 268
undulata 268
AQUIFOLIACEAE 47
Aquilegia 514
canadensis 514
chaplinei 514
chrysantha 515
var. *chaplinei* 514
var. *hinckleyana* 515
hinckleyana 515
longissima 515
Arabidopsis 185
thaliana 185
Arabis 185
canadensis 185
fendleri 185
perennans 185
petiolaris 185
Arachis 307
hypogaea 307
Aralia 48
racemosa 48
spinosa 48
ARALIACEAE 48
Arbutus 276
xalapensis 276
Arceuthobium 8
cryptopodum 8
divaricatum 611
douglasii 611
vaginatum 611
subsp. *durangense* 8
var. cryptopdum 611
var. **cryptopodum** **8**
var. **durangense** **8**
Arctostaphylos 277
pungens 277
Ardisia 452
crenata 452
Arenaria 232
benthamii 232
drummondii 232
fendleri 232
jamesiana 232
lanuginosa 233
livermorensis 233
ludens 233
muriculata 233
patula 233
serpyllifolia 233
stricta 233
var. *texana* 233
texana 233
Argemone 481
aenea 481
albiflora 481
aurantiaca 481
chisosensis 481
mexicana 482
polyanthemos 482
sanguinea 482
squarrosa 482
Argythamnia 292
aphoroides 292
argyraea 292
humilis 292
var. *laevis* 292
laevis 292
mercurialina 292
var. *pilosissima* 292
neomexicana 292
simulans 292
Ariocarpus 207
fissuratus 207
Aristolochia 49
coryi 49
erecta 49
pentandra 49
reticulata 49
serpentaria 49
tomentosa 49
wrightii 50
ARISTOLOCHIACEAE 49
Armoracia 186
lacustris 186
Arnoglossum 65
ovatum 65
plantagineum 65
Aronia 527
abutifolia 527
Artemisia 65
bigelovii 65
campestris 65
carruthii 66
dracunculoides 66
filifolia 66
ludoviciana 66
ASCLEPIADACEAE 50
Asclepias 50
amplexicaulis 50
arenaria 50
asperula 50
brachystephana 50
curassavica 50
elata 51
emoryi 51
engelmanniana 51
glaucescens 51
incarnata
var. incarnata 51
var. pulchra 51
involucrata 51
lanceolata 51
latifolia 52
linearis 52
longifolia 52
macrotis 52

nummularia 52
obovata 52
oenotheroides 53
perennis 53
prostrata 53
pumila 53
purpurascens 53
rubra 53
scaposa 54
speciosa 54
sperryi 54
stenophylla 54
subverticillata 54
texana 54
tomentosa 54
tuberosa 55
uncialis 55
variegata 55
verticillata 55
viridiflora 55
viridis 55
Asimina 31
parviflora 31
triloba 32
Asperula 542
arvensis 542
Aspicarpa 433
hirtella 433
hyssopifolia 433
Aster 157
azureus 157
drummondii 155
var. *texanus* 155
dumosus 156
ericoides 156
falcatus 156
fendleri 156
hemispherica 99
hesperius 157
laevis 156
lanceolatus 157
lateriflorus 157
oblongifolius 157
oolentangiensis 157
patens 157
var. *patens* 157
var. *patentissimus* 156
praealtus 157
pratensis 158
puniceus 158
sericeus 158
seriocarpoides 89
spinosus 80
subulatus 155, 158
var. *ligulatus* 155
tenuifolius 158
Asteraceae 60
Astragalus 307
albulus 307
allochorus 307, 312
var. *playanus* 312
austrinus 307
var. austrinus 307
var. *pleianthus* 311
var. **trichocarpus 7**, 307
bigelovii
var. bigelovii 307
var. marcidus 307
brazoensis 308
canadensis 308
ceramicus 308
var. filifolius 308
crassicarpus 308
var. berlandieri 308
var. crassicarpus 308
var. trichocalyx 308
distortus 308
var. engelmannii 308
emoryanus 308
var. emoryanus 308
var. *terlinguensis* 312
engelmannii 308
giganteus 309
gracilis 309
gypsodes 309
humistratus 309
lentiginosus 309
var. australis 309
var. higginsii 309
leptocarpus 309
lindheimeri 310
lotiflorus 310
missouriensis 310
mollissimus 310
var. coryi 310
var. earlei 310
var. mollissimus 310
nuttallianus 7, 307, 310, 311
var. *austrinus* 307
var. nuttallianus 310
var. *pleianthus* 311
var. *trichocarpus* 7
var. *trichocarpus* 307
var. zapatanus 310
pictiformis 310
plattensis 311
pleianthus 311
praelongus 311
racemosus 311
reflexus 311
soxmaniorum 311
terlinguensis 312
waterfallii 312
wootonii 312
wrightii 312
Astranthium 66
integrifolium 66
var. ciliatum 66
robustum 66
Astrophytum 207
asterias 207
Atriplex 243
acanthocarpa 243
var. acanthocarpa 243
var. coahuilensis 243
arenaria 245
argentea 244
canescens 244
confertifolia 244
elegans 244
klebergorum 244
matamorensis 244
mucronata 245
muricata 245
obovata 245
patula 245
pentandra 245
powellii 245
semibaccata 246
texana 246
wrightii 246
Aureolaria 562
flava 562
grandiflora 562
pectinata 563
Avicennia 169
germinans 169
Ayenia 591
filiformis 591
limitaris 591
microphylla 592
pilosa 592

Baccharis 67
bigelovii 67
brachyphylla 67
emoryi 67
halimifolia 67
havardii 67
neglecta 67
pteronioides 68
salicifolia 68
salicina 68
sarothroides 68
texana 68
wrightii 68
Bacopa 563
caroliniana 563
monnieri 563
repens 563
rotundifolia 563
Bahia 69
absinthifolia 69
bigelovii 69
dissecta 69
pedata 69
Baileya 69
multiradiata 69
Balsaminaceae 169
Baptisia 312
alba 312
australis 312
bracteata 313
var. *leucophaea* 313
lacta 312
leucantha 312
leucophaea 313
minor 312
var. *aberrans* 312
nuttalliana 313
sphaerocarpa 313
Bartlettia 69
scaposa 69
Bartonia 376
paniculata 376
texana 377
verna 377
Basellaceae 169
Bassia 246
hyssopifolia 246
Bastardia 437
viscosa 437
Bataceae 170
Batesimalva 437
violacea 437
Batis 170
maritima 170
Bauhinia 313
lunarioides 313
Bebbia 70
juncea 70
Bellardia 563
trixago 563
Berberidaceae 170
Berberis 170
haematocarpa 170
repens 170
swaseyi 170
trifoliata 170
Berchemia 522
scandens 522
Bergia 276
texana 276
Berlandiera 70
betonicifolia 70
lyrata 70
pumila 70
var. pumila 70
var. scabrella 70
texana 70
Bernardia 280
myricifolia 280
obovata 280
Berula 33
erecta 33
Betula 171
nigra 171
Betulaceae 171
Bidens 70
alba 618
aristosa 70
bigelovii 70
bipinnata 71
cernua 71
comosa 71
discoidea 71
frondosa 71
laevis 71
leptocephala 72
odorata 72
pilosa 72
polylepis 72
Bifora 33
americana 33

Bigelowia 72
nuttallii 72
Bignonia 172
capreolata 172
Bignoniaceae 172
Billieturnera 437
helleri 437
Boehmeria 597
cylindrica 597
anisophylla 455
Boerhavia 456
coccinea 456
diffusa 456
erecta 456
gracillima 456
intermedia 456
linearifolia 456
mathisiana 456
purpurascens 457
spicata 457
wrightii 457
Boltonia 72
asteroides 72
diffusa 73
Bonamia 264
aquatica 264
ovalifolia 255
repens 255
Boraginaceae 173
Borrichia 73
frutescens 73
Bouchea 601
linifolia 601
spathulata 601
Bouchetia 579
erecta 579
Bouvardia 542
ternifolia 542
Bowlesia 33
incana 33
Bradburia 81
hirtella 81
Brasenia 206
schreberi 206
Brassica 186
campestris 186
kaber 202
nigra 186
rapa 186
tournefortii 186
Brassicaceae 185
Brazoria 399
arenaria 399
scutellarioides 418
truncata
var. pulcherrima 400
var. truncata 400
Breweria 264
humistrata 264
Brickellia 73
baccharidea 73
brachyphylla 73
californica 73
coulteri 73
cylindracea 74
dentata 74
eupatorioides 74
var. chlorolepis 74
var. corymbulosa 74
var. eupatorioides 74
var. gracillima 74
var. texana 75
fendleri 155
hinckleyi 75
var. hinckleyi 75
var. terlinguensis 75
venosa 76
veronicifolia 76
Brickelliastrum 155
fendleri 155
Brongniartia 313
minutifolia 313
Broussonetia 451
papyrifera 451
Brunnichia 499
cirrhosa 499
ovata 499
Bryophyllum 265
fedtschenkoi 265
tubiflorum 265
Buchnera 564
americana 564
floridana 564
Buddleja 428
marrubiifolia 428
racemosa 428
var. incana 428
var. racemosa 428
scordioides 428
sessiliflora 428
Buglossoides 174
arvensis 174
Bumelia 8
lanuginosa 8
var. *ridida* 8
oblongifolia 8
Bupleurum 33
rotundifolium 33

Cabomba 206
caroliniana 206
Cabombaceae 206
Cacalia 65
ovata 65
plantaginea 65
Cactaceae 206
Caesalpinia 351
brachycarpa 351
caudata 313
drummondii 334
gilliesii 314
jamesii 351
mexicana 314
phyllanthoides 314
pulcherrima 314
wootonii 351
Cakile 186
constricta 186
fusiformis 187
geniculata 186
lanceolata 187
Calamintha 400
arkansana 400
Calibrachoa 579
parviflora 579
Callaeum 433
septentrionale 433
Calliandra 314
biflora 314
conferta 314
eriophylla 315
humilis 315
isleyi 315
media 365
Callicarpa 601
americana 601
Callirhoe 437
alcaeoides 437
involucrata 438
leiocarpa 438
papaver 438
pedata 438
scabriuscula 438
Callitrichaceae 222
Callitriche 222
heterophylla 222
nuttallii 222
peploides 222
terrestris 222
Calophanes 5
decumbens 5
schiedeanus 5
Calycanthus 618
floridus 618
Calycocarpum 449
lyonii 449
Calylophus 466
berlandieri 466
var. berlandieri 466
var. pinifolius 466
fendleri 466
filifolius 466
greggii 466
hartwegii 467
var. hartwegii 467
var. maccartii 467
var. *filifolius* 466
var. *pubescens* 466
lavandulifolius 467
serrulatus 467
tubicula 467
Calystegia 255
macounii 255
sepium 255
var. repens 255
silvatica 255
var. fraterniflora 255
Camelina 187
microcarpa 187
rumelica 187
Camissonia 467
chamaenerioides 467
Campanula 222
rapunculoides 222
reverchonii 223
rotundifolia 223
Campanulaceae 222
Campsis 172
radicans 172
Canavalia 315
rosea 315
Cannabaceae 226
Cannabis 226
sativa 226
Caperonia 281
palustris 281
Capparaceae 226
Capparis 226
incana 226
Capraria 564
mexicana 564
Caprifoliaceae 228
Capsella 187
bursa-pastoris 187
Capsicum 579
annuum 579
Cardamine 188
bulbosa 188
concatenata 187
debilis 187
hirsuta 188
macrocarpa 188
var. texana 188
parviflora 188
var. arenicola 188
pensylvanica 188
rhomboidea 188
Cardaria 188
draba 188
Cardiospermum 556
corindum 556
dissectum 556
halicacabum 556
Carduus 76
acanthoides 76
nutans 77
tenuiflorus 77
Carlowrightia 11
arizonica 11
linearifolia 11
mexicana 11
parviflora 12
parvifolia 12
serpyllifolia 12
texana 12
torreyana 12
Carminatia 77
tenuiflora 77
Carphochaete 77
bigelovii 77
Carpinus 171
caroliniana 171
Carya 397
alba 397
aquatica 396
cordiformis 396

glabra 396
illinoinensis 396
laciniosa 396
leiodermis 396
myristiciformis 397
ovata 397
texana 397
tomentosa 397
Carthamus 77
lanatus 77
tinctorius 77
CARYOPHYLLACEAE 232
Cassia 354
bauhinioides 354
calycioides 316
durangensis 355
fasciculata 316
var. *fasciculata* 316
var. *ferrisiae* 316
var. *puberula* 316
var. *rostrata* 316
leptadenia 317
lindheimeriana 355
marilandica 355
nictitans 317
obtusifolia 355
occidentalis 355
orcuttii 356
pilosior 356
pumilio 356
ripleyana 356
roemeriana 356
wislizeni 357
Castanea 366
mollissima 366
pumila 366
Castela 578
erecta 578
var. texana 578
stewartii 579
texana 578
Castilleja 565
ciliata 565
genevievana 564
indivisa 564
integra 564
lanata 565
livermorensis 565
mexicana 565
purpurea 565
var. citrina 565
var. lindheimeri 565
var. purpurea 565
rigida 565
sessiliflora 565
wootonii 565
Catalpa 172
bignonioides 172
speciosa 172
Catharanthus 45
roseus 45
Cayaponia 268
quinqueloba 268
Ceanothus 523
americanus 523
fendleri 523
greggii 523
herbaceus 523
CELASTRACEAE 241
Celastrus 241
scandens 241
Celosia 25
argentea 25
nitida 25
palmeri 25
Celtis 595
laevigata 595
lindheimeri 595
occidentalis 596
pallida 596
reticulata 596
var. *texana* 596
tenuifolia 596
Centaurea 78
americana 78
cyanus 78
melitensis 78
repens 78
solstitialis 78
Centaurium 377
arizonicum 377
beyrichii 377
breviflorum 377
calycosum 377
glanduliferum 378
maryannum 378
muhlenbergii 378
multicaule 378
nudicaule 378
pulchellum 378
texense 379
Centella 33
erecta 33
Centrosema 315
virginianum 315
Centunculus 512
minimus 512
Cephalanthus 542
occidentalis 542
salicifolius 542
Cerastium 234
axillare 234
brachypetalum 234
brachypodum 234
clawsonii 422
fontanum 234
glomeratum 234
pumilum 234
Ceratoides 250
lanata 250
CERATOPHYLLACEAE 243
Ceratophyllum 243
demersum 243
echinatum 243
Cercidium 315
texanum 315
var. texanum 315
Cercis 316
canadensis 316
var. canadensis 316
var. mexicana 316
var. texensis 316
Cercocarpus 528
breviflorus 528
glaberoides 528
var. glaber 528
montanus 528
var. argenteus 528
var. montanus 528
Cereus 221
greggii 221
pentagonus 206
poselgeri 212
Cevallia 424
sinuata 424
Chaenorhinum 566
minus 566
tainturieri 34
Chaetopappa 80
asteroides 78
bellidifolia 79
bellioides 79
effusa 79
ericoides 79
hersheyi 79
imberbis 79
keerlioides 80
parryi 80
Chamaecrista 316
calycioides 316
fasciculata 316
var. fasciculata 316
var. ferrisiae 316
var. puberula 316
var. rostrata 316
greggii 316
leptadenia 316
nictitans 317
var. *mensalis* 316
Chamaesaracha 579
coniodes 579
coronopus 579
edwardsiana 580
pallida 580
sordida 580
villosa 580
Chamaesyce 281
acuta 281
albomarginata 281
ammannioides 281
angusta 281
arizonica 281
astyla 282
capitellata 282
carunculata 282
chaetocalyx 7, 282
cinerascens 282
cordifolia 282
fendleri 7, 283
geyeri 283
glyptosperma 283
golondrina 283
hirta 283
humistrata 283
hypericifolia 284
indivisa 284
jejuna 284
laredana 284
lata 284
maculata 284
micromera 285
missurica 285
nutans 285
parryi 285
perennans 285
prostrata 285
revoluta 286
serpens 286
serpyllifolia 286
serrula 286
setiloba 286
simulans 287
spurca 7, 286
stictospora 287
theriaca 7, 287
triligulata 7, 287
velleriflora 287
villifera 287
Chamomilla 80
recutita 80
Chaptalia 80
texana 80
tomentosa 80
CHENOPODIACEAE 243
Chenopodium 246
album 246
ambrosioides 246
berlandieri 247
botrys 247
cycloides 247
fremontii 247
glaucum 247
hybridum 247
incanum 248
var. elatum 248
var. incanum 248
incisum 248
incognitum 248
leptophyllum 248
missouriense 248
murale 248
pallescens 249
pratericola 249
pumilio 249
simplex 249
standleyanum 249
Chilopsis 172
linearis 172
Chiococca 542
alba 542
Chionanthus 462
virginicus 462
Chloracantha 80
spinosa 80

Choisya 551
dumosa 551
Chorispora 189
tenella 189
Chromolaena 96
bigelovii 96
odorata 98
Chrysactinia 80
mexicana 80
Chrysopsis 111
mariana 81
pilosa 81
stenophylla 111
texana 81
villosa 111
var. *canescens* 111
Chrysothamnus 93
baileyi 81
nauseosus 93
var. *graveolens* 93
var. *texensis* 93
pulchellus 81
spathulatus 81
Cichorium 82
intybus 82
Cicuta 34
maculata 34
Cienfuegosia 438
drummondii 438
Cinnamomum 418
camphora 418
Cirsium 82
altissimum 82
carolinianum 82
discolor 82
engelmannii 82
horridulum 82
muticum 82
ochrocentrum 83
terrae-nigrae 82
texanum 83
turneri 83
undulatum 83
vulgare 83
Cissus 613
incisa 613
CISTACEAE 253
Citharexylum 601
berlandieri 601
brachyanthum 601
spathulatum 601
Citrullus 269
lanatus 269
vulgaris 269
Citrus 551
aurantium 551
Clappia 83
suaedifolia 83
Claytonia 509
virginica 509
Clematis 515
alpina 515
columbiana 515
crispa 515
drummondii 515
ligusticifolia 516
pitcheri 516
var. dictyota 516
var. pitcheri 516
reticulata 516
ternifolia 516
texensis 516
versicolor 516
virginiana 517
Cleome 226
aculeata 226
gynandra 226
multicaulis 227
serrulata 227
Cleomella 227
angustifolia 227
longipes 227
Clethra 254
alnifolia 254
CLETHRACEAE 254
Clitoria 317
mariana 317
ternatea 317
Cnidoscolus 287
texanus 287
Cocculus 449
carolinus 449
diversifolius 449
Coelophragmus 202
auriculatum 202
Coelophragmus 189
auriculatus 189
Collinsia 566
violacea 566
Cologania 371
angustifolia 317
pallida 317
Colubrina 523
greggii 523
stricta 523
texensis 524
Comandra 556
pallida 556
umbellata 556
var. *pallida* 556
Commicarpus 457
scandens 457
Condalia 524
ericoides 524
hookeri 524
spathulata 524
viridis 524
warnockii 524
Conium 34
maculatum 34
Conocarpus 254
erectus 254
Conoclinium 84
betonicifolium 84
var. betonicifolium 84
var. integrifolium 84
coelestinum 84
dissectum 84
Conopholis 479
alpina 479
Conringia 189
orientalis 189
Consolida 517
ajacis 517
ambigua 517
CONVOLVULACEAE 255
Convolvulus 255
arvensis 255
equitans 256
Conyza 84
bonariensis 84
canadensis 84
ramosissima 84
Corchorus 595
hirtus 595
Cordia 174
boissieri 174
podocephala 174
Cordylanthus 566
wrightii 566
Coreopsis 85
basalis 85
grandiflora 85
var. grandiflora 85
var. longipes 85
intermedia 85
lanceolata 85
linifolia 85
nuecensis 85
nuecensoides 86
pubescens 86
tinctoria 86
tripteris 86
wrightii 86
Coriandrum 34
sativum 34
Corispermum 249
americanum 249
orientale 250
welshii 250
CORNACEAE 264
Cornus 264
drummondii 264
florida 264
foemina 265
racemosa 265
Coronilla 354
varia 354
Coronopus 189
didymus 189
Corydalis 375
aurea 375
var. aurea 375
var. occidentalis 375
crystallina 375
curvisiliqua 375
var. curvisiliqua 375
var. grandibracteata 375
micrantha 376
var. australis 376
var. texensis 376
Coryphantha 207
chaffeyi 207
dasyacantha 207
duncanii 207
echinus 208
hesteri 208
macromeris 208
var. macromeris 208
var. runyonii 208
missouiensis 208
pottsiana 208
ramillosa 208
scheeri 209
sneedii 209
var. albicolumnaria 209
var. guadalupensis 209
var. sneedii 209
strobiliformis 209
sulcata 209
tuberculosa 209
var. tuberculosa 209
var. varicolor 209
vivipara 209
var. neomexicana 209
var. radiosa 209
var. vivipara 209
Cosmos 86
parviflorus 86
sulphureus 87
Cotinus 29
obovatus 29
Cotula 87
australis 87
Coursetia 317
axillaris 317
Cowania 537
ericifolia 537
Crassula 266
aquatica 266
connata 266
drummondii 266
longipes 266
solierii 266
CRASSULACEAE 265
Crataegus 528
anamesa 528
brachyacantha 528
crus-galli 528
dallasiana 529
engelmannii 529
greggiana 529
marshallii 529
mollis 529
nananixonii 529
opaca 530
pearsonii 530
reverchonii 530
var. *stevensiana* 530
spathulata 530
stenosepala 530
stevensiana 530
texana 531
tracyi 531
turnerorum 531

uniflora 531
viburnifolia 531
viridis 531
var. desertorum 531
var. viridis 531
Crepis 87
capillaris 87
pulchra 87
runcinata 87
setosa 87
zacintha 88
Cressa 256
depressa 256
nudicaulis 256
truxillensis 256
Croptilon 88
divaricatum 88
hookerianum 88
var. hookerianum 88
var. validum 88
rigidifolium 88
Crotalaria 318
incana 318
pumila 318
purshii 318
retusa 318
sagittalis 318
spectabilis 318
Croton 288
alabamensis 288
var. texensis 288
argenteus 288
argyranthemus 288
capitatus 288
ciliatoglandulifer 288
cortesianus 288
coryi 289
dioicus 289
fruticulosus 289
glandulosus 289
var. glandulosus 289
var. pubentissimus 289
humilis 289
incanus 289
leucophyllus 290
lindheimerianus 290
var. lindheimerianus 290
var. tharpii 290
michauxii 291
monanthogynus 290
parksii 290
pottsii 290
var. *thermophilius* 291
punctatus 290
sancti-lazari 291
soliman 291
suaveolens 291
texensis 291
thermophilus 291
torreyanus 289
willdenowii 291
Crotonopsis 291
linearis 291
Cryptantha 174
albida 174
angustifolia 174
barbigera 175
cinerea 175
var. cinerea 175
var. *laxa* 175
var. multicaulis 175
coryi 175
crassipes 175
crassisepala 175
var. crassisepala 175
var. elachantha 175
laxa 175
mexicana 176
micrantha 176
minima 176
multicaulis 175
oblata 176
palmeri 175
paysonii 176
pterocarya 176
pusilla 177
texana 177
Cryptotaenia 34
canadensis 34
Cucumis 269
anguria 269
dipsaceus 269
melo 269
Cucurbita 269
digitata 269
foetidissima 269
texana 270
CUCURBITACEAE 268
Cuminum 34
cyminum 34
Cuphea 430
carthagenensis 430
glutinosa 430
viscosissima 618
Cuscuta 271
applanata 271
attenuata 271
campestris 273
cephalanthi 271
compacta 272
cuspidata 272
decipiens 272
exaltata 272
glabrior 273
glomerata 272
gronovii 272
indecora 273
leptantha 273
longisepala 273
obtusiflora 273
pentagona 273
polygonorum 273
runyonii 273
squamata 274
suaveolens 274
umbellata 274
CUSCUTACEAE 271
Cyclanthera 270
dissecta 270
Cycloloma 250
atriplicifolium 250
Cylindropuntia 217
davisii 217
imbricata 219
kleiniae 219
leptocaulis 219
tunicata 221
Cymopteris 6
acaulis 6, 35
bulbosus 35
filifolius 6, 35
hallii 6, 35
lemmonii 35
longiradiatus 6, 36
macrorhizus 36
montanus 36
Cynanchum 56
angustifolium 56
barbigerum 56
laeve 56
maccartii 56
pringlei 56
racemosum 56
var. unifarium 56
unifarium 56
Cynoctonum 429
mitreola 429
sessilifolium 429
Cynoglossum 177
virginianum 177
zeylanicum 177
Cynosciadium 36
digitatum 36
Cyphomeris 457
crassifolia 457
gypsophiloides 457
Cyrilla 274
racemiflora 274
CYRILLACEAE 274

Dalea 319
arenicola 7, 319
argyrea 319
aurea 319
var. aurea 319
var. austrotexana 319
bartonii 319
bicolor 319
var. argyrea 319
brachystachya 319
candida 319
var. candida 319
var. oligophylla 319
compacta 320
var. compacta 320
var. pubescens 320
cylindriceps 320
drummondiana 320
emarginata 320
enneandra 320
var. *pumila* 320
formosa 320
frutescens 321
var. frutescens 321
var. laxa 321
greggii 321
grisea 321
hallii 321
jamesii 321
lachnostachys 321
lanata 322
laniceps 322
lasiathera 322
leporina 322
longipila 322
multiflora 322
nana 323
neomexicana 323
obovata 323
phleoides 323
var. *microphylla* 320
pogonathera 321
var. pogonathera 321
var. walkerae 321
polygonoides 323
purpurea 7, 324
var. arenicola 324
var. purpurea 324
reverchonii 324
sabinalis 324
scandens 324
var. pauciflora 324
tenuifolia 324
tenuis 324
terminalis 325
villosa 325
wrightii 325
var. **warnockii 7,** 325
var. wrightii 325
Dasistoma 566
macrophylla 566
Datura 580
inoxia 580
metel 580
quercifolia 581
stramonium 581
wrightii 581
Daubentonia 357
drummondii 357
Daucosma 36
laciniata 36
Daucus 36
carota 36
pusillus 37
Decodon 431
verticillatus 431
Delphinium 517
carolinianum 517
var. vimineum 517
var. virescens 517
madrense 517
virescens 517
var. *wootonii* 517
wootonii 517

Dentaria 187
laciniata 187
Descurainia 189
incana 189
pinnata 189
richardsonii 189
var. *viscosa* 189
sophia 190
Desmanthus 325
acuminatus 325
cooleyi 325
glandulosus 325
illinoensis 326
leptolobus 326
obtusus 326
reticulatus 326
tatuhyensis 326
var. brevipes 326
velutinus 326
virgatus 327
Desmodium 327
canadense 327
canescens 327
ciliare 327
cuspidatum 327
dichromum 330
glabellum 327
glutinosum 328
grahamii 328
illinoense 328
incanum 328
laevigatum 328
lindheimeri 328
lineatum 329
marilandicum 329
neomexicanum 329
nudiflorum 329
nuttallii 329
obtusum 329
paniculatum 330
var. *dillenii* 330
pauciflorum 330
psilophyllum 330
rotundifolium 330
scorpiurus 330
sessilifolium 330
tortuosum 331
tweedyi 331
viridiflorum 331
wrightii 330
Dichaetophora 88
campestris 88
Dichondra 256
argentea 256
brachypoda 256
carolinensis 256
micrantha 257
recurvata 257
Dicliptera 12
brachiata 12
sexangularis 13
Dicranocarpus 88
parviflorus 88
Dicraurus 28
leptocladus 28
Didiplis 431
diandra 431
Dimorphocarpa 190
candicans 190
palmeri 190
wislizeni 190
Diodia 542
teres 542
tricocca 550
virginiana 543
Diospyros 275
texana 275
virginiana 275
Diplotaxis 190
muralis 190
tenuifolia 190
Dipsacaceae 274
Dipsacus 274
fullonum 274
Ditaxis 292
aphoroides 292
argyraea 292
humilis 292
mercurialina 292
neomexicana 292
pilosissima 292
simulans 293
Dodecatheon 512
meadia 512
Doellingeria 89
sericocarpoides 89
Draba 190
brachycarpa 190
cuneifolia 191
platycarpa 191
reptans 191
standleyi 191
Dracopis 89
amplexicaulis 89
Drosera 275
brevifolia 275
capillaris 275
intermedia 275
tracyi 275
Droseraceae 275
Drymaria 235
laxiflora 235
leptophylla 235
molluginea 235
pachyphylla 235
Dryopetalon 191
runcinatum 191
Duchesnea 532
indica 532
Duranta 602
erecta 602
Dyschoriste 13
crenulata 13
decumbens 13
var. decumbens 13
linearis 13
var. **cinerascens 4, 13**
var. **decumbens** 5
var. linearis 13
var. *sanpatriciensis* 5, 13
var. **schiedeana** 5
schiedeana 4
var. *cinerascens* 4, 13
var. *decumbens* 13
var. *prostrata* 5
Dysodiopsis 90
tagetoides 90
Dyssodia 89
acerosa 89
aurea 89
greggii 89
micropoides 89
papposa 90
pentachaeta 90
tagetoides 90
tenuiloba 90
var. tenuiloba 90
var. wrightii 90
tephroleuca 90

Ebenaceae 275
Echeveria 266
strictiflora 266
Echinacea 90
angustifolia 90
atrorubens 91
pallida 91
var. *sanguinea* 91
purpurea 91
sanguinea 91
Echinocactus 207
asterias 207
hamatacanthus 214
horizonthalonius 209
setispinus 215
texensis 210
wislizeni 210
Echinocereus 210
berlandieri 210
chisoensis 210
coccineus 210, 213
var. paucispinus 210
var. rosei 210
dasyacanthus 210
davisii 211
enneacanthus 211
var. brevispinus 211
var. enneacanthus 211
fendleri 211
fitchii 211
lloydii 211, 212
neomexicanus 211
papillosus 212
pectinatus 212
var. neomexicanus 212
var. wenigeri 212
pentalophus 212
poselgeri 212
reichenbachii 212
roetteri 212
var. neomexicanus 212
var. roetteri 212
stramineus 213
triglochidiatus 213
var. *paucispinus* 210
viridiflorus 213
var. *canus* 213
var. chloranthus 213
var. correllii 213
var. cylindricus 213
var. neocapillus 213
var. russanthus 213
var. viridiflorus 213
var. weedinii 213
Echinomastus 213
intertextus 213
var. dasyacanthus 213
var. intertextus 213
mariposensis 214
warnockii 214
Echium 177
vulgare 177
Eclipta 91
prostrata 91
Egletes 91
viscosa 91
Ehretia 177
anacua 177
Elaeagnaceae 276
Elaeagnus 276
angustifolia 276
pungens 276
Elatinaceae 276
Elatine 276
brachysperma 276
triandra 276
Elephantopus 92
carolinianus 92
nudatus 92
tomentosus 92
Ellisia 387
nyctelea 387
Elytraria 13
bromoides 13
imbricata 13
Emex 499
spinosa 499
Emilia 92
fosbergii 92
Emorya 428
suaveolens 428
Encelia 92
scaposa 92
Enemion 518
biternatum 518
Engelmannia 92
peristenia 92
pinnatifica 92
Epifagus 479
virginiana 479
Epilobium 468
ciliatum 468
coloratum 468

Epithelantha 214
bokei 214
micromeris 214
Erechtites 93
hieraciifolia 93
Eriastrum 489
diffusum 489
ERICACEAE 276
Ericameria 168
austrotexana 168
laricifolia 93
nauseosa 93
var. graveolens 93
var. texensis 93
triantha 168
Erigeron 93
annuus 93
bellidiastrum 93
var. arenarius 93
var. bellidiastrum 93
var. robustus 93
bigelovii 93
colomexicanus 94
divergens 94
flagellaris 94
geiseri 94
modestus 94
philadelphicus 94
procumbens 95
pulchellus 95
rybius 95
strigosus 95
superbus 95
tenellus 95
tenuis 96
versicolor 96
Eriogonum 499
abertianum 499
alatum 499
var. alatum 499
var. glabriusculum 499
annuum 499
correllii 500
greggii 500
havardii 500
hemipterum 500
hieraciifolium 500
jamesii 500
lachnogynum 501
longifolium 501
var. longifolium 501
var. plantagineum 501
multiflorum 501
nealleyi 501
platyphyllum 501
polycladon 501
ramosissimum 502
rotundifolium 502
suffruticosum 502
tenellum 502
var. *platyphyllum* 501
var. *ramosissimum* 502
wrightii 502
Erodium 381
botrys 381
cicutarium 381
texanum 381
Eruca 191
sativa 191
vesicaria 191
var. sativa 191
Erucastrum 192
gallicum 192
Eryngium 37
diffusum 37
heterophyllum 37
hookeri 37
integrifolium 37
leavenworthii 37
nasturtiifolium 38
prostratum 38
yuccifolium 38
Erysimum 192
asperum 192
capitatum 192
repandum 192
Erythrina 331
herbacea 331
Eschscholtzia 482
californica 482
var. *mexicana* 482
mexicana 482
Escobaria 208
missouriensis 208
var. *caespitosa* 208
Esenbeckia 551
berlandieri 551
Eucnide 424
bartonioides 424
Eucrypta 387
micrantha 387
Euonymus 241
americana 241
atropurpurea 241
Eupatorium 96
album 96
altissimum 96
azureum 96
bigelovii 96
capillifolium 97
compositifolium 97
fistulosum 97
glaucescens 97
havanensis 61
herbaceum 61
hyssopifolium 98
incarnatum 97
ivifolium 97
leucolepis 98
mohrii 98
odoratum 98
perfoliatum 98
pinnatifidum 618
rothrockii 61
rotundifolium 98
serotinum 98
solidaginifolium 99
wrightii 61
Euphorbia 293
antisyphilitica 293
bicolor 293
bifurcata 293
bilobata 293
brachycera 293
chaetocalyx 7
corollata 294
eriantha 294
exstipulata 294
fendleri 7
var. *triligulata* 7
helleri 294
hexagona 294
innocua 294
longicruris 295
marginata 295
peplidion 295
pycnanthema 282
roemeriana 295
spathulata 295
strictior 295
tetrapora 296
texana 296
theriaca 7
var. *spurca* 7
wrightii 296
EUPHORBIACEAE 278
Eurota 250
lanata 250
Eurybia 99
hemispherica 99
Eurytaenia 38
hinckleyi 38
texana 38
Eustoma 379
exaltatum 379
grandiflorum 379
russellianum 379
Euthamia 99
camporum 99
var. *gymnospermoides* 99
graminifolia 99
gymnospermoides 99
leptocephala 99
Evax 76
candida 76
multicaulis 99
prolifera 100
Evolvulus 257
alsinoides 257
nuttallianus 257
sericeus 257
Eysenhardtia 331
spinosa 331
texana 331

FABACEAE 303
Facelis 100
retusa 100
FAGACEAE 366
Fagus 366
grandifolia 366
Fallugia 532
paradoxa 532
Fatoua 451
villosa 451
Fendlera 384
falcata 384
linearis 385
rigida 385
rupicola 385
wrightii 385
Fendlerella 385
utahensis 385
Ferocactus 214
hamatacanthus 214
var. hamatacanthus 214
setispinus 215
uncinatus 214
wislizenii 210
Filago 100
californica 100
candida 76
Firmiana 592
simplex 592
FLACOURTIACEAE 374
Flaveria 100
brownii 100
campestris 100
chlorifolia 100
trinervia 101
Fleischmannia 97
incarnata 97
Florestina 101
tripteris 101
Flourensia 101
cernua 101
Flyriella 101
parryi 101
Foeniculum 38
vulgare 38
Forestiera 462
acuminata 462
angustifolia 463
ligustrina 463
neomexicana 463
pubescens 463
reticulata 463
Forsellesia 242
planitierum 242
spinescens 242
texensis 242
Fouquiera 375
splendens 375
FOUQUIERIACEAE 375
Fragaria 532
chiloensis 532
vesca 532
virginiana 532
Frangula 525
caroliniana 525
Frankenia 375
jamesii 375
johnstonii 375

FRANKENIACEAE 375
Franseria 381
speciosa 381
Fraxinus 463
americana 463
var. americana 463
var. texensis 463
berlandieriana 464
caroliniana 464
cuspidata 464
greggii 464
papillosa 464
pennsylvanica 464
velutina 465
Froelichia 26
arizonica 26
braunii 26
drummondii 26
floridana 26
gracilis 26
Fumaria 376
officinalis 376
FUMARIACEAE 375
Funastrum 59
clausum 59
crispum 59
cynanchoides 59
torreyi 59

Gaillardia 101
aestivalis 101
var. aestivalis 101
var. austrotexana 101
var. fastigiata 101
var. winkleri 101
amblyodon 101
bicolor 6
var. *drummondii* 6
coahuilensis 102
multiceps 102
pinnatifida 102
pulchella 102
var. australis 102
var. drummondii 102
var. **drummondii 6**
var. *picta* 6
var. pulchella 102
suavis 102
Galactia 332
canescens 332
erecta 332
grayi 332
heterophylla 332
longifolia 332
marginalis 332
striata 332
texana 333
volubilis 333
wrightii 333
Galinsoga 102
parviflora 102
Galium 543
aparine 543
asperrimum 8
circaezans 543
correllii 621
frankliniense 543
hispidulum 543
mexicanum 543
subsp. *flexicum* 8
var. **asperrimum 8**
var. asperulum 543
var. flexicum 543
var. **flexicum 8**
microphyllum 544
obtusum 544
pilosum 544
proliferum 544
rothrockii 545
texense 544
tinctorium 544
trifidum 545
triflorum 545
uncinulatum 545
uniflorum 545
virgatum 545
wrightii 545
Galphimia 433
angustifolia 433
Gamochaeta 103
americana 103
falcata 103
pensylvanica 103
purpurea 103
sphacilata 103
Garrya 376
goldmanii 7, 376
ovata 7, 376
var. **goldmanii 7**, 376
wrightii 376
GARRYACEAE 376
Gaura 468
boquillensis 468
brachycarpa 468
calcicola 468
coccinea 468
demareei 469
drummondii 469
hexandra 469
var. gracilis 469
lindheimeri 469
longiflora 469
macrocarpa 469
mckelveyae 470
parviflora 470
sinuata 470
var. suffulta 470
var. terrellensis 470
triangulata 470
villosa 479
var. parksii 470
var. villosa 470
Gelsemium 428
sempervirens 428
Genistidium 333
dumosum 333
Gentiana 379
affinis 379
saponaria 379
GENTIANACEAE 376
GERANIACEAE 381
Geranium 382
caespitosum 382
carolinianum 382
dissectum 382
seemannii 382
texanum 382
Geum 532
canadense 532
Gilia 489
flavocincta 489
incisa 489
insignis 490
ludens 490
mexicana 490
rigidula 490
var. acerosa 490
var. rigidula 490
stewartii 490
Gillenia 534
stipulatus 534
Glandularia 602
bipinnatifida 602
var. bipinnatifida 602
var. ciliata 602
canadensis 602
delticola 602
polyantha 602
pulchella 602
pumila 603
quandrangulata 603
tumidula 603
verecunda 603
Glandulicactus 214
uncinatus 214
var. wrightii 214
Glaucium 483
corniculatum 483
Glechoma 618
hederacea 618
Gleditsia 333
aquatica 333
triacanthos 333
× texana 334
Glinus 450
lotoides 450
radiatus 450
Glossopetalon 242
spinescens 242
Glottidum 358
vesicarium 358
Glychrrhiza 334
lepidota 334
Gnaphalium 103
americanum 103
falcatum 103
pensylvanicum 103
purpureum 103
var. *americanum* 103
Gochnatia 103
hypoleuca 103
Gomphrena 26
caespitosa 26
decumbens 27
dispersa 27
globosa 26
haageana 27
nealleyi 27
nitida 27
serrata 27
Gonolobus 57
gonocarpos 57
GOODENIACEAE 382
Gossypianthus 27
lanuginosus 27
Gratiola 566
brevifolia 566
flava 567
neglecta 567
pilosa 567
virginiana 567
Grindelia 104
adenodonta 104
arizonica 104
ciliata 104
grandiflora 104
havardii 104
lanceolata 104
var. lanceolata 104
var. texana 104
microcephala 105
var. *adenodonta* 104
nuda 105
var. aphanactis 105
var. nuda 105
oolepis 105
oxylepis 105
pusilla 105
scabra 105
squarrosa 106
Grusonia 216
aggeria 216
emoryi 218
grahamii 218
schottii 220
Guajacum 616
angustifolium 616
Guilleminea 27
densa 27
lanuginosa 27
Gutierrezia 63
amoena 63
dracunculoides 63
microcephala 106
sarothrae 106
sphaerocephala 106
texana 106
var. glutinosa 106
var. texana 106
triflora 161
Gymnocladus 334
dioicus 334
Gymnosperma 106
glutinosum 106

Hackelia 178
besseyi 178
floribunda 178
grisea 178
pinetorum 178
virginiana 178
Halesia 593
diptera 593
Halimolobos 192
diffusa 192
Haloragaceae 383
Hamamelidaceae 384
Hamamelis 384
vernalis 384
virginiana 384
Hamatocactus 215
bicolor 215
Haploesthes 107
greggii 107
var. texana 107
Haplopappus 142
annuus 142
ciliatus 104
heterophyllus 116
pluriflorum 116
Haplophyton 46
cimicidum 46
Hedeoma 400
acinoides 400
apiculata 400
costata 400
drummondii 400
var. *reverchonii* 401
hispida 401
mollis 401
nana 401
pilosa 401
plicata 401
reverchonii 401
serpyllifolia 402
Hedera 48
helix 48
Hedyotis 546
acerosa 546
var. acerosa 546
var. polyprimoides 546
angulata 546
australis 546
boscii 549
butterwickiae 546
correllii 546
crassifolia 546
var. *micrantha* 546
croftiae 547
greenmanii 547
humifusa 547
intricata 547
longifolia 547
nigricans 547
var. austrotexana 547
var. nigricans 547
var. papillacea 547
pooleana 548
purpurea 548
rosea 548
rubra 548
subviscosa 548
uniflora 549
wrightii 548
Hedypnois 107
cretica 107
Hedysarum 334
boreale 334
Heimia 431
longipes 432
salicifolia 431
Helenium 107
amarum 107
var. amarum 107
var. badium 107
autumnale 107
drummondii 107
elegans 107
var. amphibolum 107
var. elegans 107
flexuosum 108
linifolium 108
microcephalum 108
var. microcephalum 108
var. ooclinium 108
quadridentatum 108
Helianthemum 253
carolinianum 253
georgianum 253
glomeratum 253
rosmarinifolium 253
Helianthus 108
angustifolius 108
annuus 108
argophyllus 109
ciliaris 109
debilis 109
var. cucumerifolius 109
var. silvestris 109
grosseserratus 109
hirsutus 109
laciniatus 109
maximiliani 110
mollis 110
occidentalis 110
var. plantagineus 110
paradoxus 619
petiolaris 110
var. canescens 110
var. petiolaris 110
praecox 110
subsp. *runyoni* 6
var. praecox 110
var. **runyoni 6**
var. runyonii 110
rigidus 110
var. rigidus 110
var. subrhomboideus 110
var. **subrhomdoideus 6**
salicifolius 111
subrhomboideus 6
tuberosus 111
Helietta 552
parvifolia 552
Heliopsis 111
helianthoides 111
parvifolia 111
Heliotropium 178
amplexicaule 178
angiospermum 178
confertifolium 179
convolvulaceum 179
curassavicum 179
var. curassavicum 179
var. obovatum 179
europaeum 179
glabriusculum 179
greggii 179
indicum 180
molle 180
procumbens 180
racemosum 180
tenellum 180
texanum 180
torreyi 181
Herissantia 439
crispa 439
Hermannia 592
texana 592
Heterosperma 111
pinnatum 111
Heterotheca 111
canescens 111
fulcrata 112
latifolia 112
mariana 81
pilosa 81
stenophylla 112
subaxillaris 112
var. *latifolia* 112
villosa 619
viscida 112
Heuchera 558
americana 558
leptomeria 558
rubescens 558
Hibiscus 439
aculeatus 439
coulteri 439
dasycalyx 439
denudatus 439
esculentus 439
grandiflorus 440
laevis 440
lasiocarpos 440
martianus 440
striatus 440
trionum 440
Hieracium 112
carneum 112
gronovii 112
longipilum 113
schultzii 113
Hippocastanaceae 384
Hoffmannseggia 334
brachycarpa 351
caudata 313
densiflora 335
drepanocarpa 334
drummondii 334
glauca 335
jamesii 351
melanosticta 351
oxycarpa 335
tenella 335
Holocantha 579
stewartii 579
Holodiscus 533
discolor 533
Holosteum 235
umbellatum 235
Hottonia 512
inflata 512
Houstonia 546
acerosa 546
butterwickiae 546
correllii 546
humifusa 547
intricata 547
longifolia 547
micrantha 546
nigricans 547
parviflora 547
purpurea 548
rubra 548
subviscosa 548
Hovenia 525
dulcis 525
Hunzikeria 581
texana 581
Hybanthus 609
verticillatus 609
Hydrangeaceae 384
Hydrocotyle 39
bonariensis 39
ranunculoides 39
umbellata 39
verticillata 39
Hydrolea 387
ovata 387
spinosa 387
uniflora 387
Hydrophyllaceae 387
Hygrophila 14
lacustris 14
polysperma 14
Hymenoclea 113
monogyra 113
Hymenopappus 113
artemisiifolius 113
var. artemisiifolius 113
var. riograndensis 113
biennis 113
carrizoanus 113
filifolius 114
var. cinereus 114
flavescens 114
var. canotomentosus 114
var. flavescens 114

scabiosaeus 114
var. corymbosus 114
tenuifolius 114
Hymenothrix 114
wislizeni 114
Hymenoxys 159
acaulis 159
linearifolia 160
odorata 114
richardsonii 115
scaposa 160
texana 115
turneri 160
Hypericaceae 392
Hypericum 392
cistifolium 392
crux-andreae 392
densiflorum 393
drummondii 393
fasciculatum 393
frondosum 393
galioides 393
gentianoides 393
gymnanthum 393
hypericoides 394
mutilum 394
nudiflorum 394
pauciflorum 394
perforatum 394
prolificum 394
pseudomaculatum 395
punctatum 395
setosum 395
sphaerocarpum 395
tubulosum 395
virginicum 395
Hypochaeris 115
glabra 115
microcephala 115
radicata 115
Hyptis 402
alata 402
mutabilis 619

Ibervillea 270
lindheimeri 270
tenuisecta 270
tripartita 270
Ilex 47
ambigua 47
coriacea 47
decidua 47
longipes 47
opaca 48
verticillata 48
vomitoria 48
Impatiens 169
capensis 169
Indigofera 335
lindheimeriana 335
miniata 335
var. *leptosepala* 335
var. miniata 335
var. texana 335
suffruticosa 336
Iodanthus 192
pinnatifidus 192
Ionactis 115
linariifolius 115
Ipomoea 260
acuminata 260
alba 257
amnicola 258
aristolochiifolia 258
barbatisepala 258
capillacea 258
cardiophylla 258
carnea 258
coccinea 259
costellata 259
cordatotriloba 259
var. cordatotriloba 259
var. torreyana 259
cristulata 259
dumetorum 259
fistulosa 259
hederacea 259
hederifolia 259
heptaphylla 260
imperati 260
indica 260
lacunosa 260
leptoloma 260
leptophylla 260
leucantha [hybrid] 261
lindheimeri 261
nil 261
pandurata 261
pes-caprae 261
pinnatifida 261
pubescens 262
pulchella 262
purpurea 262
quamoclit 262
rupicola 262
sagittata 262
shumardiana 263
sinuata 261
stolonifera 260
tenuiloba 263
trichocarpa 258
tricolor 263
wrightii 263
Ipomopsis 490
aggregata 490
var. *texana* 490
havardii 491
laxiflora 491
longiflora 8, 491
subsp. *australis* 8
var. **australis 8**
var. longiflora 8, 481
subsp. *neomexicana* 8
var. **neomexicana 8,** 481
polycladon 491
pumila 491
rubra 491
wrightii 492
Iresine 28
diffusa 28
heterophylla 28
leptoclada 28
palmeri 28
rhizomatosa 28
Isocarpha 116
oppositifolia 116
var. achyranthes 116
Isocoma 116
coronopifolia 116
drummondii 116
pluriflora 116
Isopyrum 518
biternatum 518
Itea 558
virginica 558
Iva 116
ambrosiifolia 116
angustifolia 116
var. angustifolia 116
var. latior 116
annua 117
axillaris 117
dealbata 117
frutescens 117
imbricata 117
xanthifolia 117

Jacquemontia 263
tamnifolia 263
Janusia 434
gracilis 434
Jatropha 296
cathartica 296
dioica 296
var. dioica 296
var. graminea 296
macrorhiza 296
Jefea 118
brevifolia 118
Juglandaceae 396
Juglans 397
major 397
microcarpa 397
nigra 398
Julocroton 288
argenteus 288
Justicia 14
americana 14
greggii 15
longii 14
ovata 14
pilosella 14
runyonii 15
turneri 15
warnockii 15
wrightii 15

Kallstroemia 616
californica 616
grandiflora 616
hirsutissima 616
maxima 616
parviflora 616
perennans 617
Karwinskia 525
humboldtiana 525
Kickxia 567
elatine 567
Kneiffia 7
sessilis 7
Koanophyllon 99
solidaginifolium 99
Kochia 250
scoparia 250
Koeberlinia 398
spinosa 398
Koeberliniaceae 398
Kosteletzkya 441
depressa 441
virginica 441
Krameria 398
erecta 398
grayi 398
lanceolata 398
ramosissima 398
Krameriaceae 398
Krapovickasia 441
physaloides 441
Krascheninnikova 250
lanata 250
Krigia 118
cespitosa 118
dandelion 118
occidentalis 118
virginica 118
wrightii 118
Kummerowia 388
stipulacea 338
striata 338

Lackeya 336
multiflora 336
Lactuca 119
canadensis 119
floridana 119
graminifolia 119
hirsuta 119
intybacea 119
ludoviciana 119
pulchella 120
saligna 120
serriola 120
Laennecia 120
coulteri 120
filaginoides 120
sophiifolia 120
turnerorum 121
Lagenaria 270
siceraria 270
Lagerstroemia 431
indica 431
Laguncularia 254
racemosa 254
Lamiaceae 399

Lamium 402
amplexicaule 402
purpureum 402
Lantana 603
achyranthifolia 603
camara 603
canescens 603
macropoda 603
montevidensis 604
urticoides 604
Lappula 181
occidentalis 181
redowskii 181
Lapsana 121
communis 121
Larrea 617
tridentata 617
Lathyrus 336
aphaca 336
graminifolius 336
hirsutus 336
lanszwertii 336
latifolius 337
polymorphus 337
pusillus 337
venosus 337
Lauraceae 418
Lavatera 441
trimestris 441
Leavenworthia 193
texana 193
Lechea 253
mensalis 253
mucronata 253
pulchella 254
san-sabeana 254
tenuifolia 254
Leitneria 419
floridana 419
Leitneriaceae 419
Lenophyllum 267
texanum 267
Lentibulariaceae 419
Leonotis 402
nepetifolia 402
Leontodon 121
hispidus 121
taraxacum 121
Leonurus 402
cardiaca 402
sibiricus 403
Lepidium 193
alyssoides 193
var. angustifolia 193
austrinum 193
densiflorum 193
lasiocarpum 193
var. orbiculare 193
var. wrigthii 193
latifolium 193
oblongum 193
perfoliatum 194
ruderale 194
sordidum 194
thurberi 194
virginicum 194
Lepidospartum 121
burgessii 121
Leptoglossis 581
texana 581
Leptopus 297
phyllanthoides 297
Lepuropetalon 559
spathulatum 559
Lespedeza 337
capitata 337
cuneata 337
hirta 338
procumbens 338
repens 338
stipulacea 338
striata 338
stuevei 338
texana 339
violacea 339
virginica 339
Lesquerella 194
angustifolia 194
argyraea 195
auriculata 195
densiflora 195
engelmannii 195
subs. *ovalifolia* 196
var. *ovalifolia* 196
fendleri 195
gordonii 195
gracilis 196
grandiflora 196
lasiocarpa 196
var. berlandiera 196
var. lasiocarpa 196
lindheimeri 196
mcvaughiana 196
ovalifolia 196
pallida 197
purpurea 197
recurvata 197
sessilis 197
thamnophila 197
valida 197
Leucaena 339
leucocephala 339
pulverulenta 339
retusa 339
Leucanthemum 121
vulgare 121
Leucophyllum 567
candidum 567
frutescens 568
minus 568
zygophyllum 568
Leucospora 568
multifida 568
Leucothoe 277
racemosa 277
Liatris 122
acidota 122
aestivalis 122
aspera 122
bracteata 122
cymosa 122
elegans 122
var. bridgesii 122
var. carizzana 122
var. elegans 122
glabrata 123
var. alabamensis 123
glandulosa 123
lancifolia 123
mucronata 6
punctata 123
var. mucronata 123
var. **mucronata 6**
var. punctata 123
pycnostachya 123
squarrosa 123
var. glabrata 123
var. squarrosa 123
tenuis 123
Ligustrum 465
japonicum 465
lucidum 465
quihoui 465
sinense 465
Lilaeopsis 39
chinensis 39
Limnophila 568
sessiliflora 568
Limnosciadium 39
pinnatum 39
pumilum 40
Limonium 489
carolinianum 489
limbatum 489
Linaceae 421
Linanthus 492
bigelovii 492
Linaria 568
anagallidea 568
canadensis 569
texana 569
vulgaris 569
Lindera 418
benzoin 418
Lindernia 569
anagallidea 569
crustacea 569
dubia 569
var. *anagallidea* 569
Lindheimera 124
texana 124
Linum 421
alatum 421
aristatum 421
australe 421
berlandieri 421
var. berlandieri 421
var. filifolium 421
compactum 421
elongatum 422
grandiflorum 422
hudsonioides 422
imbricatum 422
lewisii 422
lundellii 422
medium 423
var. texanum 423
pratense 423
var. *lewisii* 422
puberulum 423
rigidum 423
rupestre 423
schiedeanum 423
striatum 424
sulcatum 424
usitatissimum 424
vernale 424
Lippia 604
alba 604
graveolens 604
Liquidambar 384
styraciflua 384
Liriodendron 432
tulipifera 432
Lithospermum 181
arvense 181
caroliniense 181
cobrense 181
confine 182
incisum 181
matamorense 182
mirabile 182
multiflorum 182
obtusifolium 182
parksii 182
tuberosum 182
viride 182
Loasaceae 424
Lobelia 223
appendiculata 223
berlandieri 223
var. arenosa 223
var. berlandieri 223
brachypoda 223
cardinalis 223
var. cardinalis 223
var. graminea 223
fenestralis 224
flaccidifolia 224
puberula 224
reverchonii 224
siphilitica 224
spicata 224
Lobularia 198
maritima 198
Loeflingia 235
squarrosa 235
Loeselia 492
greggii 492
Loganiaceae 428
Lomatium 40
daucifolium 40
foeniculaceum 40
orientale 40

Lonicera 228
albiflora 228
arizonica 228
fragrantissima 228
japonica 229
maackii 229
sempervirens 229
Lophophora 215
williamsii 215
Lotus 340
corniculatus 340
humistratus 340
oroboides 340
plebeius 340
purshianus 340
var. helleri 340
unifoliatus 340
Ludwigia 471
alternifolia 471
decurrens 471
erecta 471
glandulosa 471
hirtella 471
leptocarpa 471
linearis 472
microcarpa 472
octovalvis 472
palustris 472
peploides 472
pilosa 472
repens 473
sphaerocarpa 473
uruguayensis 473
Luffa 270
aegyptica 270
Lupinus 340
argenteus 340
concinnus 340
havardii 341
perennis 341
plattensis 341
subcarnosus 341
texensis 341
Lycianthes 481
asarifolia 581
Lycium 481
barbarum 581
berlandieri 582
carolinianum 582
pallidum 582
puberulum 582
var. berberioides 582
var. puberulum 582
texanum 582
torreyi 582
Lycopersicon 583
esculentum 583
Lycopus 403
americanus 403
asper 403
rubellus 403
virginicus 403
Lygodesmia 136
exigua 136
juncea 124
ramosissima 124
texana 124
Lyonia 277
ligustrina 277
lucida 619
mariana 277
Lysimachia 513
lanceolata 513
radicans 513
Lythraceae 430
Lythrum 431
alatum 431
californicum 431
lanceolatum 431
lineare 619
ovalifolium 432
salicaria 432

Macfadyena 172
unguis-cati 172
Machaeranthera 142
annua 142
aurea 142
blephariphylla 124
canescens 124
gracilis 125
gypsitherma 620
gypsophila 125
parviflora 125
phyllocephala 142
pinnatifida 125
var. chihuahuana 125
var. glaberrima 125
var. pinnatifida 125
tanacetifolia 125
viscida 125
Maclura 451
pomifera 451
Macroptilium 341
atropurpureum 341
gibbosifolium 342
Macrosiphonia 46
hypoleuca 46
lanuginosa 46
macrosiphon 46
Madia 126
elegans 126
Magnolia 432
grandiflora 432
pyramidata 433
virginiana 433
Magnoliaceae 432
Malachra 441
capitata 441
Malacomeles 533
denticulata 533
Malacothrix 126
fendleri 126
sonorae 126
Malpighia 434
glabra 434
Malpighiaceae 433
Malus 533
ioensis 533
Malva 441
neglecta 441
parviflora 442
rotundifolia 442
sylvestris 442
Malvaceae 434
Malvastrum 442
americanum 442
aurantiacum 442
coromandelianum 442
Malvaviscus 443
arboreus 443
var. drummondii 443
Malvella 443
lepidota 443
leprosa 443
sagittifolia 443
Mammillaria 208
echinus 208
grahamii 215
heyderi 215
lasiacantha 215
meicantha 215
pottsii 216
prolifera 216
similis 208
sulcata 209
wrightii 216
Mancoa 198
pubens 198
Manihot 297
esculenta 297
subspicata 297
walkerae 297
Margaranthus 583
solanaceus 583
Marrubium 403
vulgare 403
Marshallia 126
caespitosa 126
var. caespitosa 126
var. signata 126
graminifolia 126
var. cynanthera 126
Martynia 485
louisianica 485
Matelea 57
biflora 57
brevicoronata 57
cynanchoides 57
decipiens 57
edwardsensis 57
gonocarpa 57
parviflora 58
parvifolia 58
producta 58
radiata 58
reticulata 58
sagittifolia 58
texensis 59
Matricaria 126
matricarioides 126
Matthiola 198
bicornis 198
Matthiola 198
longipetala 198
Maurandya 569
antirrhiniflora 569
wislizeni 570
Maytenus 242
phyllanthoides 242
texana 242
Mazus 570
japonicus 570
pumilus 570
Mecardonia 570
acuminata 570
procumbens 570
Medicago 342
arabica 342
lupulina 342
minima 342
orbicularis 342
polymorpha 342
sativa 343
Melampodium 127
cinereum 127
var. cinereum 127
var. hirtellum 127
leucanthum 127
strigosum 127
Melastomataceae 448
Melia 449
azedarach 449
Meliaceae 449
Melilotus 343
albus 343
indicus 343
officinalis 343
Melochia 592
corchorifolia 592
pyramidata 592
tomentosa 593
Melothria 271
pendula 271
Menispermaceae 449
Menispermum 449
canadense 449
Menodora 465
heterophylla 465
longiflora 466
scabra 466
Mentha 404
arvensis 404
piperita 404
rotundifolia 404
spicata 404
Mentzelia 425
albescens 425
albicaulis 425
asperula 425
decapetala 425
hispida 425
humilis 426

incisa 426
lindheimeri 426
mexicana 426
montana 426
multiflora 426
nuda 427
oligosperma 427
pachyrhiza 427
reverchonii 427
saxicola 427
strictissima 427
texana 426, 427
Menyanthceae 449
Merremia 263
dissecta 263
Meximalva 443
filipes 443
Micranthemum 470
umbrosum 570
Micromeria 404
brownei 404
Microseris 128
cuspidata 128
linearifolia 127
Mikania 127
cordifolia 127
scandens 127
Mimosa 343
aculeaticarpa 343
var. biuncifera 343
asperata 343
biuncifera 343
borealis 344
dysocarpa 344
emoryana 344
hystricina 344
latidens 344
malacophylla 344
nuttallii 345
occidentalis 345
piqua 343
var. *berlandieri* 343
quadrivalis 344
var. *hystricina* 344
var. *nuttalii* 345
var. *occidentalis* 345
var. *platycarpa* 345
roemeriana 345
rupertiana 345
strigillosa 345
texana 345
turneri 345
warnockii 346
Mimulus 570
alatus 570
dentilobus 571
glabratus 571
rubellus 571
Minuartia 232
drummondii 232
michauxii 233
var. *texana* 233
muriculata 232
muscorum 233
Mirabilis 458
albida 458
austrotexana 458
carletonii 458
cilata 458
collina 459
comata 458
decumbens 459
dumetorum 458
exaltata 458
gausapoides 459
gigantea 458
glabra 458
hirsuta 458
jalapa 459
latifolia 458
linearis 459
longiflora 459
melanotricha 459
multiflora 459
nyctaginea 459
oxybaphoides 460
texensis 460
Mitchella 549
repens 549
Mitracarpus 549
breviflorus 549
hirtus 549
villosus 549
Mitreola 429
petiolata 429
sessilifolia 429
Modiola 443
caroliniana 443
MOLLUGINACEAE 450
Mollugo 450
cerviana 450
verticillata 450
Moluccella 404
laevis 404
Monarda 405
citriodora 405
var. citriodora 405
var. parva 405
clinopodioides 405
fistulosa 405
var. fistulosa 405
var. menthifolia 405
fruticulosa 405
lindheimeri 405
maritima 406
pectinata 406
var. correllii 406
var. intermedia 406
var. lasiodonta 406
var. occidentalis 406
var. punctata 406
russeliana 406
stanfieldii 406
viridissima 407
Monolepis 250
nuttalliana 250
Monotropa 450
hypopithys 450
uniflora 451
MONOTROPACEAE 450
MORACEAE 451
Morella 452
caroliniensis 452
cerifera 452
Mortonia 242
greggii 242
scabrella 242
sempervirens 243
Morus 451
alba 451
microphylla 452
rubra 452
MYRICACEAE 452
Myagrum 198
perfoliatum 198
Myosotis 183
macrosperma 183
verna 183
Myosurus 518
minimus 518
Myrica 452
cerifera 452
heterophylla 452
pusilla 452
Myriophyllum 383
aquaticum 383
brasiliense 383
heterophyllum 383
pinnatum 383
spicatum 383
MYRSINACEAE 452
Myrtus 620
communis 620

Nama 387
berlandieri 387
carnosum 388
dichotomum 388
havardii 388
hispida 388
jamaicense 388
parvifolium 388
stenocarpum 389
stevensii 389
torynophyllum 389
undulatum 389
xylopodum 389
Nandina 170
domestica 170
Nasturtium 199
officinale 199
Nectouxia 583
formosa 583
Nelumbo 461
lutea 461
Nemophila 389
aphylla 389
phacelioides 390
sayersensis 390
triloba 389
Neolloydia 216
conoidea 216
intertexta 213
var. *dasycantha* 213
mariposensis 214
warnockii 214
Nepeta 407
cataria 407
Nephropetalum 591
pringlei 591
Neptunia 346
lutea 346
microcarpa 346
palmeri 346
plena 346
pubescens 346
var. *microcarpa* 346
Nerisyrenia 198
camporum 198
linearifolium 198
Nesaea 432
longipes 432
Nicolletia 128
edwardsii 128
Nicotiana 583
glauca 583
plumbaginifolia 583
repanda 583
trigonophylla 584
Nierembergia 584
hippomanica 584
Nissolia 346
platycalyx 346
Nomaphila 15
stricta 15
Nothocalais 128
cuspidata 128
Nuphar 461
advena 461
lutea 461
var. *advena* 461
Nuttallanthus 569
canadensis 569
texanus 569
Nyctaginia 460
capitata 460
NYCTAGINACEAE 452
Nymphaea 462
ampla 462
elegans 462
mexicana 462
odorata 462
NYMPHAEACEAE 461
Nymphoides 449
aquatica 449
peltata 450
Nyssa 265
aquatica 265
biflora 265
sylvatica 265

Obolaria 379
virginica 379

Oenothera 473
albicaulis 473
biennis 473
brachycarpa 473
caespitosa 474
canescens 474
cordata 474
coryi 474
drummondii 474
elata 474
engelmannii 475
falfurriae 475
grandis 475
havardii 475
heterophylla 475
hookeri 476
jamesii 475
kunthiana 476
laciniata 476
linifolia 476
macrocarpa 476
var. incana 476
var. macrocarpa 476
var. oklahomensis 476
mexicana 476
missouriensis 476
pallida 476
pilosella 7, 477
var. **sessilis 7,** 477
primiveris 477
pubescens 477
rhombipetala 477
rosea 477
spachiana 477
speciosa 478
tetraptera 478
texensis 478
triloba 478
villosa 478
Oleaceae 462
Oldenlandia 549
boscii 549
uniflora 549
Oligomeris 522
linifolia 522
Oligoneuron 128
nitidum 128
rigidum 128
Omphalodes 183
aliena 183
Onagraceae 466
Onopordum 128
acanthium 128
Onosmodium 183
bejariense 183
var. bejariense 183
var. occidentalis 183
helleri 183
Operculina 261
pinnatifida 261
Opuntia 216
aggeria 216
atrispina 216
aureispina 217
camanchica 217
compressa 218
cymochila 219
davisii 217
densispina 217
dulcis 217
edwardsii 217
emoryi 218
engelmannii 218
forma linguiformis 218
var. engelmanii 218
var. lindheimeri 218
ficus-indica 218
fragilis 218
grahamii 218
humifusa 218
imbricata 219
var. argentea 219
var. imbricata 219
kleiniae 219
leptocaulis 219
macrocentra 219
var. *aureispina* 217
var. macrocentra 219
var. minor 219
macrorhiza 219
var. *pottsii* 220
phaeacantha 219
polyacantha 220
var. arenaria 220
var. polyacantha 220
pottsii 220
rufida 220
schottii 220
var. *grahamii* 218
spinosibacca 220
stricta 220
strigil 221
var. flexospina 221
var. strigil 221
tortispina 221
tunicata 221
var. *davisii* 217
Orbexilum 347
pedunculatum 347
simplex 347
Orobanchaceae 479
Orobanche 479
fasciculata 479
ludoviciana 479
multicaulis 479
ramosa 479
uniflora 480
Osmorhiza 6, 40
bipatriata 6, 40
longistylis 40
mexicana 40
var. **bipatriata 6,** 40
Ostrya 171
knowltonii 171
var. chisoensis 171
var. knowltonii 171
virginiana 171
Oxalidaceae 480
Oxalis 480
amplifolia 480
corniculata 480
var. wrightii 480
corymbosa 480
dichondrifolia 480
dillenii 480
drummondii 480
lyonii 480
neaei 481
stricta 480
violacea 481
Oxypolis 40
filiformis 40
rigidior 41
Oxytropis 347
lambertii 347
sericea 347

Packera 146
fendleri 146
glabella 146
millelobata 146
obovata 146
plattensis 147
tampicanus 147
tomentosus 147
Paederia 549
foetida 549
Palafoxia 128
callosa 128
hookeriana 129
reverchonii 129
riograndensis 129
rosea 129
var. macrolepis 129
var. rosea 129
sphacelata 129
texana 129
var. ambigua 129
var. texana 129
Papaver 483
rhoeas 483
somniferum 483
Papaveraceae 481
Parentucellia 571
viscosa 571
Parietaria 597
floridana 597
obtusa 598
pensylvanica 598
var. *obtusa* 598
Parkinsonia 347
aculeata 347
Parnassia 559
asarifolia 559
grandiflora 559
Paronychia 237
chorizanthoides 237
congesta 236
drummondii 236
fastigiata 236
jamesii 236
jonesii 236
lindheimeri 236
lundelliorum 237
maccartii 237
monticola 237
nudata 237
parksii 237
sessiliflora 237
setacea 237
velutina 238
virginica 237
wilkinsonii 238
Parosela 7
warnockii 7
Parthenium 130
argentatum 130
confertum 130
var. *microcephalum* 130
hispidum 130
hysterophorus 130
incanum 130
Parthenocissus 613
heptaphylla 613
quinquefolia 613
vitacea 614
Passiflora 483
affinis 483
filipes 483
foetida 484
incarnata 484
lutea 484
suberosa 484
tenuiloba 484
Passifloraceae 483
Pavonia 444
lasiopetala 444
Pectis 130
angustifolia 130
var. angustifolia 130
var. fastigiata 130
var. tenella 130
cylindrica 131
filipes 131
papposa 131
prostrata 131
Pectocarya 183
heterocarpa 183
platycarpa 184
Pedaliaceae 484
Pedicularis 571
canadensis 571
Pediocactus 222
papyracanthus 222
Pediomelum 347
argophyllum 347
cuspidatum 348
cyphocalyx 348
digitatum 348
humile 348
hypogaeum 348
var. hypogaeum 348
var. scaposum 348
var. subulatum 348

latestipulatum 348
var. appressum 348
var. latestipulatum 348
linearifolium 349
pentaphyllum 349
reverchonii 349
rhombifolium 349
Peganum 617
harmala 617
mexicanum 617
Peniocereus 221
greggii 221
Pennellia 199
longifolia 199
micrantha 199
Penstemon 571
alamosensis 571
albidus 572
ambiguus 572
baccharifolius 572
barbatus 572
bradburii 573
buckleyi 572
cardinalis 572
cobaea 573
dasyphyllus 573
digitalis 573
fendleri 573
grandiflorus 573
guadalupensis 573
havardii 574
helleri 574
jamesii 574
lanceolatus 574
laxiflorus 574
murrayanus 574
tenuis 574
triflorus 575
var. integrifolius 575
var. triflorus 575
tubiflorus 575
wrightii 575
Penthorum 267
sedoides 267
Pentodon 549
pentandrus 549
Peperomia 486
pellucida 486
Peplis 431
diandra 431
Pereskia 221
aculeata 221
Pericome 131
caudata 131
Perilla 407
frutescens 407
Periploca 59
graeca 59
Perityle 131
aglossa 131
angustifolia 132
bisetosa 132
var. appressa 132
var. bisetosa 132
var. scalaris 132
cinerea 132
dissecta 132
fosteri 132
huecoensis 132
lindheimeri 133
var. halimifolia 133
var. lindheimeri 133
microglossa 133
parryi 133
quinqueflora 133
rupestris 133
var. albiflora 133
var. rupestris 133
vaseyi 133
vitreomontana 134
warnockii 134
Persea 419
borbonia 419
palustris 419
Petalostemon 7
arenicola 7
arenicola 319
candidus 319
var. *oligophyllus* 319
decumbens 320
emarginatum 321
glandulosum 323
grisea 321
macrostachyum 321
microphyllum 320
occidentale 319
phleoides 323
phleoides 321
var. *microphyllum* 320
pulcherrimum 320
purpureum 324
sabinale 324
stanfieldii 324
tenuifolium 324
tenuis 324
villosum 325
Peteria 349
scoparia 349
Petiveria 485
alliacea 485
Petrogenia 255
repens 255
Petrophyton 533
caespitosum 533
Petrorhagia 238
dubia 238
Petunia 584
axillaris 584
parviflora 579
Phacelia 390
coerulea 390
congesta 390
glabra 390
hirsuta 390
infundibuliformis 391
integrifolia 620
var. *texana* 620
intermedia 391
laxa 391
pallida 391
patuliflora 391
var. austrotexana 391
var. patuliflora 391
var. teucrifolia 391
popei 392
robusta 392
rupestris 392
strictiflora 392
texana 620
Phaseolus 349
acutifolius 349
angustissimus 350
filiformis 350
maculatus 350
pedicellatus 350
polystachios 350
Phaulothamnus 19
spinescens 19
Philadelphus 385
argenteus 385
crinitus 385
ernestii 385
hitchcockianus 386
mearnsii 386
microphyllus 386
occidentalis 386
pubescens 386
serpyllifolius 386
texanus 386
Philoxerus 28
vermicularis 28
Phlox 492
carolina 492
cuspidata 492
divaricata 492
drummondii 493
var. johnstonii 493
var. littoralis 493
var. mcallisteri 493
var. peregrina 493
var. tharpii 493
var. drummondii 493
glabella 493
longifolia 493
mesoleuca 493
nana 493
nivalis 493
oklahomensis 493
pilosa 494
subs. *latisepala* 494
subs. *pulcherrima* 494
subs. *vipara* 494
pulcherrima 494
roemeriana 494
stansburyi 493
triovulata 493
villosissima 494
Phoradendron 612
bolleanum 612
capitellatum 612
coryae 8
hawksworthii 612
juniperinum 612
macrophyllum 612
tomentosum 612
villosum 613
var. **coryae** **8**
Photinia 533
serratifolia 533
Phryma 485
leptostachya 485
Phrymaceae 485
Phyla 604
cuneifolia 604
fruticosa 604
lanceolata 605
nodiflora 605
strigulosa 604
Phyllanthus 297
abnormis 297
var. abnormis 297
var. riograndensis 297
caroliniensis 297
ericoides 298
fraternus 298
niruri 298
polygonoides 298
pudens 298
tenellus 298
urinaria 299
Physalis 584
acutifolia 584
angulata 584
cinerascens 585
var. cinerascens 585
var. spathuliflora 585
cordata 585
hederifolia 585
heterophylla 585
hispida 585
lobata 587
longifolia 585
var. longifolia 585
var. *subglabrata* 586
macrophysa 586
mollis 586
var. mollis 586
var. variavista 586
neomexicana 586
philadelphica 586
pubescens 586
pumila 586
texana 587
virginiana 587
var. *texana* 587
Physocarpus 533
monogynus 533
Physostegia 407
angustifolia 407
correllii 407
digitalis 407
intermedia 408
longisepala 408
micrantha 408

pulchella 408
virginiana 408
  var. arenaria 408
Phytolacca 486
americana 486
PHYTOLACCACEAE 485
Picradeniopsis 134
oppositifolia 134
woodhousei 134
Pilea 598
microphylla 598
pumila 598
Pilostyles 513
thurberi 513
Pinaropappus 134
parvus 134
roseus 134
  var. macvaughii 134
Pinguicula 419
pumila 419
PIPERACEAE 486
Pisonia 460
aculeata 460
Pistacia 30
mexicana 30
Pithecellobium 350
ebano 350
pallens 351
Pityopsis 135
graminifolia 135
Planera 596
aquatica 596
PLANTAGINACEAE 486
Plantago 486
aristata 486
coronopus 486
elongata 486
fastigiata 487
helleri 487
heterophylla 487
hookeriana 487
lanceolata 487
major 487
ovata 487
patagonica 488
rhodosperma 488
rugelii 488
virginica 488
wrightiana 488
PLATANACEAE 488
Platanus 488
occidentalis 488
Plateilema 135
palmeri 135
Pluchea 135
camphorata 135
foetida 135
odorata 135
purpurascens 134
rosea 135
sericea 136
PLUMBAGINACEAE 489
Plumbago 489
scandens 489
Podophyllum 171
peltatum 171
Poinsettia 299
cyathophora 299
davidii 299
dentata 299
heterophylla 299
radians 299
Polanisia 227
dodecandra 227
  var. riograndensis 227
  var. trachysperma 227
erosa 227
  var. breviglandulosa 227
  var. erosa 227
jamesii 228
uniglandulosa 228
POLEMONIACEAE 489
Polemonium 494
pauciflorum 494
Poliomintha 408
glabrescens 408
incana 408
Polycarpon 238
tetraphyllum 238
Polygala 494
alba 494
barbeyana 495
crenata 495
cruciata 495
glandulosa 495
hemipterocarpa 495
incarnata 495
leptocaulis 496
lindheimeri 496
  var. lindheimeri 496
  var. parviflora 496
longa 495
macradenia 496
maravillasensis 496
mariana 496
nana 496
nitida 497
  var. goliadensis 497
  var. tamaulipensis 497
nudata 497
obscura 497
ovatifolia 497
palmeri 497
polygama 497
ramosa 498
rimulicola 498
sanguinea 498
scoparioides 498
tweedyi 496
verticillata 498
watsonii 498
POLYGALACEAE 494
POLYGONACEAE 499
Polygonella 502
americana 502
parksii 503
polygama 503
Polygonum 503
amphibium 503
arenastrum 503
argyrocoleon 503
aviculare 503
buxiforme 504
caespitosum 504
  var. caespitosum 504
  var. longisetum 504
convolvulus 504
densiflorum 504
hydropiperoides 504
lapanthifolium 504
meisnerianum 505
orientale 505
pensylvanicum 505
punctatum 505
ramosissimum 505
sagittatum 505
scandens 506
setaceum 506
striatulum 506
tenue 506
texense 506
virginianum 506
Polymnia 149
uvedalia 149
Polypremum 429
procumbens 429
Polytaenia 41
nuttallii 41
Pomaria 351
austrotexana 351
brachycarpa 351
jamesii 351
melanosticta 351
Poncirus 552
trifoliata 552
Populus 553
alba 553
angustifolia 553
  × fremontii 554
arizonica 554
deltoids 554
  var. deltoides 554
  var. monilifera 554
  var. wislizenii 554
fremontii 554
hinckleyana 554
hinkleyana [hybrid] 554
nigra 554
tremuloides 554
wislizenii 554
Porlieria 616
angustifolia 616
Porophyllum 136
gracile 136
greggii 136
macrocephalum 136
scoparium 136
Porteranthus 534
stipulatus 534
Portulaca 509
halimoides 509
lanceolata 8
oleracea 509
parvula 509
pilosa 509
retusa 510
suffrutescens 510
umbraticola 510
  subsp. *lanceolata* 8
  var. **lanceolata 8,** 510
PORTULACACEAE 509
Potentilla 534
norvegica 534
paradoxa 534
pensylvanica 534
recta 534
rivalis 534
simplex 535
Prenanthella 136
exigua 136
Prenanthes 137
altissima 137
barbata 137
exigua 136
PRIMULACEAE 512
Prionopsis 104
ciliata 104
Priva 605
lappulacea 605
Proboscidea 484
althaeifolia 484
formosa 485
louisianica 485
parviflora 485
sabulosa 485
spicata 485
Proserpinaca 383
palustris 383
pectinata 383
Prosopis 352
cinerascens 352
glandulosa 351
laevigata 352
odorata 620
pubescens 352
reptans 352
  var. cinerascens 352
Prunella 409
vulgaris 409
Prunus 535
angustifolia 535
caroliniana 535
gracilis 535
havardii 535
mexicana 535
minutiflora 536
munsoniana 536
murrayana 536
persica 536
rivularis 536
serotina 536
  var. eximia 536
  var. serotina 536
  var. virens 536

texana 537
umbellata 537
virginiana 537
Psathyrotes 137
scaposa 137
Pseudabutilon 444
umbellatum 444
Pseudoclappia 137
arenaria 137
watsonii 137
Pseudocymopteris 35
filifolius 35
longiradiatus 36
longiradiatus 6
montanus 35
Pseudognaphalium 137
arizonicum 137
austrotexanum 138
canescens 138
helleri 138
jaliscense 138
luteoalbum 138
obtusifolium 138
pringlei 139
stramineum 139
viscosum 139
Pseudogynoxys 139
chenopodioides 139
Pseudostellaria 232
jamesiana 232
Psilactis 140
asteroides 140
brevilingulata 140
heterocarpa 140
tenuis 140
Psilostrophe 140
gnaphalioides 140
tagetina 140
var. cerifera 140
Psoralea 347
argophyllia 347
cuspidata 348
digitata 348
hypogaea 348
latestipulata 349
linearifolia 349
pedunculata 347
reverchonii 349
rhombifolia 349
scaposa 348
simplex 347
subulata 348
Psoralidium 352
lanceolatum 352
tenuiflorum 352
Psorothamnus 352
scoparius 352
Ptelea 552
trifoliata 552
var. angustifolia 552
var. mollis 552
Pterocaulon 141
virgatum 141
Pterospora 451
andromedea 451
Ptilimnium 41
capillaceum 41
costatum 41
nuttallii 41
Pueraria 353
lobata 353
montana 353
var. lobata 353
Purshia 537
ericifolia 537
Pycnanthemum 409
albescens 409
tenuifolium 409
Pyracantha 537
coccinea 537
Pyrrhopappus 141
carolinianus 141
grandiflorus 141
pauciflorus 141
Pyrus 537
angustifolia 537
arbutifolia 527
calleryana 538
communis 538
ioensis 533

Quercus 366
alba 366
arizonica 367
arkansana 367
boyntonii 367
buckleyi 367
carmenensis 367
chihuahuensis 367
× *gravesii* 373
depressipes 368
emoryi 368
× *gravesii* 373
falcata 368
fusiformis 368
gambelii 368
graciliformis 368
gravesii 369
grisea 369
havardii 369
hemisphaerica 369
hinckleyi 369
hypoleucoides 369
imbricaria 370
incana 370
intricata 370
laceyi 370
laurifolia 370
lyrata 370
macrocarpa 371
margarettiae 371
marilandica 371
michauxii 372
mohriana 371
muhlenbergii 371
nigra 371
oblongifolia 372
pagoda 372
phellos 372
polymorpha 372
prinus 372
pungens 372
var. *vaseyana* 372
robusta 373
rugosa 373
shumardii 373
sinuata 373
var. breviloba 373
var. sinuata 373
stellata 373
tardifolia 373
texana 374
toumeyi 374
turbinella 374
vaseyana 372
velutina 374
virginiana 374
Quincula 587
lobata 587

Rafflesiaceae 513
Rafinesquia 141
neomexicana 141
Randia 550
rhagocarpa 550
Ranunculaceae 518
Ranunculus 518
abortivus 518
carolinianus 518
cymbalaria 518
fascicularis 518
hispidus 518
laxicaulis 519
longirostris 519
macranthus 519
marginatus 519
muricatus 519
parviflorus 519
petiolaris 520
platensis 520
pusillus 520
recurvatus 520
repens 520
var. *nitidus* 518
sardous 520
scleratus 521
trilobus 521
Raphanus 199
raphanistrum 620
sativus 199
Rapistrum 199
rugosum 199
Ratibida 141
columnifera 141
peduncularis 142
tagetes 142
Rayjacksonia 142
annua 142
aurea 142
phyllocephala 142
Resedaceae 522
Reverchonia 300
arenaria 300
Rhamnaceae 522
Rhamnus 525
betulifolia 525
caroliniana 525
lanceolata 525
serrata 525
Rhexia 448
alifanus 448
lutea 448
mariana 448
var. *interior* 448
var. mariana 448
petiolata 448
virginica 448
Rhizophora 526
mangle 526
Rhizophoraceae 526
Rhododendron 277
canescens 277
coryi 277
prinophyllum 277
viscosum 277
Rhododon 409
angulatus 409
ciliatus 409
Rhus 30
aromatica 30
var. aromartica 30
var. flabelliformis 30
var. pilosissima 30
copallinum 30
glabra 30
lanceolata 30
microphylla 30
vernix 31
virens 31
var. choriophylla 31
var. virens 31
Rhynchosia 353
americana 353
difformis 353
latifolia 353
minima 353
reniformis 353
senna 354
tomentosa 620
Rhynchosida 444
physocalyx 444
Ribes 559
aureum 559
curvatum 559
leptanthum 559
mescalerum 560
odoratum 560
Richardia 550
brasiliensis 550
scabra 550
tricocca 550
Ricinus 300
communis 300

Rivina 486
humilis 486
Robinia 354
hispida 354
neomexicana 354
pseudoacacia 354
Rorippa 199
nasturtium-aquaticum 199
palustris 199
ramosa 200
sessiliflora 200
sinuata 200
sphaerocarpa 200
tenerrima 200
teres 200
truncata 201
Rosa 538
arkansana 538
bracteata 538
carolina 538
eglanteria 538
foliolosa 539
ignota 539
laevigata 539
multiflora 539
setigera 539
var. tomentosa 539
stellata 539
suffulta 538
tomentosa 539
woodsii 539
Rosaceae 526
Rotala 432
ramosior 432
Rubiaceae 542
Rubus 540
aboriginum 540
apogaeus 540
bifrons 540
flagellaris 540
louisianus 540
lucidus 540
oklahomus 541
persistens 541
saepescandens 541
trivialis 541
var. duplaris 541
var. trivialis 541
Rudbeckia 89
amplexicaulis 89
fulgida 142
grandiflora 143
hirta 143
laciniata 143
maxima 143
missouriensis 143
scabrifolia 143
subtomentosa 144
texana 144
triloba 144
Ruellia 15
brittoniana 15
caroliniensis 16
corzoi 16
drummondiana 16
humilis 16
var. depauperata 16
var. humilis 16
metziae 16
nudiflora 16
var. nudiflora 16
var. runyonii 16
occidentalis 17
parryi 17
pedunculata 17
strepens 17
yucatana 17
Rumex 507
altissimus 507
chrysocarpus 507
conglomeratus 507
crispus 507
ellipticus 507
hastatulus 507
hymenosepalus 507
langloisii 507
maritimus 508
obtusifolius 508
paraguayensis 508
pulcher 508
spiralis 508
stenophyllus 508
venosus 509
verticillatus 509
Ruta 552
chalapensis 552
graveolens 552
Rutaceae 551

Sabatia 380
angularis 380
arenicola 380
brachiata 380
calycina 380
campanulata 380
campestris 380
dodecandra 381
formosa 380
gentianoides 381
Sageretia 526
thea 526
wrightii 526
Sagina 238
decumbens 238
Salazaria 409
mexicana 409
Salicaceae 553
Salicornia 251
bigelovii 251
utahensis 251
virginica 251
Salix 556
amygdaloides 621
babylonica 554
bonplandiana 555
caroliniana 555
exigua 555
exilifolia 555
goodingii 556
humilis 555
interior 555
lasiolepis 555
nigra 556
taxifolia 555
var. *leiocarpa* 555
var. *limitanea* 555
thurberi 555
Salpichroa 587
origanifolia 587
Salsola 251
australis 251
collina 251
iberica 251
paulsenii 251
pestifer 251
tragus 251
Salvia 410
arizonica 410
azurea 410
var. azurea 410
var. grandiflora 410
ballotiflora 410
coccinea 410
dolichantha 410
dolichantha 413
engelmannii 410
farinacea 410
var. farinacea 410
var. latifolia 410
greggii 411
henryi 411
lycioides 411
lyrata 411
mexicana 409
penstemonoides 411
pinguifolia 412
reflexa 412
regla 412
reptans 412
var. glabra 412
roemeriana 412
summa 412
texana 413
tiliifolia 413
vinacea 413
whitehouseae 413
Sambucus 229
canadensis 229
canadensis 6
cerulea 6, 229
mexicana 7, 229
nigra 229
var. **canadensis 6**, 229
subsp. cerulea 6, 7
var. **cerulea 6**
Samolus 513
ebracteatus 513
var. alyssoides 513
var. cuneatus 513
parviflorus 513
valerandi 513
subsp. *parviflorus* 513
Sanguinaria 483
canadensis 483
Sanguisorba 541
annua 541
Sanicula 41
canadensis 41
odorata 42
smallii 42
Santalaceae 556
Sanvitalia 144
abertii 144
angustifolia 144
ocymoides 144
Sapindaceae 556
Sapindus 556
drummondii 556
saponaria 556
var. *drummondii* 556
Sapium 300
sebiferum 300
Saponaria 238
officinalis 238
Sapotaceae 557
Sarcocornia 251
utahensis 251
Sarcostemma 59
clausum 59
crispum 59
cynanchoides 59
var. cynanchoides 59
var. hartwegii 59
torreyi 59
Sarracenia 558
alata 558
Sarraceniaceae 558
Sartwellia 145
flaveriae 145
Sassafras 419
albidum 419
Saururaceae 558
Saururus 558
cernuus 558
Saxifraga 560
texana 560
Saxifragaceae 558
Scabiosa 274
atropurpurea 274
Scaevola 382
plumieri 382
Scandix 42
pecten-veneris 42
Schaefferia 243
cuneifolia 243
Schinus 31
molle 31
terebinthifolius 31
Schkuhria 145
multiflora 145
pinnata 145
var. wislizeni 145
Schoenocrambe 201
linearifolia 201
Schrankia 344
hystricina 344

latidens 344
nuttallii 345
occidentalis 345
quadravalvis 344
var. *latidens* 344
roemeriana 345
Sclerocarpus 145
uniserialis 145
var. austrotexanum 145
var. uniserialis 145
Scoparia 575
dulcis 575
Scorzonela 145
laciniata 145
Scrophularia 575
marilandica 575
SCROPHULARIACEAE 560
Scutellaria 413
cardiophylla 413
drummondii 413
var. drummondii 413
var. edwardsiana 413
var. runyonii 413
elliptica 414
var. elliptica 414
var. hirsuta 414
galericulata 414
integrifolia 414
laevis 414
lateriflora 414
minor 414
muriculata 415
ovata 414
var. bracteata 414
parvula 415
potosina 415
var. davisiana 415
var. tesselata 415
racemosa 415
resinosa 415
texana 416
wrightii 416
Sebastiania 300
fruticosa 300
Securigera 354
varia 354
Sedum 267
cockerellii 267
diffusum 267
havardii 267
nanifolium 267
nuttallianum 268
parvum 268
subsp. *nanifolium* 267
pulchellum 268
texanum 267
wrightii 268
Selenia 201
dissecta 201
grandis 201
jonesii 201
Selinocarpus 460
angustifolius 460
diffusus 460
subsp. *nevadensis* 8
lanceolatus 461
var. *megaphyllus* 7
maloneanus 7, 461
marshii 8
parvifolius 461
Senecio 145
ampullaceus 145
douglasii 146
fendleri 146
flaccidus 146
glabellus 146
greggii 146
millelobatus 146
obovatus 146
parryi 147
plattensis 147
quaylei 145
riddellii 147
tampicanus 147
thurberi 147
tomentosus 147
vulgaris 148
warnockii 148
wootonii 148
Senna 354
bauhinioides 354
corymbosa 355
durangensis 355
var. durangensis 355
var. iselyi 355
lindheimeriana 355
marilandica 355
obtusifolia 355
occidentalis 355
orcuttii 356
pendula 356
pilosior 356
pumilio 356
ripleyana 356
roemeriana 356
wislizeni 357
Serjania 557
brachycarpa 557
incisa 557
Sesban 357
emerus 357
Sesbania 357
drummondii 357
emerus 357
exaltata 357
exaltatus 357
herbacea 357
macrocarpa 357
punicea 357
sericea 357
vesicaria 358
Seseli 6
hallii 6
Sesuvium 20
erectum 20
maritimum 19
portulacastrum 20
sessilis 20
trianthemoides 20
verrucosum 20
Seymeria 575
scabra 575
texana 576
Sherardia 550
arvensis 550
Shinnersoseris 148
rostrata 148
Sibara 201
grisea 201
runcinata 202
virginica 202
Sicyos 271
ampelophyllus 271
angulatus 271
glaber 271
laciniatus 271
microphyllus 271
Sida 444
abutifolia 444
ciliaris 444
cordifolia 444
elliotii 445
lindheimeri 445
longipes 445
neomexicana 445
physocalyx 444
rhombifolia 445
spinosa 445
tragiifolia 446
Sidastrum 446
paniculatum 446
Sideroxylon 557
celastrinum 557
lanuginosum 557
var. albicans 557
var. **oblongifolium 8**
var. **rididum 8**, 557
lycioides 557
Silene 238
antirrhina 238
gallica 239
laciniata 239
var. greggii 239
plankii 239
stellata 239
subciliata 239
Silphium 148
albiflorum 148
asperrimum 6
gracile 148
integrifolium 149
var. asperrimum 149
var. **asperrimum 6**
var. **wrightii 6**
var. wrigthii 149
laciniatum 149
simpsonii 6
var. *wrightii* 6
Silybum 149
marianum 149
SIMAROUBACEAE 578
Simsia 149
calva 149
Sinapsis 202
arvensis 202
Siphonoglossa 14
pilosella 14
Sisymbrium *205*
altissimum 202
auriculatum 189
auriculatum 202
irio 202
officinale 203
orientale 203
shinnersii 205
Sium 42
suave 42
Smallanthus 149
uvedalius 149
SOLANACEAE 579
Solanum 587
campechiense 587
capsicastrum 587
capsicoides 588
carolinense 588
citrullifolium 588
davisense 588
dimidiatum 588
douglasii 588
elaeagnifolium 589
erianthum 589
exculentum 583
fendleri 589, 621
heterodoxum 589
interius 589
jamesii 589
leptosepalum 590
pseudocapsicum 590
ptycanthum 590
rostratum 590
sarrachoides 590
sisymbriifolium 590
tenuipes 591
var. latisectum 591
var. tenuipes 591
triflorum 591
triquetrum 591
Solidago 149
altiplanities 149
altissima 150
var. altissima 150
var. gilvocanescens 150
var. *pluricephala* 150
arguta 150
auriculata 150
caesia 150
canadensis 149
var. *gilvocanescens* 150
var. *scabra* 150
gigantea 150
var. serotina 150
graminifolia 99
gymnospermoides 99
juliae 150
ludoviciana 151

missouriensis 151
mollis 151
nemoralis 151
nitida 128
odora 151
patula 151
patula 151
petiolaris 152
radula 152
rigida 128
rugosa 152
salicina 151
sempervirens 152
speciosa 152
tortifolia 152
ulmifolia 153
velutina 153
wrightii 153
Soliva 153
anthemifolia 153
mutisii 153
pterosperma 153
Sonchus 154
asper 154
oleraceus 154
Sophora 358
affinis 358
gypsophila 358
var. guadalupensis 358
nuttalliana 358
secundiflora 358
tomentosa 358
Spartium 359
junceum 359
Spergula 239
arvensis 239
Spergularia 240
echinosperma 240
marina 240
platensis 240
salina 240
Spermacoce 551
floridana 551
glabra 550
tenuior 551
Spermolepis 42
divaricata 42
echinata 42
inermis 43
Sphaeralcea 446
angustifolia 446
coccinea 446
digitata 446
fendleri 446
hastulata 447
incana 447
leptophylla 447
lobata 621
lindheimeri 447
pedatifida 447
polychroma 621
Sphaerophysa 359
salsula 359
Sphenoclea 591
zeylanica 591
Sphenocleaceae 591
Spigelia 429
hedyotidea 429
lindheimeri 429
marilandica 429
texana 429
Spilanthes 60
americana 60
var. *repens* 60
Stachys 416
agraria 416
bigelovii 416
drummondii 416
floridana 416
grahamii 417
tenuifolia 417
Stanleya 203
pinnata 203
var. texana 203
Stellaria 240
cuspidata 240
media 240
pallida 240
parva 241
prostrata 241
Stemodia 576
multifida 576
schottii 576
tomentosa 576
Stenandrium 17
barbatum 17
dulce 18
Stenosiphon 478
linifolius 478
Stephanomeria 154
exigua 154
pauciflora 154
tenuifolia 154
Sterculiaceae 591
Stevia 154
ovata 154
serrata 155
viscida 155
Steviopsis 155
fendleri 155
Stewartia 595
malacodendron 595
Stillingia 300
sylvatica 300
texana 300
treculiana 301
Styracaceae 593
Streptanthus 203
arizonicus 203
bracteatus 203
carinatus 203
cutleri 204
hyacinthoides 204
maculatus 204
playtcarpus 204
sparsiflorus 204
Strophostyles 359
helvula 359
leiosperma 359
umbellata 359
Stylisma 264
aquatica 264
humistrata 264
pickeringii 264
villosa 264
Stylocline 155
micropoides 155
Stylodon 606
carneus 606
Stylosanthes 359
biflora 359
viscosa 360
Styphnolobium 358
affine 358
Styrax 593
americanus 593
grandifolius 593
platanifolius 593
subsp. platanifolius 593
var. platanifolius 593
var. stellatus 593
var. texanus 593
subsp. youngiae 593
var. youngiae 593
Suaeda 252
calceoliformis 252
conferta 252
depressa 252
linearis 252
suffrutescens 252
tampicensis 252
Suckleya 252
suckleyana 252
Swertia 381
speciosa 381
Symphoricarpos 229
guadalupensis 229
longiflorus 229
occidentalis 230
orbiculatus 230
oreophilus 230
palmeri 230
rotundifolius 230
Symphyotrichum 155
divaricatum 155
drummondii 155
var. texanum 155
dumosum 156
ericoides 156
expansum 156
falcatum 156
fendleri 156
laeve 156
lanceolatum 157
var. hesperium 157
var. lanceolatum 157
lateriflorum 157
oblongifolium 157
oolentangiense 157
patens 157
var. patens 157
praealtum 157
pratense 158
puniceum 158
sericeum 158
subulatum 158
tenuifolium 158
Symplocaceae 594
Symplocos 594
tinctoria 594
Synthlipsis 204
greggii 204

Taenidia 43
integerrima 43
Tagetes 158
micrantha 158
Talinopsis 510
frutescens 510
angustissimum 510
aurantiacum 510
brevicaule 511
calycinum 511
chrysanthum 511
longipes 511
paniculatum 511
parviflorum 511
rugospermum 511
Tamaricaceae 594
Tamarix 594
aphylla 594
chinensis 594
gallica 594
parviflora 594
ramosissima 594
Tamaulipa 96
azurea 96
Taraxacum 159
erythospermum 159
laevigatum 159
officinale 159
Tauschia 43
texana 43
Tecoma 173
stans 173
Telosiphonia 46
brachysiphon 46
hypoleuca 46
lanuginosa 46
macrosiphon 46
Tephrosia 360
lindheimeri 360
onobrychoides 360
potosina 360
tenella 360
virginiana 360
Tetraclea 605
coulteri 605
Tetraclea 418
viscida 418
Tetragonotheca 159
ludoviciana 159

repanda 159
texana 159
Tetramerium 18
nervosum 18
Tetraneuris 159
acaulis 159
linearifolia 160
scaposa 160
var. argyrocaulon 160
var. scaposa 160
turneri 160
Teucrium 417
canadense 417
coahuilanum 417
cubense 417
var. *densum* 417
var. *laevigatum* 417
depressum 417
laciniatum 417
Thalictrum 521
arkansanum 521
dasycarpum 521
fendleri 521
revolutum 521
texanum 522
thalictroides 522
Thamnosma 552
texanum 552
var. purpureum 552
var. texanum 552
Thaspium 43
barbinode 43
trifoliatum 43
Theaceae 595
Thelesperma 161
ambiguum 161
burridgeanum 160
filifolium 160
var. filifolium 160
var. intermedium 160
flavodiscum 160
longipes 161
megapotamicum 161
var. ambiguum 161
var. megapotamicum 161
nuecense 161
simplicifolium 161
Thelocactus 221
bicolor 221
var. bicolor 221
var. flavidispinus 221
setispinus 215
Thelypodiopsis 201
linearifolia 201
purpusii 205
shinnersii 205
vaseyi 205
Thelypodium 205
tenue 205
texanum 205
wrightii 205
Thevetia 46
peruviana 46
Thlaspi 205
arvense 205
fendleri 206
fendleri 206
montanum 296
var. *fendleri* 206
Thryallis 433
angustifolia 433
Thunbergia 18
alata 18
Thurovia 161
triflora 161
Thymelaea 595
passerina 595
Thymophylla 89
acerosa 89
aurea 89
greggii 89
micropoides 89
pentachaeta 90
tenuiloba 90
var. *tenuiloba* 90
var. *wrightii* 90
tephroleuca 90
Tidestromia 29
carnosa 29
gemmata 29
lanuginosa 29
var. *carnosa* 29
oblongifolia 29
suffruticosa 29
Tilia 595
americana 595
Tiliaceae 595
Tillaea 266
aquatica 266
connata 266
drummondii 266
longipes 266
solierii 266
Tiquilia 184
canescens 184
gossypina 184
greggii 184
hispidissima 184
mexicana 184
Tomanthera 576
auriculata 576
densiflora 576
Torilis 43
arvensis 43
nodosa 44
Toumeya 222
papyracantha 222
Tournefortia 185
volubilis 185
Tovara 506
virginiana 506
Townsendia 161
annua 161
exscapa 162
texensis 162
Toxicodendron 31
pubescens 31
radicans 31
vernix 31
Trachelospermum 47
difforme 47
Tragia 301
amblyodonta 301
betonicifolia 301
brevispica 301
cordata 301
glanduligera 301
leptophylla 302
nigricans 302
ramosa 302
smallii 302
urens 302
urticifolia 302
Tragopogon 162
dubius 162
porrifolius 162
Trepocarpus 44
aethusae 44
Triadenum 395
tubulosum 395
virginicum 395
walteri 396
Trianthema 20
portulacastrum 20
Tribulus 617
terrestris 617
Tricerma 242
phyllanthoides 242
Trichocoronis 162
rivularis 162
wrightii 162
Trichostema 418
brachiatum 418
dichotomum 418
setaceum 418
Tridax 621
procumbens 621
Trifolium 361
amphianthum 361
arvense 361
bejariense 361
campestre 361
carolinianum 361
dubium 361
incarnatum 362
lappaceum 362
polymorphum 361
pratense 362
reflexum 362
repens 362
resupinatum 362
subterraneum 363
vesiculosum 363
wormskioldii 363
var. arizonicum 363
Triodanis 225
perfoliata 225
var. *biflora* 225
biflora 225
coloradoensis 225
holzingeri 225
lamprosperma 225
leptocarpa 225
perfoliata 225
texana 226
Triosteum 230
angustifolium 230
perfoliatum 231
Tripolium 155
divaricatum 155
Tripterocalyx 461
carnea 461
Trixis 163
californica 163
inula 163
Turnera 595
diffusa 595
Turneraceae 595

Ulmaceae 595
Ulmus 596
alata 596
americana 597
crassifolia 597
pumila 597
rubra 597
Ungnadia 557
speciosa 557
Urtica 598
chamaedryoides 598
dioica 598
var. gracilis 598
gracilenta 598
urens 599
Urticaceae 597
Urvillea 557
ulmacea 557
Utricularia 420
biflora 420
cornuta 419
foliosa 420
gibba 420
juncea 420
macrorhiza 420
purpurea 420
radiata 420
subulata 421
vulgaris 420

Vaccaria 241
pyramidata 241
arboreum 278
arkansanum 278
caesium 278
corymbosum 278
elliotii 278
stamineum 278
Valeriana 599
arizonica 599
texana 599
amarella 599
florifera 599
radiata 599
stenocarpa 600

texana 600
woodsiana 600
Valerianaceae 599
Varilla 163
texana 163
Vauquelinia 541
angustifolia 541
corymbosa 541
var. angustifolia 541
Verbascum 577
blattaria 577
thapsus 577
virgatum 577
Verbena 605
bonariensis 605
bracteata 605
brasiliensis 606
canescens 606
carnea 606
cloverae 606
halei 606
hastata 606
litoralis 607
livermorensis 607
macdougalii 607
neomexicana 607
perennis 607
plicata 607
rigida 608
runyonii 608
scabra 608
simplex 608
stricta 608
urticifolia 608
xutha 609
Verbenaceae 600
Verbesina 163
alternifolia 163
encelioides 163
helianthoides 163
lindheimeri 164
microptera 164
nana 164
oreophila 164
virginica 164
walteri 164
Vernicia 303
fordii 303
Vernonia 165
baldwinii 165
var. interior 165
gigantea 165
guadalupensis 165
larsenii 165
lindheimeri 165
marginata 165
missurica 166
texana 166
Veronica 577
americana 577
anagallis-aquatica 577
arvensis 577
peregrina 578
persica 578
polita 578
Veronicastrum 578
virginicum 578
Vexibia 358
nuttalliana 358
Viburnum 231
acerifolium 231
ashei 231
australe 231
dentatum 231
nitidum 231
nudum 231
rafinesquianum 231
var. affine 231
var. rafinesquianum 231
rufidulum 232
Vicia 363
americana 363
angustifolia 363
caroliniana 363
dasycarpa 365
hirsuta 364
leavenworthii 364
ludoviciana 364
var. *laxiflora* 364
var. ludoviciana 364
var. occidentalis 364
lutea 364
minutiflora 364
sativa 364
tetrasperma 364
villosa 365
Vigna 365
luteola 365
unguiculata 365
Viguiera 166
cordifolia 166
dentata 166
longifolia 166
multiflora 166
phenax 167
stenoloba 167
Villadia 268
squamulosa 268
Vinca 47
major 47
Viola 609
bicolor 609
guadalupensis 609
lanceolata 609
missouriensis 611
palmata 610
pedata 610
primulifolia 610
pubescens 610
purpurea 610
sagittata 610
sororia 611
villosa 611
walteri 611
Violaceae 609
Viscaceae 611
Vitaceae 613
Vitex 609
agnus-castus 609
Vitis 614
acerifolia 614
aestivalis 614
arizonica 614
cinerea 614
var. cinerea 614
var. helleri 614
lincecumii 614
monticola 614
mustangensis 615
palmata 615
riparia 615
rotundifolia 615
rupestris 615
vulpina 615

Wahlenbergia 226
marginata 226
Waltheria 593
indica 593
Warnockia 418
scutellarioides 418
Wedelia 167
acapulcensis 167
var. *hispida* 167
hispida 167
Wilcoxia 212
poselgeri 212
Wislizenia 228
refracta 228
Wissadula 447
hernandioides 447
periplocifolia 448
Wisteria 365
frutescens 365
macrostachya 365
sinensis 365
Xanthisma 167
texana 167
var. drummondii 167
var. orientale 167
var. texana 167
Xanthium 167
spinosum 167
strumarium 167
Xanthocephalum 168
gymnospermoides 168
Xanthorhiza 522
simplicissima 522
Xylorhiza 168
wrightii 168
Xylosma 374
flexuosa 374
Xylothamia 168
palmeri 168
triantha 168

Yeatesia 18
platystegia 18
viridiflora 18
Youngia 168
japonica 168

Zanthoxylum 553
clava-herculis 553
fagara 553
hirsutum 553
parvum 553
Zapoteca 365
coulteri 365
Zexmenia 167
hispida 167
texana 167
Zinnia 168
acerosa 168
anomala 169
grandiflora 169
Zizia 44
aurea 44
Ziziphus 526
jujuba 526
obtusifolia 526
zizyphus 526
Zornia 366
bracteata 366
gemella 366
reticulata 366
Zygophyllaceae 616
Zygophyllum 617
fabago 617

ALPHABETICAL LIST OF FAMILIES
OF VASCULAR PLANTS KNOWN TO OCCUR IN TEXAS
(DICOTS)

Acanthaceae 11
Aceraceae 18
Aizoaceae 19
Amaranthaceae 20
Anacardiaceae 29
Annonaceae 31
Apiaceae 32
Apocynaceae 45
Aquifoliaceae 47
Araliaceae 48
Aristolochiaceae 49
Asclepiadaceae 50
Asteraceae 60
Balsaminaceae 169
Basellaceae 169
Bataceae 170
Berberidaceae 170
Betulaceae 171
Bignoniaceae 172
Boraginaceae 173
Brassicaceae 185
Cabombaceae 206
Cactaceae 206
Callitrichaceae 222
Campanulaceae 222
Cannabaceae 226
Capparaceae 226
Caprifoliaceae 228
Caryophyllaceae 232
Celastraceae 241
Ceratophyllaceae 243
Chenopodiaceae 243
Cistaceae 253
Clethraceae 254
Convolvulaceae 255
Cornaceae 264
Crassulaceae 265
Cucurbitaceae 268
Cuscutaceae 271
Cyrillaceae 274
Dipsacaceae 274
Droseraceae 275
Ebenaceae 275
Elaeagnaceae 276
Elatinaceae 276
Ericaceae 276
Euphorbiaceae 278
Fabaceae 303
Fagaceae 366
Flacourtiaceae 374
Fouquieriaceae 375
Frankeniaceae 375
Fumariaceae 375
Garryaceae 376
Gentianaceae 376
Geraniaceae 381
Goodeniaceae 382
Haloragaceae 383
Hamamelidaceae 384
Hippocastanaceae 384
Hydrangeaceae 384
Hydrophyllaceae 387
Hypericaceae 392
Juglandaceae 396
Koeberliniaceae 398
Krameriaceae 398
Lamiaceae 399
Lauraceae 418
Leitneriaceae 419
Lentibulariaceae 419
Linaceae 421
Loasaceae 424
Loganiaceae 428
Lythraceae 430
Magnoliaceae 432
Malpighiaceae 433
Malvaceae 434
Melastomataceae 448
Meliaceae 449
Menispermaceae 449
Menyanthceae 449
Molluginaceae 450
Monotropaceae 450
Moraceae 451
Myricaceae 452
Myrsinaceae 452
Nyctaginaceae 452
Nymphaeaceae 461
Oleaceae 462
Onagraceae 466
Orobanchaceae 479
Oxalidaceae 480
Papaveraceae 481
Passifloraceae 483
Pedaliaceae 484
Phrymaceae 485
Phytolaccaceae 485
Piperaceae 486
Plantaginaceae 486
Platanaceae 488
Plumbaginaceae 489
Polemoniaceae 489
Polygalaceae 494
Polygonaceae 499
Portulacaceae 509
Primulaceae 512
Rafflesiaceae 513
Ranunculaceae 518
Resedaceae 522
Rhamnaceae 522
Rhizophoraceae 526
Rosaceae 526
Rubiaceae 542
Rutaceae 551
Salicaceae 553
Santalaceae 556
Sapindaceae 556
Sapotaceae 557
Sarraceniaceae 558
Saururaceae 558
Saxifragaceae 558
Scrophulariaceae 560
Simaroubaceae 578
Solanaceae 579
Sphenocleaceae 591
Sterculiaceae 591
Styracaceae 593
Symplocaceae 594
Tamaricaceae 594
Theaceae 595
Tiliaceae 595
Turneraceae 595
Ulmaceae 595
Urticaceae 597
Valerianaceae 599
Verbenaceae 600
Violaceae 609
Viscaceae 611
Vitaceae 613
Zygophyllaceae 616